Student Solutions Manual

Felix S. Lee
The University of Western Ontario

to accompany

Introduction to Organic Chemistry
Sixth Edition

William Brown
Beloit College

Thomas Poon
Claremont McKenna College
Scripps College
Pitzer College

WILEY

Cover photo credits:© Tony Hutchings/Photographer's Choice/Getty Images, Inc.

ISBN 13   9781119106951

The inside back cover will contain printing identification and country of origin if omitted from this page. In addition, if the ISBN on the back cover differs from the ISBN on this page, the one on the back cover is correct.

Printed in the United States of America.

SKY10032776_012622

Table of Contents

Chapter 1: Covalent Bonding and Shapes of Molecules

Problems

1.1 Write and compare the ground-state electron configurations for the elements in each set. What can be said about the outermost shell of orbitals for each pair of elements?

The two elements in each pair have the same number of electrons in the outermost shell of orbitals (valence shell), which are indicated in bold.

(a) Carbon = $1s^2 2s^2 2p^2$ Silicon = $1s^2 2s^2 2p^6 3s^2 3p^2$ 4 valence electrons

(b) Oxygen = $1s^2 2s^2 2p^4$ Sulfur = $1s^2 2s^2 2p^6 3s^2 3p^4$ 6 valence electrons

(c) Nitrogen = $1s^2 2s^2 2p^3$ Phosphorus = $1s^2 2s^2 2p^6 3s^2 3p^3$ 5 valence electrons

1.2 Show how the gain of two electrons by a sulfur atom to form a sulfide ion leads to a stable octet:

$$S + 2e^- \longrightarrow S^{2-}$$

The electron configuration of the sulfur atom is $1s^2 2s^2 2p^6 3s^2 3p^4$. When sulfur gains two electrons to form S^{2-}, the electron configuration becomes $1s^2 2s^2 2p^6 3s^2 3p^6$. The valence shell has a full octet and corresponds to the configuration of the noble gas Ar.

1.3 Judging from their relative positions in the Periodic Table, which element in each pair has the larger electronegativity?

In general, electronegativity increases from left to right within a period, and from bottom to top within a group.

(a) Lithium or Potassium: Lithium, because it is higher up in the same group.

(b) Nitrogen or Phosphorus: Nitrogen, because it is higher up in the same group.

(c) Carbon or Silicon: Carbon, because it is higher up in the same group.

(d) Oxygen or Phosphorus: Oxygen, because it is higher up and further to the right.

(e) Oxygen or Silicon: Oxygen, because it is higher up and further to the right.

1.4   Classify each bond as nonpolar covalent, polar covalent, or ionic.

The classification of a bond can be determined by the difference in electronegativity between the bonded atoms. However, these classifications are very broad, and even though two bonds can be in the same category, one may be more polar than the other.

Bond	Difference in electronegativity	Type of bond
S–H	$2.5 - 2.1 = 0.4$	nonpolar covalent
P–H	$2.1 - 2.1 = 0.0$	nonpolar covalent
C–F	$4.0 - 2.5 = 1.5$	polar covalent
C–Cl	$3.0 - 2.5 = 0.5$	polar covalent

1.5   Using a bond dipole arrow and the symbols δ– and δ+, indicate the direction of polarity in these polar covalent bonds.

In a polar covalent bond, the atom that is more electronegative has a δ– charge, while the less-electronegative atom has a δ+ charge.

(a) $\overset{\delta+}{C}$–$\overset{\delta-}{N}$ N is more electronegative than C

(b) $\overset{\delta+}{N}$–$\overset{\delta-}{O}$ O is more electronegative than N

(c) $\overset{\delta+}{C}$–$\overset{\delta-}{Cl}$ Cl is more electronegative than C

1.6   Draw Lewis structures, showing all valence electrons, for these molecules:

For uncharged molecules, the total number of electrons described by the Lewis structures equals the total of the number of valence electrons contributed by each atom.

(a) C_2H_6 has a total of 14 valence electrons (4 from each C and 1 from each H). Because H can only have one bond, the only possible connectivity of the atoms is shown on the right.

(b) CS_2 has a total of 16 valence electrons (4 from each C and 6 from each S). In uncharged molecules, C generally forms 4 bonds, and S, which is in the same group as O, generally forms two bonds.

(c) HCN has a total of 10 valence electrons (1 from H, 4 from C, and 5 from N). H can form only one bond, and in uncharged molecules, C generally forms 4 bonds while N forms 3 bonds.

$$H-C\equiv N:$$

(d) HCHO has a total of 12 valence electrons (1 from each H, 4 from C, and 6 from O). The CHO portion of the formula represents an aldehyde group, in which the O forms two bonds with C.

$$\overset{\displaystyle \cdot\overset{\cdot\cdot}{O}\cdot}{\underset{\displaystyle H-\overset{\|}{C}-H}{}}$$

1.7 Draw Lewis structures for these ions, and show which atom in each bears the formal charge(s):

For positively charged species, the total number of valence electrons described by the Lewis structure is equal to the number of valence electrons contributed by each atom, less the number of electrons necessary to obtain the correct positive charge.

(a) $CH_3NH_3^+$ has a total of 14 valence electrons (1 from each H, 4 from each C, and 5 from N, minus one electron to form a positive charge). The connectivity of the atoms is suggested by the structural formula.

$$CH_3NH_3^+$$

structural formula   connectivity   Lewis structure

The formal charge of an atom is equal to the number of valence electrons in the neutral, unbonded atom, less the full number of unshared electrons (nonbonding or lone electrons) and half the number of shared electrons.

No atoms have unshared electrons. The formal charge of C is: 4 valence electrons – ½ (8 shared electrons, two from each bond) = 0. Likewise, the formal charge of N is: 5 valence electrons – ½ (8 shared electrons, two from each bond) = +1.

(b) CH_3^+ has a total of 6 valence electrons (1 from each H and 4 from C, minus one electron to form a positive charge). Carbon, with only three bonds and no unshared electrons, has a formal charge of: 4 – ½ (6 electrons, two from each bond) = +1. It can be generalized that C with three bonds and no unshared electrons has a +1 formal charge.

1.8   Predict all bond angles for these molecules:

Formula	Lewis structure	Bond angles
CH_3OH	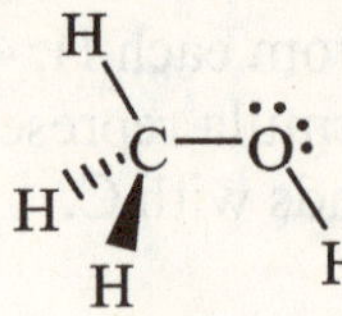	To predict bond angles, first draw the Lewis structure. Then, examine the number of regions of electron density around an atom. Here, both the C and the O atoms each have four regions of electron density, so the bond angles are all 109.5°.
CH_2Cl_2	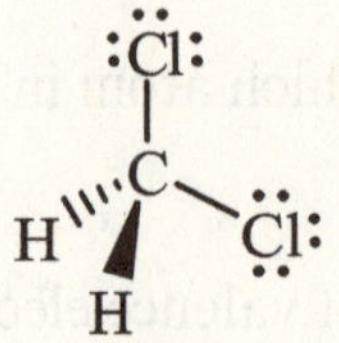	The C atom has four regions of electron density, hence a bond angle of 109.5°.
H_2CO_3	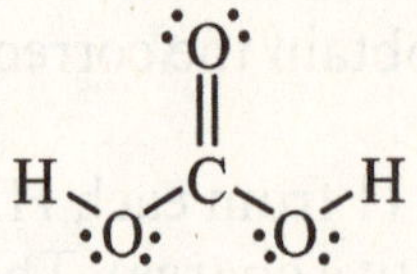	The C atom has three regions of electron density, hence a bond angle of 120°. The non-terminal O atoms each have four regions of electron density, hence bond angles of 109.5°.

1.9   Both carbon dioxide (CO_2) and sulfur dioxide (SO_2) are triatomic molecules. Account for the fact that carbon dioxide is a nonpolar molecule, whereas sulfur dioxide is a polar molecule.

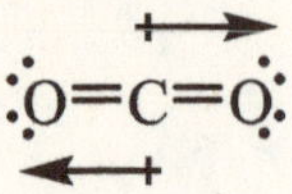

For a molecule to be polar, it requires a geometry in which the bond dipole moments do not cancel out. The linear shape of CO_2 results in a cancellation of the dipoles moments. In SO_2, which has a bent shape, the dipole moments do not cancel out.

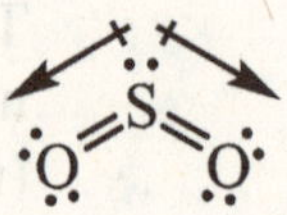

1.10   Which sets are pairs of resonance contributing structures?

(a)

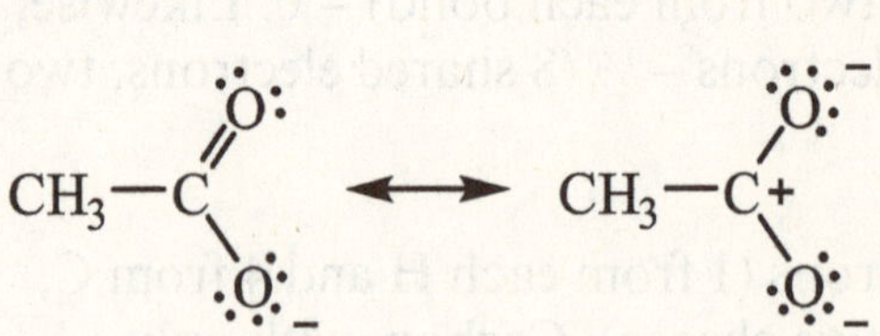

$$CH_3-C \longleftrightarrow CH_3-C+$$

(b)

$$CH_3-C \longleftrightarrow CH_3-C$$

(c)

$$H_2C=N=N \longleftrightarrow H_2C-N\equiv N$$

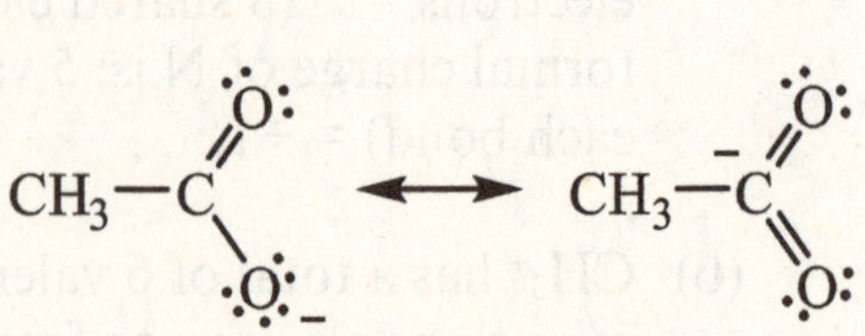

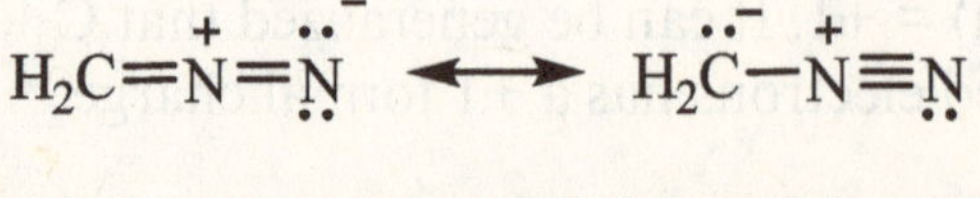

Contributing structures differ only in the distribution of the electrons, and each contributing structure itself must be a correct Lewis structure. Pairs (a) and (c) are each a set of contributing structures. Pair (b) is not a set of contributing structures; although they differ in electronic distributions, the structure on the right violates the octet rule and is an unacceptable Lewis structure.

1.11   Use curved arrows to show the redistribution of valence electrons in converting resonance contributing structure (a) to (b) and then (b) to (c). Also show, using curved arrows, how (a) can be converted to (c) without going through (b).

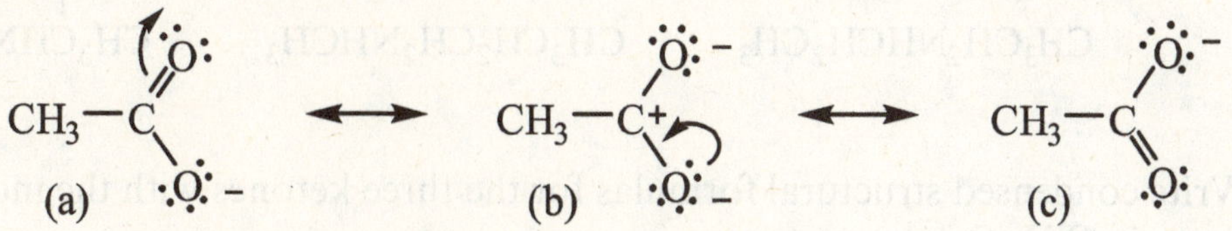

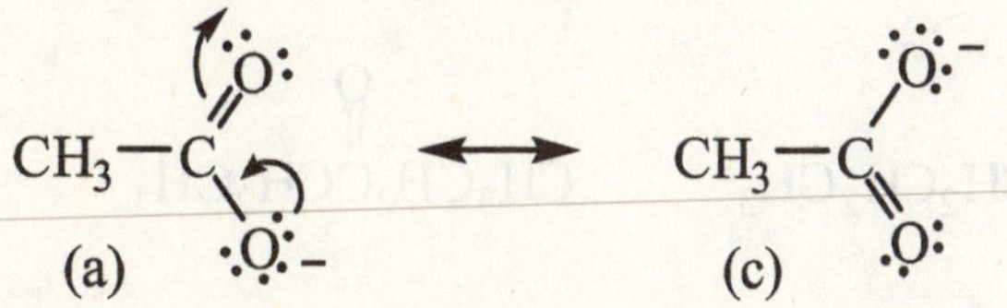

1.12   Describe the bonding in these molecules in terms of the atomic orbitals involved, and predict all bond angles.   (a) $CH_3CH=CH_2$ (b) CH_3NH_2

With the exception of hydrogen ($1s$ configuration), all other atoms are hybridized. The orbitals used to form the respective σ and π bonds are also indicated.

Hybridization	Bonding	Bond angles
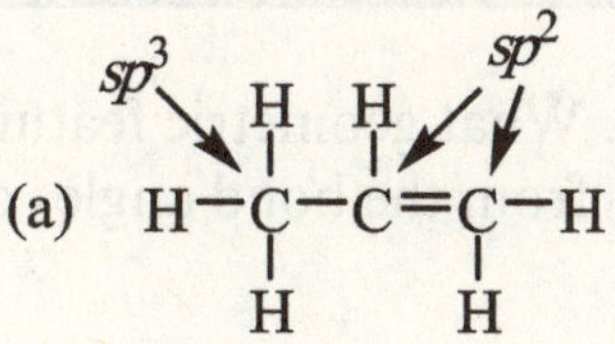 (a)	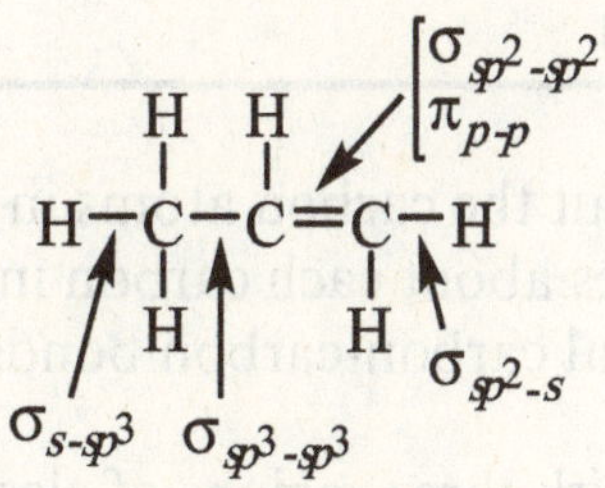	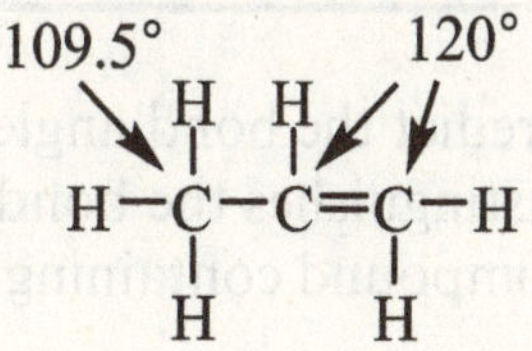
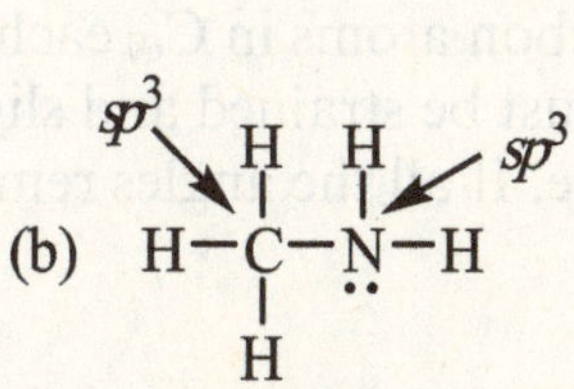 (b)	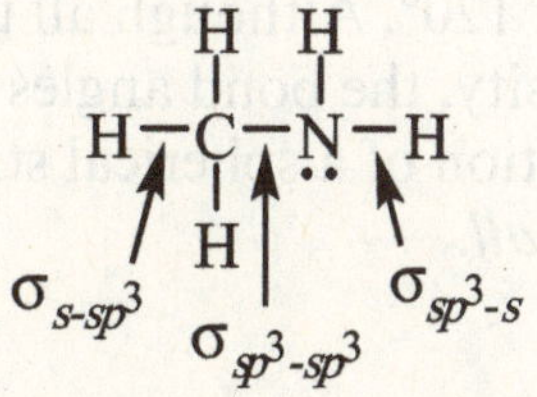	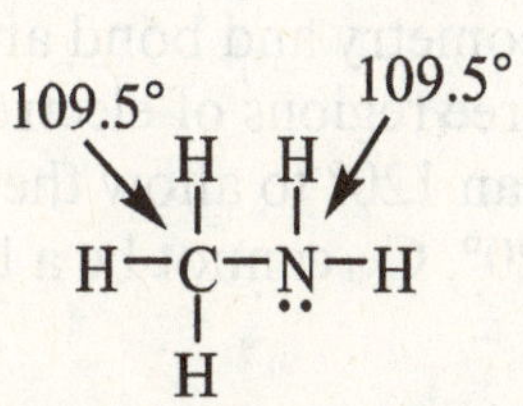

1.13   Write condensed structural formulas for the four alcohols with the molecular formula $C_4H_{10}O$. Classify each as primary, secondary, or tertiary.

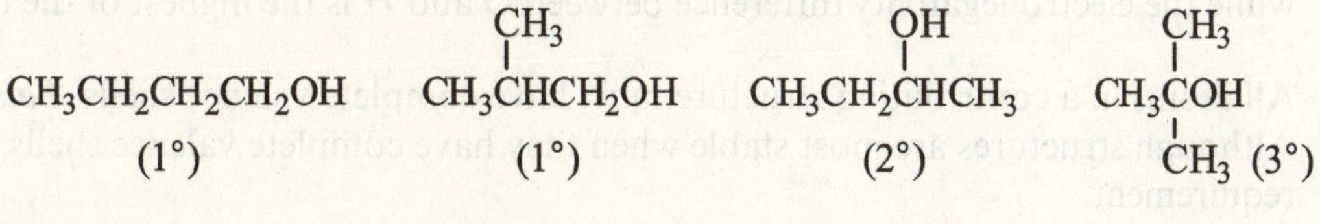

$$CH_3CH_2CH_2CH_2OH \qquad \underset{\displaystyle CH_3CHCH_2OH}{\overset{\displaystyle CH_3}{|}} \qquad \underset{\displaystyle CH_3CH_2CHCH_3}{\overset{\displaystyle OH}{|}} \qquad \overset{\displaystyle CH_3}{\underset{\displaystyle CH_3}{CH_3COH}}$$

(1°)   (1°)   (2°)   (3°)

1.14 Write condensed structural formulas for the three secondary amines with the molecular formula $C_4H_{11}N$.

$$CH_3CH_2NHCH_2CH_3 \qquad CH_3CH_2CH_2NHCH_3 \qquad \overset{\displaystyle CH_3}{\underset{\displaystyle |}{CH_3CHNHCH}}$$

1.15 Write condensed structural formulas for the three ketones with the molecular formula $C_5H_{10}O$.

$$\overset{\displaystyle O}{\overset{\displaystyle \|}{CH_3CCH_2CH_2CH_3}} \qquad \overset{\displaystyle O}{\overset{\displaystyle \|}{CH_3CH_2CCH_2CH_3}} \qquad \overset{\displaystyle O}{\overset{\displaystyle \|}{CH_3CCHCH_3}} \\ \underset{\displaystyle CH_3}{\underset{\displaystyle |}{}}$$

1.16 Write condensed structural formulas for the two carboxylic acids and four esters with the molecular formula $C_4H_8O_2$.

$$CH_3CH_2CH_2COOH \qquad (CH_3)_2CHCOOH$$

$$CH_3CH_2COOCH_3 \qquad CH_3COOCH_2CH_3 \qquad HCOOCH_2CH_2CH_3 \qquad HCOOCH(CH_3)_2$$

Chemical Connections

1A. Predict the bond angles about the carbon atoms in C_{60}. What geometric feature distinguishes the bond angles about each carbon in C_{60} from the bond angles of a compound containing typical carbon-carbon bonds?

Normally, a carbon atom with three regions of electron density has a trigonal planar geometry and bond angles of 120°. Although all the carbon atoms in C_{60} each have three regions of electron density, the bond angles C_{60} must be strained and slightly less than 120° to allow the formation of a spherical structure. If all the angles remained at 120°, C_{60} cannot be a bucky*ball*.

Quick Quiz

1. These bonds are arranged in order of increasing polarity: C−H < N−H < O−H. *True*. The difference in electronegativity between C and H is the lowest of the three, while the electronegativity difference between O and H is the highest of the three.

2. All atoms in a contributing structure must have complete valence shells. *False*. Although structures are most stable when they have complete valence shells, it is not a requirement.

3. An electron in a $1s$ orbital is held closer to the nucleus than an electron in a $2s$ orbital. *True*. The $1s$ orbital is closer to the nucleus than is the $2s$ orbital. When comparing orbitals of the same type, their distance from the nucleus increases with the principal quantum number (n).

4. A sigma bond and a pi bond have in common that each can result from the overlap of atomic orbitals. *True*. Atomic orbitals may also hybridize before bonding.

5. The molecular formula of the smallest aldehyde is C_3H_6O, and that of the smallest ketone is also C_3H_6O. *False*. While the smallest ketone is C_3H_6O (propanone), the formula of the smallest aldehyde is CH_2O (formaldehyde).

6. To predict whether a covalent molecule is polar or nonpolar, you must know both the polarity of each covalent bond and the geometry (shape) of the molecule. *True*. Whether a molecule is polar or nonpolar depends on the vector sum of the bond dipoles. Just because a molecule has polar bonds does not necessarily mean that the molecule is polar.

7. An orbital is a region of space that can hold two electrons. *True*. The maximum capacity of each orbital is two electrons.

8. In the ground-state electron configuration of an atom, only the lowest-energy orbitals are occupied. *True*. By definition, the ground-state configuration is the electron configuration of lowest energy. These electrons must be in the orbitals of lowest energy.

9. Electronegativity generally increases with atomic number. *False*. Electronegativity generally increases across a period, from left to right, and up a group, from bottom to top.

10. Paired electron spins means that the two electrons are aligned with their spins North Pole to North Pole and South Pole to South Pole. *False*. Electrons that are spin-paired must have opposite spin.

11. According to the Lewis model of bonding, atoms bond together in such a way that each atom participating in the bond acquires an outer-shell electron configuration matching that of the noble gas nearest it in atomic number. *True*. Most atoms usually have an electron configuration that consists of a full octet.

12. A primary amine contains one N−H bond, a secondary amine contains two N−H bonds, and a tertiary amine contains three N−H bonds. *False*. Amines are classified based on the number of alkyl groups, not the number of hydrogens, that are bonded to the nitrogen.

13. All bond angles in sets of contributing structures must be the same. *False*. Contributing structures differ only in the distribution of electrons. Bond angles may be different.

14. Electronegativity is a measure of an atom's attraction for electrons it shares in a chemical bond with another atom. *True*. An atom that is more electronegative will more strongly attract a bonding pair of electrons towards itself.

15. An orbital can hold a maximum of two electrons with their spins paired. *True*. Note that electrons that are spin-paired are of opposite spin.

16. Fluorine in the upper right corner of the Periodic Table is the most electronegative element; hydrogen, in the upper left corner, is the least electronegative element. *False*. While this statement is true for fluorine, it is incorrect for hydrogen. Elements with the lowest electronegativity are in the bottom-left corner of the Periodic Table.

17. A primary alcohol has one −OH group, a secondary alcohol has two −OH groups, and a tertiary alcohol has three −OH groups. *False*. Alcohols are classified not by the number of −OH groups, but by the number of carbons bonded to the carbon bearing the −OH group.

18. H_2O and NH_3 are polar molecules, but CH_4 is nonpolar. *True*. Water and ammonia have polar bonds, and their shapes (bent and trigonal pyramidal, respectively) do not allow the bond dipole moments to cancel out. Methane has nonpolar bonds and is also tetrahedral.

19. Electronegativity generally increases from top to bottom in a column of the Periodic Table. *False*. The opposite is true; electronegativity decreases from top to bottom in a group.

20. All contributing structures must have the same number of valence electrons. *True*. Contributing structures only involve the movement of electrons and not their addition or removal.

21. A carbon-carbon double bond is formed by the overlap of sp^2 hybrid orbitals, and a triple bond is formed by the overlap of sp^3 hybrid orbitals. *False*. Triple bonds involve sp hybrid orbitals. In both double and triple bonds, p atomic orbitals are also involved.

22. A covalent bond formed by sharing two electrons is called a double bond. *False*. A covalent bond consisting of two electrons is a single bond.

23. The functional groups of an alcohol, an aldehyde, and a ketone have in common the fact that each contains a single oxygen atom. *True*. However, note that the oxygen atom of an alcohol is also bonded to a hydrogen atom.

24. Electrons in atoms are confined to regions of space called principal energy levels. *True*. Within these regions of space are orbitals.

25. In a single bond, two atoms share one pair of electrons; in a double bond, they share two pairs of electrons; and in a triple bond, they share three pairs of electrons. *True*. Each bond involves the sharing of one pair of electrons.

26. The Lewis structure for ethene, C_2H_4, must show eight valence electrons. *False*. Each carbon atom has four valence electrons, and each hydrogen atom has one valence electron, giving a total of twelve valence electrons.

27. The Lewis structure for formaldehyde, CH_2O, must show eight valence electrons. *False*. It has twelve valence electrons: four from C, one from each H, and six from O.

28. The letters VSEPR stand for valence-shell electron pair repulsion. *True*. VSEPR considers only the electrons present in the valence shell.

29. In predicting bond angles about a central atom in a covalent bond, VSEPR considers only shared pairs (pairs of electrons involved in forming covalent bonds). *False*. VSEPR considers both bonding and nonbonding pairs of electrons.

30. An *sp* hybrid orbital may contain a maximum of four electrons, an sp^2 hybrid orbital may contain a maximum of six valence electrons, and an sp^3 hybrid orbital may contain a maximum of eight electrons. *False*. Each hybrid orbital is still a single orbital, which can hold a no more than of two electrons. However, there are two separate *sp*, three separate sp^2, and four separate sp^3 hybrid orbitals, which when summed, give the totals stated.

31. For a central atom surrounded by three regions of electron density, VSEPR predicts bond angles of 360°/3 = 120°. *True*. However, this equation does not always apply. For example, if there are four regions of electron density, the angle is 109.5°, not 360°/4 = 90°. Always look for the geometric shape that maximizes the separation of the regions of electron density. Do not simply divide 360° by the number of regions of electron density.

32. The three $2p$ orbitals are aligned parallel to each other. *False*. They are perpendicular (orthogonal or 90°) to each other.

33. All molecules with polar bonds are polar. *False*. Molecules with polar bonds are polar only if the dipole moments of the bonds do not cancel out. If the dipole moments of the bonds cancel out, such as in CO_2, the molecule is nonpolar.

34. Electronegativity generally increases from left to right across a period of the Periodic Table. *True*. Electronegativity also increases going up a group, from bottom to top.

35. A compound with the molecular formula C_3H_6O may be an aldehyde, a ketone, or a carboxylic acid. *False*. A carboxylic acid contains the carboxyl functional group, which has two oxygen atoms. A compound that contains both an ether and an alkene, or an alcohol and an alkene, could also have the molecular formula C_3H_6O.

36. Dichloromethane, CH_2Cl_2 is polar, but tetrachloromethane, CCl_4, is nonpolar. *True*. Although both compounds have a tetrahedral shape, the bond dipole moments in dichloromethane do not cancel out.

37. A covalent bond is formed between atoms whose difference in electronegativity is less than 1.9. *True*. When the difference in electronegativity is greater than 1.9, the bond is ionic.

38. Each principal energy level can hold two electrons. *False*. Each *orbital* can hold two electrons. Each principal energy level n can hold a maximum of $2n^2$ electrons.

39. Atoms that share electrons to achieve filled valence shells form covalent bonds. *True*. Covalent bonding involves the sharing of electrons. If the electrons were not shared, the bond would be ionic instead of covalent.

40. Contributing structures differ only in the distribution of valence electrons. *True*. Contributing structures do not involve different positions of atoms, only electrons.

41. In creating hybrid orbitals (sp, sp^2, and sp^3), the number of hybrid orbitals created is equal to the number of atomic orbitals hybridized. *True*. For example, the use of three orbitals (one s and two p) in sp^2 hybridization results in the formation of three sp^2 hybrid orbitals.

42. VSEPR treats the two electron pairs of a double bond and the three electron pairs of a triple bond as one region of electron density. *True*. A single bond, double bond, triple bond, and a nonbonding pair each count as a single region of electron density.

43. If the difference in electronegativity between two atoms is zero (they have identical electronegativities), then the two atoms will not form a covalent bond. *False*. The bond formed is still a covalent bond; it is a nonpolar covalent bond.

44. A carbon-carbon triple bond is a combination of one sigma bond and two pi bonds. *True*. The sigma bond is formed by the overlap of hybrid orbitals, while the two pi bonds are formed by the overlap of p orbitals.

45. A carbon-carbon double bond is a combination of two sigma bonds. *False*. A double bond consists of one sigma bond and one pi bond.

46. An s orbital has the shape of a sphere with the center of the sphere at the nucleus. *True*. Regardless of the principal energy level, an s orbital is always spherical.

47. A functional group is a group of atoms in an organic molecule that undergoes a predictable set of chemical reactions. *True*. The reactivity of any organic molecule is based on the functional groups present in that molecule.

48. In a polar covalent bond, the more electronegative atom has a partial negative charge ($\delta-$) and the less electronegative atom has a partial positive charge ($\delta+$). *True*. The bonding pair is closer to the atom that is more electronegative, giving the atom a partial negative charge.

49. Electronegativity depends on both the nuclear charge and the distance of the valence electrons from the nucleus. *False*. Electronegativity depends on the position of the element in the periodic table and on the oxidation state of the element.

50. There are two alcohols with the molecular formula C_3H_8O. *True*. The two alcohols are 1-propanol and 2-propanol.

51. In methanol, CH_3OH, the O−H bond is more polar than the C−O bond. *True*. The electronegativity difference between O and H is greater than that between O and C.

52. The molecular formula of the smallest carboxylic acid is $C_2H_6O_2$. *False*. The smallest carboxylic acid has a just one carbon and is HCOOH (formic acid).

53. Each $2p$ orbital has the shape of a dumbbell with the nucleus at the midpoint of the dumbbell. *True*. Note that both "ends" or "lobes" of the dumbbell together represent a single $2p$ orbital.

54. Atoms that lose electrons to achieve a filled valence shell become cations and form ionic bonds with anions. *True*. Atoms that lose electrons to achieve a filled valence shell are typically alkali and alkali earth metals, which form ionic bonds.

55. There are three amines with the molecular formula C_3H_9N. *False*. There are four amines:

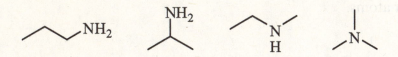

End-of-Chapter Problems

Electronic Structure of Atoms

1.17 Write the ground-state electron configuration for each element:

(a) Sodium = $1s^22s^22p^63s^1$

(b) Magnesium = $1s^22s^22p^63s^2$

(c) Oxygen = $1s^22s^22p^4$

(d) Nitrogen = $1s^22s^22p^3$

(e) Potassium = $1s^22s^22p^63s^23p^64s^1$

(f) Aluminum = $1s^22s^22p^63s^23p^1$

(g) Phosphorus = $1s^22s^22p^63s^23p^3$

(h) Argon = $1s^22s^22p^63s^23p^6$

1.18 Write the ground-state electron configuration for each ion:

(a) $Na^+ = 1s^22s^22p^6$

(b) $Cl^- = 1s^22s^22p^63s^23p^6$

(c) $Mg^{2+} = 1s^22s^22p^6$

(d) H^+ = no electrons

(e) $H^- = 1s^2$

(f) $K^+ = 1s^22s^22p^63s^23p^6$

(g) $Br^+ = 1s^22s^22p^63s^23p^64s^23d^{10}4p^4$

(h) $Li^+ = 1s^2$

1.19   Which element has the ground-state electron configuration

(a) $1s^22s^22p^63s^23p^4$ = 16 electrons = S   (b) $1s^22s^22p^4$ = 8 electrons = O

(c) [He]$2s^22p^2$ = 6 electrons = C   (d) [Ne]$3s^23p^5$ = 17 electrons = Cl

1.20   Which element or ion does not have the ground-state electron configuration $1s^22s^22p^63s^23p^6$?

(a) S^{2-} (b) Cl$^-$ (c) Ar   (d) Ca^{2+} **(e) K**

K has one more electron than all of the other atoms and has the electron configuration $1s^22s^22p^63s^23p^64s^1$. The ion K$^+$would have the same configuration as the other atoms.

1.21   Define *valence shell* and *valence electron*. Why are valence electrons more important to bonding than other types of electrons?

The *valence shell* of an atom is the outermost shell that can be occupied by electrons in the ground state. The valence shell generally has the highest principal quantum number (*n*). A *valence electron* is an electron that is situated in the valence shell. Valence electrons are more important because they are the electrons involved in bond formation.

1.22   How many electrons are in the valence shell of each element?

(a) C = $1s^2$**$2s^22p^2$** (4 valence e$^-$)   (b) N = $1s^2$**$2s^22p^3$** (5 valence e$^-$)

(c) Cl = $1s^22s^22p^4$**$3s^23p^5$** (7 valence e$^-$)   (d) Al = $1s^22s^22p^6$**$3s^23p^1$** (3 valence e$^-$)

(e) O = $1s^2$**$2s^22p^4$** (6 valence e$^-$)   (f) Si = $1s^22s^22p^6$**$3s^23p^2$** (4 valence e$^-$)

1.23   How many electrons are in the valence shell of each ion?

(a) H$^+$ = **$1s^0$** (no valence e$^-$)   (b) H$^-$ = **$1s^2$** (2 valence e$^-$)

(c) F$^-$ = $1s^2$**$2s^22p^6$** (8 valence e$^-$)   (d) Cl$^+$ = $1s^22s^22p^6$**$3s^23p^4$** (6 valence e$^-$)

(e) S^{2-} = $1s^22s^22p^6$**$3s^23p^6$** (8 valence e$^-$)

Lewis Structures

1.24 Judging from their relative positions in the Periodic Table, which element in each set is more electronegative?

(a) C or N: Both elements are in the same period, with N being more electronegative.

(b) Cl or Br: Both elements are in the same group, with Cl being more electronegative.

(c) O or S: Both elements are in the same group, with O being more electronegative.

(d) O or P: O is more electronegative than P because it is above and to the right of P.

1.25 Which compounds have nonpolar covalent bonds, which have polar covalent bonds, and which have ionic bonds?   (a)LiF   (b)CH$_3$F   (c)MgCl$_2$ (d)HCl

Bond	Difference in electronegativity	Type of bond
(a) Li–F	4.0 – 1.0 = 3.0	ionic
(b) C–F	4.0 – 2.5 = 1.5	polar covalent
C–H	2.5 – 2.1 = 0.4	nonpolar covalent
(c) Mg–Cl	3.0 – 1.2 = 1.8	polar covalent
(d) H–Cl	3.0 – 2.1 = 0.9	polar covalent

Note that the determination of bond type by electronegativity difference is only approximate and that the 0.5 and 1.9 "rules" are only guidelines. With Mg–Cl, the difference is close enough to 1.9 that many chemists consider MgCl$_2$, a salt, to be ionic.

1.26 Using the symbols δ– and δ+, indicate the direction of polarity in these polar covalent bonds.

(a) $\overset{\text{δ+}\ \ \text{δ–}}{\text{C–Cl}}$ Cl is more electronegative than C

(b) $\overset{\text{δ–}\ \ \text{δ+}}{\text{S–H}}$ S is more electronegative than H

(c) C–S   C and S have the same electronegativity

(d) P–H   P and H have the same electronegativity

1.27   Write Lewis structures for each of the following compounds, showing all valence electrons (none of the compounds contain a ring of atoms):

(a) H_2O_2 (b) N_2H_4 (c) CH_3OH (d) CH_3SH

$$H-\ddot{O}-\ddot{O}-H$$

$$H-\ddot{N}-\ddot{N}-H$$
$$\quad\;\; |\quad\; |$$
$$\quad\;\; H\quad H$$

$$\begin{array}{c} H \\ | \\ H-C-\ddot{O}-H \\ | \\ H \end{array}$$

$$\begin{array}{c} H \\ | \\ H-C-\ddot{S}-H \\ | \\ H \end{array}$$

(e) CH_3NH_2 (f) CH_3Cl (g) CH_3OCH_3 (h) CH_3CH_3

$$\begin{array}{c} H \\ | \\ H-C-\ddot{N}-H \\ | \quad\;\; | \\ H \quad H \end{array}$$

$$\begin{array}{c} H \\ | \\ H-C-\ddot{Cl}: \\ | \\ H \end{array}$$

$$\begin{array}{cc} H & H \\ | & | \\ H-C-\ddot{O}-C-H \\ | & | \\ H & H \end{array}$$

$$\begin{array}{cc} H & H \\ | & | \\ H-C-C-H \\ | & | \\ H & H \end{array}$$

(i) CH_2CH_2 (j) C_2H_2 (k) CO_2 (l) CH_2O

$$\begin{array}{cc} H & H \\ \diagdown & \diagup \\ C=C \\ \diagup & \diagdown \\ H & H \end{array}$$

$$H-C\equiv C-H$$

$$:\ddot{O}=C=\ddot{O}:$$

$$\begin{array}{c} :\ddot{O}: \\ \| \\ H-C-H \end{array}$$

(m) CH_3COCH_3 (n) H_2CO_3 (o) CH_3COOH

$$\begin{array}{ccc} H & :\ddot{O}: & H \\ | & \| & | \\ H-C-C-C-H \\ | & & | \\ H & & H \end{array}$$

$$\begin{array}{c} :\ddot{O}: \\ \| \\ H-\ddot{O}-C-\ddot{O}-H \end{array}$$

$$\begin{array}{cc} H & :\ddot{O}: \\ | & \| \\ H-C-C-\ddot{O}-H \\ | \\ H \end{array}$$

1.28   Write Lewis structures for these ions:

(a) HCO_3^- (b) CO_3^{2-} (c) CH_3COO^- (d) Cl^-

$$\begin{array}{c} :\ddot{O}: \\ \| \\ H-\ddot{O}-C-\ddot{O}:^- \end{array}$$

$$\begin{array}{c} :\ddot{O}: \\ \| \\ :^-\ddot{O}-C-\ddot{O}:^- \end{array}$$

$$\begin{array}{cc} H & :\ddot{O}: \\ | & \| \\ H-C-C-\ddot{O}:^- \\ | \\ H \end{array}$$

$$:\ddot{Cl}:^-$$

1.29   Why are the following molecular formulas impossible?

(a) CH_5 If carbon were bonded to five hydrogen atoms, the octet rule would be violated. Furthermore, each hydrogen atom can only bond with one other atom, so there is no stable connectivity of the formula CH_5.

(b) C_2H_7 Each hydrogen atom can only bond with one other atom, so a single hydrogen atom cannot be bonded to both carbons. Thus, the two carbon atoms must be bonded to each other, meaning that each carbon can only accommodate three more bonds (total of six). Obviously, it is not possible to bond seven hydrogen atoms when only six sites are available.

(c) H_2^{2+} The proton, H^+, does not contain any electrons, so it is not possible to form a bond between two protons.

(d) HN^{3-} The compound has a total of nine electrons. Because hydrogen can only form one bond with nitrogen and cannot have any nonbonding electrons, nitrogen would exceed its octet.

1.30   Following the rule that each atom of carbon, oxygen, and nitrogen reacts to achieve a complete outer shell of eight valence electrons, add unshared pairs of electrons as necessary to complete the valence shell of each atom in these ions. Then, assign formal charges as appropriate.

(a) H–Ö–C–Ö: (b) H–C–C–Ö: (c) H–C–C:⁻   (d) H–N–C–Ö:

1.31   The following Lewis structures show all valence electrons. Assign formal charges in each structure as appropriate.

With practice, some general trends will become apparent. For instance, O with one bond and six unshared electrons has a negative formal charge, O with three bonds and two unshared electrons has a positive formal charge, N with four bonds has a positive formal charge, C with three bonds and no unshared electrons has a positive formal charge, and C with three bonds and two unshared electrons has a negative formal charge.

(a) H–C–C–C–H   (b) H–N–C=C–H   (c) H–C–Ö–H   (d) H–C⁺

1.32   Each compound contains both ionic and covalent bonds. Draw a Lewis structure for each, and show by charges which bonds are ionic and by dashes which bonds are covalent.

(a) NaOH   (b) NaHCO₃   (c) NH₄Cl

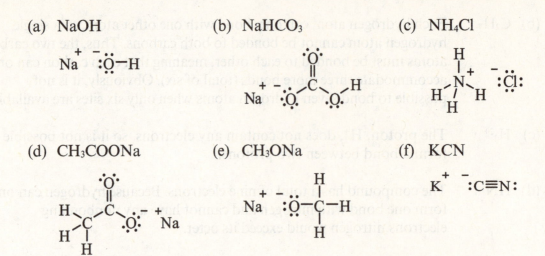

(d) CH₃COONa   (e) CH₃ONa   (f) KCN

1.33 Silver and oxygen can form a stable compound. Predict the formula of this compound and determine whether the compound consists of ionic or covalent bonds.

The most common oxidation state of silver is +1. If oxygen gained two electrons to obtain a full octet, it would have a −2 charge. The compound is silver oxide, Ag_2O. The electronegativity difference between silver and oxygen is 1.6, so the compound consists of polar covalent bonds.

1.34 Draw Lewis structures for the following molecule and ions.

(a) NH₃   (b) NH₄⁺   (c) NH₂⁻   (d) CH₃⁺

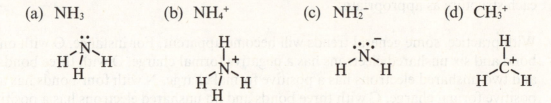

1.35 Which statement is true about electronegativity?

(a) Electronegativity increases from left to right in a period of the Periodic Table.
(b) Electronegativity increases from top to bottom in a column of the Periodic Table.
(c) Hydrogen, the element with the lowest atomic number, has the smallest electronegativity.
(d) The higher the atomic number of an element, the greater its electronegativity.

Electronegativity increases from left to right across a period, and from bottom to top within a group, so statement (a) is correct and (b) is incorrect. There is no correlation between atomic number and electronegativity, so (c) and (d) are incorrect.

1.36 Why does fluorine, the element in the upper right corner of the Periodic Table, have the largest electronegativity of any element?

Two parameters that lead to the increase of electronegativity are the increase of positive charge on the nucleus and the decrease of atomic radius (which results in a decrease of distance between the nucleus and the electrons in the valence shell). Both of these parameters exert their maximum effect in the case of fluorine.

1.37 Arrange the single covalent bonds within each set in order of increasing polarity.

Covalent bonds increase in polarity as the electronegativity difference (shown in parentheses) between the bonded atoms increases.

(a) C–H (0.4) < N–H (0.9) < O–H (1.4) (b) C–I (0) < C–H (0.4) < C–Cl (0.5)

(c) C–C (0) < C–N (0.5) < C–O (1.0) (d) C–Hg (0.6) < C–Mg (1.3) < C–Li (1.5)

1.38 Using the values of electronegativity given in Table 1.4, predict which indicated bond in each set is more polar and, using the symbols $\delta+$ and $\delta-$, show the direction of its polarity.

The bond that is more polar has the greatest difference in electronegativity between the bonded atoms.

(a) CH_3—OH or $\overset{\delta-}{\mathbf{CH_3O}}$—$\overset{\delta+}{\mathbf{H}}$ (e) H_2C=NH or $\overset{\delta+}{\mathbf{H_2C}}$=$\overset{\delta-}{\mathbf{O}}$

(b) $\overset{\delta+}{\mathbf{H}}$—$\overset{\delta-}{\mathbf{NH_2}}$ or CH_3—NH_2 (f) $\overset{\delta+}{\mathbf{H_2B}}$—$\overset{\delta-}{\mathbf{H}}$ or F_2B—F

(c) CH_3—SH or $\overset{\delta-}{\mathbf{CH_3S}}$—$\overset{\delta+}{\mathbf{H}}$ (g) $\overset{\delta+}{\mathbf{H_2C}}$=$\overset{\delta-}{\mathbf{O}}$ or H_2C=S

(d) CH_3—F or $\overset{\delta+}{\mathbf{H}}$—$\overset{\delta-}{\mathbf{F}}$ (h) CH_3—MgBr or $\overset{\delta-}{\mathbf{CH_3}}$—$\overset{\delta+}{\mathbf{Li}}$

1.39 Identify the most polar bond in each molecule.

The most polar bond is the one that has the largest difference in electronegativity between the bonded atoms.

(a) $HSCH_2CH_2OH$ The most polar bond is the O–H bond.

(b) $CHCl_2F$ The most polar bond is the C–F bond.

(c) $HOCH_2CH_2NH_2$ The most polar bond is the O–H bond.

(d) CH₃OCH₂OH   The most polar bond is the O−H bond.

(e) HOCl   The most polar bond is the O−H bond.

(f) CH₃NCHCHO   The most polar bond is the C=O bond.

1.40 Predict whether the carbon-metal bond in these organometallic compounds is nonpolar covalent, polar covalent, or ionic. For each polar covalent bond, show its direction of polarity using the symbols δ+ and δ−.

Based on electronegativity differences, all carbon-metal and metal-halogen bonds are polar covalent bonds. All three of these organometallic compounds are of exceptional relevance: tetraethyllead was used as a gasoline additive and is still used in some types of aviation fuel, methylmagnesium chloride is an important reagent in organic synthesis, and dimethylmercury is an extremely potent neurotoxin.

$$\text{(a)}\quad \overset{\delta-}{CH_3CH_2}-\overset{\delta+}{Pb}-\overset{\delta-}{CH_2CH_3} \qquad \text{(b)}\quad \overset{\delta-}{CH_3}-\overset{\delta+}{Mg}-\overset{\delta-}{Cl} \qquad \text{(c)}\quad \overset{\delta-}{CH_3}-\overset{\delta+}{Hg}-\overset{\delta-}{CH_3}$$

(with the ethyl groups $\overset{\delta-}{CH_2CH_3}$ above and below Pb in (a))

Bond Angles and Shapes of Molecules

1.41 Using VSEPR, predict bond angles about each highlighted atom:

(a) 109.5°   109.5°   (b) 120°   (c) 180°

(d) 120°   109.5°   (e) 109.5°   (f) 120°

1.42 Using VSEPR, predict bond angles about each atom of carbon, nitrogen, and oxygen in these molecules. (*Hint*: First add unshared pairs of electrons as necessary to complete the valence shell of each atom, and then make your predictions of bond angles.)

Unless otherwise indicated, all carbon, nitrogen, and oxygen bonds are 109.5°.

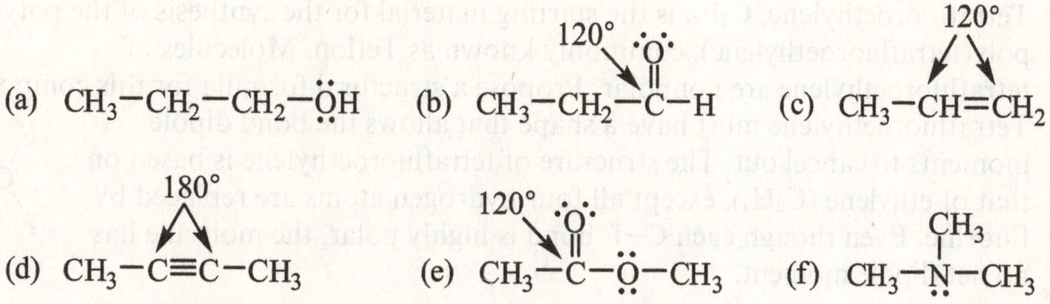

(a) $CH_3-CH_2-CH_2-\ddot{O}H$

(b) $CH_3-CH_2-\overset{\overset{\displaystyle 120°}{\underset{\displaystyle \ddot{O}\ddot{}}{\|}}}{C}-H$

(c) $CH_3-\overset{\displaystyle 120°}{CH}=CH_2$

(d) $CH_3-\overset{\displaystyle 180°}{C}\equiv C-CH_3$

(e) $CH_3-\overset{\overset{\displaystyle 120°}{\underset{\displaystyle \ddot{O}\ddot{}}{\|}}}{C}-\ddot{O}-CH_3$

(f) $CH_3-\underset{\displaystyle \ddot{}}{\overset{\displaystyle CH_3}{N}}-CH_3$

1.43 Silicon is immediately below carbon in the Periodic Table. Predict the C−Si−C bond angle in tetramethylsilane, (CH$_3$)$_4$Si.

Silicon and carbon are both in the same group of the Periodic Table. Like carbon, silicon can accommodate up to four regions of electron density, resulting in a bond angle of 109.5° at the silicon atom.

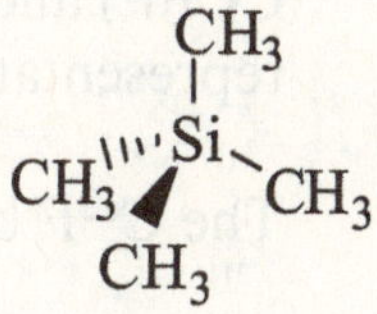

$$CH_3\overset{\displaystyle CH_3}{\underset{\displaystyle CH_3}{\overset{|}{\underset{}{Si}}}}CH_3$$

Polar and Nonpolar Molecules

1.44 Draw a three-dimensional representation for each molecule. Indicate which molecules are polar and the direction of its polarity.

In order for a molecule to be polar, it must have one or more polar bonds and also a geometric shape in which the dipole moments do not cancel out. All molecules, except CCl$_4$, have a net dipole moment and are thus polar, as shown below.

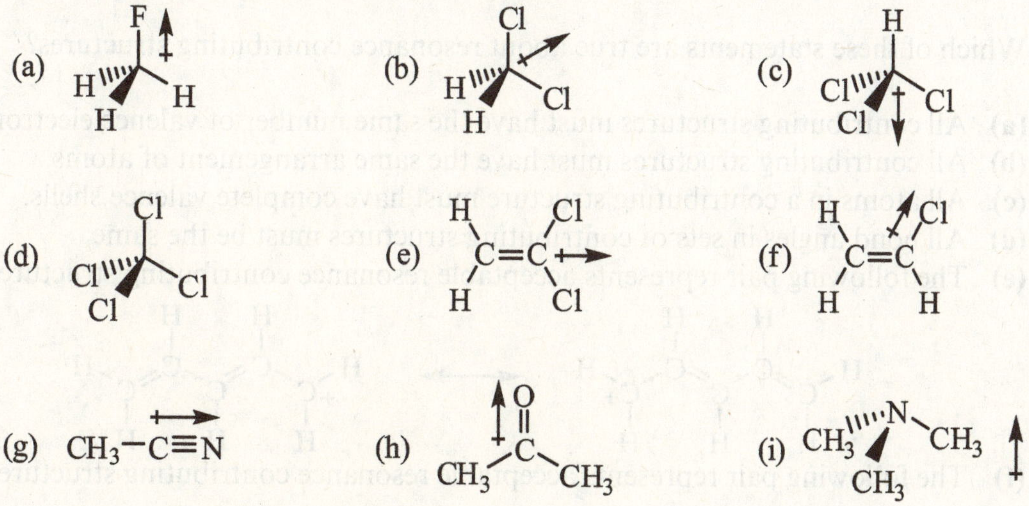

(a) H$_{\text{\tiny ////}}$ C with F (up), H, H

(b) H$_{\text{\tiny ////}}$ C with Cl (up-right), Cl, H

(c) Cl$_{\text{\tiny ////}}$ C with H (up), Cl, Cl

(d) Cl$_{\text{\tiny ////}}$ C with Cl (up), Cl, Cl

(e) C=C : H, Cl (top); H, Cl (bottom)

(f) C=C : H, Cl (top); H, H (bottom)

(g) $CH_3-C\equiv N$

(h) $CH_3\overset{\overset{\displaystyle O}{\|}}{C}CH_3$

(i) $CH_3{\text{\tiny ////}}\overset{\displaystyle N}{\underset{\displaystyle CH_3}{}}CH_3$

1.45 Tetrafluoroethylene, C_2F_4, is the starting material for the synthesis of the polymer poly(tetrafluoroethylene), commonly known as Teflon. Molecules of tetrafluoroethylene are nonpolar. Propose a structural formula for this compound. Tetrafluoroethylene must have a shape that allows the bond dipole moments to cancel out. The structure of tetrafluoroethylene is based on that of ethylene (C_2H_4), except all four hydrogen atoms are replaced by fluorine. Even though each C−F bond is highly polar, the molecule has no net dipole moment.

1.46 Until several years ago, two chlorofluorocarbons (CFCs) most widely used as heat transfer media for refrigeration systems were Freon-11 (trichlorofluoromethane, CCl_3F) and Freon-12 (dichlorodifluoromethane, CCl_2F_2). Draw a three-dimensional representation of each molecule, and indicate the direction of its polarity.

The C−F bond is more polar than the C−Cl bond. Thus, the C−F dipole is stronger.

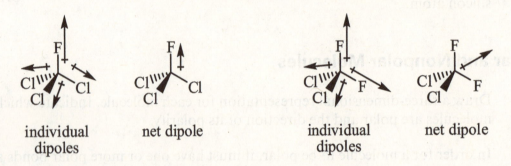

individual   net dipole   individual   net dipole
dipoles   dipoles

Contributing Structures

1.47 Which of these statements are true about resonance contributing structures?

(a) All contributing structures must have the same number of valence electrons.
(b) All contributing structures must have the same arrangement of atoms.
(c) All atoms in a contributing structure must have complete valence shells.
(d) All bond angles in sets of contributing structures must be the same.
(e) The following pair represents acceptable resonance contributing structures:

(f) The following pair represents acceptable resonance contributing structures:

(g) The following pair represents acceptable resonance contributing structures:

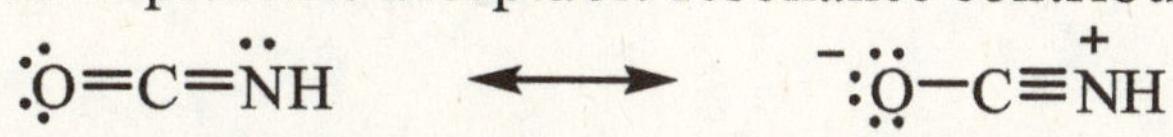

$$:\ddot{O}=C=\ddot{N}H \longleftrightarrow {}^{-}:\ddot{O}-C\equiv\overset{+}{N}H$$

Contributing structures only involve the movement of electrons, not the atomic nuclei, so (b) and (d) are true. The total number of electrons remains the same, so (a) is true. The movement of electrons often leaves one or more atoms without a filled valence shell, so (c) is false. Statements (e) through (g) are true; they involve only the movement of electrons, and each contributing structure must be a correct Lewis structure.

1.48 Draw the resonance contributing structure indicated by the curved arrow(s), and assign formal charges as appropriate:

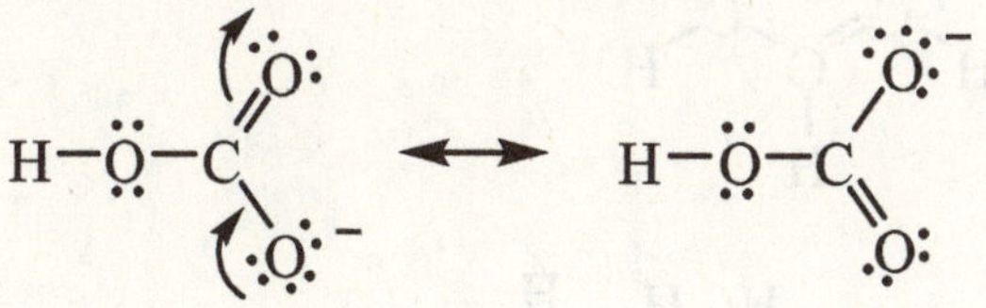

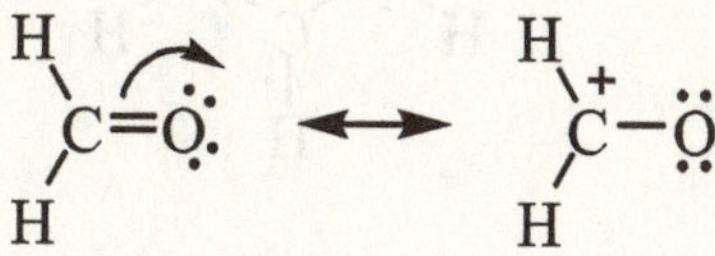

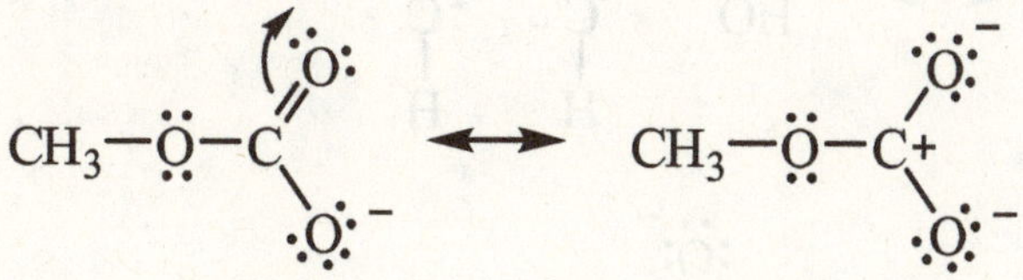

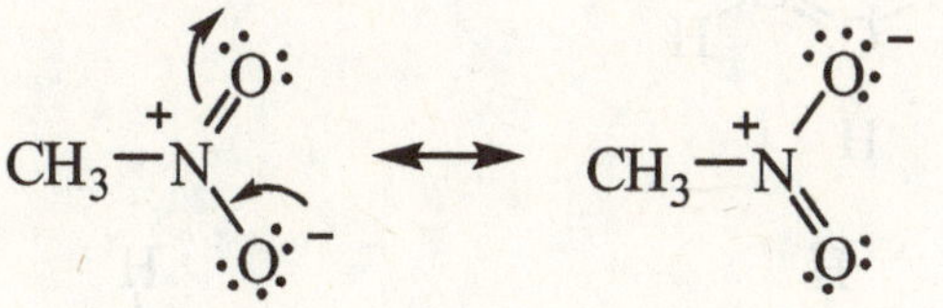

1.49 Using the VSEPR model, predict the bond angles about the carbon atom in each pair of contributing structures in Problem 1.48. In what way do the bond angles change from one contributing structure to the other?

In every pair of contributing structures in Problem 1.48, the bond angle of the carbon atom involved in resonance remains at 120° (three regions of electron density) and does not change from one contributing structure to another. The carbon atom in the CH₃ group of pairs (c) and (d) has a bond angle of 109.5° (four regions of electron density).

1.50   Draw acceptable resonance contributing structure(s) for each of the compounds shown.

(a)

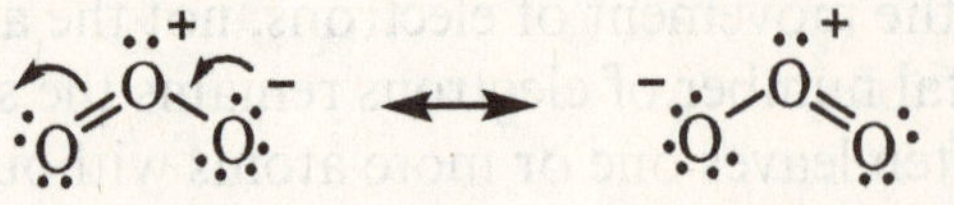

(b)

(c)

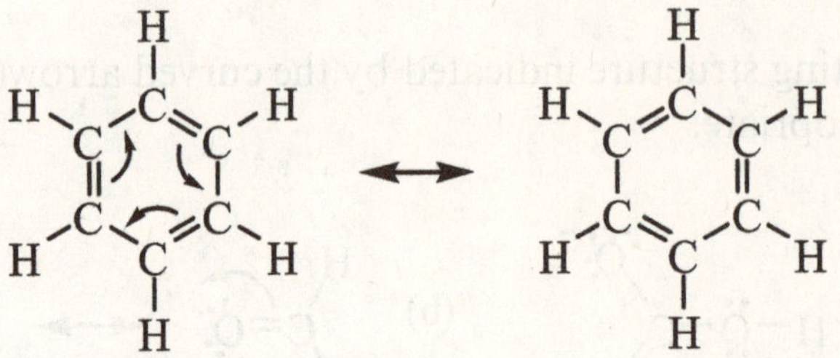

(d)

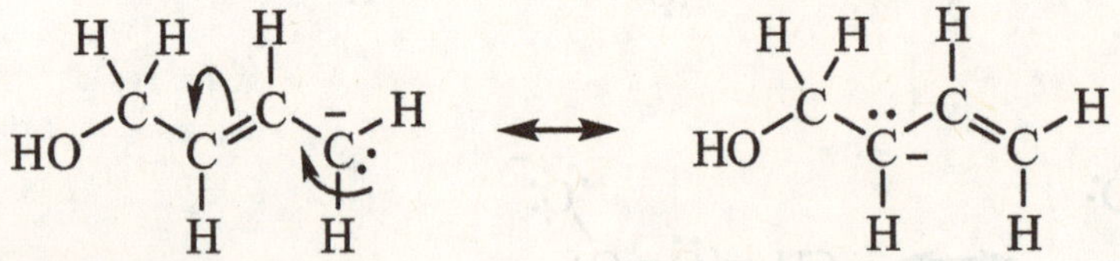

(e)

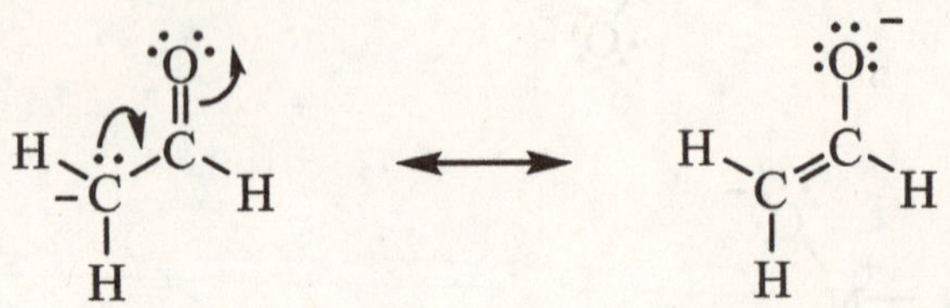

(f)

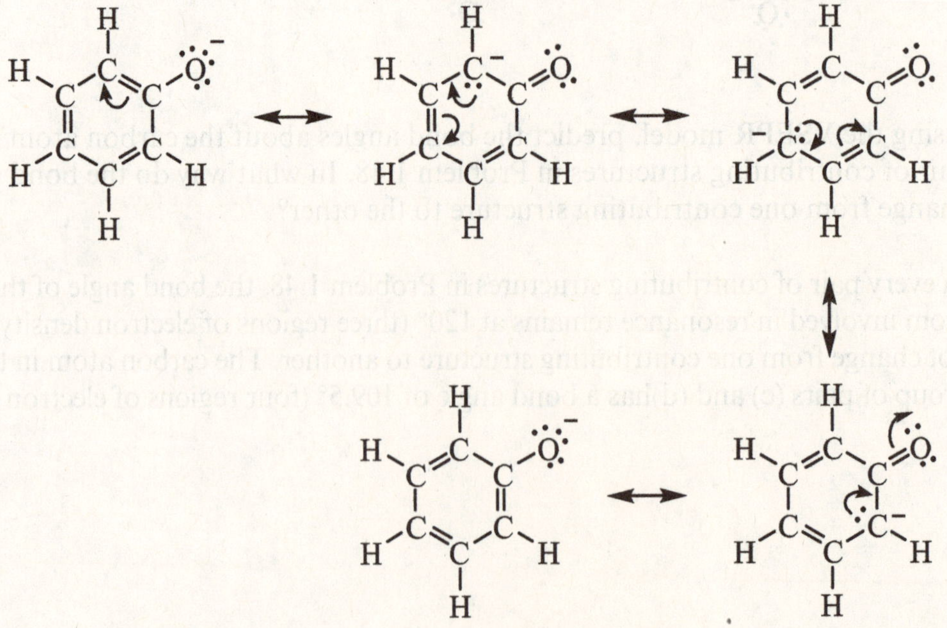

Hybridization of Atomic Orbitals

1.51   State the hybridization of each highlighted atom:

Hybridization can be assigned by examining the number of regions of electron density associated with the atom of interest. Don't forget the nonbonding electrons that are implied. Remember that a nonbonding pair, single bond, or multiple bond (whether a double or triple bond) each count as one region of electron density.

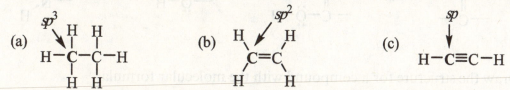

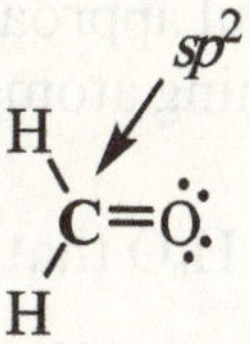

1.52   Describe each highlighted bond by indicating the type of bond(s) and the hybridization of the highlighted atoms.

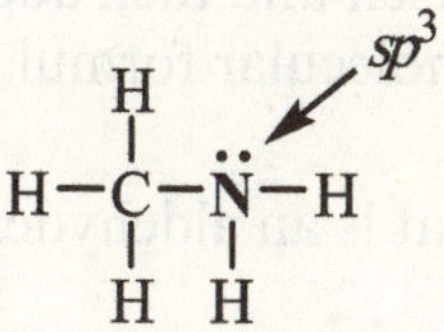

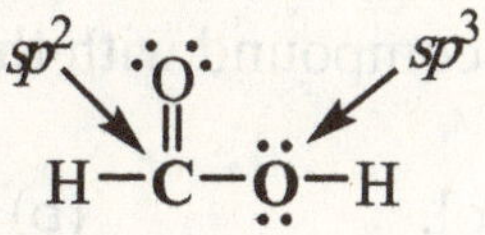

σ bonds are formed by the overlap of hybrid orbitals (with the exception of hydrogen, which uses the $1s$ atomic orbital).π bonds are formed by the overlap of two p atomic orbitals. A double bond consists of one σ bond and one π bond, while a triple bond consists of one σ bond and two π bonds.

(a)

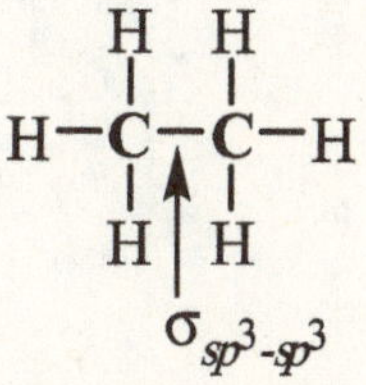

$$\begin{array}{ccc} H & H \\ | & | \\ H-C-C-H \\ | & | \\ H & H \end{array}$$

$\sigma_{sp^3-sp^3}$

(b)

$$\begin{array}{c} H \\ | \\ H-C-\ddot{O}-H \\ | \\ H \end{array}$$

$\sigma_{sp^3-sp^3}$

(c)

$$\begin{array}{cc} H & \\ | & \ddot{\,} \\ H-C-N-H \\ | & | \\ H & H \end{array}$$

$\sigma_{sp^3-sp^3}$

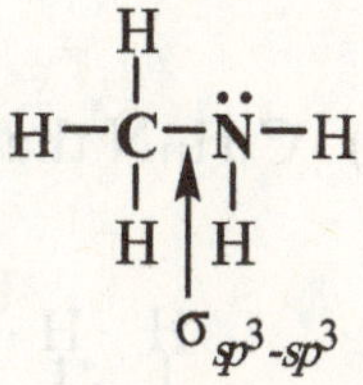

(d)

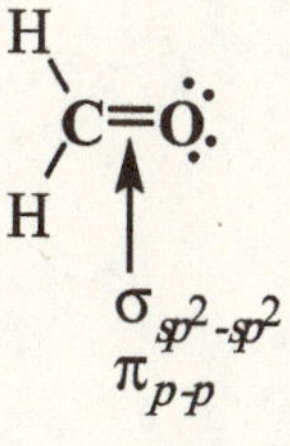

$$\begin{array}{c} H \\ \diagdown \\ C=\ddot{O}: \\ \diagup \\ H \end{array}$$

$\sigma_{sp^2-sp^2}$
π_{p-p}

(e)

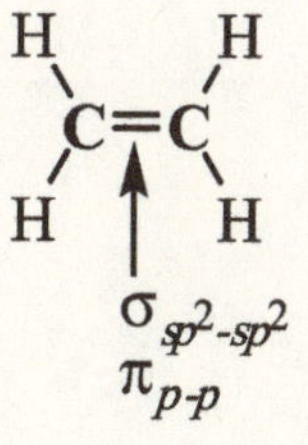

$$\begin{array}{ccc} H & & H \\ \diagdown & & \diagup \\ & C=C & \\ \diagup & & \diagdown \\ H & & H \end{array}$$

$\sigma_{sp^2-sp^2}$
π_{p-p}

(f)

$$\begin{array}{c} :\ddot{O}: \\ || \\ H-C-\ddot{O}-H \end{array}$$

$\sigma_{sp^2-sp^3}$

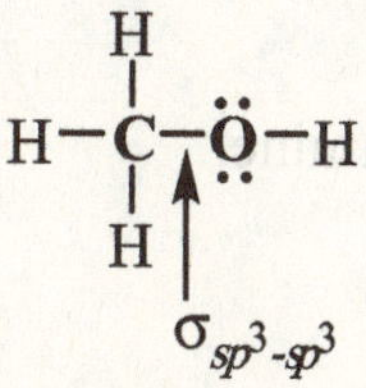

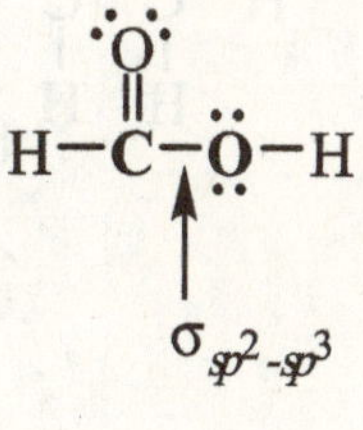

Functional Groups

1.53   Draw Lewis structures for these functional groups. Be certain to show all valence electrons on each:

   (a) Carbonyl   (b) Carboxyl 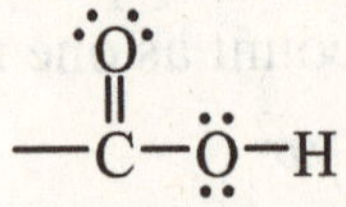   (c) Hydroxyl 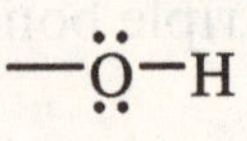   (d) Primary amino

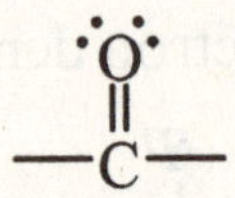

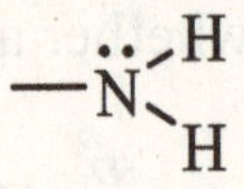

1.54   Draw the structure for a compound with the molecular formula

A useful approach is to first draw the functional group itself and then add in the remaining atoms to form a compound with the correct molecular formula.

   (a) C_2H_6O that is an alcohol.   (b) C_3H_6O that is an aldehyde.

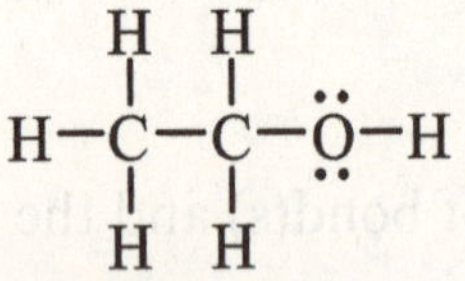

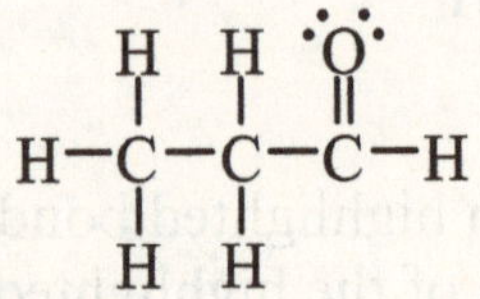

   (c) C_3H_6O that is a ketone. 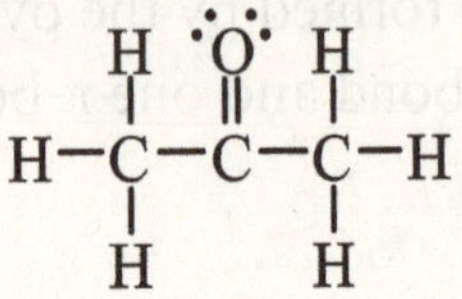   (d) $C_3H_6O_2$ that is a carboxylic acid.

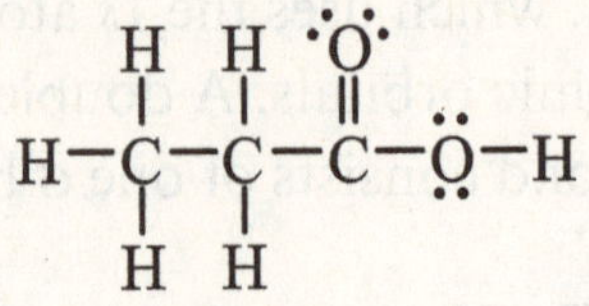

   (e) $C_4H_{11}N$ that is a tertiary amine.

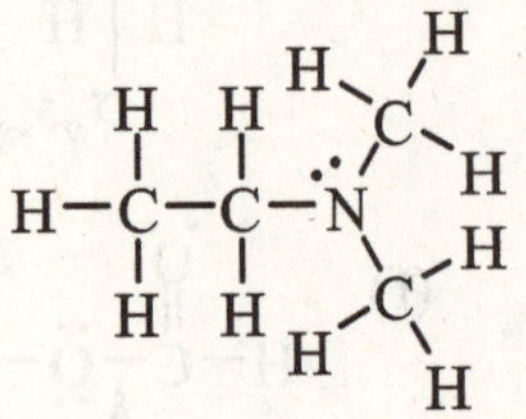

1.55 Draw condensed structural formulas for all compounds with the molecular formula C_4H_8O that contain

(a) a carbonyl group. (There are two aldehydes and one ketone.)

$$\underset{CH_3CH_2CH_2\overset{\displaystyle O}{\overset{\|}{C}}H}{} \qquad \underset{\underset{CH_3}{|}}{CH_3\overset{\displaystyle O}{\overset{\|}{C}H}CH} \qquad CH_3\overset{\displaystyle O}{\overset{\|}{C}}CH_2CH_3$$

(b) a carbon-carbon double bond and a hydroxyl group. (There are eight.)

$$CH_2=CHCH_2CH_2OH \qquad CH_3CH=CHCH_2OH \qquad CH_3CH_2CH=CHOH$$

$$\underset{CH_2=CH\overset{\displaystyle |}{\overset{OH}{C}}HCH_3}{} \qquad \underset{CH_2=\overset{\displaystyle |}{\overset{OH}{C}}CH_2CH_3}{} \qquad \underset{CH_3\overset{\displaystyle |}{\overset{OH}{C}}=CHCH_3}{}$$

$$\underset{CH_2=\overset{\displaystyle |}{\overset{CH_3}{C}}CH_2OH}{} \qquad \underset{CH_3\overset{\displaystyle |}{\overset{CH_3}{C}}=CHOH}{}$$

1.56 Draw structural formulas for

(a) the eight alcohols with the molecular formula $C_5H_{12}O$.

$$CH_3CH_2CH_2CH_2CH_2OH \qquad \underset{CH_3CH_2CH_2\overset{\displaystyle |}{\overset{OH}{C}}HCH_3}{} \qquad \underset{CH_3CH_2\overset{\displaystyle |}{\overset{OH}{C}}HCH_2CH_3}{}$$

$$\underset{CH_3\overset{\displaystyle |}{\overset{CH_3}{C}}HCH_2CH_2OH}{} \qquad \underset{\underset{OH}{|}}{\underset{CH_3\overset{\displaystyle |}{\overset{CH_3}{C}}HCHCH_3}{}} \qquad \underset{\underset{OH}{|}}{\underset{CH_3\overset{\displaystyle |}{\overset{CH_3}{C}}CH_2CH_3}{}}$$

$$\underset{CH_3CH_2\overset{\displaystyle |}{\overset{CH_3}{C}}HCH_2OH}{} \qquad \underset{\underset{CH_3}{|}}{\underset{CH_3\overset{\displaystyle |}{\overset{CH_3}{C}}CH_2OH}{}}$$

(b) the eight aldehydes with the molecular formula $C_6H_{12}O$.

$$\underset{\text{O}}{\overset{\text{O}}{\parallel}}$$

$$CH_3CH_2CH_2CH_2CH_2\overset{\text{O}}{\overset{\parallel}{CH}} \qquad CH_3\overset{CH_3}{\underset{|}{CH}}CH_2CH_2\overset{\text{O}}{\overset{\parallel}{CH}} \qquad CH_3CH_2\overset{CH_3}{\underset{|}{CH}}CH_2\overset{\text{O}}{\overset{\parallel}{CH}}$$

$$CH_3CH_2CH_2\overset{\text{O}}{\overset{\parallel}{CH}}CH \qquad CH_3\overset{CH_3}{\underset{|}{CH}}\overset{\text{O}}{\overset{\parallel}{CH}}CH \qquad CH_3\overset{CH_3}{\underset{|}{C}}CH_2\overset{\text{O}}{\overset{\parallel}{CH}}$$
$$\quad\quad\quad\underset{CH_3}{|} \qquad\qquad\qquad \underset{CH_3}{|} \qquad\qquad\qquad \underset{CH_3}{|}$$

$$CH_3CH_2\overset{CH_3}{\underset{|}{C}}\text{—}\overset{\text{O}}{\overset{\parallel}{CH}} \qquad CH_3CH_2\overset{\text{O}}{\overset{\parallel}{CH}}CH$$
$$\qquad\quad\underset{CH_3}{|} \qquad\qquad\qquad\underset{CH_2CH_3}{|}$$

(c) the six ketones with the molecular formula $C_6H_{12}O$.

$$CH_3CH_2CH_2CH_2\overset{\text{O}}{\overset{\parallel}{C}}CH_3 \qquad CH_3\overset{O}{\overset{\parallel}{CHCH_2C}}CH_3 \qquad CH_3CH_2\overset{\text{O}}{\overset{\parallel}{CH}}CCH_3$$
$$\qquad\qquad\qquad\qquad\qquad\underset{CH_3}{|} \qquad\qquad\qquad\underset{CH_3}{|}$$

$$CH_3CH_2CH_2\overset{\text{O}}{\overset{\parallel}{C}}CH_2CH_3 \qquad CH_3\overset{\text{O}}{\overset{\parallel}{CH}}CCH_2CH_3 \qquad CH_3\overset{CH_3}{\underset{|}{C}}\text{—}\overset{\text{O}}{\overset{\parallel}{C}}CH_3$$
$$\qquad\qquad\qquad\qquad\qquad\qquad\underset{CH_3}{|} \qquad\qquad\qquad\underset{CH_3}{|}$$

(d) the eight carboxylic acids with the molecular formula $C_6H_{12}O_2$.

$$CH_3CH_2CH_2CH_2CH_2\overset{\text{O}}{\overset{\parallel}{C}}OH \qquad CH_3\overset{CH_3}{\underset{|}{CH}}CH_2CH_2\overset{\text{O}}{\overset{\parallel}{C}}OH \qquad CH_3CH_2\overset{CH_3}{\underset{|}{CH}}CH_2\overset{\text{O}}{\overset{\parallel}{C}}OH$$

$$CH_3CH_2CH_2\overset{\text{O}}{\overset{\parallel}{CH}}COH \qquad CH_3\overset{CH_3}{\underset{|}{CH}}\overset{\text{O}}{\overset{\parallel}{CH}}COH \qquad CH_3\overset{CH_3}{\underset{|}{C}}CH_2\overset{\text{O}}{\overset{\parallel}{C}}OH$$
$$\qquad\underset{CH_3}{|} \qquad\qquad\qquad \underset{CH_3}{|} \qquad\qquad\qquad \underset{CH_3}{|}$$

$$CH_3CH_2\overset{CH_3}{\underset{|}{C}}\text{—}\overset{\text{O}}{\overset{\parallel}{C}}OH \qquad CH_3CH_2\overset{\text{O}}{\overset{\parallel}{CH}}COH$$
$$\qquad\quad\underset{CH_3}{|} \qquad\qquad\qquad\underset{CH_2CH_3}{|}$$

(e) the three tertiary amines with the molecular formula $C_5H_{13}N$.

A useful strategy is to first draw a 3° amino group (nitrogen atom bonded to three carbons) and then add the remaining two carbon atoms.

$$C-\underset{\underset{\displaystyle C}{|}}{N}-C \qquad\longrightarrow\qquad CH_3-\underset{\underset{\displaystyle CH_3}{|}}{N}-CH_2CH_2CH_3 \qquad CH_3-\underset{\underset{\displaystyle CH_2CH_3}{|}}{N}-CH_2CH_3$$

tertiary
amino group

$$CH_3-\underset{\underset{\displaystyle CH_3}{|}}{N}-\underset{\underset{\displaystyle CH_3}{|}}{C}HCH_3$$

1.57 Identify the functional groups in each compound (we study each compound in more detail in the indicated section):

(a)   hydroxyl group (2°)

$$CH_3-\underset{\boxed{OH}}{\underset{|}{C}}H-\underset{\boxed{\overset{\displaystyle O}{\|}}}{C}-OH$$

carboxyl group

(b)   hydroxyl groups (1°)

$$\boxed{HO}-CH_2-CH_2-\boxed{OH}$$

(c)   amino group (1°)

$$CH_3-\underset{\boxed{NH_2}}{\underset{|}{C}}H-\underset{\boxed{\overset{\displaystyle O}{\|}}}{C}-OH$$

carboxyl group

carbonyl group
(ketone)

(d)   hydroxyl groups (1° and 2°)

$$\boxed{HO}-CH_2-\underset{\boxed{OH}}{\underset{|}{C}}H-\underset{\boxed{\overset{\displaystyle O}{\|}}}{C}-H$$

carbonyl group
(aldehyde)

(e)

$$CH_3-\boxed{\overset{\displaystyle O}{\underset{\|}{C}}}-CH_2-\boxed{\overset{\displaystyle O}{\underset{\|}{C}}}-OH$$

carboxyl group

(f)   amino groups (1°)

$$\boxed{H_2N}-CH_2CH_2CH_2CH_2CH_2CH_2-\boxed{NH_2}$$

1.58   Dihydroxyacetone, $C_3H_6O_3$, the active ingredient in many sunless tanning lotions, contains two 1° hydroxyl groups, each on a different carbon, and one ketone group. Draw a structural formula for dihydroxyacetone.

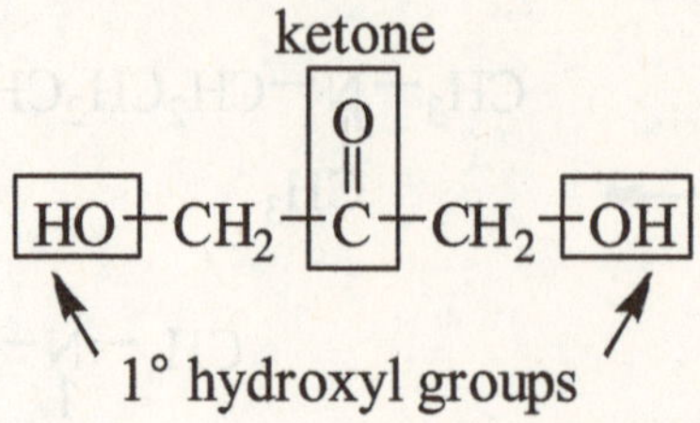

ketone

$$HO \!-\! CH_2 \!-\! C \!-\! CH_2 \!-\! OH$$

1° hydroxyl groups

1.59   Propylene glycol, $C_3H_8O_2$, commonly used in airplane deicers, contains a 1° alcohol and a 2° alcohol. Draw a structural formula for propylene glycol.

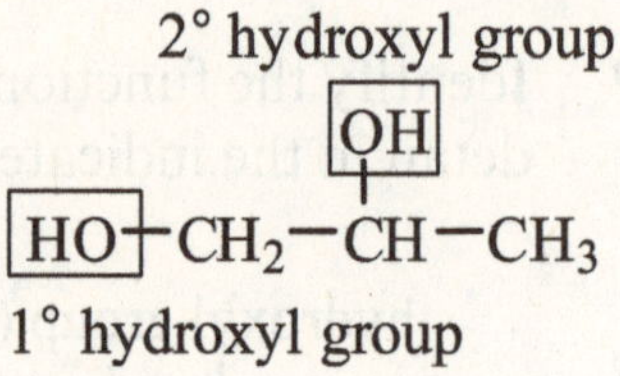

2° hydroxyl group

$$HO \!-\! CH_2 \!-\! CH \!-\! CH_3$$

1° hydroxyl group

1.60   Ephedrine is a molecule found in the dietary supplement ephedra, which has been linked to adverse health reactions such as heart attacks, strokes, and heart palpitations. The use of ephedra as dietary supplements is now banned by the FDA.

(a)   Identify at least two functional groups in ephedrine.

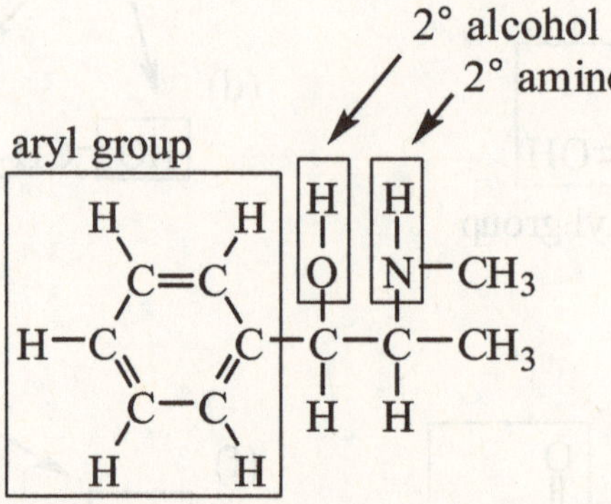

2° alcohol
2° amine
aryl group

(b)   Would you predict ephedrine to be polar or nonpolar?

Ephedrine is expected to be a polar molecule. It has polar bonds (C–O, C–N, N–H, and O–H), but the lack of symmetry in the molecule prevents the dipole moments of the individual bonds from cancelling out.

1.61 Ozone (O_3) and carbon dioxide (CO_2) are both known as greenhouse gases. Compare and contrast their shapes, and indicate the hybridization of each atom in the two molecules.

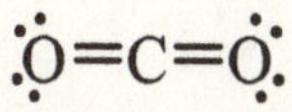

carbon dioxide 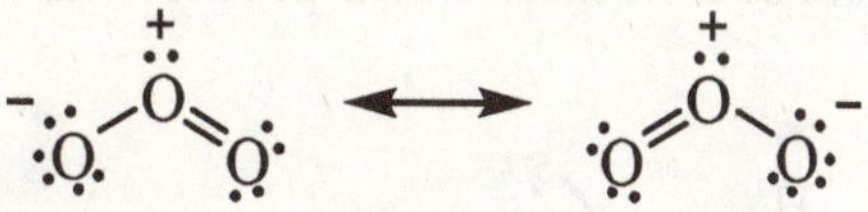   ozone

The carbon atom of CO_2 contains two regions of electron density. As a result, the carbon atom is *sp*-hybridized and CO_2 is a linear molecule. The two oxygen atoms of CO_2 each contain three regions of electron density, so they are sp^2-hybridized.

Ozone, which has three regions of electron density (one of which is a nonbonding pair) on the central oxygen atom, is a bent molecule. That oxygen atom is sp^2-hybridized. How about the hybridizations of the two terminal oxygen atoms? Looking at any one of the two contributing structures, one terminal oxygen atom has four regions of electron density while the other terminal oxygen atom has only three. Thus, for these terminal oxygen atoms, is one of them sp^3 while the other is sp^2? No. As a result of resonance, both terminal oxygen atoms are identical (sp^2). It is important to recognize that contributing structures are only Lewis diagrams that show the possible arrangement of electrons and that the actual structure of ozone is best represented by a "hybrid" of the contributing structures. In that sense, each structure contributes to the hybrid structure.

1.62 In the lower atmosphere that is also contaminated with unburned hydrocarbons, NO_2 participates in a series of reactions. One product of these reactions is peroxyacetyl nitrate (PAN). The connectivity of the atoms in PAN appears below.

(a) Determine the number of valence electrons in this molecule, and then complete its Lewis structure.

(b) Give the approximate values of the bond angles around each atom indicated with an arrow.

46 valence electrons

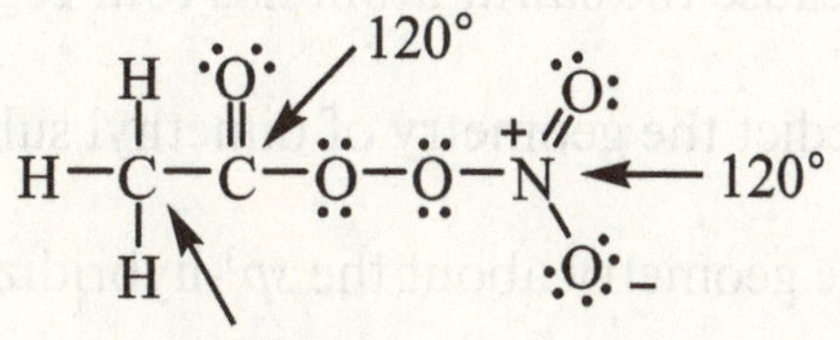

Looking Ahead

1.63   Allene, C_3H_4, has the structural formula $H_2C=C=CH_2$. Determine the hybridization of each carbon in allene and predict the shape of the molecule.

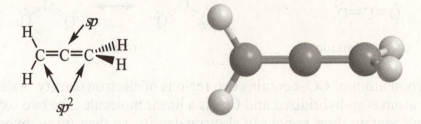

The two terminal carbon atoms are sp^2-hybridized. The central carbon atom is sp-hybridized, so the carbon chain itself is linear. One of the two remaining p orbitals on the central carbon is used to form the π bond with the carbon on the left, while the other p orbital is used to form the π bond with the carbon on the right.

Although the carbon chain is linear and the two terminal carbon atoms are trigonal planar, the molecule is actually not flat. The two p orbitals on the central carbon are perpendicular to each other and, as a result, so are the two π bonds. The plane formed by the CH_2 group on the left is perpendicular to the one on the right.

1.64   Dimethyl sulfoxide, $(CH_3)_2SO$, is a common solvent used in organic chemistry.

(a)   Write a Lewis structure for dimethyl sulfoxide.

$$\begin{array}{c} \quad\quad\ddot{\underset{\displaystyle\|}{O}}: \\ H\quad\quad\quad H \\ |\quad\ \ |\quad\ \ | \\ H-C-\ddot{S}-C-H \\ |\quad\quad\quad| \\ H\quad\quad\quad H \end{array}$$

(b)   Predict the hybridization of the sulfur atom in the molecule.

Because the sulfur atom has four regions of electron density, it is sp^3-hybridized.

(c)   Predict the geometry of dimethyl sulfoxide.

The geometry about the sp^3-hybridized carbon and sulfur atoms is tetrahedral.

(d)   Is dimethyl sulfoxide a polar or a nonpolar molecule.

Dimethyl sulfoxide contains a polar S=O bond, and the molecule is also polar.

1.65 In Chapter 5, we study a group of organic cations called carbocations. Following is the structure of one such carbocation, the *tert*-butyl cation.

tert-butyl cation
$$CH_3 \quad \overset{+}{C}-CH_3 \quad CH_3$$

(a) How many electrons are in the valence shell of the carbon bearing the positive charge?

Because this carbon has a +1 formal charge, there are six electrons in its valence shell. All six electrons are shared, and there are no unshared (nonbonding) electrons.

(b) Predict the bond angles about this carbon.

With only three regions of electron density, the bond angles are 120°.

(c) Given the bond angles you predicted in (b), what hybridization do you predict for this carbon.

Atoms with three regions of electron density, and thus bond angles of 120°, are sp^2-hybridized. Its geometry is trigonal planar, and the remaining unhybridized p orbital lies perpendicular to the trigonal plane.

1.66 We also study the isopropyl cation, $(CH_3)_2CH^+$, in Chapter 5.

(a) Write a Lewis structure for this cation. Use a plus sign to show the location of the positive charge.

isopropyl cation
$$CH_3 \quad \overset{+}{C}-H \quad CH_3$$

(b) How many electrons are in the valence shell of the carbon bearing the positive charge?

Like the *tert*-butyl cation discussed in Problem 1.65, the positively charged carbon in the isopropyl cation also has six valence electrons.

(c) Use the VSEPR model to predict all bond angles about the carbon bearing the positive charge.

With three regions of electron density, the bond angles about the positively charged carbon are 120°.

(d) Describe the hybridization of each carbon in this cation.

The positively charged carbon, which has three regions of electron density, is *sp²*-hybridized, while the carbon atoms of the two methyl groups are *sp³*-hybridized (four regions of electron density).

1.67 In Chapter 9, we study benzene, C_6H_6, and its derivatives.

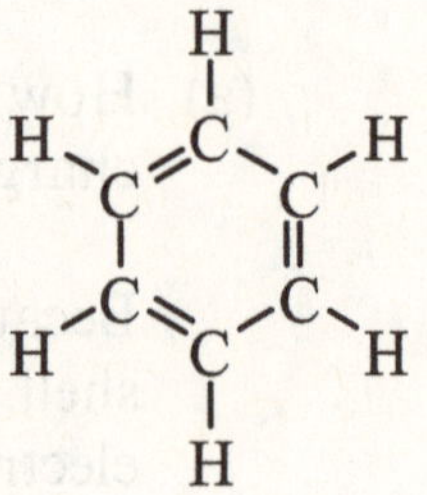

(a) Predict each H−C−C and each C−C−C bond angle on benzene.

Every carbon atom has three regions of electron density, so the bond angles are predicted to be 120° (trigonal planar).

(b) State the hybridization of each carbon in benzene.

With three regions of electron density (trigonal planar), the hybridization is *sp²*.

(c) Predict the shape of a benzene molecule.

All six carbon atoms are *sp²*-hybridized, so all carbon and hydrogen atoms lie in the same plane, resulting in a flat molecule.

1.68 Explain why *all* the carbon-carbon bonds in benzene are equal in length.

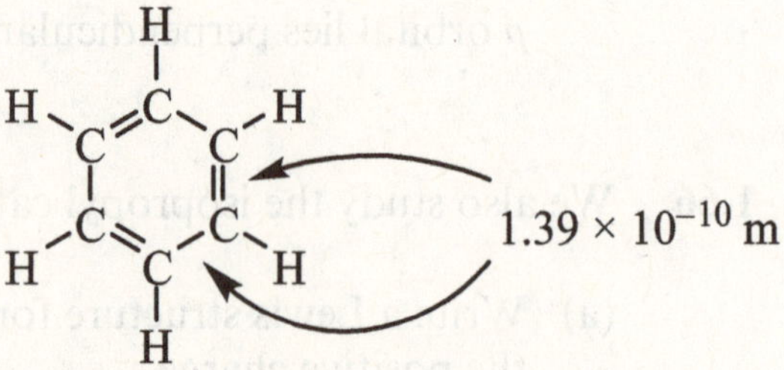

1.39×10^{-10} m

At a first glance, one may think that benzene should have three longer bonds and three shorter bonds; carbon-carbon single bonds are longer (1.54×10^{-10} m) than carbon- carbon double bonds (1.33×10^{-10} m). However, this is incorrect because we must consider resonance and the actual structure of benzene, which is a hybrid of the two energetically equivalent contributing structures shown below. In the actual structure of benzene all six carbon-carbon bonds are equivalent and equal in length (1.39×10^{-10} m) as a result of resonance.

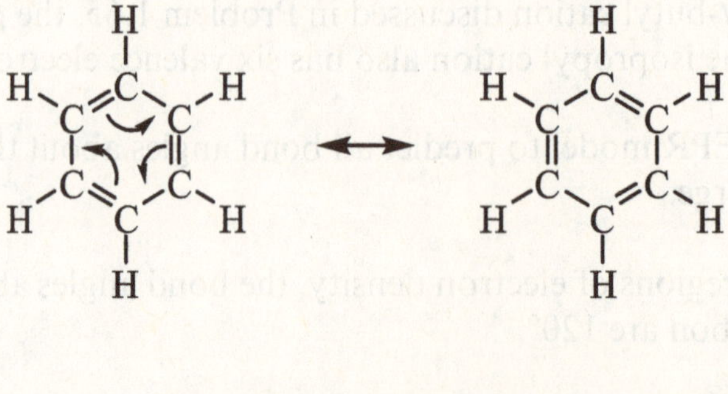

Group Learning Activities

Solutions are only provided for activities that are not open-ended.

1.70 Draw all possible contributing structures for the molecule shown. Then discuss which of these contributing structures would not contribute significantly to the resonance hybrid. Provide good reasons for the structures you eliminate.

(a)

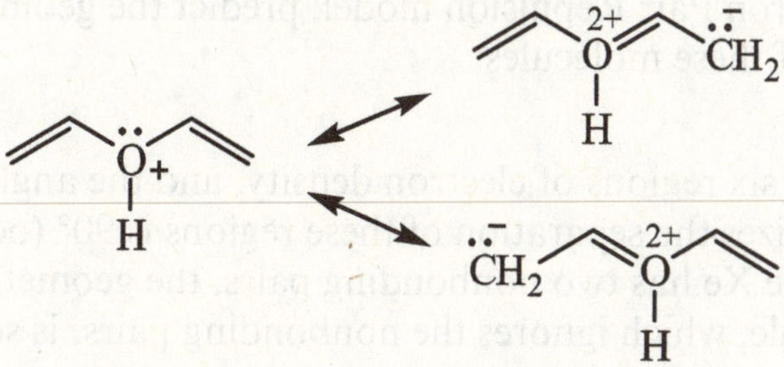

The two structures on the right are highly unstable because of the oxygen atom with two positive formal charges. As a result, they do not contribute significantly to the resonance hybrid.

(b)

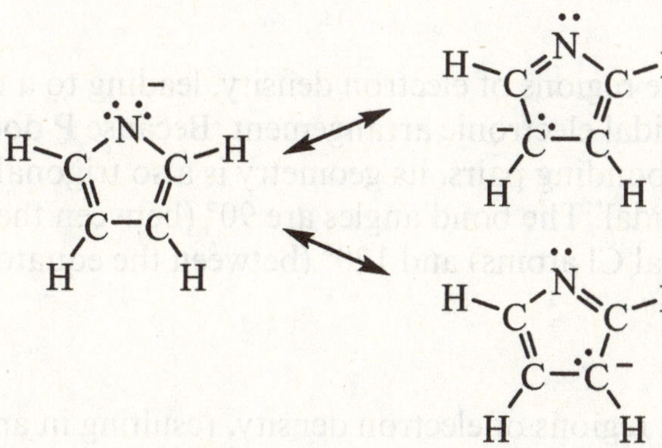

The two structures on the right each have a negatively charged carbon atom. Because the electronegativity of C is lower than that of N, a negatively charged C is less stable than a negatively charged N. These two structures do not contribute significantly to the resonance hybrid.

1.72 Rank the three types of hybridized orbitals (sp, sp^2, and sp^3) from most fitting to least fitting of the following characteristics. Provide reasons for your ranking.

(a) Most s character

An sp orbital has the most s character, because sp hybrids are made from one s and one p orbital; that is, sp hybrids have 50% s character. An sp^3 orbital has the least s character, because sp^3 hybrids are made from one s and three p atomic orbitals.

(b) Highest in energy

The sp^3 orbital has the highest energy of the three orbitals. This is because s orbitals are lower in energy than are p orbitals, and an sp^3 hybrid has the least s character. The sp orbital, which has the most s character, has the lowest energy.

(c) Forms the longest sigma bonds

The sp^3 orbital forms the longest sigma bonds. This is because s orbitals are closer to the nucleus than are p orbitals, and the sp^3 orbtial because it has the least s character. The sp orbital, which has the most s character, forms the shortest sigma bonds.

1.73   Using the Valence-Shell Electron Pair Repulsion model, predict the geometry and bond angles present in each of these molecules.

(a)

Xe has six regions of electron density, and the angle that maximizes the separation of these regions is 90° (octahedral). Because Xe has two nonbonding pairs, the geometry of the molecule, which ignores the nonbonding pairs, is square planar.

(b)

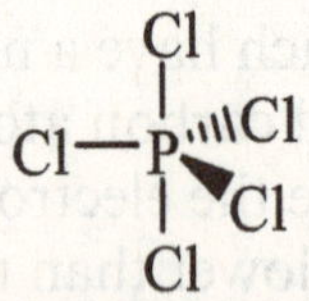

P has five regions of electron density, leading to a trigonal bipyramidal electronic arrangement. Because P does not have any nonbonding pairs, its geometry is also trigonal bipyramidal. The bond angles are 90° (between the axial and equatorial Cl atoms) and 120° (between the equatorial Cl atoms).

(c)

S has six regions of electron density, resulting in an octahedral electronic arrangement. Because S does not have any nonbonding pairs, its geometry is also octahedral. All bond angles are 90°.

Chapter 2: Acids and Bases

Problems

2.1   Write each acid-base reaction as a proton-transfer reaction. Label which reactant is the acid and which the base, as well as which product is conjugate base of the original acid and which the conjugate acid of the original base. Use curved arrows to show the flow of electrons in each reaction.

Curved arrows show the movement of electron pairs, and it is this movement of electrons that ultimately results in the transfer of a proton from the acid to the base. A common misconception is that in acid-base reactions, curved arrows show the direct movement of a proton when it is really the movement of the electrons that results in proton movement.

When looking at chemical reactions, it is very useful to examine the reactants and the products, identifying bonds that have been broken or formed. This information will be useful in determining the movement of electrons. For instance, consider the general acid-base reaction shown below, where HA is the acid and B⁻ is the base.

$$\ddot{B}^- \; + \; H{-}A \; \longrightarrow \; B{-}H \; + \; :A^-$$
$$\text{base} \qquad \text{acid} \qquad\qquad \text{conjugate} \quad \text{conjugate}$$
$$\text{acid} \qquad\quad \text{base}$$

Looking at this reaction, we see that HA has become A⁻, so not only has the H−A bond been broken, both electrons in the H−A bond have moved to A to give it a negative charge. (Why does the movement of the *two* electrons from the H−A bond to A result in only *one* negative charge? In a bond, electrons are shared. When both bonding electrons move to A, it is gaining only a net of one electron.) Likewise, B⁻ has become BH, so two of its electrons were used to make a BH bond.

In terms of electron movement, two electrons from B⁻ move to the H, forming the B−H bond, and two electrons move from the H−A bond to A, breaking the bond. This is depicted using the curved arrow notation shown below.

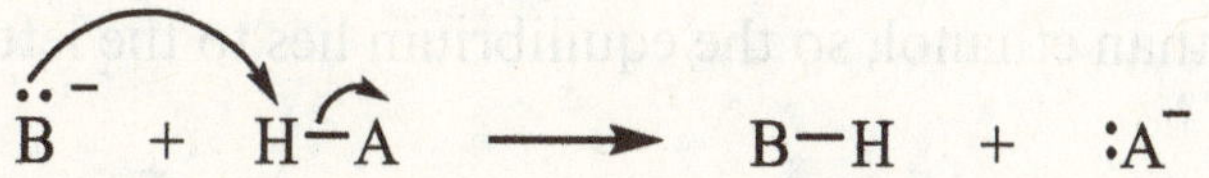

$$\ddot{B}^- \; + \; H{-}A \; \longrightarrow \; B{-}H \; + \; :A^-$$

Although we have used a negatively charged base as an example, bases can also be neutrally charged. In this case, when a neutrally charged base gains a proton, it becomes positively charged.

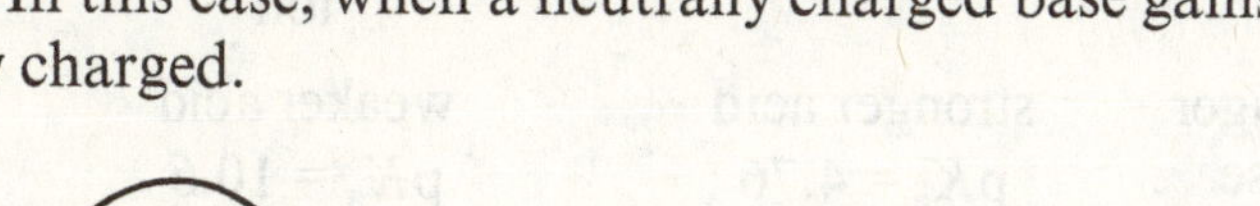

$$\ddot{B} \; + \; H{-}A \; \longrightarrow \; \overset{+}{B}{-}H \; + \; :A^-$$

(a) $CH_3\ddot{S}-H$ + $:\ddot{O}H$ ⟶ $CH_3\ddot{S}:^-$ + $H-\ddot{O}H$

     acid   base   conjugate   conjugate
                                    base   acid

(b) $CH_3\ddot{O}-H$ + $:NH_2$ ⟶ $CH_3\ddot{O}:^-$ + $H-\ddot{N}H_2$

     acid   base   conjugate   conjugate
                                    base   acid

(c) $C_6H_5\ddot{O}-H$ + $:\ddot{O}H_2$ ⟶ $C_6H_5\ddot{O}:^-$ + $H-\overset{+}{\ddot{O}}H_2$

     acid   base   conjugate   conjugate
                                    base   acid

2.2 For each value of K_a, calculate the corresponding value of pK_a. Which compound is the stronger acid?

The relationship between K_a and pK_a is: $pK_a = -\log K_a$. When K_a increases, pK_a decreases. Weak acids that are stronger lie further to the right of the equilibrium and thus have higher K_a values. Between acetic acid and water, acetic acid is a stronger acid.

(a) Acetic acid $K_a = 1.74 \times 10^{-5}$ $pK_a = 4.76$
(b) Water $K_a = 2.00 \times 10^{-16}$ $pK_a = 15.7$

2.3 For each acid-base equilibrium, label the stronger acid, the stronger base, the weaker acid, and the weaker base. Then predict whether the position of equilibrium lies toward the right or the left.

In an acid-base equilibrium, the reaction will favor the side of the equilibrium with the weaker acid and the weaker base. This is because the stronger acid will be the acid that is more likely to donate a proton, and likewise, the stronger base will be the one that is more likely to accept the proton. In (a), the methylammonium ion is a weaker acid than acetic acid, so the equilibrium lies to the right. In (b), ammonia is a weaker acid than ethanol, so the equilibrium lies to the left.

(a) CH_3NH_2 + CH_3COOH ⇌ $CH_3\overset{+}{N}H_3$ + CH_3COO^-
     Methylamine   Acetic acid   Methylammonium Acetate ion
                                          ion

     stronger   stronger acid   weaker acid   weaker
     base $pK_a = 4.76$ $pK_a = 10.6$ base

(b) $CH_3CH_2O^- + NH_3 \rightleftharpoons CH_3CH_2OH + NH_2^-$

Ethoxide ion	Ammonia	Ethanol	Amide ion
weaker base	weaker acid $pK_a = 38$	stronger acid $pK_a = 15.9$	stronger base

2.4 (a) Arrange the following compounds in the order most acidic to least acidic.

To assess the relative strengths of acids, examine the relative stabilities of the conjugate bases. When a conjugate base is more stable, more of it will be formed, the equilibrium lies further to the right, and hence, the stronger is the acid.

acids

$$CH_2=CH-\overset{H}{\underset{\cdot\cdot}{N}}-H \qquad \overset{\cdot\cdot}{O}=CH-\overset{H}{\underset{\cdot\cdot}{N}}-H \qquad CH_3CH_2-\overset{H}{\underset{\cdot\cdot}{N}}-H$$

$$\text{1} \qquad\qquad\qquad \text{2} \qquad\qquad\qquad \text{3}$$

conjugate bases

$$CH_2=CH-\overset{\cdot\cdot}{\underset{\cdot\cdot}{N}}^--H \qquad \overset{\cdot\cdot}{O}=CH-\overset{\cdot\cdot}{\underset{\cdot\cdot}{N}}^--H \qquad CH_3CH_2-\overset{\cdot\cdot}{\underset{\cdot\cdot}{N}}^--H$$

$$\overset{\cdot\cdot}{\underset{}{C}}H_2^--CH=N-H \qquad \overset{\cdot\cdot}{:\overset{\cdot\cdot}{O}}{}^--CH=N-H$$

Comparing the conjugate bases of the three acids, all are amide ions (not to be confused with the amide functional group). The conjugate base of acid **3** is not stabilized by resonance, so it is the least stable conjugate base, making acid **3** the weakest acid. How do we decide between the conjugate bases of acids **1** and **2**? The conjugate base of acid **2** has a contributing structure with a negative charge on an oxygen atom, while that of acid **1** has a negative charge on a carbon atom. Thus, there is also an electronegativity effect and the former is more stable, making acid **2** stronger than acid **1**. The order of acidity is **2 > 1 > 3**.

(b) Arrange the following compounds in the order most basic to least basic.

When assessing the relative strengths of bases, look at the bases themselves. The question implies that the compounds shown are bases, so examine them the way they are. (Do not remove protons the from bases and then look at the products because if protons are removed, the compounds would be acting as acids and not as bases!)

bases

$$CH_3CH_2-\overset{\cdot\cdot}{\underset{\cdot\cdot}{N}}^--H \qquad CH_3CH_2-\overset{\cdot\cdot}{\underset{\cdot\cdot}{O}}:^- \qquad CH_3CH_2-\overset{\cdot\cdot}{\underset{H}{C}}^--H$$

$$\text{1} \qquad\qquad\qquad \text{2} \qquad\qquad\qquad \text{3}$$

The best approach would be to compare the three bases. What is different about them? Everything else is identical except for the atom bearing the negative charge. The three atoms of interest (N, O, and C) are in the same period, so the effect of electronegativity needs to be considered. The more electronegative the atom, the better it can accommodate the negative charge, and the more stable it is. The more stable the anion, the weaker is the base. Thus, the order of basicity is **3 > 1 > 2**.

2.5   Complete this acid-base reaction. First add unshared pairs of electrons on the reacting atoms to give each a complete octet. Use curved arrows to show the redistribution of electrons in the reaction. In addition, predict whether the position of this equilibrium lies toward the left or the right.

(a)

$$CH_3\ddot{\underset{\cdot\cdot}{O}}{:}^- \;+\; CH_3-\overset{\overset{\displaystyle H}{|}}{\underset{\underset{\displaystyle CH_3}{|}}{N}}{}^+-CH_3 \;\rightleftharpoons\; CH_3\ddot{\underset{\cdot\cdot}{O}}-H \;+\; CH_3-\overset{\cdot\cdot}{\underset{\underset{\displaystyle CH_3}{|}}{N}}-CH_3$$

stronger base   stronger acid   weaker acid   weaker base
               $pK_a = 10$ $pK_a = 16$

Methanol is a weaker acid than the trimethylammonium ion, so the equilibrium lies on the right. That is, the trimethylammonium ion is more likely to donate a proton to methoxide as opposed to methanol donating a proton to trimethylamine.

(b)

$$CH_3-\overset{\overset{\displaystyle \ddot{O}}{\|}}{C}-\ddot{\underset{\cdot\cdot}{O}}-H \;+\; {:}\ddot{\underset{\cdot\cdot}{C}l}{:}^- \;\rightleftharpoons\; CH_3-\overset{\overset{\displaystyle \ddot{O}}{\|}}{C}-\ddot{\underset{\cdot\cdot}{O}}{:}^- \;+\; H-\ddot{\underset{\cdot\cdot}{C}l}{:}$$

weaker acid   weaker base   stronger base   stronger acid
   $pK_a = 5$ $pK_a = -7$

HCl is a stronger acid than acetic acid, so the equilibrium lies on the left. In other words, HCl is more likely to donate a proton to acetate as opposed to acetic acid donating a proton to the chloride ion.

2.6   Write an equation for the reaction between each Lewis acid-base pair, showing electron flow by means of curved arrows. (*Hint*: Aluminum is in Group 3A of the Periodic Table, just under boron. Aluminum in $AlCl_3$ has only six electrons in its valence shell and has an incomplete octet.)

By definition, Lewis bases are electron-pair donors and therefore must have a nonbonding pair. On the contrary, Lewis acids are electron-pair acceptors, and to be able to accept more electrons, they are usually electron-deficient species. Some Lewis acids are so reactive that they will react with a neutrally charged chlorine to generate a chloronium ion (positively charged chlorine), as seen in (b).

(a)

$$:\ddot{C}l: \quad + \quad Al-\ddot{C}l: \quad \longrightarrow \quad :\ddot{C}l-Al-\ddot{C}l:$$

Lewis base   Lewis acid

(b)

$$CH_3-\ddot{C}l: \quad + \quad Al-\ddot{C}l: \quad \longrightarrow \quad CH_3-\overset{+}{\ddot{C}l}-Al-\ddot{C}l:$$

Lewis base   Lewis acid

Quick Quiz

1. If NH_3 were to behave as an acid, its conjugate base would be NH_2^-. **True**. Acids are proton donors, and in this sense, NH_3 has donated a proton. Note that NH_3 can also act as a base and accept a proton, forming NH_4^+, the conjugate acid of NH_3.

2. Delocalization of electron density is a stabilizing factor. **True**. The delocalization, or spreading out, of electron density or charge increases stability.

3. Amide ion, NH_2^-, is a Lewis base. **True**. Lewis bases are electron-pair donors, and the amide ion has a pair of nonbonding electrons that can be donated.

4. H_3O^+ is a stronger acid than NH_4^+ and, therefore, NH_3 is a stronger base than H_2O. **True**. Within an acid-conjugate base pair, the strength of the acid is inversely related to the strength of its conjugate base.

5. Inductive effects can be used to describe electron delocalization. **True**. The presence of electronegative groups in the molecule can assist in the delocalization of electrons.

6. The direction of equilibrium in an acid-base reaction favors the side containing the stronger acid and stronger base. **False**. The side with the weaker acid and base is favored.

7. The conjugate base of CH_3CH_2OH is $CH_3CH_2O^-$. **True**. Ethanol has lost a proton, forming the conjugate base ethoxide.

8. CH_3^+ and NH_4^+ are Lewis acids. **False**. Lewis acids are electron-pair acceptors, but the ammonium ion cannot accept a pair of electrons because it already has a full octet.

9. When an acid, HA, dissolves in water, the solution becomes acidic because of the presence of H^+ ions. **False**. In water, any H^+ that is present reacts with water to form H_3O^+.

10. Between a strong acid and a weak acid, the weak acid will give rise to the stronger conjugate base. *True*. Within an acid-conjugate base pair, the weaker the acid, the stronger is its conjugate base.

11. H_2O can function as an acid (proton donor) and as a base (proton acceptor). *True*. Water is an amphoteric species.

12. The strongest base that can exist in aqueous solution is OH^-. *True*. If a base that is stronger than OH^- is added to water, it will immediately react with water (the acid) to form OH^-.

13. A strong acid is one that completely ionizes in aqueous solution. *True*. Strong acids are defined as those that completely ionize (dissociate) in aqueous solution.

14. NH_3 is a Lewis base. *True*. Lewis bases are electron-pair donors, and ammonia contains a nonbonding pair of electrons that can be donated to a Lewis acid.

15. A Brønsted-Lowry acid is a proton donor. *True*. By definition, Brønsted-Lowry acids are those that donate protons (to proton acceptors, which are Brønsted-Lowry bases).

16. When comparing the relative strength of acids, the stronger acid has the smaller value of pK_a. *True*. Acid strength increases as pK_a decreases.

17. When comparing the relative strengths of acids, the stronger acid has the smaller value of K_a. *False*. Stronger acids ionize to a greater extent and have higher K_a values.

18. The formulas of a conjugate acid-base pair differ only by a proton. *True*. An acid or base donates or accepts a proton, respectively, to form its conjugate species, so the formulas must differ by only one proton.

19. A Lewis base is an electron pair donor. *True*. A Lewis base donates a pair of electrons.

20. Acetic acid, CH_3COOH, is a stronger acid than carbonic acid, H_2CO_3, and, therefore, acetate ion, CH_3COO^-, is a stronger base than bicarbonate ion, HCO_3^-. *False*. The stronger acid gives rise to the weaker conjugate base, so bicarbonate is the stronger base.

21. The strongest acid that can exist in aqueous solution is H_3O^+. *True*. If an acid stronger than H_3O^+ is added to water, it would be neutralized by water (the base), forming H_3O^+.

22. A Lewis acid–base reaction results in the formation of a new covalent bond between the Lewis acid and the Lewis base. *True*. The Lewis base transfers a pair of electrons to the Lewis acid, forming a bond between the two species.

23. When a base accepts a proton in an acid-base reaction, it is converted into its conjugate base. *False*. When a base accepts a proton, it is converted into its conjugate *acid*.

24. When a metal hydroxide, MOH, dissolves in water, the solution becomes basic because of the presence of hydroxide ions, OH^-. *True*. The hydroxide ions are responsible for the basicity of the solution.

25. A Lewis acid is a proton acceptor. *False*. Lewis acids are electron-pair acceptors.

26. If NH_3 were to behave as a base, its conjugate acid would be NH_4^+. *True*. By acting as a base, ammonia would gain a proton to form the ammonium ion, the conjugate acid.

27. Resonance effects can be used to describe electron delocalization. *True*. Resonance involves the movement of electrons and can delocalize the electrons over multiple atoms.

28. All Lewis acid-base reactions involve transfer of a proton from the acid to the base. *False*. Proton transfer is not what defines Lewis acid-base reactions; rather, it is the transfer of a pair of electrons.

29. BF_3 is a Lewis acid. *True*. Boron is electron-deficient (does not have a full octet) and can accept a pair of electrons, thus acting as a Lewis acid.

30. When HCl dissolves in water, the major ions present are H^+ and Cl^-. *False*. No H^+ ions are actually present in water because they immediately react with H_2O to form H_3O^+.

31. According to the Arrhenius definitions, acids and bases are limited to substances that dissolve in water. *True*. Based on this definition, acids dissolve in water to produce H_3O^+ ions, and bases dissolve in water to produce OH^- ions.

32. Acid-base reactions take place only in aqueous solution. *False*. Acid-base reactions are merely proton-transfer reactions or, in the case of Lewis acids and bases, electron-pair transfer reactions, and are not limited to aqueous solution. They can occur in organic solutions and even in the gas phase.

33. The conjugate acid of HCO_3^- is H_2CO_3. *True*. If HCO_3^- (the bicarbonate ion that is found in baking soda) acts as a base and accepts a proton, the conjugate acid H_2CO_3 is formed.

End-of-Chapter Problems

Arrhenius Acids and Bases

2.7 Complete the net ionic equation for each acid when placed in water. Use curved arrows to show the flow of electron pairs in each reaction. Also for each reaction, determine the direction of equilibrium using Table 2.2 as a reference for the pK_a values of proton acids.

Note: if the question refers to the table of pK_a values, it is necessary to use the data to answer the question. In acid-base reactions, the equilibria will favor the side with the weaker acid and weaker base combination.

(a)

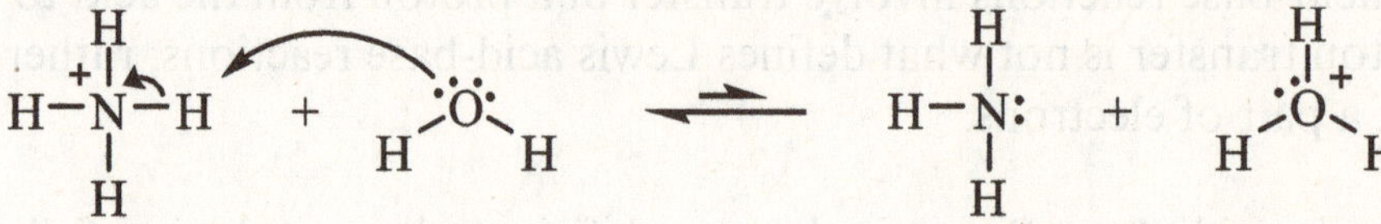

weaker acid   weaker base   stronger base   stronger acid
$pK_a = 9.24$ $pK_a = -1.74$

(b)

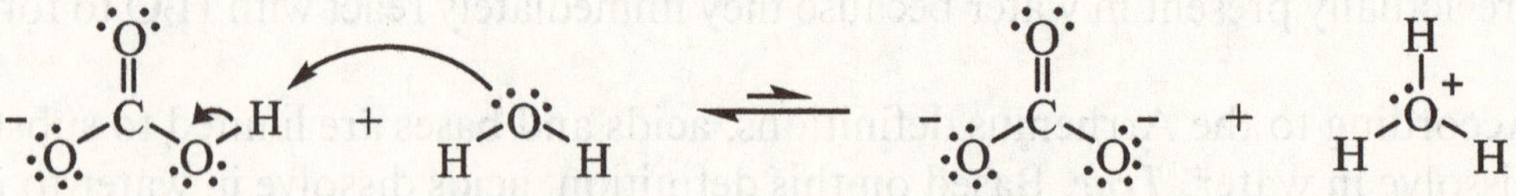

weaker acid   weaker base   stronger base   stronger acid
$pK_a = 10.33$ $pK_a = -1.74$

(c)

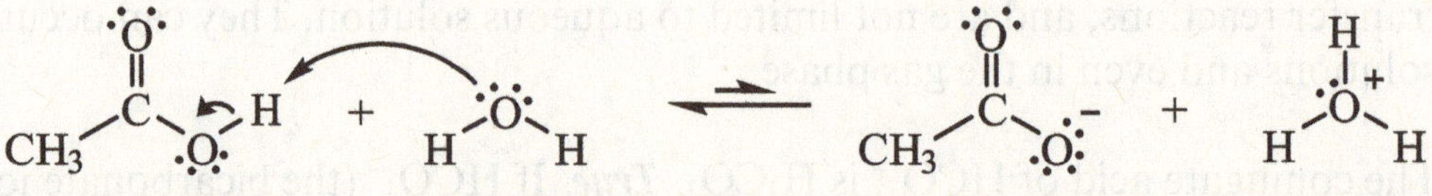

weaker acid   weaker base   stronger base   stronger acid
$pK_a = 4.76$ $pK_a = -1.74$

(d)

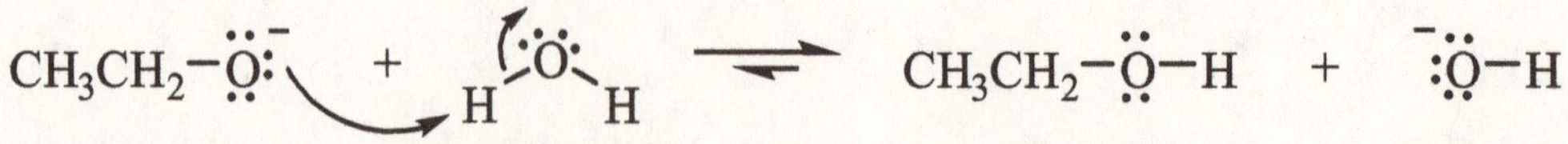

stronger base   stronger acid   weaker acid   weaker base
                $pK_a = 15.7$ $pK_a = 15.9$

This equilibrium favors the right, but only very slightly because the acid strengths of ethanol and water are very similar.

2.8   Complete the net ionic equation for each base when placed in water. Use curved arrows to show the flow of electron pairs in each reaction. Also for each reaction, determine the direction of equilibrium using Table 2.2 as a reference for the pK_a values of proton acids formed.

In this question, all compounds are to behave as bases in water, so they will accept a proton from water. All equilibria lie to the left.

(a)

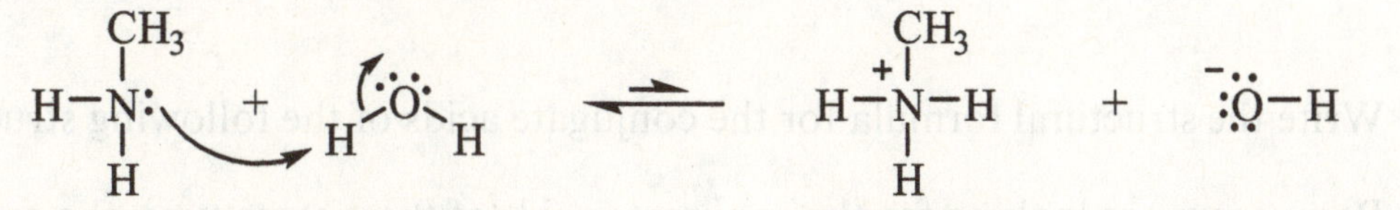

weaker base   weaker acid   stronger acid   stronger base
                  pK_a = 15.7   pK_a = 10.64

(b)

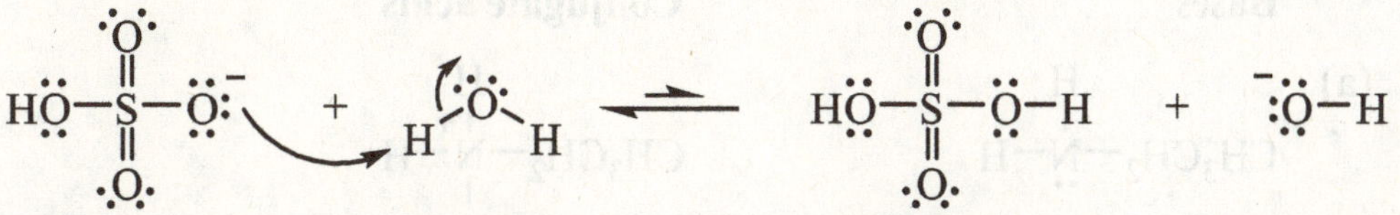

weaker base   weaker acid   stronger acid   stronger base
                  pK_a = 15.7   pK_a = –5.2

(c)

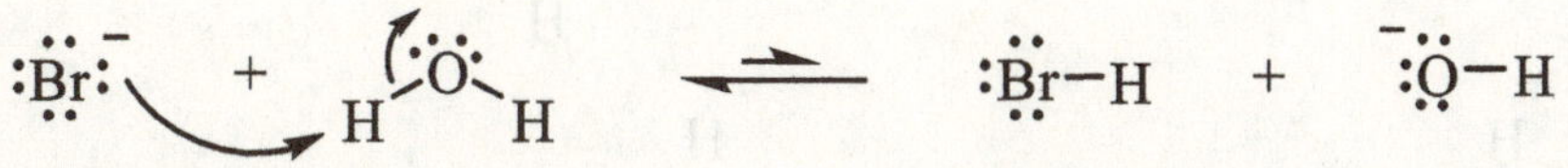

weaker base   weaker acid   stronger acid   stronger base
                  pK_a = 15.7   pK_a = –8

(d)

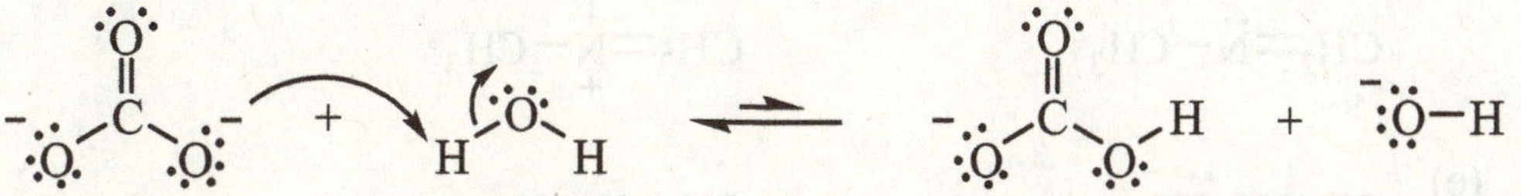

weaker base   weaker acid   stronger acid   stronger base
                  pK_a = 15.7   pK_a = 10.33

(e)

weaker base   weaker acid   stronger acid   stronger base
                  pK_a = 15.7   pK_a = 9.2

Brønsted-Lowry Acids and Bases

2.9   How are the formulas of the members of a conjugate acid-base pair related to each other? Within a pair, how can you tell which is the acid?

According to Brønsted-Lowry theory, acids are proton (H^+) donors. The formulas of a conjugate acid-base pair therefore differ by one hydrogen atom, as well as one charge. The acid has one hydrogen atom more, but one negative charge less, than the base.

2.10   Write the structural formula for the conjugate acids of the following structures.

Because we are looking for the conjugate acids of these structures, the question implies that the structures must be acting as bases (proton acceptors). Bases must have a nonbonding pair of electrons to accept the proton.

Bases	**Conjugate acids**
(a) $CH_3CH_2-\overset{\overset{\displaystyle H}{|}}{\underset{\displaystyle \cdot\cdot}{N}}-H$	$CH_3CH_2-\overset{\overset{\displaystyle H}{|}}{\underset{\displaystyle \underset{\displaystyle H}{|}}{\overset{+}{N}}}-H$
(b) $CH_3CH_2-\underset{\displaystyle \cdot\cdot}{\overset{\displaystyle \cdot\cdot}{S}}-H$	$CH_3CH_2-\underset{\displaystyle \underset{\displaystyle H}{|}}{\overset{+}{S}}-H$
(c) $H-\overset{\overset{\displaystyle H}{|}}{\underset{\displaystyle \cdot\cdot}{N}}-CH_2CH_2-\overset{\displaystyle \cdot\cdot}{\underset{\displaystyle \cdot\cdot}{O}}H$	$H-\overset{\overset{\displaystyle H}{|}}{\underset{\displaystyle \underset{\displaystyle H}{|}}{\overset{+}{N}}}-CH_2CH_2-\overset{\displaystyle \cdot\cdot}{\underset{\displaystyle \cdot\cdot}{O}}H$
(d) $CH_2=\overset{\displaystyle \cdot\cdot}{N}-CH_3$	$CH_2=\underset{\displaystyle +}{\overset{\overset{\displaystyle H}{|}}{N}}-CH_3$
(e) $CH_3CH_2\overset{\displaystyle \cdot\cdot}{\underset{\displaystyle}{C}}H_2^{-}$	$CH_3CH_2CH_3$
(f) $CH_3\underset{\displaystyle \cdot\cdot}{\overset{\displaystyle \cdot\cdot}{O}}CH_3$	$CH_3\underset{\displaystyle \cdot\cdot}{\overset{+}{\overset{\overset{\displaystyle H}{|}}{O}}}CH_3$

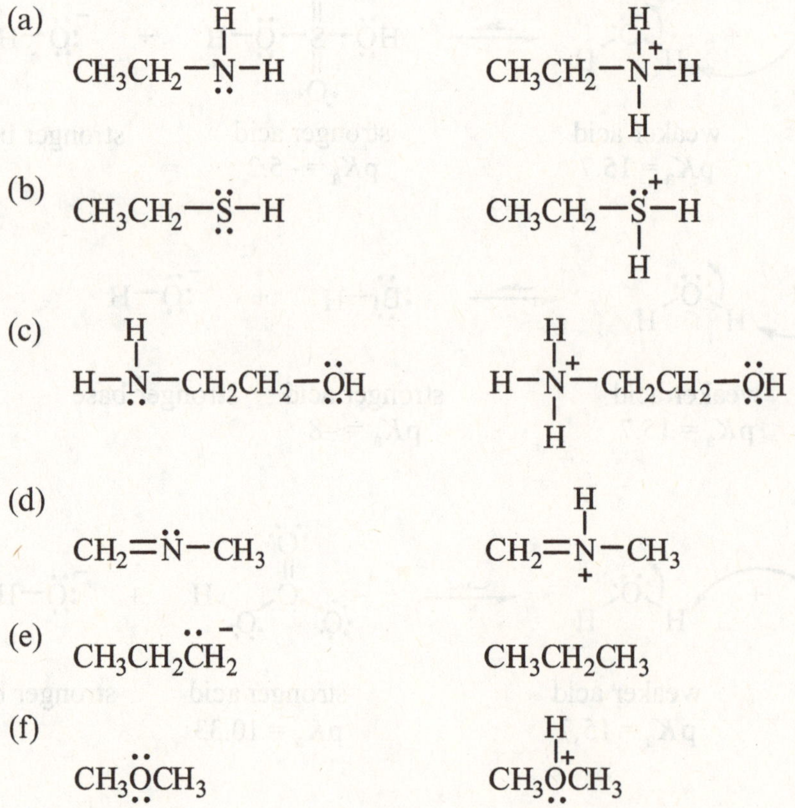

2.11   Complete a net ionic equation for each proton-transfer reaction, using curved arrows to show the flow of electron pairs in each reaction. In addition, write Lewis structures for all starting materials and products. Label the original acid and its conjugate base; label the original base and its conjugate acid. If you are uncertain about which substance in each equation is the proton donor, refer to Table 2.2 for the pK_a values of proton acids.

(a)

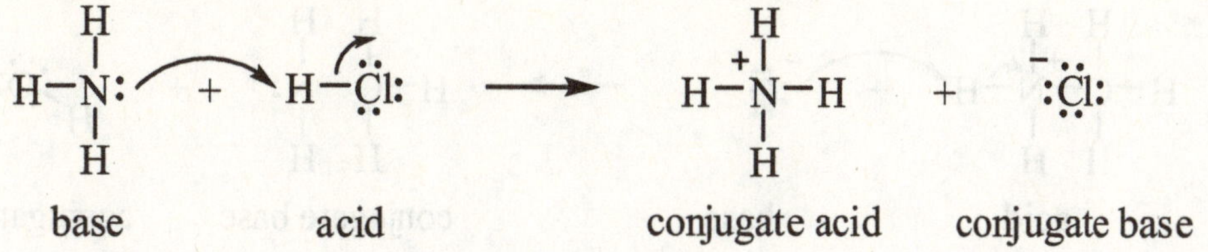

base   acid   conjugate acid   conjugate base

(b)

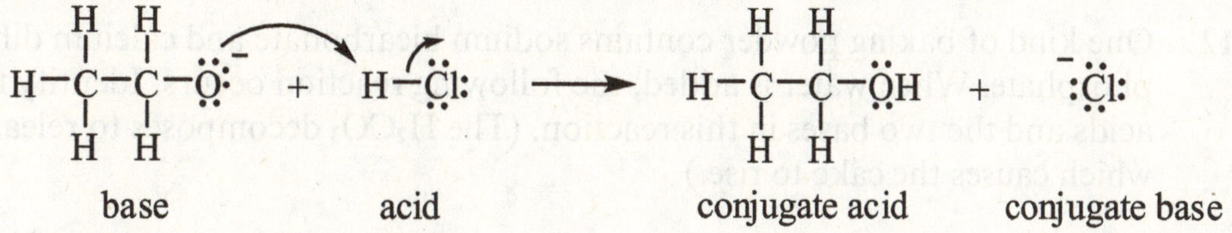

base   acid   conjugate acid   conjugate base

(c)

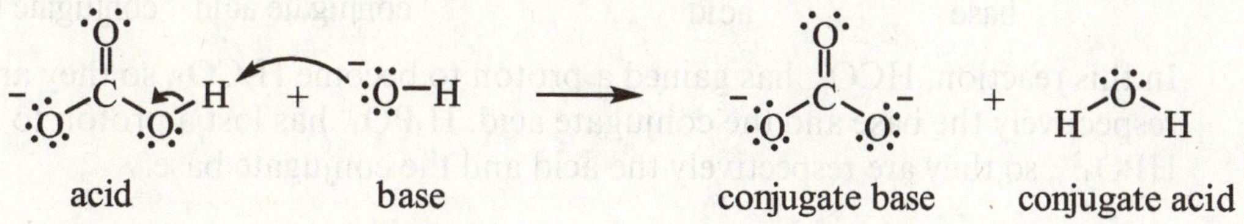

acid   base   conjugate base   conjugate acid

(d)

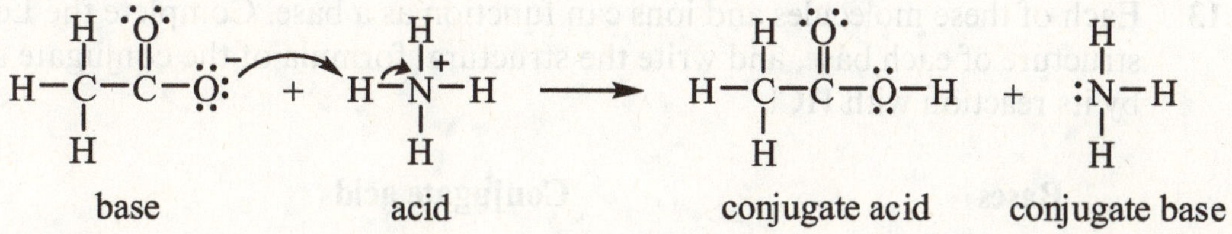

base   acid   conjugate acid   conjugate base

(e)

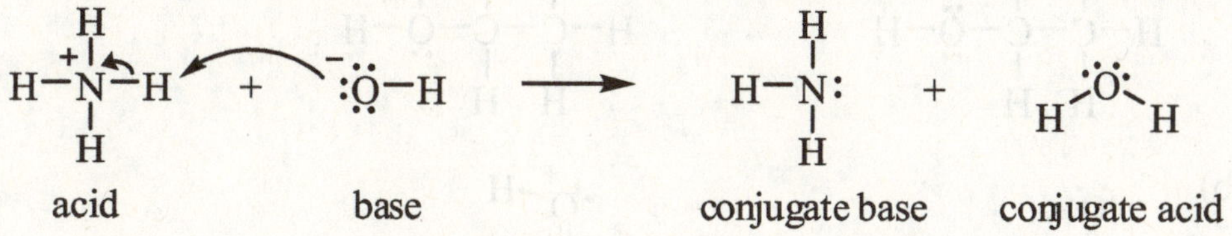

acid   base   conjugate base   conjugate acid

(f)

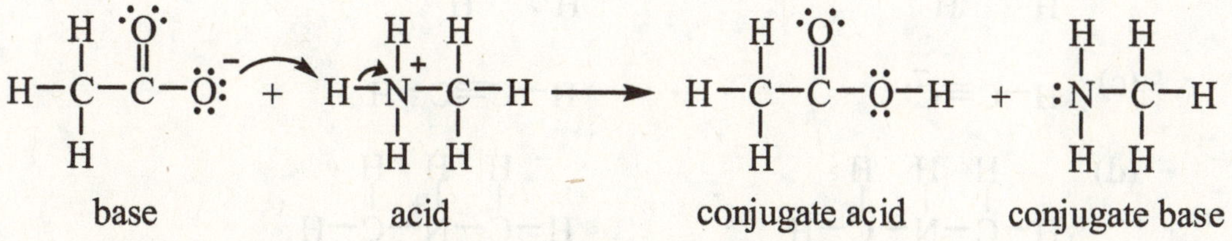

base   acid   conjugate acid   conjugate base

(g)

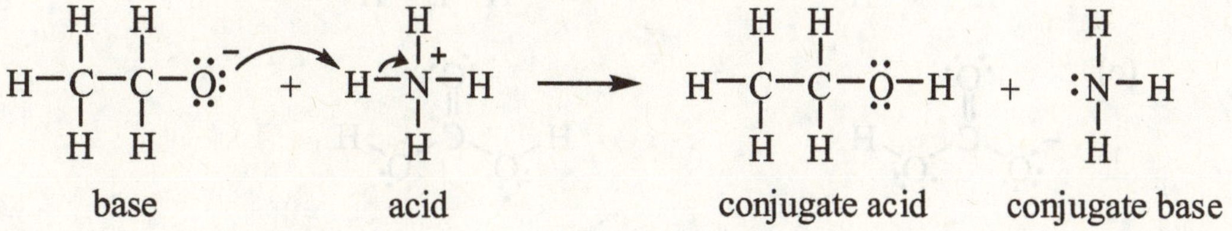

base   acid   conjugate acid   conjugate base

(h)

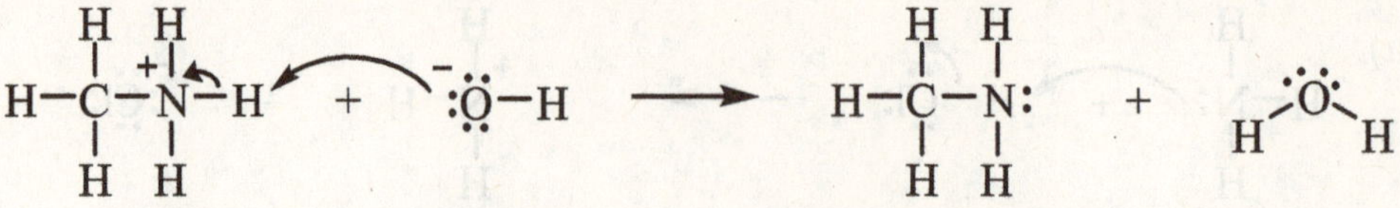

H-C-N⁺-H + :Ö⁻-H ⟶ H-C-N: + H-Ö-H

    acid   base   conjugate base   conjugate acid

2.12 One kind of baking powder contains sodium bicarbonate and calcium dihydrogen phosphate. When water is added, the following reaction occurs. Identify the two acids and the two bases in this reaction. (The H_2CO_3 decomposes to release CO_2, which causes the cake to rise.)

$$HCO_3^-(aq) + H_2PO_4^-(aq) \longrightarrow H_2CO_3(aq) + HPO_4^{2-}(aq)$$

    base   acid   conjugate acid   conjugate base

In this reaction, HCO_3^- has gained a proton to become H_2CO_3, so they are respectively the base and the conjugate acid. $H_2PO_4^-$ has lost a proton to become HPO_4^{2-}, so they are respectively the acid and the conjugate base.

2.13 Each of these molecules and ions can function as a base. Complete the Lewis structure of each base, and write the structural formula of the conjugate acid formed by its reaction with HCl.

Bases   **Conjugate acids**

(a)   H-C-C-Ö-H   H-C-C-O⁺-H

(b)   formaldehyde (O double bond C, H, H)   O⁺-H double bond C, H, H

(c)   H-C≡C:⁻   H-C≡C-H

(d) 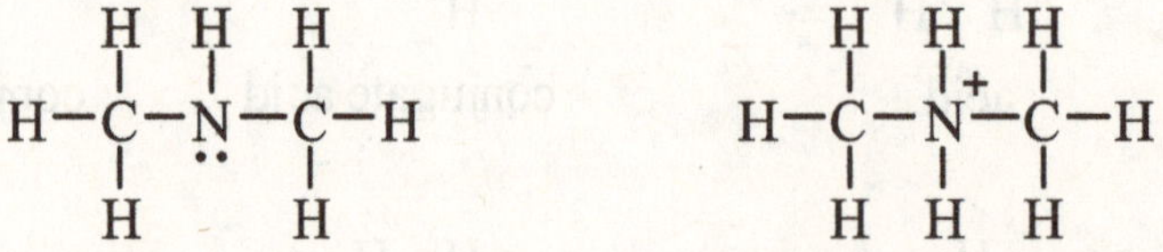   H-C-N-C-H   H-C-N⁺-C-H

(e)   ⁻O-C(=O)-O-H   H-O-C(=O)-O-H

(f)   N⁻=N⁺=N⁻   N⁻=N⁺=N-H

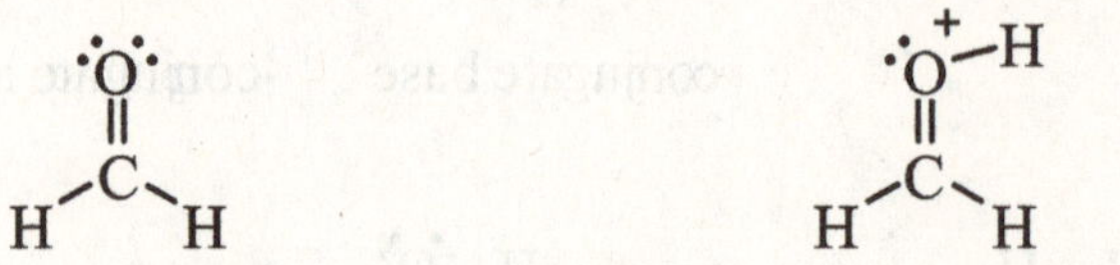

2.14 Offer an explanation for the following observations.

To assess the relative strengths of acids, remember to compare the relative stabilities of the conjugate bases. The more stable the conjugate base, the stronger the acid.

(a) H_3O^+ is a stronger acid than NH_4^+.

The respective conjugate bases are H_2O and NH_3. Due to electronegativity, the lone pair on the oxygen atom is more stable than the lone pair on the nitrogen atom.

(b) Nitric acid, HNO_3, is a stronger acid than nitrous acid, HNO_2 (pK_a 3.7).

The respective conjugate bases, nitrate (NO_3^-) and nitrite (NO_2^-), are shown below. Three contributing structures can be drawn for nitrate, but only two can be drawn for nitrite. In nitrate, the negative charge is delocalized over three oxygen atoms while in nitrite, the delocalization is over only two oxygen atoms. As a result, nitrate is more stable than nitrite, making nitric acid the stronger acid.

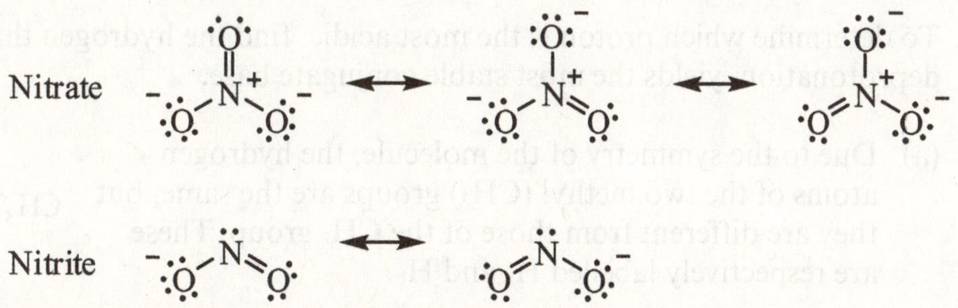

Nitrate

Nitrite

(c) Ethanol, CH_3CH_2OH, and water have approximately the same acidity.

The respective conjugate bases are ethoxide ($CH_3CH_2O^-$) and hydroxide (HO^-). Both ions have a negatively charge oxygen atom and there is no resonance stabilization in either ion. These conjugate bases are of comparable stability, so the strengths of the acids are similar.

(d) Trichloroacetic acid, CCl_3COOH (pK_a 0.64), is a stronger acid than acetic acid, CH_3COOH (pK_a 4.74).

Both conjugate bases, trichloroacetate (CCl_3COO^-) and acetate (CH_3COO^-), contain a negatively charged oxygen atom that is stabilized by resonance. However, trichloroacetate is also stabilized by the inductively withdrawing chlorine atoms. Therefore, trichloroacetic acid is a stronger acid than acetic acid.

Trichloroacetate   Acetate

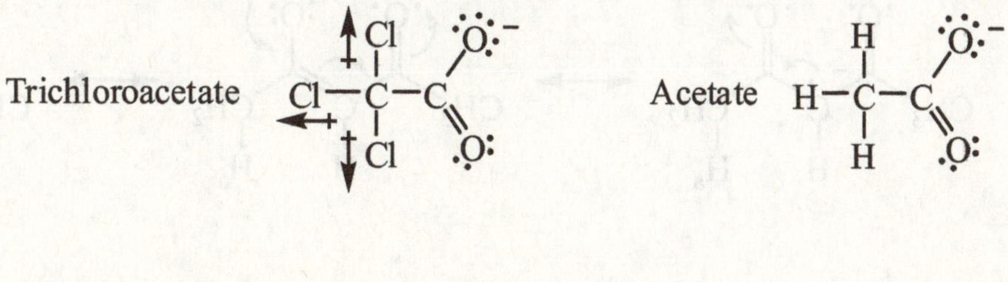

(e) Trifluoroacetic acid, CF_3COOH (pK_a 0.23), is a stronger acid than trichloroacetic acid, CCl_3COOH (pK_a 0.64).

Like (d), both conjugate bases contain a resonance-stabilized, negatively charged oxygen atom. Trifluoroacetate is more stable than trichloroacetate because the higher electronegativity of fluorine makes it a better inductively withdrawing group. Note: we are not examining the effect of atomic size (fluorine versus chlorine), as atomic size effects are used only when it involves the atom bearing the negative charge.

Trichloroacetate
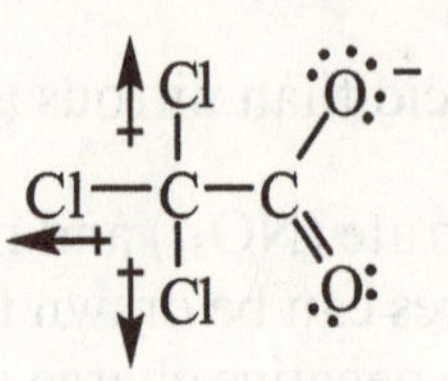
Trifluoroacetate
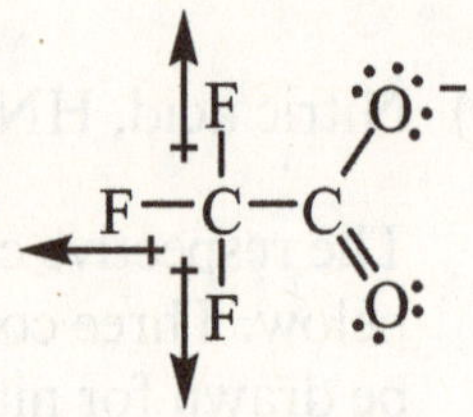

2.15 Select the most acidic proton in the following compounds:

To determine which proton is the most acidic, find the hydrogen that, after deprotonation, yields the most stable conjugate base.

(a) Due to the symmetry of the molecule, the hydrogen atoms of the two methyl (CH_3) groups are the same, but they are different from those of the CH_2 group. These are respectively labelled H_a and H_b.

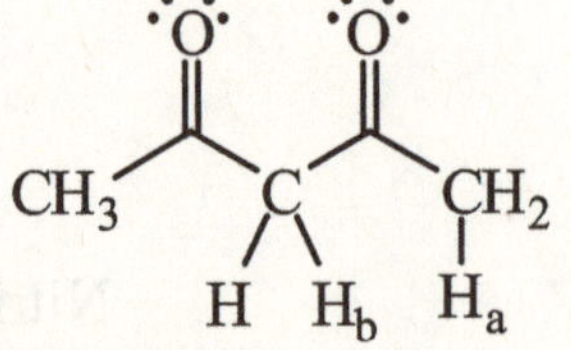

The conjugate base formed by the deprotonation of H_a is stabilized by resonance. Two contributing structures can be drawn, as shown below.

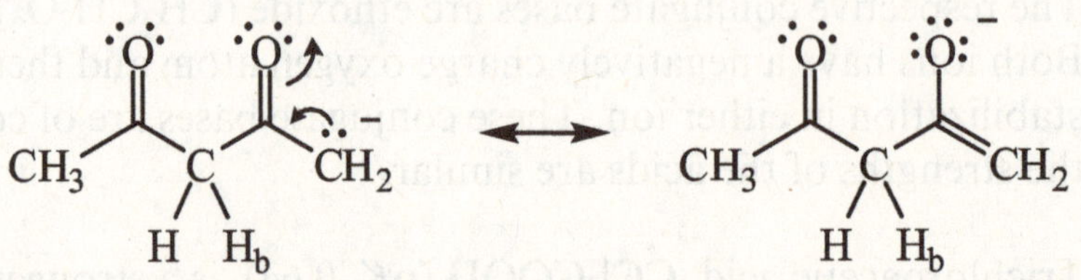

Whereas, three contributing structures can be drawn for the conjugate based that is formed by the deprotonation of H_b. The negative charge is delocalized over more atoms, so the conjugate based formed by H_b deprotonation is more stable than the conjugate base formed by H_a deprotonation. Thus, H_b is the most acidic proton.

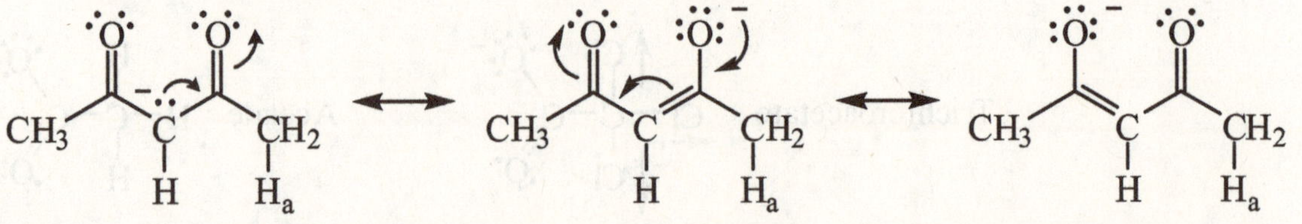

(b) The deprotonation of H_a results in a neutrally charged conjugate base while the deprotonation of the other protons would not. Furthermore, the conjugate base is also stabilized by resonance.

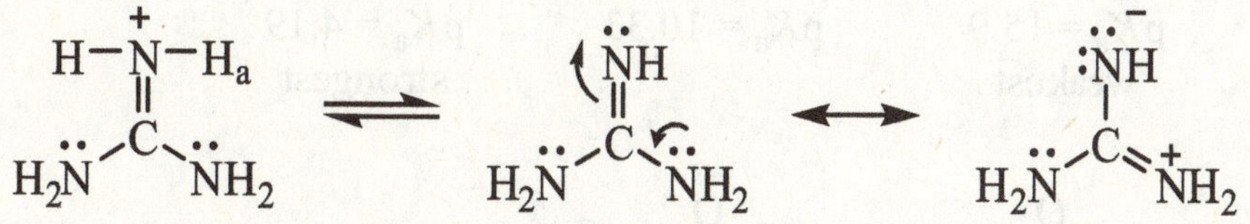

(c) When H_a is deprotonated, a conjugate base that is stabilized by resonance is formed. The deprotonation of the aldehydic H would not produce a resonance-stabilized conjugate base.

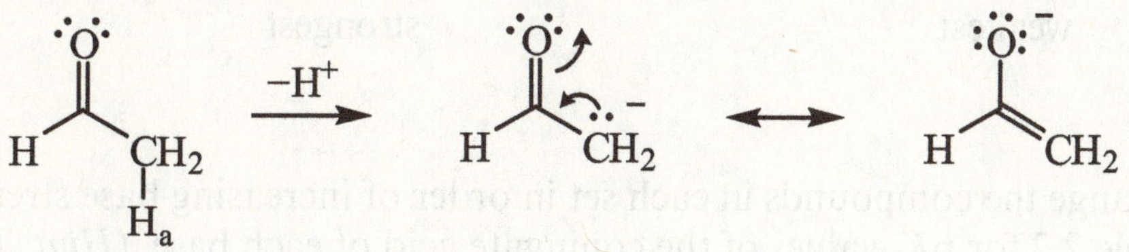

Quantitative Measure of Acid Strength

2.16 Which has the larger numerical value?

(a) The pK_a of a strong acid or the pK_a of a weak acid? As acid strength increases, K_a increases but pK_a decreases. The weak acid therefore has the larger pK_a value.

(b) The K_a of a strong acid or the K_a of a weak acid? As acid strength increases, K_a increases, giving the strong acid the larger K_a value.

2.17 In each pair, select the stronger acid:

The stronger acid has the higher K_a value, or the lower pK_a value. The stronger of the two acids in each pair is indicated in bold.

(a) **Pyruvic acid (pK_a 2.49)** or lactic acid (pK_a 3.85)
(b) Citric acid (pK_{a1} 3.08) or **phosphoric acid (pK_{a1} 2.10)**
(c) Nicotinic acid (niacin, K_a 1.4×10^{-5}) or **acetylsalicylic acid (aspirin, K_a 3.3×10^{-4})**
(d) Phenol (K_a 1.12×10^{-10}) or **acetic acid (K_a 1.74×10^{-5})**

2.18 Arrange the compounds in each set in order of increasing acid strength. Consult Table 2.2 for pK_a values of each acid.

Note that it is necessary to refer to the table of pK_a values, as the relative stabilities of some of the conjugate bases cannot be determined using structural factors alone.

(a)
$$\underset{\substack{\text{Ethanol}}}{CH_3CH_2OH} \qquad \underset{\substack{\text{Bicarbonate ion}}}{HOC\overset{O}{\overset{\|}{C}}O^-} \qquad \underset{\substack{\text{Benzoic acid}}}{C_6H_5\overset{O}{\overset{\|}{C}}OH}$$

$$\underset{\substack{\text{weakest}}}{pK_a = 15.9} \qquad pK_a = 10.33 \qquad \underset{\substack{\text{strongest}}}{pK_a = 4.19}$$

(b)
$$\underset{\substack{\text{Carbonic acid}}}{HO\overset{O}{\overset{\|}{C}}OH} \qquad \underset{\substack{\text{Acetic acid}}}{CH_3\overset{O}{\overset{\|}{C}}OH} \qquad \underset{\substack{\text{Hydrogen chloride}}}{HCl}$$

$$\underset{\substack{\text{weakest}}}{pK_a = 6.36} \qquad pK_a = 4.76 \qquad \underset{\substack{\text{strongest}}}{pK_a = -7}$$

2.19 Arrange the compounds in each set in order of increasing base strength. Consult Table 2.2 for pK_a values of the conjugate acid of each base. (*Hint*: The stronger the acid, the weaker its conjugate base, and vice versa.)

To assess the relative strengths of bases, examine the pK_a values of the conjugate acids. The stronger the conjugate acid, the weaker the base is. Note that some of the trends cannot be easily explained using the structural factors that affect acid-base strengths.

(a)
$$HO\overset{O}{\overset{\|}{C}}O^- \qquad NH_3 \qquad CH_3CH_2\overset{\cdot\cdot}{O}^-$$

$$\underset{\substack{\text{weakest base}}}{6.34} \qquad 9.24 \qquad \underset{\substack{\text{strongest base}}}{15.9} \qquad pK_a \text{ of conjugate acid}$$

(b)
$$CH_3\overset{O}{\overset{\|}{C}}O^- \qquad HO\overset{O}{\overset{\|}{C}}O^- \qquad HO^-$$

$$\underset{\substack{\text{weakest base}}}{4.76} \qquad 6.34 \qquad \underset{\substack{\text{strongest base}}}{15.7} \qquad pK_a \text{ of conjugate acid}$$

(c)
$$H_2O \qquad CH_3\overset{O}{\overset{\|}{C}}O^- \qquad NH_3$$

$$\underset{\substack{\text{weakest base}}}{-1.74} \qquad 4.76 \qquad \underset{\substack{\text{strongest base}}}{9.24} \qquad pK_a \text{ of conjugate acid}$$

(d)
$$\underset{\substack{4.76 \\ \text{weakest base}}}{CH_3\overset{\displaystyle O}{\overset{\displaystyle \|}{C}}O^-} \qquad \underset{15.7}{OH^-} \qquad \underset{\substack{38 \\ \text{strongest base}}}{NH_2^-} \qquad pK_a \text{ of conjugate acid}$$

2.20 Using only the Periodic Table, choose the stronger acid of each pair.

To assess the relative strengths of acids, compare the stabilities of the conjugate bases. Stronger acids ionize to give conjugate bases that are more stable.

(a) H_2Se or HBr

The conjugate bases are HSe^- and Br^-. Se and Br are in the same period, so Br^- is more stable due to the higher electronegativity of Br. Thus, HBr is the stronger acid.

(b) H_2Se or H_2Te

The conjugate bases are HSe^- and HTe^-. Se and Te are in the same group, so HTe^- is more stable due to the larger atomic size of Te. Thus, H_2Te is the stronger acid. (Going down a group, atomic size is more important than electronegativity.)

(c) CH_3OH or CH_3SH

The conjugate bases are CH_3O^- and CH_3S^-. O and S are in the same group. CH_3S^- is more stable due to the effect of atomic size, so CH_3SH is the stronger acid.

(d) HCl or HBr

The conjugate bases are Cl^- and Br^-. Because Cl and Br are in the same group, Br^- is more stable due to the effect of atomic size. Therefore, HBr is the stronger acid.

2.21 Explain why H_2S is a stronger acid than H_2O.

As a result of the difference in atomic size between S and O, H_2S ionizes to form a conjugate base (HS^-) that is more stable than the conjugate base of H_2O (HO^-).

2.22 Which is the stronger Brønsted-Lowry base, $CH_3CH_2O^-$ or $CH_3CH_2S^-$? What is the basis for your selection?

$CH_3CH_2O^-$ is the less stable of the two bases due to the smaller atomic size of oxygen relative to sulfur. Stronger bases are less stable than weaker bases, so $CH_3CH_2O^-$ is the stronger of the two bases.

Position of Equilibrium in Acid-Base Reactions

2.23   Unless under pressure, carbonic acid in aqueous solution breaks down into carbon
dioxide and water, and carbon dioxide is evolved as bubbles of gas. Write an
equation for the conversion of carbonic acid to carbon dioxide and water.

$$\underset{\text{HOCOH}}{\overset{\overset{\displaystyle O}{\|}}{}} \quad\longrightarrow\quad CO_2 \;+\; H_2O$$

2.24   For each of the following compounds, will carbon dioxide be evolved when sodium
bicarbonate is added to an aqueous solution of the compound?

(a) H_2SO_4 ($pK_a -5.2$)   (b) CH_3CH_2OH ($pK_a 15.9$)   (c) NH_4Cl ($pK_a 9.24$)

In order for carbon dioxide to be evolved, the bicarbonate ion (HCO_3^-) must be
protonated to form carbonic acid, H_2CO_3 ($pK_a 6.36$) The acid that protonates
bicarbonate must be a stronger acid than carbonic acid, because acid-base equilibria
favor the side with the weaker acid and base. Of the three choices, only H_2SO_4 is a
stronger acid (lower pK_a) than carbonic acid.

$$HCO_3^- \;+\; H_2SO_4 \longrightarrow H_2CO_3 \;+\; HSO_4^-$$

2.25   Acetic acid, CH_3COOH, is a weak organic acid, pK_a 4.76. Write equations for the
equilibrium reactions of acetic acid with each base. Which equilibria lie considerably
toward the left? Which lie considerably toward the right?

Acid-base equilibria favor the side with the weaker acid, the one with the higher pK_a.
Note that in (d), the right side is also favored because when comparing the two
bases, acetate is more stable than hydroxide due to resonance.

(a) $CH_3COOH \;+\; HCO_3^- \;\rightleftharpoons\; CH_3COO^- \;+\; H_2CO_3$ Right side favored
       $pK_a = 4.76$ $pK_a = 6.36$

(b) $CH_3COOH \;+\; NH_3 \;\rightleftharpoons\; CH_3COO^- \;+\; NH_4^+$ Right side favored
       $pK_a = 4.76$ $pK_a = 9.24$

(c) $CH_3COOH \;+\; H_2O \;\rightleftharpoons\; CH_3COO^- \;+\; H_3O^+$ Left side favored
       $pK_a = 4.76$ $pK_a = -1.74$

(d) $CH_3COOH \;+\; OH^- \;\rightleftharpoons\; CH_3COO^- \;+\; H_2O$ Right side favored
       $pK_a = 4.76$ $pK_a = 15.7$

2.26 The amide ion, NH_2^-, is a very strong base; it is even stronger than OH^-. Write an equation for the reaction that occurs when amide ion is placed in water. Use this equation to show why the amide ion cannot exist in aqueous solution.

The amide ion is a stronger base than hydroxide because a negatively charged nitrogen atom is less stable than a negatively charged oxygen atom (effect of electronegativity). When the amide ion is placed in water, it reacts with water to form OH^-.

$$NH_2^- + H_2O \longrightarrow NH_3 + OH^-$$

2.27 For an acid-base reaction, one way to indicate the predominant species at equilibrium is to say that the reaction arrow points to the acid with the higher value of pK_a. For example

$$NH_4^+ \quad + \quad H_2O \quad \longleftarrow \quad NH_3 \quad + \quad H_3O^+$$
$$pK_a = 9.24 \qquad\qquad\qquad\qquad\qquad pK_a = -1.74$$

$$NH_4^+ \quad + \quad OH^- \quad \longrightarrow \quad NH_3 \quad + \quad H_2O$$
$$pK_a = 9.24 \qquad\qquad\qquad\qquad\qquad pK_a = 15.7$$

Explain why this rule works.

In acid-base equilibria, the position of the equilibrium favors the reaction of the stronger acid and stronger base to give the weaker acid and weaker base. The acid with the higher pK_a is the weaker acid, the direction in which the arrow points. In the examples, NH_4^+ is a weaker acid than H_3O^+ but a stronger acid than H_2O.

Relationship Between Acidity and Basicity and Molecular Structure

2.28 For each pair of compounds, determine the stronger acid without using Table 2.2 and provide a rationale for your answer choice.

In general, stronger acids ionize to form conjugate bases that are more stable. The stronger acid in each pair is indicated in bold.

(a) **$CF_3CH_2CH_2OH$**   versus $CBr_3CH_2CH_2OH$

Both conjugate bases have a negatively charged oxygen. The conjugate base $CF_3CH_2CH_2O^-$ is more stable than $CBr_3CH_2CH_2O^-$ because F exhibits a stronger inductive effect than does Br. Both acids contain the same number of halogen atoms that are the same distance from the negatively charged oxygen atom in the conjugate base.

(b)   **$H_2C=CHOH$**   versus CH_3CH_2OH

The conjugate base of $H_2C=CHOH$ is stabilized by resonance, but that of ethanol is not.

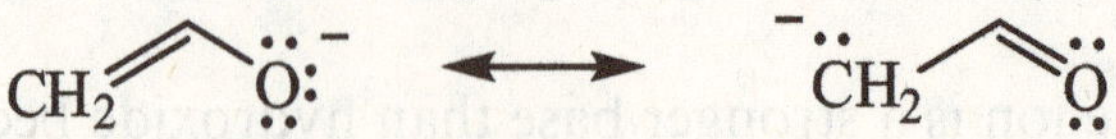

(c)   **HBr**   versus   HCl

Br^- is more stable than Cl^- because of its larger atomic size.

(d)   versus

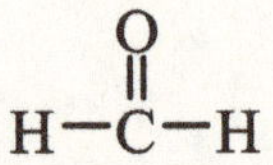

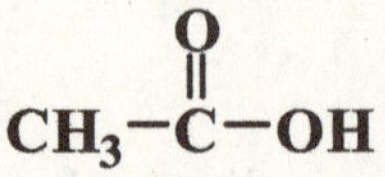

The conjugate base of acetic acid is stabilized by resonance, but that of formaldehyde is not.

(e)   **CH_3OCH_2OH**   versus CH_3SCH_2OH

Like (a), both conjugate bases have a negatively charged oxygen. The conjugate base $CH_3OCH_2O^-$ is more stable than $CH_3SCH_2O^-$ because the ether oxygen exerts a stronger inductive effect than the sulfur atom. Note that even though sulfur is a larger atom than oxygen, we are not considering the effect of atomic size because they are not the atoms that bear negative charge.

(f)   versus

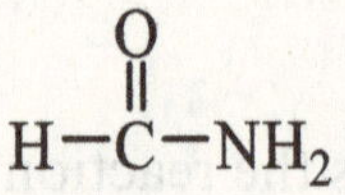

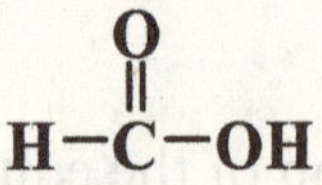

Both compounds form conjugate bases that are stabilized by resonance. The conjugate base of the carboxylic acid has a negatively charged oxygen, but that of the amide has a negatively charged nitrogen. Due to the electronegativity effect, the negatively charged oxygen is more stable than the negatively charged nitrogen.

(g) CH_3NH_2 versus   **FCH_2NH_2**

The conjugate bases are CH_3NH^- and FCH_2NH^-. The latter is more stable due to the presence of the inductively withdrawing fluorine.

2.29   For each compound, determine the more basic of the atoms highlighted in yellow and provide a rationale for your answer.

(a)

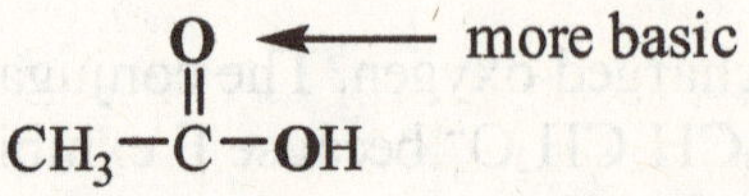

Two contributing structures for the carboxyl group are shown below. The oxygen of the carbonyl group is more basic because it has a partial negative charge and is therefore more likely to accept a proton, while the oxygen of the OH group has a partial positive charge and is much less likely to accept a proton.

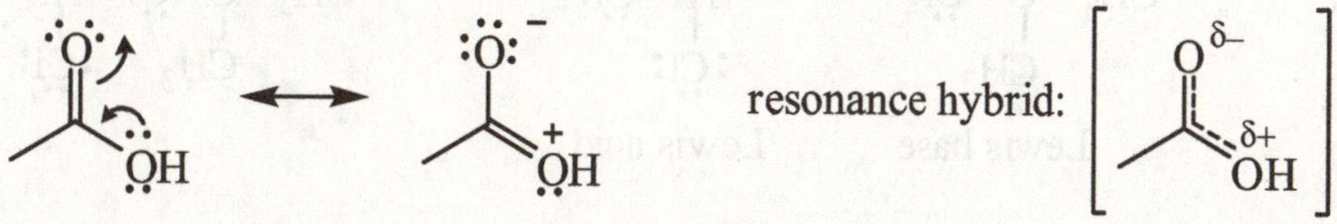

(b)

$$CH_3-\overset{\overset{\displaystyle NH}{\|}}{C}-OH \quad \longleftarrow \quad \text{more basic}$$

For the same reason as (a), the N is more basic. However, in this case, there is an additional effect: Because N is lower in electronegativity than O, the N is even more likely to accept a proton.

(c)

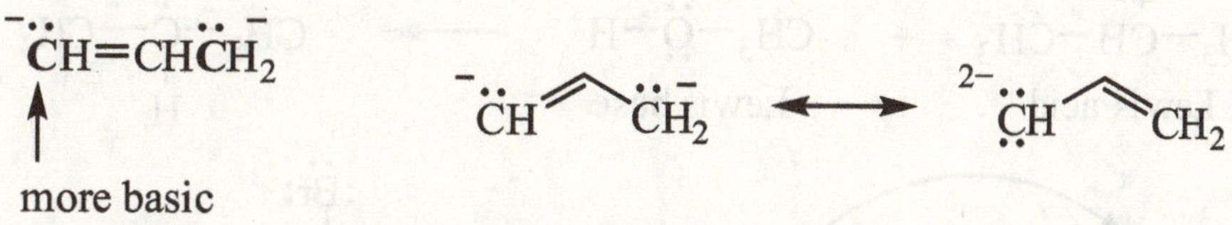

more basic

The carbon on the left is more basic because the carbon on the right is stabilized by resonance. Furthermore, the delocalization of the negative charge of the carbon on the right increases the negative charge of the carbon on the left.

(d) $HOCH_2CH_2NH_2 \quad \longleftarrow \quad \text{more basic}$

The amine is more basic than the alcohol because nitrogen has a lower electronegativity than does oxygen.

Lewis Acids and Bases

2.30 Complete the following acid-base reactions using curved arrow notation to show the flow of electron pairs. In solving these problems, it is essential that you show all valence electrons for the atoms participating directly in each reaction.

A Lewis base donates an electron pair, while a Lewis acid accepts an electron pair. Accordingly, Lewis bases must have a nonbonding pair, while Lewis acids are usually electron-deficient.

(a)

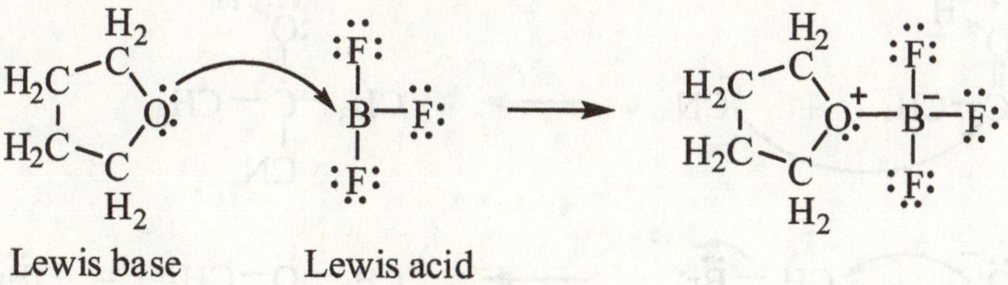

Lewis base   Lewis acid

(b)

$$CH_3-\underset{\underset{CH_3}{|}}{\overset{\overset{CH_3}{|}}{C}}-\ddot{\underset{..}{Cl}}: \quad + \quad \underset{\underset{:\ddot{Cl}:}{|}}{\overset{:\ddot{Cl}:}{Al}}-\ddot{\underset{..}{Cl}}: \quad \longrightarrow \quad CH_3-\underset{\underset{CH_3}{|}}{\overset{\overset{CH_3}{|}}{C}}-\overset{+}{\underset{..}{\ddot{Cl}}}-\underset{\underset{:\ddot{Cl}:}{|}}{\overset{:\ddot{Cl}:}{\overset{|}{Al}}}-\ddot{\underset{..}{Cl}}:$$

Lewis base   Lewis acid

2.31 Complete equations for these reactions between Lewis acid-Lewis base pairs. Label which starting material is the Lewis acid and which is the Lewis base, and use a curved arrow to show the flow of the electron pair in each reaction. In solving these problems, it is essential that you show all valence electrons for the atoms participating directly in each reaction.

(a)

$$CH_3-\overset{+}{C}H-CH_3 \quad + \quad CH_3-\overset{..}{\underset{..}{O}}-H \quad \longrightarrow \quad CH_3-\overset{\overset{H\,\ddot{O}\,\!^{+}\,CH_3}{\diagdown\diagup}}{\underset{\underset{H}{|}}{C}}-CH_3$$

Lewis acid   Lewis base

(b)

$$CH_3-\overset{+}{C}H-CH_3 \quad + \quad :\overset{..}{\underset{..}{Br}}:^{-} \quad \longrightarrow \quad CH_3-\overset{\overset{:\ddot{Br}:}{|}}{\underset{\underset{H}{|}}{C}}-CH_3$$

Lewis acid   Lewis base

(c)

$$CH_3-\overset{\overset{CH_3}{|}}{\underset{\underset{CH_3}{|}}{\overset{+}{C}}} \quad + \quad H-\overset{..}{\underset{..}{O}}-H \quad \longrightarrow \quad CH_3-\overset{\overset{CH_3\ \ H}{|\ \ \diagup}}{\underset{\underset{CH_3}{|}}{C}}-\overset{+}{\underset{\underset{H}{}}{\ddot{O}}}$$

Lewis acid   Lewis base

2.32 Use the curved arrow notation to show the flow of electron pairs in each Lewis acid-base reaction. Be certain to show all valence electron pairs on each atom participating in the reaction.

(a)

$$CH_3-\overset{\overset{\ddot{O}\cdot}{||}}{C}-CH_3 \quad + \quad :CH_3^{-} \quad \longrightarrow \quad CH_3-\overset{\overset{:\ddot{O}:^{-}}{|}}{\underset{\underset{CH_3}{|}}{C}}-CH_3$$

(b)

$$CH_3-\overset{\overset{\overset{+}{O}-H}{||}}{C}-CH_3 \quad + \quad :CN^{-} \quad \longrightarrow \quad CH_3-\overset{\overset{:\ddot{O}-H}{|}}{\underset{\underset{CN}{|}}{C}}-CH_3$$

(c)

$$CH_3-\overset{..}{\underset{..}{O}}:^{-} \quad + \quad CH_3-\overset{..}{\underset{..}{Br}}: \quad \longrightarrow \quad CH_3-\overset{..}{\underset{..}{O}}-CH_3 \quad + \quad :\overset{..}{\underset{..}{Br}}:^{-}$$

Looking Ahead

2.33   Alcohols (Chapter 8) are weak organic acids, pK_a 15–18. The pK_a of ethanol, CH_3CH_2OH, is 15.9. Write equations for the equilibrium reactions of ethanol with each base. Which equilibria lie considerably toward the right? Which lie considerably toward the left?

(a) CH_3CH_2OH + HCO_3^- $\rightleftharpoons$ $CH_3CH_2O^-$ + H_2CO_3 Left side favored

　　pK_a = 15.9   pK_a = 6.36

(b) CH_3CH_2OH + OH^- $\rightleftharpoons$ $CH_3CH_2O^-$ + H_2O Left side favored

　　pK_a = 15.9   pK_a = 15.7   very slightly

(c) CH_3CH_2OH + NH_2^- $\rightleftharpoons$ $CH_3CH_2O^-$ + NH_3 Right side favored

　　pK_a = 15.9   pK_a = 38

(d) CH_3CH_2OH + NH_3 $\rightleftharpoons$ $CH_3CH_2O^-$ + NH_4^+ Left side favored

　　pK_a = 15.9   pK_a = 9.24

2.34   Phenols (Chapter 9) are weak acids and most are insoluble in water. Phenol, C_6H_5OH (pK_a 9.95), for example, is only slightly soluble in water but its sodium salt, $C_6H_5O^-Na^+$, is quite soluble in water. In which one of these solutions will phenol dissolve?

For phenol to dissolve, it must be converted to the phenoxide ion ($C_6H_5O^-$) via an acid-base reaction. Phenol must be a stronger acid than the conjugate acid of the base used to deprotonate it. As before, acid-base equilibria favor the side of the weak acid. In other words, the base used must be a stronger base than the phenoxide ion. Of the three choices, phenol will dissolve in aqueous NaOH and Na_2CO_3.

(a) C_6H_5OH + OH^- $\rightleftharpoons$ $C_6H_5O^-$ + H_2O Right side favored

　　pK_a = 9.95   pK_a = 15.7

(b) C_6H_5OH + HCO_3^- $\rightleftharpoons$ $C_6H_5O^-$ + H_2CO_3 Left side favored

　　pK_a = 9.95   pK_a = 6.36

(c) C_6H_5OH + CO_3^{2-} $\rightleftharpoons$ $C_6H_5O^-$ + HCO_3^- Right side favored

　　pK_a = 9.95   pK_a = 10.33

2.35 Carboxylic acids (Chapter 13) of six or more carbons are insoluble in water, but their sodium salts are very soluble in water. Benzoic acid, C_6H_5COOH (pK_a 4.19), for example, is insoluble in water, but its sodium salt, $C_6H_5COO^-Na^+$, is quite soluble in water. Will benzoic acid dissolve in:

(a) Aqueous NaOH   (b) Aqueous $NaHCO_3$
(c) Aqueous Na_2CO_3 (d) Aqueous $NaCH_3CO_2$

Like phenol in Problem 2.34, benzoic acid needs to be converted to the benzoate anion ($C_6H_5COO^-$) for it to dissolve. Benzoic acid must be a stronger acid than the conjugate acid of the base used to deprotonate it. All four bases will dissolve benzoic acid, as their respective conjugate acids are H_2O (pK_a 15.7), H_2CO_3 (pK_a 6.36), HCO_3^- (pK_a 10.33), and CH_3COOH (pK_a 4.76).

2.36 As we shall see in Chapter 15, hydrogens on a carbon adjacent to a carbonyl group are far more acidic than those not adjacent to a carbonyl group. The highlighted H in propanone, for example, is more acidic than the highlighted H in ethane:

$$CH_3\overset{\overset{\displaystyle O}{\|}}{C}CH_2\!-\!H \qquad CH_3CH_2\!-\!H$$

Propanone   Ethane
pK_a = 22   pK_a = 51

Account for the greater acidity of propanone in terms of (a) the inductive effect and (b) the resonance effect.

(a) Inductive effect: Propanone contains an oxygen atom, which creates a polar carbon-oxygen bond. This polar carbonyl (C=O) group is electron-withdrawing, thereby increasing the stability of the conjugate base of propanone. Ethane does not have an electron-withdrawing group.

(b) Resonance: The conjugate base of propanone is stabilized by resonance, but that of ethane is not.

2.37 Explain why the protons in dimethyl ether, $CH_3\!-\!O\!-\!CH_3$, are not very acidic.

$$CH_3\!-\!\ddot{O}\!-\!CH_3 \;\rightleftharpoons\; CH_3\!-\!\ddot{O}\!-\!\ddot{C}H_2^- \;+\; H^+$$

The conjugate base of dimethyl ether is a highly unstable C^- ion that is stabilized by only the inductive effect of the electronegative oxygen atom. As a result, dimethyl ether is not very acidic.

2.38   Predict whether sodium hydride, NaH, will act as a base or an acid and provide a rationale for your decision.

The hydride anion (H⁻)has a lone pair of electrons and can act as a Lewis base and donate an electron pair. It can also act as a Brønsted base and accept a proton. The Na^+ ion itself is a spectator ion and is not acidic or basic.

2.39   Alanine is one of the 20 amino acids (it contains both an amino and a carboxyl group) found in proteins (Chapter 18) Is alanine better represented by the structural formula A or B? Explain.

$$CH_3-\underset{\underset{NH_2}{|}}{CH}-\overset{\overset{O}{||}}{C}-OH \qquad CH_3-\underset{\underset{\overset{+}{NH_3}}{|}}{CH}-\overset{\overset{O}{||}}{C}-\overset{-}{O}$$

(A)   (B)

Amino acids are bifunctional in that they contain both basic (amino) and acidic (carboxyl) groups. These groups undergo an intramolecular acid-base reaction to give structural formula (B), which is the better representation of alanine. In fact, at physiological pH (near pH 7), all amino acids exist in the form shown in (B).

The reaction occurs because the ammonium ion ($-NH_3^+$) is a weaker acid than the carboxyl group, and the carboxylate ion ($-COO^-$) is a weaker base than the amino group. However, the differences in acidities and basicities are not easily justified by examining the structural factors that affect acid-base strength.

2.40   Glutamic acid is another of the amino acids found in proteins (Chapter 18). Glutamic acid has two carboxyl groups, one with pK_a 2.10, the other pK_a 4.07.

(a)   Which carboxyl group has which pK_a?

$$HO-\overset{\overset{O}{||}}{C}-CH_2-CH_2-\underset{\underset{\overset{+}{NH_3}}{|}}{CH}-\overset{\overset{O}{||}}{C}-OH$$

pK_a = 4.07   pK_a = 2.10

(b)   Account for the fact that one carboxyl group is a considerably stronger acid than the other.

Deprotonation of each carboxyl group yields a resonance-stabilized carboxylate ion (the conjugate base). However, the conjugate base on the right is closer to the positively charged ammonium ion than is the conjugate base on the left. Because inductive effects diminish with distance, the conjugate base on the right experiences a greater inductive effect and is more stable, which results in the acid being stronger.

Group Learning Activities

2.41   Take turns naming one of the functional groups presented in Section 1.7. Discuss and show how each functional group can act as an acid, a base, or both.

A hydroxyl group can act as acid (a proton donor) and as a base (a proton acceptor).

As an acid: $-OH$ $\xrightarrow{\text{base}}$ $-O^-$

As a base: $-OH$ $\xrightarrow{\text{acid}}$ $-\overset{+}{O}H_2$

An amino group can also act as an acid and as a base. However, because the amide anion is not very stable, the amino group is a very weak acid and has a very high pK_a.

As an acid: $-NH_2$ $\xrightarrow{\text{base}}$ $-\overset{..}{N}H$

As a base: $-NH_2$ $\xrightarrow{\text{acid}}$ $-\overset{+}{N}H_3$

The carbonyl group of an aldehyde can only act as a base. The aldehydic proton (the hydrogen that is bonded to the carbonyl group) is not acidic.

As a base: $-\overset{\displaystyle O}{\overset{\|}{C}}-H$ $\xrightarrow{\text{acid}}$ $-\overset{\displaystyle \overset{+}{O}H}{\overset{\|}{C}}-H$

The carbonyl group of a ketone does not have any hydrogen atoms. It can only act as a base.

As a base: $-\overset{\displaystyle O}{\overset{\|}{C}}-$ $\xrightarrow{\text{acid}}$ $-\overset{\displaystyle \overset{+}{O}H}{\overset{\|}{C}}-$

A carboxyl group can act as an acid and as a base. As an acid, it can donate the carboxyl proton. As a base, the carbonyl group can accept a proton (recall from Problem 2.29 that the carbonyl oxygen is more basic than the hydroxyl oxygen).

As an acid: $-\overset{\displaystyle O}{\overset{\|}{C}}-OH$ $\xrightarrow{\text{base}}$ $-\overset{\displaystyle O}{\overset{\|}{C}}-O^-$

As a base: $-\overset{\displaystyle O}{\overset{\|}{C}}-OH$ $\xrightarrow{\text{acid}}$ $-\overset{\displaystyle \overset{+}{O}H}{\overset{\|}{C}}-OH$

2.42 Starting at the top of the following table, take turns explaining why each acid is less acidic than the acid directly below it. Which factor in Section 2.5 plays the most dominant role in your explanation?

Acid	Formula	pK_a	Explanation
ethane	CH_3CH_3	51	$CH_3CH_2^-$ is less stable than NH_2^- because of electronegativity.
ammonia	NH_3	38	NH_2^- is less stable than $CH_3CH_2O^-$ because of electronegativity.
ethanol	CH_3CH_2OH	15.9	$CH_3CH_2O^-$ is less stable than $CH_3CH_2S^-$ because of atomic size.
ethanethiol	CH_3CH_2SH	10.6	Despite the effect of atomic size, $CH_3CH_2S^-$ is less stable than $C_6H_5O^-$ because it is not stabilized by resonance.
phenol	C_6H_5OH	9.95	$C_6H_5O^-$ and CH_3COO^- are stabilized by resonance, but the latter delocalizes the charge to another oxygen atom.
acetic acid	CH_3COOH	4.76	CH_3COO^- is less stable than CF_3COO^- due to the lack of an inductive effect.
trifluoroacetic acid	CF_3COOH	0.23	

2.43 Scientists have determined that the acidity of the upper-ocean has increased an average of 30%, going from a pH value of 8.2 to 8.1 over a 250-year period. This increase in acidity is attributed to incrased CO_2 levels, which in turn affects the availability of CO_3^{2-}. With this in mind, discuss the following as a group:

(a) Explain the relationship between the pH of seawater and the availability of carbonate ion. Does the change in pH from 8.2 to 8.1 increase or decrease the availability of the carbonate ion?

The carbonate ion is a weak base that can accept a proton to form the bicarbonate ion, HCO_3^-. At a lower pH value, there is a higher proton concentration, so the formation of bicarbonate is more likely to occur. As a result, less carbonate is available at pH 8.1 than at 8.2.

(b) Using your knowledge of acid-base chemistry, complete the equations in the diagram below to show how CO_2 levels influence the availability of CO_3^{2-} in seawater.

When the concentration of CO_2 in the atmosphere is higher, the concentration of carbonic acid (H_2CO_3) in seawater is also higher. Carbonic acid ionizes to release H^+ and HCO_3^-. The H^+ reacts with the OH^- formed when CO_3^{2-} reacts with water, and this in turn causes more CO_3^{2-} to react with water. The net effect is a decrease in the concentration of CO_3^{2-} in seawater.

CO_2 (g)   Air

$H_2CO_3 \rightleftharpoons H^+ + HCO_3^-$ Ocean

$Ca^{2+} + CO_3^{2-} \overset{H_2O}{\rightleftharpoons} OH^- + HCO_3^-$

$CaCO_3$ (s)   Ocean floor

(c) How would a decrease in available carbonate ion, CO_3^{2-}, affect marine organisms' availability to build and maintain shells and other body parts from calcium carbonate?

A lower carbonate concentration would make it more difficult for such organisms to to form calcium carbonate. This is because the calcium carbonate that is formed can redissolve more easily when the carbonate concentration is low.

2.44 The pK_a of a CH_3 proton in propane is 51. The pK_a values for the CH_3 protons on propene and acetone are considerably lower. Discuss why these CH_3 protons are so much more acidic. Use drawings to inform your discussion.

The conjugate base of propane, $CH_3CH_2CH_2^-$, is not stabilized by resonance. A negatively charged carbon is highly unstable, which is why propane is such a weak acid.

On the other hand, the conjugate bases of propene and acetone are stabilized by resonance. The conjugate base of acetone is more stable than that of propene, because the conjugate base of acetone has its charge stabilized by an oxygen atom.

Chapter 3: Alkanes and Cycloalkanes

Problems

3.1 Do the structural formulas in each pair represent the same compound or constitutional isomers?

Constitutional isomers have the same molecular formula but different connectivity of atoms. To identify constitutional isomers, it is best to first identify the longest chain of carbon atoms. Then, number this chain from the end closest to a branch. Finally, compare the length of each chain, the locations of any branches, and what the branches are. If any of these are not the same for two structures, they are constitutional isomers.

(a)

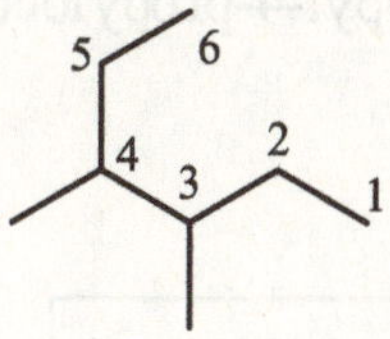

and

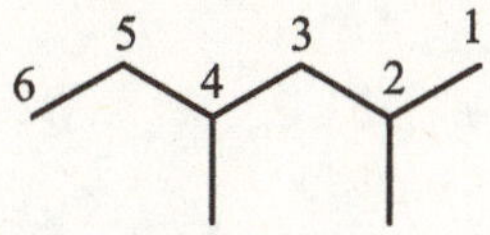

(b)

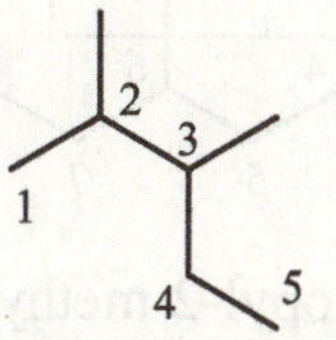

and

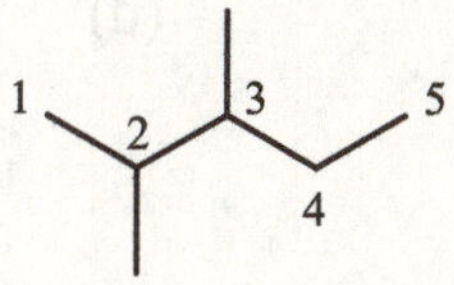

In pair (a), the longest chain in both structures is six carbons long. The structure on the left has branches of $-CH_3$ groups at positions 3 and 4, while the structure on the right has the $-CH_3$ groups on carbons 2 and 4. Pair (a) represents a pair of constitutional isomers.

In pair (b), the longest chain in both structures is five carbons long. Both structures feature $-CH_3$ branches at carbons 2 and 3. Pair (b) represents the same compound.

3.2 Draw structural formulas for the three constitutional isomers of molecular formula C_5H_{12}.

To draw constitutional isomers from a molecular formula, focus on the carbon chain and draw the line-angle diagram of the longest possible carbon chain. Then, shorten the chain by one carbon and draw all the possible branching combinations. Repeat the process until no more unique structures can be drawn. (Be attentive – structures that represent the same constitutional isomer can appear unique when they are not; for instance, the two structures shown in the middle represent the same compound.)

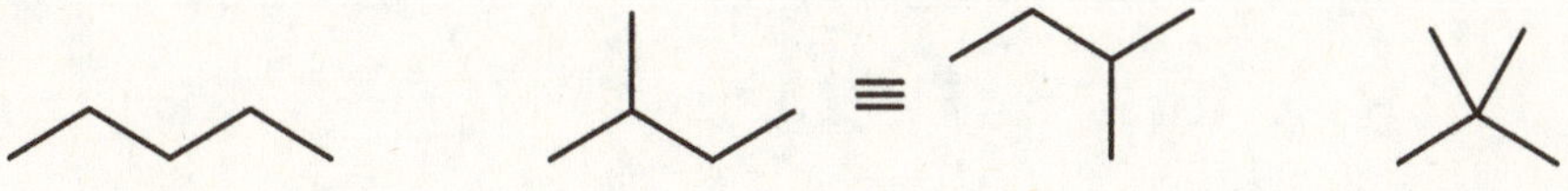

3.3   Write IUPAC names for these alkanes:

The parent name of the alkane is based on the longest chain, which is numbered from the end bearing the nearest substituent. Substituents (shown in boxes below) are named and placed in alphabetical order in front of the parent name of the alkane. Prefixes such as *di*, *tri*, *sec*, and *tert* are ignored in the alphabetizing, but *iso* (as in *isopropyl*) is included.

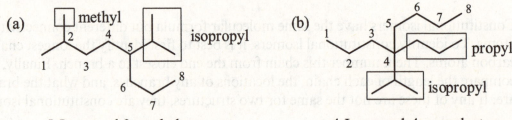

(a)   5-Isopropyl-2-methyloctane

(b)   4-Isopropyl-4-propyloctane

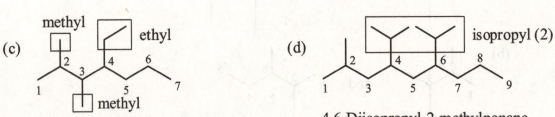

(c)   4-Ethyl-2,3-dimethylheptane

(d)   4,6-Diisopropyl-2-methylnonane

3.4   Classify each hydrogen atom in the following compounds as 1°, 2°, or 3°.

Hydrogen atoms can be classified based on the number of carbon atoms that are connected to the carbon atom bearing the hydrogen atom of interest. This is similar to how alcohols are classified. A hydrogen atom is 1° if the carbon bearing it is connected to one carbon, 2° if connected to two carbons, and 3° if connected to three carbons.

(a)
$$\overset{1°}{CH_3}$$
$$\underset{\substack{1° \quad 3° \quad 2° \quad 2° \quad 1°}}{CH_3CHCH_2CH_2CH_3}$$

(b)
$$\overset{1° \quad 1°}{CH_3 \quad CH_3}$$
$$\underset{\substack{1° \quad 2° \quad CH_3 \quad CH_3 \\ 1° \quad 1°}}{CH_3-CH_2-C-CH \; 3°}$$

3.5 Write the molecular formula and IUPAC name for each cycloalkane:

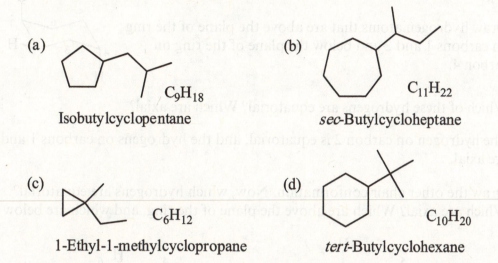

(a)

C_9H_{18}

Isobutylcyclopentane

(b)

$C_{11}H_{22}$

sec-Butylcycloheptane

(c)

C_6H_{12}

1-Ethyl-1-methylcyclopropane

(d)

$C_{10}H_{20}$

tert-Butylcyclohexane

3.6 Combine the proper prefix, infix, and suffix, and write the IUPAC name for each compound:

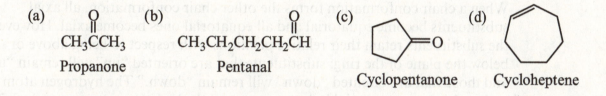

(a)

$$\overset{O}{\overset{\|}{CH_3CCH_3}}$$

Propanone

(b)

$$\overset{O}{\overset{\|}{CH_3CH_2CH_2CH_2CH}}$$

Pentanal

(c)

Cyclopentanone

(d)

Cycloheptene

3.7 Draw Newman projections for two staggered and two eclipsed conformations of 1,2-dichloroethane.

Line-angle diagrams are also provided below the Newman projections for reference.

Staggered   Eclipsed

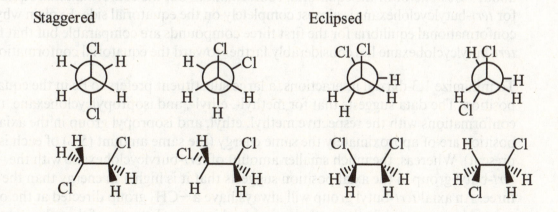

3.8 Following is a chair conformation of cyclohexane with carbon atoms numbered 1 through 6:

(a) Draw hydrogen atoms that are above the plane of the ring on carbons 1 and 2 and below the plane of the ring on carbon 4.

(b) Which of these hydrogens are equatorial? Which are axial?

The hydrogen on carbon 2 is equatorial, and the hydrogens on carbons 1 and 4 are axial.

(c) Draw the other chair conformation. Now, which hydrogens are equatorial? Which are axial? Which are above the plane of the ring, and which are below it?

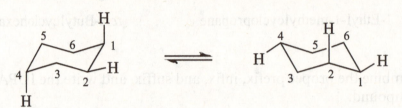

When a chair conformation forms the other chair conformation, all axial substituents become equatorial and all equatorial ones become axial. However, the substituents retain their relative positions with respect to being above or below the plane of the ring; substituents that are oriented "up" will remain "up" and those that are oriented "down" will remain "down." The hydrogen atom on carbon 1 is now equatorial but remains "up," the hydrogen atom on carbon 2 is now axial but remains "up," and the hydrogen atom on carbon 4 is now equatorial but remains "down."

3.9 The conformational equilibria for methyl-, ethyl-, and isopropylcyclohexane are all about 95% in favor of the equatorial conformation, but the conformational equilibrium for *tert*-butylcyclohexane is almost completely on the equatorial side. Explain why the conformational equilibria for the first three compounds are comparable but that for *tert*-butylcyclohexane lies considerably farther toward the equatorial conformation.

To minimize 1,3-diaxial interactions, a large substituent prefers to be in the equatorial position. The data suggests that for methyl-, ethyl-, and isopropylcyclohexane, the conformations with the respective methyl, ethyl, and isopropyl group in the axial position are of approximately the same energy (the same amount (5%) of each is present). Whereas, the much smaller amount of *tert*-butylcyclohexane with the *tert*-butyl group in the axial position suggests that it is higher in energy than the other three. An axial *tert*-butyl group will always have a $-CH_3$ group directed at the other axial substituents while the methyl, ethyl, and isopropyl groups of the other three compounds can adopt a conformation where the $-CH_3$ groups face away from the other axial positions.

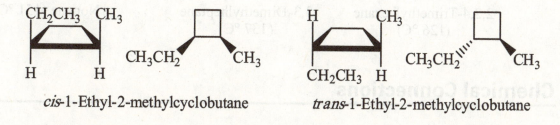

| methyl substituent | ethyl substituent | isopropyl substituent | *tert*-butyl substituent |

3.10 Which cycloalkanes show *cis-trans* isomerism? For each that does, draw both isomers.

Recall that *cis-trans* isomerism in cycloalkenes refers to the relative positions of two substituents that are bonded to different carbons of the cycloalkane ring. Therefore, *cis-trans* isomerism is only possible in (a) and (c).

(a) 1,3-Dimethylcyclopentane

cis-1,3-Dimethylcyclopentane   *trans*-1,3-Dimethylcyclopentane

(b) Ethylcyclopentane

Because there is only one substituent, *cis-trans* isomerism is not possible.

(c) 1-Ethyl-2-methylcyclobutane

cis-1-Ethyl-2-methylcyclobutane   *trans*-1-Ethyl-2-methylcyclobutane

3.11 Following is a planar hexagon representation for one isomer of 1,2,4-trimethyl-cyclohexane. Draw alternative chair conformations of this compound, and state which chair conformation is the more stable.

The conformation that is more stable is the one that places as many of the methyl substituents as possible on the equatorial positions of the cyclohexane ring. This reduces the number of 1,3-diaxial repulsive interactions. Note that when drawing conformational representations, the proper *cis-trans* relationships of the substituents must be maintained.

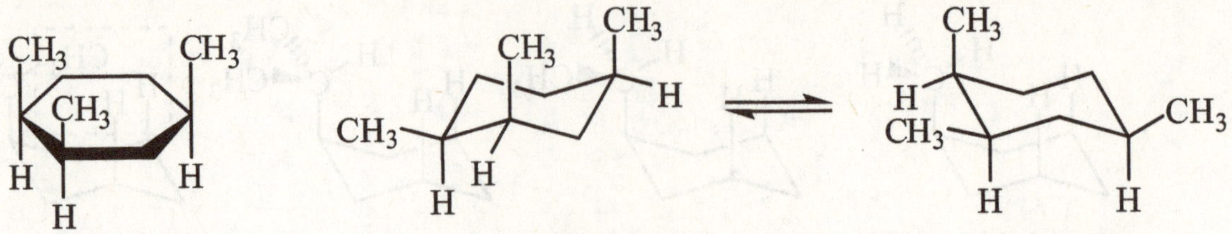

less stable   more stable

3.12 Arrange the alkanes in each set in order of increasing boiling point:

When a substance boils, the intermolecular forces between the molecules in the liquid state are overcome. The only intermolecular forces exhibited by alkanes are dispersion forces, which increase with both molecular size and surface area. For alkanes that are constitutional isomers, the surface area decreases as branching increases. This decrease of surface area decreases the amount of dispersion forces, resulting in lower boiling points.

(a) 2-Methylbutane, 2,2-dimethylpropane, and pentane

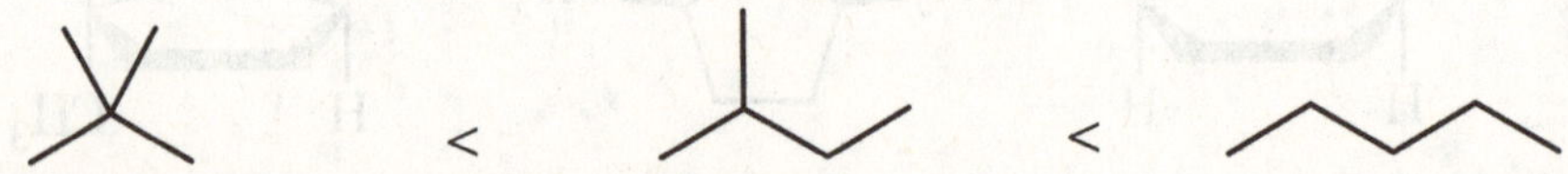

2,2-Dimethylpropane (10 °C)   2-Methylbutane (28 °C)   Pentane (36 °C)

(b) 3,3-Dimethylheptane, 2,2,4-trimethylhexane, and nonane

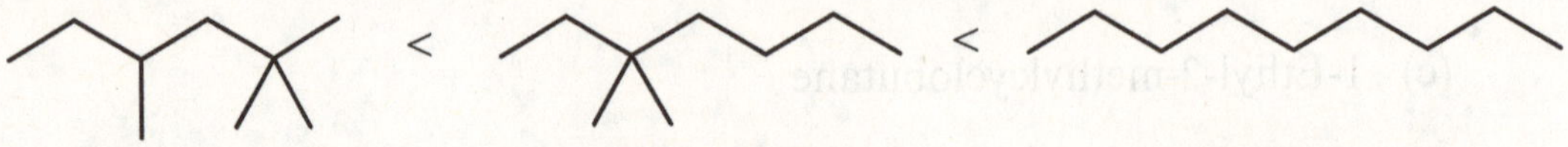

2,2,4-Trimethylhexane   3,3-Dimethylheptane   Nonane (151 °C)
(126 °C)   (137 °C)

Chemical Connections

3A. How many chair conformations are present in tetrodotoxin? Which substituents in tetrodotoxin are involved in axial-axial interactions?

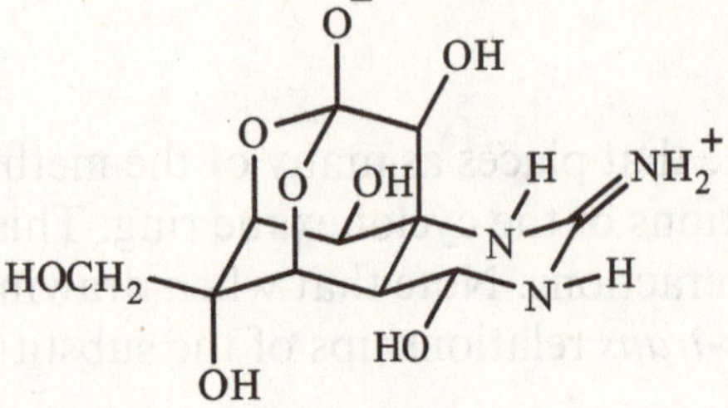

Tetrodoxin contains four chair conformations, which are indicated in bold in the following structures. Axial-axial interactions between substituents are present in three of the chairs; these substituents are highlighted and are axial with respect to the bolded chair.

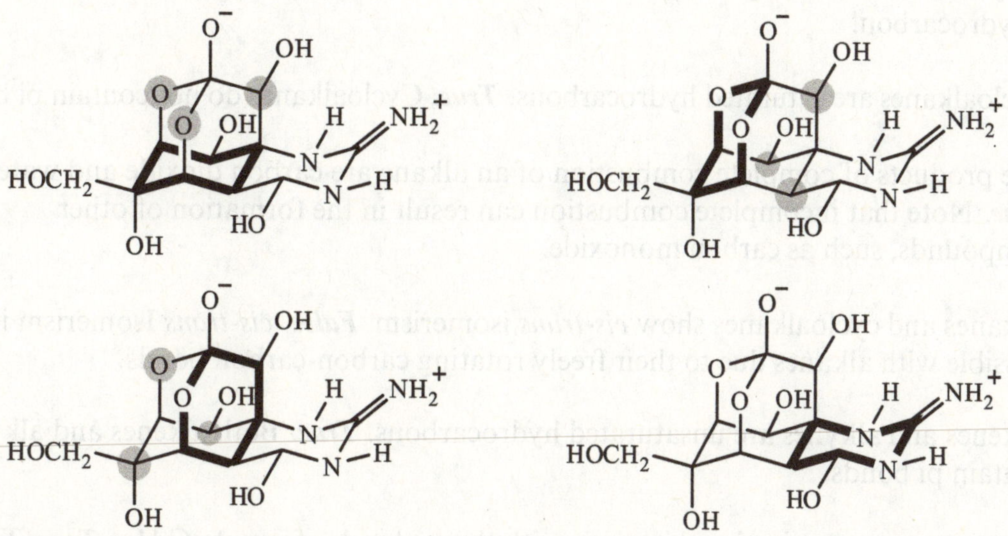

3B. Which would you expect to have a higher boiling point, octane or isooctane (2,2,4-trimethylpentane)?

Octane and isooctane are constitutional isomers; they have the same molecular formula, same molecular weight, but different connectivity of atoms. Octane, an unbranched alkane, has a higher boiling point than isooctane, a branched alkane. Compared to unbranched molecules of similar molecular weight, branched molecules have a smaller surface area and therefore a reduced amount of intermolecular forces.

Quick Quiz

1. Combustion of alkanes is an endothermic process. *False*. When alkanes burn, the reaction releases heat, so combustion is an exothermic process.

2. All alkanes that are liquid at room temperature are more dense than water. *False*. Alkanes that are liquids have a lower density than water. This is why gasoline floats on water.

3. The two main sources of alkanes the world over are petroleum and natural gas. *True*. These two fossil fuels supply the majority of our alkanes.

4. There are four alkyl groups with the molecular formula C_4H_9. *True*. They are the butyl, isobutyl, *sec*-butyl, and *tert*-butyl groups.

5. Sets of constitutional isomers have the same molecular formula and the same physical properties. *False*. Constitutional isomers have the same formula, but not necessarily the same physical properties.

6. A hydrocarbon is composed of only carbon and hydrogen. *True*. By definition, that is a hydrocarbon!

7. Cycloalkanes are saturated hydrocarbons. *True*. Cycloalkanes do not contain pi bonds.

8. The products of complete combustion of an alkane are carbon dioxide and water. *True*. Note that incomplete combustion can result in the formation of other compounds, such as carbon monoxide.

9. Alkanes and cycloalkanes show *cis-trans* isomerism. *False*. *cis-trans* Isomerism is not possible with alkanes due to their freely rotating carbon-carbon bonds.

10. Alkenes and alkynes are unsaturated hydrocarbons. *True*. Both alkenes and alkynes contain pi bonds.

11. There are two constitutional isomers with the molecular formula C_4H_{10}. *True*. They are butane and 2-methylpropane.

12. Hexane and cyclohexane are constitutional isomers. *False*. Constitutional isomers must have the same molecular formula, which is not the case with hexane and cyclohexane.

13. The propyl and isopropyl groups are constitutional isomers. *True*. They are both substituents with the formula C_3H_7.

14. There are five constitutional isomers with the molecular formula C_5H_{12}. *False*. There are only three constitutional isomers, and they are pentane, 2-methylbutane, and 2,2-dimethylpropane.

15. Boiling points among alkanes with unbranched carbon chains increase as the number of carbons in the chain increases. *True*. Increasing the number of carbons in the chain increases the surface area of the molecule, which increases dispersion forces.

16. In a cyclohexane ring, if an axial bond is above the plane of the ring on a particular carbon atom, axial bonds on the two adjacent carbons are below the plane of the ring. *True*. Recall the alternating up/down pattern of axial bonds on adjacent carbons.

17. Fractional distillation of petroleum separates hydrocarbons based on their melting points. *False*. The basis of separation by distillation is boiling point.

18. Among alkane constitutional isomers, the least branched isomer generally has the lowest boiling point. *False*. The opposite is true, and the least branched isomer usually has the highest boiling point.

19. The parent name of a cycloalkane is the name of the unbranched alkane with the same number of carbon atoms as are in the cycloalkane ring. *True*. The prefix *cyclo* is added to the parent name of the unbranched alkane.

20. Octane and 2,2,4-trimethylpentane are constitutional isomers and have the same octane number. *False*. They are constitutional isomers, and as a result, they have different chemical properties and thus do not have the same octane number.

21. Liquid alkanes and cycloalkanes are soluble in each other. *True*. Alkanes and cycloalkanes are both nonpolar, so they are soluble in each other.

22. Alkanes and cycloalkanes are insoluble in water. *True*. Hydrocarbons, which are nonpolar, will not dissolve in highly polar compounds such as water.

23. The more stable chair conformation of a substituted cyclohexane has the greater number of substituents in equatorial positions. *True*. This minimizes the amount of the diaxial repulsions.

24. The parent name of an alkane is the name of the longest chain of carbon atoms. *True*. When naming alkanes, always name it based on the longest chain.

25. Alkanes are saturated hydrocarbons. *True*. Alkanes do not contain pi bonds.

26. The general formula of an alkane is C_nH_{2n}, where n is the number of carbon atoms in the alkane. *False*. The general formula of an alkane is C_nH_{2n+2}.

27. The octane number of a particular gasoline is the number of grams of octane per liter. *False*. The octane number is a measure of how resistant is the gasoline, relative to isooctane, to engine knocking.

28. *Cis* and *trans* isomers have the same molecular formula, the same connectivity, and the same physical properties. *False*. They have the same molecular formula and the same connectivity (i.e. they are of the same constitutional isomer), but because they are stereoisomers, they have different physical properties.

29. A *cis* isomer of a disubstituted cycloalkane can be converted to a *trans* isomer by rotation about an appropriate carbon-carbon single bond. *False*. Stereoisomers do not interconvert and cannot be made to interconvert by rotation about single bonds.

30. All cycloalkanes with two substituents on the ring show *cis-trans* isomerism. *False*. Whether *cis-trans* isomerism is possible or not depends on the relative locations of the substituents. For instance, if both substituents are on the same carbon of a cycloalkane, *cis-trans* isomerism is not possible.

31. In all conformations of ethane, propane, butane, and higher alkanes, all C−C−C and C−C−H bond angles are approximately 109.5°. *True*. All carbon atoms are tetrahedral.

32. Conformations have the same molecular formula and the same connectivity, but differ in the three-dimensional arrangement of their atoms in space. *True*. Note that the different three-dimensional arrangements must also be able to interconvert.

33. Constitutional isomers have the same molecular formula and the same connectivity of their atoms. *False*. They have the same molecular formula but different atom connectivity.

End-of-Chapter Problems

Structure of Alkanes

3.13 For each condensed structural formula, write a line-angle formula:

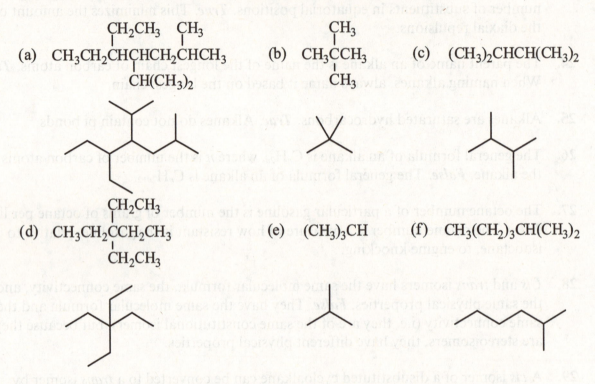

(a) $CH_3CH_2CHCHCH_2CHCH_3$ with CH_2CH_3 and CH_3 above, and $CH(CH_3)_2$ below

(b) CH_3CCH_3 with CH_3 above and CH_3 below

(c) $(CH_3)_2CHCH(CH_3)_2$

(d) $CH_3CH_2CCH_2CH_3$ with CH_2CH_3 above and CH_2CH_3 below

(e) $(CH_3)_3CH$

(f) $CH_3(CH_2)_3CH(CH_3)_2$

3.14 Write a condensed structural formula and the molecular formula of each alkane:

(a) $C_{10}H_{22}$

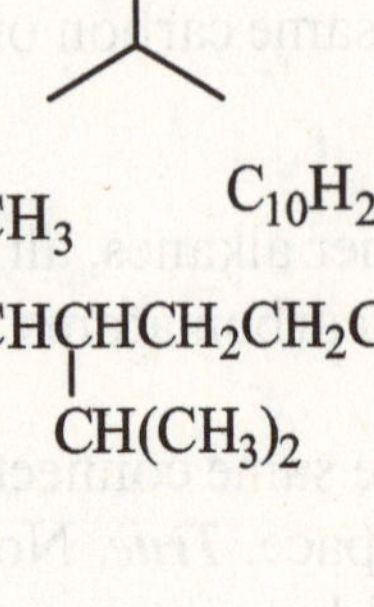

$CH_3CHCHCH_2CH_2CH_3$ with CH_3 above and $CH(CH_3)_2$ below

(b) $(CH_3)_3CC(CH_3)_3$

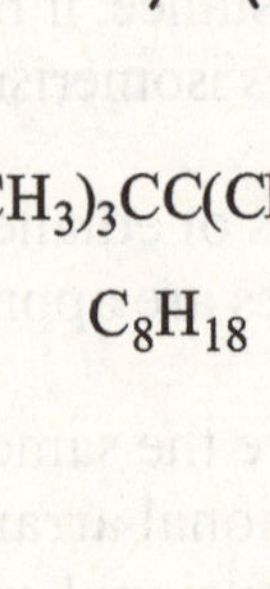

C_8H_{18}

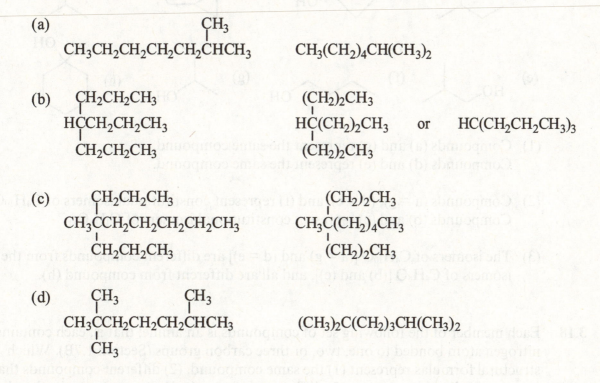

(c)

$C_{11}H_{24}$ $C(CH_3)_3$

$CH_3CH_2CH_2CHCH_2CH_2CH_3$

(d) C_6H_{14}

$CH_3CH(CH_2CH_3)_2$
or
$CH_3CH_2CH(CH_3)CH_2CH_3$

3.15 For each of the following condensed structural formulas, provide an even more abbreviated formula, using parentheses and subscripts:

(a) CH_3

$CH_3CH_2CH_2CH_2CH_2CHCH_3$ $CH_3(CH_2)_4CH(CH_3)_2$

(b) $CH_2CH_2CH_3$

$HCCH_2CH_2CH_3$

$CH_2CH_2CH_3$

$(CH_2)_2CH_3$

$HC(CH_2)_2CH_3$ or $HC(CH_2CH_2CH_3)_3$

$(CH_2)_2CH_3$

(c) $CH_2CH_2CH_3$

$CH_3CCH_2CH_2CH_2CH_2CH_3$

$CH_2CH_2CH_3$

$(CH_2)_2CH_3$

$CH_3C(CH_2)_4CH_3$

$(CH_2)_2CH_3$

(d) CH_3 CH_3

$CH_3CCH_2CH_2CH_2CHCH_3$ $(CH_3)_2C(CH_2)_3CH(CH_3)_2$

CH_3

Constitutional Isomerism

3.16 Which statements are true about constitutional isomers?

Constitutional isomers have the same molecular formula but the atoms are connected in a different sequence (connectivity). They have different properties and do not interconvert. Conformations are not constitutional isomers because conformations interconvert and have the same atom connectivity.

(a) They have the same molecular formula.   True
(b) They have the same molecular weight.   True
(c) They have the same order of attachment of atoms.   False

(d) They have the same physical properties.   False
(e) Conformations are not constitutional isomers.   True

3.17 Each member of the following set of compounds is an alcohol; that is, each contains an
−OH (hydroxyl group, Section 1.7A). Which structural formulas represent (1) the same compound, (2) different compounds that are constitutional isomers, or (3) different compounds that are not constitutional isomers?

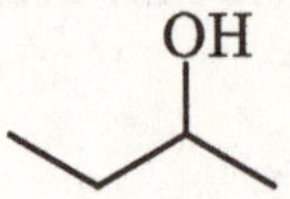

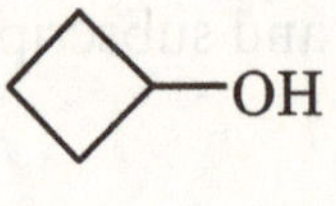

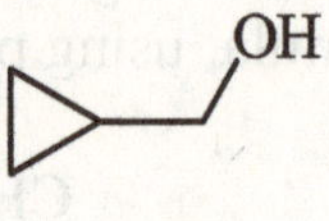

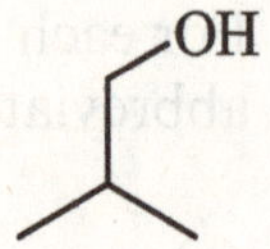

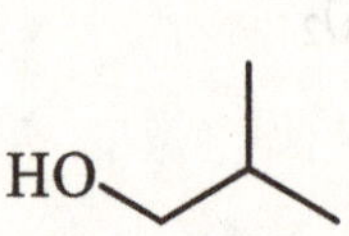

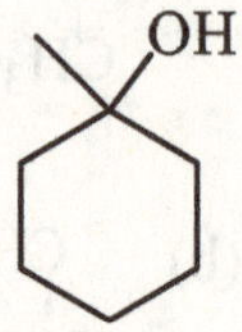

(1) Compounds (a) and (g) represent the same compound.
      Compounds (d) and (e) represent the same compound.

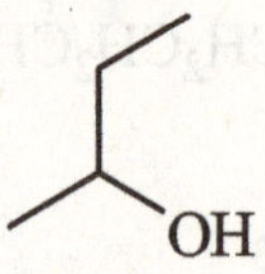

(2) Compounds (a = g), (d = e), and (f) represent constitutional isomers of $C_4H_{10}O$.
      Compounds (b) and (c) represent constitutional isomers of C_4H_8O.

(3) The isomers of $C_4H_{10}O$ [(a = g) and (d = e)] are different compounds from the isomers of C_4H_8O [(b) and (c)], and all are different from compound (h).

3.18 Each member of the following set of compounds is an amine; that is, each contains a nitrogen atom bonded to one, two, or three carbon groups (Section 1.7B). Which structural formulas represent (1) the same compound, (2) different compounds that are constitutional isomers, or (3) different compounds that are not constitutional isomers?

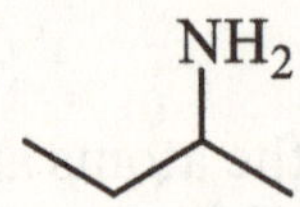

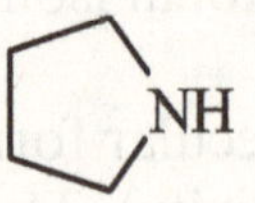

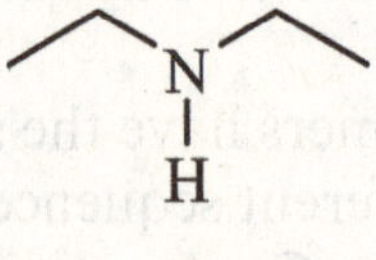

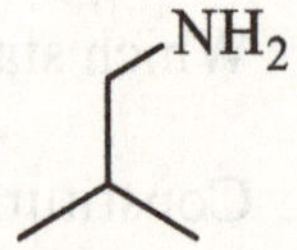

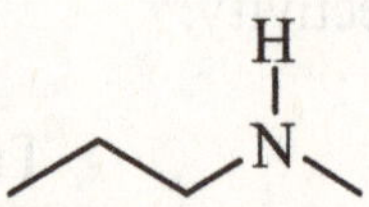

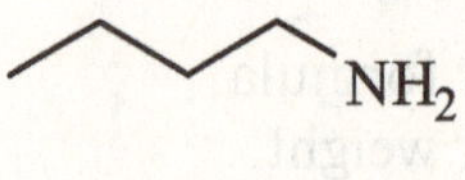

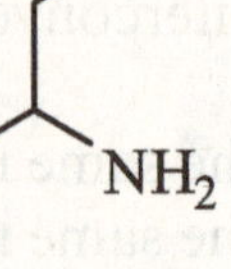

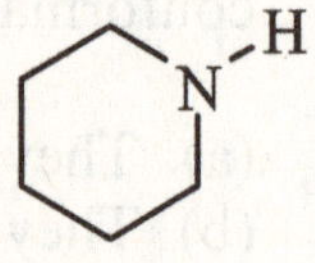

(1) Compounds (a) and (g) represent the same compound.

(2) Compounds (a = g), (c), (d), (e), and (f) represent constitutional isomers of $C_4H_{11}N$.

(3) Compounds (b) and (h) represent different compounds, and they are also
      different from the constitutional isomers of $C_4H_{11}N$ indicated in (2).

3.19 Each member of the following set of compounds is either an aldehyde or ketone
        (Section 1.7C). Which structural formulas represent (1) the same compound,
        (2) different compounds that are constitutional isomers, or (3) different compounds
        that are not constitutional isomers?

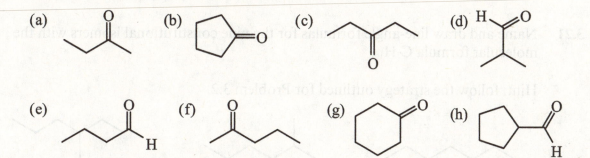

(1) None of the compounds are the same.

(2) Compounds (a), (d), and (e) represent constitutional isomers of C_4H_8O.
      Compounds (c) and (f) represent constitutional isomers of $C_5H_{10}O$.
      Compounds (g) and (h) represent constitutional isomers of $C_6H_{10}O$.

(3) Compound (b), with formula C_5H_8O, is a different compound than all of the
      isomers listed for each of the respective molecular formulas indicated in (2).

3.20 For each pair of compounds, tell whether the structural formulas shown represent
        (1) the same compound, (2) different compounds that are constitutional isomers, or
        (3) different compounds that are not constitutional isomers:

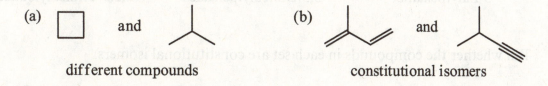

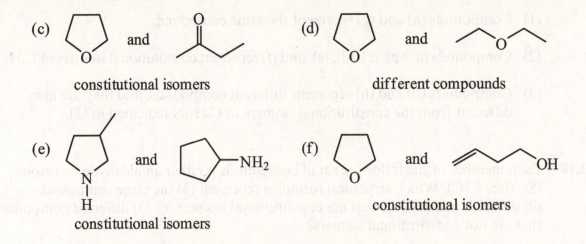

(c) and

constitutional isomers

(d) and

different compounds

(e) and —NH₂

constitutional isomers

(f) and OH

constitutional isomers

3.21 Name and draw line-angle formulas for the nine constitutional isomers with the molecular formula C_7H_{16}.

Hint: follow the strategy outlined for Problem 3.2.

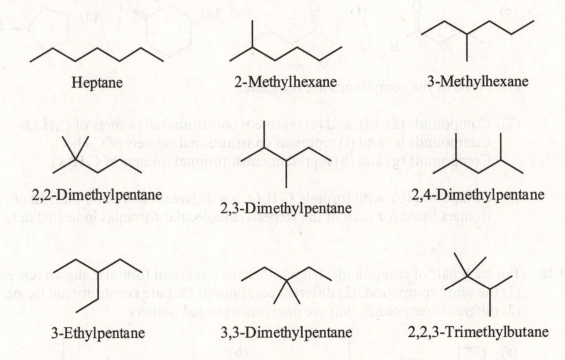

Heptane   2-Methylhexane   3-Methylhexane

2,2-Dimethylpentane   2,3-Dimethylpentane   2,4-Dimethylpentane

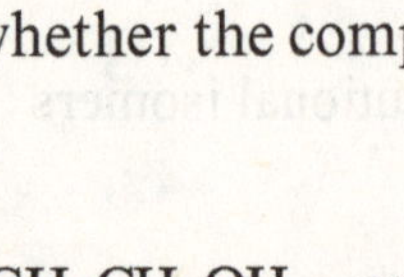

3-Ethylpentane   3,3-Dimethylpentane   2,2,3-Trimethylbutane

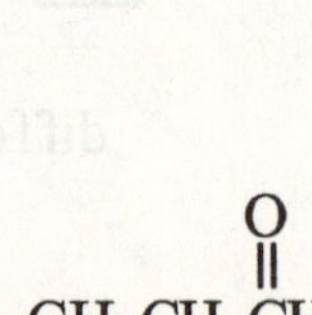

3.22 Tell whether the compounds in each set are constitutional isomers:

(a) CH_3CH_2OH and CH_3OCH_3

constitutional isomers

(b) $CH_3\overset{\overset{\displaystyle O}{\|}}{C}CH_3$ and $CH_3CH_2\overset{\overset{\displaystyle O}{\|}}{C}H$

constitutional isomers

(c) CH₃COCH₃   and   CH₃CH₂COH   (d) CH₃CHCH₂CH₃ and CH₃CCH₂CH₃

constitutional isomers   different compounds

(e) 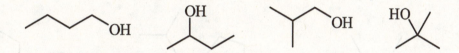 and   CH₃CH₂CH₂CH₂CH₃   (f) and   CH₂=CHCH₂CH₂CH₃

different compounds   constitutional isomers

3.23 Draw line-angle formulas for

(a) The four alcohols with the molecular formula C₄H₁₀O.

(b) The two aldehydes with the molecular formula C₄H₈O.

(c) The one ketone with the molecular formula C₄H₈O.

(d) The three ketones with the molecular formula C₅H₁₀O.

(e) The four carboxylic acids with the molecular formula C₅H₁₀O₂.

(f)   The four amines with the molecular formula C_3H_9N.

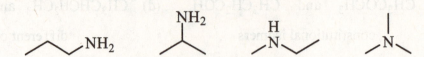

Nomenclature of Alkanes and Cycloalkanes

3.24   Write IUPAC names for these alkanes and cycloalkanes:

(a) $CH_3CHCH_2CH_2CH_3$
                |
               CH_3

2-Methylpentane

(b) $CH_3CHCH_2CH_2CHCH_3$
                | |
               CH_3 CH_3

2,5-Dimethylhexane

(c) $CH_3(CH_2)_4CHCH_2CH_3$
                            |
                         CH_2CH_3

3-Ethyloctane

(d)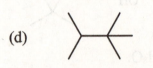

2,2,3-Trimethylbutane

(e)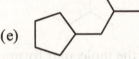

Isobutylcyclopentane

(f)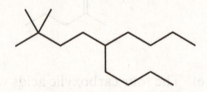

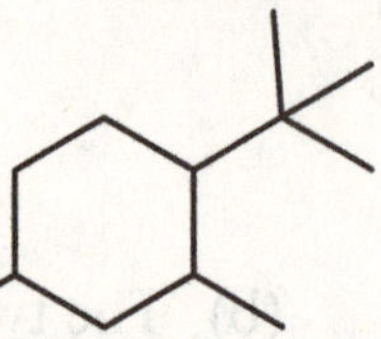

1-*tert*-Butyl-2,4-
dimethylcyclohexane

3.25   Write line-angle formulas for these alkanes:

(a)   2,2,4-Trimethylhexane

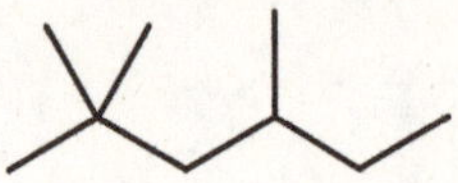

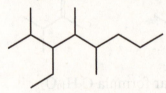

(b)   2,2-Dimethylpropane

(c)   3-Ethyl-2,4,5-trimethyloctane

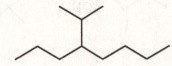

(d)   5-Butyl-2,2-dimethylnonane

(e)   4-Isopropyloctane

(f)   3,3-Dimethylpentane

(g)   *trans*-1,3-Dimethylcyclopentane   (h)   *cis*-1,2-Diethylcyclobutane

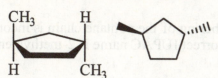   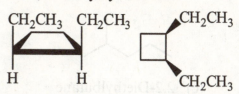

3.26   Following is the structure of limonene, the chemical component of oranges that is partly responsible for the citrus scent. Draw the hydrogens present in limonene and classify those bonded to *sp³*-hybridized carbons as 1°, 2°, or 3°.

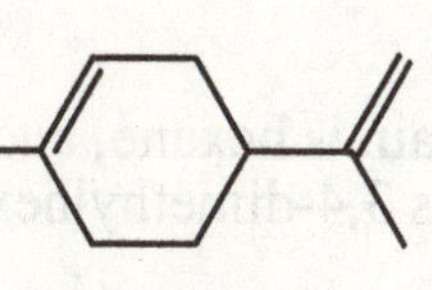

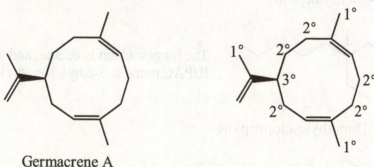

Limonene

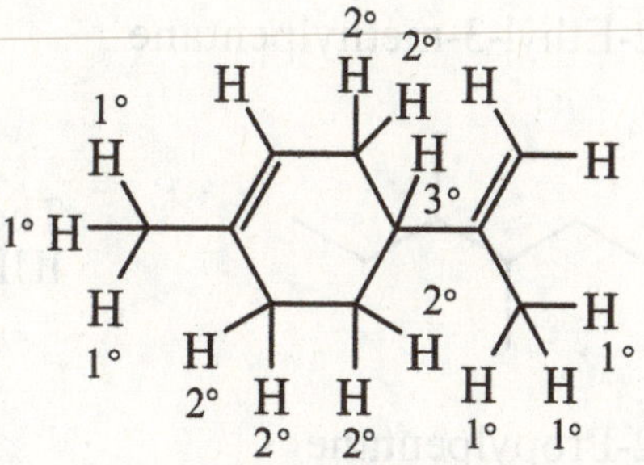

3.27   Following is the structure of Germacrene A, a hydrocarbon synthesized in plants and studied for their insecticidal properties. Classify each of the *sp³*-hybridized carbons on Germacrene A as 1°, 2°, 3° or 4°.

Germacrene A

3.28   Explain why each of the following names is an incorrect IUPAC name and write the correct IUPAC name for the intended compound:

(a)   1,3-Dimethylbutane

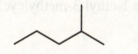

The longest chain is pentane, and the correct IUPAC name is 2-methylpentane.

(b) 4-Methylpentane

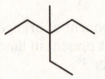

The numbering of the pentane chain is incorrect, and the correct IUPAC name is 2-methylpentane.

(c) 2,2-Diethylbutane

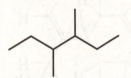

The longest chain is pentane, and the correct IUPAC name is 3-ethyl-3-methylpentane.

(d) 2-Ethyl-3-methylpentane

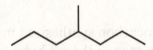

The longest chain is hexane, and the correct IUPAC name is 3,4-dimethylhexane.

(e) 2-Propylpentane

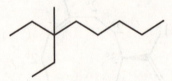

The longest chain is heptane, and the correct IUPAC name is 4-methylheptane.

(f) 2,2-Diethylheptane

The longest chain is octane, and the correct IUPAC name is 3-ethyl-3-methyloctane.

(g) 2,2-Dimethylcyclopropane

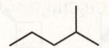

The numbering of the ring is incorrect, and the correct IUPAC name is 1,1-dimethylcyclopropane.

(h) 1-Ethyl-5-methylcyclohexane

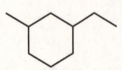

The numbering of the ring is incorrect, and the correct IUPAC name is 1-ethyl-3-methylcyclohexane.

3.29   Draw a structural formula for each compound:

(a) Ethanol

CH_3CH_2OH

(b) Ethanal

$$\overset{\displaystyle O}{\overset{\|}{CH_3CH}}$$

(c) Ethanoic acid

$$\overset{\displaystyle O}{\overset{\|}{CH_3COH}}$$

(d) Butanone

$$\overset{\displaystyle O}{\overset{\|}{CH_3CCH_2CH_3}}$$

(e) Butanal

$$\overset{\displaystyle O}{\overset{\|}{CH_3CH_2CH_2CH}}$$

(f) Butanoic acid

$$\overset{\displaystyle O}{\overset{\|}{CH_3CH_2CH_2COH}}$$

(g) Propanal

$$\overset{\displaystyle O}{\overset{\|}{CH_3CH_2CH}}$$

(h) Cyclopropanol

(i) Cyclopentanol

(j) Cyclopentene

(k) Cyclopentanone

(l) Heptanoic Acid

$$\overset{\displaystyle O}{\overset{\|}{CH_3(CH_2)_5COH}}$$

3.30   Write the IUPAC name for each compound:

(a)
$$\overset{\displaystyle O}{\overset{\|}{CH_3CCH_3}}$$
Propanone

(b)
$$\overset{\displaystyle O}{\overset{\|}{CH_3(CH_2)_3CH}}$$
Pentanal

(c)
$$\overset{\displaystyle O}{\overset{\|}{CH_3(CH_2)_8COH}}$$
Decanoic acid

(d)
Cyclohexene

(e)
Cyclohexanone

(f)
Cyclobutanol

Conformations of Alkanes and Cycloalkanes

3.31   How many *different* staggered conformations are there for 2-methylpropane? How many *different* eclipsed conformations are there?

2-Methylpropane has only one staggered conformation and one eclipsed conformation, as shown in the Newman projections. Line-angle diagrams are also shown for reference.

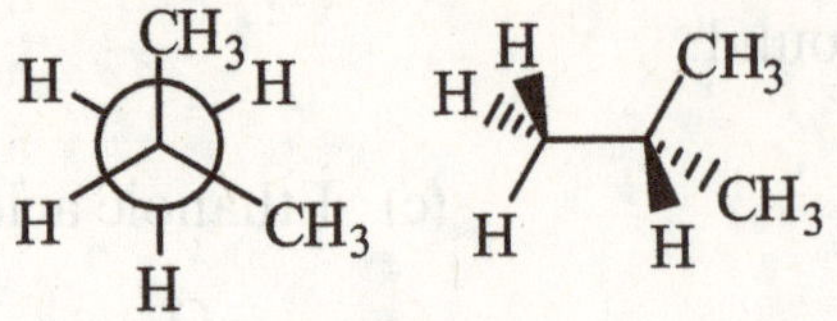

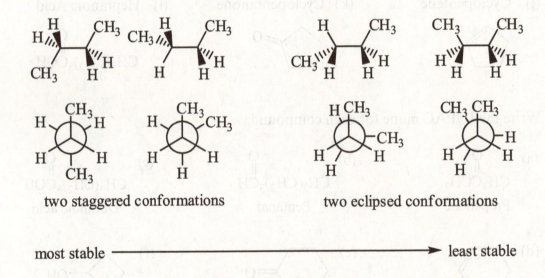

staggered 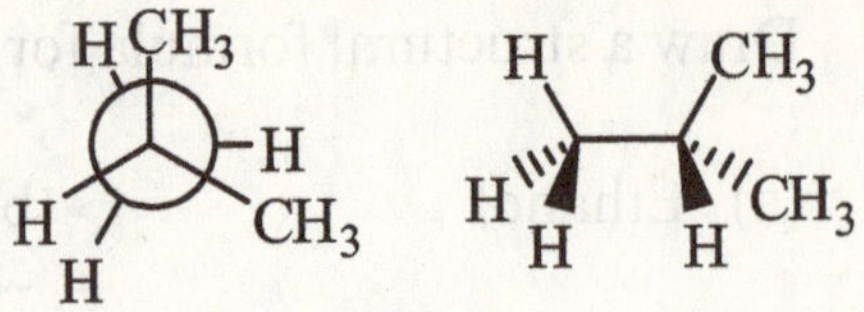   eclipsed

3.32 Looking along the bond between carbons 2 and 3 of butane, there are two different staggered conformations and two different eclipsed conformations. Draw Newman projections of each and arrange them in order from most stable conformation to least stable conformation.

Of the two staggered conformations, the most stable conformation has the two $-CH_3$ groups furthest away from each other, minimizing steric strain. Of the two eclipsed conformations, both of which are higher in energy than the staggered conformations, the least stable conformation has the two $-CH_3$ groups eclipsed. Line-angle diagrams are also provided for reference.

two staggered conformations   two eclipsed conformations

most stable ————————————————————▶ least stable

3.33 Explain why each of the following Newman projections might not represent the most stable conformation of that molecule:

(a) 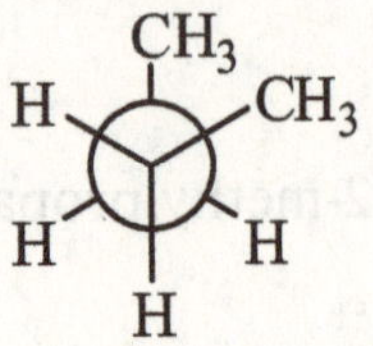   Although the molecule is in a staggered conformation, the two methyl groups are next to one another, increasing steric strain. The conformation on the right is more favorable.

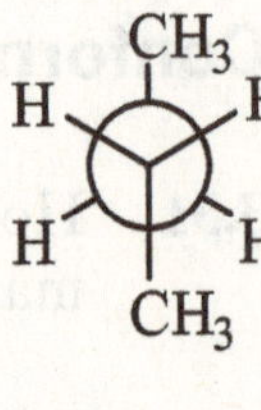

(b) 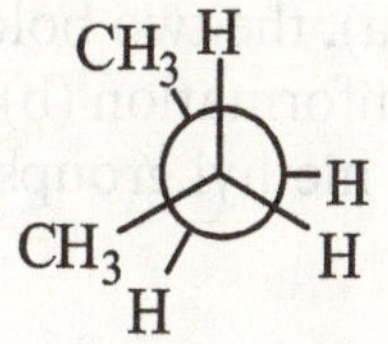 Being an eclipsed conformation, the molecule experiences high torsional strain. The staggered conformation on the right, with the two methyl groups furthest from each other, is most stable.

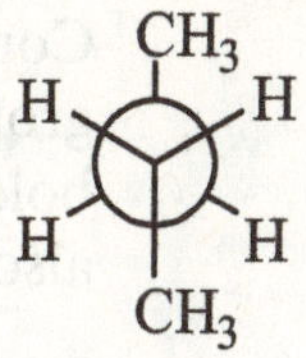

(c) 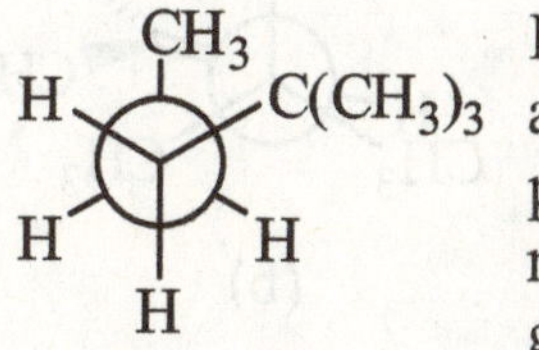 Like the molecule shown in (a), the methyl and *tert*-butyl groups are in close proximity, increasing steric strain. The most stable conformation has these two groups furthest away from each other.

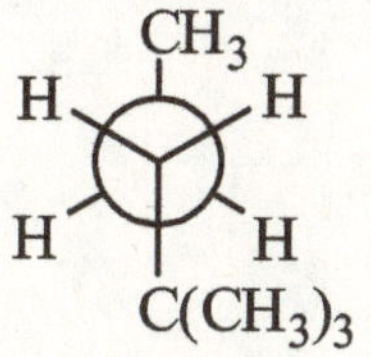

(d) 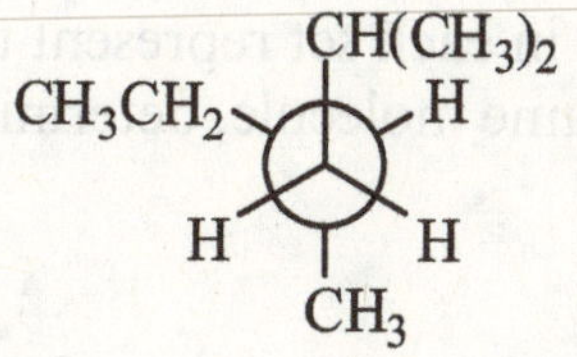 The ethyl and isopropyl groups, which are the two largest groups, are in close proximity. The staggered conformation on the right places these two groups furthest away from each other.

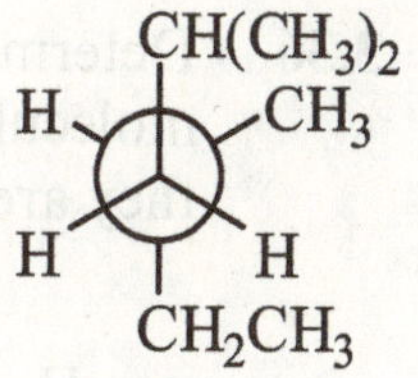

3.34 Explain why the following are not different conformations of 3-hexene:

Conformations are interconverting structures caused by rotation about a single bond. Double bonds are unable to rotate due to the sideways overlap of the p orbitals involved in π bonding. These two structures of 3-hexene differ by the spatial orientation of the two ethyl groups attached to the C=C double bond, which cannot rotate and cannot interconvert. The compounds are a pair of *cis-trans* stereoisomers (more in Chapter 4).

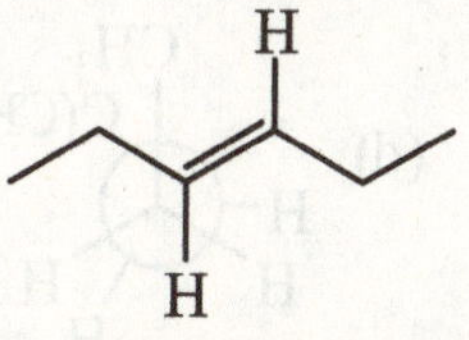

trans-3-Hexene 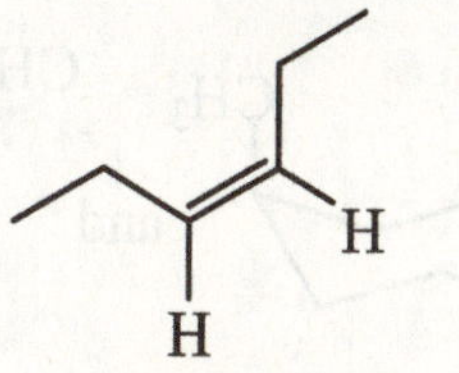   *cis*-3-Hexene

3.35 Which of the following two conformations is the more stable? (*Hint*: Use molecular models or draw Newman projections looking down the bond being rotated to compare structures):

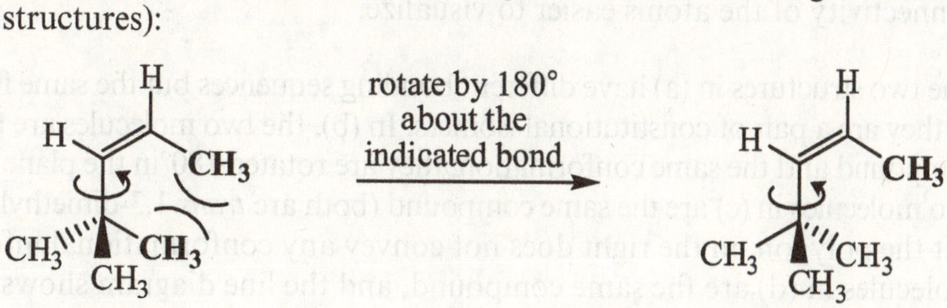

rotate by 180° about the indicated bond

(a)   (b)

Conformation (b) is more stable because in conformation (a), the two bolded methyl groups are in the same plane and interact repulsively. In conformation (b), the bolded methyl group is spaced further away from the other methyl groups. This can also be seen in the Newman projections:

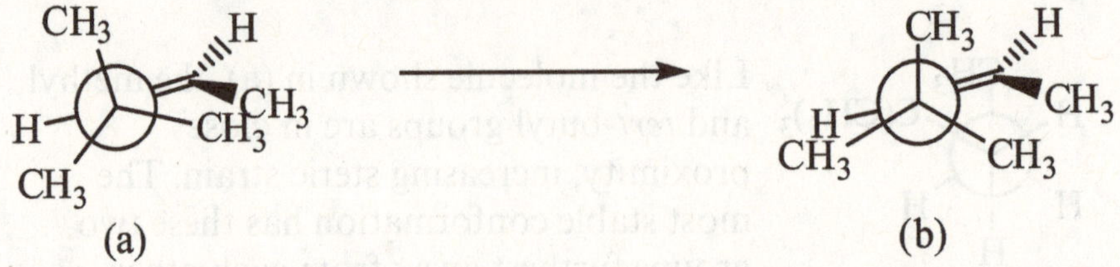

(a) → (b)

3.36 Determine whether the following pairs of structures in each set represent the same molecule or constitutional isomers. If they are the same molecule, determine whether they are in the same or different conformations:

(a) ... and ...
constitutional isomers

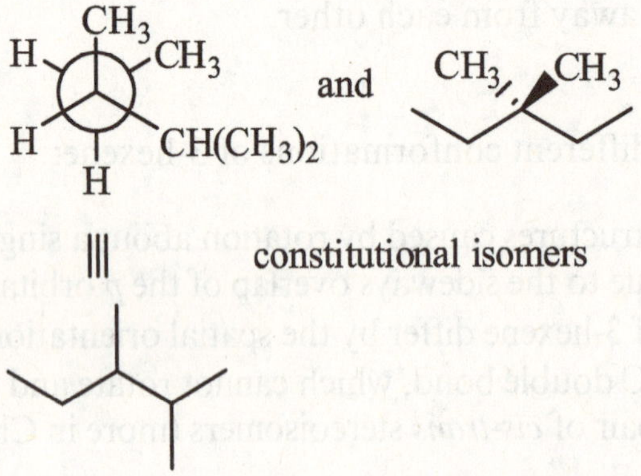

(b) H OH and HO H
same compound and conformation

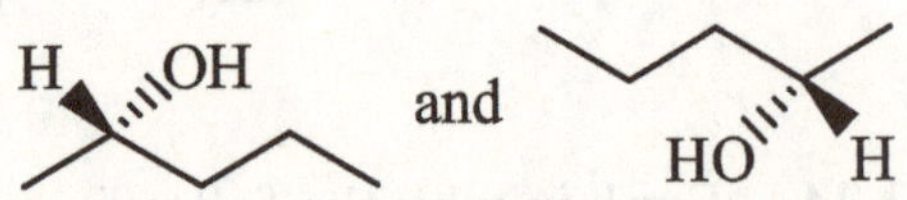

(c) CH_3 ... and ... CH_3
same compound

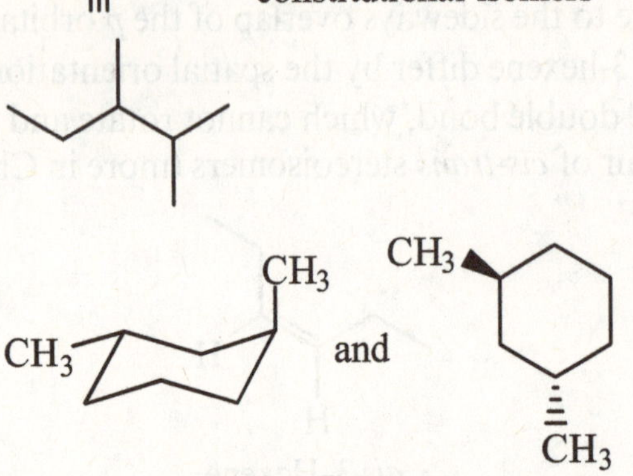

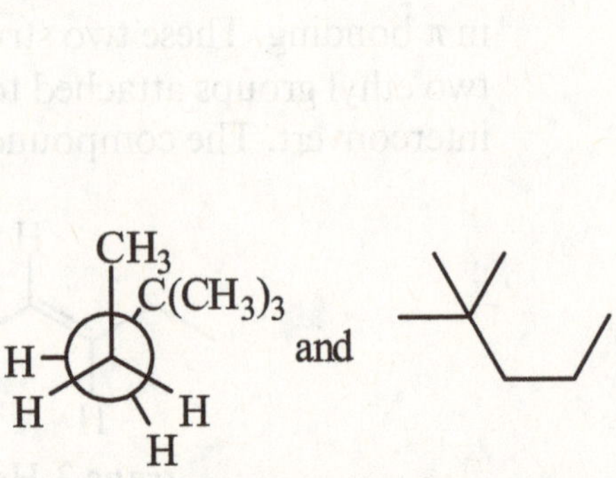

(d) ... and ...
same compound and conformation

When comparing a Newman projection to other structures, it is useful to draw the molecule depicted in the Newman projection as a line-angle diagram. This makes the connectivity of the atoms easier to visualize.

The two structures in (a) have different bonding sequences but the same formula, so they are a pair of constitutional isomers. In (b), the two molecules are the same compound and the same conformation; they are rotated 180° in the plane of the page. The two molecules in (c) are the same compound (both are *trans*-1,3-dimethylcyclohexane), but the polygon on the right does not convey any conformational information. Both molecules in (d) are the same compound, and the line diagram shows the CH_2-CH_2 bond, methyl group, and *tert*-butyl group all in the same plane.

3.37 Draw Newman projections for the most stable conformation of each of the following compounds looking down the indicated bond.

In all cases, the most stable conformation is a staggered conformation that places the largest groups as far away as possible from each other.

(a)

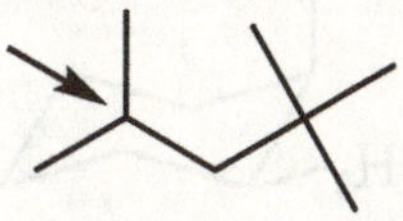

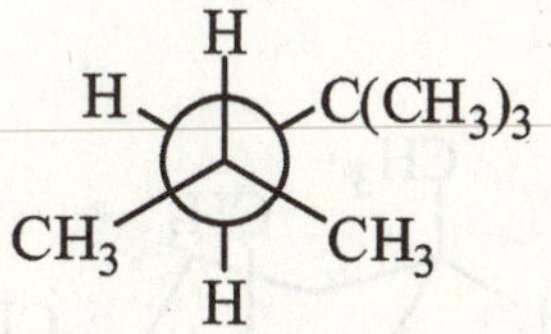

H, H on top, C(CH$_3$)$_3$ to the right, CH$_3$ lower left, CH$_3$ lower right, H at bottom

(b)

CH$_3$ on top, CH$_3$ upper left, CH$_3$ upper right, H lower left, H lower right, CH(CH$_3$)$_2$ at bottom

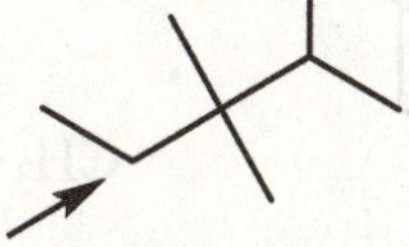

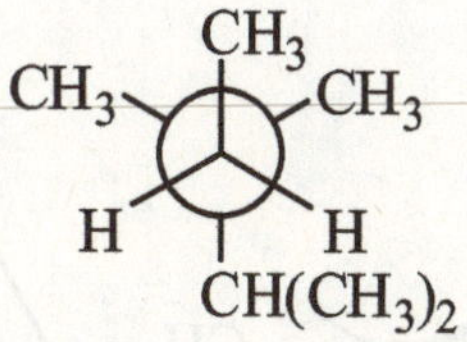

(c)

H on top, CH$_3$ upper left, CH$_3$ upper right, CH$_3$ lower left, CH$_3$ lower right, H at bottom

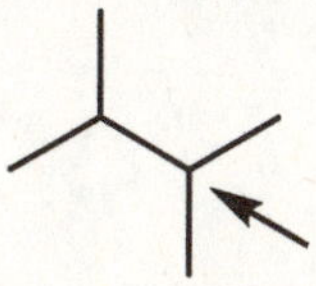

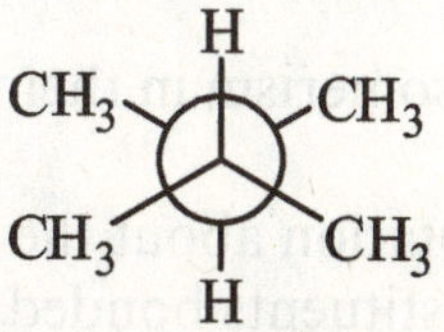

(d)

H, CH$_2$CH$_3$, H, CH$_3$

CH$_3$ on top, H upper left, CH$_3$ upper right, CH$_3$ lower left, H lower right, CH$_2$CH$_3$ at bottom

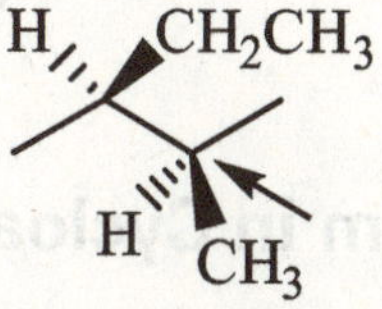

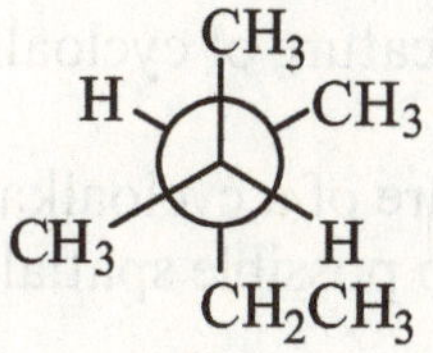

3.38 Draw both chair forms of each of the following compounds and indicate the more stable conformation.

To minimize 1,3-diaxial repulsive interactions, larger groups prefer to be in the equatorial position. In each case, the conformation on the right is more stable than the one to its left. Realize that when drawing conformational representations from polygon forms, the *cis-trans* relationships of the substituents must be maintained (that is, do not simply place of all the substituents on the equatorial positions).

(a)

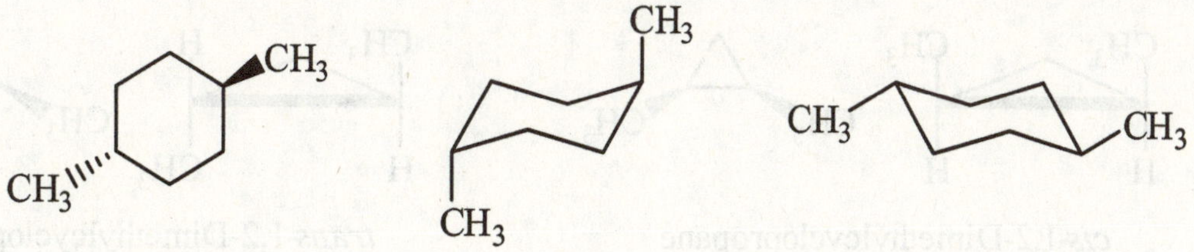

(b)

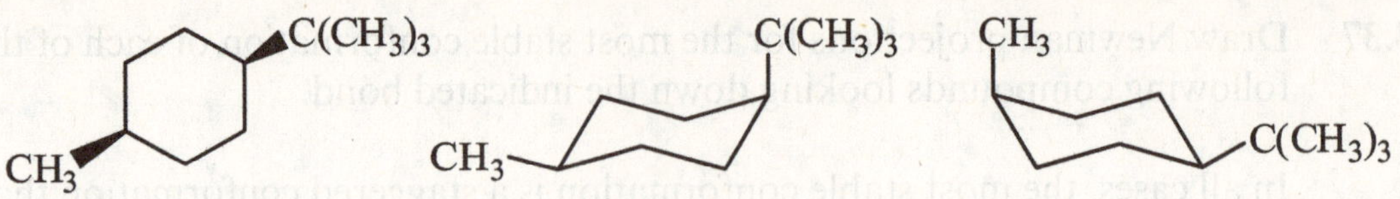

(c)

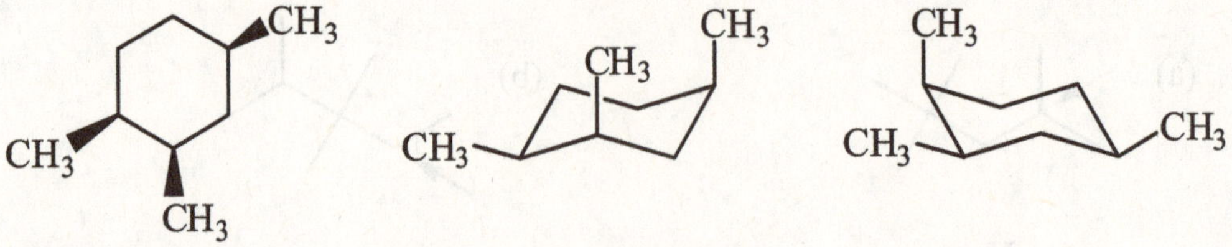

(d)

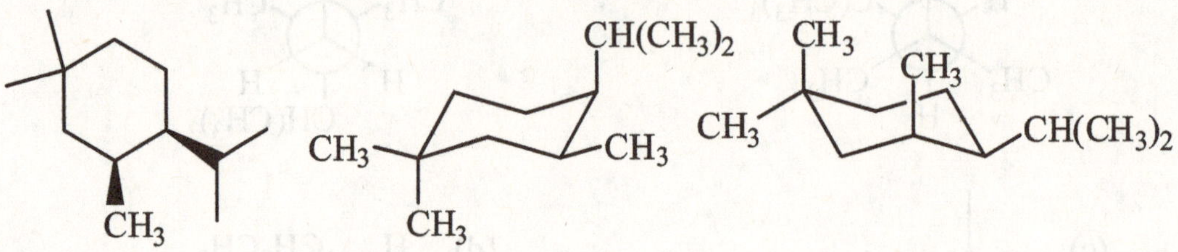

Cis-Trans Isomerism in Cycloalkanes

3.39   What structural feature of cycloalkanes makes *cis-trans* isomerism in them possible?

The cyclic structure of a cycloalkane prevents full 360° rotation about the C−C bond axis, allowing two possible spatial orientations of the substituents bonded to each sp^3 carbon.

3.40   Is *cis-trans* isomerism possible in alkanes?

No. The C−C bonds in alkanes are freely rotating, so alkanes can adopt a variety of conformations, all of which are can freely interconvert.

3.41   Name and draw structural formulas for the *cis* and *trans* isomers of 1,2-dimethylcyclopropane.

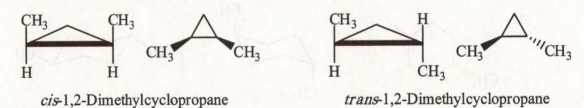

        cis-1,2-Dimethylcyclopropane   *trans*-1,2-Dimethylcyclopropane

3.42   Name and draw structural formulas for all cycloalkanes with the molecular formula
C_5H_{10}. Be certain to include *cis-trans* isomers as well as constitutional isomers.

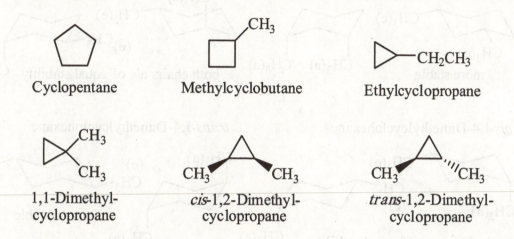

Cyclopentane   Methylcyclobutane   Ethylcyclopropane

1,1-Dimethyl-   *cis*-1,2-Dimethyl-   *trans*-1,2-Dimethyl-
cyclopropane   cyclopropane   cyclopropane

3.43   Using a planar pentagon representation for the cyclopentane ring, draw structural
formulas for the *cis* and *trans* isomers of:

(a) 1,2-Dimethylcyclopentane   (b) 1,3-Dimethylcyclopentane

cis-1,2-Dimethyl- *trans*-1,2-Dimethyl-   *cis*-1,3-Dimethyl- *trans*-1,3-Dimethyl-
cyclopentane   cyclopentane   cyclopentane   cyclopentane

3.44   Draw the alternative chair conformations for the *cis* and *trans* isomers of
1,2-dimethylcyclohexane, 1,3-dimethylcyclohexane, and 1,4-dimethylcyclohexane.

(a) Indicate by a label whether each methyl group is axial or equatorial.
(b) For which isomer(s) are the alternative chair conformations of equal stability?
(c) For which isomer(s) is one chair conformation more stable than the other?

cis-1,2-Dimethylcyclohexane   *trans*-1,2-Dimethylcyclohexane

CH₃(e)   CH₃(e)   CH₃(a)
(e)CH₃   CH₃(e)
CH₃(a)   CH₃(a)   CH₃(a)

both chairs are of equal stability   more stable

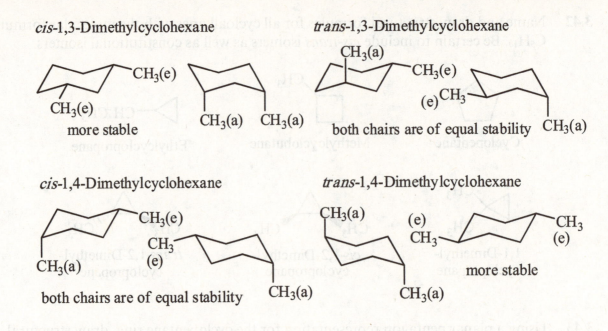

cis-1,3-Dimethylcyclohexane   *trans*-1,3-Dimethylcyclohexane

CH₃(e)

CH₃(e)

more stable   CH₃(a)   CH₃(a)

CH₃(a)

CH₃(e)

(e)CH₃

both chairs are of equal stability   CH₃(a)

cis-1,4-Dimethylcyclohexane   *trans*-1,4-Dimethylcyclohexane

CH₃(e)

CH₃
(e)

CH₃(a)

both chairs are of equal stability   CH₃(a)

CH₃(a)

(e)
CH₃

CH₃(a)

CH₃
(e)

more stable

3.45 Use your answers from Problem 3.44 to complete the table showing correlations between *cis*, *trans* and axial, equatorial for disubstituted derivatives of cyclohexane.

Position of substitution	*cis*	*trans*
1,2-	a,e or e,a	e,e or a,a
1,3-	e,e or a,a	a,e or e,a
1,4-	a,e or e,a	e,e or a,a

3.46 There are four *cis-trans* isomers of 2-isopropyl-5-methylcyclohexanol.

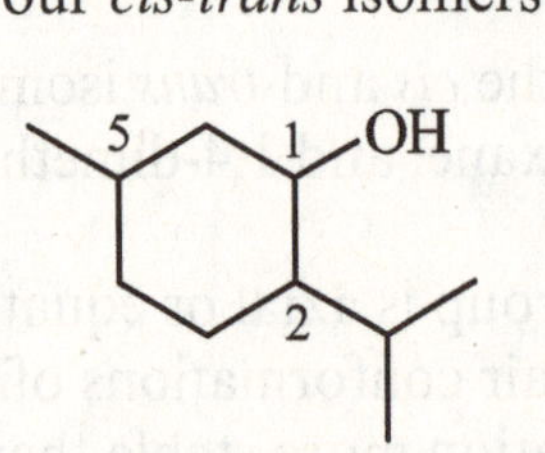

5 1 OH

2

2-Isopropyl-5-methylcyclohexanol

(a) Using a planar hexagon representation for the cyclohexane ring, draw structural formulas for these four isomers.

cis-trans Isomerism refers to the relative orientations of two substituents that are bonded to two different carbon atoms of the ring. When there are three substituents on the ring, there can be multiple *cis-trans* relationships. An effective strategy to arrive at the four stereoisomers is to pick one substituent as a reference point and then arrange the other two groups relative to it. In the following structures, the –OH group is fixed in the "up" position and used as the reference point. Note that any of the two other groups could also be used as the reference point, and regardless of which one is used, there will only be four *cis-trans* isomers for this compound.

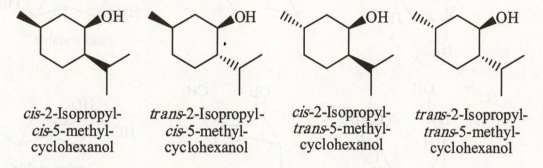

cis-2-Isopropyl-
cis-5-methyl-
cyclohexanol

trans-2-Isopropyl-
cis-5-methyl-
cyclohexanol

cis-2-Isopropyl-
trans-5-methyl-
cyclohexanol

trans-2-Isopropyl-
trans-5-methyl-
cyclohexanol

(b) Draw the more stable chair conformation for each of your answers in part (a).

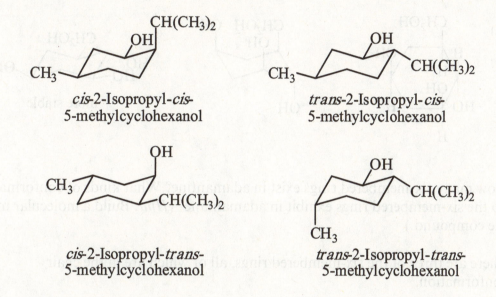

cis-2-Isopropyl-*cis*-
5-methylcyclohexanol

trans-2-Isopropyl-*cis*-
5-methylcyclohexanol

cis-2-Isopropyl-*trans*-
5-methylcyclohexanol

trans-2-Isopropyl-*trans*-
5-methylcyclohexanol

(c) For which isomer(s) is one chair conformation more stable than the other?

Of the four isomers, *trans*-2-isopropyl-*cis*-5-methylcyclohexanol is the most stable because all of its substituents are equatorial. Note that all four of these are isomers and do not interconvert with each other.

3.47 Draw alternative chair conformations for each substituted cyclohexane, and state which chair is the more stable:

(a)
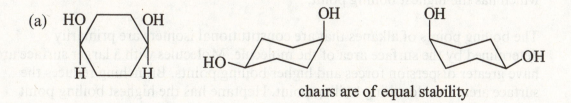

chairs are of equal stability

(b)

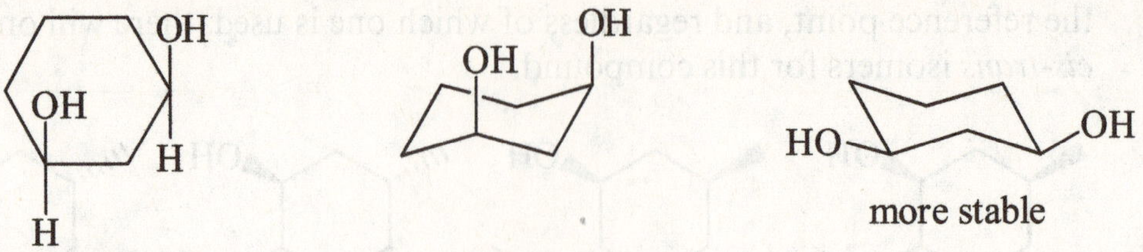

OH, OH, OH, H, H — more stable (HO—...—OH)

(c)

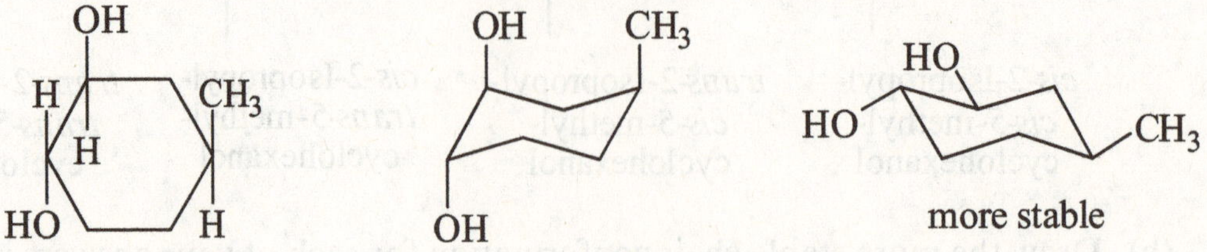

OH, H, CH₃, H, HO, H — OH, CH₃, OH — HO, HO, CH₃ — more stable

(d)

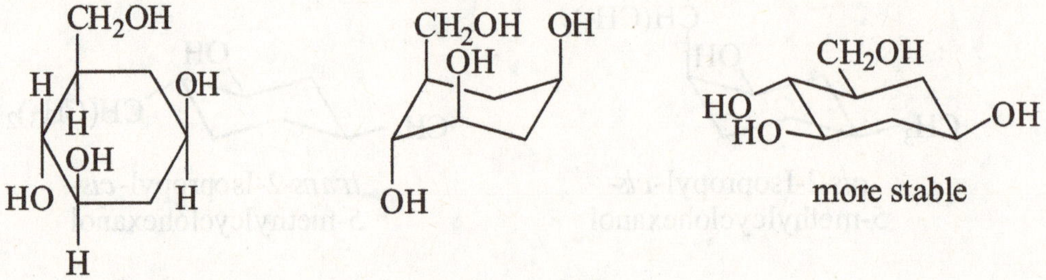

CH₂OH, H, H, OH, OH, HO, H, H — CH₂OH, OH, OH, OH — CH₂OH, HO, HO, OH — more stable

3.48 How many six-membered rings exist in adamantine? What kinds of conformations do the six-membered rings exhibit in adamantane? (*Hint*: Build a molecular model of the compound.)

There are three of the six-membered rings, all of which are in the chair conformation.

Physical Properties of Alkanes and Cycloalkanes

3.49 In Problem 3.21, you drew structural formulas for all constitutional isomers of molecular formula C_7H_{16}. Predict which isomer has the lowest boiling point and which has the highest boiling point.

The boiling points of alkanes that are constitutional isomers are primarily determined by the surface area of the molecule. Molecules with a larger surface area have greater dispersion forces and higher boiling points. Branching reduces the surface area, thus lowering boiling point. Heptane has the highest boiling point (98 °C), and the isomer with the lowest boiling point is 2,2-dimethylpentane (79 °C).

3.50 What generalizations can you make about the densities of alkanes relative to that of water?

In general, the density of an alkane is less than that of water. However, as the molecular weight of an alkane increases, its density also tends to increase.

3.51 What unbranched alkane has about the same boiling point as water (see Table 3.4)? Calculate the molecular weight of this alkane, and compare it with that of water. Explain why water, which is lower in mass than the alkane, boils at the same temperature.

Heptane has a boiling point of 98 °C. Its molecular formula is C_7H_{16}, which corresponds to a molecular weight of 100 g/mol. Although the molecular weight of water is 5.5 times lower, the relatively strong hydrogen bonding forces hold the molecules of liquid water together. On the contrary, only the relatively weak dispersion forces exist in heptane.

3.52 As you can see from Table 3.4, each CH_2 group added to the carbon chain of an alkane increases its boiling point. This increase is greater going from CH_4 to C_2H_6 and from C_2H_6 to C_3H_8 than it is from C_8H_{18} to C_9H_{20} or from C_9H_{20} to $C_{10}H_{22}$. What do you think is the reason for this?

The boiling points of unbranched alkanes are related to their surface area, and the larger the surface area, the greater the strength of the dispersion forces and the higher the boiling point. The relative increase in size per CH_2 group is greatest between CH_4 and CH_3CH_3, and this relative size increase becomes progressively smaller as molecular weight increases. When going from C_9H_{20} to $C_{10}H_{22}$, the relative increase in size is smaller, hence a smaller increase in boiling point.

3.53 Dodecane, $C_{12}H_{26}$, is an unbranched alkane. Predict the following:

(a) Will it dissolve in water?

No. Alkanes are too nonpolar to dissolve in a very polar solvent such as water.

(b) Will it dissolve in hexane?

Yes. Dodecane is nonpolar and will dissolve in nonpolar solvents such as hexane.

(c) Will it burn when ignited?

Yes. Combustion in the presence of sufficient oxygen gas to produce water and carbon dioxide is a characteristic reaction of all alkanes.

(d) Is it a liquid, solid, or gas at room temperature and atmospheric pressure?

At room temperature and atmospheric pressure, alkanes containing approximately 5-17 carbons are liquids. They are also clear and colorless.

(e) Is it more or less dense than water?

All alkanes are less dense than water. This is why gasoline, which has a high proportion of alkanes, floats on water.

3.54 As stated in Section 3.8A, the wax found in apple skins is an unbranched alkane of the molecular formula $C_{27}H_{56}$. Explain how the presence of this alkane prevents the loss of moisture from within an apple.

An alkane of this molecular formula will be a solid at room temperature. In order for moisture to be lost from an apple, the moisture (gaseous water molecules) must travel from inside the apple and through the skin to the outside. However, the presence of a solid, nonpolar compound (the wax) forms a barrier that reduces the rate at which water passes through.

Reactions of Alkanes

3.55 Write balanced equations for combustion of each hydrocarbon. Assume that each is converted completely to carbon dioxide and water.

(a) Hexane

$$2\ CH_3(CH_2)_4CH_3\ +\ 19\ O_2\ \longrightarrow\ 12\ CO_2\ +\ 14\ H_2O$$

(b) Cyclohexane

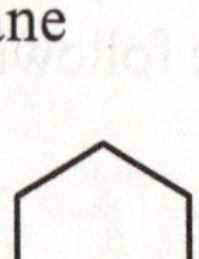

$$+\ 9\ O_2\ \longrightarrow\ 6\ CO_2\ +\ 6\ H_2O$$

(c) 2-Methylpentane

$$\underset{\overset{|}{CH_3}}{2\ CH_3CHCH_2CH_2CH_3}\ +\ 19\ O_2\ \longrightarrow\ 12\ CO_2\ +\ 14\ H_2O$$

3.56 Following are heats of combustion of methane and propane. On a gram-for-gram basis, which of these hydrocarbons is the best source of heat energy?

Hydrocarbon	Component of	$\Delta H°$ [kJ/mol (kcal/mol)]
CH_4	natural gas	−886 (−212)
$CH_3CH_2CH_3$	LPG	−2220 (−530)

The molecular weight of methane is 16.0 g/mol, and the combustion of one mole of methane yields 886 kJ of energy. Dividing this number by 16.0 g yields the amount of energy produced by one gram of methane, 55.4 kJ.

The molecular weight of propane is 44.1 g/mol, and the combustion of one mole of methane yields 2220 kJ of energy. Dividing this number by 44.1 g yields the amount of energy produced by one gram of propane, 50.3 kJ.

Therefore, on a gram-for-gram basis, methane is the best source of heat energy.

3.57 When ethanol is added to gasoline to produce gasohol, it promotes more complete combustion of the gasoline and is an octane booster (Section 3.8B). Compare the heats of combustion of 2,2,4-trimethylpentane (5460 kJ/mol) and ethanol (1369 kJ/mol). Which has the higher heat of combustion in kJ/mol? In kJ/g?

The combustion of one mole of 2,2,4-trimethylpentane releases 5460 kJ of energy, but the combustion of one mole of ethanol releases only 1369 kJ of energy. Thus, 2,2,4-trimethylpentane has the higher heat of combustion on a per mole basis.

One mole of 2,2,4-trimethylpentane corresponds to a mass of 114.2 g and 5460 kJ of energy; each gram therefore produces 47.8 kJ. One mole of ethanol weighs 46.1 g and produces 1369 kJ of energy, hence 29.7 kJ per gram. 2,2,4-Trimethylpentane therefore produces more energy on a per gram basis.

Looking Ahead

3.58 Explain why 1,2-dimethylcyclohexane can exist as *cis-trans* isomers, while 1,2-dimethylcyclododecane cannot.

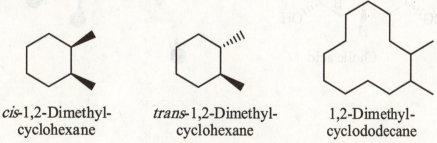

cis-1,2-Dimethyl-cyclohexane *trans*-1,2-Dimethyl-cyclohexane 1,2-Dimethyl-cyclododecane *cis-trans* Isomerism of disubstituted cyclohexane is possible because the cyclic structure restricts rotation about the carbon-carbon bonds in the ring. However, in much larger cycloalkanes, the ring is sufficiently flexible that rotation about the carbon-carbon bonds is unrestricted; the *cis-trans* isomerism of ring substituents is therefore not possible when the ring is very large. (Build a molecular model to confirm this.)

3.59 Following is a representation of the glucose molecule (we discuss the structure and chemistry of glucose in Chapter 17):

(a) Convert this representation to a planar hexagon representation.

(b) Convert this representation to a chair conformation. Which substituent groups in the chair conformation are equatorial? Which are axial?

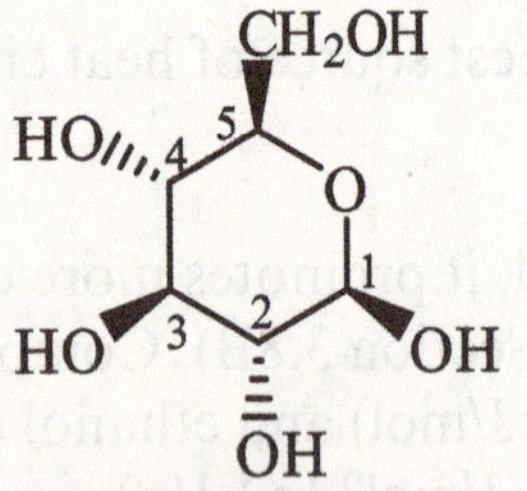

Glucose 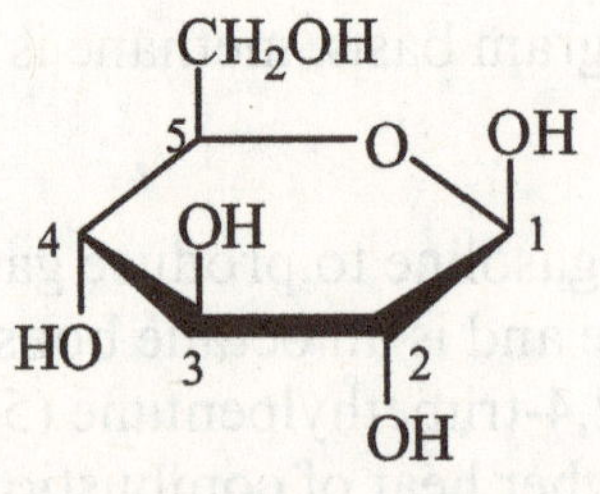   (a) 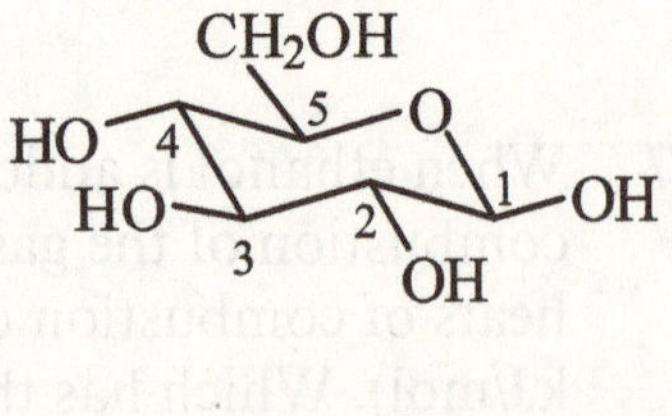   (b)

All the hydrogens are axial, and all the other substituents are equatorial. This structure of glucose is known as β-glucose.

3.60 Following is the structural formula of cholic acid (Section 19.4A), a component of human bile whose function is to aid in the absorption and digestion of dietary fats:

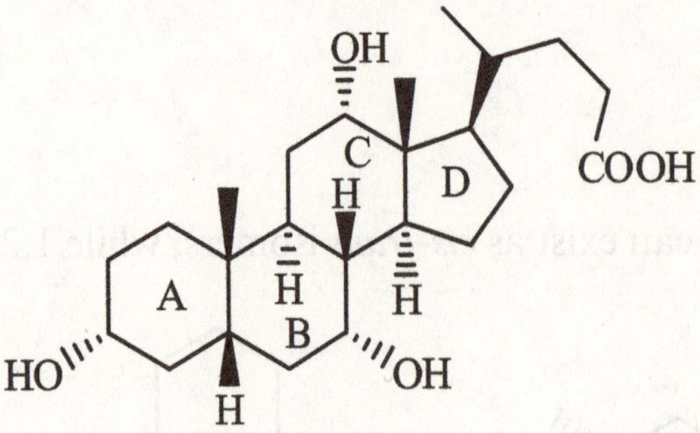

Cholic acid

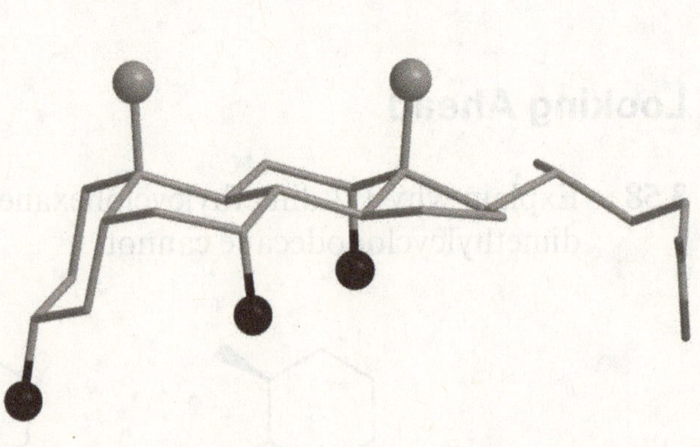

It is necessary to examine the ball-and-stick model to answer this problem because the polygonal representations of cycloalkanes do not convey conformational information. For clarity, the hydrogen atoms in the ball-and-stick model have been omitted, and the three −OH groups are colored black.

(a)   What are the conformations of rings A, B, C, and D?

    Rings A, B, and C are in the chair conformation. Ring D, a cyclopentane ring, is in an envelope conformation.

(b)   There are hydroxyl groups on ring A, B, and C. Tell whether each is axial or equatorial.

    The hydroxyl group on ring A is equatorial; notice that the ring is sideways. The hydroxyl groups on rings B and C are axial.

(c)   Is the methyl group at the junction of rings A/B axial or equatorial to ring A? Is it axial or equatorial to ring B?

    The methyl group is equatorial with respect to ring A, but axial to ring B.

(d)   Is the methyl group at the junction of rings C/D axial or equatorial to ring C?

    The methyl group at the junction of rings C/D is axial to ring C.

3.61   Following is the structural formula and ball-and-stick model of cholestanol. The only difference between this compound and cholesterol (Section 19.4A) is that cholesterol has a carbon-carbon double bond in ring B.

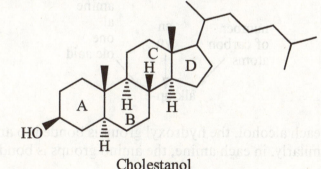

Cholestanol

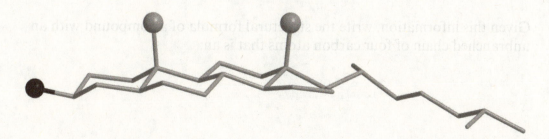

Like the previous problem, it is necessary to examine the ball-and-stick model. For clarity, the hydrogen atoms have been omitted, and the −OH group is colored black.

(a) What are the conformations of rings A, B, C, and D?

Rings A, B, and C are in the chair conformation. Ring D, a cyclopentane ring, is in an envelope conformation.

(b) Is the hydroxyl group on ring A axial or equatorial?

The hydroxyl group on ring A is equatorial.

(c) Consider the methyl group at the junction of rings A/B. Is it axial or equatorial to ring A? Is it axial or equatorial to ring B?

The methyl group at the A/B junction is axial with respect to both rings.

(d) Is the methyl group at the junction of rings C/D axial or equatorial to ring C?

The methyl group at the junction of rings C/D is axial to ring C.

3.62 As we have seen in Section 3.4, the IUPAC system divides the name of a compound into a prefix (showing the number of carbon atoms, an infix (showing the presence of carbon-carbon single, double, or triple bonds), and a suffix (showing the presence of an alcohol, amine, aldehyde, ketone, or carboxylic acid). Assume for the purposes of this problem that, to be alcohol (-ol) or amine (-amine), the hydroxyl or amino group must be bonded to a tetrahedral (sp^3-hybridized) carbon atom.

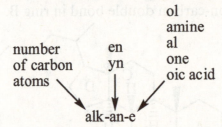

Assume that in each alcohol, the hydroxyl group is bonded to an sp^3 (tetrahedral) carbon atom; similarly, in each amine, the amino groups is bonded to an sp^3 carbon atom.

Given this information, write the structural formula of a compound with an unbranched chain of four carbon atoms that is an:

(a) Alkane

$CH_3CH_2CH_2CH_3$

(b) Alkene

$CH_2=CHCH_2CH_3$

$CH_3CH=CHCH_3$

(c) Alkyne

$CH\equiv CCH_2CH_3$

$CH_3C\equiv CCH_3$

(d) Alkanol

$CH_3CH_2CH_2CH_2OH$

$CH_3CH_2\overset{OH}{\underset{|}{C}}HCH_3$

(e) Alkenol

$CH_2=CH\overset{OH}{\underset{|}{C}}HCH_3$

$CH_2=CHCH_2CH_2OH$

$CH_3CH=CHCH_2OH$

(f) Alkynol

$CH\equiv C\overset{OH}{\underset{|}{C}}HCH_3$

$CH\equiv CCH_2CH_2OH$

$CH_3C\equiv CCH_2OH$

(g) Alkanamine

$CH_3CH_2CH_2CH_2NH_2$

$CH_3CH_2\overset{NH_2}{\underset{|}{C}}HCH_3$

(h) Alkenamine

$CH_2=CH\overset{NH_2}{\underset{|}{C}}HCH_3$

$CH_2=CHCH_2CH_2NH_2$

$CH_3CH=CHCH_2NH_2$

(i) Alkynamine

$CH\equiv C\overset{NH_2}{\underset{|}{C}}HCH_3$

$CH\equiv CCH_2CH_2NH_2$

$CH_3CH=CHCH_2NH_2$

(j) Alkanal

$CH_3CH_2CH_2\overset{O}{\overset{\|}{C}}H$

(k) Alkenal

$CH_2=CHCH_2\overset{O}{\overset{\|}{C}}H$

$CH_3CH=CH\overset{O}{\overset{\|}{C}}H$

(l) Alkynal

$CH\equiv CCH_2\overset{O}{\overset{\|}{C}}H$

$CH_3C\equiv C\overset{O}{\overset{\|}{C}}H$

(m) Alkanone

$CH_3CH_2\overset{O}{\overset{\|}{C}}CH_3$

(n) Alkenone

$CH_2=CH\overset{O}{\overset{\|}{C}}CH_3$

(o) Alkynone

$CH\equiv C\overset{O}{\overset{\|}{C}}CH_3$

(p) Alkanoic acid

$CH_3CH_2CH_2\overset{O}{\overset{\|}{C}}OH$

(q) Alkenoic acid

$CH_2=CHCH_2\overset{O}{\overset{\|}{C}}OH$

$CH_3CH=CH\overset{O}{\overset{\|}{C}}OH$

(r) Alkynoic acid

$CH\equiv CCH_2\overset{O}{\overset{\|}{C}}OH$

$CH_3C\equiv C\overset{O}{\overset{\|}{C}}OH$

Group Learning Activities

3.63   Come up with reasons for the following phenomena. You may need to refer to concepts learned from previous chapters or from general chemistry.

(a)   Gasoline is cool to the touch when spilled on bare skin.

The evaporation of any liquid into a gas is an endothermic process. The energy for this process comes from the skin, thereby causing the skin to feel cool. This is the same reason why we sweat to reduce our body temperature.

(b)   Water is more dense than methane.

As a result of hydrogen bonding, the intermolecular forces in water are very, very strong. The water molecules are pulled together very tightly, thus increasing density.

(c)   Butane is a more appropriate fuel for a disposable lighter than either propane or pentane.

A disposable lighter has a plastic chamber that holds liquid butane. When the valve is opened, the liquid butane evaporates, causing gaseous butane to be released.

Liquid butane has a certain vapor pressure at room temperature, and at this temperature, the plastic chamber is strong enough to withstand the vapor pressure of butane. This keeps the butane from boiling inside the closed chamber.

In general, the vapor pressure of a gas increases as the boiling point of the gas decreases, and as temperature increases. Propane, which has a lower boiling point than butane, has a higher vapor pressure than butane at room temperature. Plastic is not strong enough to withstand this vapor pressure, so a disposable lighter containing propane would explode. This is why propane tanks are made of metal.

Pentane, on the other hand, has a higher boiling point, and therefore a lower vapor pressure, than butane. It is not very suitable for a disposable lighter because it does not evaporate quickly enough at room temperature to generate useful amounts of gaseous fuel. The vapor pressure of pentane is low enough that at room temperature, pentane can safely be stored in glass bottles.

3.64 See who can name the following stick figure molecules the fastest:

(a)

(b)

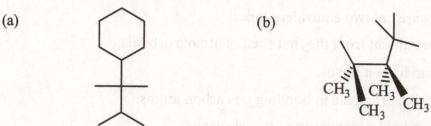

2-Cyclohexyl-2,3-dimethylbutane

2,2,3,3,4,4-Hexamethylpentane

Putting It Together

1. Which of the following molecules has a net charge of +1?

(a) CH_2CHCH_3

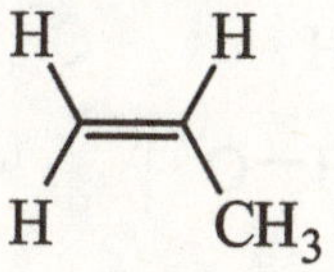

(b) CH_3CHCH_3

(c) $CHCCH_3$

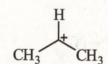

$H\!-\!\!\equiv\!\!-CH_3$

(d) $(CH_3)_3CH$

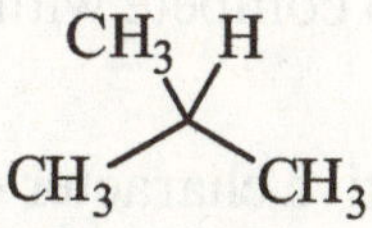

(e) CH_2CH_2

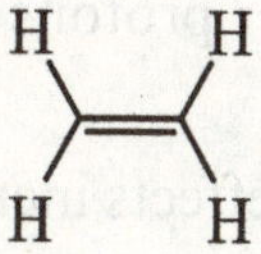

2. Which of the following statements is true concerning the compound shown below?

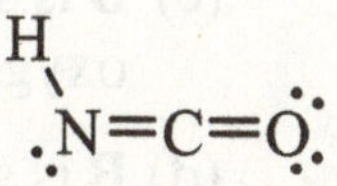

$$\overset{H}{\underset{..}{}}N=C=\overset{..}{\underset{..}{O}}$$

(a) The central carbon is sp^2-hybridized and the molecule is planar in geometry.

(b) The central carbon is sp^2-hybridized and the molecule is nonplanar in geometry.

(c) The central carbon is sp-hybridized and the molecule is planar in geometry.

(d) The central carbon is sp-hybridized and the molecule is nonplanar in geometry.

(e) None of these statements is true.

Statement (c) is correct because the central carbon has two regions of electron density. The nitrogen atom has an implied nonbonding pair, and its hybridization is sp^2. All four atoms of the molecule are in the same plane.

3.   Which of the following statements is false concerning p orbitals?

(a) They consist of two equivalent lobes.

(b) They are absent from the first shell of atomic orbitals.

(c) They can form π bonds.

(d) They only participate in bonding on carbon atoms.

(e) They can hold a maximum of two electrons.

Statement (d) is false because bonding involving p orbitals is not limited to just carbon atoms. For example, a carbonyl group (C=O) contains a π bond that involves one p orbital from each of carbon and oxygen.

4.   Which base (A or B) is stronger and why?

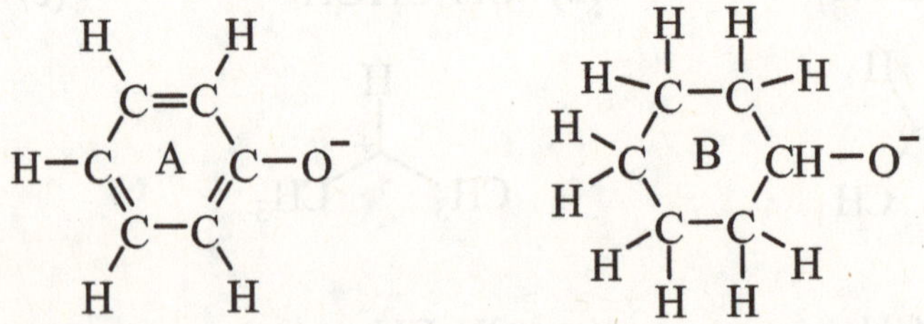

(a) **A** is stronger because it has fewer protons for the acid to compete with in acid-base reactions.

(b) **A** is stronger because inductive effects increase the negative character of its oxygen.

(c) **B** is stronger because inductive effects increase the negative character of its oxygen.

(d) **B** is stronger because resonance effects can delocalize its negative charge throughout the molecule.

(e) **B** is stronger because it has no resonance or inductive effects that can delocalize its negative charge throughout the molecule.

Statement (e) is correct because the negative charge in molecule **A** is delocalized by resonance. Compound **B** is therefore less stable than **A** and is thus the stronger base.

5. Which of the following is the initial product of the reaction between $(CH_3)_3C^+$ and CH_3OH?

(a)
$$(CH_3)_3C-\overset{\overset{\displaystyle CH_3}{|}}{\underset{\underset{\displaystyle H}{|}}{\overset{+}{O}}}$$

(b)
$$(CH_3)_3C-\overset{..}{\underset{\underset{\displaystyle H}{|}}{O}}:$$

(c) $(CH_3)_3C-\overset{..}{\underset{..}{O}}:^- + H_2$

(d) $(CH_3)_3C-CH_2 + H_2\overset{..}{O}$

(e) $(CH_3)_3C-H + CH_2-\overset{..}{\underset{..}{O}}:^-$

CH$_3$OH is a Lewis base that can attack the carbocation, $(CH_3)_3C^+$, a Lewis acid. This reaction forms the oxonium ion shown in (a).

6. Select the statement that is false concerning the following acid-base reaction.

$$CH_3-\overset{\overset{\displaystyle O}{\|}}{C}-OH + NaCl \rightleftharpoons CH_3-\overset{\overset{\displaystyle O}{\|}}{C}-ONa + HCl$$

(a) The equilibrium lies on the product side of the reaction.

(b) The carboxylic acid does not possess a positive charge.

(c) The chloride ion acts as a Lewis base.

(d) The chloride ion acts as a Brønsted-Lowry base.

(e) The carboxylic acid is a weaker acid than HCl.

Statement (a) is false because acid-base equilibria lie on the side with the weakest acid and the weakest base. Acetic acid is a weaker acid than HCl, and Cl$^-$ is a weaker base than acetate. Na$^+$ is a spectator ion. Note that statements (c) and (d) are correct because when proceeding from left to right, the chloride accepts the proton by donating an electron pair to it.

7. Which of the following statements is false?

(a) Nonbonded interaction (steric) strain contributes to the energy of butane in the eclipsed conformation.

(b) All staggered conformations possess zero strain.

(c) A Newman projection is the picture of a molecule viewed down at least one of its bonds.

(d) Bonds represented by Newman projections do not freely rotate because they must overcome an energy barrier to rotation.

(e) Ring strain contributes to the instability of cyclopropane.

A staggered conformation does not necessarily possess zero strain. Rather, a staggered conformation has less strain (but not zero) than an eclipsed conformation.

8.   Which of the following statements is true concerning the isomers *cis*-1,2-dimethylcyclohexane and *cis*-1,3-dimethylcyclohexane?

(a) They are not constitutional isomers.

(b) They are conformers.

(c) The favored conformer of the 1,3-isomer is more stable than that of the 1,2-isomer.

(d) The favored conformer of the 1,3-isomer and that of the 1,2-isomer are equal in energy.

(e) The relative stability of the two molecules cannot be predicted.

Statement (c) is true because the favored conformation of *cis*-1,3-dimethyl-cyclohexane has both of the methyl substituents in the equatorial position. With *cis*-1,2-dimethylcyclohexane, there is one equatorial and one axial methyl group.

9.   Select the correct order of stability (least stable → most stable) for the conformations shown below.

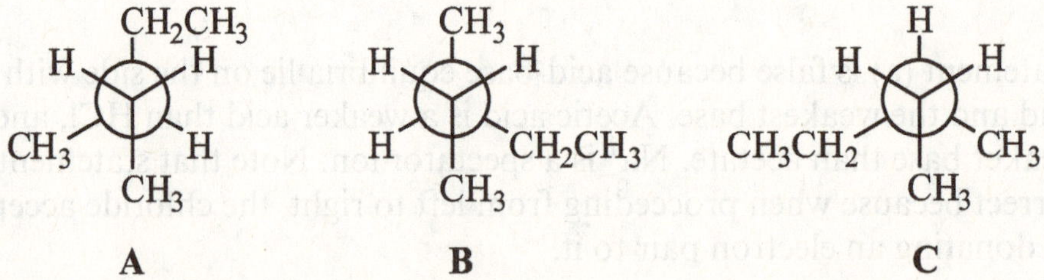

A   B   C

(a) A, B, C   (b) A, C, B   (c) B, A, C   D) C, A, B   E) C, B, A

Although all three are staggered, they differ in stability. **C** is the least stable because the three alkyl groups are in close proximity. In **B**, the ethyl group is staggered between a hydrogen and a methyl group, the methyl on the front is staggered between a hydrogen and the ethyl group, and the methyl on the back is staggered between two hydrogens. **A** is the most stable because the ethyl group is between two hydrogens, and each of the methyl groups is between the other methyl and a hydrogen.

10. Select the most stable conformation of those shown for 1-*tert*-butyl-3,5-dimethylcyclohexane.

(a)

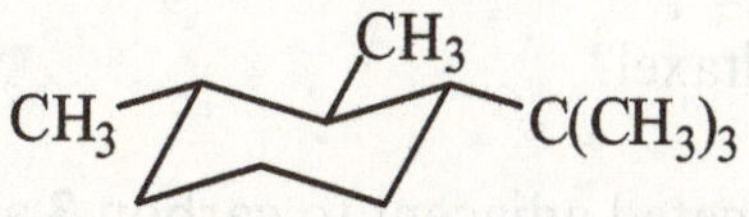

CH₃

CH₃ — C(CH₃)₃

(b)

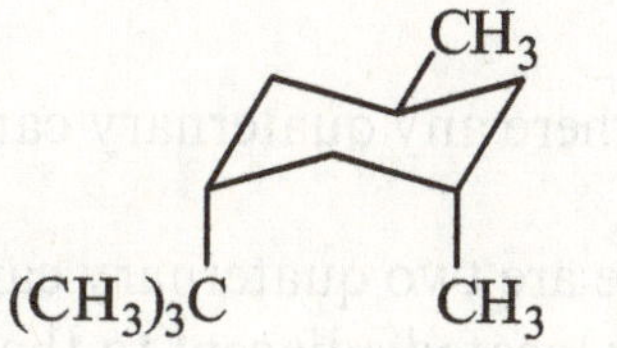

CH₃

(CH₃)₃C   CH₃

(c)

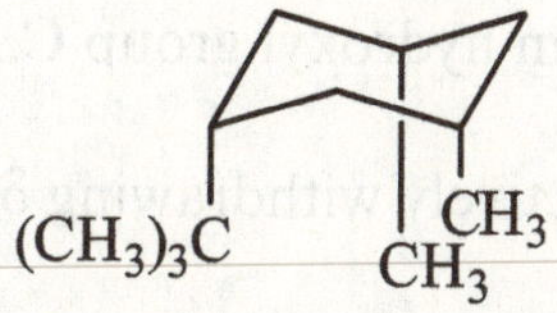

(CH₃)₃C   CH₃
CH₃

(d)

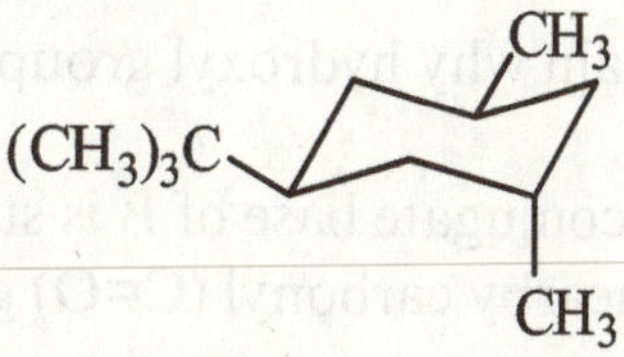

CH₃

(CH₃)₃C

CH₃

(e) (CH₃)₃C

CH₃
CH₃

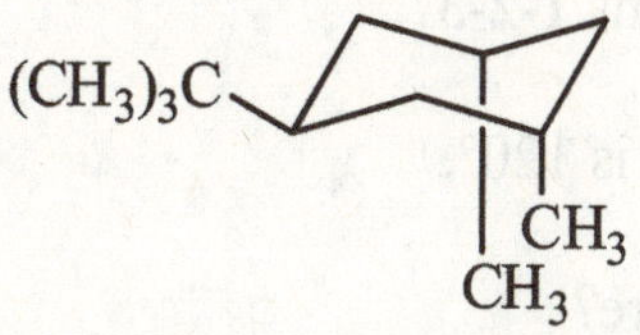

Structure (a) is not 1-*tert*-butyl-3,5-dimethylcyclohexane. Out of the rest, (d) is the most stable because the bulky *tert*-butyl substituent is in the equatorial position, as is one of the two methyl groups. Note that these compounds are not all conformations of each other.

11. Answer the questions that follow regarding the structure of paclitaxel (trade name Taxol®), a compound first isolated from the Pacific Yew tree which is now used to treat ovarian, breast and non-small cell lung cancer.

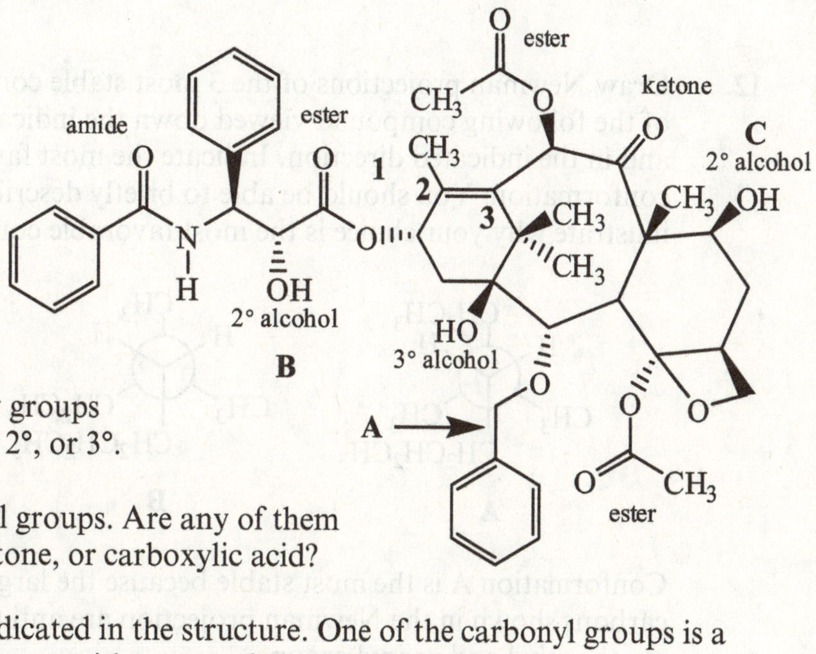

(a) Identify all the hydroxy groups and classify them as 1°, 2°, or 3°.

(b) Identify all the carbonyl groups. Are any of them part of an aldehyde, ketone, or carboxylic acid?

Carbonyl groups are indicated in the structure. One of the carbonyl groups is a part of a ketone. The rest are either part of an amideor an ester.

(c) What atomic or hybridized orbitals participate in the bond labeled **A**?

The carbon-carbon σ bond involves the overlap of sp^2 and sp^3 hybrid orbitals.

(d) Are there any quaternary carbons in paclitaxel?

There are two quaternary carbons, one located adjacent to carbon **3** and the other located adjacent to the carbon to which −OH group **C** is bonded.

(e) Explain why hydroxyl group **B** is more acidic than hydroxyl group **C**.

The conjugate base of **B** is stabilized by the inductively withdrawing δ+ charge of the nearby carbonyl (C=O) group.

(f) What is the angle of the bond containing atoms **1-2-3**?

Carbon **2** is sp^2-hybridized, so the bond angle is 120°.

(g) Locate any amide groups. How many are there?

There is one amide group.

(h) Locate any ester groups. How many are there?

There are three ester groups.

12. Draw Newman projections of the 3 most stable conformations of the following compound viewed down the indicated bond and in the indicated direction. Indicate the most favorable conformation. You should be able to briefly describe or illustrate why your choice is the most favorable conformation.

view

A   B   C

Conformation **A** is the most stable because the largest substituent on each of the two carbons shown in the Newman projection are anti to each other. These substituents are the ethyl and propyl groups.

13. Provide IUPAC names for the following compounds.

(a)

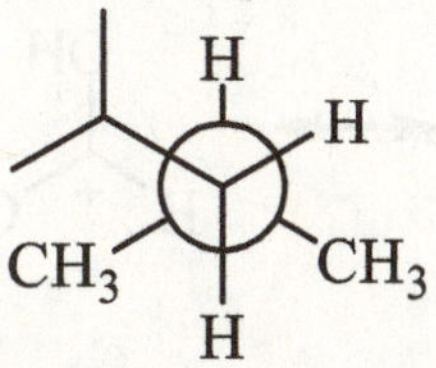

2,4-Dimethylpentane

(b)

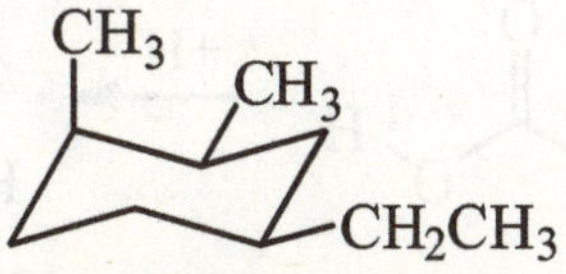

$(1R,2S,4S)$-4-Ethyl-
1,2-dimethylcyclohexane

(c) $CH_3CH_2CH(CH_3)CH(CH_2CH_3)CH_2CH(CH_3)_2$

4-Ethyl-2,5-dimethylheptane

In Chapter 6, we will present a more accurate way of naming these compounds.

14. For each pair of molecules, select the one that best fits the accompanying description. Provide a concise but thorough rationale for each of your decisions using words and/or pictures.

(a) The higher boiling point?

A   vs.   B

A has the higher boiling point because the –OH group can act as a hydrogen bonding donor and acceptor.

(b) The more acidic hydrogen?

A

$-H^+$

vs.

B

$-H^+$

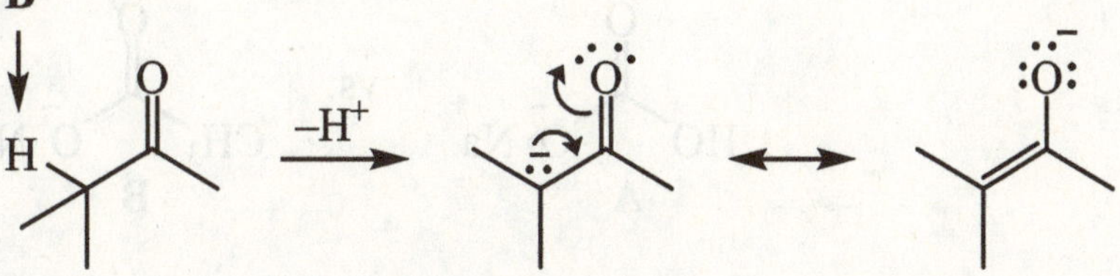

Hydrogen **B** is more acidic because the conjugate base formed by its deprotonation is more stable than the conjugate base formed by the deprotonation of **A**. The negative charge on conjugate base of **B** can be delocalized by resonance to an oxygen atom, while in **A**, the charge is delocalized to only a carbon atom.

(c) The more basic atom?

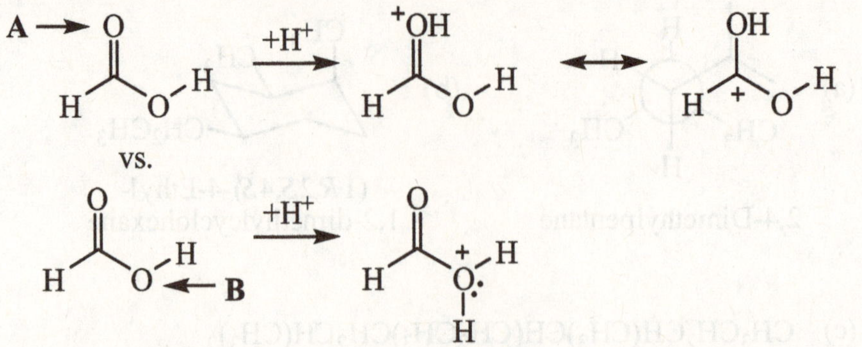

There are two reasons why atom **A** is more basic than atom **B**. The first reason is described in Problem 2.29(a). The second reason is because the positive charge formed by the protonation of **A** is stabilized by resonance, while that formed by the protonation of **B** is not.

(d) The more acidic set of protons?

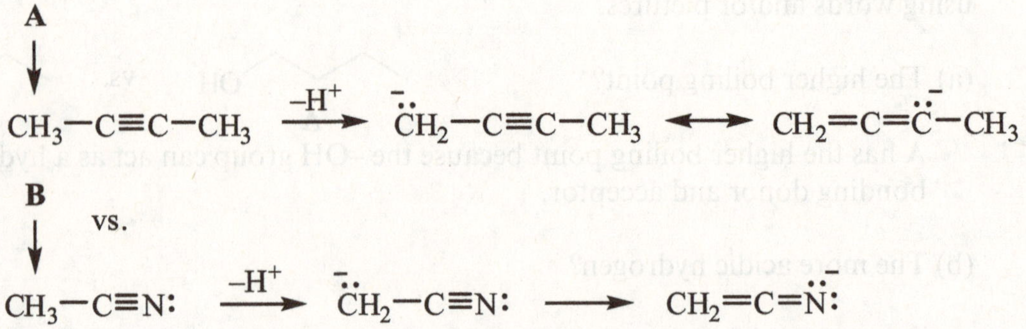

The conjugate base formed by the deprotonation of **B** has a negative charge that can be delocalized by resonance onto a nitrogen atom. Whereas, the negative charge formed by the deprotonation of **A** can only be delocalized to a carbon.

(e) The stronger base.

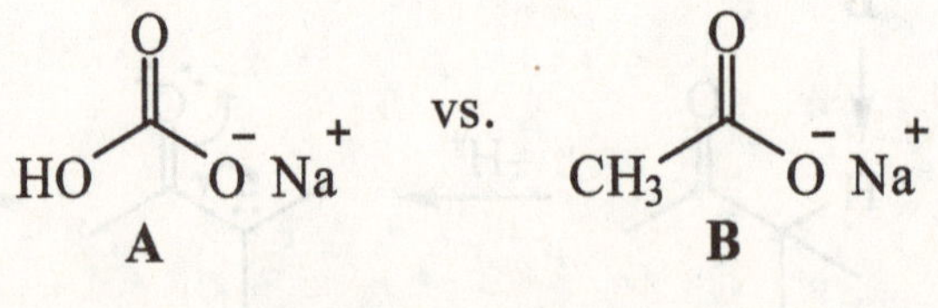

A is the stronger base. The ability of the –OH group on **A** to donate by resonance reduces the amount of positive charge located on the C=O carbon. By reducing the amount of positive charge on this carbon, the negative charge on the adjacent O⁻ becomes less stable.

(f) Possess the least nonbonding interaction (steric strain)?

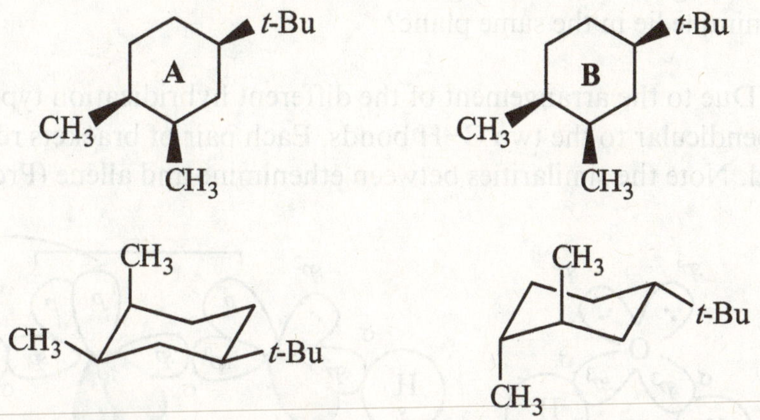

The bulky *tert*-butyl group will prefer to be in the equatorial position. When it is in the equatorial position, **A** also has a methyl group in the equatorial position. Therefore, **A** has less steric strain.

15. Glutamic acid is one of the common amino acids found in nature. Draw the predominant structure of glutamic acid when placed in a solution of pH = 3.2 and indicate its overall charge.

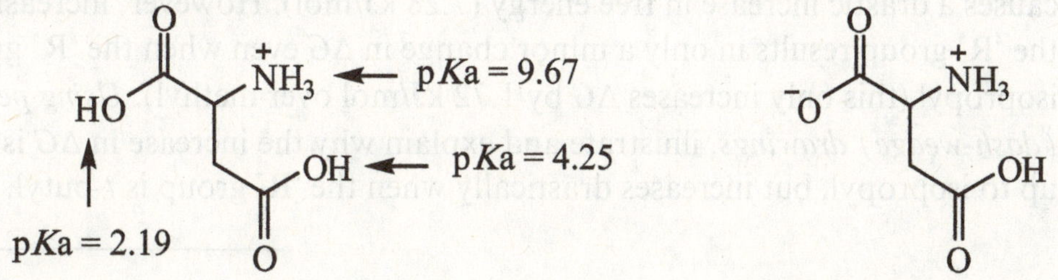

structure at pH = 3.2

At pH = 3.2, the α-carboxyl group will be ionized, because 3.2 is higher than the pK_a of that group. At this pH, the predominant form of glutamic acid has no net charge.

16. Use atomic and hybridized orbitals to illustrate (see example using H_2O) the location of bonding and nonbonding electrons in ethenimine. Do all of the atoms in ethenimine lie in the same plane?

No. Due to the arrangement of the different hybridization types, the N−H bond is perpendicular to the two C−H bonds. Each pair of brackets represents a single π bond. Note the similarities between ethenimine and allene (Problem 1.63).

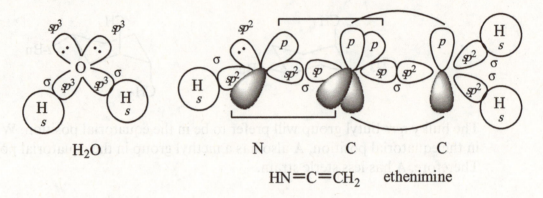

$$HN=C=CH_2 \quad \text{ethenimine}$$

17. The following values have been determined for the amount of energy it takes to place a substituent in the axial position. As shown in the table, going from H to CH_3 causes a drastic increase in free energy (7.28 kJ/mol). However, increasing the size of the 'R' group results in only a minor change in ΔG even when the 'R' group is isopropyl (this only increases ΔG by 1.72 kJ/mol over methyl). *Using perspective (dash-wedge) drawings*, illustrate and explain why the increase in ΔG is only gradual up to isopropyl, but increases drastically when the 'R' group is *t*-butyl.

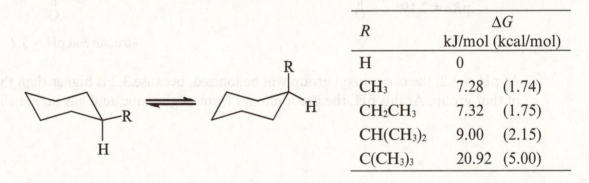

R	ΔG kJ/mol (kcal/mol)
H	0
CH_3	7.28 (1.74)
CH_2CH_3	7.32 (1.75)
$CH(CH_3)_2$	9.00 (2.15)
$C(CH_3)_3$	20.92 (5.00)

A *tert*-butyl group has a methyl group that can interact hydrogen atoms via 1,3-diaxial interactions. See Problem 3.9 for more details.

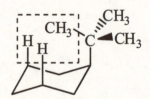

18. (a) Draw the two possible products that can form from the Lewis acid-base reaction between methyl formate and BF_3. Indicate the major product and use curved arrow notation to illustrate its formation. Show all charges & nonbonded electrons *in your products*. (b) Use pictures and words to explain why the product you indicated is favored over the other product.

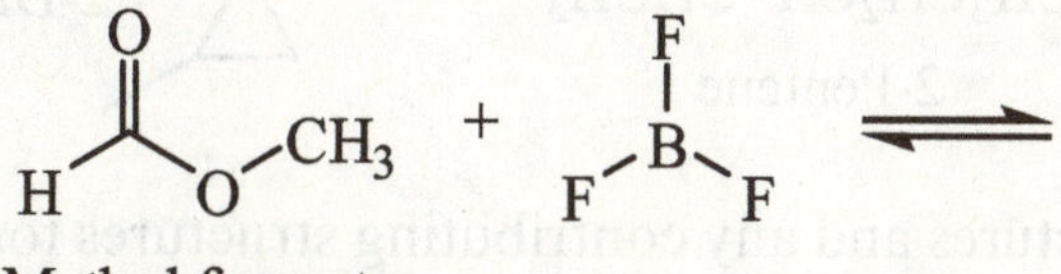

Methyl formate

Borane, which contains an electron-deficient boron atom, acts as Lewis acid. Each of the two oxygen atoms in methyl formate can act as a Lewis base. The major product is favored because the positive charge is stabilized by resonance.

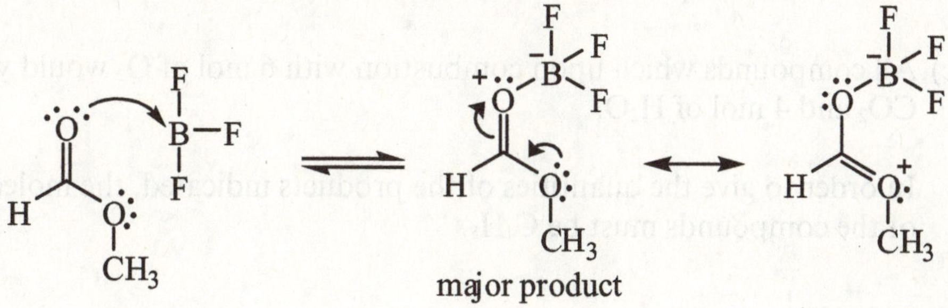

major product

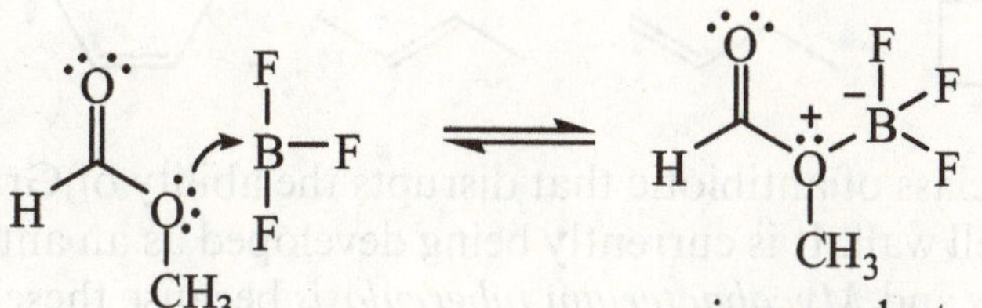

minor product

19. Use resonance theory to predict whether $[CNO]^-$ or $[NCO]^-$ is the more stable ion. Use pictures *and* words to explain your decision.

$[NCO]^-$ is more stable because the negative charge is delocalized over oxygen and nitrogen atoms. Whereas, the negative charge in $[CNO]^-$ is delocalized over oxygen and carbon atoms. The greater electronegativity of nitrogen over carbon is more favorable for the stabilization of negative charges.

20.   Provide structures as indicated:

(a) All compounds with the formula C_5H_{10} that exhibit *cis-trans* isomerism.

$CH_3CH_2CH=CHCH_3$ 1,2-Dimethylcyclopropane
2-Pentene

(b) Lewis structures and any contributing structures for the ion with formula CH_2NO_2. Show all formal charges and lone pairs of electrons.

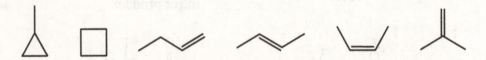

(c) All compounds which upon combustion with 6 mol of O_2 would yield 4 mol of CO_2 and 4 mol of H_2O.

In order to give the quantities of the products indicated, the molecular formula of the compounds must be C_4H_8.

21.   Teixobactin is a new class of antibiotic that disrupts the ability of Gram-positive bacteria to form its cell wall. It is currently being developed as an antibiotic for *Staphylococcus aureus* and *Mycobacterium tuberculosis* because these bacteria do not develop resistance to teixobactin as they do to other antibiotics.

(a) Locate all of the amide functional groups in teixobactin.
(b) Locate all of the amine functional groups in teixobactin and classify them as 1°, 2°, or 3°.
(c) Locate all of the ester functional groups in teixobactin.
(d) Locate all of the alcohol functional groups in teixobactin and classify them as 1°, 2°, or 3°.
(e) Label all of the 1°, 2°, or 3° carbons in teixobactin.

Note that only alkyl carbons (*sp³*-hybridized) in the structure have been classified.

(f) Draw lone pairs of nonbonded electrons for all atoms where they have been left out.

Chapter 4: Alkenes and Alkynes

Problems

4.1 Write the IUPAC name of each unsaturated hydrocarbon:

(a)

3,3-Dimethyl-1-pentene

(b)

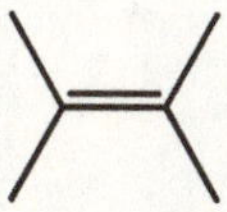

2,3-Dimethyl-1-butene

(c)

3,3-Dimethyl-1-butyne

(d)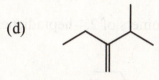

2-Isopropyl-1-butene or
2-Ethyl-3-methyl-1-butene

(e)

2,4,4-Trimethyl-2-pentene

4.2 Name each alkene, and, using the *cis-trans* system, specify its configuration:

(a)

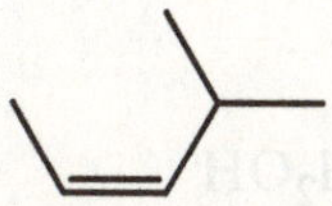

cis-4-Methyl-2-pentene

(b)

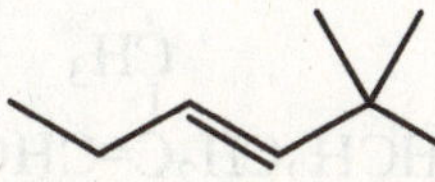

trans-2,2-Dimethyl-3-hexene

4.3 Name each alkene and specify its configuration by the E,Z system:

(a)

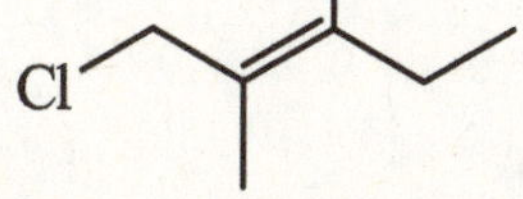

Cl

(*E*)-1-Chloro-2,3-dimethyl-2-pentene

(b)

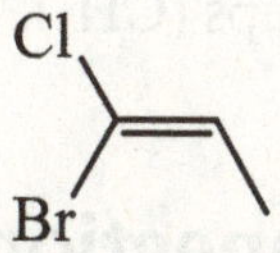

Cl

Br

(*Z*)-1-Bromo-1-chloro-1-propene

(c)

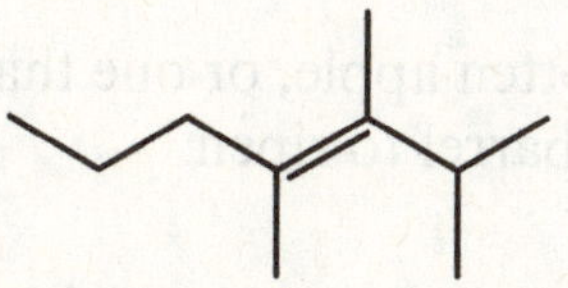

(*E*)-2,3,4-Trimethyl-3-heptene

(d)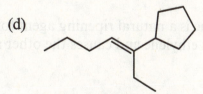

(*E*)-3-Cyclopentyl-3-heptene

4.4   Write the IUPAC name for each cycloalkene:

(a)
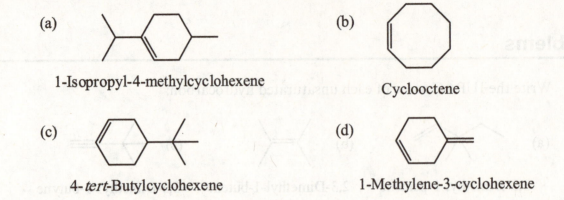

1-Isopropyl-4-methylcyclohexene

(b)

Cyclooctene

(c)

4-*tert*-Butylcyclohexene

(d)

1-Methylene-3-cyclohexene

4.5   Draw structural formulas for the other two *cis-trans* isomers of 2,4-heptadiene:

cis,trans-2,4-Heptadiene   *cis,cis*-2,4-Heptadiene

4.6   How many *E-Z* isomers are possible for the following unsaturated alcohol?

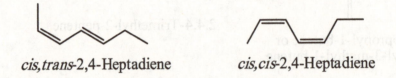

There are three carbon-carbon double bonds, but *E-Z* isomerism is only possible for the two that are indicated by arrows. *E-Z* isomerism is not possible for the alkene on the far left of the molecule because one of the two alkene carbons is bonded to two identical groups (CH_3). The number of possible *E-Z* isomers is therefore $2^2 = 4$.

Chemical Connections

4A. Explain the basis for the saying "A rotten apple can spoil the barrel."

Ethylene is a natural ripening agent for apples. A rotten apple, or one that is overripe, releases ethylene and causes the other apples in the barrel to ripen.

4B. The four *trans* double bonds in the side chain of retinal are labeled **a-d**. Double bond **c** (between carbons 11 and 12) is isomerized to its *cis* isomer by an enzyme in the body.

Which of the other three double bonds in the side chain of retinal would yield the least stable isomer of *cis*-retinal if it were to be isomerized? (*Hint:* Think steric strain.)

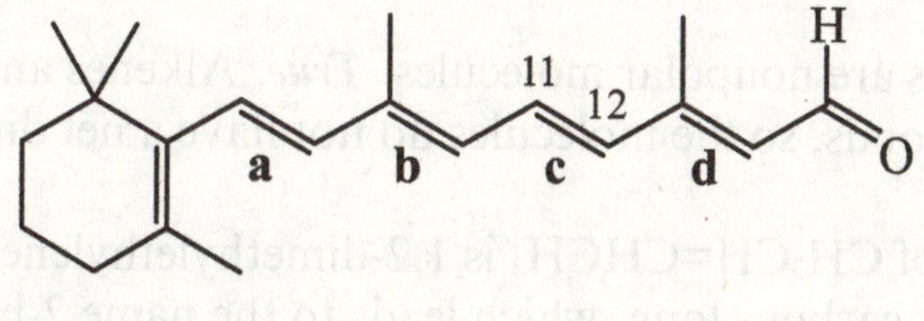

Consider the isomerization of **a**, **b**, and **d** to give the following *cis*-retinal isomers:

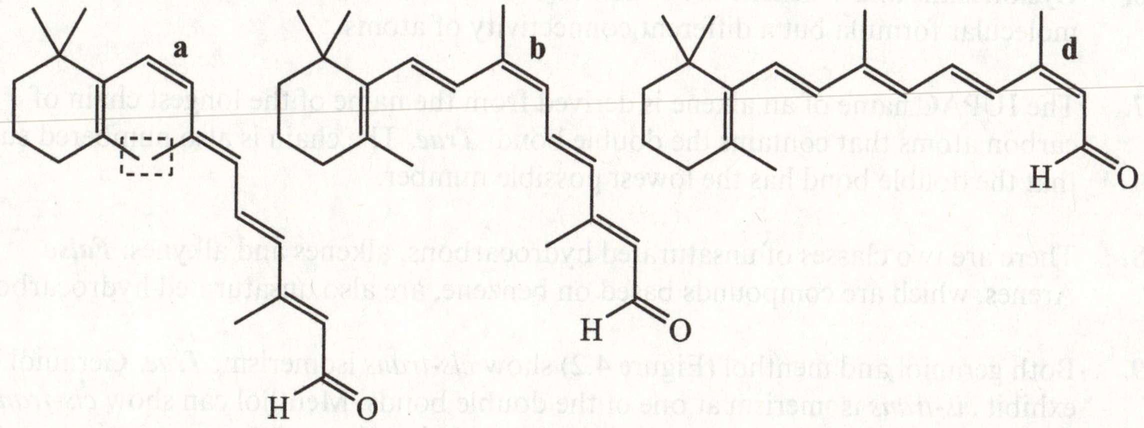

The isomerization of double bond **a** places two methyl groups in close proximity, as indicated in the box above. This isomer would have the highest energy (lowest stability) of all the possible *cis*-retinal isomers.

4C. Based on the information in this Chemical Connection, what can you deduce about the physical properties of leaf cell membranes?

Isoprene is dissolved in the leaf cell membranes before it is released into the atmosphere. Because isoprene is a nonpolar hydrocarbon, the membranes must also be relatively nonpolar. At higher temperatures, isoprene is more volatile and more easily released.

Quick Quiz

1. Ethylene and acetylene are constitutional isomers. *False.* Constitutional isomers must have the same molecular formula but different atom connectivity (bonding sequence). Ethylene and acetylene are respectively C_2H_4 and C_2H_2.

2. Alkanes that are liquid at room temperature are insoluble in water and when added to water will float on water. *True.* Alkanes are nonpolar and therefore water-insoluble, and they have a lower density than water.

3. The bulk of the ethylene used by the chemical industry worldwide is obtained from nonrenewable resources. *True*. Ethylene is derived from the cracking of hydrocarbons.

4. Alkenes and alkynes are nonpolar molecules. *True*. Alkenes and alkynes do not contain any polar bonds, so the molecules do not have a net dipole moment.

5. The IUPAC name of $CH_3CH=CHCH_3$ is 1,2-dimethylethylene. *False*. The longest carbon chain is four carbons long, which leads to the name 2-butene.

6. Cyclohexane and 1-hexene are constitutional isomers. *True*. They have the same molecular formula but a different connectivity of atoms.

7. The IUPAC name of an alkene is derived from the name of the longest chain of carbon atoms that contains the double bond. *True*. The chain is also numbered such that the double bond has the lowest possible number.

8. There are two classes of unsaturated hydrocarbons, alkenes and alkynes. *False*. Arenes, which are compounds based on benzene, are also unsaturated hydrocarbons.

9. Both geraniol and menthol (Figure 4.2) show *cis-trans* isomerism. *True*. Geraniol can exhibit *cis-trans* isomerism at one of the double bonds. Menthol can show *cis-trans* isomerism with respect to the substituents bonded to the cyclohexane ring.

10. Terpenes are identified by their carbon skeletons, namely, one that can be divided into five-carbon isoprene units. *True*. All terpenes consist of isoprene building blocks.

11. 1,2-Dimethylcyclohexene shows *cis-trans* isomerism. *False*. The double bond is a part of a six-membered ring and must be *cis* with respect to the ring. There is no *cis-trans* isomerism with respect to the methyl groups on the ring because they are both bonded to sp^2 carbons and are planar.

12. 2-Methyl-2-butene shows *cis-trans* isomerism. *False*. Carbon 2 is bonded to two identical substituents, two methyl groups.

13. Both ethylene and acetylene are planar molecules. *True*. The carbon atoms in ethylene are sp^2-hybridized, and all the carbon and hydrogen atoms lie in the same plane. Acetylene has sp-hybridized carbons, and all the atoms of acetylene lie in the same plane.

14. The physical properties of alkenes are similar to those of alkanes with the same carbon skeletons. *True*. These properties include density, melting point, and boiling point.

15. Isoprene is the common name for 2-methyl-1,3-butadiene. *True*. Realize that once this compound is incorporated into a terpene, it need not bear the same order of single and double bonds. Other chemical modifications can also be performed on the terpene. However, the five-carbon skeleton of isoprene is always retained in a terpene.

End-of-Chapter Problems

Structure of Alkenes and Alkynes

4.7   Describe what will happen when *trans*-3-heptene is added to the following compounds:

(a) Cyclohexane. Both cyclohexane and *trans*-3-heptene are nonpolar compounds. The alkene will dissolve in cyclohexane, forming a homogenous solution.

(b) Ammonia (*l*). NH₃ is a polar solvent (recall its trigonal pyramidal shape). When nonpolar *trans*-3-heptene is added to ammonia, an immiscible mixture is formed.

4.8   Each carbon atom in ethane and in ethylene is surrounded by eight valence electrons and has four bonds to it. Explain how the VSEPR model (Section 1.3) predicts a bond angle of 109.5° about each carbon in ethane but an angle of 120° about each carbon in ethylene.

Although each carbon atom in ethane and ethylene is surrounded by eight valence electrons and has four bonds to it, the shape and geometry about each carbon is based on the number of regions of electron density associated with it. In ethane, each carbon has four regions of electron density and a tetrahedral shape, which has bond angles of 109.5°. Whereas, each carbon in ethylene has three regions of electron density (remember that a double bond is counted as a single region of electron density), resulting in a trigonal planar arrangement (120°).

4.9   Explain the difference between *saturated* and *unsaturated*.

A compound that is *saturated* does not contain any carbon-carbon π bonds. Compounds that are *unsaturated* contain one or more carbon-carbon π bonds.

4.10   Use valence-shell electron-pair repulsion (VSEPR) to predict all bond angles about each of the highlighted carbon atoms.

To make these predictions, determine the number of regions of electron density associated with each carbon of interest (don't forget the implicit hydrogens). If there are two regions of electron density, the bond angle is 180°; three regions, 120°; and four regions, 109.5°. Note that these respectively correlate to *sp*, *sp²*, and *sp³* hybridization.

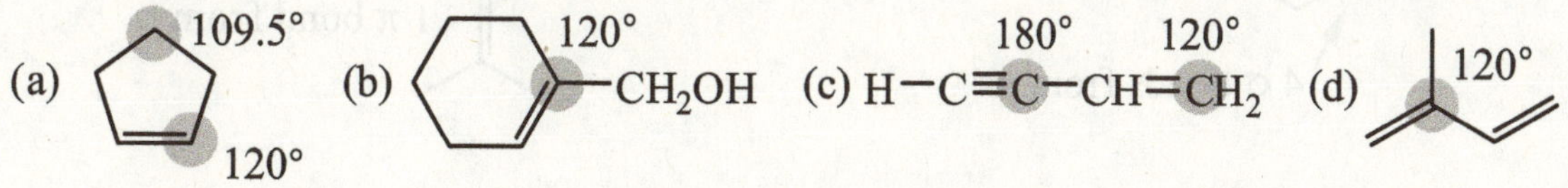

(a) [structure] 109.5° 120°   (b) [structure] 120° —CH₂OH   (c) H—C≡C—CH=CH₂ 180° 120°   (d) [structure] 120°

4.11   For each highlighted carbon atom in Problem 4.10, identify which orbitals are used to form each sigma bond and which are used to form each pi bond.

It is useful to first identify the hybridization of carbon. Hybrid orbitals are used to form σ bonds, while any remaining *p* orbitals can be used to form π bonds. Recall that a double bond consists of one σ bond and π bond, and that a triple bond consists of one σ bond and two π bonds.

(a)   4 σ bonds from sp^3

3 σ bonds from sp^2
1 π bond from *p*

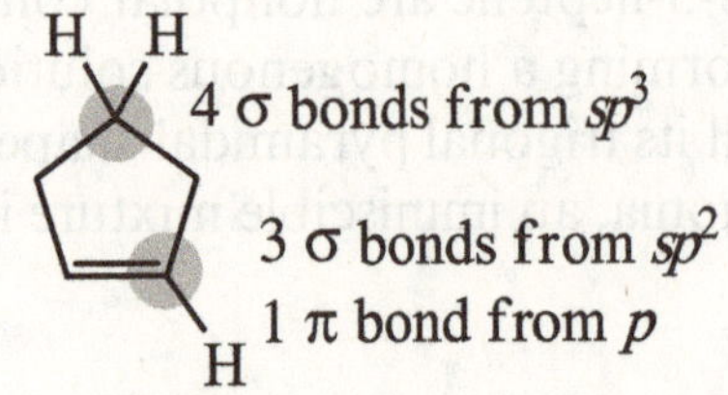

(b)   —CH₂OH
3 σ bonds from sp^2
1 π bond from *p*

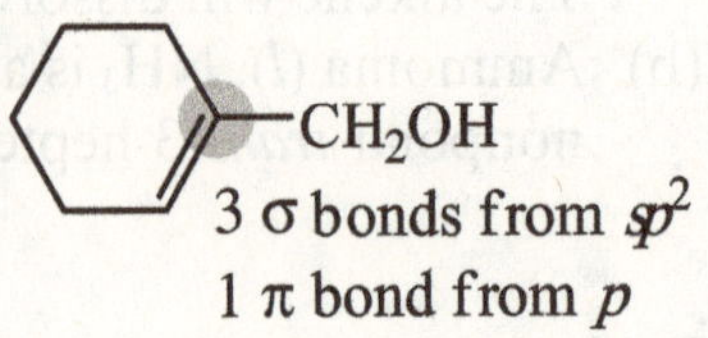

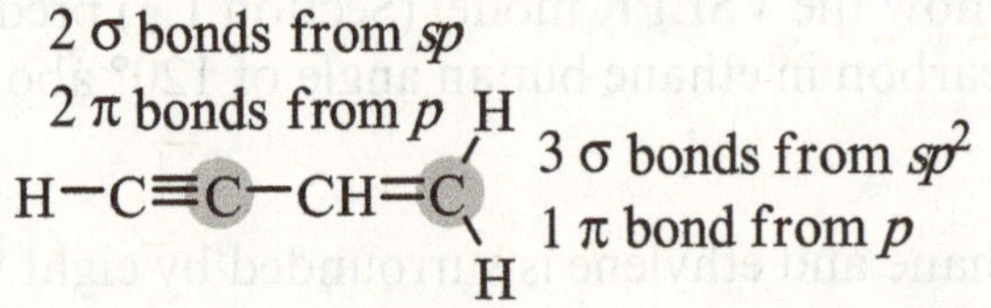

2 σ bonds from *sp*
2 π bonds from *p*
(c)   H—C≡C—CH=C
3 σ bonds from sp^2
1 π bond from *p*

(d)   3 σ bonds from sp^2
1 π bond from *p*

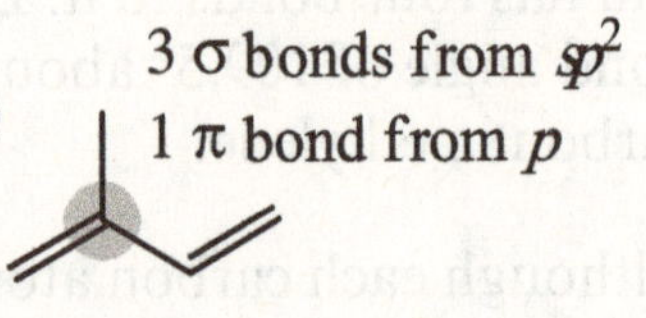

4.12   Predict all bond angles about each highlighted carbon atom:

(a)   120°
109.5°

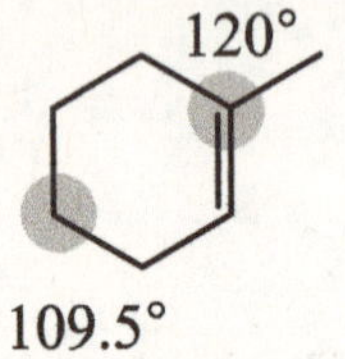

(b)   OH
120°

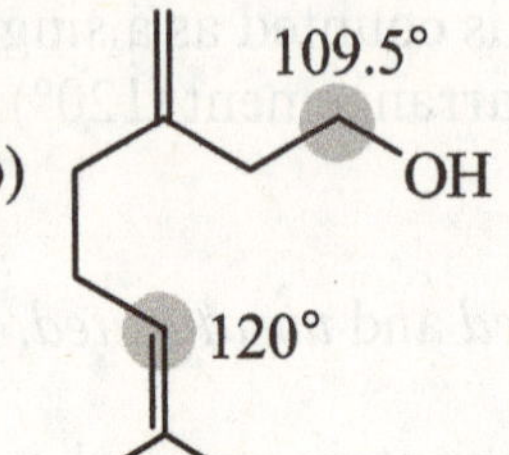

(c)   109.5°
180°

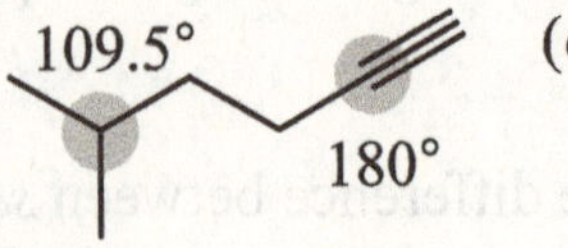

(d)   Br
109.5°
Br

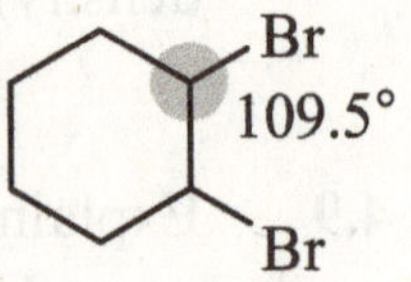

4.13   For each highlighted carbon atom in Problem 4.12, identify which orbitals are used to form each sigma bond and which are used to form each pi bond.

As with Problem 4.11, hybrid orbitals are used to form σ bonds, while remaining *p* orbitals are used to form π bonds.

(a)   3 σ bonds from sp^2
1 π bond from *p*
4 σ bonds from sp^3

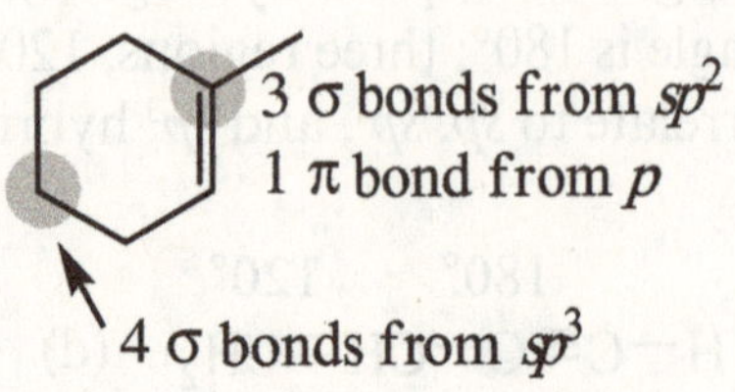

(b)   4 σ bonds from sp^3
OH
3 σ bonds from sp^2
1 π bond from *p*

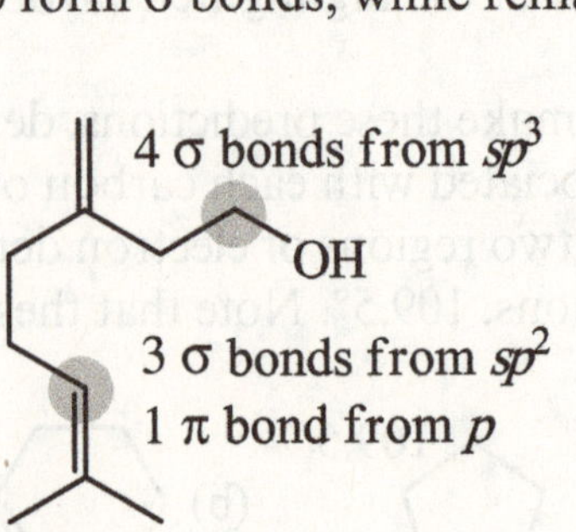

(c)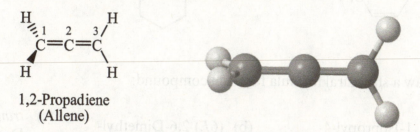

2 σ bonds from *sp*
2 π bonds from *p*

4 σ bonds from *sp³*

(d)

Br

4 σ bonds from *sp³*

Br

4.14 Following is the structure of 1,2-propadiene (allene). In it, the plane created by H−C−H of carbon 1 is perpendicular to that created by H−C−H of carbon 3.

$$\underset{\text{H}}{\overset{\text{H}}{\text{C}}}=\underset{2}{\text{C}}=\underset{\text{H}}{\overset{\text{H}}{\text{C}}}$$

1,2-Propadiene
(Allene)

(a) State the orbital hybridization of each carbon in allene.

Carbons 1 and 3 are *sp²*-hybridized, while carbon 2 is *sp*-hybridized.

(b) Account for the molecular geometry of allene in terms of the orbital overlap model. Specifically, explain why all four hydrogen atoms are not in the same plane.

Carbon 2, which is the carbon that bears two double bonds, is *sp*-hybridized. In *sp* hybridization, the two remaining *p* orbitals are perpendicular to each other. One of the two *p* orbitals forms the π bond with carbon 1 while the other *p* orbital forms the π bond with carbon 2. Accordingly, these two π bonds are perpendicular to each other. As a result, the trigonal plane formed by the CH_2 on the left is perpendicular to the one formed by the CH_2 group on the right.

Nomenclature of Alkenes and Alkynes

4.15 Draw a structural formula for each compound:

(a) *trans*-2-Methyl-3-hexene   (b) 2-Methyl-3-hexyne   (c) 2-Methyl-1-butene

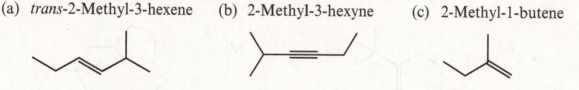

(d) 3-Ethyl-3-methyl-1-pentyne   (e) 2,3-Dimethyl-2-butene   (f) *cis*-2-Pentene

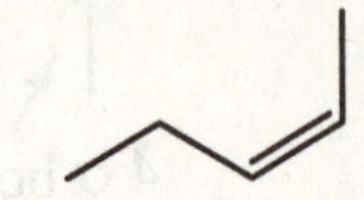

(g) (*Z*)-1-Chloropropene    (h) 3-Methylcyclohexene

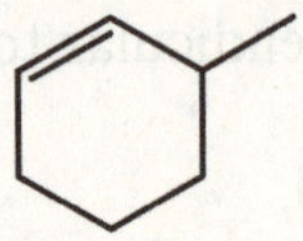

4.16 Draw a structural formula for each compound:

(a) 1-Isopropyl-4-methylcyclohexene

(b) (6*E*)-2,6-Dimethyl-2,6-octadiene

(c) *trans*-1,2-Diisopropyl-cyclopropane

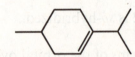   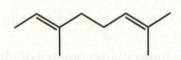

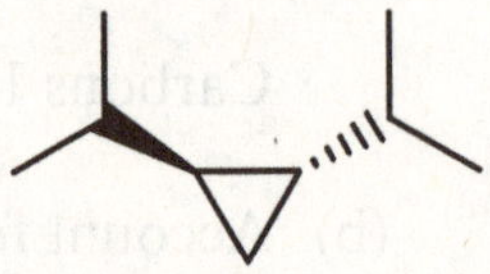

(d) 2-Methyl-3-hexyne   (e) 2-Chloropropene 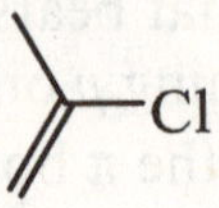   (f) Tetrachloroethylene

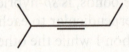

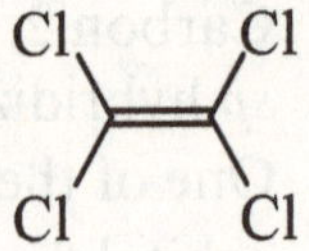

4.17 Write the IUPAC name for each compound:

(a)

1,2-Dimethylcyclohexene

(b)

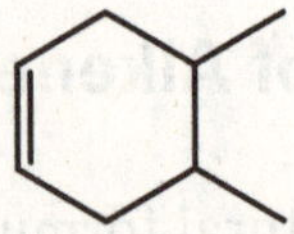

4,5-Dimethylcyclohexene

(c)

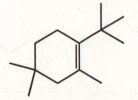

1-*tert*-Butyl-2,4,4-trimethylcyclohexene

(d)

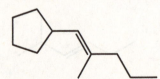

(*E*)-1-Cyclopentyl-2-methyl-1-pentene

4.18   Write the IUPAC name for each compound:

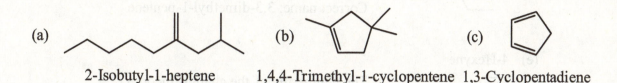

(a)   (b)   (c)

2-Isobutyl-1-heptene   1,4,4-Trimethyl-1-cyclopentene   1,3-Cyclopentadiene

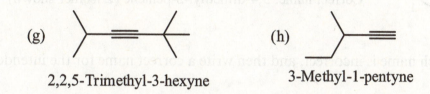

(d) 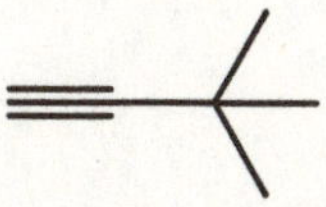   (e)   (f)

3,3-Dimethyl-1-butyne   2,4-Dimethyl-2-pentene   1-Octyne

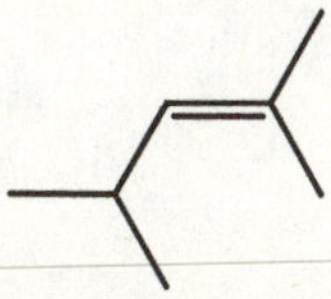

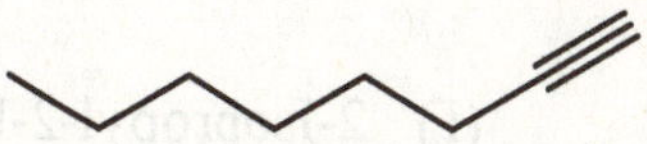

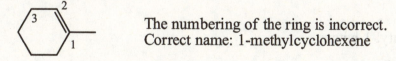

(g)   (h)

2,2,5-Trimethyl-3-hexyne   3-Methyl-1-pentyne

4.19   Explain why each name is incorrect, and then write a correct name for the intended compound:

(a) 1-Methylpropene

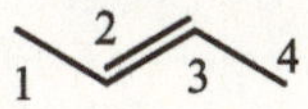

The parent chain is four carbons long, and it is also necessary to indicate the configuration (E or Z). Correct name: 2-butene (E isomer shown)

(b) 3-Pentene

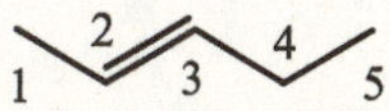

The numbering of the chain is incorrect, and it is also necessary to indicate the configuration (E or Z). Correct name: 2-pentene (E isomer shown)

(c) 2-Methylcyclohexene

The numbering of the ring is incorrect. Correct name: 1-methylcyclohexene

(d) 3,3-Dimethylpentene

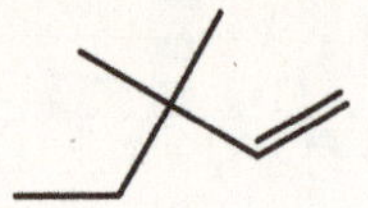

It is necessary to indicate the position of the carbon-carbon double bond.
Correct name: 3,3-dimethyl-1-pentene

(e) 4-Hexyne

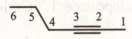

The numbering of the chain is incorrect.
Correct name: 2-hexyne

(f) 2-Isopropyl-2-butene

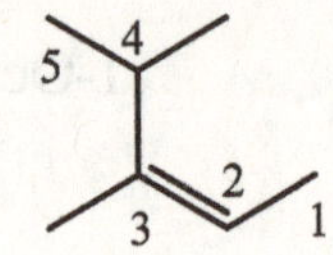

The parent chain is five carbons long, and it is also necessary to indicate the configuration (E or Z).
Correct name: 3,4-dimethyl-2-pentene (Z isomer shown)

4.20 Explain why each name is incorrect, and then write a correct name for the intended compound:

(a) 2-Ethyl-1-propene

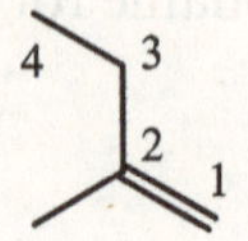

The parent chain is four carbons long.
Correct name: 2-methyl-1-butene

(b) 5-Isopropylcyclohexene

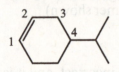

The numbering of the ring is incorrect.
Correct name: 4-isopropylcyclohexene

(c) 4-Methyl-4-hexene

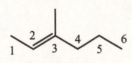

The numbering of the chain is incorrect, and it is also necessary to indicate the configuration (E or Z).
Correct name: 3-Methyl-2-hexene (E isomer shown)

(d) 2-sec-Butyl-1-butene

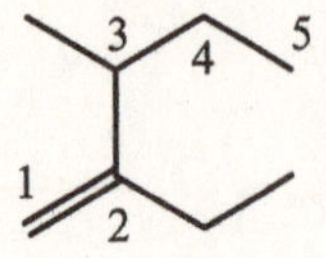

The parent chain is five carbons long.
Correct name: 2-ethyl-3-methyl-1-pentene

(e) 6,6-Dimethylcyclohexene

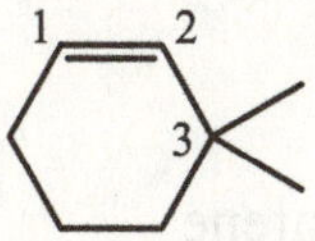

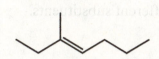

The numbering of the ring is incorrect.
Correct name: 3,3-dimethylcyclohexene

(f) 2-Ethyl-2-hexene

The parent chain is five carbons long, and it is also
necessary to indicate the configuration (E or Z).
Correct name: 3-methyl-3-heptene (E isomer shown)

Cis-Trans (E/Z) Isomerization in Alkenes and Cycloalkenes

4.21 Which of these alkenes show *cis-trans* isomerism? For each that does, draw
structural formulas for both isomers.

(a) 1-Hexene
(b) 2- Hexene
(c) 3-Hexene
(d) 2-Methyl-2-hexene
(e) 3-Methyl-2-hexene
(f) 2,3-Dimethyl-2-hexene

For *cis-trans* isomerism to exist, the two substituents bonded to each of the two
alkene carbons must be different. If either one of the alkene carbons bears two
identical substituents, *cis-trans* isomerism is not possible. Thus, (b), (c), and (e) are
the only compounds that can exhibit *cis-trans* isomerism. Note that in (e), the groups
used to determine *cis* and *trans* are those that are part of the parent chain.

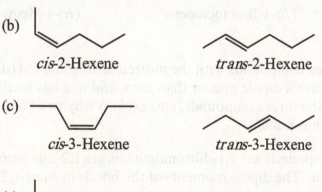

(b)

cis-2-Hexene   *trans*-2-Hexene

(c)

cis-3-Hexene   *trans*-3-Hexene

(e)

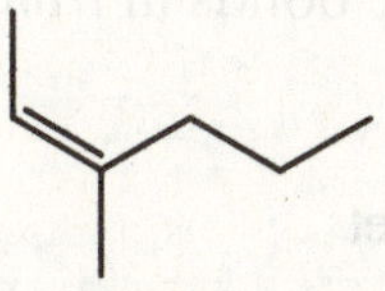

cis-3-Methyl-2-hexene 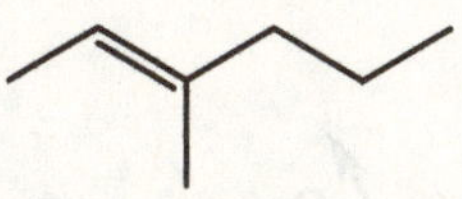   *trans*-3-Methyl-2-hexene

4.22 Which of these alkenes show *cis-trans* isomerism? For each that does, draw structural formulas for both isomers.

(a) 1-Pentene   (b) 2-Pentene
(c) 3-Ethyl-2-pentene   (d) 2,3-Dimethyl-2-pentene
(e) 2-Methyl-2-pentene   (f) 2,4-Dimethyl-2-pentene

Only 2-pentene can show *cis-trans* isomerism, because it is the only alkene where each of the two alkene carbons is bonded to two different substituents.

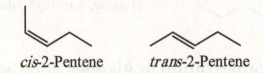

cis-2-Pentene   *trans*-2-Pentene

4.23 Which alkenes can exist as pairs of E/Z isomers? For each alkene that does, draw the E isomer.

(a) $CH_2=CHBr$ (b) $CH_3CH=CHBr$
(c) $(CH_3)_2C=CHCH_3$ (d) $(CH_3)_2CHCH=CHCH_3$

The E,Z nomenclature system can be used to name the*cis-trans* isomers of alkenes. As with *cis-trans* isomers, E/Z isomers exist only when the two substituents on each of two the alkene carbons are different, as in (b)and (d).

(b)   Br⟋⟍⟋

(*E*)-1-Bromopropene

(d)   ⟋⟍⟋⟍

(*E*)-4-Methyl-2-pentene

4.24 There are three compounds with the molecular formula $C_2H_2Br_2$. Two of these compounds have a dipole greater than zero, and one has no dipole. Draw structural formulas for the three compounds, and explain why two have dipole moments but the third one has none.

The three compounds are 1,1-dibromoethene, *cis*-1,2-dibromoethene, and *trans*-1,2-dibromoethene. The dipole moments of the bonds in *trans*-1,2-dibromoethene cancel out.

H   Br
 C=C   net
H   Br

        net
Br   Br
   C=C
 H   H

Br   H
   C=C
 H   Br

no net dipole moment

4.25   Name and draw structural formulas for all alkenes with the molecular formula C_5H_{10}. As you draw these alkenes, remember that *cis* and *trans* isomers are different compounds and must be counted separately.

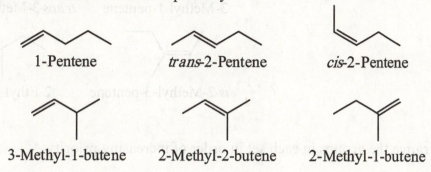

1-Pentene   *trans*-2-Pentene   *cis*-2-Pentene

3-Methyl-1-butene   2-Methyl-2-butene   2-Methyl-1-butene

4.26   Name and draw structural formulas for all alkenes with the molecular formula C_6H_{12} that have the following carbon skeletons (remember *cis* and *trans* isomers):

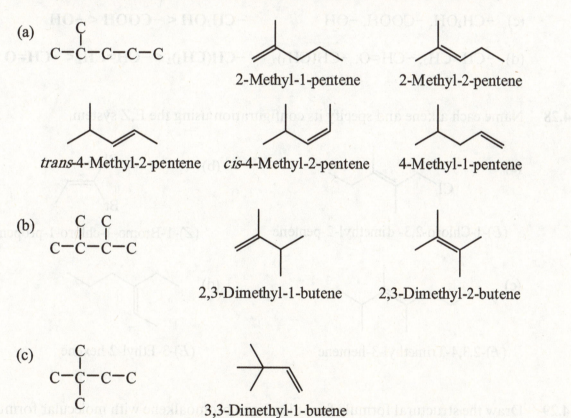

(a)   C–C–C–C–C (with C branch)

2-Methyl-1-pentene   2-Methyl-2-pentene

trans-4-Methyl-2-pentene   *cis*-4-Methyl-2-pentene   4-Methyl-1-pentene

(b)   C–C–C–C (with C, C branches)

2,3-Dimethyl-1-butene   2,3-Dimethyl-2-butene

(c)   C–C–C–C (with C, C branches)

3,3-Dimethyl-1-butene

(d)

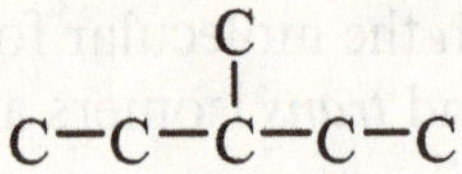

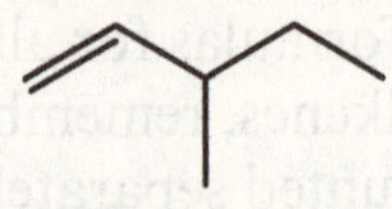

$$C-C-\overset{\overset{\displaystyle C}{|}}{\underset{}{C}}-C-C$$

3-Methyl-1-pentene 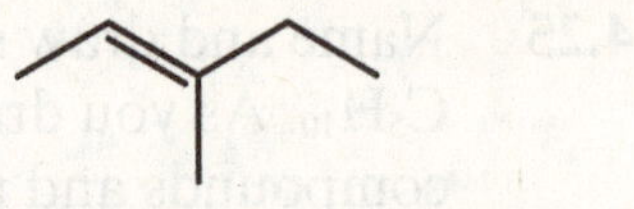   *trans*-3-Methyl-2-pentene

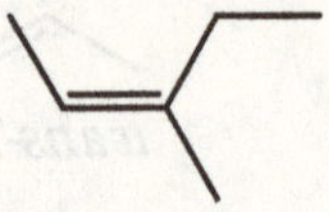

cis-2-Methyl-3-pentene 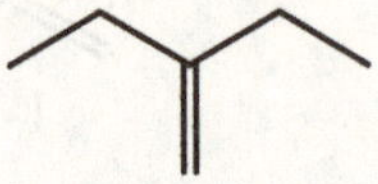   2-Ethyl-1-butene

4.27   Arrange the groups in each set in order of increasing priority:

(a) $-CH_3$, $-Br$, $-CH_2CH_3$ $-CH_3 < -CH_2CH_3 < -Br$

(b) $-OCH_3$, $-CH(CH_3)_2$, $-CH_2CH_2NH_2$ $-CH_2CH_2NH_2 < -CH(CH_3)_2 < -OCH_3$

(c) $-CH_2OH$, $-COOH$, $-OH$ $-CH_2OH < -COOH < -OH$

(d) $-CH=CH_2$, $-CH=O$, $-CH(CH_3)_2$ $-CH(CH_3)_2 < -CH=CH_2 < -CH=O$

4.28   Name each alkene and specify its configuration using the E,Z system.

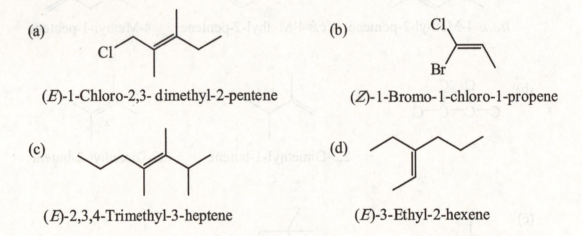

(a)

Cl

(*E*)-1-Chloro-2,3- dimethyl-2-pentene

(b)

Cl

Br

(*Z*)-1-Bromo-1-chloro-1-propene

(c)

(*E*)-2,3,4-Trimethyl-3-heptene

(d)

(*E*)-3-Ethyl-2-hexene

4.29   Draw the structural formula for at least one bromoalkene with molecular formula C_5H_9Br that:

(a) Shows E,Z isomerism

For an alkene to show E,Z isomerism, each of the two alkene carbons must be bonded to two different substituents. Thus, draw any structural isomer where each alkene carbon is bonded to two different groups. Two examples include:

(E)-2-Bromo-2-pentene   (Z)-1-Bromo-1-pentene

(b) Does not show E,Z isomerism

Draw any structural isomer where at least one of the alkene carbons is bonded to two identical substituents. Two examples include:

5-Bromo-1-pentene   1-Bromo-3-methyl-2-butene

4.30 Is *cis-trans* isomerism possible in alkanes? In alkynes? Explain.

In alkanes, the C–C bonds are freely rotating, so alkanes can adopt a variety of conformations, all of which are can freely interconvert. Whereas, in cycloalkanes, the cyclic structure of the ring restricts bond rotation. Similarly, in alkenes, the double bond cannot rotate under normal conditions.

Alkynes cannot exhibit *cis-trans* isomerism because there is only one atom or group bonded to each alkyne carbon, and alkynes have a linear geometry.

4.31 For each molecule that shows *cis-trans* isomerism, draw the *cis* isomer:

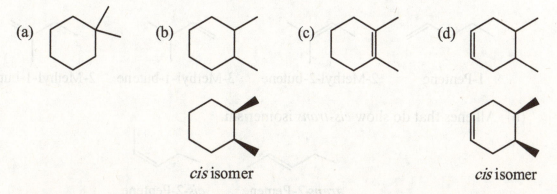

(a)   (b)   (c)   (d)

cis isomer   *cis* isomer

Recall that in cycloalkanes, *cis-trans* isomerism refers to the relative positions of two substituents bonded to two *different* tetrahedral carbons of the cycloalkane ring. Molecule (a) does not exhibit *cis-trans* isomerism because both methyl groups are bonded to the same carbon. Molecules (b) and (d) show *cis-trans* isomerism, because the two methyl groups are on different carbons and can be on either the same side (*cis*) or the opposite side (*trans*) relative to each other. The methyl groups in molecule (c) are planar due to the *sp²*-hybridized alkene carbon, so they cannot be up or down relative to each other.

4.32   Explain why each name is incorrect or incomplete, and then write a correct name:

In all four molecules, *cis-trans* isomerism (and therefore E,Z isomerism) is not possible because at least one of the two alkene carbons is bonded to two identical substituents. Correct names are indicated below each structure.

(a) (*Z*)-2-Methyl-1-pentene   (b) (*E*)-3,4-Diethyl-3-hexene

   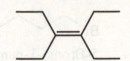

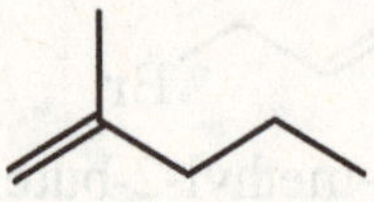

2-Methyl-1-pentene   3,4-Diethyl-3-hexene

(c) *trans*-2,3-Dimethyl-2-hexene   (d) (1*Z*,3*Z*)-2,3-Dimethyl-1,3-butadiene

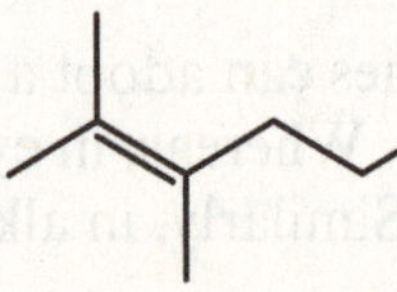

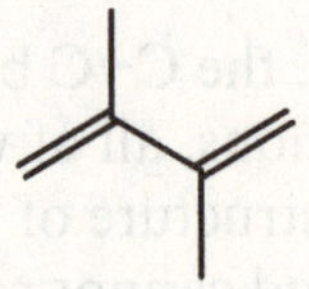

2,3-Dimethyl-2-hexene   2,3-Dimethyl-1,3-butadiene

4.33   Draw structural formulas for all compounds with the molecular formula C₅H₁₀ that are

(a) Alkenes that do not show *cis-trans* isomerism.

1-Pentene 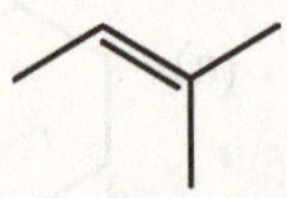   2-Methyl-2-butene 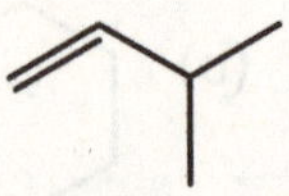   3-Methyl-1-butene   2-Methyl-1-butene

(b) Alkenes that do show *cis-trans* isomerism.

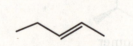

trans-2-Pentene 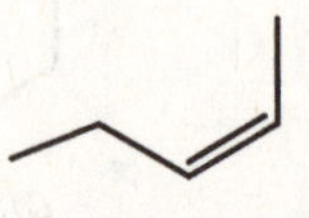   *cis*-2-Pentene

(c) Cycloalkanes that do not show *cis-trans* isomerism.

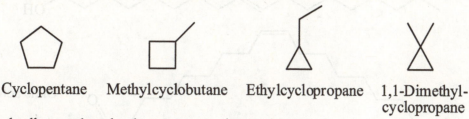

Cyclopentane   Methylcyclobutane   Ethylcyclopropane   1,1-Dimethyl-
                                                           cyclopropane

(d) Cycloalkanes that do show *cis-trans* isomerism.

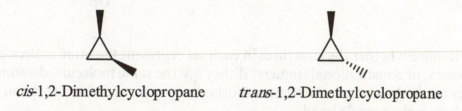

cis-1,2-Dimethylcyclopropane   *trans*-1,2-Dimethylcyclopropane

4.34 β-Ocimene, a triene found in the fragrance of cotton blossoms and several essential
      oils, has the IUPAC name (3*Z*)-3,7-dimethyl-1,3,6-octatriene. Draw a structural
      formula for β-ocimene.

(3*Z*)-3,7-Dimethyl-1,3,6-octatriene

4.35 Oleic acid and elaidic acid are, respectively, the *cis* and *trans* isomers of 9-
      octadecenoic acid. One of these fatty acids, a colorless liquid that solidifies at 4°C, is
      a major component of butterfat. The other, a white solid with a melting point of
      44−45°C, is a major component of partially hydrogenated vegetable oils. Which of
      these two fatty acids is the *cis* isomer and which is the *trans* isomer? (Hint: Think
      about the geometry of packing and the relative strengths of the resulting dispersion
      forces.)

The melting point of a compound is influenced by how well the molecules pack
together. If the molecules are closely packed together, the intermolecular forces
between them are greater. How well a molecule packs with another is highly
dependent on the three-dimensional geometry of the molecule. The *trans* isomer
(elaidic acid) is relatively linear and more compact, so it is able to pack together
more tightly than oleic acid. Elaidic acid therefore has a higher melting point than
oleic acid.

Elaidic acid

Oleic acid

4.36 Determine whether the structures in each set represent the same molecule, *cis-trans* isomers, or constitutional isomers. If they are the same molecule, determine whether they are in the same or different conformations as a result from rotation about a carbon-carbon single bond.

The two molecules in (a) are *cis-trans* isomers. The two structures in each of (b), (c), and (d) are identical, and the molecules in each of these pairs are conformations caused by rotation about the indicated carbon-carbon single bond(s).

(a)   and   *cis-trans* isomers

(b)   and   same compounds but different conformations

(c)   and   same compounds but different conformations

(d)   and   same compounds but different conformations

4.37 Following is the structural formula of lycopene, a deep-red compound that is partially responsible for the red color of ripe fruits, especially tomatoes. Approximately 20 mg of lycopene can be isolated from 1 kg of fresh, ripe tomatoes. How many of the carbon-carbon double bonds in lycopene have the possibility for *E-Z* isomerism? Use the E,Z system to assign the configuration of all applicable double bonds.

E-Z isomerism is possible for eleven of the carbon-carbon double bonds, which are indicated (*) in the structure below. All of these double bonds have the *E* configuration.

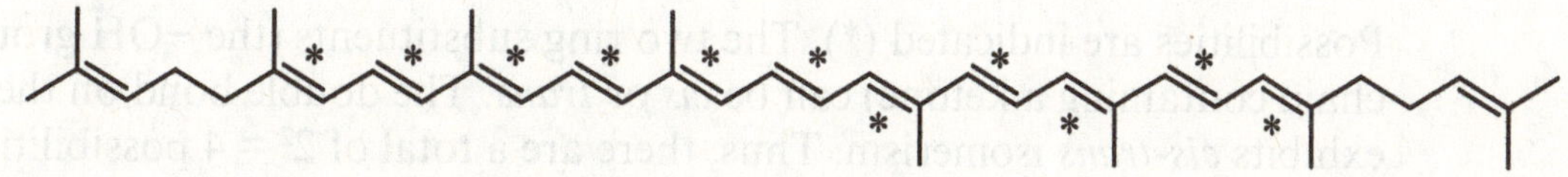

4.38 As you might suspect, β-carotene, a precursor of vitamin A, was first isolated from carrots. Dilute solutions of β-carotene are yellow – hence its use as a food coloring. In plants, it is almost always present in combination with chlorophyll to assist in the harvesting of the energy of sunlight. As tree leaves die in the fall, the green of their chlorophyll molecules is replaced by the yellows and reds of carotene and carotene-related molecules.

(a) Compare the carbon skeletons of β-carotene and lycopene. What are the similarities? What are the differences?

(b) Use the E,Z system to assign the configuration of all applicable double bonds.

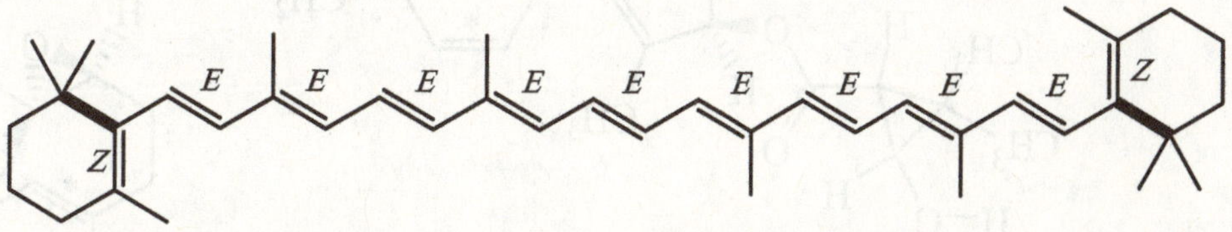

The main structural difference between β-carotene and lycopene is that β-carotene has six-membered rings at the two ends of the structure. These rings are formed by the cyclization of two sets of carbon atoms (new bonds indicated in bold).

4.39 In many parts of South America, extracts of the leaves and twigs of *Montanoa tomentosa* are used as a contraceptive, to stimulate menstruation, to facilitate labor, and as an abortifacient. The compound responsible for these effects is zoapatanol:

Zoapatanol

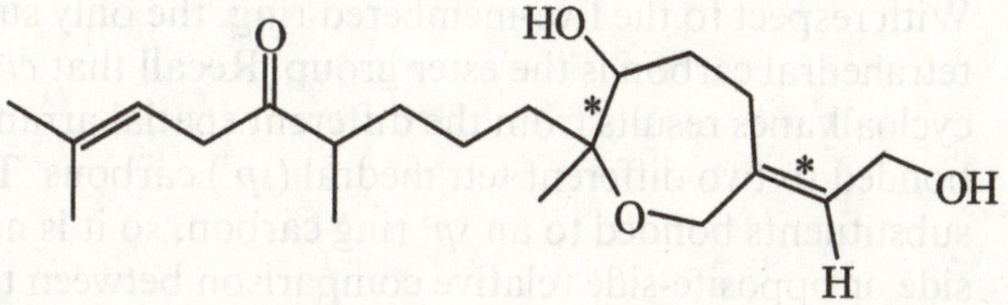

(a) Specify the configuration about the carbon-carbon double bond to the seven-membered ring, according to the E,Z system.

The double bond has the *E* configuration.

(b) How many *cis-trans* isomers are possible for zoapatanol? Consider the possibilities for *cis-trans* isomerism in cyclic compounds and about carbon-carbon double bonds.

Possibilities are indicated (*). The two ring substituents (the −OH group and the chain containing a ketone) can be *cis* or *trans*. The double bond on the right also exhibits *cis-trans* isomerism. Thus, there are a total of $2^2 = 4$ possibilities.

4.40 Pyrethrin II and pyrethrosin are natural products isolated from plants of the chrysanthemum family. Pyrethrin II is a natural insecticide and is marketed as such.

(a) Label all carbon-carbon double bonds in each about which *cis-trans* isomerism is possible.

Alkenes where *cis-trans* isomerism is possible are indicated (*).

Pyrethrin II   Pyrethrosin

(b) Why are *cis-trans* isomers possible about the three-membered ring in pyrethrin II, but not about its five-membered ring?

With respect to the five-membered ring, the only substituent bonded to a tetrahedral carbon is the ester group. Recall that *cis-trans* isomerism in cycloalkanes results from the different spatial arrangement of two substituents bonded to two different tetrahedral (sp^3) carbons. There are no other substituents bonded to an sp^3 ring carbon, so it is not possible to make a same-side or opposite-side relative comparison between the ester group and another substituent on the ring. Note that the two substituents bonded to the cycloalkene ring have a planar geometry.

Looking Ahead

4.41   Explain why the central carbon-carbon single bond in 1,3-butadiene is slightly shorter than the central carbon-carbon single bond in 1-butene.

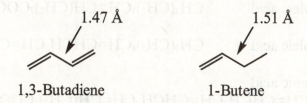

1.47 Å   1.51 Å

1,3-Butadiene   1-Butene

The indicated bond in 1,3-butadiene is formed by the overlap of two sp^2-hybridized carbons, while the indicated bond in 1-butene is formed by the overlap of one sp^3-hybridized and one sp^2-hybridized carbon. sp^2-Hybridized orbitals are smaller because they have greater *s* character (an s orbital holds its electrons closer to the nucleus of an atom), so the bond made from two sp^2 hybrids is shorter.

4.42   What effect might the ring size in the following cycloalkenes have on the reactivity of the C=C double bond in each?

The ideal bond angle for an alkene is 120°, because the alkene carbon is sp^2-hybridized. As the ring size becomes smaller, the bond angle deviates further from the ideal angle, which increases bond strain. Smaller cycloalkenes are therefore more reactive towards both ring opening (to yield an acyclic molecule) and addition reactions (Chapter 5).

4.43   What effect might each substituent have on the electron density surrounding the alkene C=C bond; that is, how does each substituent affect whether each carbon of the C−C bond is partially positive or partially negative?

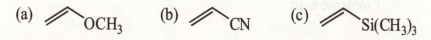

(a)   OCH$_3$ (b)   CN   (c)   Si(CH$_3$)$_3$

The electron density surrounding each alkene is affected by the electronegativity of the substituent near it. In molecules (a) and (b), the presence of oxygen and nitrogen, both of which are more electronegative than carbon, the alkenes have a reduced electron density, and the alkene carbon closest to the oxygen or nitrogen atom has a partial positive charge. Whereas, the silicon atom in (c) is lower in electronegativity than carbon, so the alkene has a higher electron density.

4.44 In Section 21.1 on the biochemistry of fatty acids, we will study the following three long-chain unsaturated carboxylic acids. Each has 18 carbons and is a component of animal fats, vegetable oils, and biological membranes. Because of their presence in animal fats, they are called fatty acids.

Oleic acid $CH_3(CH_2)_7CH=CH(CH_2)_7COOH$

Linoleic acid $CH_3(CH_2)_4CH=CHCH_2CH=CH(CH_2)_7COOH$

Linolenic acid
$CH_3CH_2CH=CHCH_2CH=CHCH_2CH=CH(CH_2)_7COOH$

(a) How many *E-Z* isomers are possible for each fatty acid?

The number of possible *E-Z* isomers corresponds to 2^n, where n is the number of carbon-carbon double bonds that can exhibit *E-Z* isomerism. Thus, oleic, linoleic, and linolenic acid respectively have 2, 4, and 8 *E-Z* isomers.

(b) These three fatty acids occur in biological membranes almost exclusively in the *cis* configuration. Draw line-angle formulas for each fatty acid, showing the *cis* configuration about each carbon-carbon double bond.

Oleic acid

Linoleic acid

Linolenic acid

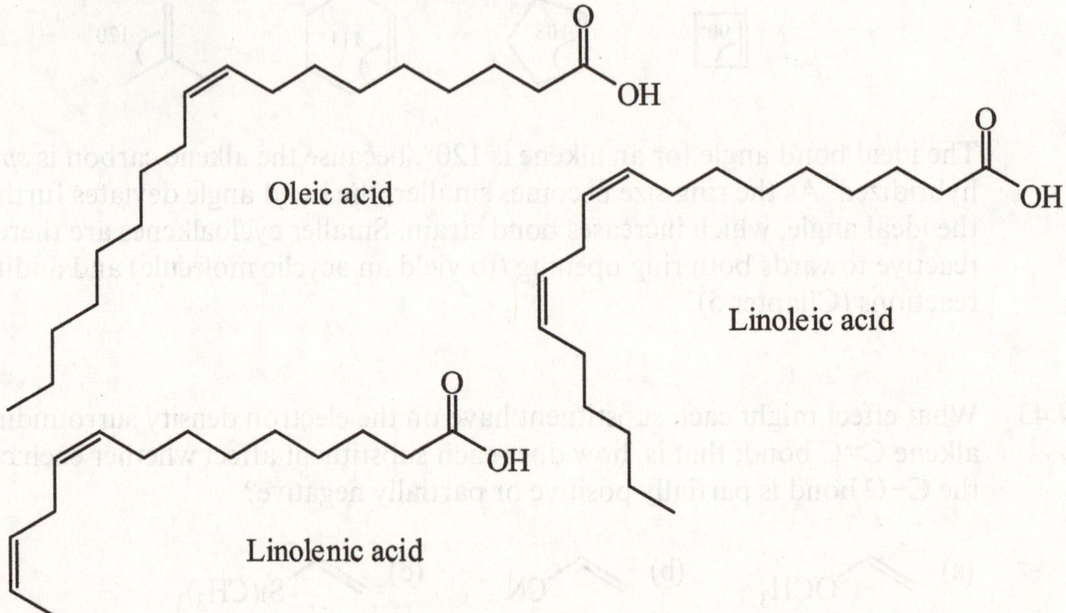

4.45 Assign an *E* or a *Z* configuration, or a *cis* or a *trans* configuration, to these carboxylic acids, each of which is an intermediate in the citric acid cycle. Under each is given its common name.

H   COOH
   C=C
HOOC   H
Fumaric acid

HOOC   COOH
   C=C
H   CH₂COOH
Aconitic acid

Fumaric acid has the *E* configuration, and it can be designated as a *trans* alkene. Aconitic acid has the *Z* configuration, and it can be designated as a *trans* alkene (note that while the two COOH groups are *cis* to each other, the C₅ parent chain is *trans*).

Group Learning Activities

4.46 Take turns coming up with structures that fit the following criteria. For each structure you come up with, explain to the group why your answer is correct.

(a) An alkene of formula C_6H_{12} that cannot be named using *cis-trans* or *E,Z*.

If an alkene cannot be named using *cis-trans* or *E,Z*, then at least one of the two alkene carbons must have two identical substituents. For example:

(b) A compound of formula C_7H_{12} that does not contain a pi bond.

In order for a compound to have that formula and not contain a pi bond, the compound must contain a ring. Your answer must have a ring, but don't forget that you can have three-, four-, five-, six-, and seven-membered rings.

(c) A compound of formula C_6H_{10} that does not contain a methylene group.

A methylene group is an alkyl –CH₂– group. An example of a compound that does not contain the methylene group is:

(d)   An alkene that uses "vinyl" in its IUPAC name.

A vinyl group is the –CH=CH₂ group. An example of such a compound is:

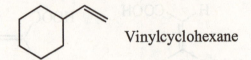

Vinylcyclohexane

(e)   A compound that can be named with the E,Z system but not with the *cis/trans* system.

This would be an alkene where the use of the *cis-trans* system would lead to ambiguity. An example would be:

(f)   A compound that can be named with the *cis-trans* system but not with the E,Z system.

Recall that the *cis-trans* system is also used to name the relative spatial positions of substituents bonded to sp^3-hybridized ring carbons. The E, Z system is used only to name alkenes. Examples would be:

(g)   A *trans* cycloalkene that has no ring or angle strain. (Hint: you may need to use a model kit to explain.)

With small cycloalkenes, because of geometric constraints, the alkene must be *cis* with respect to the carbon atoms of the ring. However, with large cycloalkenes, it is geometrically possible to have the alkene in a *trans* configuration.

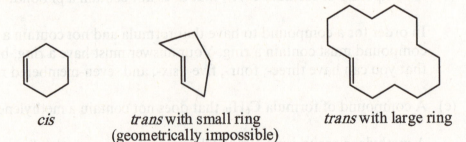

cis   *trans* with small ring   *trans* with large ring
                         (geometrically impossible)

Chapter 5: Reactions of Alkenes

Problems

5.1 In what way would the reaction energy diagram drawn in Example 5.1 change if the reaction were endothermic?

In an endothermic reaction, the products formed are higher in energy than the reactants. These reactions are thermodynamically unfavorable because they require a net input of heat.

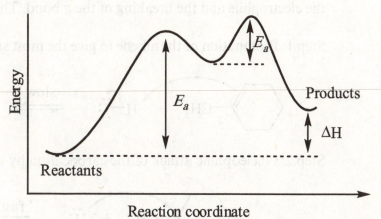

5.2 Name and draw a structural formula for the major product of each alkene addition reaction:

(a) $CH_3CH=CH_2$ + HI $\longrightarrow$ 2-Iodopropane

(b) $=CH_2$ + HI $\longrightarrow$ CH_3 / I   1-Iodo-1-methylcyclohexane

The product formed in each reaction is predicted by Markovnikov's rule; the hydrogen from HI adds to the alkene carbon that is already bonded to the most hydrogens.

5.3 Arrange these carbocations in order of increasing stability:

(a) $-CH_3$ (b) $-CH_3$ (c) $-CH_2$

most stable   least stable

Carbocations are electron-deficient and are stabilized by electron-donating groups, such as alkyl groups. Accordingly, when more alkyl groups are bonded to a carbocation, the carbocation is more stable.

5.4   Propose a mechanism for the addition of HI to 1-methylcyclohexene to give 1-iodo-1-methylcyclohexane. Which step in your mechanism is rate-determining?

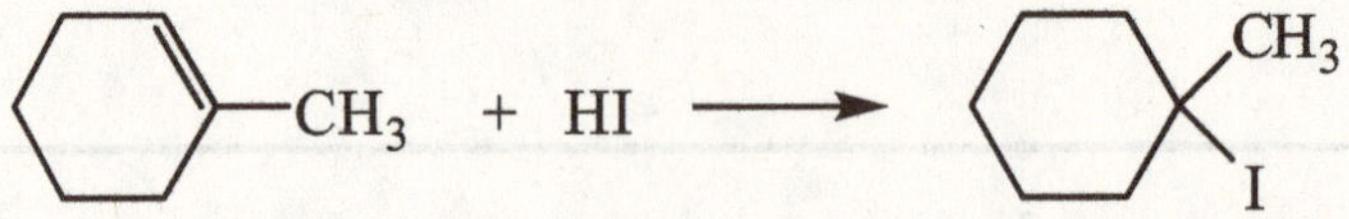

In the electrophilic addition reactions of alkenes, the first step involves the addition of the electrophile and the breaking of the π bond. This is the slow, rate-determining step.

Step 1: Protonation of the alkene to give the most stable carbocation (3°) intermediate.

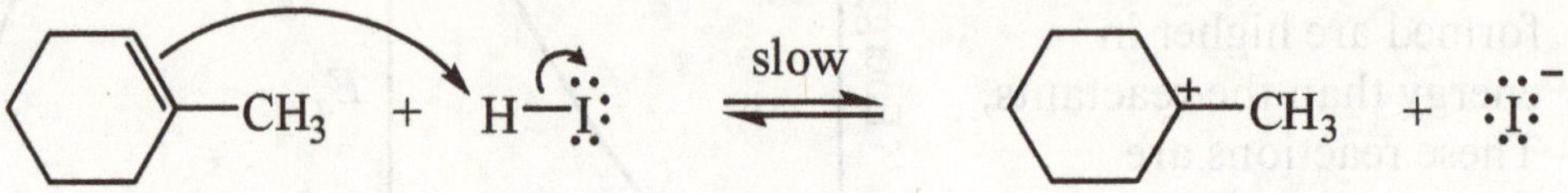

Step 2: Nucleophilic attack of the carbocation by the iodide ion, forming the product.

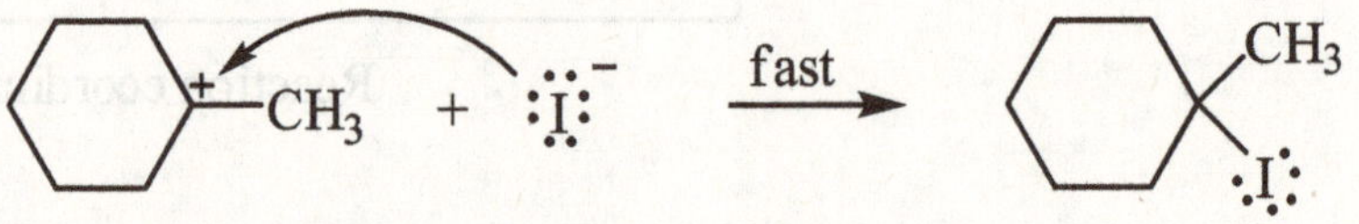

5.5   Draw a structural formula for the product of each alkene hydration reaction:

(a)

(b)

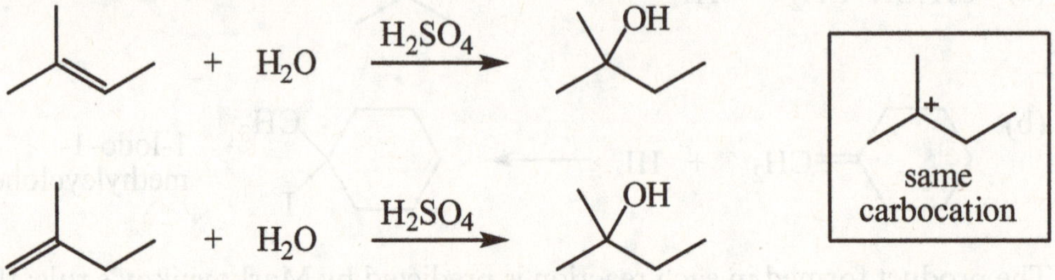

Although the two alkenes are different, the same product is obtained because both alkenes form the same 3° carbocation that is shown on the right.

5.6   Propose a mechanism for the acid-catalyzed hydration of 1-methylcyclohexene to give 1-methylcyclohexanol. Which step in your mechanism is rate-determining?

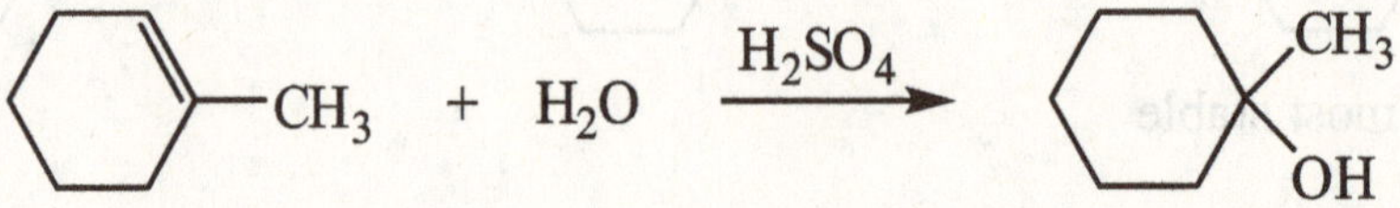

Like Problem 5.4, the slow step is the addition of the electrophile to the alkene.

Step 1: Protonation of the alkene to give the most stable carbocation (3°) intermediate:

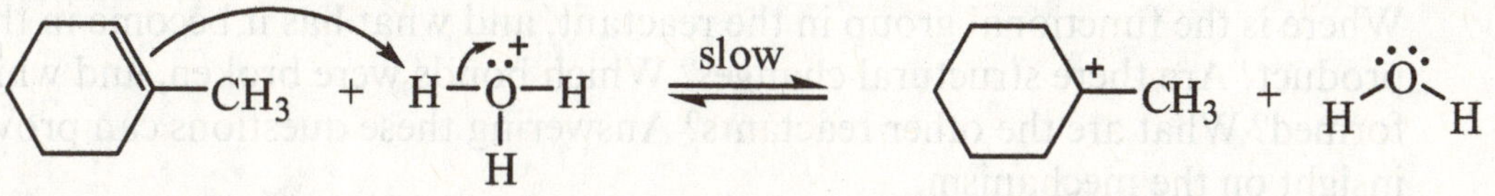

Step 2: Nucleophilic attack of the carbocation by water, forming an oxonium ion:

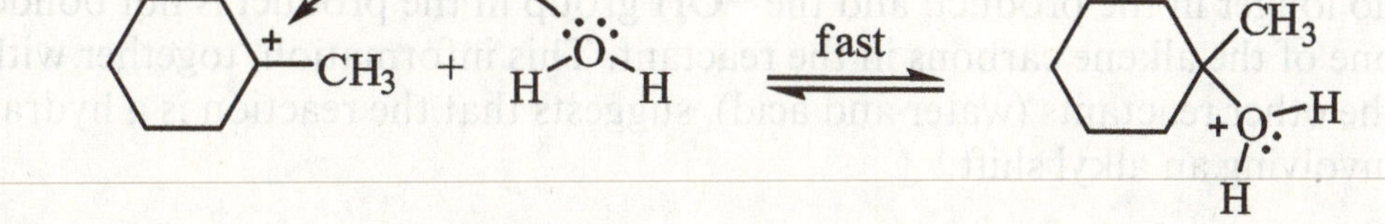

Step 3: Deprotonation of the oxonium ion to give the product and the catalyst:

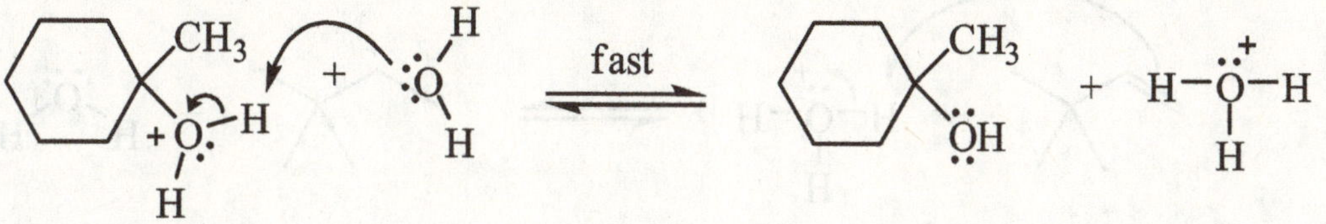

5.7 Complete these reactions:

(a)

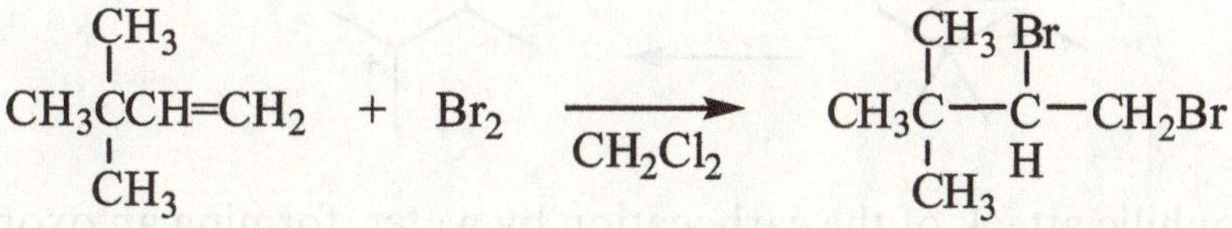

(b)

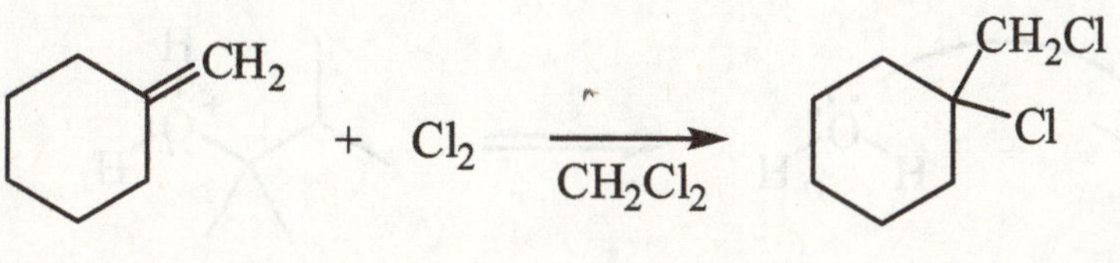

5.8 The acid-catalyzed hydration of 3,3-dimethyl-1-butene gives 2,3-dimethyl-2-butanol as the major product. Propose a mechanism for the formation of this alcohol.

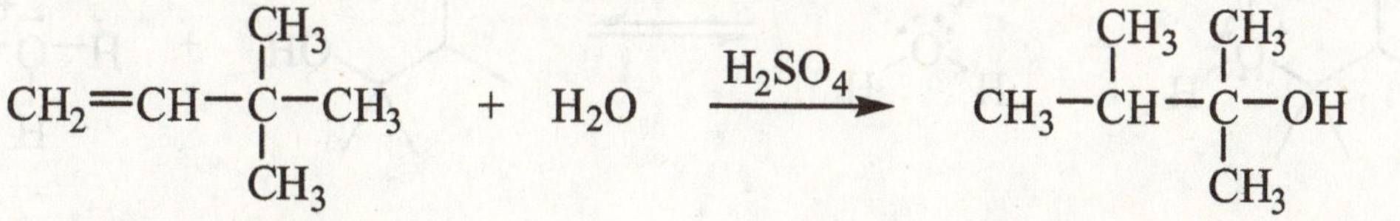

3,3-Dimethyl-1-butene   2,3-Dimethyl-2-butanol

When solving a reaction mechanism, a very useful strategy is to examine the starting materials and the products, noting the similarities and differences between them. Where is the functional group in the reactant, and what has it become in the product? Are there structural changes? Which bonds were broken, and which were formed? What are the other reactants? Answering these questions can provide insight on the mechanism.

Inspection of this reaction shows that the connectivity of the carbon atoms has changed (rearranged) from the reactant to the product. Furthermore, the alkene is no longer in the product, and the −OH group in the product is not bonded to either one of the alkene carbons in the reactant. This information, together with that about the other reactants (water and acid), suggests that the reaction is a hydration involving an alkyl shift.

Step 1: Protonation of the alkene to give the most stable carbocation (2°) intermediate:

Step 2: The 2° carbocation rearranges to form a more stable 3° carbocation:

Step 3: Nucleophilic attack of the carbocation by water, forming an oxonium ion:

Step 4: Deprotonation of the oxonium ion to give the product and the catalyst:

5.9   Draw a structural formula for the alkene that gives each alcohol on hydroboration followed by oxidation.

The hydroboration-oxidation of alkenes results in the anti-Markovinov addition of H_2O. When an alkene reacts with borane, the boron selectively adds to the alkene carbon that is least hindered. In the oxidation step of the reaction, the $-OH$ group replaces the boron.

(a)

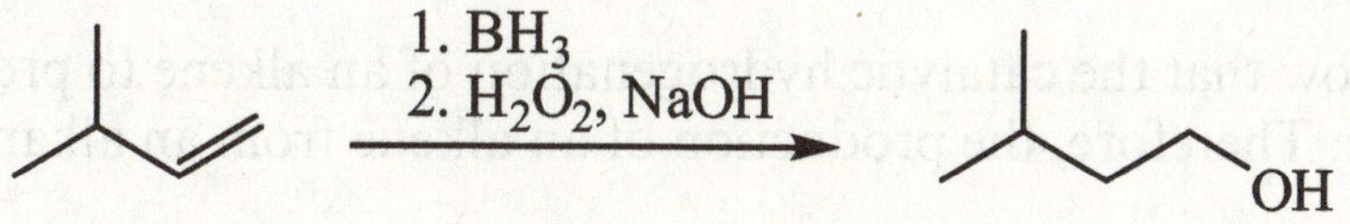

$$\text{(alkene)} \xrightarrow[\text{2. }H_2O_2\text{, NaOH}]{\text{1. }BH_3} \text{(alcohol) OH}$$

(b)

$$\text{(methylenecyclopentane)} \xrightarrow[\text{2. }H_2O_2\text{, NaOH}]{\text{1. }BH_3} \text{(cyclopentylmethanol) OH}$$

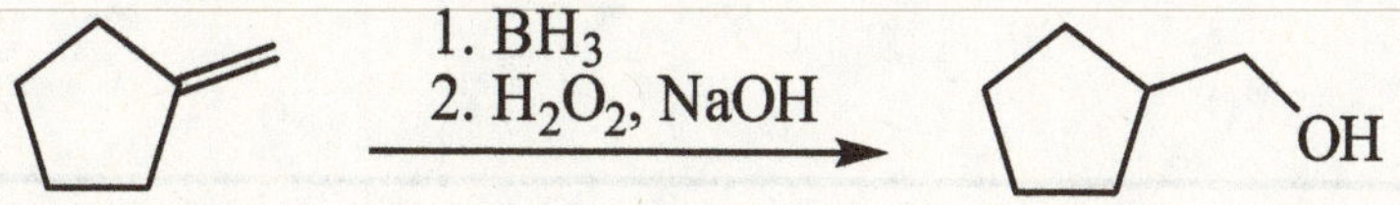

5.10   Propose a synthesis for each alkyne group starting with acetylene and any necessary organic and inorganic reagents.

Anions of acetylene (acetylides) are nucleophiles that undergo substitution reactions with haloalkanes to form new carbon-carbon bonds. A good approach is to identify the carbon atoms that originated from acetylene; any other carbons that are attached to those acetylene carbons must have been bonded to a halogen prior to the reaction.

(a) $HC \equiv CH$ $\xrightarrow[\text{2. } \triangleright - CH_2 - Br]{\text{1. NaNH}_2}$ $\triangleright - CH_2 \dagger C \equiv CH$

new bond

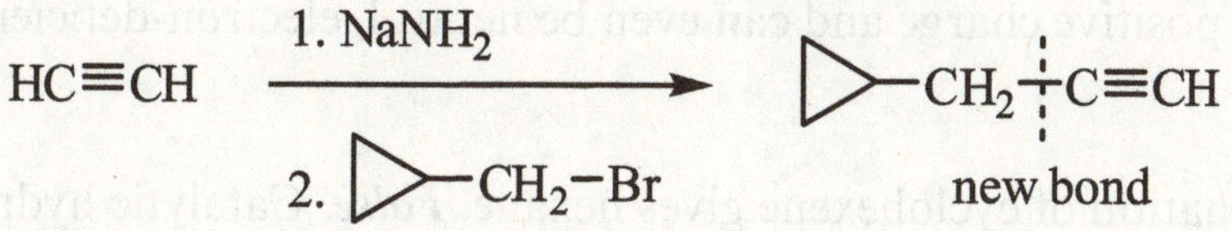

(b) $HC \equiv CH$ $\xrightarrow[\text{2. Br} \sim \sim \sim \text{Br}]{\text{1. NaNH}_2}$ Br $\sim \sim \sim$ $C \equiv CH$

$\downarrow$ NaNH$_2$

new bond

new bond

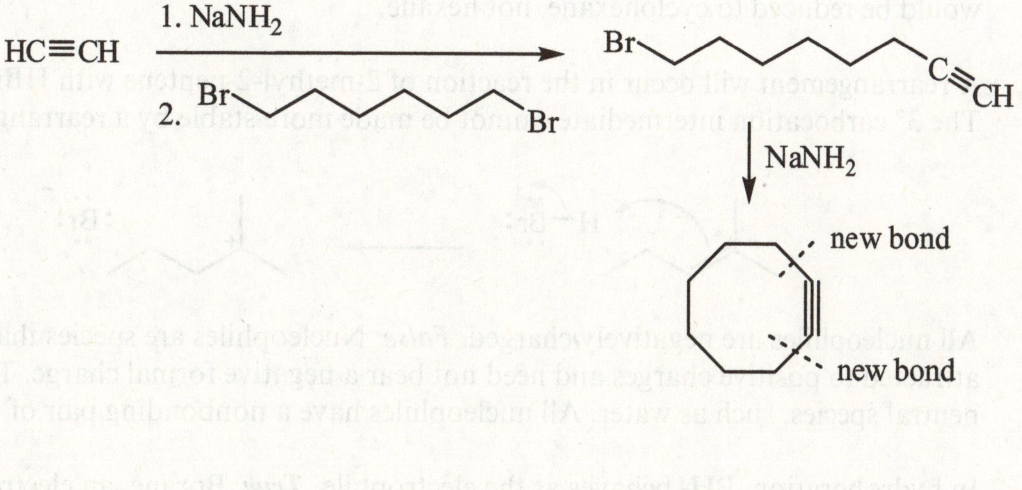

Chemical Connections

5A. Would you predict the catalytic cracking reactions to be exothermic or endothermic?

Catalytic cracking reactions, which convert an alkane to another alkane plus an alkene, occur under high temperature. The temperature requirement suggests that the reactions are endothermic.

Furthermore, we know that the catalytic hydrogenation of an alkene to produce an alkane is exothermic. Therefore, the production of an alkene from an alkane must be endothermic.

Quick Quiz

1. Catalytic reduction of an alkene is syn stereoselective. *True*. In catalytic hydrogenation, one face of the alkene binds to the surface of the catalyst, and both hydrogen atoms are added to that face of the alkene.

2. Borane, BH_3, is a Lewis acid. *True*. Lewis acids are electron-pair acceptors and are typically electron-deficient species. Borane is electron-deficient because the boron atom has only six valence electrons instead of a full octet.

3. All electrophiles are positively charged. *False*. Electrophiles are species that are attracted to electrons and do not always bear a positive formal charge. Electrophiles can have a partial positive charge and can even be neutral, electron-deficient species such as BH_3.

4. Catalytic hydrogenation of cyclohexene gives hexane. *False*. Catalytic hydrogenation reduces carbon-carbon double bonds to carbon-carbon single bonds. Cyclohexene would be reduced to cyclohexane, not hexane.

5. A rearrangement will occur in the reaction of 2-methyl-2-pentene with HBr. *False*. The 3° carbocation intermediate cannot be made more stable by a rearrangement.

6. All nucleophiles are negatively charged. *False*. Nucleophiles are species that are attracted to positive charges and need not bear a negative formal charge. They can be neutral species, such as water. All nucleophiles have a nonbonding pair of electrons.

7. In hydroboration, BH_3 behaves as the electrophile. *True*. Borane, an electron-deficient species, is attracted to the electron-rich π bond of the alkene.

8. In catalytic hydrogenation of an alkene, the reducing agent is the transition metal catalyst. *False*. The reducing agent is H_2 and not the transition metal catalyst. The catalyst only assists (catalyzes) the reaction and is not consumed in the overall reaction.

9. Alkene addition reactions involve breaking a π bond and forming two new σ bonds in its place. *True*. For example, the addition of HCl to an alkene results in the breaking of only the π bond of the C=C double bond (recall that a double bond consists of one σ and one π bond) and the formation of new C−H and C−Cl σ bonds.

10. The foundation for Markovnikov's rule is the relative stability of carbocation intermediates. *True*. Markovnikov's rule states that when HX adds to an alkene, the H is added to the alkene carbon that already has the greatest number of hydrogens bonded to it. This is because the positive charge is on the carbon adjacent to this carbon; the carbon with the positive charge must have fewer hydrogens and more alkyl groups than the carbon to which the hydrogen has added.

11. Acid-catalyzed hydration of an alkene is regioselective. *True*. The major product formed is the constitutional isomer derived from the most stable carbocation.

12. The mechanism for addition of HBr to an alkene involves one transition state and two reactive intermediates. *False*. The addition of HBr to an alkene proceeds by a two-step mechanism; there are two transition states and one reactive intermediate.

13. Hydroboration of an alkene is regioselective and stereoselective. *True*. The hydroboration of an alkene is regioselective in that the boron atom adds to the least hindered alkene carbon. The reaction is also steroselective because it proceeds via a syn addition.

14. According to the mechanism given in the text for the acid-catalyzed hydration of an alkene, the –H and –OH groups added to the double bond both arise from the same molecule of H_2O. *False*. H_2O is formed after H_3O^+ transfers a proton to the alkene, thereby generating a carbocation, but it is not necessarily this same H_2O molecule that acts as a nucleophile and attacks the carbocation. There are many, many other water molecules in the solution.

15. Acid-catalyzed addition of H_2O to an alkene is called *hydration*. *True*. Hydration refers to the addition of water to an alkene.

16. If a compound fails to react with Br_2, it is unlikely that the compound contains a carbon-carbon double bond. *True*. Alkenes usually react with bromine; this reaction is visually observed as a disappearance of the orange-red color of Br_2 and is a good test for the presence of an alkene.

17. Addition of Br_2 and Cl_2 to cyclohexene is anti stereoselective. *True*. One halogen atom is added to one face of the alkene, and the other halogen is added to the other face of the alkene. When halogens add to cycloalkenes, *trans*-1,2-dihalocycloalkanes are formed.

18. A carbocation is a carbon that has four bonds to it and bears a positive charge. *False*. Although a carbocation bears a positive formal charge, it only has three bonds to it. It also does not have any nonbonding pairs; it is an electron-deficient species.

19. The geometry about the positively charged carbon of a carbocation is best described as trigonal planar. *True*. Carbocations have three regions of electron density (three bonding regions and no nonbonding electrons) and are sp^2-hybridized. The remaining p orbital is perpendicular to the trigonal plane.

20. The carbocation derived by proton transfer to ethylene is $CH_3CH_2^+$. *True*. The proton adds to one of the alkene carbons, and the other carbon becomes a carbocation.

21. Alkyl carbocations are stabilized by the electron-withdrawing inductive effect of the positively charged carbon of the carbocation. *True*. The positively charged carbon pulls the electrons between it and the inductively donating alkyl groups closer towards itself.

22. The oxygen atom of an oxonium ion obeys the octet rule. *True*. Although the oxygen atom is positively charged, it has a full valence shell (three bonding pairs and one lone pair).

23. Markovnikov's rule refers to the regioselectivity of addition reactions to carbon-carbon double bonds. *True*. The rule specifies that when HX adds to alkenes, the hydrogen is added to the alkene carbon that already bears the most hydrogen atoms.

24. A rearrangement, in which a hydride ion shifts, will occur in the reaction of 3-methyl-1-pentene with HCl. *True*. A rearrangement forms a 3° carbocation, as shown below.

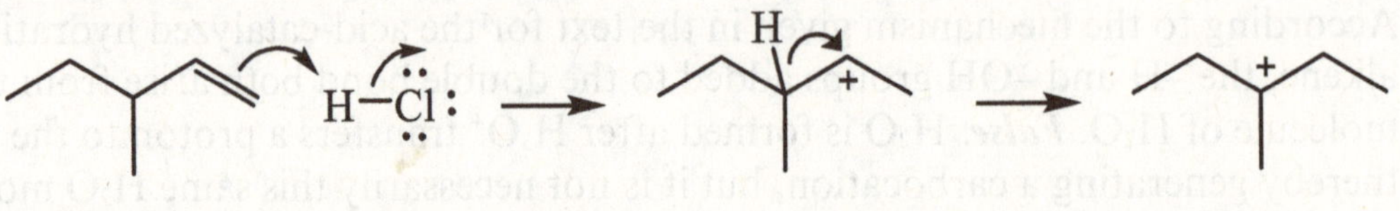

25. Acid-catalyzed hydration of 1-butene gives 1-butanol, and acid-catalyzed hydration of 2-butene gives 2-butanol. *False*. The acid-catalyzed hydration of both 1- and 2-butene gives 2-butanol. A 2° carbocation is formed during the hydration of 1-butene.

26. Alkenes are good starting materials for reactions in which it is necessary to form a C−C bond. *False*. It is very difficult to form a new C−C bond using an alkene. Rather, alkynes are much better starting materials due to the lower acidity of the alkyne hydrogen.

27. Alkynes can be reduced to *cis* alkenes. *True*. Alkynes can be stereoselectively reduced to *cis* alkenes by catalytic hydrogenation with the Lindlar catalyst.

End-of-Chapter Problems

Energy Diagrams

5.11   Describe the differences between a transition state and a reaction intermediate.

A transition state is a point on the reaction coordinate where the energy is at a maximum. Because a transition state is at an energy maximum, it cannot be isolated and its structure can often only be postulated. A reaction intermediate corresponds to an energy minimum between two transition states, but the energy of the intermediate is usually higher than the energies of the products or the reactants.

5.12   Sketch an energy diagram for a one-step reaction that is very slow and only slightly exothermic. How many transition states are present in this reaction? How many intermediates are present?

A one-step reaction that is very slow has a relatively high activation energy. Reactions that occur in one step have only one transition state, so there are no intermediates.

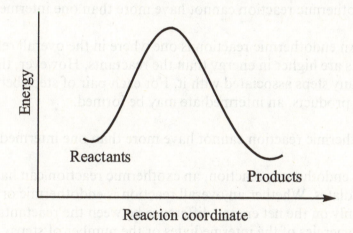

5.13   Sketch an energy diagram for a two-step reaction that is endothermic in the first step, exothermic in the second step, and exothermic overall. How many transition states are present in this two-step reaction? How many intermediates are present?

A two-step reaction has two transition states and one intermediate. The intermediate corresponds to the product of the first step and the reactant for the second step.

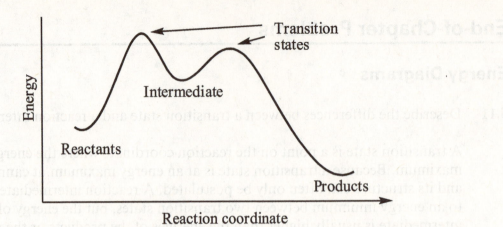

Reaction coordinate

5.14   Determine whether each of the following statements is true or false and provide a rationale for your decision:

(a)   A transition state can never be lower in energy than the reactants from which it was formed.

True. Transition states are always energy maxima on reaction energy diagrams.

(b)   An endothermic reaction cannot have more than one intermediate.

False. An endothermic reaction is one where in the overall reaction, the products are higher in energy than the reactants. However, the reaction can have many steps associated with it. For each pair of steps between the reactants and the products, an intermediate may be formed.

(c)   An exothermic reaction cannot have more than one intermediate.

Like an endothermic reaction, an exothermic reaction can have many steps and intermediates. Whether an overall reaction is endothermic or exothermic is based only on the net energy difference between the reactants and the products, not the energies of the intermediates or the number of steps.

(d)   The rate-determining step is the step with the largest difference in energy between products and reactants.

False. The rate-determining step is the step with the highest barrier. The difference in energy between the products and reactants is the ΔH of the reaction.

(e)   Transition states exist for long periods of time, but can be readily isolated.

False. Transition states are very short-lived and cannot be isolated.

Electrophilic Additions to Alkenes, Rearrangements, and Hydroboration-Oxidation

5.15 From each pair, select the more stable carbocation:

Carbocations that are bonded to more alkyl groups, which are electron-donating by induction, are more stable. In general, tertiary carbocations are more stable than secondary carbocations, which are more stable than primary carbocations.

(a) $CH_3CH_2\overset{+}{C}H_2$ or $CH_3\overset{+}{C}HCH_3$ (b) $CH_3\overset{\underset{|}{CH_3}}{C}H\overset{+}{C}HCH_3$ or $CH_3\overset{\underset{|}{CH_3}}{\overset{+}{C}}CH_2CH_3$

less stable (1°)   more stable (2°)   less stable (2°)   more stable (3°)

5.16 From each pair, select the more stable carbocation:

(a) 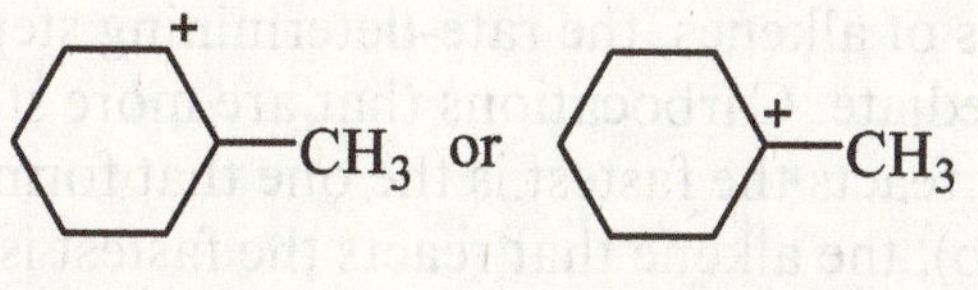 [ring]—CH_3 or [ring with +]—CH_3 (b) [ring with +]—CH_3 or [ring]—$\overset{+}{C}H_2$

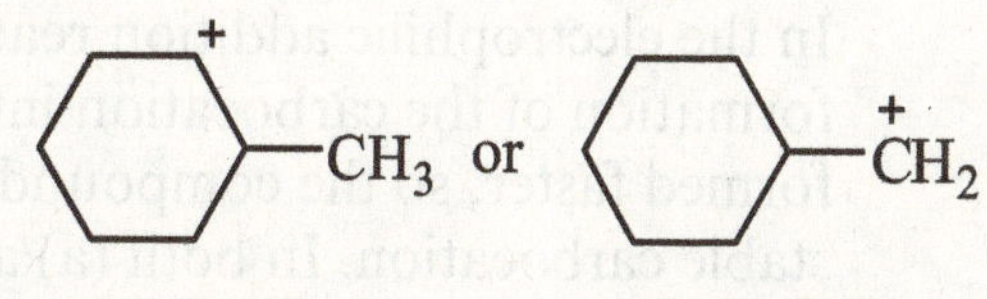

less stable (2°)   more stable (3°)   more stable (2°)   less stable (1°)

5.17 Draw structural formulas for the isomeric carbocation intermediates formed by the reaction of each alkene with HCl. Label each carbocation as primary, secondary, or tertiary, and state which, if either, of the isomeric carbocations is formed more readily.

The rate at which a carbocation is formed depends on its stability, because the formation of carbocations that are more stable requires less activation energy. Carbocation stability is affected by the number of alkyl groups bonded to the positively charged carbon.

(a)

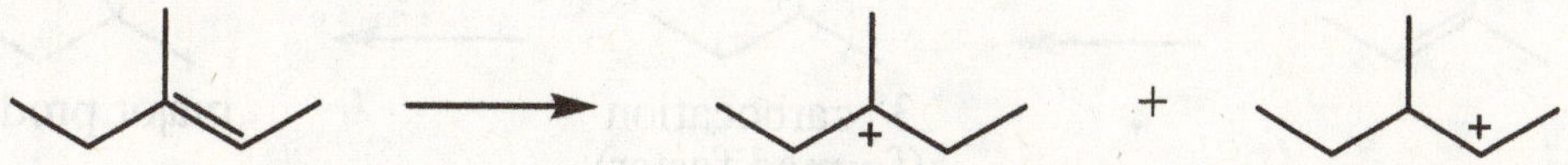

                                    3° carbocation   2° carbocation

                               (formed more readily)

(b)

Both are 2° carbocations and are of equal stability, so they are both formed at approximately equal rates

(c)

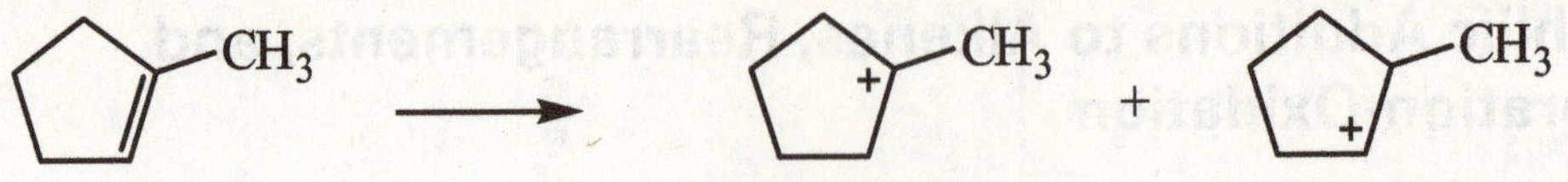

3° carbocation
(formed more readily)

2° carbocation

(d)

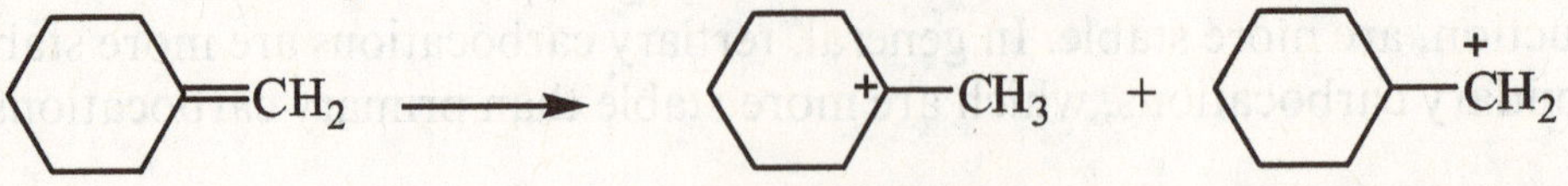

3° carbocation
(formed more readily)

1° carbocation

5.18 From each pair of compounds, select the one that reacts more rapidly with HI, draw the structural formula of the major product formed in each case, and explain the basis for your ranking.

In the electrophilic addition reactions of alkenes, the rate-determining step is the formation of the carbocation intermediate. Carbocations that are more stable are formed faster, so the compound that reacts the fastest is the one that forms the most stable carbocation. In both (a) and (b), the alkene that reacts the fastest is the one that involves a tertiary carbocation, and the major product is derived from that carbocation.

(a) and

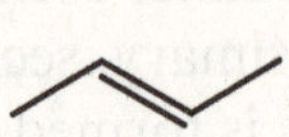

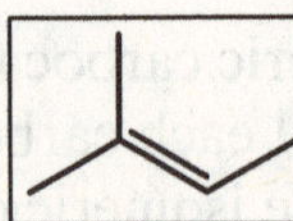

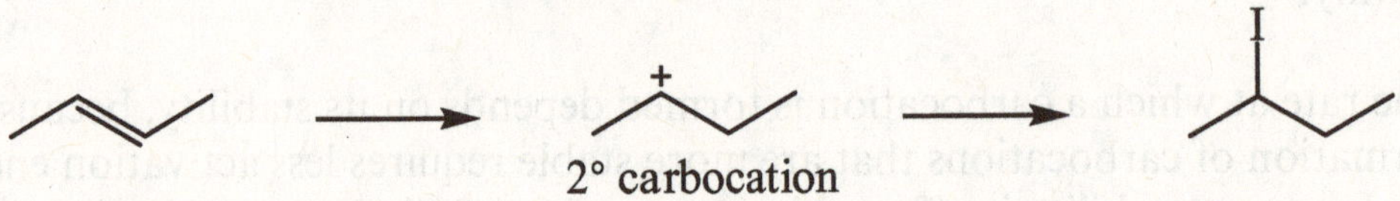

2° carbocation

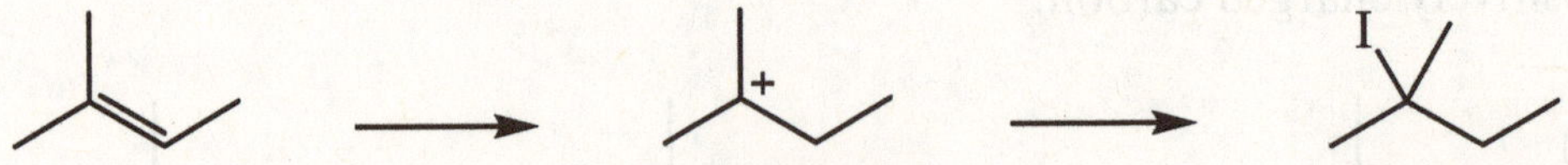

3° carbocation
(formed faster)

major product

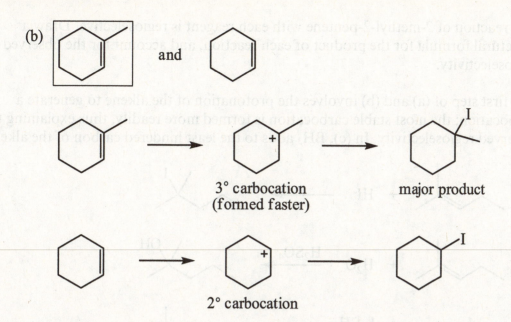

(b) [structure] and [structure]

[structure] ——→ [structure] ——→ [structure]

3° carbocation   major product
(formed faster)

[structure] ——→ [structure] ——→ [structure]

2° carbocation

5.19 Complete these equations by predicting the major product formed in each reaction:

(a) [structure] + HCl ——→ [structure]

(b) [structure] + H_2O $\xrightarrow{H_2SO_4}$ [structure]

(c) [structure] + HI ——→ [structure]

(d) [structure] + HCl ——→ [structure]

(e) [structure] + H_2O $\xrightarrow{H_2SO_4}$ [structure]

(f) [structure] + H_2O $\xrightarrow{H_2SO_4}$ [structure]

5.20   The reaction of 2-methyl-2-pentene with each reagent is regioselective. Draw a structural formula for the product of each reaction, and account for the observed regioselectivity.

The first step of (a) and (b) involves the protonation of the alkene to generate a carbocation; the most stable carbocation is formed more readily, thus explaining the observed regioselectivity. In (c), BH_3 adds to the least hindered carbon of the alkene.

(a)

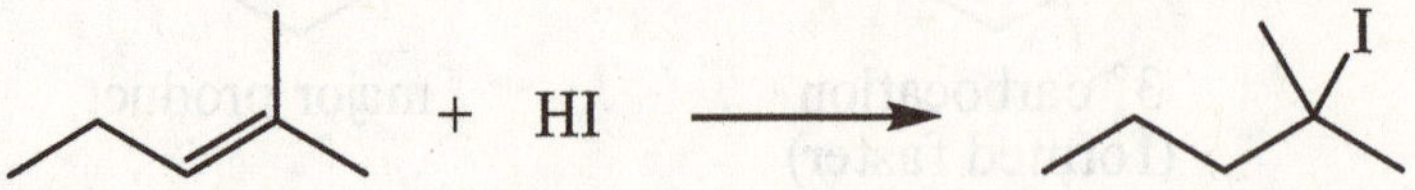

(b)

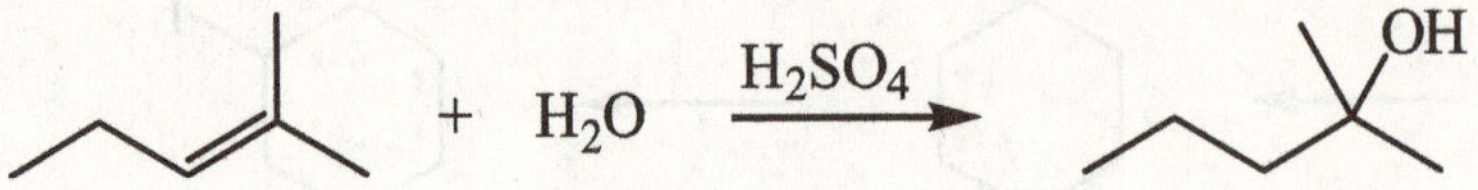

(c)

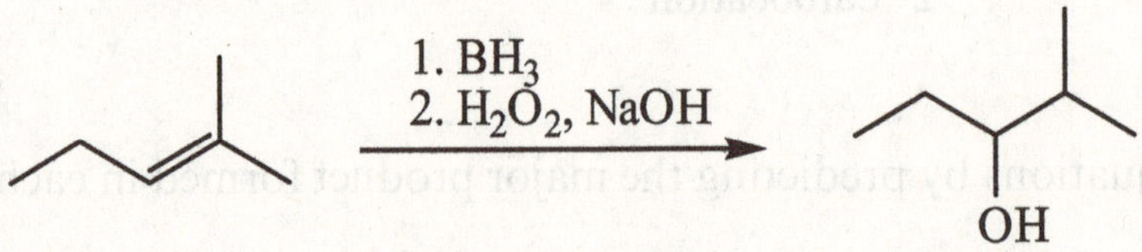

5.21   The addition of bromine and chlorine to cycloalkenes is stereoselective. Predict the stereochemistry of the product formed in each reaction.

The addition of halogens to alkenes is stereoselective in that the mechanism proceeds via anti addition; one halogen adds to one face of the planar double bond, while the other halogen adds to the other face. In cycloalkenes, anti addition gives 1,2-*trans* dihalides.

(a)

(b)

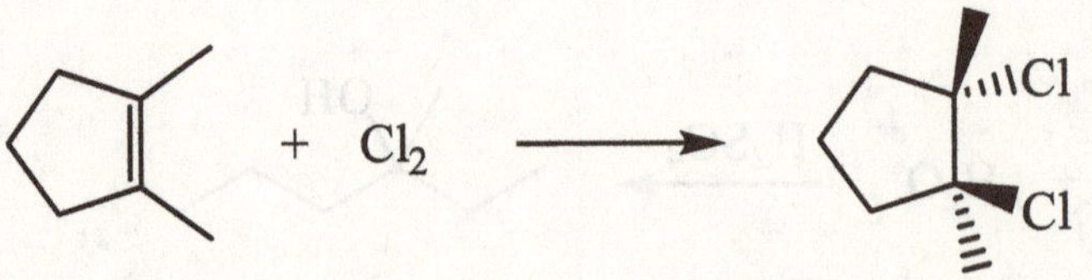

5.22 Draw a structural formula for an alkene with the indicated molecular formula that gives the compound shown as the major product. Note that more than one alkene may give the same compound as the major product.

Examine the products and work backwards. With (a) and (c), the carbon bearing the −OH or −Cl group, respectively, corresponds to the location of the most stable carbocation that was formed during the reaction. Then, determine the possible alkene(s) that could form the corresponding carbocation most readily. With (b), the alkene must have been between the two carbon atoms bearing the two −Br groups.

(a) C_5H_{10} + H_2O $\xrightarrow{H_2SO_4}$ [structure] or [structures]

(b) C_5H_{10} + Br_2 ⟶ [structures]

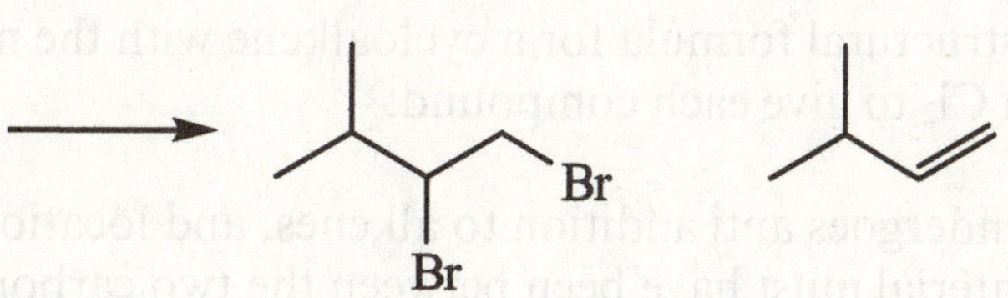

(c) C_7H_{12} + HCl ⟶ [structures] or [structure]

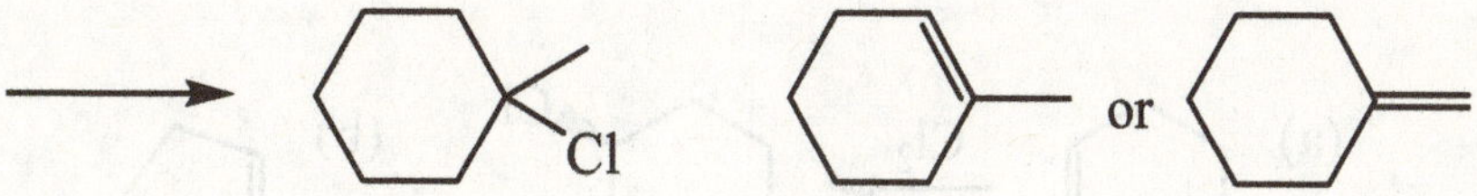

5.23 Draw the structural formula for an alkene with the molecular formula C_5H_{10} that reacts with Br_2 to give each product:

Working backwards from the product, the carbon-carbon double bond in the starting material must have been between the two carbon atoms bearing the two −Br groups.

(a) 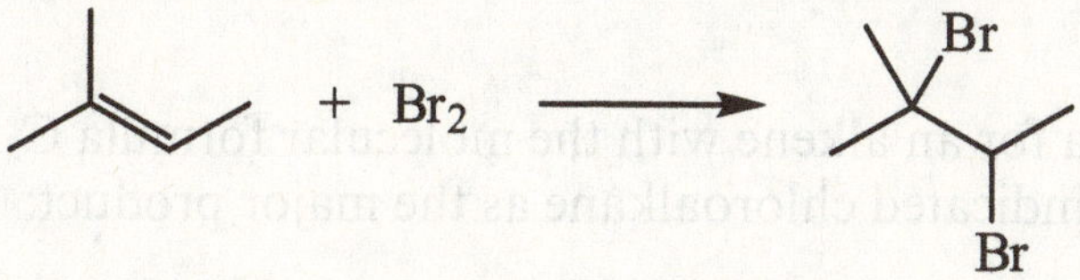 [structure] + Br_2 ⟶ [structure]

(b) 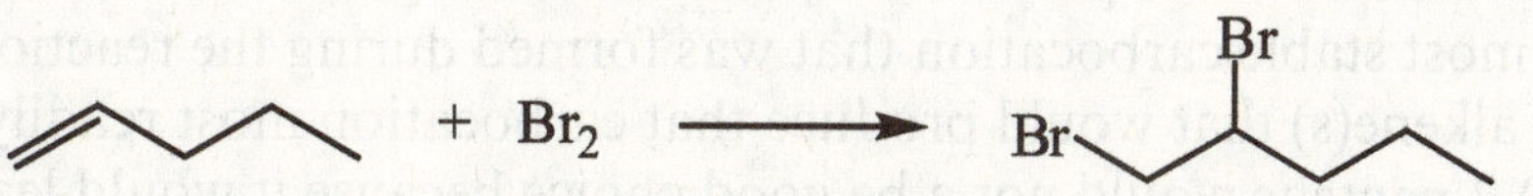 [structure] + Br_2 ⟶ [structure]

(c) 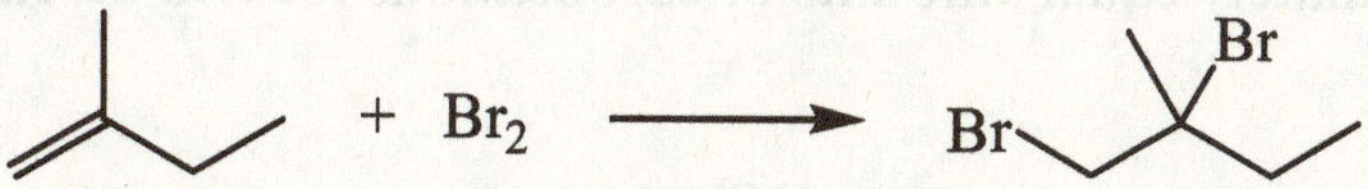 [structure] + Br_2 ⟶ [structure]

(d) Notice that in the product, the two −Br groups are both pointing out of the page. The addition of Br_2 is anti, so we would expect one Br to be pointing out of the page and the other to be behind the page. It is important to realize that after the addition of Br_2 to the alkene, the resulting carbon-carbon single bond can freely rotate. To make the alkene easier to identify, "undo" the rotation by turning the single bond until the two −Br groups are anti.

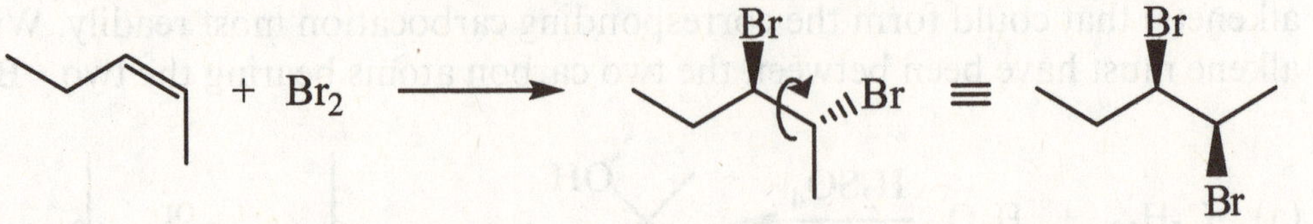

5.24 Draw the structural formula for a cycloalkene with the molecular formula C_6H_{10} that reacts with Cl_2 to give each compound:

Chlorine undergoes anti addition to alkenes, and location of the alkene in the starting material must have been between the two carbon atoms bearing the −Cl substituents.

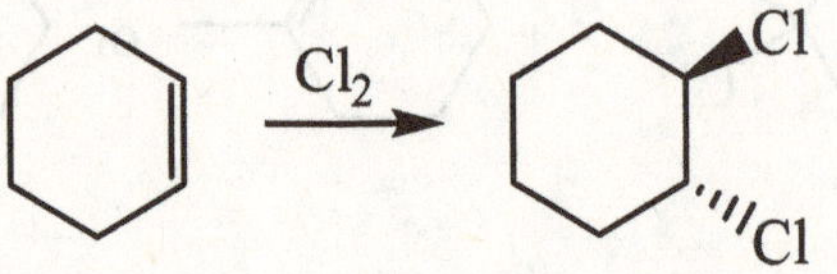

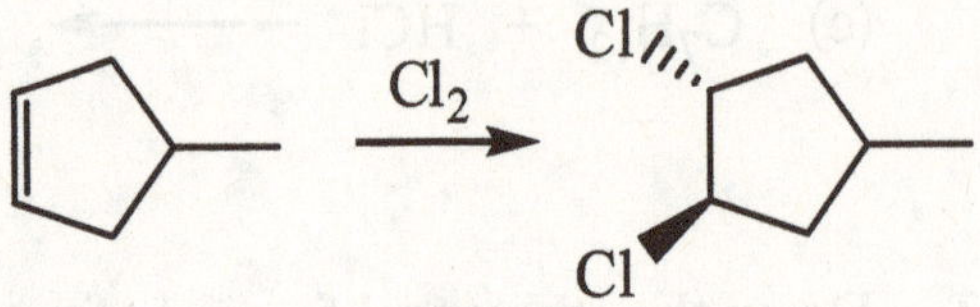

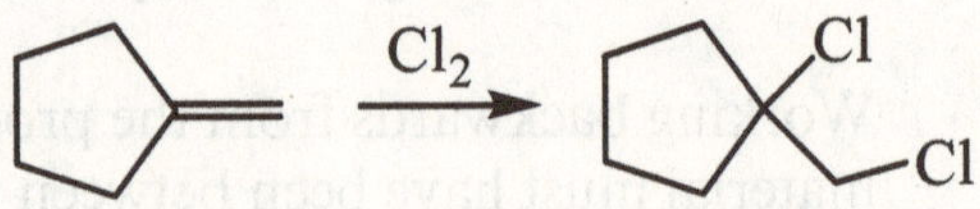

5.25 Draw the structural formula for an alkene with the molecular formula C_5H_{10} that reacts with HCl to give the indicated chloroalkane as the major product:

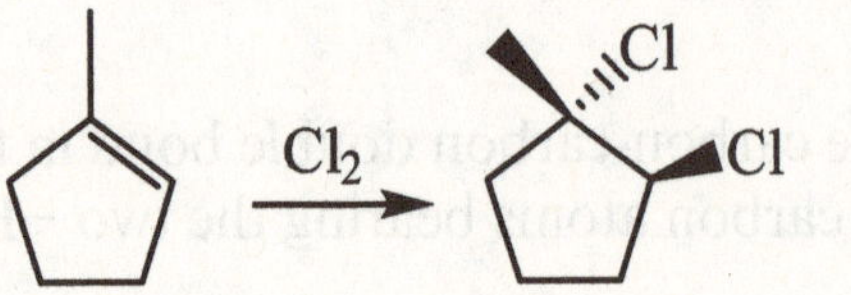

Working backwards from the product, the carbon bearing the −Cl substituent must have been the most stable carbocation that was formed during the reaction; from there, find the alkene(s) that would produce that carbocation most readily. For example, in (c), 2-pentene would not a be good choice because it would lead to the formation of approximately equal amounts of carbocations located on carbons 2 and 3.

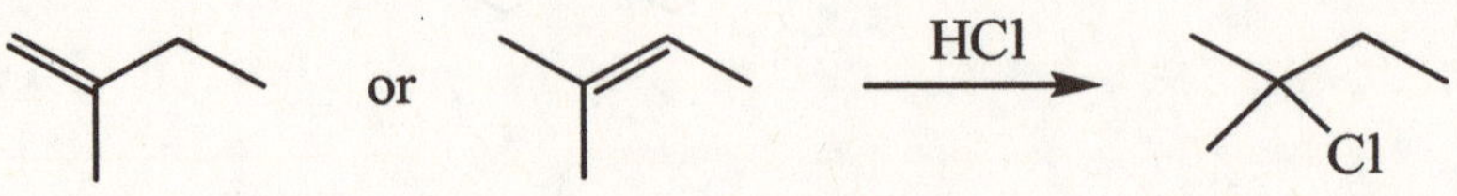

(b)

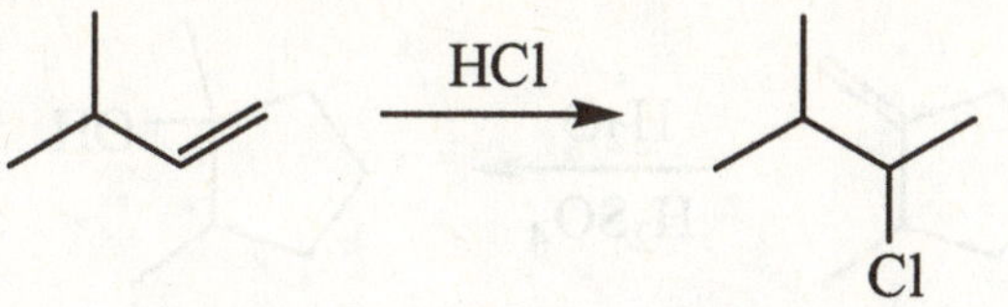

HCl

(c)

HCl

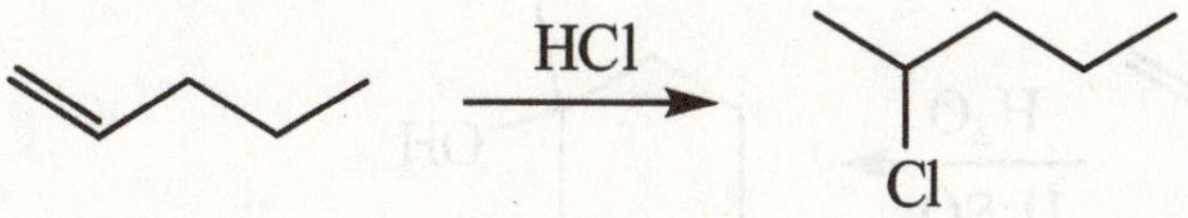

5.26 Draw the structural formula of an alkene that undergoes acid-catalyzed hydration to give the indicated alcohol as the major product. More than one alkene may give each compound as the major product.

Note that we are looking for the alkene that gives the largest amount of the indicated product. For example, with (a), only 3-hexene gives the highest yield of 3-hexanol; using 2-hexene gives approximately equal amounts of 2-hexanol and 3-hexanol because the two carbocations derived from 2-hexene are of equal stability.

(a)

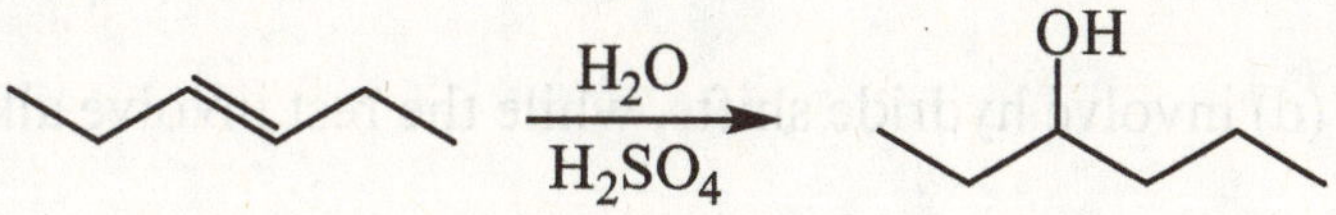

H_2O
H_2SO_4

(b)

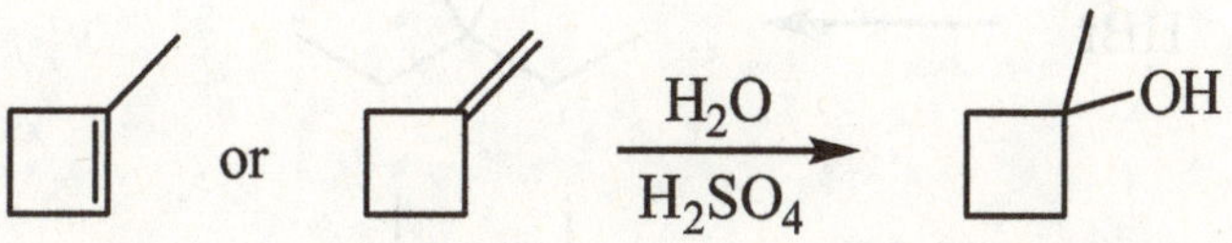

or

H_2O
H_2SO_4

(c)

or

H_2O
H_2SO_4

(d)

H_2O
H_2SO_4

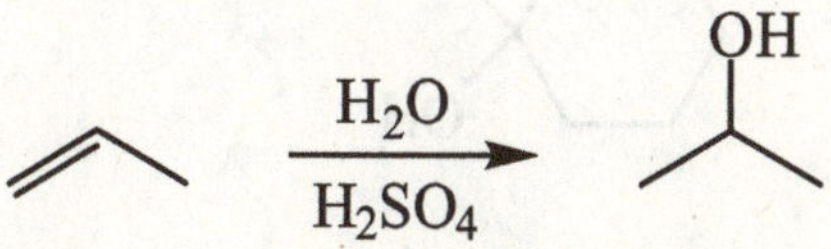

5.27 Draw the structural formula of an alkene that undergoes acid-catalyzed hydration to give each alcohol as the major product. More than one alkene may give each compound as the major product.

(a)

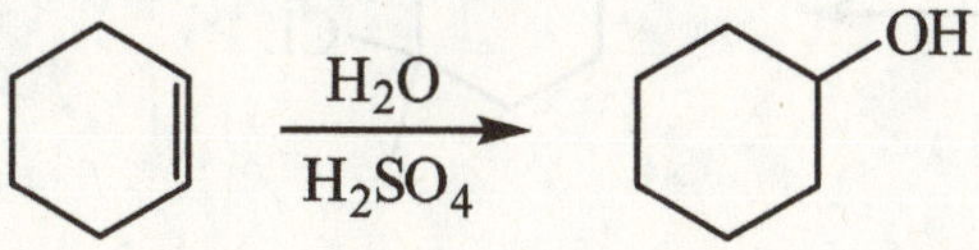

H_2O
H_2SO_4

(b)

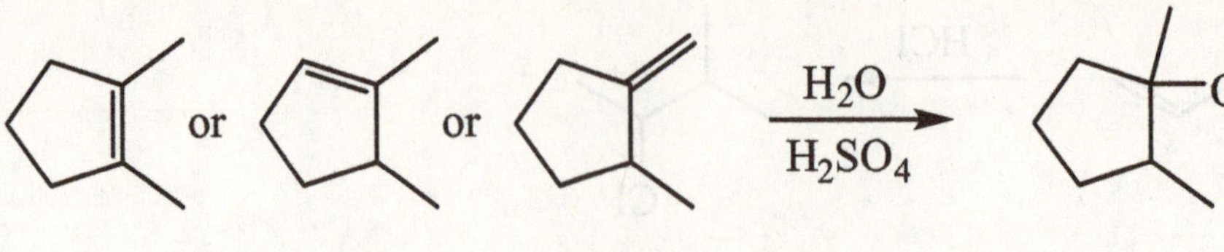

(c)

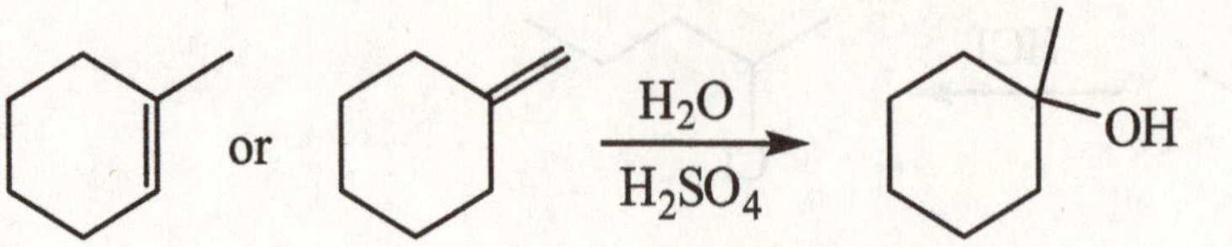

(d)

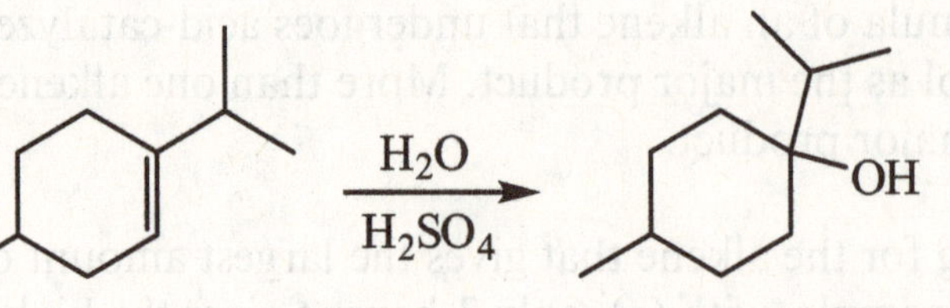

5.28 Complete these equations by predicting the major product formed in each reaction. Note that some of these reactions involve rearrangements.

Reactions (a), (c), and (d) involve hydride shifts, while the rest involve alkyl shifts.

(a)

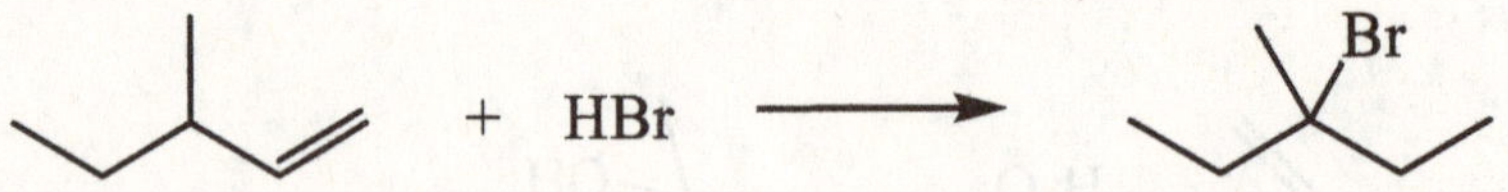

(b)

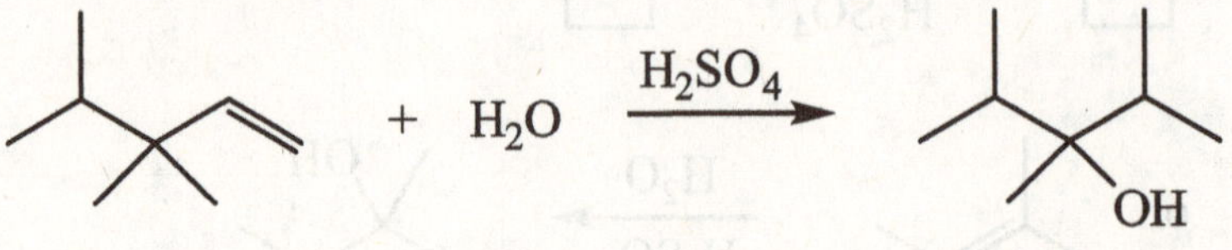

(c)

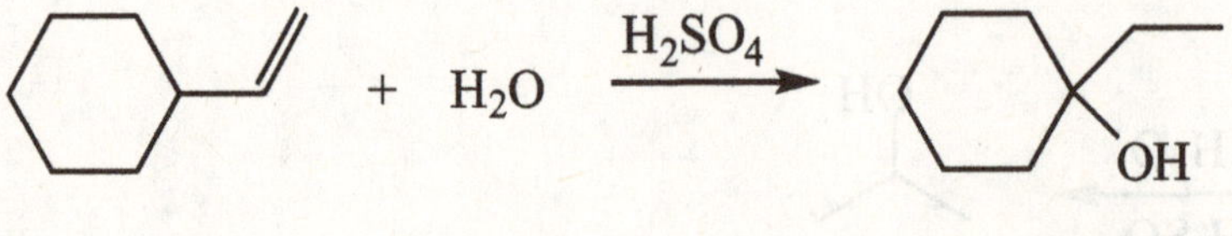

(d)

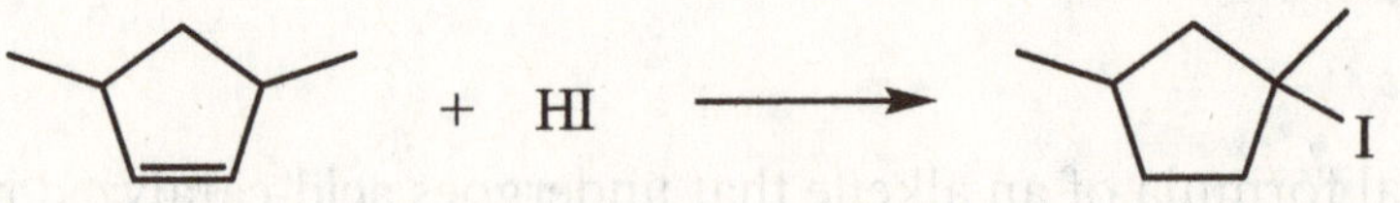

(e)

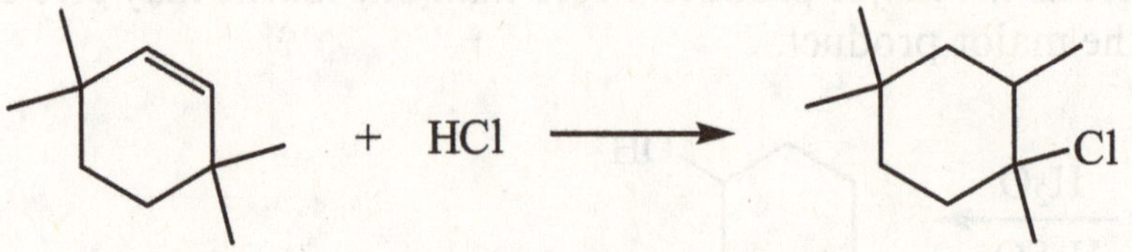

(f)

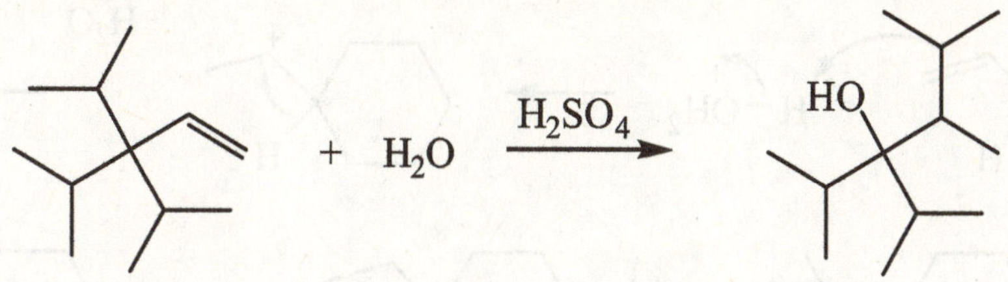

5.29 Propose a mechanism for each reaction in Problem 5.28.

Notice that in the mechanisms below, nonbonding electrons are not shown. This reiterates an important point: in Lewis structures, nonbonding electrons are implied to be present unless otherwise specified. Of course, the number of nonbonding electrons present on each atom must be consistent with the element and its formal charge. You should be able to add the proper number of nonbonding electrons to each species and recognize trends (number of bonds, charge, number of nonbonding electrons). If you encounter any difficulties, now would be a good opportunity to review Chapter 1.

(a)

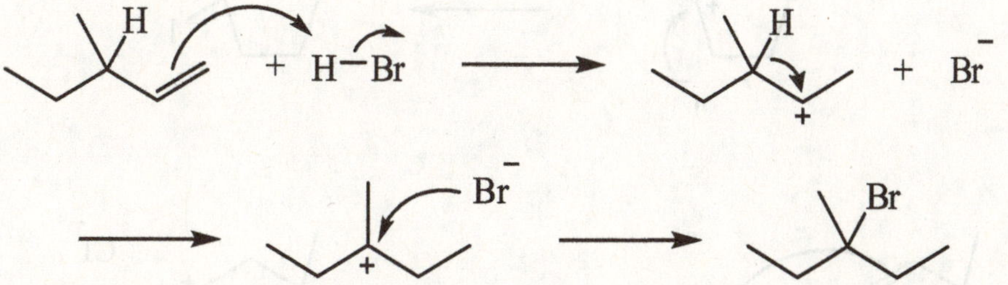

(b)

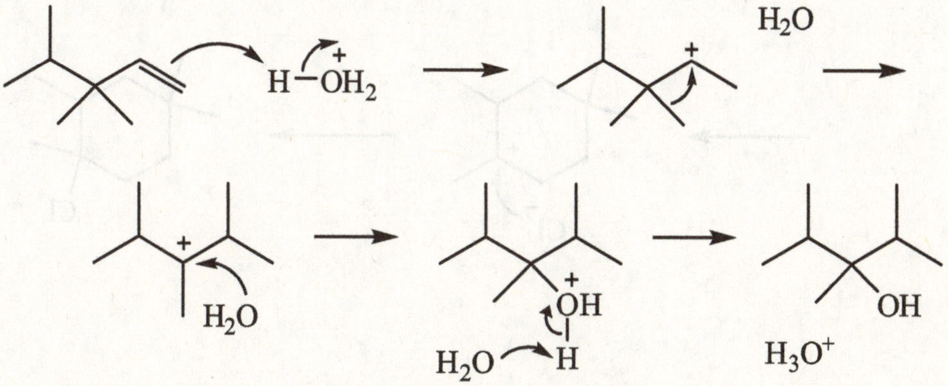

(c)

(d)

(e)

(f)

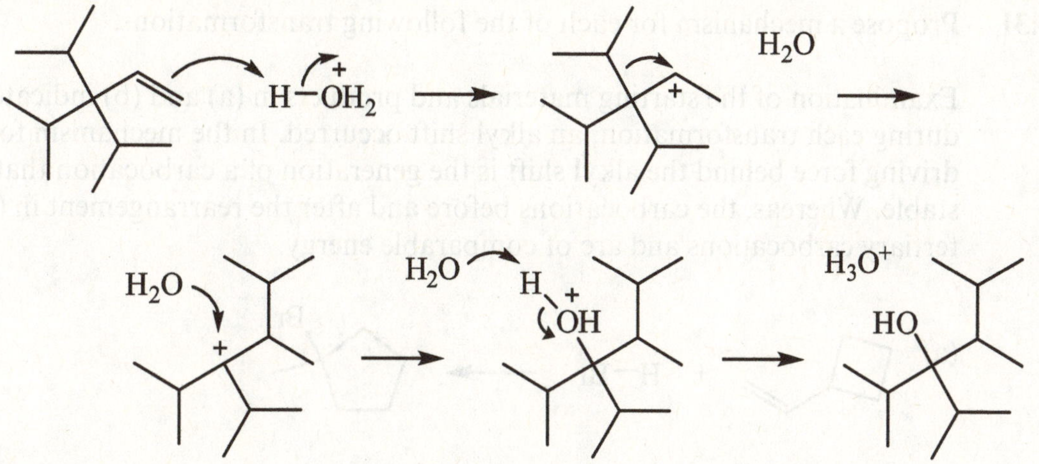

H_2O

5.30 Propose a mechanism for the following acid-catalyzed dehydration.

$$\xrightarrow{H_2SO_4}$$

$+ H_2O$

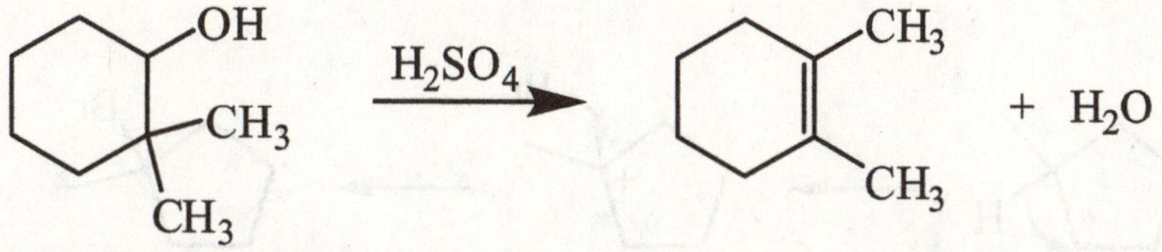

Inspection of the starting material and the product reveals that during the dehydration of the alcohol to form an alkene, a 1,2-methyl shift occurred. Mechanistically, an acid-catalyzed dehydration proceeds in the reverse direction of a hydration reaction, and the same intermediates and steps are involved (except of course in the reverse direction).

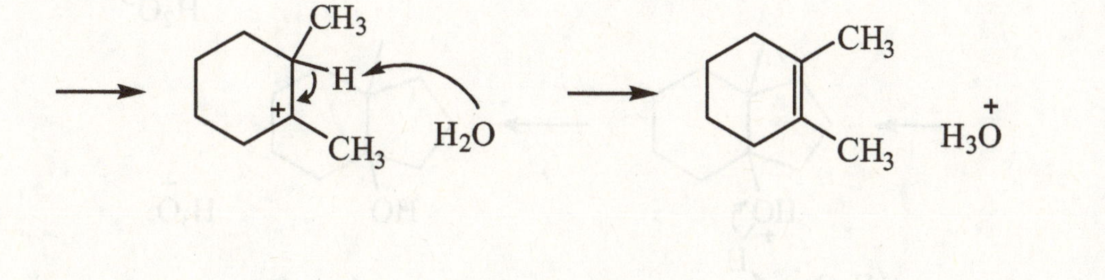

5.31 Propose a mechanism for each of the following transformations.

Examination of the starting materials and products in (a) and (b) indicates that during each transformation, an alkyl shift occurred. In the mechanism for (a), the driving force behind the alkyl shift is the generation of a carbocation that is more stable. Whereas, the carbocations before and after the rearrangement in (b) are both tertiary carbocations and are of comparable energy.

(a)

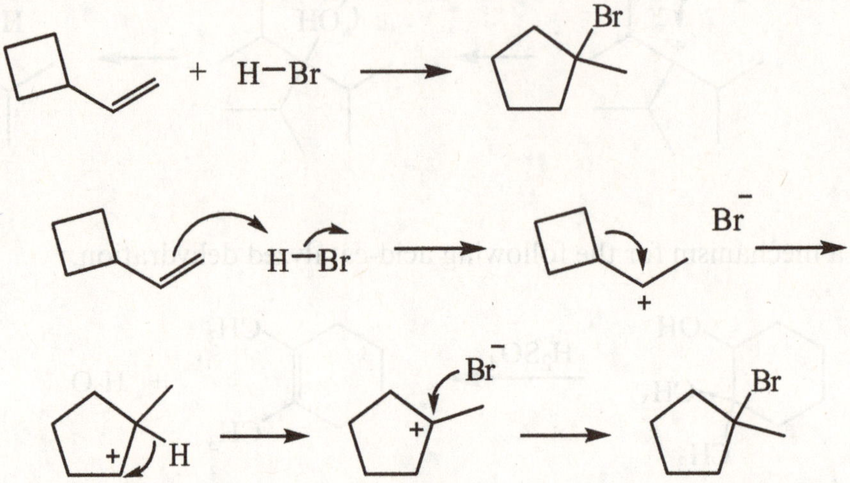

(b)

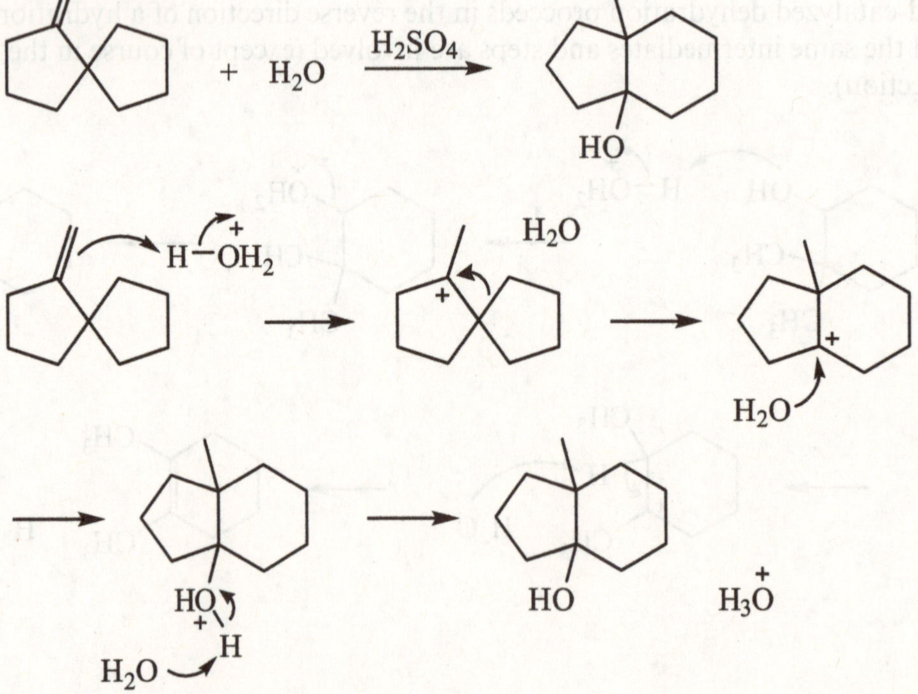

5.32   Terpin is prepared commercially by the acid-catalyzed hydration of limonene:

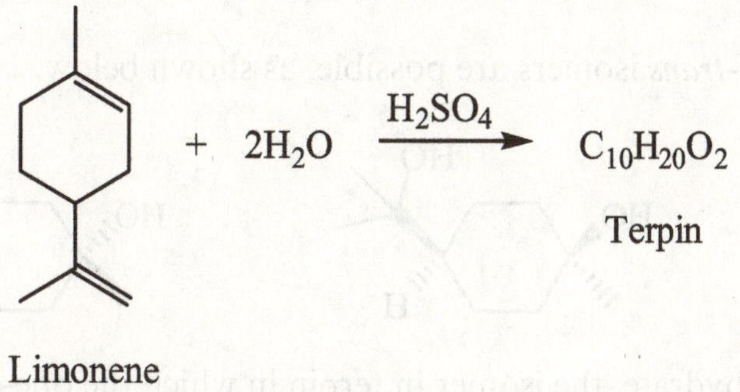

Limonene

(a)   Propose a structural formula for terpin and a mechanism for its formation.

Limonene contains two alkenes. With both alkenes, protonation yields the most stable carbocation (3°); the addition of water is therefore regioselective. Because both alkenes give carbocations of similar stability, both alkenes will react at about the same rate. The mechanism shows the cycloalkene as the first alkene undergoing the addition of water, but it is just as likely for the other alkene to react first.

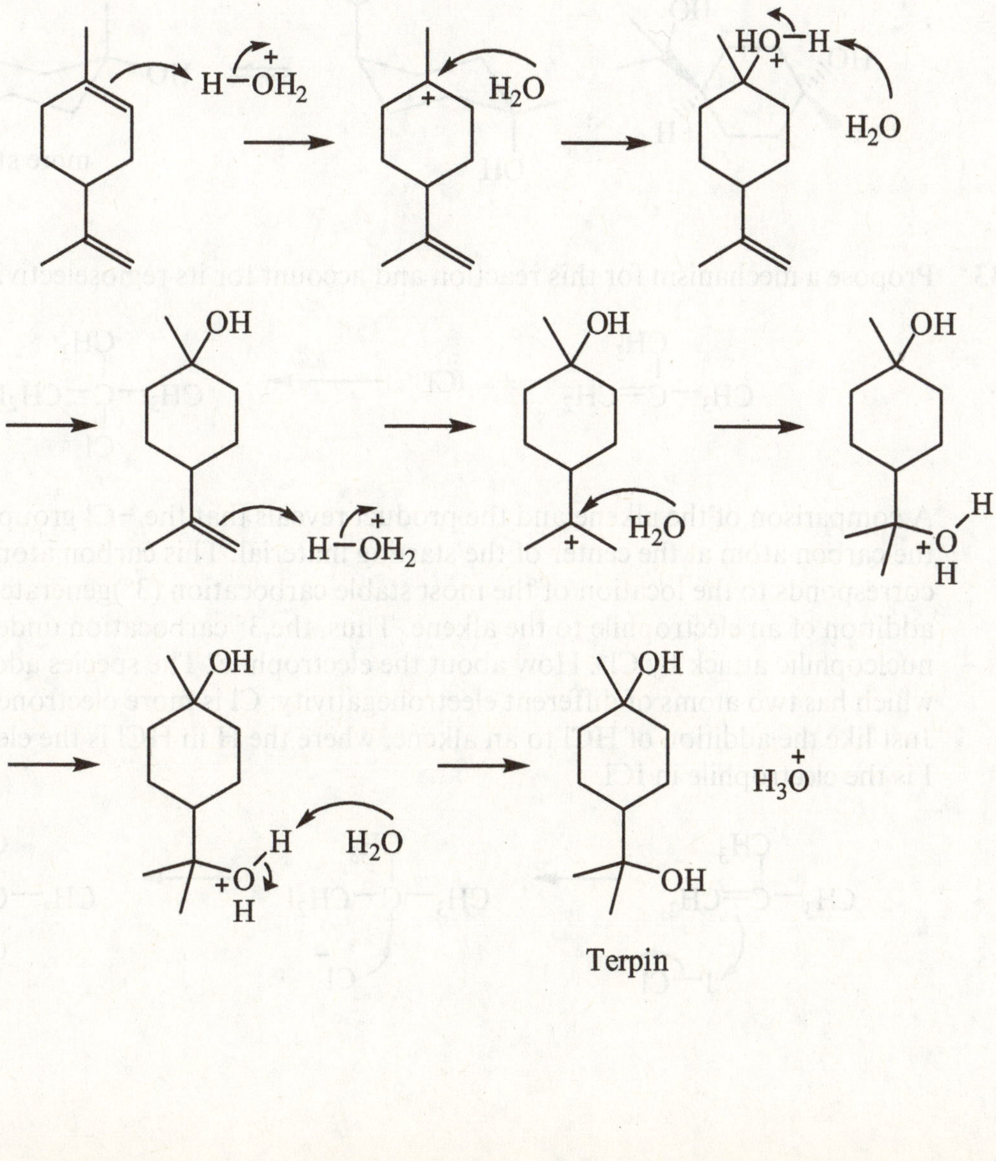

Terpin

(b) How many *cis-trans* isomers are possible for the structural formula you propose?

Two *cis-trans* isomers are possible, as shown below.

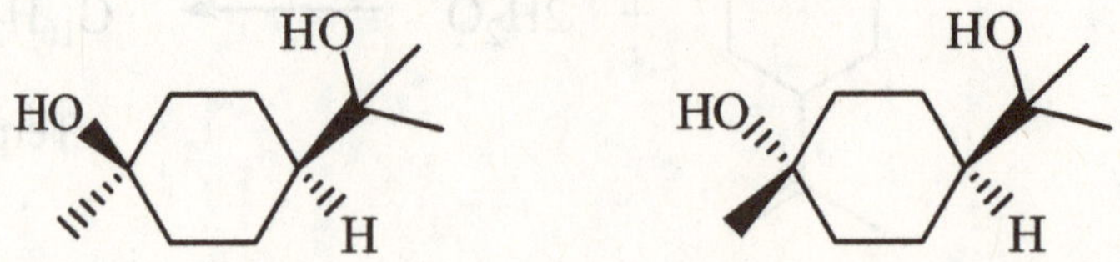

(c) Terpin hydrate, the isomer in terpin in which the one-carbon and three-carbon substituent are *cis* to each other, is used as an expectorant in cough medicines. Draw alternative chair conformations for terpin hydrate, and state which of the two is the more stable?

The three-carbon substituent is bulky (about the same size as a *tert*-butyl group) and will have a preference for the equatorial position.

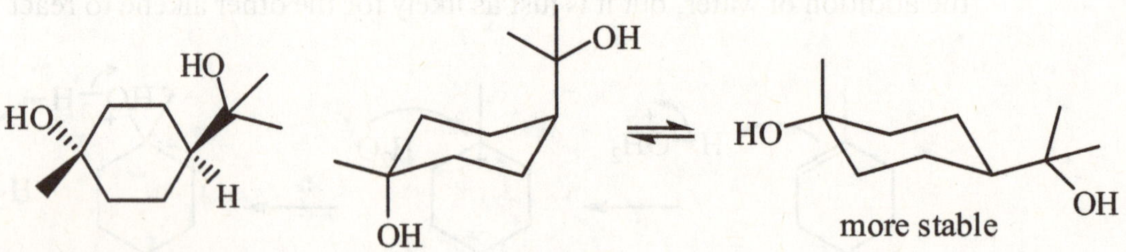

more stable

5.33 Propose a mechanism for this reaction and account for its regioselectivity.

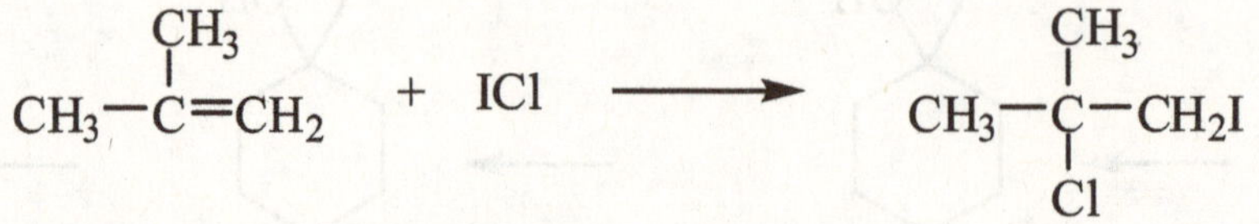

A comparison of the alkene and the product reveals that the −Cl group has added to the carbon atom at the center of the starting material. This carbon atom also corresponds to the location of the most stable carbocation (3°)generated by the addition of an electrophile to the alkene. Thus, the 3° carbocation underwent nucleophilic attack by Cl⁻. How about the electrophile? The species added is ICl, which has two atoms of different electronegativity; Cl is more electronegative than I. Just like the addition of HCl to an alkene, where the H in HCl is the electrophile, the I is the electrophile in ICl.

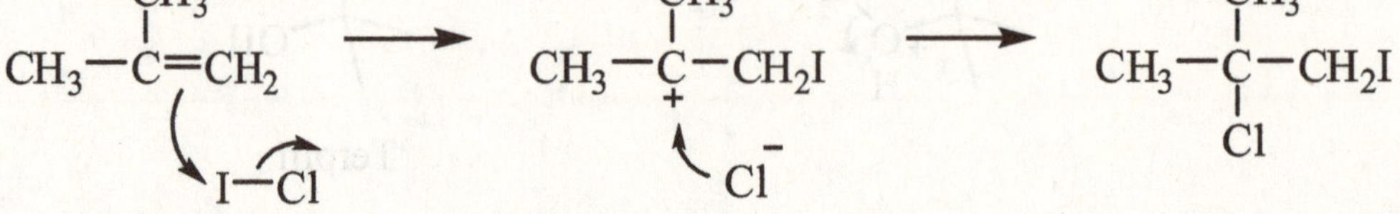

5.34   Treatment of 2-methylpropene with methanol in the presence of a sulfuric acid catalyst gives *tert*-butyl methyl ether. Propose a mechanism for the formation of this ether.

$$CH_3-\underset{\underset{CH_3}{|}}{C}=CH_2 \quad + \quad CH_3OH \quad \xrightarrow{H_2SO_4} \quad CH_3-\underset{\underset{CH_3}{|}}{\overset{\overset{CH_3}{|}}{C}}-OCH_3$$

2-Methylpropene   Methanol   *tert*-Butyl methyl ether

The acid-catalyzed reaction between an alkene and an alcohol is mechanistically the same as that of alkene hydration, except the nucleophile is an alcohol instead of water. The reaction is regioselective because the tertiary carbocation is favored over the primary.

$$CH_3-\underset{\underset{CH_3}{|}}{C}=CH_2 \quad \rightleftharpoons \quad CH_3-\underset{\underset{CH_3}{|}}{\overset{+}{C}} \quad \rightleftharpoons \quad CH_3-\underset{\underset{CH_3}{|}}{C}-\overset{+}{O}$$

$$\underset{\underset{H}{|}}{H-\overset{+}{O}CH_3} \qquad CH_3OH \qquad CH_3OH$$

$$\rightleftharpoons \quad CH_3-\underset{\underset{CH_3}{|}}{\overset{\overset{CH_3}{|}}{C}}-OCH_3 \quad + \quad CH_3\overset{+}{O}H_2$$

5.35   Treating cyclohexene with HBr in the presence of acetic acid gives a mixture of bromocyclohexane and cyclohexyl acetate. Account for the formation of each product but do not be concerned with the relative percentages of each.

$$\bigcirc \quad + \quad HBr \quad \xrightarrow{CH_3-\overset{\overset{O}{||}}{C}-OH} \quad \text{(bromocyclohexane)} \quad + \quad \text{(cyclohexyl acetate)}$$

In the presence of two acids, HBr and acetic acid, the stronger acid (HBr) is more likely to protonate cyclohexene.

$$\bigcirc \quad \underset{H-Br}{} \quad \rightleftharpoons \quad \bigcirc^+$$

Bromocyclohexane is formed from the nucleophlic attack of the carbocation by Br⁻.

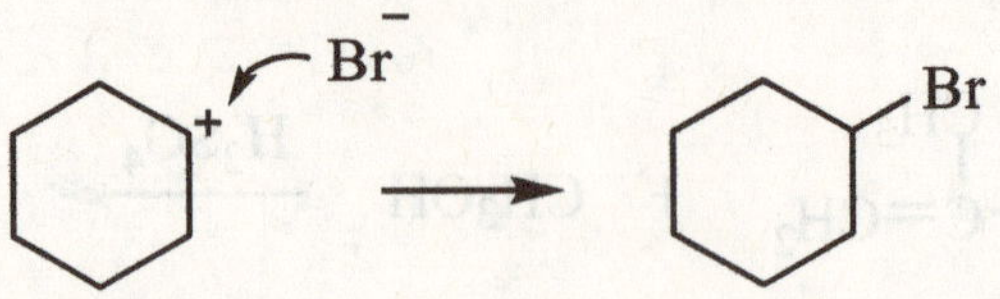

Whereas, nucleophilic attack by acetic acid followed by deprotonation results in the formation of cyclohexyl acetate. Note that the carbonyl oxygen is the nucleophile.

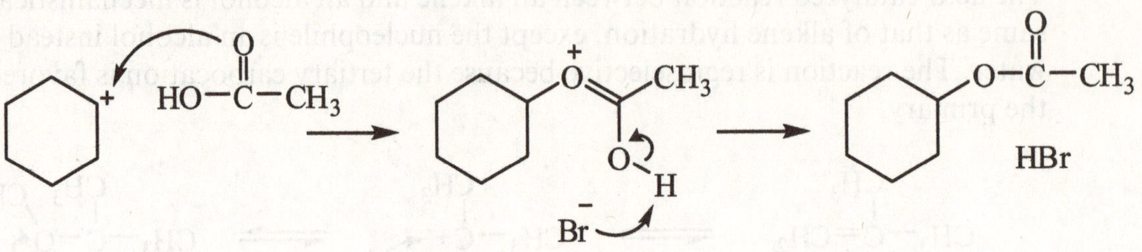

5.36 Draw a structural formula for the alcohol formed by treating each alkene with borane in tetrahydrofuran (THF) followed by hydrogen peroxide in aqueous sodium hydroxide, and specify the stereochemistry where appropriate.

The reagents are characteristic of hydroboration-oxidation, which results in the syn and anti-Markovnikov addition of water to alkenes. Anti-Markovnikov addition occurs because the boron atom of borane forms a bond with the least sterically hindered carbon, which is the one with the most hydrogens (fewest carbons) bonded to it. The resulting borane is subsequently oxidized to the alcohol.

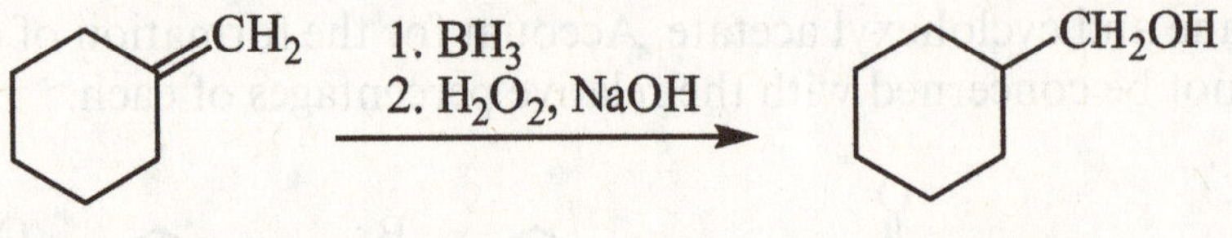

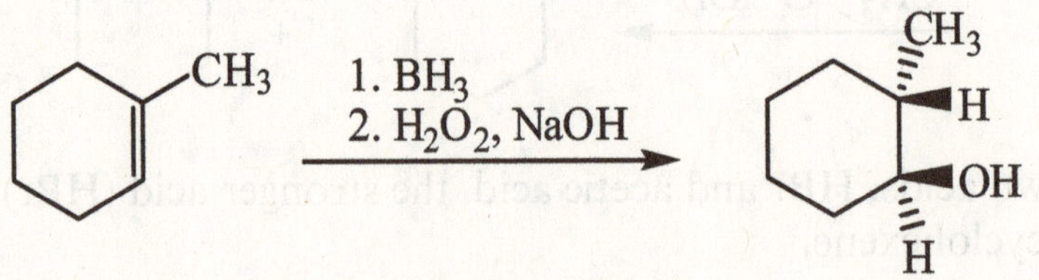

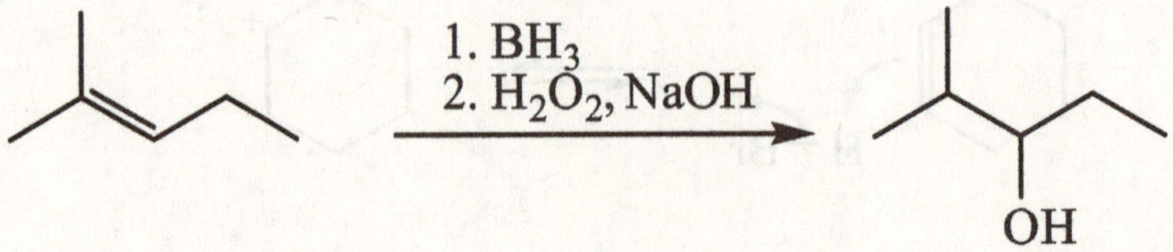

(d)

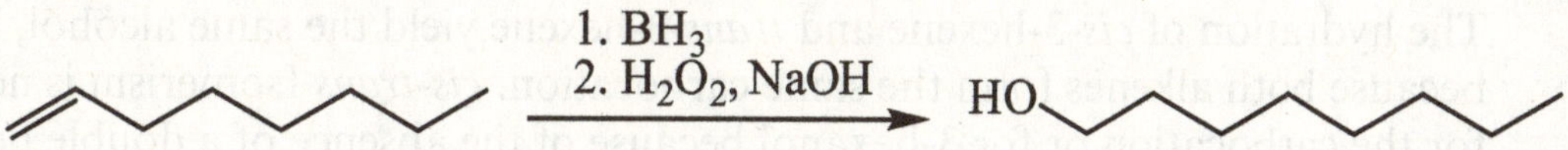

1. BH_3
2. H_2O_2, NaOH

(e)

1. BH_3
2. H_2O_2, NaOH

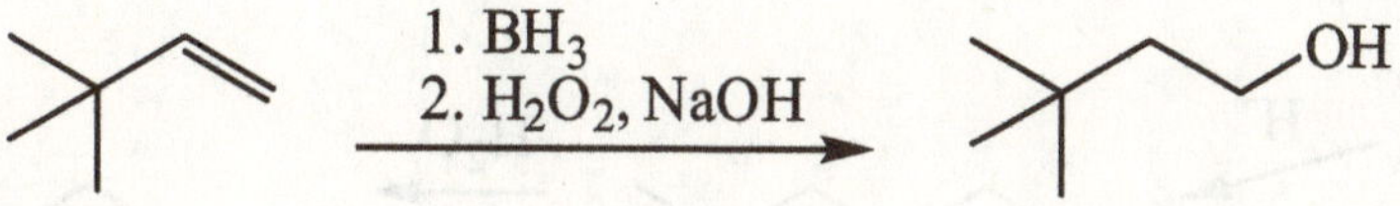

5.37 Treatment of 1-methylcyclohexene with methanol in the presence of a sulfuric acid catalyst gives a compound with the molecular formula $C_8H_{16}O$. Propose a structural formula for this compound and a mechanism for its formation.

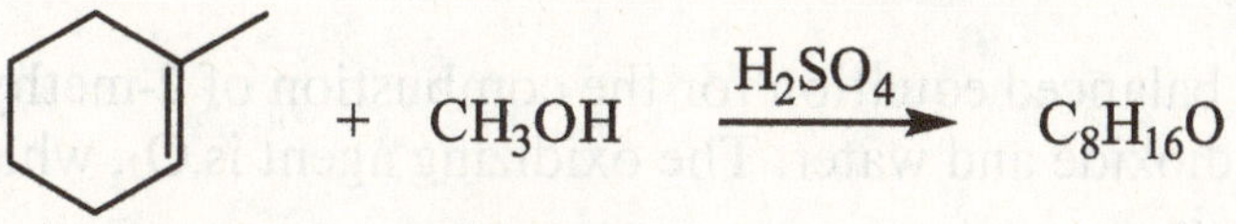

$+ \quad CH_3OH \quad \xrightarrow{H_2SO_4} \quad C_8H_{16}O$

1-Methylcyclohexene   Methanol

The molecular formula of the product suggests that methanol has added to 1-methylcyclohexne (C_7H_{12}). The mechanism by which this proceeds is similar to the one proposed in Problem 5.34, except the alkene is different. The reaction is regioselective and the ether formed is derived from the most stable carbocation.

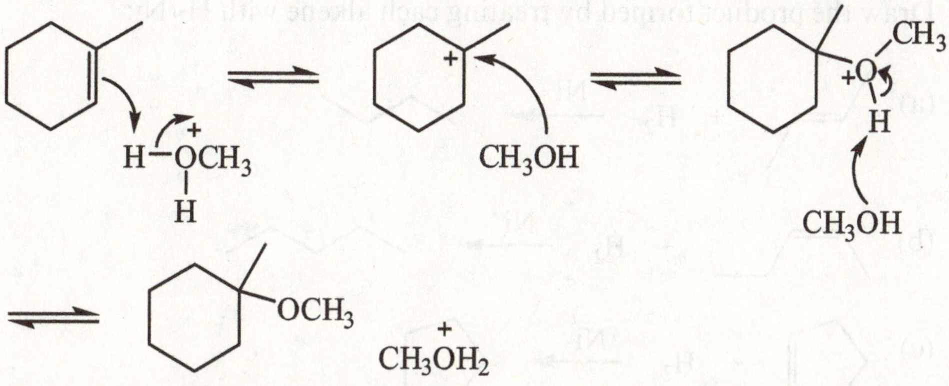

$H{-}\overset{+}{O}CH_3$
$\overset{|}{H}$

CH_3OH

CH_3OH

$\overset{+}{CH_3OH_2}$

$C_8H_{16}O$

5.38 *cis*-3-Hexene and *trans*-3-hexene are different compounds and have different physical and chemical properties. Yet, when treated with H_2O/H_2SO_4, each gives the same alcohol. What is the alcohol, and how do you account for the fact that each alkene gives the same one?

The hydration of *cis*-3-hexene and *trans*-3-hexene yield the same alcohol, 3-hexanol, because both alkenes form the same carbocation. *cis-trans* Isomerism is not possible for the carbocation or for 3-hexanol because of the absence of a double bond; single bonds are able to freely rotate and adopt a variety of conformations.

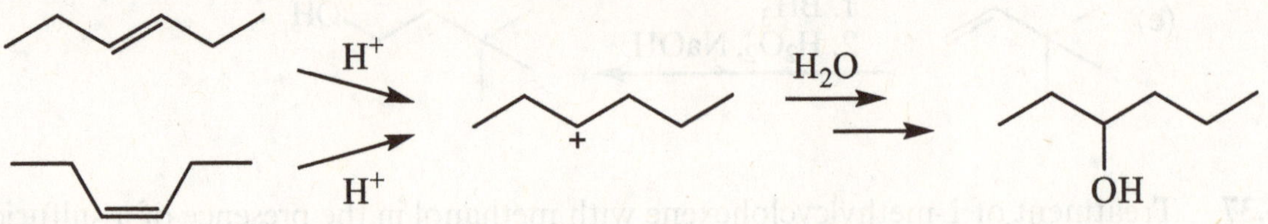

Oxidation-Reduction

5.39 Write a balanced equation for the combustion of 2-methylpropene in air to give carbon dioxide and water. The oxidizing agent is O_2, which makes up approximately 20% of air.

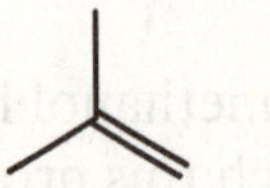

$$+ \; 6\,O_2 \longrightarrow 4\,CO_2 \; + \; 4\,H_2O$$

$$C_4H_8$$

5.40 Draw the product formed by treating each alkene with H_2/Ni:

(a)

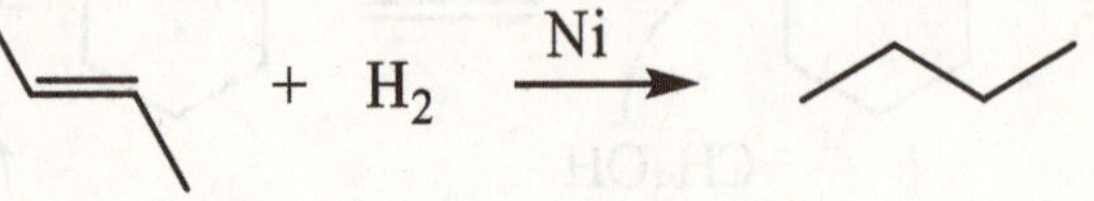

(b)

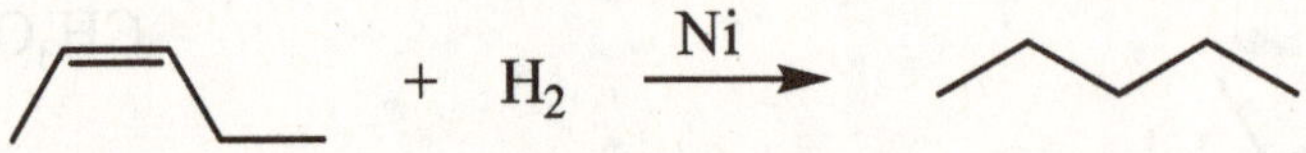

(c)

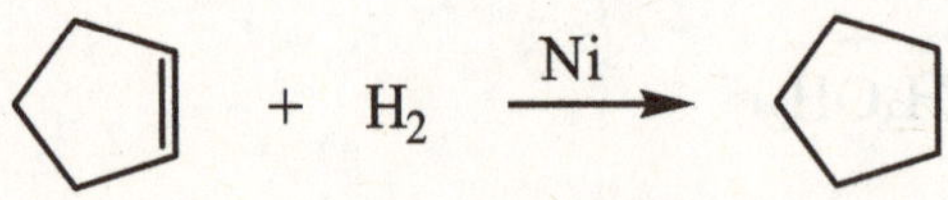

(d)

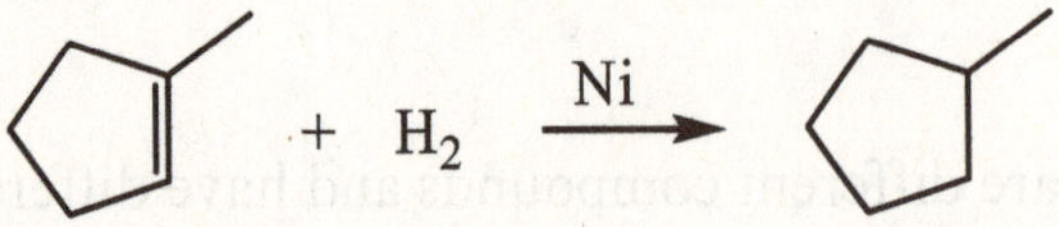

5.41 Hydrocarbon A, C_5H_8, reacts with 2 moles of Br_2 to give 1,2,3,4-tetrabromo-2-methylbutane. What is the structure of hydrocarbon A?

In a bromination reaction, Br_2 adds across the carbon-carbon double bond. Work backwards from the product to determine the structure of hydrocarbon A. Arrows in the product indicate where the C=C bonds were found in the starting material.

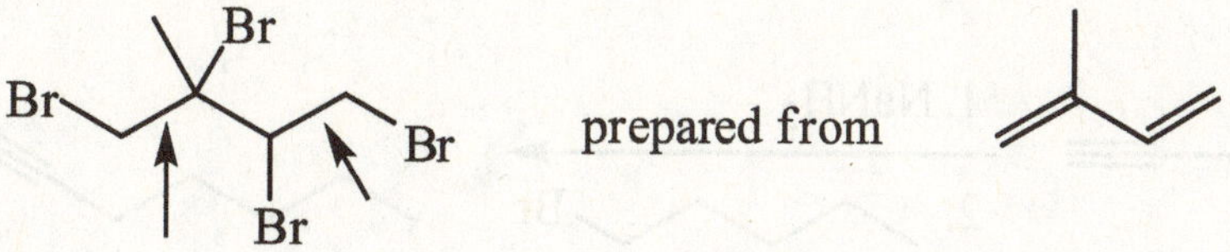
Br
Br
Br
Br
prepared from

5.42 Two alkenes, A and B, each have the formula C_5H_{10}. Both react with H_2/Pt and with HBr to give identical products. What are the structures of A and B?

Recognize that in order for the two alkenes to give the same products, their carbon skeletons must identical. Furthermore, the location of the double bond must be such that both A and B form the same carbocation upon reaction with HBr, and for this to be the case, there must be one common alkene carbon in both A and B.

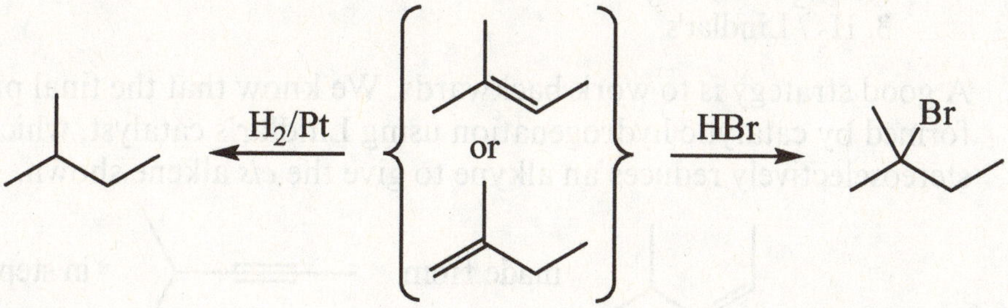
H2/Pt
or
HBr
Br

Reactions of Alkynes

5.43 Complete these equations by predicting the major products formed in each reaction. If more than one product is equally likely, draw both products.

(a)

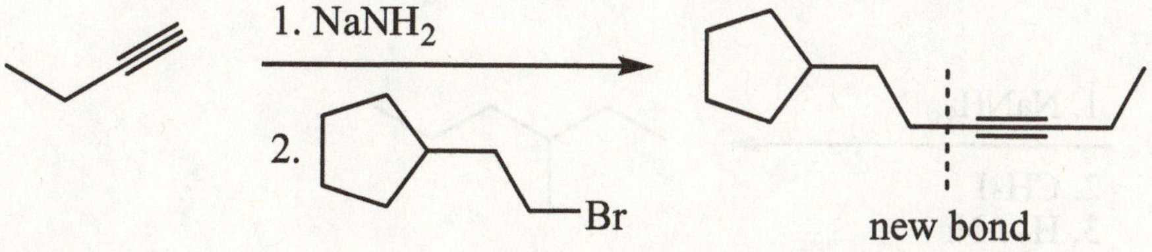
1. NaNH2
2.
Br
new bond

(b)

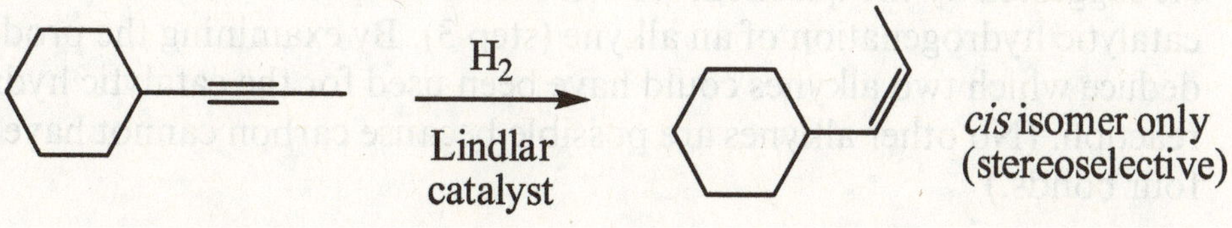
H2
Lindlar
catalyst
cis isomer only
(stereoselective)

(c)

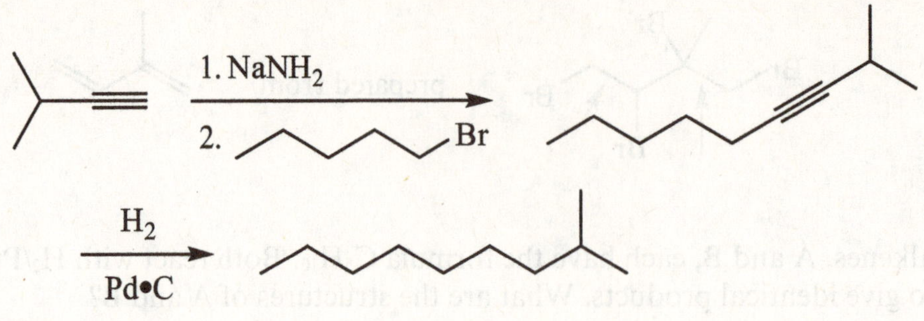

$$\xrightarrow[\text{Pd}\bullet\text{C}]{\text{H}_2}$$

5.44   Determine the alkyne that would be required in the following sequences of reactions.

(a) ? $\xrightarrow[\begin{array}{l}\text{2. CH}_3\text{CHBrCH}_3 \\ \text{3. H}_2 \text{ / Lindlar's}\end{array}]{\text{1. NaNH}_2}$

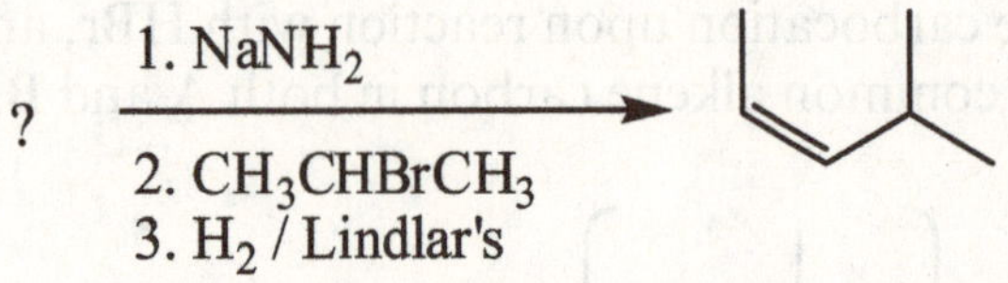

A good strategy is to work backwards. We know that the final product is formed by catalytic hydrogenation using Lindlar's catalyst, which stereoselectively reduces an alkyne to give the *cis* alkene shown.

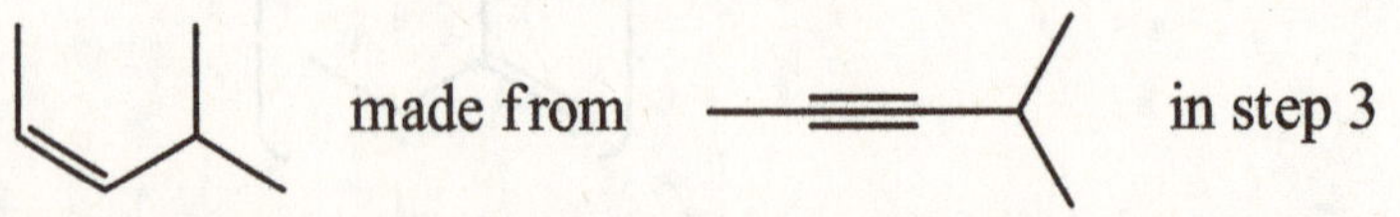

made from   in step 3

The alkyne used in step 3 can be made from 2-bromopropane and propyne, which is the original alkyne required for the reaction sequence.

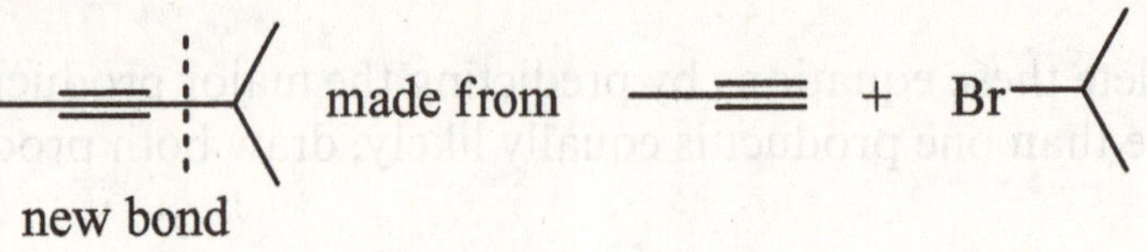

made from

new bond

(b) ? $\xrightarrow[\begin{array}{l}\text{2. CH}_3\text{I} \\ \text{3. H}_2 \text{ / Ni}\end{array}]{\text{1. NaNH}_2}$

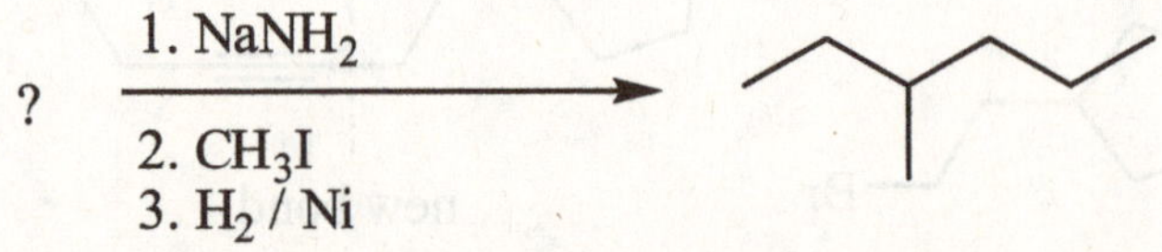

As suggested by the question, we know that the product was formed by the catalytic hydrogenation of an alkyne (step 3). By examining the product, we can deduce which two alkynes could have been used for the catalytic hydrogenation reaction. (No other alkynes are possible because carbon cannot have more than four bonds.)

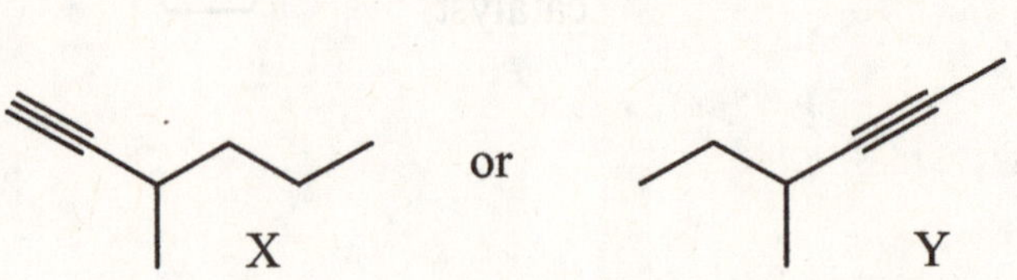

X   or   Y

Because step 2 involves CH_3I, which has only one single carbon, it can be concluded that alkyne Y must be made after step 2. It is not possible to make alkene X from CH_3I and the anion of an alkyne.

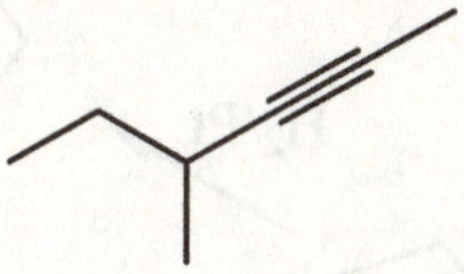

made from CH_3I and

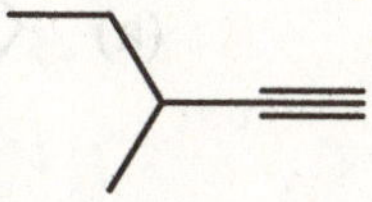

(c)

1. $NaNH_2$
2. CH_3I

? $\longrightarrow$

3. $NaNH_2$
4. $(CH_3)_2CHCH_2Br$
5. H_2/Pd

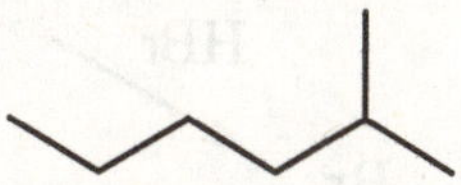

By examining the product and the organic starting materials (found in steps 2 and 4), it can be concluded that the bonds indicated below were formed. The bond on the left is made by subjecting acetylene to steps 1 and 2, the product of which is then subjected to steps 3 and 4 to make the other bond. Complete reduction of the alkyne forms the alkane product.

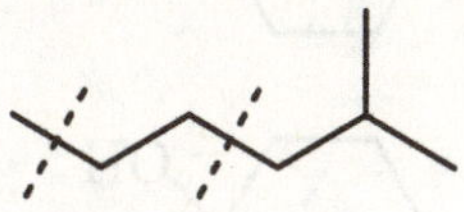

made from CH_3I + $HC{\equiv}CH$ + Br

new bonds

Synthesis

5.45   Show how to convert ethylene into these compounds:

(a) $CH_2{=}CH_2 \xrightarrow[Ni]{H_2} CH_3CH_3$

(b) $CH_2{=}CH_2 \xrightarrow[H_2SO_4]{H_2O} CH_3CH_2OH$

(c) $CH_2{=}CH_2 \xrightarrow{HBr} CH_3CH_2Br$

(d) $CH_2{=}CH_2 \xrightarrow{Br_2} BrCH_2CH_2Br$

(e) $CH_2{=}CH_2 \xrightarrow{HCl} CH_3CH_2Cl$

(f)
$CH_2{=}CH_2 \xrightarrow[\substack{3.\ H_2/Lindlar's}]{\substack{1.\ HBr \\ 2.\ NaC{\equiv}CH}} CH_3CH_2CH{=}CH_2$

5.46   Show how to convert cyclopentene into these compounds:

(a) — Br₂ ← cyclopentene → H₂/Pt — (b)

(c) — HBr ← cyclopentene → H₂O/H₂SO₄ — (d)

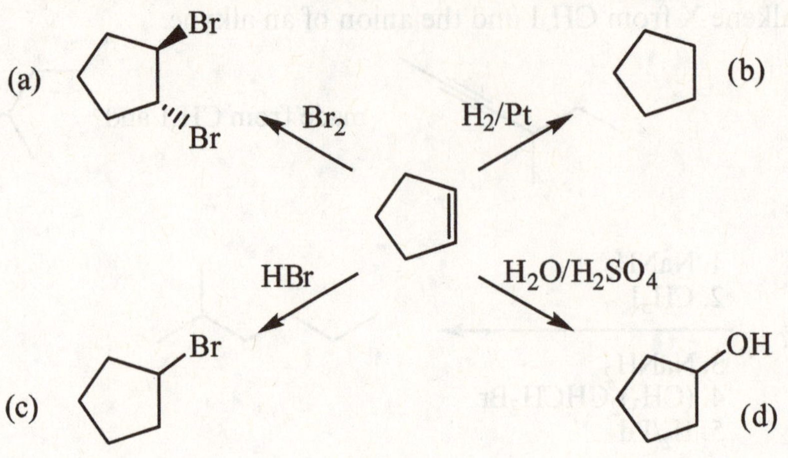

5.47   Show how to convert methylenecyclohexane into these compounds:

1. BH₃
2. H₂O₂, NaOH → —CH₂OH   (a)

=CH₂   H₂O/H₂SO₄ → OH / CH₃   (b)

Methylenecyclohexane   H₂/Ni → (c)

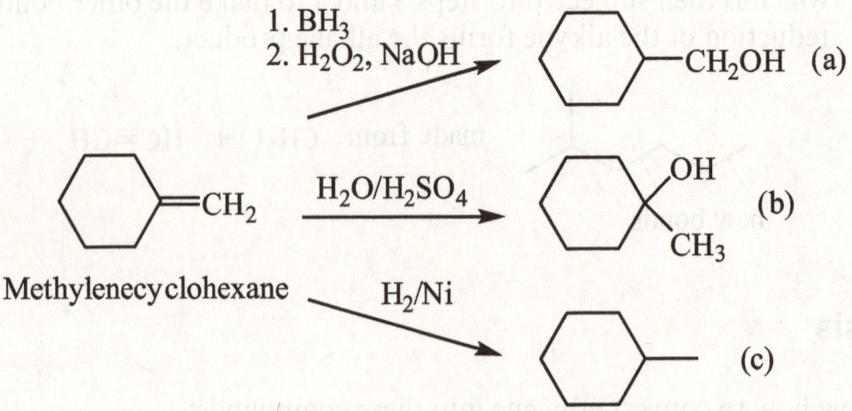

5.48   Show how to convert 1-butene into these compounds:

(a) — $\dfrac{H_2}{Ni}$ →

(b) — $\dfrac{H_2O}{H_2SO_4}$ → OH

(c) — 1. BH₃ 2. H₂O₂, NaOH → OH

(d) — HBr → Br

(e) — Br₂ → Br / Br

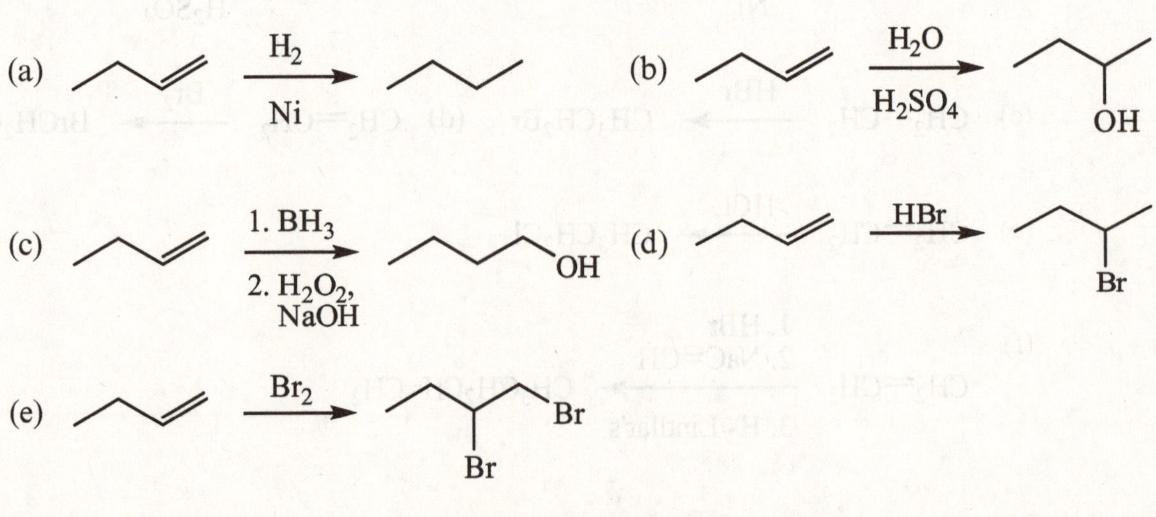

(f)   By comparing the structures of 3-methylpentane and 1-butene, the four-carbon
      fragment in 3-methylpentane that originated from 1-butene can be located.

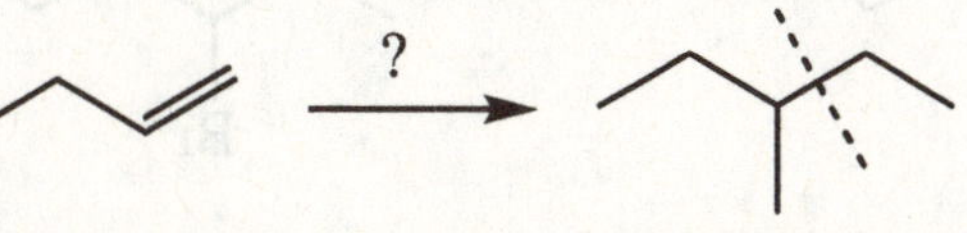

      The carbon skeleton of the product can be made by combining the four-
      carbon fragment together with a two-carbon fragment. This can be done by
      converting 1-butene into 2-bromobutane, followed by treatment with sodium
      acetylide, which can be made from acetylene and $NaNH_2$.

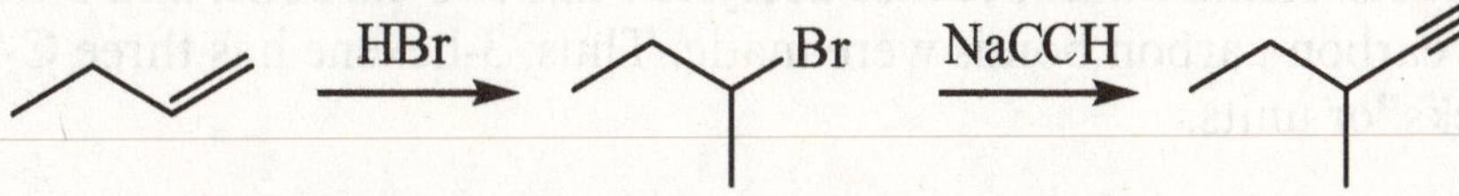

      Catalytic hydrogenation of the alkyne in the presence of excess hydrogen gas
      produces the final product.

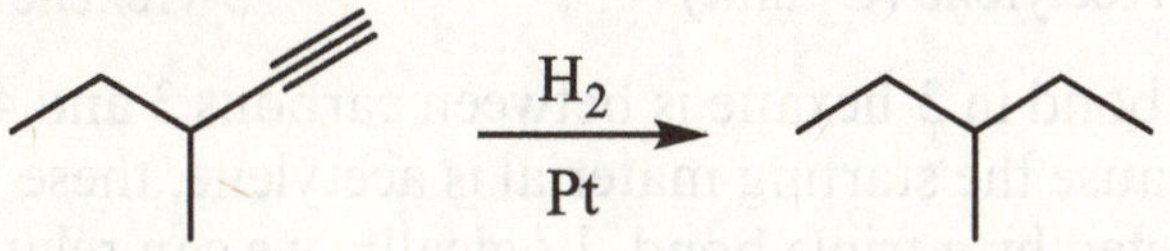

5.49   Show how the following compounds can be synthesized in good yields from an alkene.

Starting by examining the product to determine where the double bond could have
been in the alkene. If the reagent used is regioselective, the location of the double bond
in the alkene must be such that it leads to the product shown as the major product.

(a)

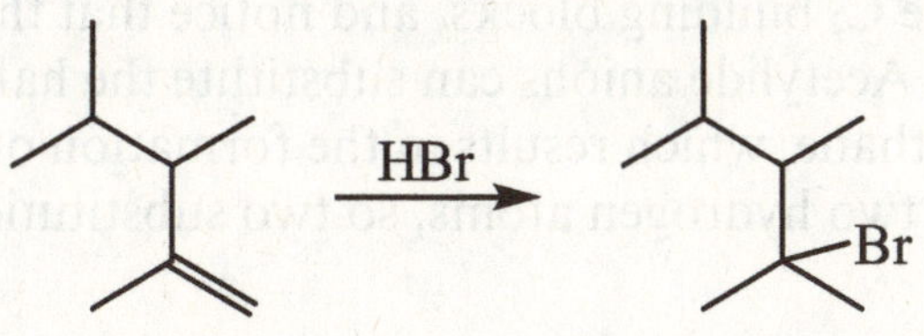

(b)

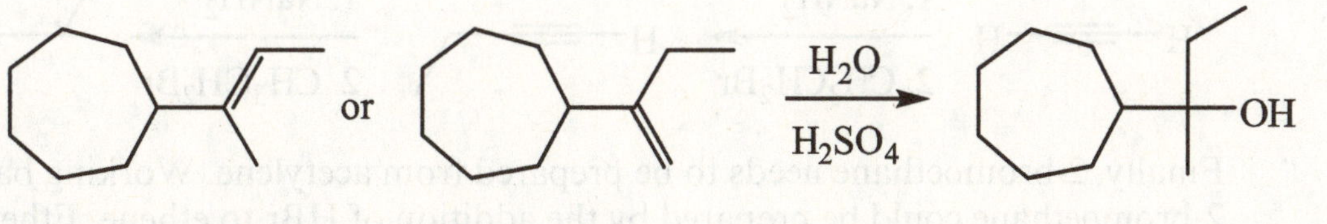

(c)

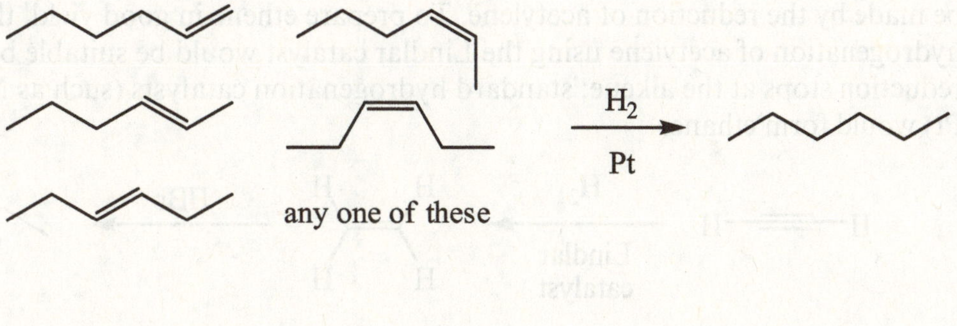

any one of these

(d)

$$\xrightarrow{Br_2}$$

5.50 How would you prepare *cis*-3-hexene using only acetylene as the source of carbon atoms, and using any necessary inorganic reagents.

To devise a synthetic scheme, start by comparing the starting materials and the products. Realize that because acetylene has two carbons, and 3-hexene has six, two new carbon-carbon bonds were made. Thus, 3-hexene has three C_2"building blocks"or units.

$$H-C \equiv C-H \qquad\qquad CH_3CH_2 \vdots CH=CH \vdots CH_2CH_3$$
$$\text{Acetylene (C}_2 \text{ unit)} \qquad\qquad \text{3-Hexene}$$

The double bond in 3-hexane is between carbons 3 and 4, which belong to the same C_2 unit. Because the starting material is acetylene, these two carbons were, at one time, connected by a triple bond. Logically, we can relate 3-hexene back to 3-hexyne, the stereoselective reduction of which results in the *cis* isomer of 3-hexene.

$$\xrightarrow[\text{catalyst}]{\underset{\text{Lindlar}}{H_2}}$$

The next question is, how do we prepare 3-hexyne? Recall that two carbon-carbon bonds were needed to link the three C_2 building blocks, and notice that the two terminal C_2 units are alkyl groups. Acetylide anions can substitute the halogen atom of a haloalkane, such as 2-bromoethane, which results in the formation of a carbon-carbon single bond. Acetylene has two hydrogen atoms, so two substitution reactions could be performed.

$$H \!\!-\!\!\!\equiv\!\!\!-\!\! H \xrightarrow[\text{2. CH}_3\text{CH}_2\text{Br}]{\text{1. NaNH}_2} H \!\!-\!\!\!\equiv\!\!\!\diagdown \xrightarrow[\text{2. CH}_3\text{CH}_2\text{Br}]{\text{1. NaNH}_2} \diagdown\!\!\!\equiv\!\!\!\diagdown$$

Finally, 2-bromoethane needs to be prepared from acetylene. Working backwards, 2-bromoethane could be prepared by the addition of HBr to ethene. Ethene could be made by the reduction of acetylene. To prepare ethene in good yield, the hydrogenation of acetylene using the Lindlar catalyst would be suitable because the reduction stops at the alkene; standard hydrogenation catalysts (such as Ni, Pd, and Pt) would form ethane.

$$H \!\!-\!\!\!\equiv\!\!\!-\!\! H \xrightarrow[\substack{\text{Lindlar}\\\text{catalyst}}]{H_2} \underset{H \quad\quad H}{\overset{H \quad\quad H}{\diagup\!\!=\!\!\diagdown}} \xrightarrow{\text{HBr}} \diagup\!\!\diagdown_{Br}$$

Chemical Transformations

5.51 Test your cumulative knowledge of the reactions learned thus far by completing the following chemical transformations. Note that some will require more than one step.

(a)

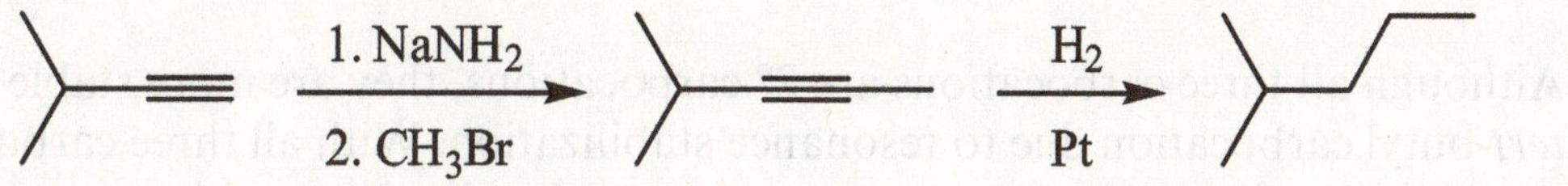

(b)

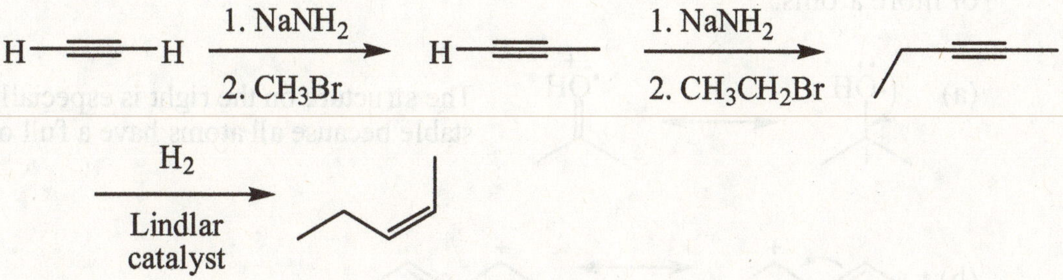

(c)

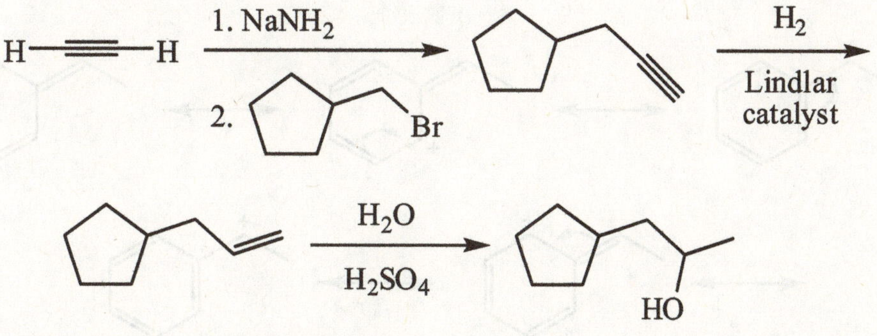

(d)

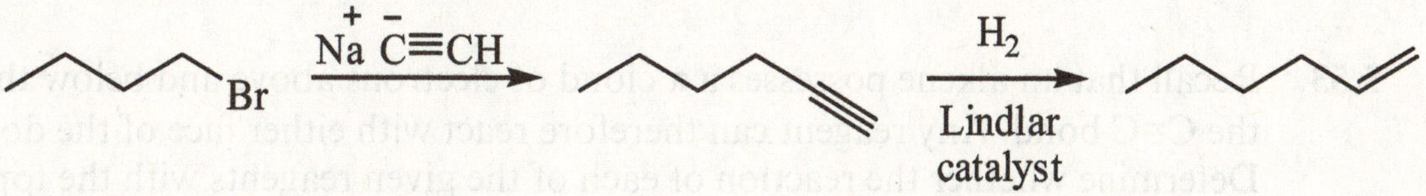

(e)

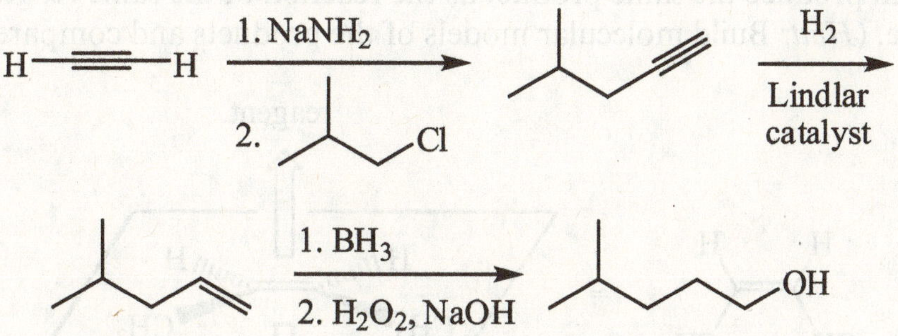

(f)

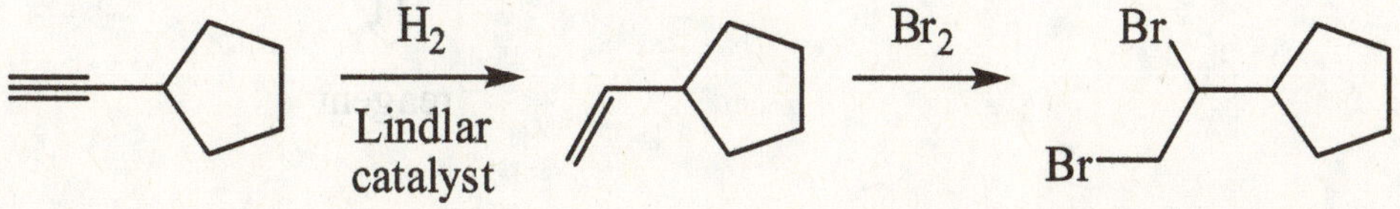

Looking Ahead

5.52   Each of the following 2° carbocations is more stable than the tertiary-butyl carbocation shown. Provide an explanation for each cation's enhanced stability.

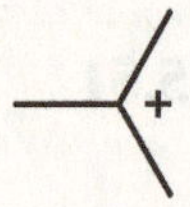

Although all three carbocations are 2° carbocations, they are more stable than the *tert*-butyl carbocation due to resonance stabilization. With all three carbocations, it is possible to draw resonance structures that delocalize the positive charge over two or more atoms.

(a) 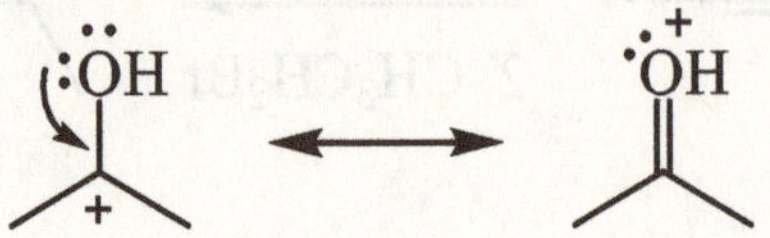   The structure on the right is especially stable because all atoms have a full octet.

(b)

(c)

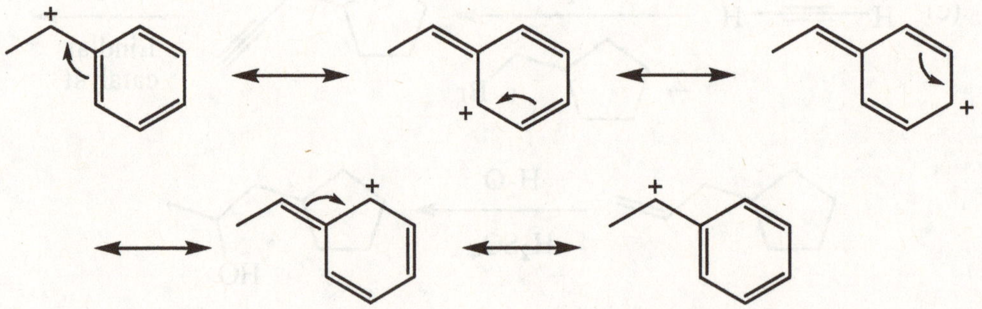

5.53   Recall that an alkene possesses a π cloud of electrons above and below the plane of the C=C bond. Any reagent can therefore react with either face of the double bond. Determine whether the reaction of each of the given reagents with the top face of *cis*-2-butene will produce the same product as the reaction of the same reagent with the bottom face. (*Hint:* Build molecular models of the products and compare them.)

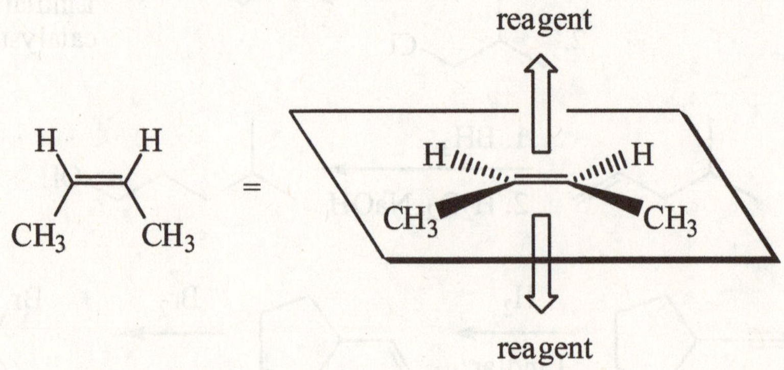

The hydrogenation reaction in (a) proceeds by syn addition, and the addition of two H atoms to either face yields the same product. Hydroboration-oxidation (b) also proceeds by syn addition, this time adding water in an anti-Markovnikov fashion. The bromination reaction in (c) proceeds by anti addition, where if one Br adds to the bottom face, the other Br must add to the top face, and vice versa. The hydroboration-oxidation and bromination of *cis*-2-butene each forms two different products that are mirror images of each other, which are known as *enantiomers* (more in Chapter 6).

(a)

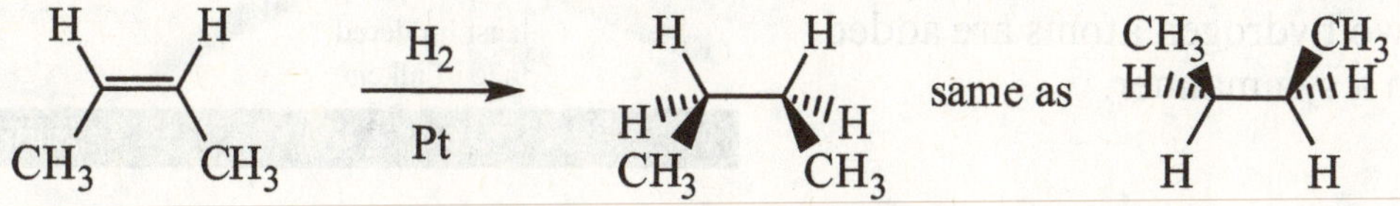

same as

(b)

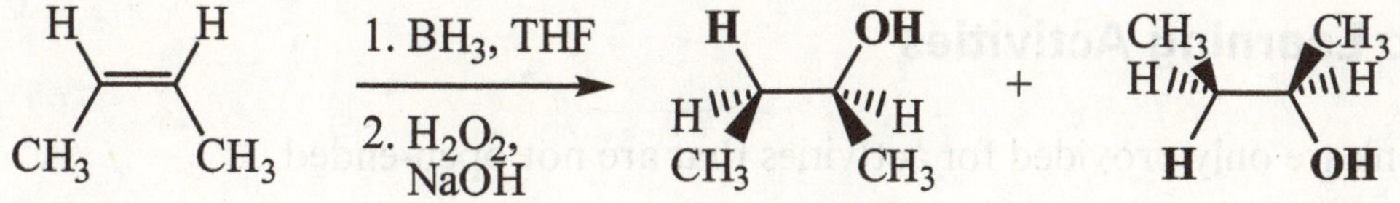

(c)

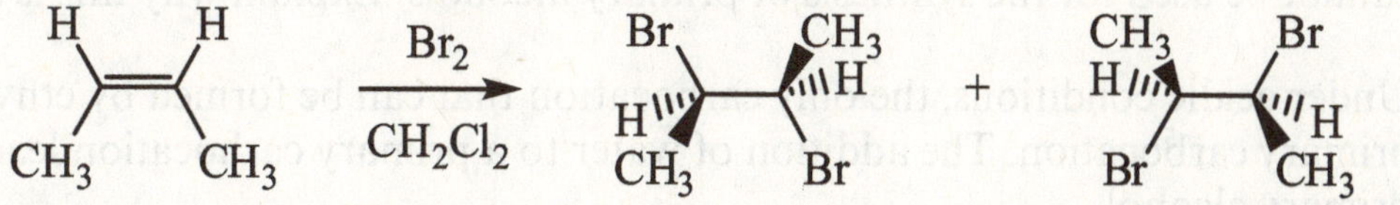

5.54 This reaction yields two products in differing amounts. Draw the two products and predict which product is favored.

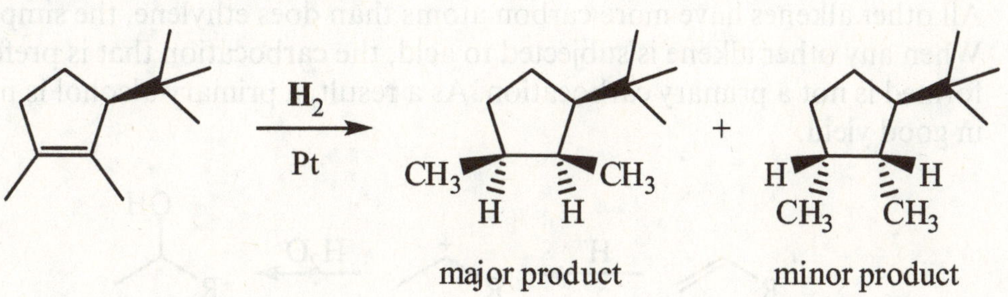

major product   minor product

In catalytic hydrogenation, the least hindered face of the alkene preferentially binds to the surface of the transition metal catalyst. In this case, the least hindered face is the face opposite to that of the bulky *tert*-butyl group. The face that binds to the catalyst is also the face to which two hydrogen atoms are added in a syn manner.

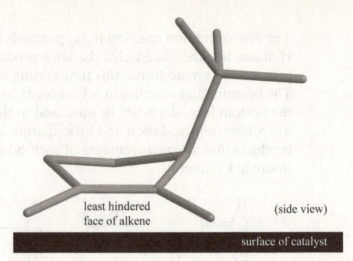

least hindered face of alkene

(side view)

surface of catalyst

Group Learning Activities

Solutions are only provided for activities that are not open-ended.

5.57   With the exception of ethylene to ethanol, the acid-catalyzed hydration of alkenes cannot be used for the synthesis of primary alcohols. Explain why this is so.

Under acidic conditions, the only carbocation that can be formed by ethylene is a primary carbocation. The addition of water to a primary carbocation leads to primary alcohol.

$$CH_2{=}CH_2 \xrightarrow{\ H^+\ } CH_3{-}\overset{+}{C}H_2 \xrightarrow{\ H_2O\ } CH_3{-}CH_2OH$$

All other alkenes have more carbon atoms than does ethylene, the simplest alkene. When any other alkene is subjected to acid, the carbocation that is preferentially formed is not a primary carbocation. As a result, a primary alcohol is not produced in good yield.

$$R\diagup\diagdown \xrightarrow{\ H^+\ } R\diagup\overset{+}{\diagdown} \xrightarrow{\ H_2O\ } R\overset{OH}{\diagup\diagdown}$$

5.58   Discuss the following word problem by proposing structures and debating possible outcomes. Use a whiteboard or chalkboard if possible.

Consider alkenes A, B, and C, each with molecular formula C_6H_{12}. A-C undergo catalytic reduction to give hexane as the only product. Acid-catalyzed hydration of A gives only one alcohol. Acid-catalyzed hydration of B gives two alcohols. Acid-catalyzed hydration of C gives only one alcohol. All alcohols are constitutional isomers with the molecular formula $C_6H_{14}O$.

(a) Propose structural formulas and names for A-C.
(b) Propose structural formulas for the product alcohols that are consistent with these experimental results.
(c) Propose structural formulas for two other alkenes, each with the molecular formula C_6H_{12} that will also give only one alcohol upon acid-catalyzed hydration.

Because the only product made upon catalytic hydrogenation is hexane, it can be concluded that alkenes A, B, and C are unbranched. There are three constitutional isomers of unbranched alkenes with the correct molecular formula.

$$CH_3CH_2CH=CHCH_2CH_3 \qquad CH_3CH_2CH_2CH=CHCH_3 \qquad CH_3CH_2CH_2CH_2CH=CH_2$$

3-hexene 2-hexene 1-hexene

The acid-catalyzed hydration of 3-hexene gives only one product because the starting material is symmetrical. The hydration of 2-hexene gives two products, because the two possible carbocations involved in the reaction are equally stable. The hydration of 1-hexene forms only one product.

$$CH_3CH_2CH=CHCH_2CH_3 \longrightarrow CH_3CH_2CH_2\overset{\overset{\displaystyle OH}{|}}{C}HCH_2CH_3$$

3-hexene

$$CH_3CH_2CH_2CH=CHCH_3 \longrightarrow \begin{bmatrix} CH_3CH_2CH_2\overset{\overset{\displaystyle OH}{|}}{C}HCH_2CH_3 \\ CH_3CH_2CH_2CH_2\overset{\overset{\displaystyle OH}{|}}{C}HCH_3 \end{bmatrix}$$

2-hexene

$$CH_3CH_2CH_2CH_2CH=CH_2 \longrightarrow CH_3CH_2CH_2CH_2\overset{\overset{\displaystyle OH}{|}}{C}HCH_3$$

1-hexene

Other alkenes with the formula C_6H_{12} that will each form one alcohol upon acid-catalyzed hydration are:

$$(CH_3)_3CCH=CH_2 \qquad CH_3CH_2CH(CH_3)CH=CH_2$$

$$(CH_3)_2CHCH_2CH=CH_2 \qquad (CH_3)_2C=C(CH_3)_2$$

Chapter 6: Chirality

Problems

6.1 Each molecule has one stereocenter. Identify the stereocenter in each and draw stereorepresentations of the enantiomers of each.

(a)

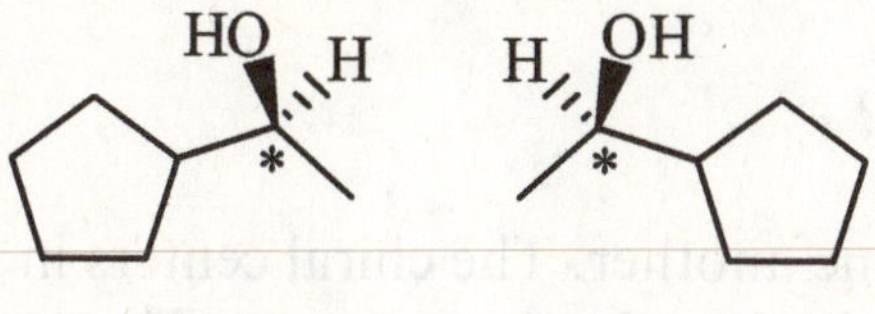

(b)

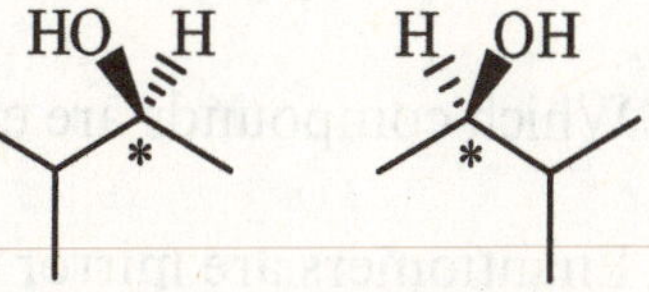

(c)

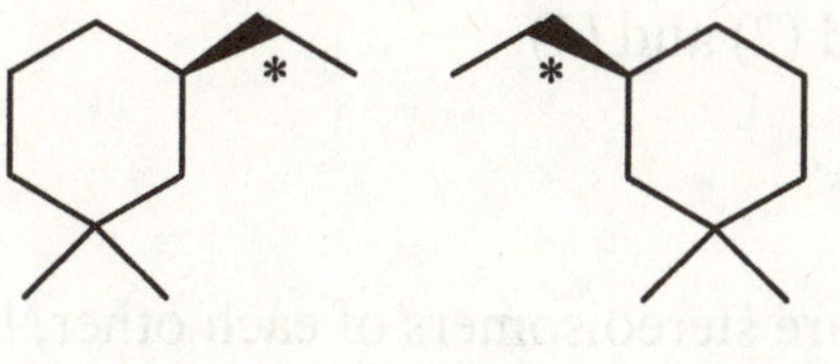

(d)

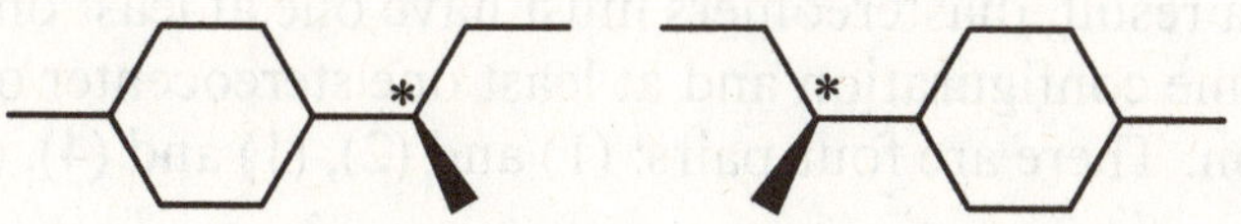

6.2 Assign an *R* or *S* configuration to each stereocenter:

When determining R,S configuration, the group of lowest priority must be pointing towards the back. That is, view down the C–H bond from the C. Your molecular model kit is very useful for this purpose – simply build the molecule and place #4 in the back.

(a)

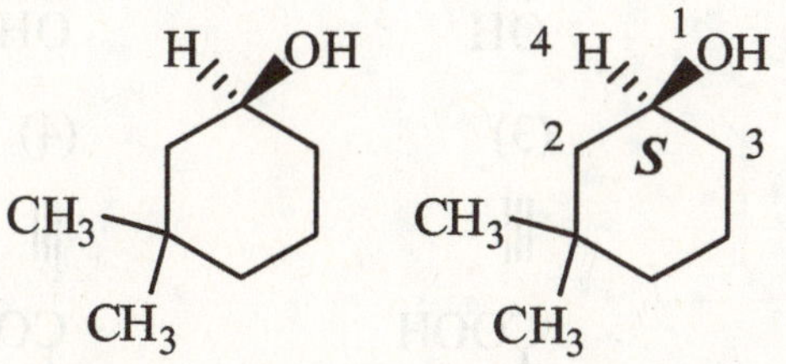

(b)

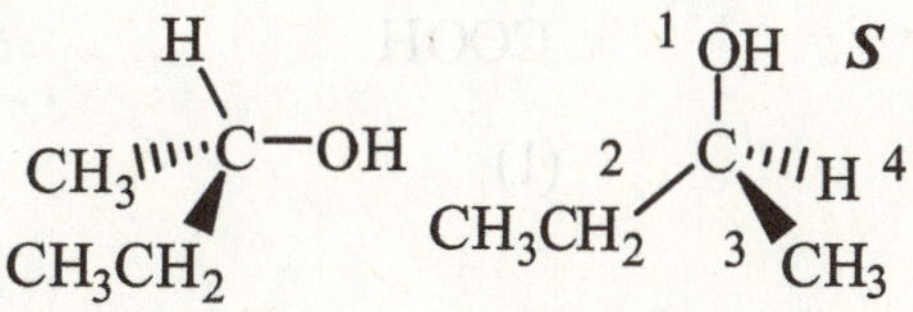

(c)

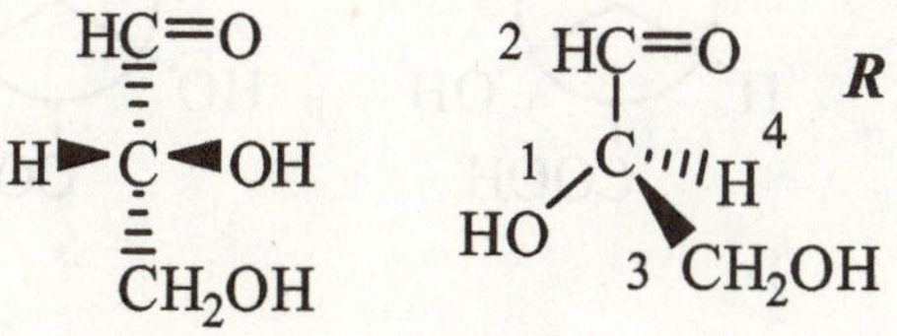

(d)

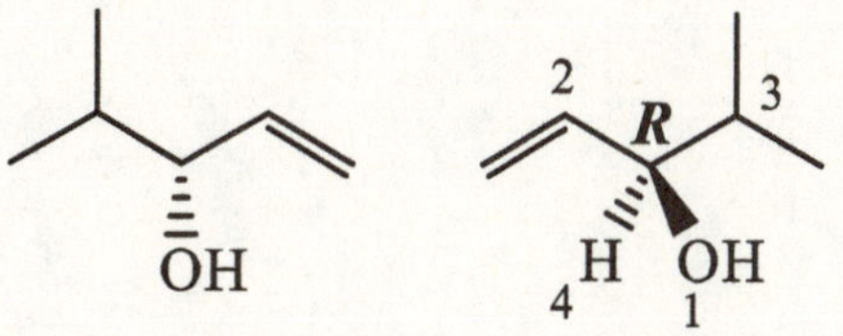

6.3 Following are stereo-representations of the four stereoisomers of 3-chloro-2-butanol:

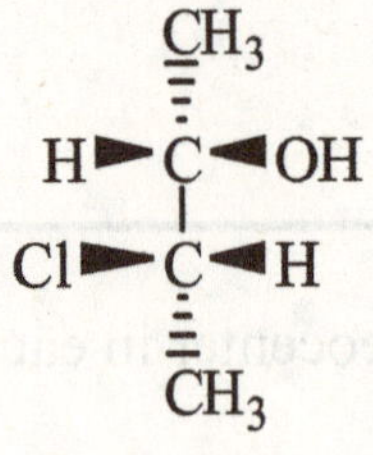

$$\begin{array}{cccc}
CH_3 & CH_3 & CH_3 & CH_3 \\
H \blacktriangleright C \blacktriangleleft OH & H \blacktriangleright C \blacktriangleleft OH & HO \blacktriangleright C \blacktriangleleft H & HO \blacktriangleright C \blacktriangleleft H \\
Cl \blacktriangleright C \blacktriangleleft H & H \blacktriangleright C \blacktriangleleft Cl & H \blacktriangleright C \blacktriangleleft Cl & Cl \blacktriangleright C \blacktriangleleft H \\
CH_3 & CH_3 & CH_3 & CH_3 \\
(1) & (2) & (3) & (4)
\end{array}$$

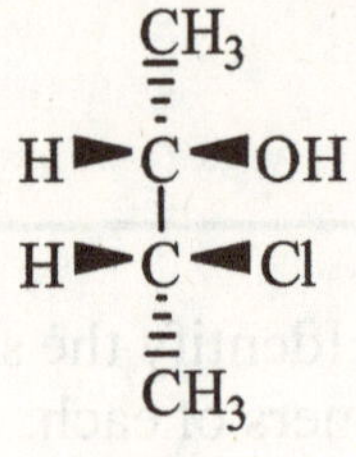

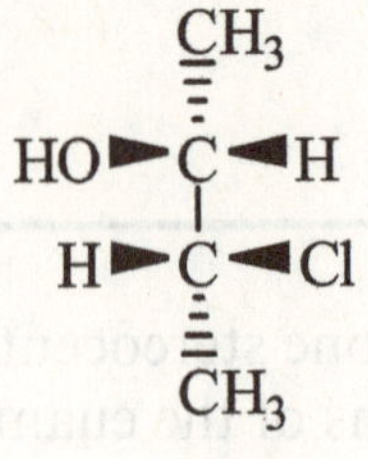

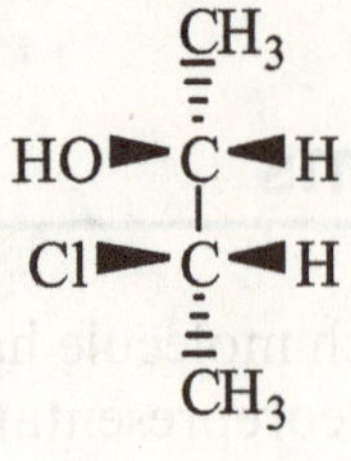

(a) Which compounds are enantiomers?

Enantiomers are mirror images of one another. The chiral centers in one enantiomer are all opposite of those in the other enantiomer. There are two pairs of enantiomers, (1) and (3), and (2) and (4).

(b) Which compounds are diastereomers?

Diastereomers are compounds that are stereoisomers of each other, but they are not enantiomers. As a result, diastereomers must have one at least one stereocenter of the same configuration and at least one stereocenter of a different configuration. There are four pairs: (1) and (2), (1) and (4), (2) and (3), and (3) and (4).

6.4 Following are four Newman projection formulas for tartaric acid:

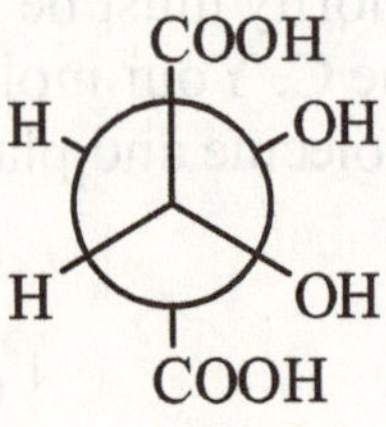

(1) 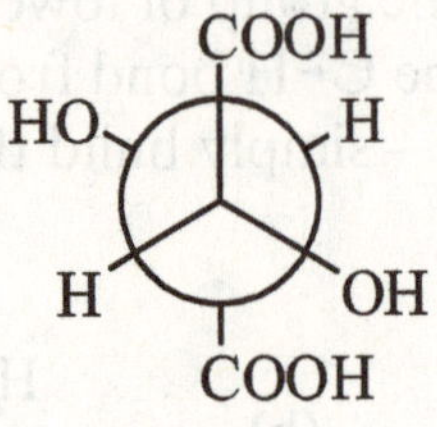 (2) 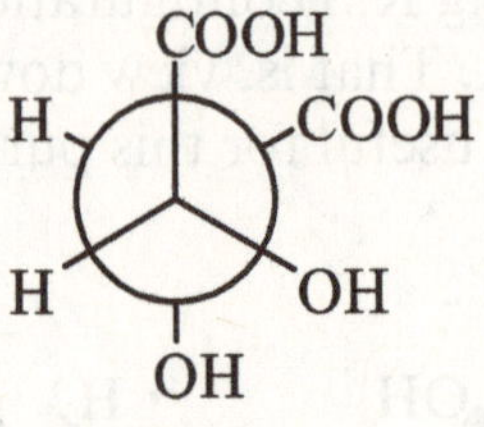 (3) 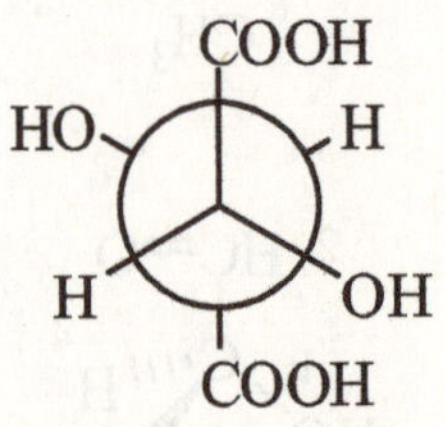 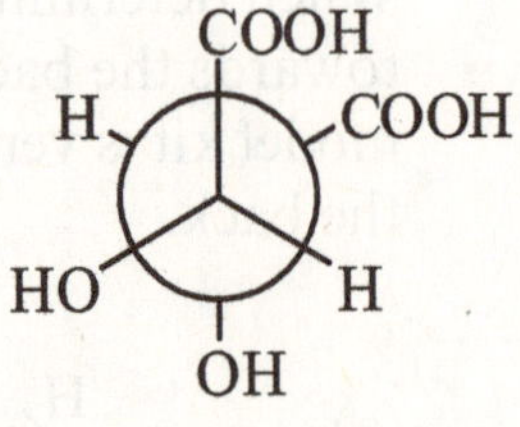 (4)

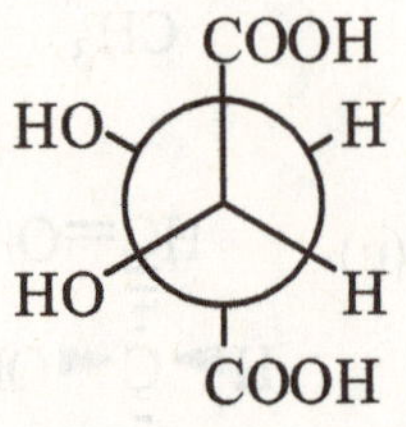

(a) Which represent the same compound?

When comparing Newman projections, it is often helpful to rotate the carbon-carbon bond to obtain a useful reference point. In the structures above, (3) and (4) have been rotated so that the two −COOH groups are anti to each other, as in (1) and (2). This allows us to see that (2) and (3) are identical.

(b) Which represent enantiomers?

When compounds (1) and (4) are lined up side-by-side, it can be seen that they are mirror images of each other.

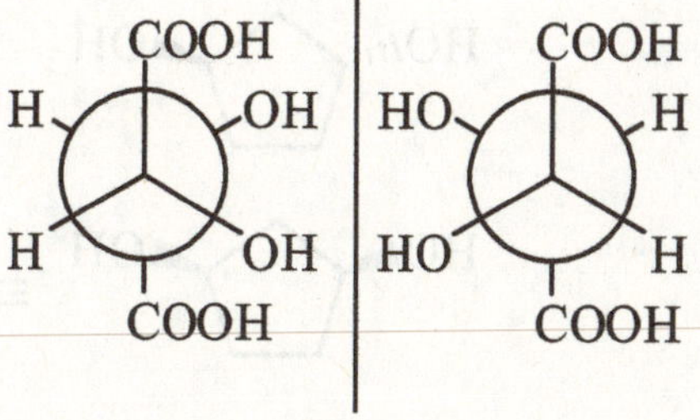

(c) Which represent(s) meso tartaric acid?

(2) and (3), which are the same, are meso because they have an internal plane of symmetry. It is helpful to draw an eclipsed conformation; in meso compounds, the substituents that are eclipsing each other are identical. The internal plane of symmetry lies parallel to the page and is between the front and rear carbon atoms of the Newman projection. Drawing the structure using the representation used in Problem 6.3 also reveals an internal plane of symmetry.

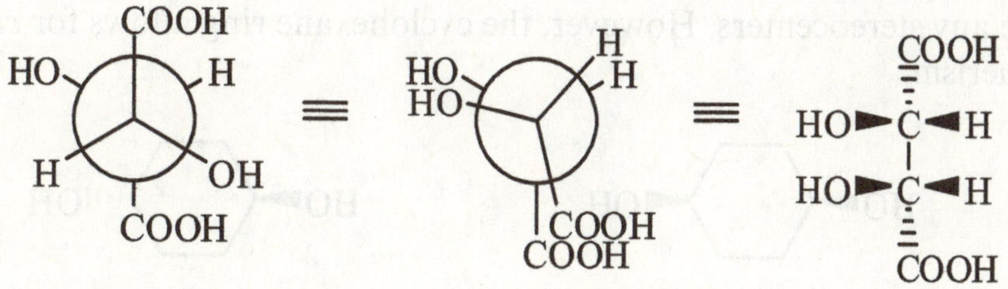

(d) Which are diastereomers?

When compounds (3) and (4) are compared side-by-side, it can be seen that the rear stereocenters have the same configuration, but the front stereocenters have a different configuration. (Notice the positions of the H and OH groups). These two are a pair of diasteromers.

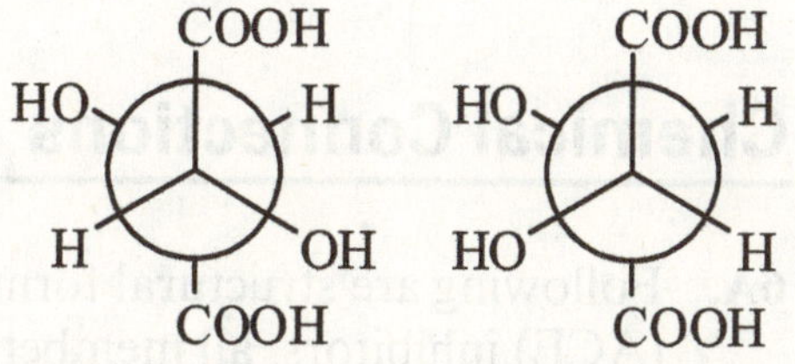

There also exists another pair of diastereomers: compounds (1) and (3). In this case, the two rear stereocenters have the same configuration, while the front stereocenters have different configuration.

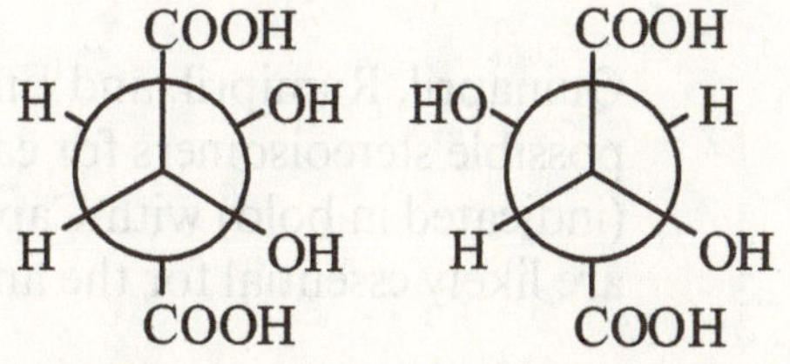

6.5   How many stereoisomers exist for 1,3-cyclopentanediol?

Because 1,3-cyclopentanediol has two stereocenters, a maximum of four stereoisomers are possible. The two −OH groups can be configured up/up, up/down, down/down, and down/up. However, the up/up and down/down configurations represent the same meso compound, so there are only three possible stereoisomers.

HO//,,   OH   HO   ,,\\OH   *trans*-1,3-Cyclopentanediol
                                            (pair of enantiomers)

HO   OH   ≡   HO//,,   ,,\\OH   *cis*-1,3-Cyclopentanediol
                                            (meso)

This problem reiterates an important point: the name *trans*-1,3-cyclopentanediol does not distinguish between the two *trans* enantiomers. The *trans* designation simply means that the two groups are on opposite sides, *relative to each other*. Thus, there are only two *cis-trans* isomers of 1,3-cyclopentanediol, but there are a total of three stereoisomers.

6.6   How many stereoisomers exist for 1,4-cyclohexanediol?

The compound is achiral because it has an internal plane of symmetry. It does not have any stereocenters. However, the cyclohexane ring allows for *cis-trans* isomerism.

HO   OH   HO   ,,\\OH

cis-1,4-Cyclohexanediol   *trans*-1,4-Cyclohexanediol

Chemical Connections

6A. Following are structural formulas for three other angiotensin-converting enzyme (ACE) inhibitors, all members of the "pril" family. Which are chiral? For each that is chiral, determine the number of stereoisomers possible for it. List the similarities in structure among each of these four drugs.

Quinapril, Ramipril, and Enalapril each contain three chiral centers, giving rise to $2^3 = 8$ possible stereoisomers for each compound. These drugs share structural similarities (indicated in bold) with Captopril, which is redrawn below. The indicated similarities are likely essential for the angiotensin-converting enzyme to bind to these drugs.

Captopril   Quinapril (Accupril)

Ramipril (Altace)   Enalapril (Vasotec)

Quick Quiz

1. Enantiomers are always chiral. *True*. In order for an object or a molecule to be chiral, it cannot have an internal plane of symmetry. As a result, the object or molecule would not be superimposable on its mirror image.

2. An unmarked cube is chiral. *False*. An unmarked cube has a plane of symmetry – in fact, it has planes of symmetry through each face of the cube and through the diagonal corners.

3. Stereocenters can be designated using E and Z. *False*. The E/Z system is used to identify *cis-trans* isomers of alkenes. Stereocenters are designated using R and S.

4. A chiral molecule will always have a diastereomer. *False*. A chiral molecule will always have an enantiomer, but not necessarily a diastereomer. Molecules with two or more stereocenters may have diastereomers.

5. Every object in nature has a mirror image. *True*. It is possible to have a mirror image of every object – it's a matter of if the mirror images are the same or different!

6. A molecule that possesses an internal plane of symmetry can never be chiral. *True*. The mirror image of such a molecule would be superimposable and thus identical.

7. Pairs of enantiomers have the same connectivity. *True*. Stereoisomers have the same connectivity but different, noninterconverting spatial positionings of the atoms and groups. Compounds that are stereoisomers must all be of the same constitutional isomer.

8.   Enantiomers, like gloves, occur in pairs. ***True***. Enantiomers are always mirror images of each other, so they can only exist as pairs.

9.   A cyclic molecule with two stereocenters will always have only three stereoisomers. ***False***. The number of possible stereoisomers will depend on whether the substituents bonded to the ring are identical or not. For example, 1-chloro-2-fluorocyclohexane has four stereosiomers, but 1,2-dichlorocyclohexane has only three stereoisomers.

10.   An achiral molecule will always have a diastereomer. ***False***. Only achiral molecules that have two or more stereocenters (meso compounds) will have a diastereomer.

11.   The *cis* and *trans* isomers of 2-butene are chiral. ***False***. Although *cis-trans* isomerism is a type of stereoisomerism, it is not the same as chirality. The *cis-trans* isomers of 2-butene are not mirror images of each other.

12.   A human foot is chiral. ***True***. Your left and right feet are mirror images of each other, and this is why we need left and right shoes, which are also mirror images. Also recall enantiomers behave differently in a chiral environment, which is why your left foot does not properly fit in your right shoe.

13.   A compound with n stereocenters will always have 2^n stereoisomers. ***False***. It is incorrect that it will *always* have 2^n stereoisomers. This number is simply a mathematical *maximum* of the number of stereoisomers that may exist.

14.   A molecule with three or more stereocenters cannot be meso. ***False***. As long as a molecule has at least two stereocenters and an internal plane of symmetry, it is meso.

15.   A molecule with three or more stereocenters must be chiral. ***False***. Meso compounds, which are achiral, may have three or more stereocenters.

16.   Each member of a pair of enantiomers will have the same boiling point. ***True***. Enantiomers have the same physical properties, which includes boiling point.

17.   If a molecule is not superposable on its mirror image, it is chiral. ***True***. Chiral objects have nonsuperimposable mirror images.

18.   For a molecule with two tetrahedral stereocenters, four stereoisomers are possible. ***True***. Keep in mind that this is the maximum number that is possible and that it can be reduced if a meso stereoisomer exists.

19.   Constitutional isomers have the same connectivity. ***False***. Constitutional isomers have the same molecular formula, but they have different connectivity.

20.   Enantiomers can be separated by interacting them with the same chiral environment or chemical agent. ***True***. For example, chiral substrate chromatography and enzymes can be used to separate enantiomers.

21. Enzymes are achiral molecules that can differentiate chiral molecules. *False*. Enzymes are chiral, which is why they can differentiate chiral molecules.

22. *Cis* and *trans* stereoisomers of a cyclic compound can be classified as diastereomers. *True*. *cis-trans* isomers of cyclic compounds are not mirror images of each other, but they are stereoisomers. Note that in this context, *cis-trans* is in reference to the substituents bonded to the ring and not to an alkene.

23. 3-Pentanol is the mirror image of 2-pentanol. *False*. These two compounds have the same molecular formula but different connectivity, so they are constitutional isomers.

24. Diastereomers do not have a mirror image. *False*. Unless one of the diastereomers is meso, each of the diastereomers will have its own mirror image.

25. The most common cause of chirality in organic molecules is the presence of a tetrahedral carbon atom with four different groups bonded to it. *True*. A carbon bearing four different groups is a stereocenter.

26. Each member of a pair of enantiomers will have the same density. *True*. Enantiomers have the same physical properties.

27. The carbonyl carbon of an aldehyde or ketone cannot be a stereocenter. *True*. The carbon of a carbonyl group is only bonded to three substituents.

28. For a molecule with three stereocenters, $3^2 = 9$ stereoisomers are possible. *False*. The maximum number of possible stereoisomers is $2^3 = 8$.

29. Diastereomers can be resolved using traditional methods such as distillation. *True*. Unlike enantiomers, diastereomers have different physical and chemical properties.

30. A racemic mixture is optically inactive. *True*. A 1:1 mixture of two enantiomers does not rotate plane polarized light.

31. 2-Pentanol and 3-pentanol are chiral and show enantiomerism. *False*. Only 2-pentanol is chiral; 3-pentanol is achiral.

32. A diastereomer of a chiral molecule must also be chiral. *False*. The diastereomer of a chiral molecule may be meso and thus achiral.

33. In order to designate the configuration of a stereocenter, the priority of groups must be read in a clockwise or counterclockwise fashion after the lowest priority group is placed facing towards the viewer. *False*. The group of lowest priority must be placed facing *away* from the viewer.

34. A compound with *n* stereocenters will always be one of the 2^n stereoisomers of that compound. *True*. The compound shown or drawn is one of the possible stereoisomers.

35. Each member of a pair of enantiomers could react differently in a chiral environment. *True*. In a chiral environment, enantiomers have different chemical properties. This is why enantiomers of drugs often behave differently in the body.

36. A chiral molecule will always have an enantiomer. *True*. In order for a molecule to be chiral, it must have a nonsuperimposable mirror image, which is the enantiomer.

37. Each member of a pair of diastereomers will have the same melting point. *False*. Diastereomers have different physical and chemical properties.

38. If a chiral compound is dextrorotatory, its enantiomer is levorotatory by the same number of degrees. *True*. Enantiomers rotate plane polarized light to the same magnitude but in opposite directions.

39. All stereoisomers are optically active. *False*. Examples of stereoisomers that are optically inactive are meso compounds and *cis-trans* isomers of alkenes.

40. There are usually equal amounts of each enantiomer of a chiral biological molecule in a living organism. *False*. In living organisms, usually one enantiomer of a chiral biological molecule is favored over the other.

End-of-Chapter Problems

Chirality

6.7 Define the term *stereoisomer*. Name four types of stereoisomers.

Stereoisomers are compounds that have the same molecular formula and the same atom connectivity (i.e. they are of the same constitutional isomer) but have different, noninterconverting orientations of their atoms or groups in three-dimensional space.

Four types of stereoisomers include:
- Enantiomers, which are nonsuperimposable mirror images
- Diastereomers, which are stereoisomers that are not enantiomers
- *cis-trans* isomers, which differ by the relative positions of the substituents bonded to a cycloalkane ring or an alkene
- Meso isomers, which are achiral compounds with two or more stereocenters

6.8   In what way are constitutional isomers different from stereoisomers? In what way
are they the same?

Constitutional isomers and stereoisomers have the same molecular formula. They
are different in that constitutional isomers have a different connectivity of atoms,
but stereoisomers have the same atom connectivity and only a different,
noninterconverting spatial orientation of the atoms.

6.9   Compare and contrast the meaning of the terms *conformation* and *configuration*.

Conformation refers to the different, interconverting arrangements of atoms, in
three-dimensional space, that are the result of rotation about single bonds.
Configuration also refers to the different spatial arrangement of atoms, but these
spatial arrangements are noninterconverting and cannot be made the same by
rotation about single bonds.

6.10   Which of these objects are chiral (assume that there is no label or other identifying
mark)?

Any object is chiral if it cannot be superimposed on its mirror image. This happens
when the object does not have an internal plane of symmetry.

(a)   A pair of scissors. Scissors are chiral because they do not have a plane of
symmetry. This is why there are left- and right-handed scissors.

(b)   A tennis ball. Achiral. A tennis ball actually has two planes of symmetry.

(c)   A paper clip. Achiral. There is a plane of symmetry that slices through it.

(d)   A beaker. Achiral. There is a vertical plane of symmetry that runs through the
spout.

(e)   The swirl created in water as it drains out of a sink or bathtub. Chiral. Spirals
can be clockwise and counterclockwise, and they do not have mirror planes.

6.11   Think about the helical coil of a telephone cord or the spiral binding on a notebook,
and suppose that you view the spiral from one end and find that it has a left-handed
twist. If you view the same spiral from the other end, does it have a right-handed
twist or a left-handed twist from that end as well?

A spiral with a left-handed twist will have a left-handed twist when viewed from
either end. For the same reason, viewing a chiral molecule from one direction and in
another direction does not mean that you are viewing its mirror image – it is still the
same molecule.

6.12   Next time you have the opportunity to view a collection of augers or other seashells that have a helical twist, study the chirality of their twists. Do you find an equal number of left-handed and right-handed augers, or, for example, do they all have the same handedness? What about the handedness of augers compared with that of other spiral shells?

This question is meant to encourage you to think about chirality in everyday objects. Please share your observations with other members of the class and come to a conclusion.

6.13   Next time you have an opportunity to examine any of the seemingly endless varieties of spiral pasta (rotini, fusilli, radiatori, tortiglione), examine their twist. Do the twists of any one kind all have a right-handed twist, do they all have a left-handed twist, or are they a racemic mixture?

This of course depends on the manufacturer of the pasta! However, in any given box of spiral pasta, usually every piece of pasta is of the same twist because the entire box was manufactured using the same machine. So, the next time you are having spiral pasta, you will most likely be consuming something that is enantiomerically pure!

6.14   One reason we can be sure that sp^3-hybridized carbon atoms are tetrahedral is the number of stereoisomers that can exist for different organic compounds.

(a)   How many stereoisomers are possible for $CHCl_3$, CH_2Cl_2, and $CHBrClF$ if the four bonds to carbon have a tetrahedral geometry?

A tetrahedral carbon can only be a stereocenter if there are four different substituents bonded to it, and only $CHBrClF$ fulfills this requirement. There are two stereoisomers for $CHBrClF$, and they represent a pair of enantiomers.

(b)   How many stereoisomers are possible for each compound if the four bonds to the carbon have a square planar geometry?

cis-trans isomerism is possible in a square planar geometry if it has at least two or more different groups bonded to the central atom. CH_2Cl_2 (left pair) and $CHBrClF$ (right pair) satisfy this requirement; note that for the latter, Br and Cl were arbitrarily chosen for the comparision of *cis* and *trans*.

Enantiomers

6.15 Which compounds contain stereocenters?

(a)

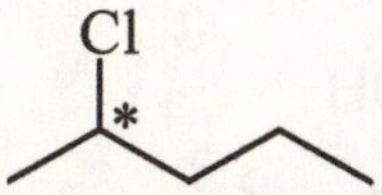

2-Chloropentane

(b)

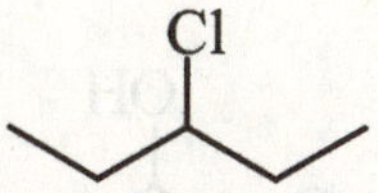

3-Chloropentane

(c)

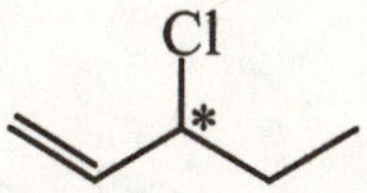

3-Chloro-1-pentene

(d)

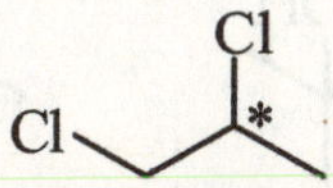

1,2-Dichloropropane

All compounds except for 3-chloropropane each contain one stereocenter. 3-Chloropropane has an internal plane of symmetry and is therefore achiral.

6.16 Using only C, H, and O, write structural formulas for the lowest-molecular-weight chiral molecule of each of the following compounds:

(a) Alkane

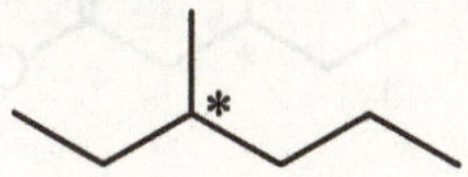

or

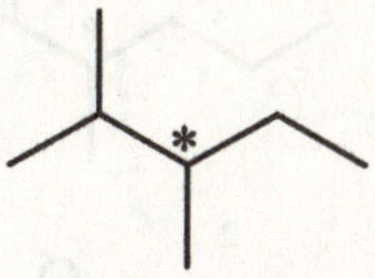

(b) Alcohol

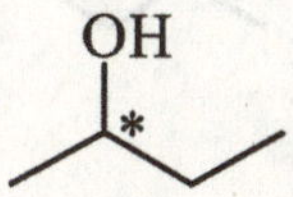

(c) Aldehyde

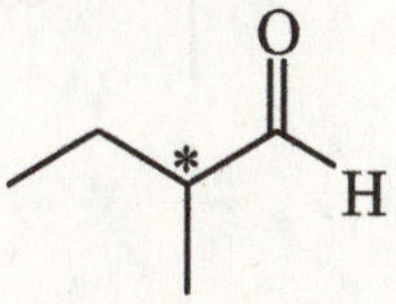

(d) Ketone

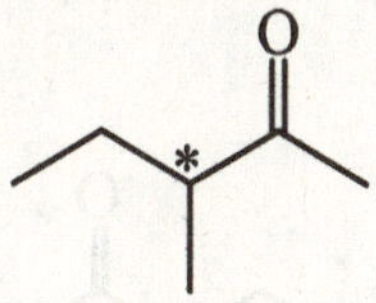

(e) Carboxylic acid

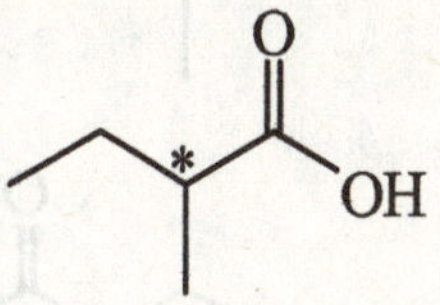

(f) Ester

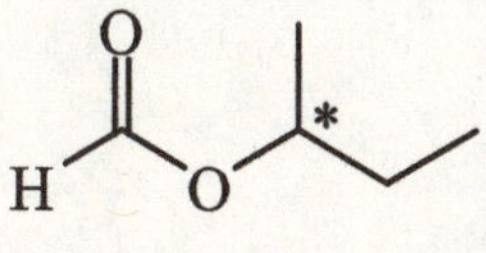

6.17 Which alcohols with the molecular formula $C_5H_{12}O$ are chiral?

There are eight alcohols with the molecular formula $C_5H_{12}O$, and only three of these alcohols are chiral. Stereocenters are indicated by an asterisk.

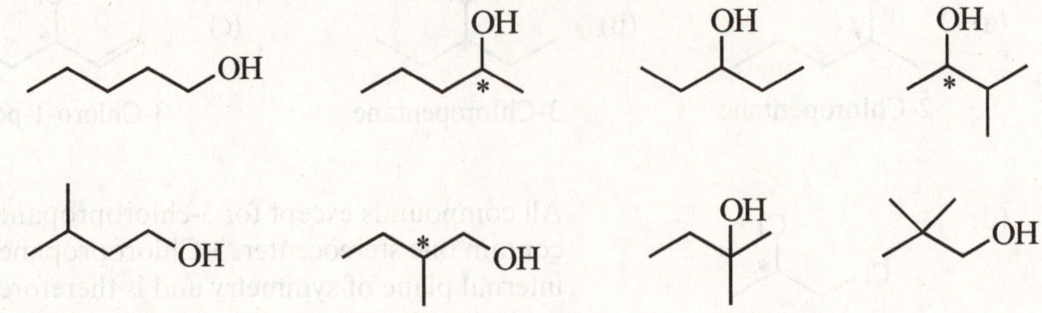

6.18 Which carboxylic acids with the molecular formula $C_6H_{12}O_2$ are chiral?

There are eight carboxylic acids with this molecular formula. Three of them are chiral.

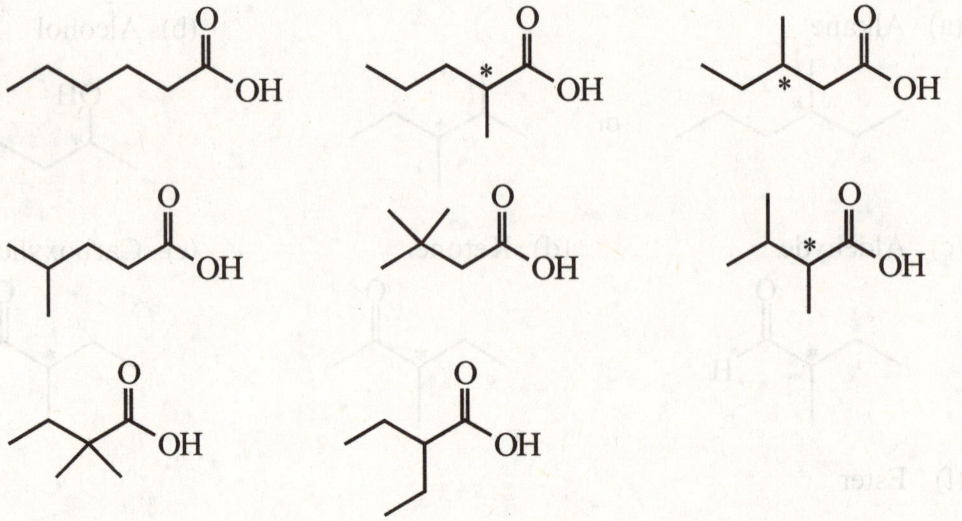

6.19   Draw the enantiomer for each molecule:

(a)

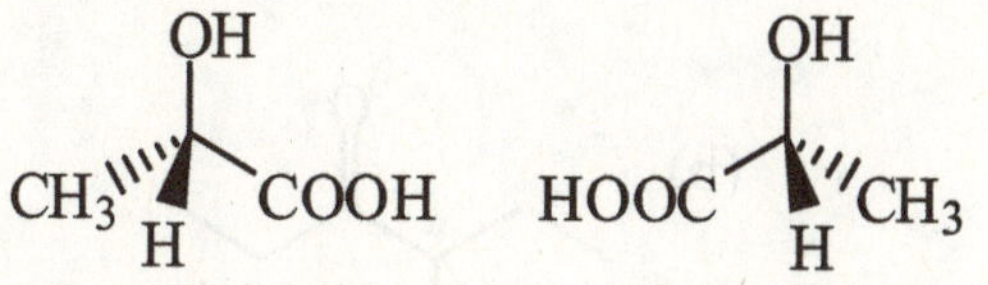

(b)

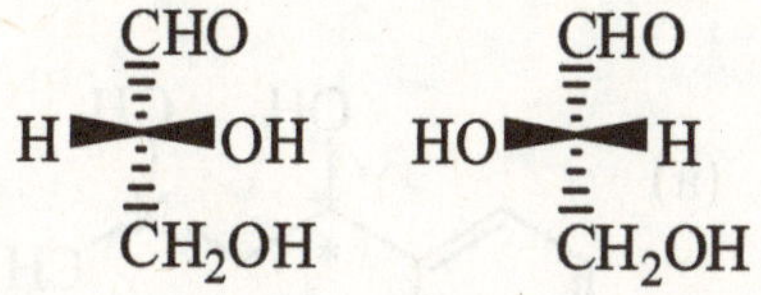

(c)

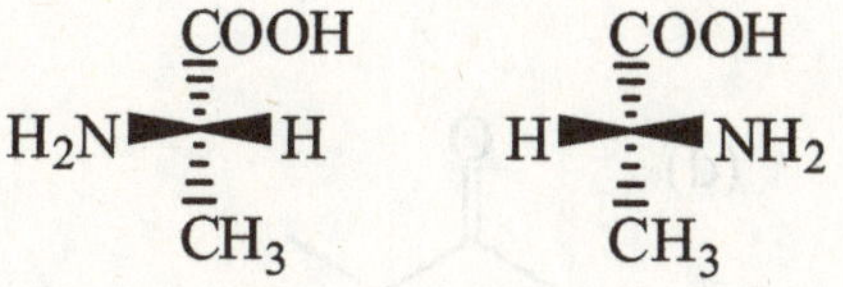

(d)

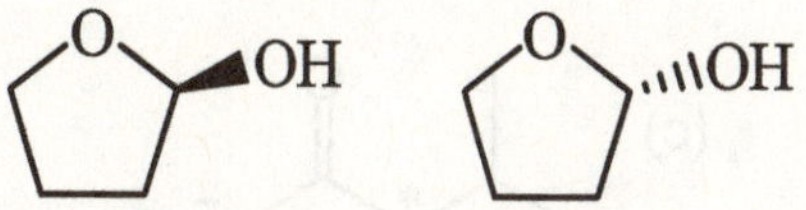

(e)

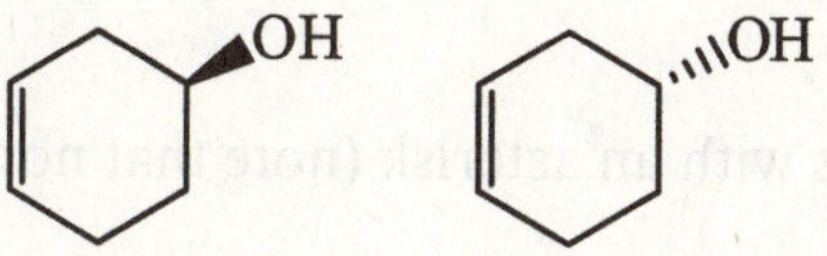

(f)

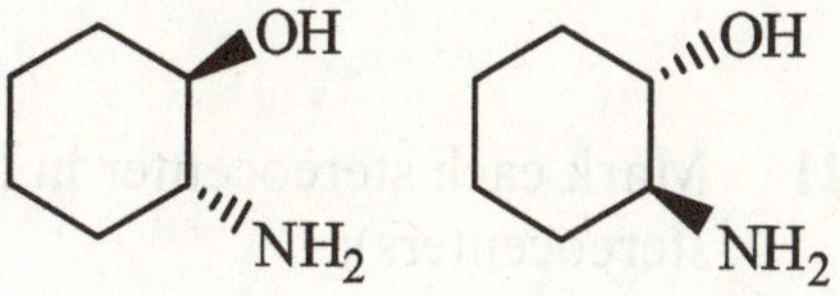

(g)

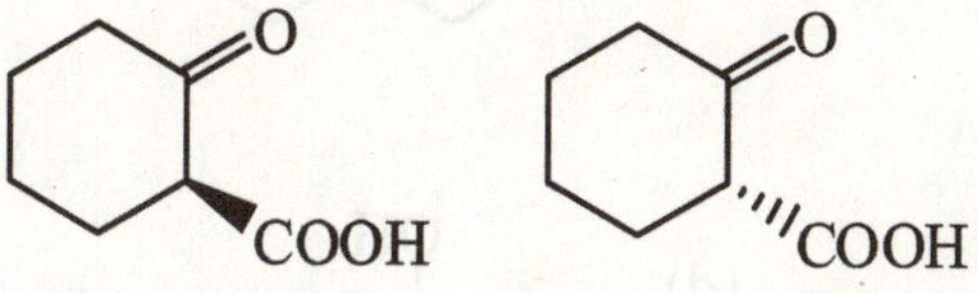

(h)

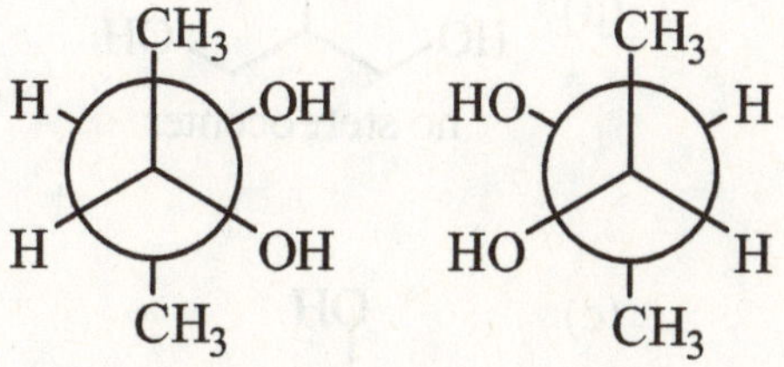

(i)

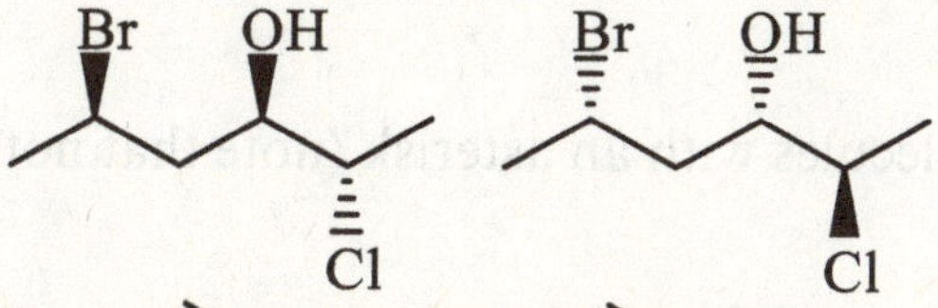

(j)

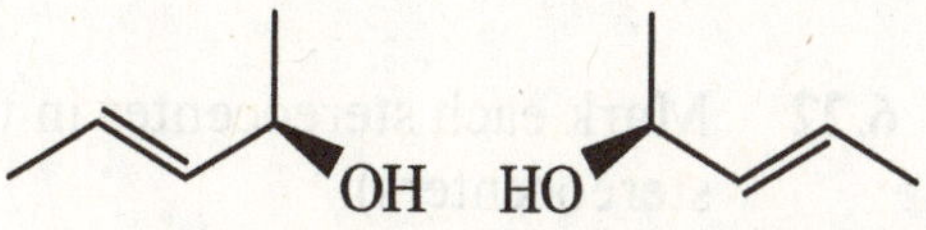

(k)

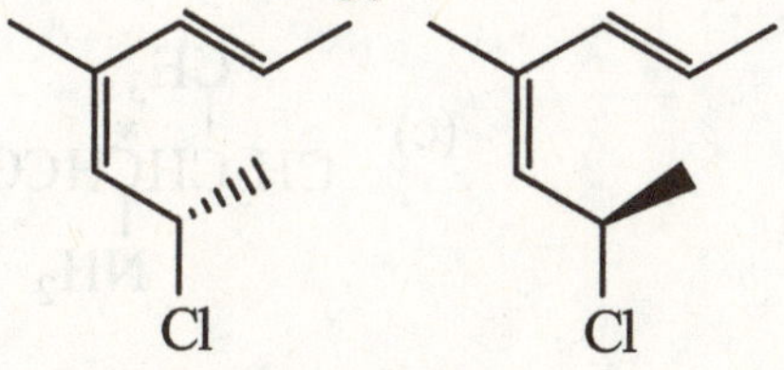

(l)

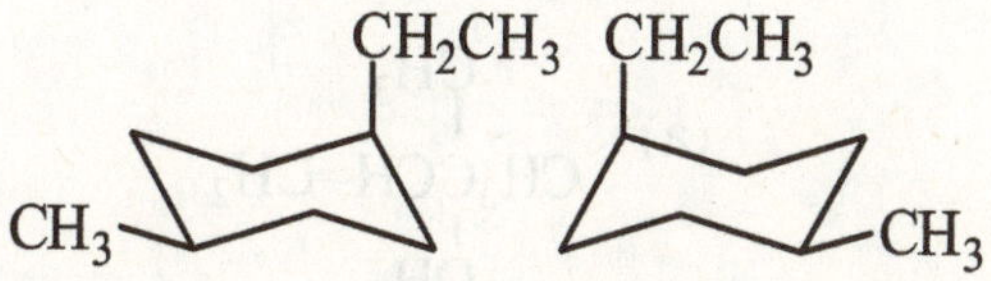

Note that enantiomers can be drawn a variety of different ways. For example, (d):

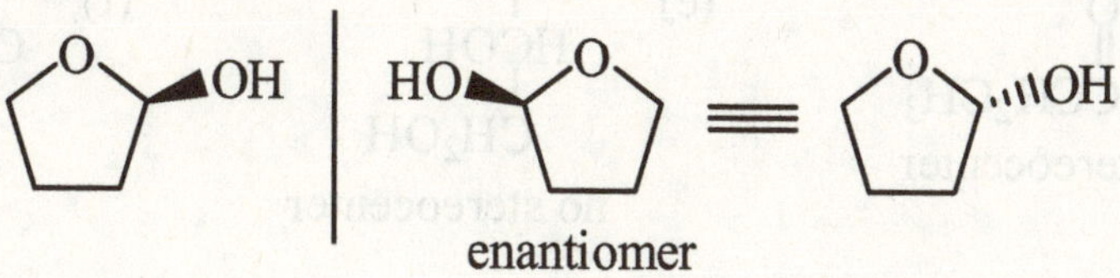

enantiomer

6.20 Mark each stereocenter in these molecules with an asterisk (note that not all contain stereocenters):

(a)

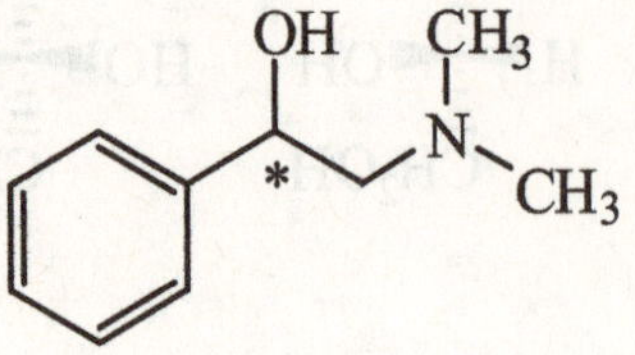

OH   CH₃
| |
   * N
      |
      CH₃

(b)

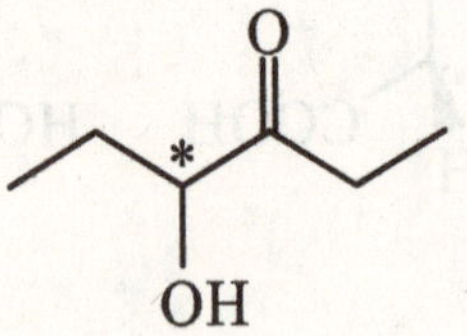

O
||

OH

(c)

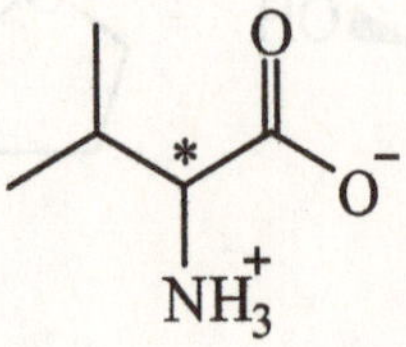

O
||

O⁻

NH₃⁺

(d)

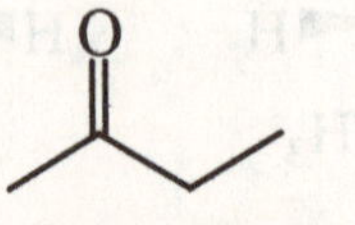

O
||

no stereocenter

6.21 Mark each stereocenter in these molecules with an asterisk (note that not all contain stereocenters):

(a)

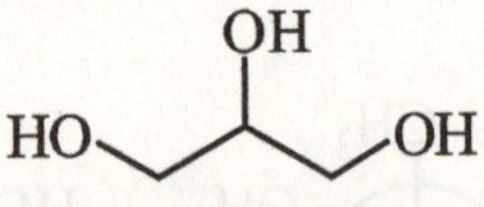

OH
|
HO   OH

no stereocenter

(b)

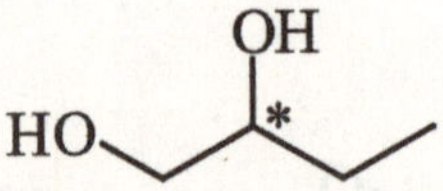

OH
|*
HO

(c)

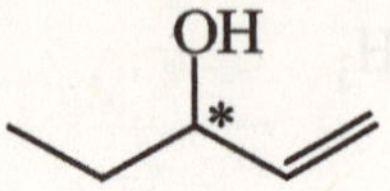

OH
|*

(d)

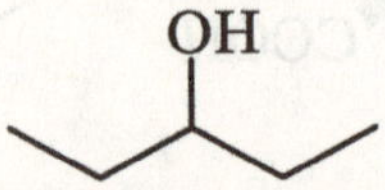

OH
|

no stereocenter

6.22 Mark each stereocenter in these molecules with an asterisk (note that not all contain stereocenters):

(a)

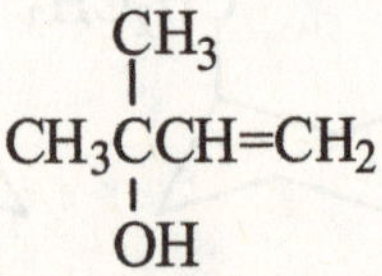

CH₃
|
CH₃CCH=CH₂
|
OH

no stereocenter

(b)

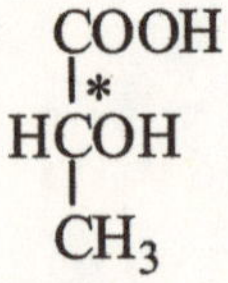

COOH
|*
HCOH
|
CH₃

(c)

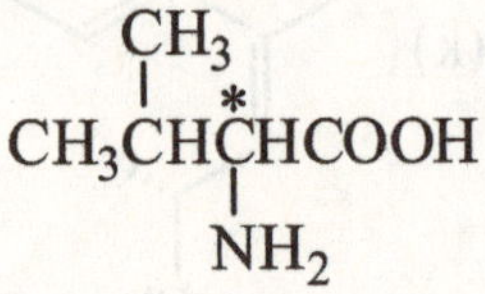

CH₃
|*
CH₃CHCHCOOH
|
NH₂

(d)

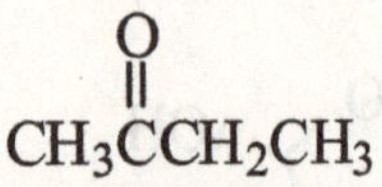

O
||
CH₃CCH₂CH₃

no stereocenter

(e)

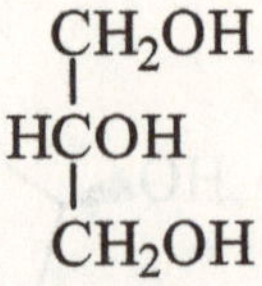

CH₂OH
|
HCOH
|
CH₂OH

no stereocenter

(f)

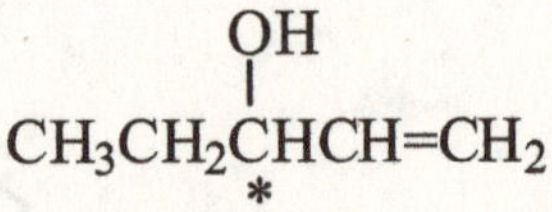

OH
|
CH₃CH₂CHCH=CH₂
*

(g)
$$
\begin{array}{c}
COOH \\
| \\
HOCCOOH \\
| \\
COOH
\end{array}
$$

(h)
$$
\begin{array}{c}
CH_3 \\
| \quad * \\
CH_3CCH_2CHCH_3 \\
| \quad | \\
CH_3 \quad OH
\end{array}
$$

no stereocenter

6.23 Following are eight stereorepresentations of lactic acid. Take (a) as a reference structure. Which stereorepresentations are identical with (a) and which are mirror images of (a)?

(a)
$$
\begin{array}{c}
COOH \\
S \\
H^{\prime\prime\prime}{-}OH \\
CH_3
\end{array}
$$

(b)
$$
\begin{array}{c}
CH_3 \\
S \\
HO^{\prime\prime\prime}{-}H \\
HOOC
\end{array}
$$

(c)
$$
\begin{array}{c}
COOH \\
S \\
HO^{\prime\prime\prime}{-}CH_3 \\
H
\end{array}
$$

(d)
$$
\begin{array}{c}
CH_3 \\
S \\
H^{\prime\prime\prime}{-}COOH \\
HO
\end{array}
$$

(e)
$$
\begin{array}{c}
COOH \\
H{-}R{-}OH \\
CH_3
\end{array}
$$

(f)
$$
\begin{array}{c}
CH_3 \\
H{-}S{-}OH \\
COOH
\end{array}
$$

(g)
$$
\begin{array}{c}
OH \\
CH_3{-}R{-}COOH \\
H
\end{array}
$$

(h)
$$
\begin{array}{c}
CH_3 \\
H{-}R{-}COOH \\
OH
\end{array}
$$

A good way to approach this type of problem is to take advantage of your molecular model kit and let it do some of the thinking! Build a model of (a) and see how it compares to the other structures. Use colored balls to represent each of the four substituents. Structures (b), (c), (d), and (f) are identical to (a). Structures (e), (g), and (h) are mirror images of (a).

It is also completely acceptable to solve this problem by determining the R,S configuration of each structure, but the process is time-consuming.

Designation of Configuration: The R,S Convention

6.24   Assign priorities to the groups in each set:

(a)   —H   —CH₃   —OH   —CH₂OH
       4   3   1   2

(b)   —CH₂CH=CH₂   —CH=CH₂   —CH₃   —CH₂COOH
          3   1   4   2

(c)   —CH₃   —H   —COO⁻   —NH₃⁺
        3   4   2   1

(d)   —CH₃   —CH₂SH   —NH₃⁺   —COO⁻
        4   2   1   3

(e)   —CH(CH₃)₂   —CH=CH₂   —C(CH₃)₃   —C≡CH
          4   3   2   1

6.25   Which molecules have *R* configurations?

(a)
          3
          CH₃
      1   S   4
     Br | '''H
          CH₂OH
          2

#4 already
in the back

(b)
                    4
                    H
               2 S   3
     HOCH₂ | '''CH₃
               Br 1
                ⇧
            look here

#4 in the plane
of the page; look
from the bottom
of the page

(c)
          2
          CH₂OH
     3 R   1
    CH₃ '''Br
          H
          4

#4 pointing towards
you; look from the
back of the page

(d)
                    look here
              1   ⬋
             Br
     4 R   2
     H '''CH₂OH
          CH₃
          3

#4 in the plane
of the page; look
from the top-right
corner of the page

When determining R,S configuration, the group of lowest priority must be pointing
towards the back. However, group #4 is not pointing towards the back in structures
(b), (c), and (d). The best way to solve these structures would be to either use
molecular models or imagine that you are looking at the three-dimensional
structures from a perspective that places group #4 at the back. The latter method
takes practice and requires spatial reasoning, but it works!

6.26 Following are structural formulas for the enantiomers of carvone. Each enantiomer has a distinctive odor characteristic of the source from which it can be isolated. Assign an *R* or *S* configuration to the stereocenter in each. How can they have such different properties when they are so similar in structure?

(−)-Carvone
(Spearmint oil)

(+)-Carvone
(Caraway and dillseed oil)

Recall that in achiral environments, enantiomers have the same physical and chemical properties. However, in a chiral environment, such as that of the chiral odor receptors in our nose, enantiomers behave differently. The chiral odor receptors in our nose are able to differentiate between the two enantiomers.

6.27 Following is a staggered conformation of one of the stereoisomers of 2-butanol:

(a) Is this (*R*)-2-butanol or (*S*)-2-butanol?

(*S*)-2-butanol

(b) Draw a Newman projection for this staggered conformation, viewed along the bond between carbons 2 and 3.

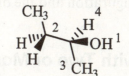 View along bond between carbons 2 and 3

(c) Draw a Newman projection for one more staggered conformation of this molecule. Which of your conformations is the more stable? Assume that −OH and −CH₃ are comparable in size.

There are two more staggered conformations. By assuming that the hydroxyl and methyl groups are of the same size, the Newman projection in part (b) and the one on the far right are of equal stability. The one in the middle is less stable because the three large groups are close to each other, increasing steric strain.

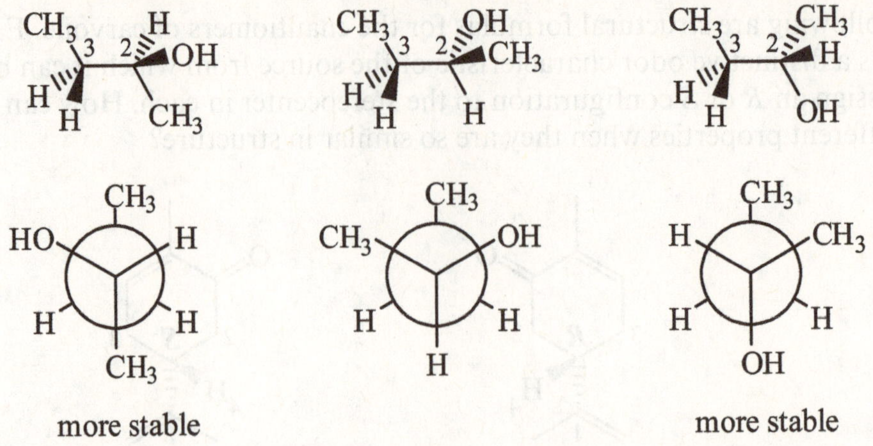

more stable   more stable

(d) Do diastereomers exist for 2-butanol? If so, draw them.

No. In molecules where chirality is caused by the presence of one or more stereocenters, it is necessary to have at least two stereocenters in order to have diasteromers. 2-Butanol only has one stereocenter.

(e) Is this the dextrorotatory or levorotatory form of 2-butanol?

This cannot be predicted, because there is no correlation between R,S configuration and the direction of rotation of plane-polarized light.

Molecules with Two or More Stereocenters

6.28 Write the structural formula of an alcohol with molecular formula $C_6H_{14}O$ that contains two stereocenters.

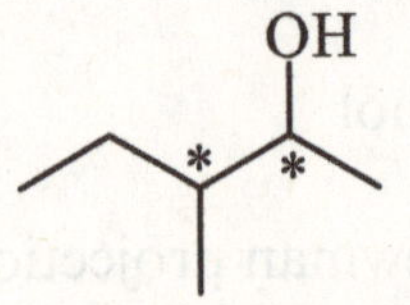

2-Methyl-2-pentanol

6.29 For centuries, Chinese herbal medicine has used extracts of *Ephedra sinica* to treat asthma. Investigation of this plant resulted in the isolation of ephedrine, a potent dilator of the air passages of the lungs. The naturally occurring stereoisomer is levorotatory and has the following structure. Assign R or S configuration to each stereocenter.

When there are multiple stereocenters, treat each stereocenter separately.

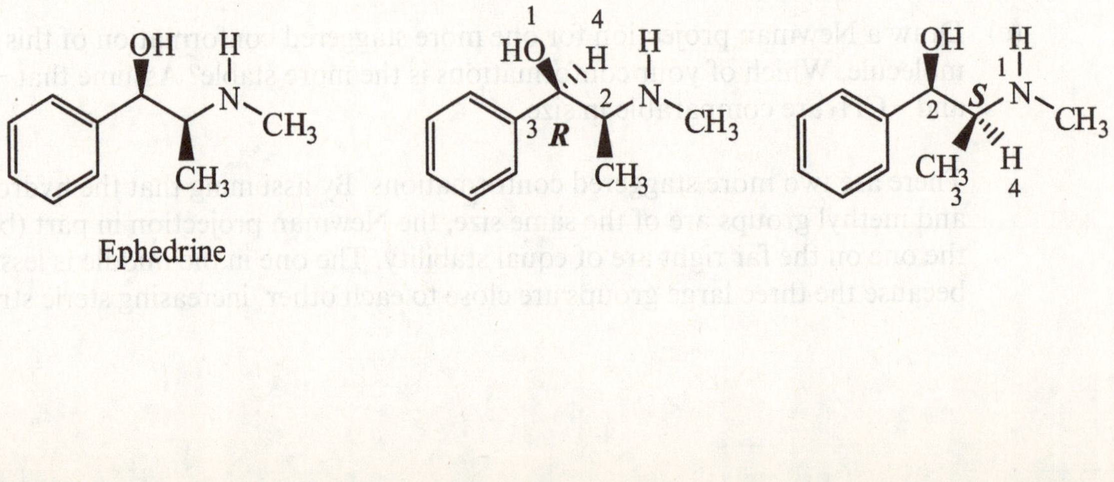

Ephedrine

6.30 The specific rotation of naturally occurring ephedrine, shown in Problem 6.29, is −41°. What is the specific rotation of its enantiomer?

Enantiomers rotate plane-polarized light to the same magnitude but in opposite directions, so the specific rotation of the enantiomer would be +41°.

6.31 Label each stereocenter in these molecules with an asterisk and tell how many stereoisomers are possible for each.

(a) $CH_3\overset{*}{C}H\overset{*}{C}HCOOH$
$\quad\quad\;\; | \;\;\; |$
$\quad\quad\; HO \;\; OH$

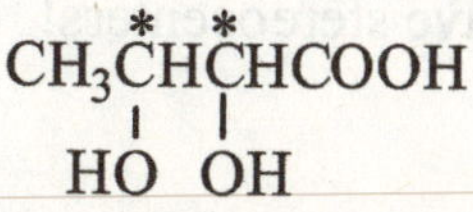

four stereoisomers

(b) CH_2-COOH
$\quad\; |$
$\quad\; \overset{*}{C}H-COOH$
$\quad\quad\; |$
$HO-\overset{*}{C}H-COOH$

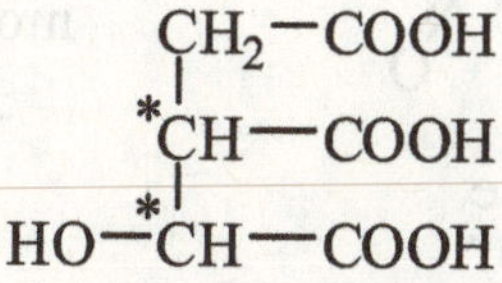

four stereoisomers

(c)

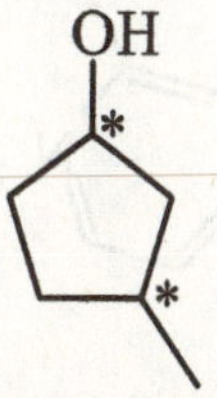

four stereoisomers

(d)

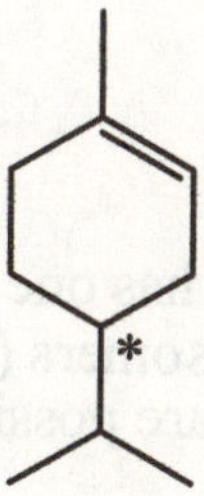

two stereoisomers

(e)

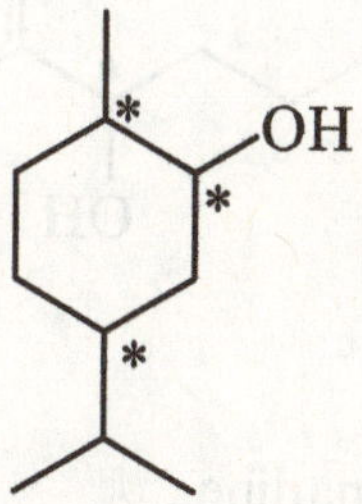

eight stereoisomers

(f)

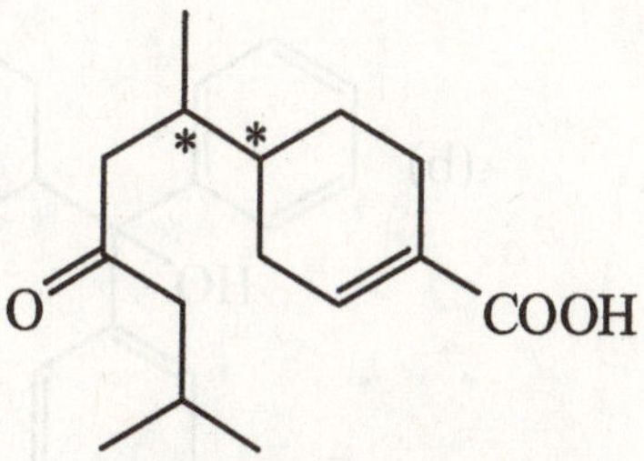

four stereoisomers

(g)

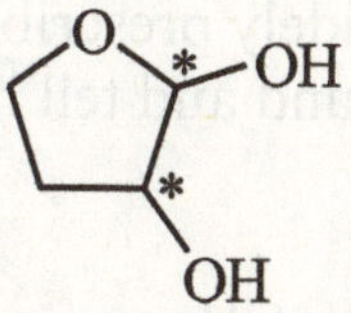

four stereoisomers

(h)

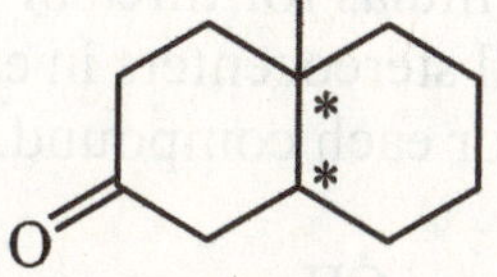

four stereoisomers

6.32 Label the four stereocenters in amoxicillin, which belongs to the family of semisynthetic penicillins:

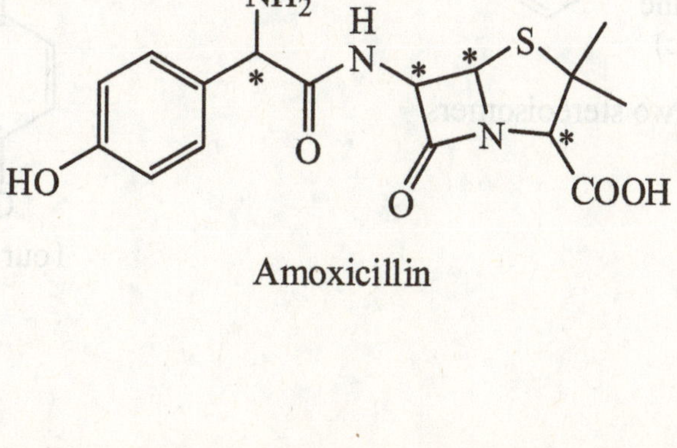

Amoxicillin

6.33   Label all stereocenters in loratadine (Claritin) and fexofenadine (Allegra), now the top-selling antihistamines in the United States. Tell how many stereoisomers are possible for each.

(a)

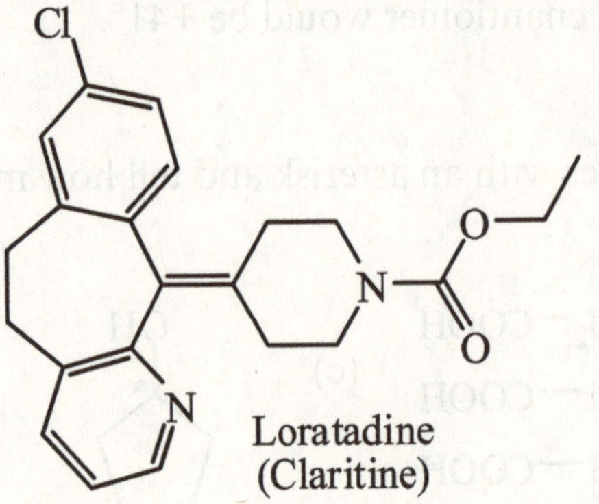

Loratadine
(Claritine)

Loratadine has no stereocenters and is achiral. Not all large molecules have stereocenters!

(b)

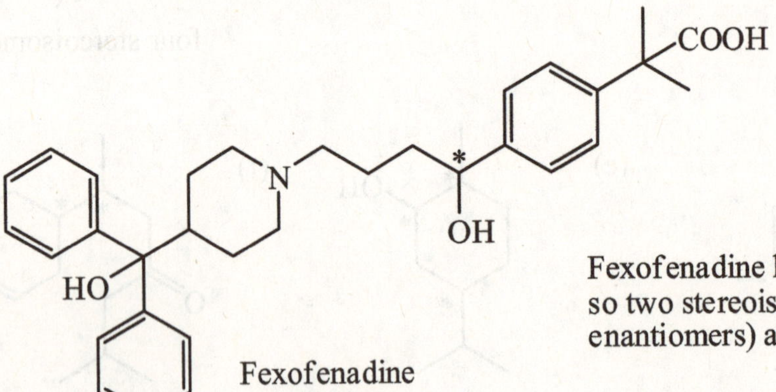

Fexofenadine
(Allegra)

Fexofenadine has one stereocenter, so two stereoisomers (a pair of enantiomers) are possible.

6.34   Following are structural formulas for three of the most widely prescribed drugs used to treat depression. Label all stereocenters in each compound and tell how many stereoisomers are possible for each compound.

(a)

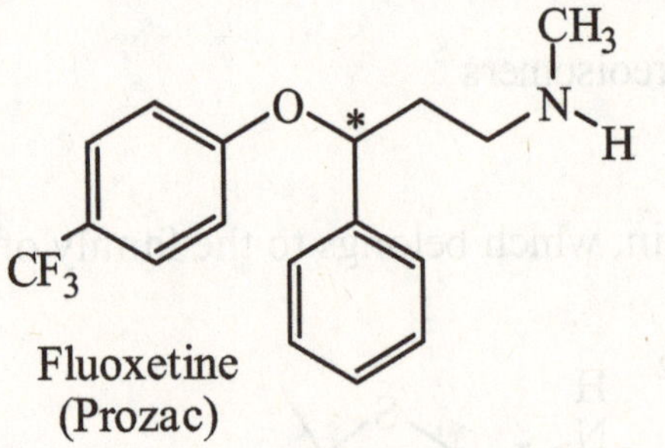

Fluoxetine
(Prozac)

two stereoisomers

(b)

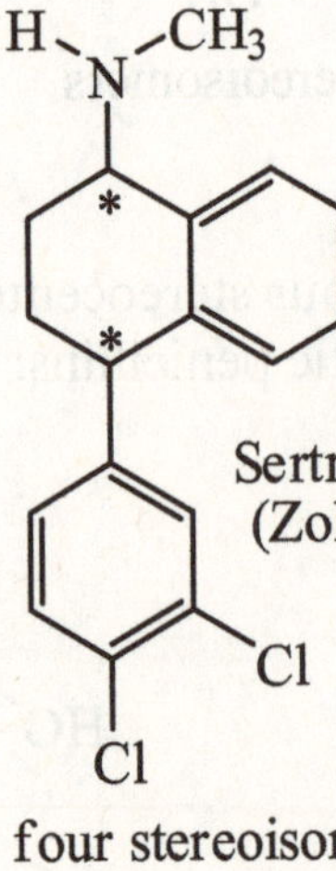

Sertraline
(Zoloft)

four stereoisomers

(c)

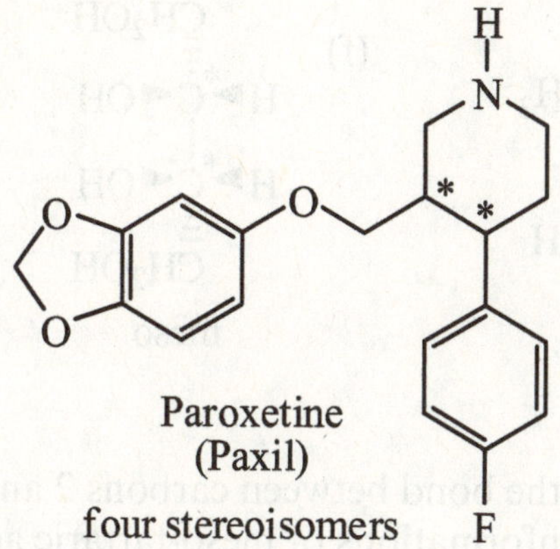

Paroxetine
(Paxil)
four stereoisomers

6.35 Triamcinolone acetonide, the active ingredient in Azmacort Inhalation Aerosol, is a steroid used to treat bronchial asthma:

(a) Label the eight stereocenters in this molecule.

Triamcinolone acetonide

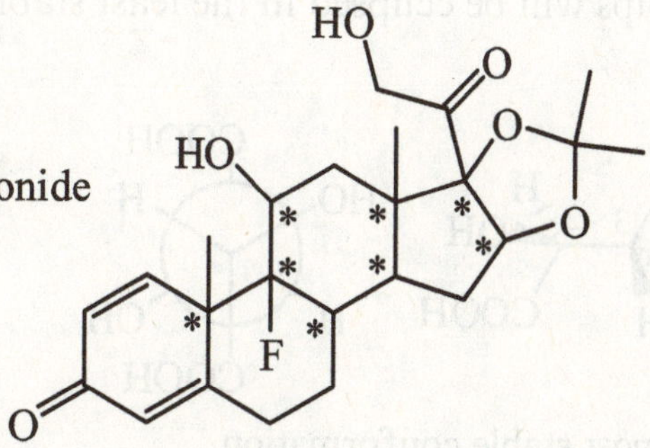

(b) How many stereoisomers are possible for the molecule? (Of this number, only one is the active ingredient in Azmacort.)

There are eight stereocenters, hence $2^8 = 256$ possible stereoisomers.

6.36 Which of these structural formulas represent meso compounds?

A compound is meso if it contains stereocenters *and* has an internal plane of symmetry; both of these conditions must be satisfied.

(a)

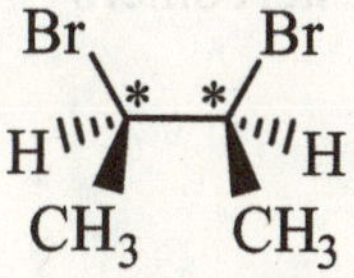

Br   Br
H\\\\   \\\\H
CH₃   CH₃
meso

(b)

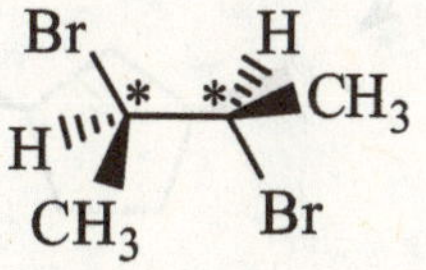

Br   H
H\\\\   \\\\CH₃
CH₃   Br

(c)

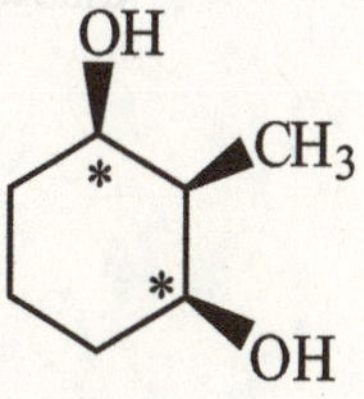

OH

CH₃

OH
meso

(d) meso

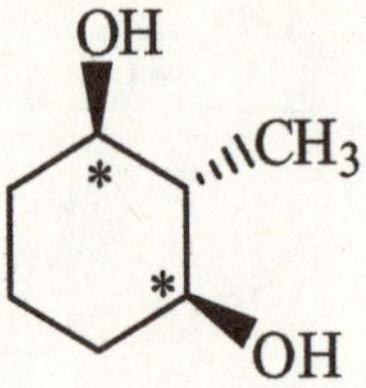

(e)

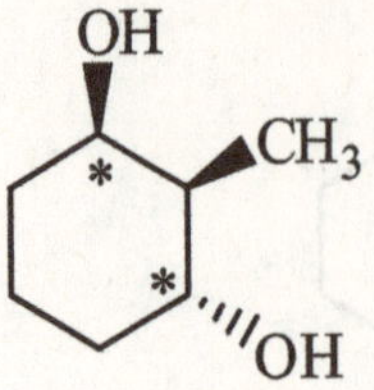

(f) 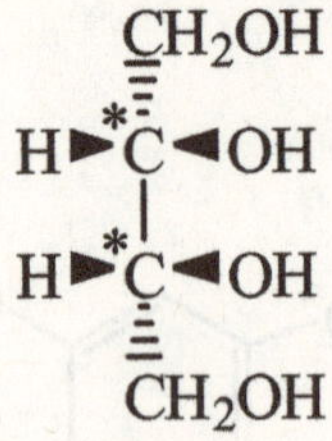 meso

6.37 Draw a Newman projection, viewed along the bond between carbons 2 and 3, for both the most stable and the least stable conformations of meso-tartaric acid:

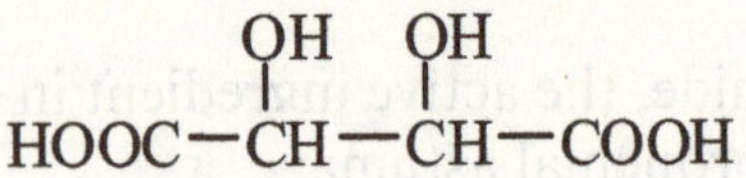

$$HOOC-\overset{\overset{\displaystyle OH}{|}}{CH}-\overset{\overset{\displaystyle OH}{|}}{CH}-COOH$$

Because the carboxyl groups are the largest groups in the molecule, the most stable conformation will have them on sides opposite (anti) to each other. The carboxylic acid groups will be eclipsed in the least stable conformation.

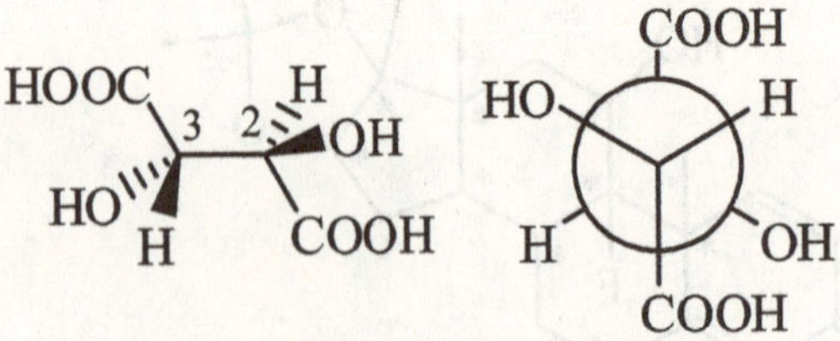

most stable conformation 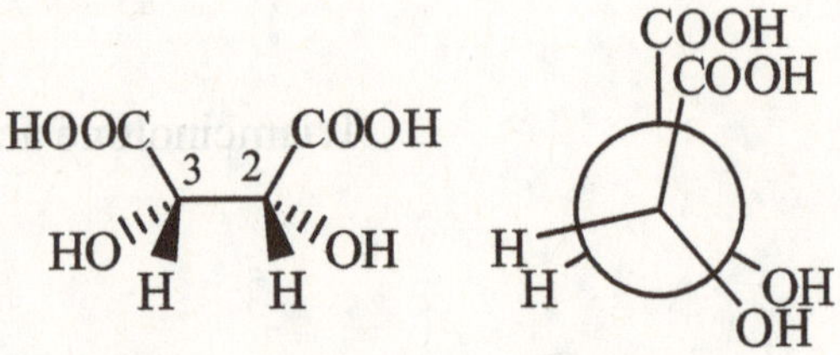   least stable conformation

6.38 How many stereoisomers are possible for 1,3-dimethylcyclopentane? Which are pairs of enantiomers? Which are meso compounds?

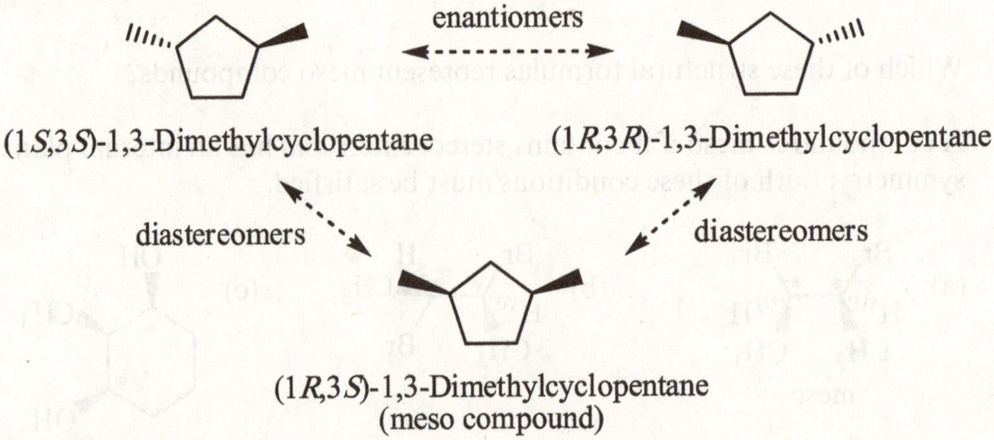

enantiomers

(1*S*,3*S*)-1,3-Dimethylcyclopentane   (1*R*,3*R*)-1,3-Dimethylcyclopentane

diastereomers   diastereomers

(1*R*,3*S*)-1,3-Dimethylcyclopentane
(meso compound)

6.39 In Problem 3.59, you were asked to draw the more stable chair conformation of glucose, a molecule in which all groups on the six-membered ring are equatorial.

(a) Identify all stereocenters in this molecule.

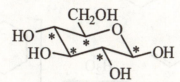

(b) How many stereoisomers are possible for the molecule?

There are $2^5 = 32$ possible stereoisomers.

(c) How many pairs of enantiomers are possible?

Normally, there are half as many pairs of enantiomers as there are number of stereoisomers. So, there are 16 pairs of enantiomers.

(d) What is the configuration (R or S) at carbons 1 and 5 in the stereoisomer shown?

Carbons 1 and 5 both have the R configuration. (To help you determine the configuration, add the axial hydrogens and build a molecular model.)

6.40 What is a racemic mixture? Is a racemic mixture optically active? That is, will it rotate the plane of polarized light?

A racemic mixture is a 50:50 mixture (1:1 mixture) of both enantiomers. Enantiomers rotate plane polarized light to the same magnitude but in opposite directions, so racemic mixtures are optically inactive. The rotation caused by one enantiomer is cancelled by that of the other enantiomer.

Chemical Transformations

6.41 Test your cumulative knowledge of the reactions learned so far by completing the following chemical transformations. Pay particular attention to the stereochemistry in the product. Where more than one stereoisomer is possible, show each stereoisomer. *Note that some transformations will require more than one step.*

(a)

(b)

(c)

(d)

(e)

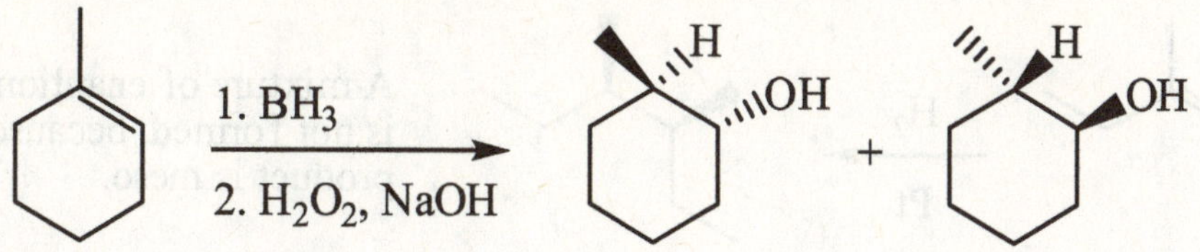

1. BH_3
2. H_2O_2, NaOH

(f)

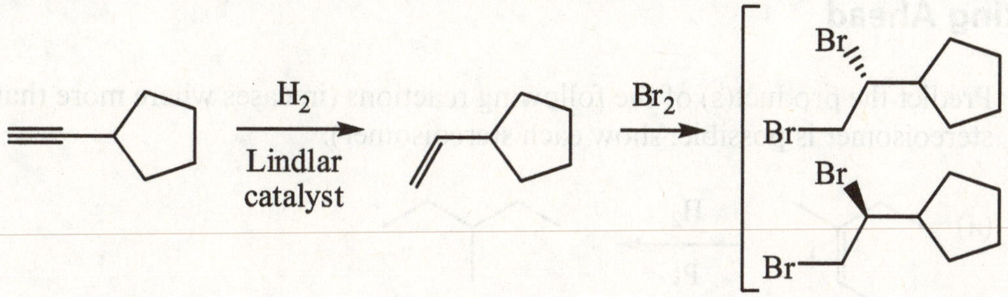

H_2

Lindlar
catalyst

Br_2

(g)

CH_3

HCl

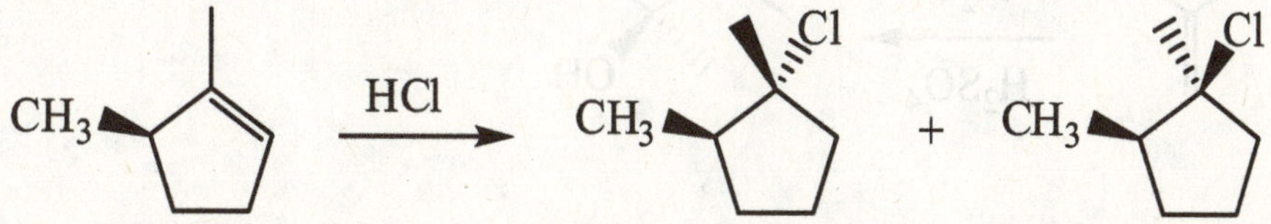

(h)

Cl_2

mixture of *trans*
enantiomers

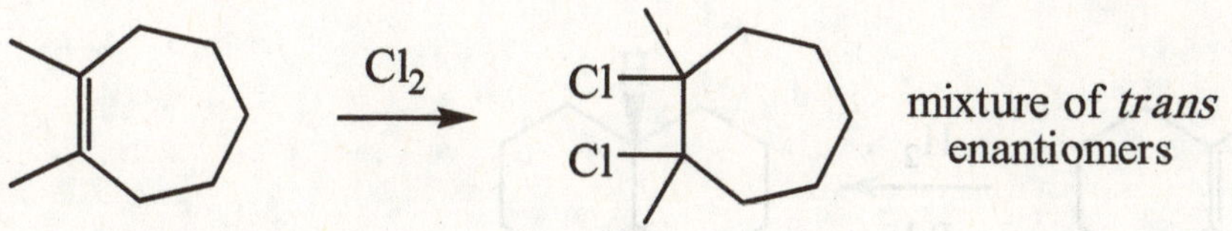

(i)

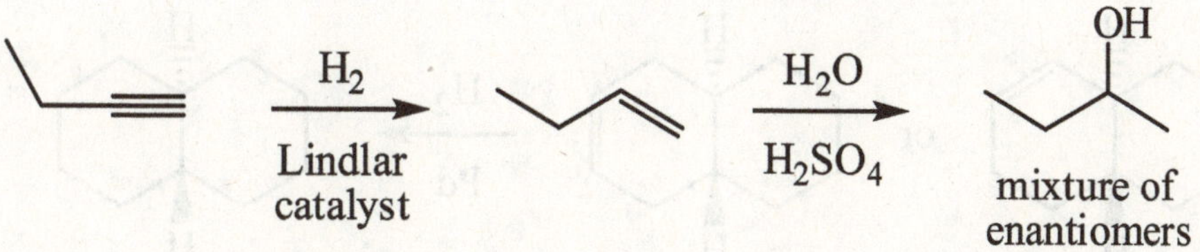

H_2

Lindlar
catalyst

H_2O

H_2SO_4

mixture of
enantiomers

(j) $CH_2=CH_2$

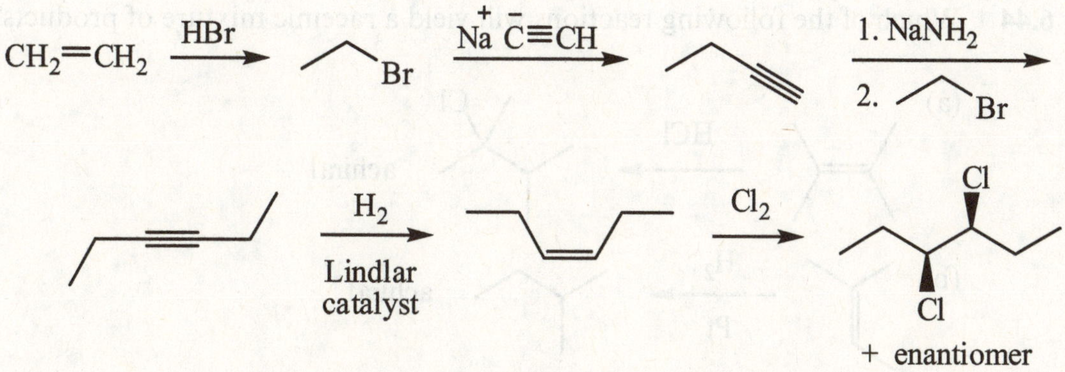

HBr

$\overset{+}{Na}\ \overset{-}{C}\equiv CH$

1. $NaNH_2$

2.

H_2

Lindlar
catalyst

Cl_2

+ enantiomer

(k)

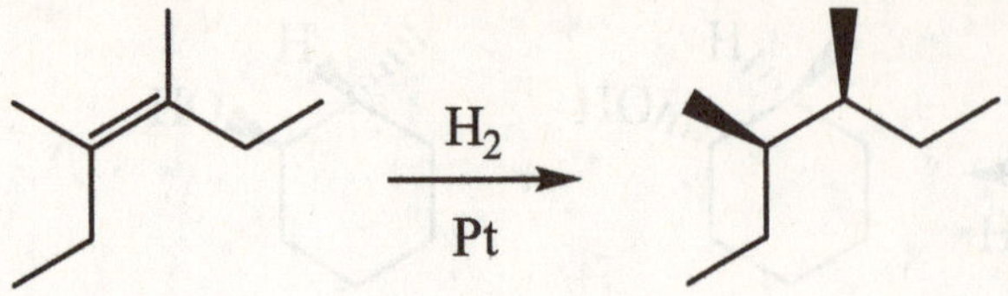

A mixture of enantiomers is not formed, because the product is meso.

Looking Ahead

6.42 Predict the product(s) of the following reactions (in cases where more than one stereoisomer is possible, show each stereoisomer):

(a)

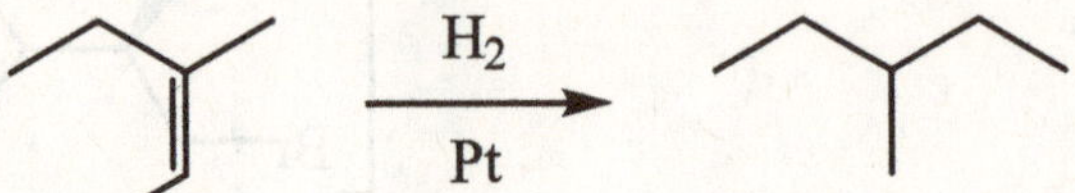

(b)

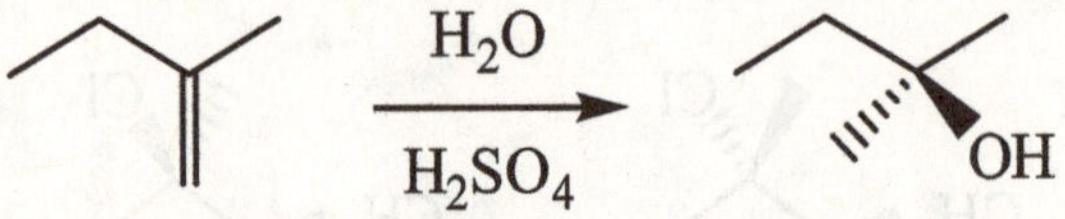

6.43 What alkene, when treated with H_2/Pd, will ensure a 100% yield of the stereoisomer shown?

(a)

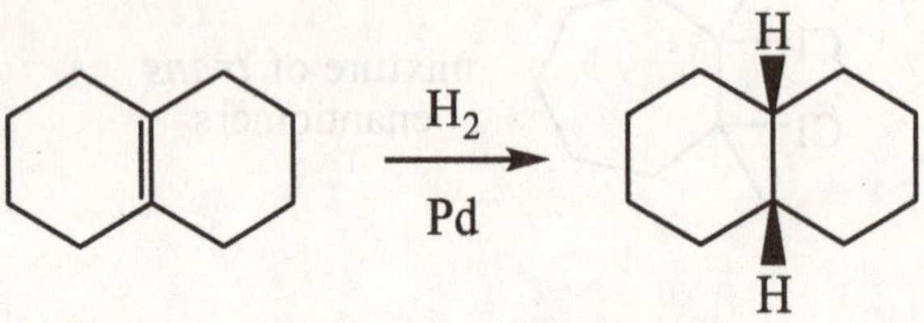

(b)

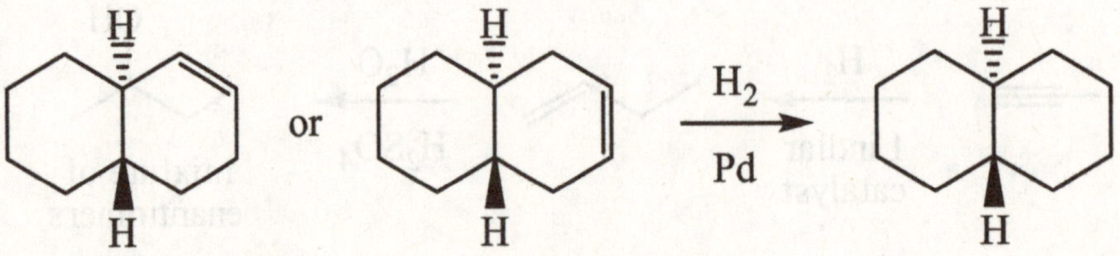

6.44 Which of the following reactions will yield a racemic mixture of products?

(a)

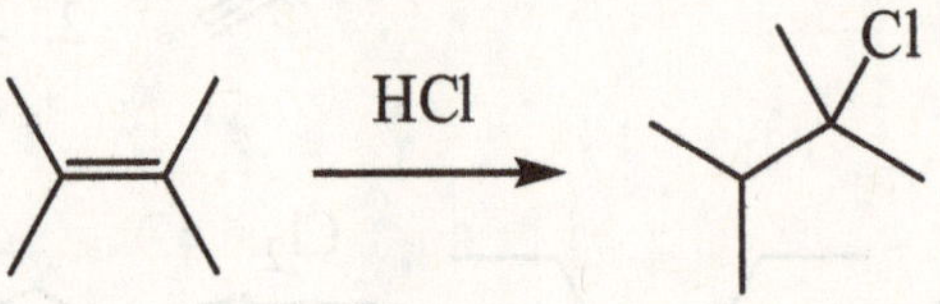

achiral

(b)

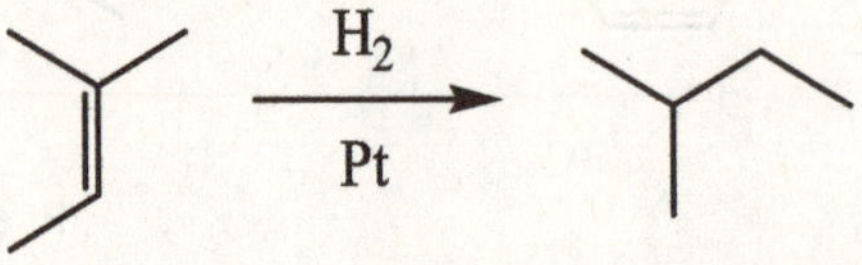

achiral

(c)

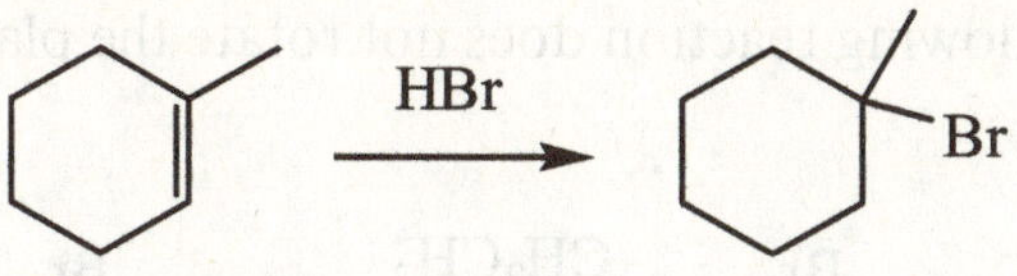

HBr

Br   achiral

(d)

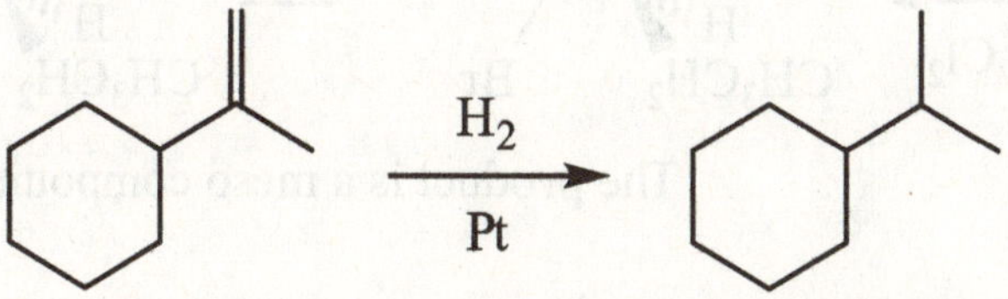

H₂
⟶
Pt

achiral

(e)

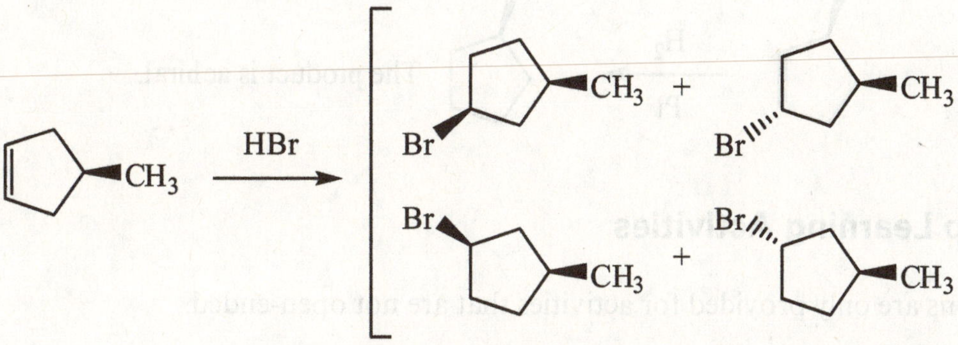

HBr

two sets of enantiomers (two racemic mixtures)

(f)

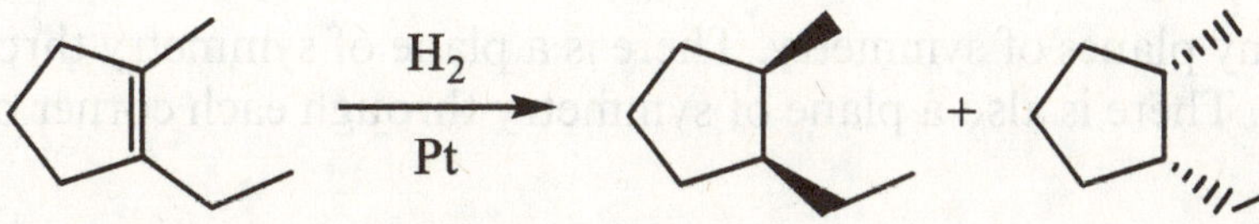

H₂
⟶
Pt

racemic
mixture

6.45 Draw all the stereoisomers that can be formed in the following reaction. Comment
on the utility of this particular reaction as a synthetic method.

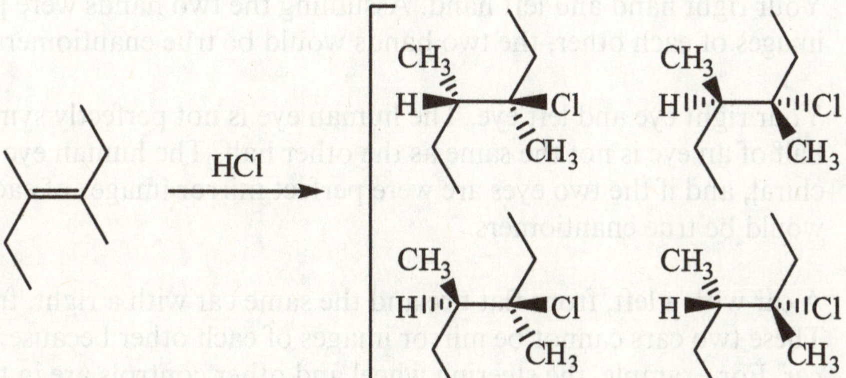

HCl

The reaction produces four different stereoisomers. If the goal of the synthesis is to
selectively synthesize one of the four stereoisomers, this method is not very useful.

6.46 Explain why the product of the following reaction does not rotate the plane of polarized light:

(a)

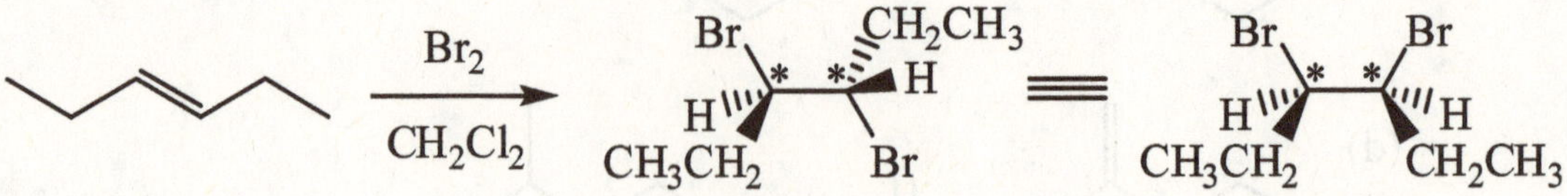

$$\text{(alkene)} \xrightarrow[\text{CH}_2\text{Cl}_2]{\text{Br}_2} \qquad \text{(product)} \equiv \text{(product)}$$

The product is a meso compound.

(b)

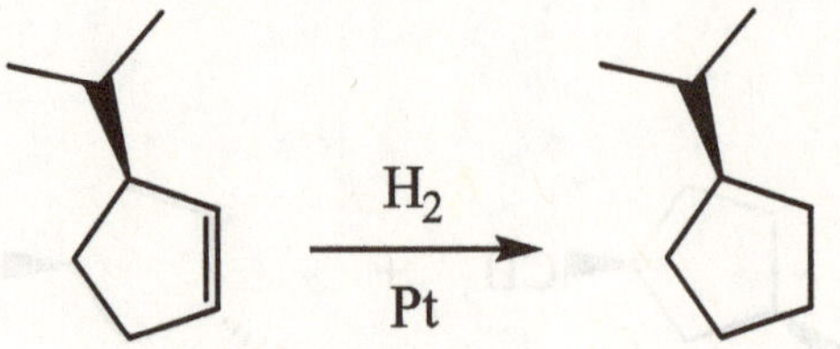

$$\xrightarrow[\text{Pt}]{\text{H}_2}$$

The product is achiral.

Group Learning Activities

Solutions are only provided for activities that are not open-ended.

6.48 Take turns identifying the planes of symmetry in cubane (note: the hydrogen atoms are not shown).

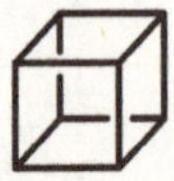

Cubane has many planes of symmetry. There is a plane of symmetry through each face of the cube. There is also a plane of symmetry through each corner of the cube.

6.49 Discuss whether the following pairs of objects are true enantiomers of each other. For those that are not true enantiomers, decide what it would take for them to be true enantiomers.

(a) Your right hand and left hand. Assuming the two hands were perfect mirror images of each other, the two hands would be true enantiomers.

(b) Your right eye and left eye. The human eye is not perfectly symmetrical: one-half of an eye is not the same as the other half. The human eye is therefore chiral, and if the two eyes are were perfect mirror images of each other, they would be true enantiomers.

(c) A car with a left, front flat tire and the same car with a right, front flat tire. These two cars cannot be mirror images of each other because they are the *same car*. For example, the steering wheel and other controls are in the exact same positions and are not mirror images of each other. *Hint:* Draw a front view of each of the two cars.

6.50 Compound **A** (C₅H₈) is not optically active and cannot be resolved. It reacts with Br₂ in CCl₄ to give compound **B** (C₅H₈Br₂). When compound **A** is treated with H₂/Pt, it is converted to compound **C** (C₅H₁₀). When treated with HBr, compound **A** is converted to compound **D** (C₅H₉Br). Given this information, propose structural formulas for **A**, **B**, **C**, and **D**. There are at least three possibilities for compound **A** and, in turn, three possibilities for compounds **B**, **C**, and **D**. As a group, try to come up with all the possibilities.

The first statement provides much useful information about compound **A**. First, it has four fewer hydrogens than a C₅ acyclic alkane, so it must have either two rings or two π bonds, or one of each. Second, it cannot be resolved, so the compound is not a racemic mixture. Finally, because the compound is optically inactive, it must be achiral.

The reactions show that only one mole of Br₂, H₂, or HBr can add to compound **A**. Therefore, the compound must contain exactly one alkene and, in turn, also one ring.

There are a variety of achiral C₅H₈ compounds that contain one alkene and one ring. Each of these compounds could react to form compounds **B**, **C**, and **D**.

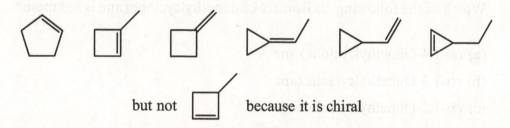

but not ☐ because it is chiral

6.51 In Section 5.2, we learned that a reaction mechanism must be consistent with all experimental observations, and in Section 5.3 we proposed a mechanism for the acid-catalyzed hydration of 1-butene to give 2-butanol. In this chapter, we learned that 2-butanol is chiral, with a maximum of two stereoisomers (a pair of enantiomers) possible. Experimentally, the acid-catalyzed hydration of 1-butene gives a racemic mixture of (R)-2-butanol and (S)-2-butanol. Show, using illustrations, that the formation of a racemic product is consistent with the mechanism proposed in Section 5.3.

$$\diagup\!\!\!\diagdown\!\!\!\diagup \xrightarrow[\text{H}_2\text{O}]{\text{H}_2\text{SO}_4} \text{(racemic)}$$

1-butene 2-butanol

The acid-catalyzed hydration of an alkene involves a carbocation intermediate. Because carbocations are sp^2-hybridized, they have trigonal planar geometry and are flat. When the nucleophile (water) attacks the carbocation, it can do so from either face of the carbocation, leading to the formation of two stereoisomers.

$$\text{CH}_3\text{CH}_2 \overset{+}{\underset{\text{H}}{\diagup\!\!\!\text{\scriptsize{,}}\text{\scriptsize{\textbackslash}}}}\text{CH}_3$$

side view of carbocation

Putting It Together

1. Which of the following will *not* rotate the plane of polarized light?

    (a) A 50:50 ratio of (*R*)-2-butanol and *cis*-2-butene.

    (b) A 70:20 ratio of (*R*)-2-butanol and (*S*)-2-butanol.

    (c) A 50:25:25 ratio of (*S*)-2-butanol, *cis*-2-butene, and *trans*-2-butene.

    (d) A 20:70 ratio of *trans*-2-butene and *cis*-2-butene.

    (e) None of the above (i.e., all of them will rotate plane polarized light)

    cis-trans Isomers of alkenes do not rotate plane polarized light and do not affect the rotation of plane polarized light by other molecules. Mixtures (a) and (b) can be treated as though they contain only (*R*)-2-butanol, and (c) as though it contains only (*S*)-2-butanol.

2. Which of the following *cis* isomers of dimethylcyclohexane is *not* meso?

    (a) *cis*-1,4-Dimethylcyclohexane

    (b) *cis*-1,3-Dimethylcyclohexane

    (c) *cis*-1,2-Dimethylcyclohexane

    (d) All of the above (i.e. none of them is meso)

    (e) None of the above (i.e. all of them are meso)

    Meso compounds are those that contain chiral centers *and* are achiral because they contain an internal plane of symmetry. Although (a) is achiral and has an internal plane of symmetry, it does not have any chiral centers.

3. How many products are possible in the following Lewis acid-base reaction?

    (a) 1   **(b)** 2   (c) 3   (d) 4   (e) None (no reaction wil take place)

    The boron atom is electron-deficient and *sp²*-hybridized. Nucleophilic attack by OH⁻ results in the formation of a new chiral center and a pair of enantiomers. Realize that whenever an *sp²* atom becomes *sp³*, it has the potential to become a stereocenter.

4. What is the relationship between the following two molecules?

(a) They are identical.

(b) They are enantiomers.

(c) They are diastereomers.

(d) They are constitutional isomers.

(e) They are non-isomers.

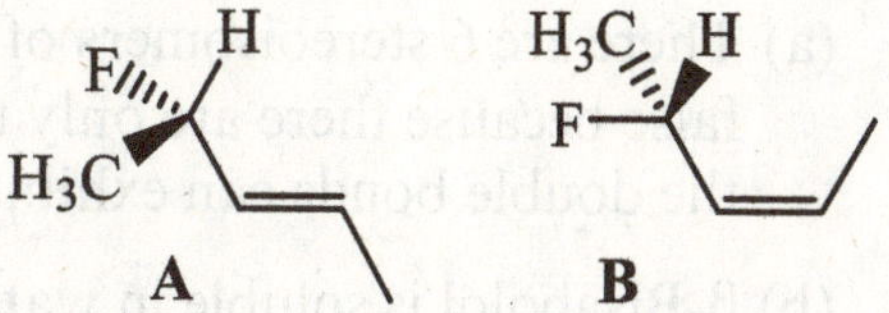

The two compounds have the same molecular formula and the same connectivity, so they are not constitutional isomers or non-isomers (compounds with different formula). The compounds differ in the configuration of both the chiral center and the double bond, so they are not identical. They are not mirror images either. Accordingly, they are best described as diastereomers.

5. Which stereoisomer of 2,4-hexadiene is the *least* stable?

(a) *Z,Z*-2,4-hexadiene   (b) *Z,E*-2,4-hexadiene

(c) *E,Z*-2,4-hexadiene   (d) *E,E*-2,4-hexadiene

(e) All are equal in stability.

Both double bonds in *Z,Z*-2,4-hexadiene are *cis* with respect to the main carbon chain, resulting in high steric hindrance.

6. Select the shortest C-C single bond in the following molecule.

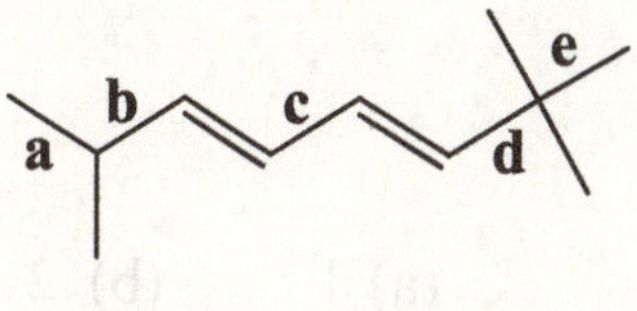

Bond **c** is the shortest bond because it is formed by the overlap of two sp^2-hybridized carbon atoms. Compared to sp^3 orbitals, sp^2 orbitals are closer to the nucleus due to their increased s character, which results in shorter bonds.

7.   Which of the following statements is true of β-bisabolol?

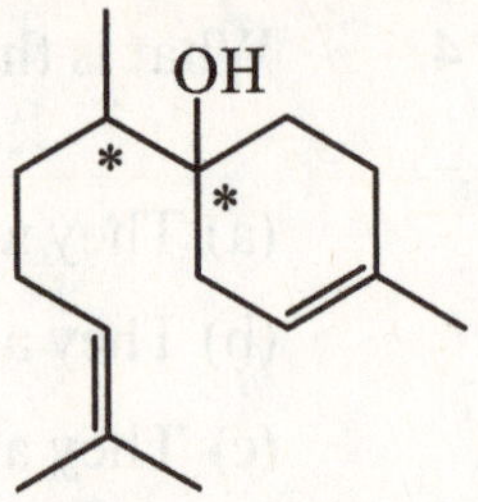

(a) There are 6 stereoisomers of β-bisabolol. This statement is
    false because there are only two chiral centers, and none of
    the double bonds can exhibit *cis-trans* isomerism.

(b) β-Bisabolol is soluble in water. The compound has only one
    hydroxyl group and a large nonpolar region, so it is expected   β-Bisabolol
    to be water-insoluble.

(c) β-Bisabolol is achiral. The compound has two chiral centers, and it is not
    possible to have a combination of stereocenter configurations that leads to a
    compound with an internal plane of symmetry. That is, a meso stereoismer is not
    possible.

(d) β-Bisabolol has a meso stereoisomer.

(e) None of the above.

8.   How many products are formed in the following reaction?

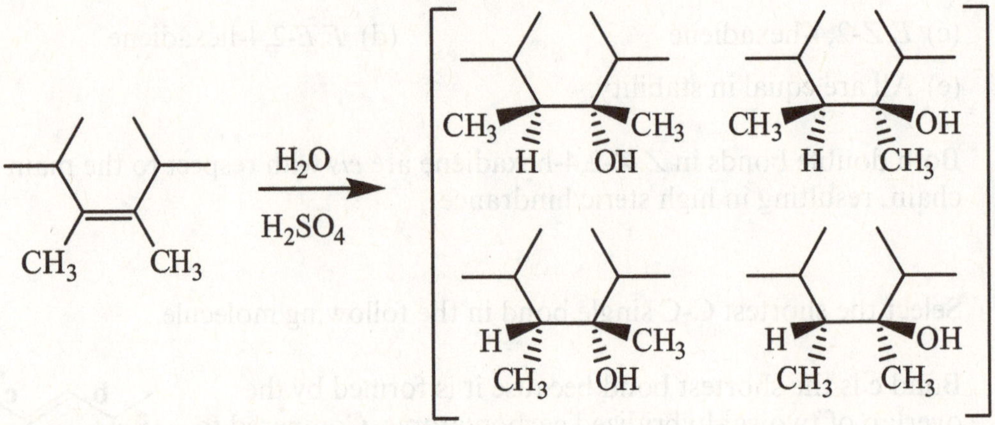

(a) 1   (b) 2   (c) 3   **(d) 4**   (e) 5

The reaction is a hydration reaction, which involves protonation of the alkene,
generation of a carbocation, and nucleophilic attack by water. Because the left-half
of the alkene is same as the right-half, it does not matter which alkene carbon is
protonated. However, we need to consider that protonation could occur from either
face (top or bottom) of the alkene. Likewise, the nucleophlic attack of the
carbocation could also occur from either face, regardless of the face of protonation
(the addition of water is not stereoselective). As a result, four products are formed.

9.   Which of the following is *true* when two isomeric alkenes are treated with H_2 / Pt?

(a) The alkene that releases more energy in the reaction is the more stable alkene.

(b) The alkene with the lower melting point will release less energy in the reaction.

(c) The alkene with the lower boiling point will release less energy in the reaction.

(d) Both alkenes will release equal amounts of energy in the reaction.

(e) None of these statements is true.

The amount of energy releaed in a hydrogenation reaction depends on the difference in energy between the alkene and the corresponding alkane product. Therefore, the alkene that is less stable will release more energy upon hydrogenation; the opposite of statement (a) would be correct. There is no correlation between the amount of energy released in a hydrogenation reaction and the melting or boiling point of the alkene.

10.   An unknown compound reacts with two equivalents of H_2 catalyzed by Ni. The unknown also yields 5 CO_2 and 4 H_2O upon combustion. Which of the following could be the unknown compound?

(a)    (b) 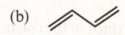   (c)

(d)   (e) 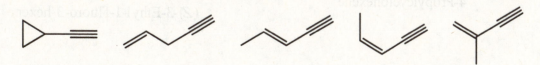   (f) $CH_3CH=CHCH_2CH_3$

The data indicates that the compound contains two π bonds (either two alkenes or one alkyne) and has the molecular formula C_5H_8. Only (d) matches this data. Note that the ring of a cycloalkane is not "opened" by catalytic hydrogenation.

11.   Provide structures for all possible compounds of formula C_5H_6 that would react quantitatively with $NaNH_2$.

Hydrocarbons that react with $NaNH_2$ are terminal alkynes. The molecular formula C_5H_6 corresponds to a compound that has six fewer hydrogens than a fully saturated C_5 alkane (C_5H_{12}), so C_5H_6 must have a combined total of three π bonds and rings.

12. Answer the questions that follow regarding the following compound, which has been found in herbal preparations of *Echinacea*, the genus name for a variety of plants that are marketed for their immunostimulant properties.

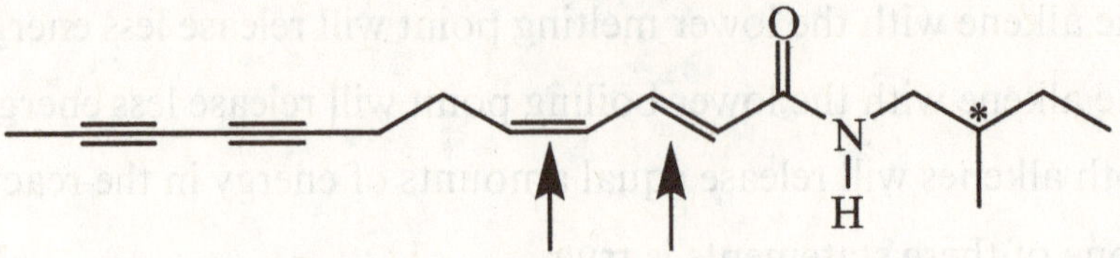

(a) How many stereoisomers exist for the compound shown?

The compound contains two double bonds that can exhibit *cis-trans* isomerism and one chiral center. There are $2^3 = 8$ possible stereoisomers.

(b) Would you expect the compound to be soluble in water?

The compound is not expected to be water-soluble. It has large, nonpolar hydrocarbon groups and only one relatively small polar amide group.

(c) Is the molecule chiral?

Yes. It contains one chiral center.

(d) What would be the product formed in the reaction of this compound with an excess amount of H_2 / Pt.

All alkenes and alkynes would be reduced to alkanes. However, the carbonyl group (C=O) of the amide would not be reduced by catalytic hydrogenation.

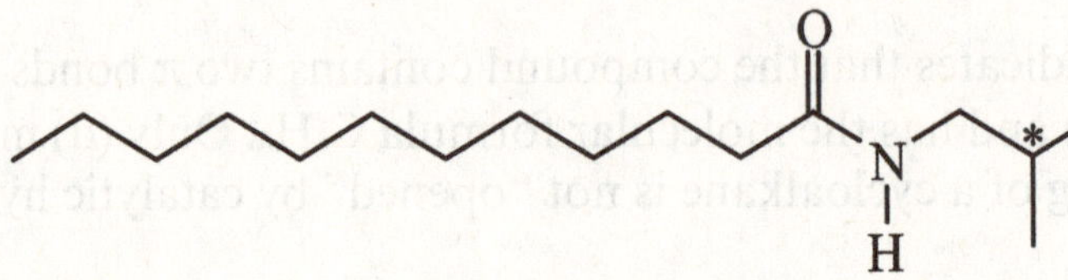

13. Provide IUPAC names for the following compounds.

(a)

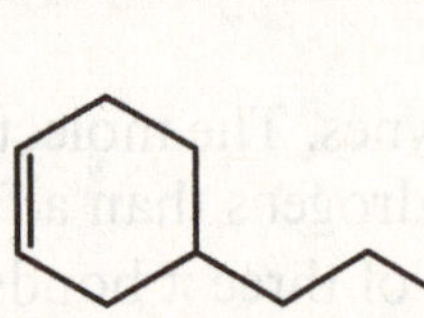

4-Propylcyclohexene

(b)

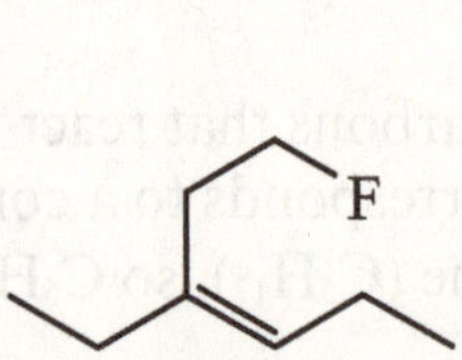

(*Z*)-3-Ethyl-1-fluoro-3-hexene

(c)

(S)-4-Butyl-7-methyl-2-octyne

(d)

(1S,2S)-1,2-Dichloro-
1-fluorocyclohexane

14. Compound **A** is an optically inactive compound with a molecular formula of C_5H_8. Catalytic hydrogenation of **A** gives an optically inactive compound, **B** (C_5H_{10}), as the sole product. Furthermore, reaction of **A** with HBr results in a single compound, **C**, with a molecular formula of C_5H_9Br. Provide structures for **A**, **B**, and **C**.

Compound **A** is achiral and its molecular formula has four fewer hydrogens than a C_5 alkane (C_5H_{12}). Because the catalytic hydrogenation of **A** forms only C_5H_{10} (**B**), it can be concluded that **A** contains one π bond and one ring. The presence of only one π bond is further suggested its reaction with only one equivalent of HBr, forming **C**. Compound **C** is a single compound, so the reaction of **A** with HBr must not give a mixture of constitutional isomers or stereoisomers.

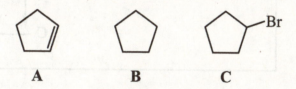

A   **B**   **C**

15. An optically active compound, **A**, has a molecular formula of C_6H_{12}. Hydroboration-oxidation of **A** yields an optically active product, **B**, with a molecular formula of $C_6H_{14}O$. Catalytic hydrogenation of **A** yields an optically inactive product, **C**, with a molecular formula of C_6H_{14}. Propose structures for **A**, **B**, and **C**.

Compound **A** is chiral and its formula has two fewer hydrogens than a C_6 alkane (C_6H_{12}). The reactivity of **A** towards hydroboration-oxidation and hydrogenation is attributed to the presence of a carbon-carbon π bond. The only chiral alkene that becomes achiral upon reduction is 2-methylpentene.

A   **B**   **C**

16. Based on the following hydrogenation data, which is more stable, the alkene (**A**) with the double bond outside of the ring or the alkene (**B**) with the double bond inside the ring? Use a reaction energy diagram to illustrate your point.

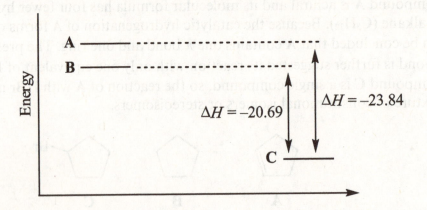

$\Delta H = -23.84$ kcal/mol $\Delta H = -20.69$ kcal/mol

The same product (**C**) is formed by the reduction of the two alkenes, but the reduction of **A** is more exothermic (releases more energy) than the reduction of **B**. That is, **A** is higher in energy (less stable) than **B** to start with. This is consistent with the fact alkenes that are more substituted are more stable.

17. Briefly explain how you would separate the following pairs of compounds. If separation is not possible, indicate so and explain your answer.

(a)

The compounds are enantiomers therefore and have the same physical properties. It is necessary to use a chiral agent to separate them.

(b)

You can always *try* to separate them, but you won't get anywhere! They may appear to be look like enantiomers, but they are identical (the compound is meso).

(c)

The two compounds are *cis-trans* isomers and have different properties. For instance, they can be separated by distillation.

18. Predict whether solutions containing equal amounts of each pair of the structures shown would rotate the plane of polarized light (PPL).

(a)

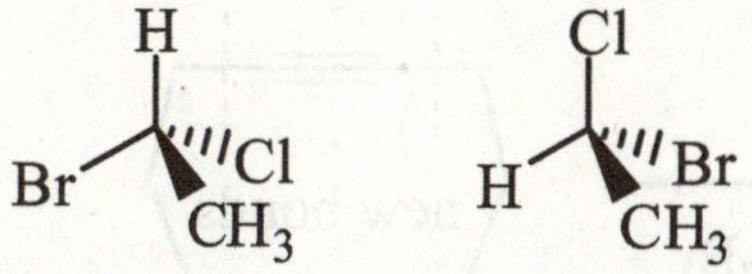

The two structures are identical (both are *S*), so the solution would be enantiomerically pure and will rotate PPL.

(b)

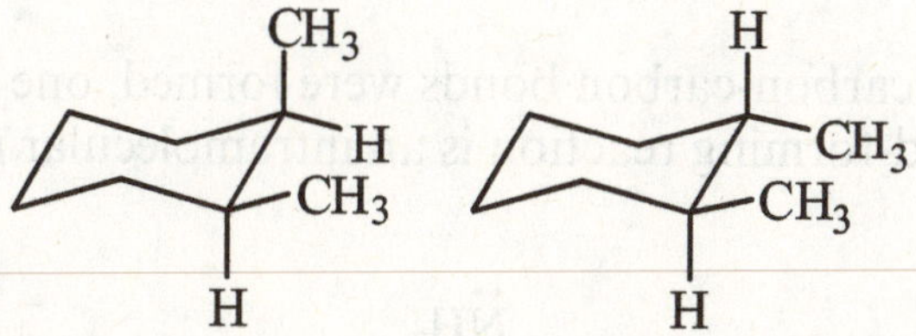

The compound on the left is meso and does not rotate PPL. The compound on the right is chiral and rotates PPL. As a whole, the solution rotates PPL.

(c)

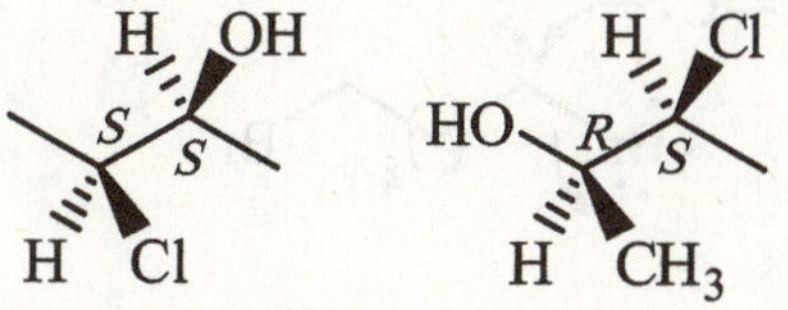

The compounds are diastereomers, which do not have equal and opposite optical activites. Overall, the solution rotates PPL.

19. Complete the following chemical transformations.

(a)

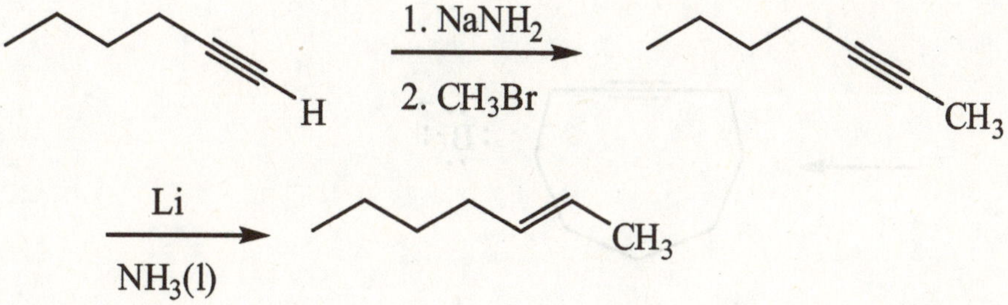

(b)

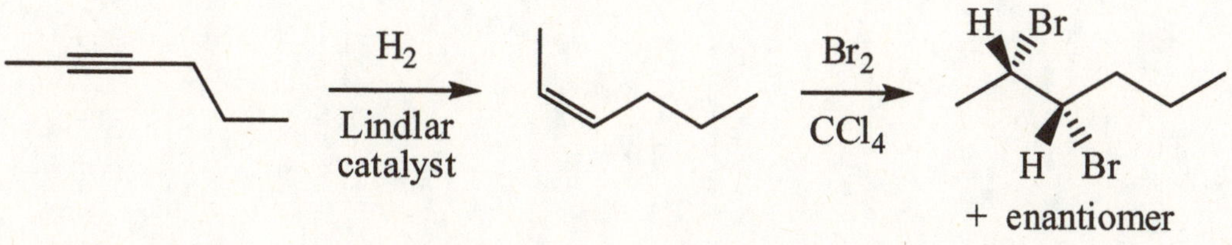

(c)

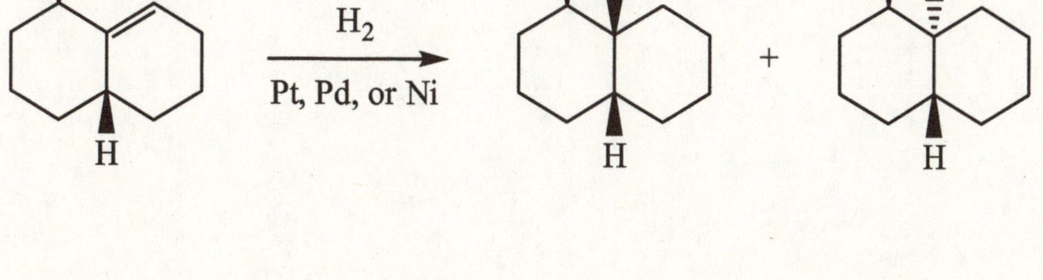

20. Provide a mechanism for the following series of reactions. Show all charges and lone pairs of electrons in your structures as well as the structures of all intermediates.

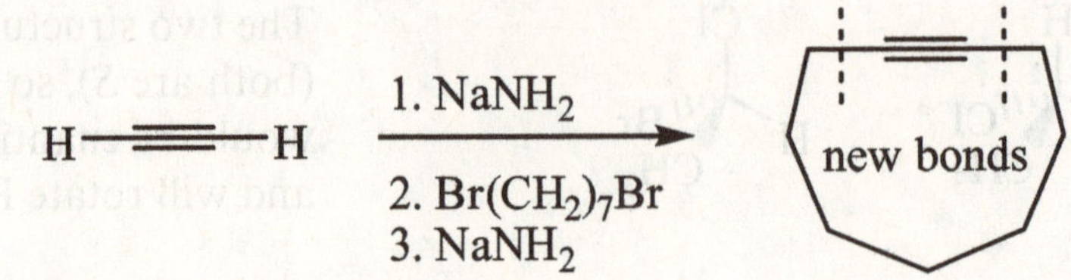

$$H \!-\!\!\equiv\!\!-\! H \quad \xrightarrow[\substack{\text{2. Br(CH}_2)_7\text{Br} \\ \text{3. NaNH}_2}]{\text{1. NaNH}_2} \quad \boxed{\text{new bonds}}$$

The reaction shows that two new carbon–carbon bonds were formed, one at each end of acetylene. The second bond-forming reaction is an intramolecular reaction.

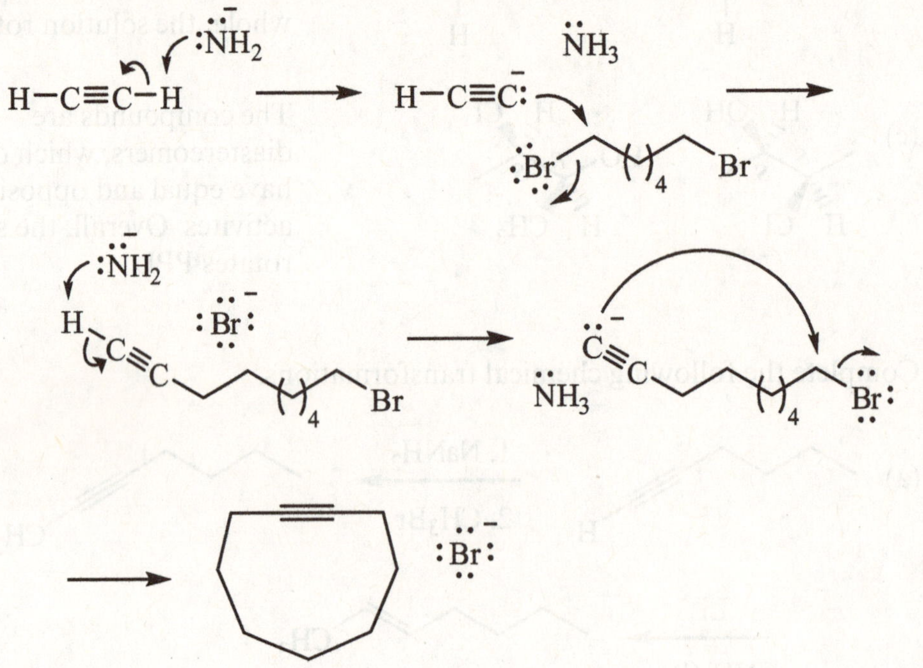

21. Predict the major product or products of each of the following reactions. Be sure to consider stereochemistry in your answers.

(a)

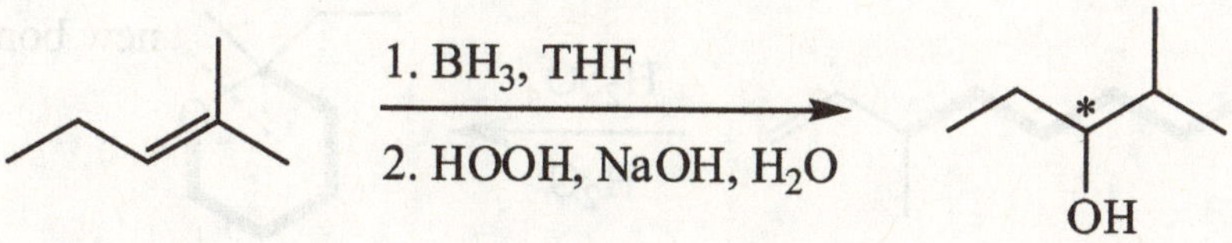

1. BH_3, THF

2. HOOH, NaOH, H_2O

racemic mixture

(b)

H CH_2CH_3

H_2

Pt

H CH_2CH_3 H

H CH_2CH_3 H

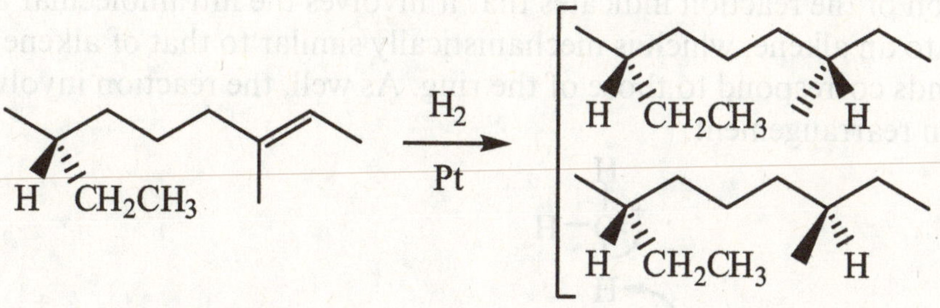

mixture of diastereomers

(c)

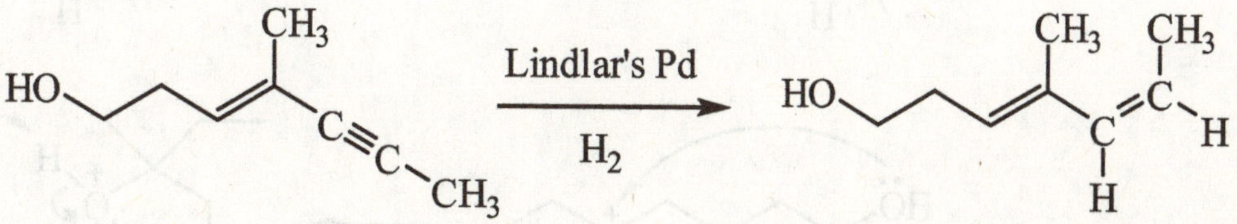

HO

CH_3

C≡C

CH_3

Lindlar's Pd

H_2

HO

CH_3 CH_3

C

H

H

(d)

HI

I

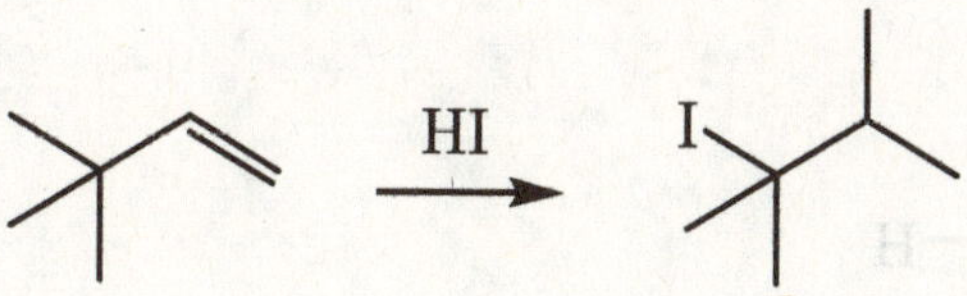

1,2-alkyl shift occurs to form a more stable carboation

(e)

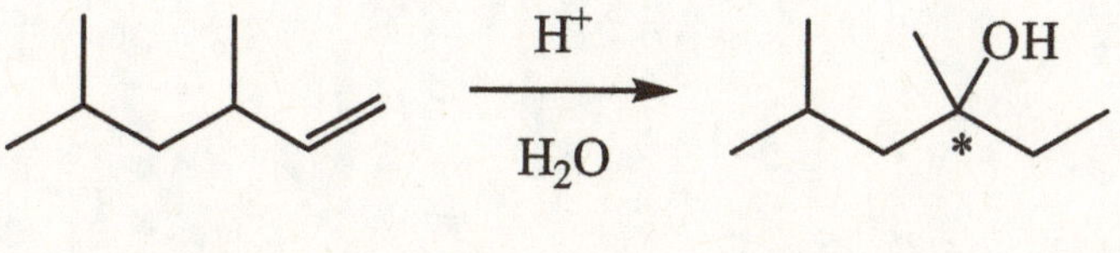

H^+

H_2O

OH

racemic mixture

(f)

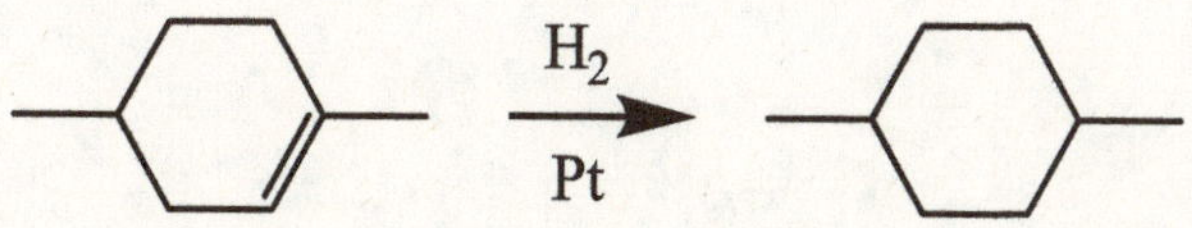

H_2

Pt

mixture of *cis-trans* isomers

22.   Provide a mechanism for the following reaction. Show all charges and lone pairs of electrons in your structures as well as the structures of all intermediates.

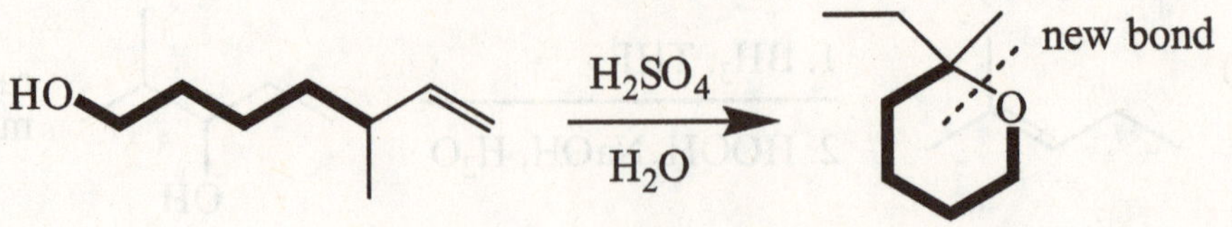

Examination of the reaction indicates that it involves the intramolecular addition of an alcohol to an alkene, which is mechanistically similar to that of alkene hydration. Bolded bonds correspond to those of the ring. As well, the reaction involves a carbocation rearrangement.

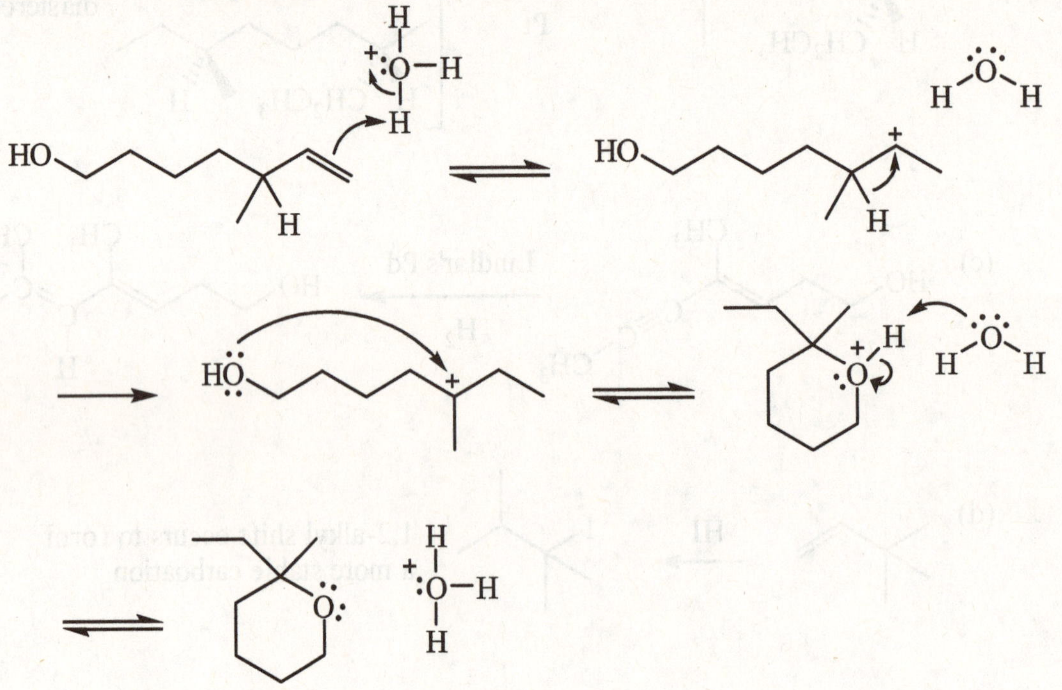

Chapter 7: Haloalkanes

Problems

7.1   Write the IUPAC name for each compound:

(a)

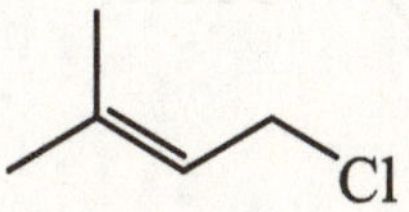

1-Chloro-3-
methyl-2-butene

(b)

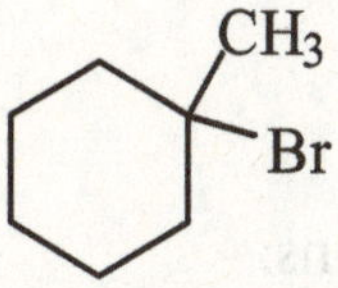

1-Bromo-1-
methylcyclohexane

(c)

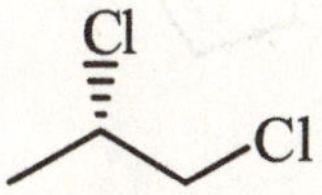

(S)-1,2-
Dichloropropane

(d)

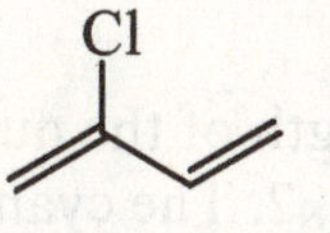

2-Chloro-1,3-butadiene

(e)

(S)-3-Chloro-3-methyl-1-pentene

7.2   Determine whether the following haloalkanes underwent substitution, elimination, or both substitution and elimination:

(a)

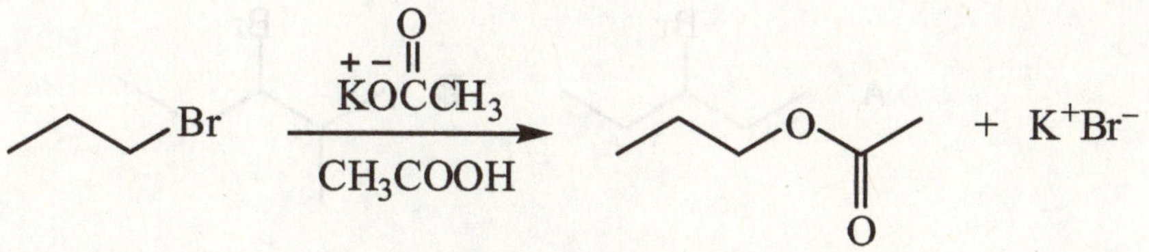

$$\text{CH}_3\text{CH}_2\text{CH}_2\text{Br} \xrightarrow[\text{CH}_3\text{COOH}]{\overset{\text{O}}{\overset{\|}{\text{KOCCH}_3}}} \text{product} + \text{K}^+\text{Br}^-$$

In this reaction, the haloalkane underwent substitution. The bromine was replaced by an acetate (CH₃COO−) group.

(b)

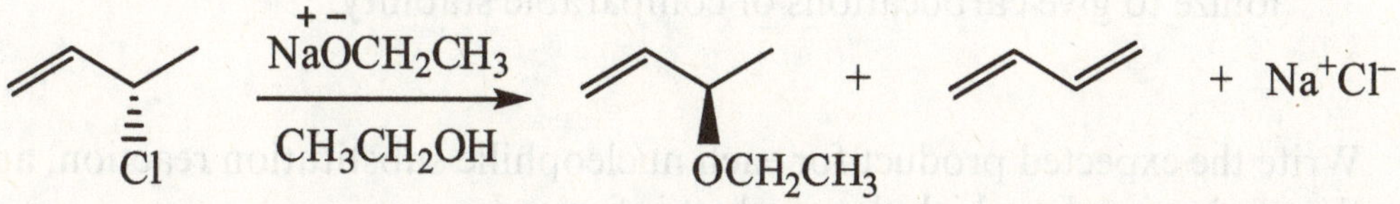

$$\xrightarrow[\text{CH}_3\text{CH}_2\text{OH}]{\overset{+-}{\text{NaOCH}_2\text{CH}_3}} \text{product} + \text{product} + \text{Na}^+\text{Cl}^-$$

In this reaction, the haloalkane underwent both substitution and elimination. In the substitution, the chlorine was replaced by an ethoxy (CH₃CH₂O−) group.

7.3   Complete these nucleophilic substitution reactions:

(a) [cyclopentyl]–Br + $CH_3CH_2S^-Na^+$ $\longrightarrow$ [cyclopentyl]–SCH_2CH_3 + Na^+Br^-

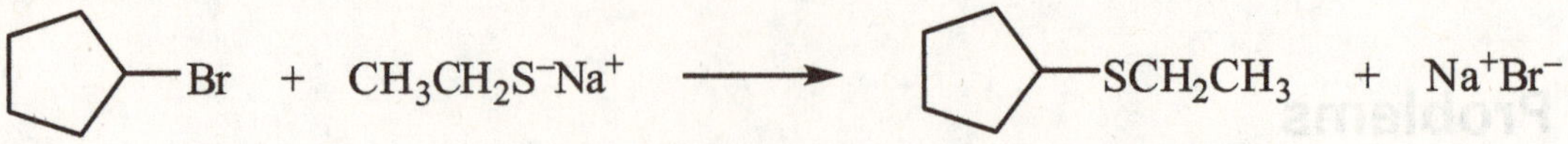

(b) [cyclopentyl]–Br + $CH_3\overset{\displaystyle O}{\overset{\|}{C}}O^-Na^+$ $\longrightarrow$ [cyclopentyl]–$O\overset{\displaystyle O}{\overset{\|}{C}}CH_3$ + Na^+Br^-

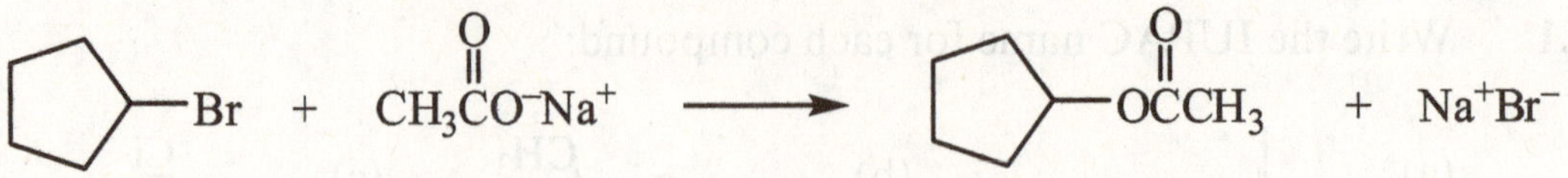

7.4   Answer the following questions:

(a) Potassium cyanide, KCN, reacts faster than trimethylamine, $(CH_3)_3N$, with 1-chloropentane. What type of substitution mechanism does this haloalkane likely undergo?

The rate of this reaction is influenced by the strength of the nucleophile, suggesting that the mechanism of substitution is S_N2. The cyanide ion is a much better nucleophile than is an amine.

(b) Compound A reacts faster with dimethylamine, $(CH_3)_2NH$, than compound B. What does this reveal about the relative ability of each haloalkane to undergo S_N2? S_N1?

A [structure with Br on secondary carbon of butane chain]   B [structure with Br on secondary carbon, branched]

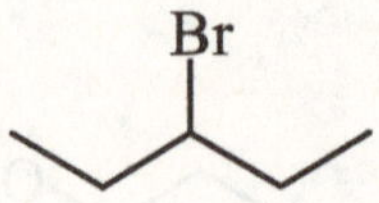

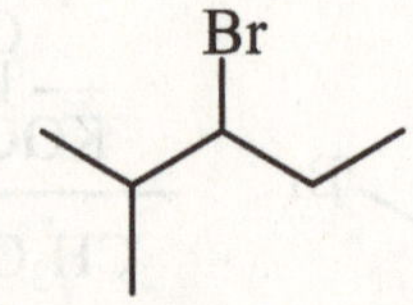

Compound A is more able to undergo S_N2, which involves backside attack by the nucleophile, because it is sterically less hindered than B. Both compounds A and B have about the same reactivity towards S_N1, because both A and B can ionize to give carbocations of comparable stability.

7.5   Write the expected product for each nucleophilic substitution reaction, and predict the mechanism by which the product is formed:

(a) [cyclohexane ring with Br and tert-butyl substituent] + Na^+SH^- $\xrightarrow[\text{acetone}]{S_N2}$ [cyclohexane ring with SH and tert-butyl substituent] + Na^+Br^-

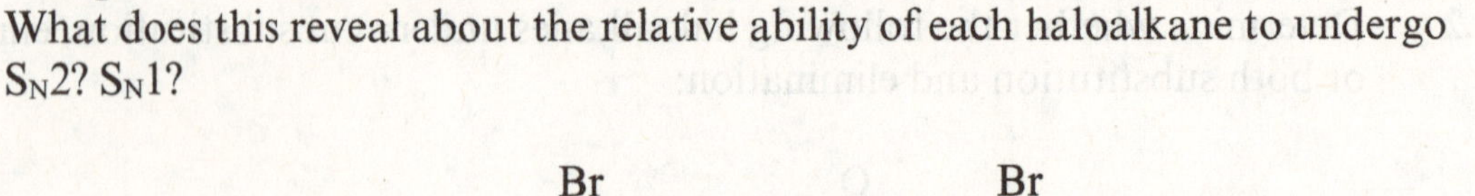

SH^- is a good nucleophile, and the haloalkane is secondary. Acetone is a polar aprotic solvent. These factors favor an S_N2 reaction. Because the mechanism of reaction is S_N2, an inversion of stereochemistry at the carbon occurs.

(b)

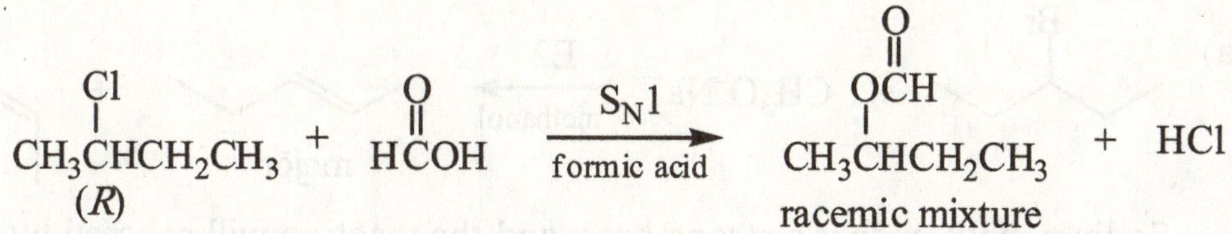
Cl
CH3CHCH2CH3
(R)
+
O
HCOH
SN1
formic acid
O
OCH
CH3CHCH2CH3
racemic mixture
+ HCl

The solvent, formic acid, is highly ionizing and is also a weak nucleophile. The reaction therefore proceeds by SN1.

7.6 Predict the β-elimination product formed when each chloroalkane is treated with sodium ethoxide in ethanol (if two products might be formed, predict which is the major product):

(a)

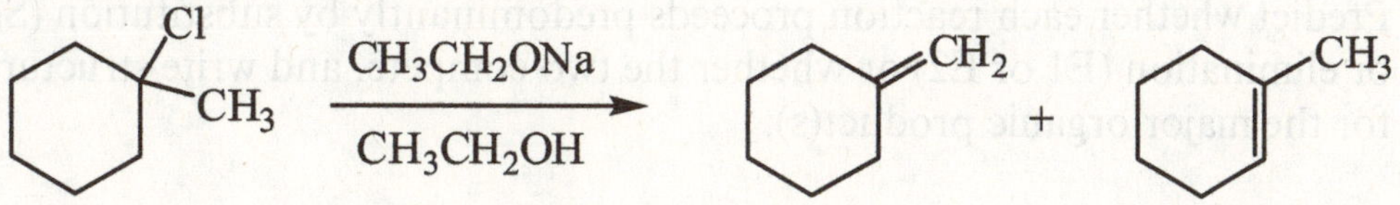
Cl
CH3
CH3CH2ONa
CH3CH2OH
CH2
+
CH3

minor major

(b)

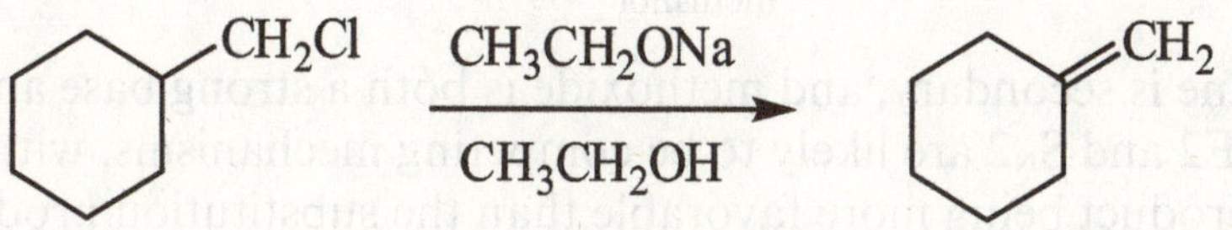
CH2Cl
CH3CH2ONa
CH3CH2OH
CH2

(c)

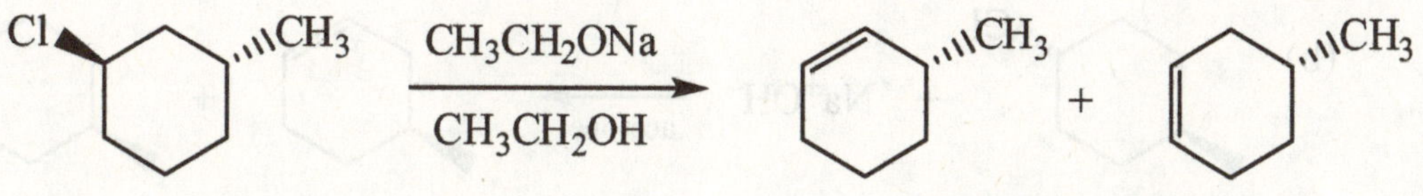
Cl
CH3
CH3CH2ONa
CH3CH2OH
CH3
+
CH3

equal amounts

(d)

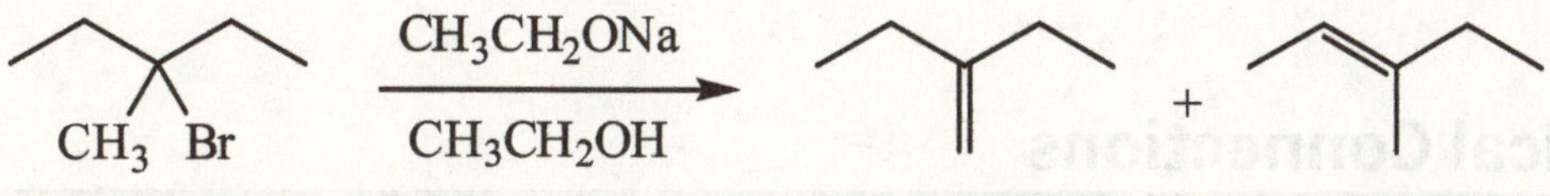
CH3 Br
CH3CH2ONa
CH3CH2OH
+

minor major

7.7 Predict whether each elimination reaction proceeds predominantly by an E1 or E2 mechanism, and write a structural formula for the major organic product:

Note that if there are two or more possible products, the major product is the one that is the most stable. Alkenes that are more substituted are more stable (Zaitsev's Rule). *trans* Alkenes are also more stable than *cis* alkenes.

(a)

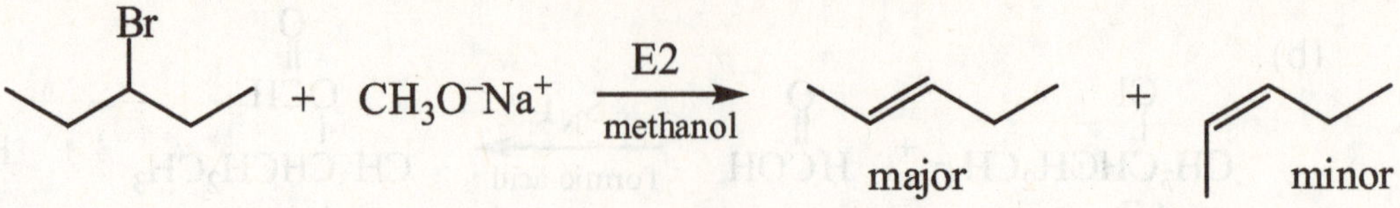

Sodium methoxide is a strong base, and the reaction will proceed by E2.

(b)

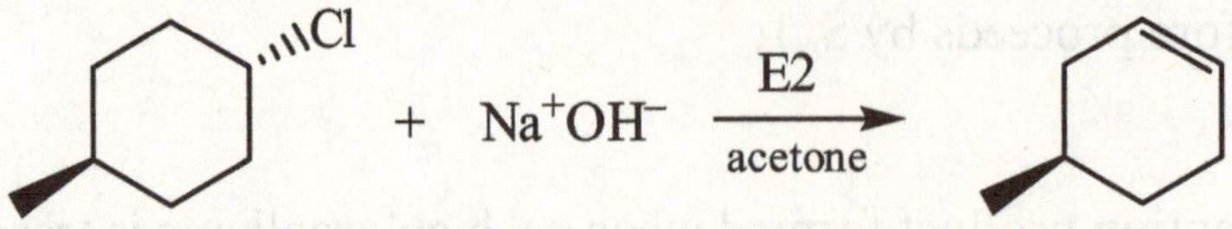

The hydroxide ion is a strong base, and the reaction will proceed by E2.

7.8   Predict whether each reaction proceeds predominantly by substitution (S_N1 or S_N2) or elimination (E1 or E2) or whether the two compete, and write structural formulas for the major organic product(s):

(a)

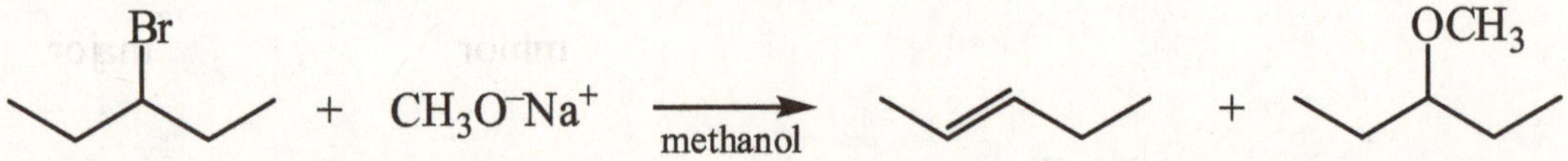

The haloalkane is secondary, and methoxide is both a strong base and a good nucleophile. E2 and S_N2 are likely to be competing mechanisms, with the elimination product being more favorable than the substitution product.

(b)

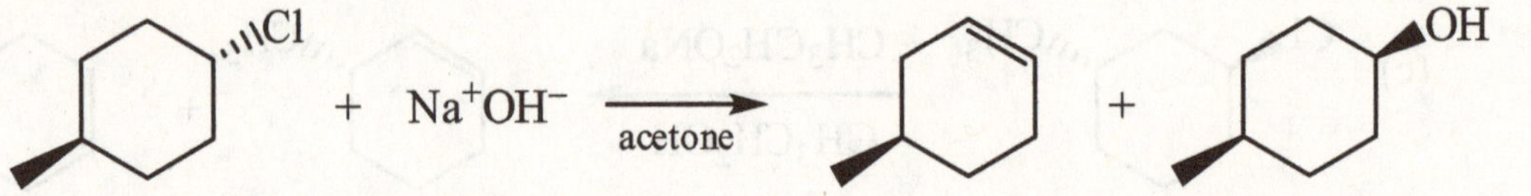

The haloalkane is secondary, and hydroxide is both a strong base and a good nucleophile. E2 and S_N2 are likely to be competing mechanisms.

Chemical Connections

7A.   Provide IUPAC names for HFC-134a and HCFC-141b.

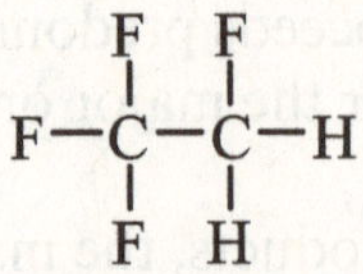

HFC-134a

HCFC-141b

1,1,1,2-Tetrafluoroethane   1,1-Dichloro-1-fluoroethane

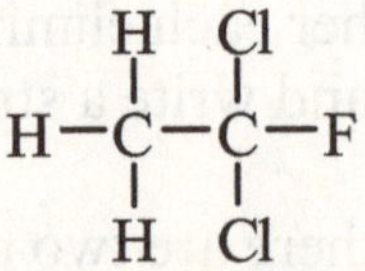

7B. Would you expect HFA-134a or HFA-227 to undergo an S_N1 reaction? An S_N2 reaction? Why or why not?

$$\underset{\text{HFA-134a}}{\overset{\begin{array}{cc} F & F \\ | & | \end{array}}{F-\overset{|}{\underset{|}{C}}-\overset{|}{\underset{|}{C}}-H}} \qquad\qquad \underset{\text{HFA-227}}{\overset{\begin{array}{ccc} F & F & F \\ | & | & | \end{array}}{F-\overset{|}{\underset{|}{C}}-\overset{|}{\underset{|}{C}}-\overset{|}{\underset{|}{C}}-F}}$$

Neither HFA-134a nor HFA-227 is expected to undergo an S_N1 or S_N2 reaction. First, fluoride is a very poor leaving group. Second, the carbocation generated in an S_N1 reaction would be extremely unstable; the other fluorine atoms would act as strong electron- withdrawing groups and destabilize the carbocation.

Quick Quiz

1. An S_N1 reaction can result in two products that are stereoisomers. *True*. An S_N1 reaction involves a trigonal planar carbocation intermediate that can be attacked by the nucleophile from either face of the trigonal plane.

2. In naming halogenated compounds, "haloalkane" is the IUPAC form of the name while "alkyl halide" is the common form of the name. *True*. In IUPAC nomenclature, the halogen is named as a substituent bonded to a parent alkane.

3. A substitution reaction results in the formation of an alkene. *False*. The formation of an alkene is indicative of an elimination reaction.

4. Sodium ethoxide ($CH_3CH_2O^- Na^+$) can act as a base and as a nucleophile in its reaction with bromocyclohexane. *True*. Ethoxide is both a strong base and a strong nucleophile, and bromocyclohexane can eliminate (E2) or substitute (S_N2). However, elimination is more likely than substitution because the haloalkane is secondary.

5. The rate law of the E2 reaction is dependent on just the haloalkane concentration. *False*. An E2 reaction is bimolecular at the rate-determining step, so the rate law will also depend on the concentration of the base.

6. The mechanism of the S_N1 reaction involves the formation of a carbocation intermediate. *True*. The formation of the carbocation is the rate-determining step of the reaction.

7. Polar protic solvents are required for E1 or S_N1 reactions to occur. *True*. Polar protic sovents stabilize the carbocation intermediates involved in these two reactions.

8. OH^- is a better leaving group than Cl^-. *False*. As a result of its atomic size, chloride is a much more stable anion than is hydroxide.

9. When naming haloalkanes with more than one type of halogen, numbering priority is given to the halogen with the higher mass. *False*. Regular IUPAC convention applies, and the numbering priority is given to the halogen first encoutered in the chain.

10. S$_N$2 reactions prefer good nucleophiles, while S$_N$1 reactions proceed with almost any nucleophile. *True*. The nucleophile is not involved in the rate-determing step of an S$_N$1 reaction, so the strength of the nucleophile is much less important.

11. The stronger the base, the better is the leaving group. *False*. Better leaving groups are groups that are more stable, meaning that they are weaker bases.

12. S$_N$2 reactions are more likely to occur with 2° haloalkanes than with 1° haloalkanes. *False*. Haloalkanes that are sterically more hindered will react slower in an S$_N$2 reaction.

13. A solvolysis reaction is a reaction performed without solvent. *False*. In the context of substitution, a solvolysis reaction is where the solvent itself acts as the nucleophile.

14. The degree of substitution at the reaction center affects the rate of an S$_N$1 reaction but not an S$_N$2 reaction. *False*. Both S$_N$1 and S$_N$2 reactions are affected by the degree of substitution at the carbon center. Substitution affects the stability of the carbocation intermediate involved in an S$_N$1 reaction, but it also affects steric hindrance, which is important in an S$_N$2 reaction.

15. A reagent must possess a negative charge to react as a nucleophile. *False*. Nucleophiles can be neutrally charged, but they be able to donate a pair of electrons.

16. Elimination reactions favor the formation of the more substituted alkene. *True*. Alkenes that are more substituted are more stable (Zaitsev's rule).

17. The best leaving group is one that is unstable as an anion. *False*. The opposite is true: the best leaving group is one that is stable.

18. In the S$_N$2 reaction, the nucleophile attacks the carbon from the side opposite that of the leaving group. *True*. Backside attack also results in stereochemical inversion.

19. Only haloalkanes can undergo substitution reactions. *False*. Other compounds containing good leaving groups can also undergo substitutioon reactions.

20. All of the following are polar aprotic solvents: acetone, DMSO, ethanol. *False*. Ethanol, CH_3CH_2OH, is polar but is capable of hydrogen bonding; it is a polar protic solvent.

End-of-Chapter Problems

Nomenclature

7.9 Write the IUPAC name for each compound:

(a) $CH_2=CF_2$

1,1-Difluoroethene

(b)

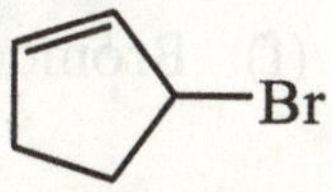

3-Bromocyclopentene

(c)

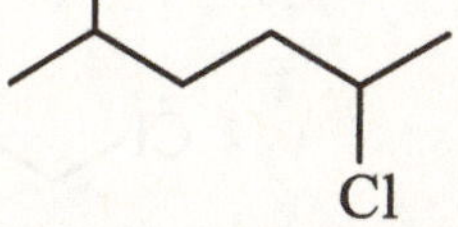

2-Chloro-5-methylhexane

(d) $Cl(CH_2)_6Cl$

1,6-Dichlorohexane

(e) CF_2Cl_2

Dichlorodi-
fluoromethane

(f)

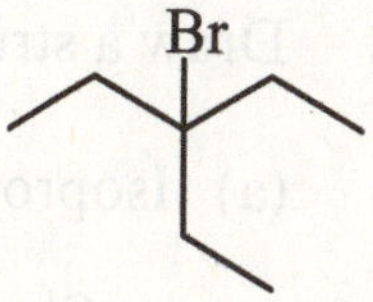

3-Bromo-3-ethylpentane

7.10 Write the IUPAC name for each compound (be certain to include a designation of configuration, where appropriate, in your answer):

(a)

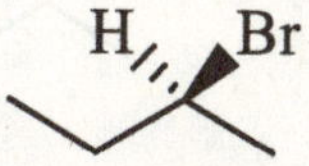

(S)-2-Bromobutane

(b) 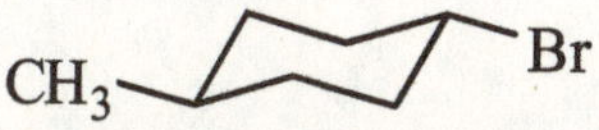 CH_3

trans-1-Bromo-4-
methylcyclohexane

(c)

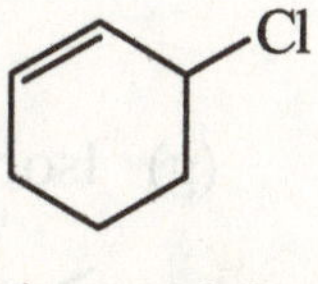

3-Chlorocyclohexene

(d)

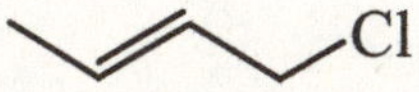

(E)-1-Chloro-2-butene

(e)

(R)-2-Bromo-2-
chlorobutane

(f)

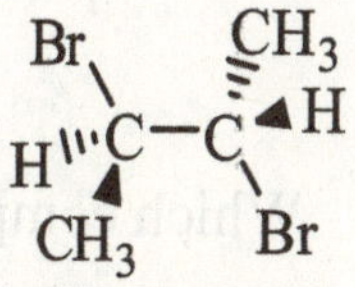

meso-2,3-
Dibromobutane

7.11 Draw a structural formula for each compound (given are IUPAC names):

(a) 3-Bromopropene

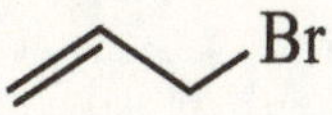

(b) (R)-2-Chloropentane

(c) meso-3,4-
Dibromohexane

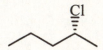

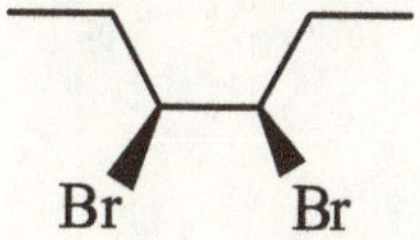

(d) *trans*-1-Bromo-3-isopropylcyclohexane

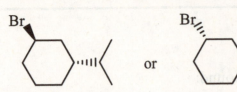   or   There are two enantiomers of *trans*-1-bromo-3-isopropylcyclohexane.

(e) 1,2-Dichloroethane   (f) Bromocyclobutane

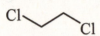

7.12 Draw a structural formula for each compound (given are common names):

(a) Isopropyl chloride   (b) *sec*-Butyl bromide   (c) Allyl iodide

(d) Methylene chloride   (e) Chloroform   (f) *tert*-Butyl chloride

CH_2Cl_2 $CHCl_3$

(g) Isobutyl chloride

7.13 Which compounds are 2° alkyl halides?

(a) Isobutyl chloride   (b) 2-Iodooctane

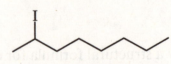

1° alkyl halide   2° alkyl halide

(c) *trans*-1-Chloro-4-methylcyclohexane

2° alkyl halide

(d) 3-Bromo-3-methylpentane

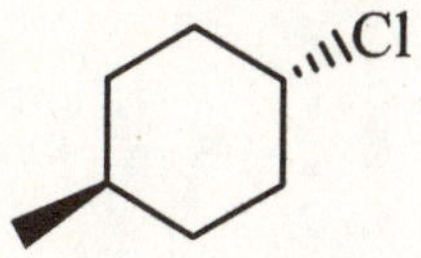

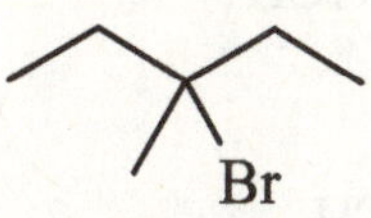

3° alkyl halide

Synthesis of Alkyl Halides

7.14 What alkene or alkenes and reaction conditions give each alkyl halide in good yield? (*Hint*: Review Chapter 5.)

All of these alkyl halides can be prepared by the Markovnikov addition of a hydrogen halide to an alkene.

(a) ⬠ $\xrightarrow{\text{HBr}}$ ⬠—Br

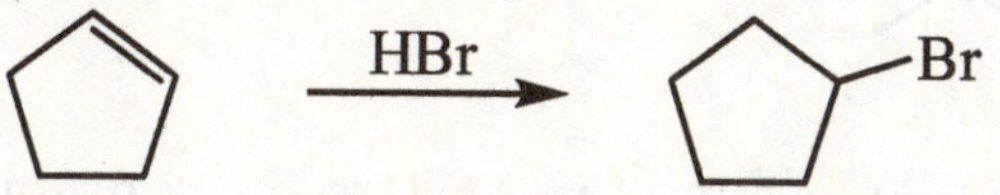

(b) [alkene] or [alkene] $\xrightarrow{\text{HBr}}$ [product with Br]

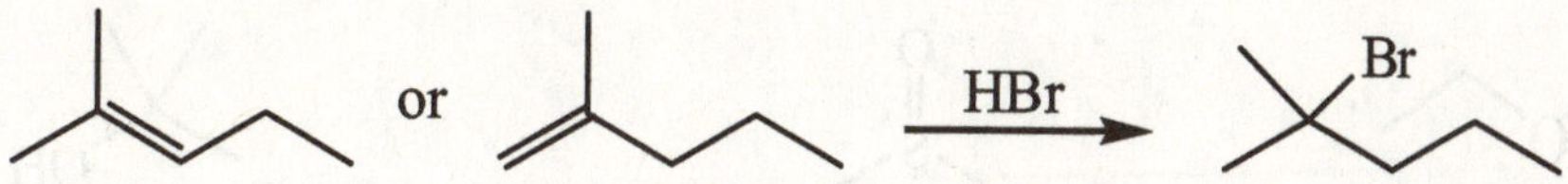

(c) [cyclohexane structure] or [cyclohexene structure] $\xrightarrow{\text{HCl}}$ [cyclohexane with CH₃ and Cl]

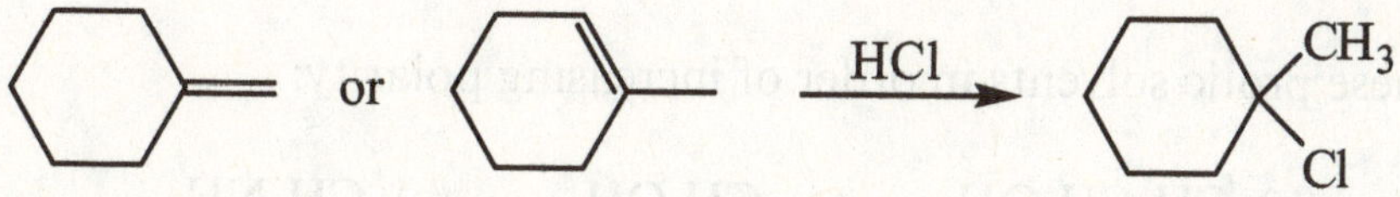

(d) [alkene] $\xrightarrow{\text{HCl}}$ [product with Cl]   racemic

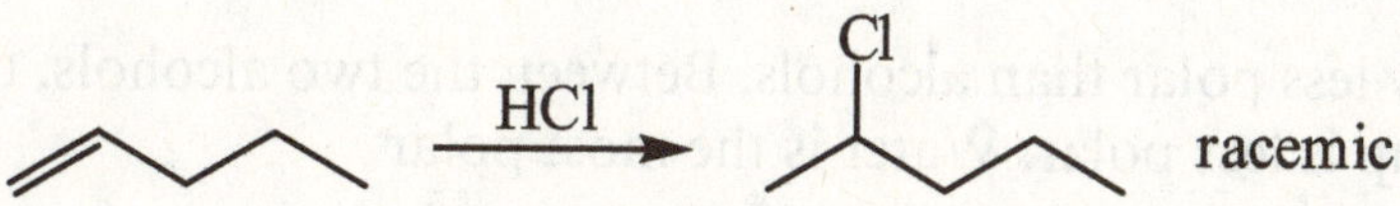

7.15 Show reagents and conditions that bring about these conversions:

(a) [alkene] $\xrightarrow{\text{HCl}}$ [product with Cl]

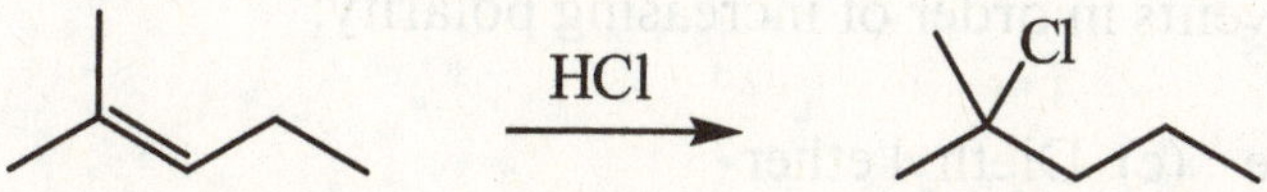

(b) $CH_3CH_2CH=CH_2$ $\xrightarrow{\text{HI}}$ $CH_3CH_2\overset{\text{I}}{\underset{|}{C}}HCH_3$

(c) $CH_3CH=CHCH_3$ $\xrightarrow{HCl}$ $CH_3\overset{\overset{\displaystyle Cl}{|}}{C}HCH_2CH_3$

(d) 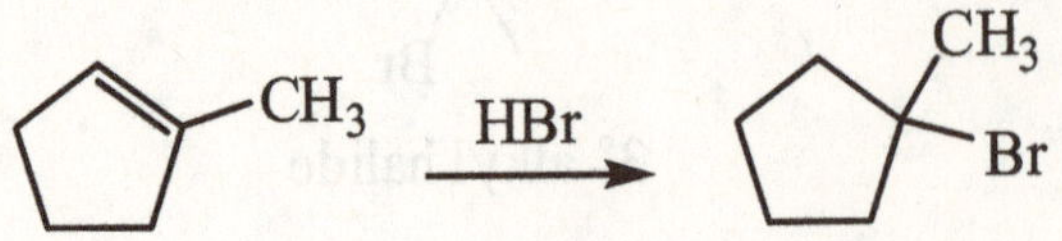 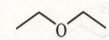 $\xrightarrow{HBr}$

Nucleophilic Aliphatic Substitution

7.16 Write structural formulas for these common organic solvents:

(a) Dichloromethane   (b) Acetone 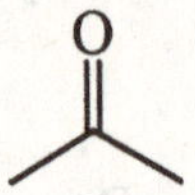   (c) Ethanol

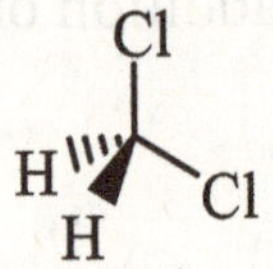

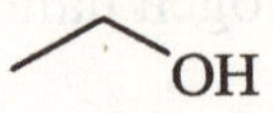

(d) Diethyl ether   (e) Dimethyl sulfoxide 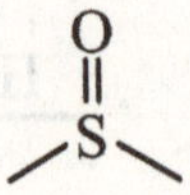   (f) *tert*-Butyl alcohol

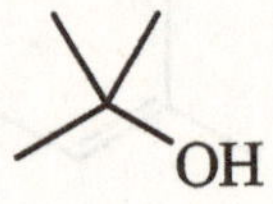

7.17 Arrange these protic solvents in order of increasing polarity:

(a) H_2O (b) CH_3CH_2OH (c) CH_3OH (c) CH_3NH_2

Amines are typically less polar than alcohols. Between the two alcohols, the one with the larger alkyl group is less polar. Water is the most polar.

$$CH_3NH_2 < CH_3CH_2OH < CH_3OH < H_2O$$

7.18 Arrange these aprotic solvents in order of increasing polarity:

(a) Acetone   (b) Pentane   (c) Diethyl ether

Hydrocarbons, such as pentane, are nonpolar. Between acetone and diethyl ether, diethyl ether is less polar because the bent geometry at the oxygen atom allows the bond dipole moments to partially cancel out. In the order of increasing polarity, they are:

$$Pentane < diethyl\ ether < acetone$$

7.19 From each pair, select the better nucleophile:

(a) H_2O or **OH^-**

When comparing species with the same nucleophilic atom, the strength of a nucleophile is related to base strength. Hydroxide is a stronger base than water, so it is a better nucleophile.

(b) CH_3COO^- or **OH^-**

Both are oxygen nucleophiles, but acetate is stabilized by resonance and less basic than hydroxide. Hydroxide is therefore the better nucleophile.

(c) CH_3SH or **CH_3S^-**

Similar to (a), these two species have the same nucleophilic atom. The negatively charged species (CH_3S^-) is the better nucleophile.

7.20 Which statements are true for S_N2 reactions of haloalkanes?

S_N2 reactions are favored by a good leaving group (required for both S_N1 and S_N2), a good nucleophile, a haloalkane with low steric hindrance, and a weakly ionizing (polar aprotic) solvent.

(a) Both the haloalkane and the nucleophile are involved in the transition state.

True. Both species are reacting together at the rate-determining step.

(b) The reaction proceeds with inversion of configuration at the substitution center.

True. S_N2 reactions occur by backside attack.

(c) The reaction proceeds with retention of optical activity.

True. Because an S_N2 reaction always occurs by backside attack, the product will always show inversion of stereochemistry and be enantiomerically pure.

(d) The order of reactivity is 3° > 2° > 1° > methyl.

False. Rather, the opposite trend would be correct. A methyl halide has the least steric hindrance and would react the fastest by an S_N2 reaction.

(e) The nucleophile must have an unshared pair of electrons and bear a negative charge.

False. Although it must be able to donate a pair of electrons, a negative charge is not necessary. Ammonia, for example, does not have a negative charge but is a reasonably suitable nucleophile for S_N2 reactions.

(f) The greater the nucleophilicity of the nucleophile, the greater the rate of reaction.

True. Because the nucleophile is involved in the rate-determining step of an S_N2 reaction, it only follows that the better the nucleophile, the faster the reaction.

7.21 Complete these S_N2 reactions:

(a) $Na^+I^- + CH_3CH_2CH_2Cl \xrightarrow{\text{acetone}} Na^+Cl^- + CH_3CH_2CH_2I$

(b) 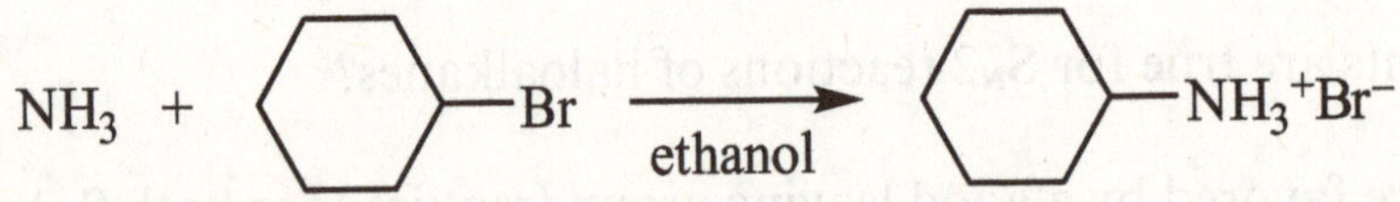 $NH_3 + $ $\xrightarrow{\text{ethanol}}$

(c) $CH_3CH_2O^-Na^+ + CH_2=CHCH_2Cl \xrightarrow{\text{ethanol}} Na^+Cl^- + CH_2=CHCH_2OCH_2CH_3$

7.22 Complete these S_N2 reactions:

(a) 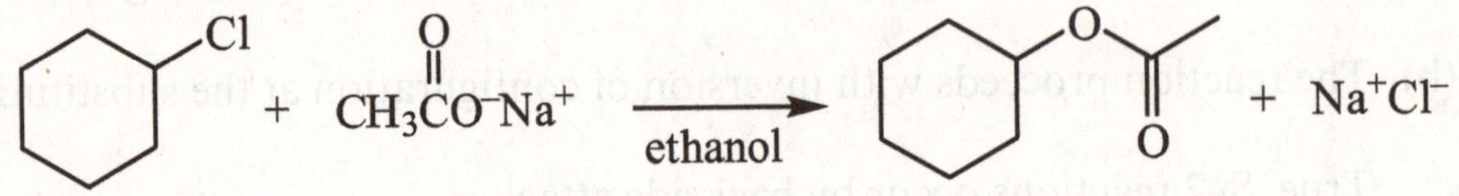 $+ CH_3\overset{O}{\overset{\|}{C}}O^-Na^+ \xrightarrow{\text{ethanol}}$ $+ Na^+Cl^-$

(b) 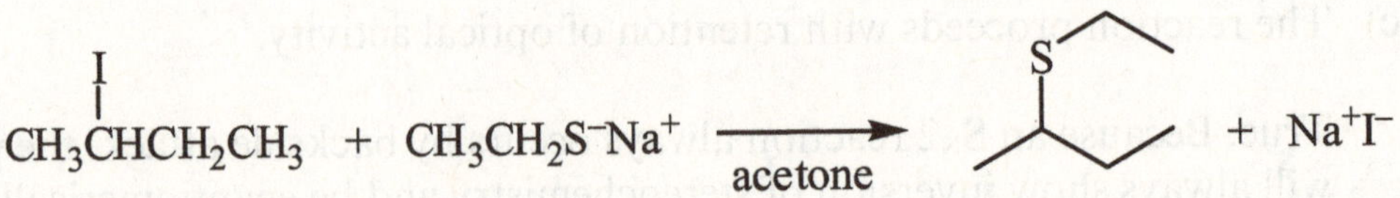 $CH_3\overset{I}{\underset{|}{C}}HCH_2CH_3 + CH_3CH_2S^-Na^+ \xrightarrow{\text{acetone}}$ $+ Na^+I^-$

(c) $CH_3\overset{CH_3}{\underset{|}{C}}HCH_2CH_2Br + Na^+I^- \xrightarrow{\text{acetone}}$ $+ Na^+Br^-$

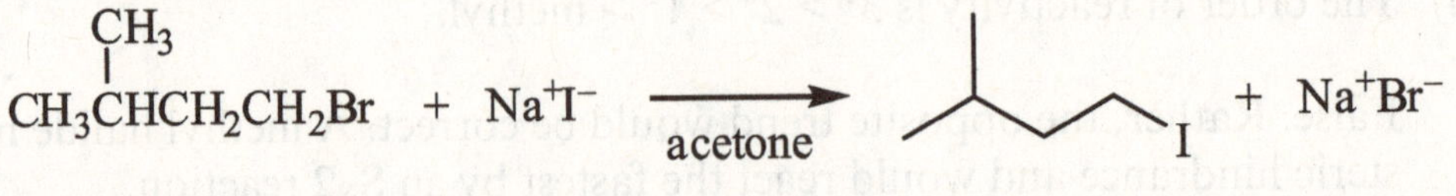

(d) $(CH_3)_3N + CH_3I \xrightarrow{\text{acetone}} (CH_3)_4N^+I^-$

(e) $\bigcirc$—CH_2Br + $CH_3O^-Na^+$ $\xrightarrow{\text{methanol}}$ $\bigcirc$—CH_2OCH_3 + Na^+Br^-

(f)

CH_3⌁$\bigcirc$—Cl + $CH_3S^-Na^+$ $\xrightarrow{\text{ethanol}}$ CH_3⌁$\bigcirc$^{SCH_3} + Na^+Cl^-

(stereochemical inversion)

(g) $\bigcirc$NH + $CH_3(CH_2)_6CH_2Cl$ $\xrightarrow{\text{ethanol}}$ $\bigcirc$$^+N\overset{H}{\underset{(CH_2)_7CH_3}{}}$ + Cl^-

(h) $\bigcirc$—CH_2Cl + NH_3 $\xrightarrow{\text{ethanol}}$ $\bigcirc$—$CH_2NH_3^+$ + Cl^-

7.23   You were told that each reaction in Problem 7.22 proceeds by an S_N2 mechanism. Suppose you were not told the mechanism. Describe how you could conclude, from the structure of the haloalkane, the nucleophile, and the solvent, that each reaction is in fact an S_N2 reaction.

The conditions that are ideal for an S_N2 mechanism are a haloalkane that is not very sterically hindered, a good nucleophile, and a weakly ionizing solvent. However, note that these are only the optimal conditions and that many reactions that do not meet these criteria can still proceed by S_N2.

(a)   The haloalkane is 2°, acetate is a moderate nucleophile that is a weak base, and ethanol is a moderately ionizing solvent; these favor an S_N2 mechanism.

(b)   The haloalkane is 2°, ethyl thiolate is a good nucleophile that is a weak base, and acetone is a weakly ionizing solvent; these favor an S_N2 mechanism.

(c)   The haloalkane is 1°, and when combined with an excellent nucleophile such as iodide, the reaction will proceed by an S_N2 mechanism.

(d)   Although the nucleophile is only a moderate nucleophile, the reaction can only proceed by S_N2 because the haloalkane is a methyl halide. Because methyl carbocations are extremely unstable, the reaction cannot proceed by an S_N1 mechanism.

(e)   The haloalkane is 1°, and when combined with a good nucleophile such as methoxide, the reaction will proceed by an S_N2 mechanism. Although methoxide is also a good base, 1° haloalkanes will favor substitution over elimination.

(f) The haloalkane is 2°, methyl thiolate is a good nucleophile that is a weak base, and ethanol is a moderately ionizing solvent; these favor an S_N2 mechanism.

(g) Similar to (d), the amine is a moderate nucleophile. The haloalkane is 1°, which will not react by an S_N1 mechanism. Thus, the favored mechanism is S_N2.

(h) As with (d) and (g), the amine nucleophile (ammonia) is moderate. When combined with a 1° haloalkane, which cannot proceed by S_N1, the favored mechanism is S_N2.

7.24 In the following reactions, a haloalkane is treated with a compound that has two nucleophilic sites. Select the more nucleophilic site in each part, and show the product of each S_N2 reaction.

When comparing nucleophilic atoms that are on the same row of the periodic table, as in reactions (a) and (b), atoms that are less electronegative (more basic) are better nucleophiles; nitrogen is therefore a better nucleophile than oxygen. For reaction (c), which involve nucleophilic atoms of the same group of the periodic table, nucleophilicity increases from top to bottom, so sulfur is a better nucleophile than is oxygen.

(a) $HOCH_2CH_2NH_2$ + CH_3I $\xrightarrow{\text{ethanol}}$ $HOCH_2CH_2\overset{+}{N}H_2CH_3$ + I^-

(b) 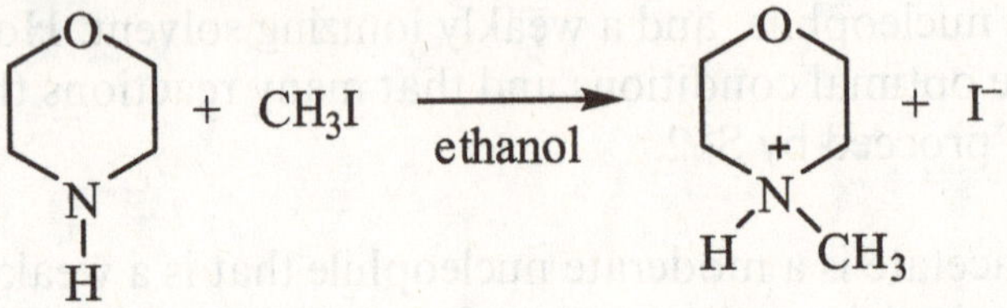 + CH_3I $\xrightarrow{\text{ethanol}}$ + I^-

(c) $HOCH_2CH_2SH$ + CH_3I $\xrightarrow{\text{ethanol}}$ $HOCH_2CH_2SCH_3$ + HI

7.25 Which statements are true for S_N1 reactions of haloalkanes?

The ideal conditions for an S_N1 mechanism are a haloalkane that forms a stable carbocation, a poor nucleophile, and an ionizing solvent.

(a) Both the haloalkane and the nucleophile are involved in the transition state of the rate-determining step.

False. In an S_N1 reaction, the rate-determining step is the first step, the ionization of the alkyl halide. The nucleophile is only involved in the second step of the reaction.

(b) The reaction at a stereocenter proceeds with retention of configuration.

False. In an S$_N$1 reaction, a carbocation, which is sp^2-hybridized and has trigonal planar geometry, is involved. The nucleophile can add to either face of the carbocation, resulting in the formation of a mixture of stereoisomers.

(c) The reaction at a stereocenter proceeds with loss of optical activity.

True. Because a carbocation intermediate is involved, a 1:1 mixture of enantiomers will be formed (assuming that the product has only one stereocenter). Racemic mixtures are optically inactive.

(d) The order of reactivity is 3° > 2° > 1° > methyl.

True. The order of reactivity corresponds to the stability of the carbocation intermediate. Tertiary carbocations are stabilized by three inductively donating alkyl groups.

(e) The greater the steric crowding around the reactive center, the lower the rate of reaction.

False. This would be true for an S$_N$2 reaction, but in an S$_N$1 reaction, the nucleophilic attack of the carbocation is not the rate-determining step. As a result, steric crowding around the reactive center has no effect on the rate of reaction.

(f) The rate of reaction is greater with good nucleophiles as compared with poor nucleophiles.

False. The rate-determining step of an S$_N$1 reaction is the ionization of the haloalkane and does not involve the nucleophile. (In the presence of a good nucleophile, a reaction is more likely to proceed by S$_N$2.)

7.26 Draw a structural formula for the product of each S$_N$1 reaction:

(a)

$$CH_3CHClCH_2CH_3 + CH_3CH_2OH \xrightarrow{ethanol} CH_3CH(OCH_2CH_3)CH_2CH_3 + CH_3CH(OCH_2CH_3)CH_2CH_3$$

S enantiomer → racemic mixture

(b)

(c)

$$\underset{\underset{CH_3}{|}}{\overset{\overset{CH_3}{|}}{CH_3CCl}} + CH_3\overset{O}{\overset{||}{C}}OH \xrightarrow{\text{acetic acid}} \underset{\underset{CH_3}{|}}{\overset{\overset{CH_3}{|}}{CH_3C}}-O\overset{O}{\overset{||}{C}}CH_3$$

(d)

racemic mixture

(e)

$+ \ CH_3CH_2OH \xrightarrow{\text{ethanol}}$

(f)

$+ \ CH_3\overset{O}{\overset{||}{C}}OH \xrightarrow{\text{acetic acid}}$

Four stereoisomers are generated in this reaction.

7.27 You were told that each substitution reaction in Problem 7.26 proceeds by an S_N1 mechanism. Suppose that you were not told the mechanism. Describe how you could conclude, from the structure of the haloalkane, the nucleophile, and the solvent, that each reaction is in fact an S_N1 reaction.

Recall that the ideal conditions for an S_N1 mechanism are a haloalkane that forms a stable carbocation, a poor nucleophile, and an ionizing solvent. All four reactions involve poor nucleophiles, so they are most likely to proceed by S_N1.

(a) The 2° haloalkane forms a relatively stable carbocation, ethanol is a weak nucleophile, and the solvent (also ethanol) is moderately ionizing.

(b) The 3° haloalkane forms a very stable carbocation, methanol is a weak nucleophile, and the solvent (also methanol) is moderately ionizing.

(c) The 3° haloalkane forms a very stable carbocation, acetic acid is a weak nucleophile, and the solvent (also acetic acid) is strongly ionizing.

(d) The 2° haloalkane forms a carbocation that is stabilized by resonance, methanol is a weak nucleophile, and the solvent (also methanol) is moderately ionizing.

(e) The 2° haloalkane forms a carbocation that can rearrange (hydride shift) to form a 3° carbocation, and like (a), the nucleophile and solvent favor an S_N1 mechanism.

(f)   The 2° haloalkane forms a carbocation that can rearrange (alkyl shift) to form a 3° carbocation, and like (c), the nucleophile and solvent favor an S_N1 mechanism.

7.28   Select the member of each pair that undergoes nucleophilic substitution in aqueous ethanol more rapidly:

The conditions used (aqueous ethanol) are suggestive of an S_N1 reaction. In the absence of any resonance stabilization of the carbocation intermediate, the order of reactivity is 3° > 2° > 1° > methyl.

(a)

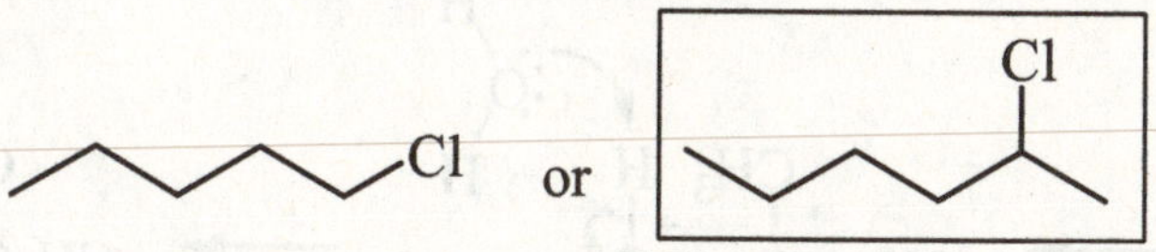

(b)

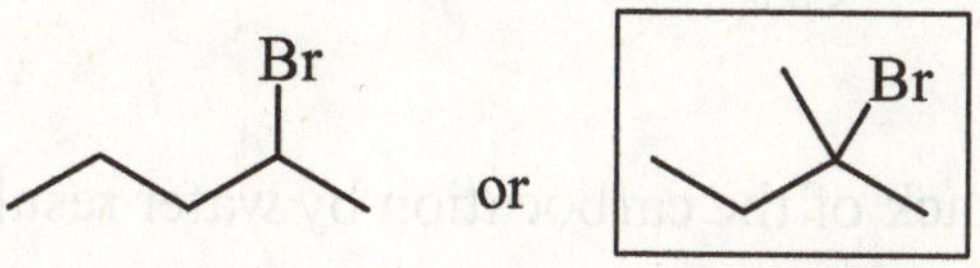

(c)

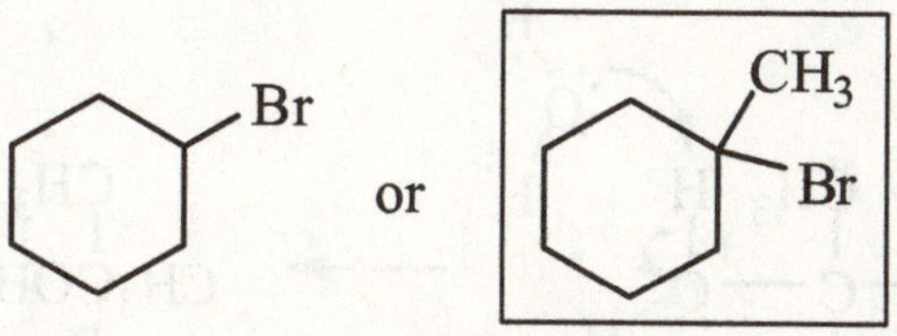

7.29   Propose a mechanism for the formation of the products (but not their relative percentages) in this reaction.

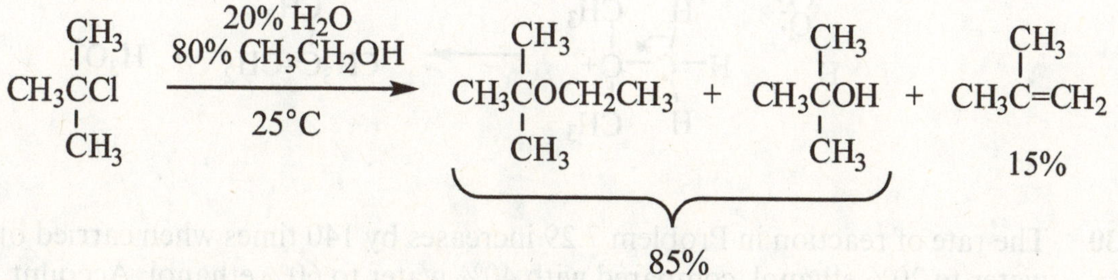

A combination of substitution (85%) and elimination (15%) products are obtained. Under the reaction conditions used, the only mechanisms that can occur are S_N1 and E1. H_2O and CH_3CH_2OH are poor nucleophiles and weak bases, so S_N2 and E2 can be precluded.

Both the S_N1 and E1 mechanisms involve the highly stable *tert*-butyl carbocation, which is generated in the first step of the reaction:

$$CH_3-\underset{\underset{CH_3}{|}}{\overset{\overset{CH_3}{|}}{C}}-Cl \xrightarrow{\text{slow}} CH_3-\underset{\underset{CH_3}{|}}{\overset{\overset{CH_3}{|}}{C}}+ \;+\; Cl^-$$

Attack of the carbocation by ethanol followed by deprotonation gives the ether product:

$$CH_3-\underset{\underset{CH_3}{|}}{\overset{\overset{CH_3}{|}}{C}}+ \quad \overset{..}{\underset{..}{O}} \longrightarrow CH_3-\overset{\overset{CH_3}{|}}{\underset{\underset{CH_3}{|}}{C}}-\overset{+}{\overset{..}{O}}\quad \xrightarrow{H_3O^+} \quad CH_3\underset{\underset{CH_3}{|}}{\overset{}{C}}OCH_2CH_3$$

In a similar fashion, nucleophilic attack of the carbocation by water results in the alcohol:

$$CH_3-\underset{\underset{CH_3}{|}}{\overset{\overset{CH_3}{|}}{C}}+ \quad \overset{..}{\underset{..}{O}} \longrightarrow CH_3-\overset{\overset{CH_3}{|}}{\underset{\underset{CH_3}{|}}{C}}-\overset{+}{\overset{..}{O}}\quad \longrightarrow \quad CH_3\underset{\underset{CH_3}{|}}{\overset{}{C}}OH \;+\; H_3O^+$$

Deprotonation of the carbocation by the solvent (water or ethanol) gives the alkene:

$$\overset{H}{\underset{H}{O:}} \quad H-\overset{\overset{H}{|}}{\underset{\underset{H}{|}}{C}}-\overset{\overset{CH_3}{|}}{\underset{\underset{CH_3}{|}}{C}}+ \longrightarrow CH_3\overset{\overset{CH_3}{|}}{C}=CH_2 \;+\; H_3O^+$$

7.30 The rate of reaction in Problem 7.29 increases by 140 times when carried out in 80% water to 20% ethanol, compared with 40% water to 60% ethanol. Account for this difference.

In S_N1 and E1 reactions, the formation of the carbocation intermediate is the rate-determining step. Water is more polar than is ethanol, so the solvent mixture that has more water is better at stabilizing the carbocation intermediate. The stabilization of the carbocation intermediate increases the rate of the reaction.

7.31 Select the member of each pair that shows the greater rate of S_N2 reaction with KI in acetone:

Larger alkyl groups that are bonded to the carbon at which substitution occurs present greater steric hindrance to the attacking nucleophile.

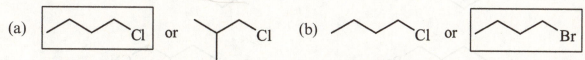

(a)

Both are 1°, but the compound on the left is less sterically hindered.

(b)

The structures are the same, but bromide is a better leaving group.

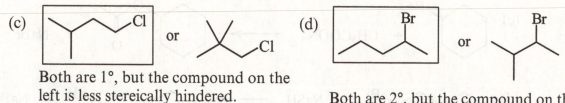

(c)

Both are 1°, but the compound on the left is less sterically hindered.

(d)

Both are 2°, but the compound on the left is less sterically hindered.

7.32 What hybridization best describes the reacting carbon in the S_N2 transition state?

At the transition state of an S_N2 reaction, the reacting carbon is best described as being sp^2-hybridized. The three substitutents on the carbon atom are trigonal planar relative to each other, while the leaving group and the nucleophile are on opposite sides of the trigonal plane. The incoming nucleophile and the departing leaving group are each partially bonded to an unhybridized p orbital.

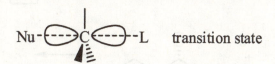

Nu $\cdots$ C $\cdots$ L   transition state

7.33 Haloalkenes such as vinyl bromide, $CH_2=CHBr$, undergo neither S_N1 nor S_N2 reactions. What factors account for this lack of reactivity?

Haloalkenes fail to undergo S_N1 reactions because the alkenyl carbocations produced through ionization of the carbon-halogen bond are too unstable. They fail to undergo S_N2 reactions because the planar geometry of the alkene does not allow backside attack, and the electron-rich nature of the alkene does not attract nucleophiles.

7.34 Show how you might synthesize the following compounds from a haloalkane and a nucleophile:

(a) [cyclohexyl bromide] + NaCN $\longrightarrow$ [cyclohexyl CN] + NaBr

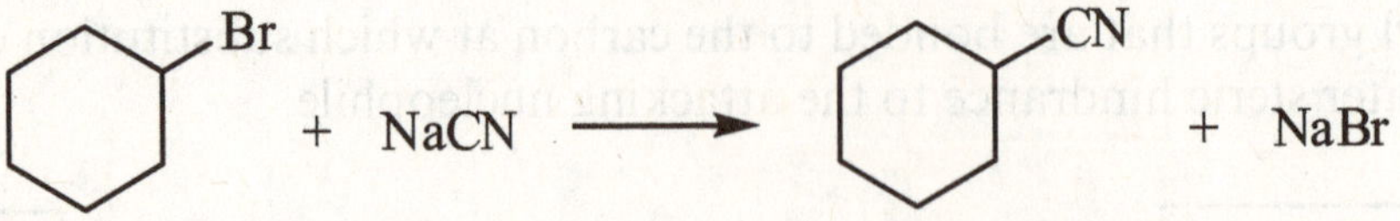

(b) [benzyl-type CH₂Br on ring] + NaCN $\longrightarrow$ [CH₂CN on ring] + NaBr

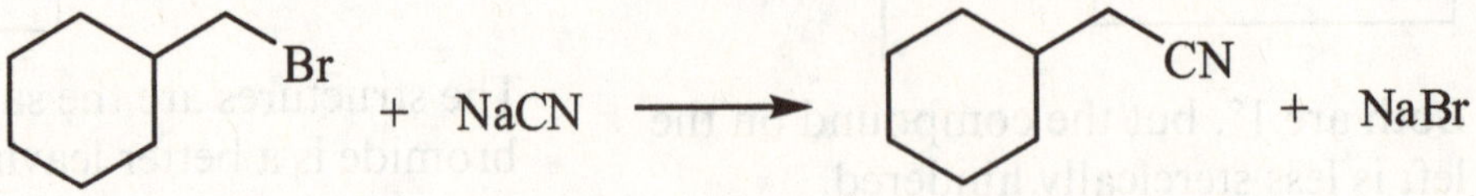

(c) [cyclohexyl Br] + CH₃COONa $\longrightarrow$ [cyclohexyl O–C(=O)CH₃] + NaBr

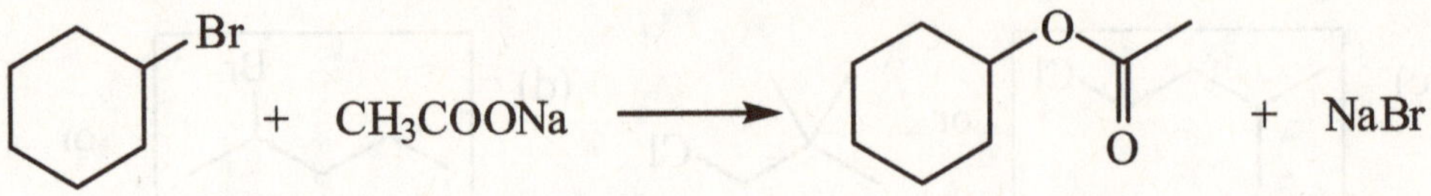

(d) [alkyl Br chain] + NaSH $\longrightarrow$ [alkyl SH chain] + NaBr

(e) [alkyl Br chain] + NaOCH₃ $\longrightarrow$ [alkyl OCH₃ chain] + NaBr

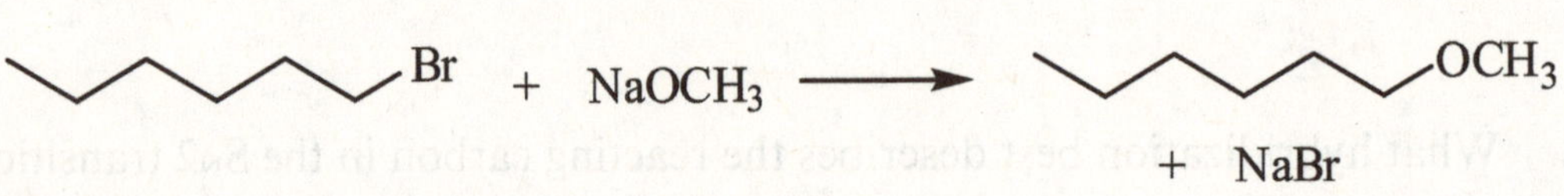

(f) [alkyl Br] + NaOCH₂CH₃ $\longrightarrow$ [ether] + NaBr

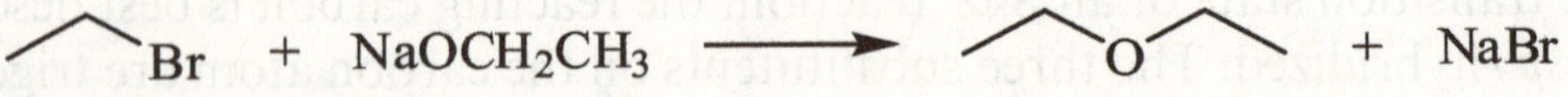

(g) [cyclopentane with Br and CH₃ substituents] + NaSH $\longrightarrow$ [cyclopentane with SH and CH₃] + NaBr

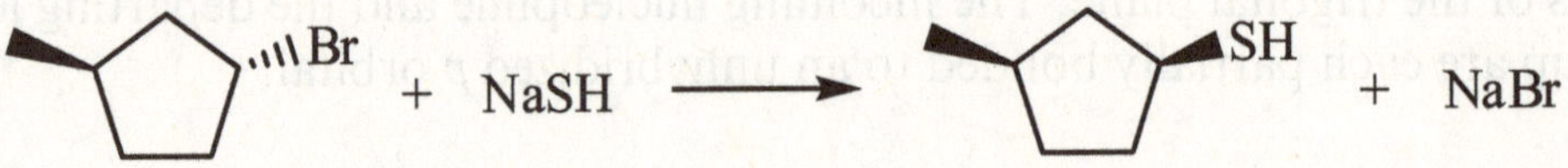

(h) [alkyl Br] + [NaO–CH with chain] $\longrightarrow$ [CH₃CH₂O–CH with chain]

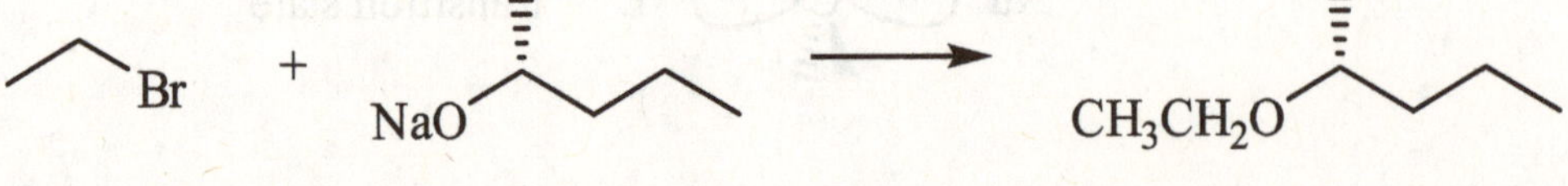

7.35 Show how you might synthesize each compound from a haloalkane and a nucleophile:

(a) [cyclohexyl]—Br + 2NH₃ $\longrightarrow$ [cyclohexyl]—NH₂ + NH₄Br

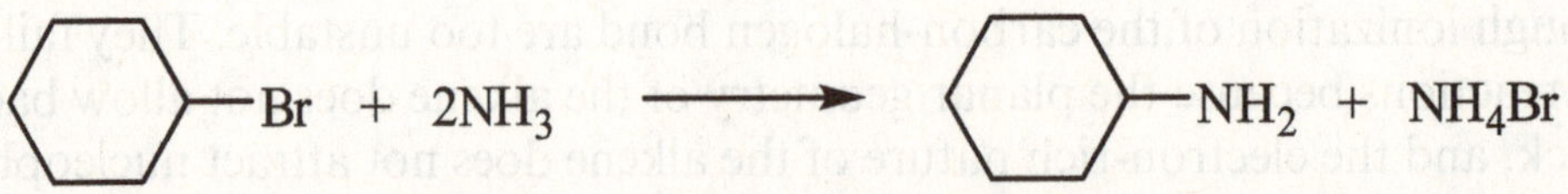

The reaction requires two equivalents of NH₃ because the product that is initially formed is the salt cyclohexylammonium bromide. The second equivalent of NH₃ neutralizes the salt to give cyclohexylamine and ammonium bromide.

(b) 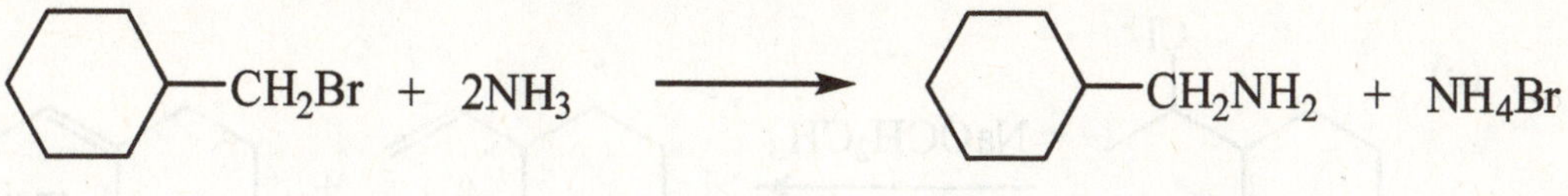 [cyclohexyl]–CH$_2$Br + 2NH$_3$ ⟶ [cyclohexyl]–CH$_2$NH$_2$ + NH$_4$Br

(c) 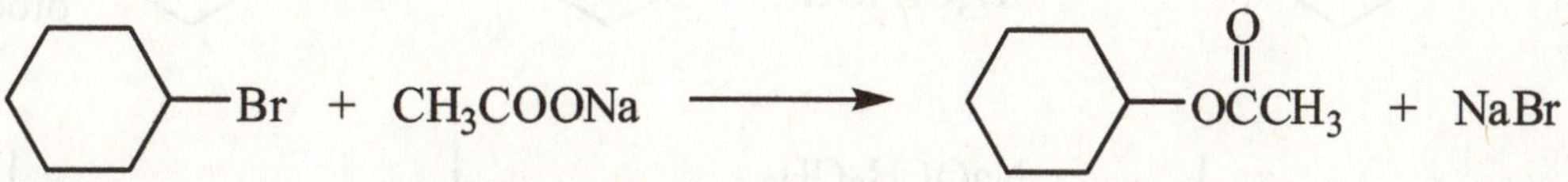 [cyclohexyl]–Br + CH$_3$COONa ⟶ [cyclohexyl]–OCCH$_3$ (C=O) + NaBr

(d) 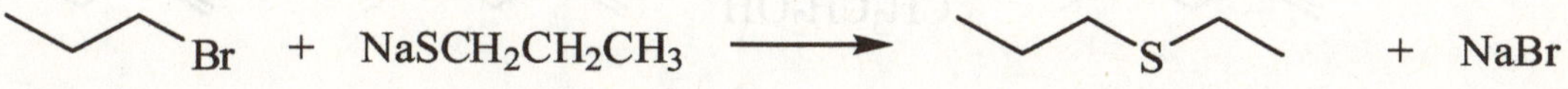 [propyl]–Br + NaSCH$_2$CH$_2$CH$_3$ ⟶ [structure with S] + NaBr

(e) 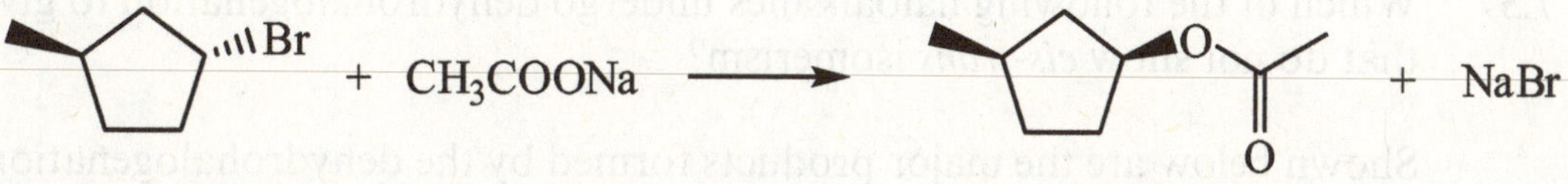 [cyclopentane with Br and CH$_3$ wedges] + CH$_3$COONa ⟶ [cyclopentane with ester O] + NaBr

(f) CH$_3$(CH$_2$)$_2$CH$_2$Br + NaOCH$_2$(CH$_2$)$_2$CH$_3$ ⟶ (CH$_3$CH$_2$CH$_2$CH$_2$)$_2$O + NaBr

β-Eliminations

7.36 Draw structural formulas for the alkene(s) formed by treating each of the following haloalkanes with sodium ethoxide in ethanol. Assume that elimination is by an E2 mechanism. Where two alkenes are possible, use Zaitsev's rule to predict which alkene is the major product:

In an elimination reaction, the most stable alkene is the major product. Alkene stability increases as the number of alkyl substituents bonded to the C=C group increases. Note however, that elimination can only occur at a β-carbon that has a hydrogen atom.

(a) 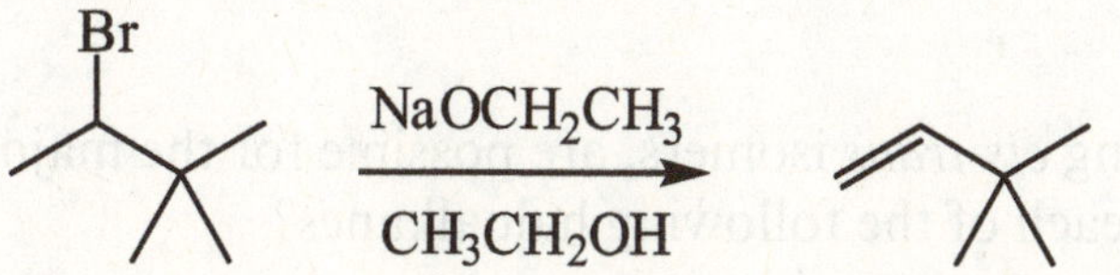 [structure with Br]
NaOCH$_2$CH$_3$ / CH$_3$CH$_2$OH ⟶ [alkene structure]

(b) 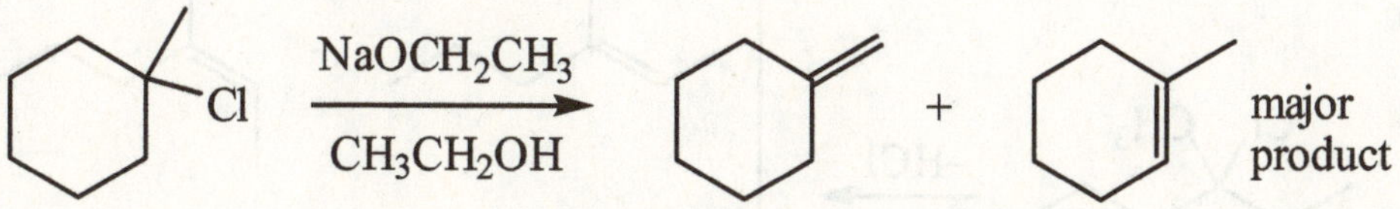 [cyclohexane with CH$_3$ and Cl]
NaOCH$_2$CH$_3$ / CH$_3$CH$_2$OH ⟶ [methylenecyclohexane] + [methylcyclohexene] major product

238 Chapter 7: Haloalkanes

(c)

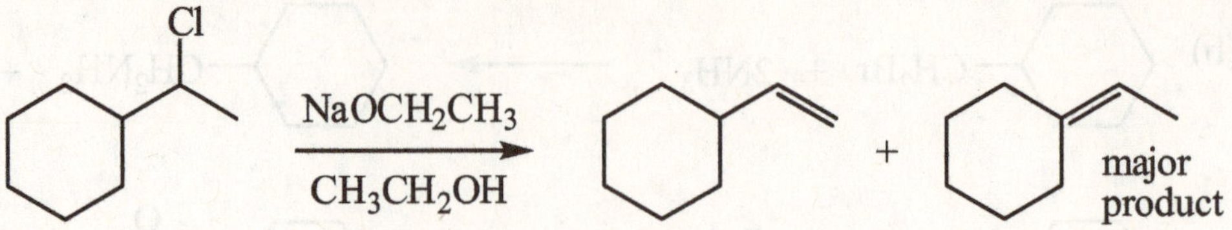

major product

(d)

major product

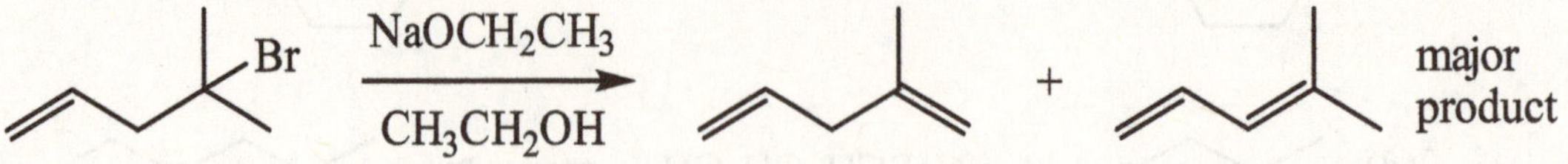

7.37 Which of the following haloalkanes undergo dehydrohalogenation to give alkenes that do not show *cis-trans* isomerism?

Shown below are the major products formed by the dehydrohalogenation (removal of HX via an elimination reaction) of the corresponding compound. No *cis-trans* isomerism is possible for the products formed from chlorocyclohexane and isobutyl chloride.

(a) 2-Chloropentane

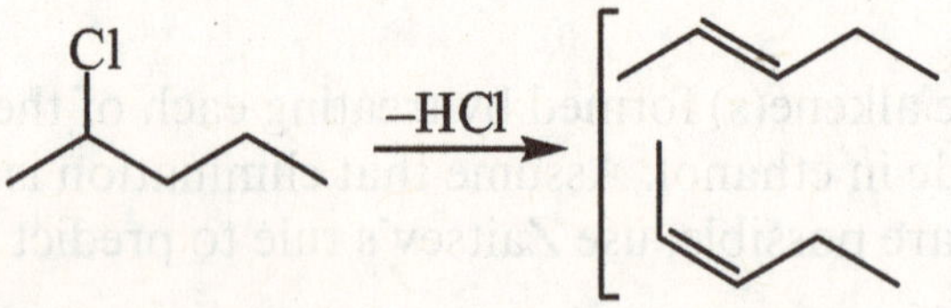

(b) 2-Chlorobutane

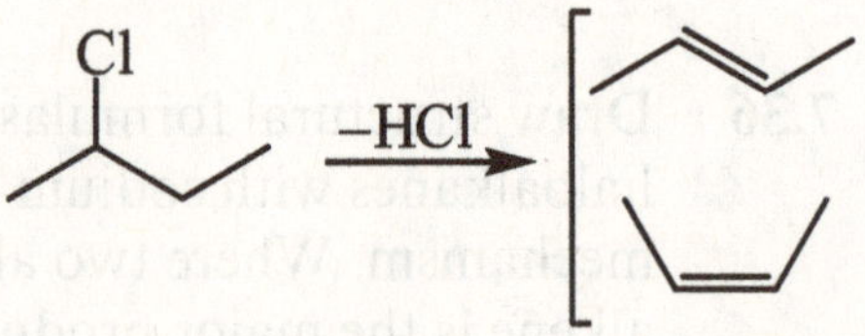

(c) Chlorocyclohexane

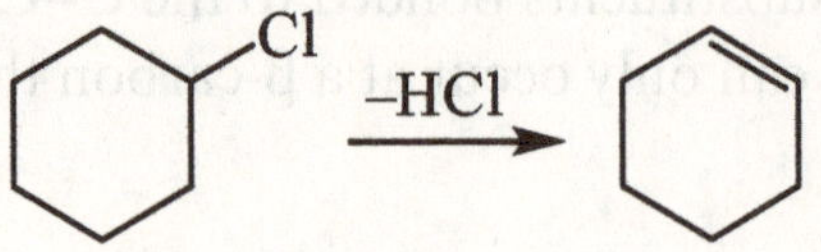

(d) Isobutyl chloride

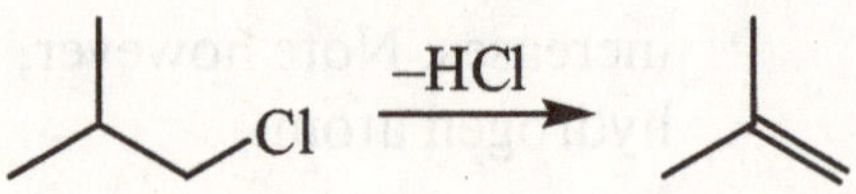

7.38 How many isomers, including *cis-trans* isomers, are possible for the major product of dehydrohalogenation of each of the following haloalkanes?

(a)

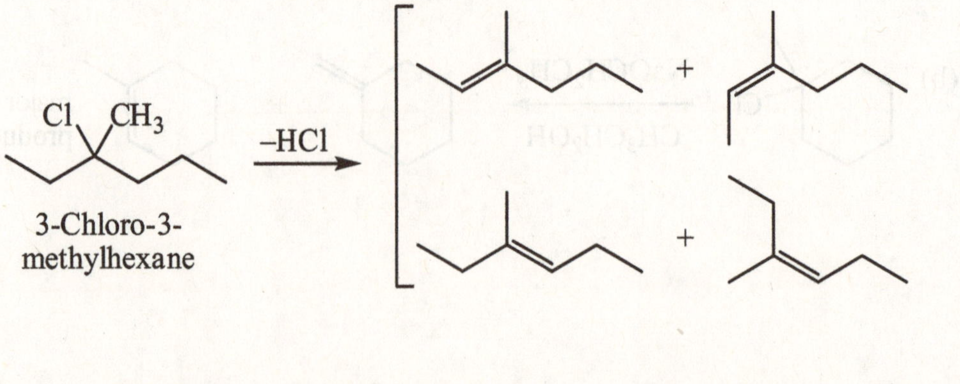

3-Chloro-3-methylhexane

(b) 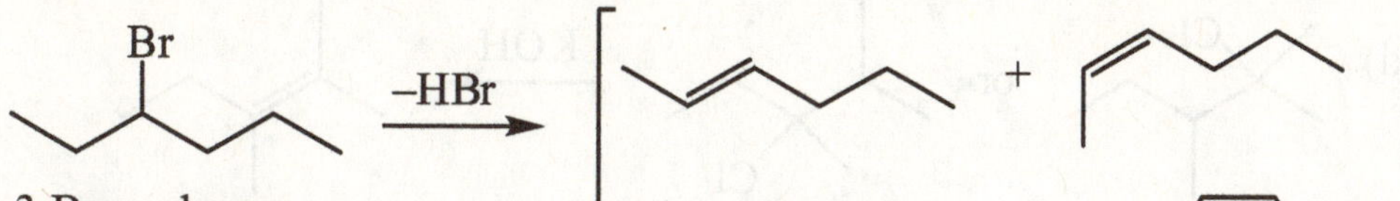

3-Bromohexane

7.39 What haloalkane might you use as a starting material to produce each of the following alkenes in high yield and uncontaminated by isomeric alkenes?

Because the alkenes were prepared from haloalkanes, we can conclude that a halogen was elminated from one of the alkene carbon atoms and that one hydrogen was eliminated from the other alkene carbon atom. By placing the halogen on the terminal carbon atom of each haloalkane, the corresponding alkene can be obtained in high yield.

(a) 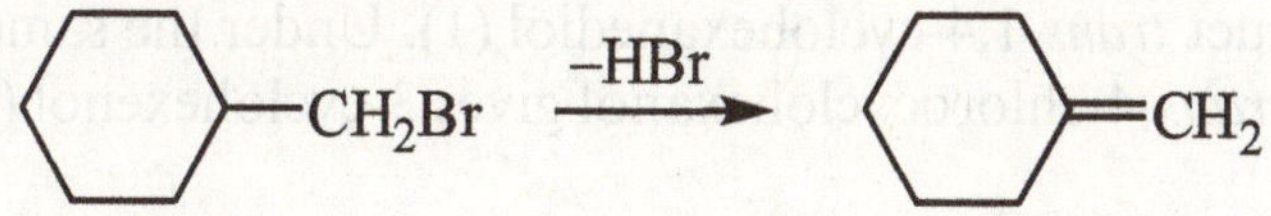

(b) $CH_3CH(CH_3)CH_2CH_2CH_2Br \xrightarrow{-HBr} CH_3CH(CH_3)CH_2CH{=}CH_2$

7.40 For each of the following alkenes, draw structural formulas of all chloroalkanes that undergo dehydrohalogenation when treated with KOH to give that alkene as the major product (for some parts, only one chloroalkane gives the desired alkene as the major product; for other parts, two chloroalkanes may work):

(a) 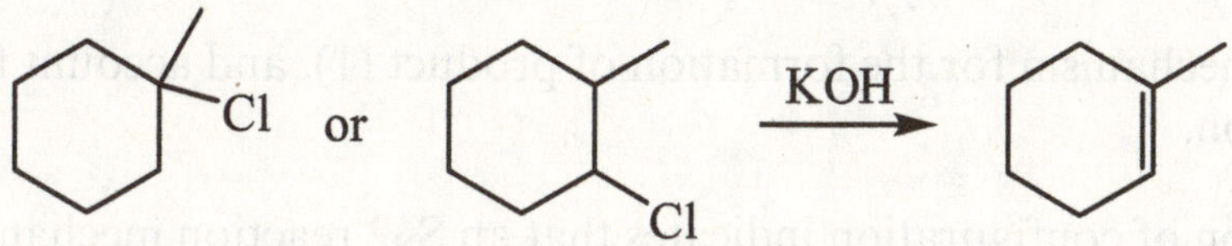

(b) 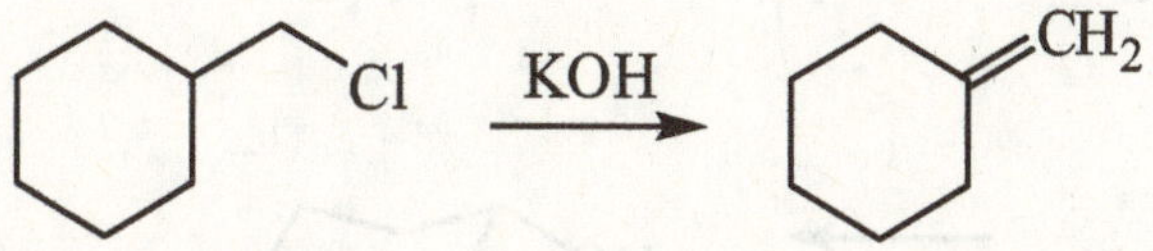

(c) 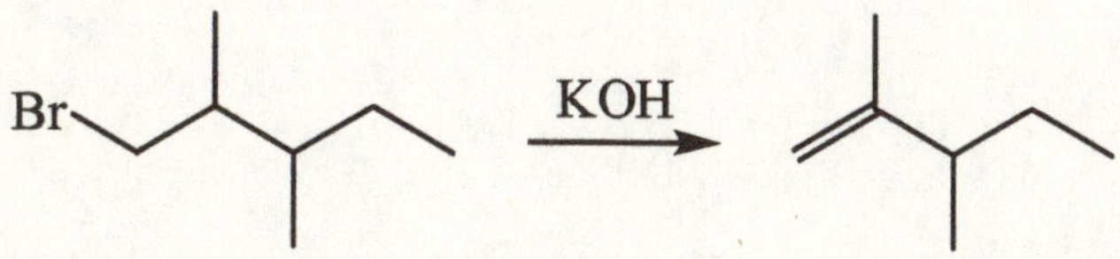

(d)

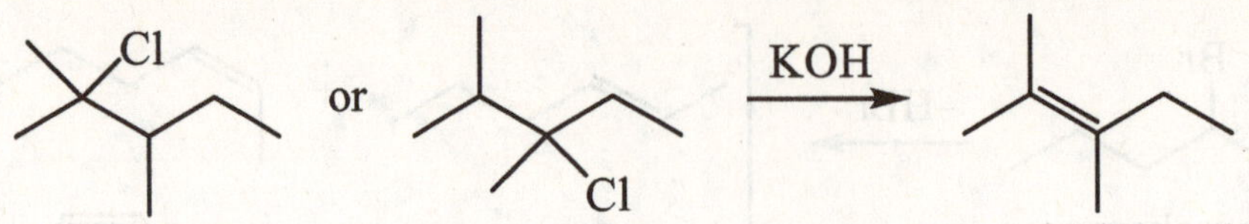

(e)

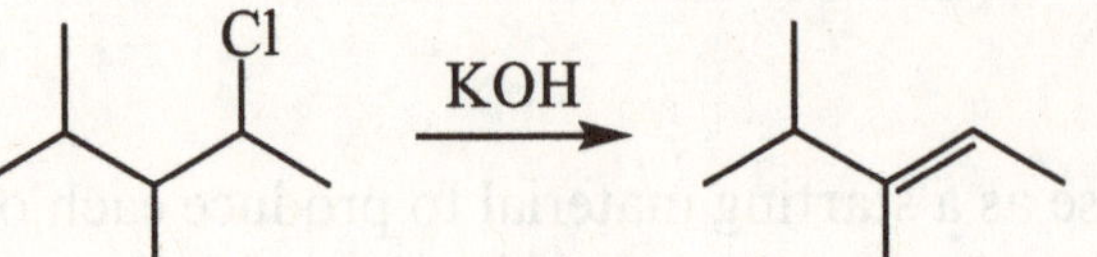

(f)

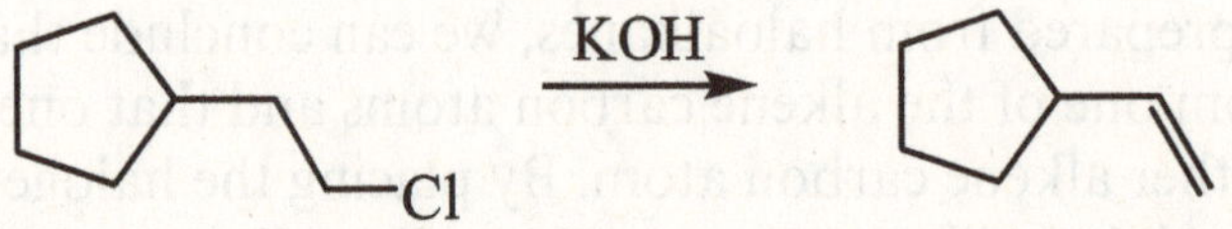

7.41 When *cis*-4-chlorocyclohexanol is treated with sodium hydroxide in ethanol, it gives only the substitution product *trans*-1,4-cyclohexanediol (1). Under the same experimental conditions, *trans*-4-chlorocyclohexanol gives 3-cyclohexenol (2) and the product (3):

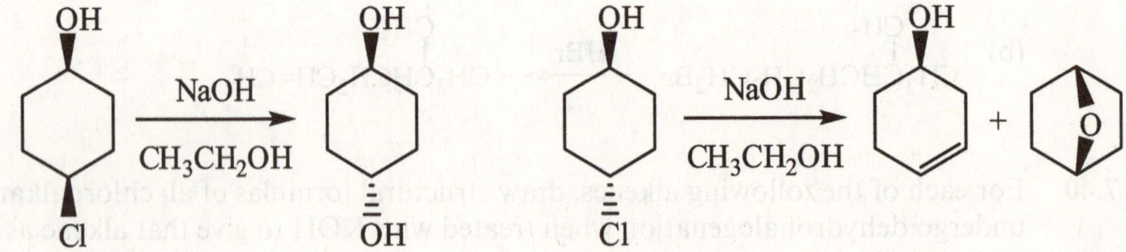

cis-4-Chloro-
cyclohexanol

(1)

trans-4-Chloro-
cyclohexanol

(2)   (3)

(a) Propose a mechanism for the formation of product (1), and account for its configuration.

The inversion of configuration indicates that an S_N2 reaction mechanism occurred. Hydroxide is a good nucleophile, thus favoring S_N2.

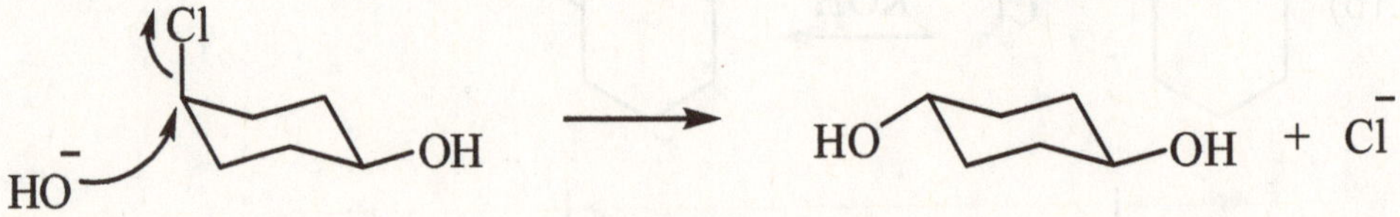

(b) Propose a mechanism for the formation of product (2).

Product (2) is formed via an elimination reaction. Because hydroxide is a strong base, the favored mechanism of elimination is E2.

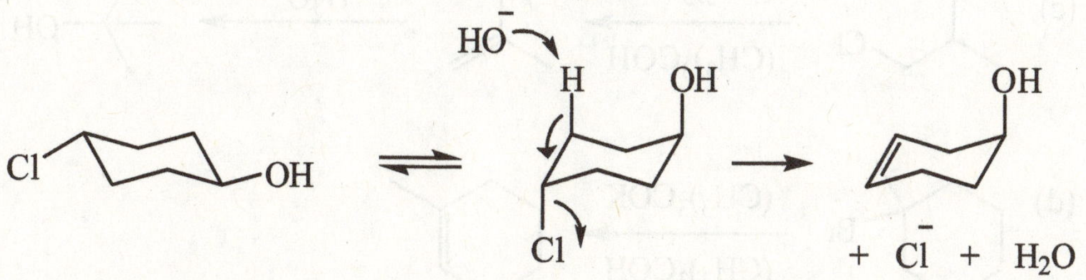

(c) Account for the fact that the product (3) is formed from the *trans* isomer, but not from the *cis* isomer.

In an S_N2 reaction, the nucleophile undergoes backside attack of the carbon bearing the leaving group. With *trans*-4-chlorocyclohexanol, the nucleophilic atom is created by deprotonating the hydroxyl group. The resulting alkoxide is properly oriented to initiate a backside attack. Whereas, the alkoxide nucleophile generated from *cis*-4-chlorocyclohexanol is situated on the same side as the leaving group and cannot initiate a backside attack.

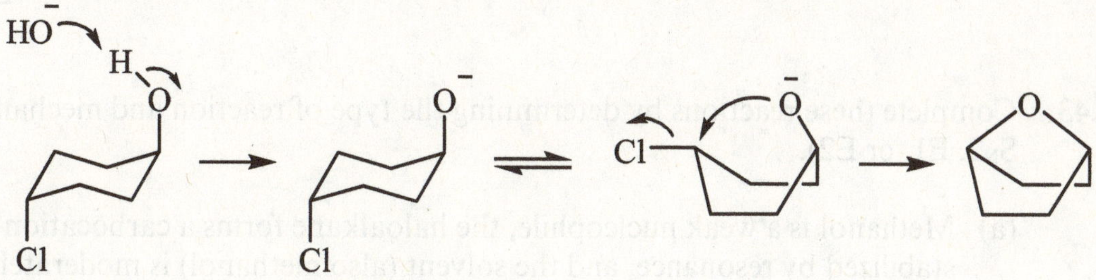

Synthesis and Predict the Product

7.42 Show how to convert the given starting material into the desired product (note that some syntheses require only one step, whereas others require two or more steps):

For all reactions that involve an elimination step, it is best to choose a bulky strong base, such as potassium *tert*-butoxide dissolved in *t*-butanol, to ensure high yield of the elimination product. The use of a smaller strong base, such as sodium methoxide, may also produce some substitution product, especially with 1° and 2° haloalkanes.

(a)

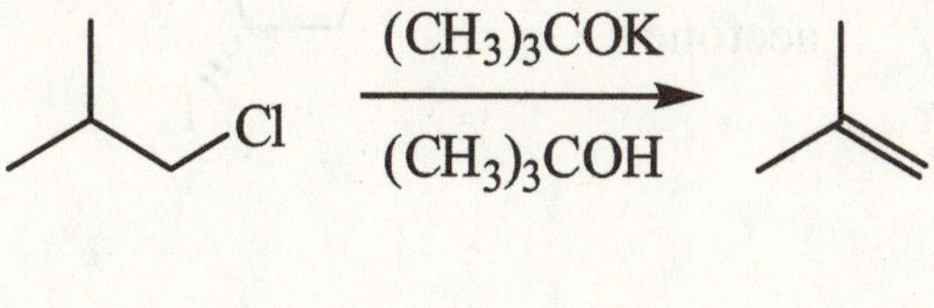

$(CH_3)_3COK$

$(CH_3)_3COH$

(b)

$$\diagdown\diagup = \diagdown \xrightarrow{\text{HBr}} \diagdown\diagup\diagdown\text{Br}$$

(c)

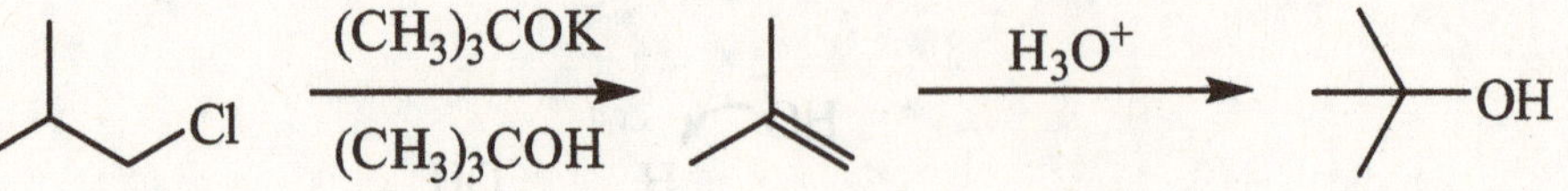

$$\diagup\diagdown\diagdown\text{Cl} \xrightarrow[\text{(CH}_3)_3\text{COH}]{\text{(CH}_3)_3\text{COK}} \diagdown\diagup = \diagdown \xrightarrow{\text{H}_3\text{O}^+} \diagdown\diagup\diagdown\text{OH}$$

(d)

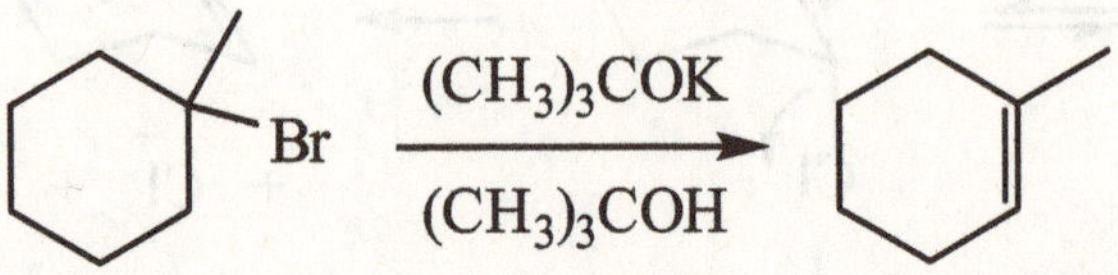

$$\xrightarrow[\text{(CH}_3)_3\text{COH}]{\text{(CH}_3)_3\text{COK}}$$

(e)

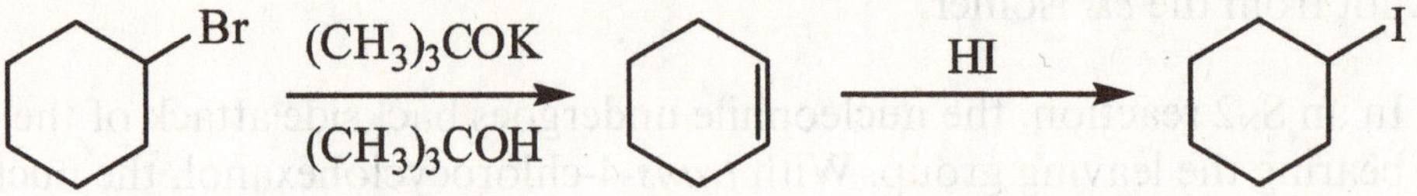

$$\xrightarrow[\text{(CH}_3)_3\text{COH}]{\text{(CH}_3)_3\text{COK}} \xrightarrow{\text{HI}}$$

(f)

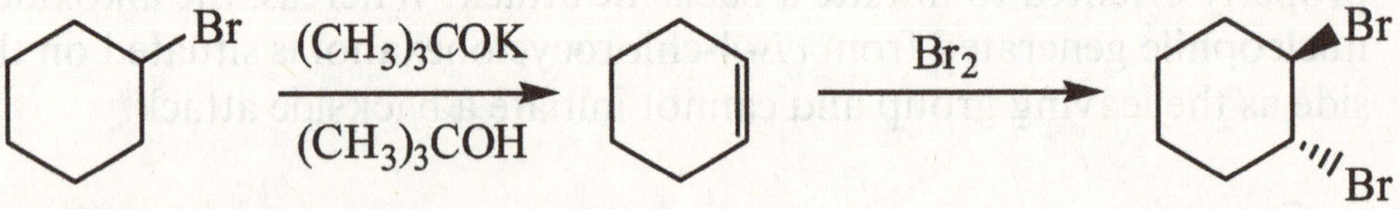

$$\xrightarrow[\text{(CH}_3)_3\text{COH}]{\text{(CH}_3)_3\text{COK}} \xrightarrow{\text{Br}_2} \quad \text{racemic}$$

7.43 Complete these reactions by determining the type of reaction and mechanism (S_N1, S_N2, E1, or E2).

(a) Methanol is a weak nucleophile, the haloalkane forms a carbocation that is stabilized by resonance, and the solvent (also methanol) is moderately ionizing. S_N1 occurs, and a racemic mixture is formed.

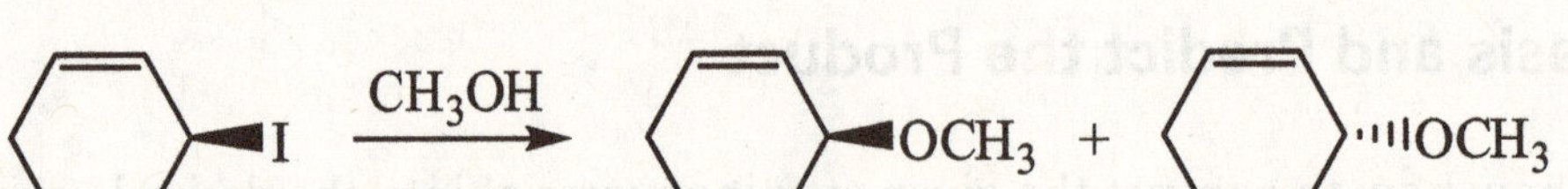

$$\xrightarrow{\text{CH}_3\text{OH}} \quad \text{OCH}_3 \quad + \quad \text{OCH}_3$$

(b) Iodide is an excellent nucleophile, and the solvent is weakly ionizing. Because the haloalkane is 2°, the reaction can proceed by S_N2, giving a single enantiomer.

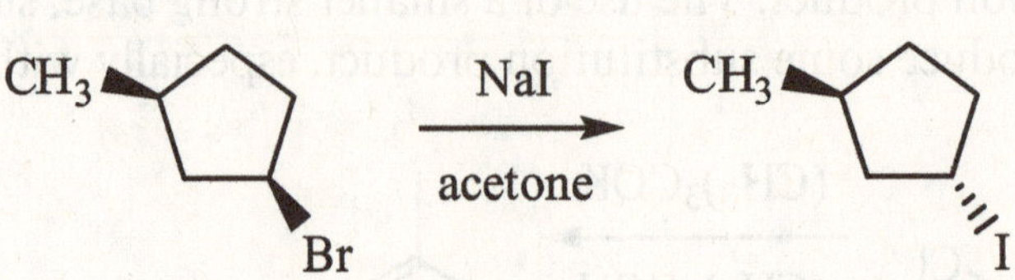

$$\text{CH}_3 \cdots \text{Br} \xrightarrow[\text{acetone}]{\text{NaI}} \text{CH}_3 \cdots \text{I}$$

(c) Hydroxide is both a strong nucleophile and a strong base, suggesting S_N2 or E2. The haloalkane, however, is 3° and cannot proceed by S_N2. Therefore, E2 occurs, giving the most stable alkene as the major product.

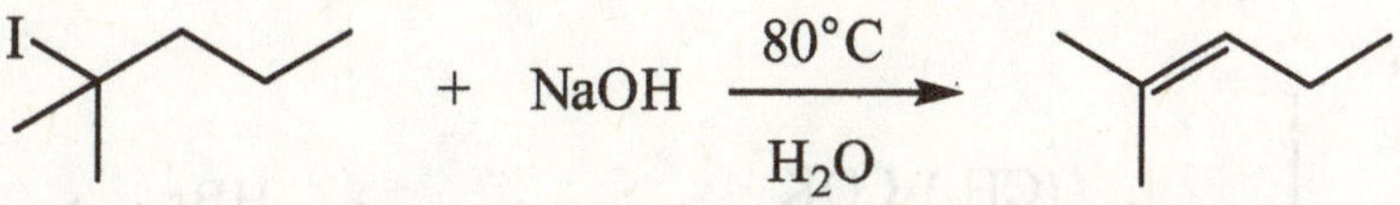

(d) Methoxide is both a strong nucleophile and a strong base. When combined with a 2° haloalkane, E2 elimination is more probable than S_N2 substitution. Note that elimination can only occur at a carbon atom bearing a β-hydrogen.

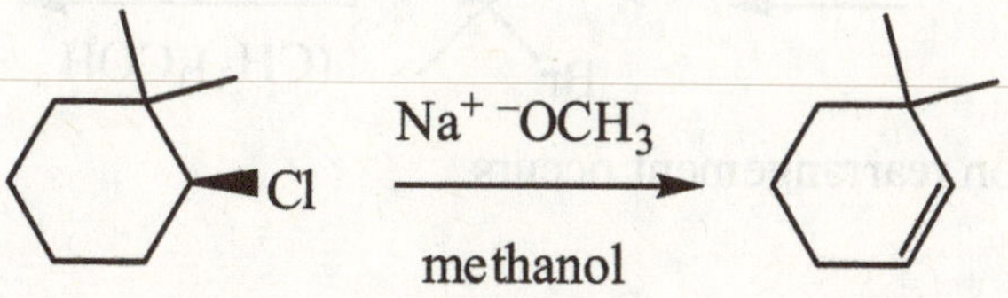

(e) The 2° haloalkane can form a relatively stable carbocation, and acetic acid is both a weak nuclophile and a very ionizing solvent. The reaction proceeds S_N1, forming a mixture of enantiomers.

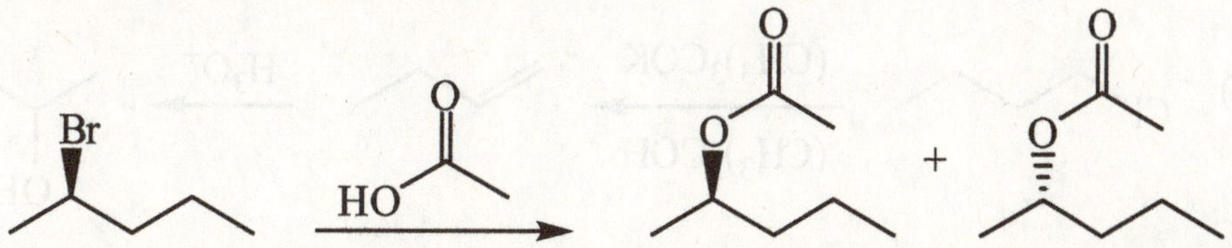

(f) The haloalkane is 1°, cyanide is a good nucleophile, and dimethylsulfoxide is a polar, aprotic solvent. These conditions are ideal for an S_N2 reaction.

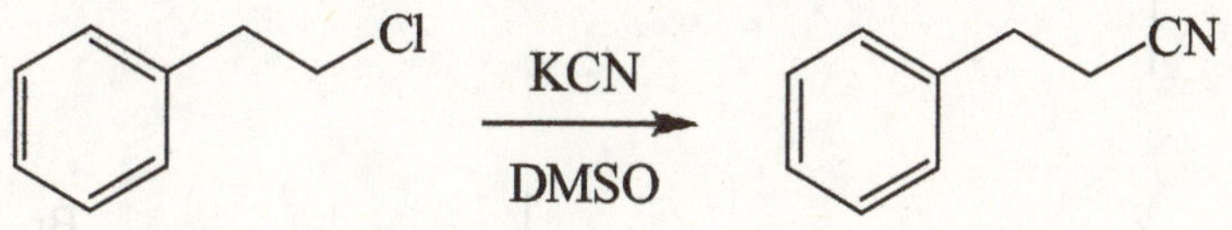

Chemical Transformations

7.44 Test your cumulative knowledge of the reactions learned thus far by completing the following chemical transformations. *Note: Some will require more than one step.*

(a)

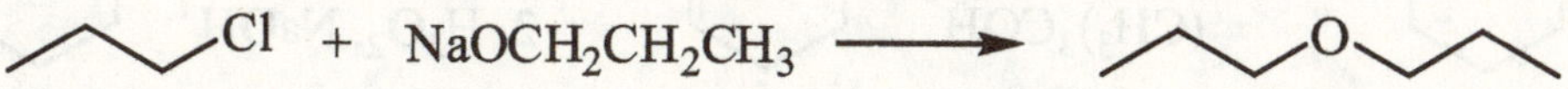

(b)

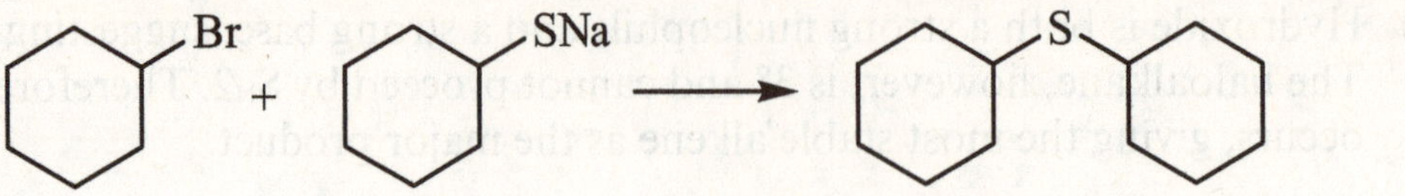

(c)

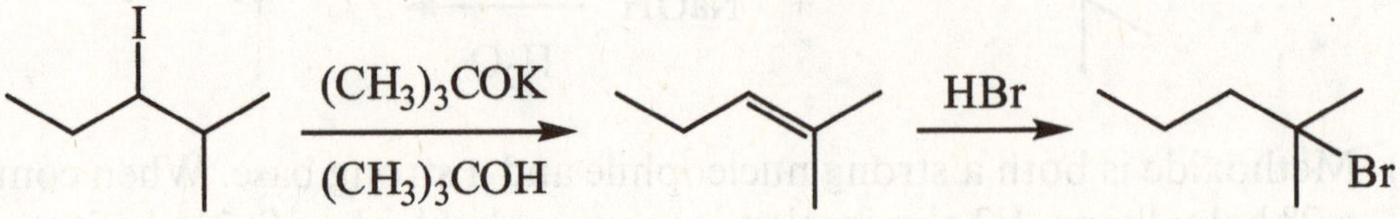

(d)

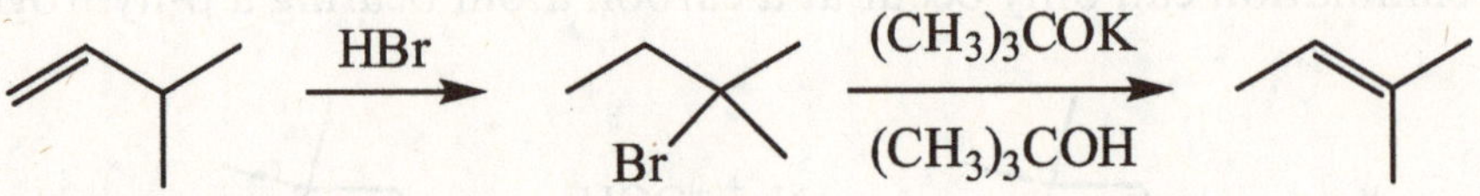

carbocation rearrangement occurs

(e)

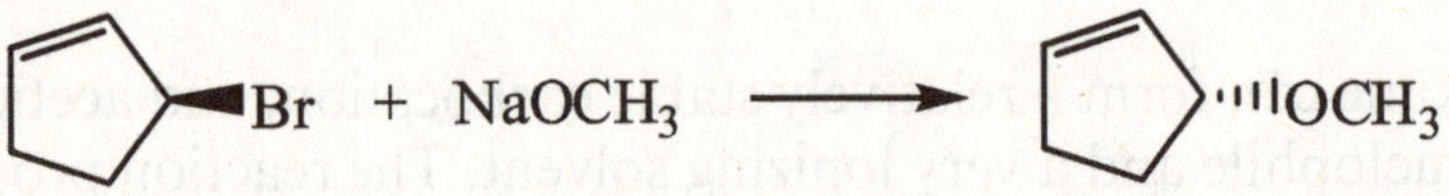

(f)

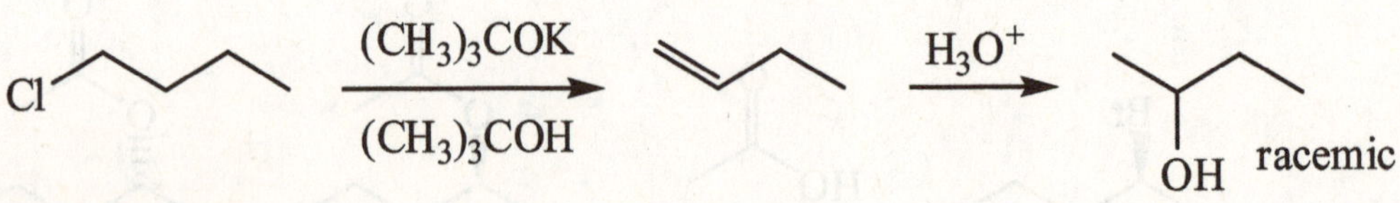

racemic

(g)

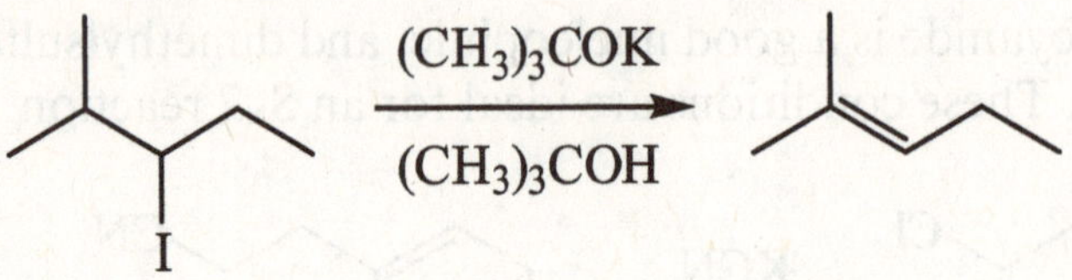

(h)

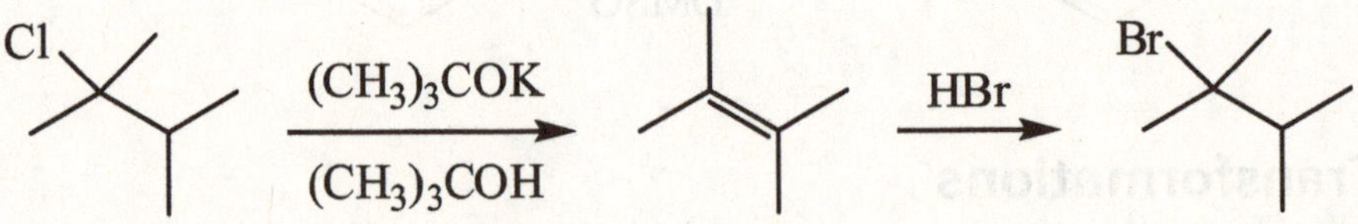

(i)

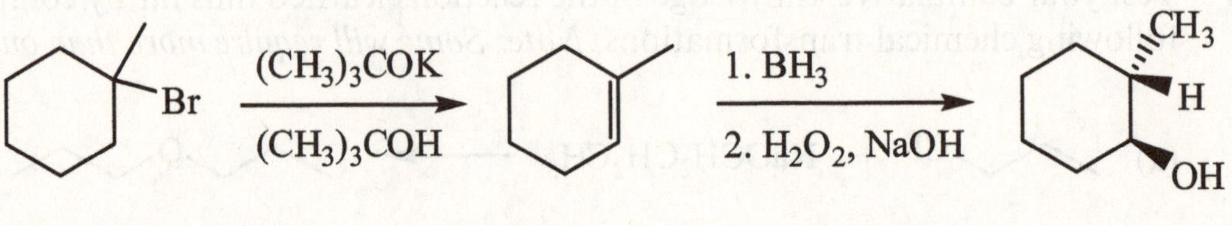

(j)

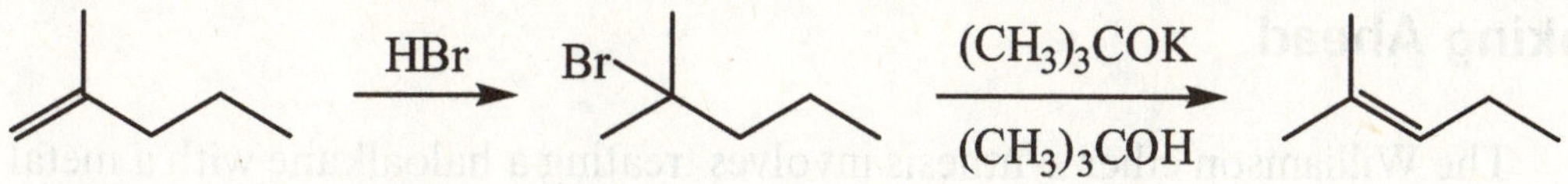

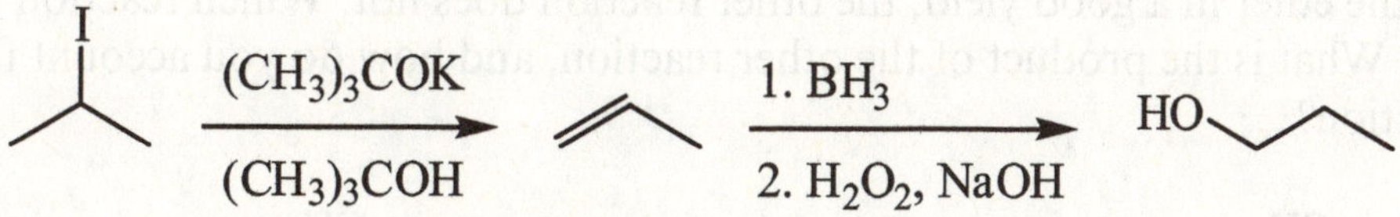

(k)

(l)

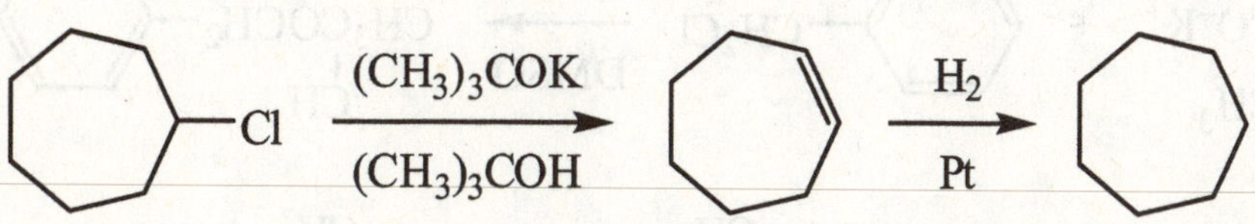

(m

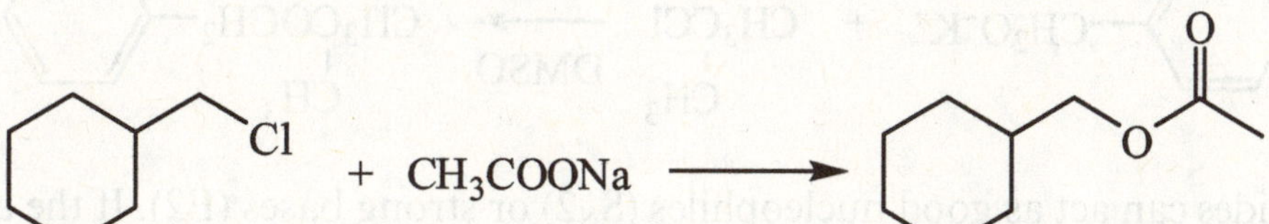

(n)

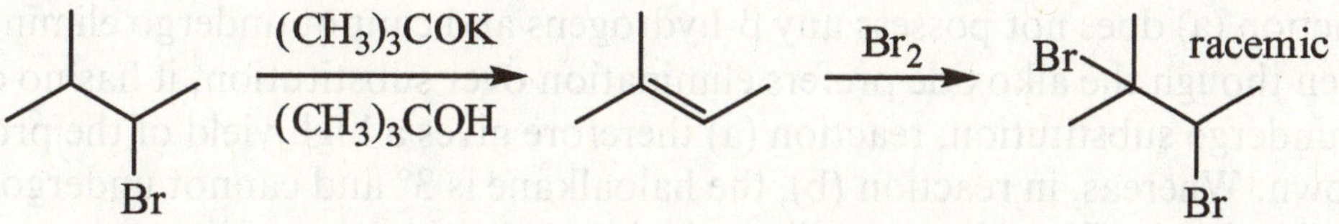

(o)

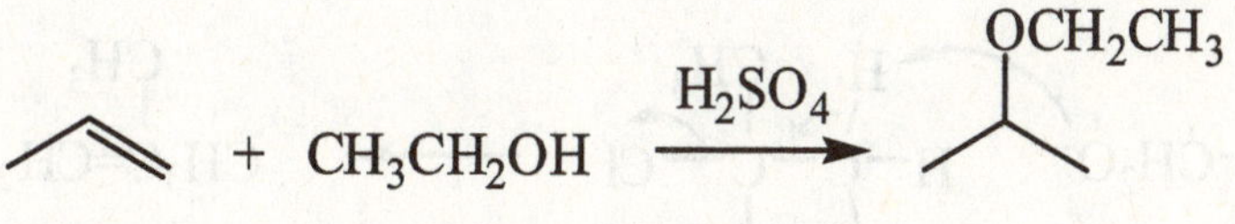

(p)

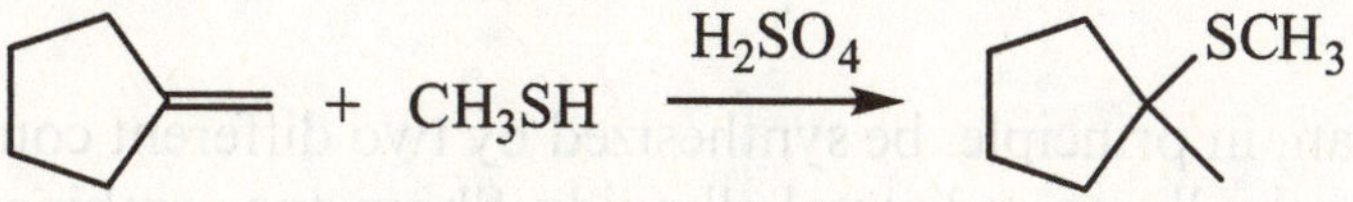

(q)

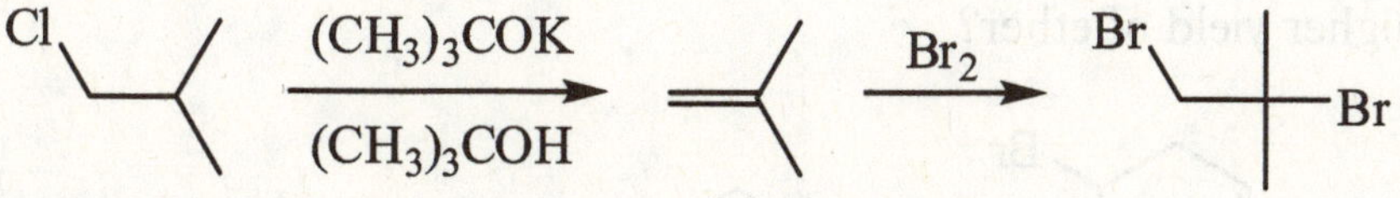

(r)

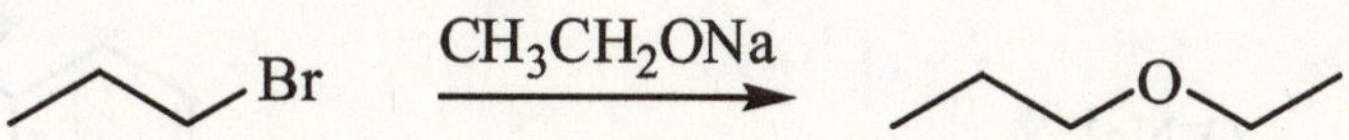

Looking Ahead

7.45   The Williamson ether synthesis involves treating a haloalkane with a metal alkoxide. Following are two reactions intended to give benzyl *tert*-butyl ether. One reaction gives the ether in a good yield, the other reaction does not. Which reaction gives the ether? What is the product of the other reaction, and how do you account for its formation?

(a)

$$CH_3CO^-K^+ \ (with\ CH_3,\ CH_3) \ + \ \langle C_6H_5 \rangle-CH_2Cl \ \xrightarrow{\ DMSO\ } \ CH_3COCH_2-\langle C_6H_5 \rangle \ (with\ CH_3,\ CH_3) \ + \ KCl$$

(b)

$$\langle C_6H_5 \rangle-CH_2O^-K^+ \ + \ CH_3CCl \ (with\ CH_3,\ CH_3) \ \xrightarrow{\ DMSO\ } \ CH_3COCH_2-\langle C_6H_5 \rangle \ (with\ CH_3,\ CH_3) \ + \ KCl$$

Alkoxides can act as good nucleophiles (S$_N$2) or strong bases (E2). If the alkoxide is large and bulky, as in the case of *tert*-butoxide in reaction (a), it will be more likely to act as a base, causing elimination, than as a nucleophile. Yet, the haloalkane in reaction (a) does not possess any β-hydrogens and cannot undergo elimination. Even though the alkoxide prefers elimination over substitution, it has no choice but to undergo substitution; reaction (a) therefore gives a high yield of the product shown. Whereas, in reaction (b), the haloalkane is 3° and cannot undergo S$_N$2, so elimination by E2 to form an alkene is the predominant reaction.

$$\langle C_6H_5 \rangle-CH_2O^- \qquad H-\overset{H}{\underset{H}{C}}-\overset{CH_3}{\underset{CH_3}{C}}-Cl \ \longrightarrow \ CH_3\overset{CH_3}{C}=CH_2$$

7.46   The following ethers can, in principle, be synthesized by two different combinations of haloalkane or halocycloalkane and metal alkoxide. Show one combination that forms ether bond (1) and another that forms ether bond (2). Which combination gives the higher yield of ether?

(a)

Bond (1):   [cyclohexenyl bromide]   +   [CH₃CH₂O⁻Na⁺]   ⟶

Bond (2):   [cyclohexenyl O⁻Na⁺]   +   [CH₃CH₂Br]   ⟶   [product with O bonds labeled (1) and (2)]

The formation of bond (1) involves the use of a secondary alkyl halide, which is much more likely to undergo an elimination reaction than is a primary alkyl halide. The formation of bond (2) gives the higher yield of ether.

(b) Bond (1):   [structure] —O⁻Na⁺ + CH₃Br ⟶   (1)   (2)

Bond (2):   [structure] —Br + CH₃O⁻Na⁺ ⟶

The alkyl halide used in the formation of bond (2) is a tertiary alkyl halide; it will react with the alkoxide by E2 instead of S$_N$2. The alkyl halide used in the formation of bond (1) contains only one carbon atom and cannot eliminate even though the alkoxide is bulky. The formation of bond (1) gives the higher yield of ether.

(c) Bond (1):   [structure] Br + [structure] O⁻Na⁺ ⟶   (1)   (2)

Bond (2):   [structure] O⁻Na⁺ + [structure] Br ⟶

The secondary alkyl halide used in the formation of bond (2) is much more likely to eliminate than the primary halide used in the formation of bond (1). Thus, the formation of bond (1) gives the highest yield of ether.

7.47 Propose a mechanism for this reaction:

$$Cl—CH_2—CH_2—OH \xrightarrow[H_2O]{Na_2CO_3} H_2C\overset{O}{—}CH_2$$

2-Chloroethanol   Ethylene oxide

In this reaction, an intramolecular S$_N$2 reaction occurs. An aqueous solution of sodium carbonate is basic, which results in the deprotonation of 2-chloroethanol. The alkoxide subsequently attacks the carbon atom bearing the halogen.

$$Cl—CH_2—CH_2—O—H \quad {}^-OH \longrightarrow Cl—CH_2—CH_2—O^-$$

$$\longrightarrow H_2C\overset{O}{—}CH_2$$

7.48   An OH group is a poor leaving group, and yet substitution occurs readily in the following reaction. Propose a mechanism for this reaction that shows how OH overcomes its limitation of being a poor leaving group.

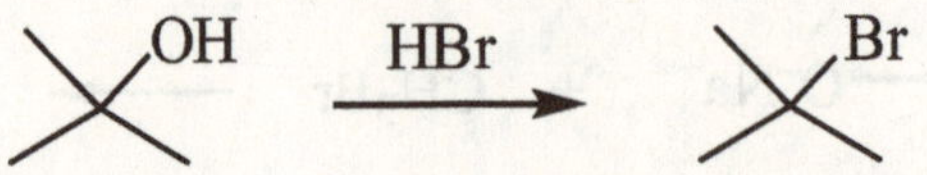

The OH group is a poor leaving group because it would leave as OH⁻, a relatively unstable species. In this reaction, the OH group is protonated by the strong acid to form an oxonium ion that leaves as water, which is a much better (more stable) leaving group.

First, HBr protonates the hydroxyl group to form an oxonium ion.

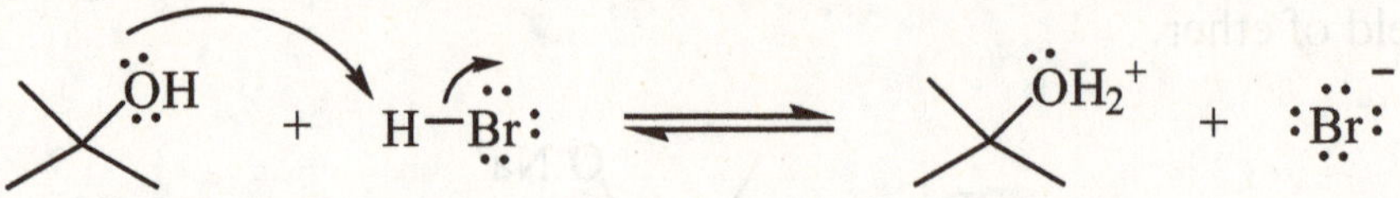

Loss of the leaving group generates a tertiary carbocation.

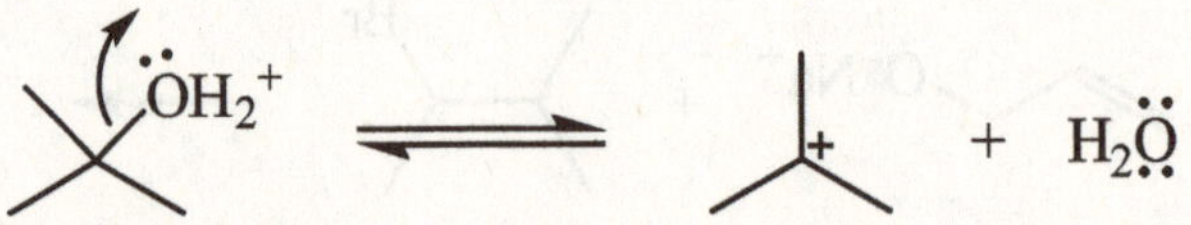

Nucleophilic attack of the carbocation by bromide forms the alkyl halide.

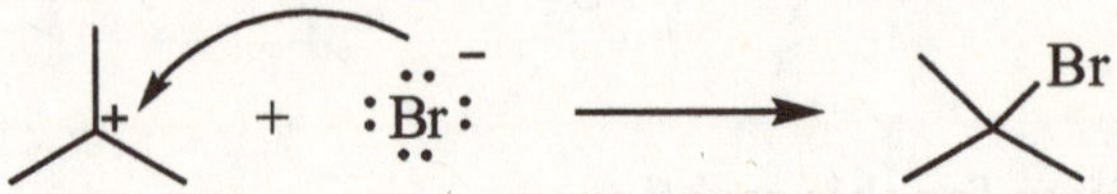

7.49   Explain why (*S*)-2-bromobutane becomes optically inactive when treated with sodium bromide in dimethylsulfoxide (DMSO):

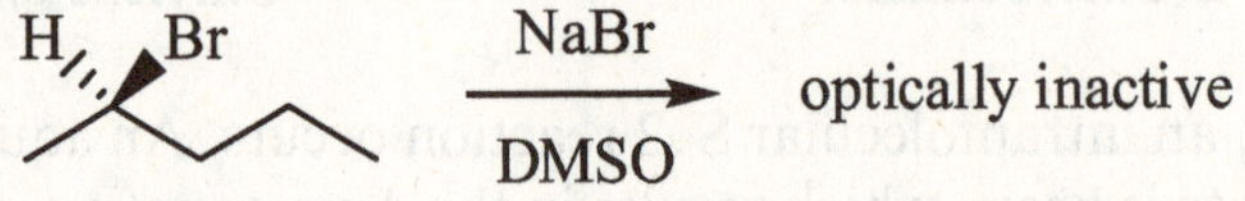

Bromide is both a good leaving group and a good nucleophile. In the presence of a polar, aprotic solvent such as DMSO, bromide displaces the bromide on the haloalkane via an S_N2 reaction, resulting in stereochemical inversion. The product that is formed can also react with bromide to regenerate the starting material. Over time, an equilibrium mixture consisting of equal amounts of both enantiomers is formed.

7.50   Explain why phenoxide is a much poorer nucleophile and weaker base than cyclohexoxide:

Sodium phenoxide   Sodium cyclohexoxide

The negative charge on phenoxide is stabilized by resonance and delocalized over multiple atoms. As a result, phenoxide is less basic, and the electron pairs are less available for nucleophilic reactions,compared to cyclohexoxide, which has no electron delocalization by resonance.

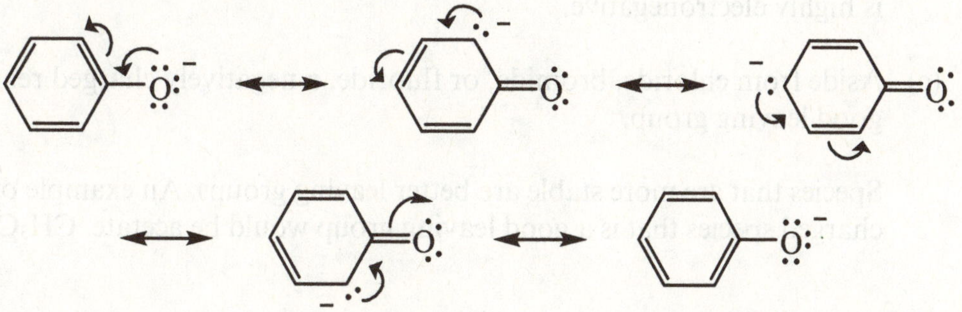

7.51   In ethers, each side of the oxygen is essentially an OR group and is thus a poor leaving group. Epoxides are three-membered ring ethers. Explain why an epoxide reacts readily with a nucleophile despite being an ether. Would you expect the five-membered and six-membered ring ethers shown to react with nucleophiles the same way that epoxides react with nucleophiles? Why or why not?

$$R-O-R \ + \ :Nu^- \longrightarrow \text{no reaction}$$

An ether

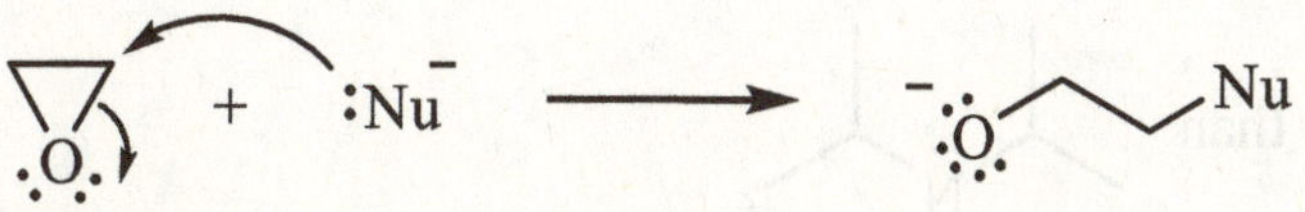

An epoxide

Although alkoxides are poor leaving groups, the opening of the highly strained three-membered ring is the driving force behind the opening of epoxides by nucleophiles. Five- or six-membered ring ethers are not highly strained and will not react the same way.

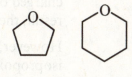

Group Learning Activities

7.52   Discuss and come up with examples of the following:

(a) A negatively charged reagent that is a weak base.

Weak bases are species with electron pairs that are more stable. An example would be the iodide ion, I^-, which is very stable due to its atomic size.

(b) A negatively charged reagent that is a poor nucleophile.

Although most negatively charged species tend to be good nucleophiles, that is not always the case. The fluoride ion, F^-, is a weak nucleophile because fluorine is highly electronegative.

(c) Aside from chloride, bromide, or fluoride, a negatively charged reagent that is a good leaving group.

Species that are more stable are better leaving groups. An example of a negatively charged species that is a good leaving group would be acetate, CH_3COO^-.

7.53   Discuss reasons why the following statements are true:

(a) Although hexane is an aprotic solvent, it is a poor solvent for an S_N2 reaction.

The nucleophiles used in S_N2 reactions tend to be negatively charged and originate from salts (such as NaBr). Because salts are not very soluble in aprotic solvents, the concentration of nucleophile in the solution would be very low, leading to a very slow reaction.

(b)

$CH_3\overset{..}{\underset{-}{N}}H$ is a better nucleophile than $CH_3\overset{..}{\underset{..}{O}}\!:^-$, but

is better than

Negatively charged nitrogen atoms are more nucleophilic than negatively charged oxygen atoms because nitrogen is less electronegative than oxygen. As a result, the methylamide anion is a better nucleophile than the methoxide anion.

However, the diisopropylamide anion is worse of a nucleophile than the isopropoxide anion because the diisopropylamide anion is much bulkier. As a result, in an S_N2 reaction, it cannot attack a carbon atom as effectively.

7.54 Discuss ways that you could speed up the following reactions without changing the products formed.

(a)

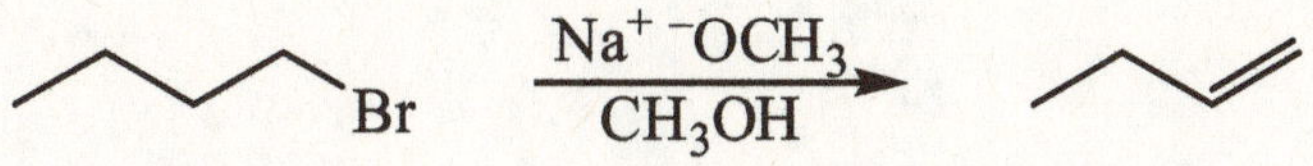

$$\text{CH}_3\text{CH}_2\text{CH}_2\text{CH}_2\text{Br} \xrightarrow[\text{CH}_3\text{OH}]{\text{Na}^+ {}^-\text{OCH}_3} \text{CH}_2=\text{CHCH}_2\text{CH}_3$$

(b)

$$\xrightarrow{\text{NH}_2\text{CH}_3}$$

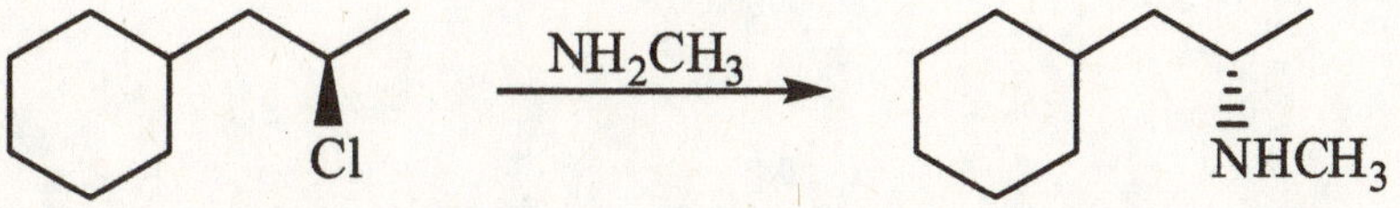

The speed of each of these reactions can be increased by using a starting material that has a better leaving group. For example, an alkyl iodide can be used instead of an alkyl bromide or chloride.

The reactions can also be performed on a hot plate. Chemical reactions occur faster at higher temperatures.

Chapter 8: Alcohols, Ethers, and Thiols

Problems

8.1 Write the IUPAC name for each alcohol:

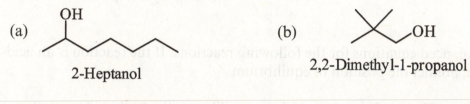

(a)

2-Heptanol

(b)

2,2-Dimethyl-1-propanol

(c)

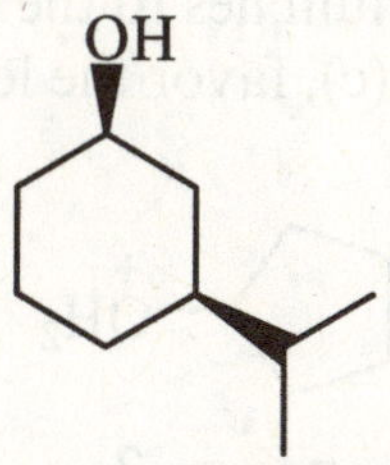

(1R,3S)-3-Isopropylcyclohexanol

(d)

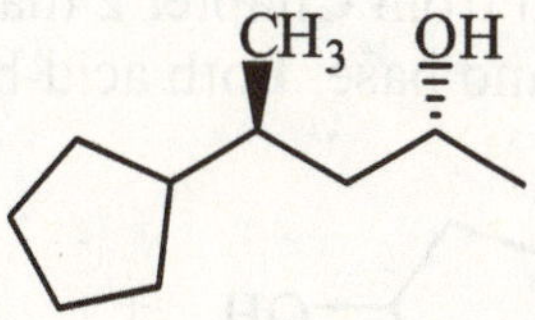

(2R,4S)-4-Cyclopentyl-2-pentanol

8.2 Classify each alcohol as primary, secondary, or tertiary:

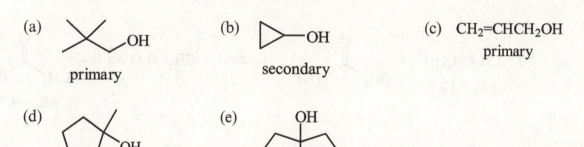

(a)

primary

(b)

secondary

(c) CH$_2$=CHCH$_2$OH
primary

(d)

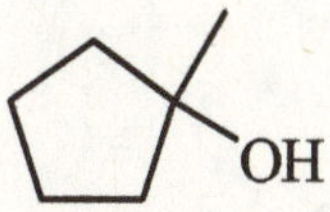

tertiary

(e)

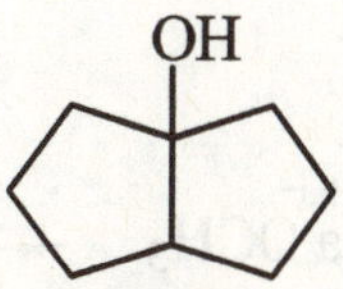

tertiary

8.3 Write the IUPAC name for each alcohol:

(a)

trans-3-Penten-1-ol

(b)

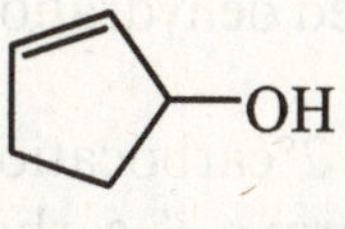

2-Cyclopentenol

(c)

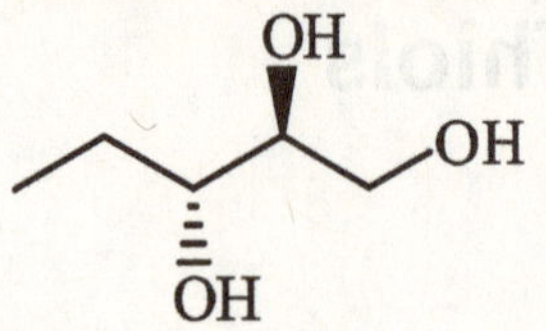

(2S,3R)-1,2,3-Pentanetriol

(d)

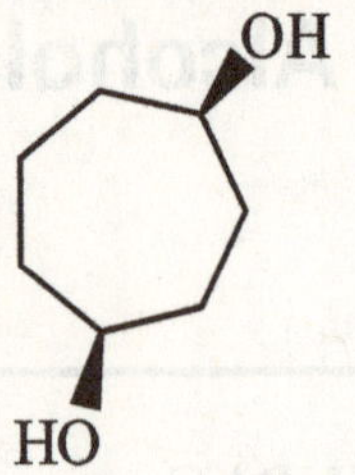

(1R,4S)-Cycloheptane-1,4-diol

8.4 Write balanced equations for the following reactions. If the reaction is an acid-base reaction, predict the position of equilibrium.

Recall from Chapter 2 that an acid-base equilibrium lies to the side with the weakest acid and base. Both acid-base reactions, (a) and (c), favor the left side.

(a)

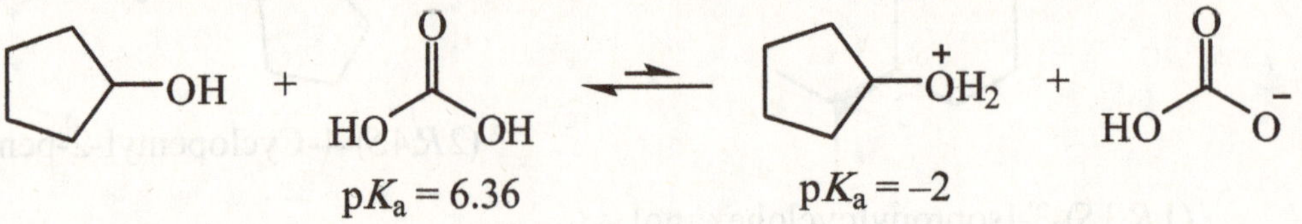

(b)

(c)

(d)

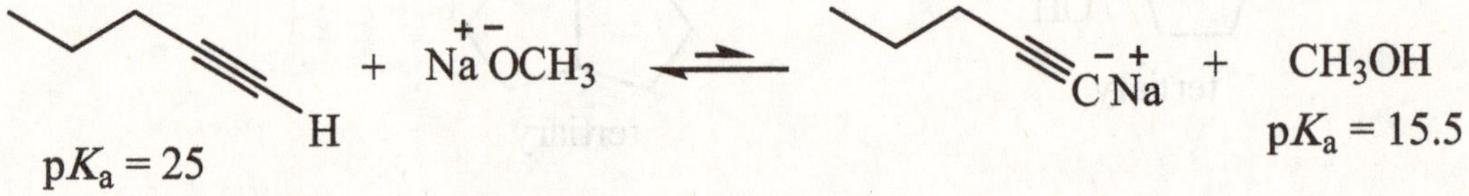

8.5 For each of the following alcohols, draw structural formulas for the alkenes that form upon acid-catalyzed dehydration, and predict which alkene is the major product:

In (a) and (b), the 2°carbocation intermediate undergoes rearrangement by a 1,2-hydride shift to form a 3° carbocation. The major product formed from all three reactions is the one bonded to more alkyl substituents, because alkenes with more alkyl substituents are more stable (Zaitsev's Rule). However, in (c), the *trans* product is more favourable than the *cis* product because there is less steric hindreance.

(a)

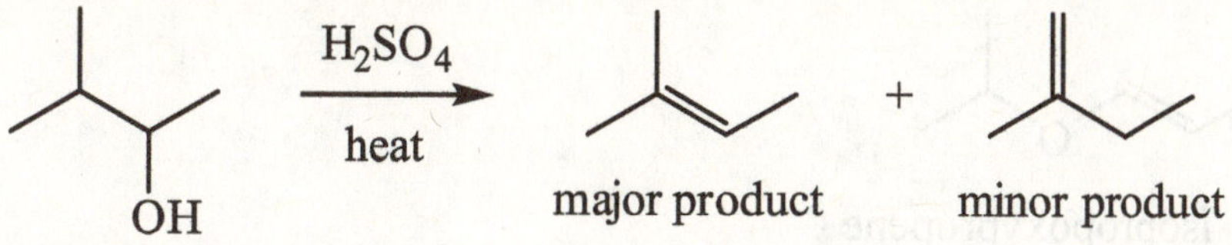

(b)

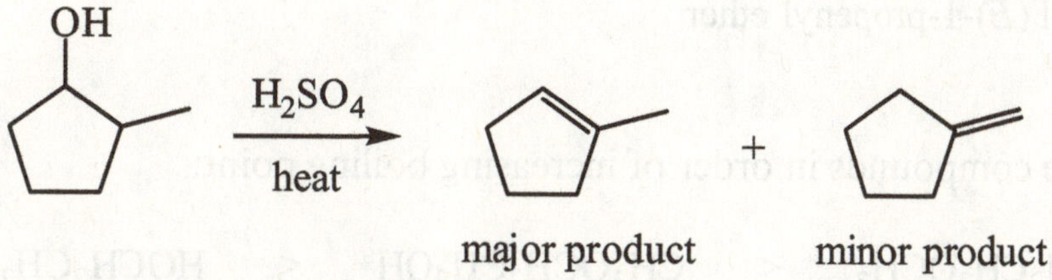

(c)

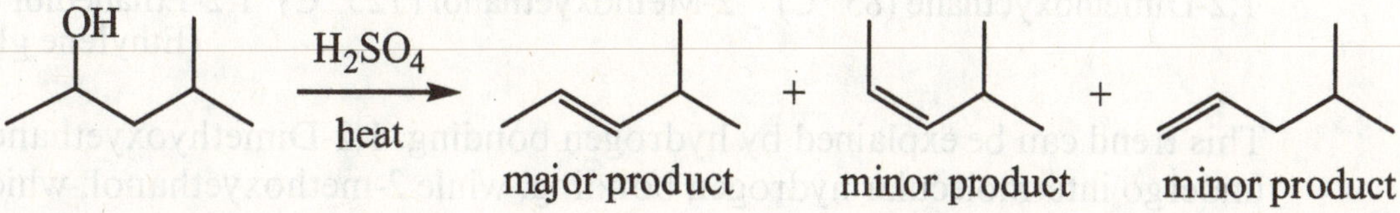

8.6 Draw the product formed by treating each alcohol in Example 8.6 with chromic acid.

Chromic acid is a strong oxidizing agent that oxidizes primary alcohols to carboxylic acids (via an aldehyde intermediate). Secondary alcohols are oxidized to ketones, which cannot be oxidized further due to the lack of a hydrogen atom on the ketone carbon.

(a)

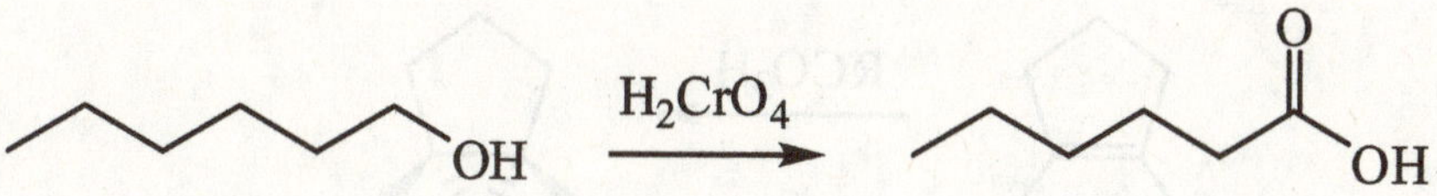

(b)

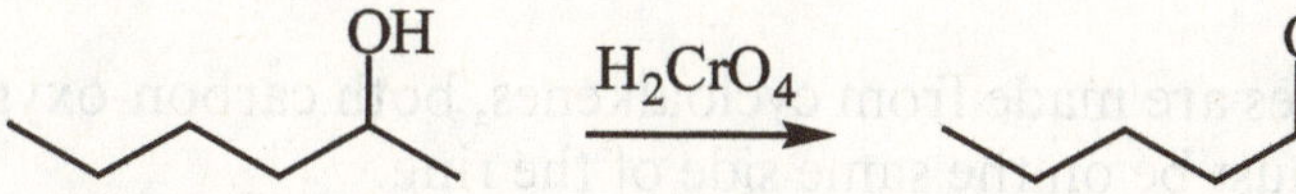

(c)

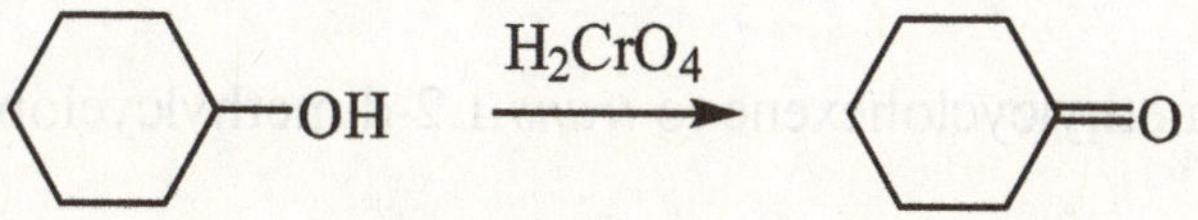

8.7 Write the IUPAC and common names for each ether:

(a)
$$CH_3CHCH_2OCH_2CH_3$$
with CH_3 substituent

1-Ethoxy-2-methylpropane

Ethyl isobutyl ether

(b)

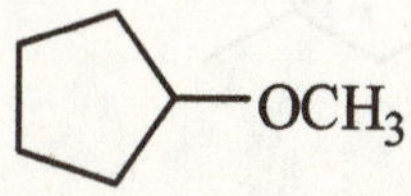

Methoxycyclopentane

Cyclopentyl methyl ether

(c)

(*E*)-1-Isopropoxypropene

Isopropyl (*E*)-1-propenyl ether

8.8   Arrange these compounds in order of increasing boiling point:

$$CH_3OCH_2CH_2OCH_3 \quad < \quad CH_3OCH_2CH_2OH \quad < \quad HOCH_2CH_2OH$$

1,2-Dimethoxyethane (85 °C)   2-Methoxyethanol (125 °C) 1,2-Ethanediol (197 °C)
                                                                    (Ethylene glycol)

This trend can be explained by hydrogen bonding. 1,2-Dimethyoxyethane cannot undergo intermolecular hydrogen bonding, while 2-methoxyethanol, which contains one −OH group, is capable of intermolecular hydrogen bonding. Ethylene glycol, which has two −OH groups and thus more hydrogen-bonding sites, has the highest boiling point. (A mixture of water and ethylene glycol is used as automotive coolant due to its ability to resist boiling at high temperatures.)

8.9   Draw the structural formula of the epoxide formed by treating 1,2-dimethylcyclopentene with a peroxycarboxylic acid.

Recall that when epoxides are made from cycloalkenes, both carbon-oxygen bonds to the epoxide oxygen must be on the same side of the ring.

8.10   Show how to convert 1,2-dimethylcyclohexene to *trans*-1,2-dimethylcyclohexane-1,2-diol.

The conversion of a cycloalkene to a *trans*-1,2-diol (glycol) can be accomplished by epoxidation followed by acid-catalyzed ring opening, which is stereoselective. (Compare the mechanism of ring opening to the addition of Cl_2 or Br_2 to alkenes.)

8.11 Write the IUPAC name for each thiol:

(a)

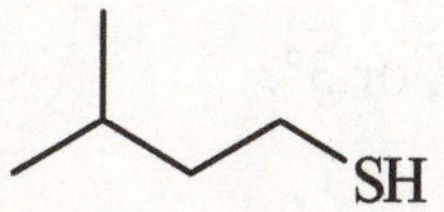

3-Methyl-1-butanethiol

(b)

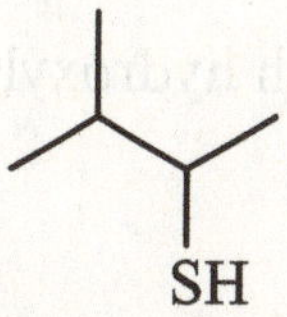

SH

3-Methyl-2-butanethiol

(c)

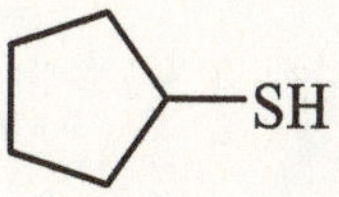

—SH

Cyclopentanethiol

8.12 Predict the products of the following reactions. If the reaction is an acid-base reaction, predict its position of equilibrium.

Thiols are stronger acids ($pK_a \approx 9$) than alcohols ($pK_a \approx 16$) but are weaker acids than carboxylic acids ($pK_a \approx 5$). Reaction (a) lies to the right because acetate is more stable than the 3-pentylsulfide ion; however, this is not easily predicted from the structural effects learned in Chapter 2 due to the presence of multiple effects. Reaction (b) lies to the right because, as a result of atomic size, the sulfide ion is more stable than hydroxide.

(a)

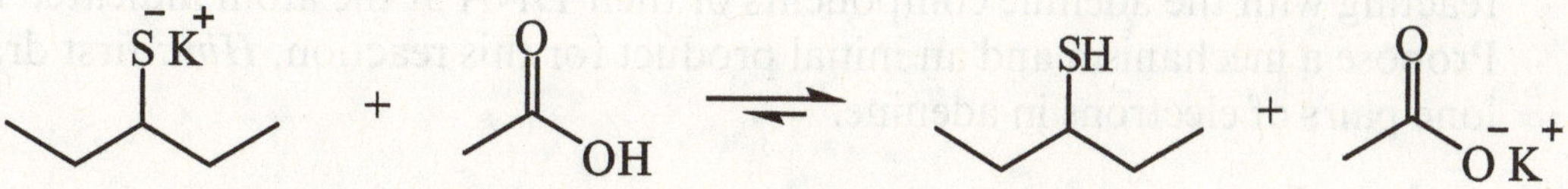

$$\bar{S}\,K^{+}$$

$+$

$$pK_a = 4.76$$

$\rightleftharpoons$

SH

$$pK_a = 9$$

$+$

$$\bar{O}\,K^{+}$$

(b) HS⌒⌒OH + NaOH $\rightleftharpoons$ Na$^{+}\bar{S}$⌒⌒OH + H_2O

pK_a of thiol $= 9$

$pK_a = 15.7$

(c)

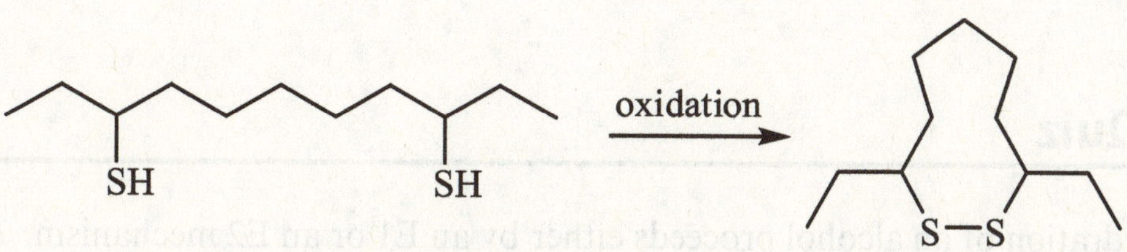

SH   SH

oxidation $\longrightarrow$

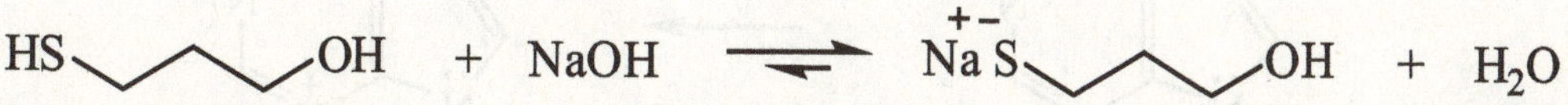

S—S

(d)

$\overset{}{\underset{S}{}}$S

reduction $\longrightarrow$

SH + HS

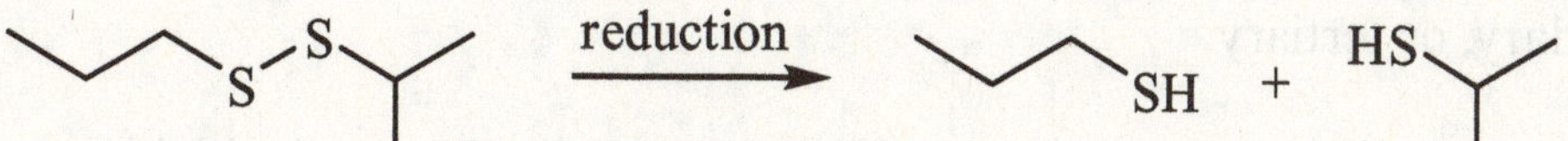

Chemical Connections

8A. Classify each hydroxyl group in glycerol as either 1°, 2°, or 3°.

$$CH_2{-}OH \quad 1°$$
$$CH{-}OH \quad 2°$$
$$CH_2{-}OH \quad 1°$$

8B. Although methanol and isopropyl alcohol are much more toxic than ethanol and would rarely be found in one's breath, would these two compounds also give a positive alcohol screening test? If so, what would be the products of these reactions?

A positive alcohol screening test is indicated by a color change caused by the reduction of the reddish-orange dichromate ion to green Cr(III). Concomitant with the reduction of dichromate is the oxidation of the alcohol. Both methanol and isopropyl alcohol will give a positive test. Methanol would be oxidized to methanal (formaldehyde) and subsequently to methanoic acid (formic acid). Isopropyl alcohol, a secondary alcohol, would only be oxidized to 2-propanone (acetone).

8C. One of the ways that ethylene oxide has been found to kill microorganisms is by reacting with the adenine components of their DNA at the atom indicated in red. Propose a mechanism and an initial product for this reaction. *Hint:* first draw in any lone pairs of electrons in adenine.

Quick Quiz

1. Dehydration of an alcohol proceeds either by an E1 or an E2 mechanism. *True*. The mechanism by which dehydration proceeds depends whether the alcohol is primary, secondary, or tertiary.

2. Epoxides are more reactive than acyclic ethers. *True*. Due to the highly strained three-membered ring, epoxides are much more reactive than acyclic ethers, which are relatively unreactive except towards combustion.

3. Attack of an electrophile on the carbon of an epoxide ring results in opening of the ring. *False*. Ring opening is caused by *nucleophilic* attack.

4. A hydrogen bond is a form of dipole-dipole interaction. *True*. Hydrogen bonds arise from interactions between the dipoles formed by hydrogen bonded to an electronegative atom.

5. Alcohols have higher boiling points than thiols with the same molecular weight. *True*. Alcohols are capable of intermolecular hydrogen bonding, but thiols are not.

6. Thiols are more acidic than alcohols. *True*. The conjugate base of a thiol (a sulfur atom with a negative charge) is more stable than that of an alcohol (an oxygen atom with a negative charge) due to the larger atomic size of sulfur.

7. Alcohols can act as hydrogen-bond donors but not as hydrogen-bond acceptors. *False*. Oxygen has lone pairs that can act as hydrogen-bond acceptors.

8. Alcohols can function as both acids and bases. *True*. Just like water, an alcohol can accept or donate a proton.

9. Ethers can act as hydrogen-bond donors but not as hydrogen-bond acceptors. *False*. Rather, the opposite is true. Without any O−H bonds, ethers cannot act as hydrogen-bond donors.

10. Reduction of a thiol produces a disulfide. *False*. The process involves the removal of hydrogen, so it is an oxidation.

11. Ethers are more reactive than alcohols. *False*. Except towards combustion, ethers (acyclic) are generally very unreactive.

12. $(CH_3CH_2)_2CHOH$ is classified as a 3° alcohol. *False*. The carbon bearing the hydroxyl group is bonded to two alkyl groups (in this case, ethyl), so it is a secondary alcohol.

13. PCC will oxidize a secondary alcohol to a ketone. *True*. PCC will also oxidize a primary alcohol to an aldehyde.

14. PCC will oxidize a primary alcohol to a carboxylic acid. *False*. The oxidation of a primary alcohol will stop at the aldehyde when PCC is used.

15. Alcohols have higher boiling points than ethers with the same molecular weight. *True*. No intermolecular hydrogen bonding is possible with ethers.

16. A dehydration reaction yields an epoxide as the product. *False*. The dehydration of an alcohol yields an alkene. The alkene can then be oxidized to an epoxide using a peroxycarboxylic acid. A balanced half-reaction will show that the transformation is a two-electron oxidation.

17.   Alcohols can be converted to alkenes. *True*. Alkenes can be prepared from the dehydration of alcohols.

18.   Alcohols can be converted to haloalkanes. *True*. The hydroxyl group is a poor leaving group, but it can be converted to a good leaving group by protonating it to generate an oxonium ion. Substitution of the leaving group with a halide forms a haloalkane.

19.   In naming alcohols, "alkyl alcohol" is the IUPAC form of the name, while "alkanol" is the common form of the name. *False*. The exact opposite is true.

20.   −OH is a poor leaving group. *True*. It can be converted into a good leaving group by protonation, which generates an oxonium ion that leaves as water.

21.   A glycol is any alcohol with at least two hydroxyl groups. *True*. Glycols contain two hydroxyl groups on different carbons.

End-of-Chapter Problems

Structure and Nomenclature

8.13   Classify the alcohols as primary, secondary, or tertiary.

The classification of alcohols is based on the number of carbon atoms connected to the carbon atom bearing the hydroxyl (−OH) group.

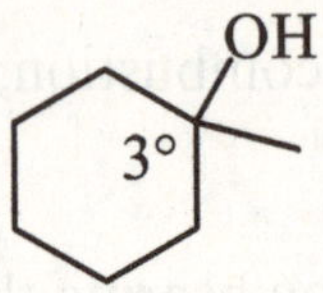

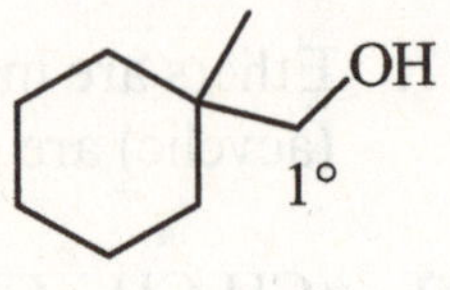

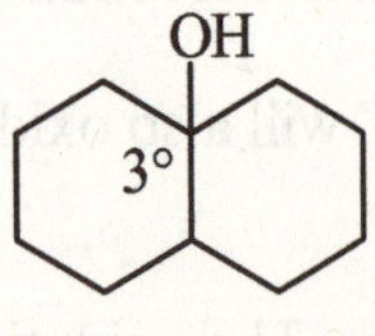

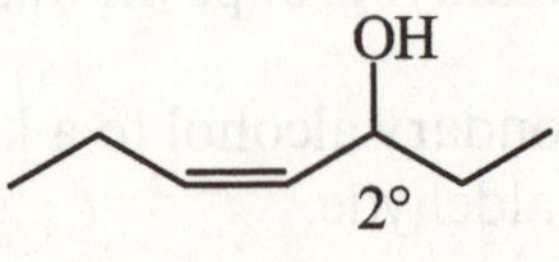

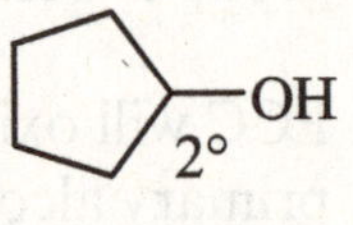

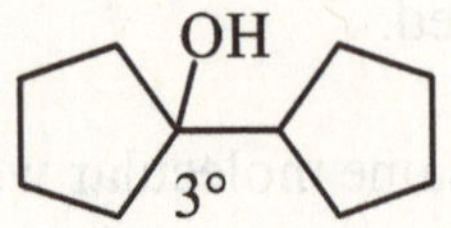

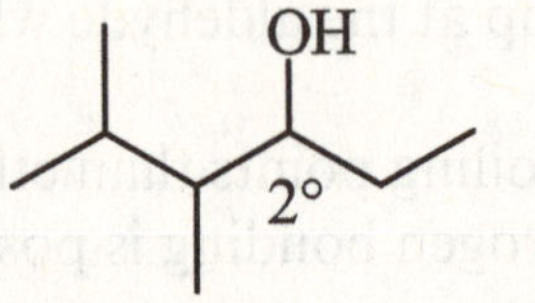

8.14   Provide an IUPAC or common name for these compounds:

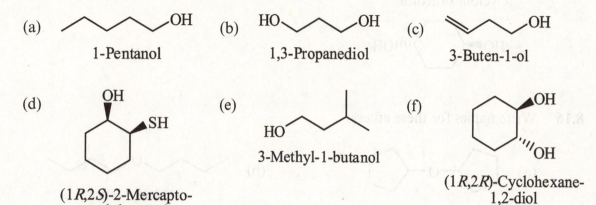

(a)

1-Pentanol

(b)

1,3-Propanediol

(c)

3-Buten-1-ol

(d)

(1R,2S)-2-Mercapto-
cyclohexanol

(e)

3-Methyl-1-butanol

(f)

(1R,2R)-Cyclohexane-
1,2-diol

(g)

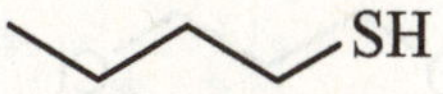

1-Butanethiol

(h)

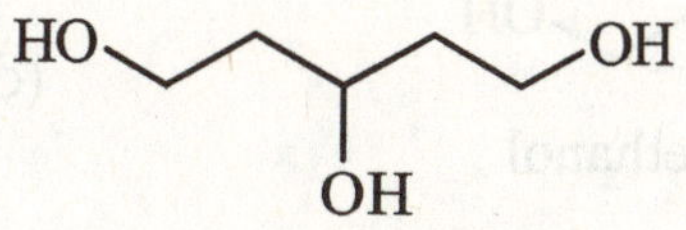

1,3,5-Pentanetriol

8.15   Draw a structural formula for each alcohol:

(a) Isopropyl alcohol

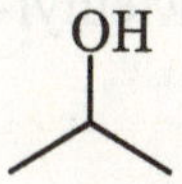

(b) Propylene glycol

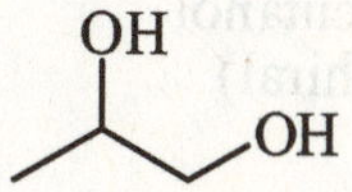

(c) (R)-5-Methyl-2-
hexanol

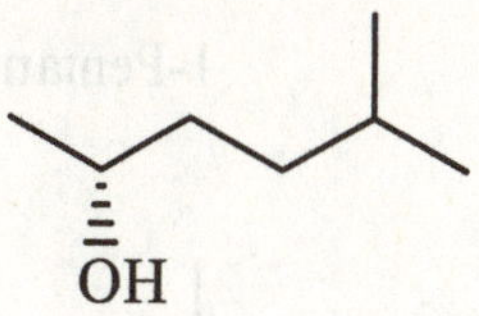

(d) 2-Methyl-2-propyl-
1,3-propanediol

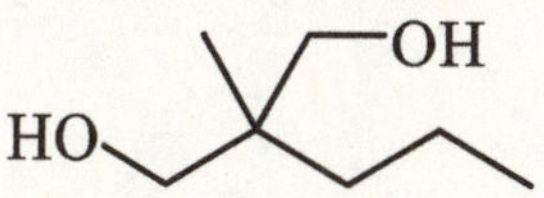

(e) 2,2-Dimethyl-1-
propanol

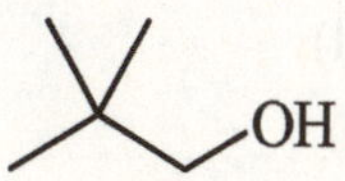

(f) 2-Mercaptoethanol

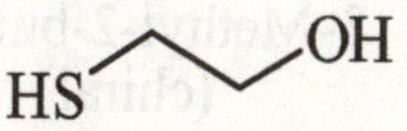

(g) 1,4-Butanediol

(h) (Z)-5-Methyl-
2-hexen-1-ol

(i) cis-3-Penten-1-ol

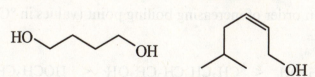

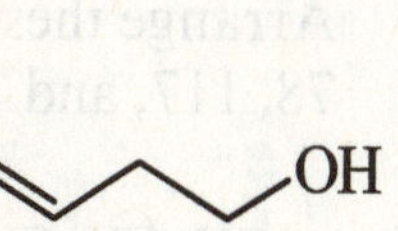

(j)   *trans*-1,4-
      Cyclohexanediol

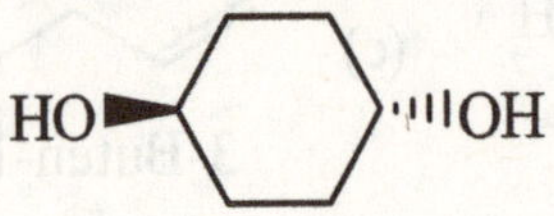

HO—⬛⟨cyclohexane⟩ⁱⁱⁱⁱOH

8.16   Write names for these ethers:

(a)   ⟨cyclopentyl⟩—O—⟨cyclopentyl⟩

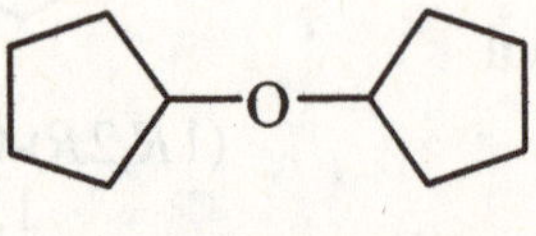

Dicyclopentyl ether

(b)   ⟨butyl⟩—O—⟨butyl⟩

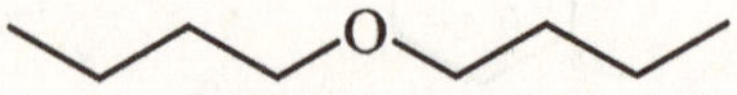

Dibutyl ether

(c)   ⟨ethyl⟩—O—⟨ethyl⟩—OH

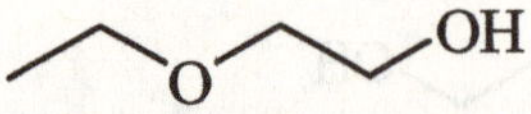

2-Ethoxyethanol

(d)   ⟨ethyl⟩—O—⟨propyl⟩—Cl

1-Chloro-3-ethoxypropane

8.17   Name and draw structural formulas for the eight isomeric alcohols with the molecular formula $C_5H_{12}O$. Which are chiral?

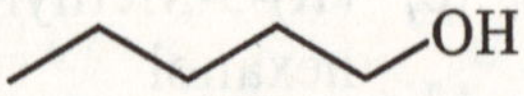

⟨structure⟩—OH

1-Pentanol

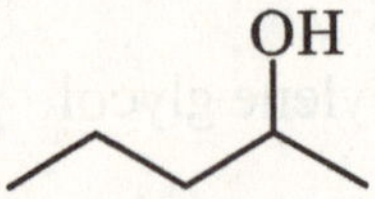

⟨structure with OH⟩

2-Pentanol
(chiral)

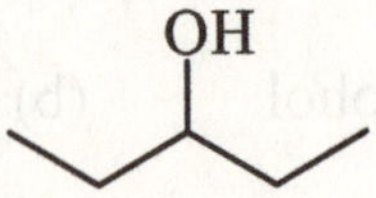

⟨structure with OH⟩

3-Pentanol

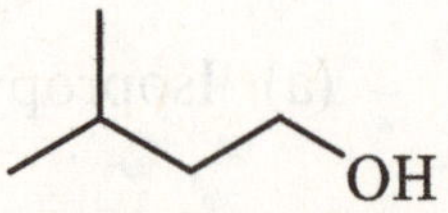

⟨structure⟩—OH

3-Methyl-1-butanol

⟨structure⟩—OH

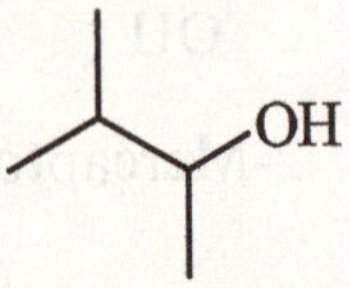

3-Methyl-2-butanol
(chiral)

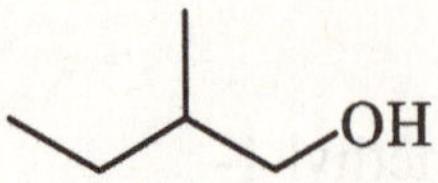

⟨structure⟩—OH

2-Methyl-1-butanol
(chiral)

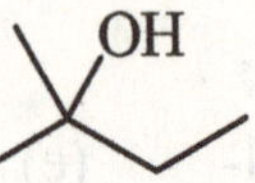

⟨structure with OH⟩

2-Methyl-2-butanol

⟨structure⟩—OH

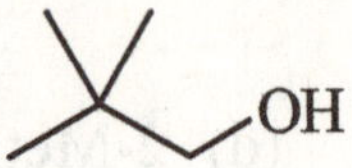

2,2-Dimethyl-1-propanol

Physical Properties

8.18   Arrange these compounds in order of increasing boiling point (values in °C are −42, 78, 117, and 198):

$$CH_3CH_2CH_3 \;<\; CH_3CH_2OH \;<\; CH_3CH_2CH_2CH_2OH \;<\; HOCH_2CH_2OH$$

propane (−42°C)   ethanol (78°C)   1-butanol (117°C)   ethylene glycol (198°C)

The only intermolecular forces between propane molecules in the liquid state are the relatively weak dispersion forces. Ethanol, which is approximately the same size as propane, exhibits intermolecular hydrogen bonding due to the −OH group and thus has a higher boiling point than propane. 1-Butanol also has an −OH group, but the larger size of the molecule allows for more dispersion forces. Ethylene glycol, though smaller than 1-butanol, has two −OH groups (two sites of hydrogen bonding) and has the highest boiling point.

8.19 Arrange these compounds in order of increasing boiling point (values in °C are −42, −24, 78, and 118):

$$CH_3CH_2CH_3 \quad < \quad CH_3OCH_3 \quad < \quad CH_3CH_2OH \quad < \quad CH_3COOH$$

propane (−42°C)   dimethyl   ethanol (78°C)   acetic acid
                       ether (−24°C)   (118°C)

Dimethyl ether, which has intermolecular dipole-dipole interactions (but no hydrogen bonding), has a higher boiling point than propane, which only has dispersion forces. The hydrogen bonding interactions in ethanol are stronger than the dipole-dipole interactions in dimethyl ether. Both acetic acid and ethanol have an −OH group, but acetic acid has a higher boiling point than ethanol because the oxygen of the carbonyl (C=O) group can act as an additional hydrogen-bond acceptor.

8.20 Propanoic acid and methyl acetate are constitutional isomers, and both are liquids at room temperature. One of these compounds has a boiling point of 141°C; the other has a boiling point of 57°C. Which compound has which boiling point?

$$CH_3CH_2\overset{\displaystyle O}{\overset{\|}{C}}OH \qquad\qquad CH_3\overset{\displaystyle O}{\overset{\|}{C}}OCH_3$$

Propanoic acid   Methyl acetate

Methyl acetate is not capable of intermolecular hydrogen bonding because it does not contain any hydrogen-bond donors. It has the lower boiling point, while propanoic acid has the higher boiling point.

8.21 Draw all possible staggered conformations of ethylene glycol (HOCH₂CH₂OH). Can you explain why the conformation in which the −OH groups are closest to each other is more stable than the conformation in which the −OH groups are farthest apart by approximately 4.2 kJ/mol (1 kcal/mol)?

intramolecular
hydrogen bonding

Normally, the conformation with the larger groups closest to each other is higher in energy than the conformation with the larger groups farthest apart. This is because the close proximity of the two larger groups increases torsional strain. However, in the case of ethylene glycol, *intramolecular* hydrogen bonding can occur between the hydroxyl groups. The strength of the hydrogen bond is greater than 4.2 kJ/mol, so the net result is that the conformation with the larger groups closest to each other is more stable.

8.22 Following are structural formulas for 1-butanol and 1-butanethiol. One of these compounds has a boiling point of 98.5°C; the other has a boiling point of 117°C. Which compound has which boiling point?

1-Butanol   1-Butanethiol

The intermolecular forces in 1-butanethiol are much lower than those in 1-butanol. Unlike −OH groups, −SH groups are not capable of hydrogen bonding due to the relatively small difference between the electronegativities of sulfur and hydrogen. 1-Butanethiol has a lower boiling point than does 1-butanol.

8.23 From each pair of compounds, select the one that is more soluble in water:

(a) CH_2Cl_2 or **CH_3OH**

Methanol is more soluble in water than is dichloromethane (CH_2Cl_2). The polar −OH group in methanol participates as both a hydrogen-bond donor and acceptor, thus allowing methanol to strongly interact with water.

(b)

$$\underset{CH_3CCH_3}{\overset{O}{\underset{\parallel}{}}} \quad or \quad \underset{CH_3CCH_3}{\overset{CH_2}{\underset{\parallel}{}}}$$

Propanone (acetone) is a polar molecule because it has a net dipole moment, while 2-methylpropene is a nonpolar molecule. As a result, propanone is more soluble in water, a polar solvent. In addition, the oxygen atom in propanone can accept hydrogen bonds.

(c) CH_3CH_2Cl or   **NaCl**

Sodium chloride, an ionic compound, is much more soluble in water than is chloroethane. Ionic compounds dissociate in water, and the ions are surrounded by water molecules that stabilize the ions. This stabilization arises from the energy of hydration.

(d) $CH_3CH_2CH_2SH$ or   **$CH_3CH_2CH_2OH$**

1-Propanol is more polar than 1-propanethiol due to the higher electronegativity of oxygen compared to sulfur. The $-OH$ group in 1-propanol can also participate in hydrogen bonding. As a result, 1-propanol is more soluble in water.

(e)

$$\underset{CH_3CH_2CHCH_2CH_3}{\overset{\overset{\displaystyle OH}{|}}{}}$$ or $$\underset{CH_3CH_2CCH_2CH_3}{\overset{\overset{\displaystyle O}{\|}}{}}$$

Although both 3-pentanol and 3-pentanone are polar molecules, the $-OH$ group in the alcohol is both a hydrogen-bond donor and acceptor. The oxygen atom in the ketone is only a hydrogen-bond acceptor. The alcohol therefore interacts more strongly with water and is more soluble in it.

8.24   Arrange the compounds in each set in order of decreasing solubility in water:

(a)   Ethanol > diethyl ether > butane

The order of decreasing solubility in water is related to the ability of these compounds to hydrogen bond. Ethanol can act as both a hydrogen-bond donor and acceptor, diethyl ether is only a hydrogen-bond acceptor, and butane is a nonpolar hydrocarbon.

(b)   1,2-Hexanediol > 1-hexanol > hexane

1,2-Hexanediol is the most water-soluble of the three compounds because it contains two hydroxyl groups, which interact favorably with water by hydrogen bonding.

8.25   Each of the following compounds is a common organic solvent. From each pair of compounds, select the solvent with the greater solubility in water:

(a) CH_2Cl_2 or CH_3CH_2OH

Ethanol is more soluble in water than is dichloromethane (CH_2Cl_2). The polar
−OH group can also participate as both a hydrogen-bond donor and acceptor,
thus allowing methanol to strongly interact with water.

(b) $CH_3CH_2OCH_2CH_3$ or CH_3CH_2OH

The polarity of an ether is much lower than that of an alcohol. The −OH group
of an alcohol also acts as a hydrogen-bond donor and acceptor. As a result,
ethanol is more soluble in water than is diethyl ether.

(c)

$$CH_3\overset{\displaystyle O}{\overset{\|}{C}}CH_3 \quad \text{or} \quad CH_3CH_2OCH_2CH_3$$

Ketones are more polar than ethers because they have a higher net dipole
moment. There is no cancelling out of the C=O dipole in a ketone, but the two
C−O dipoles in an ether partially cancel out due to the bent geometry at the
ether oxygen.

(d) $CH_3CH_2OCH_2CH_3$ or $CH_3(CH_2)_3CH_3$

Although ethers are relatively low in polarity, they are nonetheless more polar
than alkanes, which are nonpolar. The oxygen atom in an ether can also act as
a hydrogen bond acceptor. Ethers are more soluble in water than are alkanes.

Synthesis of Alcohols

8.26 Give the structural formula of an alkene or alkenes from which each alcohol or
glycol can be prepared:

(a) Acid-catalyzed hydration

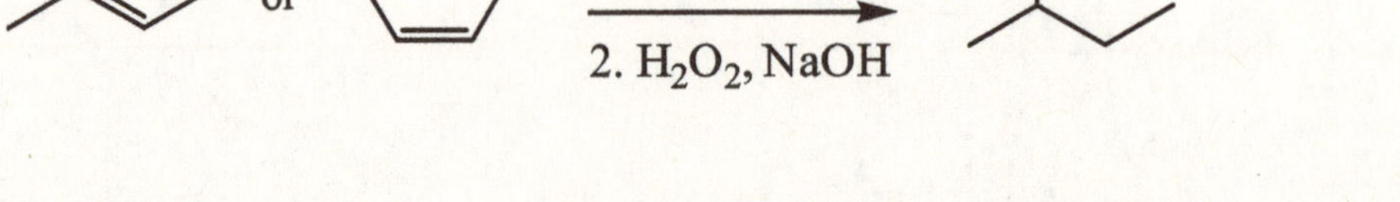

Hydroboration-oxidation

(b)

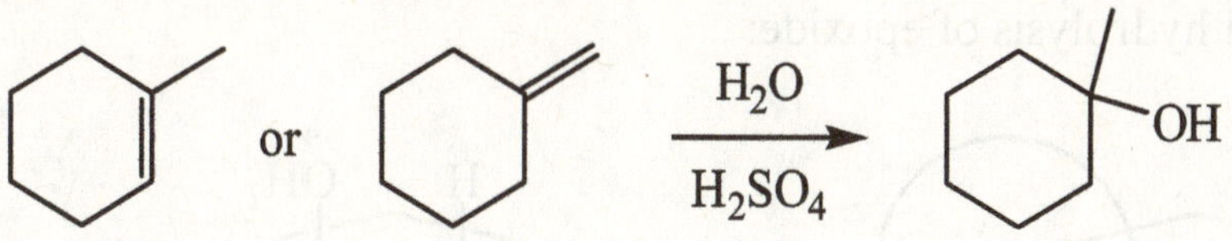

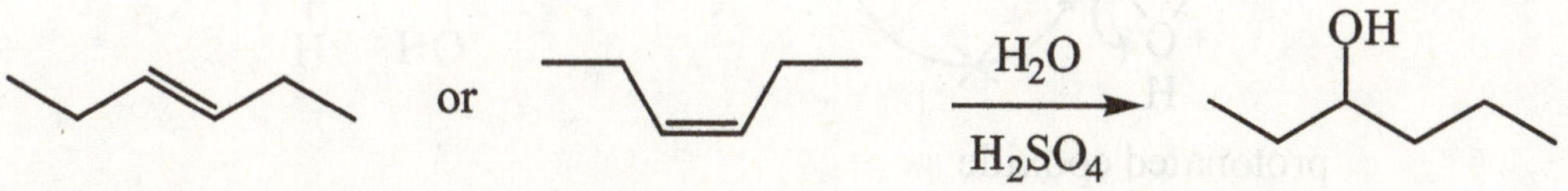

3-Hexanol could also be made from the hydroboration-oxidation of the two alkenes.

(d)

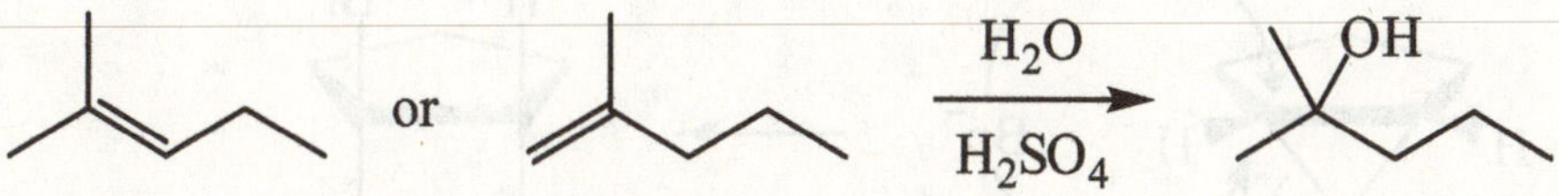

(e)

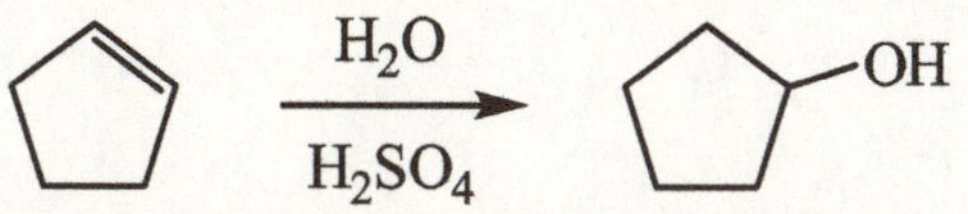

The hydroboration-oxidation of cyclopentene would also give cyclopentanol.

(f)

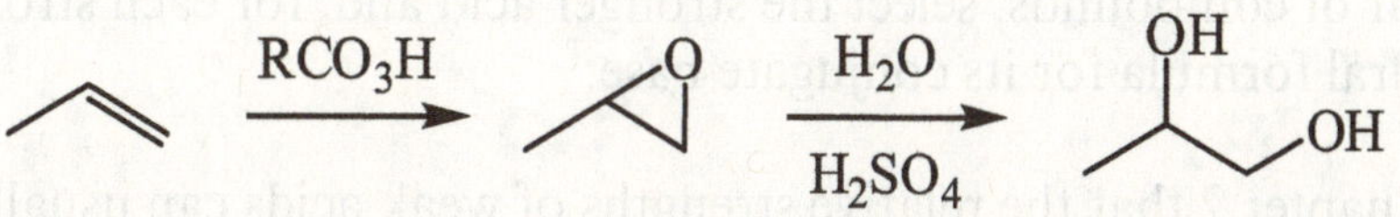

8.27 The addition of bromine to cyclopentene and the acid-catalyzed hydrolysis of cyclopentene oxide are both stereoselective; each gives a *trans* product. Compare the mechanisms of these two reactions, and show how each mechanism accounts for the formation of the *trans* product.

The two reactions are stereoselective and result in *trans* products because they both involve an intermediate consisting of a three-membered ring. To open the ring, the nucleophile preferentially attacks from the least-hindered side, which is the side opposite (anti) to that of the three-membered ring. In cyclic compounds, anti addition to the ring results in *trans* products. The key ring-opening steps are shown below.

Acid-catalyzed hydrolysis of epoxide:

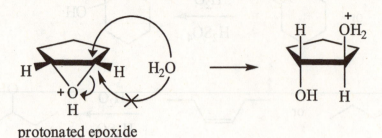

protonated epoxide

Bromination:

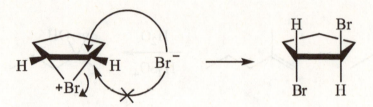

cyclic brominium ion

Acidity of Alcohols and Thiols

8.28 From each pair of compounds, select the stronger acid and, for each stronger acid, write a structural formula for its conjugate base:

Recall from Chapter 2 that the relative strengths of weak acids can usually (but not always) be determined by comparing the relative stabilities of the conjugate bases.

(a) H_2O or **H_2CO_3**

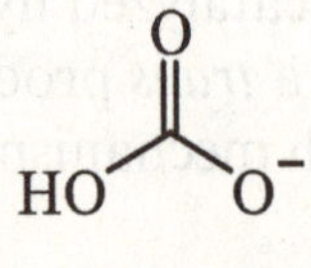

H_2CO_3 is the stronger acid because its conjugate base, the bicarbonate ion, is more stable than the conjugate base of water, the hydroxide ion. Both conjugate bases have a negatively charged oxygen, but bicarbonate is stabilized by resonance.

(b) CH_3OH or **CH_3COOH**

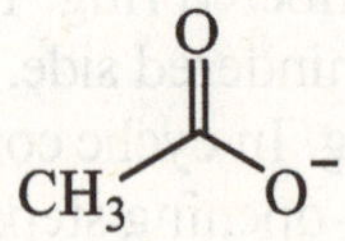

For the same reasons as (a), acetic acid is a stronger acid than methanol. The acetate ion is more stable than the methoxide ion.

(c) **CH₃COOH** or CH₃CH₂SH

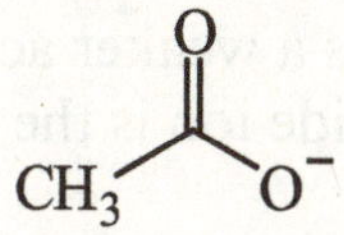

The conjugate base of acetic acid, the acetate ion, is more stable than that of ethanethiol, the ethylsulfide ion. Sulfur has a larger atomic size than oxygen, but the negatively charged oxygen in acetate is also stabilized by resonance. The relative stabilities of these conjugate bases are also suggested by the pK_a of acetic acid (4.76) and ethanethiol (8.5).

8.29 Arrange these compounds in order of increasing acidity (from weakest to strongest):

$$CH_3CH_2CH_2OH \quad < \quad CH_3CH_2CH_2SH \quad < \quad CH_3CH_2COH$$

$$pK_a = 16 \qquad\qquad pK_a = 9 \qquad\qquad pK_a = 5$$

$$CH_3CH_2CH_2O^- \qquad CH_3CH_2CH_2S^- \qquad CH_3CH_2CO^-$$

least stable   most stable
conjugate base   conjugate base

Propylsulfide is more stable than propoxide due to the larger atomic size of sulfur. However, propanoate, which has a negatively charged oxygen that is resonance-stabilized, is more stable than the sulfide.

8.30 From each pair of compounds, select the stronger base and, for each stronger base, write the structural formula of its conjugate acid:

The strongest base is the one with the least-stable pair of electrons. The strongest base is also derived from the weakest parent acid.

(a) HO⁻ or **CH₃O⁻**

CH₃OH

Methoxide is a stronger base than hydroxide because the electron-donating properties of the alkyl group reduce the stability of the negatively charged oxygen. CH₃OH (pK_a = 15.9) is a weaker acid than H₂O (pK_a = 15.7), so methoxide is the stronger base.

(b) CH₃CH₂S⁻ or **CH₃CH₂O⁻**

CH₃CH₂OH

Sulfur has a larger atomic size than oxygen, so ethylsulfide is more stable than ethoxide. Both atoms are in the same group, so the effect of atomic size predominates over the effect of electronegativity. CH₃CH₂OH (pK_a = 15.9) is a weaker acid than CH₃CH₂SH (pK_a = 8.5), so ethoxide is the stronger base.

(c) $CH_3CH_2O^-$ or NH_2^-

NH_3

Nitrogen and oxygen are in the same period, so ethoxide is more stable than the amide ion. CH_3CH_2OH ($pK_a = 15.9$) is a weaker acid than NH_3 ($pK_a = 38$), so the amide ion is the stronger base.

8.31 Label the stronger acid, stronger base, weaker acid, and weaker base in each of the following equilibria, and predict the position of each equilibrium; that is, does each lie considerably to the left, does it lie considerably to the right, or are the concentrations evenly balanced?

(a) $CH_3CH_2O^-$ + HCl $\rightleftharpoons$ CH_3CH_2OH + Cl^-

    stronger base   stronger acid   weaker acid   weaker base

The equilibrium lies far to the right because the chloride ion is much more stable than the alkoxide ion. HCl is a much stronger acid ($pK_a = -7$) than ethanol($pK_a = 15.9$).

(b)

$$\overset{O}{\overset{\|}{CH_3COH}} + CH_3CH_2O^- \rightleftharpoons \overset{O}{\overset{\|}{CH_3CO^-}} + CH_3CH_2OH$$

   stronger acid   stronger base   weaker base   weaker acid

The equilibrium lies far to the right because the acetate ion, as a result of resonance, is much more stable than the ethoxide anion. Acetic acid is a stronger acid ($pK_a = 4.76$) than ethanol ($pK_a = 15.9$).

8.32 Predict the position of equilibrium for each acid-base reaction; that is, does each lie considerably to the left, does it lie considerably to the right, or are the concentrations evenly balanced?

(a) CH_3CH_2OH + Na^+OH^- $\rightleftharpoons$ $CH_3CH_2O^-Na^+$ + H_2O

The equilibrium is evenly balanced, although it very slightly favors the left side. The hydroxide ion is more stable than the ethoxide ion because alkyl groups are electron-donating and reduce the stability of the ethoxide ion, making ethoxide the stronger base. Ethanol ($pK_a = 15.9$) is weaker acid than water ($pK_a = 15.7$). In acid-base equilibria, the side with the weaker acid and the weaker base is favored.

(b) CH_3CH_2SH + Na^+OH^- $\rightleftharpoons$ $CH_3CH_2S^-Na^+$ + H_2O

The equilibrium lies far to the right because the sulfide ion is much more stable due to the atomic size of sulfur, than the hydroxide ion. Ethanethiol ($pK_a = 8.5$) is a much stronger acid than water ($pK_a = 15.7$).

(c) CH_3CH_2OH + $CH_3CH_2S^-Na^+$ $\rightleftharpoons$ $CH_3CH_2O^-Na^+$ + CH_3CH_2SH

The sulfide ion is more stable than the alkoxide ion due to atomic size. Ethanethiol is a is a stronger acid ($pK_a = 8.5$) than ethanol ($pK_a = 15.9$).

(d) $CH_3CH_2S^-Na^+$ + $CH_3\overset{O}{\overset{\|}{C}}OH$ $\rightleftharpoons$ CH_3CH_2SH + $CH_3\overset{O}{\overset{\|}{C}}O^-Na^+$

As seen in Problem 8.28, a carboxylate ion is more stable than a sulfide ion. Acetic acid is a stronger acid ($pK_a = 4.76$) than ethanethiol ($pK_a = 8.5$).

Reactions of Alcohols

8.33   Show how to distinguish between cyclohexanol and cyclohexene by a simple chemical test. (*Hint:* Treat each with Br_2 in CCl_4 and watch what happens.)

Both cyclohexanol and cyclohexene are colorless. However, cyclohexene will decolorize an orange-red solution of bromine dissolved in CCl_4 (the solvent), while cyclohexanol does not react with bromine.

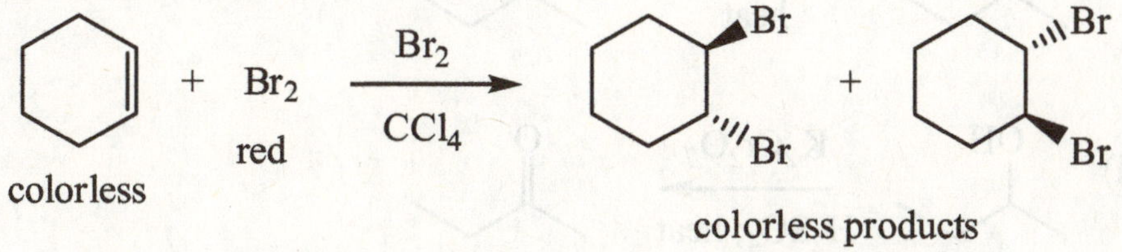

8.34   Write equations for the reaction of 1-butanol, a primary alcohol, with these reagents:

(a)

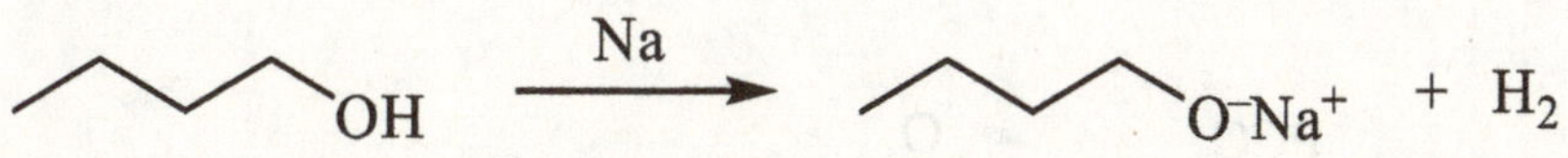

(b)

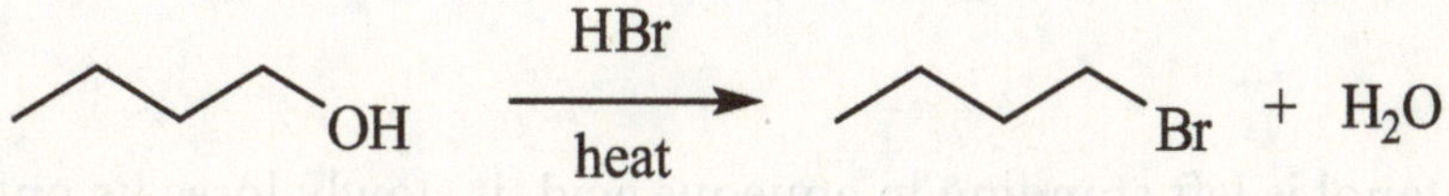

(c)

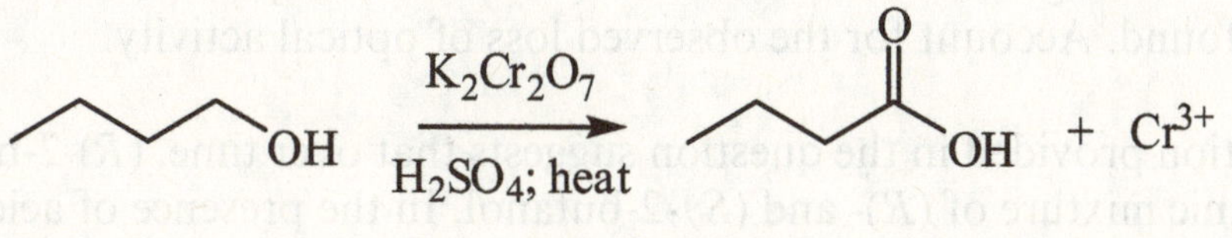

(d) 
$$\text{\raisebox{0pt}{\includegraphics}} \xrightarrow{\text{SOCl}_2} \text{\raisebox{0pt}{\includegraphics}} + SO_2 + HCl$$

(e) 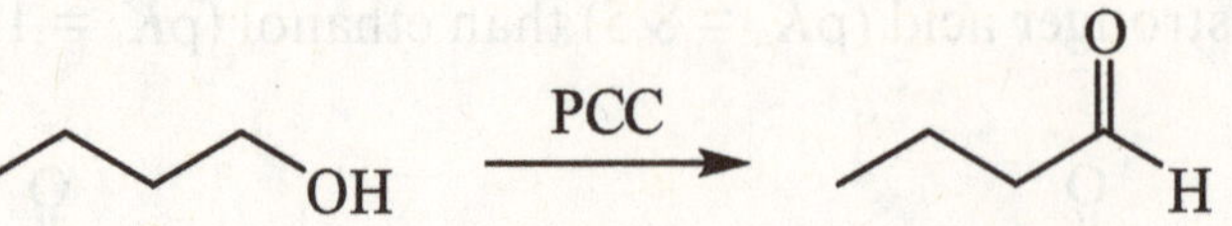
$$\xrightarrow{\text{PCC}}$$

8.35 Write equations for the reaction of 2-butanol, a secondary alcohol, with these reagents:

(a) 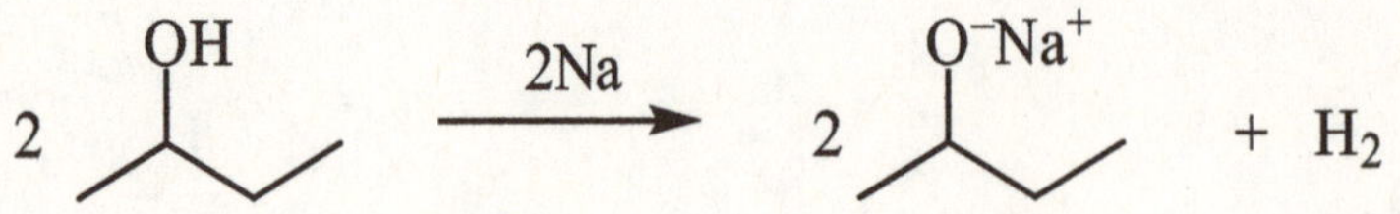
$$2 \text{ \raisebox{0pt}{OH}} \xrightarrow{2Na} 2 \text{ \raisebox{0pt}{O}^-\text{Na}^+} + H_2$$

(b) 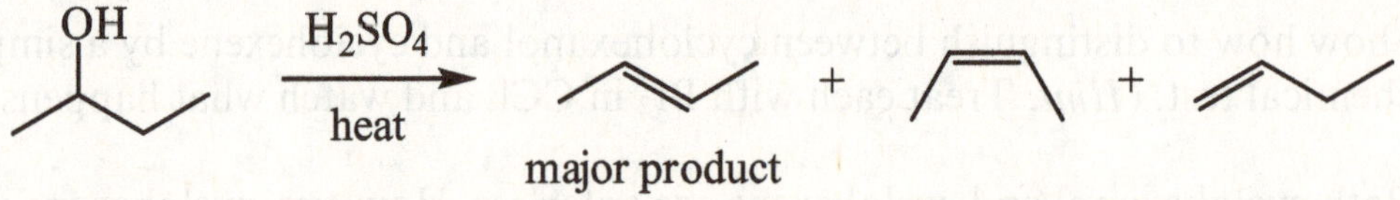
$$\text{OH} \xrightarrow[\text{heat}]{\text{H}_2\text{SO}_4} \quad + \quad + \quad$$

major product

(c) 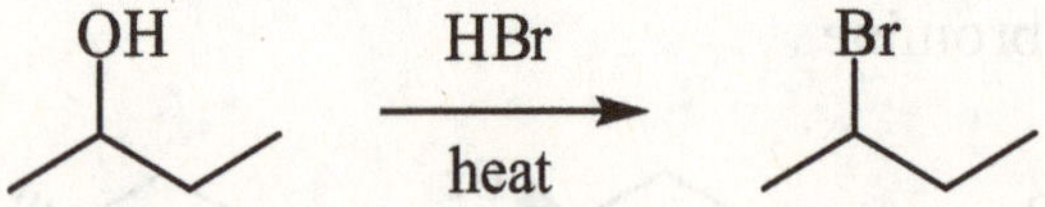
$$\text{OH} \xrightarrow[\text{heat}]{\text{HBr}} \text{Br}$$

(d) 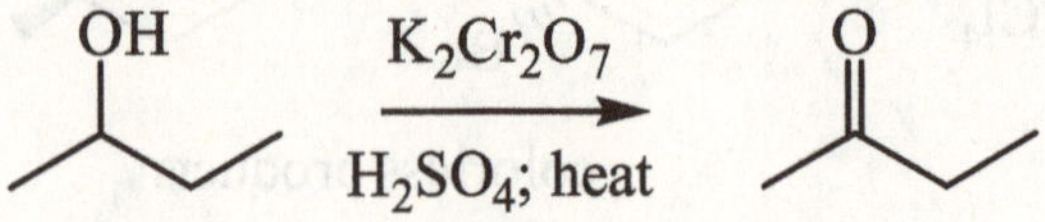
$$\text{OH} \xrightarrow[\text{H}_2\text{SO}_4;\ \text{heat}]{\text{K}_2\text{Cr}_2\text{O}_7} \text{O}$$

(e) 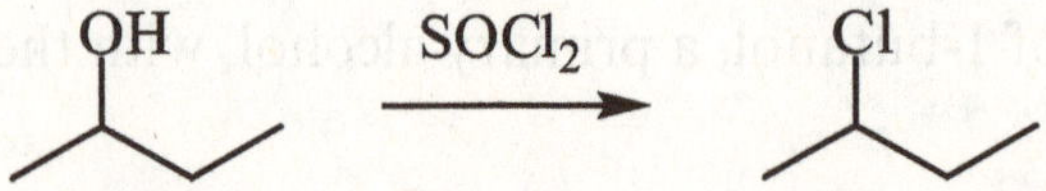
$$\text{OH} \xrightarrow{\text{SOCl}_2} \text{Cl}$$

(f) 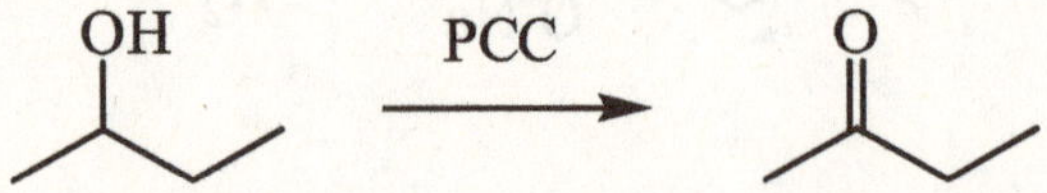
$$\text{OH} \xrightarrow{\text{PCC}} \text{O}$$

8.36 When (R)-2-butanol is left standing in aqueous acid, it slowly loses its optical activity. When the organic material is recovered from the aqueous solution, only 2-butanol is found. Account for the observed loss of optical activity.

The information provided in the question suggests that over time, (R)-2-butanol forms a racemic mixture of (R)- and (S)-2-butanol. In the presence of acid, the hydroxyl group is converted into a good leaving group. Loss of the leaving group generates a carbocation intermediate, which is planar due to its sp^2 hybridization. The nucleophile, water, can attack either face of the carbocation to regenerate the alcohol. Over time, a 1:1 equilibrium mixture of the two stereoisomers is formed.

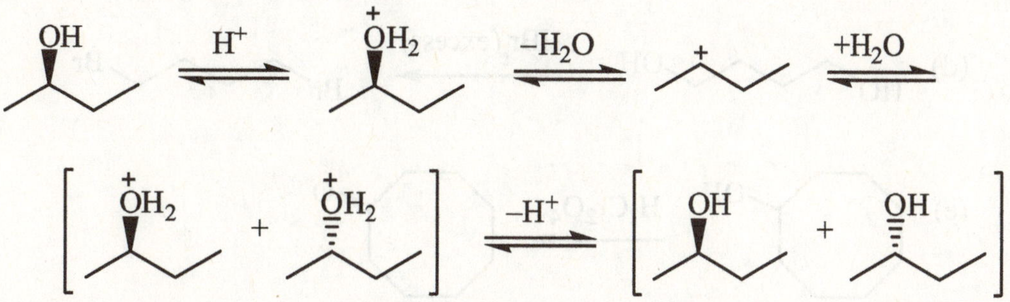

8.37 What is the most likely mechanism of the following reaction? Draw a structural formula for the intermediate(s) formed during the reaction.

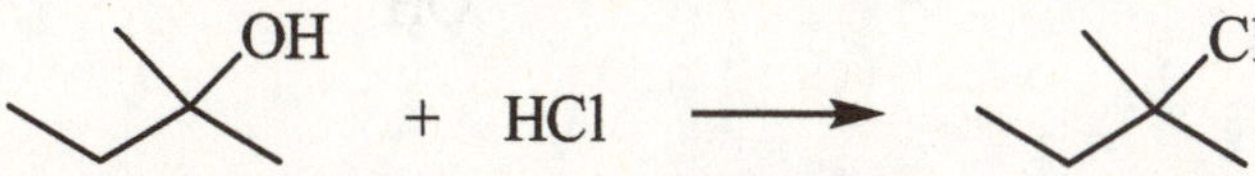

The reaction most likely proceeds by an S_N1 mechanism. Under acidic conditions, the hydroxyl group is protonated to generate a good leaving group. Departure of the leaving group results in a stable 3° carbocation that is subsequently attacked by chloride.

oxonium ion intermediate   carbocation intermediate

8.38 Complete the equations for these reactions:

(a)

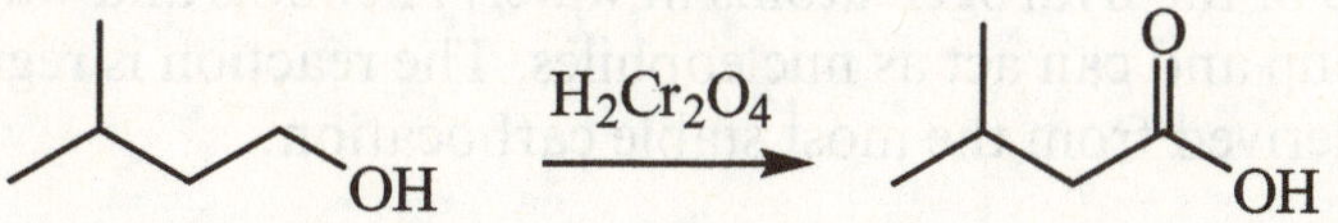

(b)

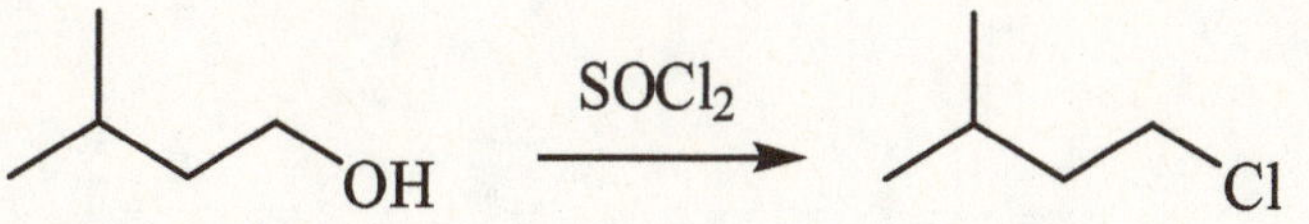

(c)

$$\text{1-methylcyclohexanol} \xrightarrow{\text{HCl}} \text{1-chloro-1-methylcyclohexane}$$

(d)

$$HO\text{-}(\text{CH}_2)_4\text{-}OH \xrightarrow{\text{HBr (excess)}} Br\text{-}(\text{CH}_2)_4\text{-}Br$$

(e)

$$\text{cyclooctanol} \xrightarrow{\text{H}_2\text{Cr}_2\text{O}_4} \text{cyclooctanone}$$

(f)

$$\text{cyclohexene} \xrightarrow[\text{2. H}^+/\text{H}_2\text{O}]{\text{1. RCO}_3\text{H}} \text{trans-1,2-cyclohexanediol} + \text{trans-1,2-cyclohexanediol}$$

8.39   In the commercial synthesis of methyl *tert*-butyl ether (MTBE), once used as an antiknock, octane-improving gasoline additive, 2-methylpropene and methanol are passed over an acid catalyst to give the ether. Propose a mechanism for this reaction.

$$
\begin{array}{c}
\underset{\substack{|\\ \text{CH}_3}}{\text{CH}_3\text{C}}\text{=CH}_2 \quad + \quad \text{CH}_3\text{OH} \xrightarrow[\text{catalyst}]{\text{acid}} \underset{\substack{|\\ \text{CH}_3}}{\underset{\substack{|\\ }}{\text{CH}_3\text{C}}}\overset{\substack{\text{CH}_3\\ |}}{\text{COCH}_3}
\end{array}
$$

2-Methylpropene   Methanol

(Isobutylene)

2-Methoxy-2-methylpropane

(Methyl *tert*-butyl ether, MTBE)

The mechanism for the addition of an alcohol to an alkene is analogous to the addition of water. The only difference between an alcohol and water is that an alkyl group has replaced one of the hydrogen atoms in water. Alcohols and water both contain a hydroxyl group and can act as nucleophiles. The reaction is regioselective in that the product is derived from the most stable carbocation.

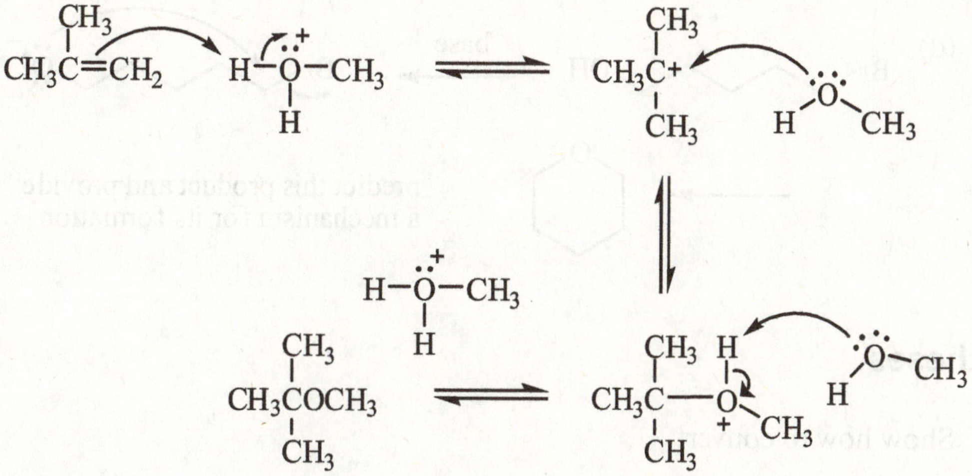

8.40 Cyclic bromoalcohols, upon treatment with base, can sometimes undergo intramolecular reactions to form the ethers shown in reactions (a) and (b). Provide a mechanism for reactions (a) and (b). Indicate why equation (c) does not yield a similar reaction.

When treated with a base, the hydroxyl group is deprotonated to form an alkoxide ion, which is a good nucleophile and can undergo S_N2 reactions. Reaction (c) does not occur because in an S_N2 reaction, the nucleophile must attack from the side opposite to that of the leaving group (see Problem 7.41 for more details). The starting material in (c) has the hydroxyl and bromo groups on the same side.

(a)

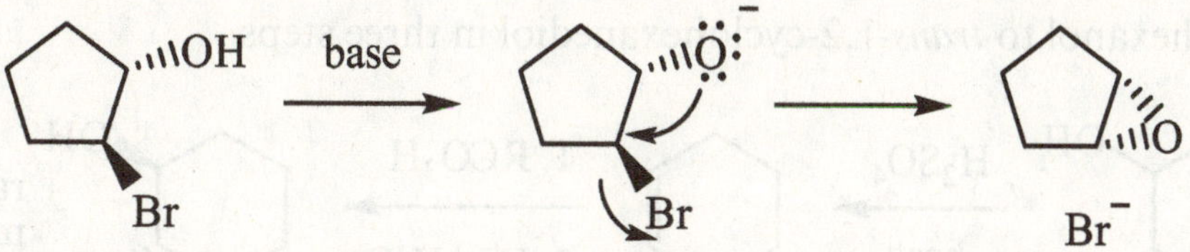

(b)

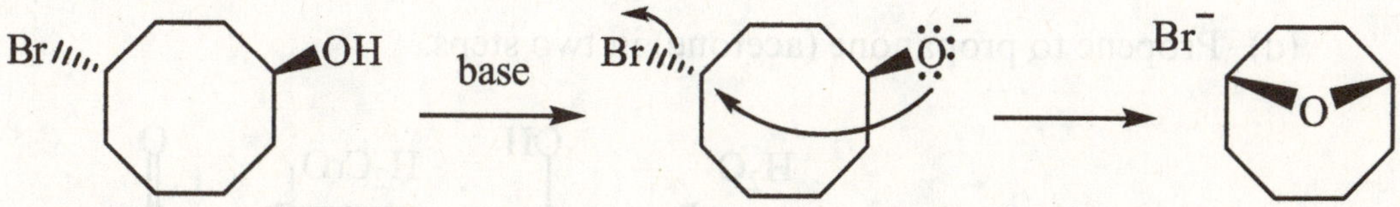

(c)

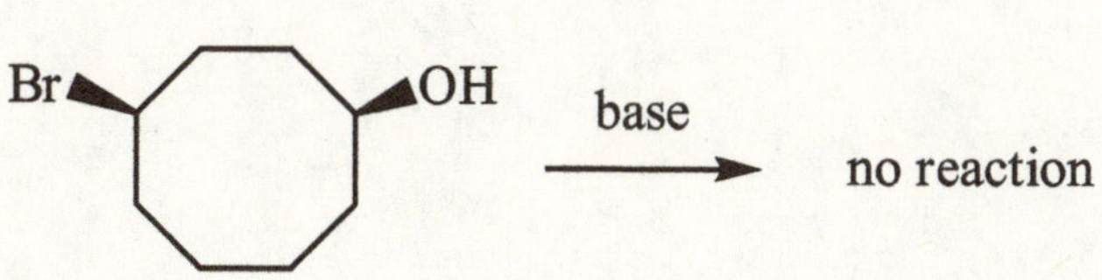

(d)

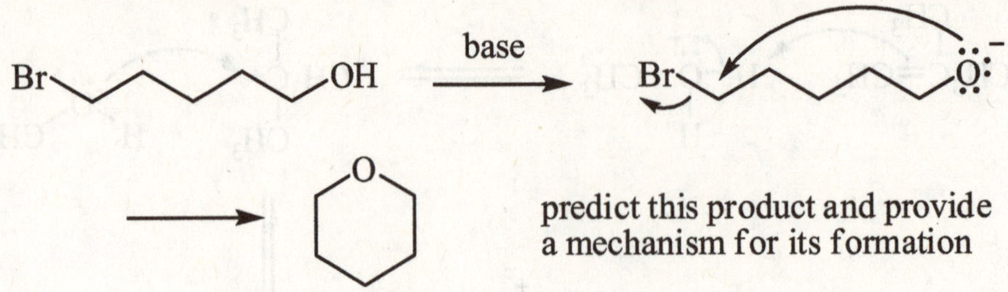

Br~~~~~~OH $\xrightarrow{\text{base}}$ Br~~~~~~$\ddot{\text{O}}:^-$

$\longrightarrow$ [pyran ring with O]   predict this product and provide
a mechanism for its formation

Syntheses

8.41   Show how to convert

(a)   1-Propanol to 2-propanol in two steps.

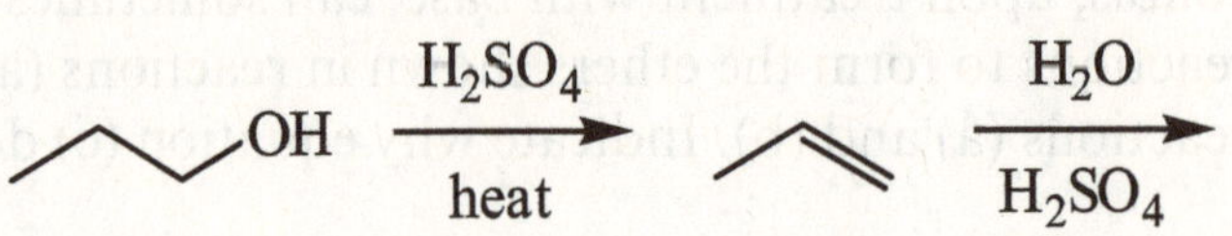

~~~OH $\xrightarrow[\text{heat}]{\text{H}_2\text{SO}_4}$ ~~~ $\xrightarrow[\text{H}_2\text{SO}_4]{\text{H}_2\text{O}}$ [isopropanol with OH]

(b)   Cyclohexene to cyclohexanone in two steps.

[cyclohexene] $\xrightarrow[\text{H}_2\text{SO}_4]{\text{H}_2\text{O}}$ [cyclohexanol, OH] $\xrightarrow[\text{or PCC}]{\text{H}_2\text{CrO}_4}$ [cyclohexanone, O]

(c)   Cyclohexanol to *trans*-1,2-cyclohexanediol in three steps.

[cyclohexanol, OH] $\xrightarrow[\text{heat}]{\text{H}_2\text{SO}_4}$ [cyclohexene] $\xrightarrow[\text{2. H}^+ / \text{H}_2\text{O}]{\text{1. RCO}_3\text{H}}$ [trans-1,2-cyclohexanediol, OH and ''''OH]   racemic
mixture

(d)   Propene to propanone (acetone) in two steps.

~~~ $\xrightarrow[\text{H}_2\text{SO}_4]{\text{H}_2\text{O}}$ [isopropanol, OH] $\xrightarrow[\text{or PCC}]{\text{H}_2\text{CrO}_4}$ [acetone, O]

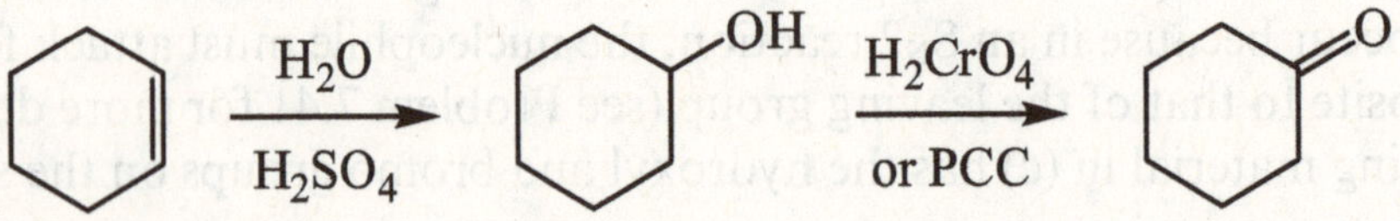

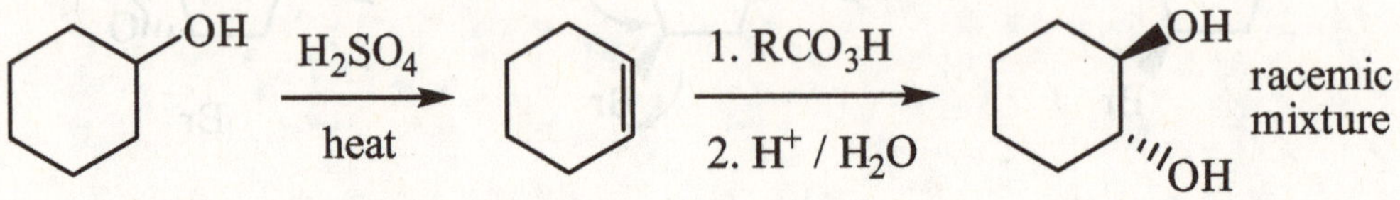

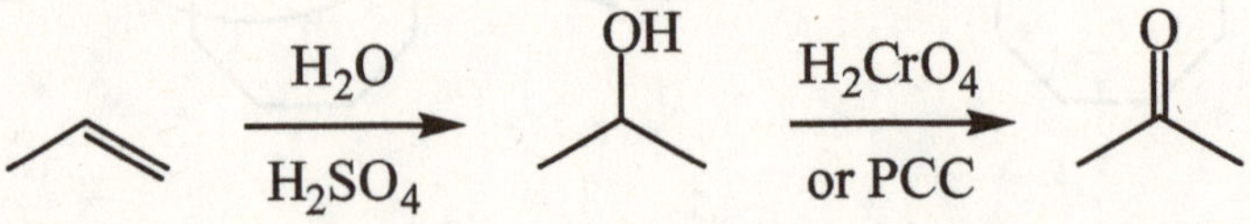

8.42 Show how to convert cyclohexanol to these compounds:

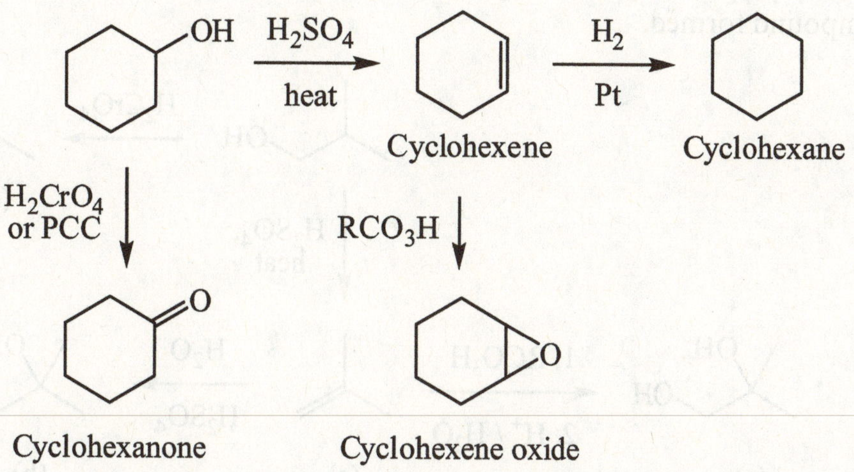

8.43 Show reagents and experimental conditions that can be used to synthesize these compounds from 1-propanol (any derivative of 1-propanol prepared in an earlier part of this problem may be used for a later synthesis):

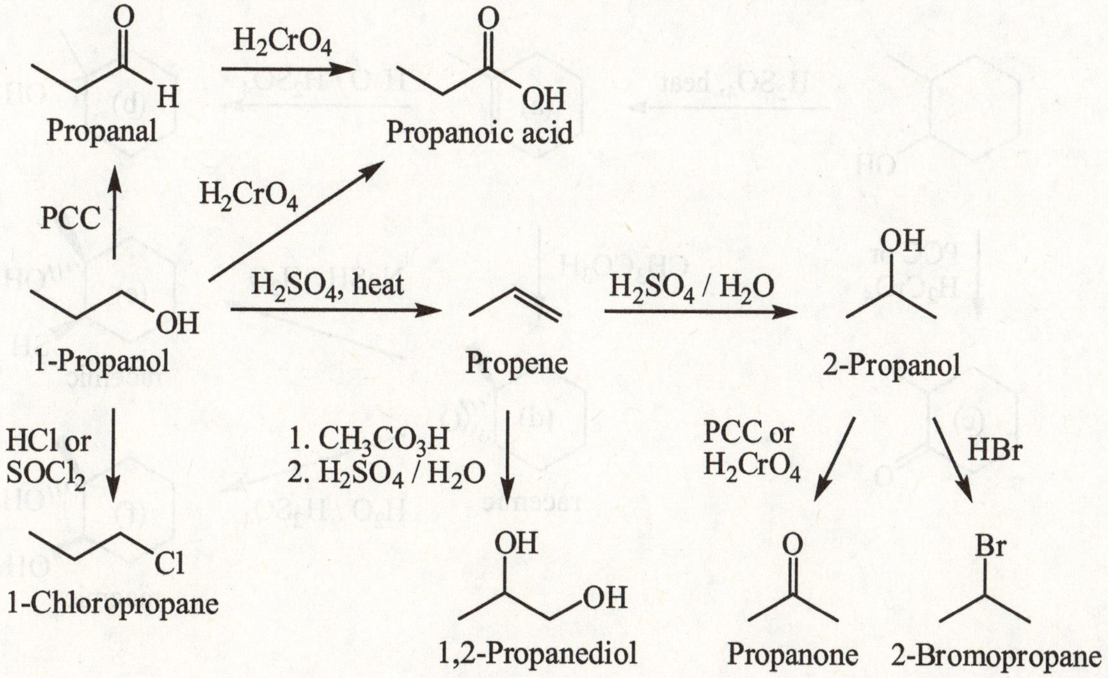

8.44 Show how to prepare each compound from 2-methyl-1-propanol (isobutyl alcohol). For any preparation involving more than one step, show each intermediate compound formed.

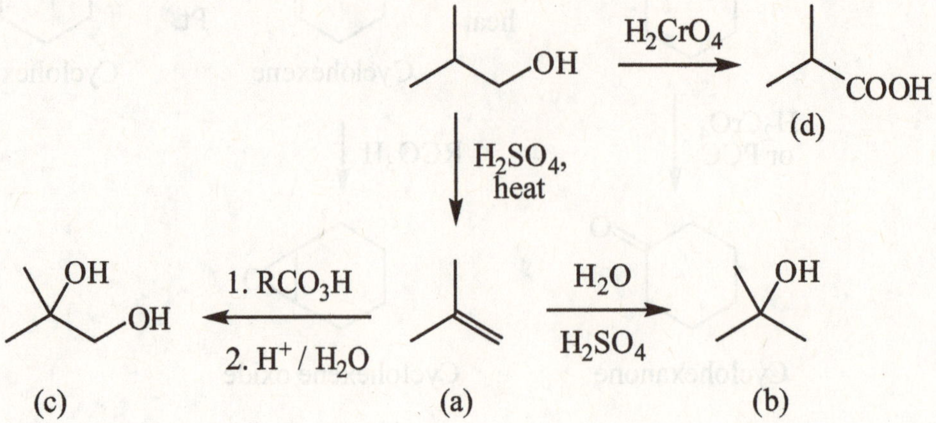

8.45 Show how to prepare each compound from 2-methylcyclohexanol. For any preparation involving more than one step, show each intermediate compound formed.

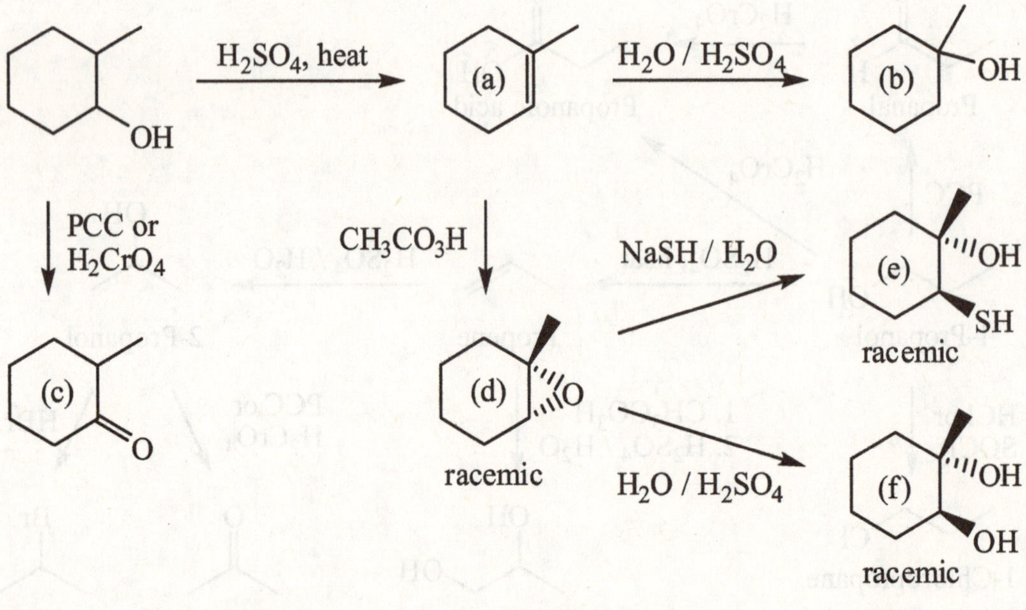

8.46 Show how to convert the alcohol on the left to compounds (a), (b), and (c).

$SOCl_2$ → (a) [structure: cyclopentane with methyl and CH_2Cl]

CH_2OH [cyclopentane structure with methyl]

PCC → (b) [structure: cyclopentane with methyl and CHO group]

H_2CrO_4 → (c) [structure: cyclopentane with methyl and COOH group]

8.47 Disparlure, a sex attractant of the gypsy moth (*Porthetria dispar*), has been synthesized in the laboratory from the following (Z)-alkene:

[structure: (Z)-2-Methyl-7-octadecene] → [structure: Disparlure epoxide]

(Z)-2-Methyl-7-octadecene   **Disparlure**

(a) How might the (Z)-alkene be converted to disparlure?

A peroxycarboxylic acid, such as CH_3CO_3H, will convert the alkene into an epoxide.

(b) How many stereoisomers are possible for disparlure? How many are formed in the sequence you chose?

There are two stereocenters, giving rise to a maximum of $2^2 = 4$ stereoisomers. However, only two are formed due to the stereoselectivity of the epoxidation reaction. Both carbon-oxygen bonds of the epoxide are formed on the same face of the alkene, but there are two faces to choose from, with equal probability.

[structures: two epoxide stereoisomers]

8.48   The chemical name for bombykol, the sex pheromone secreted by the female silkworm moth to attract male silkworm moths, is *trans*-10-*cis*-12-hexadecadien-1-ol. (The compound has one hydroxyl group and two carbon-carbon double bonds in a 16-carbon chain.)

(a)   Draw a structural formula for bombykol, showing the correct configuration about each carbon-carbon double bond.

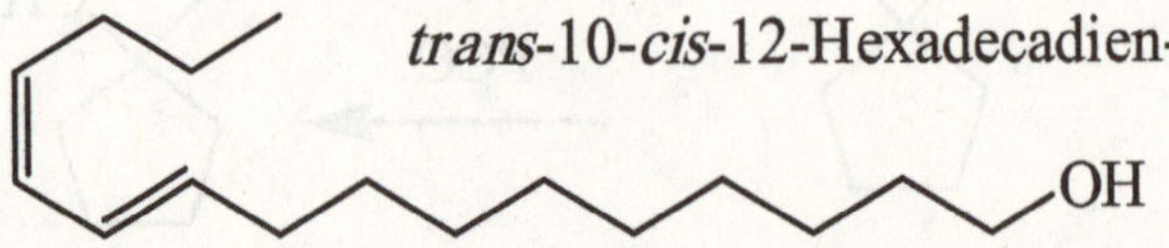

trans-10-*cis*-12-Hexadecadien-1-ol

(b)   How many *cis-trans* isomers are possible for the structural formula you drew in part (a)? All possible *cis-trans* isomers have been synthesized in the laboratory, but only the one named bombykol is produced by the female silkworm moth, and only it attracts male silkworm moths.

Each carbon-carbon double bond can exhibit *cis-trans* isomerism, giving rise to a total of four possible combinations of *cis-trans* isomers.

Chemical Transformations

8.49   Test your cumulative knowledge of the reactions learned so far by completing the following chemical transformations. Pay particular attention to the stereochemistry in the product. Where more than one stereoisomer is possible, show each stereoisomer. Note that transformations will require more than one step.

(a)

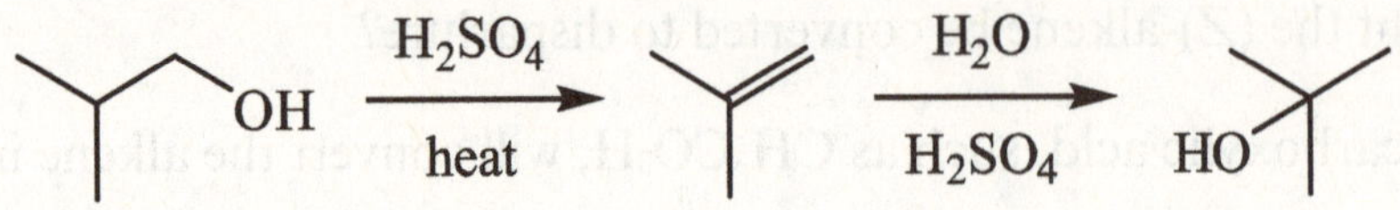

(b)

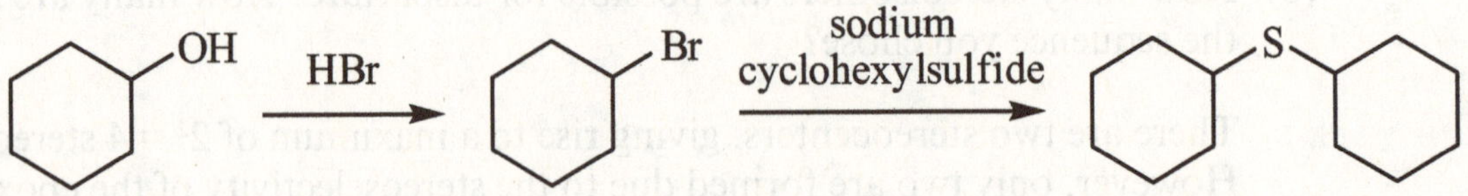

(c)

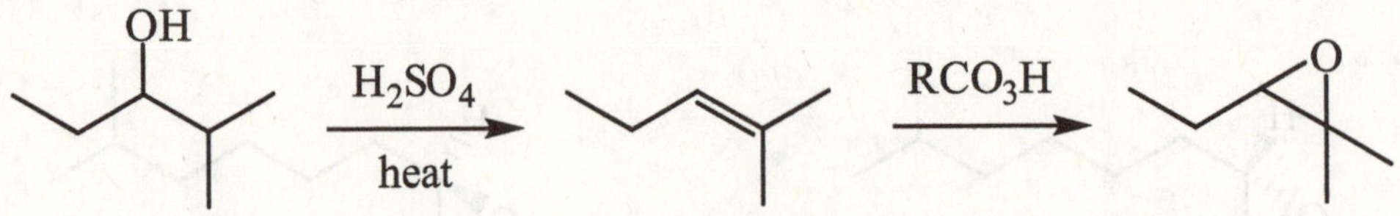

(d)

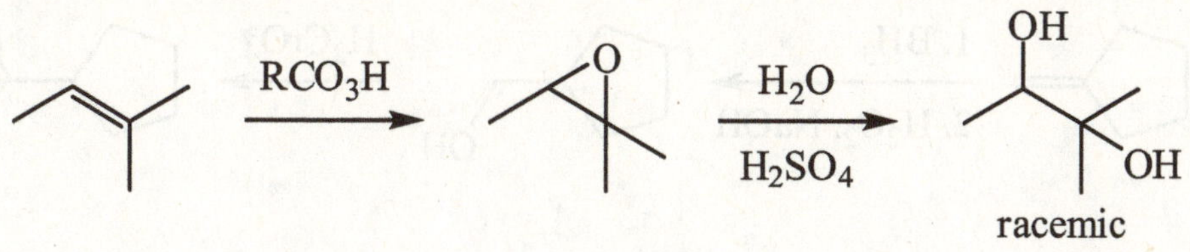

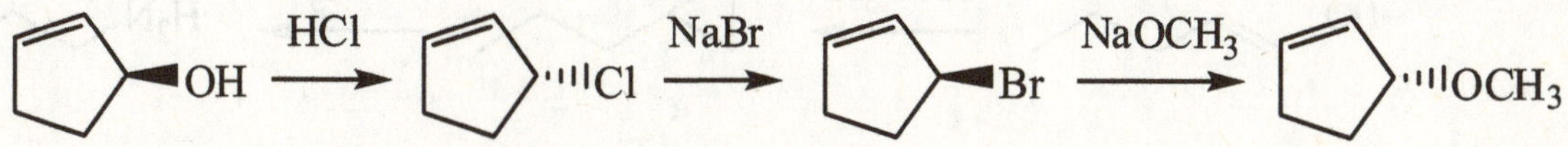

(e)

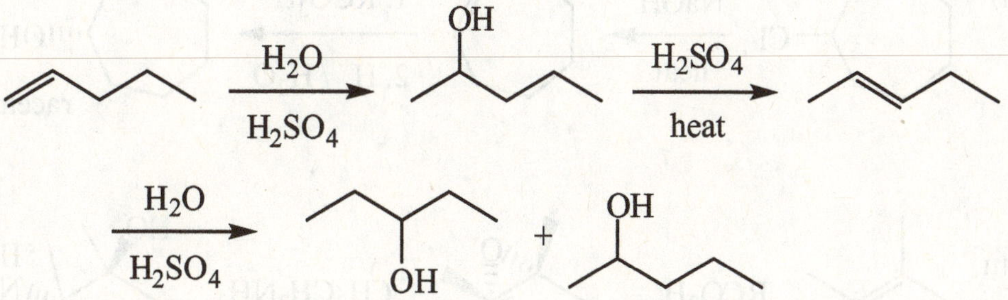

(f)

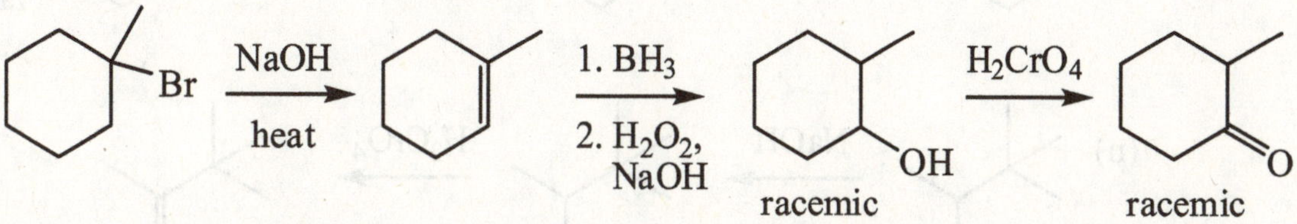

(g)

(h)

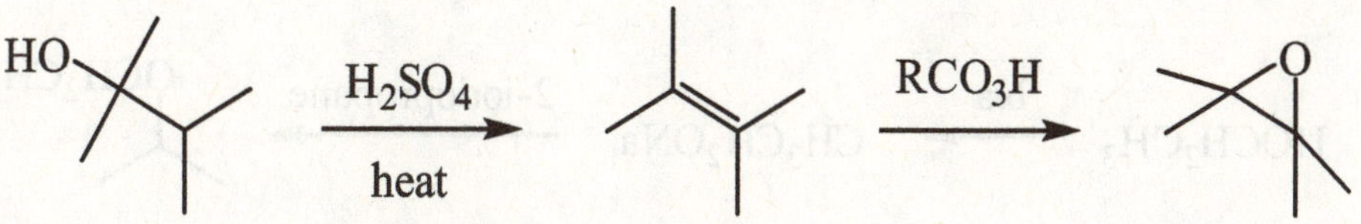

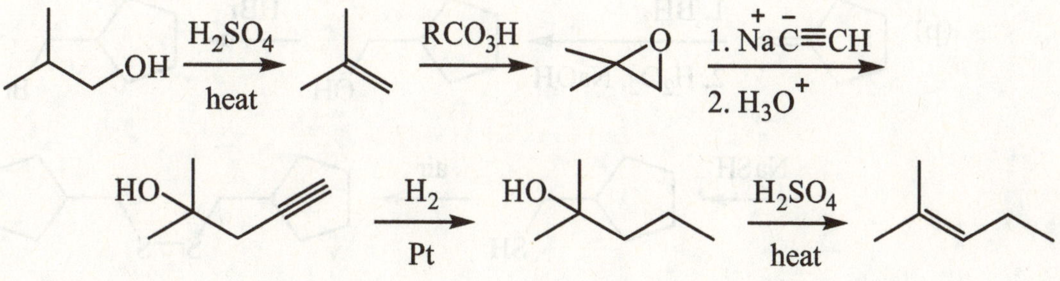

(i)

(j)

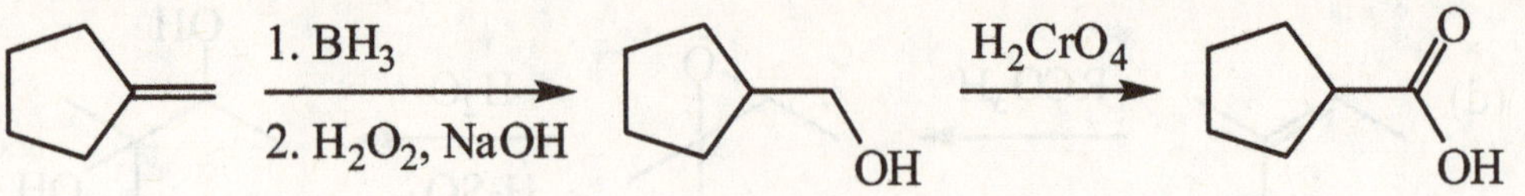

$$\text{methylenecyclopentane} \xrightarrow[\text{2. } H_2O_2, NaOH]{\text{1. } BH_3} \text{(cyclopentylmethanol) } \xrightarrow{H_2CrO_4} \text{cyclopentanecarboxylic acid}$$

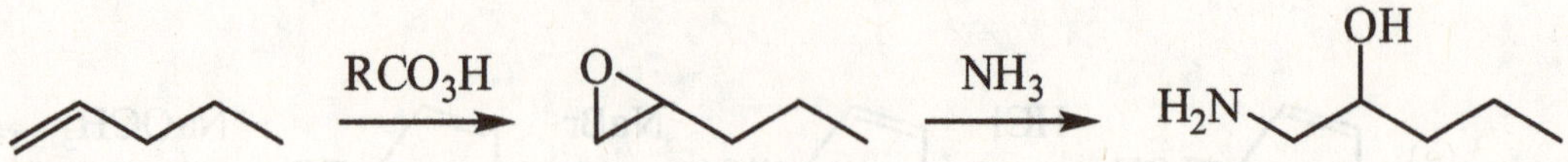

(k)

$$\text{1-pentene} \xrightarrow{RCO_3H} \text{epoxide} \xrightarrow{NH_3} \text{1-amino-2-pentanol}$$

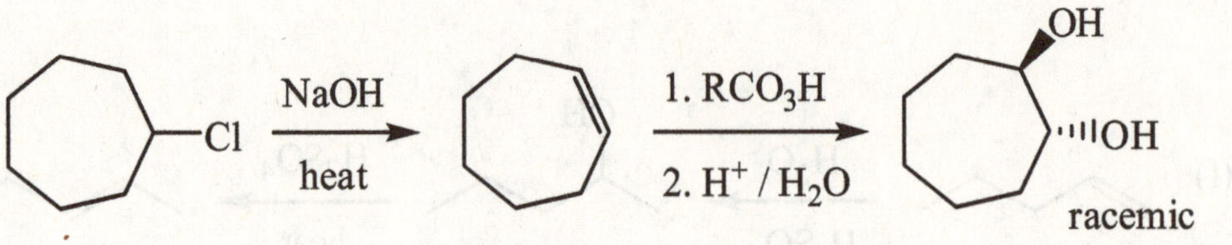

(l)

$$\text{chlorocycloheptane} \xrightarrow[\text{heat}]{NaOH} \text{cycloheptene} \xrightarrow[\text{2. } H^+/H_2O]{\text{1. } RCO_3H} \text{trans-1,2-cycloheptanediol}$$

racemic

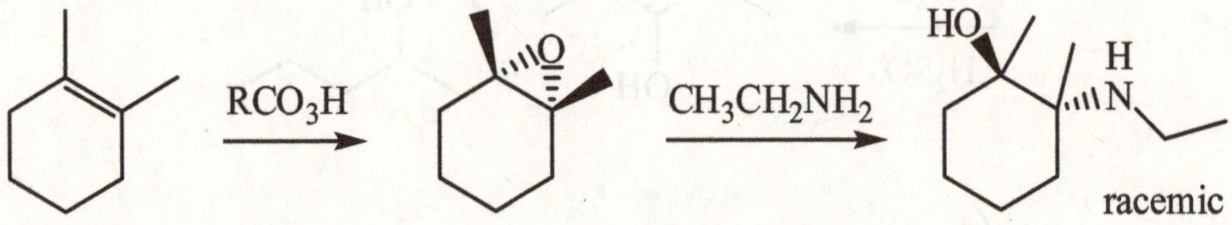

(m)

$$\text{1,2-dimethylcyclohexene} \xrightarrow{RCO_3H} \text{epoxide} \xrightarrow{CH_3CH_2NH_2} \text{amino alcohol}$$

racemic

(n)

$$\text{2-bromo-3,3-dimethylbutane} \xrightarrow{NaOH} \text{3,3-dimethyl-2-butanol} \xrightarrow{H_2CrO_4} \text{ketone}$$

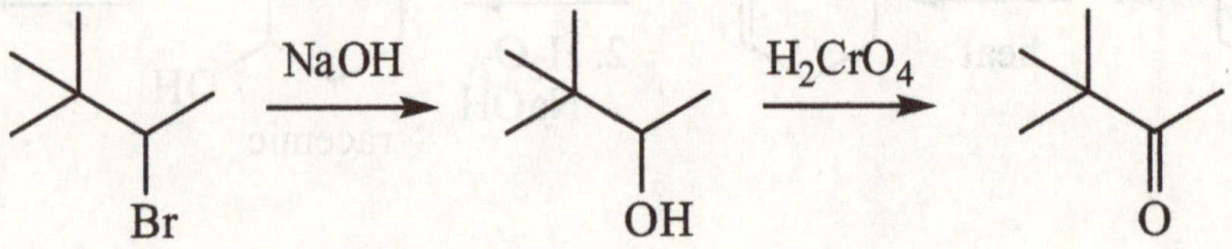

(o)

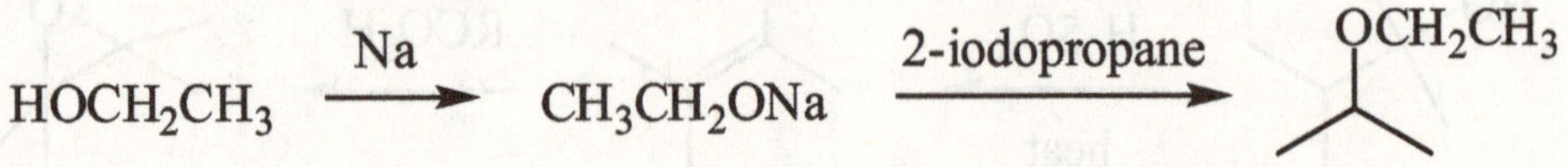

$$HOCH_2CH_3 \xrightarrow{Na} CH_3CH_2ONa \xrightarrow{\text{2-iodopropane}} \text{isopropyl ethyl ether (} OCH_2CH_3 \text{)}$$

(p)

$$\text{methylenecyclopentane} \xrightarrow[\text{2. } H_2O_2, NaOH]{\text{1. } BH_3} \text{cyclopentylmethanol (} OH \text{)} \xrightarrow{HBr} \text{cyclopentylmethyl bromide (} Br \text{)}$$

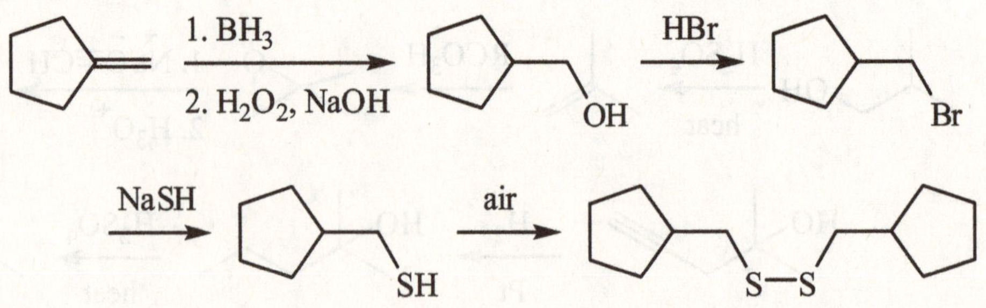

$$\xrightarrow{NaSH} \text{cyclopentylmethanethiol (} SH \text{)} \xrightarrow{\text{air}} \text{disulfide (} S-S \text{)}$$

(q)

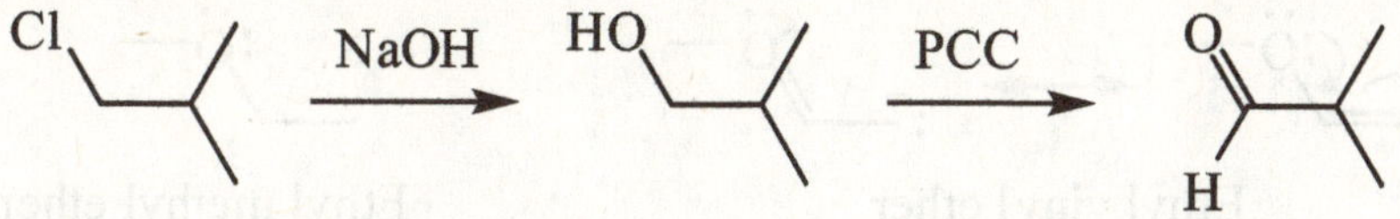

(r)

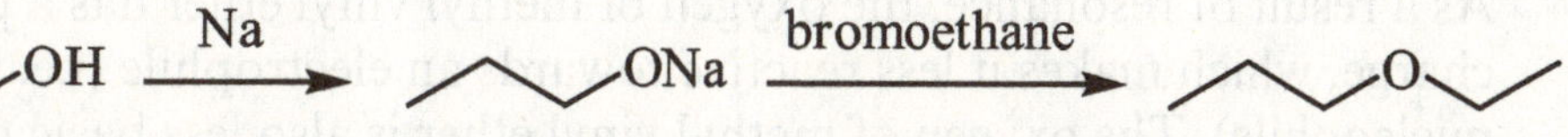

(s)

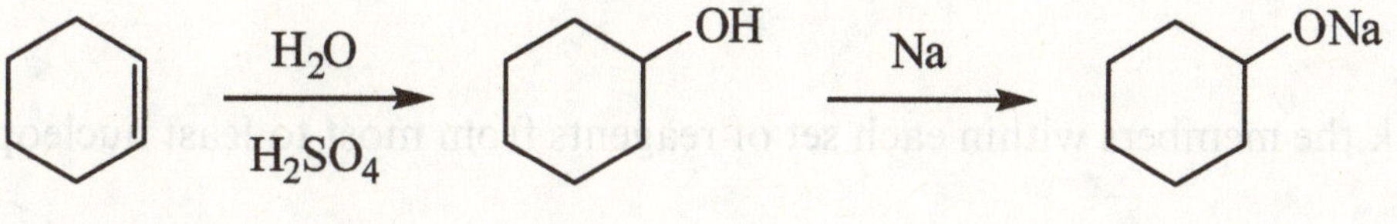

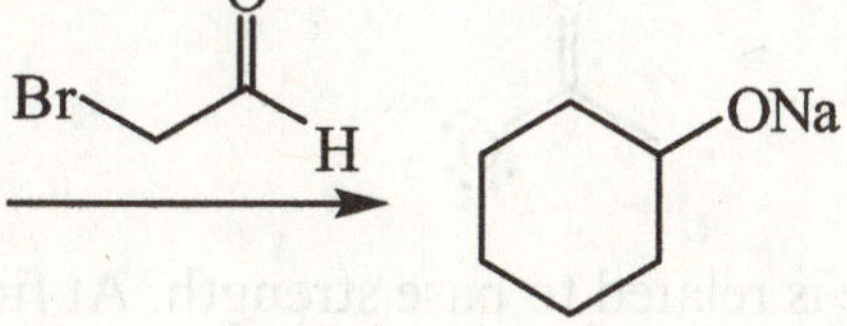

Looking Ahead

8.50 Compounds that contain an N−H group associate by hydrogen bonding.

(a) Do you expect this association to be stronger or weaker than that between compounds containing an O−H group?

Hydrogen bonding arises as a result of the polar bond formed between hydrogen and an electronegative atom, such as N or O. The N−H bond is less polar than the O−H bond, so the association involving the former will be weaker.

(b) Based on your answer to part (a), which would you predict to have the higher boiling point, 1-butanol or 1-butanamine?

1-Butanol   1-Butanamine

1-Butanol will have a higher boiling point because the strength of the intermolecular association will be stronger. A higher temperature is therefore required to overcome this intermolecular association.

8.51 Draw a contributing structure for methyl vinyl ether in which the oxygen is positively charged. Compared with ethyl methyl ether, how does the contributing structure for methyl vinyl ether influence the reactivity of its oxygen toward an electrophile?

Ethyl vinyl ether   Ethyl methyl ether

As a result of resonance, the oxygen of methyl vinyl ether has a partial positive charge, which makes it less reactive towards an electrophile (i.e. it is a weaker nucleophile). The oxygen of methyl vinyl ether is also less basic than the oxygen of ethyl methyl ether.

8.52 Rank the members within each set of reagents from most to least nucleophilic:

(a)

The strength of a nucleophile is related to base strength. At first, it appears that all three anions are stabilized by resonance and that their differences lie only in the identities of the negatively charged atoms (N, S, and O). A negatively charged nitrogen is less stable than a negatively charged oxygen, so the latter is a stronger base and better nucleophile. A negatively charged sulfur is normally more stable than a negatively charged oxygen (effect of atomic size), but in this case, the compound with the sulfur atom is actually less stable. Sulfur, a third-period element with valence orbitals of principal quantum number $n = 3$, cannot undergo effective orbital overlap with the $2p$ orbital of the adjacent carbon atom; p orbitals of different principal quantum numbers differ greatly in size. This poor orbital overlap results in only a little resonance stabilization of the negatively charged sulfur.

(b) $R-\overset{..}{\underset{..}{C}}H_2$ > $R-\overset{..}{N}H$ > $R-\overset{..}{\underset{..}{O}}:^-$

The relative stabilities, and hence the relative basicities and nucleophilic strengths, of these reagents can be be explained by the effect of electronegativity.

8.53 In Chapter 14 we will see that the reactivity of the following carbonyl compounds is directly proportional to the stability of the leaving group. Rank the order of reactivity of these carbonyl compounds from most reactive to least reactive based on the stability of the leaving group.

The best leaving group is the most stable leaving group. Due to the effect of atomic size, chloride is much more stable than methoxide or the amide ion.

Group Learning Activities

8.54   Discuss why primary alcohols (with the exception of ethanol) cannot be prepared by the acid-catalyzed hydration of alkenes. You'll want to discuss the mechanism of acid-catalyzed hydration and consider the reactive species involved.

To prepare a primary alcohol from an alkene, a primary carbocation must be made by protonating the alkene. However, the protonation of any alkene, except for ethylene, will preferentially give a carbocation other than a primary carbocation. For example:

most-stable   least-stable
carbocation   carbocation

8.55   Discuss why sodium hydroxide is not a good reagent for the synthesis of alkoxides from alcohols. Similarly, could sodium hydroxide be used to synthesize alkylsulfides from thiols? Why or why not? You'll want to discuss the type of reactions that would occur in both reactions as well as what makes a reaction synthetically useful.

Both of these reactions are acid-base reactions.

$$ROH + HO^- \rightleftharpoons RO^- + H_2O$$

$$RSH + HO^- \rightleftharpoons RS^- + H_2O$$

Sodium hydroxide is not a good reagent for the preparation of alkoxides from alcohols. Hydroxide is a slightly weaker base than an alkoxide, so the equilibrium for the reaction lies slightly to the left. To prepare an alkoxide, a large excess of hydroxide would be required to give a reasonable yield of the alkoxide. From a synthetic standpoint, this is not very efficient.

On the other hand, sodium hydroxide is a good reagent for the preparation of alkylsulfides from thiols. Hydroxide is a much stronger base than an alkylsulfide, so the equilibrium lies far to the right. Only a stoichiometric amount of hydroxide would be required for the reaction to give a good yield of the alkylsulfide.

8.56   One of the following alcohols is used to de-ice airplanes during extreme cold weather. As a group, decide which of the three alcohols would be most suitable for the job by discussing factors that one must consider for such an application.

$$CH_3OH \qquad CH_3CH_2OH \qquad (CH_3)_3COH$$

Methanol   Ethanol   *tert*-Butanol

For a substance to be a good de-icing agent, it must dissolve the ice and form an aqueous solution that is resistant to freezing. Out of the three alcohols, methanol has the lowest melting point (freezing point). Consequently, the aqueous solution that is formed by methanol also has the lowest freezing point. Methanol is the most suitable alcohol for de-icing purposes.

8.57 Shown is the structure of Erythromycin A, an important antibiotic that is used to treat a number of bacterial infections. In your study group,

(a) Locate all the hydroxyl groups in Erythromycin A and classify each as 1°, 2°, or 3°.

(b) Four of these hydroxyl groups are involved in intramolecular hydrogen bonding. One of these is pointed out on the structural formula, and it creates a five-membered ring. Locate the other three alcohols that have the potential for intramolecular hydrogen bonding. Note that nitrogen can also form hydrogen bonds.

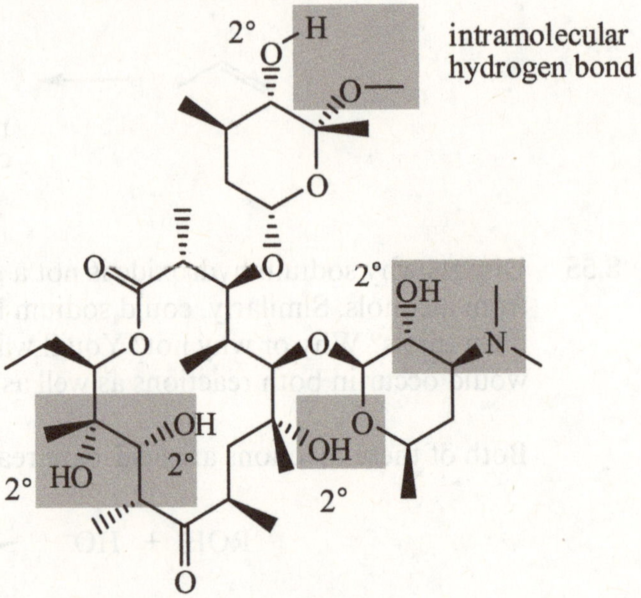

(c) Identify the other functional groups in Erythromycin A.

Note that the acetal functional group will be covered in Chapter 10.

(d) Locate all of the stereocenters, determine their R/S configuration, and predict the maximum number of stereoisomers possible for Erythromycin A.

Out of the $2^{18} = 262,144$ possible stereoisomers, only one of them is Erythromycin A.

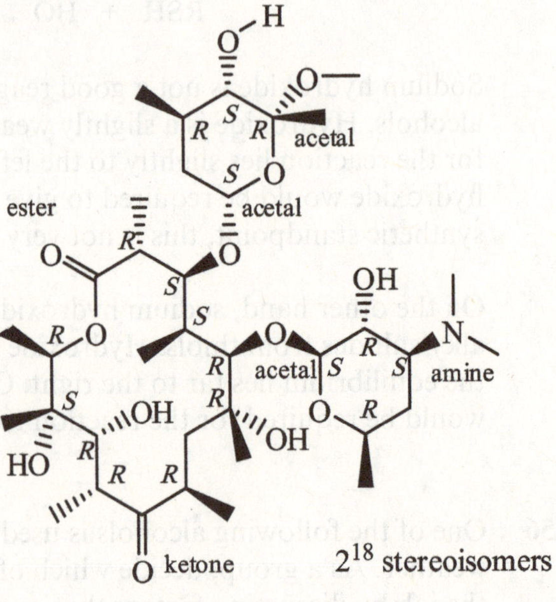

8.58 Practice your arrow pushing skills by proposing mechanisms for the following reactions:

Because the starting materials for both reactions allow the initial formation of relatively stable carbocations, the dehydration reactions proceeds by E1 elimination. After the carbocations have been formed, they undergo rearrangement.

(a)

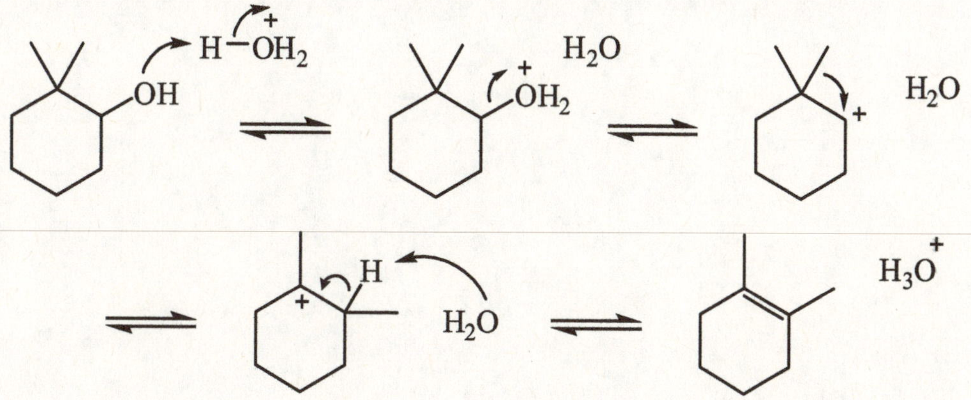

(b)

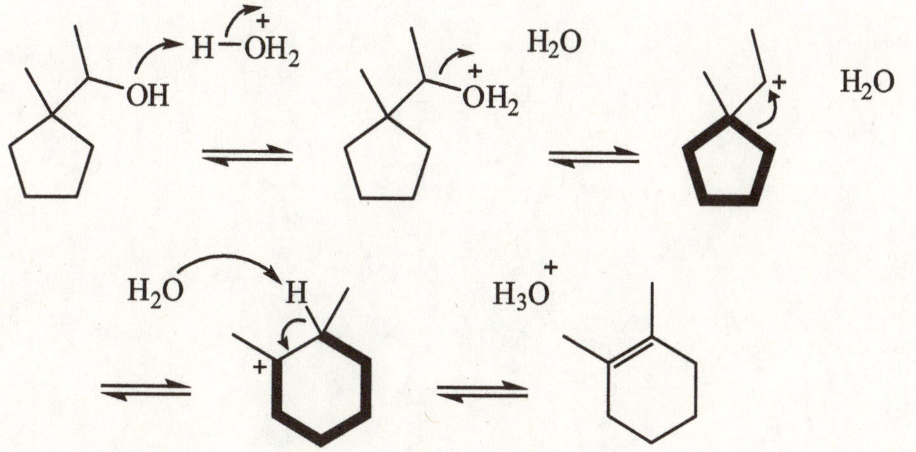

Chapter 9: Benzene and Its Derivatives

Problems

9.1 Which of the following compounds are aromatic?

(a)   (b)   (c)   (d)

In order for a compound to be aromatic, each atom of the ring must contain a $2p$ orbital, the ring must be planar, and there must be a total of 2, 6, 10, 14, etc. π electrons in the p orbitals of the ring atoms. Compound (a) is aromatic because all the ring atoms are sp^2-hybridized, so each atom has a p orbital, and there are a total of 2 π electrons. Compound (b) is not aromatic because even though all the ring atoms are sp^2 and the ring is planar, there are only 4 π electrons present. Compounds (c) and (d) are aromatic because there are 6 π electrons, all the ring atoms are sp^2-hybridized, and the ring is planar. Note that in (d), the lone pair is placed into a p orbital, making the hybridization of that carbon sp^2.

9.2 Write names for these compounds:

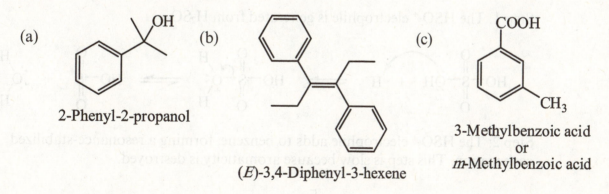

(a)

2-Phenyl-2-propanol

(b)

(*E*)-3,4-Diphenyl-3-hexene

(c)

3-Methylbenzoic acid
or
m-Methylbenzoic acid

(d)

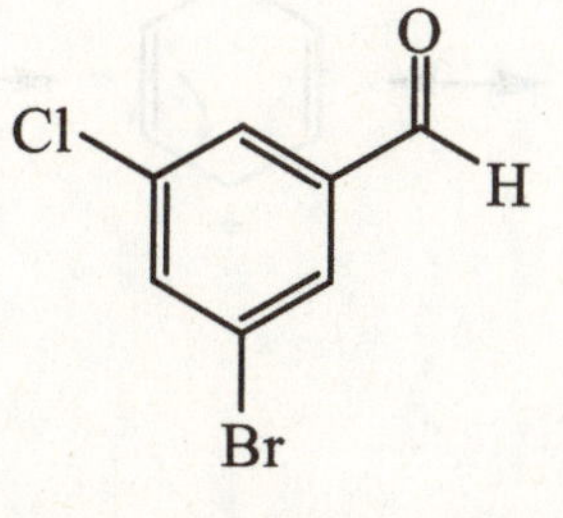

3-Bromo-5-chlorobenzaldehyde

(e)

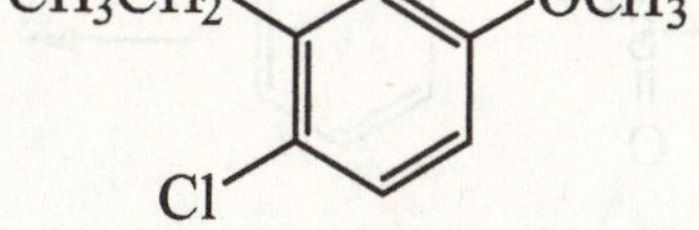

4-Chloro-3-ethylanisole

9.3 Predict the products resulting from vigorous oxidation of each compound by H_2CrO_4:

When aromatic compounds are treated with a strong oxidizing agent, the benzylic carbon is oxidized to a carboxylic acid if it bears one or more hydrogen atoms. The oxidation of a benzylic carbon results in the removal of all other carbons that are bonded to it. Aldehydes are oxidized to carboxyl groups in the presence of an oxidizing agent.

(a)

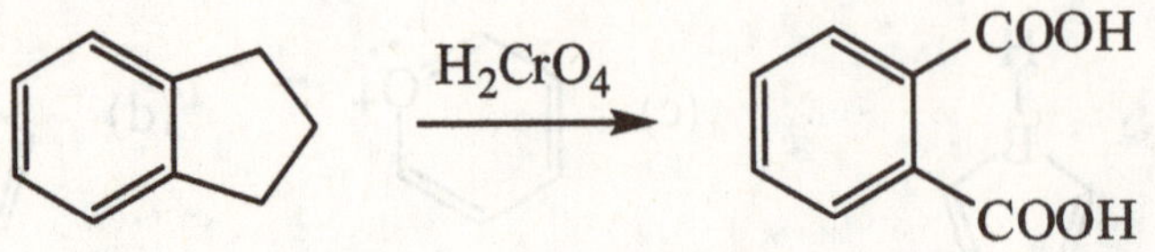

(b)

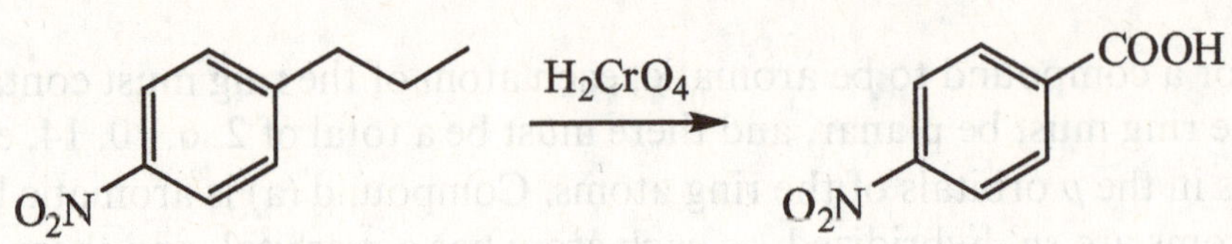

(c)

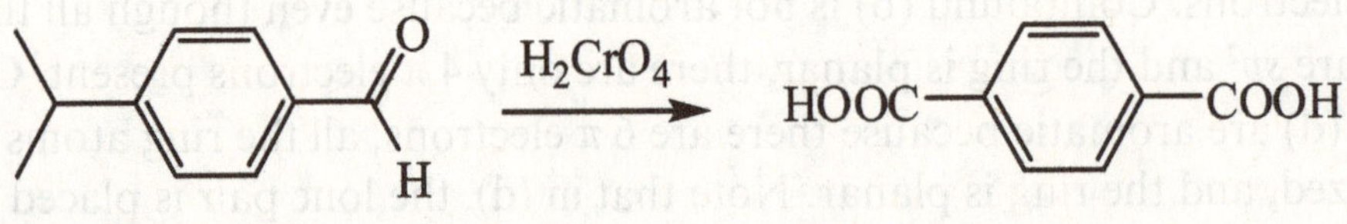

9.4 Write a stepwise mechanism for the sulfonation of benzene. Use HSO_3^+ as the electrophile.

Step 1: The HSO_3^+ electrophile is generated from H_2SO_4.

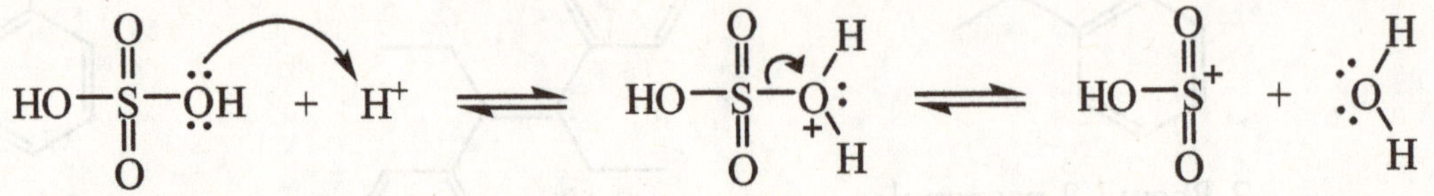

Step 2: The HSO_3^+ electrophile adds to benzene, forming a resonance-stabilized carbocation. This step is slow because aromaticity is destroyed.

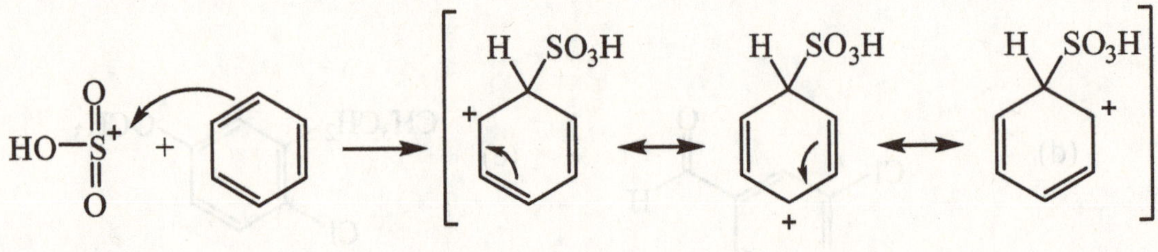

Step 3: Deprotonation of the carbocation regenerates the benzene ring. This step is fast because aromaticity is regenerated. Note that the proton is removed from the same carbon to which the electrophile (HSO$_3$$^+$) was added.

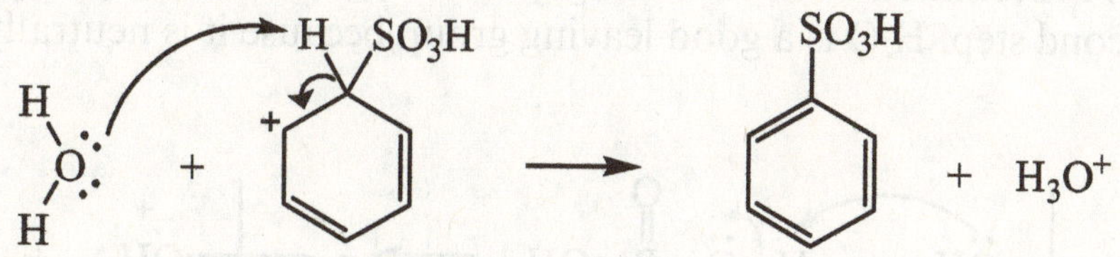

9.5   Write a structural formula for the product formed from Friedel-Crafts alkylation or acylation of benzene with:

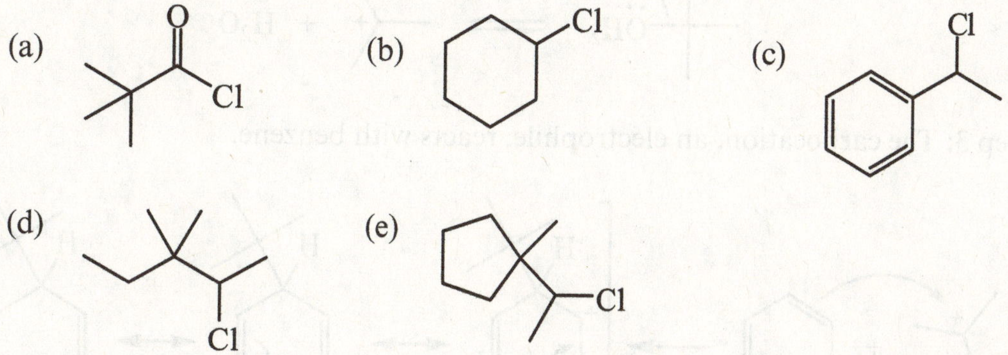

(a)

(b)

(c)

(d)

(e)

In a Friedel-Crafts alkylation or acylation, the catalyst (AlCl$_3$) removes the chloride ion from the alkyl halide or the acid halide, respectively. The resulting carbocations are electrophiles that react with benzene, forming the corresponding products below. The formation of (d) and (e) involves a 1,2-methyl shift to make the carbocation more stable.

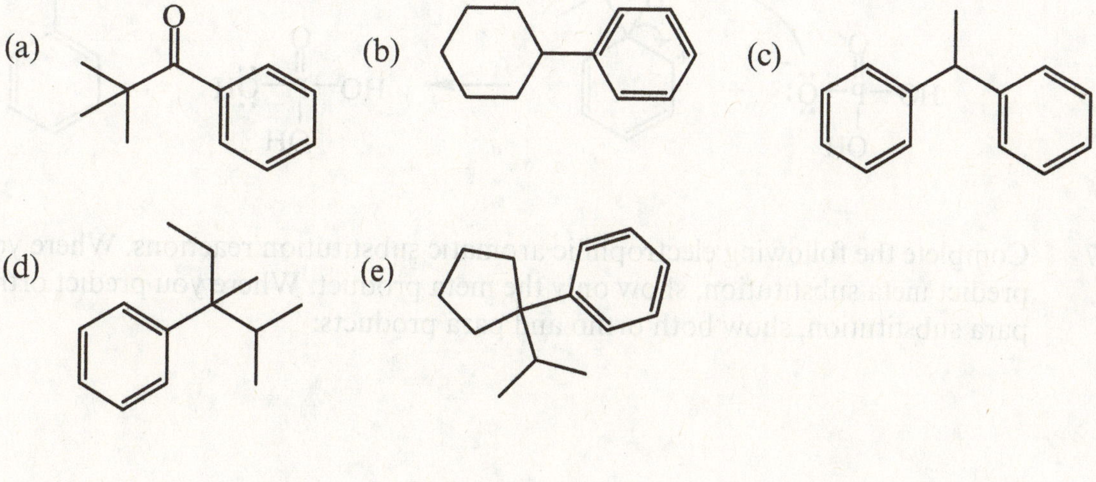

(a)

(b)

(c)

(d)

(e)

9.6   Write a mechanism for the formation of *tert*-butylbenzene from benzene and *tert*-butyl alcohol in the presence of phosphoric acid.

Step 1: Protonation of the alcohol generates a good leaving group, which departs in the second step. H_2O is a good leaving group because it is neutrally charged and stable.

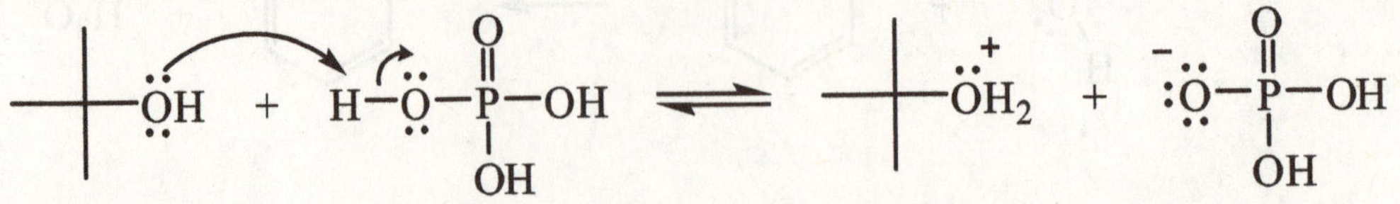

Step 2: The leaving group leaves as water, generating the *tert*-butyl carbocation.

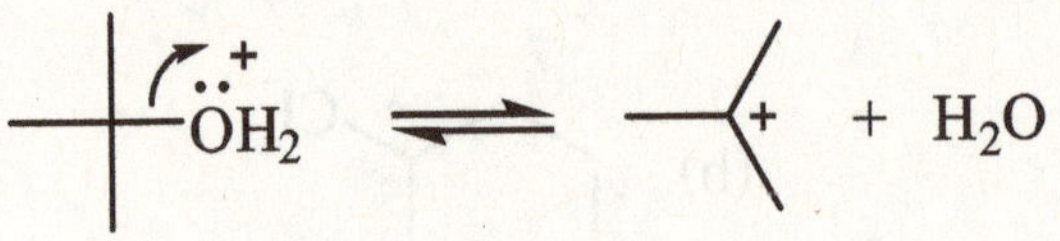

Step 3: The carbocation, an electrophile, reacts with benzene.

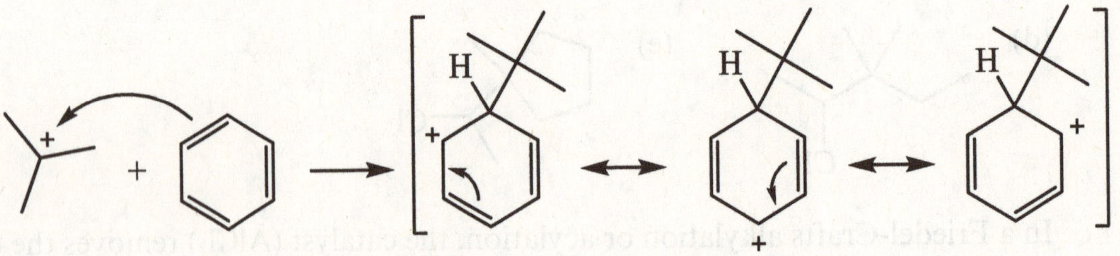

Step 4: Deprotonation regenerates the aromatic ring.

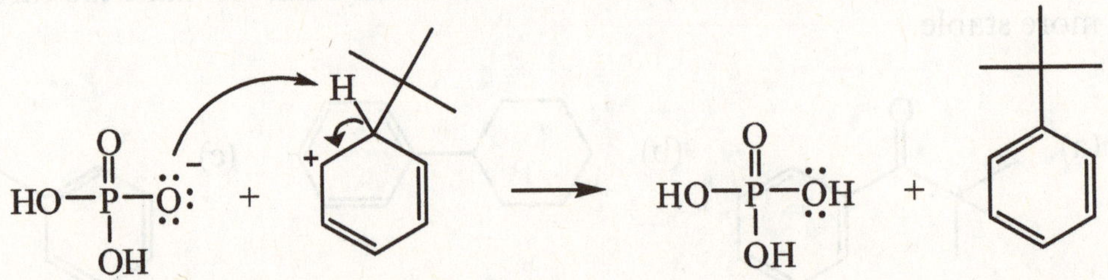

9.7   Complete the following electrophilic aromatic substitution reactions. Where you predict meta substitution, show only the meta product. Where you predict ortho-para substitution, show both ortho and para products:

(a)

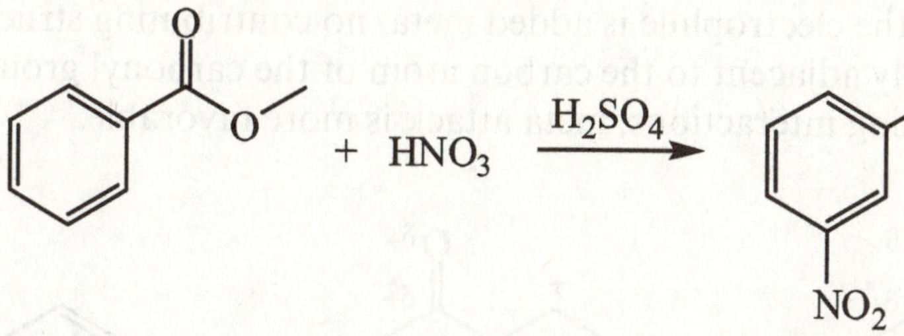

(b)

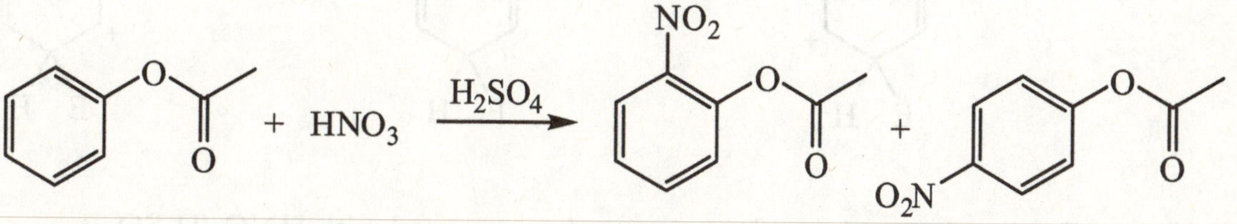

In reaction (a), the carbonyl group of the ester is bonded to the aromatic ring, and the δ+ charge on the carbon acts as an electron-withdrawing group (meta director). In reaction (b), the oxygen atom bonded to the aromatic ring can stabilize, by resonance, the carbocation that is formed during the reaction and is therefore an ortho-para director.

9.8 Because the electronegativity of oxygen is greater than that of carbon, the carbon of a carbonyl group bears a partial positive charge, and its oxygen bears a partial negative charge. Using this information, show that a carbonyl group is meta directing:

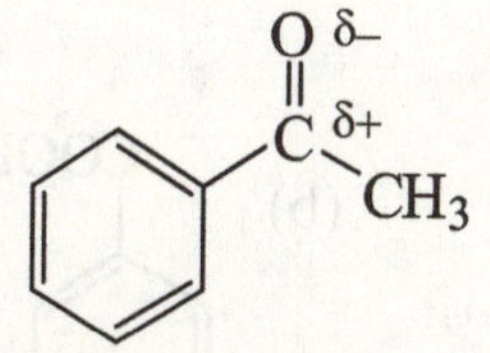

If the electrophile is added ortho or para, it is possible to draw a contributing structure that places the positive charge directly adjacent to the partially positive carbon atom of the carbonyl group. This interaction destabilizes the carbocation, which in turn disfavors ortho-para attack of the electrophile.

Para

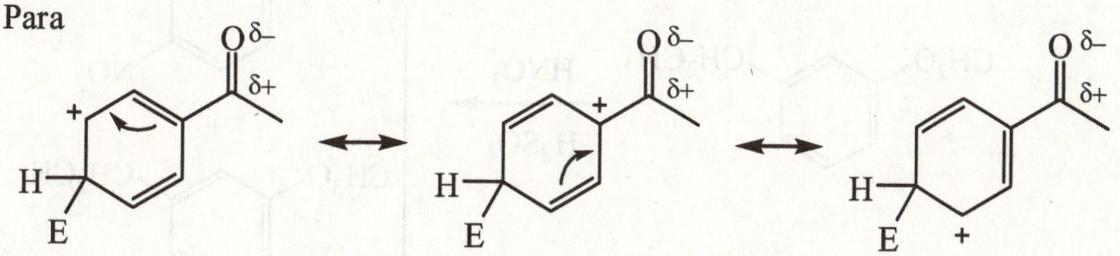

Ortho

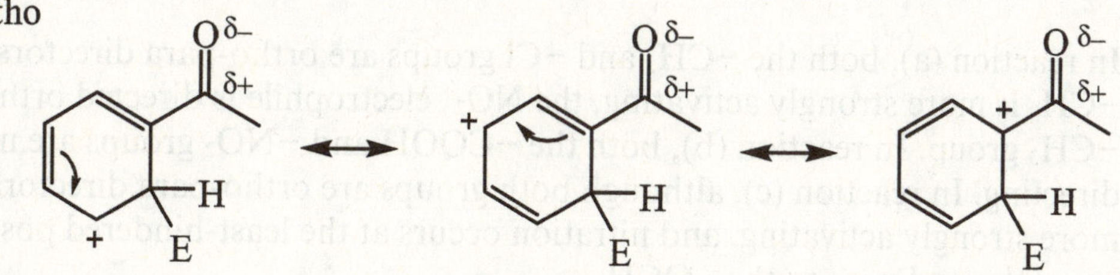

On the other hand, if the electrophile is added meta, no contributing structure places the carbocation directly adjacent to the carbon atom of the carbonyl group. Because there are no destabilizing interactions, meta attack is more favorable.

Meta

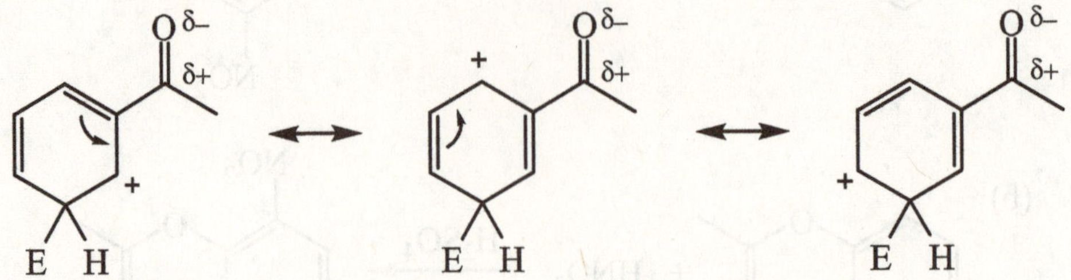

9.9   Predict the product of treating each compound with HNO_3/H_2SO_4.

(a)

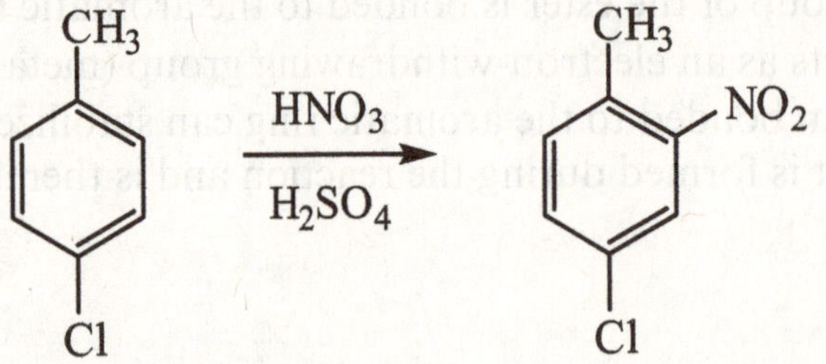

(b)

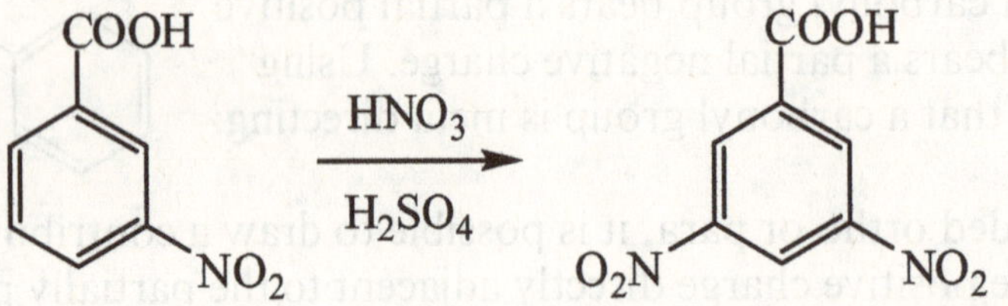

(c)

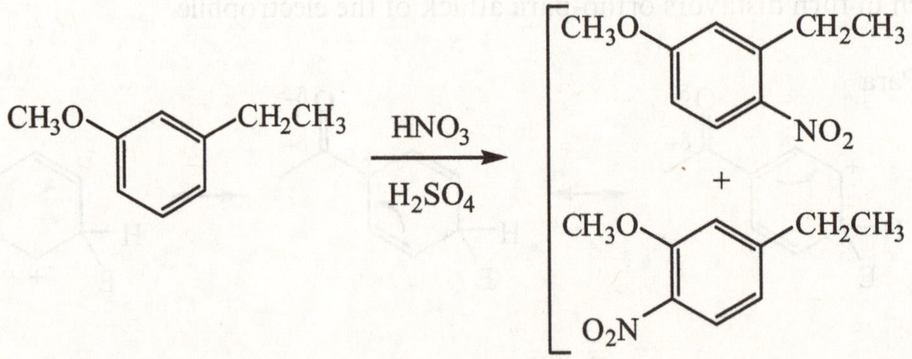

In reaction (a), both the $-CH_3$ and $-Cl$ groups are ortho-para directors; because $-CH_3$ is more strongly activating, the NO_2^+ electrophile is directed ortho to the $-CH_3$ group. In reaction (b), both the $-COOH$ and $-NO_2$ groups are meta-directing. In reaction (c), although both groups are ortho-para directors, $-OCH_3$ is more strongly activating, and nitration occurs at the least-hindered positions that are ortho and para to the $-OCH_3$.

9.10   Arrange these compounds in order of increasing acidity: 2,4-dichlorophenol, phenol, cyclohexanol.

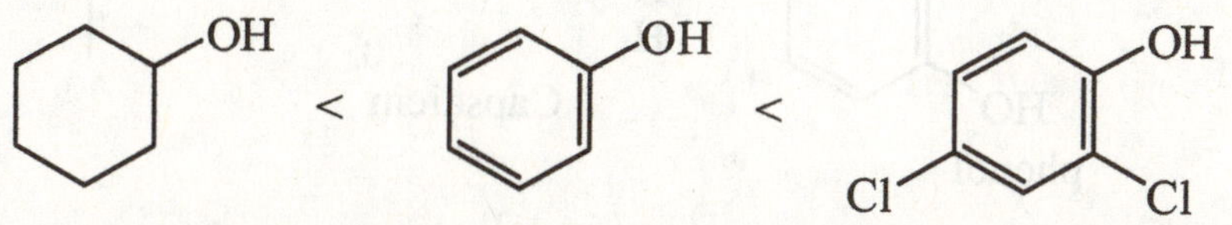

2,4-dichlorophenol is the most acidic because its conjugate base is stabilized by both resonance and the inductive effect. The conjugate base of phenol is stabilized only by resonance, and the conjugate base of cyclohexanol is not stabilized by either effect.

Chemical Connections

9A.   Show how the outer perimeter of benzo[a]pyrene satisfies Hückel's criteria for aromaticity. Is the outer perimeter of the highlighted portion of the diol epoxide product of benzo[a]pyrene also aromatic?

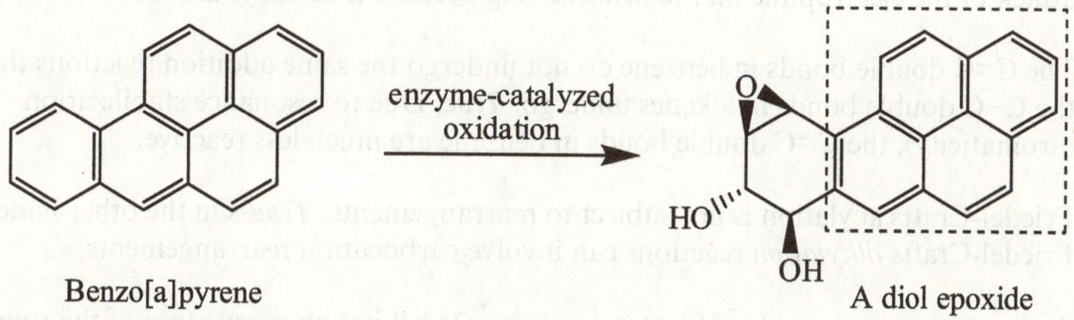

Benzo[a]pyrene   A diol epoxide

The outer perimeter of benzo[a]pyrene is an eighteen-membered ring. All of the eighteen ring carbons are sp^2-hybridized and each member contributes one p electron to the aromatic system.

The highlighted portion of the diol epoxide is also aromatic. The perimeter of the highlighted portion is a fourteen-membered ring, and each of the fourteen sp^2-hybridized members contributes one p electron to the aromatic system.

9B.   Would you predict capsaicin to be more soluble in water or more soluble in 1-octanol? Would your prediction remain the same if capsaicin were first treated with a molar equivalent of NaOH?

Capsaicin is relatively nonpolar, so it will be more soluble in 1-octanol, which is less polar than water. If capsaicin were treated with NaOH, the phenol would be deprotonated to form the sodium salt of capsaicin. The salt, being ionic, would be more soluble in water.

Quick Quiz

1. The mechanism of electrophilic aromatic substitution involves three steps: generation of the electrophile, attack of the electrophile on the benzene ring, and proton transfer to regenerate the ring. *True*. Note that the slow step of the overall reaction is the attack of the electrophile on the benzene ring because it destroys aromaticity.

2. The C=C double bonds in benzene do not undergo the same addition reactions that the C−C double bonds in alkenes undergo. *True*. Due to resonance stabilization (aromaticity), the C=C double bonds in benzene are much less reactive.

3. Friedel-Crafts acylation is not subject to rearrangements. *True*. On the other hand, Friedel-Crafts *alkylation* reactions can involve carbocation rearrangements.

4. An aromatic compound is planar, possesses a *2p* orbital on every atom of the ring, and contains either 4, 8, 12, 16, and so on, pi electrons. *False*. The number of pi electrons required are 2, 6, 10, 14, 18, and so on.

5. When naming disubstituted benzenes, the locators para, meta, and ortho refer to substituents that are 1,2, 1,3, and 1,4, respectively. *False*. Para refers to substituents that are 1,4, meta refers to 1,3, and ortho refers to 1,2.

6. The electrophile in the chlorination or bromination of benzene is an ion pair containing a chloronium or bromonium ion. *True*. These halonium ions are exceptionally reactive electrophiles, and they are needed to react with the relatively stable benzene ring.

7. An ammonium group ($-NH_3^+$) on a benzene ring will direct an attacking electrophile to a meta position. *True*. The ammonium group is an electron-withdrawing group and therefore a meta director.

8. Reaction of chromic acid, H_2CrO_4, with a substituted benzene always oxidizes every alkyl group at the benzylic position to a carboxyl group. *False*. Oxidation occurs only if the benzylic carbon contains one or more hydrogen atoms.

9. Benzene consists of two contributing structures that rapidly interconvert between each other. *False*. Recall that contributing structures are not species that are in equilibrium and interconverting. There is only one structure of benzene, the resonance hybrid.

10. The electrophile in the nitration of benzene is the nitrate ion. *False*. The electrophile is the nitronium ion (NO_2^+) and not nitrate (NO_3^-).

11. A benzene ring with an −OH bonded to it is referred to as "phenyl." *False*. *Phenyl* refers to the benzene ring as a substituent. A benzene ring with a hydroxyl group is a *phenol*.

12. Friedel-Crafts alkylation of a primary haloalkane with benzene will always result in a new bond between benzene and the carbon that was bonded to the halogen. *False*. The carbocation formed from a primary haloalkane may undergo rearrangement.

13. Resonance energy is the energy a ring contains due to the stability afforded it by its contributing structures. *False*. It is not the contributing structures themselves that afford stability, but rather the delocalization of the π electrons throughout the molecule.

14. A phenol will react quantitatively with NaOH. *True*. Phenols are weak acids and can be neutralized by strong bases such as NaOH.

15. The use of a haloalkane and $AlCl_3$ is the only way to synthesize an alkylbenzene. *False*. When a haloalkane is combined with $AlCl_3$, a carbocation intermediate is generated. Carbocations could also be generated using an alkene and a strong acid.

16. Phenols are more acidic than alcohols. *True*. The conjugate base of a phenol is more stable than the conjugate base of an alcohol due to resonance stabilization of the negative charge.

17. Substituents of polysubstituted benzene rings can be numbered according to their distance from the substituent that imparts a special name to the compound. *True*. The substituent that imparts a special name to the compound is automatically assigned position one.

18. If a benzene ring contains both a weakly activating group and a strongly deactivating group, the strongly deactivating group will direct the attack of an electrophile. *False*. In the presence of both activating and deactivating groups, the former will direct the attack.

19. Oxygen, O_2, can be considered a diradical. *True*. Many of the reactive properties of molecular oxygen are due to its diradical nature.

20. The contributing structures for the attack of an electrophile to the ortho position of aniline are more stable than those for the attack at the meta position. *True*. The electron-donating amino group is an ortho-para director.

21. A deactivating group will cause its benzene ring to react slower than benzene itself. *True*. Deactivating groups withdraw electron density from the ring, reducing its reactivity.

22. Friedel-Crafts alkylation is promoted by the presence of electron-withdrawing groups. *False.* Friedel-Crafts alkylation reactions are inhibited by electron-withdrawing groups.

23. Autoxidation takes place at allylic carbons. *True.* Allylic carbons (carbon atoms directly adjacent to an alkene) are oxidized by oxygen alone.

24. The contributing structures for the attack of an electrophile to the meta position of nitrobenzene are more stable than those for the attack at the ortho or para position. *True.* For this reason, the electron-withdrawing nitro group is a meta director.

End-of-Chapter Problems

Aromaticity

9.11 Which of the following compounds or chemical entities are aromatic?

Recall that in order for a compound to be aromatic, it must be planar, have all ring atoms sp^2-hybridized ($2p$ orbital on each atom), and have a total of 2, 6, 10, 14, etc. π electrons. If an atom with one or more lone pairs is normally sp^3 but changing its hybridization to sp^2 (and placing a lone pair in the p orbital) allows the criteria for aromaticity to be satisfied, this change of hybridization will occur.

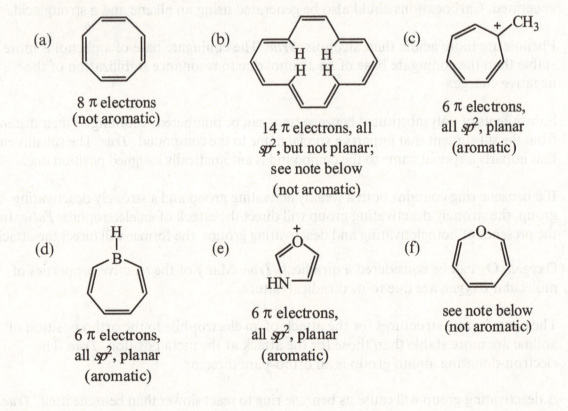

(a)

8 π electrons
(not aromatic)

(b)

14 π electrons, all
sp^2, but not planar;
see note below
(not aromatic)

(c)

6 π electrons,
all sp^2, planar
(aromatic)

(d)

6 π electrons,
all sp^2, planar
(aromatic)

(e)

6 π electrons,
all sp^2, planar
(aromatic)

(f)

see note below
(not aromatic)

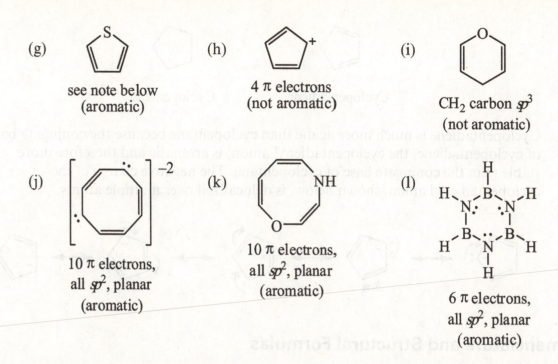

(g) see note below
(aromatic)

(h) 4 π electrons
(not aromatic)

(i) CH_2 carbon sp^3
(not aromatic)

(j) 10 π electrons,
all sp^2, planar
(aromatic)

(k) 10 π electrons,
all sp^2, planar
(aromatic)

(l) 6 π electrons,
all sp^2, planar
(aromatic)

Although compound (b) has 14 π electrons and all of the ring atoms are sp^2, it is actually not planar due to the steric interactions between the four hydrogens atoms shown in the line diagram. Due to the nonplanar structure, the carbon framework of which is shown in the model below, the compound is not aromatic.

Compound (f) is not aromatic. It has 6 π electrons, but the oxygen atom is sp^3. If the hybridization of oxygen became sp^2, two electrons would be placed in the *p* orbital. Yet, a compound with 8 π electrons is not aromatic. (Compare (f) to (k), where nitrogen and oxygen are both sp^2 due to the requirements for aromaticity being fulfilled.)

Compound (g) contains sulfur, an element in the third period. The sp^2 hybridization of sulfur would result in the placement of two electrons in a 3*p* orbital. Aromaticity requires, at least in theory, 2*p* orbitals, but computational and experimental studies have shown that this compound indeed exhibits aromatic character.

9.12 Explain why cyclopentadiene (pK_a 16) is many orders of magnitude more acidic than cyclopentane (pK_a > 50). (*Hint*: Draw the structural formula for the anion formed by removing one of the protons on the $-CH_2-$ group, and then apply the Hückel criteria for aromaticity.)

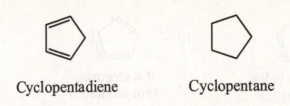

Cyclopentadiene   Cyclopentane

Cyclopentadiene is much more acidic than cyclopentane because the conjugate base of cyclopentadiene, the cyclopentadienyl anion, is aromatic and therefore more stable than the conjugate base of cyclopentane. The negative charge of the cyclopentadienyl anion, shown below, is delocalized over multiple atoms.

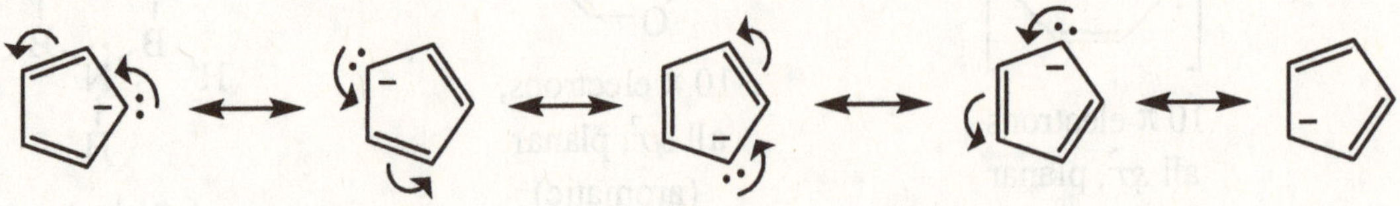

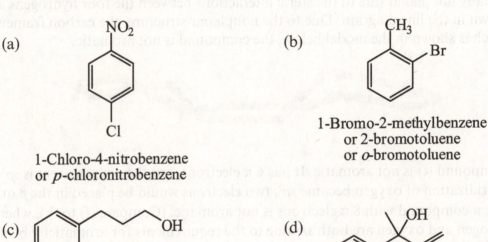

Nomenclature and Structural Formulas

9.13   Name these compounds:

(a)

NO_2

Cl

1-Chloro-4-nitrobenzene
or *p*-chloronitrobenzene

(b)

CH_3
Br

1-Bromo-2-methylbenzene
or 2-bromotoluene
or *o*-bromotoluene

(c)

OH

3-Phenyl-1-propanol

(d)

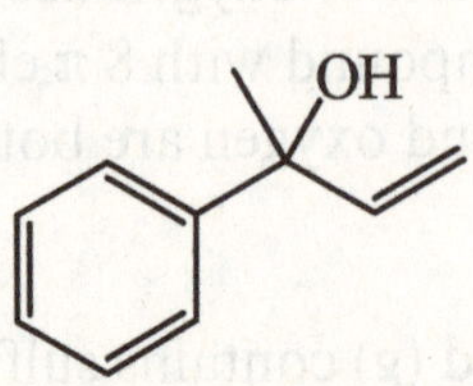

OH

2-Phenyl-3-buten-2-ol

(e)

COOH

NO_2

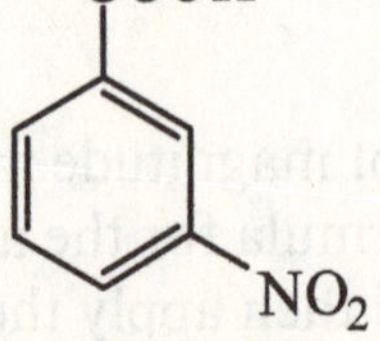

3-Nitrobenzoic acid
or *m*-nitrobenzoic acid

(f)

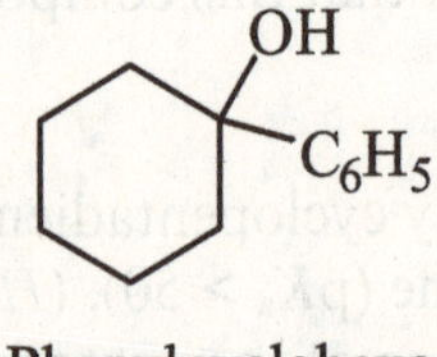

OH
C_6H_5

1-Phenylcyclohexanol

(g) 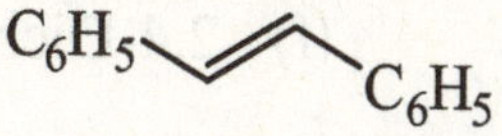 C_6H_5 $\diagdown$ C_6H_5

(E)-1,2-Diphenylethene
or *trans*-1,2-diphenylethene

(h)

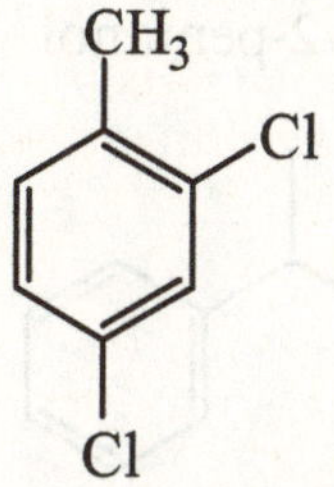

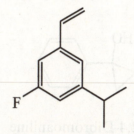

2,4-Dichlorotoluene

(i)

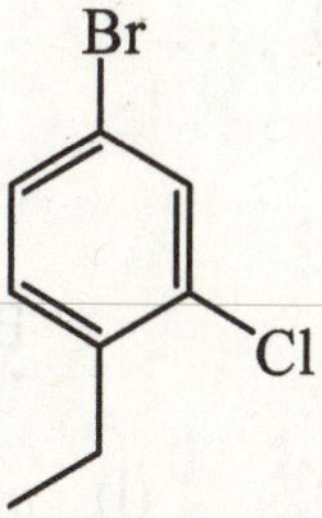

4-Bromo-2-chloro-1-ethylbenzene

(j)

1-Fluoro-3-isopropyl-
5-vinylbenzene
or 3-fluoro-5-isopropylstyrene

(k)

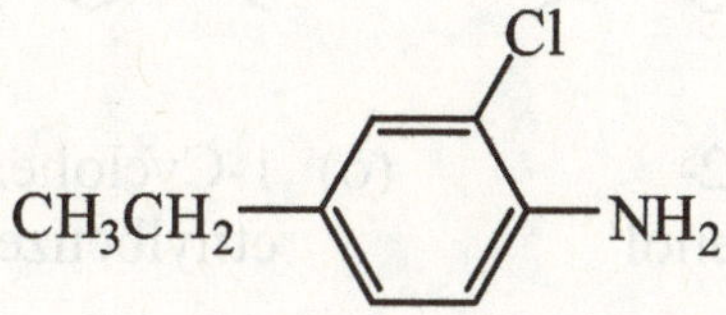

1-Amino-2-chloro-4-ethylbenzene
or 2-chloro-4-ethylaniline

(l) 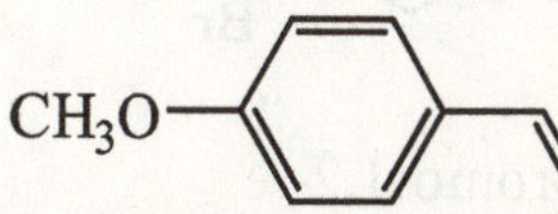 CH_3O-

1-Methoxy-4-vinylbenzene
or *p*-methoxystyrene

9.14 Draw structural formulas for these compounds:

(a) 1-Bromo-2-chloro-
4-ethylbenzene

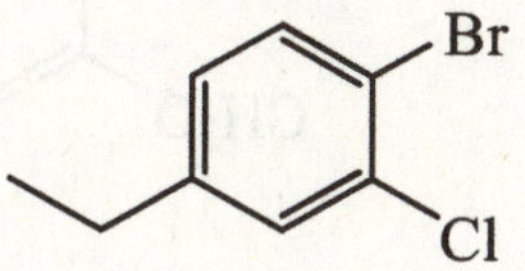

(b) 4-Iodo-1,2-
dimethylbenzene

(c) 2,4,6-Trinitrotoluene
(TNT)

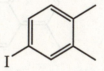

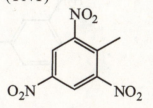

(d) 4-Phenyl-2-pentanol

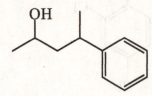

(e) *p*-Cresol

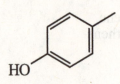

(f) 2,4-Dichlorophenol

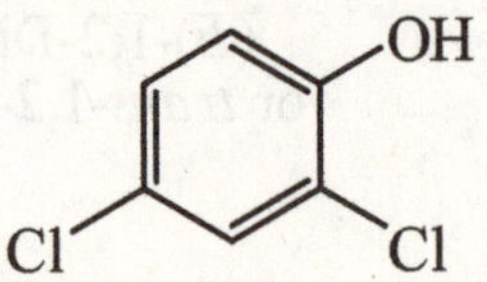

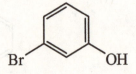

(g) 1-Phenylcyclopropanol

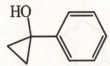

(h) Styrene (phenylethylene)

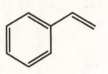

(i) *m*-Bromophenol

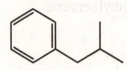

(j) 2,4-Dibromoaniline

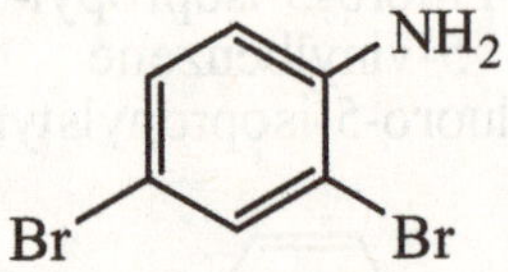

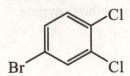

(k) Isobutylbenzene

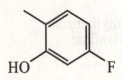

(l) *m*-Xylene

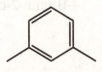

(m) 4-Bromo-1,2-dichlorobenzene

(n) 5-Fluoro-2-methylphenol

(o) 1-Cyclohexyl-3-ethylbenzene

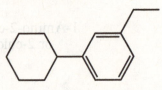

(p) *m*-Phenylaniline

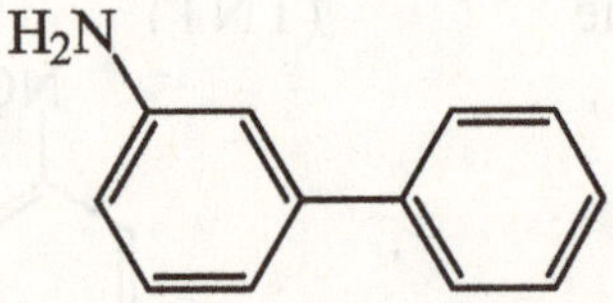

(q) 3-Methyl-2-vinylbenzoic acid

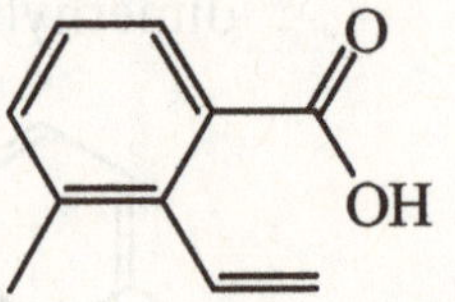

(r) 2,5-Dimethylanisole

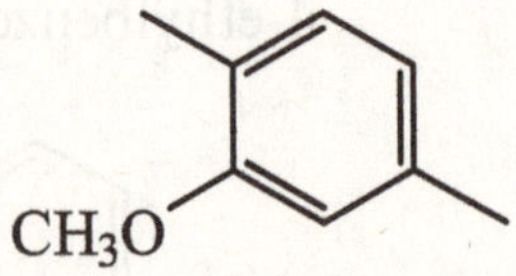

9.15 Show that pyridine can be represented as a hybrid of two equivalent contributing structures.

9.16 Show that naphthalene can be represented as a hybrid of three contributing structures. Show also, by the use of curved arrows, how one contributing structure is converted to the next.

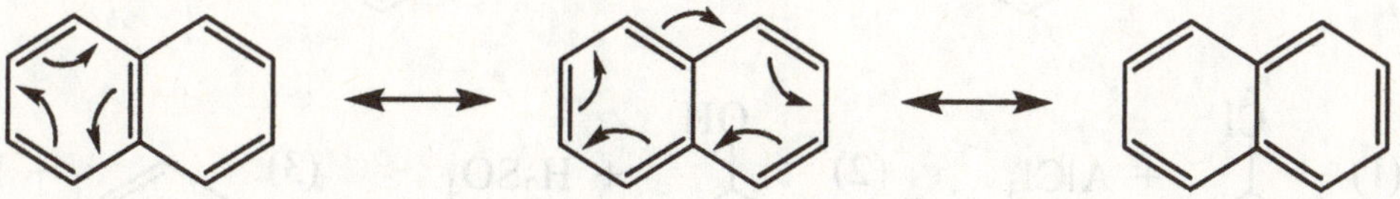

9.17 Draw four contributing structures for anthracene.

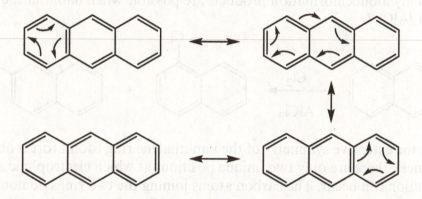

Electrophilic Aromatic Substitution: Monosubstitution

9.18 Draw a structural formula for the compound formed by treating benzene with each of the following combinations of reagents:

(a) $CH_3CH_2Cl / AlCl_3$
(b) $CH_2=CH_2 / H_2SO_4$
(c) CH_3CH_2OH / H_2SO_4
(d) CH_3OCH_3 / H_2SO_4

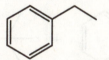

Ethylbenzene

Toluene

Reagent combinations (a), (b), and (c) lead to the formation of ethylbenzene. These reactions are alkylation reactions that involve the ethyl carbocation.

Reagent combination (d) leads to the formation of toluene. In this reaction, acid protonates dimethyl ether, and the subsequent loss of methanol generates the methyl carbocation.

9.19 Show three different combinations of reagents you might use to convert benzene to isopropylbenzene.

To prepare isopropylbenzene from benzene, it is necessary to generate the isopropyl carbocation. Any of the three combinations below produce the same carbocation.

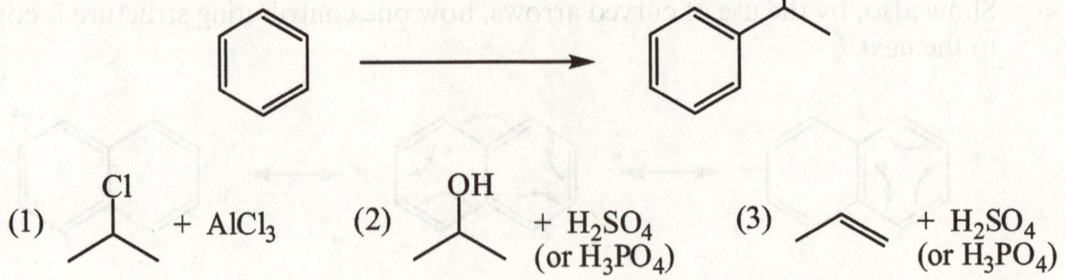
(1) Cl + AlCl3
(2) OH + H2SO4 (or H3PO4)
(3) + H2SO4 (or H3PO4)

9.20 How many monochlorination products are possible when naphthalene is treated with $Cl_2/AlCl_3$?

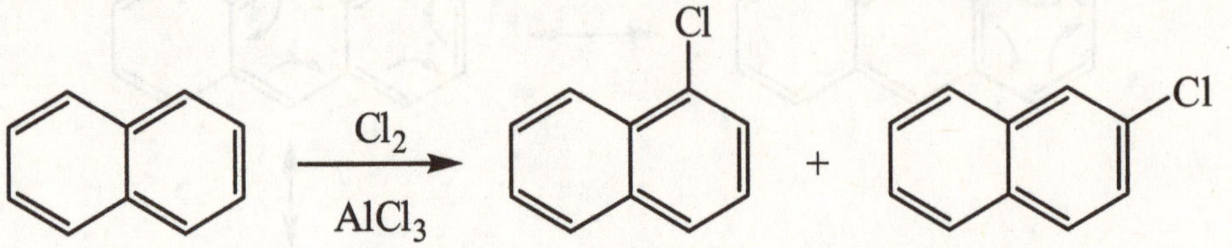
Cl2
AlCl3
Cl
Cl
+

Due to the extensive symmetry of the naphthalene ring (don't forget about resonance), there are only two unique positions at which electrophilic aromatic substitution can occur. The carbon atoms joining the two rings do not bear hydrogen atoms and cannot undergo electrophilic aromatic substitution.

9.21 Write a stepwise mechanism for the following reaction, using curved arrows to show the flow of electrons in each step:

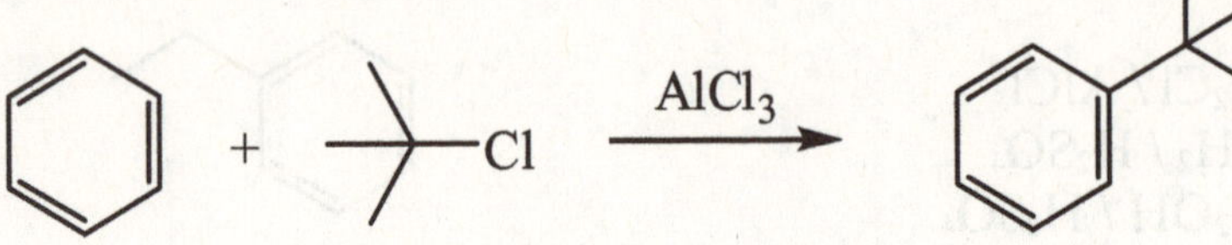
+
Cl
AlCl3

Step 1: Reaction of *tert*-butyl chloride (Lewis base) with $AlCl_3$ (Lewis acid).

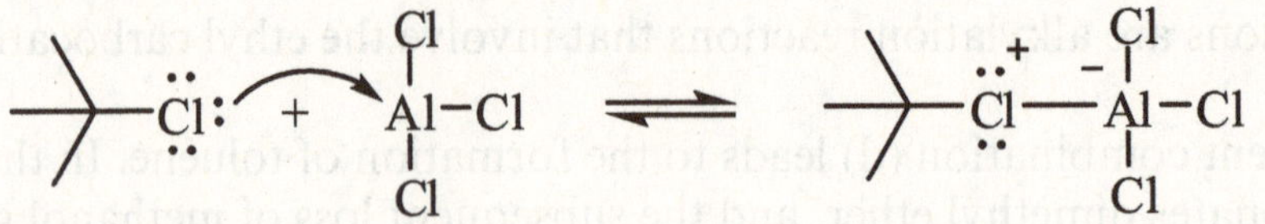

Step 2: Formation of the *tert*-butyl carbocation.

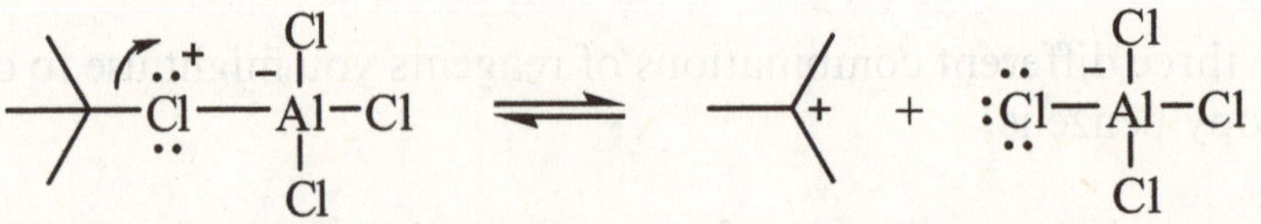

Step 3: Electrophilic attack on the benzene ring by the *tert*-butyl carbocation.

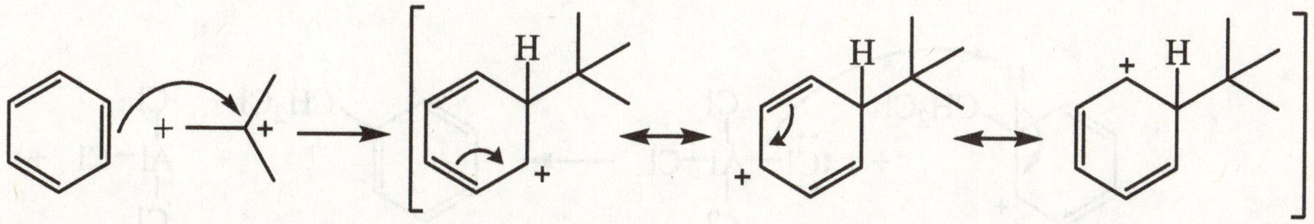

Step 4: Deprotonation to regenerate the aromatic ring and the catalyst.

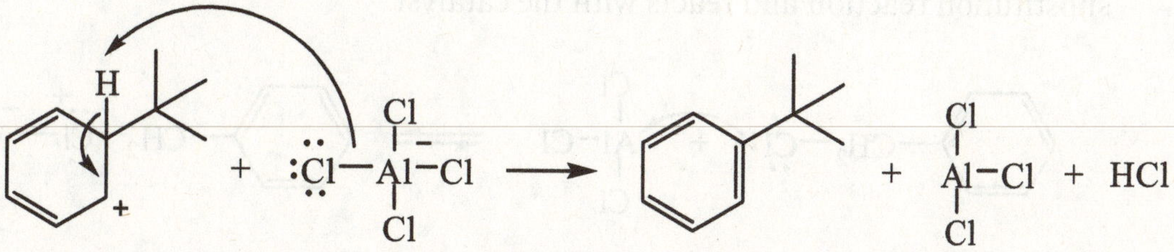

9.22   Write a stepwise mechanism for the preparation of diphenylmethane by treating benzene with dichloromethane in the presence of an aluminum chloride catalyst.

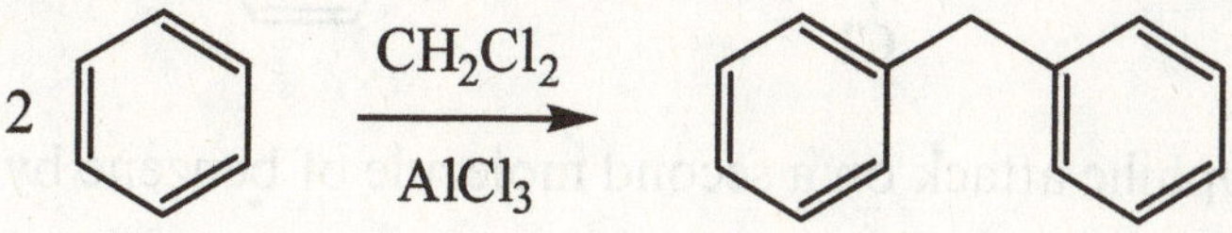

Step 1: Reaction of dichloromethane (Lewis base) with $AlCl_3$ (Lewis acid).

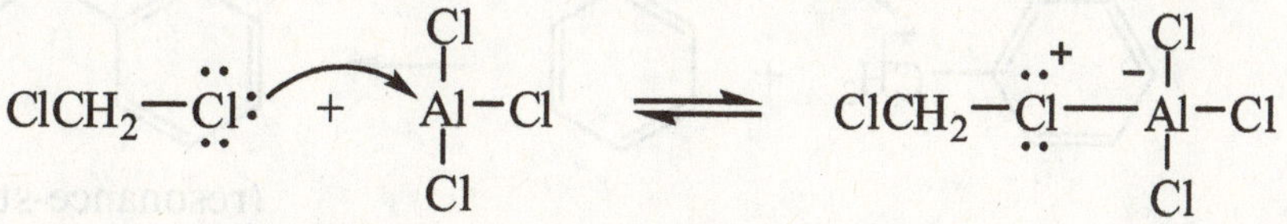

Step 2: Electrophilic attack on the benzene ring by the Lewis acid-base complex. Unlike Problem 9.21, a free $ClCH_2^+$ carbocation is not formed because it would be very unstable. This step is an S_N2-like reaction.

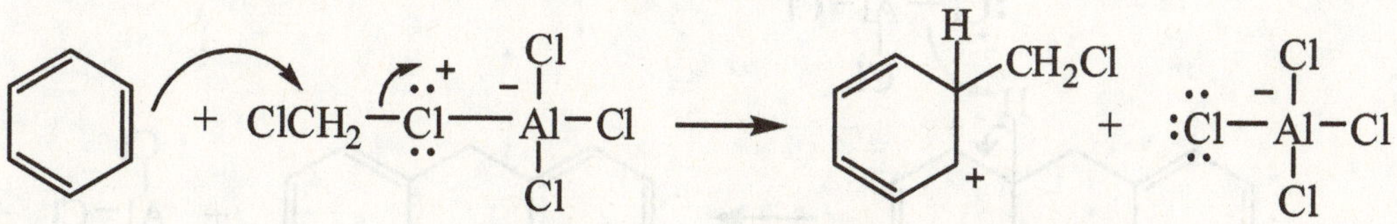

(resonance-stabilized)

Step 3: Deprotonation to regenerate the aromatic ring and the catalyst. The aromatic product formed is benzyl chloride.

Step 4: Benzyl chloride is the alkyl halide for the second electrophilic aromatic substitution reaction and reacts with the catalyst.

Step 5: Formation of the resonance-stabilized benzyl carbocation.

Step 6: Electrophilic attack on a second molecule of benzene by the benzylic carbocation.

(resonance-stabilized)

Step 7: Deprotonation to give the final product and regenerate the catalyst.

9.23 The following alkylation reactions do not yield the compounds shown as the major product. Predict the major product for each reaction and provide a mechanism for their formation.

(a)

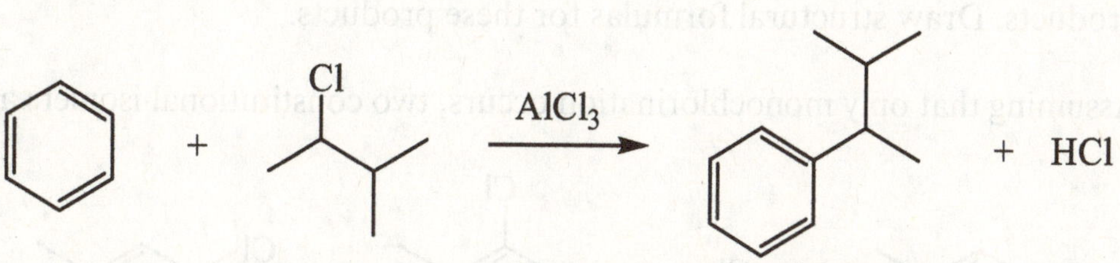

not major product

(b)

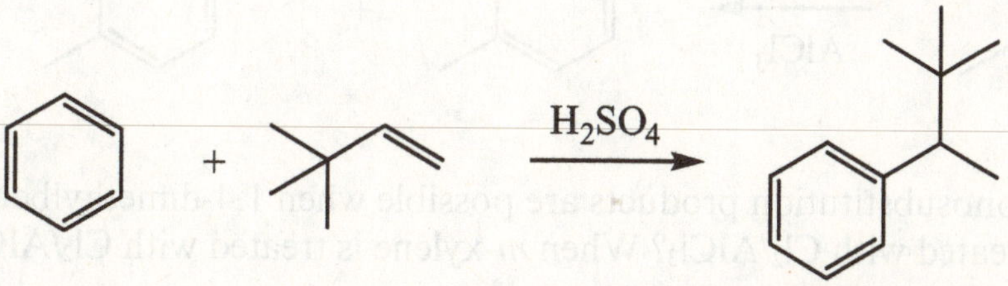

not major product

(c)

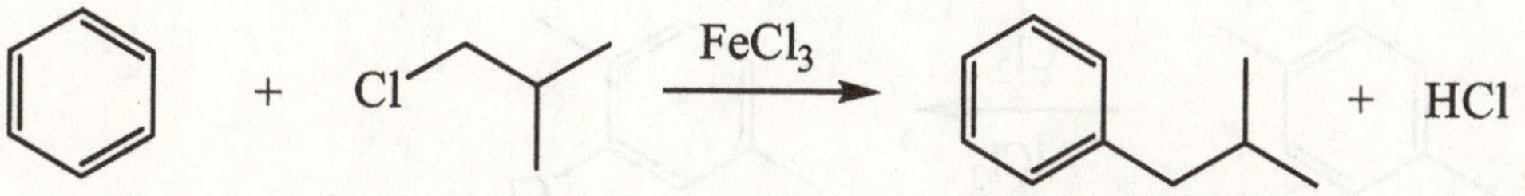

not major product

All three reactions proceed by electrophilic aromatic substitution. The electrophile (the carbocation) can be generated using an alkyl chloride and a Lewis acid, as in (a) and (c), or by the addition of H^+ to an alkene, as in (b). However, the carbocations made in these reactions rearrange (hydride and alkyl shifts) to more stable carbocations. These carbocations then react with benzene as usual (see Problem 9.21).

(a)

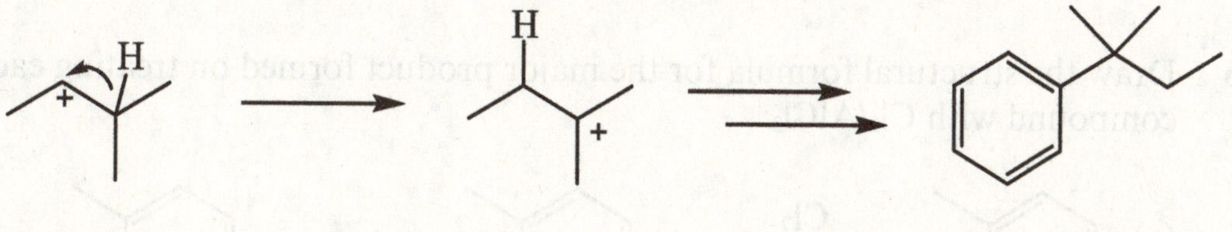

(b)

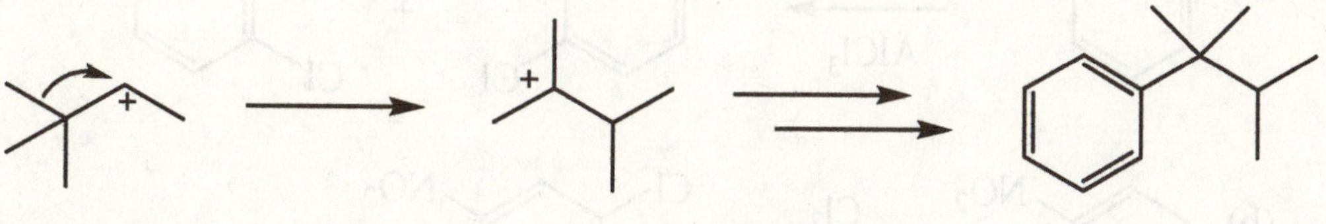

(c)

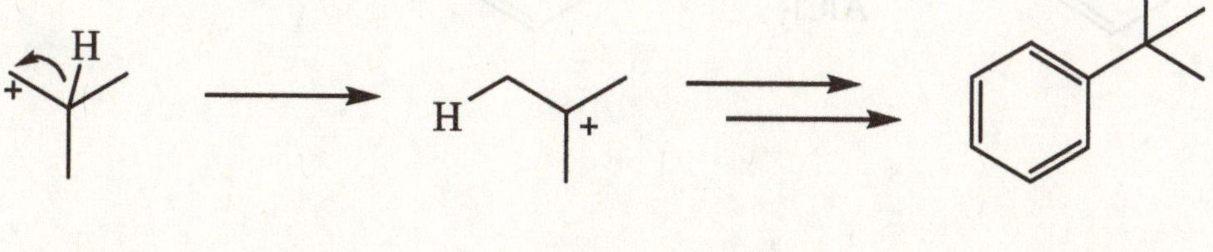

Electrophilic Aromatic Substitution: Substituent Effects

9.24 When treated with $Cl_2/AlCl_3$, 1,2-dimethylbenzene (*o*-xylene) gives a mixture of two products. Draw structural formulas for these products.

Assuming that only monochlorination occurs, two constitutional isomers are formed.

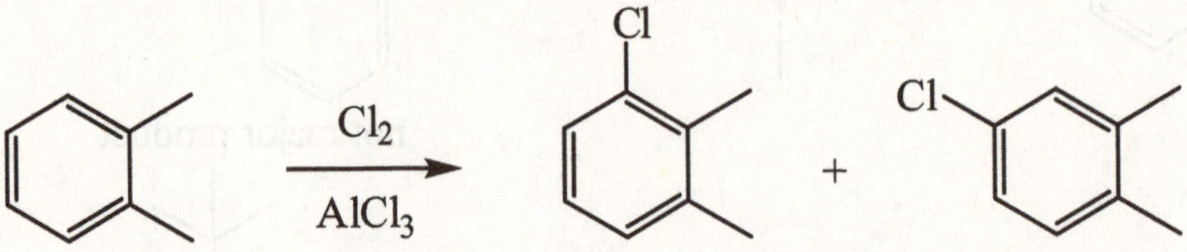

9.25 How many monosubstitution products are possible when 1,4-dimethylbenzene (*p*-xylene) is treated with $Cl_2/AlCl_3$? When *m*-xylene is treated with $Cl_2/AlCl_3$?

One product is formed from *p*-xylene, but *m*-xylene forms two products.

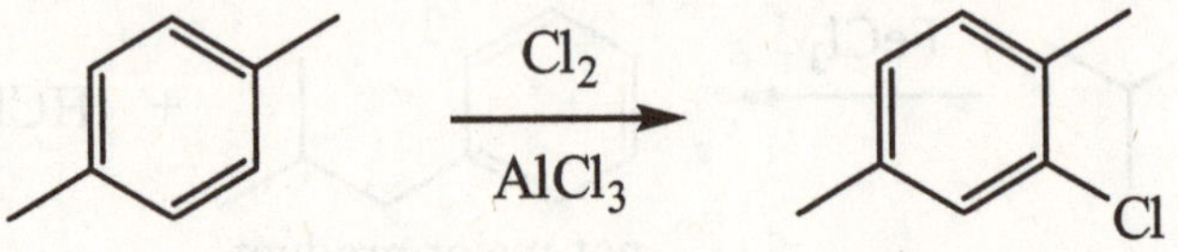

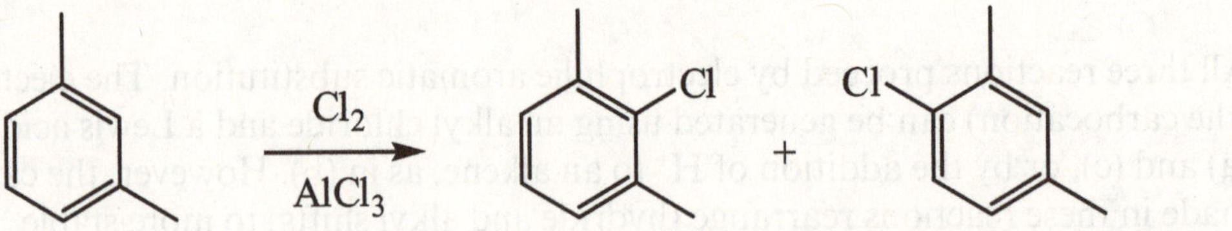

sterically more favorable

9.26 Draw the structural formula for the major product formed on treating each compound with $Cl_2/AlCl_3$:

(a)

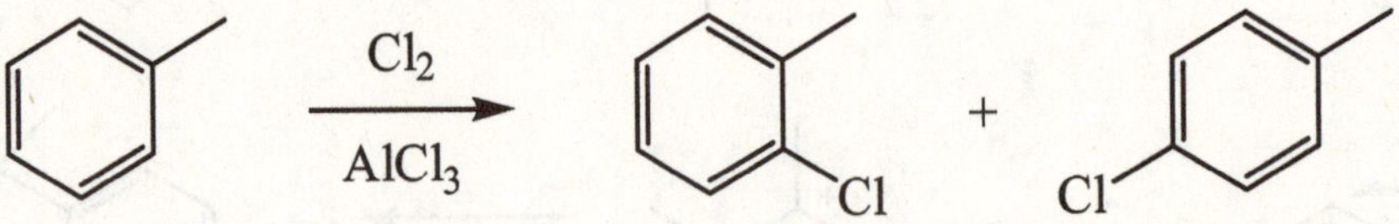

(b)

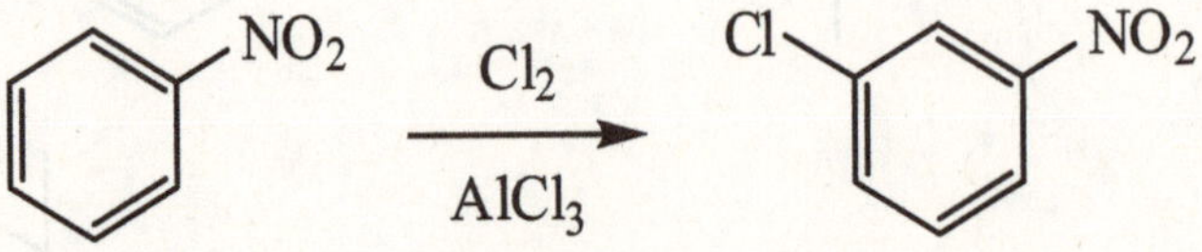

(c)

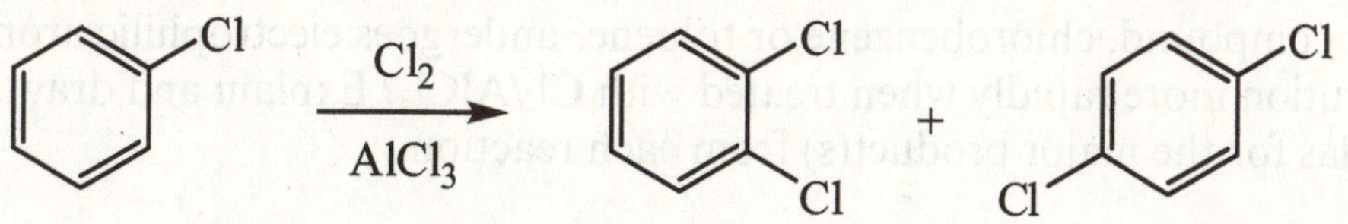

(d)

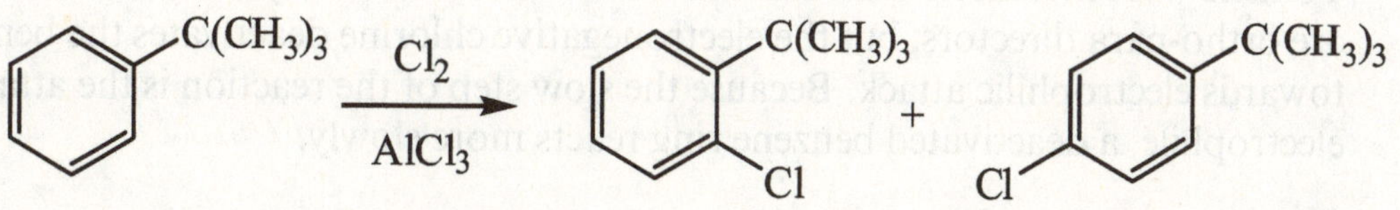

sterically more favorable

(e)

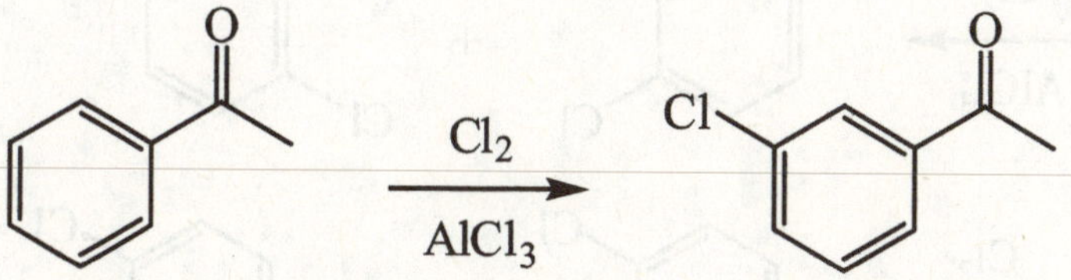

(f)

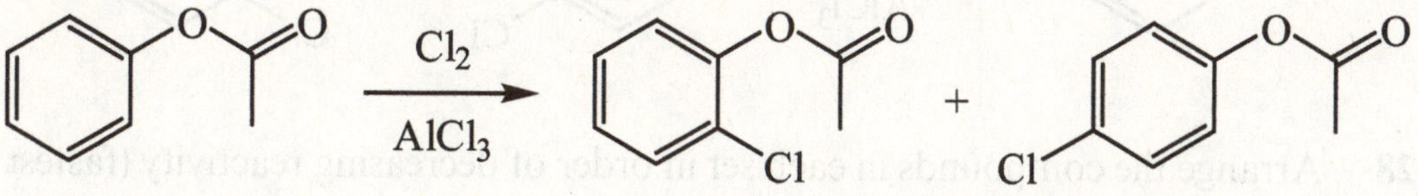

(g)

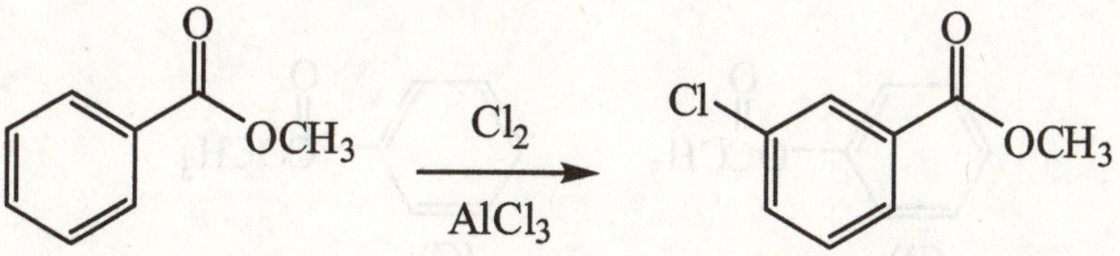

(h)

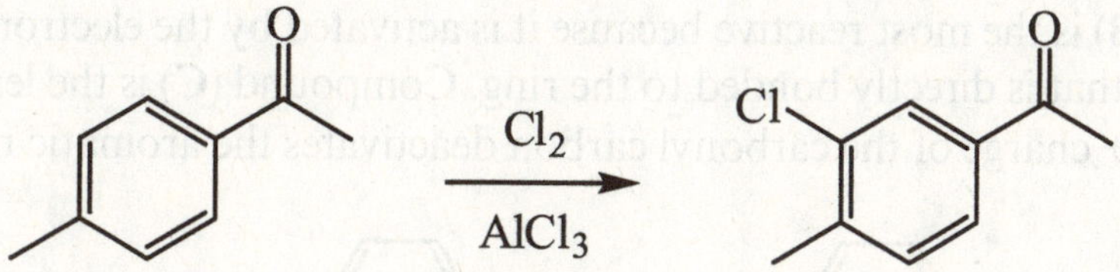

(i)

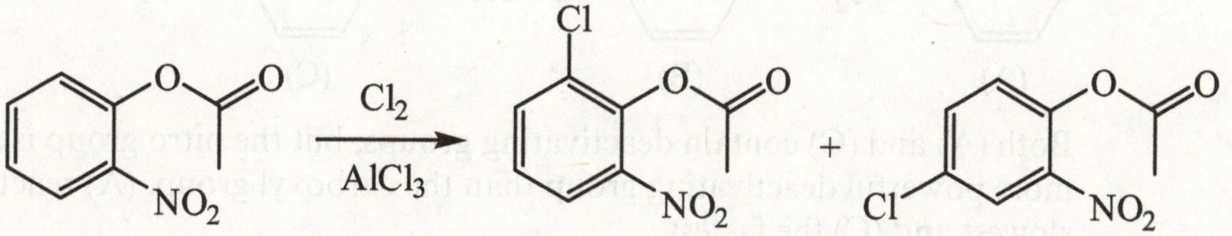

(j)

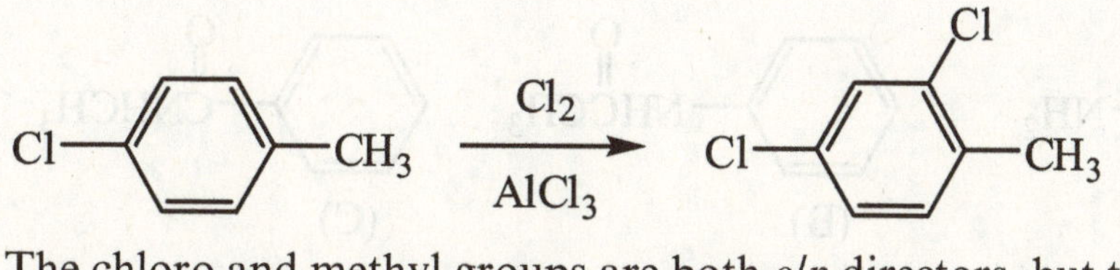

The chloro and methyl groups are both *o/p* directors, but the methyl group is more activating than the chloro group.

9.27 Which compound, chlorobenzene or toluene, undergoes electrophilic aromatic substitution more rapidly when treated with $Cl_2/AlCl_3$? Explain and draw structural formulas for the major product(s) from each reaction.

Toluene will react faster than chlorobenzene. Both the methyl and chlorine substituents are ortho-para directors, but the electronegative chlorine deactivates the benzene ring towards electrophilic attack. Because the slow step of the reaction is the attack of the electrophile, a deactivated benzene ring reacts more slowly.

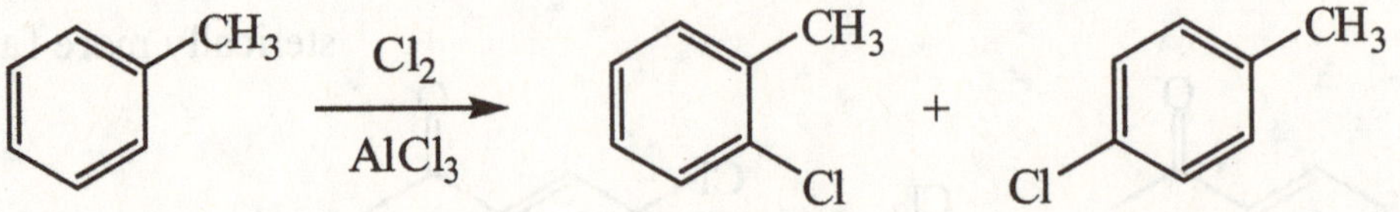

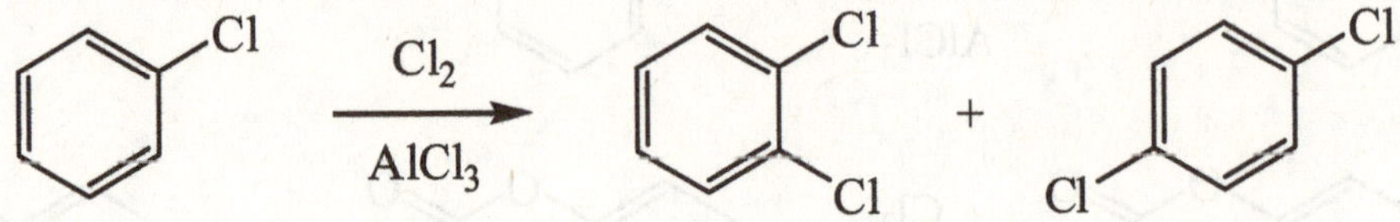

9.28 Arrange the compounds in each set in order of decreasing reactivity (fastest to slowest) toward electrophilic aromatic substitution:

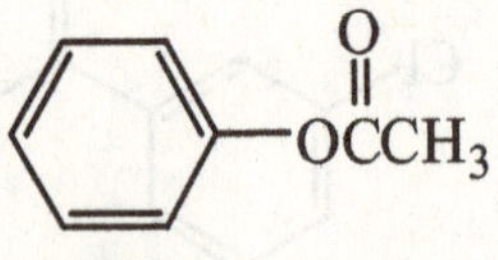

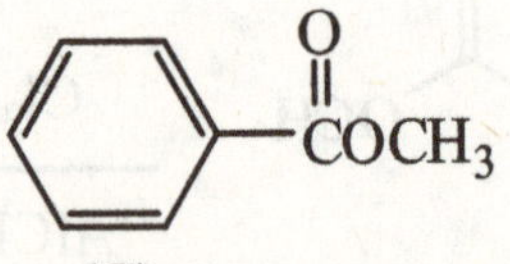

(a)

Compound (B) is the most reactive because it is activated by the electron-donating oxygen atom that is directly bonded to the ring. Compound (C) is the least reactive because the $\delta+$ charge of the carbonyl carbon deactivates the aromatic ring.

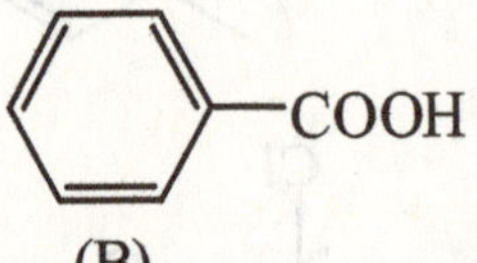

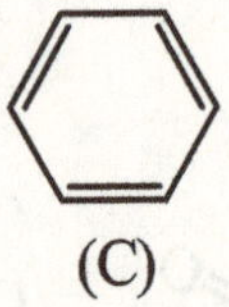

(b)

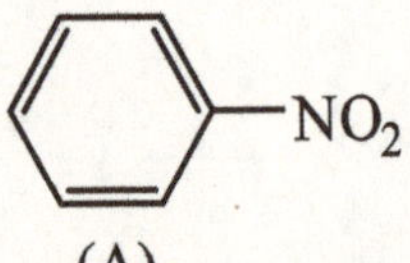

Both (A) and (C) contain deactivating groups, but the nitro group is a much more powerful deactivating group than the carboxyl group. (A) reacts the slowest and (C) the fastest.

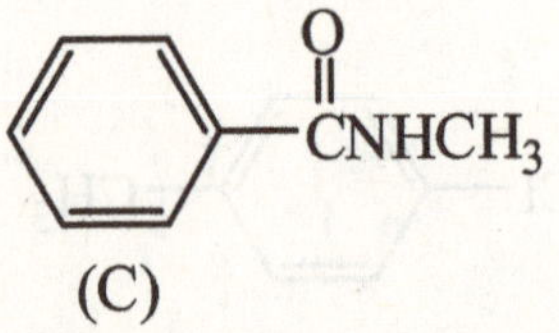

(c)

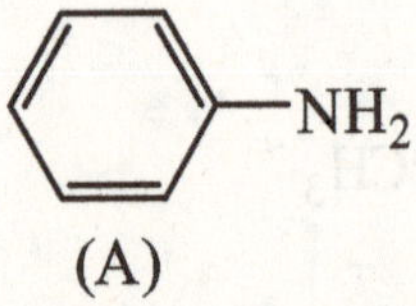

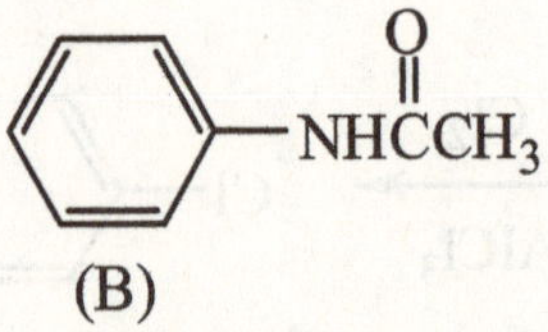

Both (A) and (B) contain activating groups, but the amino group is a much better activating group than the amide group (the nitrogen lone pair of the amide is delocalized into the carbonyl group). The carbonyl group of (C) is deactivating. (A) reacts the fastest and (C) the slowest.

(d)

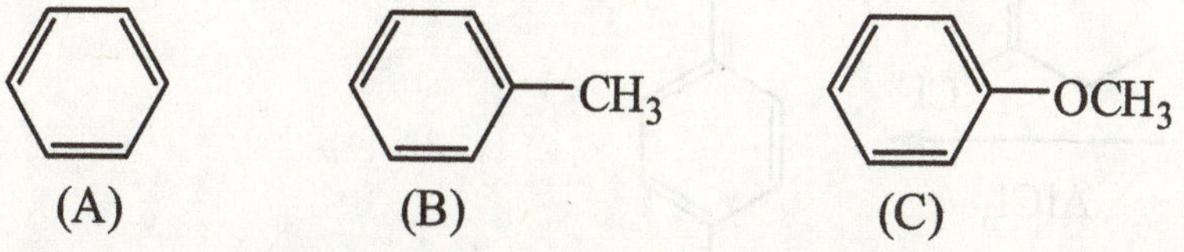

(A)   (B)   (C)

Both (B) and (C) contain activating groups. The methyl group in (B) is an inductively donating group, which is less activating than the methoxy group in (C), which can donate by resonance. (C) reacts the fastest and (A) the slowest.

9.29   Account for the observation that the trifluoromethyl group is meta directing, as shown in the following example:

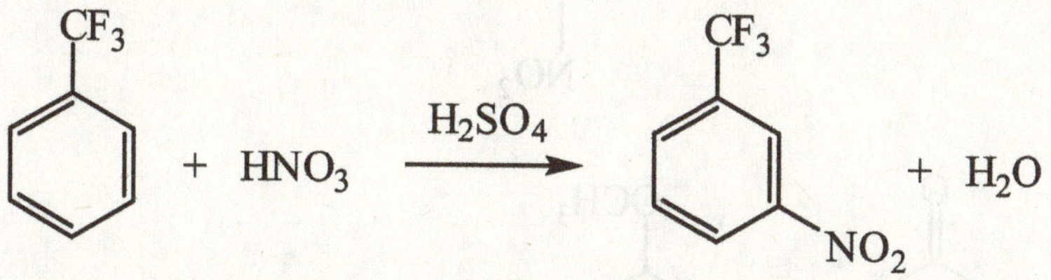

The trifluoromethyl group is a strong electron-withdrawing group and therefore a meta director. The three fluorine atoms make the carbon atom of the trifluoromethyl group, which is bonded directly to the ring, highly $\delta+$.

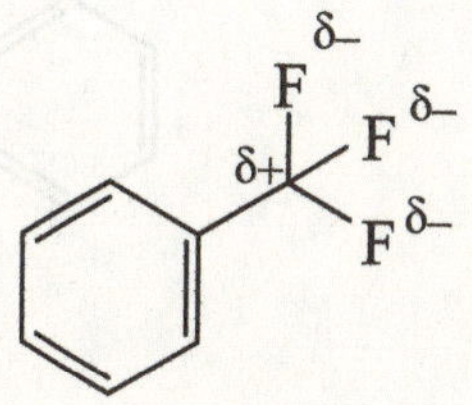

9.30   Show how to convert toluene to these carboxylic acids:

(a)

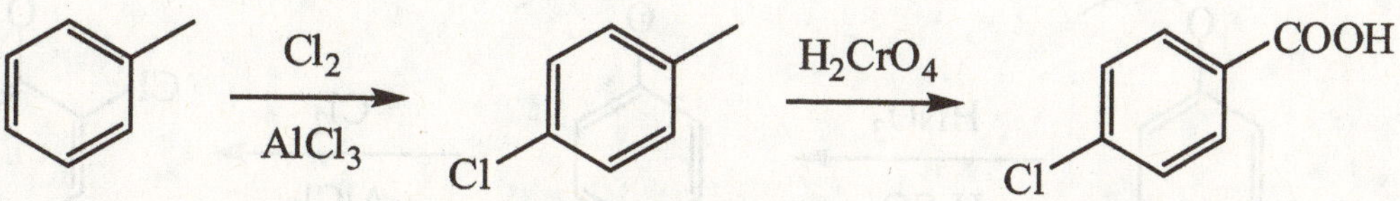

The preparation of 4-chlorobenzoic acid requires the use of the methyl group to direct the chloro group to the para position.

(b)

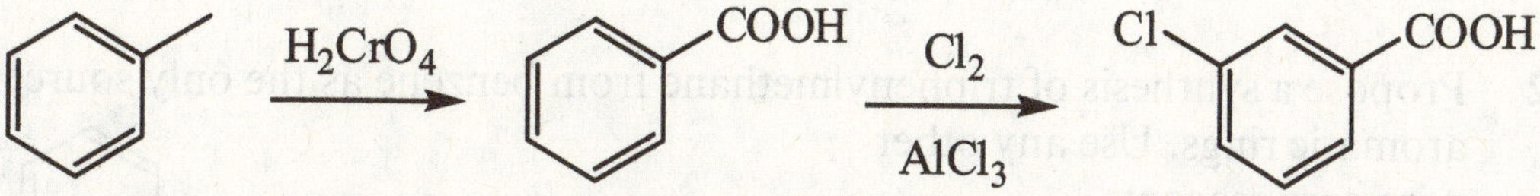

Reversing the order of the reactions in (a) allows the synthesis of 3-chlorobenzoic acid. The carboxyl group directs the chloro group to the meta position.

9.31 Show reagents and conditions that can be used to bring about these conversions:

(a)

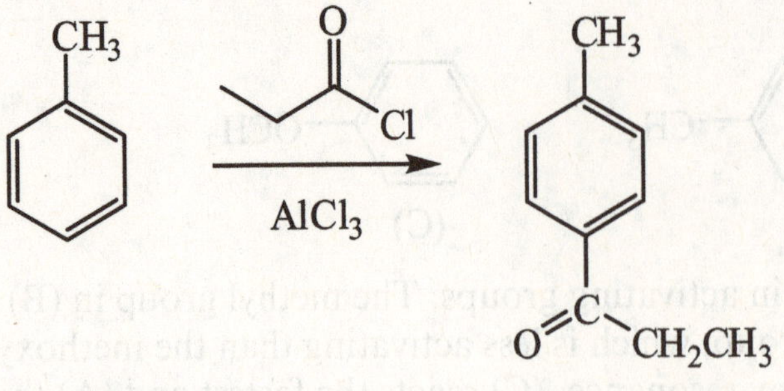

(b)

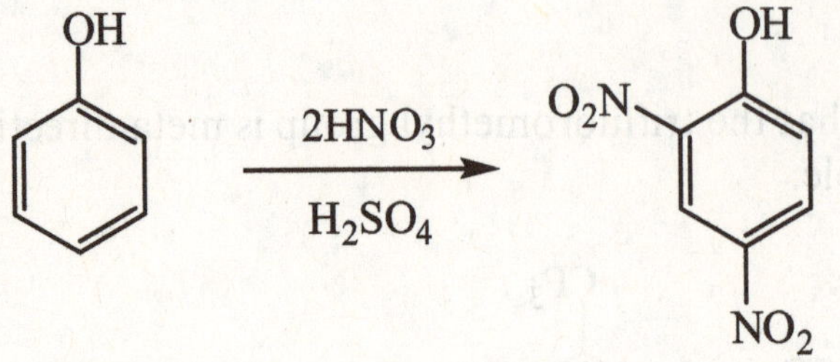

(c)

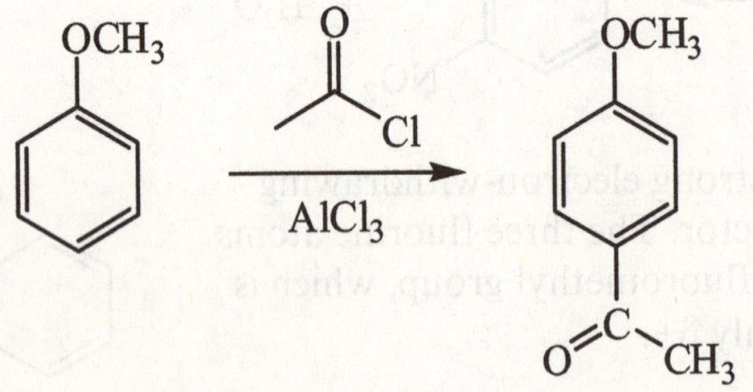

(d)

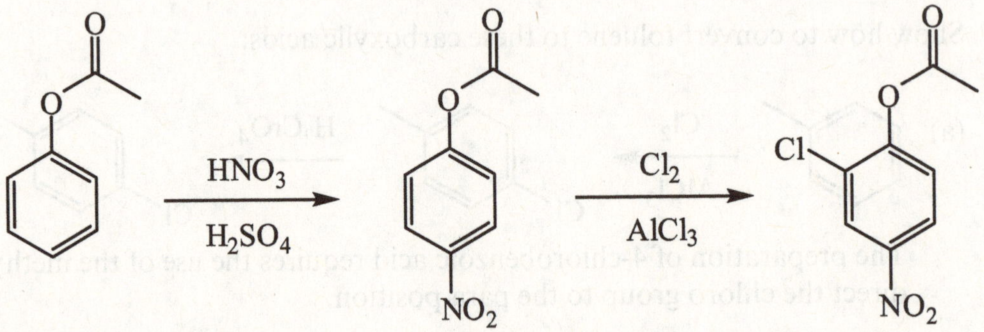

9.32 Propose a synthesis of triphenylmethane from benzene as the only source of aromatic rings. Use any other necessary reagents.

The synthesis can be performed similar to Problem 9.22 except using $CHCl_3$ and three equivalents of benzene.

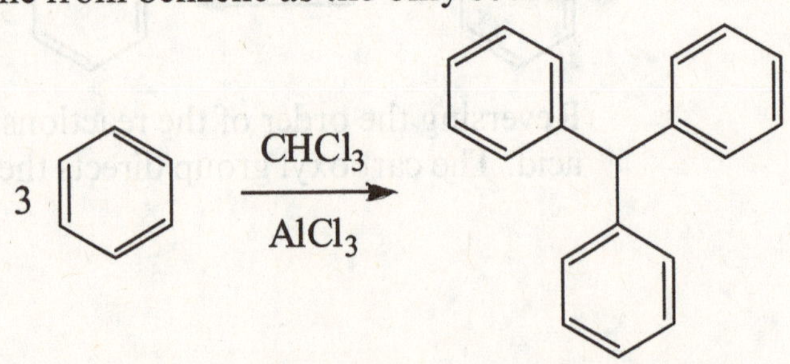

9.33 Reaction of phenol with acetone in the presence of an acid catalyst gives bisphenol A, a compound used in the production of polycarbonate and epoxy resins (Sections 16.4C and 16.4E). Propose a mechanism for the formation of bisphenol A. (*Hint:* The first step is a proton transfer from phosphoric acid to the oxygen of the carbonyl group of acetone.)

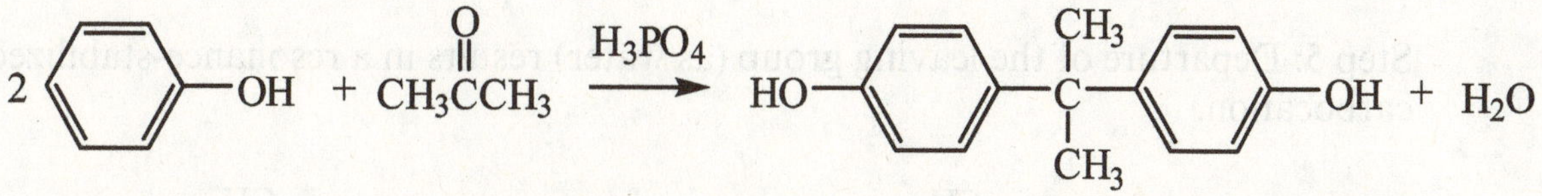

A comparison of the starting materials and the product reveals that the center carbon of acetone has made two new carbon-carbon bonds to the benzene rings. This suggests that the respective acetone carbon acted as an electrophile (a carbocation), which could be generated by the protonation of acetone.

Step 1: Protonation of acetone, forming a resonance-stabilized cation.

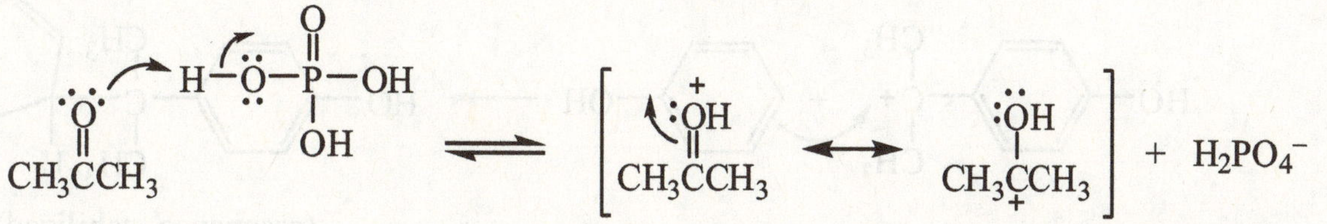

Step 2: Electrophilic attack of the benzene ring by the carbocation. The −OH group is an ortho-para director, and the carbocation adds to the least-hindered position (para).

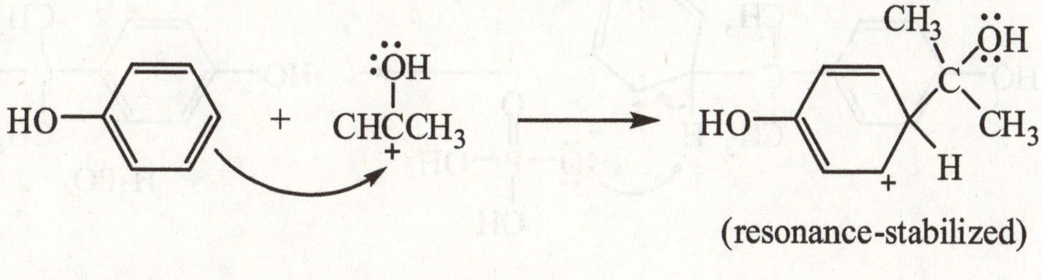

(resonance-stabilized)

Step 3: Deprotonation results in an alcohol.

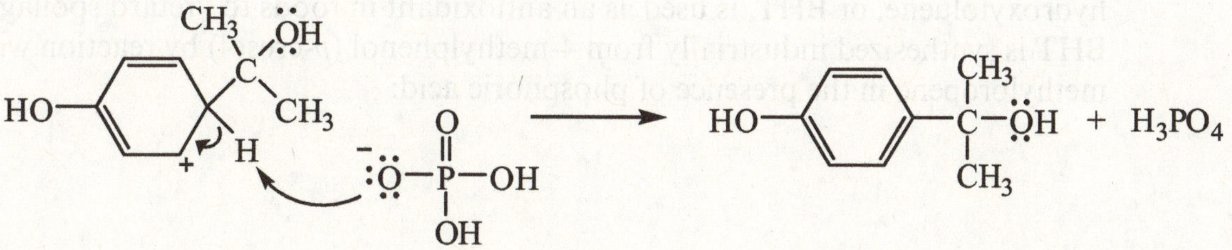

Step 4: In this step, the alcohol is converted into a good leaving group via protonation.

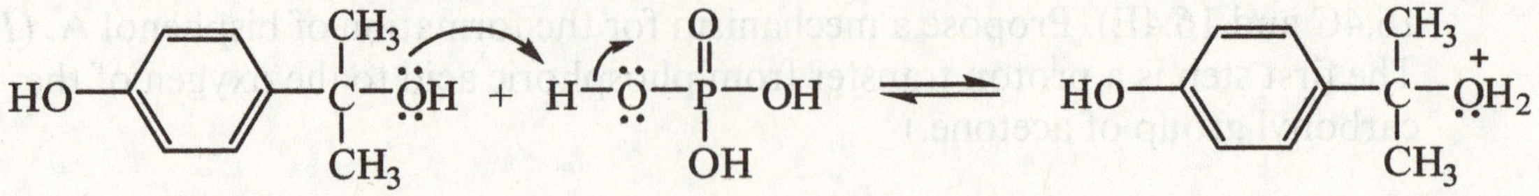

Step 5: Departure of the leaving group (as water) results in a resonance-stabilized carbocation.

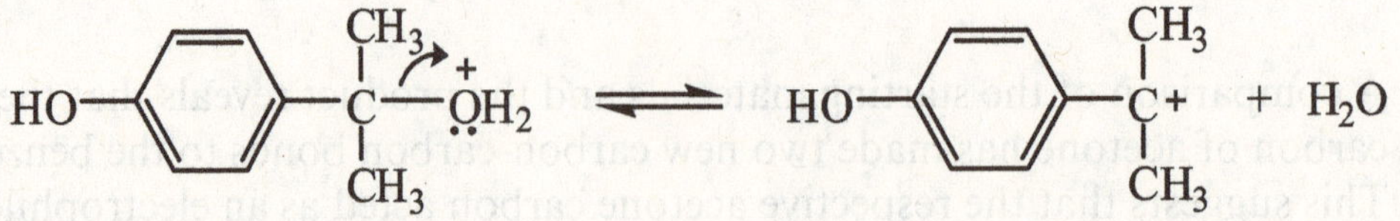

Step 6: This carbocation undergoes an electrophilic aromatic substitution with a second molecule of phenol. First, the carbocation attacks the benzene ring.

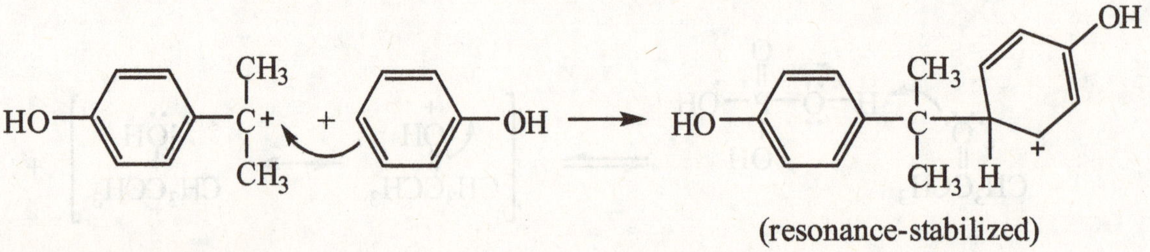

(resonance-stabilized)

Step 7: Deprotonation regenerates the aromatic ring and yields bisphenol A.

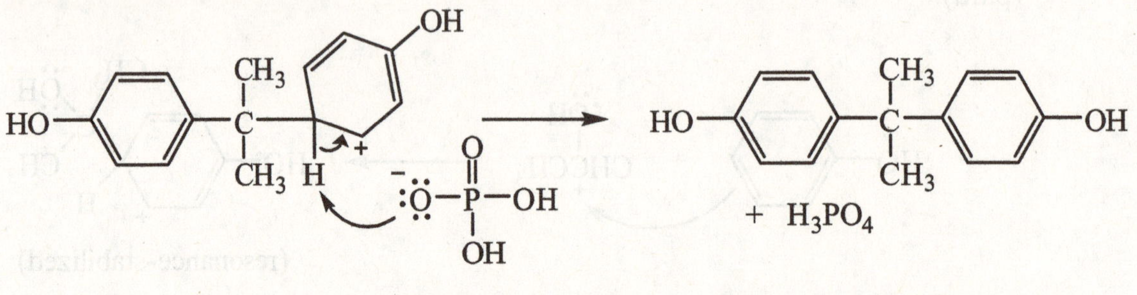

9.34   2,6-Di-*tert*-butyl-4-methylphenol, more commonly known as butylated hydroxytoluene, or BHT, is used as an antioxidant in foods to "retard spoilage." BHT is synthesized industrially from 4-methylphenol (*p*-cresol) by reaction with 2-methylpropene in the presence of phosphoric acid:

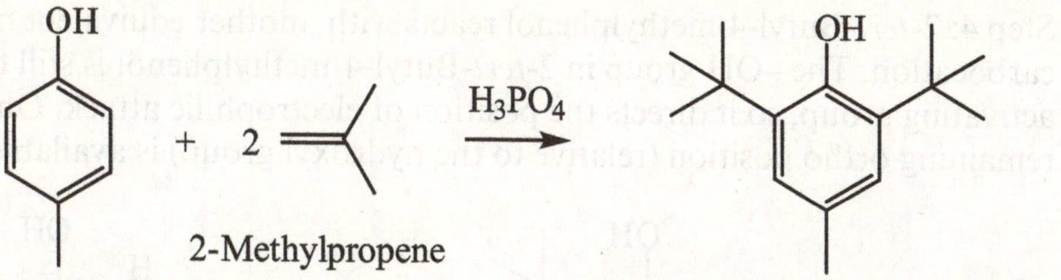

2-Methylpropene

4-Methylphenol

2,4-Di-*tert*-butyl-4-methylphenol
(Butylated hydroxytoluene, BHT)

In the preparation of BHT, two electrophilic aromatic substitution reactions occur. The hydroxyl group of 4-methylphenol is more strongly activating than the alkyl group, so the hydroxyl group directs the position of electrophilic attack.

Step 1: Protonation of 2-methylpropene to form the *tert*-butyl carbocation. Note that for every equivalent of BHT made, two equivalents of 2-methylpropene are required.

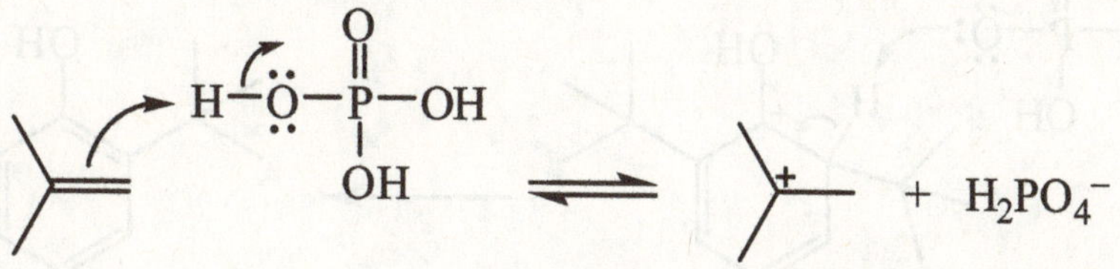

Step 2: Electrophilic attack of the benzene ring by the carbocation. The −OH group is an ortho-para director, but only the ortho position is available for electrophilic attack.

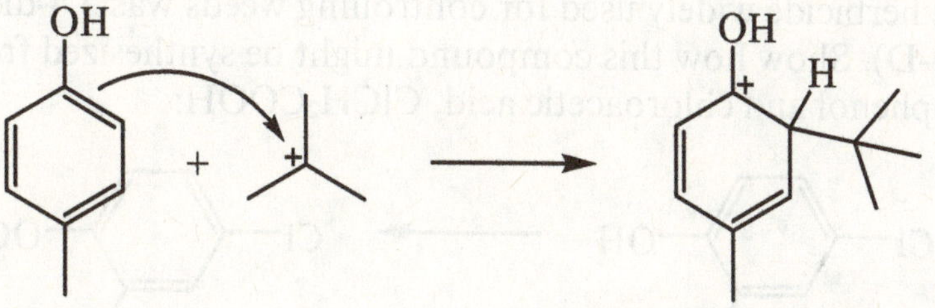

(resonance-stabilized)

Step 3: Deprotonation completes the first electrophilic aromatic substitution to give 2-*tert*-butyl-4-methylphenol. We're halfway there!

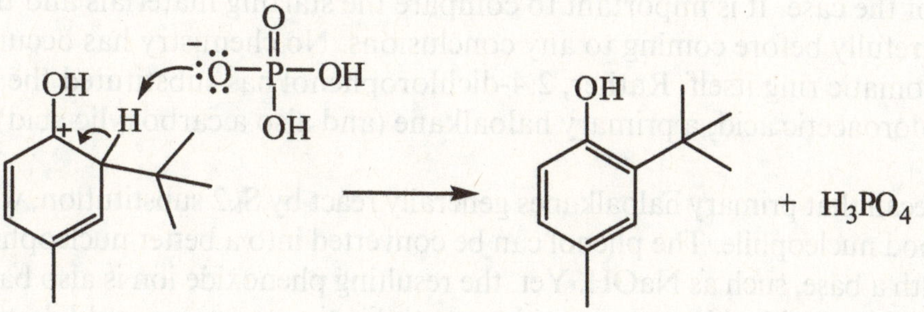

Step 4: 2-*tert*-Butyl-4-methylphenol reacts with another equivalent of the *tert*-butyl carbocation. The –OH group in 2-*tert*-Butyl-4-methylphenol is still the strongest activating group, so it directs the position of electrophilic attack. Only one remaining ortho position (relative to the hydroxyl group) is available.

(resonance-stabilized)

Step 5: Deprotonation completes the second electrophilic aromatic substitution to give the final product.

9.35 The first herbicide widely used for controlling weeds was 2,4-dichlorophenoxyacetic acid (2,4-D). Show how this compound might be synthesized from 2,4-dichlorophenol and chloroacetic acid, $ClCH_2COOH$:

2,4-Dichlorophenol   2,4-Dichlorophenoxyacetic acid (2,4-D)

Although the reaction may appear to be a reaction involving the aromatic ring, it is not the case. It is important to compare the starting materials and the product very carefully before coming to any conclusions. No chemistry has occurred on the aromatic ring itself. Rather, 2,4-dichlorophenol has substituted the chlorine atom of chloroacetic acid, a primary haloalkane (and also a carboxylic acid).

Recall that primary haloalkanes generally react by S_N2 substitution, which requires a good nucleophile. The phenol can be converted into a better nucleophile by treating it with a base, such as NaOH. Yet, the resulting phenoxide ion is also basic, and if combined with chloroacetic acid, a neutralization reaction would destroy the phenoxide ion and regenerate the phenol. It is therefore necessary to first convert chloroacetic to its conjugate base (by neutralization with a non-nucleophilic base, such as $NaHCO_3$) prior to performing the substitution reaction. After the reaction, the carboxylate ion can be protonated via the addition of acid.

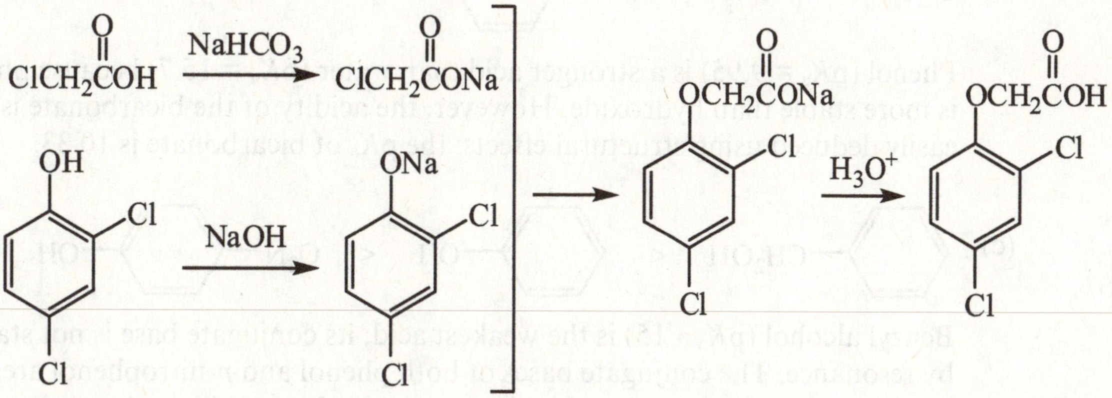

Acidity of Phenols

9.36 Use resonance theory to account for the fact that phenol (pK_a 9.95) is a stronger acid than cyclohexanol (pK_a 18).

Phenol is a stronger acid than cyclohexanol because the conjugate base of phenol, the phenoxide ion, is stabilized by resonance, while that of cyclohexanol is not. Recall from Chapter 2 that when a conjugate base is more stable, the acid becomes stronger.

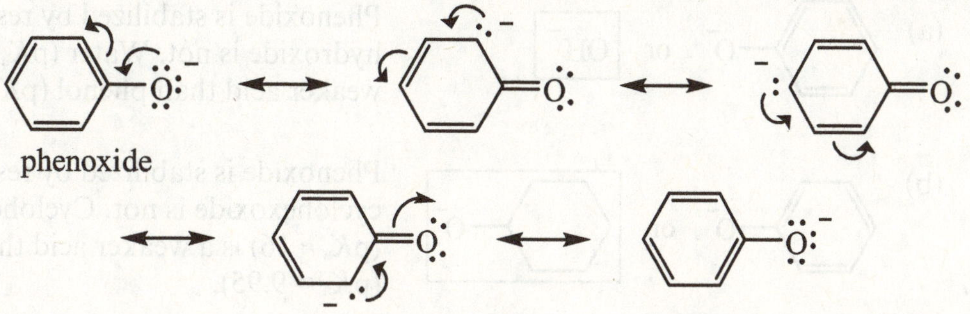

phenoxide

9.37 Arrange the compounds in each set in order of increasing acidity (from least acidic to most acidic):

(a) 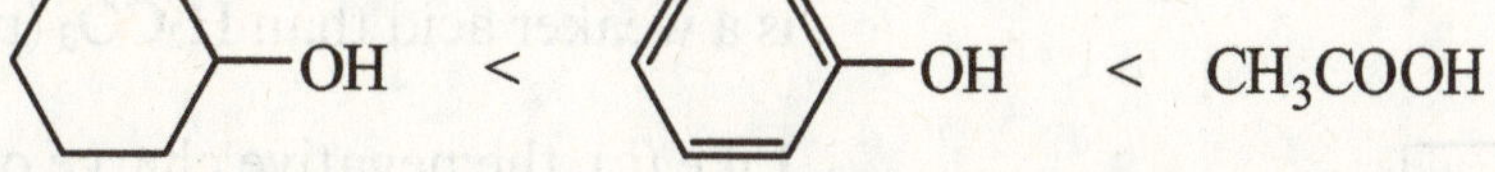 $<$ $<$ CH_3COOH

Acetic acid ($pK_a = 4.76$) is a stronger acid than phenol ($pK_a = 9.95$) because the acetate ion is more stable than the phenoxide ion. The negative charge of acetate, which is delocalized over two oxygen atoms, is more stable than that of phenoxide, which is delocalized over a single oxygen and three less-electronegative carbons. Cyclohexanol ($pK_a \approx 16$) is the weakest acid because cyclohexoxide is not resonance-stabilized.

(b) H_2O < $NaHCO_3$ <   [benzene ring]—OH

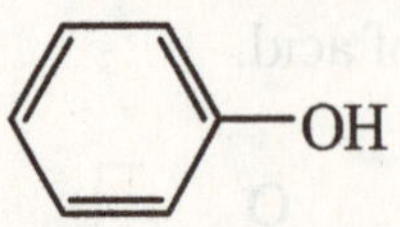

Phenol ($pK_a = 9.95$) is a stronger acid than water ($pK_a = 15.7$) because phenoxide is more stable than hydroxide. However, the acidity of the bicarbonate is not easily deduced using structural effects; the pK_a of bicarbonate is 10.33.

(c) 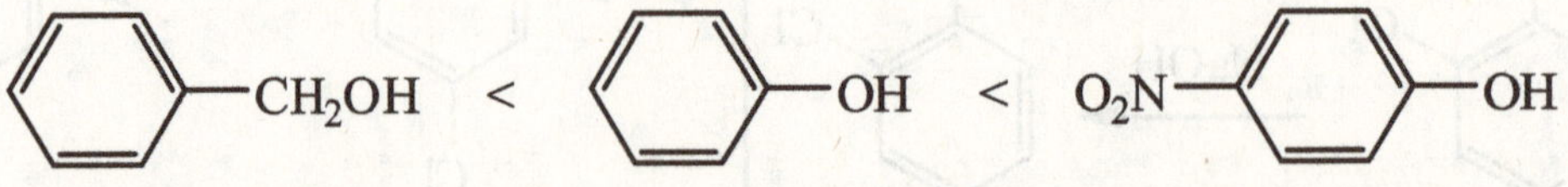   [benzene ring]—CH_2OH <   [benzene ring]—OH   < O_2N—[benzene ring]—OH

Benzyl alcohol ($pK_a \approx 15$) is the weakest acid; its conjugate base is not stabilized by resonance. The conjugate bases of both phenol and p-nitrophenol are stabilized by resonance, but the nitro group further stabilizes the conjugate base by resonance and induction. p-Nitrophenol ($pK_a = 7.15$) is a stronger acid than phenol ($pK_a = 9.95$).

9.38   From each pair, select the stronger base:

The stronger base can be identified by determining which base is the least stable, or by determining which parent acid is the weakest acid (recall that the weaker the acid, the stronger the conjugate base).

(a) 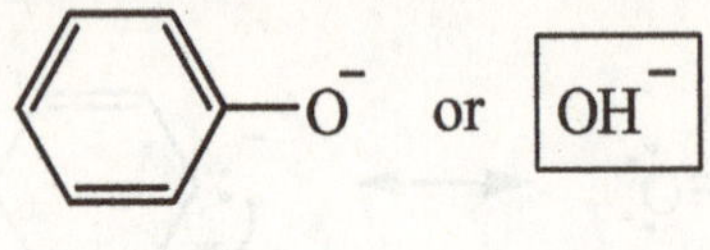   [benzene ring]—O^- or OH^- Phenoxide is stabilized by resonance, but hydroxide is not. Water ($pK_a = 15.7$) is a weaker acid than phenol ($pK_a = 9.95$).

(b) 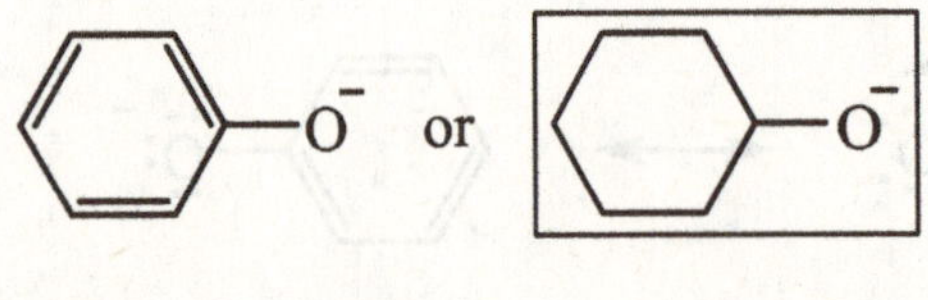   [benzene ring]—O^- or   [cyclohexane ring]—O^- Phenoxide is stabilized by resonance, but cyclohexoxide is not. Cyclohexanol ($pK_a \approx 16$) is a weaker acid than phenol ($pK_a = 9.95$).

(c) 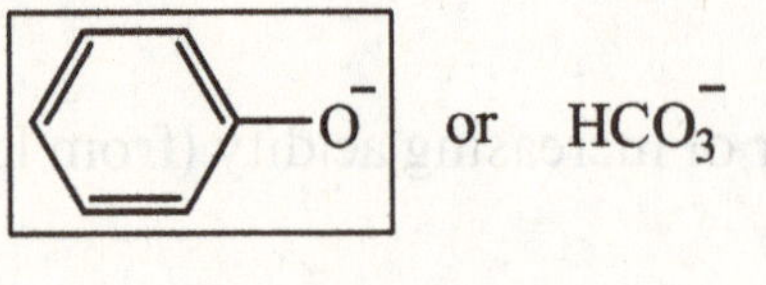   [benzene ring]—O^- or HCO_3^- Both species are stabilized by resonance, but the negative charge of bicarbonate is delocalized over two oxygen atoms (more preferable than one oxygen and three carbons in phenoxide). Phenol ($pK_a = 9.95$) is a weaker acid than H_2CO_3 ($pK_a = 6.36$).

(d) 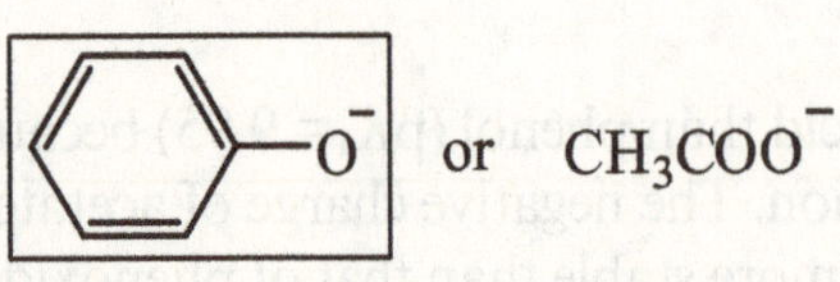   [benzene ring]—O^- or CH_3COO^- Like (c), the negative charge of acetate is delocalized over two oxygen atoms. Phenol ($pK_a = 9.95$) is a weaker acid than acetic acid ($pK_a = 4.76$).

9.39 Account for the fact that water-insoluble carboxylic acids (pK_a 4–5) dissolve in 10% sodium bicarbonate with the evolution of a gas, but water-insoluble phenols (pK_a 9.5–10.5) do not show this chemical behavior.

The gas formed is carbon dioxide, which arises from the decomposition of carbonic acid.

$$H_2CO_3 \longrightarrow CO_2 + H_2O$$

Carbonic acid ($pK_a = 6.36$) is formed when HCO_3^- acts as a base. A carboxylic acid is a stronger acid than carbonic acid, so the bicarbonate ion is strong enough of a base to deprotonate a carboxylic acid; that is, the carboxylate ion is more stable than the bicarbonate ion. However, because a phenol is a weaker acid than carbonic acid, bicarbonate is not strong enough of a base to deprotonate phenol; that is, the bicarbonate ion is more stable than the phenoxide ion. These phenomena also show that acid-base equilibria favor the side of the weaker acid.

9.40 Describe a procedure for separating a mixture of 1-hexanol and 2-methylphenol (*o*-cresol) and recovering each in pure form. Each is insoluble in water but soluble in diethyl ether.

The proposed procedure relies on the fact that phenolic −OH groups are weakly acidic ($pK_a \approx 10$) and can be completely deprotonated by an aqueous strong base, such as NaOH. Alcohols, such as 1-hexanol, are not acidic enough to be deprotonated by NaOH.

The mixture can first be dissolved in diethyl ether. Extraction of the ethereal solution with aqueous NaOH extracts 2-methylphenol into the aqueous layer as the sodium salt. 1-Hexanol remains dissolved in diethyl ether, the evaporation of which would leave 1-hexanol. The aqueous solution, which contains sodium 2-methylphenoxide, can be acidified with concentrated HCl to precipitate 2-methyphenol.

Syntheses

9.41   Using styrene, $C_6H_5CH=CH_2$, as the only aromatic starting material, show how to synthesize these compounds. In addition to styrene, use any other necessary organic or inorganic chemicals. Any compound synthesized in one part of this problem may be used to make any other compound in the problem:

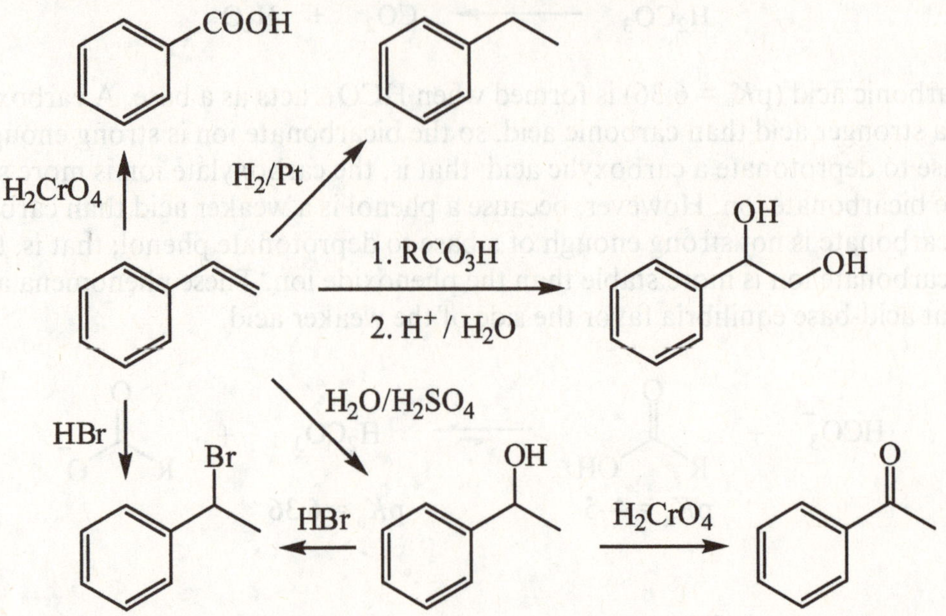

9.42   Show how to synthesize these compounds, starting with benzene, toluene, or phenol as the only sources of aromatic rings. Assume that, in all syntheses, you can separate mixtures of ortho-para products to give the desired isomer in pure form:

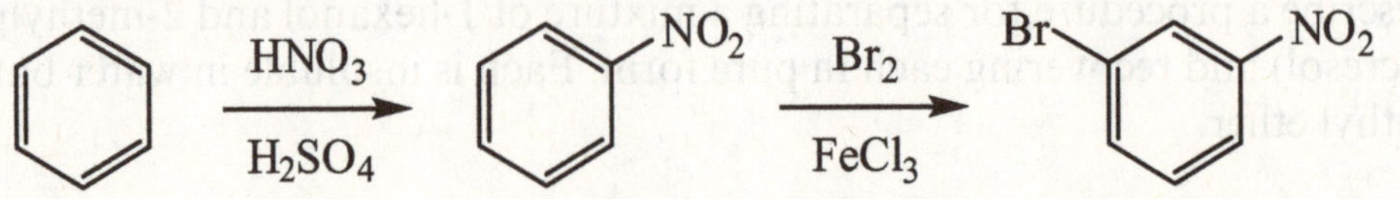

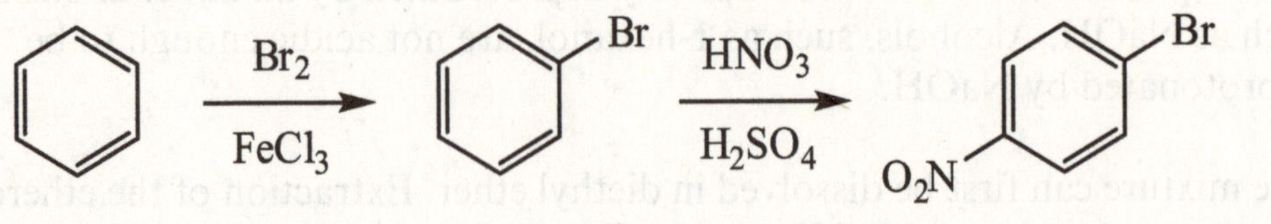

(c)

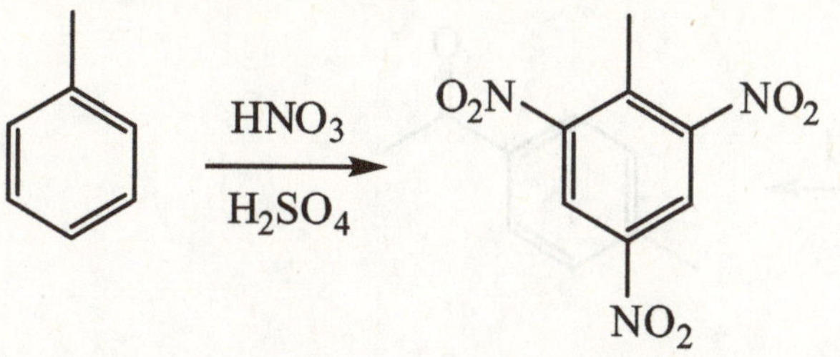

The product is the explosive trinitrotoluene (TNT). Don't try this reaction at home!

(d)

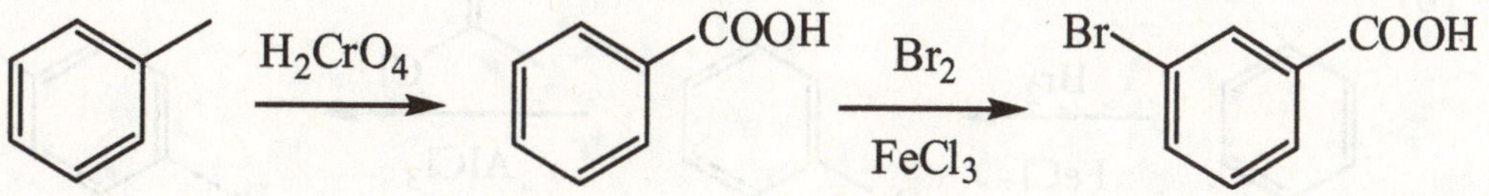

(e)

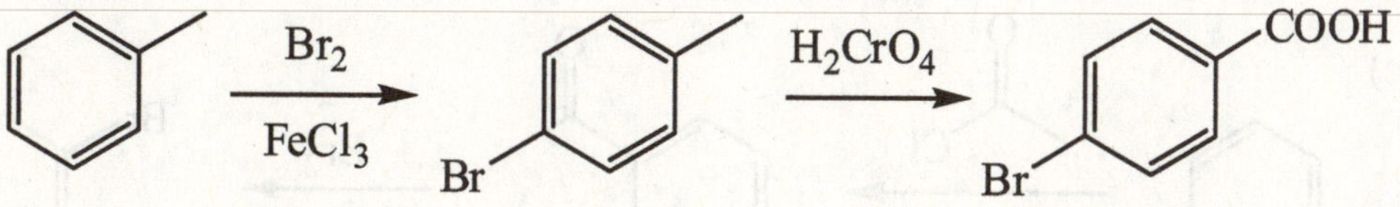

(f)

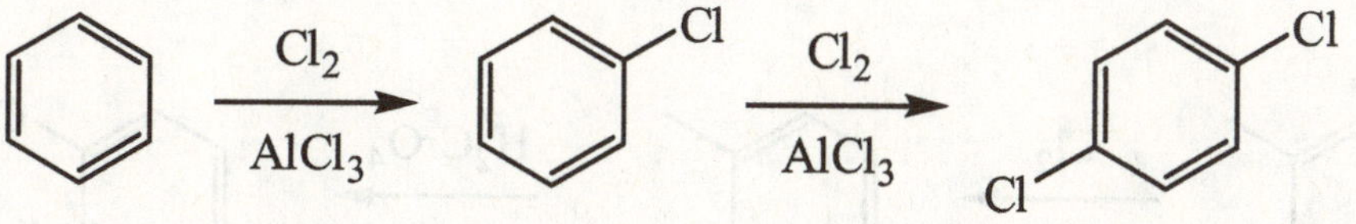

(g)

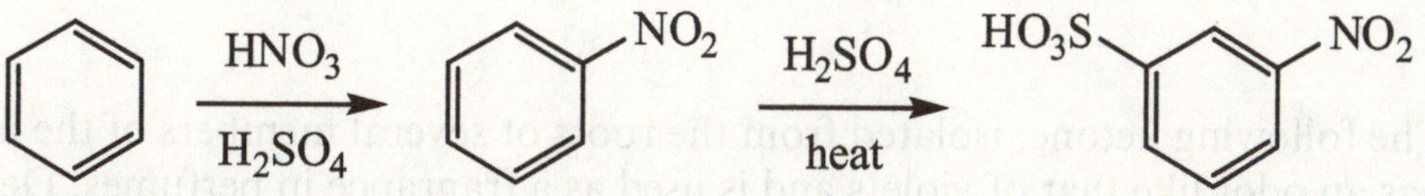

Performing the reaction in the reverse order (sulfonation followed by nitration) would yield the same product.

(h)

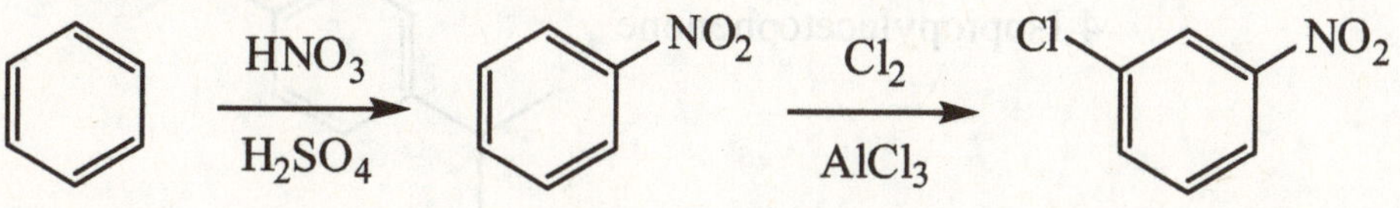

9.43 Show how to synthesize these aromatic ketones starting with benzene or toluene as the only sources of aromatic rings. Assume in all syntheses that mixtures of ortho-para products can be separated to give the desired isomer in pure form:

(a)

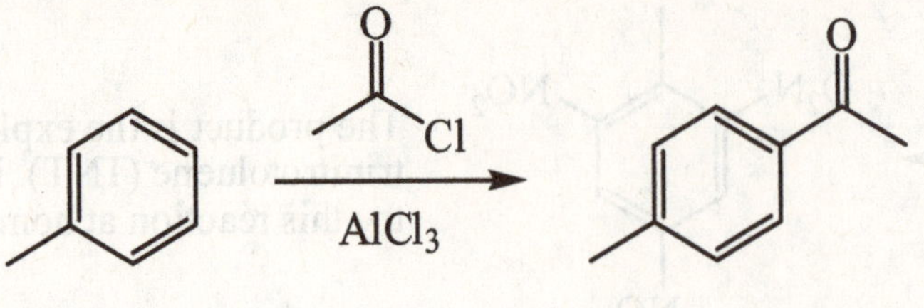

(b)

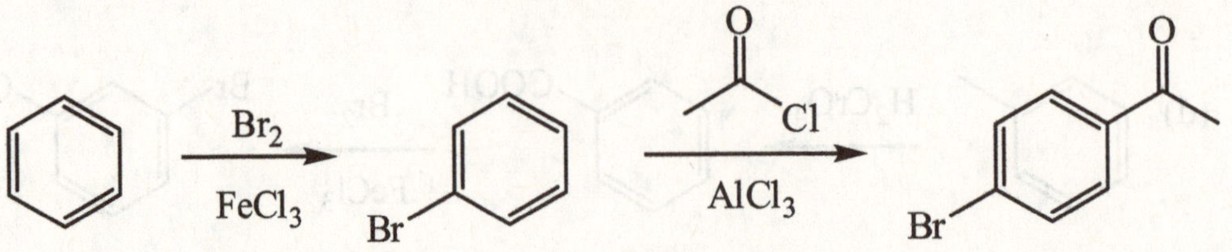

(c)

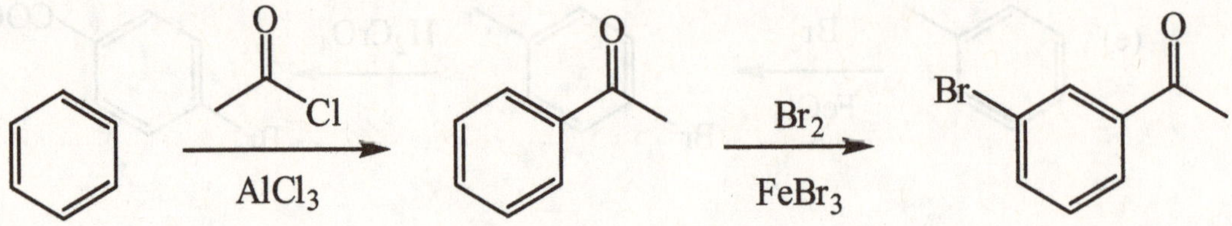

(d)

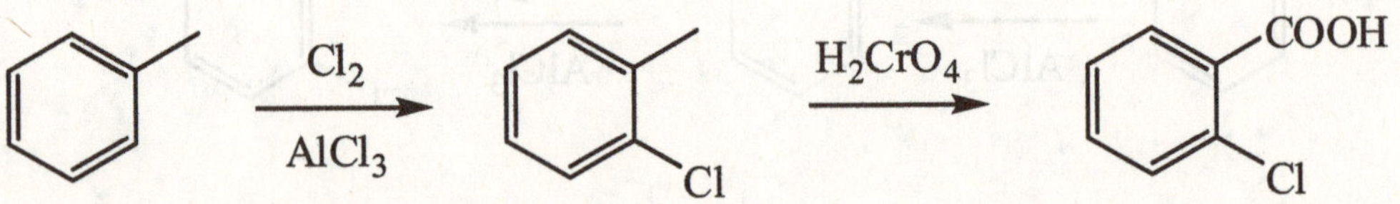

9.44 The following ketone, isolated from the roots of several members of the iris family, has an odor like that of violets and is used as a fragrance in perfumes. Describe the synthesis of this ketone from benzene.

4-Isopropylacetophenone

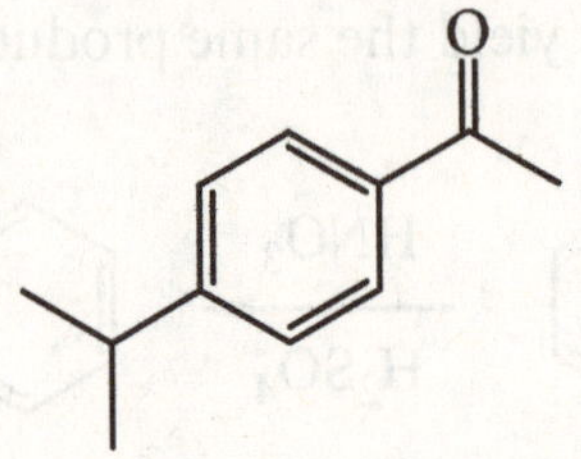

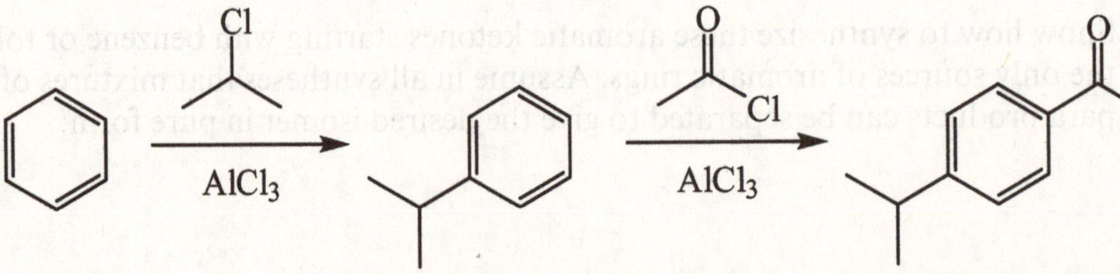

4-Isopropylacetophenone

9.45 The bombardier beetle generates *p*-quinone, an irritating chemical, by the enzyme-catalyzed oxidation of hydroquinone using hydrogen peroxide as the oxidizing agent. Heat generated in this oxidation produces superheated steam, which is ejected, along with *p*-quinone, with explosive force.

(a) Balance the equation.

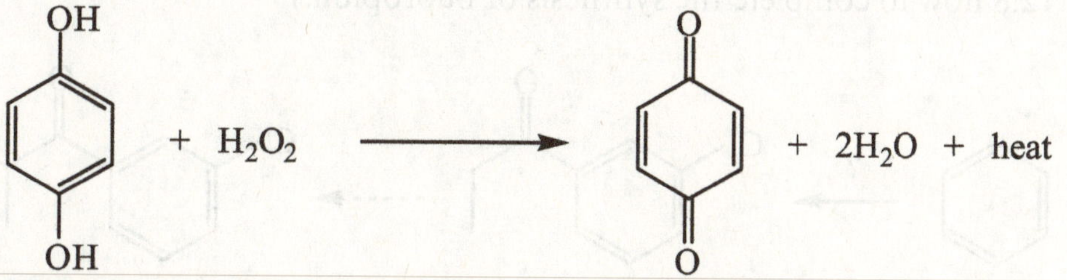

Hydroquinone   *p*-Quinone

(b) Show that this reaction of hydroquinone is an oxidation.

Two hydrogen atoms have been removed from hydroquinone.

9.46 Following is a structural formula for musk ambrette, a synthetic musk used in perfumes to enhance and retain fragrance. Propose a synthesis for musk ambrette from *m*-cresol.

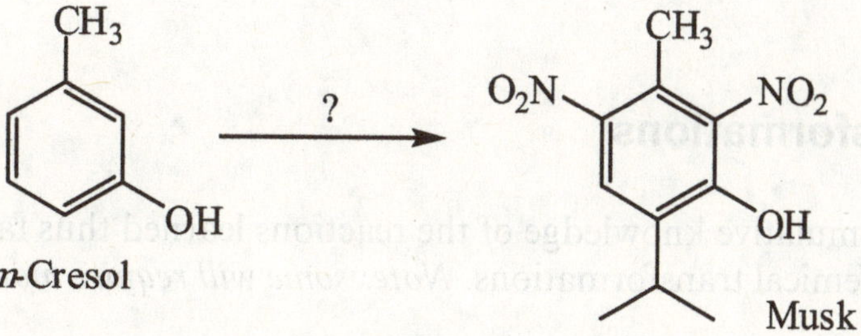

m-Cresol   Musk ambrette

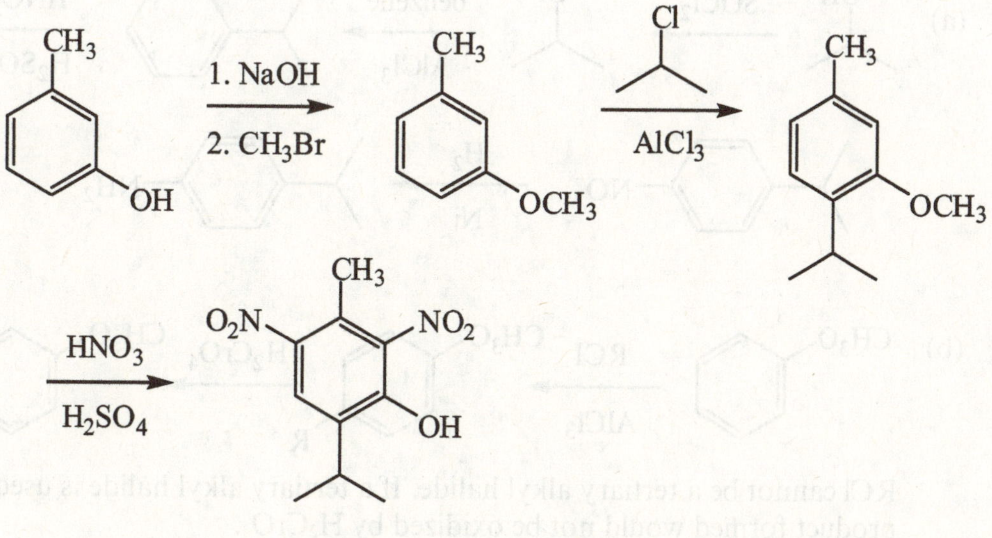

9.47 (3-Chlorophenyl)-1-propanone is a building block in the synthesis of bupropion, the hydrochloride salt of which is the antidepressant Wellbutrin. During clinical trials, researchers discovered that smokers reported a diminished craving for tobacco after one to two weeks on the drug. Further clinical trials confirmed this finding, and the drug is also marketed under the trade name Zyban as an aid in smoking cessation. Propose a synthesis for this building block from benzene. (We will see in Section 12.8 how to complete the synthesis of bupropion.)

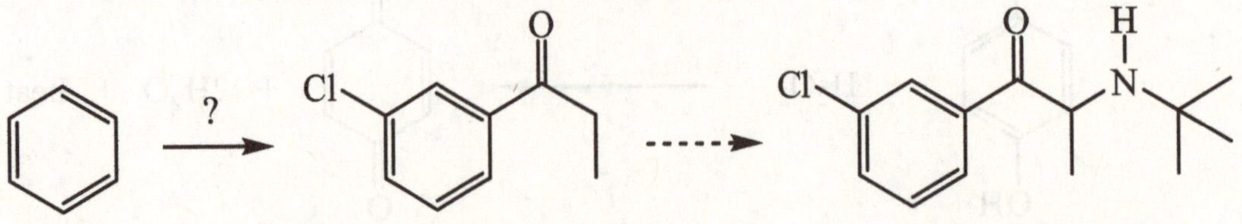

Benzene   (3-Chlorophenyl)-1-propanone   Bupropion
(Wellbutrin, Zyban)

Synthesis of 1-(3-chlorophenyl)-1-propanone building block:

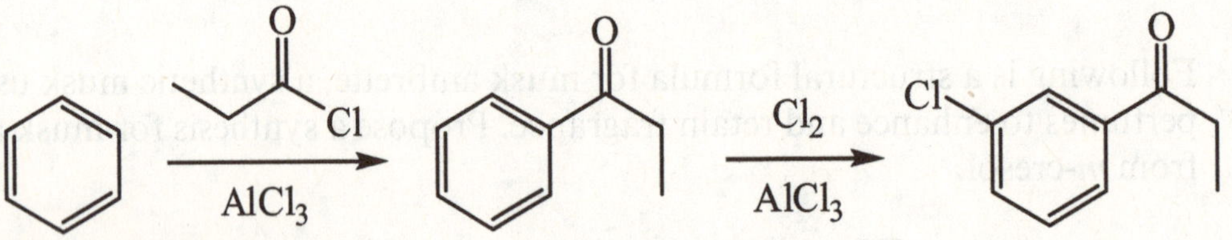

Chemical Transformations

9.48 Test your cumulative knowledge of the reactions learned thus far by completing the following chemical transformations. *Note: some will require more than one step.*

(a)

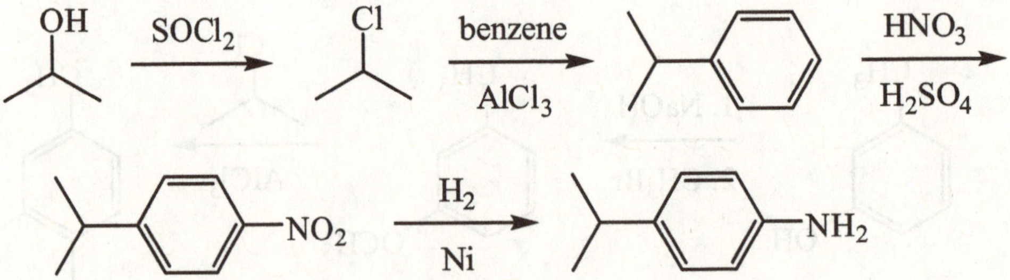

(b)

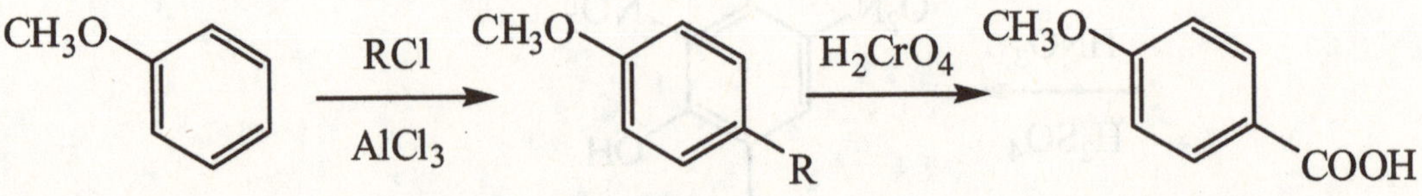

RCl cannot be a tertiary alkyl halide. If a tertiary alkyl halide is used, the product formed would not be oxidized by H_2CrO_4.

(c)

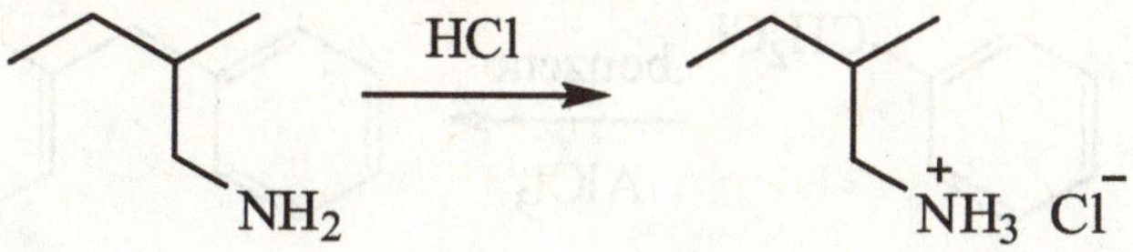

$$\text{(2-methylbutyl)aniline} \xrightarrow{\text{HCl}} \overset{+}{N}H_3 \; Cl^-$$

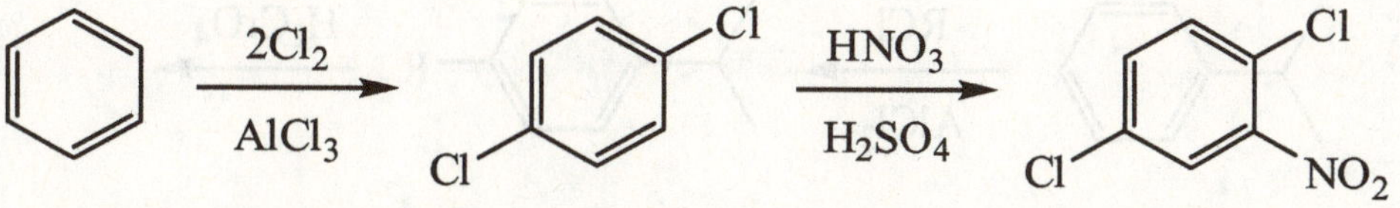

(d)

$$\text{benzene} \xrightarrow[\text{AlCl}_3]{2Cl_2} \text{1,4-dichlorobenzene} \xrightarrow[\text{H}_2\text{SO}_4]{\text{HNO}_3} \text{product (Cl, Cl, NO}_2\text{)}$$

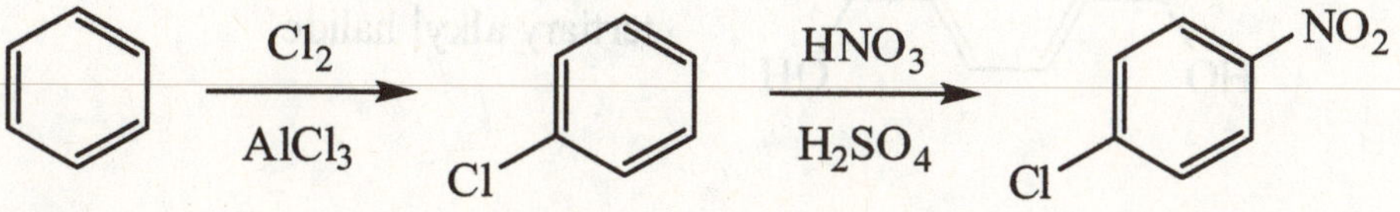

(e)

$$\text{benzene} \xrightarrow[\text{AlCl}_3]{Cl_2} \text{chlorobenzene} \xrightarrow[\text{H}_2\text{SO}_4]{\text{HNO}_3} \text{4-chloronitrobenzene}$$

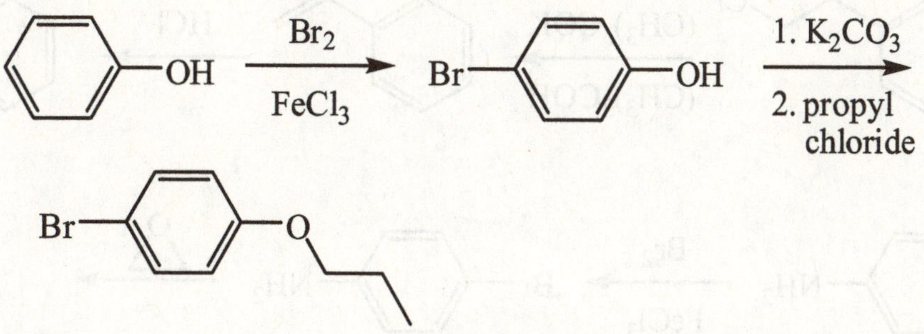

(f)

$$\text{phenol} \xrightarrow[\text{FeCl}_3]{\text{Br}_2} \text{4-bromophenol} \xrightarrow[\text{2. propyl chloride}]{\text{1. K}_2\text{CO}_3} $$

$$Br{-}\!\!\!\bigcirc\!\!\!{-}O{-}CH_2CH_2CH_3$$

(g)

$$\text{chlorobenzene} \xrightarrow[\text{AlCl}_3]{CH_3Cl} \text{4-chlorotoluene} \xrightarrow[\text{AlCl}_3]{Cl_2} \text{product (Cl, CH}_3\text{, Cl)}$$

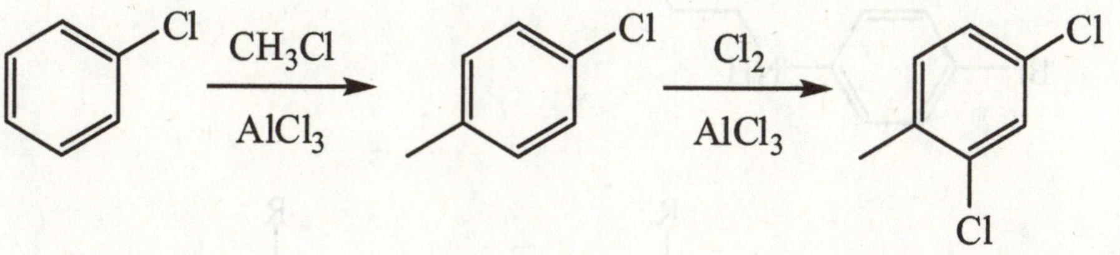

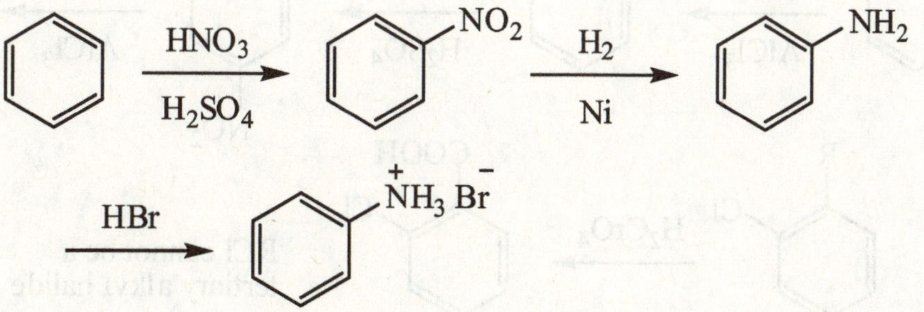

(h)

$$\text{benzene} \xrightarrow[\text{H}_2\text{SO}_4]{\text{HNO}_3} \text{nitrobenzene} \xrightarrow[\text{Ni}]{H_2} \text{aniline (NH}_2\text{)}$$

$$\xrightarrow{\text{HBr}} \overset{+}{N}H_3 \; Br^-$$

(i)

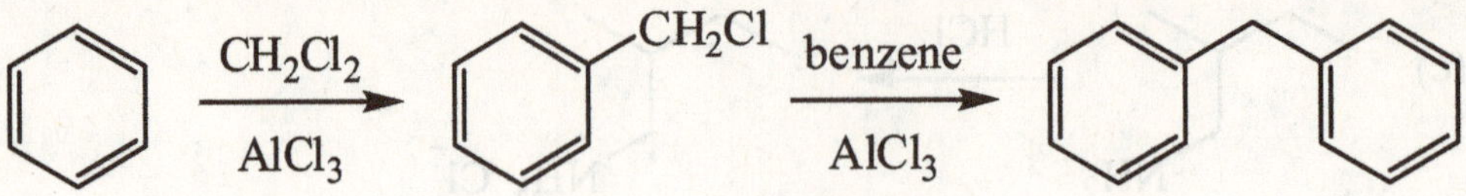

$$\underset{\text{AlCl}_3}{\xrightarrow{\text{CH}_2\text{Cl}_2}}$$

$$\underset{\text{AlCl}_3}{\xrightarrow{\text{benzene}}}$$

(j)

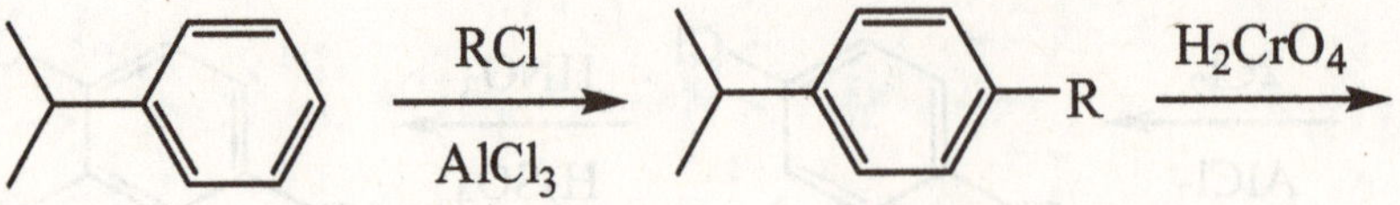

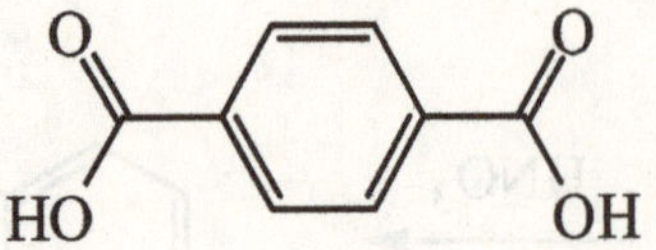

$$\underset{\text{AlCl}_3}{\xrightarrow{\text{RCl}}}$$

$$\xrightarrow{\text{H}_2\text{CrO}_4}$$

RCl cannot be a
tertiary alkyl halide

(k)

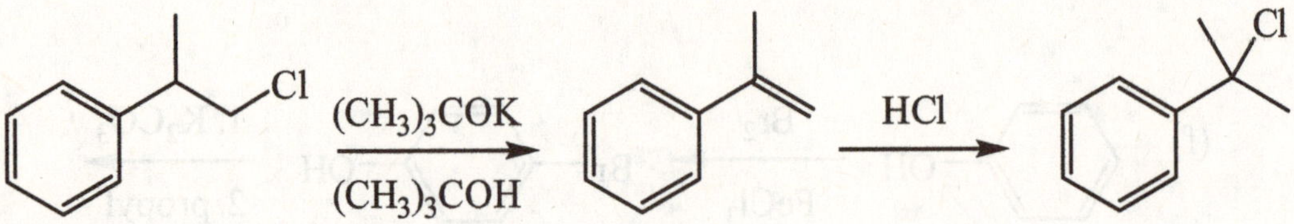

$$\xrightarrow[(\text{CH}_3)_3\text{COH}]{(\text{CH}_3)_3\text{COK}}$$

$$\xrightarrow{\text{HCl}}$$

(l)

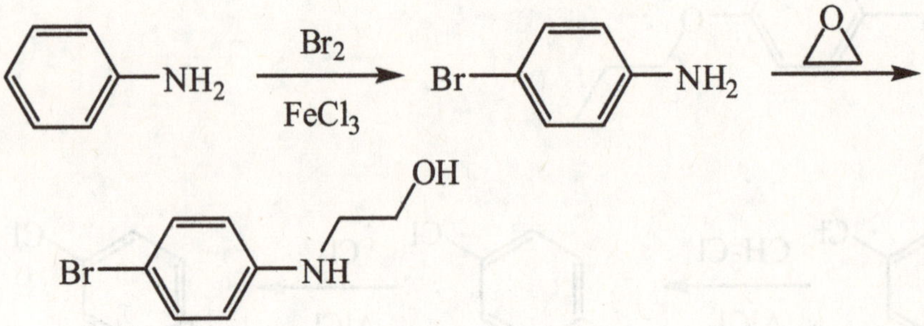

$$\underset{\text{FeCl}_3}{\xrightarrow{\text{Br}_2}}$$

(m)

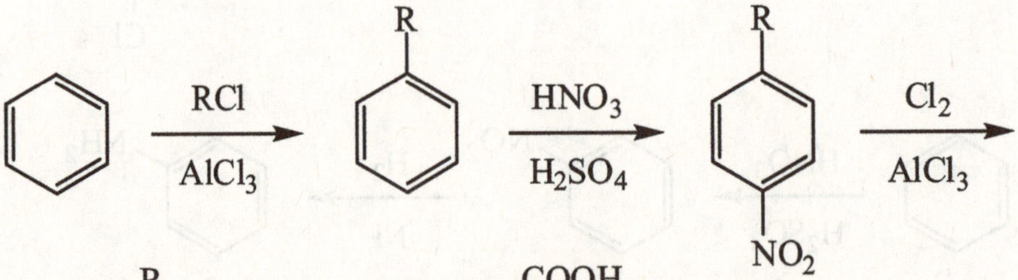

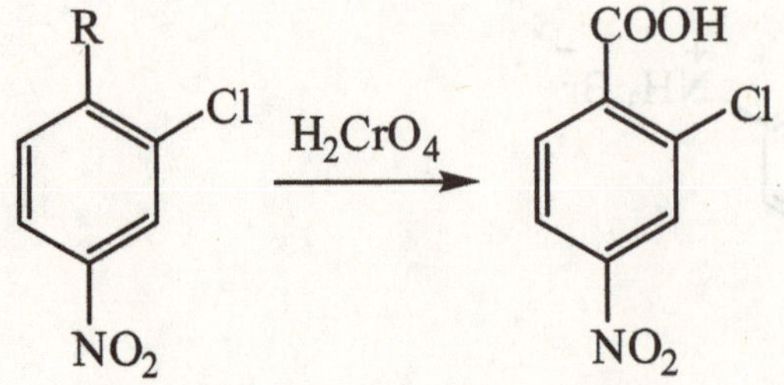

$$\underset{\text{AlCl}_3}{\xrightarrow{\text{RCl}}}$$

$$\underset{\text{H}_2\text{SO}_4}{\xrightarrow{\text{HNO}_3}}$$

$$\underset{\text{AlCl}_3}{\xrightarrow{\text{Cl}_2}}$$

$$\xrightarrow{\text{H}_2\text{CrO}_4}$$

RCl cannot be a
tertiary alkyl halide

(n)

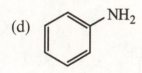

Looking Ahead

9.49 Which of the following compounds can be made directly by using an electrophilic aromatic substitution reaction?

None of these compounds can be made directly. In compounds that are made directly from electrophilic aromatic substitution, the substituent bonded to the benzene ring must have acted as an electrophile during the reaction.

(a) Making this compound would require the 1° propyl carbocation, which rearranges via a hydride shift to the more stable 2° isopropyl carbocation.

(b) Making this compound would require the vinyl carbocation ($CH_2=CH^+$), which is too unstable to form.

(c) Making this compound would require the OH^+ cation, which is too unstable to form.

(d) Making this compound would require the NH_2^+ cation, which is too unstable to form.

9.50 Which compound is a better nucleophile?

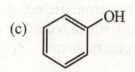

Aniline   or   Cyclohexanamine

Cyclohexanamine is a better nucleophile because the nitrogen lone pair of aniline is delocalized into the aromatic ring by resonance, making it less available for a nucleophilic reaction.

9.51   Suggest a reason that the following arenes do not undergo electrophilic aromatic substitution when $AlCl_3$ is used in the reaction.

(a) 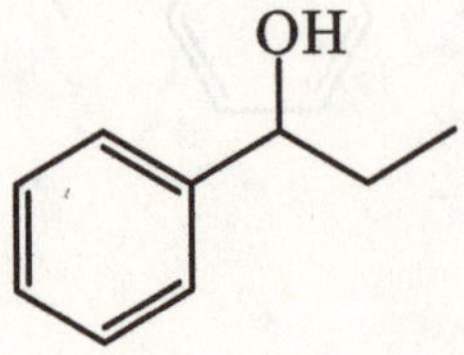   (b) 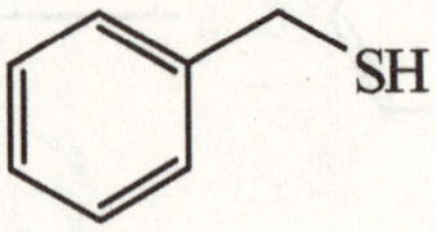   (c) 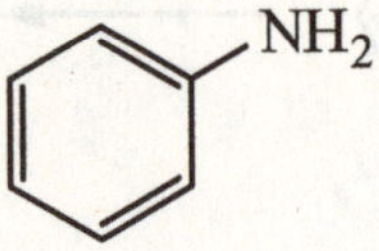   (d)

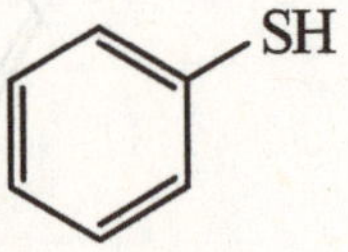

There are a couple of reasons why these arenes do not undergo electrophilic aromatic substitution when $AlCl_3$ is used. First, $AlCl_3$ decomposes in the presence of protic acids (ionizable hydrogens) to form HCl. Second, $AlCl_3$ (a Lewis acid) reacts with −OH, −SH, and −NH_2 groups, which are Lewis bases. These two reactions destroy the catalyst.

9.52   Predict the product of the following acid-base reaction:

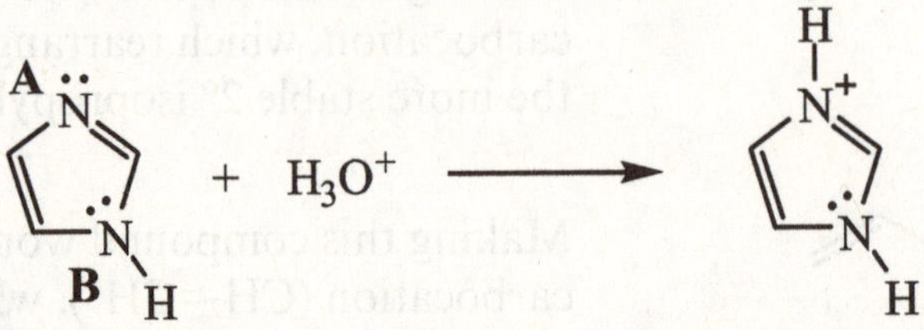

When the compound is treated with acid, it is protonated at nitrogen **A** instead of **B** because **A** is the stronger base. The lone pair of **B** is part of the aromatic sextet. The protonation of **B** would lead to the destruction of aromaticity, which would be an unfavorable process. The lone pair of **A** is not part of the aromatic sextet.

9.53   Which haloalkane reacts faster in an S_N1 reaction?

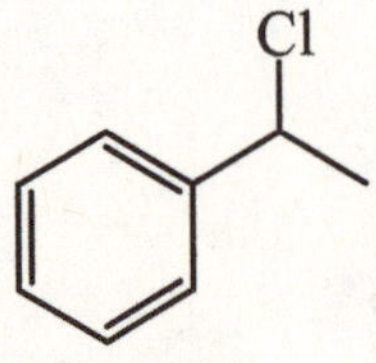

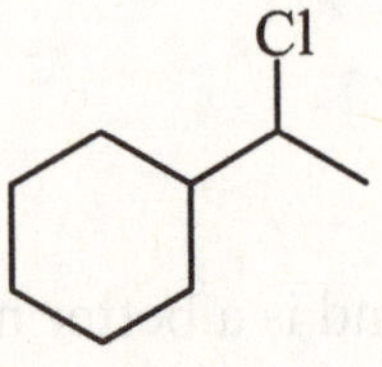

faster   slower

An S_N1 reaction involves a carbocation intermediate. Both of these compounds lose chloride to form 2° carbocations, but the carbocation derived from the compound on the left is stabilized by resonance.

9.54 Which of the following compounds is more basic?

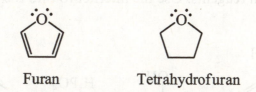

Furan   Tetrahydrofuran

Furan is aromatic, and the hybridization of its oxygen atom is sp^2. One of the lone pairs of oxygen is part of the aromatic sextet, while the other is not. The lone pair that can accept a proton is more tightly held by the sp^2 oxygen in furan compared to the sp^3 oxygen in tetrahydrofuran. Tetrahydrofuran is therefore more basic than furan.

Group Learning Activities

Solutions are only provided for activities that are not open-ended.

9.55 Following are benzene compounds with substituents we have yet to encounter. As a group, decide whether each ring will be activated or deactivated. Then determine whether each substituent is ortho–para or meta directing by analyzing their intermediates in an electrophilic aromatic substitution reaction.

(a)

(b) $Si(CH_3)_3$

The substituent in compound (a) is an electron-withdrawing group (meta director), so the ring will be deactivated. A contributing structure that places the positive charge on the substituent's carbon atom, which is directly attached to the ring, can be drawn. The addition of an electrophile to the meta position makes it impossible to place the positive charge of the ring carbon on the same carbon that bears the substituent.

The $Si(CH_3)_3$ substituent in compound (b) is similar to an alkyl substituent, which is an electron-donating group (ortho-para director) and a ring activator. The $Si(CH_3)_3$ group is a better ring activator than an alkyl group because of the lower electronegativity of Si. The addition of the electrophile to the ortho or para position allows the positive charge of the ring carbon to be placed next to the donating group, where it can be stabilized.

9.58   Propose a synthesis of 2,6-diisopropylphenol (Propofol) starting with phenol and
any three-carbon reagents. Use the internet to find the uses and cultural significance
of Propofol.

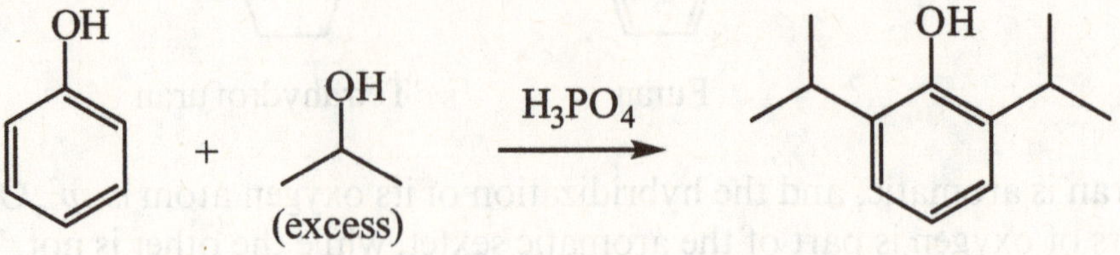

(excess)

Under acidic conditions, isopropanol forms the isopropyl carbocation, which acts as
electrophile in the electrophilic aromatic substitution reaction. Propene could also be
used in place of isopropanol.

Note that it is not possible to use isopropyl chloride and AlCl₃, because phenol is
acidic and would destroy the Lewis acid catalyst.

This reaction also produces a significant amount of para-substituted product, so the
para position is actually temporarily "hidden" to prevent alkylation at that position
(see Problem 10.43).

Chapter 10: Amines

Problems

10.1 Identify all carbon stereocenters in coniine, nicotine, and cocaine. Assign R/S configurations for all stereocenters in cocaine.

(a) Coniine   (b) Nicotine   (c) Cocaine (hint: build a model)

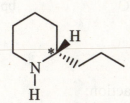

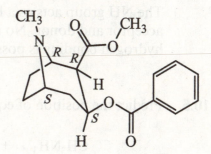

10.2 Write a structural formula for each amine:

(a) 2-Methyl-1-propanamine (b) Cyclohexanamine   (c) (*R*)-2-Butanamine

(d) (2*S*,4*S*)-2,4-Hexanediamine

10.3 Write a structural formula for each amine:

(a) Isobutylamine   (b) Triphenylamine   (c) Diisopropylamine

(d) Butylcyclohexylamine

10.4 Account for the fact that diethylamine has
a higher boiling point than diethyl ether.

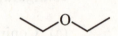

Diethylamine
bp 55°C

Diethyl ether
bp 34.6°C

The NH group acts as a hydrogen-bond
acceptor and donor. No intermolecular
hydrogen bonding is possible with the ether.

10.5 Predict the position of equilibrium for this acid-base reaction:

$$CH_3NH_3^+ \; + \; H_2O \; \rightleftharpoons \; CH_3NH_2 \; + \; H_3O^+$$

Amines are stronger bases than water because the electron pair on the nitrogen atom
is less stable than the electron pair on the oxygen atom (recall the effect of
electronegativity in Chapter 2). Therefore, the equilibrium lies to the left, the more
stable side.

10.6 Select the stronger acid from each pair of ions:

(a) O_2N—⬡—NH_3^+ or CH_3—⬡—NH_3^+

                A   B

When comparing the strengths of acids, examine the conjugate bases. (A) is the
stronger acid because its conjugate base, *p*-nitroaniline, is more stable, as a
result of the inductively withdrawing nitro group, than the conjugate base of
(B), *p*-methylaniline.

(b) ⬡NH^+ or   ⬡—NH_3^+

                 C   D

(C) is the stronger acid because its conjugate base, pyridine, is more stable, as a
result of the sp^2-hybridized nitrogen atom, than the conjugate base of (D),
cyclohexanamine. The nitrogen atom of cyclohexanamine is sp^3-hybridized.

10.7 Complete each acid-base reaction and name the salt formed:

(a) $(CH_3CH_2)_3N$ + HCl $\longrightarrow$ $(CH_3CH_2)_3NH^+Cl^-$ Triethylammonium chloride

(b) [piperidine structure] NH + CH_3COOH $\longrightarrow$ [piperidinium structure] NH_2^+ CH_3COO^- Piperidinium acetate

(c) [benzene] $-NH_2$ + H_2SO_4 $\longrightarrow$ $\left([\text{benzene}]-NH_3^+\right)_2 SO_4^{2-}$ Anilinium sulfate

10.8 As shown in Example 10.8, alanine is better represented as an internal salt. Suppose that the internal salt is dissolved in water.

$$\underset{\underset{NH_2}{|}}{CH_3CHCOH} \rightleftharpoons \underset{\underset{NH_3^+}{|}}{CH_3CHCO^-}$$

(a) In what way would you expect the structure of alanine in aqueous solution to change if concentrated HCl were added to adjust the pH of the solution to 2.0?

At a pH of 2, the carboxylate will be protonated. This is because the carboxylate ion is a weak base, and it will be protonated under acidic conditions. The amino acid has a net positive overall charge.

$$\underset{\underset{NH_3^+}{|}}{CH_3CHCOH}$$

(b) In what way would you expect the structure of alanine in aqueous solution to change if concentrated NaOH were added to bring the pH of the solution to 12.0?

At a pH of 12, the ammonium ion will be deprotonated. The ammonium ion is a weak acid, and it will be deprotonated under basic conditions. The amino acid has a net negative overall charge.

$$\underset{\underset{NH_2}{|}}{CH_3CHCO^-}$$

10.9 Show how you can use the same set of steps in Example 10.9, but in a different order, to convert toluene to 3-aminobenzoic acid.

To prepare 3-hydroxybenzoic acid (*m*-hydroxybenzoic acid) from toluene, toluene must first be converted to benzoic acid, which contains a meta-directing group, prior to the nitration reaction.

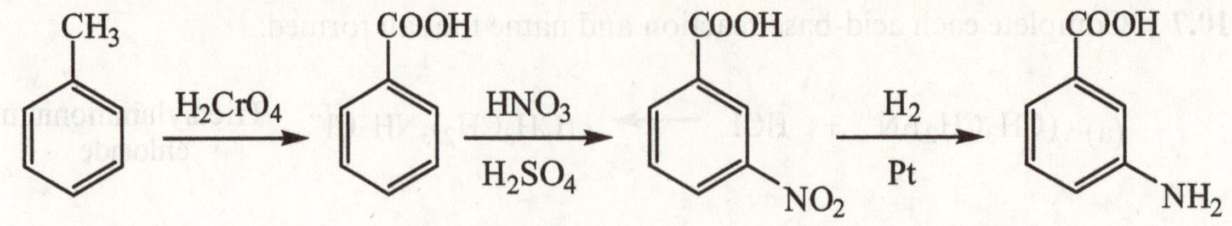

10.10 Determine all possible nitrogen-based products that can be formed in the following reaction:

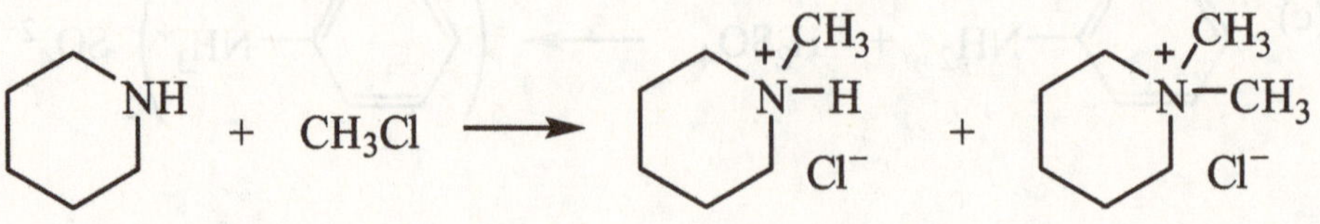

Chemical Connections

10A. Identify the functional groups in morphine and meperidine. Classify the amino group in these opiates as primary, secondary, or tertiary (Section 1.8).

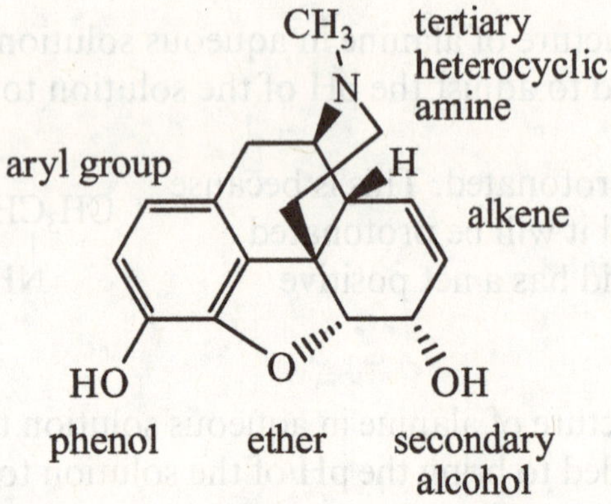

Morphine

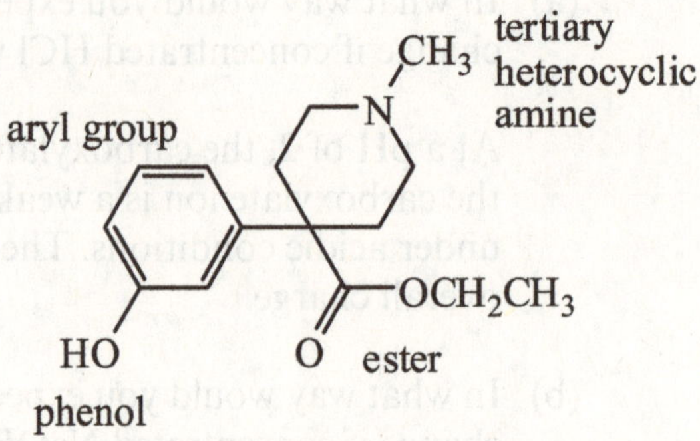

Meperidine

10B. Would you expect batrachotoxin or batrachotoxin A to be more soluble in water? Why?

Predict the product formed from the reaction of batrachotoxin with one equivalent of a weak acid such as acetic acid, CH_3COOH.

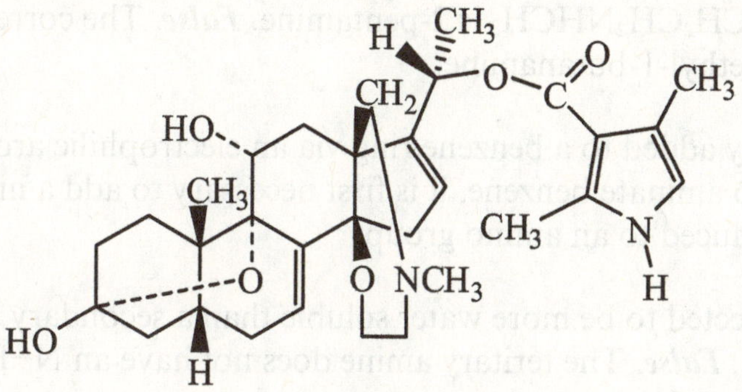

Batrachotoxin

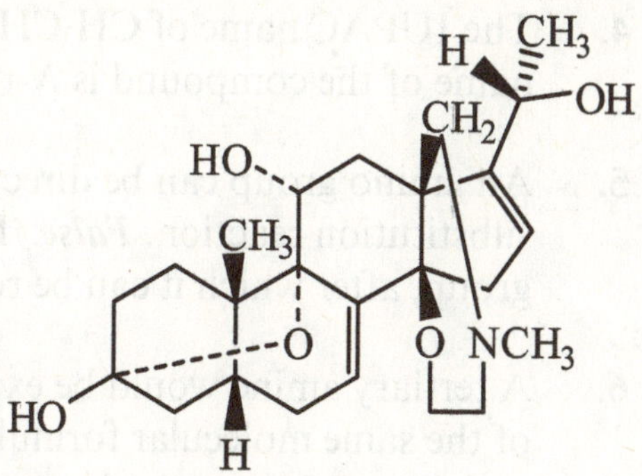

Batrachotoxin A

Batrachotoxin A contains an alcohol that is replaced by an ester and a substitututed pyrrole in batrachotoxin. The much higher polarity of the hydroxyl group, and its ability to hydrogen bond with water, is expected to make batrachotoxin more soluble in water.

When batrachotoxin is treated with one equivalent of acetic acid, the most basic atom in batrachotoxin is protonated first. This would be the nitrogen corresponding to the tertiary amine and not the nitrogen of the pyrrole ring. The lone pair of the pyrrole nitrogen is part of the aromatic sextet and therefore not very basic.

Batrachotoxin acetate

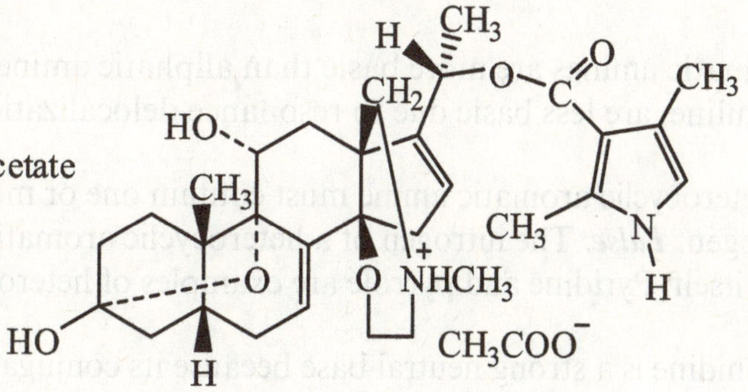

Quick Quiz

1. An amine with an $-NH_2$ group bonded to a tertiary carbon is classified as a tertiary amine. *False*. While this would be true for the classification of alcohols, amines are classified based on the number of carbons bonded directly to the nitrogen atom.

2. The reaction of an amine with a haloalkane initially results in an ammonium halide salt. *True*. After the salt has been made, it can be neutralized, as long as the salt does not contain a quaternary ammonium group, by the addition of a base.

3. An efficient way to make diethylamine is to react ammonia with two equivalents of chloroethane. *False*. Such a reaction would give a mixture of ethylamine, diethylamine, triethylamine, and possibly even tetraethylammonium chloride.

4. The IUPAC name of $CH_3CH_2CH_2CH_2NHCH_3$ is 2-pentamine. *False*. The correct name of the compound is *N*-methyl-1-butanamine.

5. An amino group can be directly added to a benzene ring via an electrophilic aromatic substitution reaction. *False*. To aminate benzene, it is first necessary to add a nitro group, after which it can be reduced to an amino group.

6. A tertiary amine would be expected to be more water soluble than a secondary amine of the same molecular formula. *False*. The teritary amine does not have an N−H bond and cannot act as a hydrogen-bond donor.

7. The pK_b of an amine can be determined from the pK_a of its conjugate acid. *True*. For any acid-conjugate base (or base-conjugate acid) pair, $pK_a + pK_b = 14$.

8. The lower the value of pK_b, the stronger the base. *True*. Stronger bases have higher K_b values and therefore lower pK_b values.

9. The basicity of amines and the solubility of amine salts in water can be used to separate amines from water-insoluble, nonbasic compounds. *True*. The mixture of compounds can be dissolved in an organic solvent, such as diethyl ether. Subsequent extraction of the ethereal solution with acid allows the isolation of the amine as the amine salt.

10. Aromatic amines are more basic than aliphatic amines. *False*. Aromatic amines, such as aniline, are less basic due to resonance delocalization of the nitrogen lone pair.

11. A heterocyclic aromatic amine must contain one or more aryl groups directly bonded to nitrogen. *False*. The nitrogen of a heterocyclic aromatic amine is part of the aromatic ring itself. Pyridine and pyrrole are examples of heterocyclic aromatic amines.

12. Guanidine is a strong neutral base because its conjugate acid is resonance stabilized. *True*. Guanidine is an example of where it is the resonance stabilization of the conjugate acid, and not the base itself, that affects the strength of the base. Guanidine is found in the side chain of the amino acid arginine, one of the 20 essential amino acids.

13. Ammonia is a slightly weaker base than most aliphatic amines. *True*. The electron-donating nature of alkyl groups increases the basicity of aliphatic amines.

14. An amino group forms stronger hydrogen bonds than a hydroxy group. *False*. Oxygen is more electronegative than nitrogen, so a hydroxyl group forms stronger hydrogen bonds due to the higher polarity of the OH bond.

15. A heterocyclic amine must contain a ring and a nitrogen atom as a member of the ring. *True*. *Heterocyclic amine* implies that the nitrogen atom is a part of the cyclic structure.

16. An electron-withdrawing group in an amine decreases its basicity. *True*. Electron-withdrawing groups stabilize the lone pair of the nitrogen atom.

End-of-Chapter Problems

Structure and Nomenclature

10.11 Draw a structural formula for each amine:

(a) (*R*)-2-Butanamine

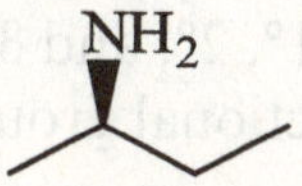

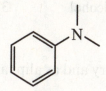

(b) 1-Octanamine

CH$_3$(CH$_2$)$_6$CH$_2$NH$_2$

(c) 2,2-Dimethyl-1-propanamine

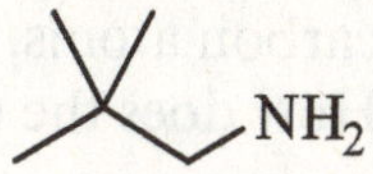

(d) 1,5-Pentanediamine

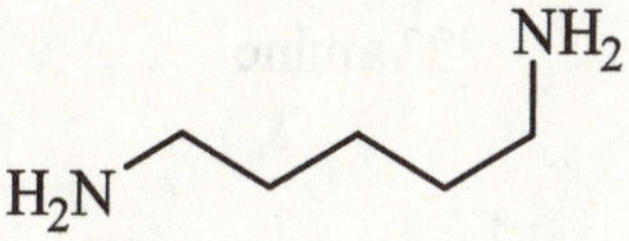

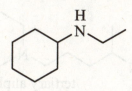

(e) 2-Bromoaniline

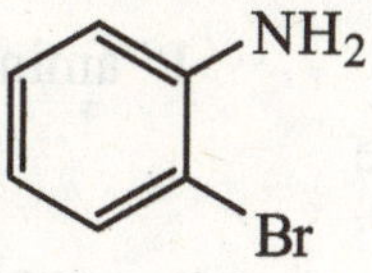

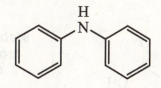

(f) Tributylamine

(CH$_3$CH$_2$CH$_2$CH$_2$)$_3$N

(g) *N*,*N*-Dimethylaniline

(h) Benzylamine

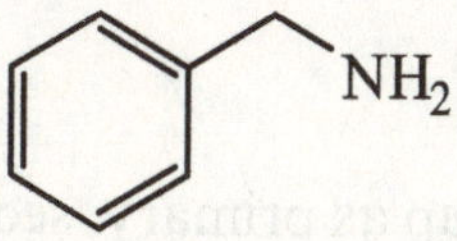

(i) *tert*-Butylamine

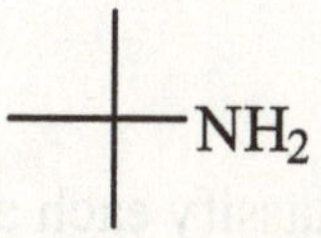

(j) *N*-Ethylcyclohexanamine

(k) Diphenylamine

(l) Isobutylamine

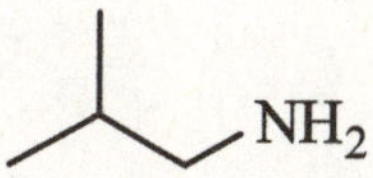

10.12 Draw a structural formula for each amine:

(a) 4-Aminobutanoic acid

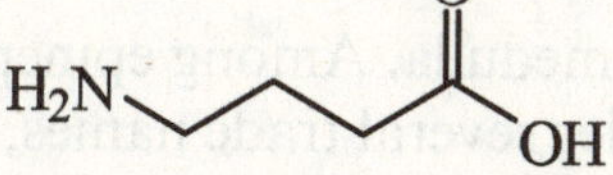

(b) 2-Aminoethanol (ethanolamine)

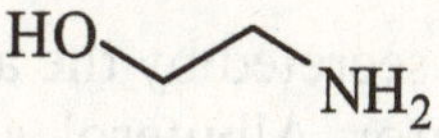

(c) 2-Aminobenzoic acid

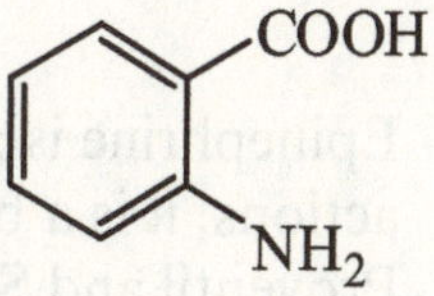

(d) (*S*)-2-Aminopropanoic   (e) 4-Aminobutanal   (f) 4-Amino-2-butanone
    acid (alanine)

10.13 Draw examples of 1°, 2°, and 3° amines that contain at least four *sp*³-hybridized carbon atoms. Using the same criterion, provide examples of 1°, 2°, and 3° alcohols. How does the classification system differ between the two functional groups?

Amines are classified by the number of carbon substituents bonded to the nitrogen atom. Whereas, alcohols are classified by the number of carbon substituents bonded to the carbon bearing the −OH group.

1° amine   2° amine   3° amine

1° alcohol   2° alcohol   3° alcohol

10.14 Classify each amino group as primary, secondary, or tertiary and as aliphatic or aromatic:

(a) Benzocaine   (b) Chloroquine

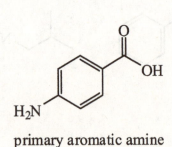

primary aromatic amine

secondary aromatic amine

tertiary aliphatic amine

heterocyclic aromatic amine

10.15 Epinephrine is a hormone secreted by the adrenal medulla. Among epinephrine's actions, it is a bronchodilator. Albuterol, sold under several trade names, including Proventil and Salbumol, is one of the most effective and widely prescribed antiasthma drugs. The R enantiomer of albuterol is 68 times more effective in the treatment of asthma than the S enantiomer.

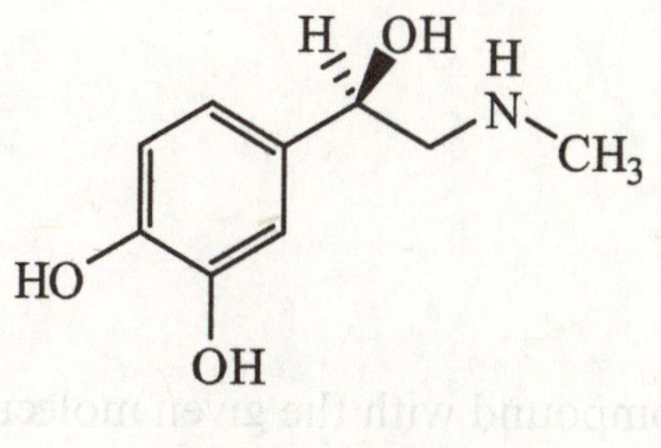

(R)-Epinephrine
(Adrenaline)

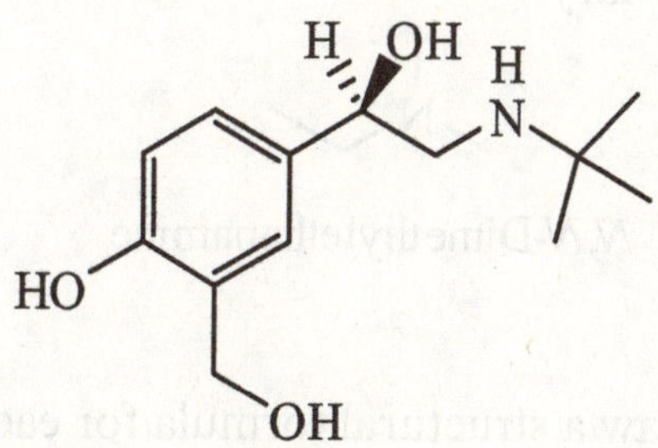

(R)-Albuterol

(a) Classify each amino group as primary, secondary, or tertiary.

Both amino groups are secondary.

(b) List the similarities and differences between the structural formulas of these compounds.

Both compounds contain the same basic framework and have the same configuration at the stereocenter. Differences are the *tert*-butyl and $-CH_2OH$ groups in (R)-albuterol versus the methyl and $-OH$ groups in (R)-epinephrine, respectively.

10.16 There are eight constitutional isomers with the molecular formula $C_4H_{11}N$. Name and draw structural formulas for each. Classify each amine as primary, secondary, or tertiary.

Primary

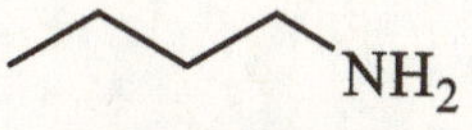

1-Butanamine
(butylamine)

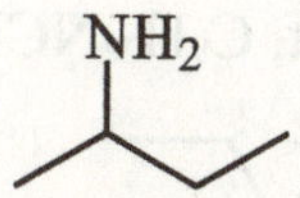

2-Butanamine
(*sec*-Butylamine)

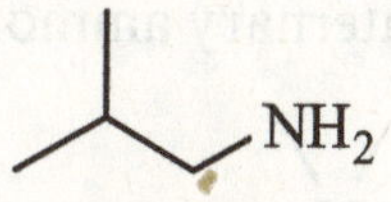

2-Methyl-
1-propanamine
(Isobutylamine)

2-Methyl-
2-propanamine
(*tert*-Butylamine)

Secondary

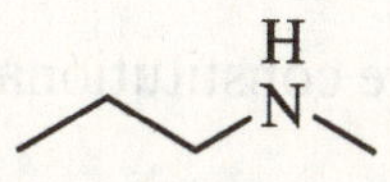

N-Methyl-1-propanamine

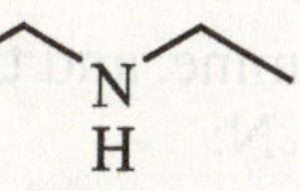

N-Ethylethanamine
(Eiethylamine)

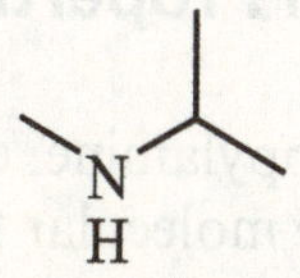

N-Methyl-2-methyl-
1-Ethanamine

Tertiary

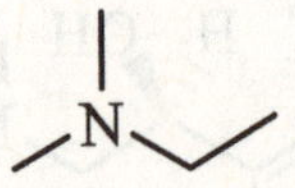

N,N-Dimethylethanamine

10.17 Draw a structural formula for each compound with the given molecular formula:

(a) A 2° arylamine, C_7H_9N

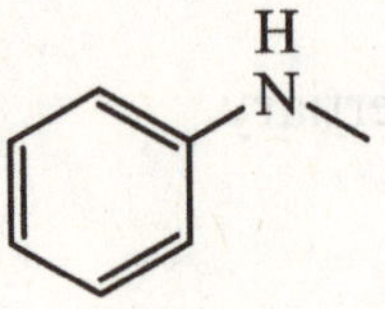

(b) A 3° arylamine, $C_8H_{11}N$

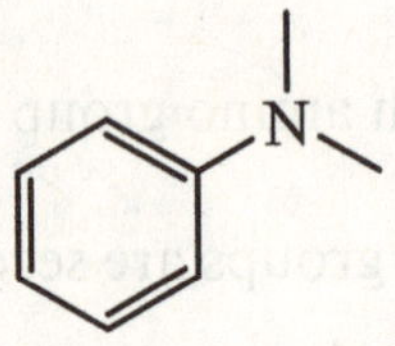

(c) A 1° aliphatic amine, C_7H_9N

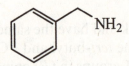

(d) A chiral 1° amine, $C_4H_{11}N$

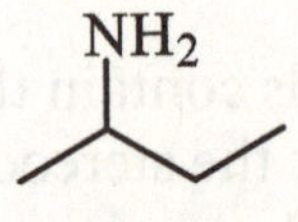

(e) A 3° heterocyclic amine, $C_5H_{11}N$

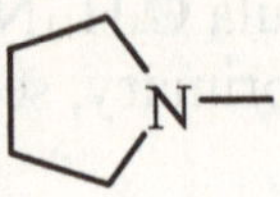

(f) A trisubstituted 1° arylamine, $C_9H_{13}N$

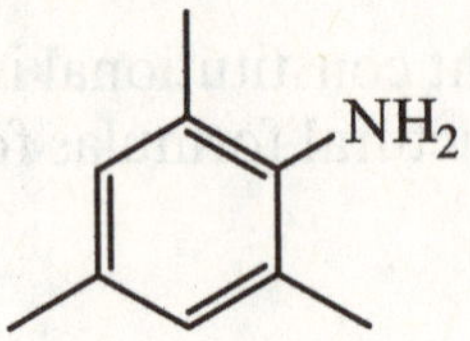

+ isomers

(g) A chiral quaternary ammonium salt, $C_9H_{22}NCl$

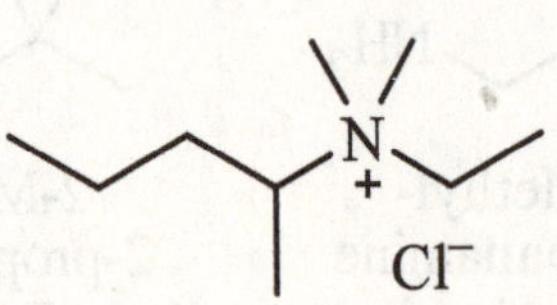

or

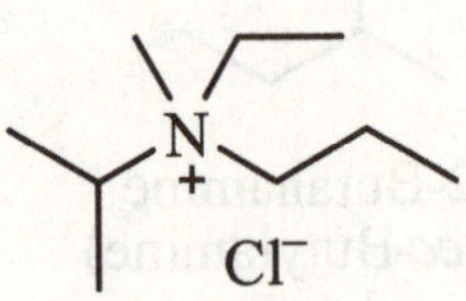

Physical Properties

10.18 Propylamine, ethylmethylamine, and trimethylamine are constitutional isomers with the molecular formula C_3H_9N:

$CH_3CH_2CH_2NH_2$ $CH_3CH_2NHCH_3$ $(CH_3)_3N$

bp 48°C   bp 37°C   bp 3°C

Propylamine   Ethylmethylamine   Trimethylamine

Trimethylamine does not have any N−H bonds and therefore cannot form intermolecular hydrogen bonds. The other two amines have N−H bonds and can participate in intermolecular hydrogen bonding. Propylamine has the highest boiling point because it can act as a hydrogen-bond donor to two other molecules, whereas ethylmethylamine can act as a hydrogen-bond donor to only one other molecule.

In addition to hydrogen bonding, dispersion forces also play a role in the boiling points of these three amines. Propylamine, which is unbranched, has the largest surface area and the greatest dispersion forces, hence the highest boiling point.

10.19 Account for the fact that 1-butanamine has a lower boiling point than 1-butanol:

bp 78°C   bp 117°C
1-Butanamine   1-Butanol

1-Butanol has a higher boiling point because an O−H---O hydrogen bond is stronger than an N−H---N hydrogen bond as a result of the higher polarity of the OH bond. (See Problem 8.50 for more details.) The intermolecular forces are greater in 1-butanol compared to 1-butanamine, hence the higher boiling point of 1-butanol.

10.20 Account for the fact that putrescine, a foul-smelling compound produced by rotting flesh, ceases to smell upon treatment with two equivalents of HCl:

1,4-Butanediamine
(Putrescine)

In order for us to be able to smell any compound, it needs to be volatile and be detectable by our nose. When putrescine is treated with HCl, it is converted to its salt form. Due to strong electrostatic interactions, ionic compounds are not very volatile.

Basicity of Amines

10.21 Account for the fact that amines are more basic than alcohols.

The relative strengths of bases can be assessed by comparing the stabilities of the lone pair of electrons on the bases. Nitrogen is less electronegative than oxygen, so the lone pair on the nitrogen atom of an amine is less stable (and less tightly held) compared to that on the oxygen atom of an alcohol. As a result, amines are more basic than alcohols.

10.22 From each pair of compounds, select the stronger base:

(a)

stronger base

Piperidine is a stronger base than pyridine because the nitrogen atom of piperidine is sp^3-hybridized, while that of pyridine is sp^2. As a result, the nitrogen lone pair of pyridine is more stable (held more tightly), making pyridine a weaker base than piperidine.

(b)

stronger base

The nitrogen lone pair of the aromatic amine is delocalized by resonance, making the nitrogen less basic compared to that of the aliphatic amine.

(c)

stronger base

The lone pair of the nitrogen in 4-ethylaniline, an aromatic amine, is delocalized by resonance and therefore less basic than the nitrogen of benzylamine.

(d)

stronger base

The nitrogen lone pair of both compounds is delocalized by resonance, but p-nitroaniline has a withdrawing nitro group that further reduces electron density about the nitrogen.

10.23 Account for the fact that substitution of a nitro group makes an aromatic amine a weaker base, but makes a phenol a stronger acid. For example, 4-nitroaniline is a weaker base than aniline, but 4-nitrophenol is a stronger acid than phenol.

The nitro group is an electron-withdrawing group. In addition, it is able to stabilize the lone pair of electrons on the nitrogen atom in 4-nitroaniline by resonance. Bases that are more stable are weaker bases. Therefore, 4-nitroaniline is a weaker base than aniline.

Likewise, the nitro group can stabilize the conjugate base of 4-nitrophenol, 4-nitrophenoxide, by both induction and resonance. The more stable is the conjugate base, the stronger is the acid. Therefore, 4-nitrophenol is a stronger acid than phenol.

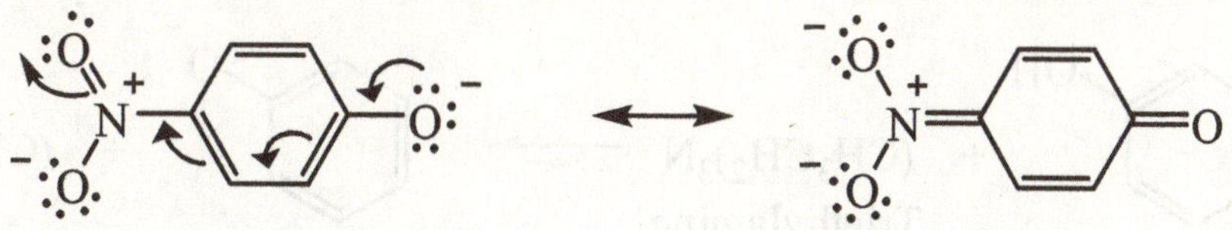

10.24 Select the stronger base in this pair of compounds:

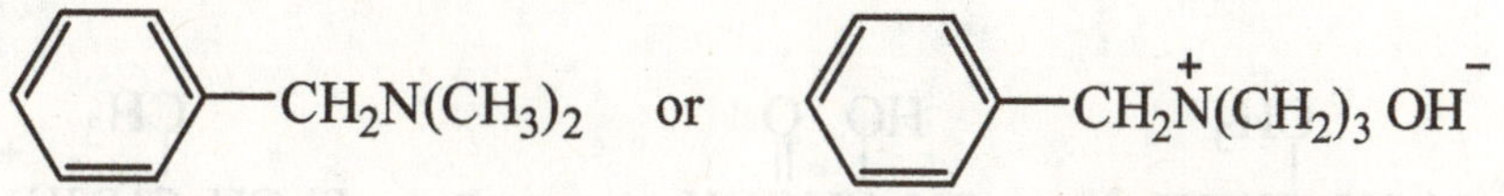

The compound on the right, benzyltrimethylammonium hydroxide, is a stronger base than the compound on the left, benzyldimethylamine, because the former contains a hydroxide ion. The quaternary ammonium cation is simply a counterion, much like how Na^+ is a counterion for the OH^- anion in NaOH, a strong base.

10.25 Complete the following acid-base reactions and predict the position of equilibrium for each. Justify your prediction by citing values of pK_a for the stronger and weaker acid in each equilibrium. For values of acid ionization constants, consult Table 2.2 (pK_a's of some inorganic and organic acids), Table 8.3 (pK_a's of alcohols), Section 9.8B (acidity of phenols), and Table 10.2 (base strengths of amines). Where no ionization constants are given, make the best estimate from the aforementioned tables and section.

Recall that in acid-base reactions, the equilibrium favors the side with the weaker acid, which is the side with the acid of the highest pK_a. For any acid-conjugate base or base-conjugate acid pair, the value of $pK_a + pK_b$ is 14. The stronger the acid, the weaker (and more stable) is its conjugate base. The stronger (and less stable) the base, the weaker is its conjugate acid. Note that the acid and base strengths of these compounds cannot be easily predicted from trends and effects, so it is necessary to use the quantitative data.

(a)

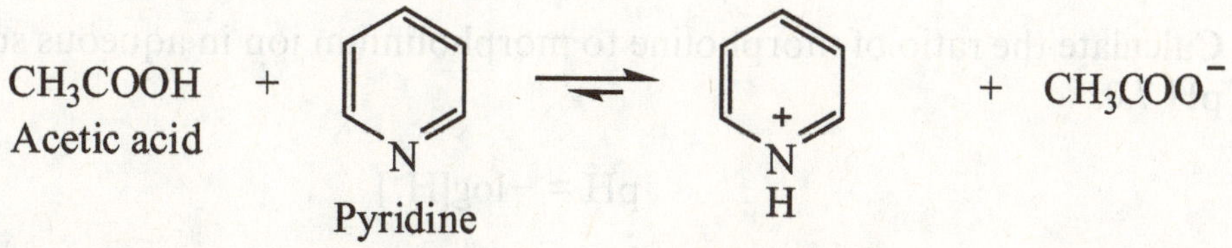

CH₃COOH	Pyridine		
Acetic acid			
stronger acid	stronger base	weaker acid	weaker base
$pK_a = 4.76$		$pK_a = 5.25$	

(b)

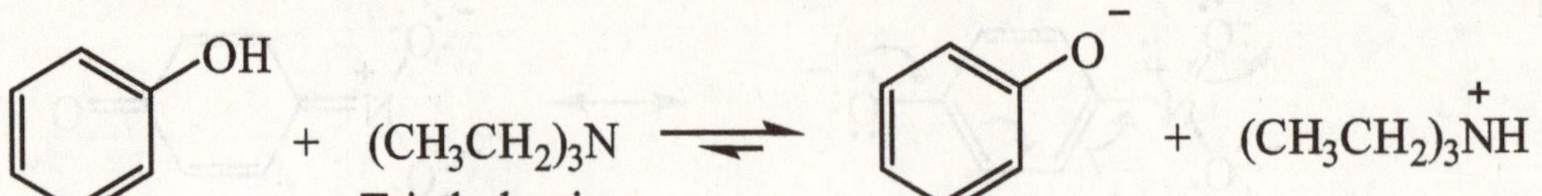

Phenol

stronger acid	stronger	weaker	weaker acid
$pK_a = 9.95$	base	base	$pK_a = 10.75$

(c)

1-Phenyl-2-propanamine (Amphetamine)   2-Hydroxypropanoic acid (Lactic acid)

stronger base	stronger acid $pK_a = 3.08$	weaker acid $pK_a = 11$	weaker base

(d)

Methamphetamine   Acetic acid

stronger base	stronger acid $pK_a = 4.76$	weaker acid $pK_a = 11$	weaker base

10.26 The pK_a of the morpholinium ion is 8.33:

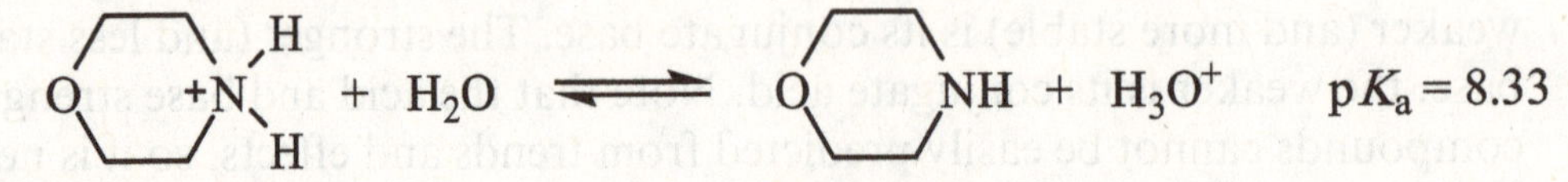

Morpholinium ion   Morpholine $pK_a = 8.33$

(a) Calculate the ratio of morpholine to morpholinium ion in aqueous solution at pH 7.0.

$$pH = -\log[H^+]$$

$$[H^+] = 10^{-pH} = 10^{-7.0} = 1 \times 10^{-7.0} M$$

$$K_a = 10^{-pK_a} = 10^{-8.33} = \frac{[\text{Morpholine}][H^+]}{[\text{Morpholinium ion}]} = 4.68 \times 10^{-9}$$

$$4.68 \times 10^{-9} = \frac{[\text{Morpholine}][1 \times 10^{-7}]}{[\text{Morpholinium ion}]}$$

$$\frac{[\text{Morpholine}]}{[\text{Morpholinium ion}]} = 0.047$$

The ratio indicates that there are about 21 morpholinium ions present for every molecule of morpholine. This is as expected, given that morpholine is basic and would predominantly exist as its conjugate acid, the morpholinium ion, at pH 7.0.

(b) At what pH are the concentrations of morpholine and morpholinium ion equal?

When the two concentrations are equal (a ratio of 1:1), the $[\text{H}^+]$ must be equal to the K_a. Thus, the concentrations are equal at a pH of 8.33.

10.27 The pK_b of amphetamine (Example 10.2) is approximately 3.2. Calculate the ratio of amphetamine to its conjugate acid at pH 7.4, the pH of blood plasma.

Because the question provides a pK_b value instead of pK_a, there are two possible approaches to this question. The value of pK_b can be used to find the pK_a of the conjugate acid. Or, we could use pK_b but use the pH to find pOH (recall that pH + pOH = 14). The former approach uses calculations similar to those of Problem 10.26, so the latter approach is presented for this problem.

$$\text{pOH} = 14 - \text{pH} = 14 - 7.4 = 6.6$$

$$[\text{OH}^-] = 10^{-\text{pOH}} = 10^{-6.60} = 2.51 \times 10^{-7} \text{M}$$

$$K_b = 10^{-pK_b} = 10^{-3.2} = \frac{[\text{Conjugate acid}][\text{OH}^-]}{[\text{Amphetamine}]} = 6.31 \times 10^{-4}$$

$$6.31 \times 10^{-4} = \frac{[\text{Conjugate acid}][2.51 \times 10^{-7}]}{[\text{Amphetamine}]}$$

$$\frac{[\text{Conjugate acid}]}{[\text{Amphetamine}]} = 2.5 \times 10^3$$

The ratio of amphetamine to its conjugate acid is therefore about 1:2500.

10.28 Calculate the ratio of amphetamine to its conjugate acid at pH 1.0, such as might be present in stomach acid.

Use the same approach as shown for Problem 10.27, except using a pH of 1.0 instead of 7.4. At pH 1.0, the ratio of amphetamine to its conjugate acid about one in 6.3×10^9. As expected, at low pH, bases are predominantly in the protonated form.

10.29 Following is a structural formula of pyridoxamine, one form of vitamin B₆:

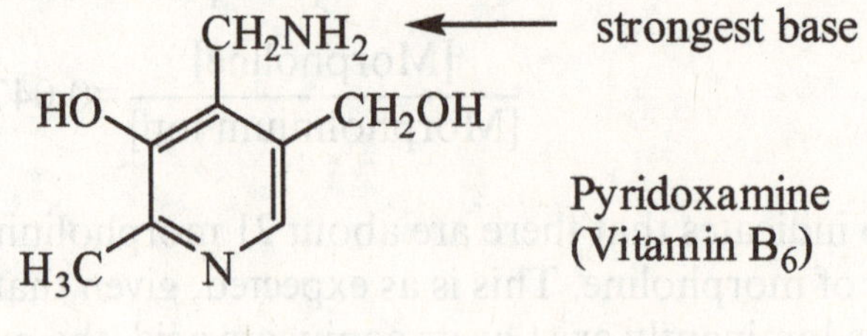

CH₂NH₂ ← strongest base

HO — CH₂OH

H₃C — N

Pyridoxamine
(Vitamin B₆)

(a) Which nitrogen atom of pyridoxamine is the stronger base?

The indicated nitrogen atom is the stronger base because it is sp^3-hybridized. It is therefore less stable than the pyridine nitrogen atom, which is sp^2-hybridized.

(b) Draw the structural formula of the hydrochloride salt formed when pyridoxamine is treated with one equivalent of HCl.

When a compound containing more than one basic group is treated with one equivalent of a strong acid, the most basic site is protonated first.

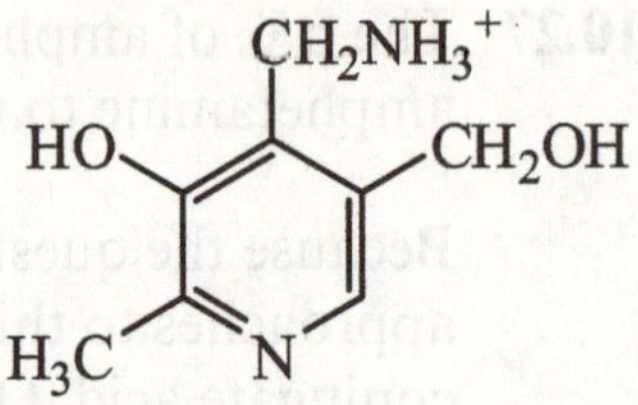

CH₂NH₃⁺

HO — CH₂OH

H₃C — N

10.30 Epibatidine, a colorless oil isolated from the skin of the Ecuadorian poison frog *Epipedobates tricolor*, has several times the analgesic potency of morphine. It is the first chlorine-containing, non-opioid (nonmorphine-like in structure) analgesic ever isolated from a natural source.

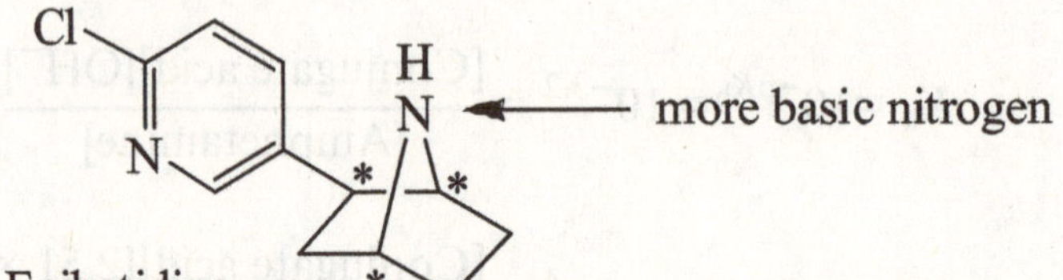

Cl

H
N ← more basic nitrogen

N

*   *

Epibatidine   *

(a) Which of the two nitrogen atoms of epibatidine is the more basic?

The nitrogen atom of the aliphatic amine (sp^3-hybridized nitrogen) is more basic than that of the pyridine ring (sp^2-hybridized nitrogen)

(b) Mark all stereocenters in this molecule.

Epibatidine has a total of three stereocenters.

10.31 Procaine was one of the first local anesthetics for infiltration and regional anesthesia. Its hydrochloride salt is marketed as Novocaine.

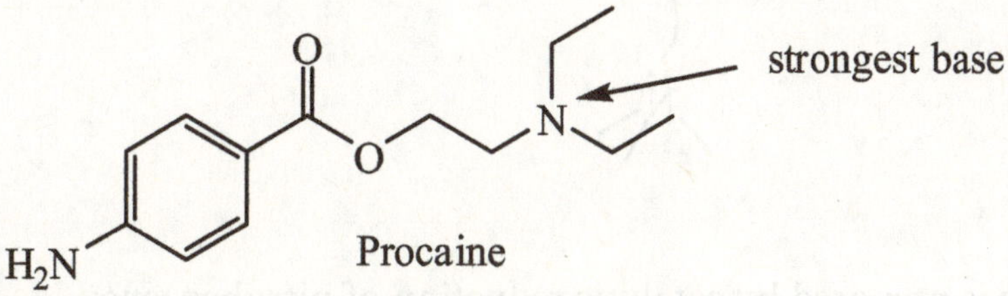

Procaine

(a) Which nitrogen atom of procaine is the stronger base?

The indicated nitrogen atom is the stronger base. The lone pair of the aniline nitrogen is delocalized into the aromatic ring and stabilized by resonance, so it is a weaker base than the indicated introgen atom.

(b) Draw the formula of the salt formed by treating procaine with one mole of HCl.

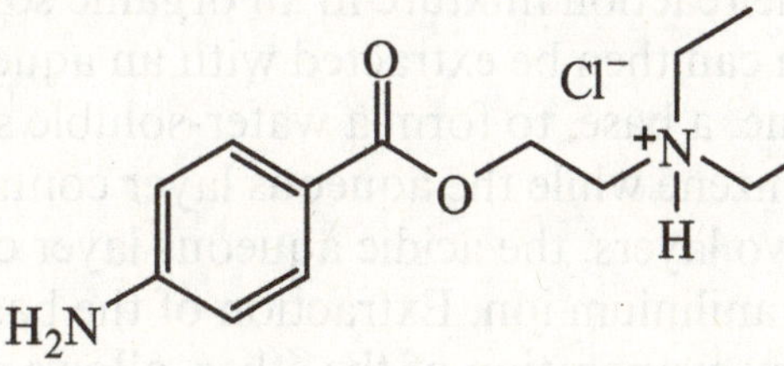

(c) Is procaine chiral? Would a solution of Novocaine in water be optically active or optically inactive?

Neither procaine nor Novocaine is chiral. A solution of Novocaine would be optically inactive. Achiral compounds do not rotate plane-polarized light.

10.32 Treatment of trimethylamine with 2-chloroethyl acetate gives the neurotransmitter acetylcholine as its chloride salt:

$$(CH_3)_3N \ + \ CH_3COCH_2CH_2Cl \ \longrightarrow \ C_7H_{16}ClNO_2$$

Acetylcholine chloride

Propose a structural formula for this quaternary ammonium salt and a mechanism for its formation.

The information given indicates that the product is a chloride salt, suggesting that the −Cl group of 2-chloroethyl acetate must have departed as a leaving group. The nitrogen atom of trimethylamine is good enough of a nucleophile to react by an S_N2 reaciton.

10.33 Aniline is prepared by catalytic reduction of nitrobenzene:

Devise a chemical procedure based on the basicity of aniline to separate it from any unreacted nitrobenzene.

First, dissolve the reaction mixture in an organic solvent such as diethyl ether. The ethereal solution can then be extracted with an aqueous acid, such as HCl, which reacts with aniline, a base, to form a water-soluble salt. Thus, the ether layer contains nitrobenzene while the aqueous layer contains the anilinium salt. After separating the two layers, the acidic aqueous layer can be basified with NaOH to deprotonate the anilinium ion. Extraction of the basified aqueous layer with diethyl ether, followed by evaporation of the ether, allows aniline to be isolated.

10.34 Suppose that you have a mixture of the following three compounds:

4-Nitrotoluene	4-Methylaniline	4-Methylphenol
(*p*-Nitrotoluene)	(*p*-Toluidine)	(*p*-Cresol)

Devise a chemical procedure based on their relative acidity or basicity to separate and isolate each in pure form.

The three compounds have different acid base properties: 4-nitrotoluene is neither acidic nor basic, 4-methylaniline is a weak base, and 4-methylphenol is a weak acid (recall that phenols are mildly acidic).

First, dissolve the reaction mixture in an organic solvent in which all three compounds are soluble. Extraction of the organic solution with dilute aqueous HCl extracts 4-methylaniline as the chloride salt (subsequent basification of the acidic solution regenerates the water-insoluble 4-methylaniline). The organic solvent, which now contains the two remaining compounds, can be extracted with dilute aqueous NaOH to extract 4-methylpenol as the sodium salt (subsequent acidification of this basic solution regenerates the water-insoluble 4-methylphenol). Evaporation of the remaining organic solution gives 4-nitrotoluene.

10.35 Following is a structural formula for metformin, the hydrochloride salt of which is marketed as the antidiabetic Glucophage. Metformin was introduced into clinical medicine in the United States in 1995 for the treatment of type 2 diabetes. More than 25 million prescriptions for this drug were written in 2000, making it the most commonly prescribed brand-name diabetes medication in the nation.

Metformin

(a) Draw the structural formula for Glucophage.

Metformin contains two guanidine groups, and the two most basic nitrogen atoms are those with the double bond. Protonation of either one or both of these nitrogen atoms results in a conjugate acid that is stabilized by resonance. Glucophage is a hydrochloride salt (not a dihydrochloride) and only one of these is protonated.

Glucophage

(b) Would you predict Glucophage to be soluble or insoluble in water? Soluble or insoluble in blood plasma? Would you predict it to be soluble or insoluble in diethyl ether? In dichloromethane? Explain your reasoning.

As the hydrochloride salt of an amine, Glucophage is ionic and is soluble in water. It will also be soluble in blood plasma; the pK_a of the guanidinium group is much higher than the pH of blood (7.4), so the guanidinium nitrogen will be protonated and carry a positive charge. Because Glucophage is ionic, it is insoluble in solvents of much lower polarity, such as diethyl ether or dichloromethane.

Synthesis

10.36 4-Aminophenol is a building block in the synthesis of the analgesic acetaminophen. Show how this building block can be synthesized in two steps from phenol (in Chapter 15, we will see how to complete the synthesis of acetaminophen):

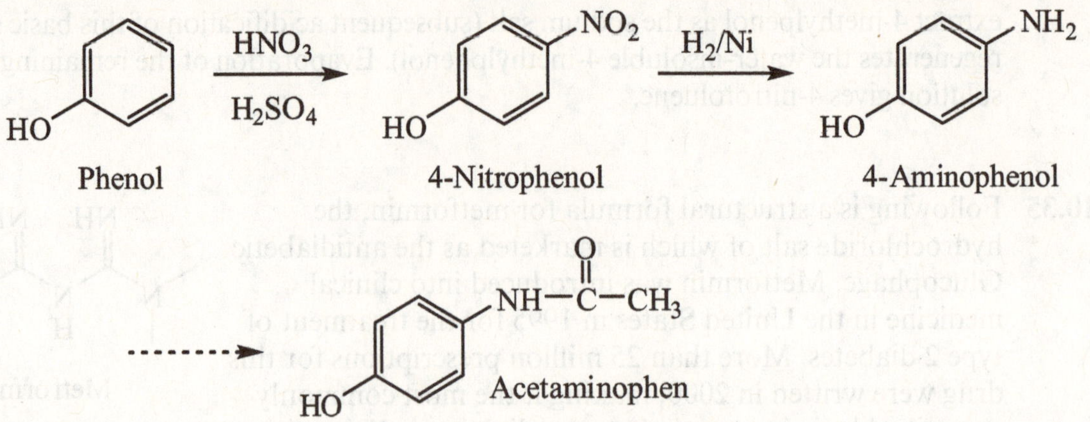

Phenol → 4-Nitrophenol → 4-Aminophenol → Acetaminophen

10.37 4-Aminobenzoic acid is a building block in the synthesis of the topical anesthetic benzocaine. Show how this building block can be synthesized in three steps from toluene (in Chapter 14, we will see how to complete the synthesis of benzocaine):

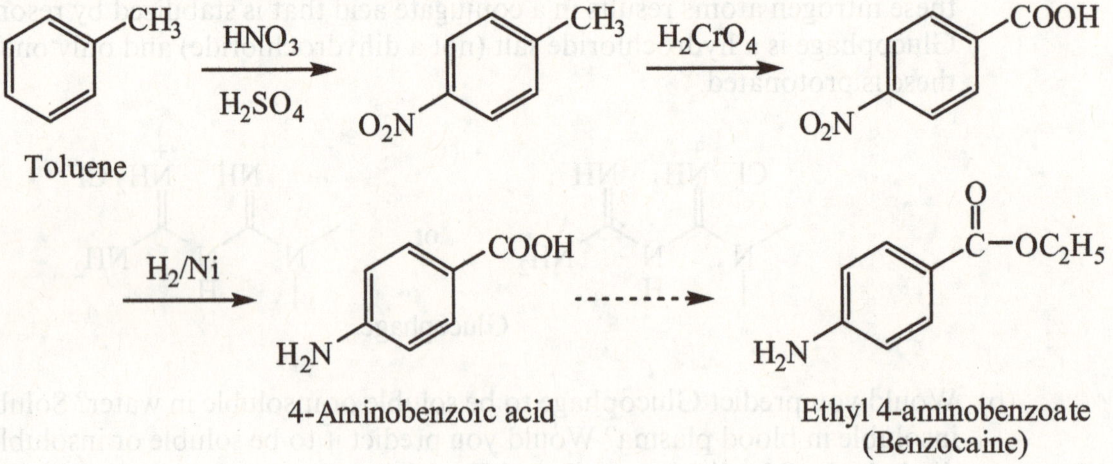

Toluene

4-Aminobenzoic acid

Ethyl 4-aminobenzoate
(Benzocaine)

10.38 The compound 4-amino-5-nitrosalicylic acid is one of the building blocks needed for the synthesis of propoxycaine, one of the family of "caine" anesthetics. Some other members of this family of local anesthetics are procaine (Novocaine), lidocaine (Xylocaine), and mepivicaine (Carbocaine). 4-Amino-5-nitrosalicylic acid is synthesized from salicylic acid in three steps. Show reagents that will bring about the synthesis of 4-amino-5-nitrosalicylic acid.

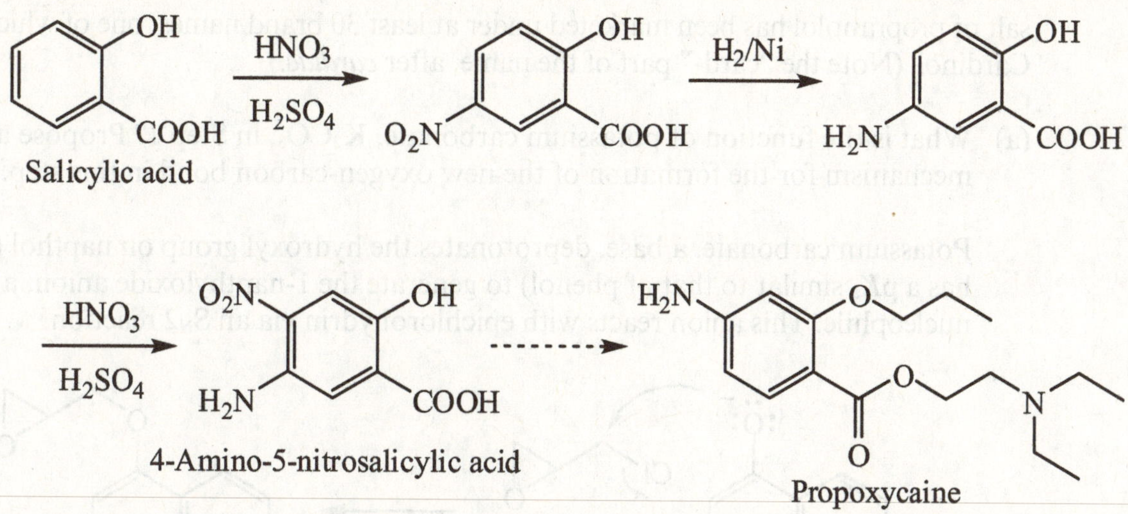

Salicylic acid

$\xrightarrow[\text{H}_2\text{SO}_4]{\text{HNO}_3}$

$\xrightarrow{\text{H}_2/\text{Ni}}$

$\xrightarrow[\text{H}_2\text{SO}_4]{\text{HNO}_3}$

4-Amino-5-nitrosalicylic acid

- - - - - →

Propoxycaine

10.39 A second building block for the synthesis of propoxycaine is 2-diethylaminoethanol. Show how this compound can be prepared from ethylene oxide and diethylamine.

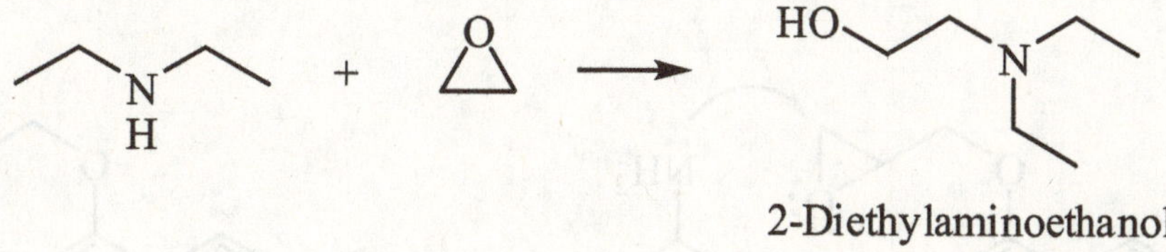

2-Diethylaminoethanol

10.40 Following is a two-step synthesis of the antihypertensive drug propranolol, a so-called beta blocker with vasodilating action:

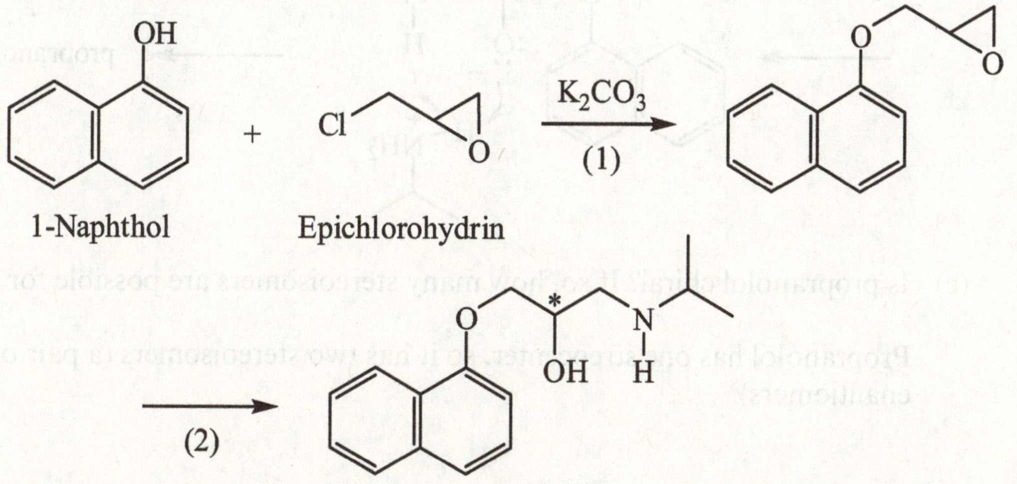

1-Naphthol   Epichlorohydrin

$\xrightarrow[(1)]{\text{K}_2\text{CO}_3}$

$\xrightarrow{(2)}$

Propranolol and other beta blockers have received enormous clinical attention because of their effectiveness in treating hypertension (high blood pressure), migraine headaches, glaucoma, ischemic heart disease, and certain cardiac arrhythmias. The hydrochloride salt of propranolol has been marketed under at least 30 brand names, one of which is Cardinol. (Note the "card-" part of the name, after *cardiac*.)

(a) What is the function of potassium carbonate, K_2CO_3, in Step 1? Propose a mechanism for the formation of the new oxygen-carbon bond in this step.

Potassium carbonate, a base, deprotonates the hydroxyl group on napthol (which has a pK_a similar to that of phenol) to generate the 1-napthyloxide anion, a good nucleophile. This anion reacts with epichlorohydrin via an S_N2 reaction.

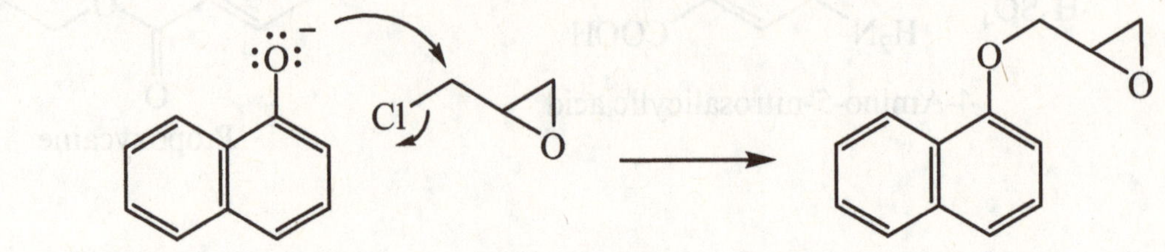

(b) Name the amine used to bring about Step 2, and propose a mechanism for this step.

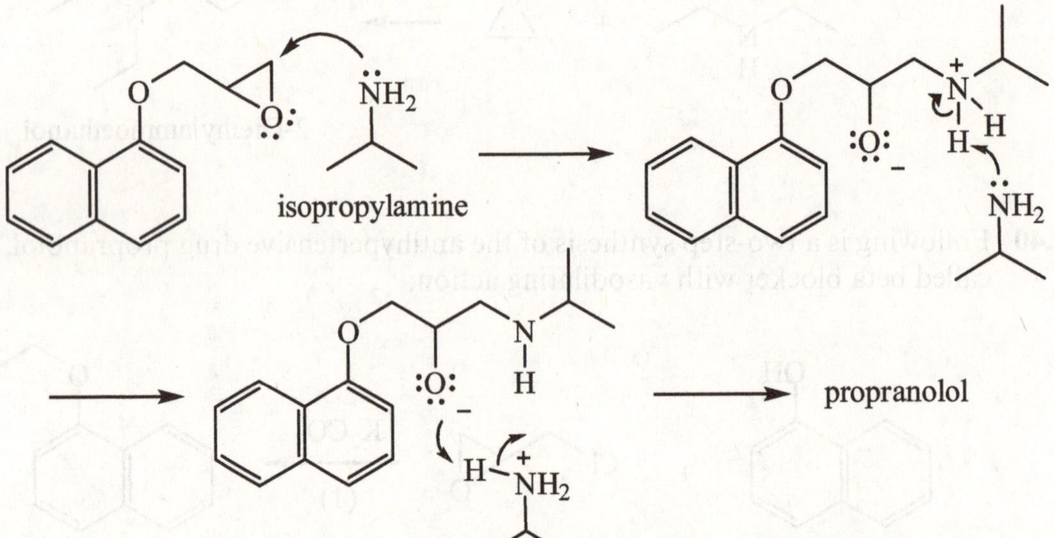

(c) Is propranolol chiral? If so, how many stereoisomers are possible for it?

Propranolol has one streocenter, so it has two stereoisomers (a pair of enantiomers).

10.41 The compound 4-ethoxyaniline, a building block of the over-the-counter analgesic phenacetin, is synthesized in three steps from phenol. Show reagents for each step in the synthesis of 4-ethoxyaniline. (In Chapter 14, we will see how to complete this synthesis.)

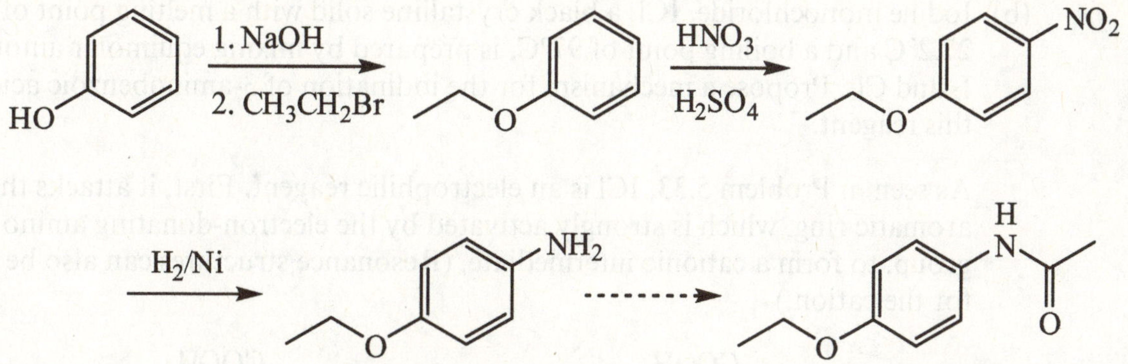

4-Ethoxyaniline   Phenacetin

10.42 Radiopaque imaging agents are substances administered either orally or intravenously that absorb X rays more strongly than body material does. One of the best known of these agents is barium sulfate, the key ingredient in the "barium cocktail" used for imaging of the gastrointestinal tract. Among other X-ray imaging agents are the so-called triiodoaromatics. You can get some idea of the kinds of imaging for which they are used from the following selection of trade names: Angiografin, Gastrografin, Cardiografin, Cholografin, Renografin, and Urografin. The most common of the triiodoaromatics are derivatives of these three triiodobenzenecarboxylic acids:

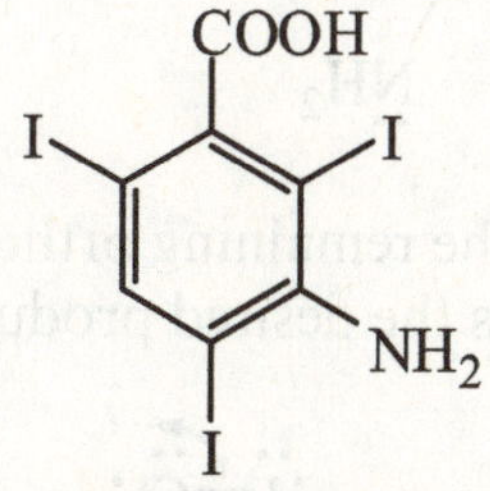

3-Amino-2,4,6-   3,5-Diamino-2,4,6-   5-Amino-2,4,6-
triiodobenzoic acid   triiodobenzoic acid   triiodoisophthalic acid

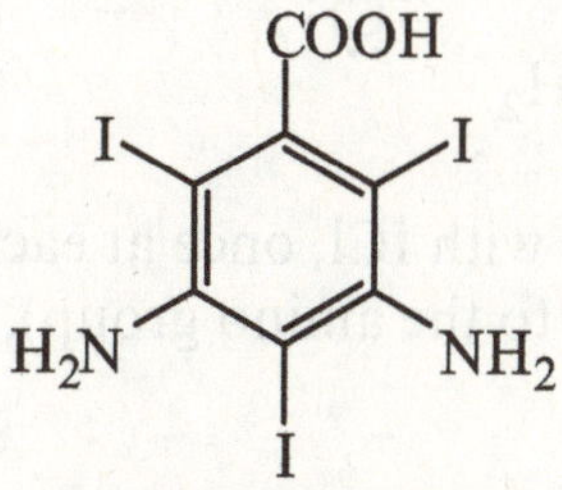

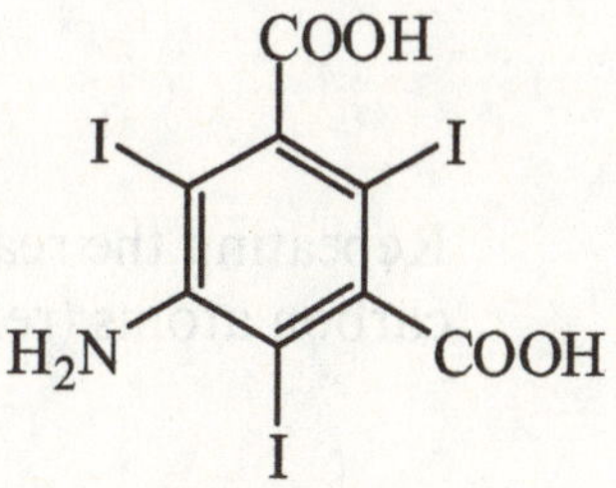

3-Amino-2,4,6-triiodobenzoic acid is synthesized from benzoic acid in three steps:

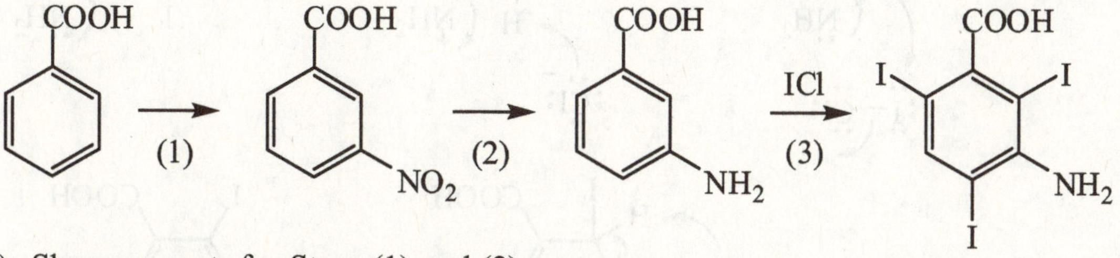

(a) Show reagents for Steps (1) and (2).

Step (1) is a nitration reaction and can be performed using HNO_3/H_2SO_4. Step (2) reduces the nitro group to an amino group and can be performed using H_2/Ni.

(b) Iodine monochloride, ICl, a black crystalline solid with a melting point of
27.2°C and a boiling point of 97°C, is prepared by mixing equimolar amounts of
I₂ and Cl₂. Propose a mechanism for the iodination of 3-aminobenzoic acid by
this reagent.

As seen in Problem 5.33, ICl is an electrophilic reagent. First, it attacks the
aromatic ring, which is strongly activated by the electron-donating amino
group, to form a cationic intermediate. (Resonance structures can also be drawn
for the cation.)

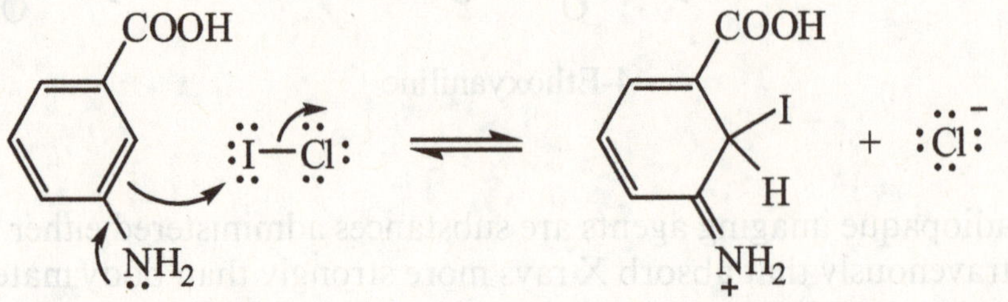

Deprotonation regenerates the aromatic ring, completing the electrophilic
aromatic substitution reaction.

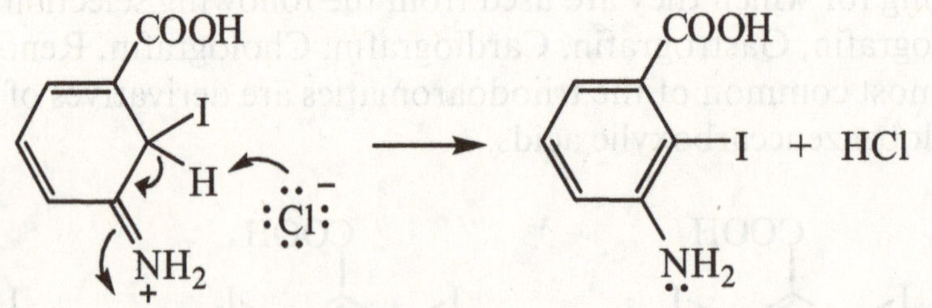

Repeating the reaction with ICl, once at each of the remaining ortho and para
carbon atoms (relative to the amino group), yields the desired product.

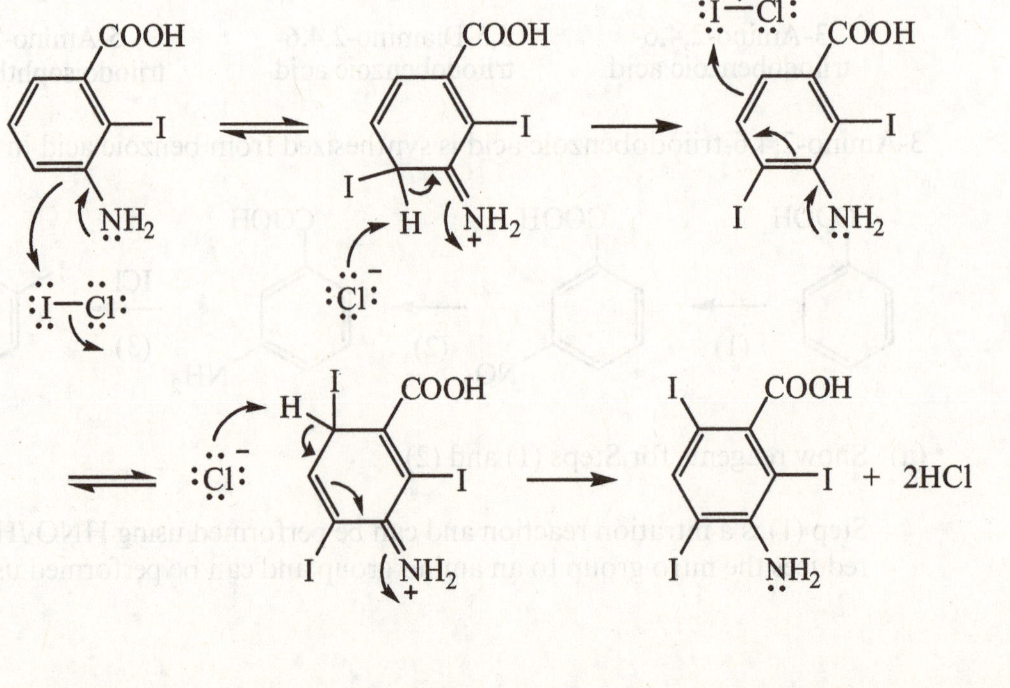

(c) Show how to prepare 3,5-diamino-2,4,6-triiodobenzoic acid from benzoic acid.

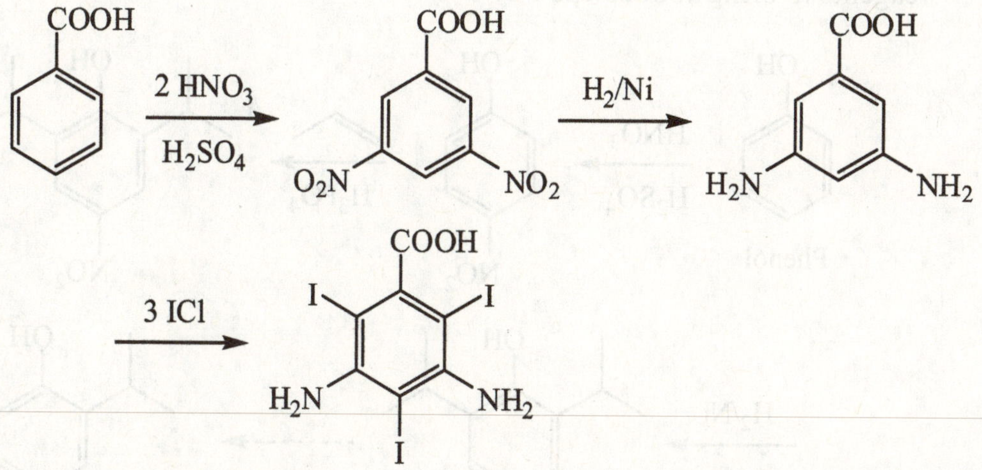

(d) Show how to prepare 5-amino-2,4,6-triiodoisophthalic acid from isophthalic acid (1,3-benzenedicarboxylic acid).

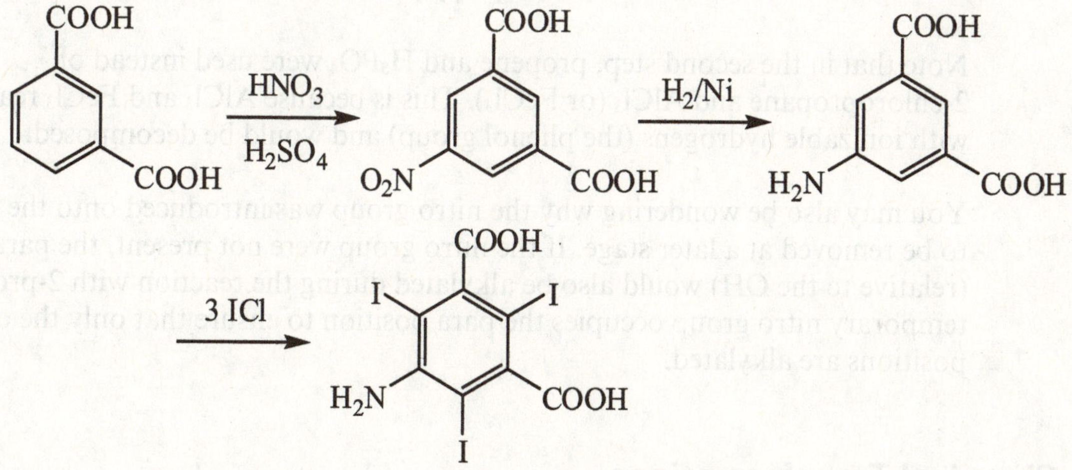

10.43 The intravenous anesthetic propofol is synthesized in four steps from phenol. Show reagents to bring about steps 1-3.

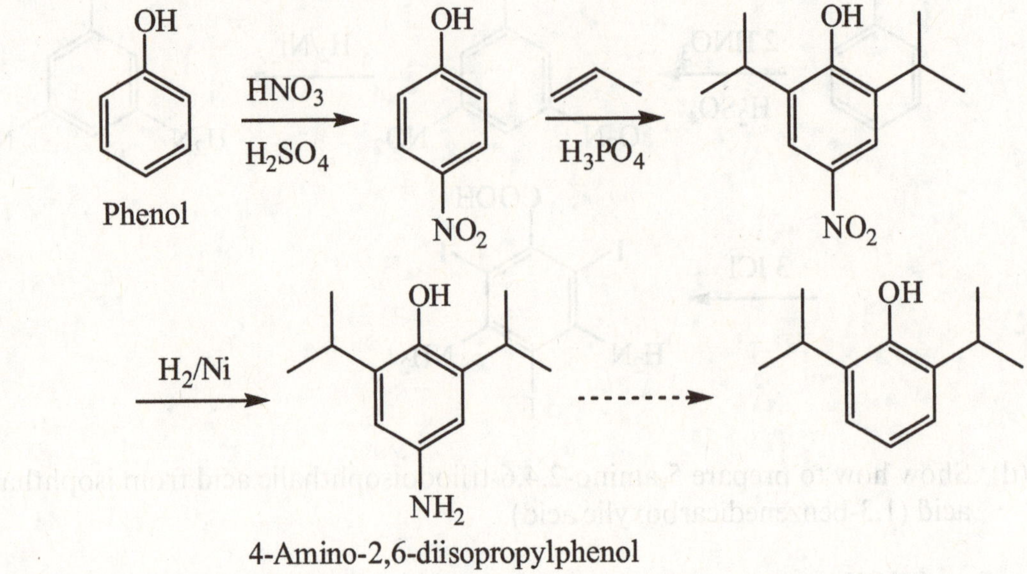

Phenol

4-Amino-2,6-diisopropylphenol

Note that in the second step, propene and H_3PO_4 were used instead of 2-chloropropane and $AlCl_3$ (or $FeCl_3$). This is because $AlCl_3$ and $FeCl_3$ react with ionizable hydrogens (the phenol group) and would be decomposed.

You may also be wondering why the nitro group was introduced onto the ring only to be removed at a later stage. If the nitro group were not present, the para position (relative to the OH) would also be alkylated during the reaction with 2-propene. The temporary nitro group occupies the para position to ensure that only the ortho positions are alkylated.

Chemical Transformations

10.44 Test your cumulative knowledge of the reactions learned thus far by completing the following chemical transformations. *Note: some will require more than one step.*

(a)

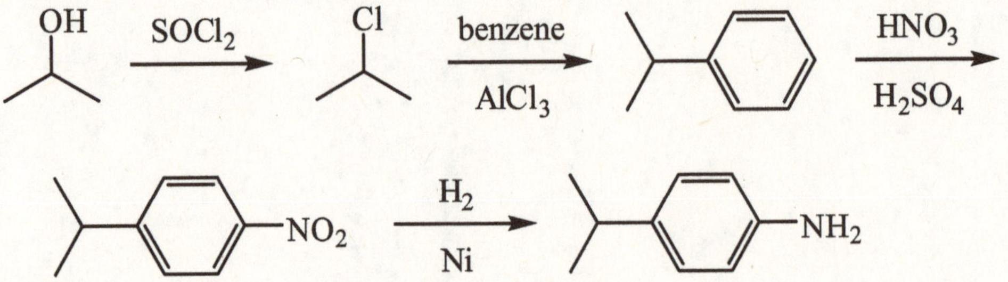

(b)

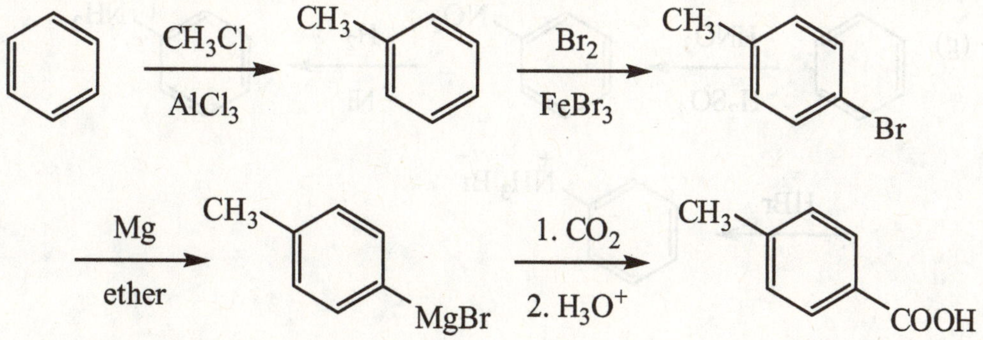

(c)

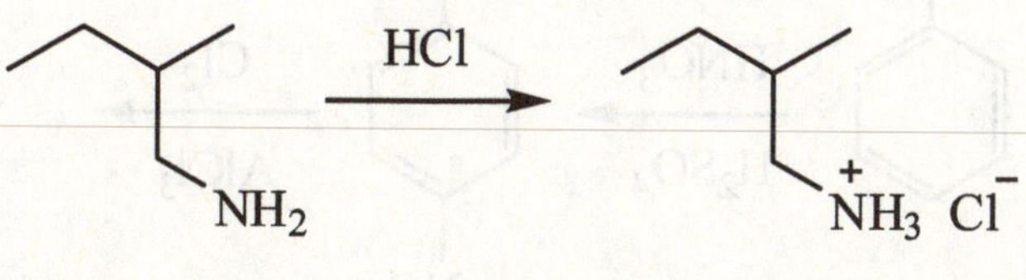

(d)

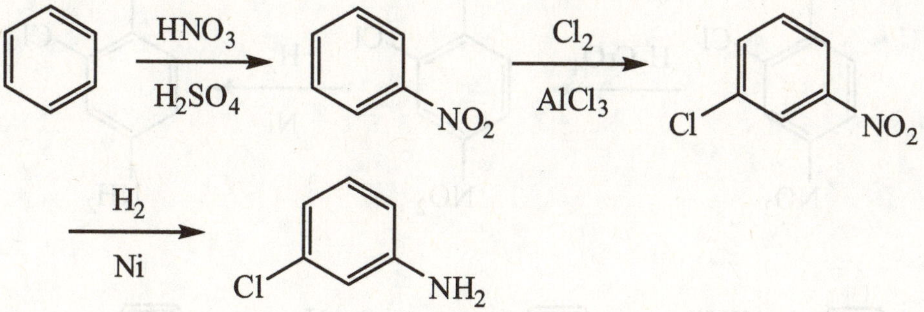

(e)

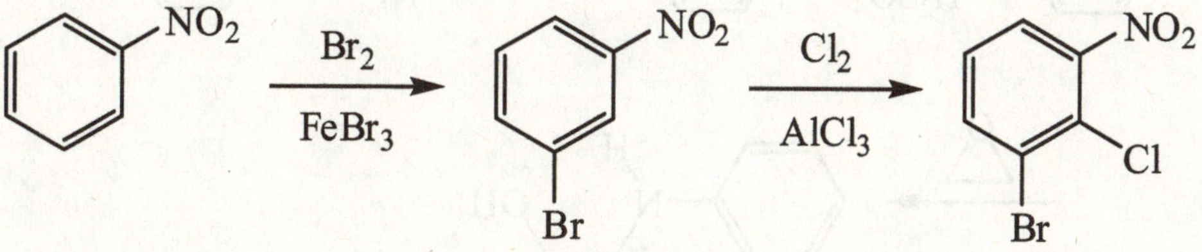

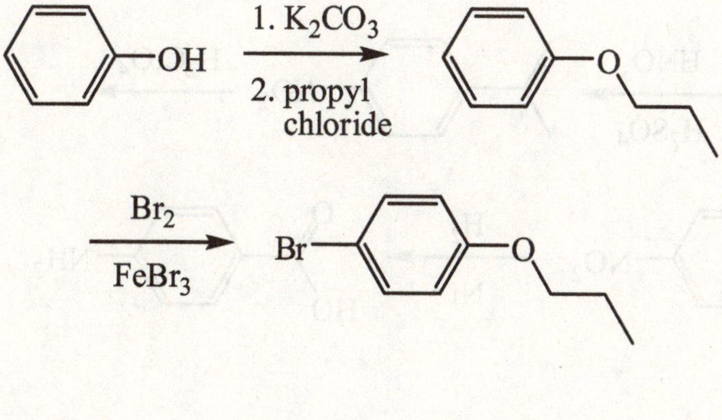

(f)

(g) Benzene $\xrightarrow[H_2SO_4]{HNO_3}$ nitrobenzene (NO_2) $\xrightarrow[Ni]{H_2}$ aniline (NH_2)

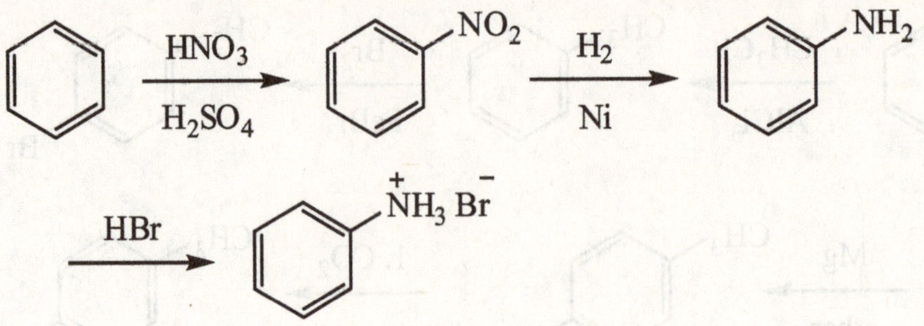

$\xrightarrow{HBr}$ anilinium bromide ($\overset{+}{N}H_3 \, Br^-$)

(h) Benzene $\xrightarrow[AlCl_3]{RCl}$ R-benzene (R) $\xrightarrow[H_2SO_4]{HNO_3}$ R, NO_2-benzene (R, NO_2) $\xrightarrow[AlCl_3]{Cl_2}$

R, Cl, NO_2-benzene $\xrightarrow{H_2CrO_4}$ COOH, Cl, NO_2-benzene $\xrightarrow[Ni]{H_2}$ COOH, Cl, NH_2-benzene

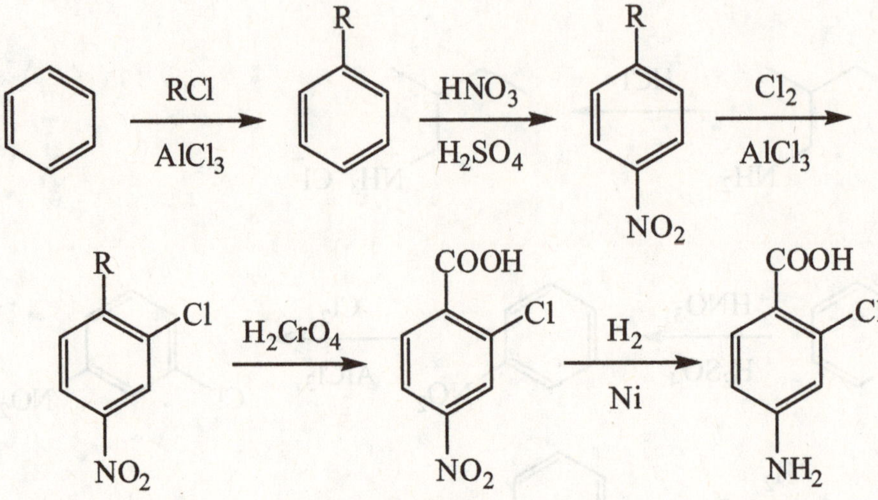

(i) Benzene $\xrightarrow[H_2SO_4]{HNO_3}$ NO_2-benzene $\xrightarrow[Ni]{H_2}$ NH_2-benzene

$\xrightarrow{\text{(epoxide, O)}}$ $C_6H_5-\overset{H}{N}-CH_2CH_2OH$ (N, OH)

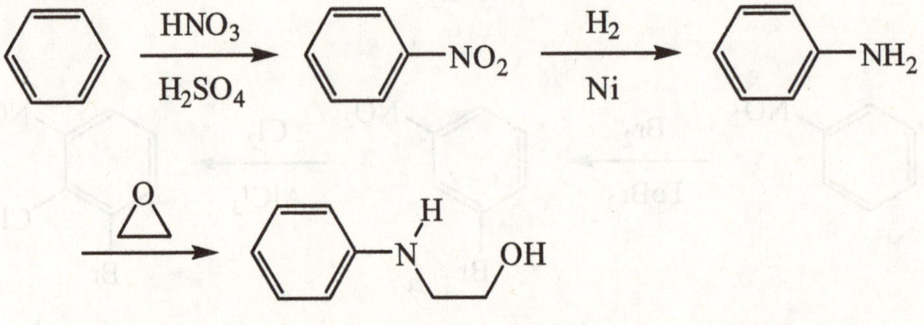

(j) isopropylbenzene $\xrightarrow[H_2SO_4]{HNO_3}$ 4-isopropyl-NO_2-benzene $\xrightarrow{H_2CrO_4}$

HOOC-C_6H_4-NO_2 $\xrightarrow[Ni]{H_2}$ HOOC-C_6H_4-NH_2

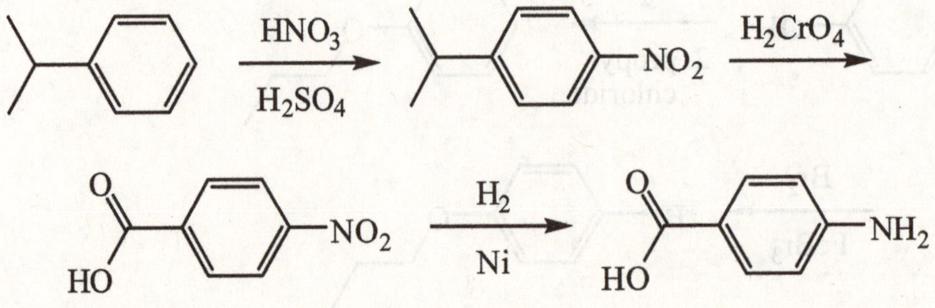

(k)

$$\underset{}{\bigcirc} \xrightarrow[\text{H}_2\text{SO}_4]{\text{HNO}_3} \underset{\text{NO}_2}{\bigcirc} \xrightarrow[\text{AlCl}_3]{\text{Cl}_2} \underset{\text{NO}_2}{\overset{\text{Cl}}{\bigcirc}} \xrightarrow[\text{Ni}]{\text{H}_2}$$

$$\underset{\text{NH}_2}{\overset{\text{Cl}}{\bigcirc}} \xrightarrow{\text{CH}_3\text{Cl}} \underset{+\text{N(CH}_3)_3 \ \ \text{Cl}^-}{\overset{\text{Cl}}{\bigcirc}}$$

(l)

$$\underset{}{\bigcirc} \xrightarrow[\text{FeBr}_3]{\text{Br}_2} \text{Br}{-}\underset{}{\bigcirc} \xrightarrow[\text{H}_2\text{SO}_4]{\text{HNO}_3} \text{Br}{-}\underset{}{\bigcirc}{-}\text{NO}_2$$

$$\xrightarrow[\text{Ni}]{\text{H}_2} \text{Br}{-}\underset{}{\bigcirc}{-}\text{NH}_2 \xrightarrow{\overset{O}{\triangle}} \text{Br}{-}\underset{}{\bigcirc}{-}\text{NH}\overset{\text{OH}}{\diagdown}$$

Looking Ahead

10.45 State the hybridization of the nitrogen atom in each of the following compounds:

(a) pyridine with N

(b) pyrrole with N–H

(c) aniline with NH$_2$

(d) N,N-dimethylacetamide, H$_3$C–C(=O)–N(CH$_3$)–CH$_3$

All of the nitrogen atoms are sp^2. The nitrogen atom in (a) has three regions of electron density; note that the lone pair is not part of the aromatic π system. The nitrogen atom in (b) is sp^2 because the lone pair is located in a $2p$ orbital, thus allowing the compound to be aromatic. The nitrogen atom in (c) is sp^2 to allow the lone pair to be located in a $2p$ orbital, which interacts with the π system of the ring. The nitrogen atom in (d) is sp^2 because of the resonance stabilization of the amide bond; this structural feature explains the planarity of the peptide bond in proteins.

$$\text{H}_3\text{C}{-}\overset{\overset{\overset{O}{\|}}{}}{\underset{\underset{\text{CH}_3}{|}}{\text{N}}}{-}\text{CH}_3 \longleftrightarrow \text{H}_3\text{C}{-}\overset{\overset{\overset{:\ddot{O}:^-}{\|}}{}}{\underset{\underset{\text{CH}_3}{|}}{\overset{+}{\text{N}}}}{-}\text{CH}_3$$

10.46 Amines can act as nucleophiles. For each of the following molecules, circle the most likely atom that would be attacked by the nitrogen of an amine:

(a) 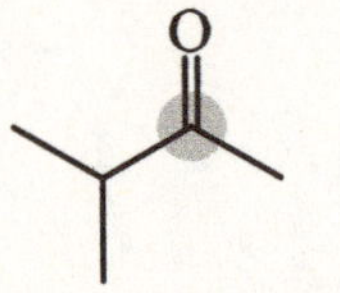   (b) 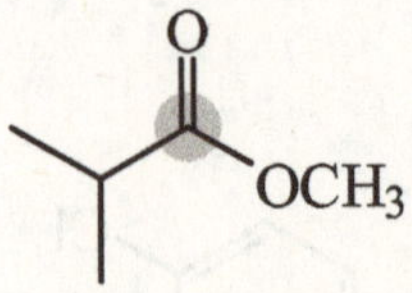   (c)

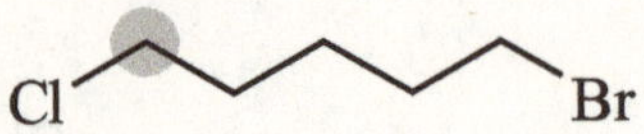

In each of these molecules, the most likely atom that would be attacked by the nitrogen of an amine is the one that has the greatest δ+ charge (the most electrophilic). However, realize that according to kinetic molecular theory, not every attack (collision) is successful. Although the indicated carbon in (c) is more likely to be attacked than is the carbon atom bearing the Br, the latter is more likely to yield a successful collision due to −Br being a better leaving group than −Cl.

10.47 Draw a Lewis structure for a molecule with the formula C_3H_7N that does not contain a ring or an alkene (a carbon-carbon double bond).

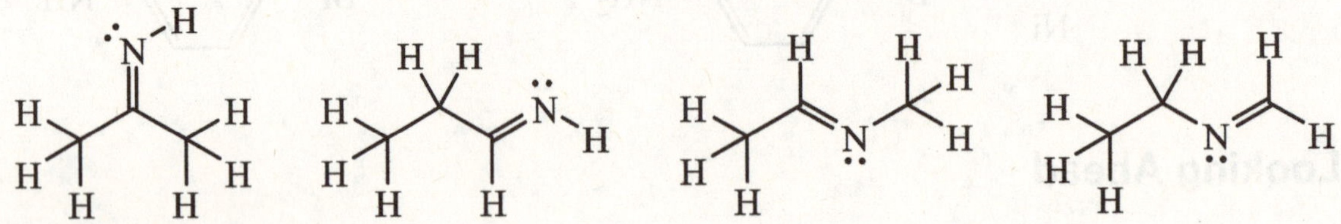

10.48 Rank the following leaving groups in order from best to worst:

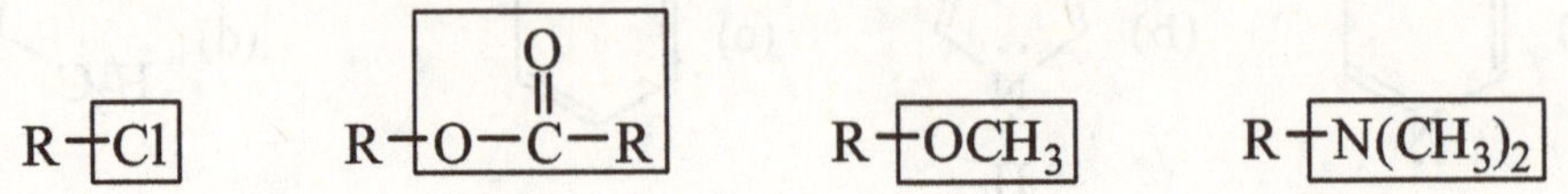

R—Cl   R—O—C—R   R—OCH₃   R—N(CH₃)₂

leaving   Cl⁻   ⁻O—C—R   ⁻OCH₃   ⁻N(CH₃)₂
groups:

best leaving group   worst leaving group
(most stable; weakest base)   (least stable; strongest base)

Recall from Chapter 7 that the best leaving groups are those that are most stable. Leaving group ability is therefore inversely related to the strength of the leaving group as a base.

Group Learning Activities

Solutions are only provided for activities that are not open-ended.

10.49 Discuss why ⁻NH₂ is a stronger base than ⁻OH. Are both bases strong enough to quantitatively abstract the hydrogen from a terminal alkyne? Why or why not?

The amide ion is a stronger base than the hydroxide ion due to the effect of electronegativity. The negatively charged nitrogen is less stable than the negatively charged oxygen. The respective pK_a values for ammonia and water are 38 and 16.

Only the amide ion is strong enough to quantitatively deprotonate a terminal alkyne. The amine ion is a stronger base than the acetylide ion, but the hydroxide ion is a weaker base than the acetylide ion. The pK_a of an alkyne is approximately 25.

10.51 Compare the basicity of amide nitrogens to that of amines.

As seen in Problem 10.45, the lone pair of the nitrogen atom of an amide is delocalized and stabilized by resonance. Amide nitrogens are therefore much less basic than the nitrogens of amines.

Putting It Together

1. Arrange the following amines from lowest to highest boiling point.

$$\underset{\textbf{A}}{\underset{CH_3}{\overset{CH_3CH_2}{\diagdown}}N-H} \qquad \underset{\textbf{B}}{CH_3CH_2CH_2-NH_2} \qquad \underset{\textbf{C}}{\underset{CH_3}{\overset{CH_3}{\diagdown}}N-CH_3}$$

(a) **A, B, C**   (b) **C, B, A**   (c) **B, C, A**   (d) **B, A, C**   (e) **C, A, B**

The boiling point of a compound is influenced by the strength of the intermolecular forces. Compoound **C** does not contain an N−H bond and therefore cannot be a hydrogen-bond donor. Between **A** and **B**, the latter has two N−H bonds and is unbranched (more surface area), so it has a higher boiling point than **A**.

2.   Which of the following statements is true regarding the following two molecules?

(a) Both **A** and **B** are aromatic.

(b) Both **A** and **B** are aliphatic amines.

(c) The nitrogen atoms in **A** and **B** are both sp^3-hybridized.

(d) B is more basic than **A**.

(e) Both **A** and **B** are planar molecules.

Compound **A** is aromatic, and it is a heterocyclic aromatic (not aliphatic) amine. All of its ring atoms are sp^2-hybridized, so it is planar. The nitrogen atom of **B** is more basic than that of **A** because it is sp^3-hybridized.

3.   Which series of reagents can be used to achieve the following transformation?

(a) 1. HBr   2. H_2SO_4

(b) 1. H_2SO_4, H_2O 2. PCC

(c) 1. HCl   2. $SOCl_2$

(d) 1. H_3PO_4, H_2O 2. H_2CrO_4

(e) More than one of these will achieve this transformation

The transformation involves the conversion of an alkene to a ketone. A intermediate common to these two functional groups would be an alcohol. Thus, the conversion is a hydration followed by an oxidation. Both (b) and (d) will work.

4.   Arrange the following from strongest to weakest base.

(a) **A, B, C**   (b) **B, C, A**   (c)**C, A, B**   (d) **A, C, B**   (e) **B, A, C**

The three anilines differ in the nature of the para substituent. Compound **B** contains an inductively electron-withdrawing cyano group, which increases the stability of the lone pair on the nitrogen atom and decreases base strength relative to **A**. Compound **C** contains a methyl group, which is inductively electron-donating and destabilizes the lone pair on the nitrogen atom, thus increasing its basicity.

5. How many products are possible from the following elimination reaction?

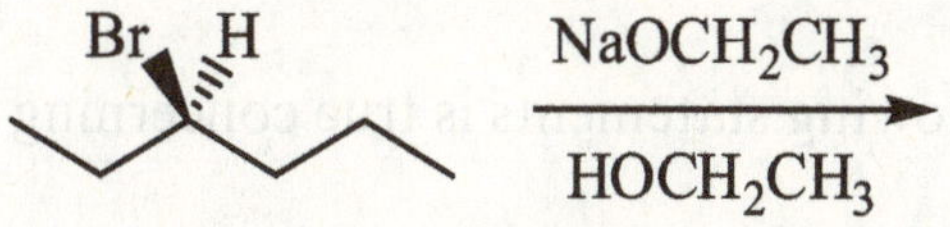

(a) one   (b) two   (c) three   **(d)** four   (e) six

Don't forget about *cis-trans* isomerism; *cis-trans* isomers of each of 2-hexene and 3-hexene are possible.

6. Which series of reagents can be used to achieve the following transformation?

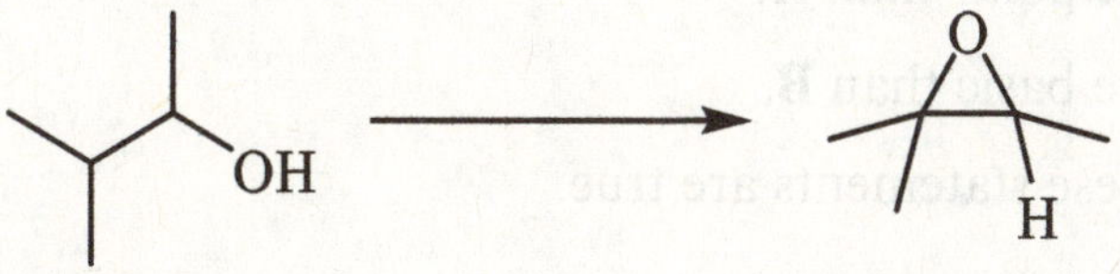

(a) 1. HCl   2. RCO_3H

(b) 1. $SOCl_2$ 2. RCO_3H

(c) 1. Na   2. RCO_3H

(d) 1. H_3PO_4 2. RCO_3H

(e) 1. H_2CrO_4 2. RCO_3H

In the transformation, an alcohol is converted to an epoxide. An intermediate common these two compounds would be an alkene. The transformation therefore involves a dehydration followed by an epoxidation.

7. Consider the following situation: An ether solution containing phenol and a neutral compound is extracted with 30% sodium bicarbonate. Next the ether solution is extracted with 30% NaOH. Finally, the ether solution is extracted with distilled water. Which solution contains the phenol?

(a) The 30% sodium bicarbonate solution.

(b) The 30% NaOH solution.

(c) The ether.

(d) The distilled water.

(e) Not enough information to determine.

Phenols are weakly acidic ($pK_a = 10$). The pK_a of carbonic acid, which is the conjugate acid of the bicarbonate ion, is approximately 6.4. Accordingly, bicarbonate is too weak of a base to deprotonate phenol. However, sodium hydroxide is strong enough of a base (pK_a of water = 16) to deprotonate phenol.

8.   Which of the following statements is true concerning the following two molecules?

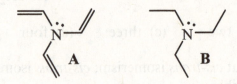

(a) Both are aromatic.

(b) Only one molecule is an amine.

(c) B is more polar than A.

(d) **A** is more basic than **B**.

(e) All of these statements are true.

The two compounds differ by the three double bonds in **A**. In this compound, the lone pair on the nitrogen atom is stabilized by resonance, thereby reducing its basicity. A direct consequence is the hybridization of the nitrogen atom of **A**: instead of being sp^3-hybridized, it is sp^2 and has a trigonal planar geometry. This geometry allows the bond dipole moments to cancel out, resulting in a nonpolar molecule. Recall that the most correct structure is best represented by the resonance hybrid.

9.   Which combination of reagents would be most likely to undergo an S_N2 reaction?

(c)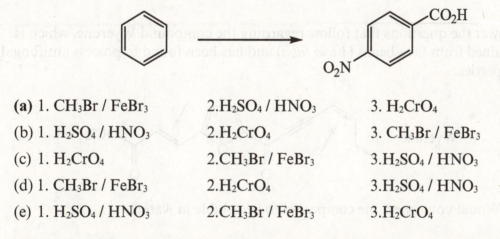

NaI
———→
DMSO

(d)

HI
———→
DMSO

(e)

HI
———→
DMSO

Recall that the ideal conditions for an S_N2 reaction are low steric hindrance, a good nucleophile, a good leaving group, and a polar, aprotic solvent. All compounds are secondary haloalkanes, but (c), (d), and (e) are less hindered. Comparing these three compounds, (c) and (d) have a better leaving group (bromide) than does (e). The difference is now down to the source of the iodide ion: it originates from NaI in (c) and HI in (d). In a nonaqueous solvent such as DMSO, HI is actually not a very strong acid, and the amount of iodide that dissociates from HI is relatively small.

10. Which series of reagents can be used to achieve the following transformation?

(a) 1. CH_3Br / $FeBr_3$ 2. H_2SO_4 / HNO_3 3. H_2CrO_4

(b) 1. H_2SO_4 / HNO_3 2. H_2CrO_4 3. CH_3Br / $FeBr_3$

(c) 1. H_2CrO_4 2. CH_3Br / $FeBr_3$ 3. H_2SO_4 / HNO_3

(d) 1. CH_3Br / $FeBr_3$ 2. H_2CrO_4 3. H_2SO_4 / HNO_3

(e) 1. H_2SO_4 / HNO_3 2. CH_3Br / $FeBr_3$ 3. H_2CrO_4

The two substituents in the product are para to each other, so at some point during the transformation, a para director must be involved. Friedel-Crafts methylation of benzene gives toluene, which contains a para director. The nitration of toluene forms *p*-nitrotoluene, which can then be oxidized to the carboxylic acid. If the nitration step were performed either first (e) or last (d), *m*-nitrobenzoic acid would be formed. Recall that nitro and carboxyl groups are meta directors.

11. Determine which aryl amine (**A** or **B**) is more basic and provide a rationale for your determination.

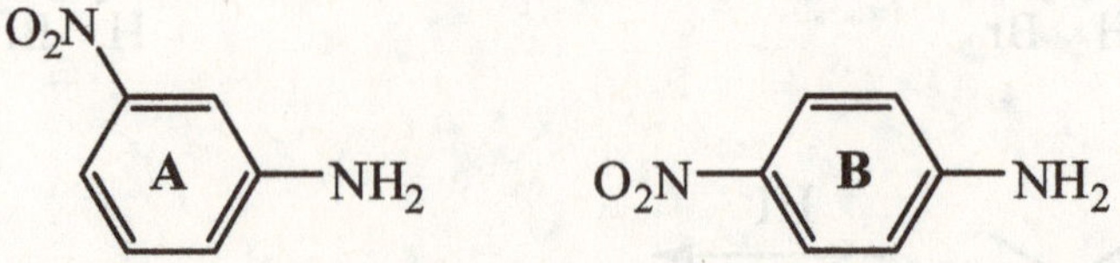

At a first glance, one may think that **B** is the less stable base, and hence the more basic compound, because the inductively withdrawing nitro group is furthest away from the lone pair of the amino group. However, we also need to take resonance effects into account; resonance structures that delocalize the lone pair of the amino group to the *p*-nitro group, but not the *m*-nitro group, can be drawn. The resonance effect more than compensates for the increased distance of the nitro group, and **B** is in fact the weaker base of the two. **A** is therefore more basic.

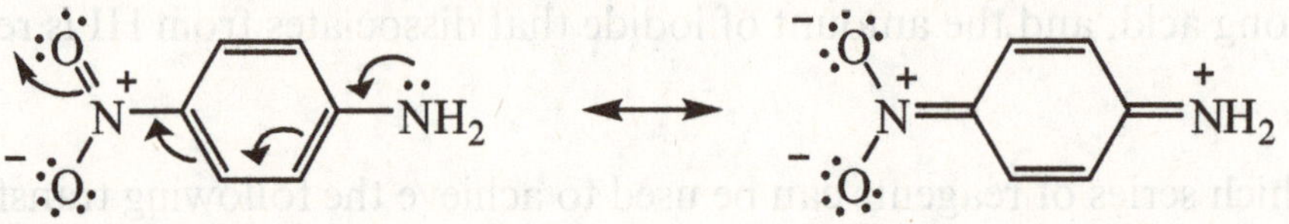

12. Answer the questions that follow regarding the compound Wyerone, which is obtained from fava beans (*Vicia faba*) and has been found to possess antifungal properties.

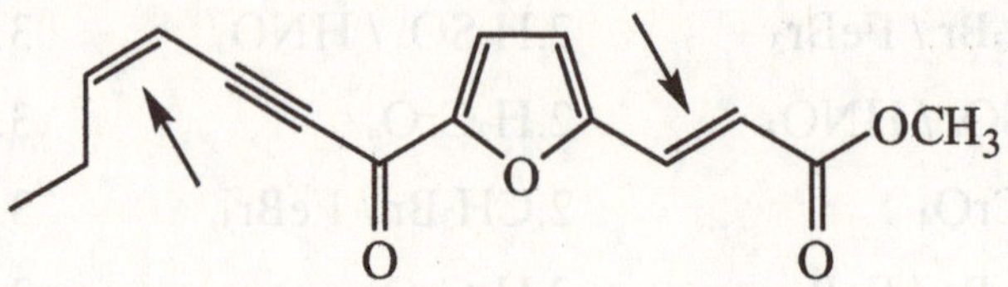

(a) Would you expect the compound to be soluble in water?

No. The compound is relatively nonpolar.

(b) How many stereoisomers exist for the compound shown?

Two double bonds exhibit *cis-trans* isomerism, and there is one chiral center. There are $2^3 = 8$ possible stereoisomers.

(c) Is the molecule chiral?

Yes. The chiral center is indicated in the structure.

(d) How many equivalents of Br₂ in CH₂Cl₂ would Wyerone be expected to react with?

Both alkenes and alkynes react with Br₂. However, the alkyne will react with two equivalents of bromine. Therefore, Wyerone is expected to react with five equivalents of bromine.

13. Provide IUPAC names for the following compounds.

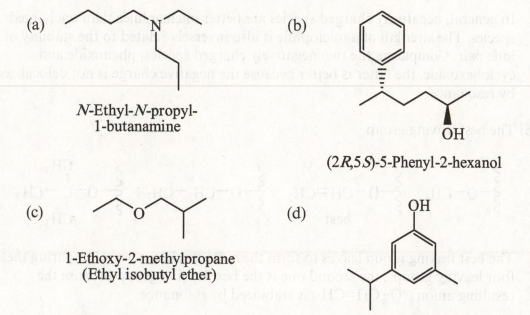

(a)

N-Ethyl-*N*-propyl-1-butanamine

(b)

(2*R*,5*S*)-5-Phenyl-2-hexanol

(c)

1-Ethoxy-2-methylpropane
(Ethyl isobutyl ether)

(d)

3-Isopropyl-5-methylphenol

14. Determine whether highlighted proton **A** or **B** is more acidic and provide a rationale for your selection.

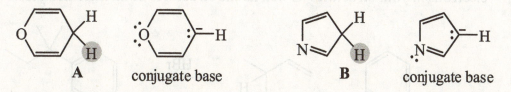

A   conjugate base   **B**   conjugate base

Recall that the strength of an acid is affected by the stability of its conjugate base. In this case, the conjugate base of **B** is aromatic, but that of **A** is not. As a result, proton **B** is more acidic than proton **A**.

15. Select the answer that best fits each description and provide an explanation for your decision.

(a) The best nucleophile

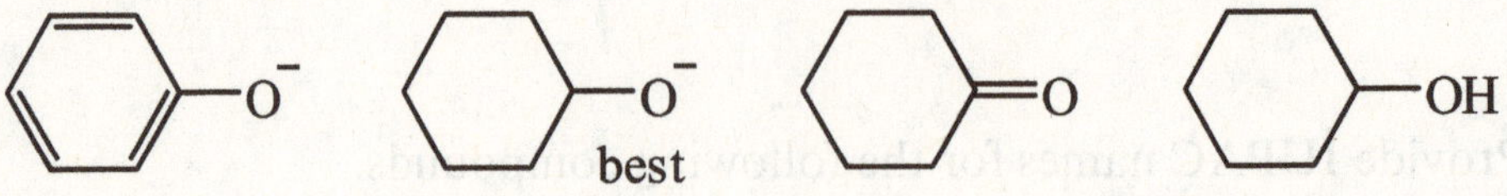

In general, negatively charged species are better nucleophiles than uncharged species. The strength of a nuclophile is also inversely related to the stability of the lone pair. Comparing the two negatively charged species, phenoxide and cyclohexoxide, the latter is better because the negative charge is not delocalized by resonance.

(b) The best leaving group

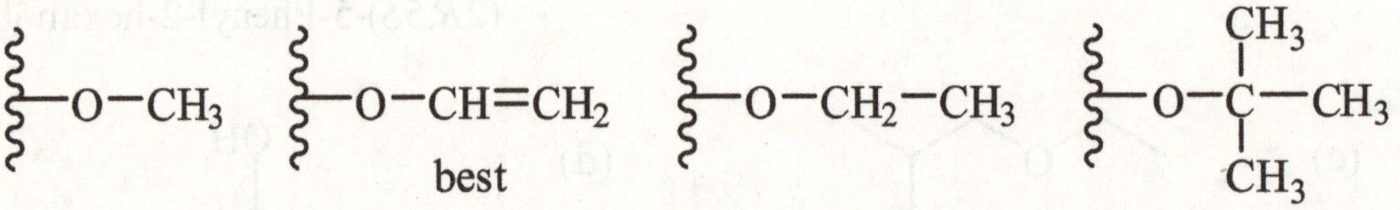

The best leaving group leaves to form the most stable species. Comparing these four leaving groups, the second one is the best leaving group because the resulting anion ($^-$O–CH=CH$_2$) is stabilized by resonance.

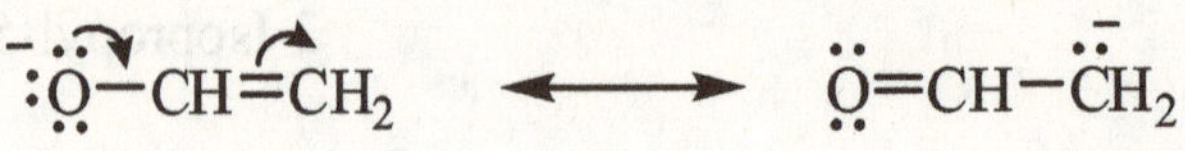

16. Provide a mechanism for the following reaction. Show all charges and lone pairs of electrons in your structures as well as the structures of all intermediates.

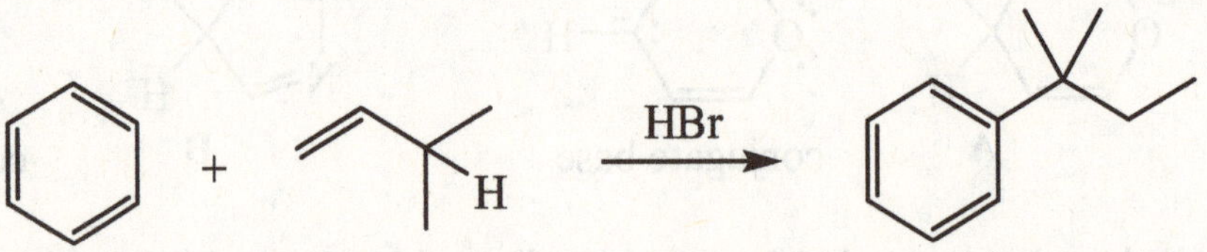

Close examination of the reaction reveals that this is an alkylation reaction involving a carbocation rearrangement – look closely at where the hydrogen atoms are found in the alkyl chain of the product.

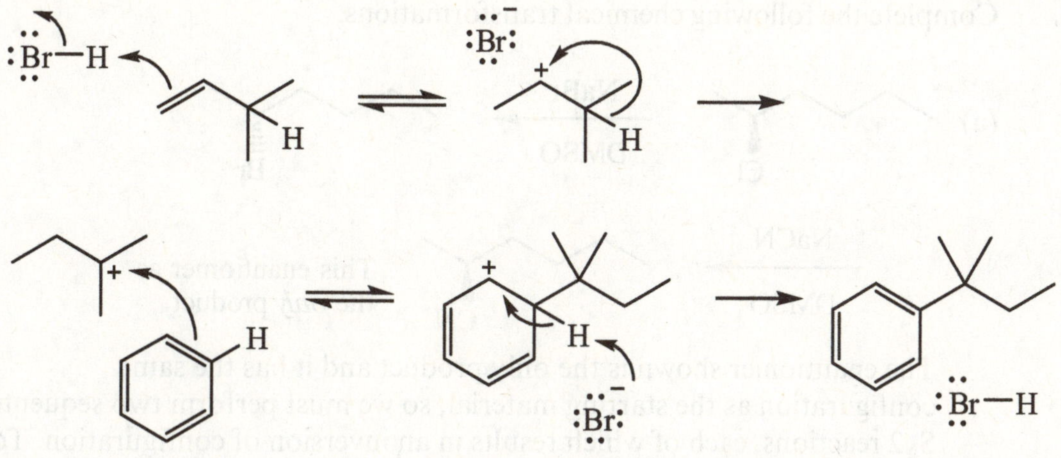

17. When the following nucleophilic substitution reaction was performed, the major product was found to possess the molecular formula $C_{13}H_{30}N$ rather than $C_5H_{13}N$, the formula of the desired product shown below. Provide the structure of the major product and explain why it is formed over the desired product.

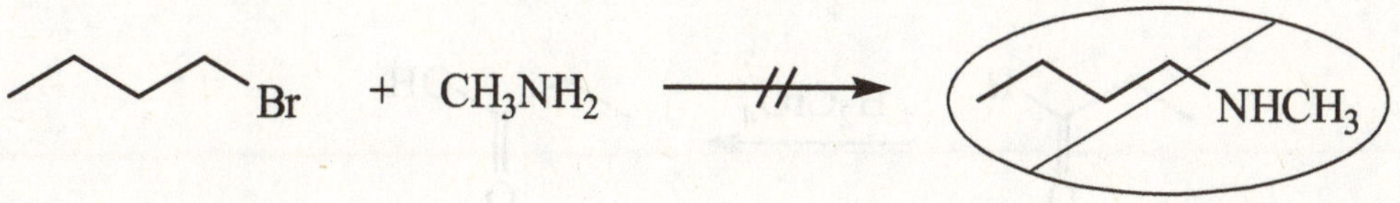

The molecular formula of the major product suggests that it has two more butyl groups than the desired product. That is, the desired product reacted with two more equivalents of 1-bromobutane. Yet, why is the desired product more reactive than methylamine? It is because alkyl groups are electron-donating, so amines with more alkyl groups are more basic and hence better nucleophiles.

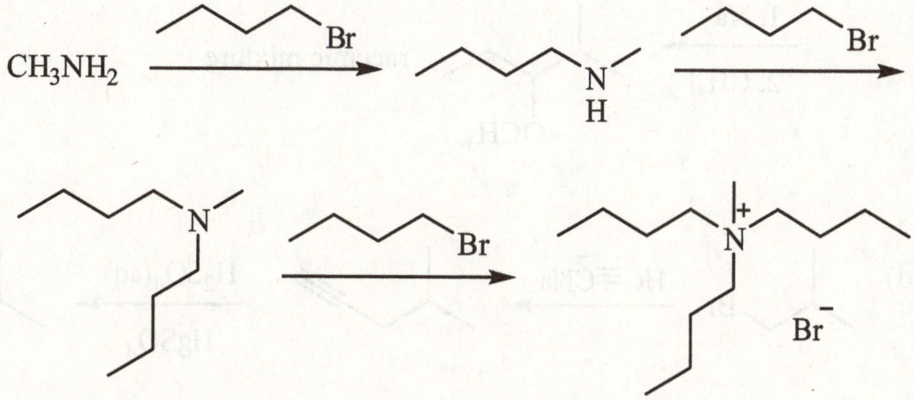

18. Complete the following chemical transformations.

(a)

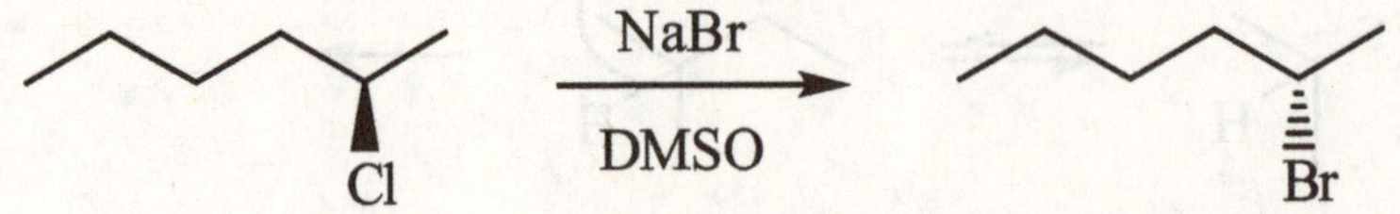

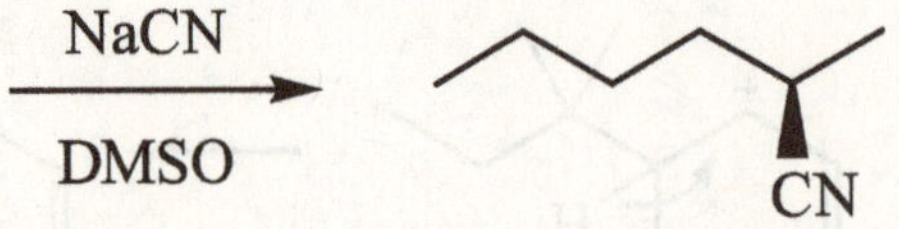

This enantiomer as the *only* product

The enantiomer shown is the only product and it has the same configuration as the starting material, so we must perform two sequential S_N2 reactions, each of which results in an inversion of configuration. To ensure the highest possible yield, the use of DMSO, a polar, aprotic solvent, encourages the reaction to proceed by S_N2 instead of S_N1.

(b)

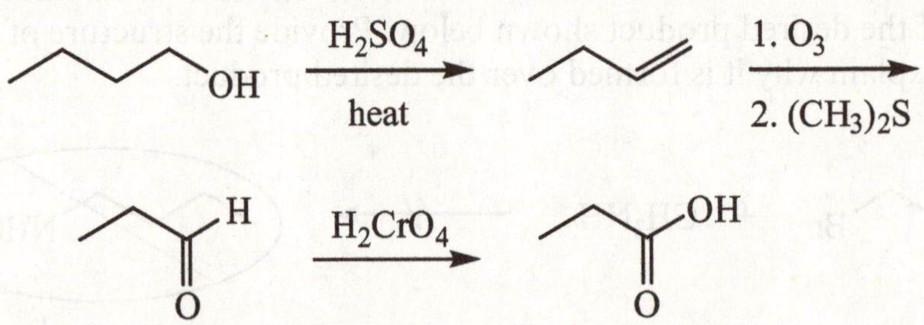

(c)

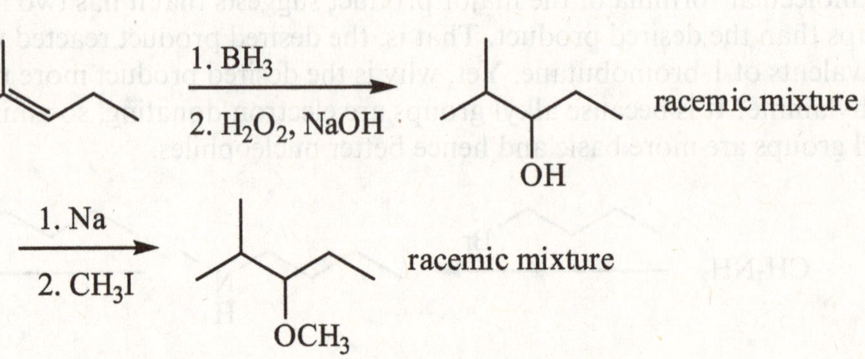

(d)

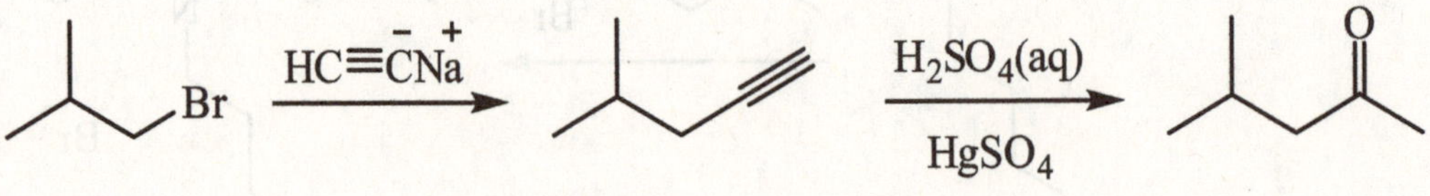

19.   Provide a mechanism for the following reaction. Show all charges and lone pairs of electrons in your structures as well as the structures of all intermediates.

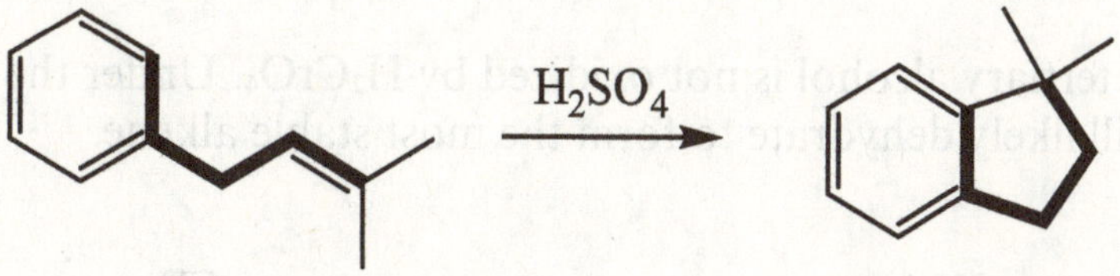

Like Problem 16, this reaction is also an alkylation reaction except that it is intramolecular and there is no carbocation rearrangement. The bonds involved in the formation of the five-membered ring are indicated in bold.

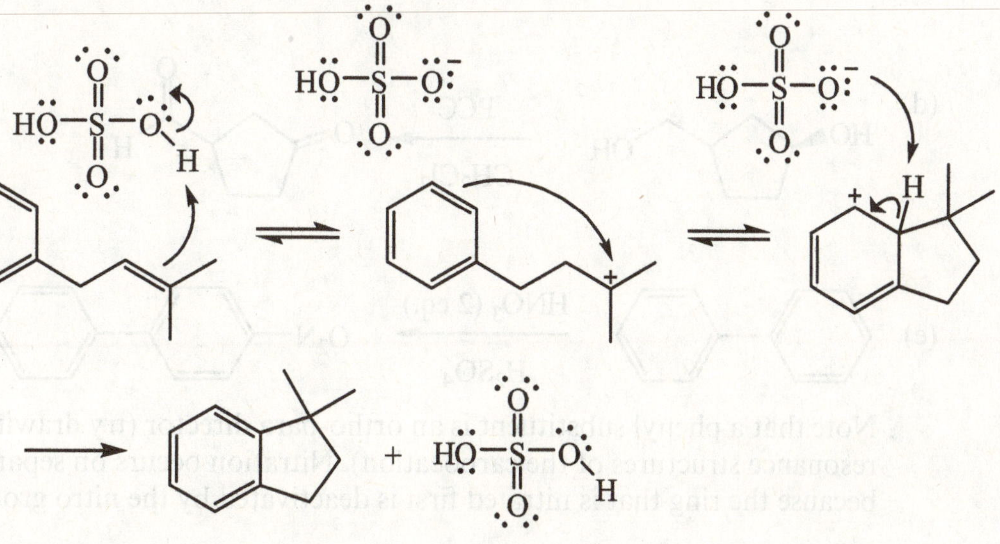

20.   Predict the major product or products of each of the following reactions. Be sure to consider stereochemistry in your answers.

(a)

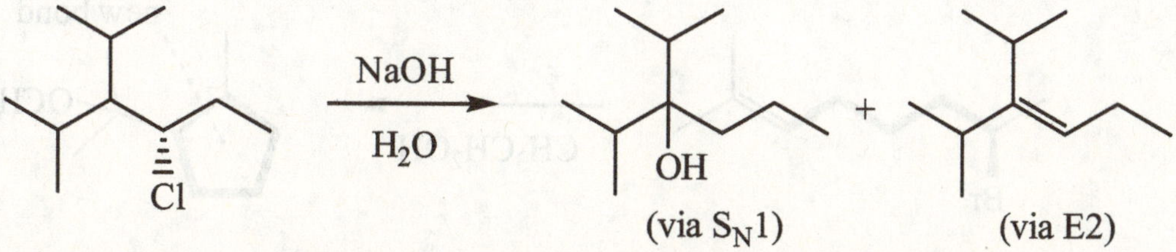

The high steric hindrance around the carbon bearing the chlorine atom will make S_N2 substitution difficult. Any substitution product formed would be via an S_N1 reaction involving a carbocation rearrangement.

(b) 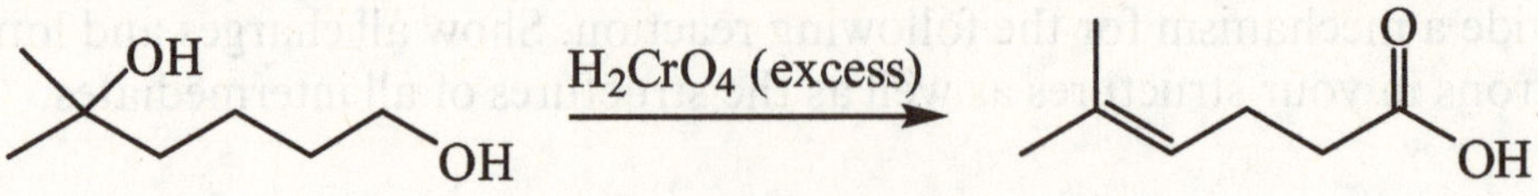
OH
OH
H_2CrO_4 (excess)
O
OH

The tertiary alcohol is not oxidized by H_2CrO_4. Under the acidic conditions, it will likely dehydrate to form the most stable alkene.

(c) 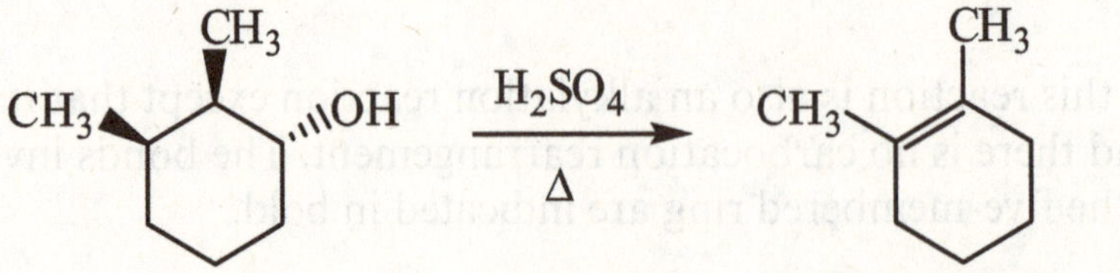
CH_3
CH_3
OH
H_2SO_4
Δ
CH_3
CH_3

A carbocation rearrangement occurs during the reaction.

(d) 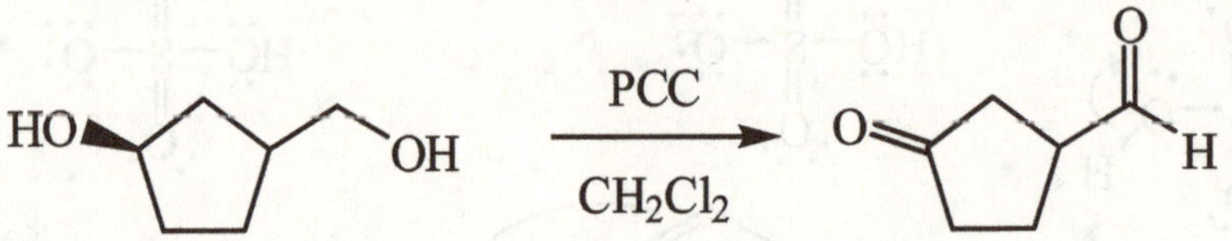
HO
OH
PCC
CH_2Cl_2
O
O
H

(e) 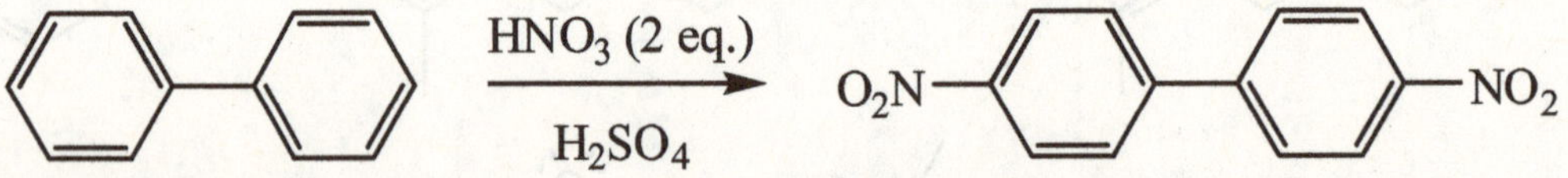
HNO_3 (2 eq.)
H_2SO_4
O_2N
NO_2

Note that a phenyl substituent is an ortho-para director (try drawing the resonance structures of the carbocation). Nitration occurs on separate rings because the ring that is nitrated first is deactivated by the nitro group.

21. Provide a mechanism for the following reaction. Show all charges and lone pairs of electrons in your structures as well as the structures of all intermediates.

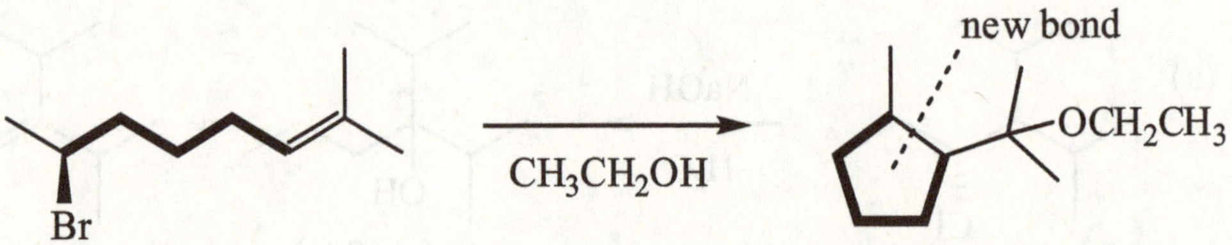
Br
CH_3CH_2OH
new bond
OCH_2CH_3

Inspection of the starting materials and the product reveals that both a substitution and a cyclization have occurred. Because ethanol is a polar protic solvent and also a weak nucleophile, the reaction proceeds by an S_N1 mechanism. The 2° carbocation generated by the departure of bromide acts as an electrophile and adds to the alkene, forming a 3° carbocation. The carbon atoms used to form the five-membered ring are shown in bold.

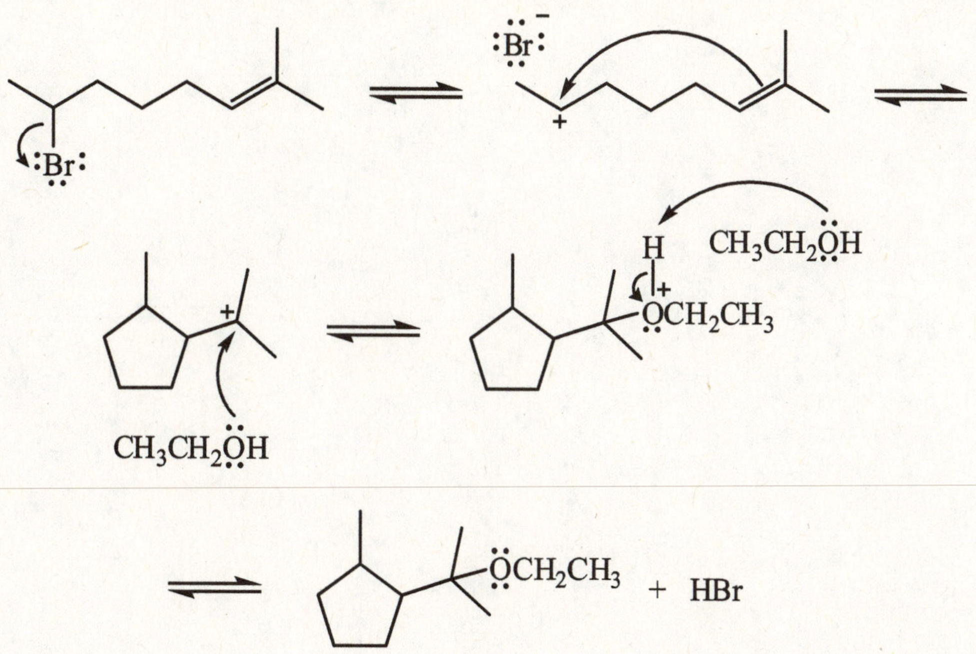
:Br:
:Br:⁻
+
CH3CH2ÖH
H
+
ÖCH2CH3
CH3CH2ÖH
ÖCH2CH3
+ HBr

Chapter 11: Spectroscopy

Problems

11.1 Calculate the energy of red light (680 nm) in kilocalories per mole. Which form of radiation carries more energy, infrared radiation of wavelength 2.50 μm or red light of wavelength 680 nm?

The energy of red light (680 nm) is 176 kJ/mol (42.1 kcal/mol) and is higher than the energy from infrared radiation of 2500 nm (47.7 kJ/mol or 11.4 kcal/mol).

11.2 A compound shows strong, very broad IR absorption in the region from 3200 to 3500 cm^{-1} and strong absorption at 1715 cm^{-1}. What functional group accounts for both of these absorptions?

Carboxyl group. The absorption at 1715 cm^{-1} is attributed to the carboxyl C=O group, and the absorption in the region from 3200 to 3500 cm^{-1} is attributed to the −OH group of a carboxyl group.

11.3 Propanoic acid and methyl ethanoate are constitutional isomers. Show how to distinguish between these two compounds by IR spectroscopy.

$$
\begin{array}{cc}
\overset{\displaystyle O}{\underset{\displaystyle \|}{}} & \overset{\displaystyle O}{\underset{\displaystyle \|}{}} \\
CH_3CH_2COH & CH_3COCH_3 \\
\text{Propanoic acid} & \text{Methyl ethanoate} \\
& \text{(Methyl acetate)}
\end{array}
$$

Propanoic acid is a carboxylic acid while methyl ethanoate is an ester. Both compounds have a strong absorption around 1700−1725 cm^{-1} due to the C=O group, but only propanoic acid will have a strong, broad OH absorption between 3200−3500 cm^{-1}.

11.4 What does the value of the wavenumber of the stretching frequency for a particular functional group indicate about the relative strength of the bond in that functional group?

The wavenumber is directly related to the energy of the infrared radiation required to stretch the bond. Bonds that require more energy to stretch are stronger bonds.

11.5   Calculate the IHD of cyclohexene, C_6H_{10}, and account for this deficiency by reference to the structural formula of the compound.

Cyclohexene has one ring and one π bond, so it has an IHD of two.

11.6   The IHD of niacin is 5. Account for this value by reference to the structural formula of niacin.

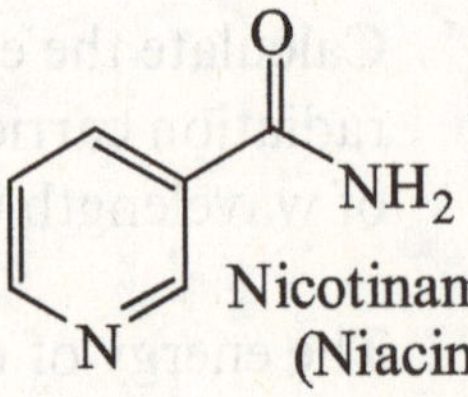

Each IHD could be a ring or a π bond. Niacin contains one ring (the pyridine ring) and four π bonds.

Nicotinamide
(Niacin)

11.7   Determine possible structures for the same spectrum (above) for a compound with molecular formula $C_8H_{10}O$. What does example 11.7 and this problem tell you about the effectiveness of IR spectroscopy for determining the structure of an unknown compound?

The $C_8H_{10}O$ compounds that would have the same spectrum as the C_7H_8O compounds are shown below. IR spectroscopy only provides functional group information and is not effective for the determination of the actual structure of a compound.

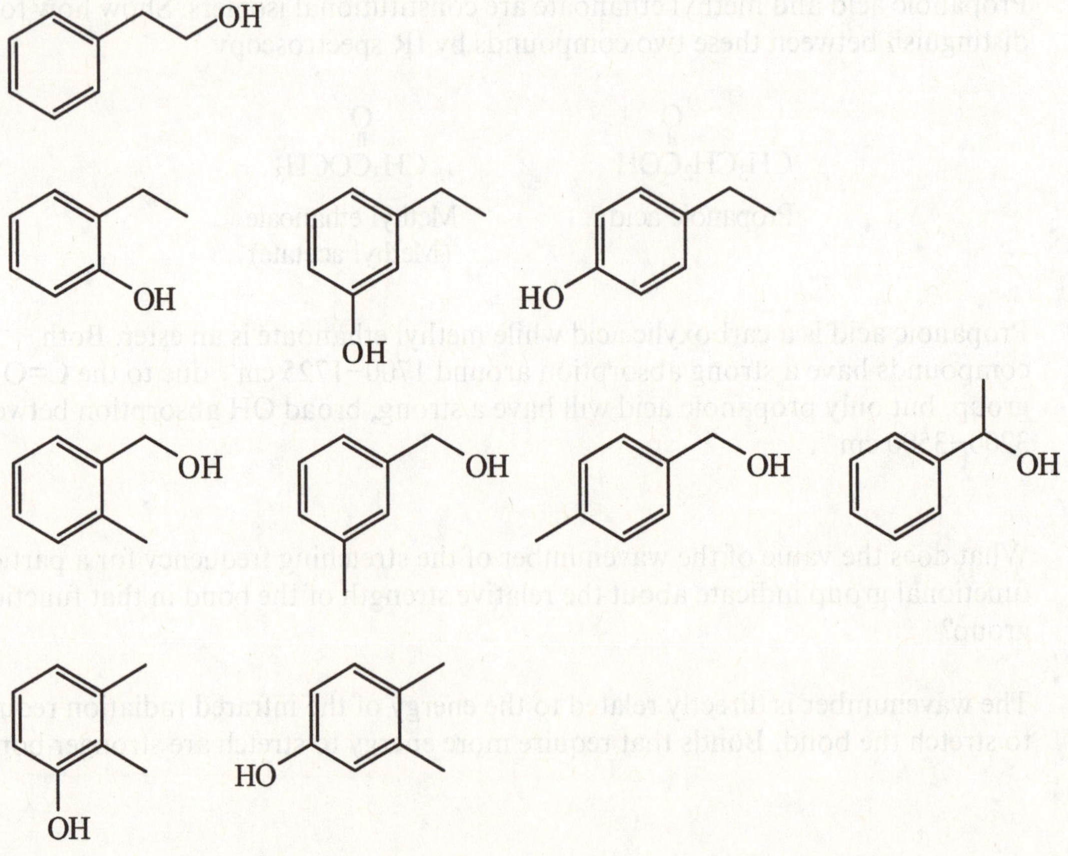

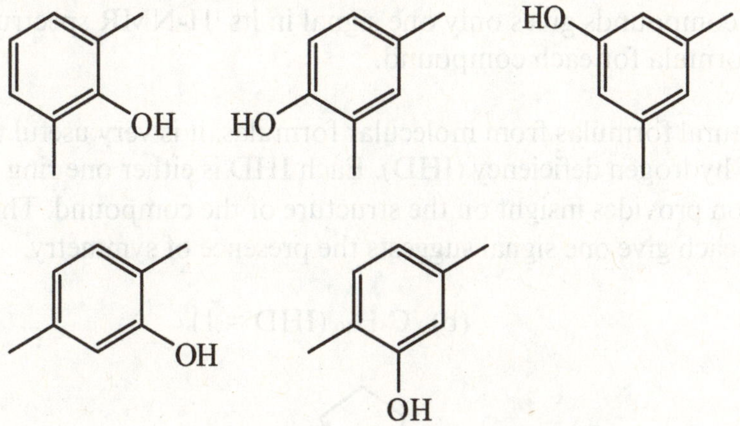

11.8 Which of the following nuclei are capable of behaving like tiny bar magnets?

(a) $^{31}_{15}P$ (b) $^{195}_{78}Pt$

Nuclei with an odd mass number or an odd atomic number can behave as tiny bar magnets. Therefore, both ^{31}P and ^{195}Pt can behave as tiny bar magnets and can be detected by NMR spectroscopy.

11.9 State the number of sets of equivalent hydrogens in each compound and the number of hydrogens in each set.

To identify the number of sets of equivalent hydrogens, look for any symmetry (mirror planes) that may be present in the molecule. In addition, all hydrogens that are bonded to the same carbon atom are usually equivalent.

(a) 3-Methylpentane has four sets of equivalent hydrogens. The number of hydrogens in each set are 6, 4, 3, and 1, which are respectively labelled a, b, c, and d.

(b) 2,2,4-Trimethylpentane has four sets of equivalent hydrogens. The number of hydrogens in each set are 9, 6, 2, and 1, which are respectively labelled a, b, c, and d.

(c) 1,4-Dichloro-2,5-dimethylbenzene has two sets of equivalent hydrogens. The number of hydrogens in each set are 6 and 2, which are respectively labelled a and b.

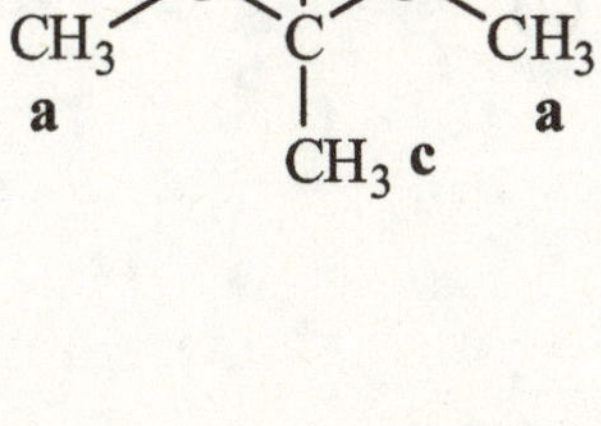

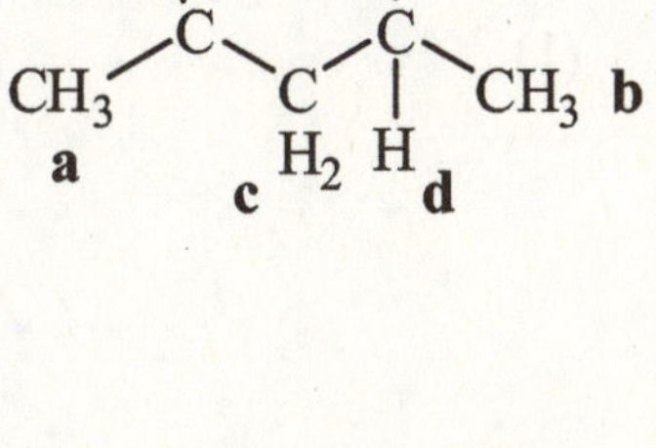

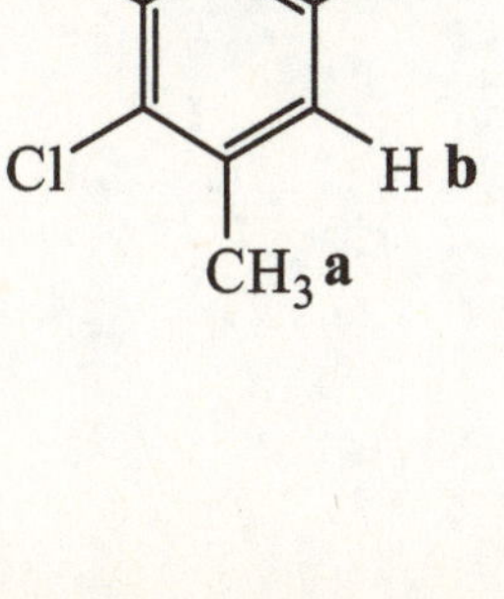

11.10 Each of the following compounds gives only one signal in its ¹H-NMR spectrum. Propose a structural formula for each compound.

When proposing structural formulas from molecular formulas, it is very useful to first determine the index of hydrogen deficiency (IHD). Each IHD is either one ring or one π bond; this information provides insight on the structure of the compound. The fact that these compounds each give one signal suggests the presence of symmetry.

(a) C_3H_6O (IHD = 1)

(b) C_5H_{10} (IHD = 1)

$$\underset{CH_3}{\overset{O}{\underset{\quad}{\|}}}\underset{\;}{\overset{\|}{C}}\underset{CH_3}{}$$

(c) C_5H_{12} (IHD = 0)

(d) $C_4H_6Cl_4$ (IHD = 0)

$$CH_3-\overset{\overset{CH_3}{|}}{\underset{\underset{CH_3}{|}}{C}}-CH_3$$

$$CH_3-\overset{\overset{Cl}{|}}{\underset{\underset{Cl}{|}}{C}}-\overset{\overset{Cl}{|}}{\underset{\underset{Cl}{|}}{C}}-CH_3$$

11.11 The line of integration of the two signals in the ¹H-NMR spectrum of a ketone with the molecular formula $C_7H_{14}O$ shows a vertical rise of 62 and 10 chart divisions. Calculate the number of hydrogens giving rise to each signal, and propose a structural formula for this ketone.

The presence of two signals indicates the presence of two sets of equivalent hydrogens. The integration of each signal, which is associated with the vertical rise of chart divisions, is proportional to the number of hydrogens. The ratio of the chart divisions (integrations) is approximately 6:1, and because there are 14 hydrogens in the molecule, the number of hydrogens in each equivalent set must be 12 (a) and 2 (b). The compound that is consistent with the data is 2,4-dimethyl-3-pentanone.

11.12 Following are two constitutional isomers with the molecular formula $C_4H_8O_2$:

$$\underset{(1)}{CH_3CH_2O\overset{O}{\overset{\|}{C}}CH_3} \qquad \underset{(2)}{CH_3CH_2\overset{O}{\overset{\|}{C}}OCH_3}$$

(a) Predict the number of signals in the ^{1}H-NMR spectrum of each isomer.

There is no symmetry in either of the two compounds. Both compounds will have three signals (one for each of the two $-CH_3$ groups, and one for the CH_2).

(b) Predict the ratio of areas of the signals in each spectrum.

The ratio of the areas of the signals will be 3:3:2.

(c) Show how to distinguish between these isomers on the basis of chemical shift.

In compound (1), the CH_2 group is directly bonded to oxygen and due to the electron-withdrawing effects of oxygen, the chemical shift of the CH_2 hydrogens will be more downfield than that of the CH_2 hydrogens in compound (2).

11.13 Following are pairs of constitutional isomers. Predict the number of signals and the splitting pattern of each signal in the ^{1}H-NMR spectrum of each isomer.

(a) Both compounds will each have three signals.

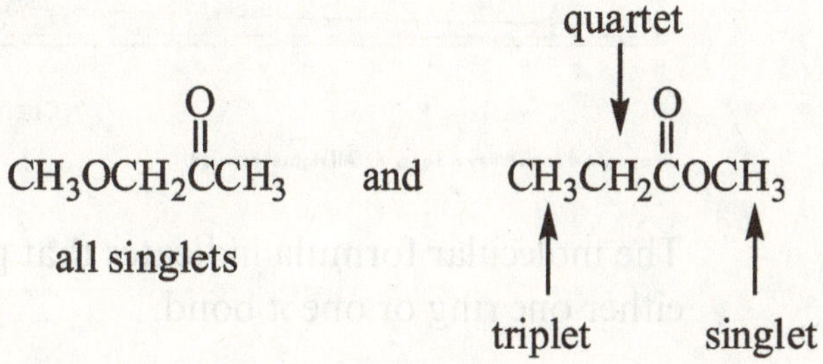

$$CH_3OCH_2\overset{\overset{\displaystyle O}{\|}}{C}CH_3 \quad \text{and} \quad \overset{\text{quartet}}{CH_3}CH_2\overset{\overset{\displaystyle O}{\|}}{C}CH_3$$

all singlets

triplet   singlet

(b) The compound on the left has one signal; the compound on the right has two signals. Note that in the compound on the right, the hydrogens bonded to the middle carbon are split by four neighboring hydrogens; all four of these are equivalent.

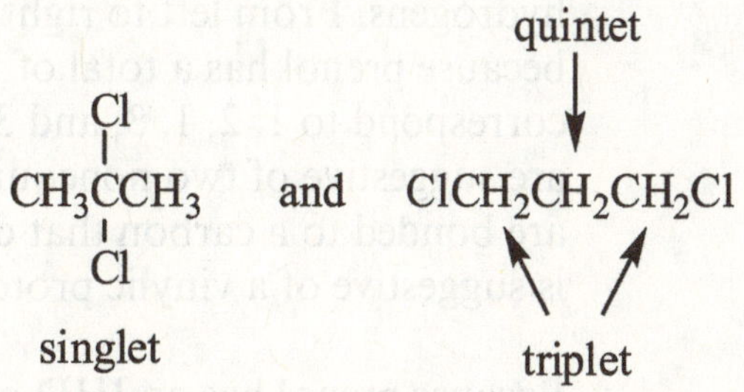

$$CH_3\overset{\overset{\displaystyle Cl}{|}}{\underset{\underset{\displaystyle Cl}{|}}{C}}CH_3 \quad \text{and} \quad \overset{\text{quintet}}{ClCH_2}CH_2CH_2Cl$$

singlet   triplet

11.14 Explain how to distinguish between the members of each pair of constitutional isomers, on the basis of the number of signals in the ^{13}C-NMR spectrum of each isomer:

(a) The compound on the left has a plane of symmetry and will have five ^{13}C signals, while the compound on the right will have seven signals.

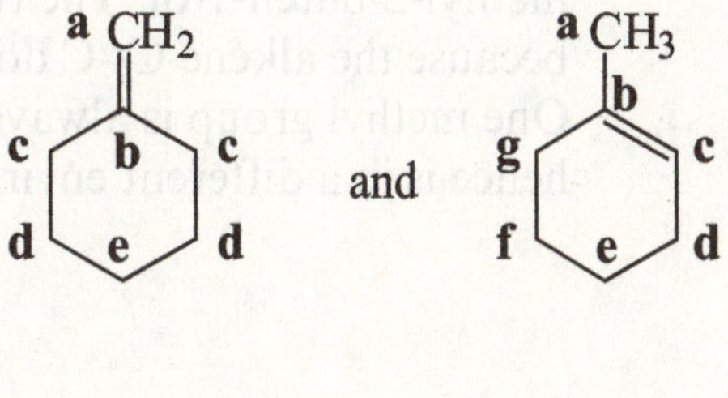

a CH_2
c   b   c
d   e   d

and

a CH_3
b
g   c
f   e   d

(b) The compound on the left will have six ^{13}C signals, while the compound on the right, which has symmetry, will have three signals.

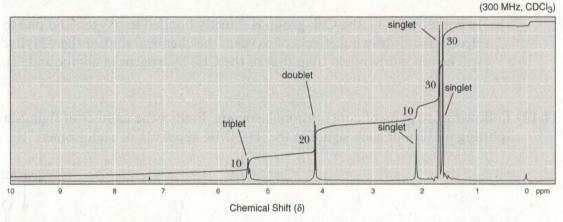

11.15 Following is a ^{1}H-NMR spectrum for prenol, a compound that possesses a fruity odor and that is commonly used in perfumes. Prenol has the molecular formula $C_5H_{10}O$. Propose a structural formula for prenol.

(300 MHz, CDCl$_3$)

singlet
30
doublet
30 singlet
triplet   10
20   singlet
10

10   9   8   7   6   5   4   3   2   1   0 ppm
Chemical Shift (δ)

The molecular formula indicates that prenol has an IHD of one, so prenol must have either one ring or one π bond.

The ^{1}H-NMR spectrum has five signals, which correspond to five sets of equivalent hydrogens. From left to right, the integration ratios are 10, 20,10, 30, and 30, and because prenol has a total of ten hydrogens, these signals must respectively correspond to 1, 2, 1, 3, and 3 hydrogen atoms. The two 3H singlets near 1.65 ppm are suggestive of two nonequivalent methyl groups, and they are not split (i.e. they are bonded to a carbon that does not bear a hydrogen). The 1H triplet near 5.4 ppm is suggestive of a vinylic proton.

Because prenol has an IHD of only one and we have a vinylic proton, hence a carbon-carbon double bond, there cannot be a carbonyl group. The 1H singlet near 2.1 ppm is therefore attributed to an –OH group. The 2H signal near 4.2 ppm is suggestive of a CH_2 adjacent to an –OH group.

The structure that is consistent with the ^{1}H-NMR data is 3-methyl-2-buten-1-ol. The two methyl groups are nonequivalent because the alkene C=C double bond is unable to freely rotate. One methyl group is always closer to the $–CH_2OH$ group, and hence is in a different environment, than the other methyl group.

11.16 Following is a ¹H-NMR spectrum for a compound that is a colorless liquid with the molecular formula $C_7H_{14}O$. Propose a structural formula for the compound.

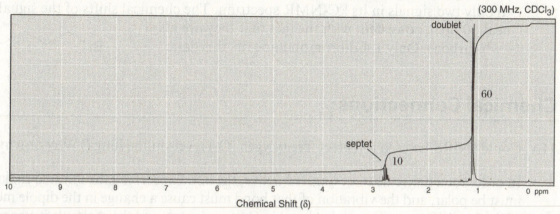

(300 MHz, CDCl₃)

doublet

60

septet

10

Chemical Shift (δ)

The molecular formula indicates that the compound has an IHD of one, so it must have either one ring or one π bond.

The ¹H-NMR spectrum has just two signals, which indicates that the seven-carbon compound has an extensive amount of symmetry. From left to right, the integration ratios are 10 and 60, and because the compound has a total of fourteen hydrogens, the two signals correspond to 2 and 12 hydrogens, respectively. The 12H doublet near 1.2 ppm is suggestive of four methyl groups, while the 2H septet near 3.8 ppm is suggestive of two equivalent CH groups, each of which is bonded to two equivalent methyl groups. The chemical shift of 3.8 ppm suggests that the CH groups are adjacent to a carbonyl group. Because of the absence of singlets, we can conclude that no alcohol is present.

The compound consistent with the data is 2,4-dimethyl-3-pentanone.

11.17 Following is a ¹³C-NMR spectrum for a compound that is a colorless liquid with the molecular formula $C_4H_8Br_2$. Propose a structural formula for the compound

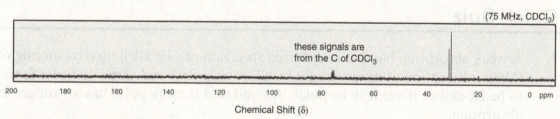

(75 MHz, CDCl₃)

these signals are from the C of CDCl₃

Chemical Shift (δ)

The molecular formula indicates that the compound has an IHD of zero, so it does not have any rings or π bonds. Therefore, it must be a dibromobutane that gives rise to only two signals in its ^{13}C-NMR spectrum. The chemical shifts of the signals are near 30 ppm, consistent with the fact that they are alkyl (sp^3) carbons. Only 1,4-dibromobutane fits this data.

Chemical Connections

11A. Could fNIRS be used to detect free oxygen (O_2) levels in the lungs? Why or why not?

In order for a molecule to be infrared-active, the bond absorbing the infrared radiation must be polar, and the vibration of that bond must cause a change in the dipole moment of the bond. The oxygen-oxygen bond in O_2 is nonpolar, and the molecule is symmetrical. As a result, free oxygen is not infrared-active and cannot be detected by fNIRS.

11B. Besides atmospheric lifetime and GWP, what other factor(s) might affect a compound's potential harm to the atmosphere?

The chemical reactivity of the compound, both in the presence and absence of light, also needs to be taken into consideration. For example, does the compound react with ozone in the upper atmosphere? Does the compound react with ground-level pollutants and exacerbate the formation of smog?

11C. In ^1H-NMR spectroscopy, the chemical sample is set spinning on its long axis to ensure that all parts of the sample experience a homogenous applied field. Homogeneity is also required in MRI. Keeping in mind that the "sample" in MRI is a human being, how do you suppose this is achieved?

In MRI imaging, the human being is usually larger than the magnet, and the magnet's magnetic field is most homogenous at the center. To ensure that the entire area being imaged is subjected to a homogeneous field, the patient lays on a moving table that slides within the magnet to position the area being imaged in a homogenous field.

Quick Quiz

1. A weak absorption band in an infrared spectrum can be attributed to, among other things, absorption of infrared light by a low polarity bond. *True.* In order for a bond to be IR-active, it needs to be polar. A bond that is more polar has a stronger infrared absorption.

2. Integration reveals the number of neighboring hydrogens in a ^1H-NMR spectrum. *False.* It is splitting, not integration, that reveals the number of neighboring hydrogens.

3. Wavelength and frequency are directly proportional. That is, as wavelength increases, frequency increases. *False*. Wavelength and frequency are *inversely* proportional, and as wavelength increases, frequency decreases.

4. An alkene (vinylic) hydrogen can be distinguished from a benzene ring hydrogen via ^{1}H-NMR spectroscopy. *True*. Vinylic and aromatic hydrogens have different chemical shifts.

5. IR spectroscopy can be used to distinguish between a terminal alkyne and an internal alkyne. *True*. A terminal alkyne contains a $\equiv$C$-$H bond, which absorbs near 3300 cm^{-1}.

6. The NMR signal of a shielded nucleus appears more upfield than the signal for a deshielded nucleus. *True*. Nuclei that are more deshielded are further downfield.

7. A transition between two energy states, E_1 and E_2, can be made to occur using light equal to or greater than the energy difference between E_1 and E_2. *False*. Energy levels in molecules are quantized, and a transition occurs only if the energy of the light is exactly equal to the energy difference between the two states.

8. The chemical shift of a nucleus depends on its resonance frequency. *True*. Chemical shift, reported in units of ppm, is related to resonance frequency.

9. A compound with the molecular formula $C_5H_{10}O$ could contain a C$-$C triple bond, two C=O bonds, or two rings. *False*. The formula $C_5H_{10}O$ has an IHD of one. Each IHD could be a π bond or a ring.

10. A ketone can be distinguished from an aldehyde via ^{13}C-NMR spectroscopy. *False*. Aldehydes and ketones have very similar ^{13}C-NMR chemical shifts. However, if a compound is suspected to be an aldehyde, ^{1}H-NMR spectroscopy will reveal the presence of an aldehydic proton.

11. A compound with the molecular formula $C_7H_{12}O$ has an IHD of 2. *True*. It has four fewer hydrogens than an alkane with seven carbons, C_7H_{16}.

12. A ^{1}H-NMR spectrum with an integration ratio of 3 : 1 : 2 could represent a compound with the molecular formula C_5H_9O. *False*. A compound with this integration ratio cannot have nine hydrogens in its molecular formula.

13. Electromagnetic radiation can be described as a wave, as a particle, and in terms of energy. *True*. Electromagnetic radiation has particle-wave duality and a certain energy that is dependent on the wavelength (or frequency) of the electromagnetic radiation.

14. A set of hydrogens are equivalent if replacing each of them with a halogen results in compounds of the same name. *True*. This is a very useful method for verifying hydrogen equivalency!

15. The collection of absorption peaks in the 1000−400 cm⁻¹ region of an IR spectrum is unique to a particular compound (i.e., no two compounds will yield the same spectrum in this region). *True*. This region of an IR spectrum is known as the fingerprint region.

16. The area under each peak in a ¹H-NMR spectrum can be determined using a technique known as integration. *True*. Integration refers to the area under any peak.

17. All atomic nuclei have a spin, which allows them to be analyzed by NMR spectroscopy. *False*. Only nuclei with an odd mass or odd atomic number are NMR-active.

18. C−H stretching vibrations occur at higher wavenumbers than C−C stretching vibrations. *True*. C−H stretching requires more energy than C−C stretching.

19. The resonance frequency of a nucleus depends on its amount of shielding. *True*. It is this shielding, which affects resonance frequency, that makes NMR spectroscopy useful for the elucidation of chemical structure.

20. It is not possible to use IR spectroscopy to distinguish between a ketone and a carboxylic acid. *False*. Although both functional groups will show an absorption corresponding to the C=O group, the carboxylic acid also has an −OH group that is very infrared-active.

21. A carboxylic acid can be distinguished from an aldehyde via ¹H-NMR spectroscopy. *True*. The hydrogen of a carboxylic acid is usually more deshielded (by about 2−3 ppm) than the hydrogen of an aldehyde.

22. A wavenumber, $\bar{\nu}$, is directly proportional to frequency. *True*. Wavenumber increases as frequency increases, and it is also directly proportional to energy (but inversely proportional to wavelength).

23. Resonance is the excitation of a magnetic nucleus in one spin state to a higher spin state. *True*. It is the energy of this spin flip that is detected by an NMR spectrometer.

24. IR spectroscopy cannot be used to distinguish between an alcohol and an ether. *False*. The alcohol will contain a very broad and strong IR absorption corresponding to the −OH group.

25. A compound with an IHD deficiency of 1 can contain either one ring, one double bond, or one triple bond. *False*. While it can contain either one ring or one double bond, it cannot contain a triple bond, which has an IHD of 2.

26. Infrared spectroscopy measures transitions between electronic energy levels. *False*. IR spectroscopy measures transitions between the *vibrational* energy levels of bonds.

27. A set of hydrogens represented by a doublet indicates that there are two neighboring equivalent hydrogens. *False*. A doublet is indicative of one neighboring hydrogen.

28. The IHD can reveal the possible number of rings, double bonds, or triple bonds in a compound based solely on its molecular formula. ***True***. Each ring or π bond in the compound will affect the number of hydrogens in the molecular formula.

29. TMS, tetramethylsilane, is a type of solvent used in NMR spectroscopy. ***False***. TMS is not a solvent, but rather a compound that has been arbitrary given a chemical shift of 0 ppm. The chemical shifts of hydrogen and carbon atoms are measured relative to TMS.

30. Light of wavelength 400 nm is higher in energy than light of wavelength 600 nm. ***True***. Wavelength and energy are inversely proportional. That is, as wavelength increases (becomes longer), energy decreases.

31. The methyl carbon of 1-chlorobutane will yield a ^{1}H-NMR signal that appears as a triplet. ***True***. The methyl group is bonded to a CH_2 group, which has two equivalent hydrogens.

32. A compound with the molecular formula $C_6H_{14}FN$ has an IHD of one. ***False***. A compound with that formula has an IHD of zero. The fluorine atom is treated as a single hydrogen, while the presence of one nitrogen requires the total number of hydrogens to be reduced by one.

33. IR spectroscopy can be used to distinguish between 1°, 2°, and 3° amines. ***True***. The number of signals corresponding to the N−H vibrations depend on the number of N−H bonds in the molecule. 1° amines will generally show two N−H signals, 2° amines will show just one signal, and 3° amines will not show any N−H signals.

End-of-Chapter Problems

Electromagnetic Radiation

11.18 Which puts out light of a higher energy, a green laser pointer or a red laser pointer?

Green light is of a shorter wavelength than red light. Because wavelength and energy are inversely proportional, green light is of a higher energy.

11.19 Calculate the energy, in kilocalories per mole of radiation, of a wave with a wavelength of 2 m. What type of radiant energy is this?

Electromagnetic radiation with a wavelength of 2 m corresponds to microwave radiation, and the corresponding energy is 5.9×10^{-5} kJ/mol (1.4×10^{-5} kcal/mol).

11.20 A molecule possesses molecular orbitals that differ in energy by 343 kJ/mol. What wavelength of light would be required to cause a transition between these two energy levels? What region of the electromagnetic spectrum does this energy correspond to?

The wavelength of light with an energy of 343 kJ/mol (82 kcal/mol) is 349 nm, which corresponds to light in the UV region of the electromagnetic spectrum.

Interpreting Infrared Spectra

11.21 Calculate the IHD of each compound:

Compound	Molecular Formula	Reference Compound	Reference Hydrocarbon	Index
Aspirin	$C_9H_8O_4$	C_9H_8	C_9H_{20}	6
Ascorbic acid (vitamin C)	$C_6H_8O_6$	C_6H_8	C_6H_{14}	3
Pyridine	C_5H_5N	C_5H_4	C_5H_{12}	4
Urea	CH_4N_2O	CH_2	CH_4	1
Cholesterol	$C_{27}H_{46}O$	$C_{27}H_{46}$	$C_{27}H_{56}$	5
Trichloroacetic acid	$C_2HCl_3O_2$	C_2H_4	C_2H_6	1

11.22 Compound A, with the molecular formula C_6H_{10}, reacts with H_2/Ni to give compound B, with the molecular formula C_6H_{12}. The IR spectrum of compound A is provided.

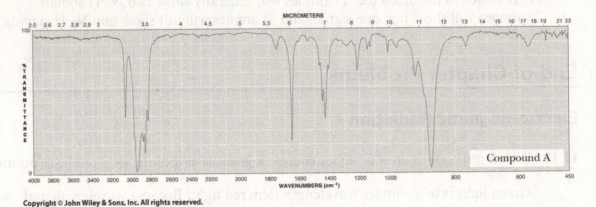

From this information about compound A tell

(a) Its IHD.

Compound A has an IHD of 2.

(b) The number of rings or pi bonds (or both) in compound A.

An IHD of 2 indicates that the compound could have two rings, two π bonds, or one ring and one π bond. However, catalytic hydrogenation only adds hydrogen to a π bond and does not cause ring opening. Because the product formed from catalytic hydrogenation, C_6H_{12}, still has one IHD, a ring must be present. Compound A has one π bond and one ring.

(c) What structural feature(s) would account for compound A's IHD.

Compound A contains one cycloalkane ring and one double bond. It is important to realize that IR spectroscopy cannot deduce the structure of the compound; there are many constitutional isomers of C_6H_{10} with one ring and one double bond. Yet, we do know from the IR absorption at 3150 cm^{-1} that a vinylic =C−H is present.

11.23 Compound C, with the molecular formula C_6H_{12}, reacts with H_2/Ni to give compound D, with the molecular formula C_6H_{14}. The IR spectrum of compound C is provided.

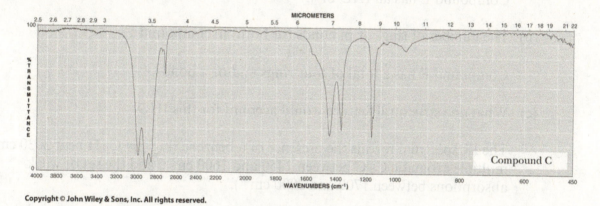

From this information about compound C, tell

(a) Its IHD.

Compound C has an IHD of 1.

(b) The number of rings or pi bonds (or both) in compound C.

An IHD of 1 indicates that the compound could have one ring or one π bond. Because the compound can undergo catalytic hydrogenation, which only adds hydrogen to a π bond and does not cause ring opening, compound C must have one π bond.

(c) What structural feature(s) would account for compound C's IHD.

Compound C contains a carbon-carbon double bond that is not bonded to any hydrogen atoms (no IR absorption corresponding to the vinylic =C−H is present in the spectrum).

11.24 Examine the following IR spectrum and the molecular formula of compound E, $C_9H_{12}O$, and tell:

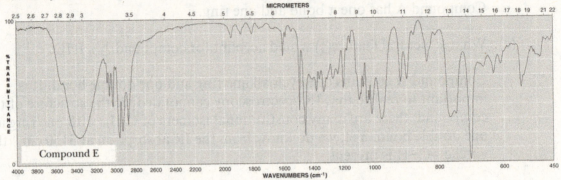

(a) Its IHD.

Compound E has an IHD of 4.

(b) The number of rings or pi bonds (or both) in compound E.

Compound E has a total of four rings and/or π bonds.

(c) What one structural feature would account for this IHD.

The IR spectrum reveals the presence of a benzene ring (sp^2 C–H near 3030 cm^{-1}, multiple aromatic C=C between 1450 and 1600 cm^{-1}, and the set of weak absorptions between 1700 and 2000 cm^{-1}).

(d) What oxygen-containing functional group compound E contains.

The strong, broad absorption near 3400 cm^{-1} is indicative of an –OH group.

11.25 Examine the following IR spectrum and the molecular formula of compound F, $C_5H_{13}N$, and tell:

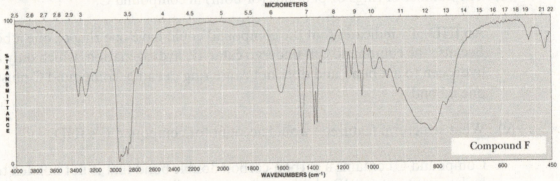

(a) Its IHD.

Compound F has an IHD of zero.

(b) The number of rings or pi bonds (or both) in compound F.

With an IHD of zero, it does not have any rings or π bonds.

(c) The nitrogen-containing functional group(s) compound F might contain.

The IR spectrum indicates the presence of an −NH₂ group, as evidenced by the two broad N−H stretches near 3300 and 3380 cm⁻¹.

11.26 Examine the following IR spectrum and the molecular formula of compound G, $C_6H_{12}O$, and tell:

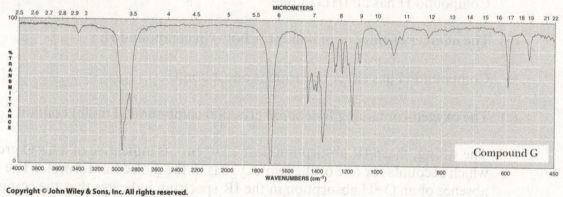

(a) Its IHD.

Compound G has an IHD of one.

(b) The number of rings or pi bonds (or both) in compound G.

Compound G has either one ring or one π bond.

(c) What structural features would account for this IHD.

The strong, sharp IR absorption near 1720 cm⁻¹ is indicative of a C=O group. Because the molecular formula only contains one oxygen atom, and only carbons and hydrogens, the C=O is due to an aldehyde or a ketone.

11.27 Examine the following IR spectrum and the molecular formula of compound H, $C_6H_{12}O_2$, and tell:

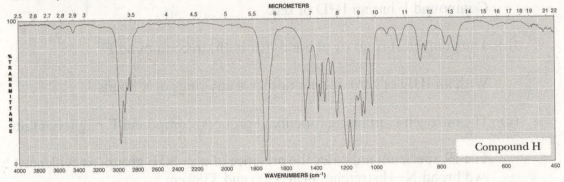

(a) Its IHD.

Compound H has an IHD of one.

(b) The number of rings or pi bonds (or both) in compound H.

Compound H has either one ring or one π bond.

(c) The oxygen-containing functional group(s) compound H might contain.

The strong, sharp IR absorption near 1730 cm^{-1} is indicative of a C=O group, which accounts for one of the two oxygens in the molecular formula. The absence of an O−H absorption in the IR spectrum indicates that the compound cannot be an alcohol or a carboxylic acid. The remaining oxygen atom must be attributed to either an ether or an ester, both of which would give the C−O stretching between $1100-1200 \text{ cm}^{-1}$. Therefore, compound L might contain either an ester alone or an aldehyde/ketone in conjunction with an ether.

11.28 Examine the following IR spectrum and the molecular formula of compound I, C_3H_7NO, and tell:

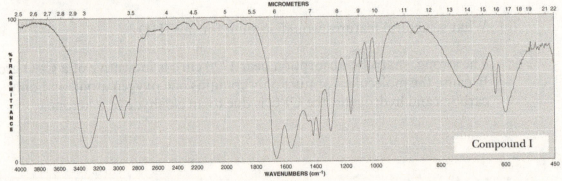

Chapter 11: Spectroscopy — 391

(a) Its IHD.

Compound I has an IHD of one.

(b) The number of rings or pi bonds (or both) in compound I.

Compound I has either one ring or one π bond.

(c) The oxygen- and nitrogen-containing functional group(s) in compound I.

The strong, sharp C=O absorption near 1680 cm^{-1}, along with the broad N−H absorption near 3300 cm^{-1}, is indicative of an amide.

11.29 Show how IR spectroscopy can be used to distinguish between the compounds in each of the following pairs:

(a) 1-Butanol and diethyl ether

1-Butanol will have a strong, broad O−H absorption between 3200 and 3400 cm^{-1}.

(b) Butanoic acid and 1-butanol

Butanoic acid will have a strong C=O absorption near 1700 cm^{-1}.

(c) Butanoic acid and 2-butanone

Butanoic acid will have a broad O−H absorption between 2400 and 3400 cm^{-1}.

(d) Butanal and 1-butene

Butanal will have a strong C=O absorption near 1700 and 1725 cm^{-1}. 1-Butene will have a vinylic =C−H absorption near 3100 cm^{-1}.

(e) 2-Butanone and 2-butanol

Butanone will have a strong C=O absorption near 1725 cm^{-1}. 2-Butanol will have a strong, broad O−H absorption between 3200 and 3400 cm^{-1}.

(f) Butane and 2-butene

2-Butene will have a vinylic =C−H absorption near 3100 cm^{-1}.

11.30 For each pair of compounds that follows, list one major feature that appears in the IR spectrum of one compound, but not the other. In your answer, state what type of bond vibration is responsible for the spectral feature you list, and give its approximate position in the IR spectrum.

(a) 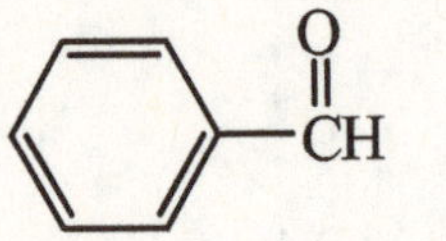 and 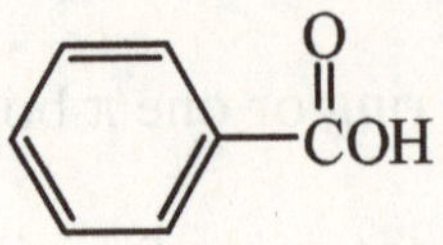

Benzoic acid has a strong, broad O–H stretching absorption between 2400–3400 cm^{-1}.

(b) 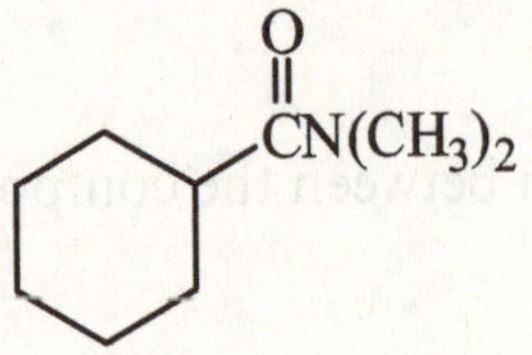 and 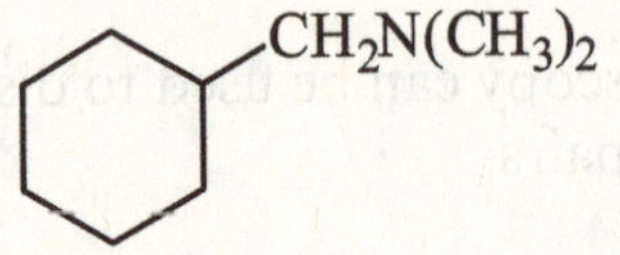

The amide has a strong C=O stretching absorption between 1630–1680 cm^{-1}.

(c) 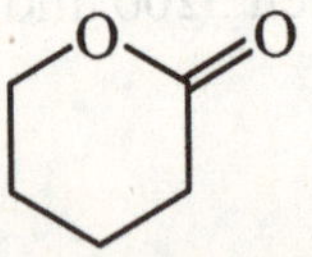 and $HO(CH_2)_4\overset{O}{\overset{\|}{C}}OH$

The acid has a strong, broad O–H stretching absorption between 2400–3400 cm^{-1}. In addition to the carboxylic acid –OH group, this compound also contains an alcohol –OH group, which has a stretching absorption between 3200–3400 cm^{-1}.

(d) 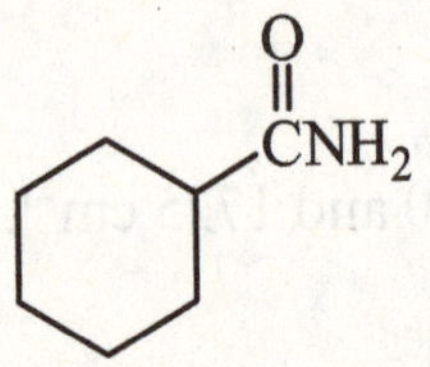 and 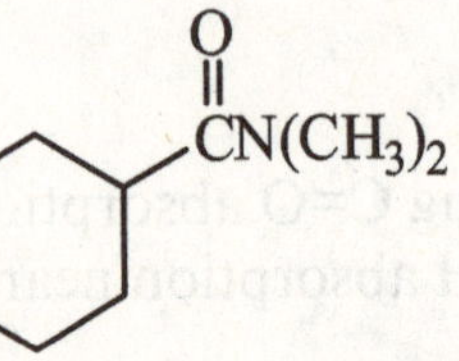

The primary amide has two broad N–H stretching absorptions between 3200–3400 cm^{-1}.

(e) $CH_3C{\equiv}CH$ and $CH_3C{\equiv}CCH_3$

The terminal alkyne (left) has a stretching absorption at 3300 cm^{-1} due to the bond between an H bonded to an *sp*-hybridized C.

11.31 Following are an infrared spectrum and a structural formula for methyl salicylate, the fragrant component of oil of wintergreen. On this spectrum, locate the absorption peak(s) due to:

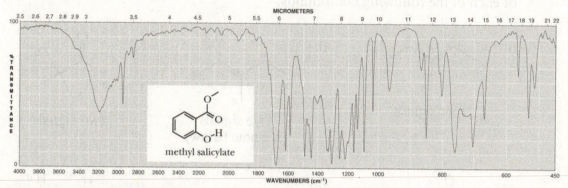

methyl salicylate

(a) O—H stretching of the hydrogen-bonded —OH group (very broad and of medium intensity).

The O—H stretching absorption is centered at 3200 cm^{-1}.

(b) C—H stretching of the aromatic ring (sharp and of weak intensity).

The aromatic C—H stretching is a hard-to-see weak signal near 3050 cm^{-1}; it is partially obscured by the broad O—H stretch at 3200 cm^{-1}. The sharp peak at 2975 cm^{-1} is due to the C—H stretching of the —OCH$_3$ group.

(c) C=O stretching of the ester group (sharp and of strong intensity).

The C=O stretching absorption is located at 1680 cm^{-1}. Note that ester C=O absorptions are usually between 1735 and 1750 cm^{-1}, but the absorption of this ester is lower due because (1) the C=O is conjugated to the aromatic ring and (2) the C=O oxygen can undergo intramolecular hydrogen bonding with the adjacent —OH group.

(d) C=C stretching of the aromatic ring (sharp and of medium intensity).

The aromatic C=C stretching absorptions are between 1450 and 1620 cm^{-1}.

Equivalency of Hydrogens and Carbons

11.32 Determine the number of signals you would expect to see in the ¹H-NMR spectrum of each of the following compounds.

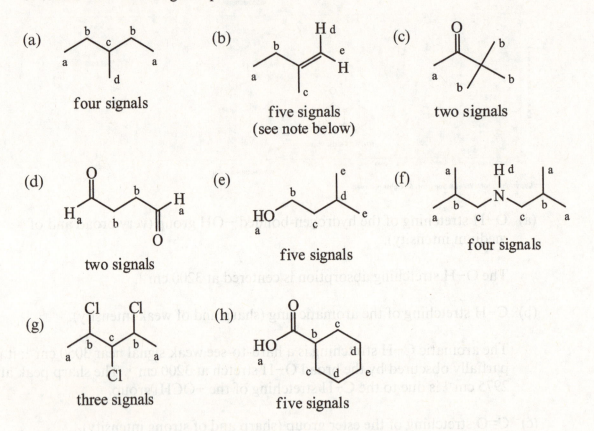

(a)

four signals

(b)

five signals
(see note below)

(c)

two signals

(d)

two signals

(e)

five signals

(f)

four signals

(g)

three signals

(h)

five signals

Compound (b) has five ¹H-NMR signals because protons H$_d$ and H$_e$, although both bonded to the same carbon, are not equivalent. The double bond cannot rotate, and as a result, H$_d$ is closer to the ethyl group while H$_e$ is closer to the methyl group.

11.33 Determine the number of signals you would expect to see in the ¹C-NMR spectrum of each of the compounds in Problem 11.32.

(a)

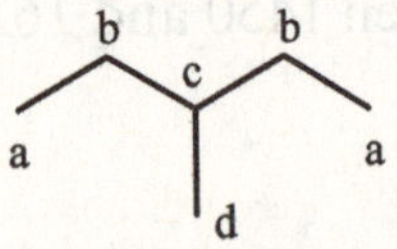

four signals

(b)

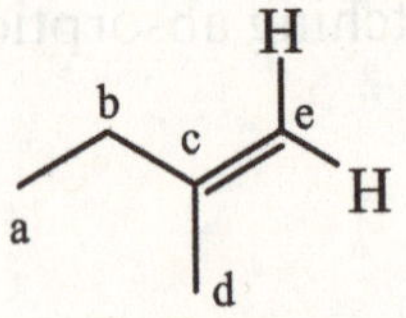

five signals

(c)

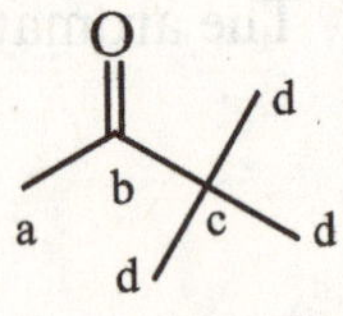

four signals

(d)

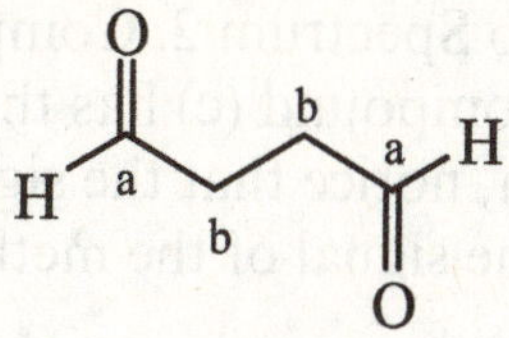

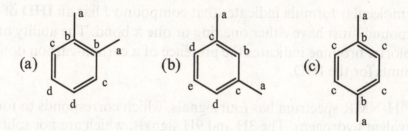

two signals

(e)

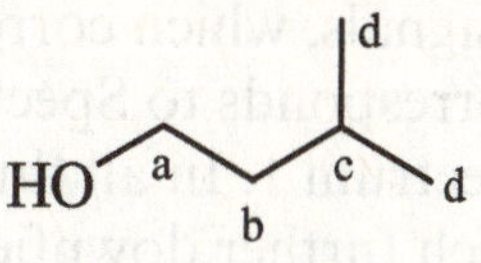

four signals

(f)

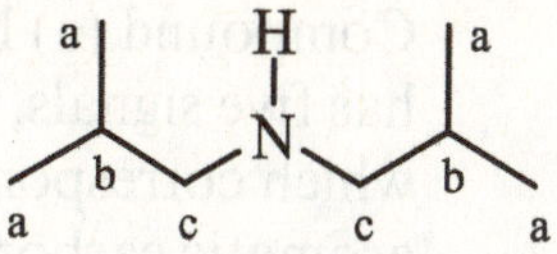

three signals

(g)

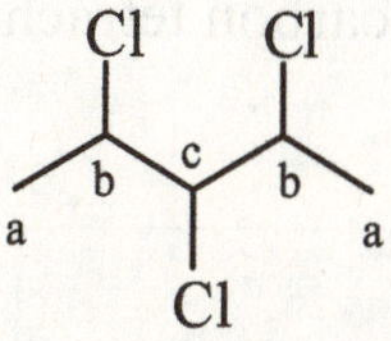

three signals

(h)

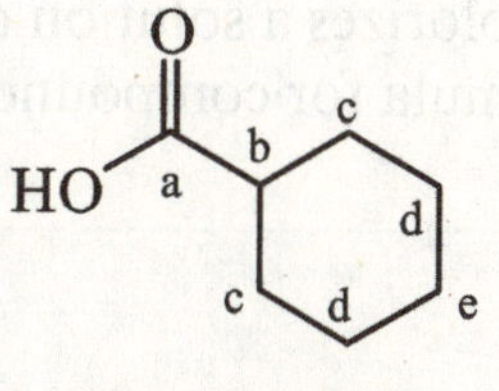

five signals

Interpreting ¹H-NMR and ¹³C-NMR Spectra

11.34 Following are structural formulas for the constitutional isomers of xylene and three sets of ¹³C-NMR spectra. Assign each constitutional isomer its correct spectrum.

(a)   (b)   (c)

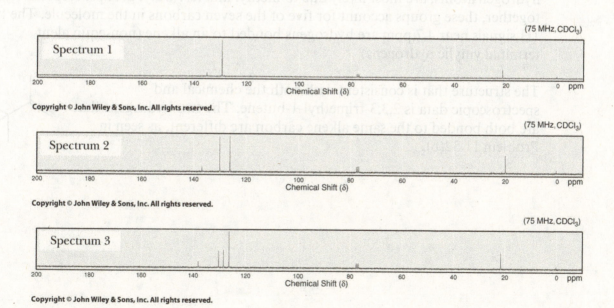

Spectrum 1   (75 MHz, CDCl₃)

200 180 160 140 120 100 80 60 40 20 0 ppm
Chemical Shift (δ)

Spectrum 2   (75 MHz, CDCl₃)

200 180 160 140 120 100 80 60 40 20 0 ppm
Chemical Shift (δ)

Spectrum 3   (75 MHz, CDCl₃)

200 180 160 140 120 100 80 60 40 20 0 ppm
Chemical Shift (δ)

Compound (a) has four signals, which corresponds to Spectrum 2. Compound (b) has five signals, which corresponds to Spectrum 3. Compound (c) has three signals, which corresponds to Spectrum 1. In all three spectra, notice that the signals for the aromatic carbons are much further downfield than the signal of the methyl carbon.

11.35 Following is a ¹H-NMR spectrum for compound J, with the molecular formula C_7H_{14}. Compound J decolorizes a solution of bromine in carbon tetrachloride. Propose a structural formula for compound J.

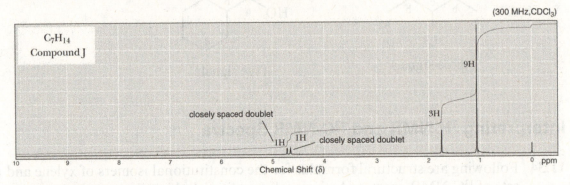

The molecular formula indicates that compound J has an IHD of one, so the compound must have either one ring or one π bond. The ability of the compound to decolorize bromine indicates the presence of a carbon-carbon double bond, which accounts for the IHD.

The ¹H-NMR spectrum has four signals, which corresponds to four sets of equivalent hydrogens. The 3H and 9H signals, which are not split by any other hydrogen atoms, are most likely due to methyl and *tert*-butyl groups, respectively; together, these groups account for five of the seven carbons in the molecule. The two 1H signals near 4.7 ppm are hydrogens bonded to an alkene (non-equivalent terminal vinylic hydrogens).

The structure that is consistent with both the chemical and spectroscopic data is 2,3,3-trimethyl-1-butene. The two hydrogens that are both bonded to the same alkene carbon are different, as seen in Problem 11.32(b).

11.36 Following is a ^{1}H-NMR spectrum of compound K, with the molecular formula C_8H_{16}. Compound K decolorizes a solution of Br_2 in CCl_4. Propose a structural formula for compound K.

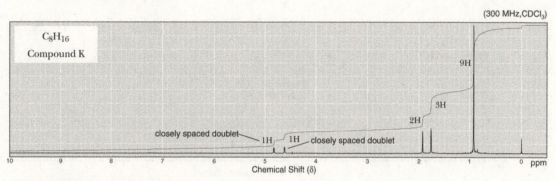

Compound K has an IHD of one and decolorizes a bromine solution, so it must contain a carbon-carbon double bond.

The ^{1}H-NMR spectrum has five signals, none of which are split. The 3H and 9H signals, are most likely due to methyl and *tert*-butyl groups. The 2H singlet is a CH_2 that does not have any adjacent hydrogens. The two 1H signals near 4.7 ppm are non-equivalent terminal vinylic hydrogens.

The structure that is consistent with both the chemical and spectroscopic data is 2,4,4-trimethyl-1-butene.

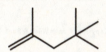

11.37 Following are the ^{1}H-NMR spectra of compounds L and M, each with the molecular formula C_4H_7Cl. Each compound decolorizes a solution of Br_2 in CCl_4. Propose structural formulas for compounds L and M.

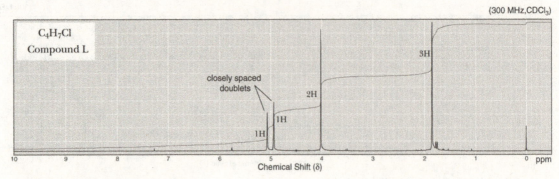

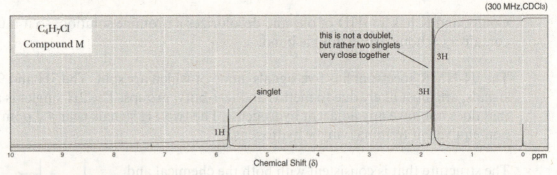

Both compounds L and M have an IHD of one and decolorize a solution of bromine. Both of these compounds must have a carbon-carbon double bond.

The ^{1}H-NMR spectrum of compound L has four signals, none of which are split. The 3H singlet is most likely a methyl group with no adjacent hydrogens, the 2H singlet is most likely a CH_2 with no adjacent hydrogens, and the two signals near 5.0 ppm are most likely due to the two non-equivalent terminal vinylic hydrogens. The chemical shift of 4.0 ppm for the 2H signal is relatively downfield for a CH_2, which suggests that the group is bonded to the electronegative chlorine. Compound L is 3-chloro-2-methyl-1-propene.

The ^{1}H-NMR spectrum of compound M has three signals, none of which are split. The two 3H singlets likely represent two non-equivalent CH_3 groups with no adjacent hydrogens. The ^{1}H singlet near 5.8 ppm is likely a vinylic proton, but the relatively downfield position of the signal suggests that it is close to the electronegative chlorine. Compound M is 1-chloro-2-methyl-1-propene. The two CH_3 groups are non-equivalent due to the non-rotating double bond.

compound L   compound M

11.38 Following are the structural formulas for three alcohols with the molecular formula $C_7H_{16}O$ and three sets of ^{13}C-NMR spectral data. Assign each constitutional isomer to its correct spectral data.

(a) $\overset{g}{C}H_3\overset{f}{C}H_2\overset{e}{C}H_2\overset{d}{C}H_2\overset{c}{C}H_2\overset{b}{C}H_2\overset{a}{C}H_2OH$

	Spectrum 1	Spectrum 2	Spectrum 3
	74.66	70.97	62.93
	30.54	43.74	32.79
	7.73	29.21	31.86
		26.60	29.14
		23.27	25.75
		14.09	22.63
			14.08

(b)
$$\overset{f}{C}H_3\overset{e}{\underset{\underset{f}{|}}{C}}\overset{}{}\overset{d}{C}H_2\overset{c}{C}H_2\overset{b}{C}H_2\overset{a}{C}H_3$$
with OH above the e carbon and CH_3 (f) below.

(c)
$$\overset{a}{C}H_3\overset{b}{C}H_2\overset{c}{\underset{\underset{b}{|}}{C}}\overset{b}{C}H_2\overset{a}{C}H_3$$
with OH above the c carbon and $\overset{b}{C}H_2\overset{a}{C}H_3$ below.

Each of the constitutional isomers has a different number of non-equivalent carbons and therefore give a different number of ^{13}C signals. Compound (a) has seven signals and corresponds to spectrum 3. Compound (b) has six signals and corresponds to spectrum 2. Compound (c) has three signals and corresponds to spectrum 1. Note that with all three compounds, carbons that are closer to the electronegative oxygen are more downfield.

11.39 Alcohol N, with the molecular formula $C_6H_{14}O$, undergoes acid-catalyzed dehydration when it is warmed with phosphoric acid, giving compound O, with the molecular formula C_6H_{12}, as the major product. A ^1H-NMR spectrum of compound N shows peaks at δ 0.89 (t, 6H), 1.12 (s, 3H), 1.38 (s, 1H), and 1.48 (q, 4H). The ^{13}C-NMR spectrum of compound N shows peaks at δ 72.98, 33.72, 25.85, and 8.16. Propose structural formulas for compounds N and O.

Alcohol N has an IHD is zero, so there are no rings or π bonds. The alcohol dehydrates to form compound O, an alkene, which has an IHD of one.

The ^{13}C-NMR spectrum of compound N reveals that there are four different sets of non-equivalent carbons. Likewise, the ^1H-NMR spectrum indicates that there are four sets of non-equivalent hydrogens. The absence of a ^1H signal between 3.3 and 4.0 ppm is significant; hydrogens on a carbon that is directly bonded to an alcohol OH are typically in that range (i.e. the OH is *not* bonded to a carbon bearing an H). The (t, 6H) and (q, 4H) splitting and integration patterns are suggestive of two equivalent ethyl groups. The structure of compound N that best matches the data is shown below along with the ^1H chemical shifts. The major product, O, is a *trans* alkene (recall that *trans* alkenes are more stable than *cis* alkenes).

1.48   1.48
0.89   0.89
          OH
1.12   1.38

compound N

H₃PO₄
heat
→

compound O

11.40 Compound P, $C_6H_{14}O$, does not react with sodium metal and does not discharge the color of Br_2 in CCl_4. The ^1H-NMR spectrum of compound P consists of only two signals, a 12H doublet at δ 1.1 and a 2H septet at δ 3.6. Propose a structural formula for compound P.

The compound has an IHD of zero, so it does not have any rings or π bonds. It is therefore no surprise that it does not discharge the color of bromine. However, the fact that it does not react with sodium metal is significant because it indicates that the compound is *not* an alcohol. If it contains an oxygen but is not an alcohol, what could it be? It cannot be an aldehyde or a ketone, because they contain a π bond. From this information, it can be concluded that the compound is an ether.

A compound with this molecular formula and only two sets of equivalent hydrogens must be very symmetrical. The septet suggests the presence of six neighboring hydrogens, all of which are equivalent. The structure that fits the data is diisopropyl ether.

1.1   O   1.1
   3.6   3.6
1.1   1.1

11.41 Propose a structural formula for each haloalkane:

(a) $C_2H_4Br_2$
    δ2.5 (d, 3H) and 5.9 (q, 1H)

2.5
5.9
Br   Br

(b) $C_4H_8Cl_2$
    δ1.67 (d, 6H) and 2.15 (q, 2H)

Cl
2.15   1.67
1.67   2.15
Cl

(c) $C_5H_8Br_4$
    δ3.6 (s, 8H)

CH₂Br
BrCH₂—⟨ ⟩—CH₂Br
CH₂Br

(d) C_4H_9Br
    δ1.1 (d, 6H), 1.9 (m, 1H), and 3.4 (d, 2H)

1.1
3.4   Br
1.1   1.9

(e) C₅H₁₁Br
1.1 (s, 9H) and 3.2 (s, 2H)

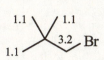

(f) C₇H₁₅Cl
δ 1.1 (s, 9H) and 1.6 (s, 6H)

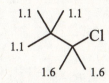

Notice that only adjacent, non-equivalent hydrogens cause splitting. For instance, the CH hydrogen in (b) is split only by the CH₃ hydrogens,

11.42 Following are structural formulas for esters (1), (2), and (3) and three ¹H-NMR spectra. Assign each compound its correct spectrum (Q, R, or S) and assign all signals to their corresponding hydrogens.

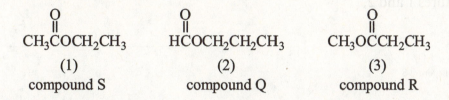

$$CH_3COCH_2CH_3 \quad HCOCH_2CH_2CH_3 \quad CH_3OCCH_2CH_3$$

(1)   (2)   (3)

compound S   compound Q   compound R

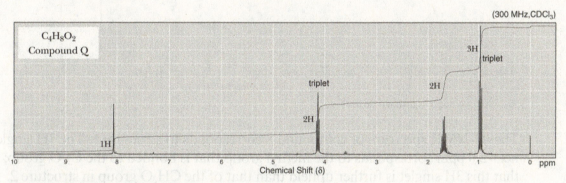

The ¹H-NMR spectrum of compound Q is distinctive in that there is a signal near 8.1 ppm. This highly downfield signal is attributed to the hydrogen that is bonded to the C=O group of the formate ester in structure 2. Also present in this spectrum is a 2H triplet near 4.2 ppm, which corresponds to a CH₂ group bonded directly to an oxygen.

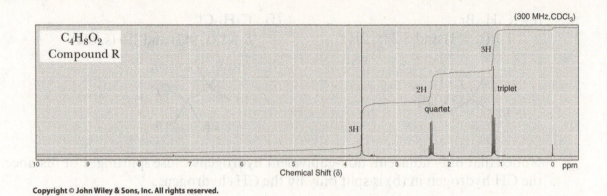

The ¹H-NMR spectrum of compound R corresponds to structure 3. The most obvious feature of this spectrum is the 3H singlet near 3.7 ppm, which corresponds to the CH_3 group bonded directly to the oxygen. This signal, with respect to its chemical shift and splitting pattern, is not expected for the methyl group in structures 1 and 2.

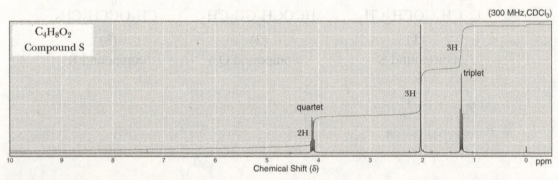

The ¹H-NMR spectrum of compound S corresponds to structure 1. The 3H singlet near 2.0 ppm corresponds to the methyl group that is bonded to the C=O group; notice that this 3H singlet is further upfield than that of the CH_3O group in structure 2. Compound S also has a 2H triplet near 4.2 ppm, which corresponds to the CH_2 of the CH_3CH_2O group.

11.43 Compound T, $C_{10}H_{10}O_2$, is insoluble in water, 10% NaOH, and 10% HCl. A ¹H-NMR spectrum of compound T shows signals at δ 2.55 (s, 6H) and 7.97 (s, 4H). A ¹³C-NMR spectrum of compound T shows four signals. From this information, propose a structural formula for T.

The molecular formula of compound T reveals that it has an IHD of six, so the compound has a combined total of six rings and/or π bonds. The chemical data suggests that the compound is neither acidic nor basic, because it does not dissolve NaOH or HCl. The compound therefore cannot be a carboxylic acid or a phenol, and from the molecular formula, it is clearly not an amine.

The 4H singlet at 7.97 ppm of the ¹H-NMR spectrum indicates the presence of an aromatic ring. The 6H singlet at 2.55 ppm suggests the presence of two CH₃ groups, and the chemical shift indicates that it is somewhat near an electron-withdrawing atom.

Along with the data from the ¹³C-NMR spectrum, which shows four signals for a C₁₀ formula, we conclude that the compound has a very high degree of symmetry. The structure that fits the chemical and spectroscopic data is shown on the right.

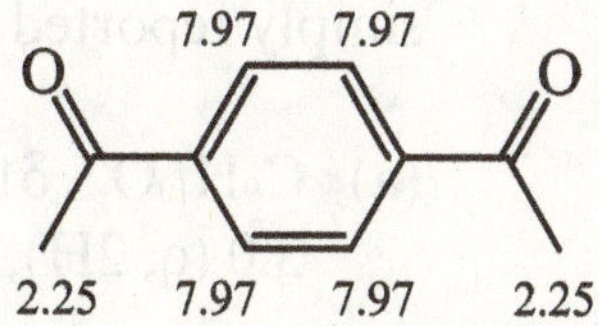

11.44 Compound U, C₁₅H₂₄O, is used as an antioxidant in many commercial food products, synthetic rubbers, and petroleum products. Propose a structural formula for compound U based on its ¹H-NMR and ¹³C-NMR spectra.

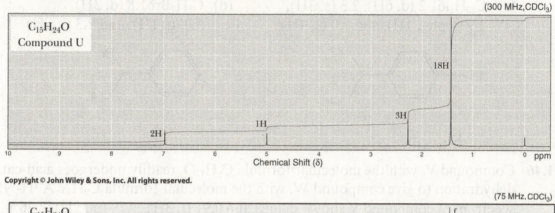

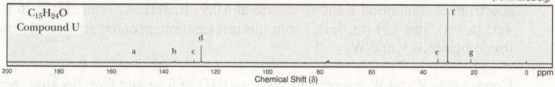

The compound has an IHD of four. The signals in the aromatic regions of the two spectra indicate the presence of a benzene ring, which accounts for the IHD. The 3H singlet near 2.3 ppm in the ¹H-NMR spectrum is indicative of a methyl group bonded to a benzene ring, the 18H singlet near 1.3 ppm suggests the presence of two *tert*-butyl groups, and the 1H singlet near 5.0 ppm is likely an alcohol. The presence of four aromatic signals (one of which is very weak) in the ¹³C-NMR spectrum indicates a symmetrically substituted benzene ring.

The compound that supports the data is the preservative BHT, short for *butylated hydroxytoluene*. Its antioxidant properties are attributed to the phenolic −OH group, which is the same group is found in vitamin E and many other naturally occurring antioxidants.

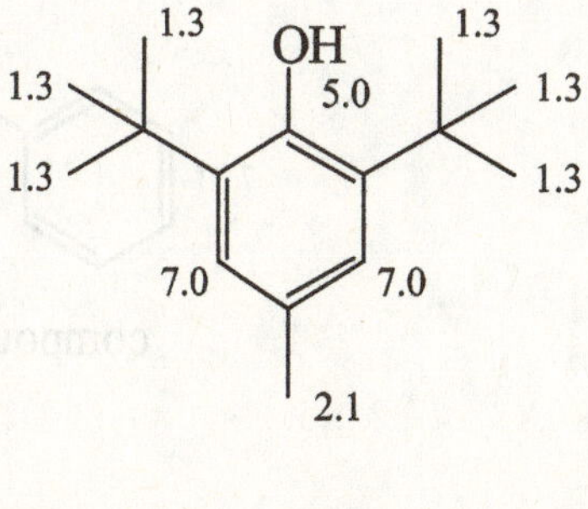

11.45 Propose a structural formula for these compounds, each of which contains an aromatic ring:

Note: It is important to emphasize at this point that splitting patterns in aromatic rings are usually difficult to resolve and describe. Accordingly, they are sometimes simply reported as singlets or multiplets.

(a) $C_9H_{10}O$ δ1.2 (t, 3H), 3.0 (q, 2H), and 7.4-8.0 (m, 5H)

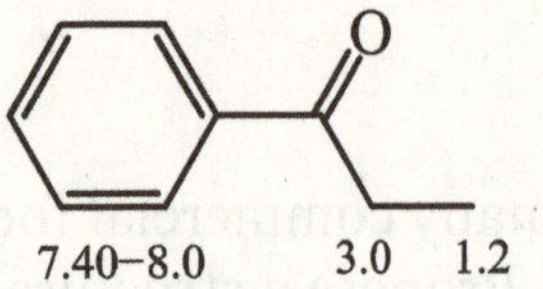

(b) $C_{10}H_{12}O_2$δ2.2 (s, 3H), 2.9 (t, 2H), 4.3 (t, 2H), and 7.3 (s, 5H)

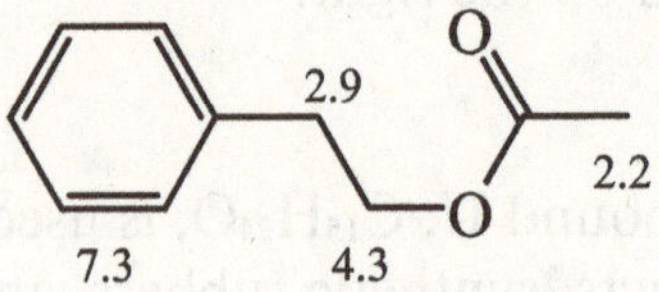

7.40–8.0   3.0   1.2

7.3   4.3   2.9   2.2

(c) $C_{10}H_{14}$δ1.2 (d, 6H), 2.3 (s, 3H), 2.9 (septet, 1H), and 7.0 (s, 4H)

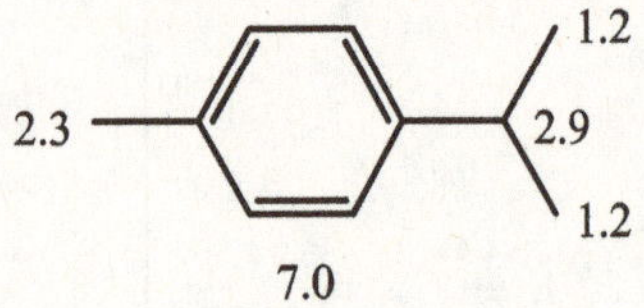

(d) C_8H_9Brδ1.8 (d, 3H), 5.0 (q, 1H), and 7.3 (s, 5H)

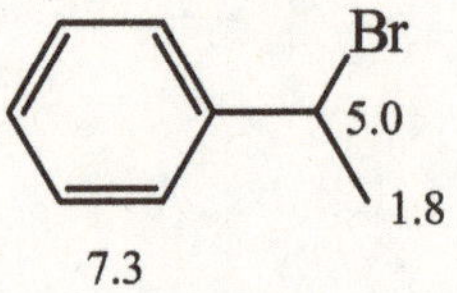

2.3   2.9   1.2   1.2   7.0

Br   5.0   1.8   7.3

11.46 Compound V, with the molecular formula $C_9H_{12}O$, readily undergoes acid-catalyzed dehydration to give compound W, with the molecular formula C_9H_{10}. A ^{1}H-NMR spectrum of compound V shows signals at δ 0.91 (t, 3H), 1.78 (m, 2H), 2.26 (s, 1H), 4.55 (t, 1H), and 7.31 (m, 5H). From this information, propose structural formulas for compounds V and W.

Compounds V and W respectively have an IHD of four and five. Because compound V undergoes dehydration under acidic conditions, it is an alcohol; dehydration forms compound W, an alkene.

The 5H multiplet at 7.31 ppm indicates the presence of a monosubstituted benzene ring, which accounts for the IHD of four. The 3H triplet at 0.91 ppm is attributed to a CH_3 group bonded to a CH_2 group, which is in turn bonded to a CH group. The structure of compound V is 1-phenyl-1-propanol. (Recall that the hydrogen of an −OH group is not split and does not cause splitting.)

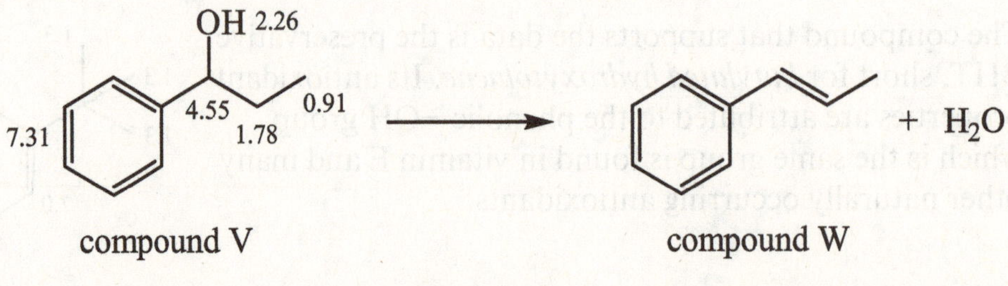

OH 2.26   4.55   0.91   7.31   1.78

compound V   compound W   + H_2O

11.47 Propose a structural formula for each ketone:

(a) C_4H_8O δ 1.0 (t, 3H),
2.1 (s 3H), and 2.4 (q, 2H)

(b) $C_7H_{14}O$ δ 0.9 (t, 6H),
1.6 (sextet, 4H), and 2.4 (t, 4H)

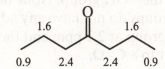

11.48 Propose a structural formula for compound X, a ketone with the molecular formula $C_{10}H_{12}O$:

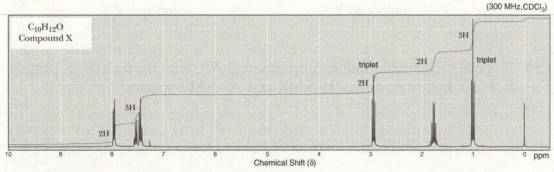

A total integration of 5 for the signals between 7–8 ppm indicates the presence of a monosubstituted benzene ring. The benzene ring and the ketone together account for the IHD of five. The 3H triplet near 1.0 ppm, 2H multiplet near 2.8 ppm, and the 2H triplet near 2.9 ppm are indicative of a propyl group; the signal at triplet at 2.9 ppm is attributed to a CH_2 group adjacent to a carbonyl group.

11.49 Following is a ¹H-NMR spectrum for compound Y, with the molecular formula $C_6H_{12}O_2$. Compound Y undergoes acid-catalyzed dehydration to give compound Z, $C_6H_{10}O$. Propose structural formulas for compounds Y and Z.

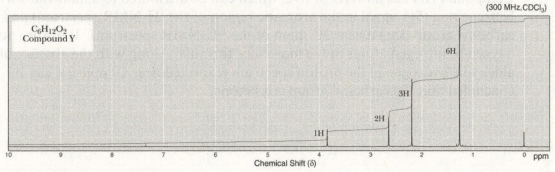

Compound Y has an IHD of one and must contain an alcohol, as it undergoes acid-catalyzed dehydration to form Q, which therefore must contain an alkene. The 6H signal at 1.2 ppm is attributed to two methyl groups, the 3H signal at 2.2 ppm to one methyl group, the 2H signal at 2.6 ppm to one CH_2 group, and the 1H signal at 3.8 ppm to the −OH group. None of the signals in the spectrum show any splitting, so those groups do not have any adjacent C−H hydrogens. The chemical shifts of the methyl group at 2.2 ppm and the CH_2 group at 2.6 ppm suggest that they are adjacent to a C=O.

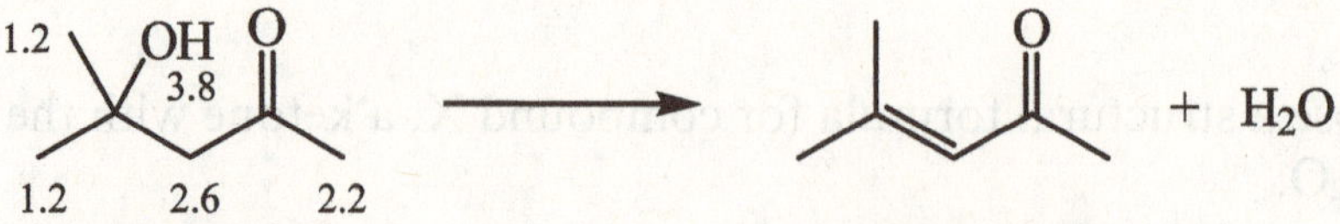

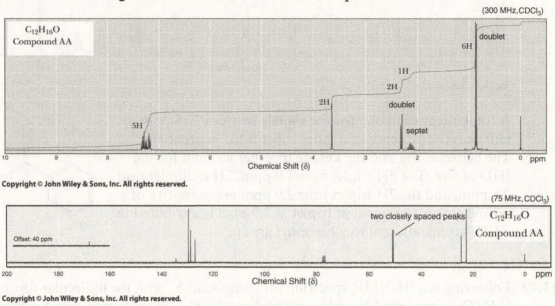

11.50 Propose a structural formula for compound AA, with the molecular formula $C_{12}H_{16}O$. Following are its ^{1}H-NMR and ^{13}C-NMR spectra:

Compound AA has an IHD of five, which can be attributed to a monosubstituted benzene ring (5H signal in the aromatic region of the ^{1}H-NMR spectrum) and a carbonyl group (signal near 207 ppm of the ^{13}C-NMR spectrum). There are no other oxygen or nitrogen atoms in the molecular formula. Along with the absence of an aldehydic hydrogen in the proton spectrum (expected near 10 ppm), it can be concluded that the carbonyl group is a ketone.

The 6H doublet near 0.9 ppm of ^{1}H-NMR spectrum is suggestive of two CH$_3$ groups bonded to a CH group. There are also two CH$_2$ groups; the one near 2.3 ppm is a doublet must be bonded to a CH group, while the one near 3.6 ppm is a singlet. The 1H multiplet near 2.1 ppm can be attributed to the CH that is bonded to the two methyl groups. The structure consistent with this data is 4-methyl-1-phenyl-2-pentanone.

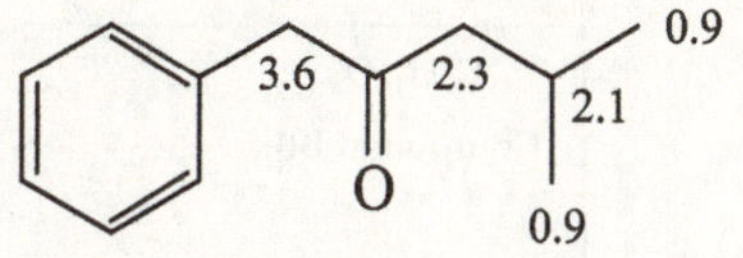

11.51 Propose a structural formula for each carboxylic acid:

(a) C$_5$H$_{10}$O$_2$

^{1}H-NMR	^{13}C-NMR
0.94 (t, 3H)	180.7
1.39 (m, 2H)	33.89
1.62 (m, 2H)	26.76
2.35 (t, 2H)	22.21
12.0 (s, 1H)	13.69

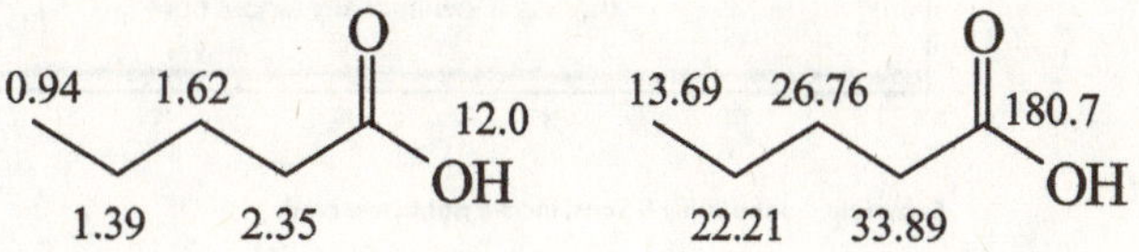

(b) C$_6$H$_{12}$O$_2$

^{1}H-NMR	^{13}C-NMR
1.08 (s, 9H)	179.29
2.23 (s, 2H)	46.82
12.1 (s, 1H)	30.62
	29.57

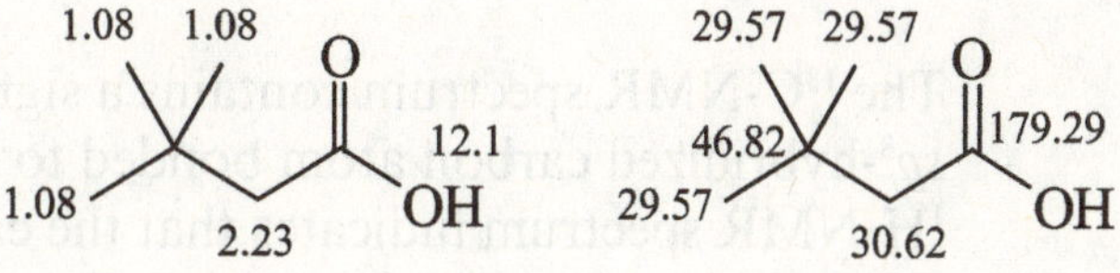

(c) C$_5$H$_8$O$_4$

^{1}H-NMR	^{13}C-NMR
0.93 (t, 3H)	170.94
1.80 (m, 2H)	53.28
3.10 (t, 1H)	21.90
12.7 (s, 2H)	11.18

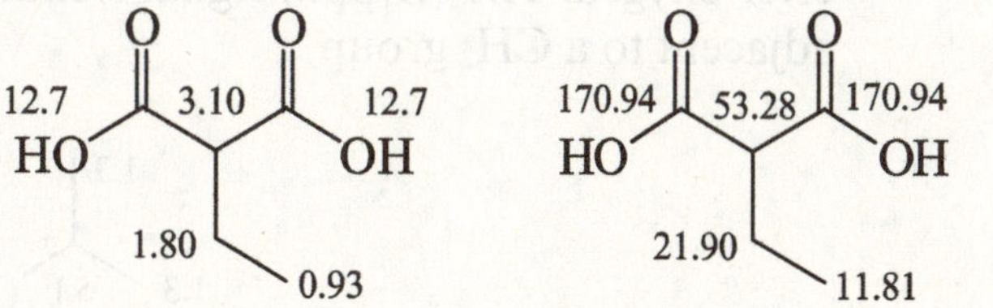

11.52 Following are ^{1}H-NMR and ^{13}C-NMR spectra of compound BB, with the molecular formula $C_7H_{14}O_2$. Propose a structural formula for compound BB.

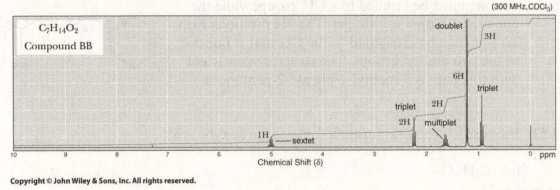

Compound BB has an IHD of one, which can be attributed to a carbonyl group (signal near 175 ppm in the ^{13}C-NMR spectrum). The chemical shift of the carbonyl group is indicative of a carboxylic acid or an ester, both of which are also supported by the fact that the molecular formula has two oxygens. However, no proton corresponding to the COOH group is observed in the ^{1}H-NMR spectrum, so the compound is likely an ester.

The ^{13}C-NMR spectrum contains a signal near 67 ppm, consistent with an sp^3-hybridized carbon atom bonded to an oxygen. The signal at 5.1 ppm in the ^{1}H-NMR spectrum indicates that the carbon atom bonded to the ester oxygen has one hydrogen, but it is bonded to five adjacent hydrogens. The signal at 2.3 ppm (t, 2H) is a CH_2 adjacent to a carbonyl group, and it is also adjacent to another CH_2 group (1.7 ppm). The 6H doublet at 1.3 ppm is consistent with two CH_3 groups bonded to a CH group, and this correlates to the CH group that is bonded to the ester oxygen. The 0.9 ppm signal is indicative of another methyl group that is adjacent to a CH_2 group.

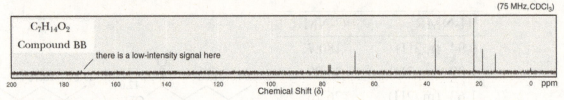

11.53 Propose a structural formula for each ester:

(a) $C_6H_{12}O_2$

^{1}H-NMR	^{13}C-NMR
1.18 (d, 6H)	117.16
1.26 (t, 3H)	60.17
2.51 (m, 1H)	34.04
4.13 (q, 2H)	19.01
	14.25

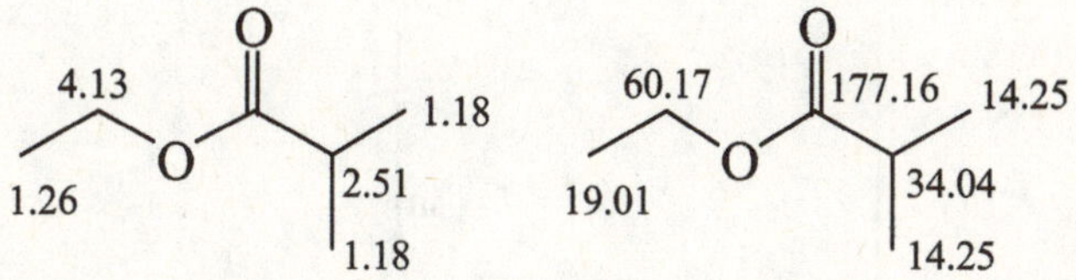

(b) $C_7H_{12}O_4$

^{1}H-NMR	^{13}C-NMR
1.28 (t, 6H)	166.52
3.36 (s, 2H)	61.43
4.21 (q, 4H)	41.69
	14.07

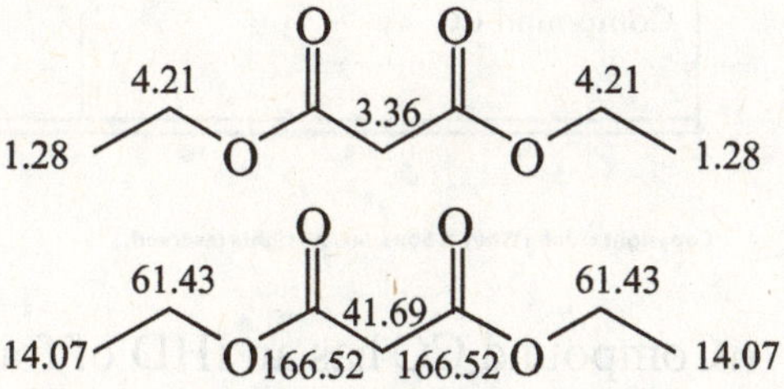

(c) $C_7H_{14}O_2$

^{1}H-NMR	^{13}C-NMR
0.92 (d, 6H)	171.15
1.52 (m, 2H)	63.12
1.70 (m, 1H)	37.31
2.09 (s, 3H)	25.05
4.10 (t, 2H)	22.45
	21.06

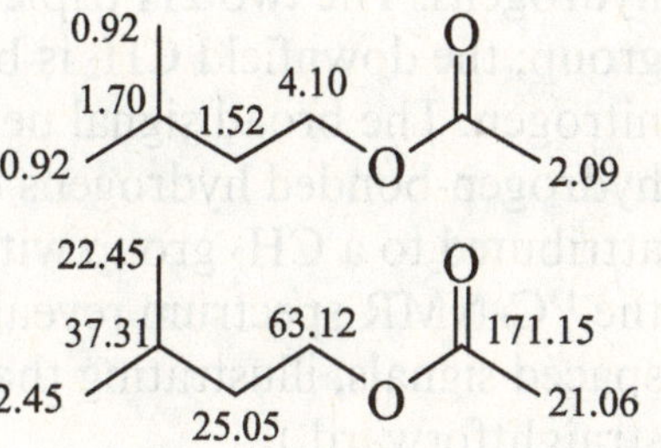

11.54 Following are ^{1}H-NMR and ^{13}C-NMR spectra of compound CC, with the molecular formula $C_{10}H_{15}NO$. Propose a structural formula for compound CC.

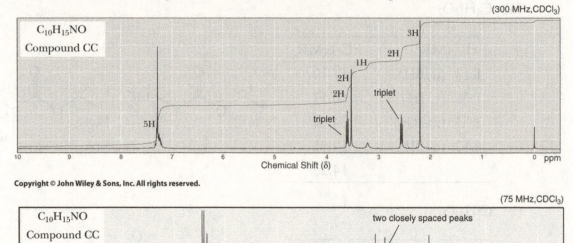

Compound CC has an IHD of four, all of which are attributed to a monosubstituted benzene ring (5H signal near 7.3 ppm of the ^{1}H-NMR spectrum). Other than the benzene ring, there cannot be any other rings or π bonds ion the structure.

The 3H singlet near 2.2 ppm in the ^{1}H-NMR spectrum is a CH_3 with no neighboring hydrogens. The two 2H triplets (near 2.5 and 3.7 ppm) are consistent with a CH_2CH_2 group; the downfield CH_2 is bonded to oxygen while the upfield CH_2 is bonded to nitrogen. The broad signal near 3.2 ppm is indicative of an −OH group (signals of hydrogen-bonded hydrogens often appear broad). The singlet near 3.5 ppm is attributed to a CH_2 group with no neighboring protons. (Note: Careful inspection of the ^{13}C-NMR spectrum reveals that the signal at 58 ppm is actually two closely spaced signals, illustrating that the interpretation of NMR spectra is not always straightforward.)

11.55 Propose a structural formula for amide DD, with molecular formula $C_6H_{13}NO$:

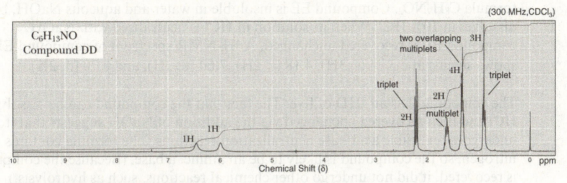

The compound has an IHD of one, which is attributed to the carbonyl group of the amide. The signals at between 0.9 and 2.2 ppm are consistent with a $CH_3CH_2CH_2CH_2CH_2$ structure. Note that the absence of 1H signals in the alkyl region of the spectrum is a good indicator of the absence of branching. The last CH_2 group, which is adjacent to the carbonyl group of the amide, is at 2.2 ppm. The two 1H signals near 6.0 and 6.6 ppm are each due to an amide hydrogen. Realize that although both amide hydrogens are bonded to the same nitrogen, the two hydrogens are not equivalent. This is because the amide group is stabilized by resonance and the carbon-nitrogen bond has double-bond character.

11.56 Propose a structural formula for the analgesic phenacetin, with the molecular formula $C_{10}H_{13}NO_2$, based on its ^1H-NMR spectrum:

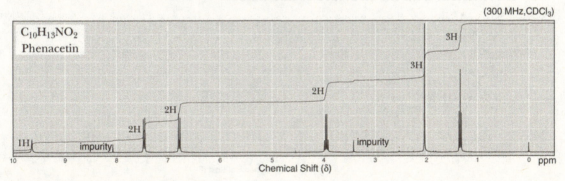

Phenacetin has an IHD of five. The two 2H doublets near 6.8 and 7.5 ppm in the spectrum are noteworthy in that the integration and splitting are characteristic of a para-disubstituteted benzene ring. The 3H triplet near 1.3 ppm and the 2H quartet near 4.0 ppm suggest the presence of an ethyl group that is bonded to oxygen; recall that a CH_2 group bonded to an oxygen is near 4 ppm. The 3H singlet near 2.0 ppm is typical of a CH_3 bonded to a carbonyl group. Finally, the signal near 9.7 ppm is attributed to the hydrogen of an amide.

11.57 Propose a structural formula for compound EE, an oily liquid with the molecular formula $C_8H_9NO_2$. Compound EE is insoluble in water and aqueous NaOH, but dissolves in 10% HCl. When its solution in HCl is neutralized with NaOH, compound EE is recovered unchanged. A ^1H-NMR spectrum of compound EE shows signals at δ 3.84 (s, 3H), 4.18 (s, 2H), 7.60 (d, 2H), and 8.70 (d, 2H).

The compound has an IHD of five. The fact that the compound can be dissolved in HCl and then recovered unchanged via the addition of NaOH suggests that it undergoes reversible acid-base neutralization reactions. The formula contains nitrogen, so the compound is likely to be an amine, a base. (Because the compound is recovered, it did not undergo other chemical reactions, such as hydrolysis.) Referring to the ^1H-NMR data, the pair of 2H doublets in the aromatic region (7.60 and 8.70 ppm) indicates a para-disubstituted benzene ring. The 3H singlet at 3.84 ppm is a methyl group bonded to oxygen (methoxy group). Knowing that the compound is an amine, we can assign the 2H singlet at 4.18 ppm to an NH_2 group. The disubstituted benzene ring and the methoxy and amino groups account for all atoms in the molecular formula except for one C and one O. These two atoms must be a carbonyl group because the compound has an IHD of five, four of which are allocated to the ring. The only structure consistent with all the data is methyl 4-aminobenzoate.

11.58 Following is a ^1H-NMR spectrum and a structural formula for anethole, $C_{10}H_{12}O$, a fragrant natural product obtained from anise. Using the line of integration, determine the number of protons giving rise to each signal. Show that this spectrum is consistent with the structure of anethole.

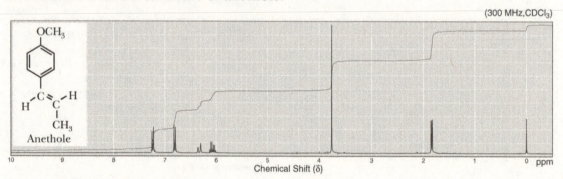

(300 MHz, CDCl₃)

Chemical Shift (δ)

The approximate chemical shift of the signals, the splitting patterns, and the number of hydrogen atoms corresponding to each signal are: δ 1.8 (d, 3H), 3.8 (s, 3H), 6.5 (m, 1H), 6.3 (d, 1H), 6.8 (d, 2H), 7.2 (d, 2H). The pair of 2H doublets near 6.8 and 7.2 ppm are attributed to the aromatic ring. The methyl group bonded to oxygen is more downfield than the methyl group bonded to the alkene. The vinylic hydrogen near 6.5 ppm has four adjacent hydrogens, but the vinylic hydrogen near 6.3 ppm has only one adjacent hydrogen.

11.59 Propose a structural formula for compound FF, with the molecular formula C_4H_6O, based on the following IR and ^{1}H-NMR spectra:

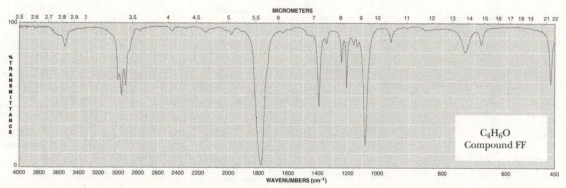

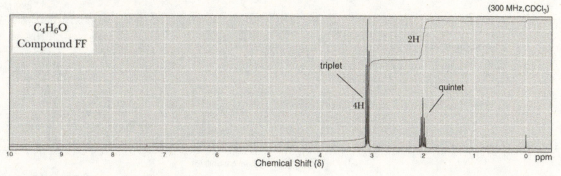

Compound FF has an IHD of two. A notable feature in the IR spectrum is the strong absorption near 1780 cm^{-1}, which indicative of a carbonyl group, accounting for one IHD. How can we account for the other IHD? The spectra are useful because they provide us with not only an idea of the functional groups that are present, but also which groups are absent. In this case, the IR spectrum shows an absence of both C=C and =C−H absorptions, and the NMR spectrum also confirms the absence of the vinylic hydrogens. Therefore, the second IHD is attributed to a ring. Cyclobutanone is consistent with the data.

11.60 Propose a structural formula for compound GG, with the molecular formula $C_5H_{10}O_2$, based on the following IR and ^1H-NMR spectra:

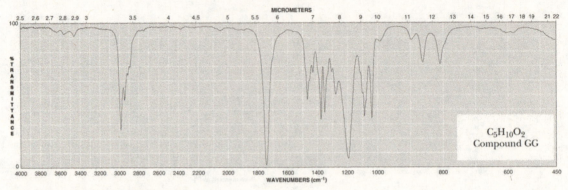

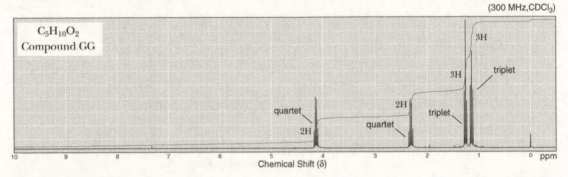

Compound GG has an IHD of one. The IR spectrum reveals a a carbonyl group (1750 cm^{-1}), which accounts for the IHD. A carbonyl group near 1750 cm^{-1} is likely an ester or a carboxylic acid, but we can rule out the latter due to the absence of an O–H stretch between 2400 and 3400 cm^{-1}.

The ^1H-NMR spectrum has two characteristic triplet-quartet pairs, suggesting two ethyl groups. The signals near 1.3 (t, 3H) and 4.1 (q, 2H) come from an ethoxy (CH$_3$CH$_2$O) group, while those near 1.1 (t, 3H) and 2.3 (q, 2H) come from an ethyl group. The compound matching all the data is ethyl propanoate.

11.61 Propose a structural formula for compound HH, with the molecular formula $C_5H_9ClO_2$, based on the following IR and ^{1}H-NMR spectra:

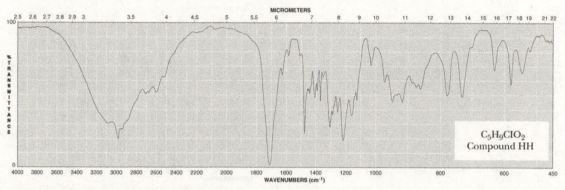

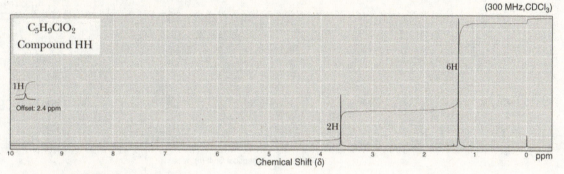

Compound HH has an IHD of one. The IR spectrum reveals a carbonyl group (sharp absorption near 1700 cm^{-1}) and an −OH group (strong, broad absorption spanning the region between 2400 and 3400 cm^{-1}). The compound therefore contains a carboxylic acid, which accounts for the IHD.

The ^{1}H-NMR spectrum confirms the presence of a carboxylic acid proton (1H singlet at 12.4 ppm). The 6H singlet near 1.3 ppm is indicative of two methyl groups that do not have any adjacent hydrogens. The 2H singlet near 3.6 ppm arises from a CH$_2$ group without any adjacent hydrogens. The only possible structure fitting this data is 3-chloro-3-methylbutanoic acid.

11.62 Propose a structural formula for compound II, with the molecular formula $C_6H_{14}O$, based on the following IR and ^1H-NMR spectra:

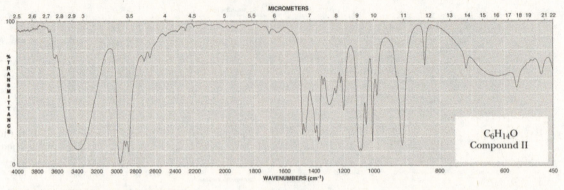

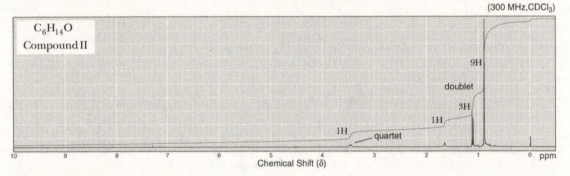

Compound II has an IHD of zero. The IR spectrum reveals an alcohol −OH group (strong, broad absorption near 3400 cm⁻¹); notice the difference in wavenumber between the peak of an alcohol −OH and that of a carboxylic acid −OH, as seen in the previous problem.

In the ^1H-NMR spectrum, the 9H singlet near 0.9 ppm indicates a *tert*-butyl group, −C(CH₃)₃, which accounts for four of the six carbons in the molecule. The 3H doublet near 1.1 ppm indicates a methyl group bonded to a CH. This CH (1H quartet near 3.5 ppm) is also attached to an −OH, evidenced by downfield chemical shift. The signal corresponding to the −OH hydrogen is near 1.7 ppm. The compound matching all of the data is 3,3-dimethyl-2-butanol.

Looking Ahead

11.63 Predict the position of the C=O stretching absorption in acetate ion relative to that in acetic acid:

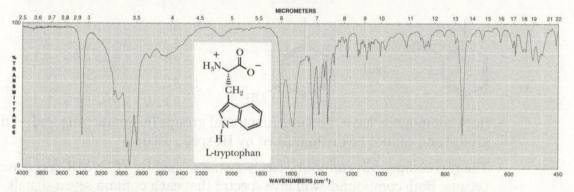

Acetic acid   **Acetate ion**

The C=O stretching absorption for acetate will be at a lower wavenumber. This is because resonance structures can be drawn for acetate, keeping in mind that each resonance structure contributes to the actual structure of acetate. Both carbon-oxygen bonds in acetate are neither single bonds nor double bonds, but rather somewhere in between. The carbon-oxygen bonds in acetate are therefore weaker than the true C=O double bond in acetate. Weaker bonds require less energy for stretching to occur.

11.64 Following is the IR spectrum of L-tryptophan, a naturally occurring amino acid that is abundant in foods like turkey. For many years, the L-tryptophan in turkey was believed to make people drowsy after Thanksgiving dinner. Scientists now know that consumption of L-tryptophan only makes one drowsy if taken on an empty stomach. Therefore it is unlikely that one's Thanksgiving day turkey is the cause of drowsiness. Notice that L-tryptophan contains one stereocenter. Its enantiomer, D-tryptophan, does not occur in nature but can be synthesized in the laboratory. What would the IR spectrum of D-tryptophan look like?

The IR spectrum of D-tryptophan will look identical to that of L-tryptophan. D- and L-tryptophan are enantiomers, which have the same physical and chemical properties in an achiral environment, such as that used in a simple IR experiment.

Group Learning Activities

11.65 Discuss whether IR or NMR spectroscopy could be used to distinguish between the following pairs of molecules. Be very specific in describing the spectral data that would allow you to identify each compound. Assume that you do not have the reference spectra of either molecule.

(a)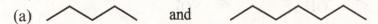

Both compounds are alkanes, and with no other functional groups, they cannot be distinguished by IR spectroscopy.

However, they can be distinguished by either ^{1}H-NMR or ^{13}C-NMR spectroscopy. Both the ^{1}H-NMR and ^{13}C-NMR spectra of the compound on the left will each have three signals, while the spectra of the compound on the right will each have four signals.

(b)

Both compounds contain a vinylic =C–H and cannot be distinguished by IR spectroscopy.

However, they can easily be distinguished by ^{13}C-NMR spectroscopy. The ^{13}C-NMR spectrum of the compound on the left will contain three signals, while that of the compound on the right will have six signals.

^{1}H-NMR spectroscopy could also be used to distinguish between the two compounds. The ^{1}H-NMR spectrum of the compound on the left will contain three signals, while that of the compound on the right will have six signals.

(c)

Both compounds have the same functional groups (aromatic ring and –OH group) and cannot be distinguished by IR spectroscopy.

The two compounds cannot be easily distinguished by ^{13}C-NMR spectroscopy because both compounds will give spectra that each contain seven signals.

However, they can be distinguished by ^{1}H-NMR spectroscopy. The compound on the right will contain a singlet in the aromatic region of the spectrum.

(d)

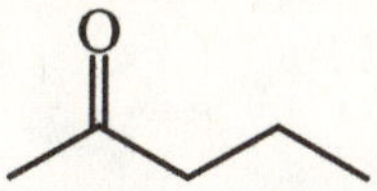

and

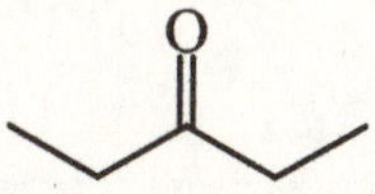

Both compounds are ketones and cannot be distinguished by IR spectroscopy.

Like the two compounds in (b), they can easily be distinguished by ^{13}C-NMR spectroscopy. The ^{13}C-NMR spectrum of the compound on the left will contain five signals, while that of the compound on the right will have three signals.

^{1}H-NMR spectroscopy could also be used to distinguish between the two compounds. The ^{1}H-NMR spectrum of the compound on the left will contain four signals, while that of the compound on the right will have two signals.

(e)

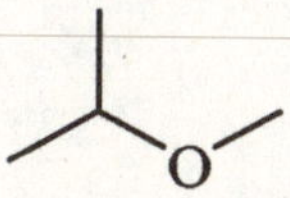

and

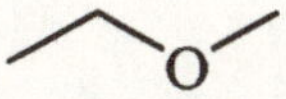

Both compounds are ethers and cannot be distinguished by IR spectroscopy.

They also cannot be distinguished by ^{13}C-NMR spectroscopy because the ^{13}C-NMR spectra of both compounds will each contain three signals.

However, they can be distinguished by ^{1}H-NMR spectroscopy. The compound on the left would give three signals (one doublet, one multiplet, and one singlet). The compound on the right would also have three signals, but the signals would have a different splitting pattern (one triplet, one quartet, and one singlet).

Chapter 12: Aldehydes and Ketones

Problems

12.1 Write the IUPAC name for each compound:

(a)

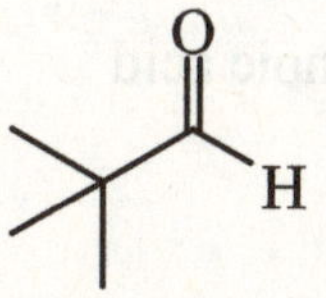

2,2-Dimethylpropanal

(b)

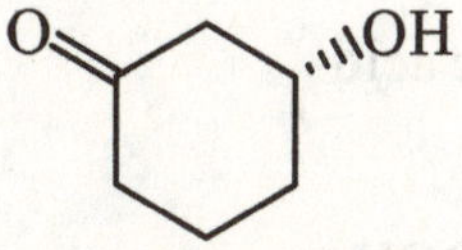

(R)-3-Hydroxy-
cyclohexanone

(c)

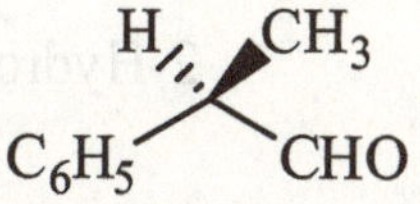

(R)-2-Phenylpropanal

(d)

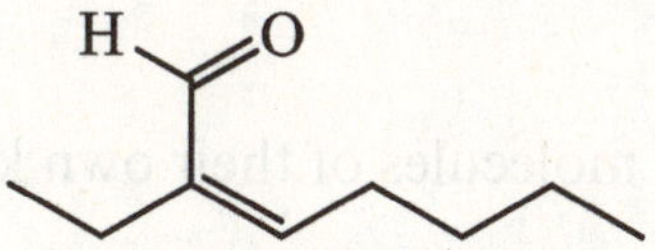

(Z)-2-Ethyl-2-heptenal

12.2 Write structural formulas for all aldehydes with molecular formula $C_6H_{12}O$, and give each its IUPAC name. Which of these aldehydes are chiral?

The stereocenter in each chiral aldehyde is indicated by an asterisk.

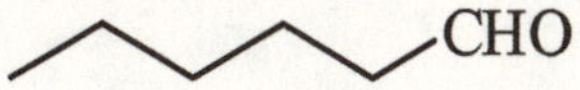

Hexanal

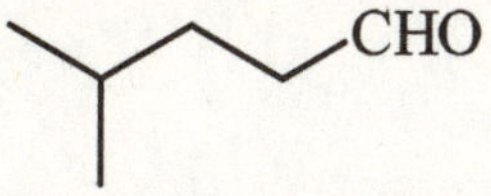

4-Methylpentanal

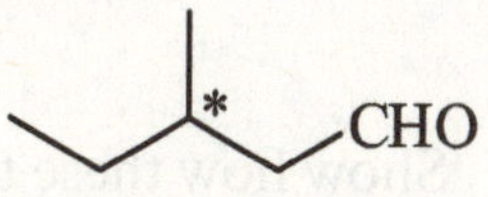

3-Methylpentanal

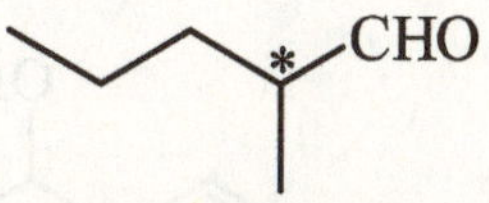

2-Methylpentanal

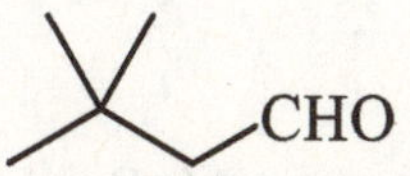

3,3-Dimethylbutanal

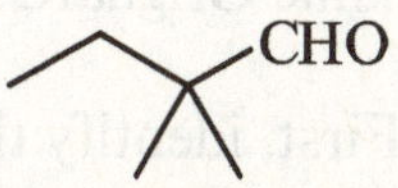

2,2-Dimethylbutanal

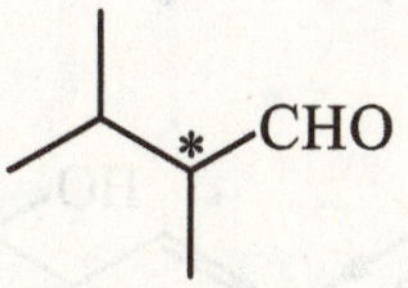

2,3-Dimethylbutanal

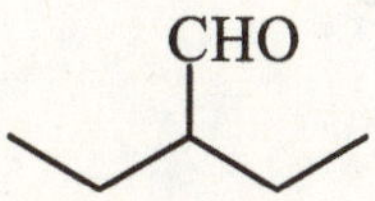

2-Ethylbutanal

12.3   Write IUPAC names for each compounds, each of which is important in intermediary metabolism. The name shown is the one by which the compound is more commonly known in the biological sciences.

(a)

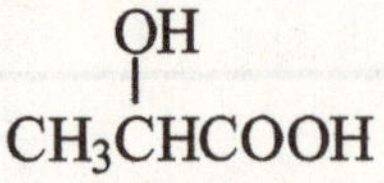

$$\underset{\displaystyle \text{CH}_3\text{CHCOOH}}{\overset{\displaystyle \overset{\text{OH}}{|}}{}}$$

Lactic acid

2-Hydroxypropanoic acid

(b)

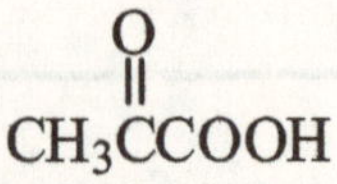

$$\underset{\displaystyle \text{CH}_3\text{CCOOH}}{\overset{\displaystyle \overset{\text{O}}{\|}}{}}$$

Pyruvic acid

2-Oxopropanoic acid

(c) $\text{H}_2\text{NCH}_2\text{CH}_2\text{CH}_2\text{COOH}$

γ-Aminobutyric acid

4-Aminobutanoic acid

12.4   Explain how these Grignard reagents react with molecules of their own kind to "self-destruct":

(a) 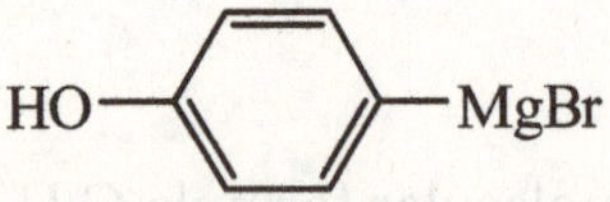   HO—⬡—MgBr

(b)

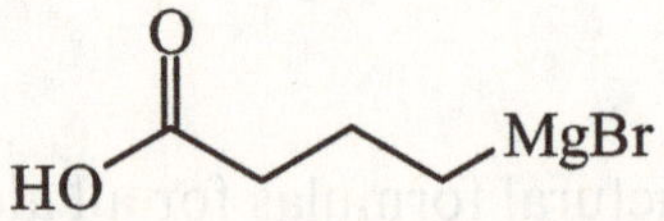

Grignard reagents, which behave as carbanion equivalents, are highly basic and are destroyed by acids. Both of these reagents contain ionizable hydrogens; (a) contains a hydroxyl group and (b) contains a carboxyl group.

12.5   Show how these three compounds can be synthesized from the same Grignard reagent:

First, identify the common structural fragment in all three compounds; this would be the cyclohexene ring. In each product, the carbon bearing the −OH group was the carbonyl group in the starting material.

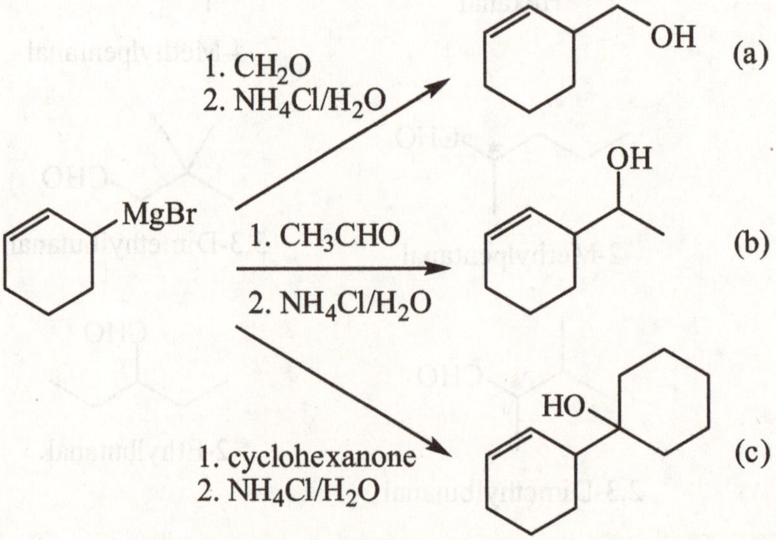

1. CH_2O
2. $\text{NH}_4\text{Cl}/\text{H}_2\text{O}$

(a)

1. CH_3CHO
2. $\text{NH}_4\text{Cl}/\text{H}_2\text{O}$

(b)

1. cyclohexanone
2. $\text{NH}_4\text{Cl}/\text{H}_2\text{O}$

(c)

12.6 The hydrolysis of an acetal forms an aldehyde or ketone and two alcohols. Following are structural formulas for four acetals. Draw the structural formulas for the products of hydrolysis of each in aqueous acid (i.e., provide the carbonyl compound and alcohol(s) from which each acetal was derived).

In the hydrolysis of an acetal, the acetal carbon (the carbon bonded to two −OR groups) reverts to either an aldehyde or a ketone. If the acetal carbon bears an H, it becomes an aldehyde. The acetal carbon in each compound is highlighted.

(a)

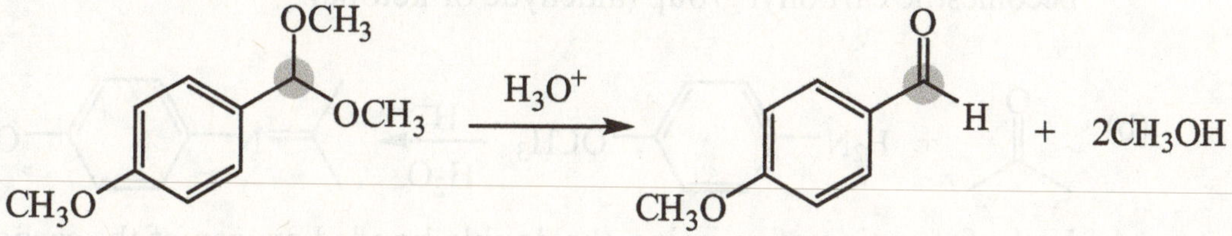

$$\text{(aryl acetal)} \xrightarrow{\text{H}_3\text{O}^+} \text{(aryl aldehyde)} + 2\text{CH}_3\text{OH}$$

(b)

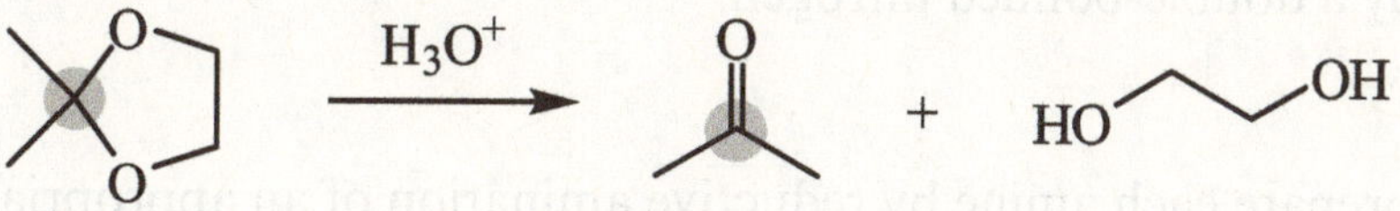

$$\xrightarrow{\text{H}_3\text{O}^+} + \text{HO} \diagdown \text{OH}$$

(c)

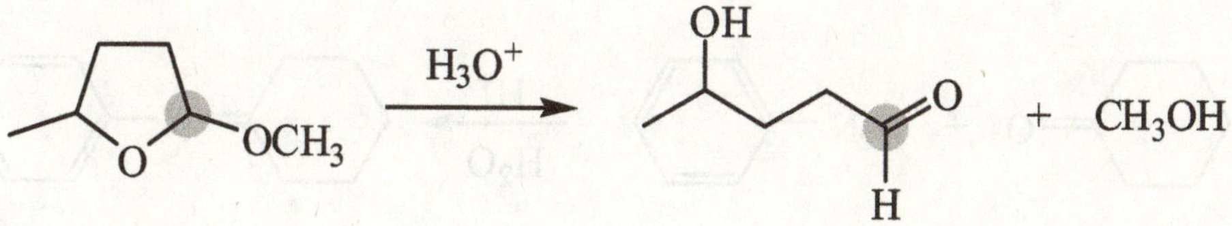

$$\xrightarrow{\text{H}_3\text{O}^+} + \text{CH}_3\text{OH}$$

(d)

$$\xrightarrow{\text{H}_3\text{O}^+}$$

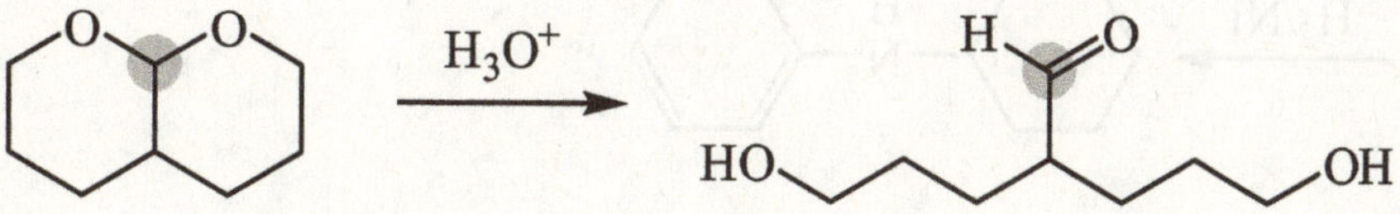

12.7 Propose a method for the following transformation:

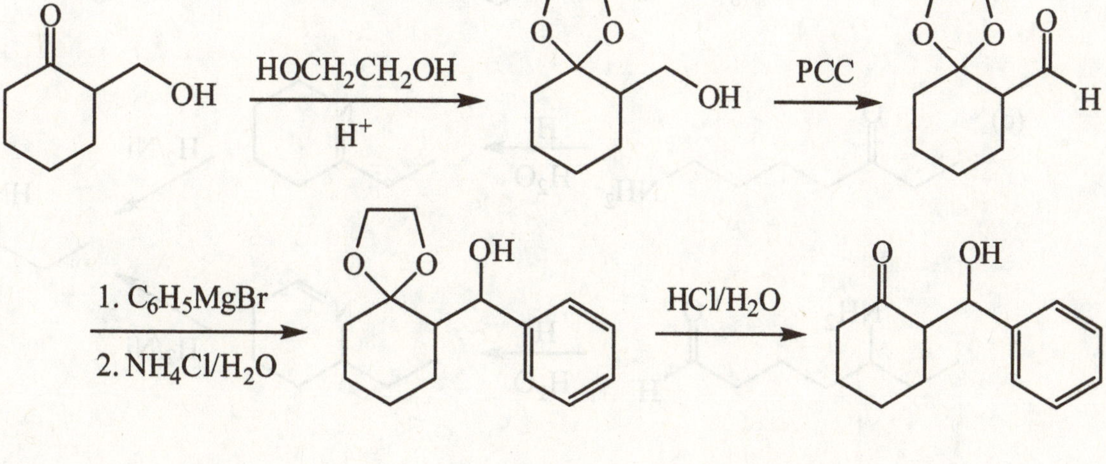

$$\xrightarrow[\text{H}^+]{\text{HOCH}_2\text{CH}_2\text{OH}} \xrightarrow{\text{PCC}}$$

$$\xrightarrow[\text{2. NH}_4\text{Cl/H}_2\text{O}]{\text{1. C}_6\text{H}_5\text{MgBr}} \xrightarrow{\text{HCl/H}_2\text{O}}$$

12.8   Predict the products formed in each reaction. Note: Acid-catalyzed hydrolysis of an imine gives an amine and an aldehyde or a ketone. When one equivalent or more of acid is used, the amine is converted to its ammonium salt.

(a) C_6H_5—CH=NCH$_2$CH$_3$ + H$_2$O $\xrightarrow{\text{HCl}}$ C_6H_5—CHO + H$_3\overset{+}{N}$CH$_2$CH$_3$

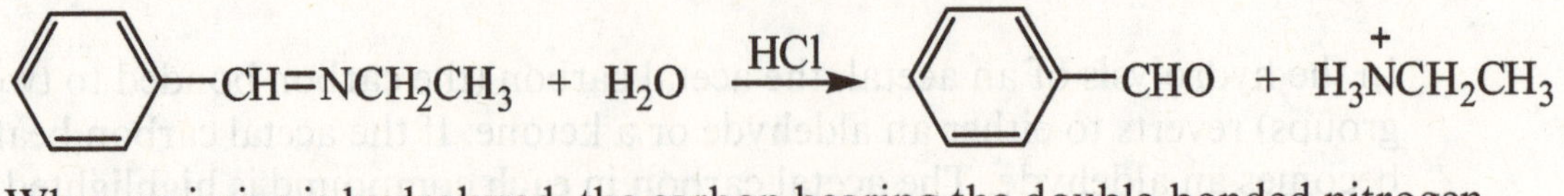

When an imine is hydrolyzed, the carbon bearing the double-bonded nitrogen becomes the carbonyl group (aldehyde or ketone).

(b)   (acetone) + H$_2$N—C_6H_4—OCH$_3$ $\xrightarrow[\text{H}_2\text{O}]{\text{H}^+}$ (imine)=N—C_6H_4—OCH$_3$ + H$_2$O

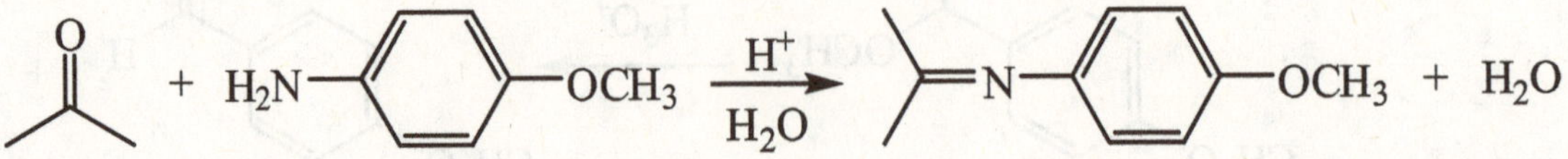

In the formation of an imine, the double-bonded oxygen of the carbonyl group is replaced by a double-bonded nitrogen.

12.9   Show how to prepare each amine by reductive amination of an appropriate aldehyde or ketone:

(a)   (cyclohexanone)=O + H$_2$N—C_6H_5 $\xrightarrow[\text{H}_2\text{O}]{\text{H}^+}$ (cyclohexylidene)=N—C_6H_5

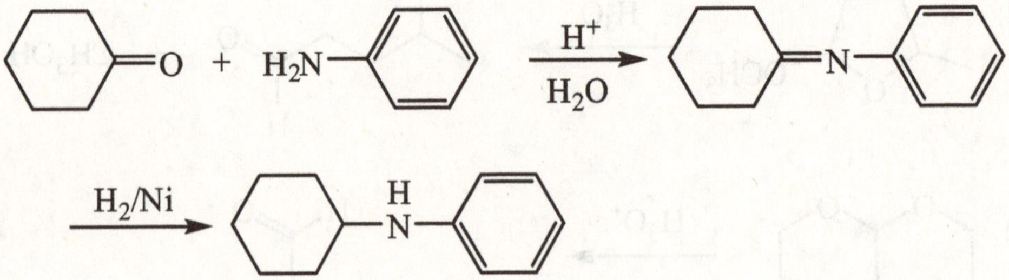

$\xrightarrow{\text{H}_2/\text{Ni}}$ (cyclohexyl)—$\overset{\text{H}}{\text{N}}$—$C_6H_5$

(b)   (propiophenone) + NH$_3$ $\xrightarrow[\text{H}_2\text{O}]{\text{H}^+}$ (NH imine) $\xrightarrow{\text{H}_2/\text{Ni}}$ (NH$_2$ amine)

(c)   (keto-amine) $\xrightarrow[\text{H}_2\text{O}]{\text{H}^+}$ (cyclic imine) $\xrightarrow{\text{H}_2/\text{Ni}}$ (piperidine, HN)

(amino-aldehyde) $\xrightarrow[\text{H}_2\text{O}]{\text{H}^+}$ (cyclic imine) $\xrightarrow{\text{H}_2/\text{Ni}}$

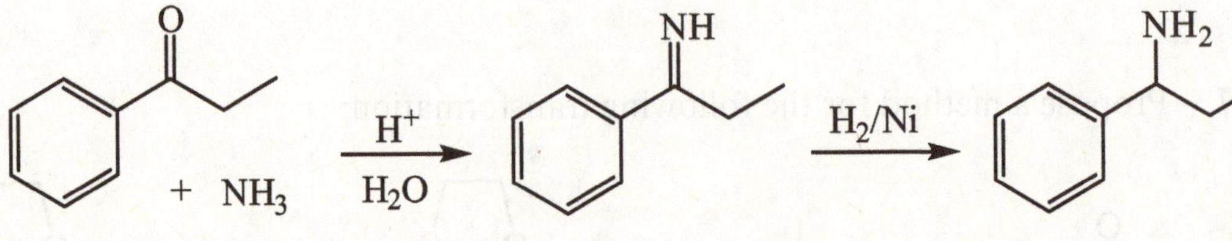

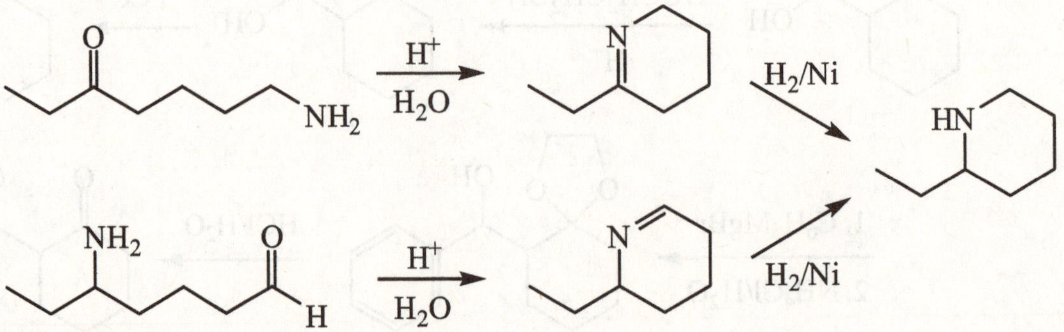

12.10 Draw the structural formula for the keto form of each enol:

(a)

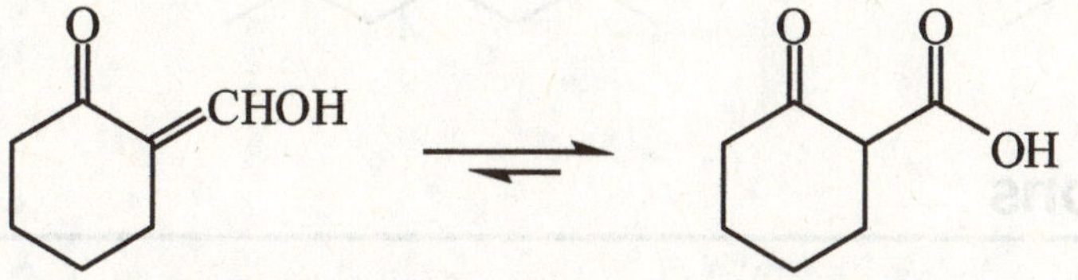

(b)

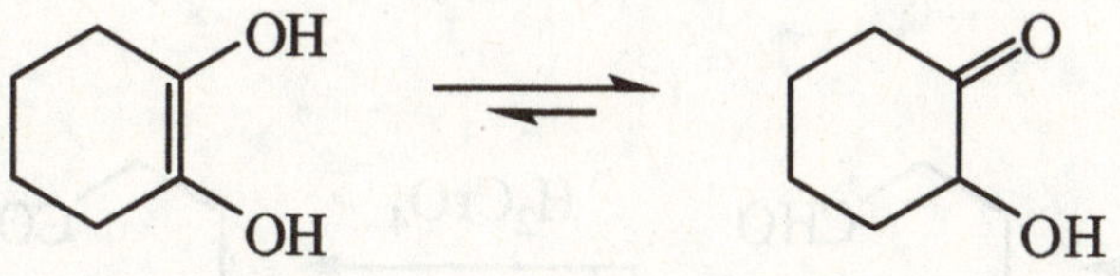

(c)

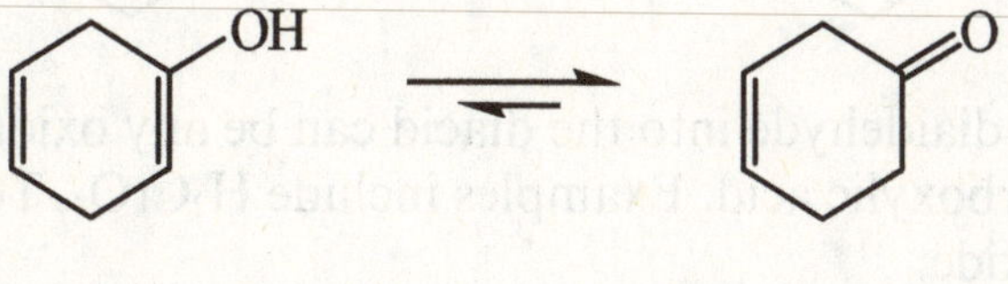

(d)

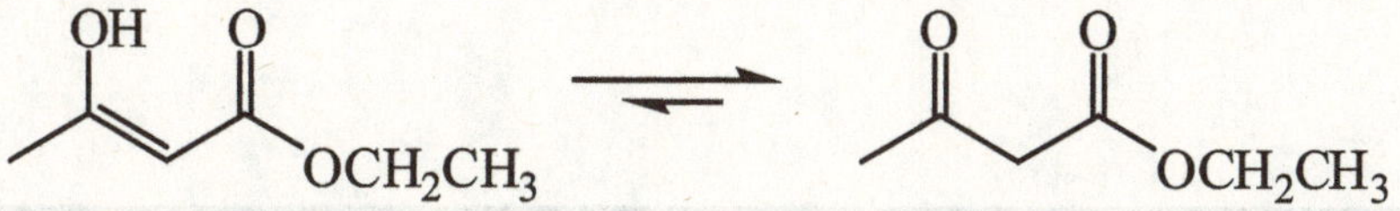

12.11 Complete these oxidations:

(a)

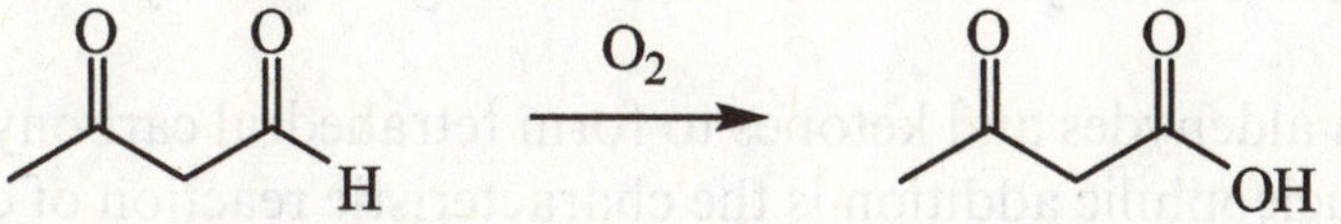

(b)

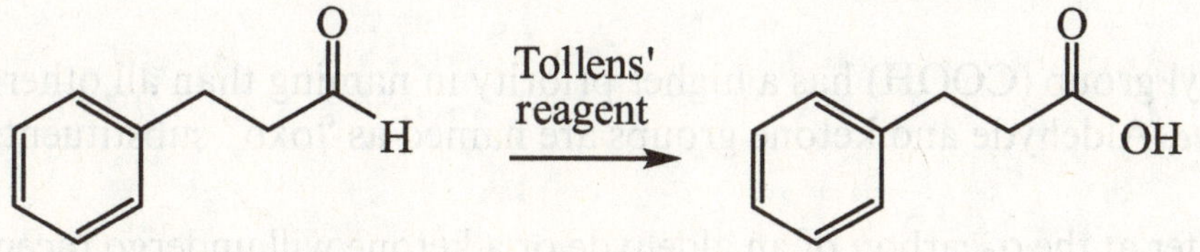

12.12 What aldehyde or ketone gives each alcohol on reduction by $NaBH_4$?

(a)

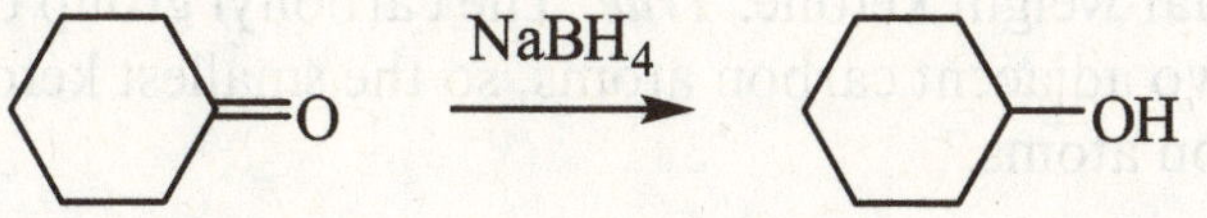

(b)

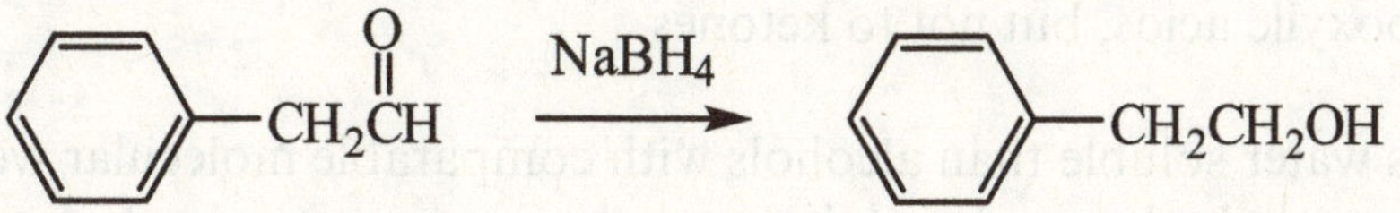

(c)

Chemical Connections

12A. Using chemistry presented in this and previous chapters, propose a synthesis for adipic acid from cyclohexene.

The oxidant used to convert the dialdehyde into the diacid can be any oxidant capable of oxidizing an aldehyde to a carboxylic acid. Examples include H_2CrO_4, Tollens' reagent, O_2, and hydrogen peroxide.

Quick Quiz

1. In a compound that contains both an aldehyde and a C−C double bond, each functional group can be reduced exclusive of the other. *True*. Alkenes are can be reduced by catalytic hydrogenation, and aldehydes can be reduced using a metal hydride.

2. Nucleophiles react with aldehydes and ketones to form tetrahedral carbonyl addition intermediates. *True*. Nucleophilic addition is the characteristic reaction of carbonyl compounds.

3. The carboxyl group (COOH) has a higher priority in naming than all other functional groups. *True*. Aldehyde and ketone groups are named as "oxo" substituents.

4. A stereocenter at the α-carbon of an aldehyde or a ketone will undergo racemization over time in the presence of an acid or a base. *True*. An enol intermediate, the formation of which is accelerated in acid or base, is involved in the racemization process.

5. Acetone is the lowest molecular weight ketone. *True*. The carbonyl group of a ketone must be directly bonded to two adjacent carbon atoms, so the smallest ketone must have at least three total carbon atoms.

6. Aldehydes can be oxidized to ketones and carboxylic acids. *False*. Aldehydes can be oxidized to carboxylic acids, but not to ketones.

7. Ketones are less water soluble than alcohols with comparable molecular weight. *True*. Ketones cannot act as hydrogen-bond donors.

8. A Grignard reagent cannot be formed in the presence of an NH, OH, or SH group. *True*. Grignard reagents are highly basic and would be destroyed by these acidic groups.

9. Ketones have higher boiling points than alkanes with comparable molecular weight. *True*. The intermolecular forces present in ketones, which are polar compounds, are higher.

10. An aldehyde has a higher priority in naming than a ketone. *True*. If a compound contains both an aldehyde and a ketone, the ketone is named as an "oxo" substituent.

11. A Grignard reagent is a good electrophile. *False*. Grignard reagents are good *nucleophiles*.

12. Any reaction that oxidizes an aldehyde to a carboxylic acid will also oxidize a ketone to a carboxylic acid. *False*. Ketones are much more resistant to oxidation and require different conditions for oxidation.

13. Aldehydes are more water soluble than ethers with comparable molecular weight. *True*. Aldehydes are more polar than ethers.

14. Aldehydes react with Grignard reagents (followed by acid workup) to form 1° alcohols. *False*. A new carbon-carbon bond is formed, resulting in a secondary alcohol.

15. An imine can be reduced to an amine through catalytic hydrogenation. *True*. The formation of an imine, followed by its reduction to an amine, is known as reductive amination.

16. Sodium borohydride, $NaBH_4$, is more reactive and less selective than lithium aluminum hydride, $LiAlH_4$. *False*. The exact opposite is true.

17. An acetal can only result from the base-catalyzed addition of an alcohol to a hemiacetal. *False*. The formation of acetals from hemiacetals requires *acid* catalysis.

18. A Grignard reagent is a strong base. *True*. Grignard reagents behave as carbanions and are therefore good nucleophiles and strong bases.

19. Acetal formation is reversible. *True*. The hydrolysis of an acetal results in a hemiacetal and an alcohol.

20. An imine is the result of the reaction of a 2° amine with an aldehyde or a ketone. *False*. Imines are formed using ammonia or primary amines.

21. Ketones react with Grignard reagents (followed by acid workup) to form 2° alcohols. *False*. Tertiary alcohols would be formed.

22. Aldehydes and ketones can undergo tautomerism. *True*. Both aldehydes and ketones can tautomerize to enols.

23. Acetaldehyde is the lowest molecular weight aldehyde. *False*. The lowest molecular weight aldehyde is formaldehyde (methanal).

24. A ketone that possesses an α-hydrogen can undergo α-halogenation. *True*. The α-hydrogen of a ketone or an aldehyde is relatively acidic.

25. A carbonyl group is polarized such that the oxygen atom is partially positive and the carbon atom is partially negative. *False*. The reverse is true because oxygen is the more electronegative of the two elements.

26. Acetals are stable to bases, nucleophiles, and reducing agents. *True*. However, acetals are not stable under acidic conditions.

27. A "carbaldehyde" is an aldehyde in which the carbonyl group is adjacent to a C—C double bond. *False*. "Carbaldehyde" refers to an aldehyde group that is bonded to a ring.

28. A hemiacetal can result from the acid-catalyzed or base-catalyzed addition of an alcohol to an aldehyde or a ketone. *True*. Regardless of whether an aldehyde or a ketone is used, the resulting product is called a hemiacetal.

End-of-Chapter Problems

Preparation of Aldehydes and Ketones

12.13 Complete these reactions:

(a)

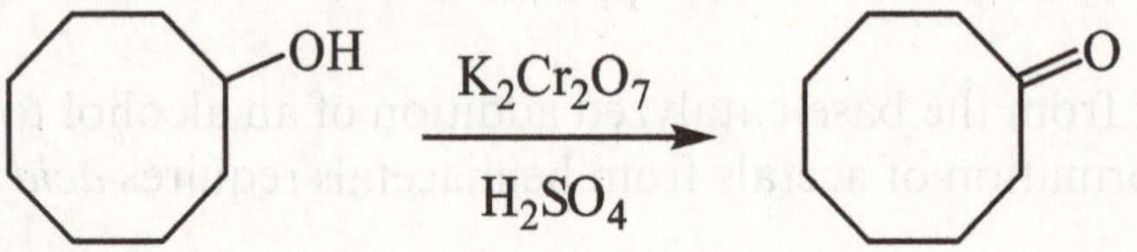

$$\xrightarrow[\text{H}_2\text{SO}_4]{\text{K}_2\text{Cr}_2\text{O}_7}$$

(b)

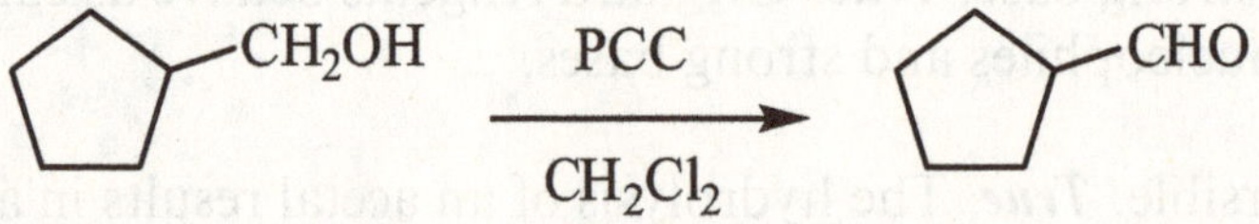

$$\xrightarrow[\text{CH}_2\text{Cl}_2]{\text{PCC}}$$

(c)

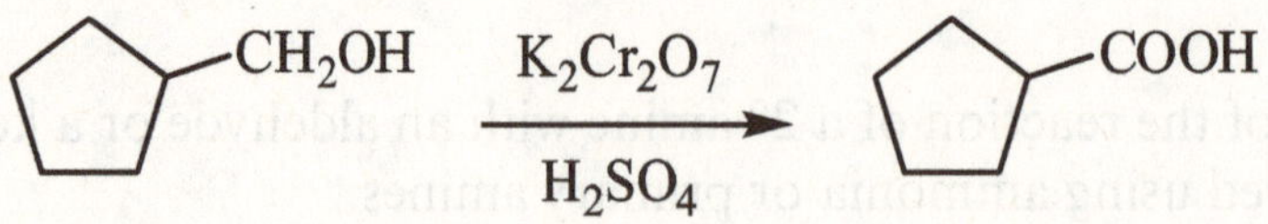

$$\xrightarrow[\text{H}_2\text{SO}_4]{\text{K}_2\text{Cr}_2\text{O}_7}$$

(d)

$$\xrightarrow{\text{AlCl}_3}$$

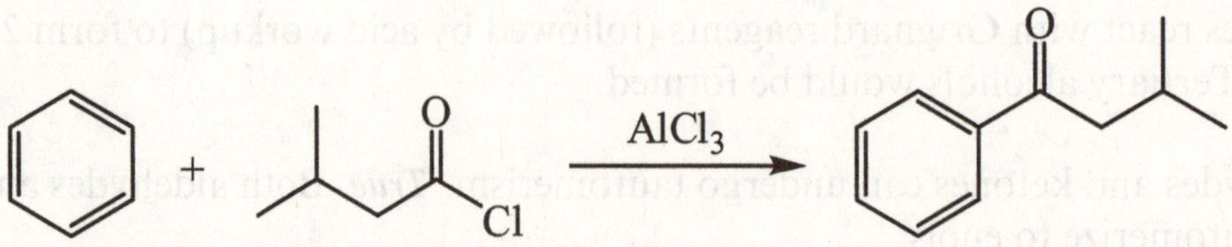

12.14 Show how you would bring about these conversions:

In all conversions involving the oxidation of a secondary alcohol to a ketone, both $K_2Cr_2O_7/H_2SO_4$ and PCC/CH_2Cl_2 are suitable reagents.

(a)

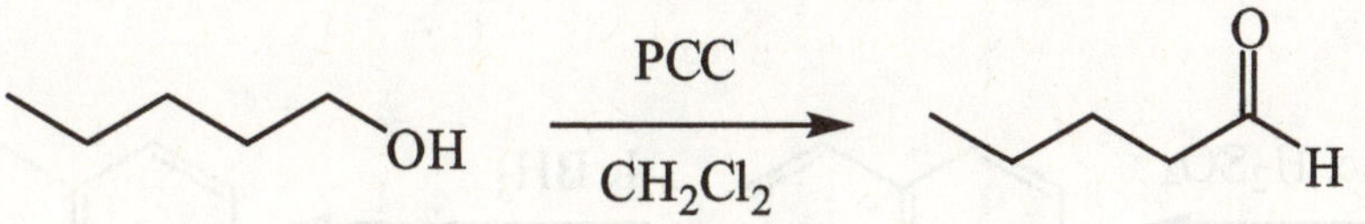

(b)

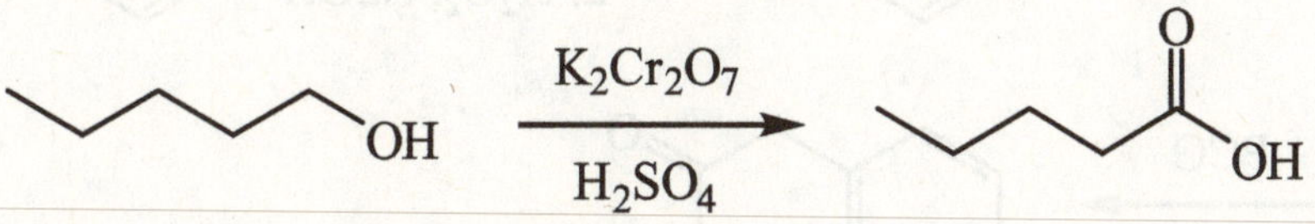

(c)

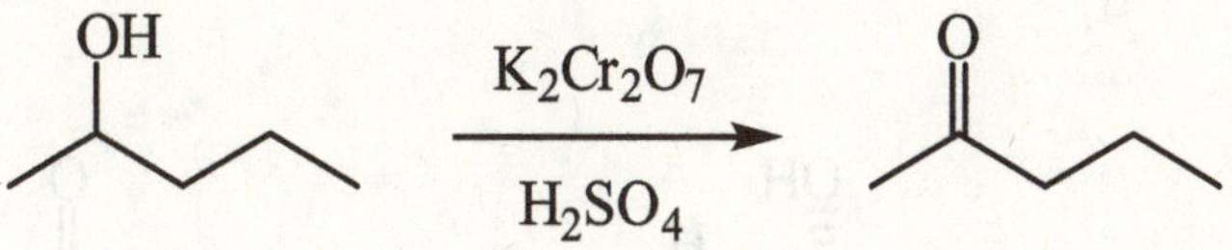

(d)

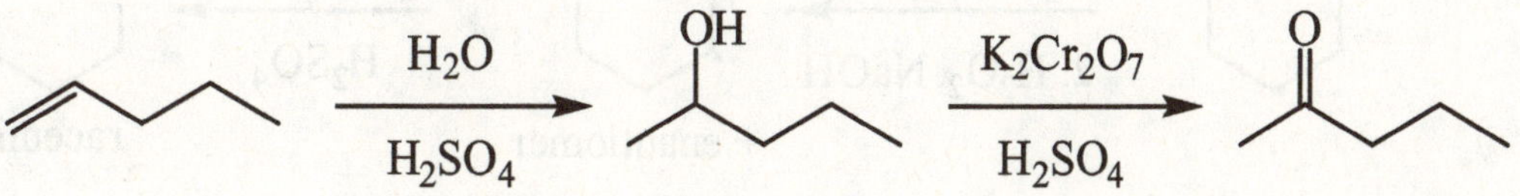

(e)

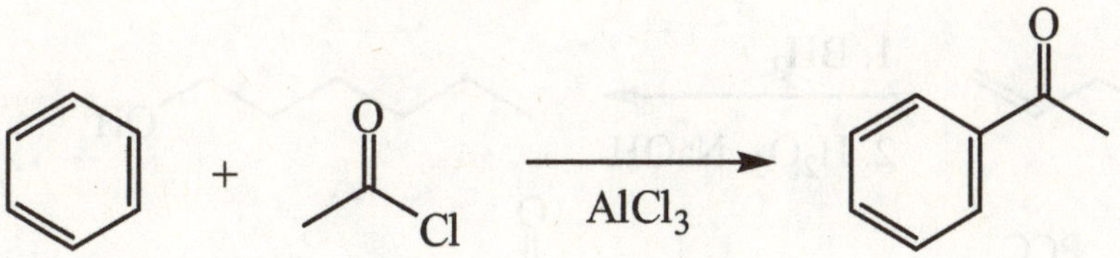

(f)

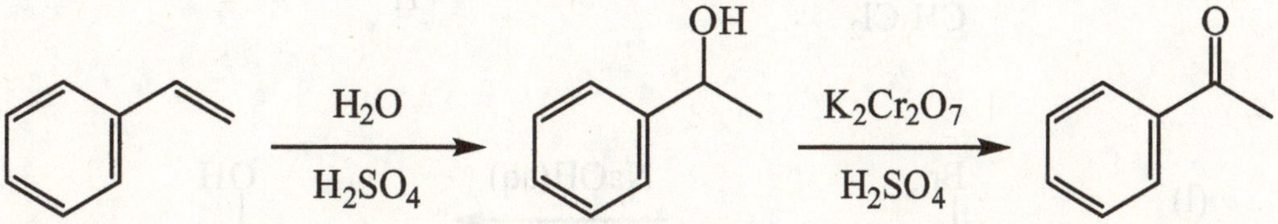

(g)

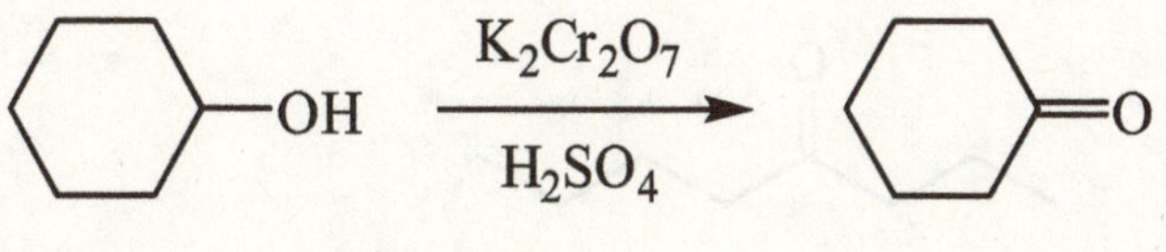

(h)

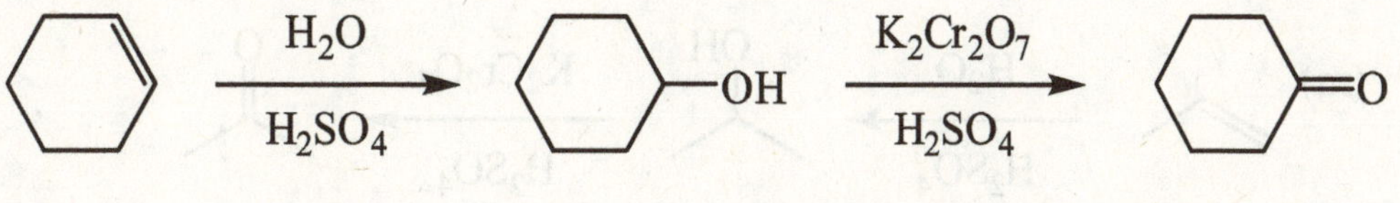

(i)

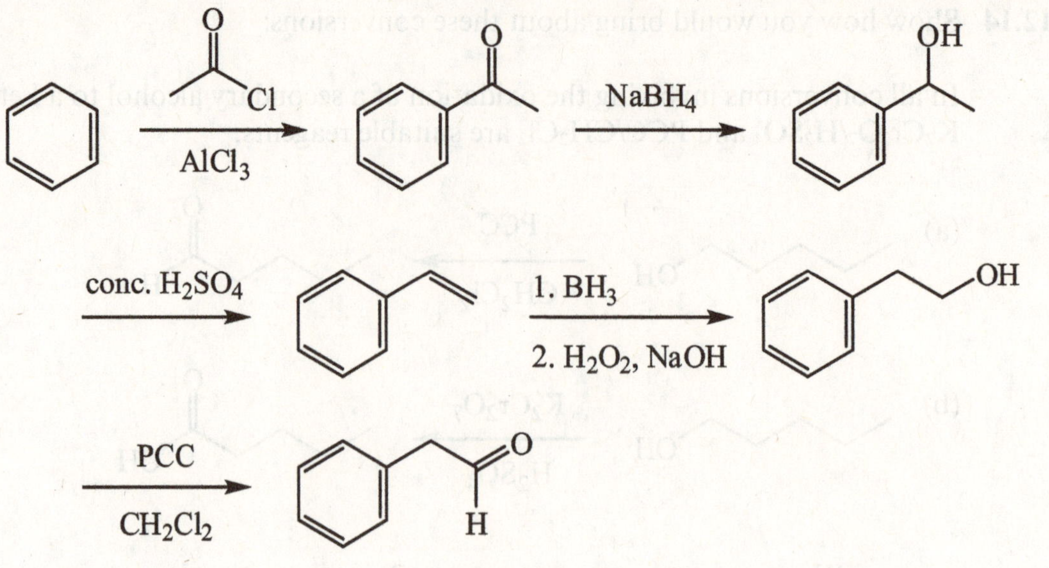

(j)

+ enantiomer   racemic

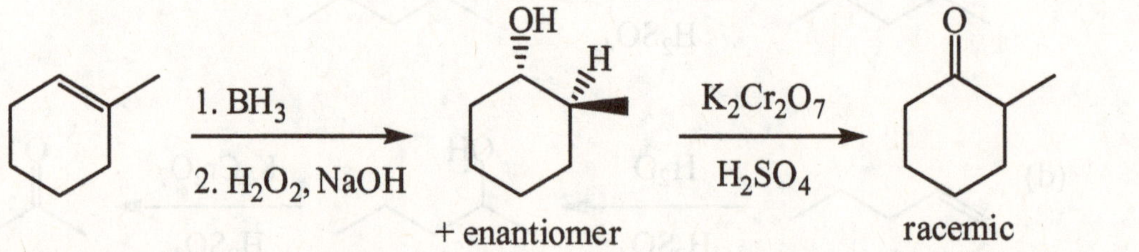

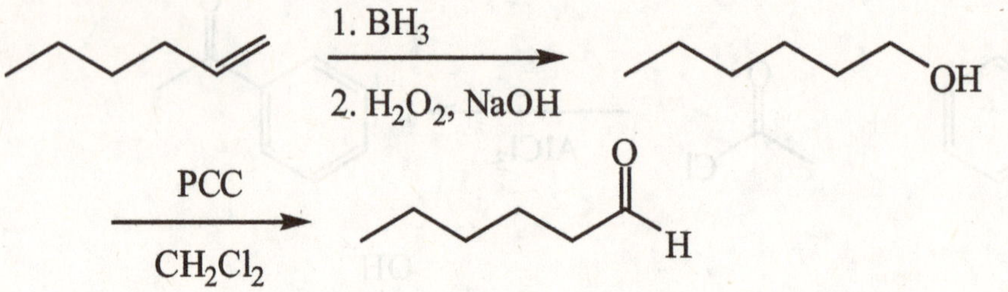

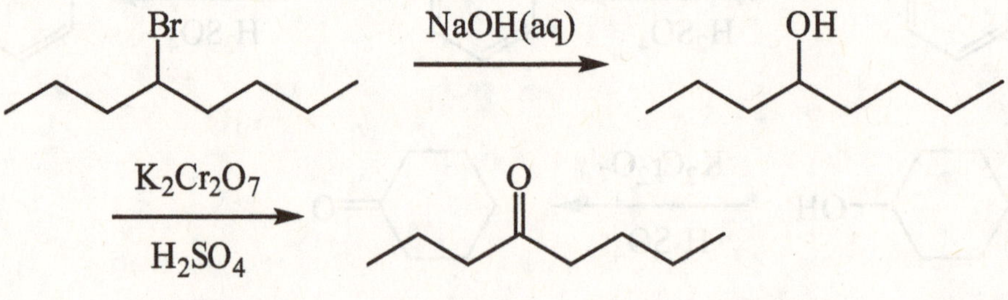

(k)

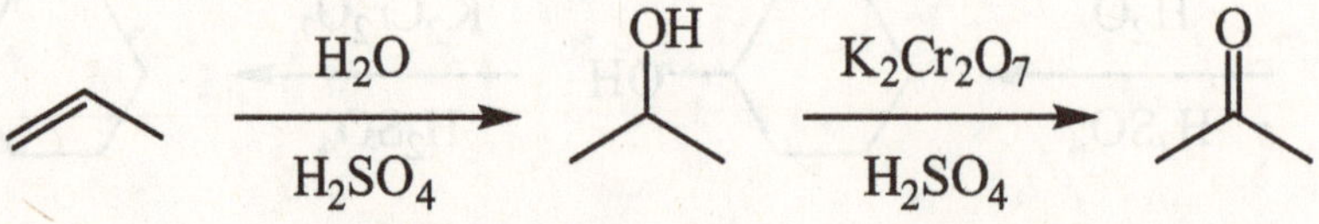

(l)

(m)

Structure and Nomenclature

12.15 Draw a structural formula for the one ketone with the molecular formula C_4H_8O and for the two aldehydes with the molecular formula C_4H_8O.

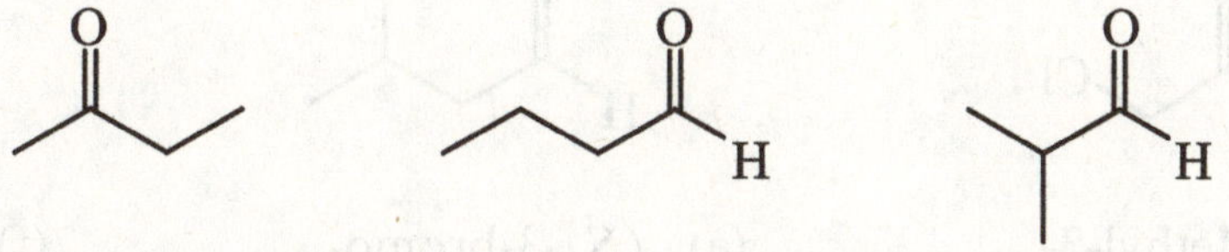

12.16 Draw structural formulas for the four aldehydes with the molecular formula $C_5H_{10}O$. Which of these aldehydes are chiral?

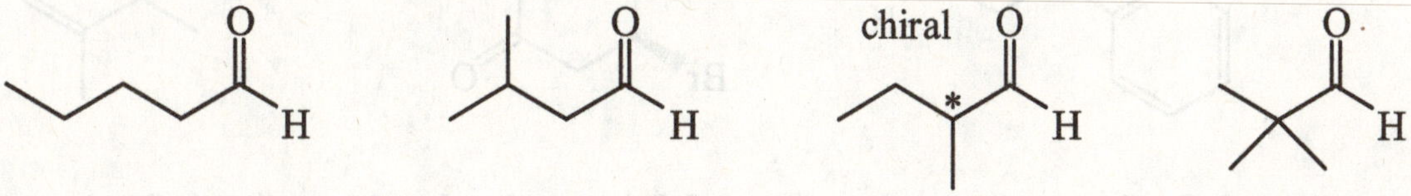

chiral

12.17 Name these compounds:

(a)

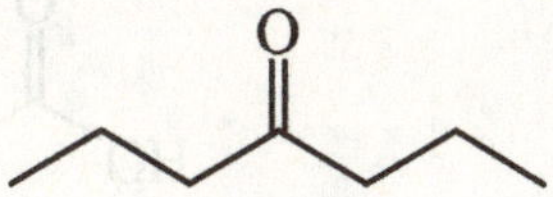

4-Heptanone

(b)

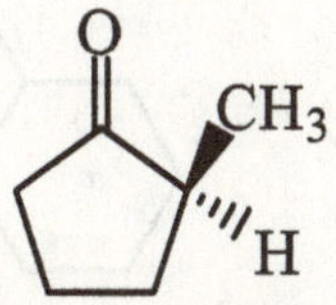

(S)-2-Methyl-cyclopentanone

(c)

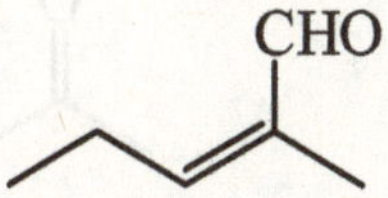

(Z)-2-Methyl-2-pentenal

(d)

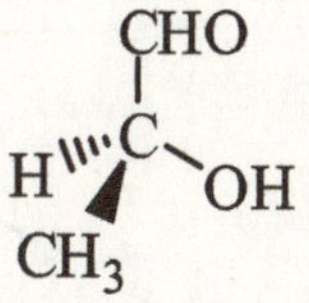

(S)-2-Hydroxypropanal

(e)

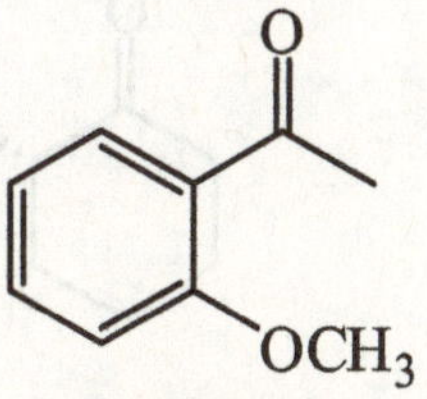

2-Methoxy-acetophenone

(f)

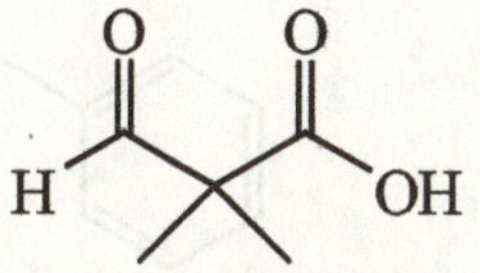

2,2-Dimethyl-3-oxopropanoic acid

(g)

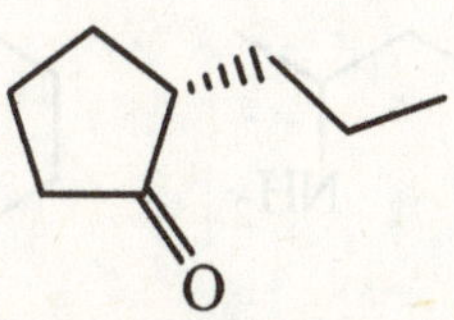

(S)-2-Propylcyclopentanone

(h)

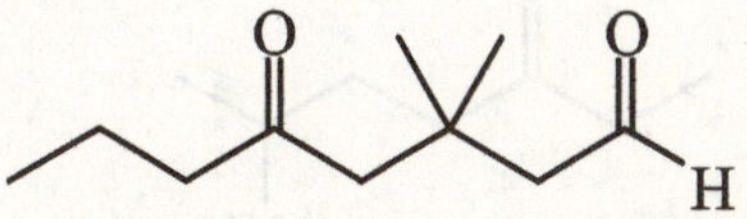

3,3-Dimethyl-5-oxooctanal

12.18 Draw structural formulas for these compounds:

(a) 1-Chloro-2-propanone (b) 3-Hydroxybutanal (c) 4-Hydroxy-4-methyl-2-pentanone

(d) 3-Methyl-3-phenylbutanal (e) (S)-3-bromo-cyclohexanone (f) 3-Methyl-3-buten-2-one

(g) 5-Oxohexanal (h) 2,2-Dimethylcyclo-hexanecarbaldehyde (i) 3-Oxobutanoic acid

(j) 2-Phenylethanal (k) (R)-2-Methyl-cyclohexanone (l) 2,4-Pentadione

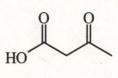

(m) 6-Amino-3-heptanone (n) 6-Amino-4-oxo-heptanal (o) (S)-2-Ethoxycyclo-hexanone

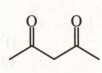

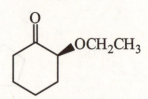

Addition of Carbon Nucleophiles

12.19 Write an equation for the acid-base reaction between phenylmagnesium iodide and a carboxylic acid. Use curved arrows to show the flow of electrons in this reaction. In addition, show that this reaction is an example of a stronger acid and stronger base reacting to form a weaker acid and weaker base.

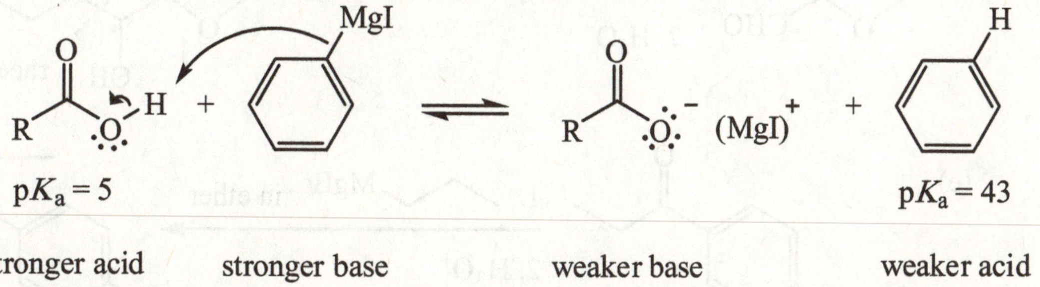

$pK_a = 5$ $pK_a = 43$

stronger acid   stronger base   weaker base   weaker acid

12.20 Diethyl ether is prepared on an industrial scale by the acid-catalyzed dehydration of ethanol. Explain why diethyl ether used in the preparation of Grignard reagents must be carefully purified to remove all traces of ethanol and water.

$$2CH_3CH_2OH \xrightarrow[180\,°C]{H_2SO_4} CH_3CH_2OCH_2CH_3 + H_2O$$

Grignard reagents, which behave as carbon nucleophiles, are also strong bases and would be neutralized by the acidic OH protons present in ethanol and water.

12.21 Draw structural formulas for the product formed by treating each compound with propylmagnesium bromide, followed by hydrolysis in aqueous acid:

The new carbon-carbon bond formed in each reaction is indicated by the dashed line.

(a) CH_2O $\xrightarrow[\text{2. H}_3O^+]{\text{1. } \diagup\!\!\diagdown\!\!\diagup MgBr \text{ in ether}}$

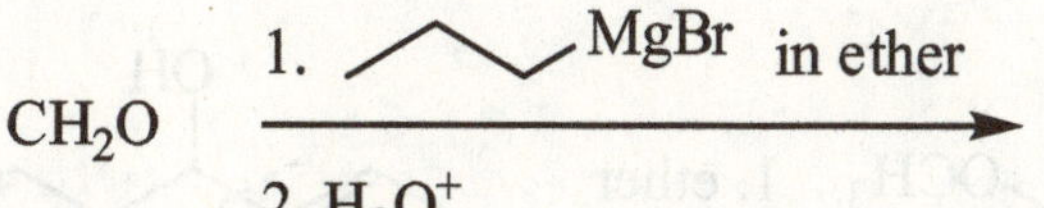

(b)

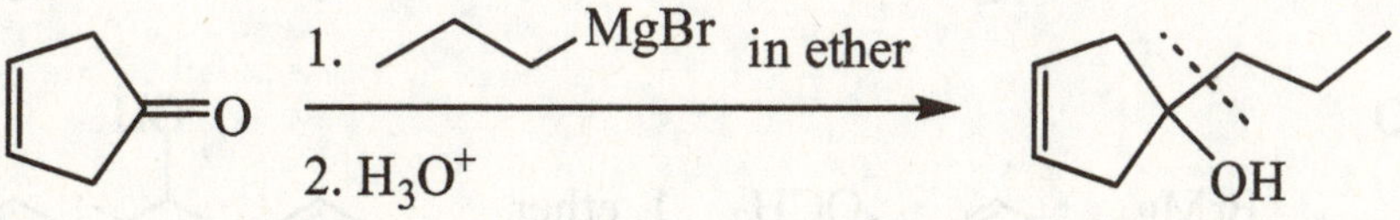

(c)

$$\text{(ketone)} \xrightarrow[\text{2. }H_3O^+]{\text{1. } \bigwedge\!\!\!\!\diagup\!\!\!\!\text{MgBr in ether}} \text{HO}\cdots \quad \text{racemic}$$

(d)

$$\underset{O}{\diagdown}\!\!\diagup\!\!\text{CHO} \xrightarrow[\text{2. }H_3O^+]{\text{1. } \bigwedge\!\!\!\!\diagup\!\!\!\!\text{MgBr in ether}} \quad \text{OH} \quad \text{racemic}$$

(e)

$$CH_3O\!-\!\!\bigcirc\!\!-\!\!\underset{O}{\overset{\parallel}{C}}\!\!\diagup \xrightarrow[\text{2. }H_3O^+]{\text{1. } \bigwedge\!\!\!\!\diagup\!\!\!\!\text{MgBr in ether}} CH_3O\!-\!\!\bigcirc\!\!-\!\!\overset{OH}{\underset{\cdots}{C}} \quad \text{racemic}$$

12.22 Suggest a synthesis for each alcohol, starting from an aldehyde or a ketone and an appropriate Grignard reagent (the number of combinations of Grignard reagent and aldehyde or ketone that might be used is shown in parentheses below each target molecule):

The new carbon-carbon bond formed in each reaction is indicated by the dashed line.

(a)

$$\diagdown\!\!\diagup\!\!\diagdown\!\!\diagup\!\!\underset{O}{\overset{\parallel}{C}}\!\!H \xrightarrow[\text{2. }H_3O^+]{\text{1. }CH_3MgBr \text{ in ether}} \diagdown\!\!\diagup\!\!\diagdown\!\!\diagup\!\!\overset{OH}{\underset{\cdots}{C}}$$

$$CH_3\!-\!\!\underset{O}{\overset{\parallel}{C}}\!\!H \xrightarrow[\text{2. }H_3O^+]{\text{1. } \bigwedge\!\!\!\!\diagup\!\!\!\!\text{MgBr in ether}} \quad \text{OH}$$

(b)

$$\bigcirc\!\!-\!MgBr \;+\; H\!-\!\!\underset{O}{\overset{\parallel}{C}}\!\!-\!\!\bigcirc\!\!-\!OCH_3 \xrightarrow[\text{2. }H_3O^+]{\text{1. ether}} \bigcirc\!\!-\!\!\underset{\cdots}{C}\!\!-\!\!\bigcirc\!\!-\!OCH_3$$

$$\bigcirc\!\!-\!CHO \;+\; BrMg\!-\!\!\bigcirc\!\!-\!OCH_3 \xrightarrow[\text{2. }H_3O^+]{\text{1. ether}} \bigcirc\!\!-\!\!\overset{OH}{\underset{\cdots}{C}}\!\!-\!\!\bigcirc\!\!-\!OCH_3$$

(c)

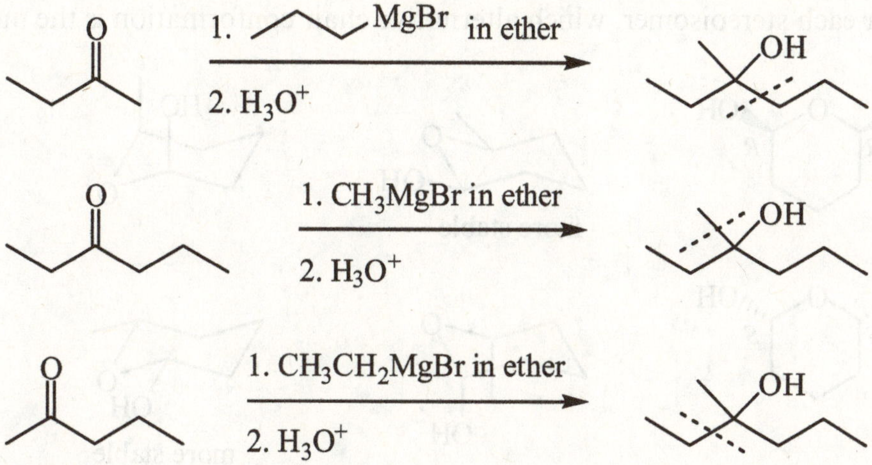

1. ⌁⌁⌁MgBr in ether

2. H_3O^+

1. CH_3MgBr in ether

2. H_3O^+

1. CH_3CH_2MgBr in ether

2. H_3O^+

Addition of Oxygen Nucleophiles

12.23 5-Hydroxyhexanal forms a six-membered cyclic hemiacetal that predominates at equilibrium in aqueous solution.

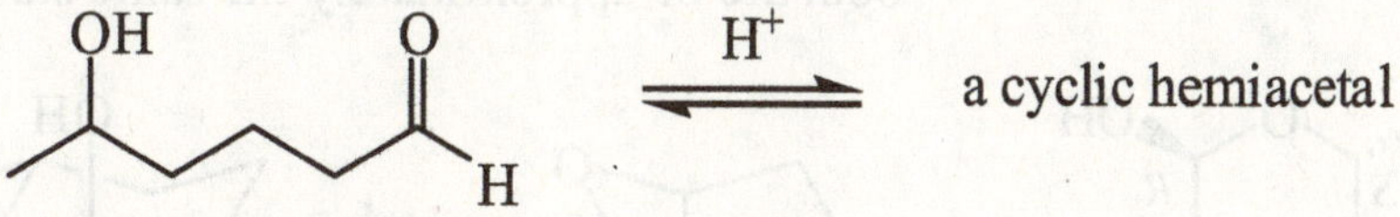

5-Hydroxyhexanal   ⇌ H^+ a cyclic hemiacetal

(a) Draw a structural formula for this cyclic hemiacetal.

carbon atom from aldehyde

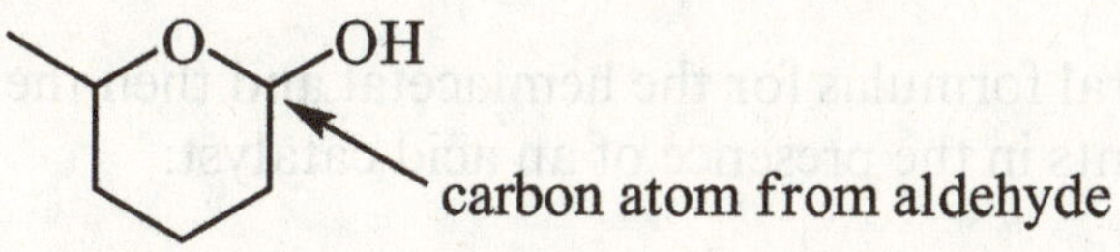

(b) How many stereoisomers are possible for 5-hydroxyhexanal?

5-Hydroxyhexanal contains one chiral center, so two stereoisomers are possible.

(c) How many stereoisomers are possible for this cyclic hemiacetal?

The cyclic hemiacetal contains two chiral centers, so there are four stereoisomers.

(d) Draw alternative chair conformations for each stereoisomer.

(e) For each stereoisomer, which alternative chair conformation is the more stable?

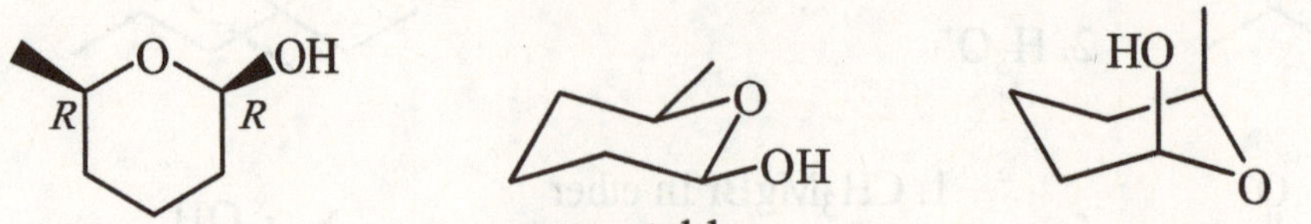

more stable

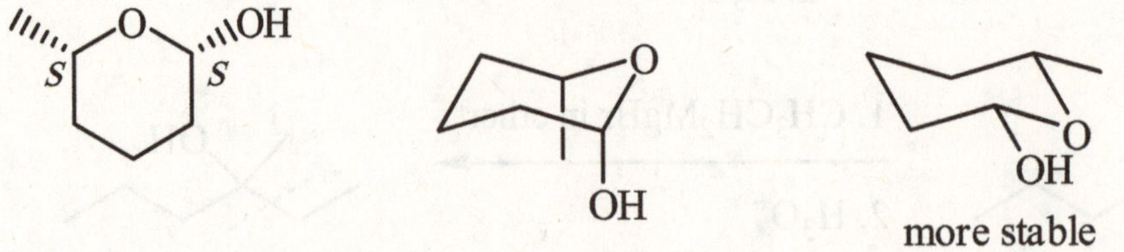

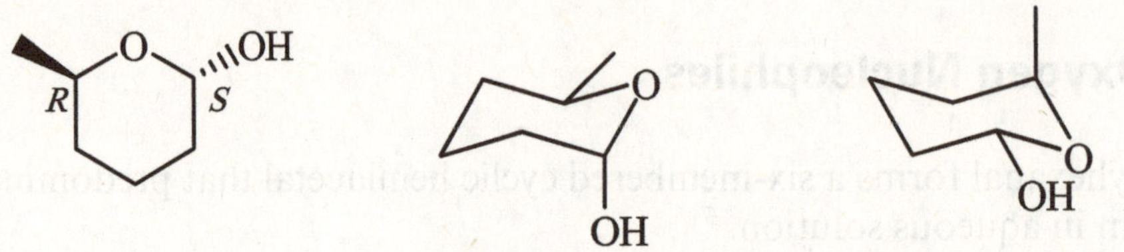

more stable

both are of approximately the same stability

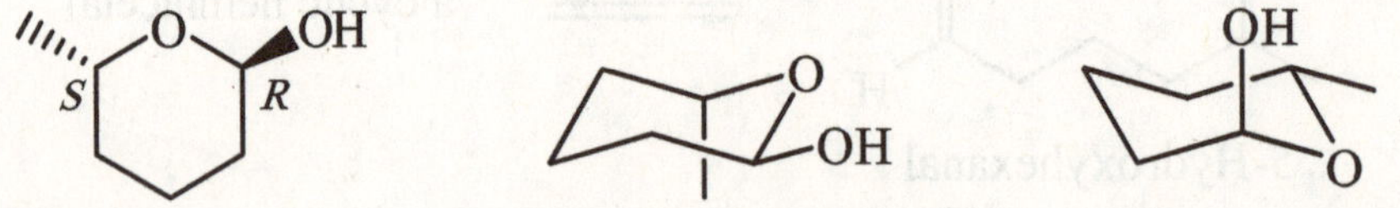

both are of approximately the same stability

12.24 Draw structural formulas for the hemiacetal and then the acetal formed from each pair of reactants in the presence of an acid catalyst:

(a)

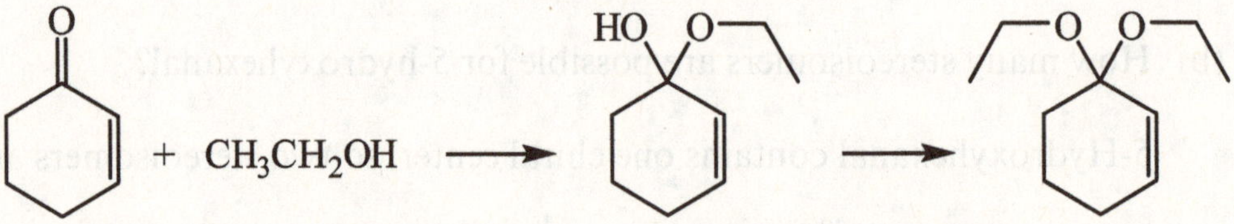

$+ \ CH_3CH_2OH \longrightarrow$

(b)

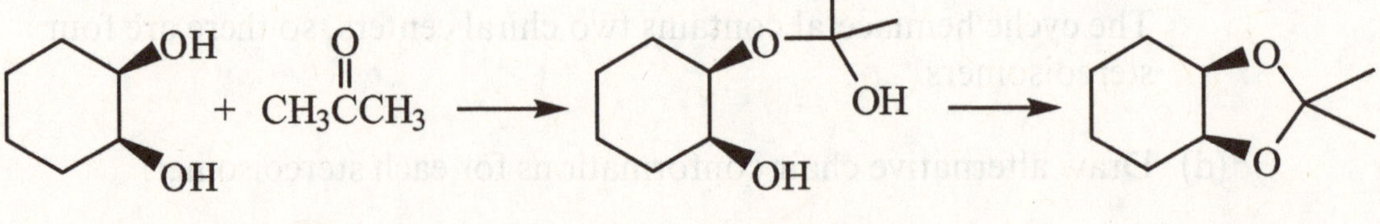

$+ \ CH_3CCH_3 \longrightarrow$

(c)

$CHO + CH_3OH \longrightarrow$

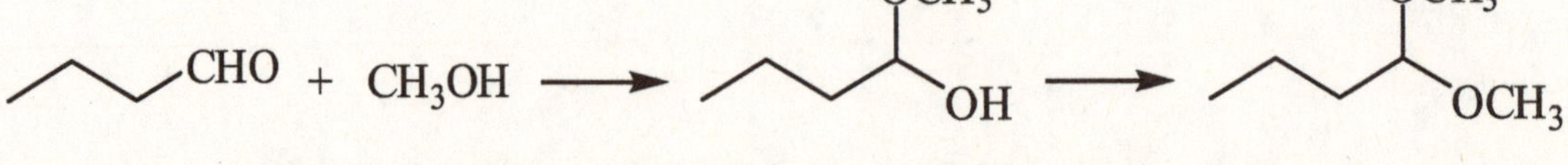

12.25 Draw structural formulas for the products of hydrolysis of each acetal in aqueous acid:

(a)

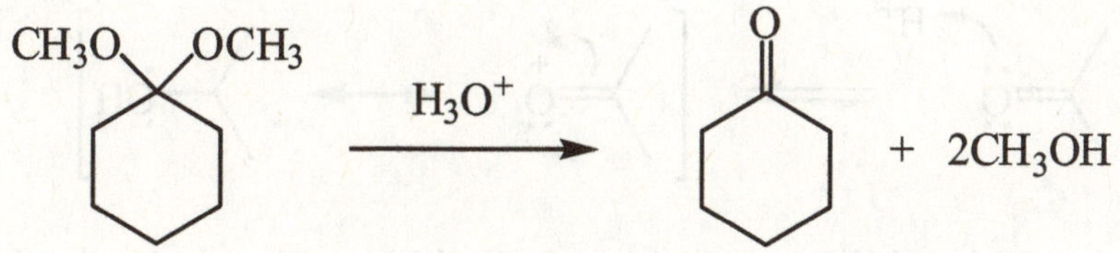

(b)

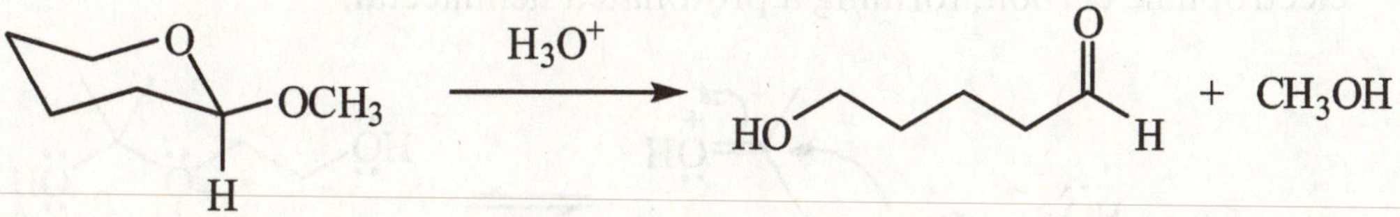

(c)

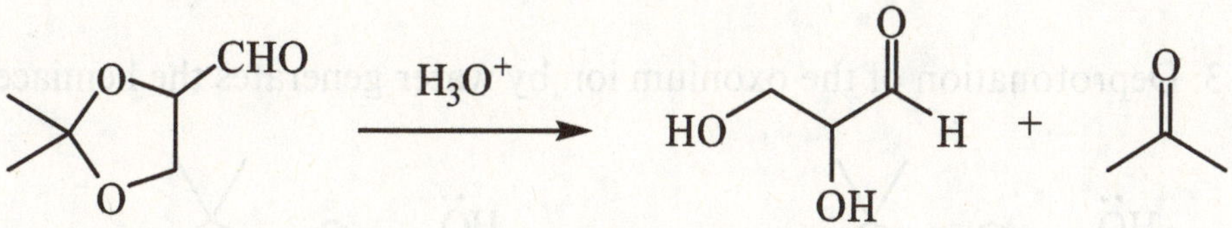

(d)

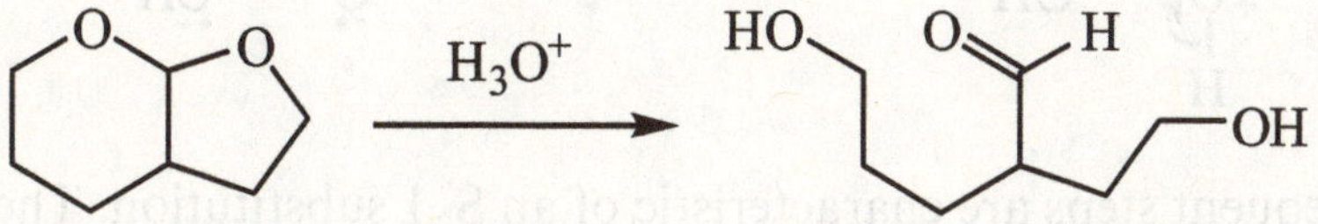

12.26 The following compound is a component of the fragrance of jasmine: From what carbonyl-containing compound and alcohol is the compound derived?

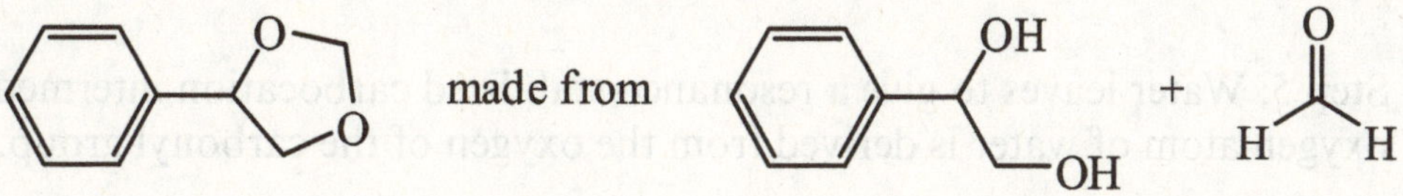

12.27 Propose a mechanism for formation of the cyclic acetal by treating acetone with ethylene glycol in the presence of an acid catalyst. Make sure that your mechanism is consistent with the fact that the oxygen atom of the water molecule is derived from the carbonyl oxygen of acetone.

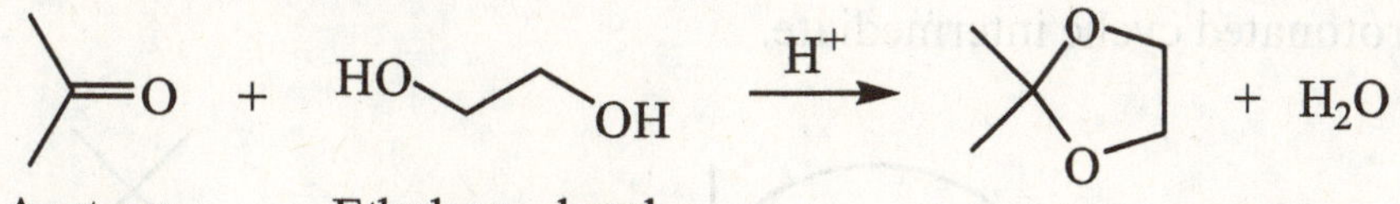

Acetone   Ethylene glycol

Step 1: Protonation of the carbonyl group to form a resonance-stabilized electrophilic intermediate (a resonance-stabilized carbocation).

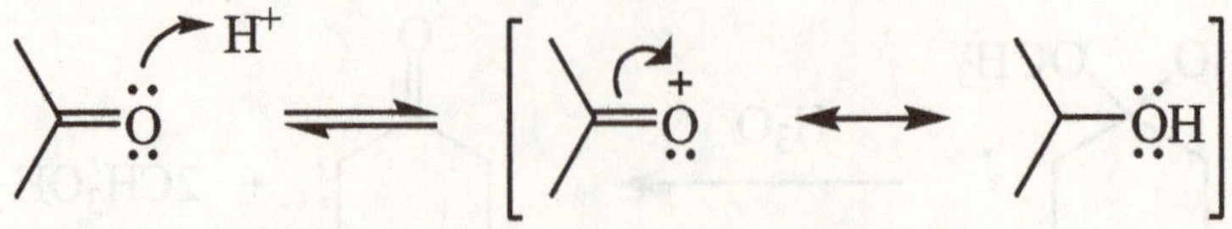

Step 2: One of the nucleophilic oxygen atoms of ethylene glycol attacks the electrophilic carbon, forming a protonated hemiacetal.

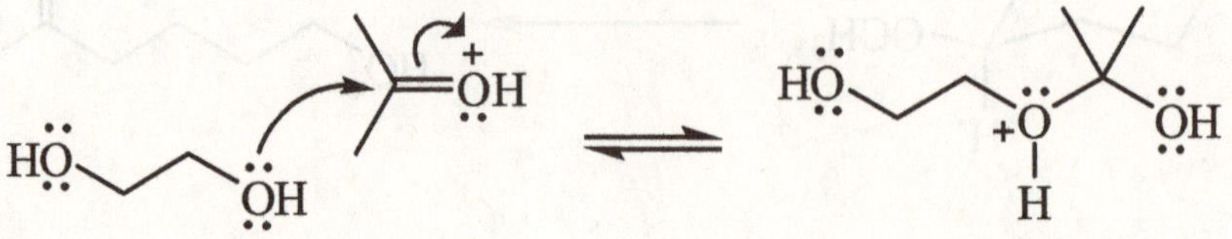

Step 3: Deprotonation of the oxonium ion by water generates the hemiacetal.

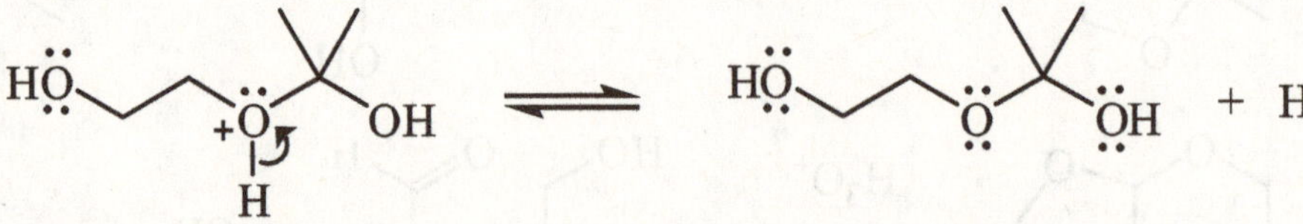

Step 4: The subsequent steps are characteristic of an S_N1 substitution. The hydroxyl group the hemiacetal is protonated, forming a good leaving group.

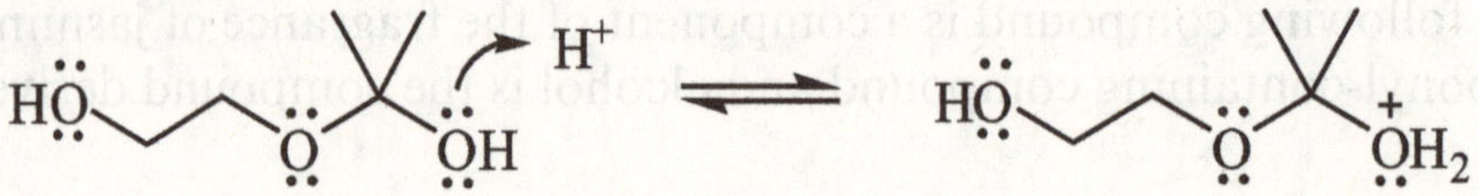

Step 5: Water leaves to give a resonance-stabilized carbocation intermediate. The oxygen atom of water is derived from the oxygen of the carbonyl group.

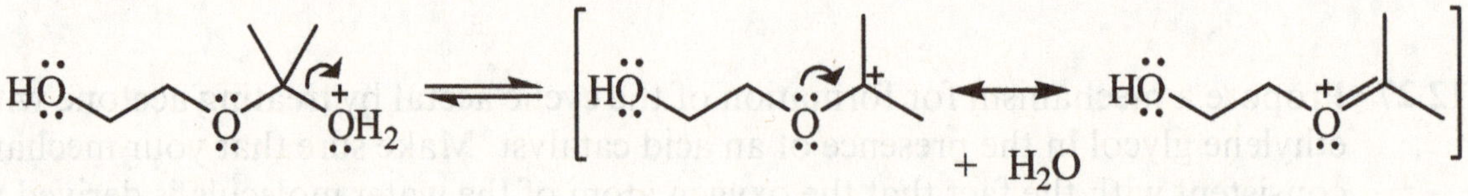

Step 6: The remaining hydroxyl oxygen atom attacks the electrophilic carbon to give a protonated cyclic intermediate.

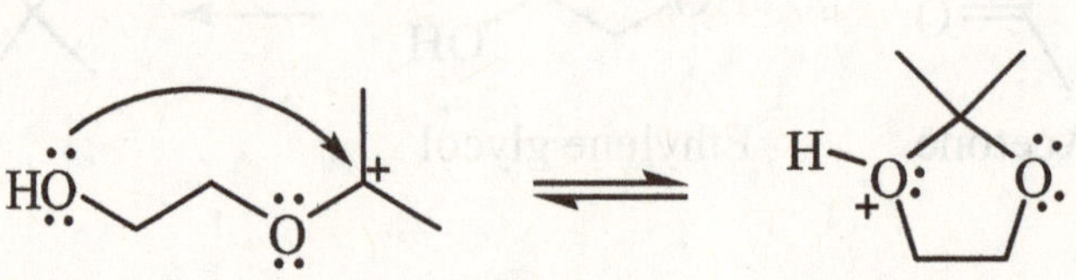

Step 7: Deprotonation of the oxonium ion by water results in the final product.

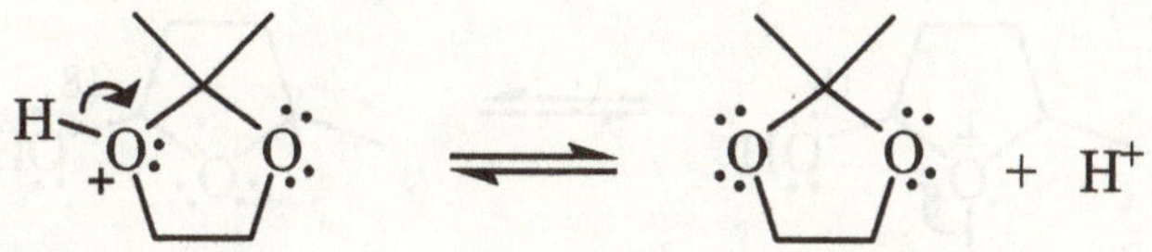

12.28 Propose a mechanism for the formation of a cyclic acetal from 4-hydroxypentanal and one equivalent of methanol: If the carbonyl oxygen of 4-hydroxypentanal is enriched with oxygen-18, does your mechanism predict that the oxygen label appears in the cyclic acetal or in the water? Explain.

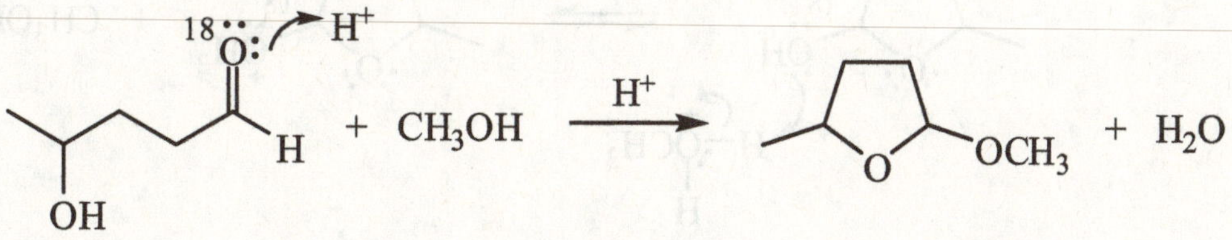

Oxygen-18 will be found in water because during the formation of the acetal, the oxygen of the carbonyl group is lost as water.

Step 1: Protonation of the carbonyl group to form a resonance-stabilized intermediate.

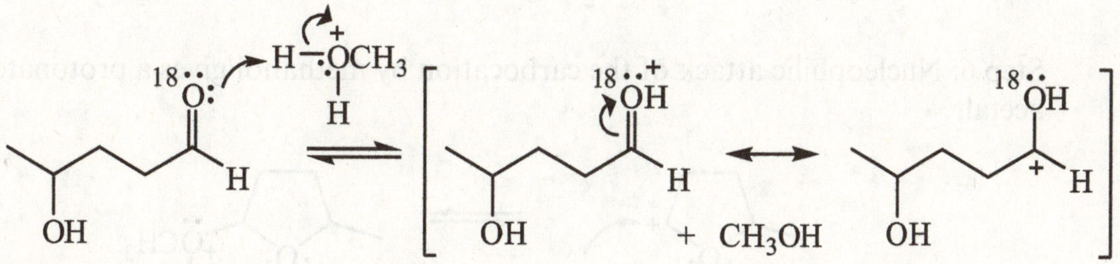

Step 2: Nucleophilic attack by the hydroxyl group forms a protonated cyclic hemiacetal.

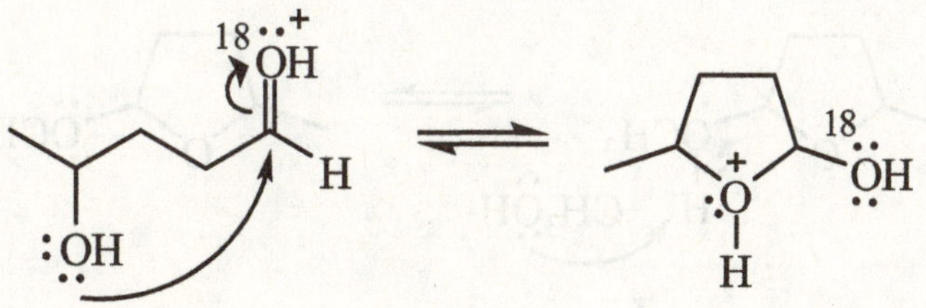

Step 3: Deprotonation generates the hemiacetal.

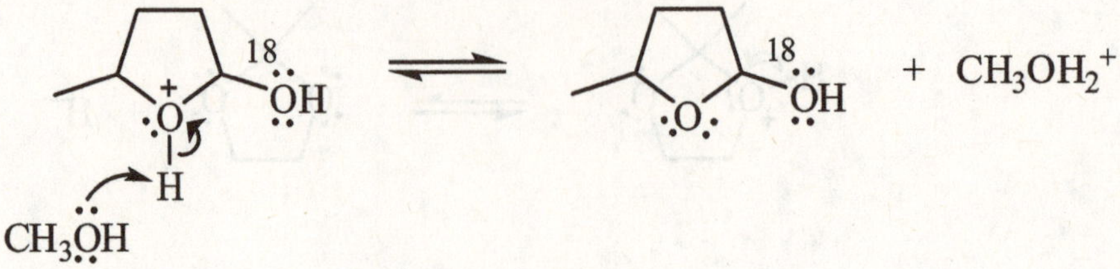

Step 4: The hydroxyl group of the hemiacetal is protonated to form a good leaving group.

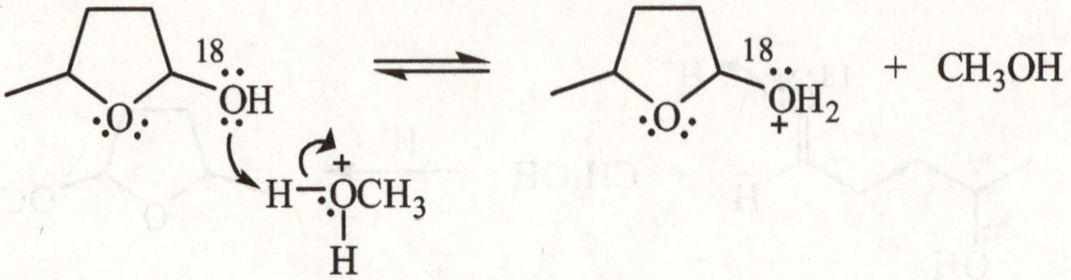

Step 5: Water, containing oxygen-18, leaves to give a resonance-stabilized carbocation.

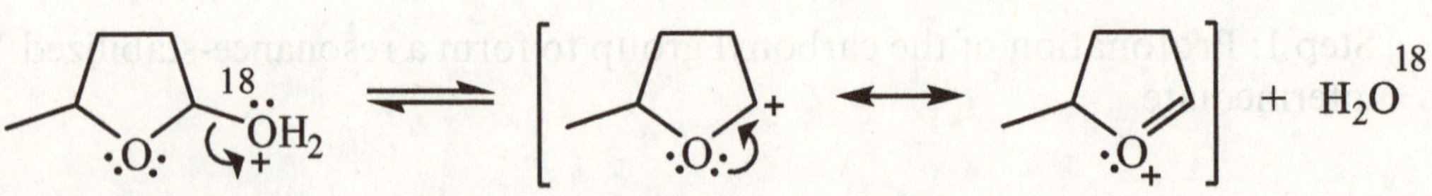

Step 6: Nucleophilic attack of the carbocation by methanol gives a protonated acetal.

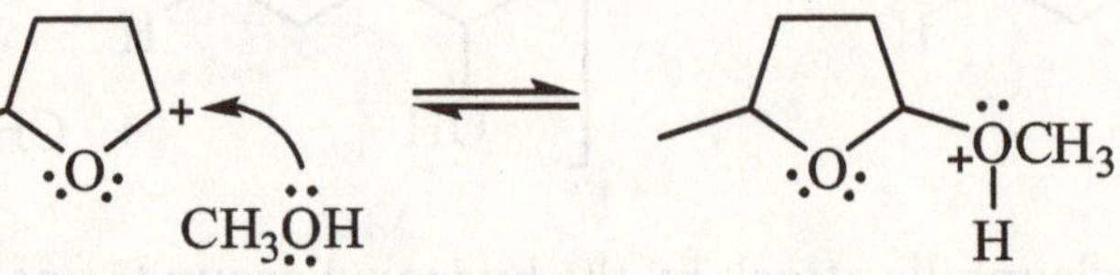

Step 7: Deprotonation results in the final product.

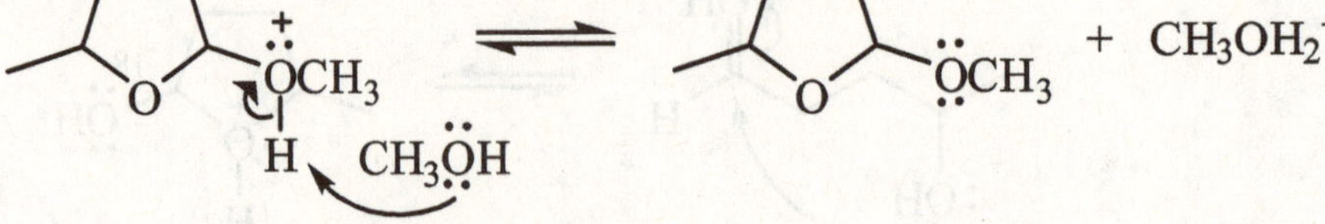

Addition of Nitrogen Nucleophiles

12.29 Show how this secondary amine can be prepared by two successive reductive aminations:

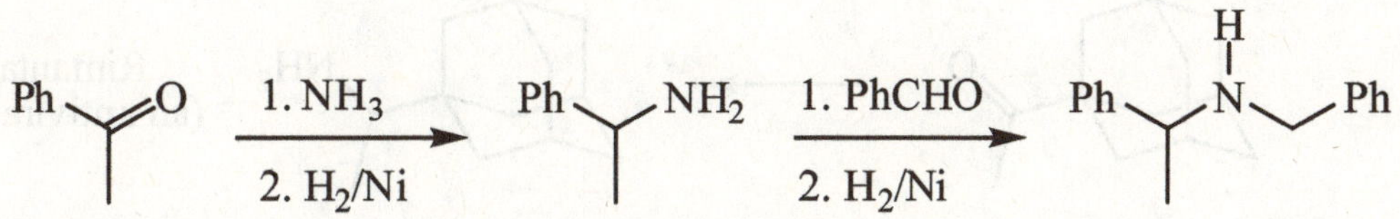

12.30 Show how to convert cyclohexanone to each of the following amines:

(a)
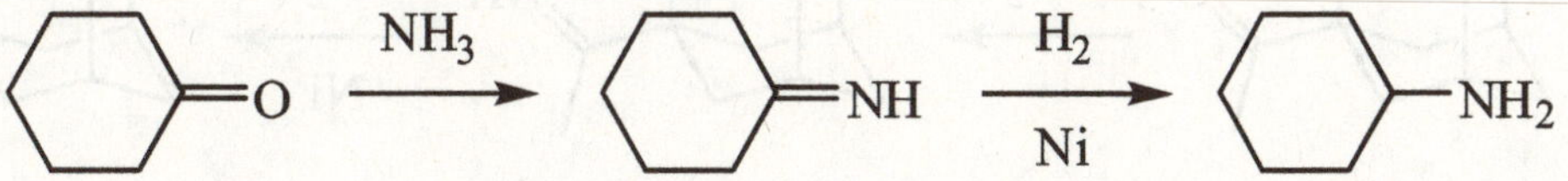

(b)
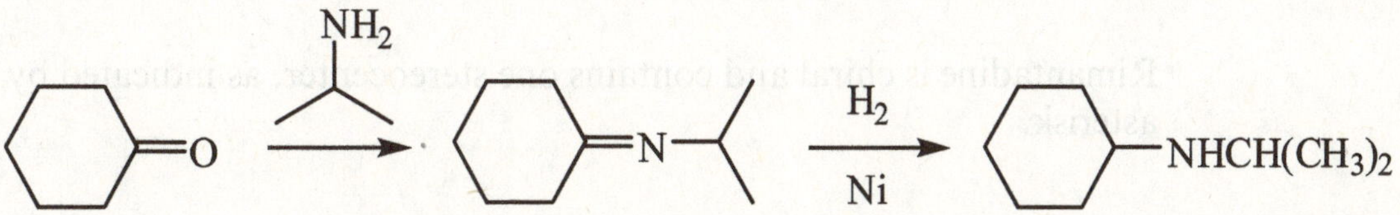

(c)
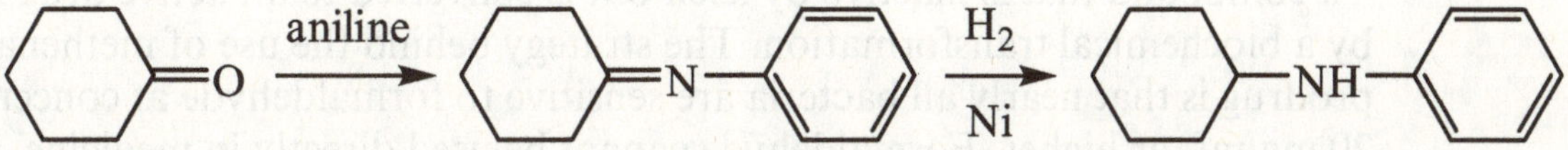

12.31 Following are structural formulas for amphetamine and methamphetamine. The major central nervous system effects of amphetamine and amphetamine-like drugs are locomotor stimulation, euphoria and excitement, stereotyped behavior, and anorexia. Show how each drug can be synthesized by the reductive amination of an appropriate aldehyde or ketone.

(a)
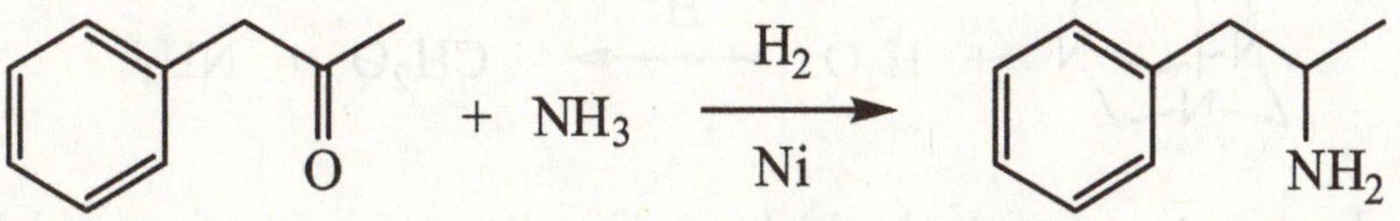

Amphetamine

(b)
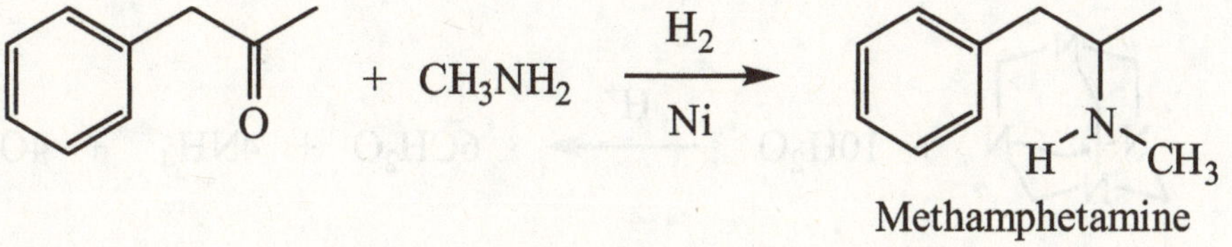

Methamphetamine

12.32 Rimantadine was once used to prevent infections caused by the influenza A virus, but virus strains have since acquired immunity to the drug. It has, however, been used to some success in treating Parkinson's disease. Following is the final step in the synthesis of rimantadine:

Rimantadine
(an antiviral agent)

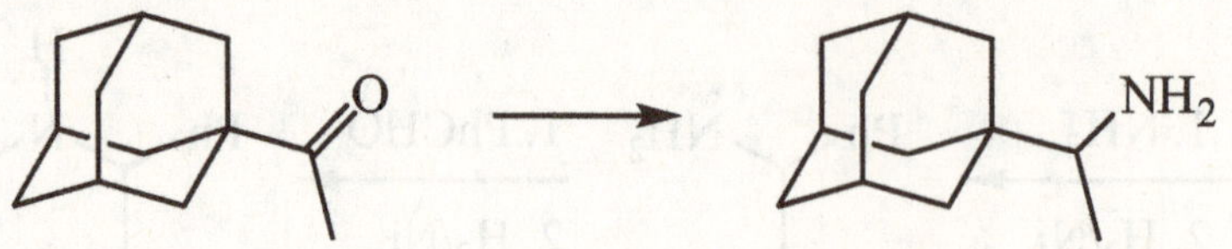

(a) Describe experimental conditions to bring about this conversion.

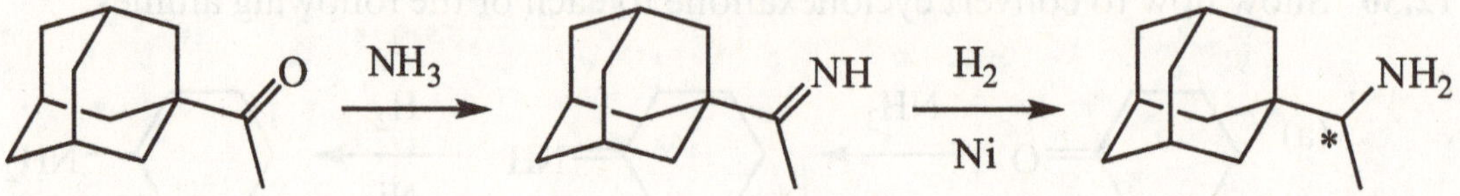

(b) Is rimantadine chiral?

Rimantadine is chiral and contains one stereocenter, as indicated by the asterisk.

12.33 Methenamine, a product of the reaction of formaldehyde and ammonia, is a *prodrug* – a compound that is inactive by itself but is converted to an active drug in the body by a biochemical transformation. The strategy behind the use of methenamine as a prodrug is that nearly all bacteria are sensitive to formaldehyde at concentrations of 20 mg/mL or higher. Formaldehyde cannot be used directly in medicine, however, because an effective concentration in plasma cannot be achieved with safe doses. Methenamine is stable at pH 7.4 (the pH of blood plasma) but undergoes acid-catalyzed hydrolysis to formaldehyde and ammonium ion under the acidic conditions of the kidneys and the urinary tract. Thus, methenamine can be used as a site-specific drug to treat urinary infections.

$$+ H_2O \xrightarrow{H^+} CH_2O + NH_4^+$$

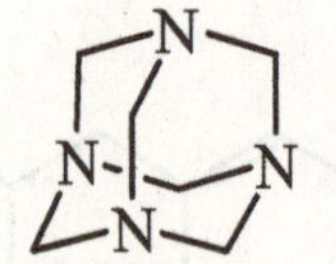

(a) Balance the equation for the hydrolysis of methenamine to formaldehyde and ammonium ion.

$$+ 10H_2O \xrightarrow{H^+} 6CH_2O + 4NH_4^+ + 4OH^-$$

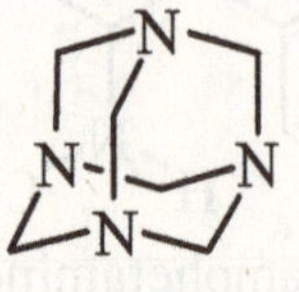

(b) Does the pH of an aqueous solution of methenamine increase, remain the same, or decrease as a result of the hydrolysis of the compound? Explain.

When methenamine is hydrolyzed, ammonium hydroxide is formed. Ammonium hydroxide is a base, so the pH will increase.

(c) Explain the meaning of the following statement: The functional group in methenamine is the nitrogen analog of an acetal.

Acetals are 1,1-diethers, where a single carbon atom is bonded to two –OR groups. In the case of methenamine, each carbon atom is bonded to two amine functional groups.

(d) Account for the observation that methenamine is stable in blood plasma but undergoes hydrolysis in the urinary tract (pH 6.0).

The pH of blood plasma is slightly basic (pH of 7.4), but the pH of the urinary tract is acidic. Acetals (and methenamine, which is a nitrogen analog of an acetal) require acidic conditions for hydrolysis to occur, but they are stable under basic conditions.

Keto–Enol Tautomerism

12.34 The following molecule belongs to a class of compounds called enediols. Each carbon of the double bond carries an group. Draw structural formulas for the α-hydroxyketone and the α-hydroxyaldehyde with which this enediol is in equilibrium.

$$
\begin{array}{ccc}
\mathrm{HC{=}O} & \mathrm{HC{-}OH} & \mathrm{H_2C{-}OH} \\
| & \| & | \\
\mathrm{HC{-}OH} \rightleftharpoons & \mathrm{C{-}OH} \rightleftharpoons & \mathrm{C{=}O} \\
| & | & | \\
\mathrm{CH_3} & \mathrm{CH_3} & \mathrm{CH_3}
\end{array}
$$

α-hydroxyaldehyde   An enediol α-hydroxyketone

12.35 In dilute aqueous acid, (R)-glyceraldehyde is converted into an equilibrium mixture of (R,S)-glyceraldehyde and dihydroxyacetone. Propose a mechanism for this isomerization.

$$
\begin{array}{ccc}
\mathrm{CHO} & \mathrm{CHO} & \mathrm{CH_2OH} \\
| & | & | \\
\mathrm{CHOH} \xrightarrow{\mathrm{H_2O,\ HCl}} & \mathrm{CHOH} \quad + & \mathrm{C{=}O} \\
| & | & | \\
\mathrm{CH_2OH} & \mathrm{CH_2OH} & \mathrm{CH_2OH}
\end{array}
$$

(R)-Glyceraldehyde   (R,S)-Glyceraldehyde   Dihydroxyacetone

In the presence of acid, the aldehyde group of glyceraldehyde tautomerizes, forming an enediol. The enediol is the enol form of both an aldehyde and a ketone.

Step 1: Protonation of the carbonyl group to form a resonance-stabilized intermediate.

$$
\begin{array}{c}
\text{H} \overset{\ddots}{\underset{\text{C}}{\text{C}}} = \ddot{\text{O}}: \quad \text{H} - \overset{+}{\underset{\ddot{\text{O}}}{\text{O}}}\text{H}_2 \\
\text{H} \blacktriangleright \text{C} \blacktriangleleft \text{OH} \\
\text{CH}_2\text{OH}
\end{array}
\quad \rightleftharpoons \quad
\left[
\begin{array}{c}
\text{H} \underset{\text{C}}{\text{C}} = \overset{+}{\underset{\ddot{\text{O}}}{\text{O}}} \text{H} \\
\text{H} \blacktriangleright \text{C} \blacktriangleleft \text{OH} \\
\text{CH}_2\text{OH}
\end{array}
\quad \longleftrightarrow \quad
\begin{array}{c}
\text{H} \underset{\text{C}}{\overset{+}{\text{C}}} \; \ddot{\text{O}} \; \text{H} \\
\text{H} \blacktriangleright \text{C} \blacktriangleleft \text{OH} \\
\text{CH}_2\text{OH}
\end{array}
\right]
\; + \; \text{H}_2\text{O}
$$

(*R*)-Glyceraldehyde

Step 2: Deprotonation of the resonance-stabilized intermediate at the α-carbon (carbon 2) leads to the enediol intermediate. The Z isomer is shown, but the E isomer can be formed as well. Note that the chiral center is destroyed and that the enediol is achiral.

$$
\begin{array}{c}
\text{H} \underset{\text{C}}{\text{C}} = \overset{+}{\underset{\ddot{\text{O}}}{\text{O}}} \text{H} \\
\text{H} \blacktriangleright \text{C} \blacktriangleleft \text{OH} \\
\text{CH}_2\text{OH}
\end{array}
\quad
\overset{\text{H}_2\ddot{\text{O}}:}{}
\quad \rightleftharpoons \quad
\begin{array}{c}
\text{H} \underset{\text{C}}{\text{C}} \; \ddot{\text{O}} \; \text{H} \\
\parallel \\
\text{HOCH}_2 \underset{\text{C}}{\text{C}} \text{OH}
\end{array}
\quad + \; \text{H}_3\text{O}^+
$$

achiral

Step 3: Reprotonation of the same carbon (carbon 2) can occur from either face of the alkene with equal probability. As a result, a racemic mixture is formed.

$$
\begin{array}{c}
\text{H}_2\overset{+}{\underset{\ddot{\text{O}}}{\text{O}}} - \text{H} \\
\\
\begin{array}{c}
\text{H} \underset{\text{C}}{\text{C}} \; \ddot{\text{O}} \; \text{H} \\
\parallel \\
\text{HOCH}_2 \underset{\text{C}}{\text{C}} \text{OH}
\end{array}
\end{array}
\quad \rightleftharpoons \quad
\begin{array}{c}
\text{H} \underset{\text{C}}{\text{C}} = \overset{+}{\underset{\ddot{\text{O}}}{\text{O}}} \text{H} \\
\text{H} \blacktriangleright \text{C} \blacktriangleleft \text{OH} \\
\text{CH}_2\text{OH}
\end{array}
\quad + \quad
\begin{array}{c}
\text{H} \underset{\text{C}}{\text{C}} = \overset{+}{\underset{\ddot{\text{O}}}{\text{O}}} \text{H} \\
\text{HO} \blacktriangleright \text{C} \blacktriangleleft \text{H} \\
\text{CH}_2\text{OH}
\end{array}
$$

$$
\Big\downarrow -\text{H}^+ \qquad\qquad\qquad \Big\downarrow -\text{H}^+
$$

$$
\begin{array}{c}
\text{H} \underset{\text{C}}{\text{C}} = \text{O} \\
\text{H} \blacktriangleright \text{C} \blacktriangleleft \text{OH} \\
\text{CH}_2\text{OH}
\end{array}
\quad + \quad
\begin{array}{c}
\text{H} \underset{\text{C}}{\text{C}} = \text{O} \\
\text{HO} \blacktriangleright \text{C} \blacktriangleleft \text{H} \\
\text{CH}_2\text{OH}
\end{array}
$$

(*R*)-Glyceraldehyde   (*S*)-Glyceraldehyde

Step 4: If the enediol formed in Step 2 is protonated at carbon 1 instead of carbon 2, the intermediate that is formed can subsequently lead to the formation of dihydroxyacetone.

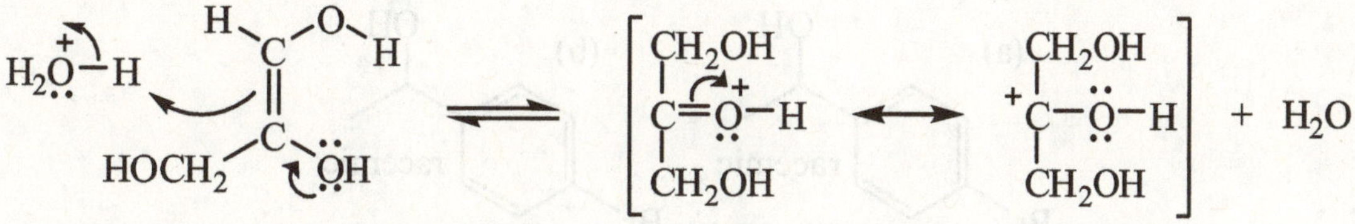

Step 5: Deprotonation results in the formation of dihydroxyacetone.

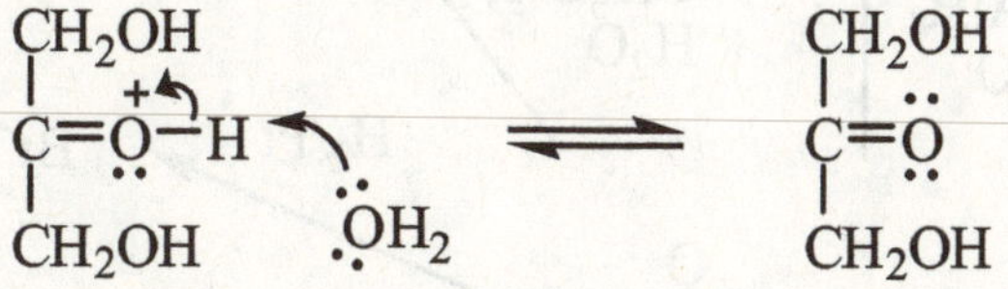

Oxidation/Reduction of Aldehydes and Ketones

12.36 Draw a structural formula for the product formed by treating butanal with each of the following sets of reagents:

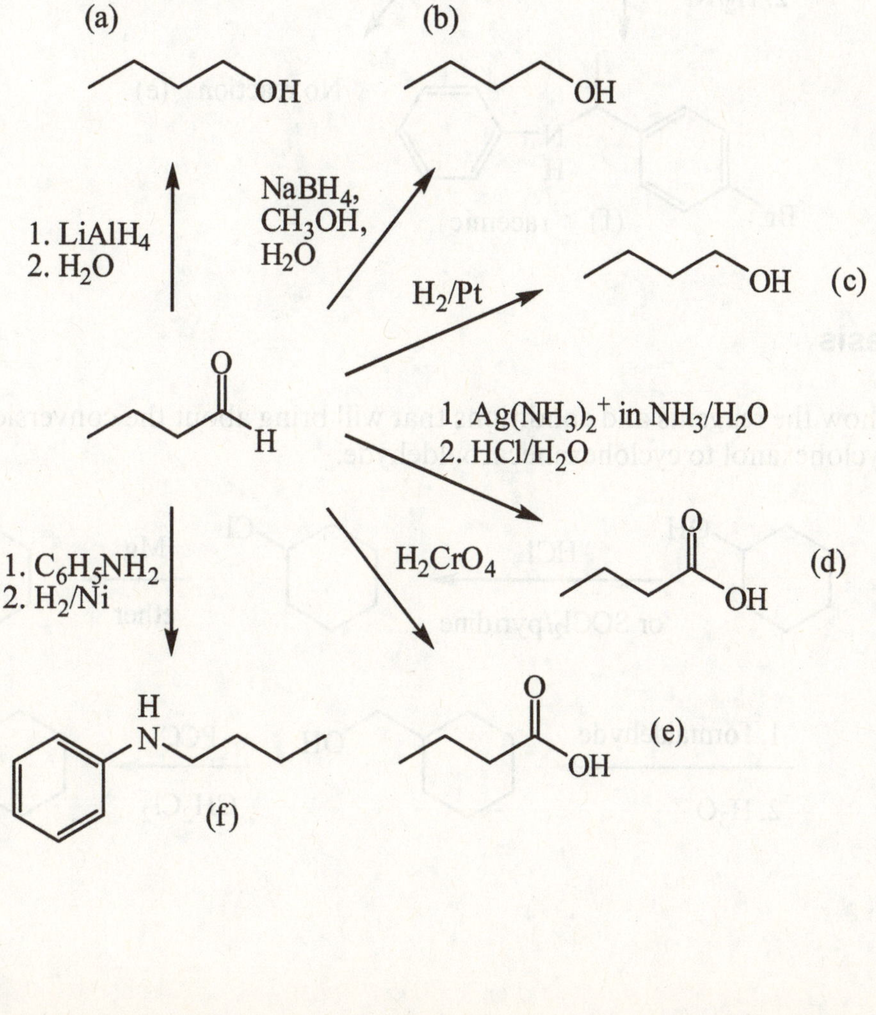

12.37 Draw a structural formula for the product of the reaction of *p*-bromoacetophenone with each set of reagents in Problem 12.36.

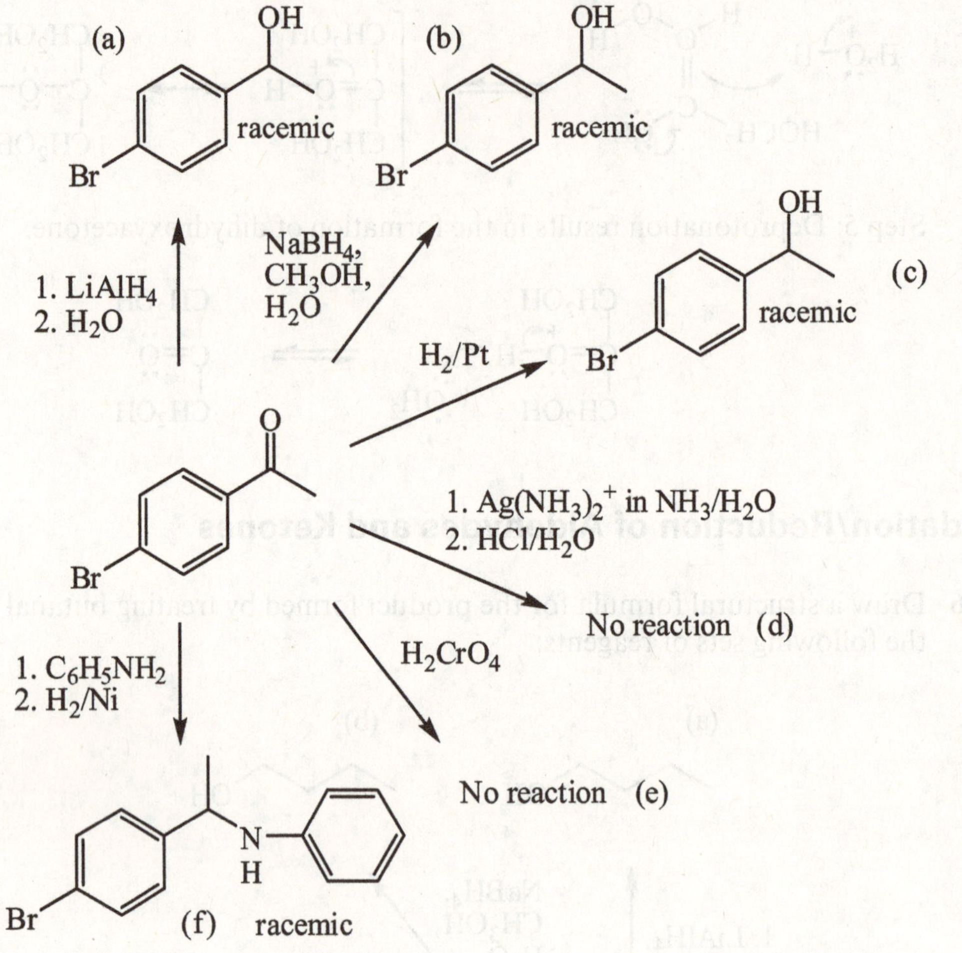

Synthesis

12.38 Show the reagents and conditions that will bring about the conversion of cyclohexanol to cyclohexanecarbaldehyde:

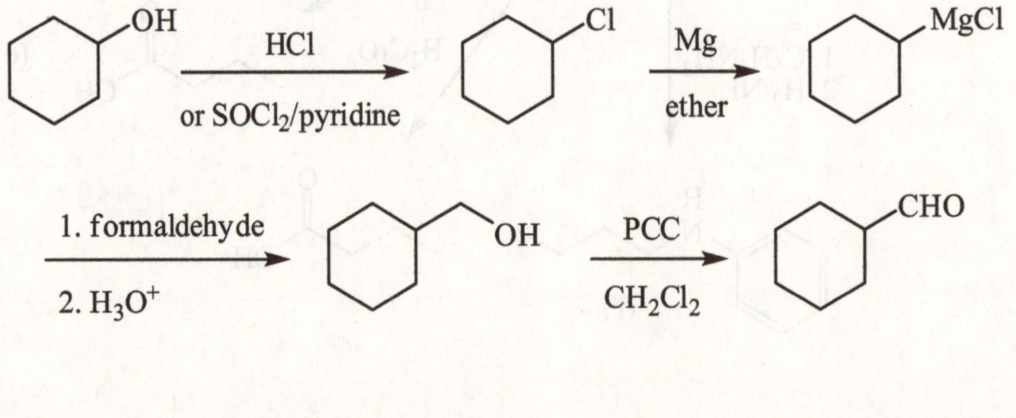

12.39 Starting with cyclohexanone, show how to prepare these compounds (in addition to the given starting material, use any other organic or inorganic reagents, as necessary):

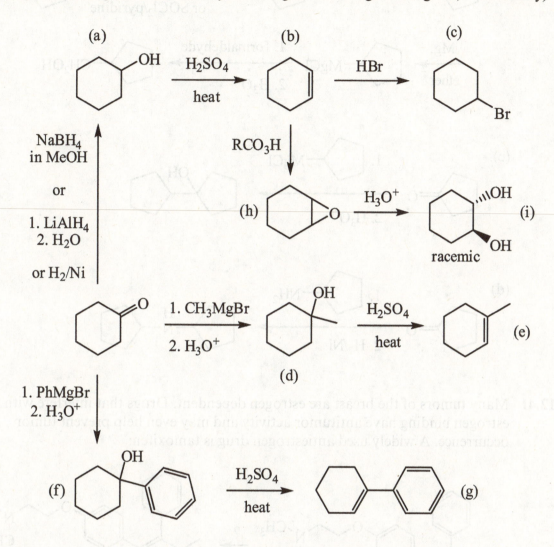

12.40 Show how to bring about these conversions (in addition to the given starting material, use any other organic or inorganic reagents, as necessary):

In reactions (a) and (b), the ketone can be reduced to the alcohol either by using a hydride reagent (NaBH₄ or LiAlH₄) or by catalytic hydrogenation.

(a)

$$C_6H_5\overset{O}{\underset{\|}{C}}CH_2CH_3 \longrightarrow C_6H_5\overset{OH}{\underset{|}{C}H}CH_2CH_3 \xrightarrow[\text{heat}]{H_2SO_4} C_6H_5CH{=}CHCH_3$$

(b)

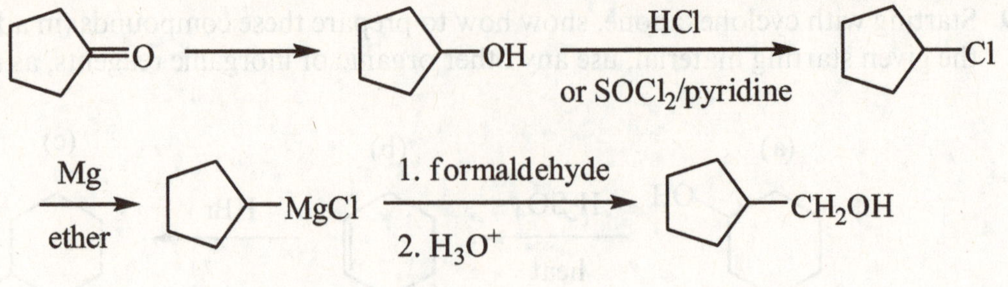

(c)

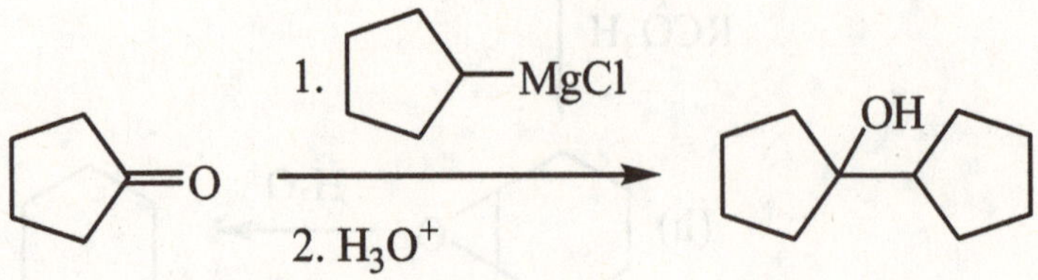

(d)

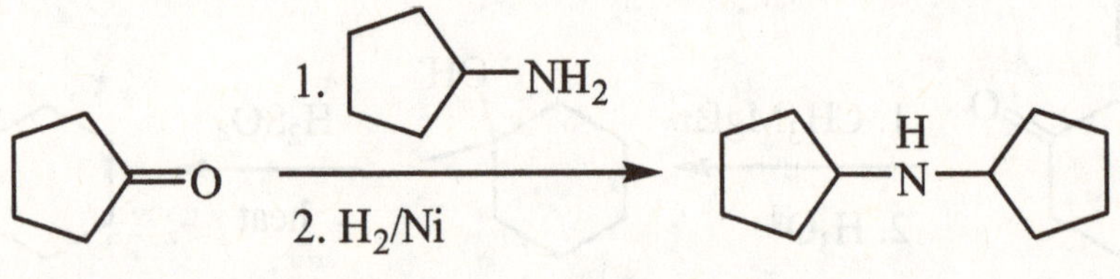

12.41 Many tumors of the breast are estrogen dependent. Drugs that interfere with estrogen binding have antitumor activity and may even help prevent tumor occurrence. A widely used antiestrogen drug is tamoxifen:

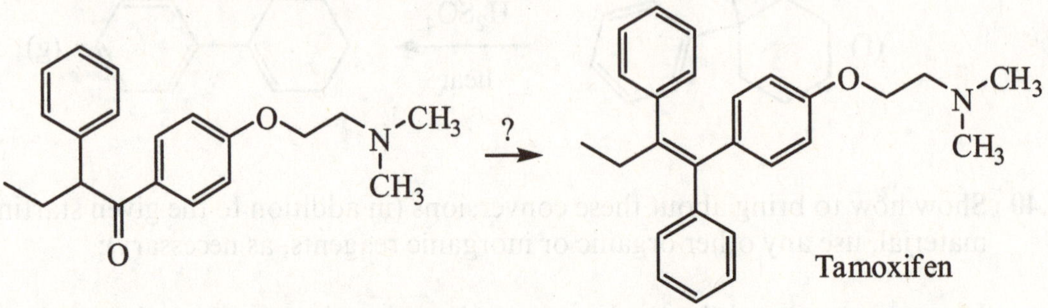

Tamoxifen

(a) How many stereoisomers are possible for tamoxifen?

Tamoxifen has one double bond for which *cis-trans* isomerism is possible, and there are no stereocenters. Therefore, two stereoisomers are possible.

(b) Specify the configuration of the stereoisomer shown here.

The stereoisomer shown has the Z configuration.

(c) Show how tamoxifen can be synthesized from the given ketone using a Grignard reaction, followed by dehydration.

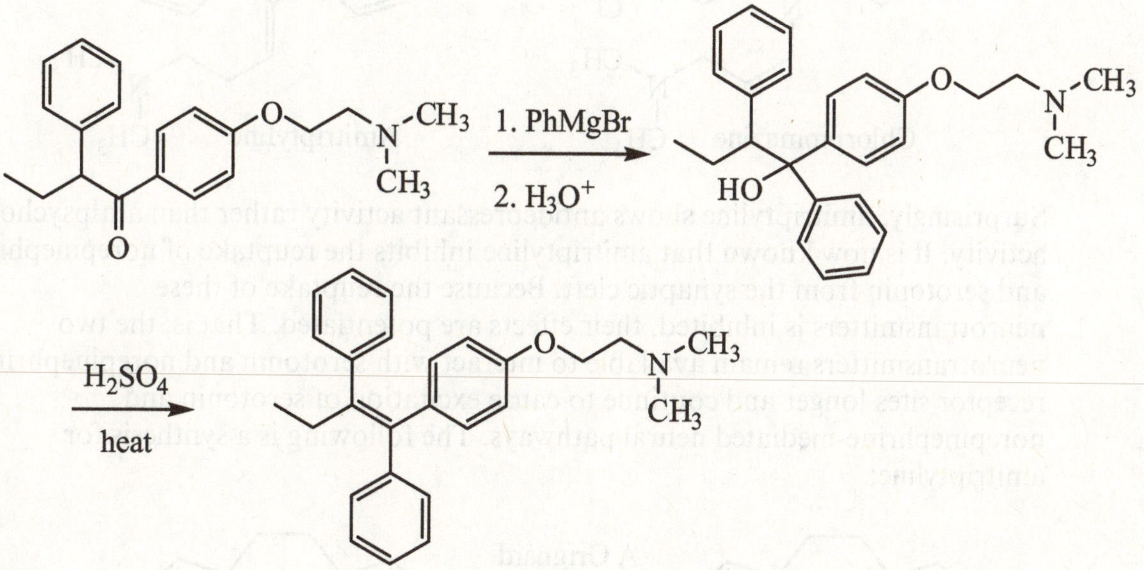

12.42 Following is a possible synthesis of the antidepressant bupropion (Wellbutrin). Show reagents to bring about each step in this synthesis.

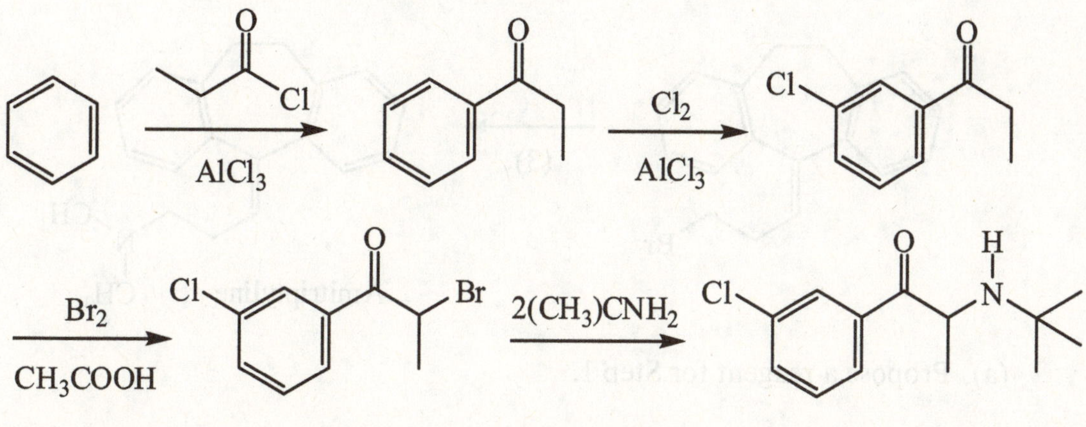

12.43 The synthesis of chlorpromazine in the 1950s and the discovery soon thereafter of the drug's antipsychotic activity opened the modern era of biochemical investigations into the pharmacology of the central nervous system. One of the compounds prepared in the search for more effective antipsychotics was amitriptyline.

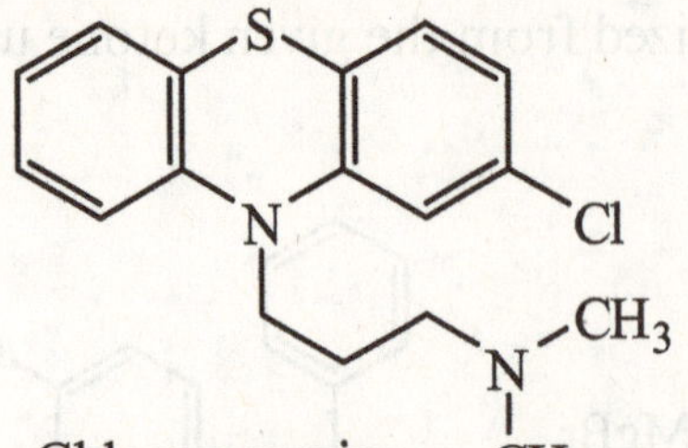

Chlorpromazine   CH₃   Amitriptyline   CH₃

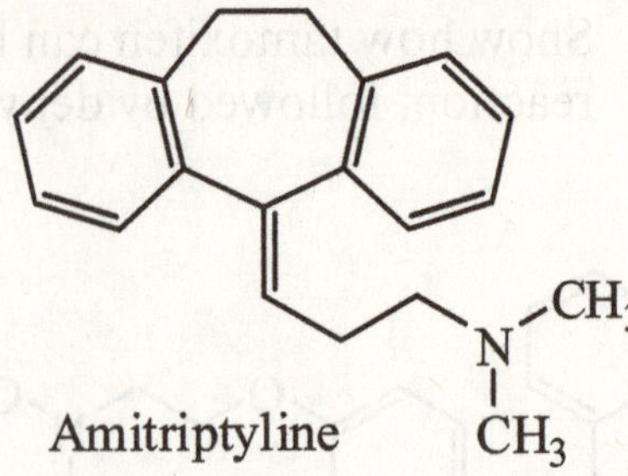

Surprisingly, amitriptyline shows antidepressant activity rather than antipsychotic activity. It is now known that amitriptyline inhibits the reuptake of norepinephrine and serotonin from the synaptic cleft. Because the reuptake of these neurotransmitters is inhibited, their effects are potentiated. That is, the two neurotransmitters remain available to interact with serotonin and norepinephrine receptor sites longer and continue to cause excitation of serotonin and norepinephrine-mediated neural pathways. The following is a synthesis for amitriptyline:

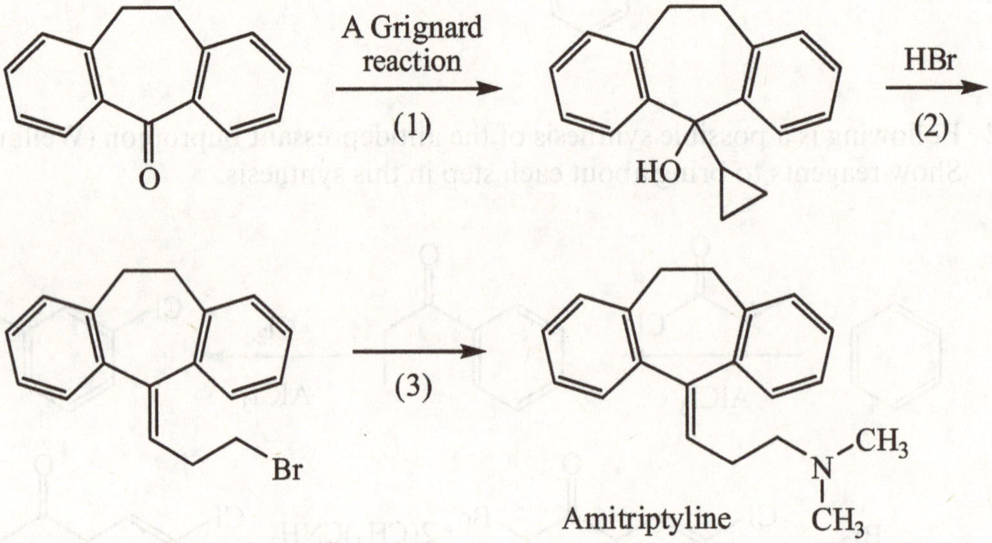

(a) Propose a reagent for Step 1.

   Cyclopropylmagnesium bromide followed by acid treatment will form the alcohol.

(b) Propose a mechanism for Step 2. (*Note*: It is not acceptable to propose a primary carbocation as an intermediate.)

   In Step 2, the hydroxyl group has been lost, the cyclopropane ring has opened, and bromine has been added. The acidic conditions (HBr) suggest that the hydroxyl is turned into a good leaving group via protonation and that the −Br group must have been a nucleophile.

First, the hydroxyl group is protonated to form a good leaving group.

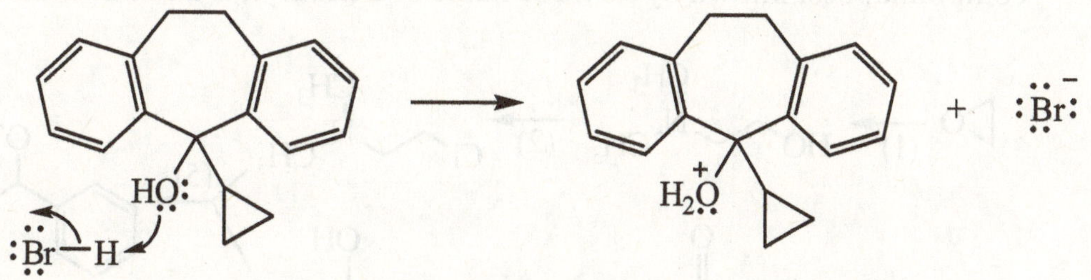

Next, the loss of water generates a resonance-stabilized carbocation.

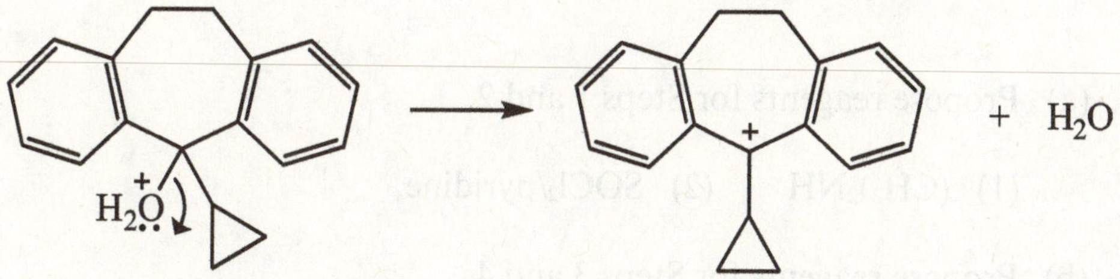

Finally, attack of the cyclopropane ring by bromide forms the product of Step 2. (Recall that three-membered rings are highly strained and prone to opening.)

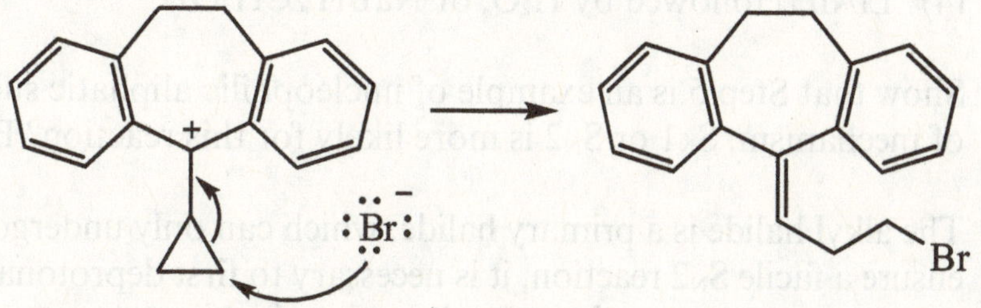

(c) Propose a reagent for Step (3).

In this step, dimethylamine has substituted the Br. Thus, two equivalents of dimethylamine will produce the final product. The first equivalent undergoes an S_N2 reaction to substitute the Br, resulting in an ammonium salt, while the second equivalent acts as a base to deprotonate the ammonium salt, forming the amine.

12.44 Following is a synthesis for diphenhydramine. The hydrochloride salt of this compound, best known by its trade name of Benadryl, is an antihistamine.

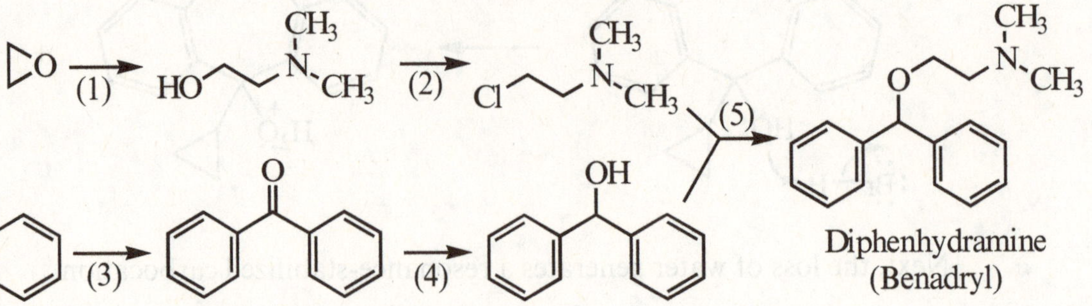

Diphenhydramine
(Benadryl)

(a) Propose reagents for Steps 1 and 2.

(1) $(CH_3)_2NH$ (2) $SOCl_2$/pyridine

(b) Propose reagents for Steps 3 and 4.

(3) $PhCOCl/AlCl_3$

(4) $LiAlH_4$ followed by H_2O, or $NaBH_4/CH_3OH$

(c) Show that Step 5 is an example of nucleophilic aliphatic substitution. What type of mechanism, S_N1 or S_N2 is more likely for this reaction? Explain.

The alkyl halide is a primary halide, which can only undergo an S_N2 reaction. To ensure a facile S_N2 reaction, it is necessary to first deprotonate diphenylmethanol using a strong base to form an alkoxide, which is a strong nucleophile.

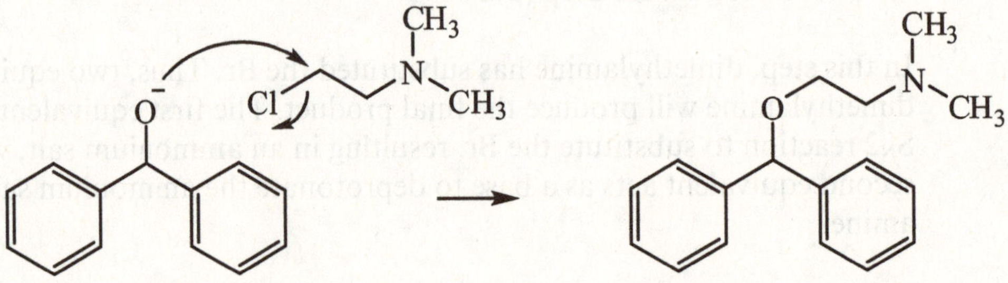

12.45 Following is a synthesis for the antidepressant venlafaxine:

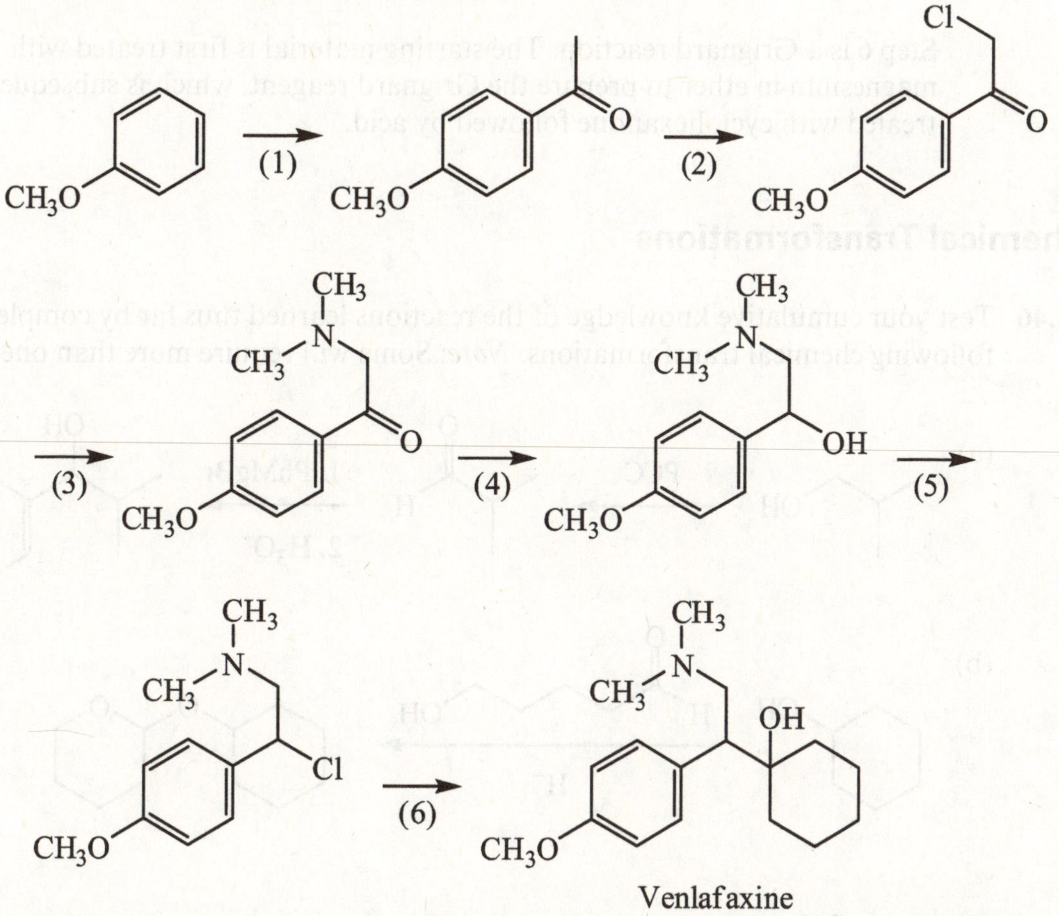

Venlafaxine

(a) Propose a reagent for Step 1 and name the type of reaction that takes place.

Step 1 is a Friedel-Crafts acylation and involves the use of acetyl chloride (CH_3COCl) and a Lewis acid catalyst, such as $AlCl_3$.

(b) Propose reagents for Steps 2 and 3.

Step 2 involves the chlorination of the α-carbon, and Cl_2 in acetic acid can be used.

Step 3 involves the substitution of the Cl with dimethylamine. Two equivalents of dimethylamine are necessary.

(c) Propose reagents for Steps 4 and 5.

Step 4 involves the reduction of the ketone, so $NaBH_4$ in an alcohol, or $LiAlH_4$ followed by water, can be used. Step 5 involves the substitution of the alcohol with a halogen, so $SOCl_2$ in pyridine can be used.

(d) Propose a reagent for Step 6, and name the type of reaction that takes place.

Step 6 is a Grignard reaction. The starting material is first treated with magnesium in ether to prepare the Grignard reagent, which is subsequently treated with cyclohexanone followed by acid.

Chemical Transformations

12.46 Test your cumulative knowledge of the reactions learned thus far by completing the following chemical transformations. *Note*: Some will require more than one step.

(a)

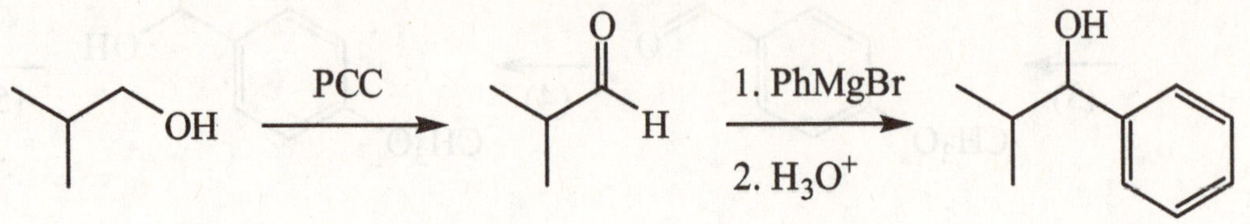

(b)

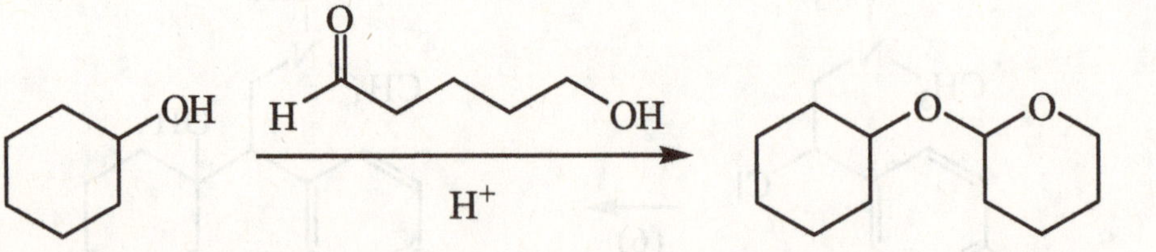

(c)

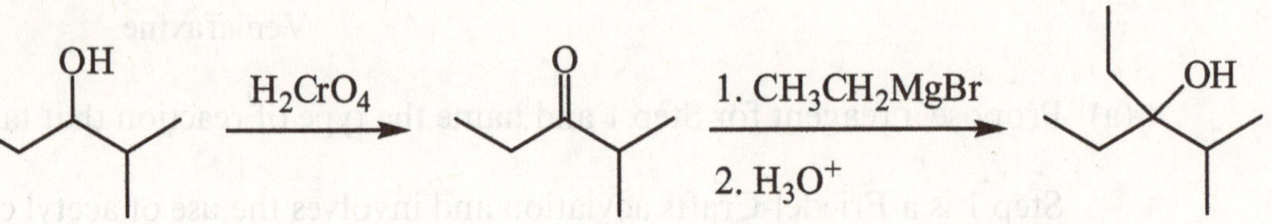

(d)

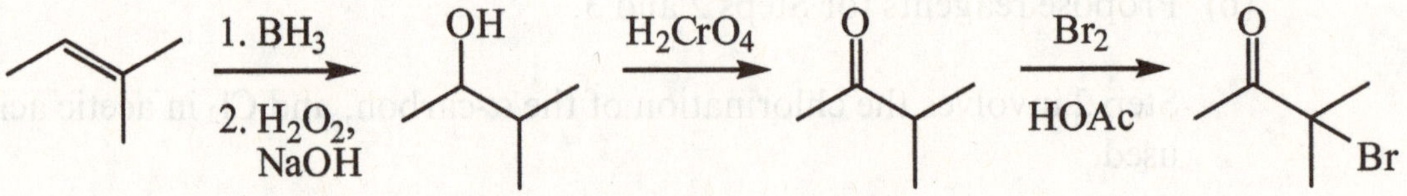

(e)

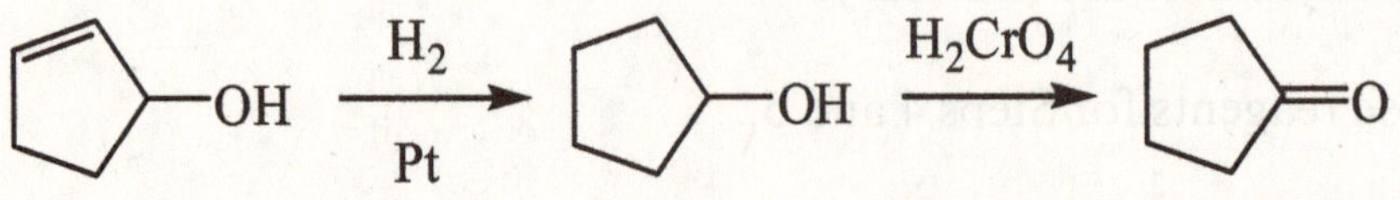

(f)

(g)

(h)

(i)

(j)

(k)

(l)

(m)

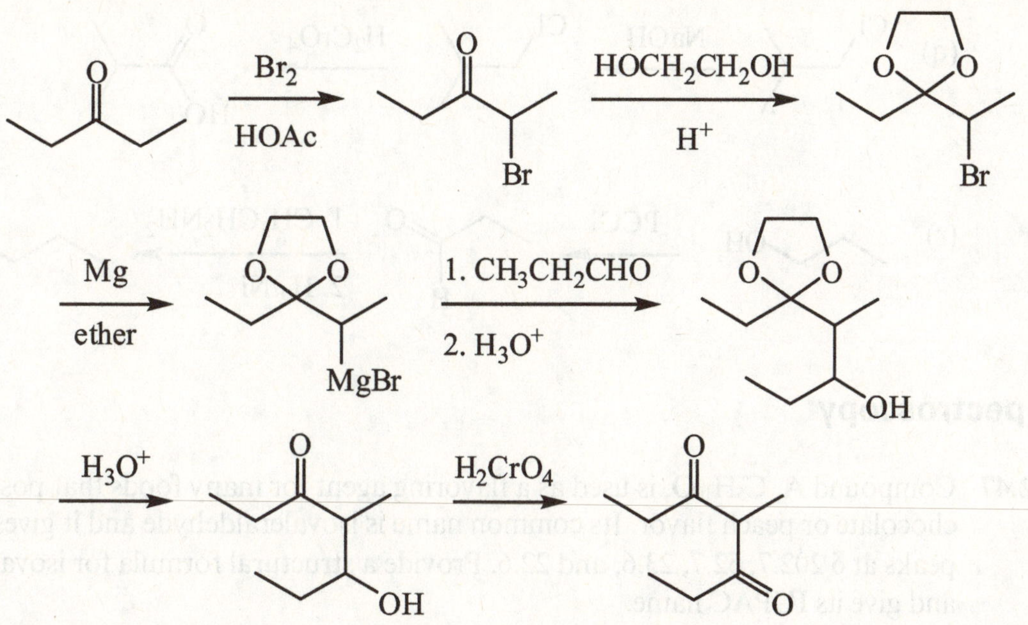

$$\xrightarrow{\text{Br}_2 \atop \text{HOAc}}$$

$$\xrightarrow{\text{HOCH}_2\text{CH}_2\text{OH} \atop \text{H}^+}$$

$$\xrightarrow{\text{Mg} \atop \text{ether}}$$

$$\xrightarrow[\text{2. H}_3\text{O}^+]{\text{1. CH}_3\text{CH}_2\text{CHO}}$$

$$\xrightarrow{\text{H}_3\text{O}^+}$$

$$\xrightarrow{\text{H}_2\text{CrO}_4}$$

(n)

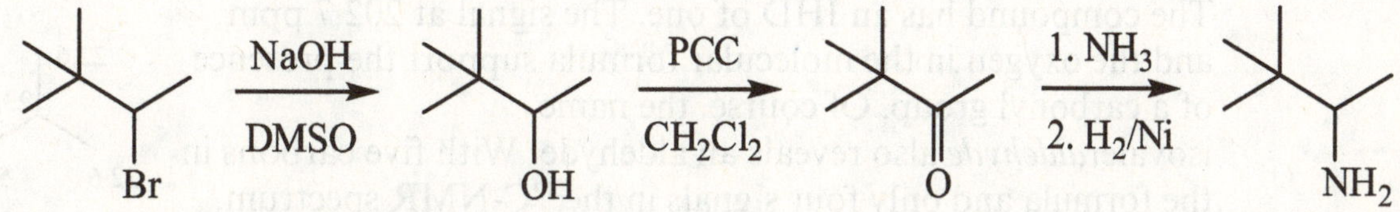

$$\xrightarrow{\text{NaOH} \atop \text{DMSO}}$$

$$\xrightarrow{\text{PCC} \atop \text{CH}_2\text{Cl}_2}$$

$$\xrightarrow[\text{2. H}_2/\text{Ni}]{\text{1. NH}_3}$$

(o) $CH_3-C\equiv C-H$

$$\xrightarrow[\text{2. CH}_3\text{Br}]{\text{1. NaNH}_2}$$

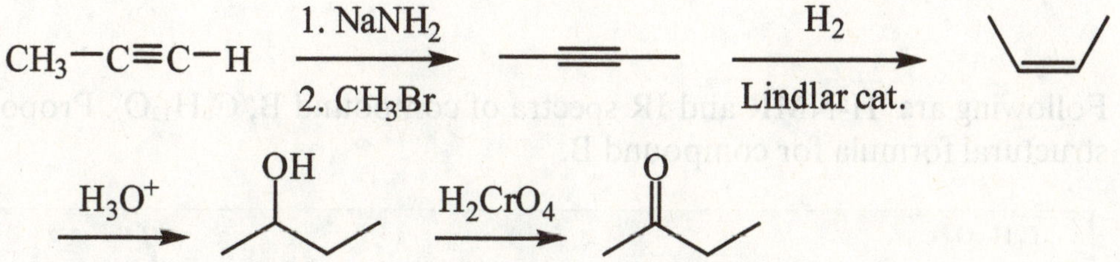

$$\xrightarrow[\text{Lindlar cat.}]{\text{H}_2}$$

$$\xrightarrow{\text{H}_3\text{O}^+}$$

$$\xrightarrow{\text{H}_2\text{CrO}_4}$$

(p)

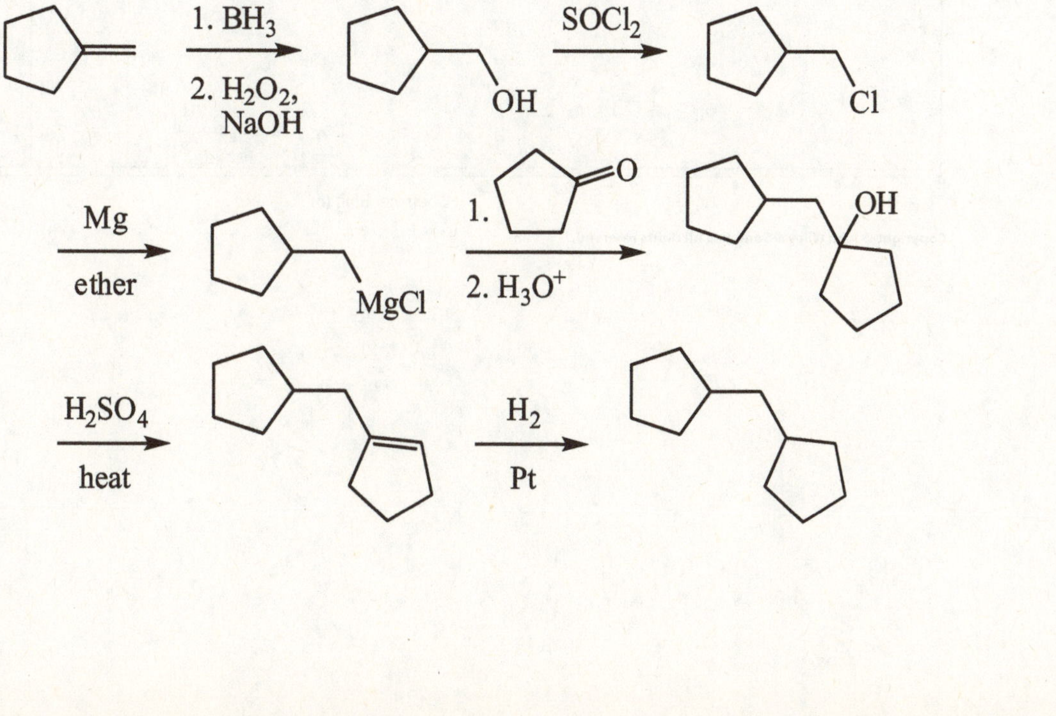

$$\xrightarrow[\substack{\text{2. H}_2\text{O}_2, \\ \text{NaOH}}]{\text{1. BH}_3}$$

$$\xrightarrow{\text{SOCl}_2}$$

$$\xrightarrow{\text{Mg} \atop \text{ether}}$$

$$\xrightarrow[\text{2. H}_3\text{O}^+]{\text{1.}}$$

$$\xrightarrow{\text{H}_2\text{SO}_4 \atop \text{heat}}$$

$$\xrightarrow{\text{H}_2 \atop \text{Pt}}$$

(q) 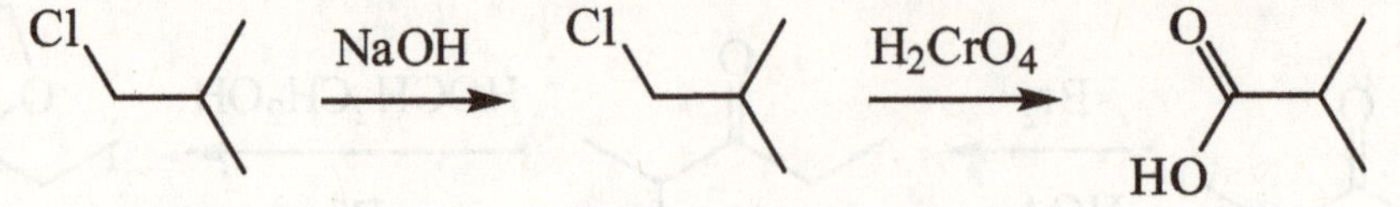 Cl — $\overset{\text{NaOH}}{\longrightarrow}$ — Cl — $\overset{\text{H}_2\text{CrO}_4}{\longrightarrow}$ — HO—C(=O)

(r) 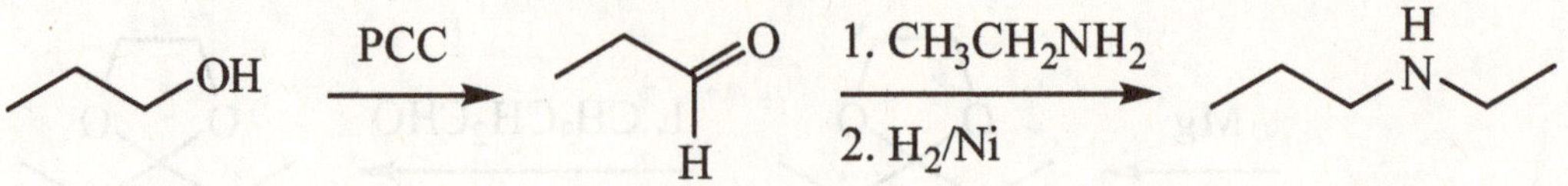 OH $\xrightarrow{\text{PCC}}$ —CHO $\xrightarrow[\text{2. H}_2/\text{Ni}]{\text{1. CH}_3\text{CH}_2\text{NH}_2}$ —N(H)—

Spectroscopy

12.47 Compound A, $C_5H_{10}O$, is used as a flavoring agent for many foods that possess a chocolate or peach flavor. Its common name is isovaleraldehyde and it gives ^{13}C-NMR peaks at δ 202.7, 52.7, 23.6, and 22.6. Provide a structural formula for isovaleraldehyde and give its IUPAC name.

The compound has an IHD of one. The signal at 202.7 ppm and the oxygen in the molecular formula support the presence of a carbonyl group. Of course, the name isovaler*aldehyde* also reveals an aldehyde! With five carbons in the formula and only four signals in the ^{13}C-NMR spectrum, the aldehyde must be 3-methylbutanal.

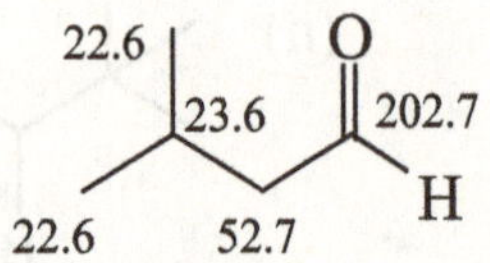

22.6 │
        ├ 23.6   O ‖ 202.7
22.6   52.7   H

12.48 Following are ^{1}H-NMR and IR spectra of compound B, $C_6H_{12}O_2$. Propose a structural formula for compound B.

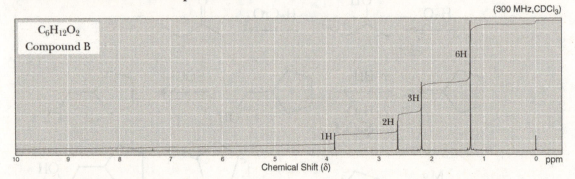

(300 MHz,CDCl₃)

$C_6H_{12}O_2$
Compound B

6H

3H

2H

1H

10   9   8   7   6   5   4   3   2   1   0   ppm
Chemical Shift (δ)

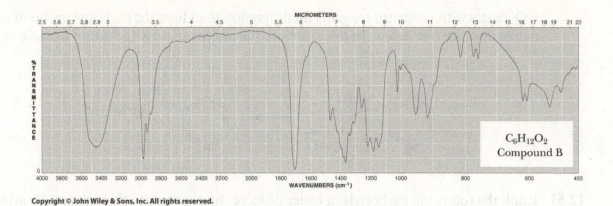

MICROMETERS

%TRANSMITTANCE

WAVENUMBERS (cm⁻¹)

$C_6H_{12}O_2$
Compound B

The compound has an IHD of one, which can be attributed to a carbonyl group (IR absorption near 1700 cm⁻¹). The strong, broad absorption near 3450 cm⁻¹ in the IR spectrum indicates the presence of an alcohol. The compound cannot be a carboxylic acid for two reasons: there is no carboxylic acid proton in the NMR spectrum, and the IR absorption for a carboxylic acid −OH group occurs at a lower wavenumber.

12.49 Compound C, $C_9H_{18}O$, is used in the automotive industry to retard the flow of solvent and thus improve the application of paints and coatings. It yields ¹³C NMR peaks at δ 210.5, 52.4, 24.5, and 22.6. Provide a structure and IUPAC name for C.

The compound has an IHD of one. The signal at 210.5 ppm and the oxygen in the molecular formula support the presence of a carbonyl group. The formula contains nine carbons, but there are only for signals in the spectrum, indicating the presence of equivalent carbons. The compound is 2,6-dimethyl-4-heptanone.

22.6   O   22.6
24.5   210.5   24.5

22.6   52.4   52.4   22.6

Looking Ahead

12.50 Reaction of a Grignard reagent with carbon dioxide, followed by treatment with aqueous HCl, gives a carboxylic acid. Propose a structural formula for the bracketed intermediate formed by the reaction of phenylmagnesium bromide with CO_2 and propose a mechanism for the formation of this intermediate:

MgBr

+ CO_2 →

$\begin{bmatrix} \text{intermediate} \\ \text{(not isolated)} \end{bmatrix}$ →

O
‖
C−OH

CO_2 reacts with Grignard reagents no differently than how the carbonyl group of an aldehyde or ketone reacts. In this reaction, the intermediate is the salt of benzoic acid.

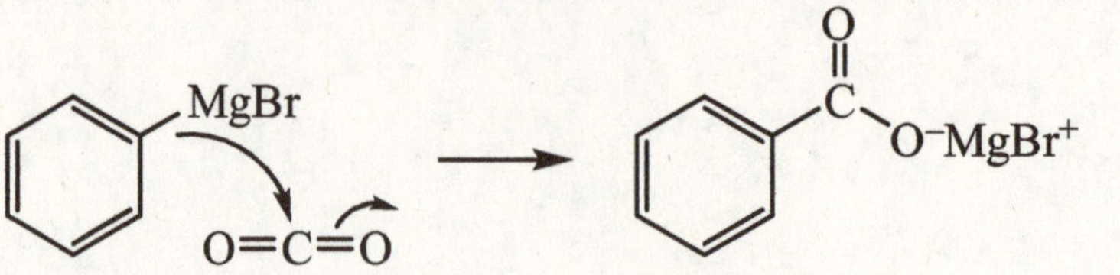

12.51 Rank the following carbonyls in order of increasing reactivity to nucleophilic attack, and explain your reasoning.

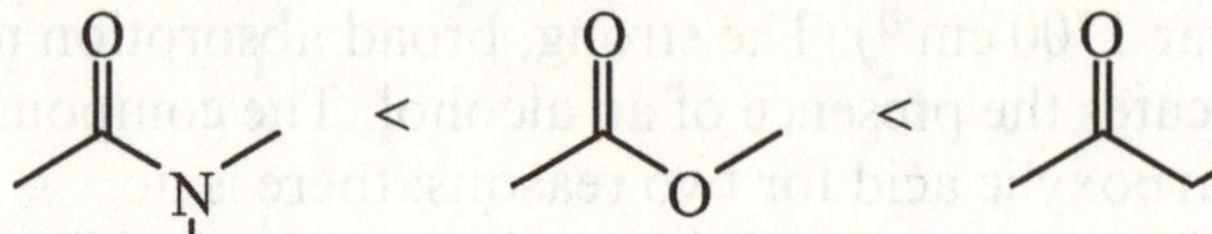

Each carbonyl compound can be described as a resonance hybrid composed of contributing structures. The ketone has an electron-deficient carbon and is thus susceptible to nucleophilic attack.

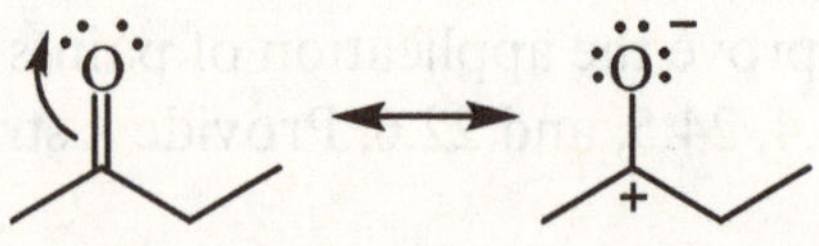

The reactivity of these carbonyl compounds to nucleophilic attack depends on the magnitude of positive charge on the carbonyl carbon. The two other compounds, an ester and an amide, have additional resonance structures that involve the participation of an electron pair from the oxygen and nitrogen atom, respectively. This reduces the magnitude of the positive charge on the carbonyl carbon, making it less electrophilic and,therefore, less reactive towards nucleophilic attack. Yet, is the ester oxygen or the amide nitrogen better at stabilizing the positive charge of the carbonyl carbon? Nitrogen is less electronegative than oxygen, so the lone pair of nitrogen is more available for resonance donation compared to that of oxygen. As a result, the carbonyl carbon of an amide is less positive than that of an ester and, therefore, less reactive towards nucleophilic attack.

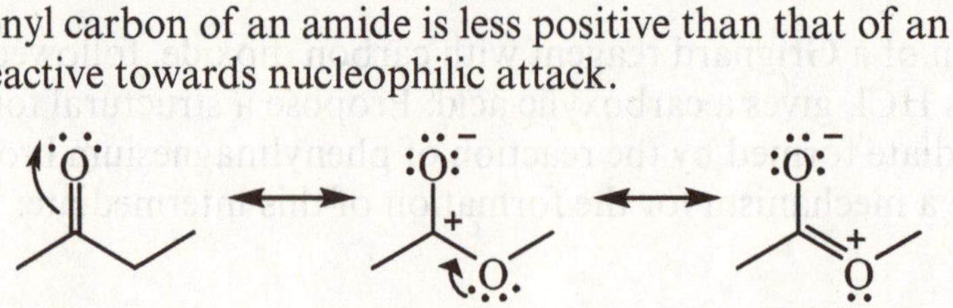

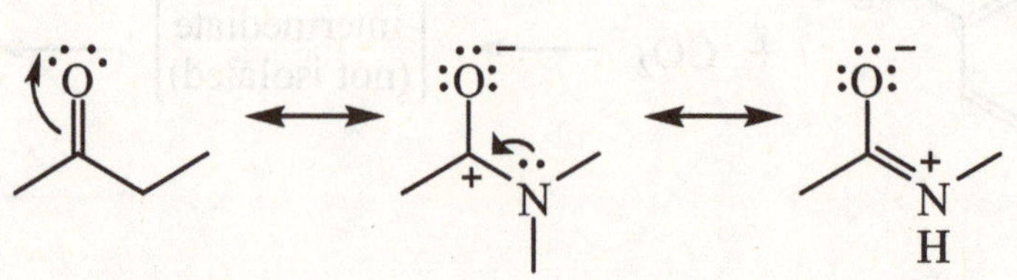

12.52 Provide the enol form of this ketone and predict the direction of equilibrium.

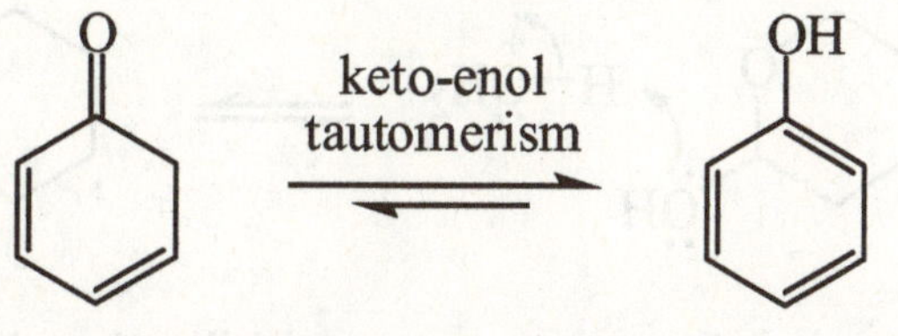

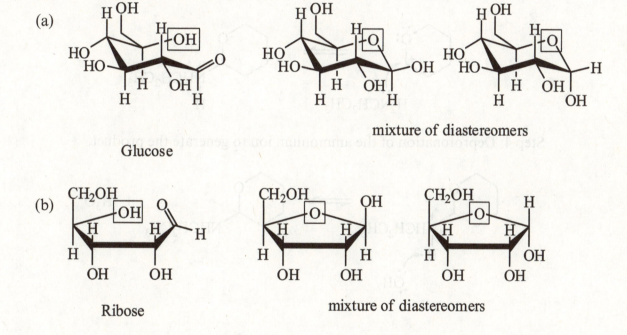

Although the keto tautomer is normally more favorable than the enol tautomer, the enol is more favorable in this case because the product is aromatic and more stable.

12.53 Draw the cyclic hemiacetal formed by reaction of the highlighted −OH group with the aldehyde group:

(a)

Glucose   mixture of diastereomers

(b)

Ribose   mixture of diastereomers

12.54 Propose a mechanism for the acid-catalyzed reaction of the following hemiacetal, with an amine acting as a nucleophile:

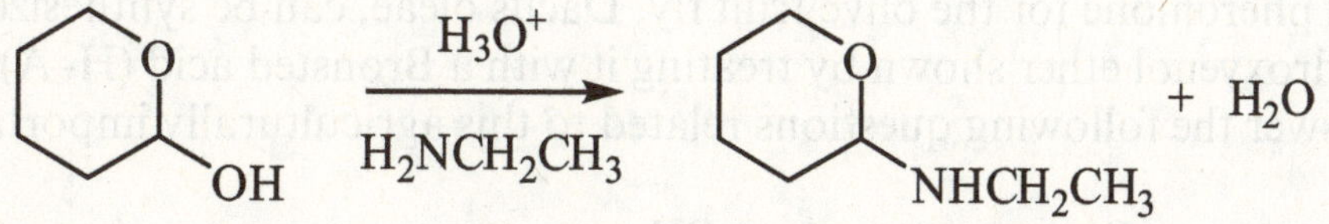

$$\text{(structure)} \quad \xrightarrow[H_2NCH_2CH_3]{H_3O^+} \quad \text{(structure)} \quad + \ H_2O$$

The mechanism of this reaction is analogous to the conversion of a hemiacetal to an acetal, except the nucleophile is an amine instead of an alcohol.

Step 1: Protonation of the hydroxyl group to form a good leaving group.

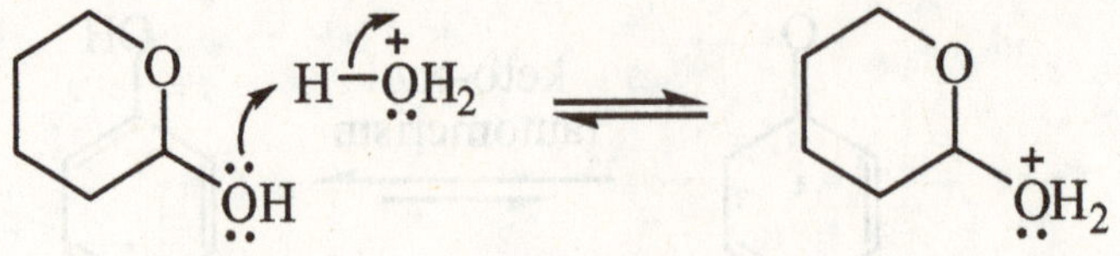

Step 2: Loss of water gives a resonance-stabilized carbocation.

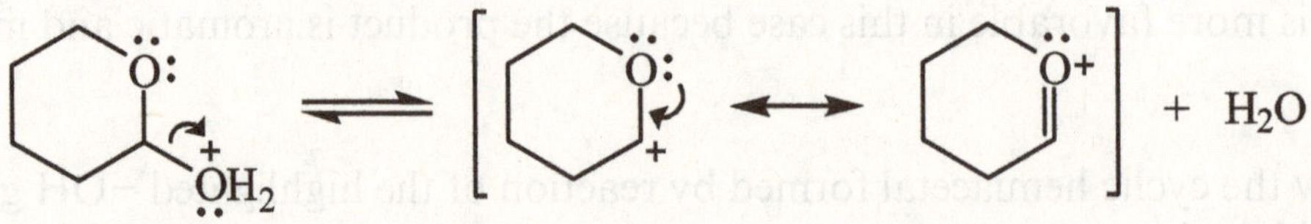

$+ \ H_2O$

Step 3: Nucleophilic attack of the carbocation by the amine.

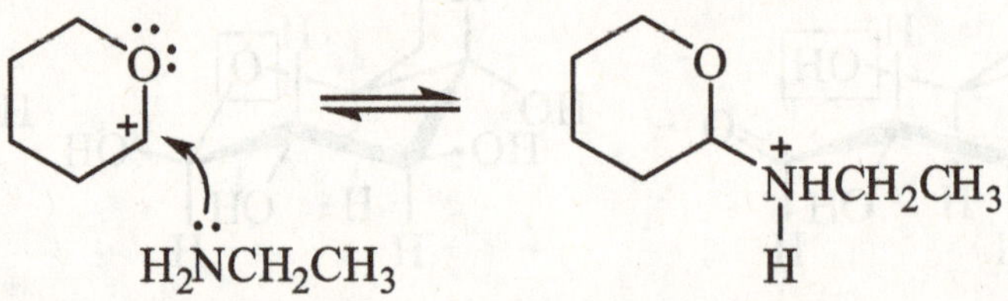

Step 4: Deprotonation of the ammonium ion to generate the product.

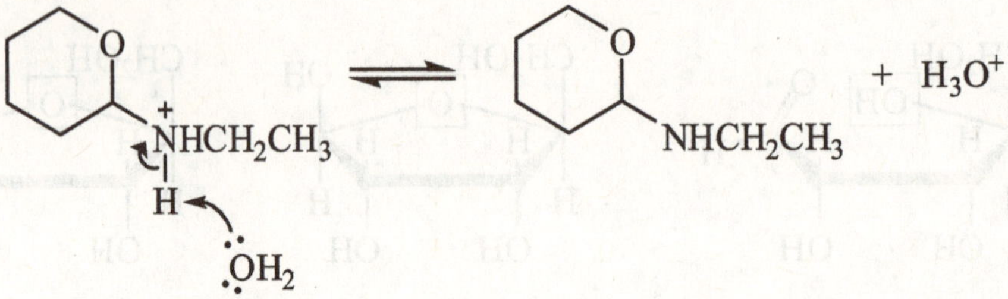

$+ \ H_3O^+$

Group Learning Activities

12.55 Pheromones are important organic compounds in agriculture because they represent one means of baiting and trapping insects that may be harmful to crops. Olean, the sex pheromone for the olive fruit fly, *Dacus oleae*, can be synthesized from the hydroxyenol ether shown by treating it with a Brønsted acid (H–A). As a group, answer the following questions related to this agriculturally important product:

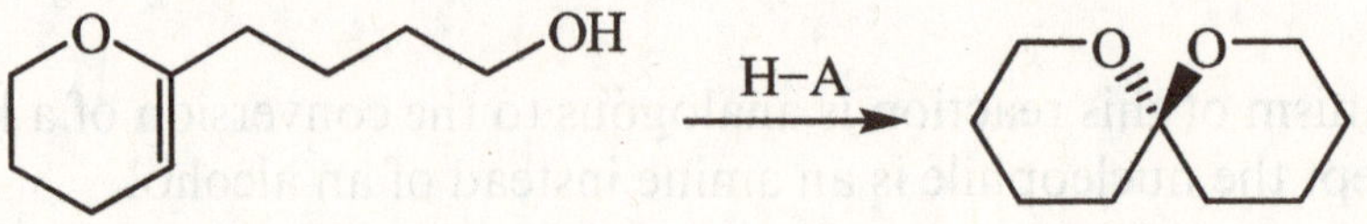

Olean

(a) Name the functional group in Olean.

There is an acetal present in Olean.

(b) Propose a mechanism for the reaction. Hint: The mechanism consists of the following patterns: (1) add a proton, (2) reaction of an electrophile and a nucleophile to form a new covalent bond, and (3) take a proton away.

Inspection of the starting material and the product shows that a new carbon-oxygen bond has been made. This new bond is made at the alkene carbon to which the other oxygen atom is already bonded. The reaction is mechanistically similar to the acid-catalyzed addition of an alcohol to an alkene.

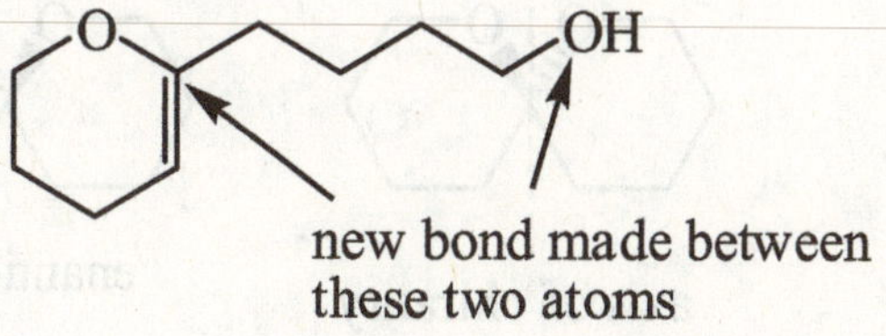

new bond made between these two atoms

The first step involves the protonation of the alkene. The proton preferentially adds to the lower alkene carbon because it gives the more-stable carbocation.

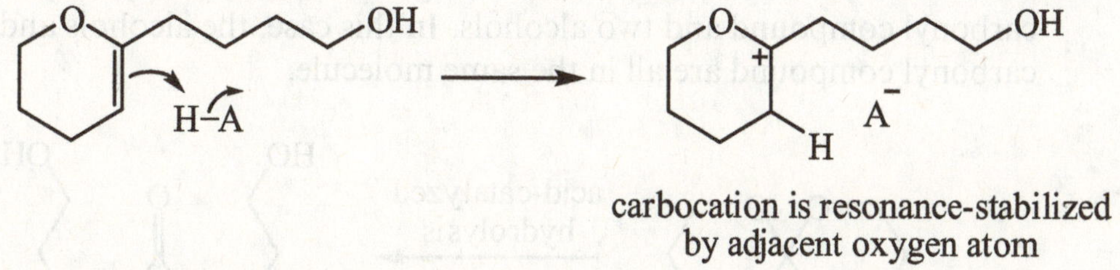

carbocation is resonance-stabilized by adjacent oxygen atom

The carbocation is then attacked by the oxygen of the alcohol.

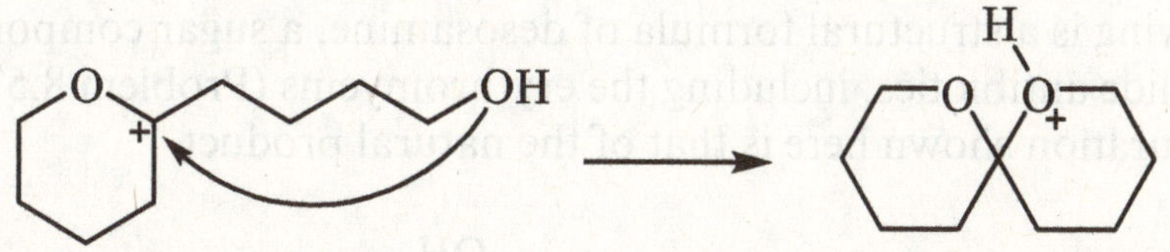

Finally, the oxonium ion is deprotonated to give the product and regenerate the acid.

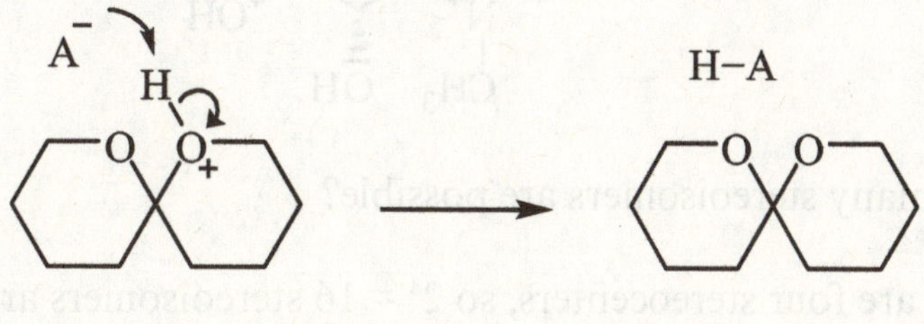

(c) Is Olean chiral? If so how many stereoisomers are possible? Hint: Build a model of olean. Then build a second model in which the two central C–O bonds are swapped.

Although Olean does not have a stereocenter (a carbon atom with four different substituents attached), Olean is a chiral molecule with two stereoisomers. The source of chirality is not a stereocentre but rather an axis of chirality. The two mirror images of Olean are shown below. Note that they are mirror images and not identical structures because the rotation of the structure on the left about the indicated axis does not give the same structure as the one on the right. (This is similar to how a left-handed spiral is the mirror image of a right-handed spiral.)

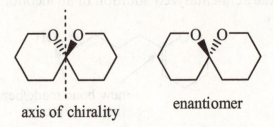

axis of chirality   enantiomer

(d) Predict the product formed by acid-catalyzed hydrolysis of Olean.

Because Olean is an acetal, it will undergo acid-catalyzed hydrolysis to form a carbonyl compound and two alcohols. In this case, the alcohols and the carbonyl compound are all in the same molecule.

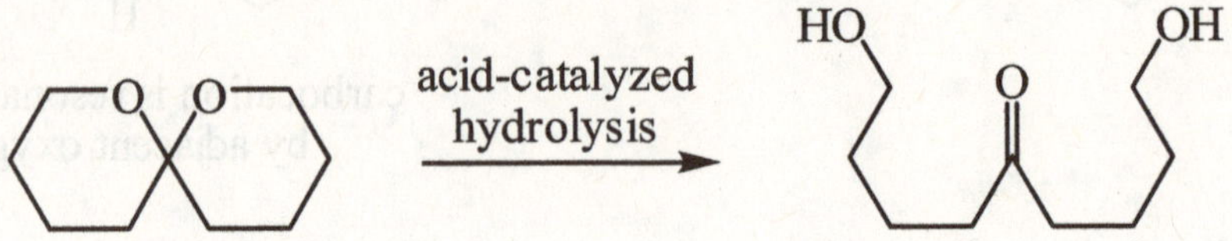

12.56 Following is a structural formula of desosamine, a sugar component of several macrolide antibiotics, including the erythromycins (Problem 8.57). The configuration shown here is that of the natural product.

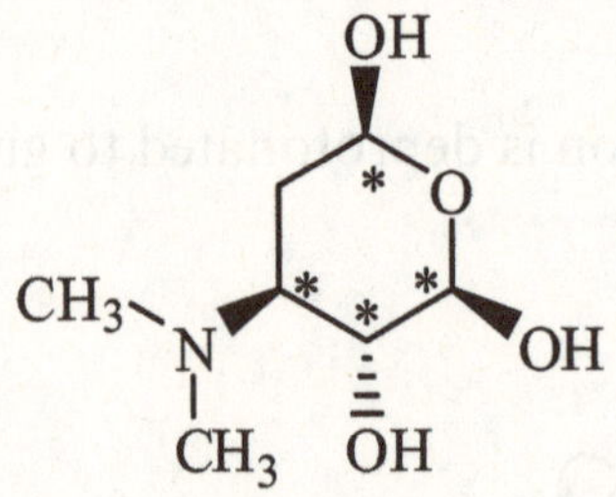

(a) How many stereoisomers are possible?

There are four stereocenters, so $2^4 = 16$ stereoisomers are possible.

(b) Can you spot the hemiacetals or acetals? For each that you spot, show the structure of the ketone or aldehyde and the alcohols from which it can be derived.

There are two hemiacetals, and they share a common oxygen atom. They can be made from:

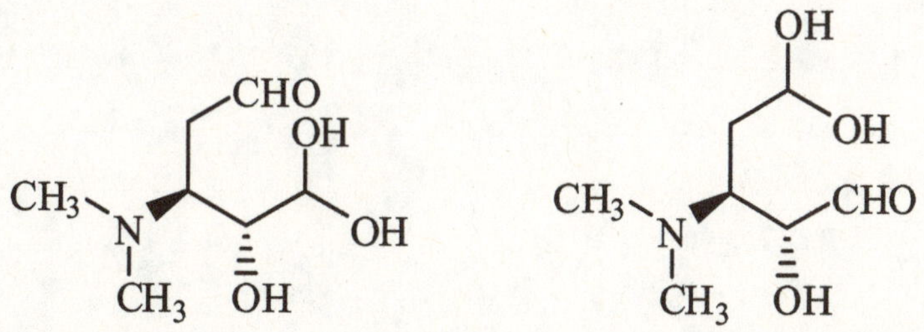

(c) Draw alternative chair conformations for desosamine and label the groups on the ring as either axial or equatorial.

(d) Which of the alternative chair conformations is more stable?

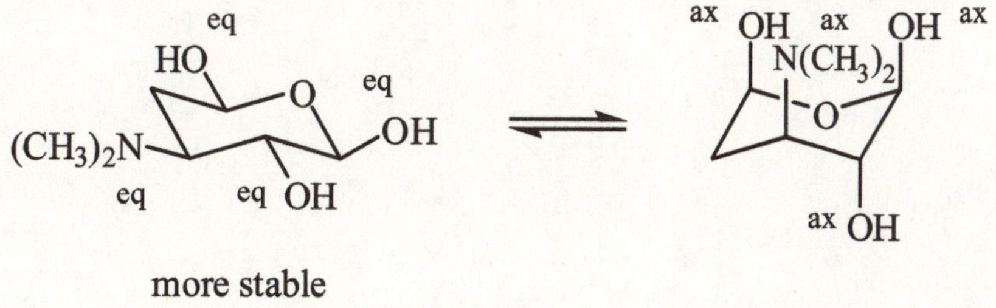

more stable

Chapter 13: Carboxylic Acids

Problems

13.1 Each of these compounds has a well-recognized common name. A derivative of glyceric acid is an intermediate in glycolysis (Section 21.3). Maleic acid is an intermediate in the tricarboxylic acid (TCA) cycle. Mevalonic acid is an intermediate in the biosynthesis of steroids (Section 19.4B). Lactic acid is a product of fermentation in animals (Section 21.4A).

Write the IUPAC name for each compound. Be certain to show the configuration of each.

(a)

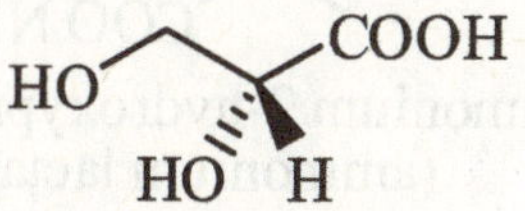

Glyceric acid

(*R*)-2,3-Dihydroxypropanoic acid

(b)

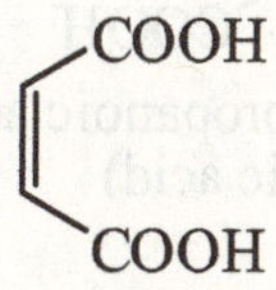

Maleic acid

cis-2-Butenedioic acid
or (*Z*)-butenedioic acid

(c)

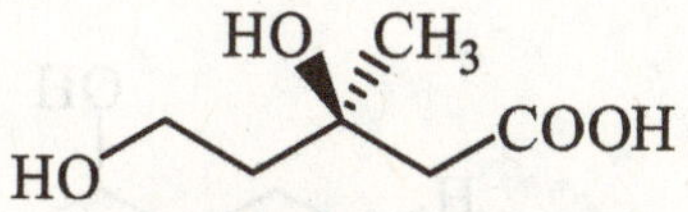

Mevalonic acid

(*R*)-3,5-Dihydroxy-
3-methylpentanoic acid

(d)

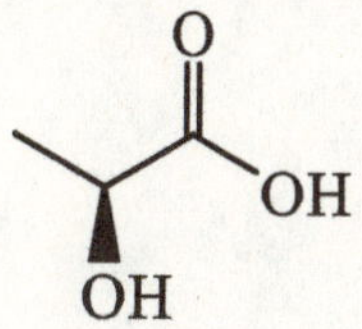

Lactic acid

(*S*)-2-Hydroxypropanoic acid

13.2 Match each compound with its appropriate pK_a value:

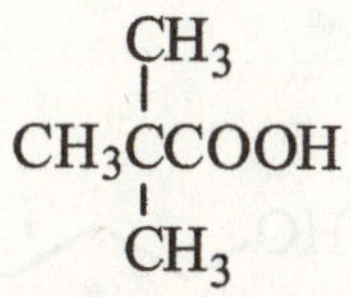

CH_3CCOOH with CH_3 groups

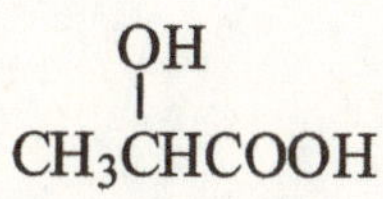

2,2-Dimethylpropanoic acid

CF_3COOH

Trifluoroacetic acid

$CH_3CHCOOH$ with OH

2-Hydroxypropanoic acid
(Lactic acid)

pK_a	5.03	0.22	3.08

Recall from Chapter 2 that the strength of an acid is related to the stability of its conjugate base. All three conjugate bases (carboxylate ions) are stabilized by resonance. Trifluoroacetic acid is the strongest acid because the trifluoroacetate ion is also stabilized by three inductively withdrawing fluorine atoms. Lactic acid also has an −OH group that stabilizes the lactate ion, but it is not as inductively withdrawing as three fluorines. 2,2-Dimethylpropanoic acid does not have any inductively withdrawing groups that can stabilize its conjugate base.

13.3   Write an equation for the reaction of each acid in Example 13.3 with ammonia, and name the salt formed.

(a)   ~~~COOH + NH₃ ⟶ ~~~COO⁻ NH₄⁺

Butanoic acid   Ammonium butanoate

(b)

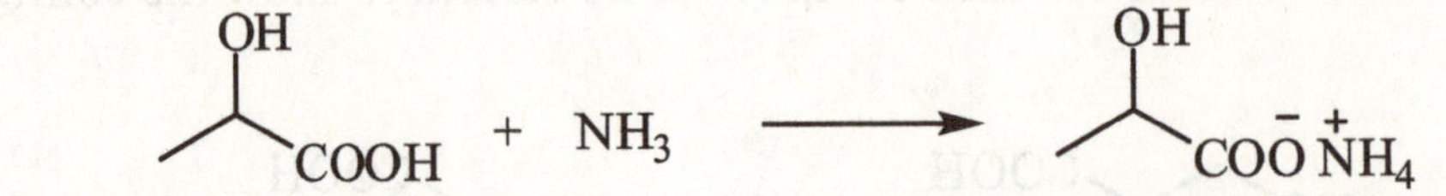

2-Hydroxypropanoic acid   Ammonium 2-hydroxypropanoate
(lactic acid)   (ammonium lactate)

13.4   Provide the product formed when each of the following is treated with: (i) H₂/Pd, (ii) LiAlH₄ in ether followed by H₂O, (iii) NaBH₄ in EtOH followed by H₂O. Presume that an excess of reagent is available for each reaction.

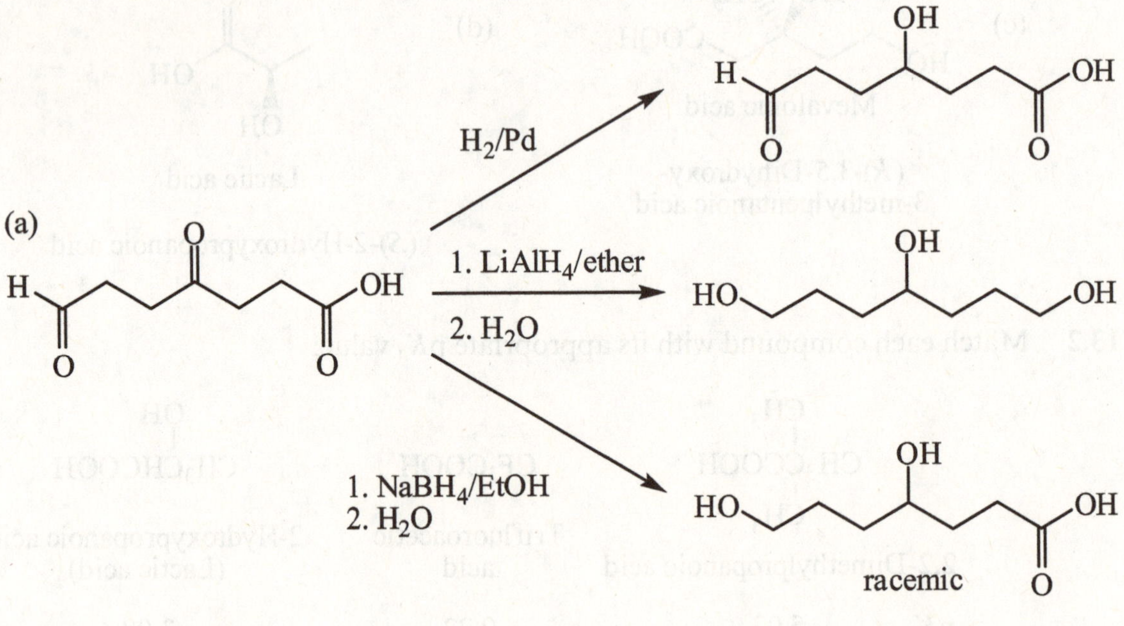

(b)

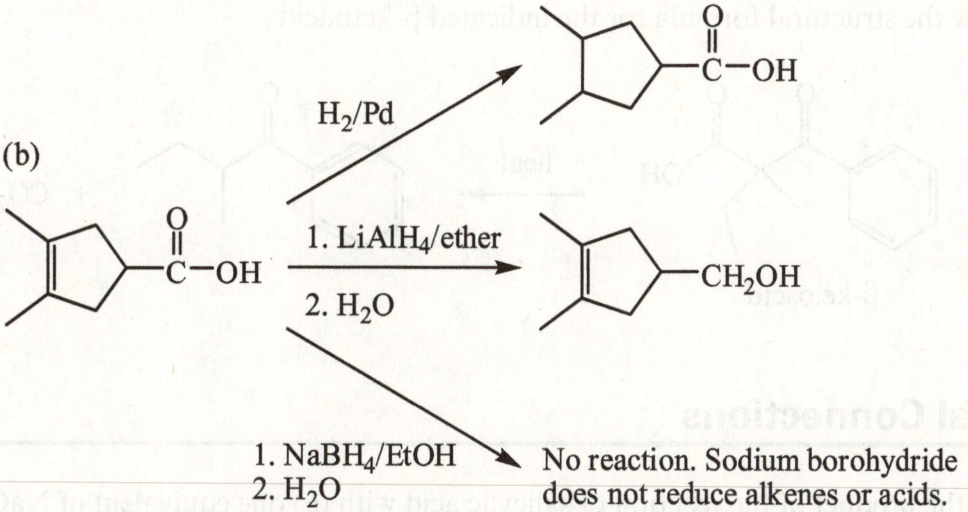
H2/Pd
C−OH
1. LiAlH4/ether
2. H2O
CH2OH
1. NaBH4/EtOH
2. H2O
No reaction. Sodium borohydride
does not reduce alkenes or acids.

13.5 Complete these Fischer esterification reactions:

(a)

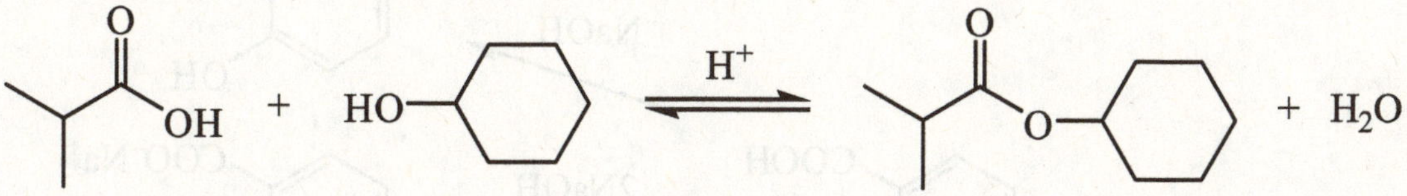
OH
+
HO
H+
O
+ H2O

(b)

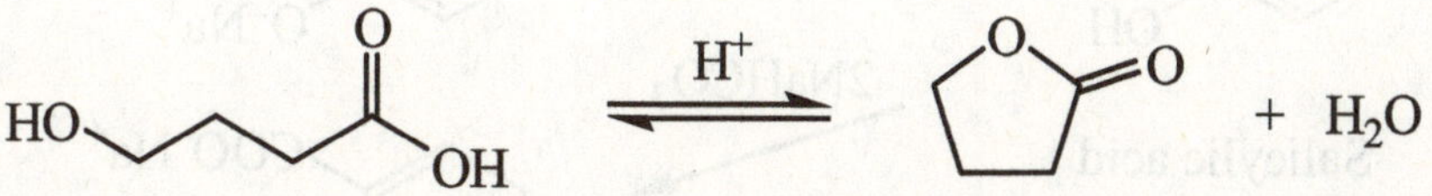
HO
OH
H+
O
+ H2O

13.6 Complete each equation:

(a)

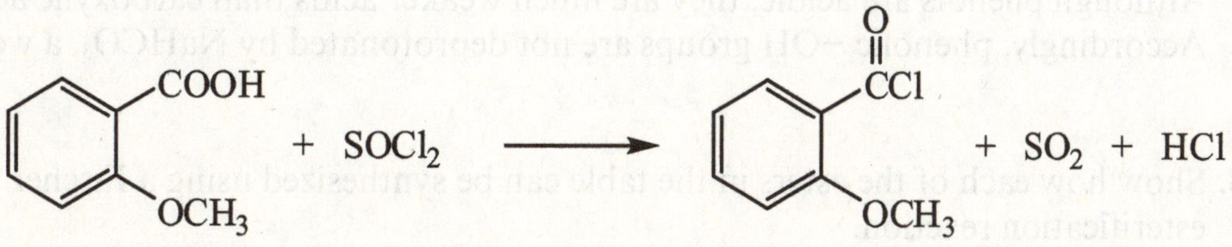
COOH
OCH3
+ SOCl2
CCl
OCH3
+ SO2 + HCl

(b)

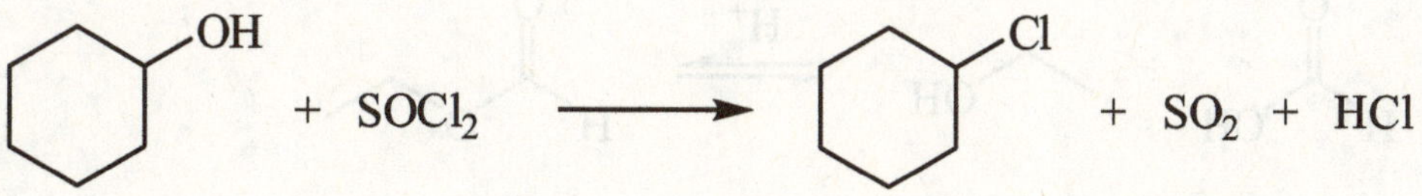
OH
+ SOCl2
Cl
+ SO2 + HCl

13.7   Draw the structural formula for the indicated β-ketoacid:

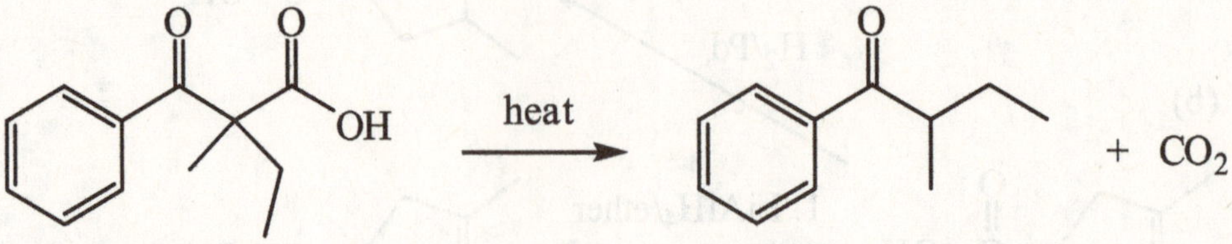

β-ketoacid

heat → + CO_2

Chemical Connections

13A. Draw the product of the reaction of salicylic acid with (a) one equivalent of NaOH, (b) two equivalents of NaOH, and (c) two equivalents of $NaHCO_3$.

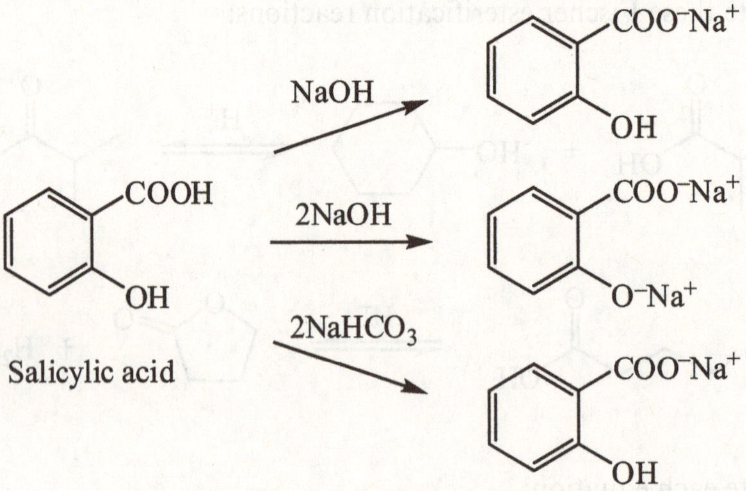

Salicylic acid

NaOH

2NaOH

$2NaHCO_3$

Although phenols are acidic, they are much weaker acids than carboxylic acids. Accordingly, phenolic –OH groups are not deprotonated by $NaHCO_3$, a weak base.

13B. Show how each of the esters in the table can be synthesized using a Fischer esterification reaction.

(a)

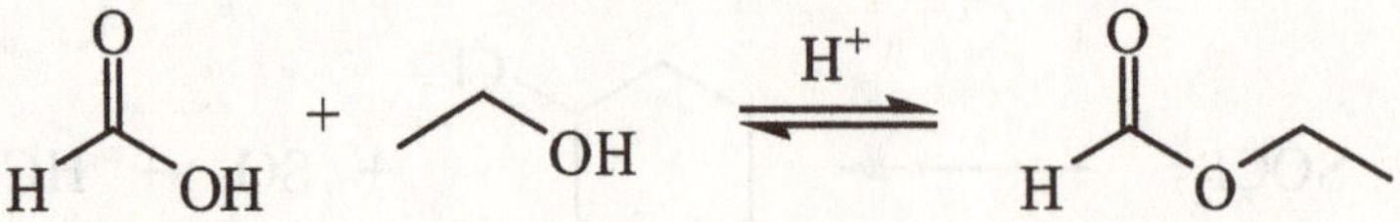

(b)

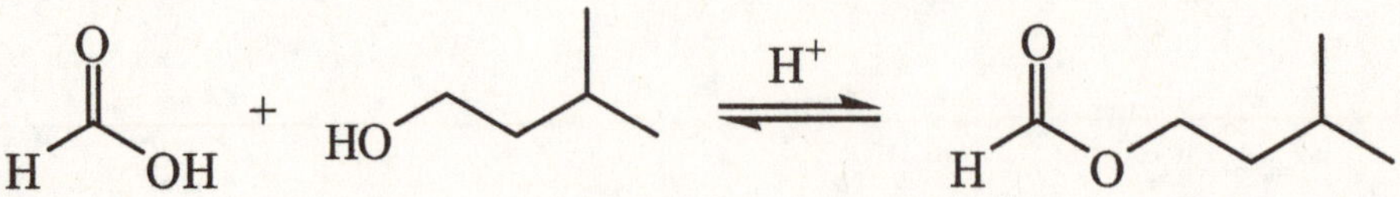

(c)

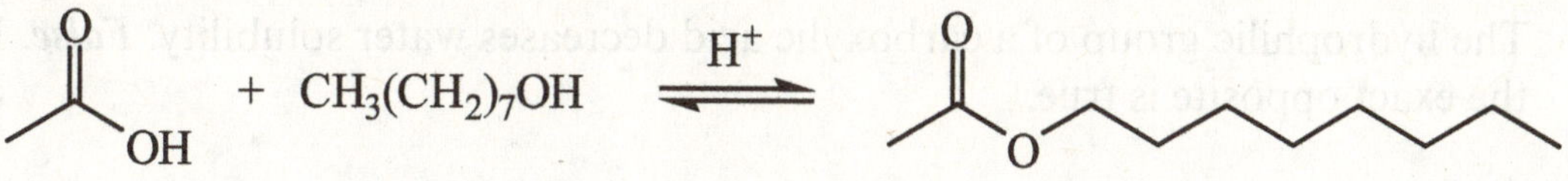

(d)

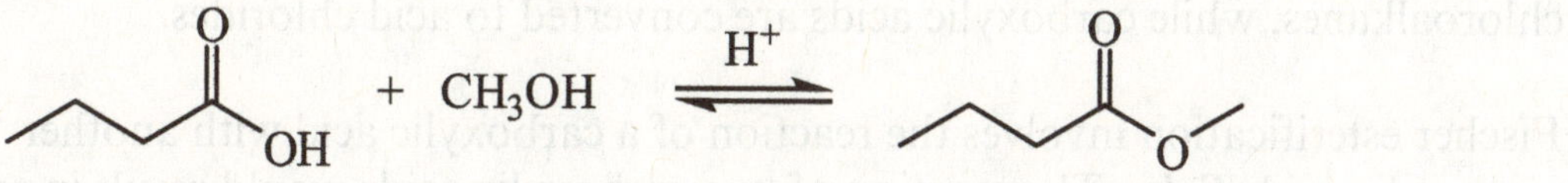

(e)

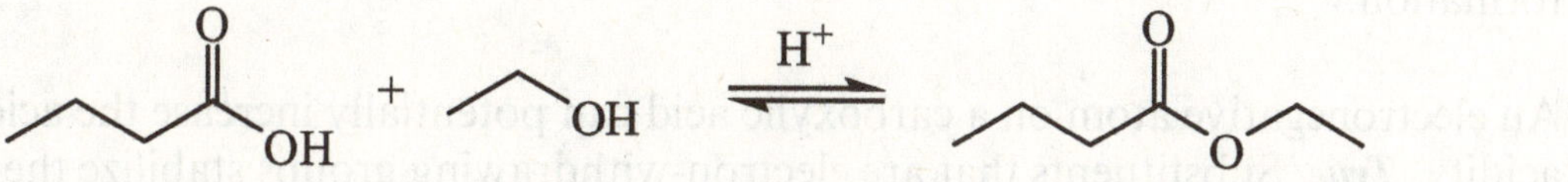

(f)

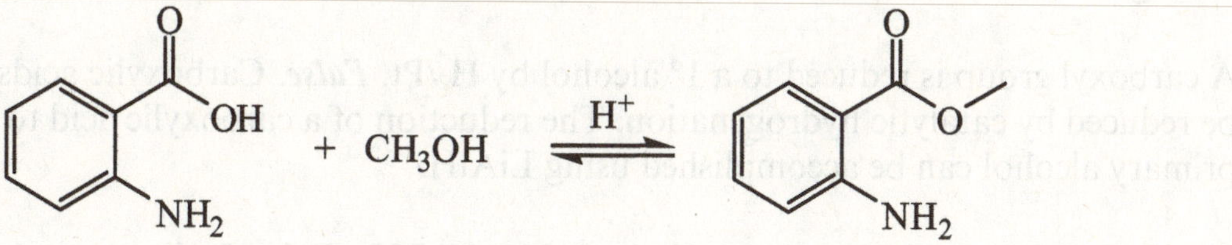

13C. Show the mechanism for the decarboxylation of acetoacetic acid. Explain why 3-hydroxybutanoic acid cannot undergo decarboxylation.

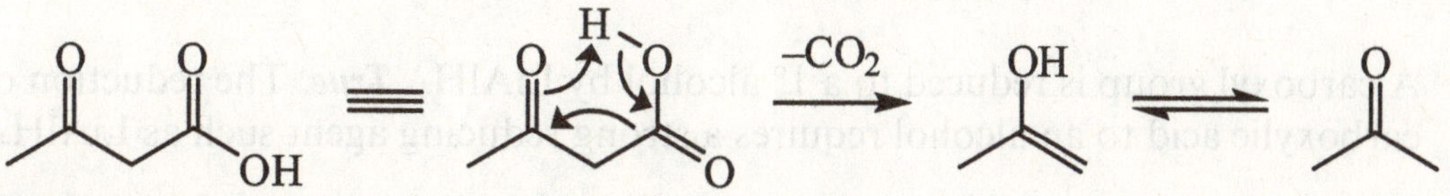

Acetoacetic acid

3-Hydroxybutanoic acid cannot undergo decarboxylation because it has a β-hydroxy group and not a β-keto group, which is necessary for a carboxylic acid to decarboxylate.

Quick Quiz

1. In naming carboxylic acids, it is always necessary to indicate the position at which the carboxyl group occurs. *False*. In acyclic compounds, the carbon of the carboxyl group is always assigned position one.

2. 2-Propylpropanedioic acid can undergo decarboxylation at relatively moderate temperatures. *True*. Each carboxylic acid in the compound has a β-keto group relative to it.

3. Fischer esterification is reversible. *True*. In a Fischer esterification, the starting materials (alcohol and carboxylic acid) are in equilibrium with the products (ester and water).

4.   The hydrophilic group of a carboxylic acid decreases water solubility. *False*. Rather, the exact opposite is true.

5.   Both alcohols and carboxylic acids react with $SOCl_2$. *True*. Alcohols are converted to chloroalkanes, while carboxylic acids are converted to acid chlorides.

6.   Fischer esterification involves the reaction of a carboxylic acid with another carboxylic acid. *False*. The reaction of two carboxylic acids would result in anhydride formation.

7.   An electronegative atom on a carboxylic acid can potentially increase the acid's acidity. *True*. Substituents that are electron-withdrawing groups stabilize the conjugate base.

8.   A carboxyl group is reduced to a 1° alcohol by H_2/Pt. *False*. Carboxylic acids cannot be reduced by catalytic hydrogenation. The reduction of a carboxylic acid to a primary alcohol can be accomplished using $LiAlH_4$.

9.   A carboxyl group is reduced to a 1° alcohol by $NaBH_4$. *False*. Sodium borohydride is too weak of a reducing agent to reduce a carboxylic acid. $LiAlH_4$ is required.

10.   A carboxyl group that has been deprotonated is called a carboxylate group. *True*. A carboxylate is the conjugate base of a carboxylic acid.

11.   A carboxyl group is reduced to a 1° alcohol by $LiAlH_4$. *True*. The reduction of a carboxylic acid to an alcohol requires a strong reducing agent such as $LiAlH_4$.

12.   The conjugate base of a carboxylic acid is resonance-stabilized. *True*. The negative charge the carboxylate group is delocalized over both oxygen atoms.

13.   Carboxylic acids possess both a region of polarity and a region of nonpolarity. *True*. The carboxyl group is polar, while the alkyl group is nonpolar.

14.   Carboxylic acids are less acidic than phenols. *False*. Although the conjugate bases of both are stabilized by resonance, the delocalization of negative charge over two oxygen atoms results in a conjugate base that is more stable than the conjugate base of phenol.

15.   4-Oxopentanoic acid can undergo decarboxylation at relatively moderate temperatures. *False*. The compound does not have a β-keto group.

16.   The γ position of a carboxylic acid refers to carbon-4 of the chain. *True*. The α position is carbon-2 (the carbon adjacent to the carboxyl group), β is carbon-3, γ is carbon-4, δ is carbon-5, and so forth.

End-of-Chapter Problems

Structure and Nomenclature

13.8 Name and draw structural formulas for the four carboxylic acids with the molecular formula $C_5H_{10}O_2$. Which of these carboxylic acids is chiral?

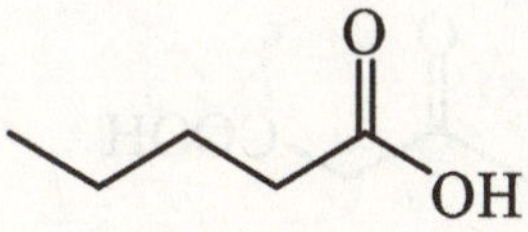

Pentanoic acid

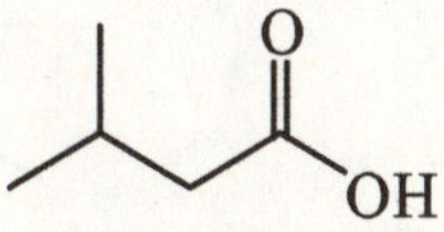

3-Methylbutanoic acid

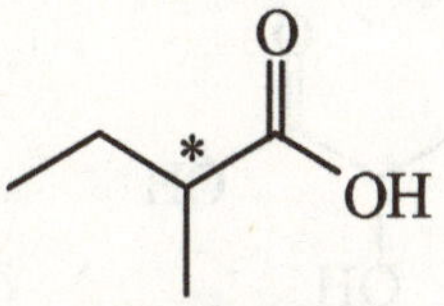

2-Methylbutanoic acid (chiral)

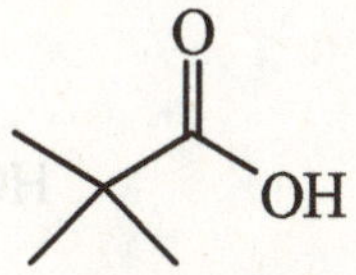

2,2-Dimethyl-propanoic acid

13.9 Write the IUPAC name for each compound:

(a) —COOH

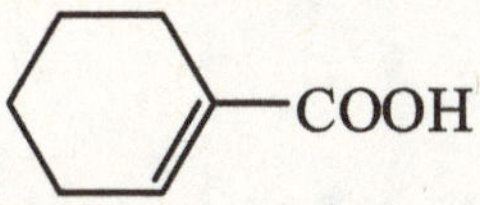

1-Cyclohexenecarboxylic acid

(b)

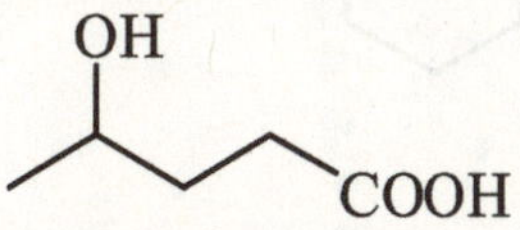

4-Hydroxypentanoic acid

(c)

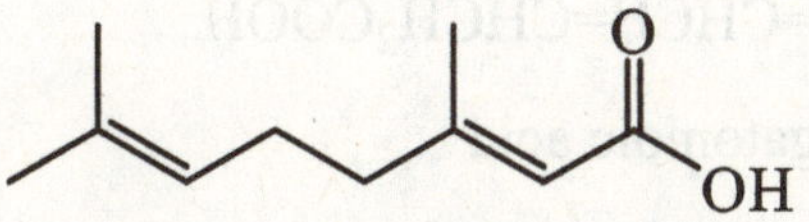

(E)-3,7-Dimethyl-2,6-octadienoic acid

(d)

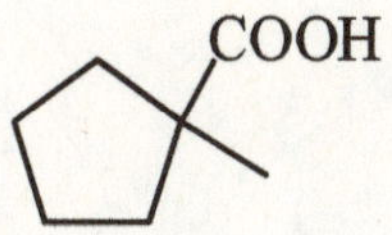

1-Methylcyclopentanecarboxylic acid

(e)

$COO^- NH_4^+$

Ammonium hexanoate

(f)

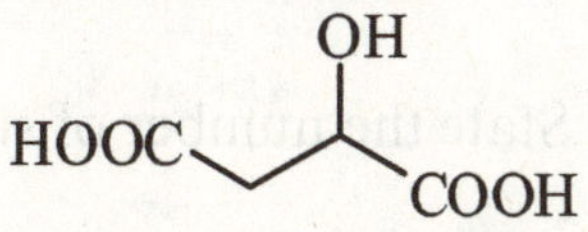

2-Hydroxybutanedioic acid

13.10 Draw a structural formula for each carboxylic acid:

(a) 4-Nitrophenylacetic acid

(b) 4-Aminopentanoic acid

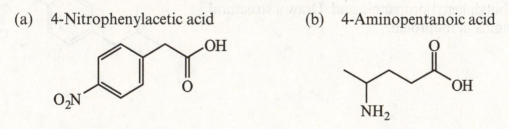

(c)   3-Chloro-4-phenylbutanoic acid

(d)   *cis*-3-Hexenedioic acid

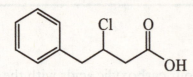

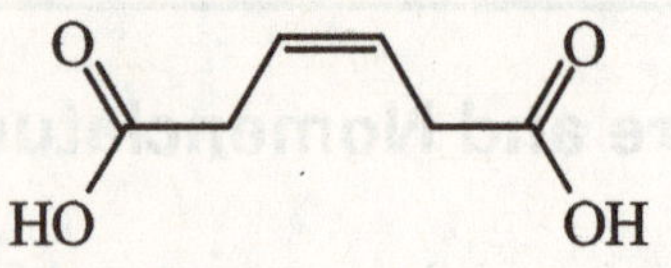

(e)   2,3-Dihydroxypropanoic acid

(f)   3-Oxohexanoic acid

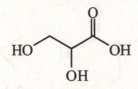

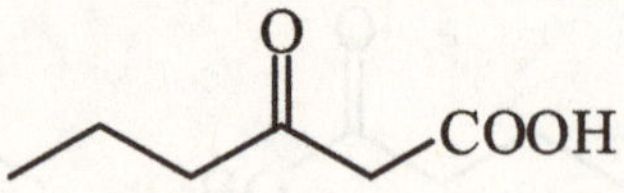

(g)   2-Oxocyclohexanecarboxylic acid

(h)   2,2-Dimethylpropanoic acid

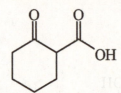

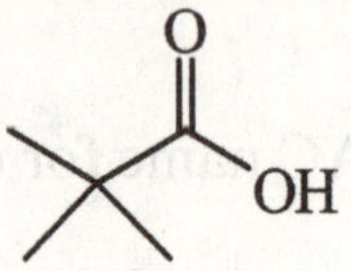

13.11 Megatomoic acid, the sex attractant of the female black carpet beetle, has the structure

$$CH_3(CH_2)_7CH=CHCH=CHCH_2COOH$$

Megatomoic acid

(a)   What is the IUPAC name of megatomoic acid?

3,5-Tetradecadienoic acid

(b)   State the number of stereoisomers possible for this compound.

The compound has two alkenes, and *cis-trans* isomerism is possible for each of them. Therefore, it has four possible stereoisomers.

13.12 The IUPAC name of ibuprofen is 2-(4-isobutylphenyl)propanoic acid. Draw a structural formula of ibuprofen.

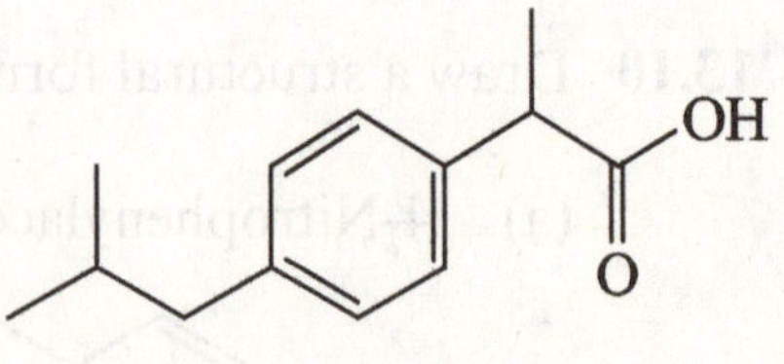

13.13 Draw structural formulas for these salts:

  (a) Sodium benzoate   (b) Lithium acetate   (c) Ammonium acetate

  (d) Disodium adipate   (e) Sodium salicylate   (f) Calcium butanoate

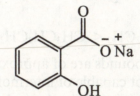

13.14 The monopotassium salt of oxalic acid is present in certain leafy vegetables, including rhubarb. Both oxalic acid and its salts are poisonous in high concentrations. Draw a structural formula of monopotassium oxalate.

13.15 Potassium sorbate is added as a preservative to certain foods to prevent bacteria and molds from causing food spoilage and to extend the foods' shelf life. The IUPAC name of potassium sorbate is potassium (2*E*,4*E*)-2,4-hexadienoate. Draw a structural formula of potassium sorbate.

13.16 Zinc 10-undecenoate, the zinc salt of 10-undecenoic acid, is used to treat certain fungal infections, particularly *tinea pedis* (athlete's foot). Draw a structural formula of this zinc salt.

Physical Properties

13.17 Arrange the compounds in each set in order of increasing boiling point:

(a) $CH_3(CH_2)_6CHO$ < $CH_3(CH_2)_6CH_2OH$ < $CH_3(CH_2)_5COOH$

All three compounds are of approximately the same molecular weight and size. The alcohol and the carboxylic acid are capable of intermolecular hydrogen bonding, so they have a higher boiling point than the aldehyde. The boiling point of the carboxylic acid is higher than that of the alcohol because of the highly polar carboxyl group and the ability of the carbonyl oxygen to act as a hydrogen-bonding acceptor.

(b) $CH_3CH_2OCH_2CH_3$ < $CH_3CH_2CH_2CH_2OH$ < CH_3CH_2COOH

All three compounds are of approximately the same molecular weight and size. The ether is not capable of intermolecular hydrogen bonding, so it has the lowest boiling point of the three. As in (a), the carboxylic acid has a higher boiling point than the alcohol.

Preparation of Carboxylic Acids

13.18 Draw a structural formula for the product formed by treating each compound with warm chromic acid, H_2CrO_4:

(a)
$$CH_3(CH_2)_4CH_2OH \xrightarrow{H_2CrO_4} CH_3(CH_2)_4COOH$$

(b)

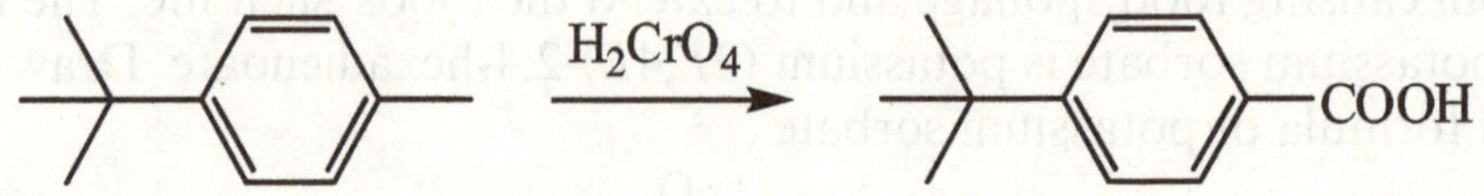

(c)

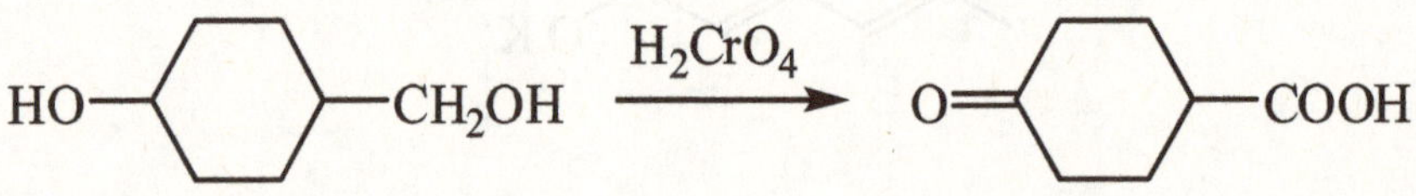

13.19 Draw a structural formula for a compound of the given molecular formula that, on oxidation by chromic acid, gives the carboxylic acid or dicarboxylic acid shown:

(a)

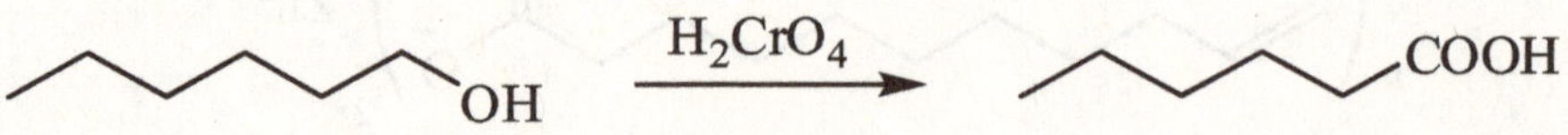

(b)

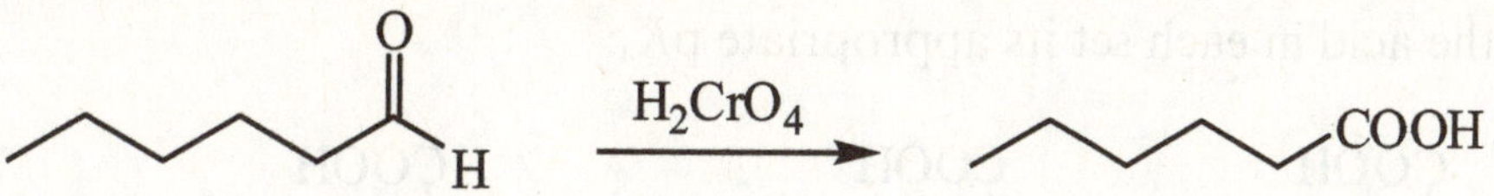

$$\text{pentanal} \xrightarrow{\text{H}_2\text{CrO}_4} \text{pentanoic acid (COOH)}$$

(c)

$$\text{HO}\dots\text{OH} \xrightarrow{\text{H}_2\text{CrO}_4} \text{HOOC}\dots\text{COOH}$$

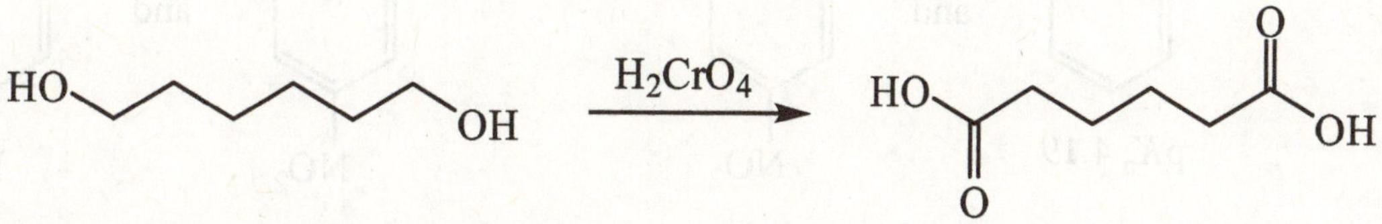

Acidity of Carboxylic Acids

13.20 Which is the stronger acid in each pair?

(a) Phenol (pK_a 9.95) or **benzoic acid (pK_a 4.17)**. Lower pK_a = more acidic.

(b) **Lactic acid (K_a 1.4 × 10^{-4})** or ascorbic acid (K_a 6.8 × 10^{-5}). Higher K_a = more acidic.

13.21 Arrange these compounds in order of increasing acidity: benzoic acid, benzyl alcohol, and phenol.

Relative strengths of acids

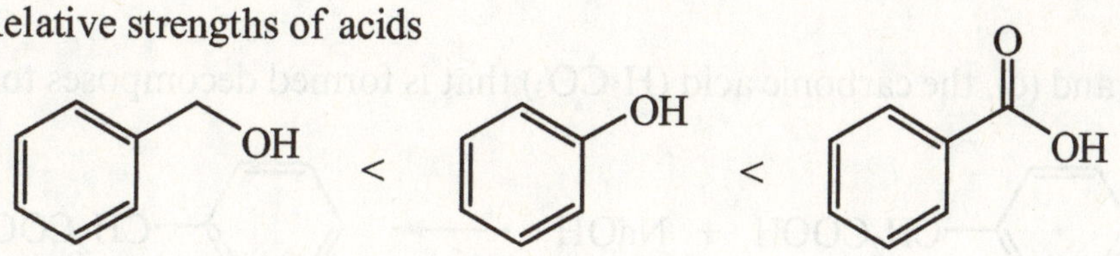

Relative stabilities of conjugate bases

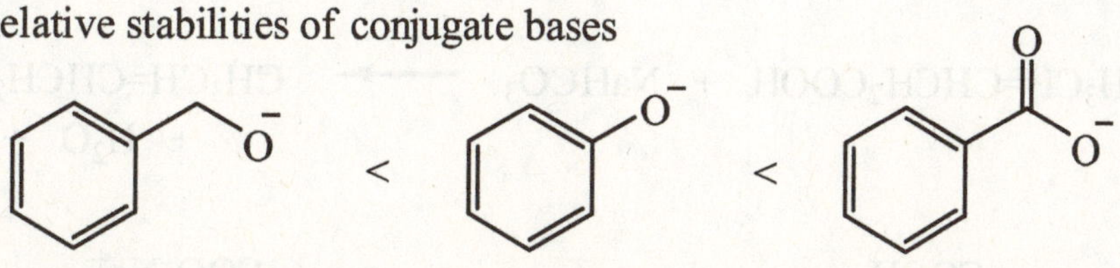

Recall that the strength of an acid is related to the stability of its conjugate base. Here, all three conjugate bases feature a negatively charged oxygen atom. The conjugate base of benzyl alcohol is not stabilized by resonance, so it is the least stable of the three. The phenoxide and benzoate ions are both stabilized by resonance, but benzoate is more stable than phenoxide. In benzoate, the negative charged is delocalized over two oxygen atoms, while in phenoxide, the negative charge is delocalized over an oxygen and carbon atoms (less favorable than two oxygen atoms).

13.22 Assign the acid in each set its appropriate pK_a:

(a)

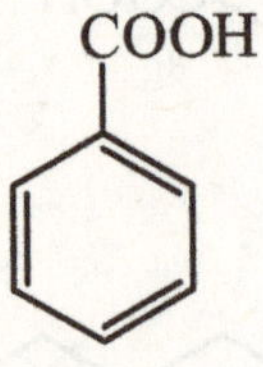

COOH

and

COOH

NO$_2$

pK_a 4.19

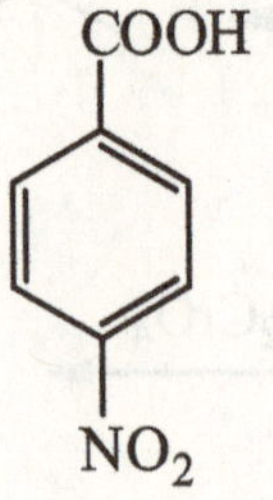

pK_a 3.14

The inductively withdrawing nitro group makes *p*-nitrobenzoic acid a stronger acid than benzoic acid.

(b)

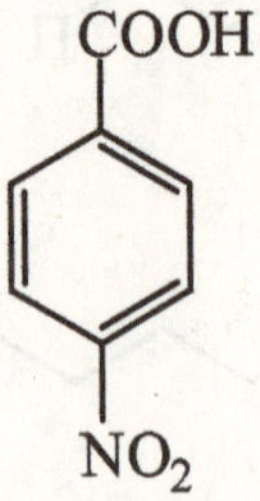

COOH

and

COOH

NO$_2$

NH$_2$

pK_a 3.14

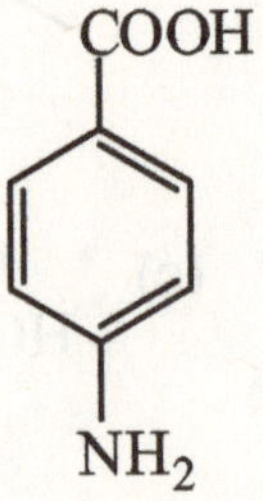

pK_a 4.92

The nitro group is an electron-withdrawing group, while the amino group is electron-donating.

(c) CH_3CCH_2COOH and CH_3CCOOH

pK_a 3.58   pK_a 2.49

The carbonyl group, which has a δ+ carbon, is inductively withdrawing.

(d) $CH_3CHCOOH$ and CH_3CH_2COOH

OH

pK_a 3.85   pK_a 4.78

The hydroxyl group is inductively withdrawing.

13.23 Complete these acid-base reactions:

In (b) and (c), the carbonic acid (H_2CO_3) that is formed decomposes to H_2O and CO_2.

(a) $\bigcirc$—CH_2COOH + NaOH ⟶ $\bigcirc$—$CH_2COO^-Na^+$ + H_2O

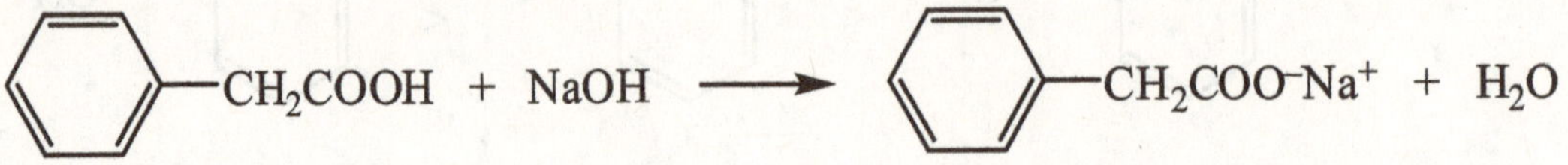

(b) $CH_3CH=CHCH_2COOH$ + $NaHCO_3$ ⟶ $CH_3CH=CHCH_2COO^-Na^+$ + H_2O + CO_2

(c) COOH / OH + $NaHCO_3$ ⟶ COO$^-$Na$^+$ / OH + H_2O + CO_2

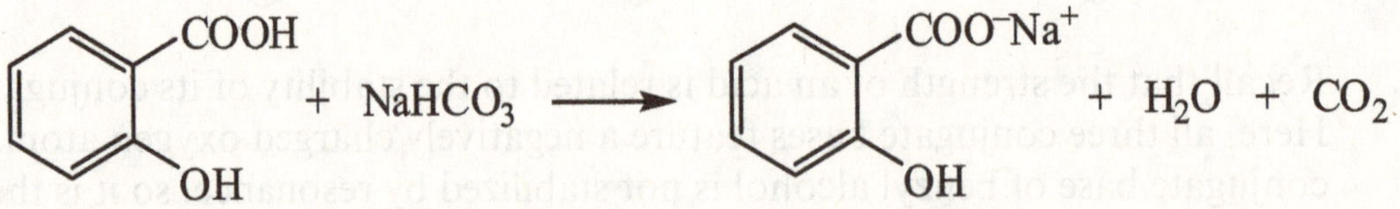

(d) $CH_3CHCOOH$ (OH) + $H_2NCH_2CH_2OH$ ⟶ CH_3CHCOO^- $H_3\overset{+}{N}CH_2CH_2OH$

(e) $CH_3CH=CHCH_2COO^-Na^+$ + HCl ⟶ $CH_3CH=CHCH_2COOH$ + NaCl

13.24 The normal pH range for blood plasma is 7.35–7.45. Under these conditions, would you expect the carboxyl group of lactic acid (pK_a 3.85) to exist primarily as a carboxyl group or as a carboxylate anion? Explain.

When the pH is higher than the pK_a of an acid, the acid will be deprotonated (see Problem 10.26 for more details). Lactic acid exists primarily a carboxylate anion.

$$pH = -\log[H^+]$$

$$[H^+] = 10^{-pH} = 10^{-7.4} = 4.0 \times 10^{-8}M$$

$$K_a = 10^{-pK_a} = 10^{-7.4} = \frac{[\text{Lactate}][H^+]}{[\text{Lactic acid}]} = 8.4 \times 10^{-4}$$

$$8.4 \times 10^{-4} = \frac{[\text{Lactate}]\left[4.0 \times 10^{-8}\right]}{[\text{Lactic acid}]}$$

$$\frac{[\text{Lactate}]}{[\text{Lactic acid}]} = 21000$$

13.25 The pK_a of salicylic acid (Section 13.2) is 2.97. Would you expect salicylic acid dissolved in blood plasma (pH 7.35–7.45) to exist primarily as salicylic acid or as salicylate anion? Explain.

Similar to Problem 13.24, when the pH is greater than the pK_a of the acid, the acid will be deprotonated. Salicylic acid in blood plasma exists primarily as the salicylate anion.

13.26 Vanillylmandelic acid (pK_a 3.42) is a metabolite found in urine, the pH of which is normally in the range from 4.8 to 8.4. Provide the structure of vanillylmandelic acid that you would expect to find in urine with pH 5.8?

Vanillylmandelic acid will be deprotonated because the pH of urine is higher than the pK_a of the acid. Note that it is the carboxyl group, which is the most acidic functional group in the molecule, that is deprotonated.

13.27 The pH of human gastric juice is normally in the range from 1.0 to 3.0. What form of lactic acid (pK_a 4.07), lactic acid itself or its anion, would you expect to be present in the stomach?

When the pH is less than the pK_a of an acid, the acid will not be deprotonated. Lactic acid will therefore exist as lactic acid itself.

13.28 Following are two structural formulas for the amino acid alanine (Section 18.2). Is alanine better represented by structural formula A or B? Explain.

$$CH_3-\overset{\displaystyle NH_2}{\underset{\displaystyle |}{CH}}-\overset{\displaystyle O}{\overset{\displaystyle ||}{C}}-OH \qquad\qquad CH_3-\overset{\displaystyle \overset{+}{N}H_3}{\underset{\displaystyle |}{CH}}-\overset{\displaystyle O}{\overset{\displaystyle ||}{C}}-O^-$$

(A)   (B)

Alanine is best represented by structural formula (B). See Problem 2.39 for more details.

13.29 In Chapter 18, we discuss a class of compounds called amino acids, so named because they contain both an amino group and a carboxyl group. Following is a structural formula for the amino acid alanine in the form of an internal salt:

$$CH_3\overset{\displaystyle \overset{+}{N}H_3}{\underset{\displaystyle |}{CH}}\overset{\displaystyle O}{\overset{\displaystyle ||}{C}}O^-$$

What would you expect to be the major form of alanine present in aqueous solution at:

(a) pH 2.0

The ammonium group remains protonated because the pH of the solution is less than the pK_a of the group (approximately 10). The pH is also less than the pK_a of carboxyl group (approximately 2), so the carboxylate group will be protonated to form a carboxyl group.

$$CH_3\overset{\displaystyle \overset{+}{N}H_3}{\underset{\displaystyle |}{CH}}\overset{\displaystyle O}{\overset{\displaystyle ||}{C}}OH$$

(b) pH 5–6

At a pH that is above the pK_a of the carboxyl group, the carboxyl group will be ionized and found as the carboxylate ion. The ammonium group is still protonated because the pK_a of the ammonium group is higher than the pH of the solution.

$$CH_3\overset{\displaystyle \overset{+}{N}H_3}{\underset{\displaystyle |}{CH}}\overset{\displaystyle O}{\overset{\displaystyle ||}{C}}O^-$$

(c) pH 11.0

The ammonium ion will be deprotonated because the pH of the solution is now higher than the pK_a of the ammonium group.

$$CH_3\overset{\displaystyle NH_2}{\underset{\displaystyle |}{CH}}\overset{\displaystyle O}{\overset{\displaystyle ||}{C}}O^-$$

Reactions of Carboxylic Acids

13.30 Give the expected organic products formed when phenylacetic acid, $PhCH_2COOH$, is treated with each of the following reagents:

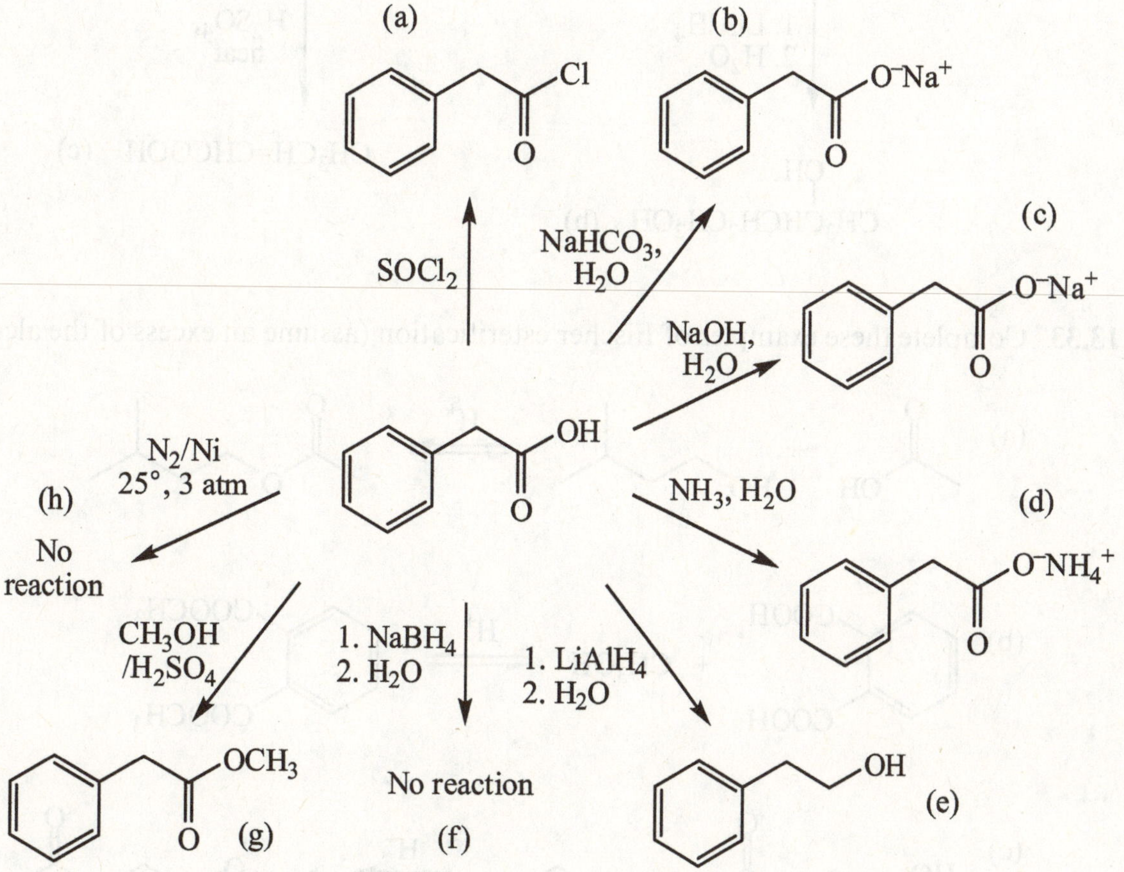
(a)
Cl
O
(b)
O⁻Na⁺
O
SOCl₂
NaHCO₃,
H₂O
(c)
O⁻Na⁺
NaOH,
H₂O
O
OH
O
N₂/Ni
25°, 3 atm
(h)
No
reaction
NH₃, H₂O
(d)
O⁻NH₄⁺
O
CH₃OH
/H₂SO₄
1. NaBH₄
2. H₂O
1. LiAlH₄
2. H₂O
OCH₃
O
(g)
No reaction
(f)
OH
(e)

13.31 Show how to convert *trans*-3-phenyl-2-propenoic acid (cinnamic acid) to these compounds:

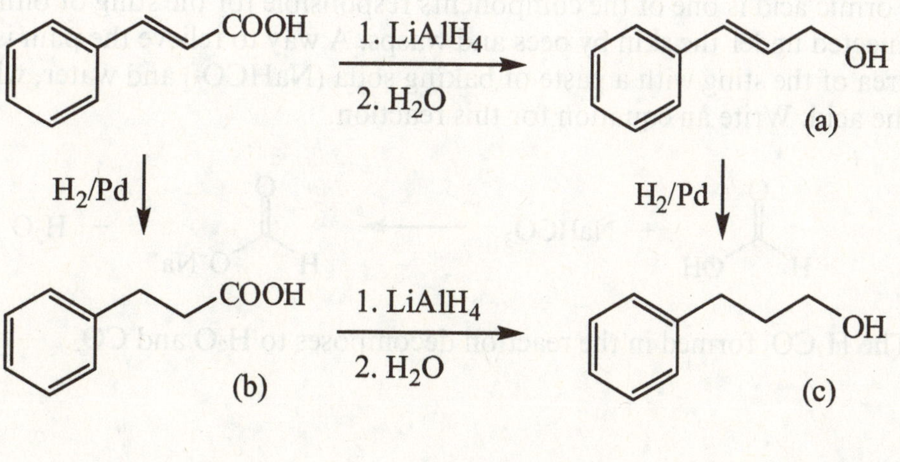
COOH
1. LiAlH₄
2. H₂O
OH
(a)
H₂/Pd
H₂/Pd
COOH
(b)
1. LiAlH₄
2. H₂O
OH
(c)

13.32 Show how to convert 3-oxobutanoic acid (acetoacetic acid) to these compounds:

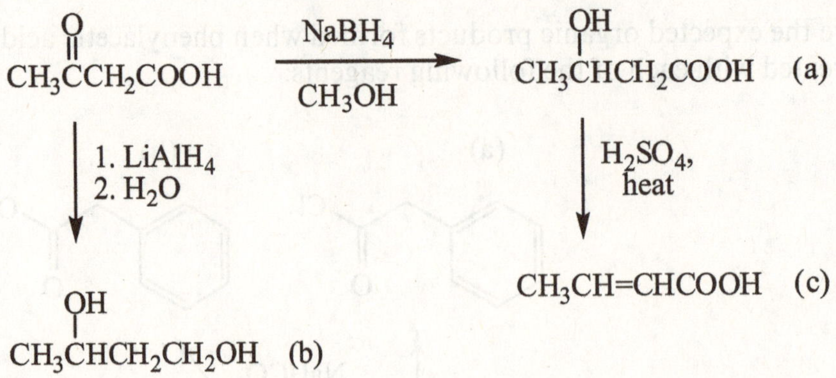

$$\underset{\substack{\text{O}\\\|\\}}{\text{CH}_3\text{CCH}_2\text{COOH}} \xrightarrow[\text{CH}_3\text{OH}]{\text{NaBH}_4} \underset{\substack{\text{OH}\\|\\}}{\text{CH}_3\text{CHCH}_2\text{COOH}} \quad \text{(a)}$$

$$\downarrow \begin{array}{l}\text{1. LiAlH}_4\\\text{2. H}_2\text{O}\end{array} \qquad\qquad\qquad \downarrow \begin{array}{l}\text{H}_2\text{SO}_4,\\\text{heat}\end{array}$$

$$\text{CH}_3\text{CH=CHCOOH} \quad \text{(c)}$$

$$\underset{\substack{\text{OH}\\|\\}}{\text{CH}_3\text{CHCH}_2\text{CH}_2\text{OH}} \quad \text{(b)}$$

13.33 Complete these examples of Fischer esterification (assume an excess of the alcohol):

(a)

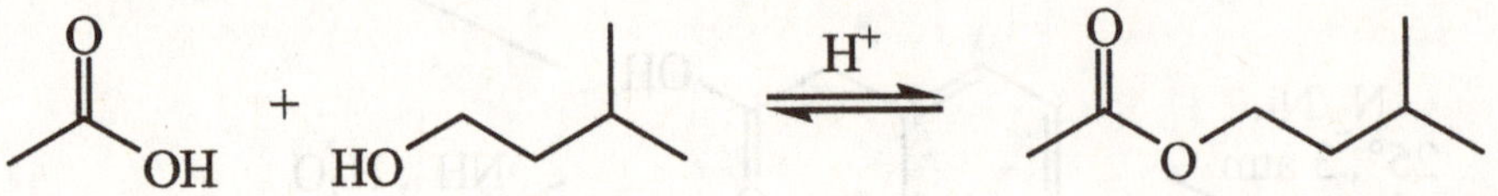

(b)

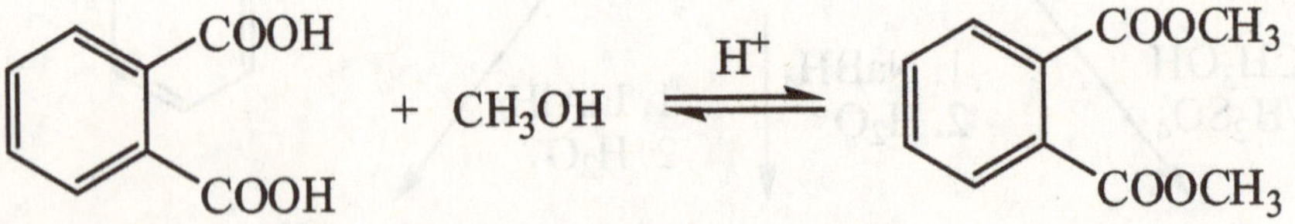

(c)

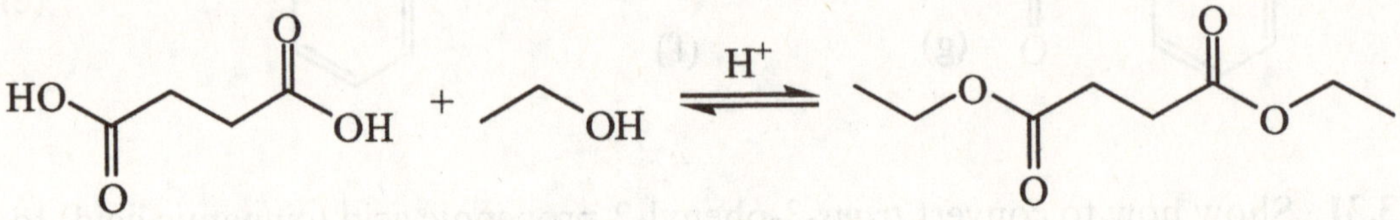

13.34 Formic acid is one of the components responsible for the sting of biting ants and is injected under the skin by bees and wasps. A way to relieve the pain is to rub the area of the sting with a paste of baking soda (NaHCO₃) and water, which neutralizes the acid. Write an equation for this reaction.

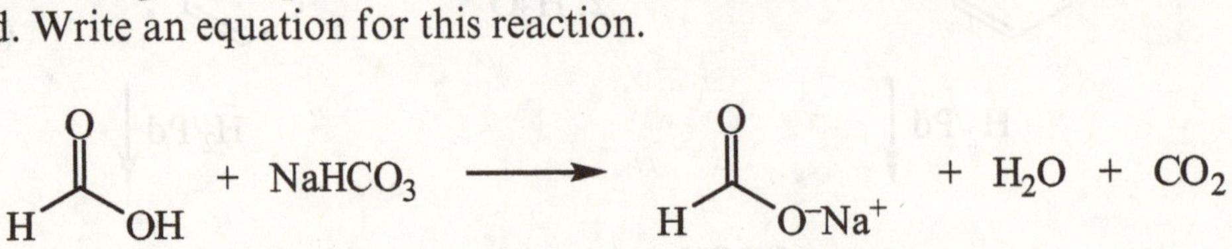

The H_2CO_3 formed in the reaction decomposes to H_2O and CO_2.

13.35 Methyl 2-hydroxybenzoate (methyl salicylate) has the odor of oil of wintergreen. This ester is prepared by the Fischer esterification of 2-hydroxybenzoic acid (salicylic acid) with methanol. Draw a structural formula of methyl 2-hydroxybenzoate.

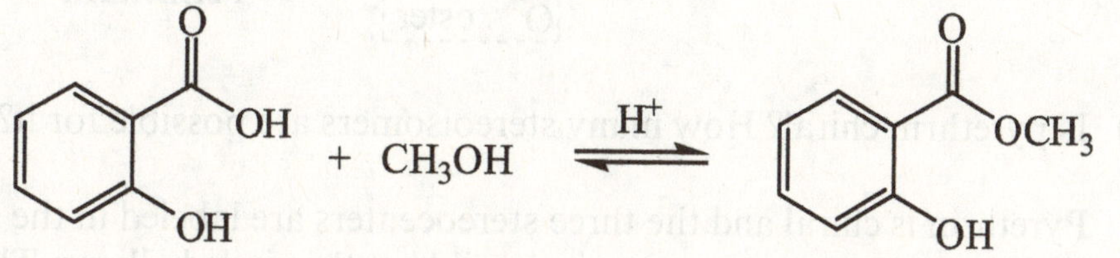

13.36 Benzocaine, a topical anesthetic, is prepared by treating 4-aminobenzoic acid with ethanol in the presence of an acid catalyst, followed by neutralization. Draw a structural formula of benzocaine.

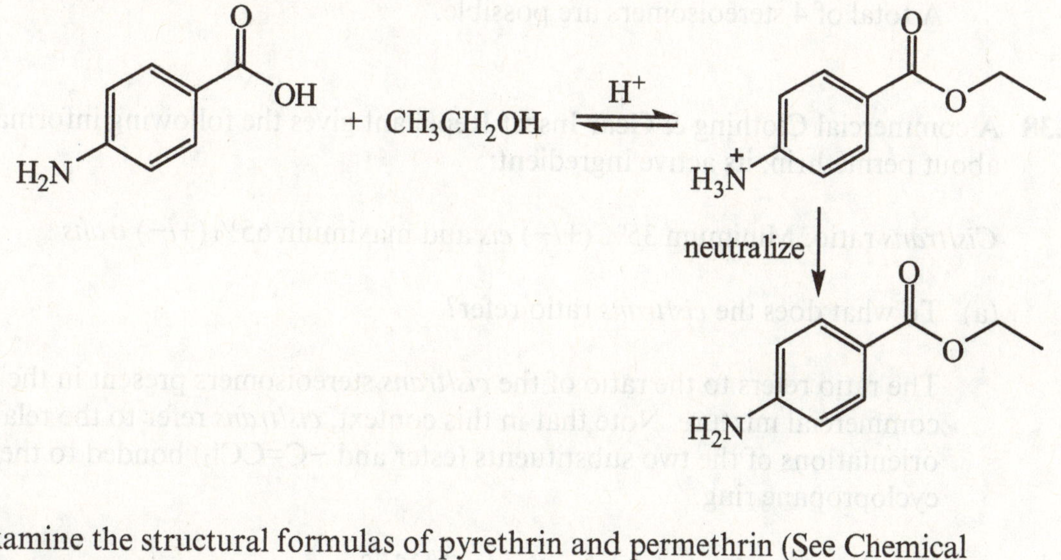

13.37 Examine the structural formulas of pyrethrin and permethrin (See Chemical Connections 14D.)

(a) Locate the ester groups in each compound.

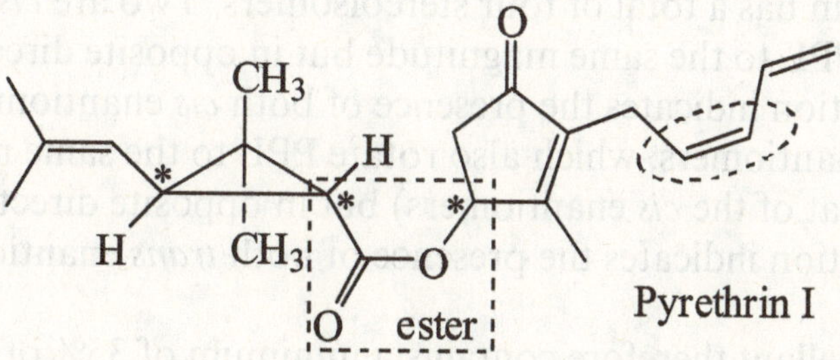

Pyrethrin I

Permethrin

(b) Is pyrethrin chiral? How many stereoisomers are possible for it?

Pyrethrin is chiral and the three stereocenters are labeled in the structure. In
addition, *cis-trans* isomerism is possible at the circled alkene. Thus, a total of
$2^4 = 16$ stereoisomers are possible.

(c) Is permethrin chiral? How many stereoisomers are possible for it?

Permethrin is chiral and the two stereocenters are labeled in the structure.
A total of 4 stereoisomers are possible.

13.38 A commercial Clothing & Gear Insect Repellant gives the following information
about permethrin, its active ingredient:

Cis/trans ratio: Minimum 35% (+/−) *cis* and maximum 65% (+/−) *trans*

(a) To what does the *cis/trans* ratio refer?

The ratio refers to the ratio of the *cis/trans* stereoisomers present in the
commercial mixture. Note that in this context, *cis/trans* refer to the relative
orientations of the two substituents (ester and −C=CCl₂) bonded to the
cyclopropane ring.

(b) To what does the designation "(+/−)" refer?

The + and − refer to the direction of rotation of plane-polarized light (PPL).

Pyrethrin has a total of four stereoisomers. Two are *cis* enantiomers, which
rotate PPL to the same magnitude but in opposite directions, and the (+/−)
designation indicates the presence of both *cis* enantiomers. The other two are
trans enantiomers, which also rotate PPL to the same magnitude (yet different
from that of the *cis* enantiomers) but in opposite directions, and the (+/−)
designation indicates the presence of both *trans* enantiomers.

The repellant therefore contains a minimum of 35% of a racemic mixture of *cis*
enantiomers and a maximum of 65% of a racemic mixture of *trans* enantiomers.

13.39 From what carboxylic acid and alcohol is each of the following esters derived?

(a)

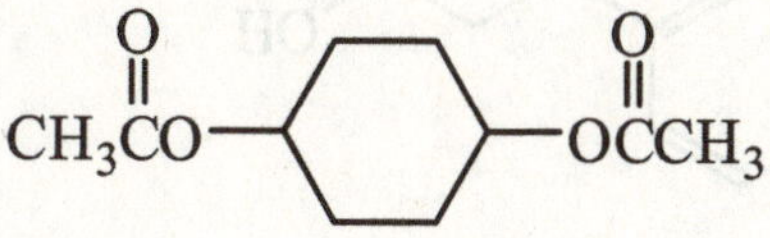

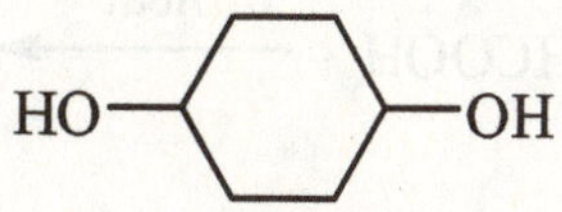

$$CH_3CO-\!\!\!\bigcirc\!\!\!-OCCH_3 \qquad HO-\!\!\!\bigcirc\!\!\!-OH \qquad CH_3COH$$

alcohol   carboxylic acid

(b)

$$CH_3OCCH_2CH_2COCH_3 \qquad CH_3OH \qquad HOCCH_2CH_2COH$$

alcohol   carboxylic acid

(c)

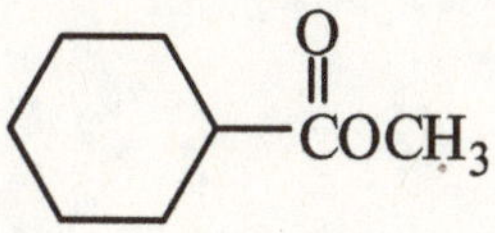

$$\bigcirc\!\!\!-COCH_3 \qquad CH_3OH \qquad \bigcirc\!\!\!-COH$$

alcohol   carboxylic acid

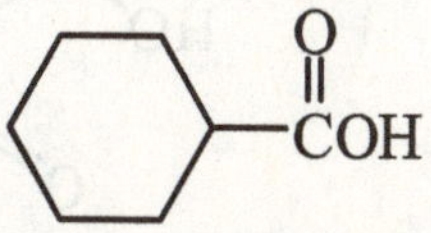

(d)

$$CH_3CH_2CH=CHCOCH(CH_3)_2 \qquad (CH_3)_2CHOH \qquad CH_3CH_2CH=CHCOH$$

alcohol   carboxylic acid

13.40 When treated with an acid catalyst, 4-hydroxybutanoic acid forms a cyclic ester (a lactone). Draw the structural formula of this lactone.

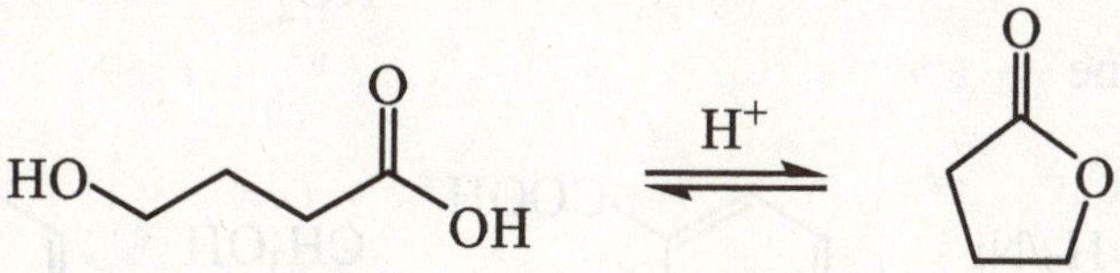

$$HO\diagdown\diagup\diagdown COOH \quad \xrightleftharpoons{H^+} \quad \bigcirc$$

13.41 Draw a structural for the product formed on thermal decarboxylation of each of the following compounds:

(a)

$$C_6H_5CCH_2COOH \quad \xrightarrow{\text{heat}}$$

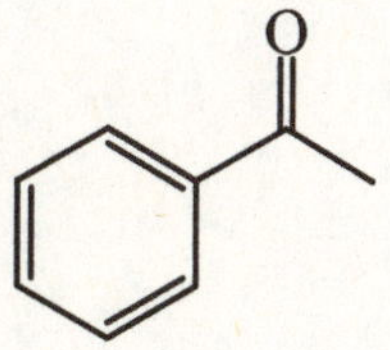

(b)

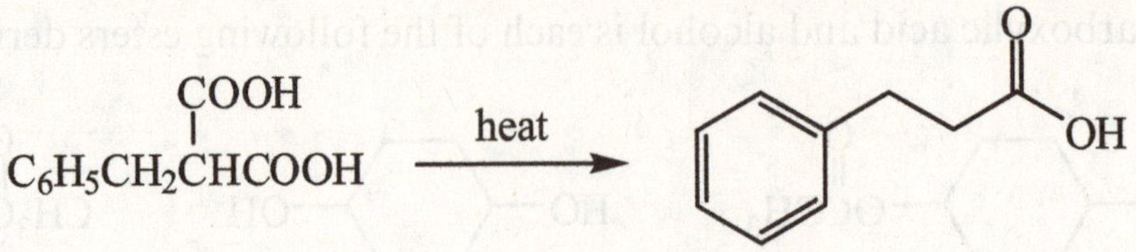

$$C_6H_5CH_2CHCOOH \xrightarrow{\text{heat}}$$

with COOH substituent

(c)

cyclopentane with CCH₃ and COOH groups $\xrightarrow{\text{heat}}$ cyclopentane with CCH₃

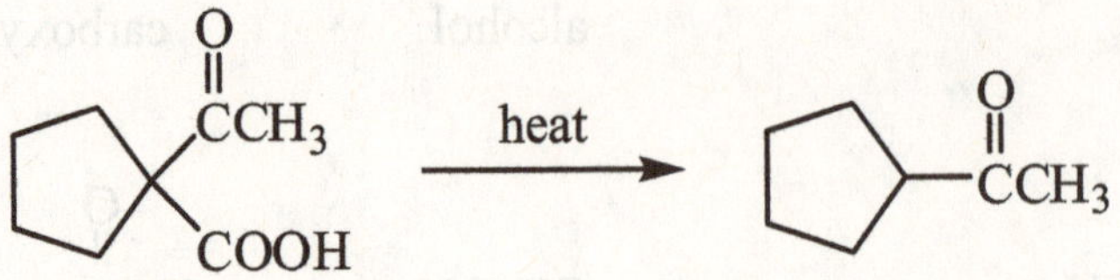

(d)

malonic-type acid with ethyl group $\xrightarrow{\text{heat}}$ ketone

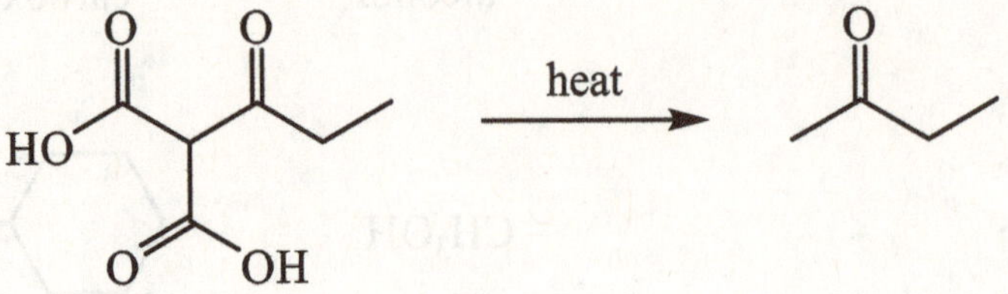

Synthesis

13.42 Methyl 2-aminobenzoate, a flavoring agent with the taste of grapes (see Chemical Connections 13B), can be prepared from toluene by the following series of steps. Show how you might bring about each step in this synthesis.

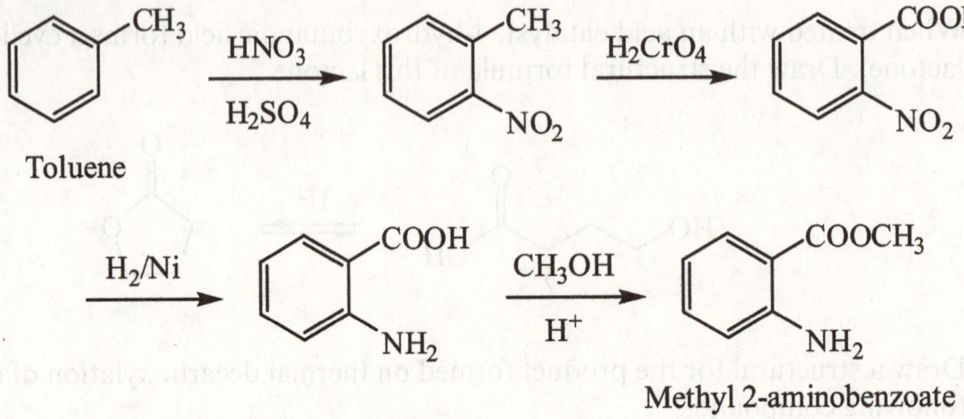

Toluene $\xrightarrow[\text{H}_2\text{SO}_4]{\text{HNO}_3}$ o-nitrotoluene $\xrightarrow{\text{H}_2\text{CrO}_4}$ 2-nitrobenzoic acid

$\xrightarrow{\text{H}_2/\text{Ni}}$ 2-aminobenzoic acid $\xrightarrow[\text{H}^+]{\text{CH}_3\text{OH}}$ Methyl 2-aminobenzoate

13.43 Methylparaben and propylparaben are used as preservatives in foods, beverages, and cosmetics. Show how the synthetic scheme in Problem 13.42 can be modified given each of these compounds.

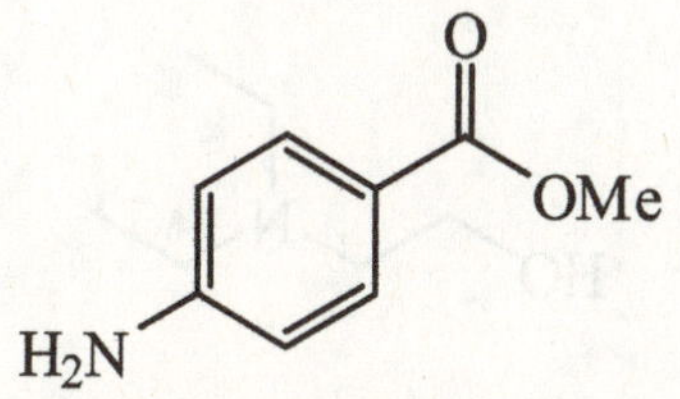

Methyl 4-aminobenzoate
(Methylparaben)

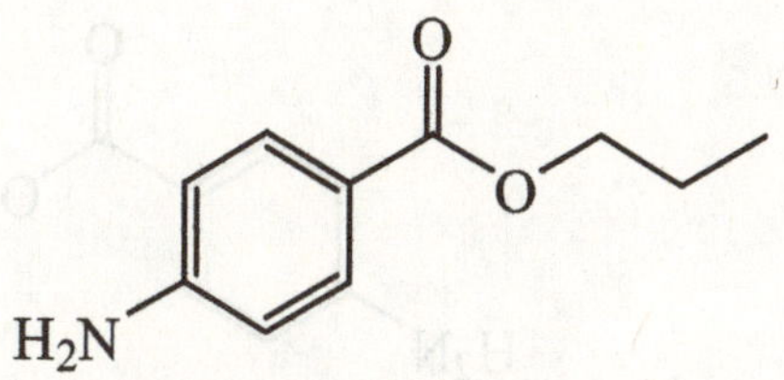

Propyl 4-aminobenzoate
(Propylparaben)

The first three steps are the same, except *p*-nitrotoluene from the nitration reaction (the first step) is used instead of *o*-nitrotoluene. Acid-catalyzed Fischer esterification of *p*-aminobenzoic acid with methanol or propanol yields the desired product.

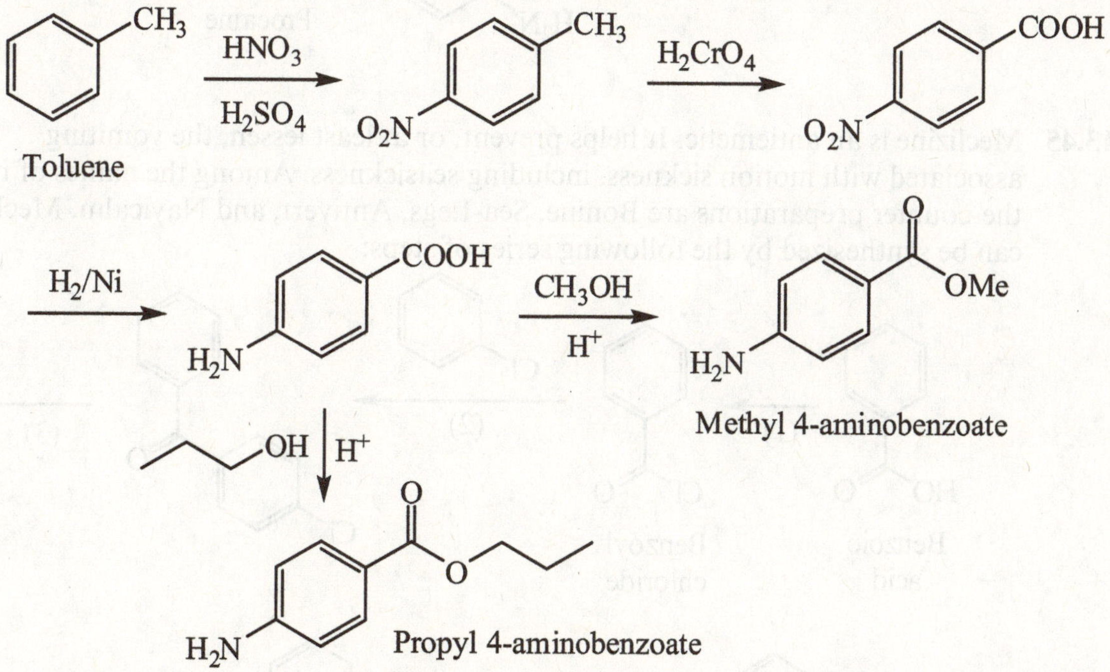

Toluene

HNO_3

H_2SO_4

H_2CrO_4

H_2/Ni

CH_3OH

H^+

Methyl 4-aminobenzoate

OH | H^+

Propyl 4-aminobenzoate

13.44 Procaine (its hydrochloride is marketed as Novocaine) was one of the first local anesthetics developed for infiltration and regional anesthesia. It is synthesized by the following Fischer esterification. Draw a structural formula for procaine.

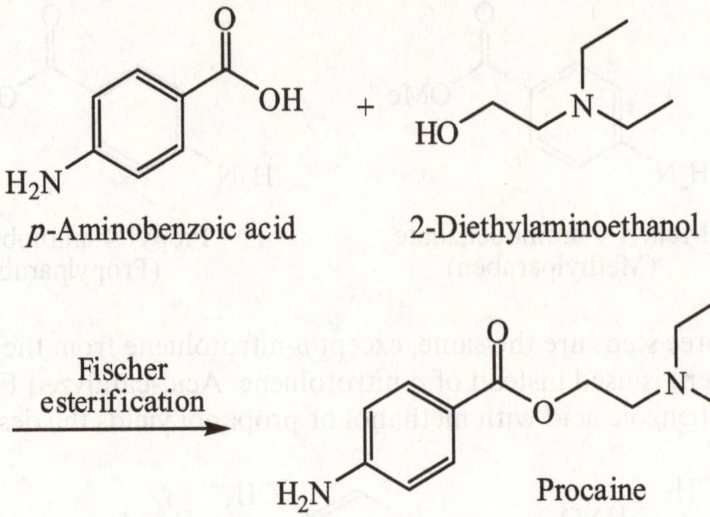
OH
HO
N
H_2N
p-Aminobenzoic acid
2-Diethylaminoethanol
Fischer esterification
Procaine

13.45 Meclizine is an antiemetic: It helps prevent, or at least lessen, the vomiting associated with motion sickness, including seasickness. Among the names of its over-the-counter preparations are Bonine, Sea-Legs, Antivert, and Navicalm. Meclizine can be synthesized by the following series of steps:

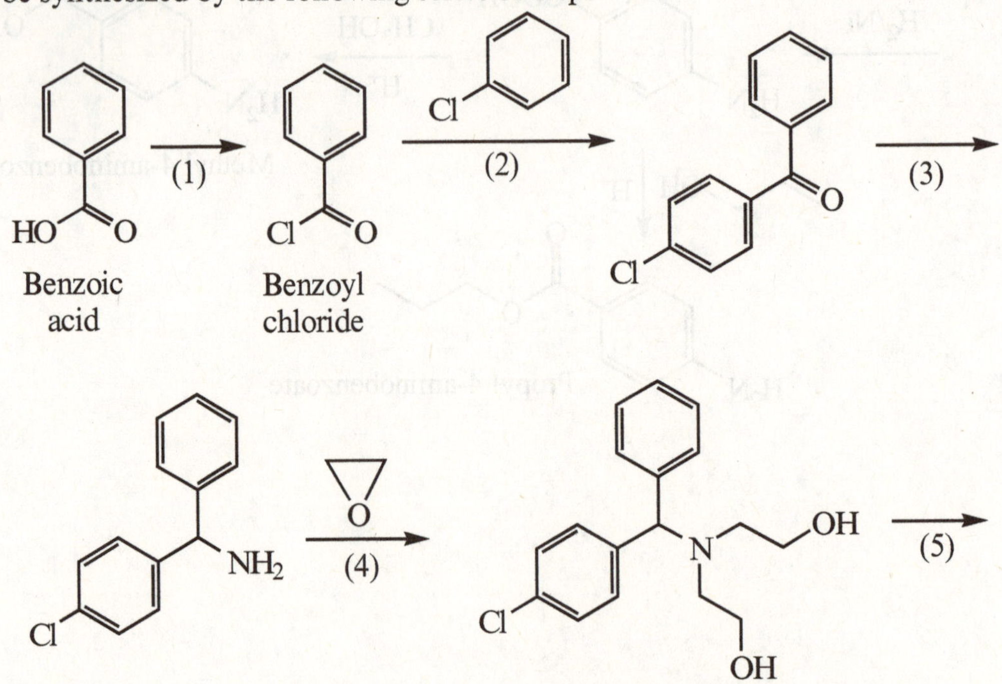
(1)
(2)
(3)
HO
Cl
Benzoic acid
Benzoyl chloride
NH_2
(4)
N
OH
(5)

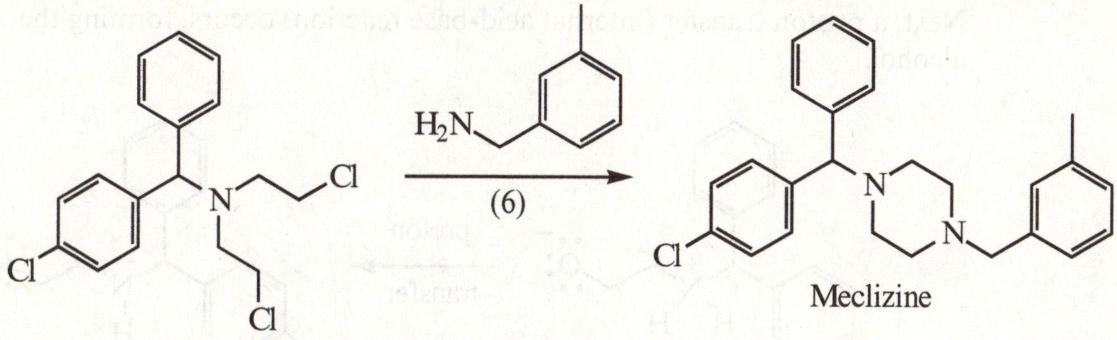

(a) Propose a reagent for Step 1.

SOCl$_2$ in the presence of a base, such as pyridine.

(b) The catalyst for Step 2 is AlCl$_3$. Name the type of reaction that occurs in Step 2.

Friedel-Crafts acylation

(c) Propose reagents for Step 3.

NH$_3$ followed by catalytic hydrogenation, such as H$_2$/Pt.

(d) Propose a mechanism for Step 4, and show that it is an example of nucleophilic aliphatic substitution.

In this reaction, the nitrogen atom of the amino group substitutes the oxygen atom that is bonded to the carbon atoms of the epoxide. The driving force behind this reaction is the opening of the highly strained epoxide ring. First, the amine attacks the epoxide:

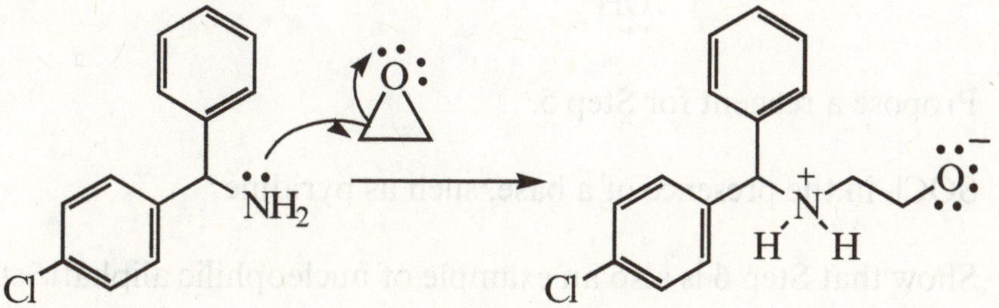

Next, a proton transfer (internal acid-base reaction) occurs, forming the alcohol.

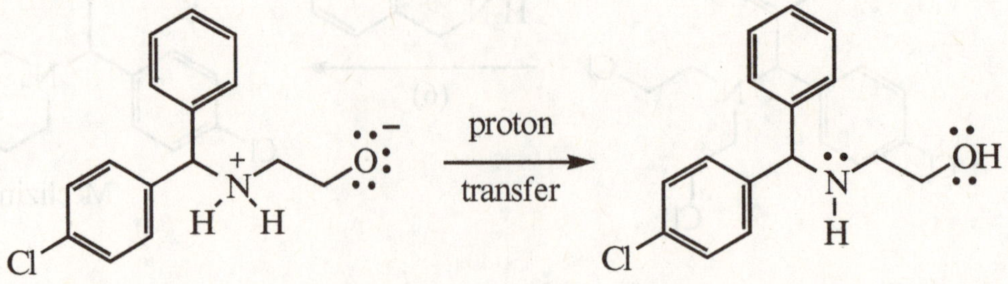

The mechanism repeats, using another equivalent of the epoxide.

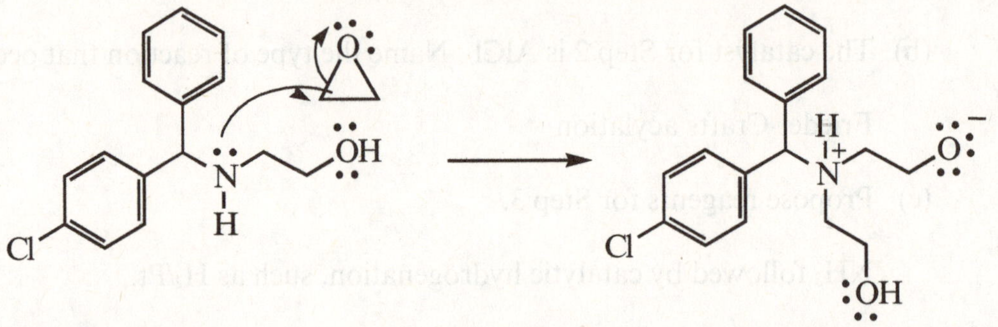

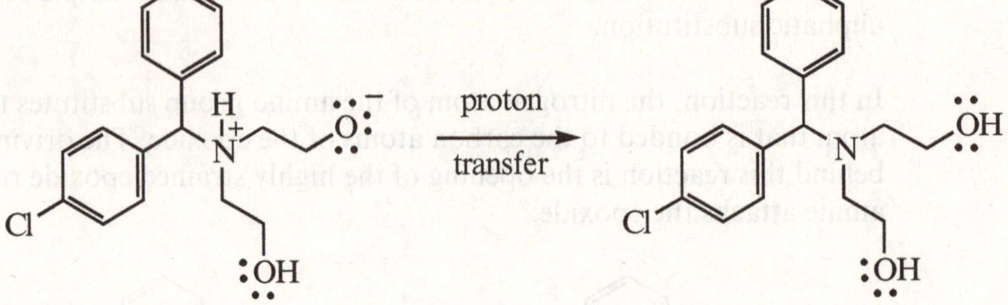

(e) Propose a reagent for Step 5.

$SOCl_2$ in the presence of a base, such as pyridine.

(f) Show that Step 6 is also an example of nucleophilic aliphatic substitution.

In this reaction, the nitrogen atom of the amino group substitutes the chlorine atoms that are bonded to their respective carbon atoms.

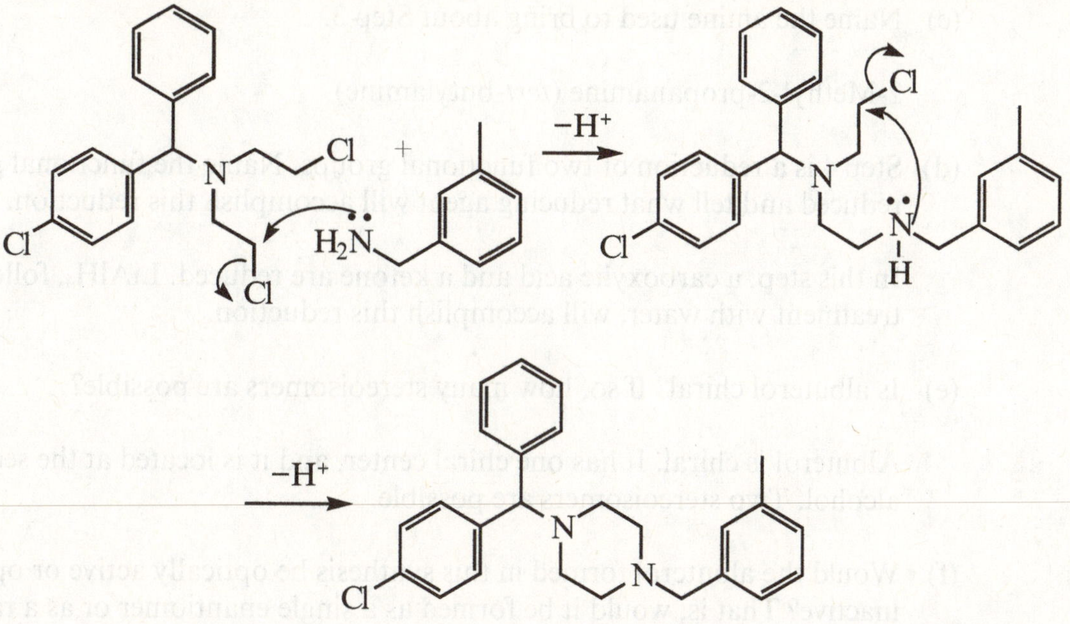

13.46 Chemists have developed several syntheses for the anti-asthmatic drug albuterol (Proventil). One of these syntheses starts with salicylic acid, the same acid that is the starting material for the synthesis of aspirin.

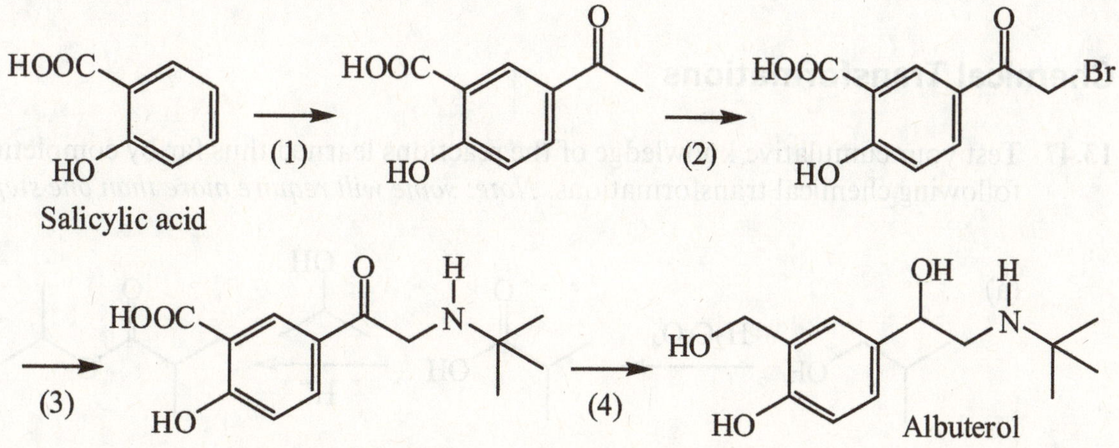

Salicylic acid

Albuterol

(a) Propose a reagent and a catalyst for Step 1. What name is given to this type of reaction?

The reaction is a Friedel-Crafts acylation and uses CH_3COCl and $AlCl_3$.

(b) Propose a reagent for Step 2.

In this reaction, bromination of an α-carbon occurs. Br_2 in acetic acid can be used.

(c) Name the amine used to bring about Step 3.

2-Methyl-2-propanamine (*tert*-butylamine)

(d) Step 4 is a reduction of two functional groups. Name the functional groups reduced and tell what reducing agent will accomplish this reduction.

In this step, a carboxylic acid and a ketone are reduced. $LiAlH_4$, followed by treatment with water, will accomplish this reduction.

(e) Is albuterol chiral? If so, how many stereoisomers are possible?

Albuterol is chiral. It has one chiral center, and it is located at the secondary alcohol. Two stereoisomers are possible.

(f) Would the albuterol formed in this synthesis be optically active or optically inactive? That is, would it be formed as a single enantiomer or as a racemic mixture?

The synthesis forms a racemic mixture, so the albuterol would be optically inactive. A racemic mixture is formed because when the ketone is reduced to the alcohol, the hydride can add to either face of the planar, sp^2-hybridized carbonyl group.

Chemical Transformations

13.47 Test your cumulative knowledge of the reactions learned thus far by completing the following chemical transformations. *Note: some will require more than one step.*

(a)

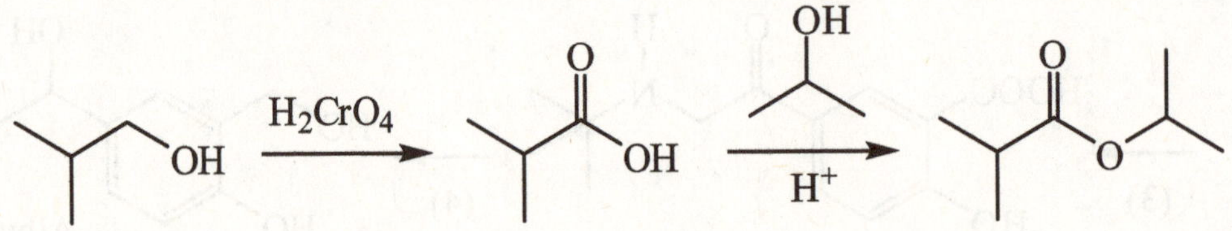

(b)

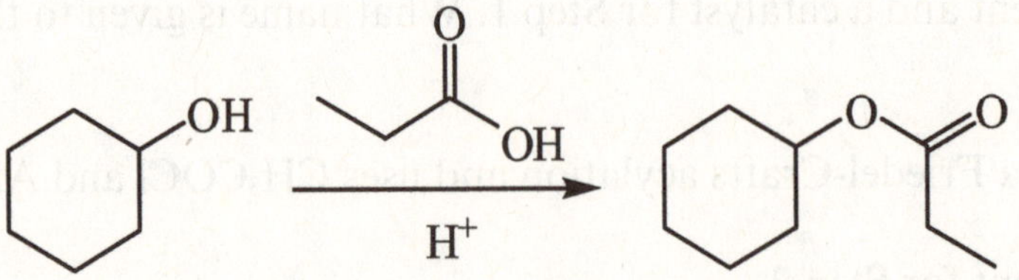

(c)

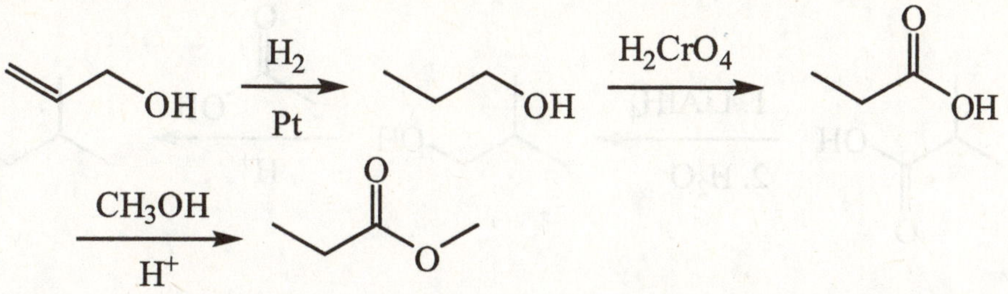

(d)

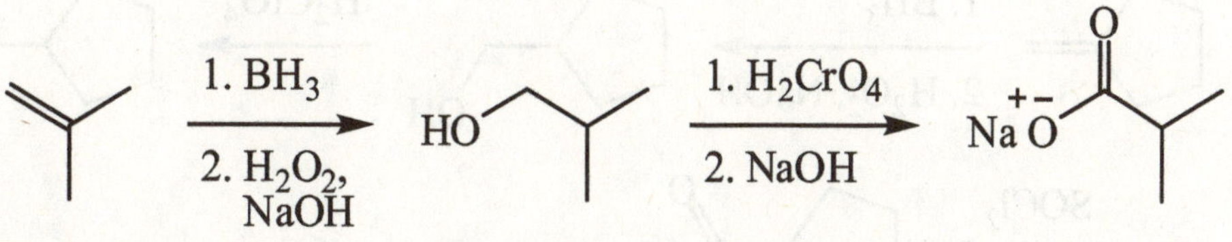

(e)

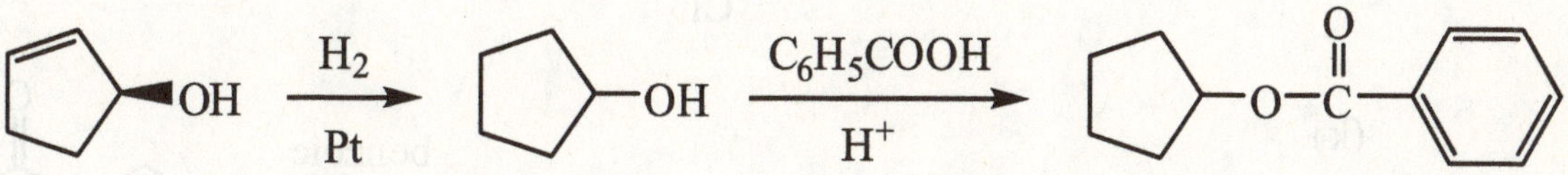

(f)

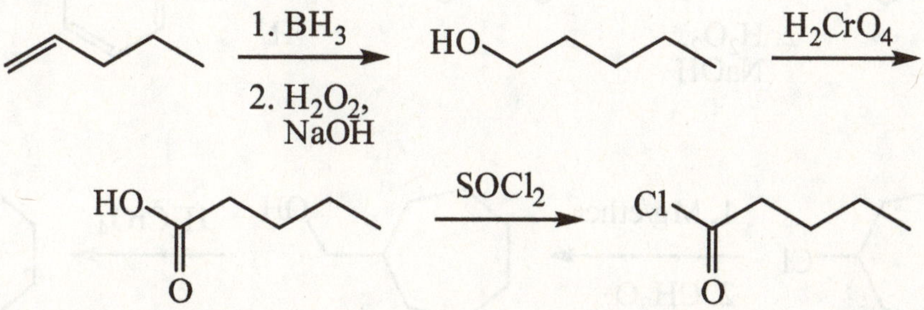

(g)

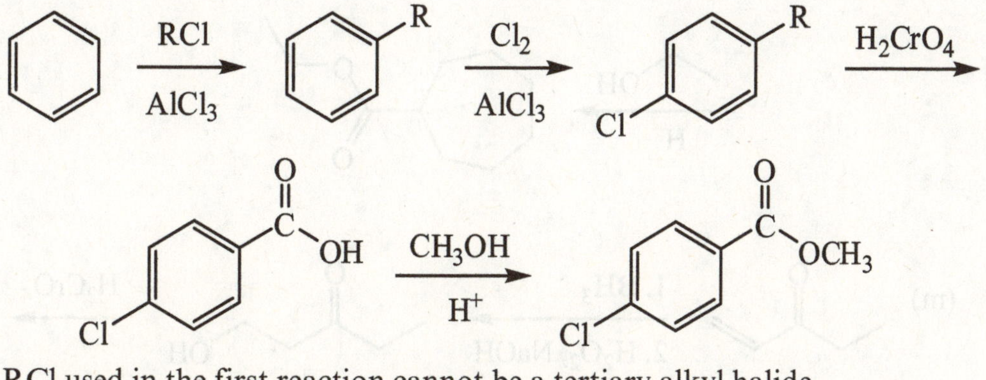

RCl used in the first reaction cannot be a tertiary alkyl halide.

(h)

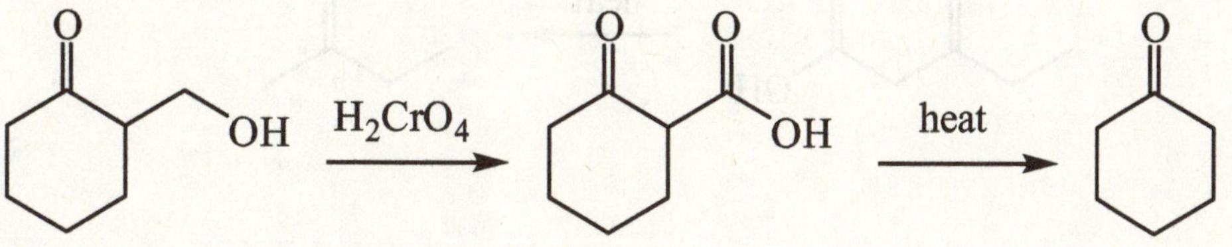

(i)

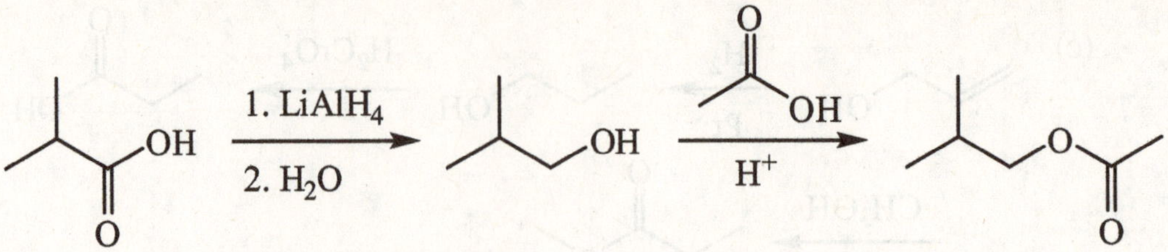

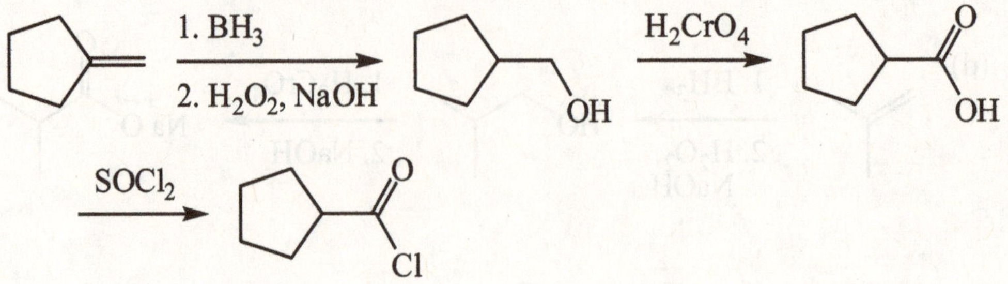

(j)

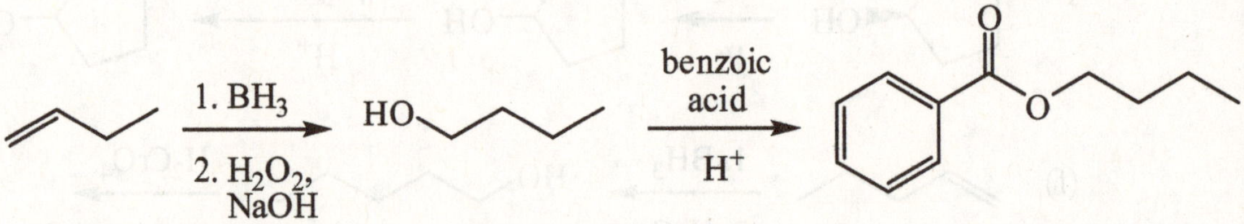

(k)

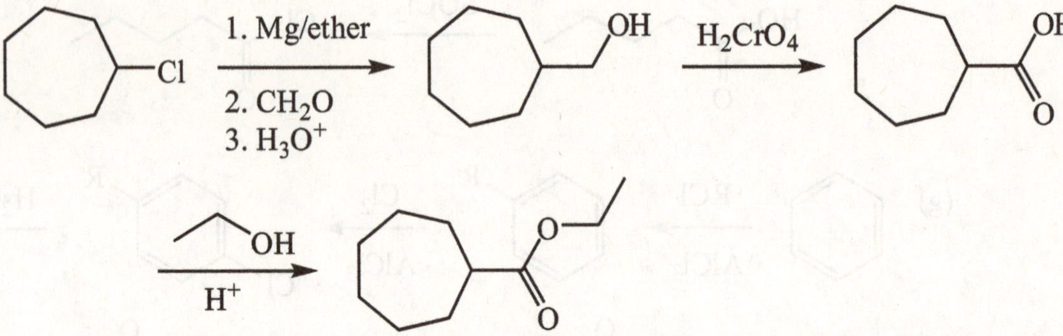

(l)

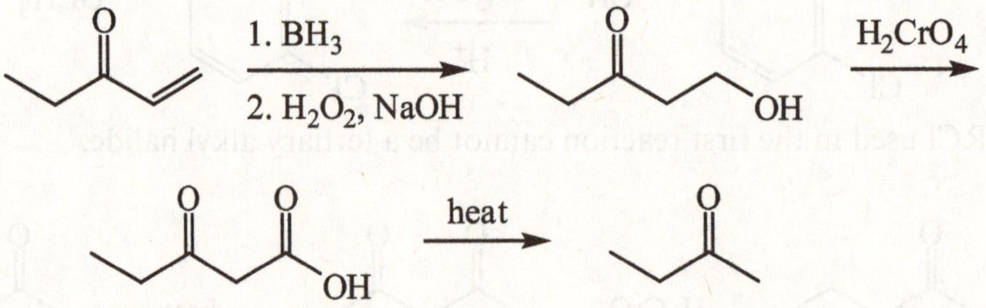

(m)

(n)

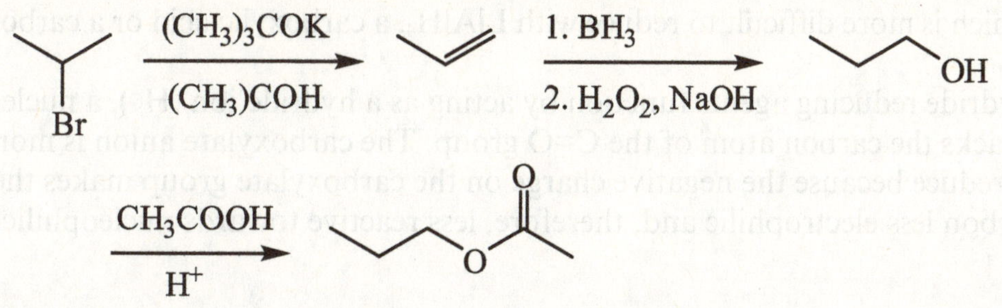

(o)

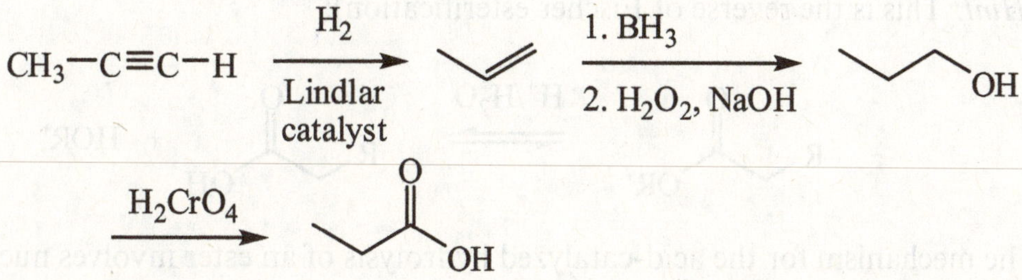

(p)

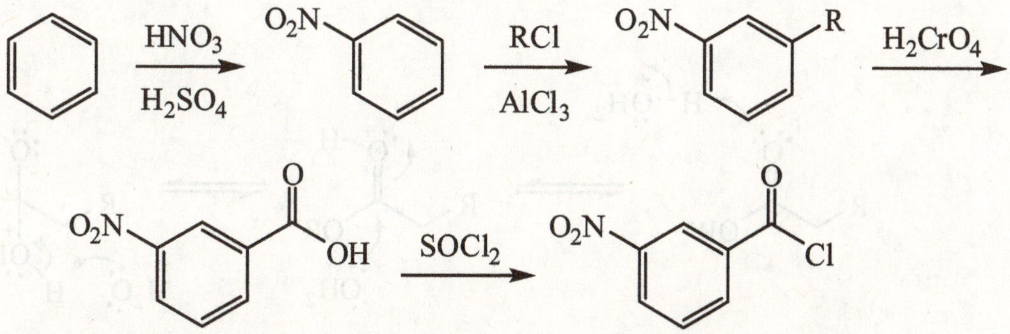

RCl used in the second reaction cannot be a tertiary alkyl halide.

Looking Ahead

13.48 Explain why α-amino acids, the building blocks of proteins (Chapter 18), are nearly a thousand times more acidic than aliphatic carboxylic acids:

The positively charged ammonium group is a strong inductively withdrawing group. It increases the strength of the carboxylic acid by stabilizing its conjugate base, the carboxylate ion.

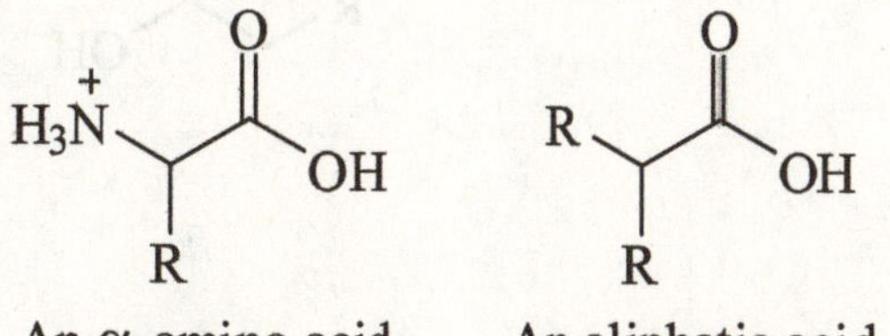

An α-amino acid
$pK_a \approx 2$

An aliphatic acid
$pK_a \approx 5$

13.49 Which is more difficult to reduce with LiAlH₄, a carboxylic acid or a carboxylate ion?

Hydride reducing agents function by acting as a hydride ion (H⁻), a nucleophile that attacks the carbon atom of the C=O group. The carboxylate anion is more difficult to reduce because the negative charge on the carboxylate group makes the carbonyl carbon less electrophilic and, therefore, less reactive towards nucleophilic attack.

13.50 Show how an ester can react with H⁺/H₂O to give a carboxylic acid and an alcohol (*Hint:* This is the reverse of Fischer esterification):

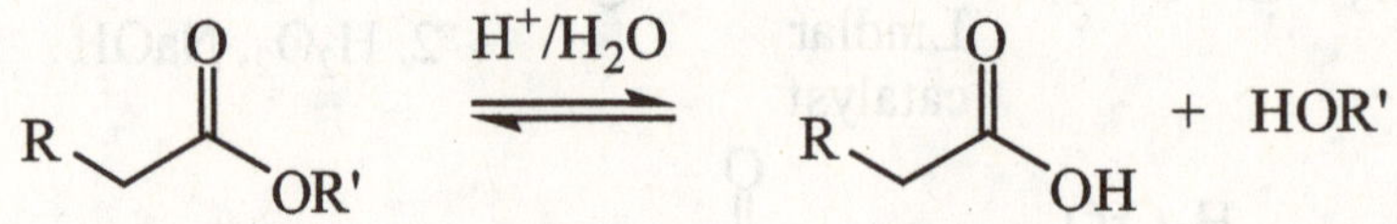

The mechanism for the acid-catalyzed hydrolysis of an ester involves nucleophilic attack by water and the departure of the alcohol as the leaving group. A tetrahedral carbonyl addition intermediate is involved in the reaction.

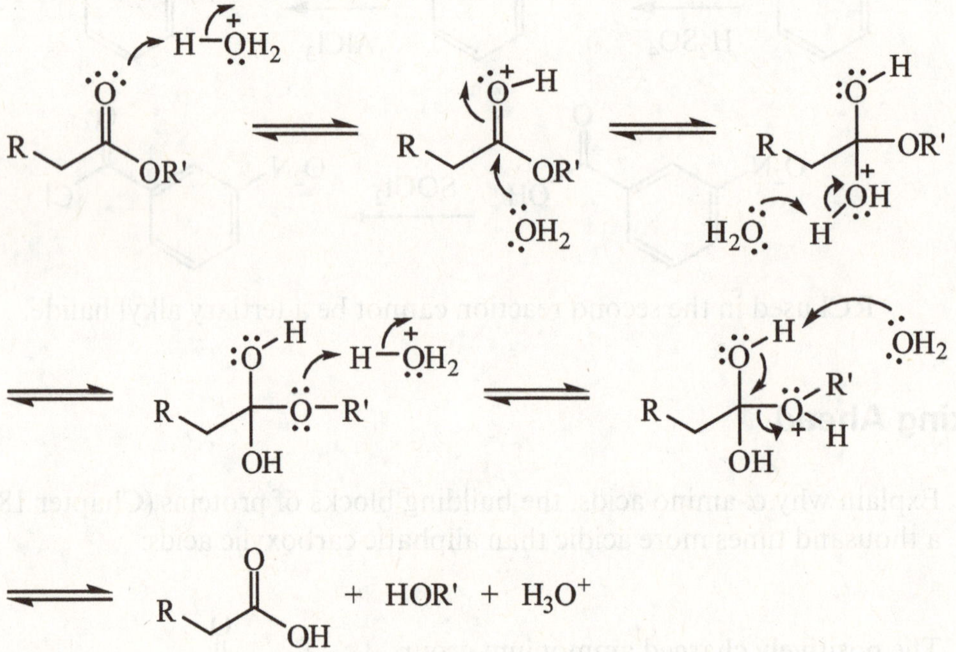

13.51 In Chapter 12, we saw how Grignard reagents readily attack the carbonyl carbon of ketones and aldehydes. Should the same process occur with Grignards and carboxylic acids? With esters?

The carbonyl carbon of both carboxylic acids and esters is electrophilic. However, Grignard reagents are base-sensitive and are destroyed by the acidic proton found in carboxylic acids. On the other hand, esters will react with Grignard reagents.

13.52 In Section 13.6, it was suggested that the mechanism for the Fischer esterification of carboxylic acids would be a model for many of the reactions of the functional derivatives of carboxylic acids. One such reaction, the reaction of an acid halide with water, is the following:

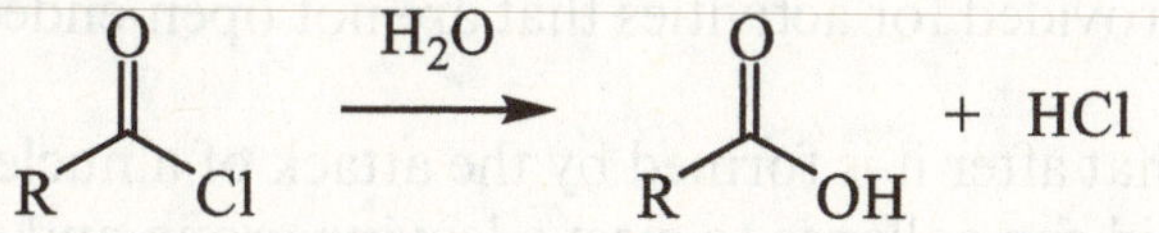

Suggest a mechanism for this reaction.

Although there is no acid catalyst present in the reaction mixture, it is not needed because the carbon atom of the carbonyl group has a relatively high $\delta+$ charge due to the inductively withdrawing chlorine atom. (In a Fischer esterification, an acid catalyst was required to protonate the carbonyl group, thereby increasing its electrophilicity.)

Step 1: Nucleophilic attack by water forms a tetrahedral carbonyl addition intermediate.

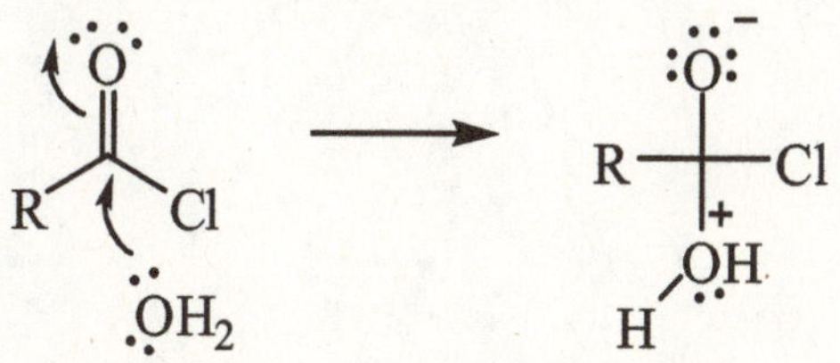

Step 2: The tetrahedral intermediate collapses, displacing chloride. Note that unlike a Fischer esterification, protonation of the leaving group is not necessary because chloride is already a good leaving group..

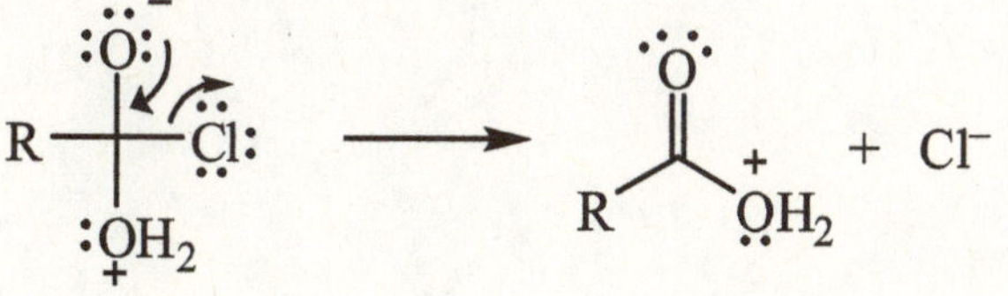

Step 3: Deprotonation forms the carboxylic acid. Together, H_3O^+ and Cl^- represent HCl ionized in water.

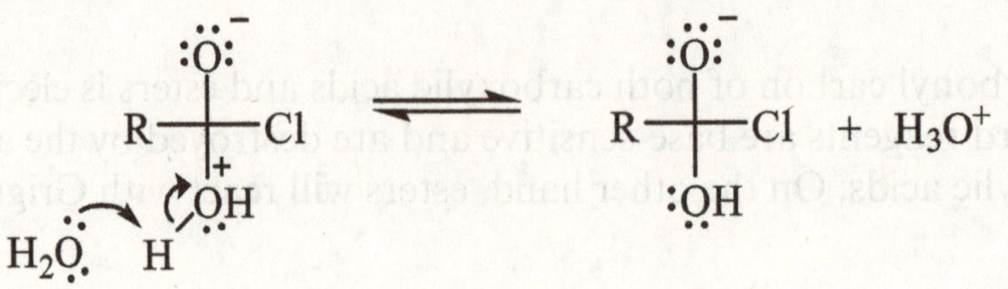

Group Learning Activities

Solutions are only provided for activities that are not open-ended.

13.54 We learned that after it is formed by the attack of a nucleophile, the TCAI of a carboxylic acid can collapse to eject a leaving group and regenerate a carbonyl group. Discuss why the TCAI's of ketones and aldehydes don't undergo this same collapse.

Ketones and aldehydes do not undergo the same process because the leaving groups that would be ejected are respectively carbanions and hydrides. These are very poor leaving groups because they are unstable. It is not possible to protonate these groups, thus turning them into better leaving groups, prior to ejecting them because they do not have any lone pairs.

Chapter 14: Functional Derivatives of Carboxylic Acids

Problems

14.1 Draw a structural formula for each compound:

(a) *N*-Cyclohexylacetamide   (b) *sec*-Butyl acetate 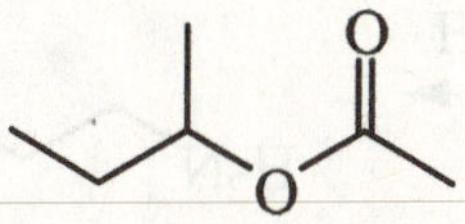   (c) Cyclobutyl butanoate

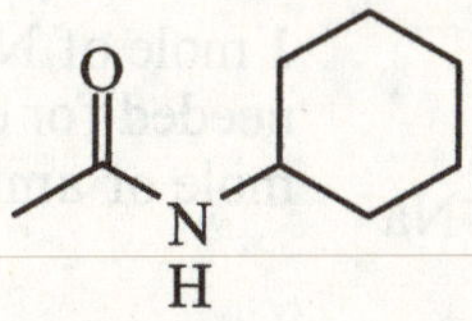

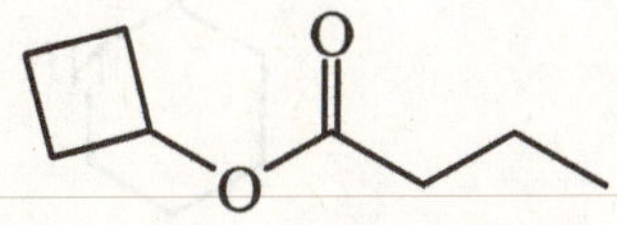

(d) *N*-(2-Octyl)benzamide 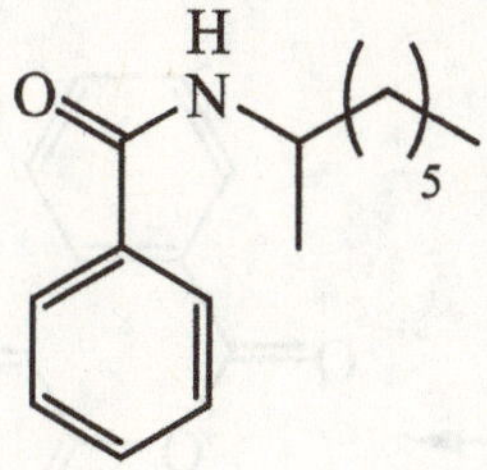   (e) Diethyl adipate 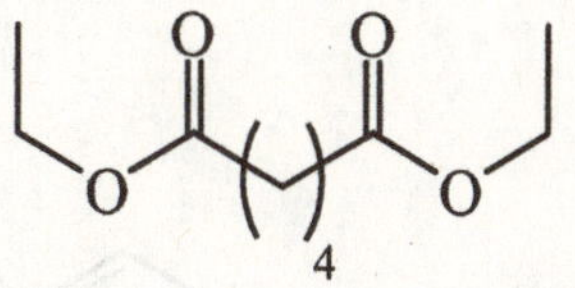   (f) Propanoic anhydride

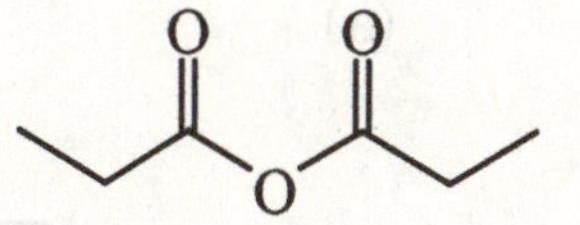

14.2 Complete and balance equations for hydrolysis of each ester in aqueous solution, showing each product as it is ionized under the given experimental conditions:

Under basic conditions, the carboxylic acid will be ionized and found in the carboxylate form, but under acidic conditions, the carboxylic acid will be in the acid form.

(a)

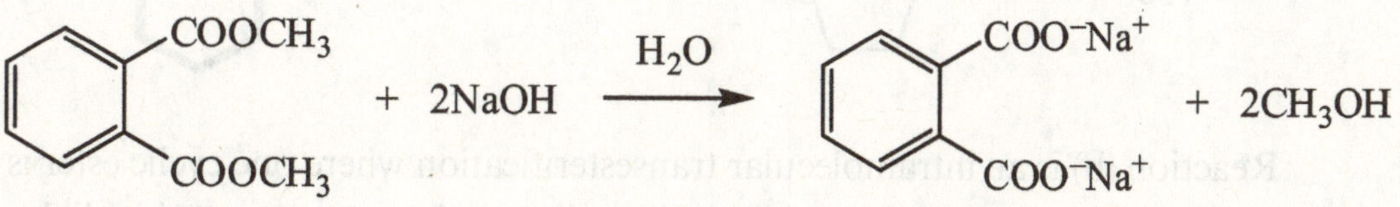

$$\text{(benzene with COOCH}_3\text{, COOCH}_3\text{)} + 2\text{NaOH} \xrightarrow{\text{H}_2\text{O}} \text{(benzene with COO}^-\text{Na}^+\text{, COO}^-\text{Na}^+\text{)} + 2\text{CH}_3\text{OH}$$

(b)

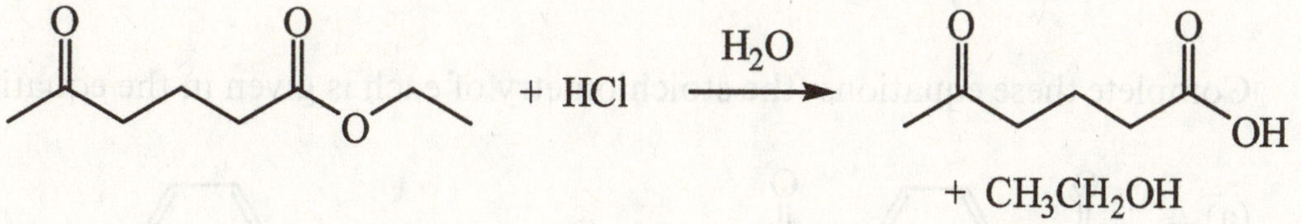

$$\text{(ester)} + \text{HCl} \xrightarrow{\text{H}_2\text{O}} \text{(keto acid)} + \text{CH}_3\text{CH}_2\text{OH}$$

14.3 Complete equations for the hydrolysis of the amides in Example 14.3 in concentrated aqueous NaOH. Show all products as they exist in aqueous NaOH, and show the number of moles of NaOH required for the hydrolysis of each amide.

(a)

$$CH_3CN(CH_3)_2 \xrightarrow[H_2O]{NaOH} CH_3CO^-Na^+ + (CH_3)_2NH$$

1 mole of NaOH needed for every mole of amide

(b)

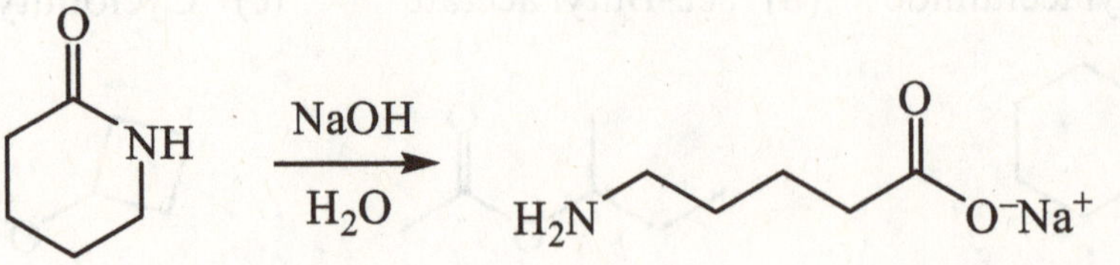

1 mole of NaOH needed for every mole of amide

14.4 Complete these equations (the stoichiometry of each is given in the equation):

(a)

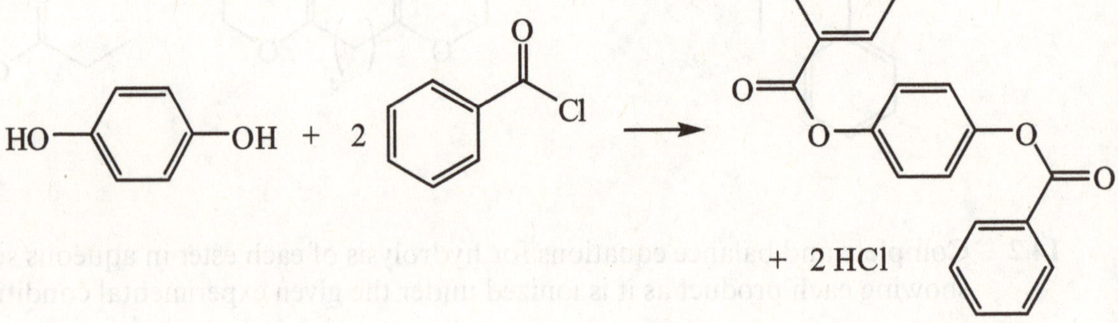

(b)

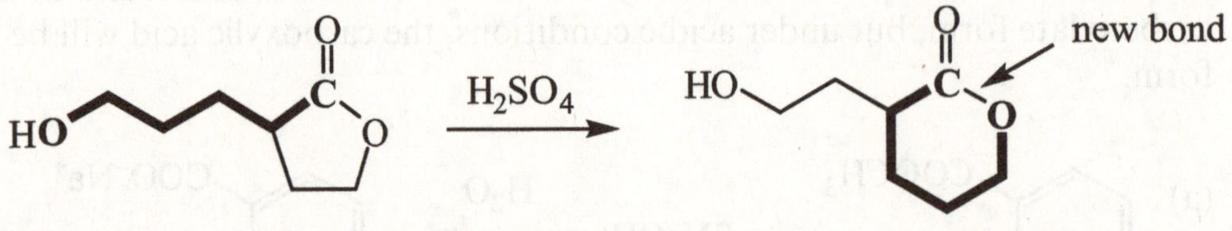

Reaction (b) is an intramolecular transesterification where one cyclic ester is replaced by another. The atoms and bonds corresponding to the new ester are highlighted in bold.

14.5 Complete these equations (the stoichiometry of each is given in the equation):

(a)

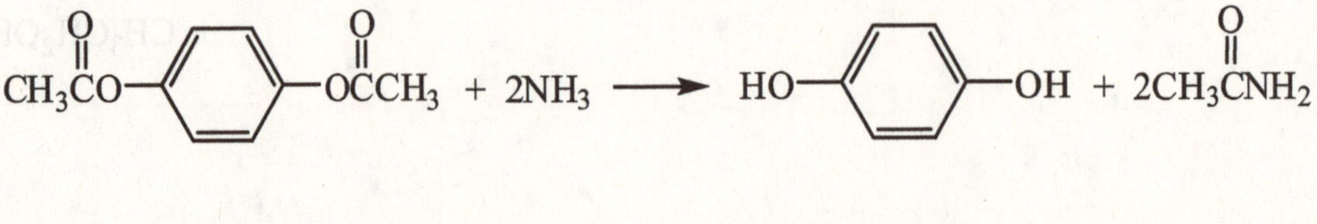

(b)

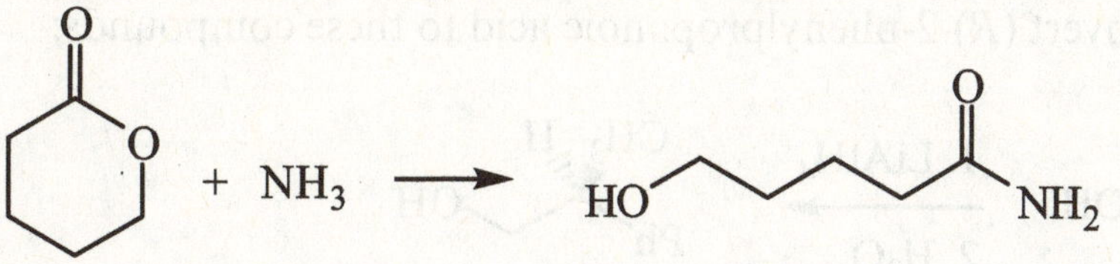

14.6 Show how to prepare each alcohol by treating an ester with a Grignard reagent:

(a)

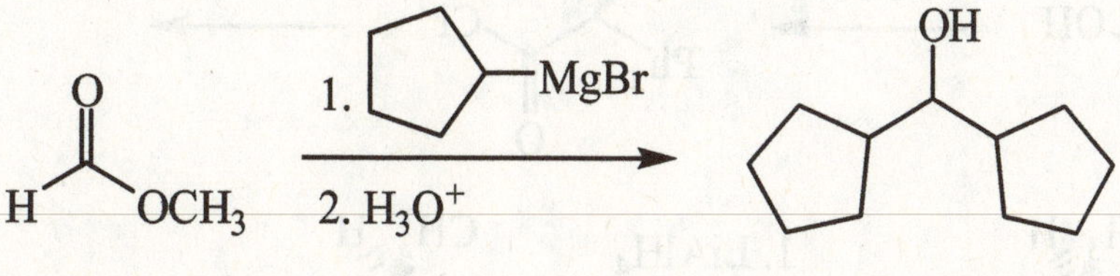

(b)

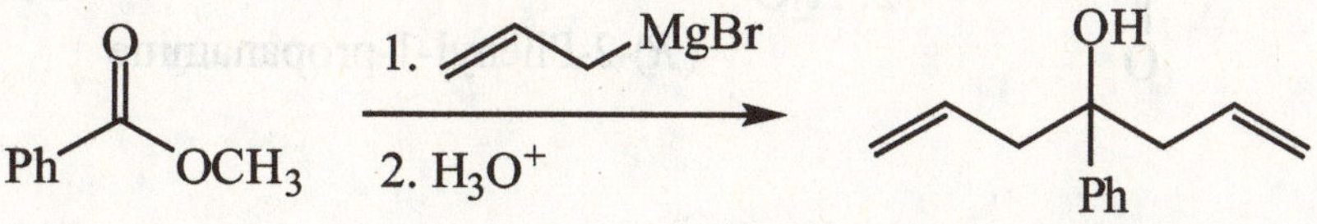

14.7 Show how to convert hexanoic acid to each amine in good yield:

(a)

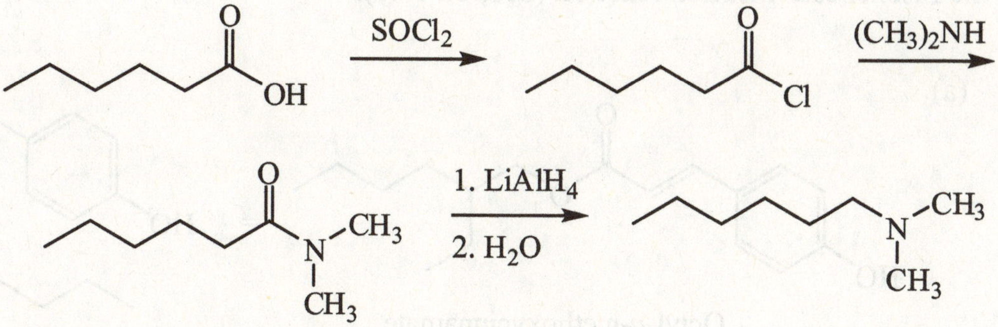

(b)

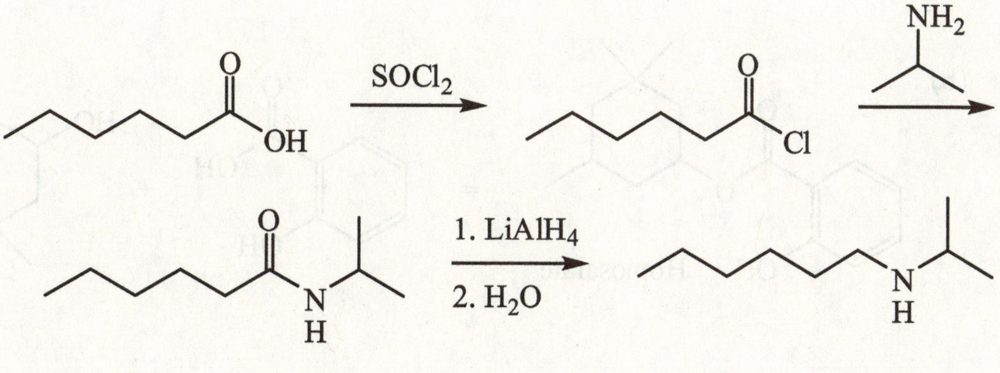

14.8   Show how to convert (*R*)-2-phenylpropanoic acid to these compounds:

(a)

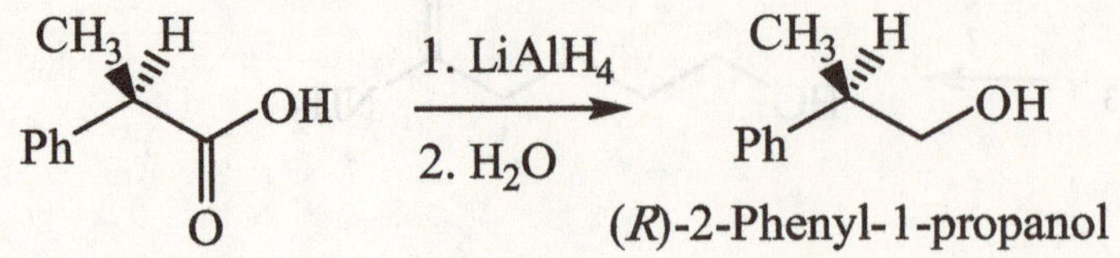

(*R*)-2-Phenyl-1-propanol

(b)

(*R*)-2-Phenyl-1-propanamine

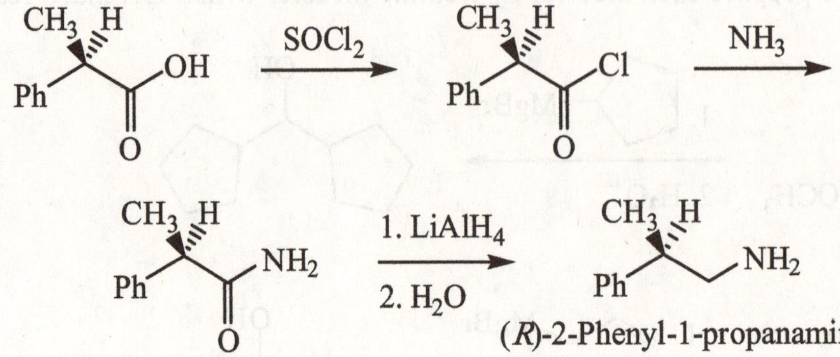

Chemical Connections

14A. Show how each sunscreen can be synthesized from a carboxylic acid and alcohol using the Fischer esterification reaction (Section 13.6).

(a)

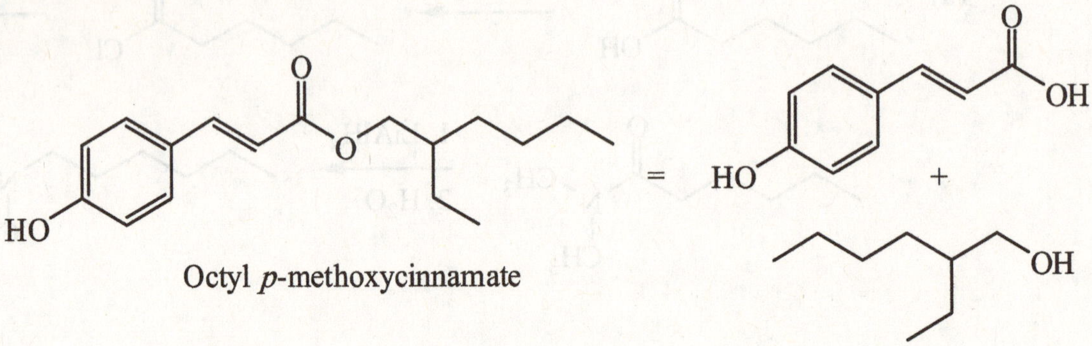

Octyl *p*-methoxycinnamate

(b)

Homosalate

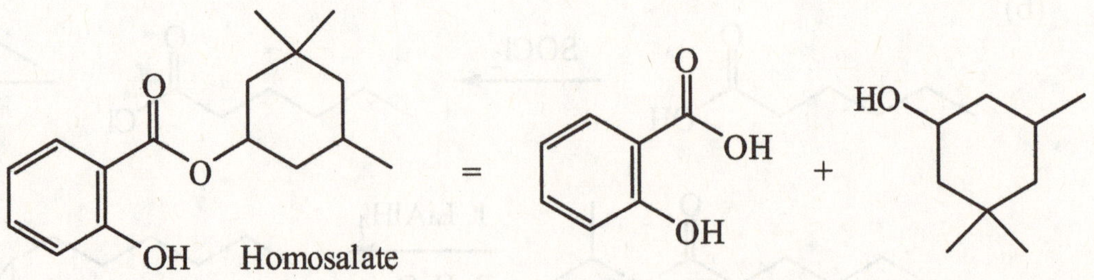

(c)

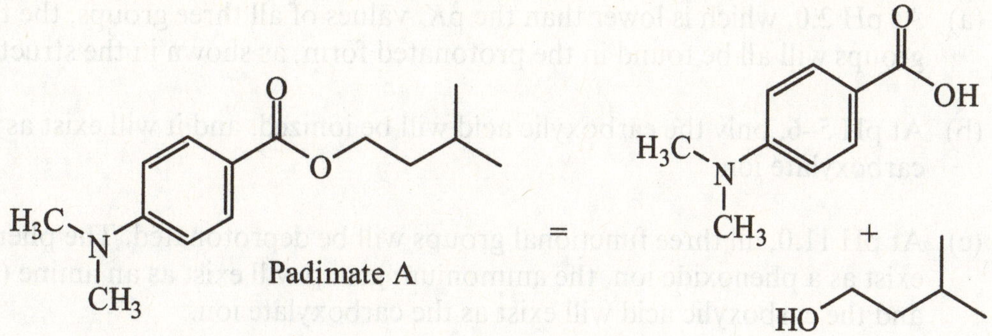

Padimate A

$=$

14B. Identify warfarin as an α, β, γ, etc., lactone. Identify each part of warfarin that can undergo keto-enol tautomerisation and show the tautomer at that position.

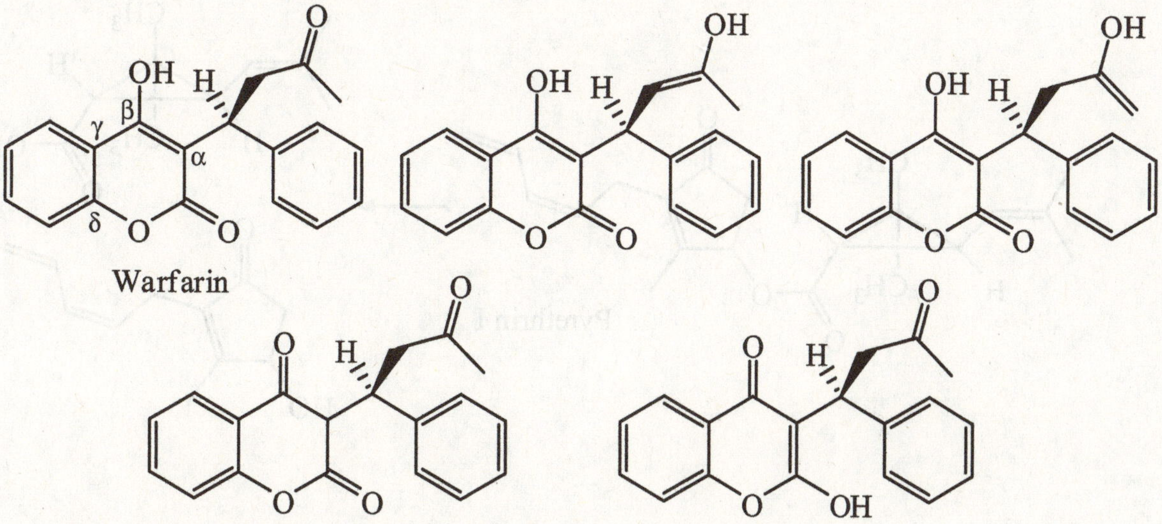

Warfarin

Warfarin is a δ lactone and contains multiple groups that can undergo tautomerism.

14C. What would you except to be the major form of amoxicillin present in aqueous solution at (a) pH 2.0, (b) at pH 5–6, and (c) at pH 11.0? Explain.

There are three pK_a values that need to be considered. The pK_a of the phenol is near 10, that of the protonated amine is near 9, and that of the carboxylic acid is near 4.

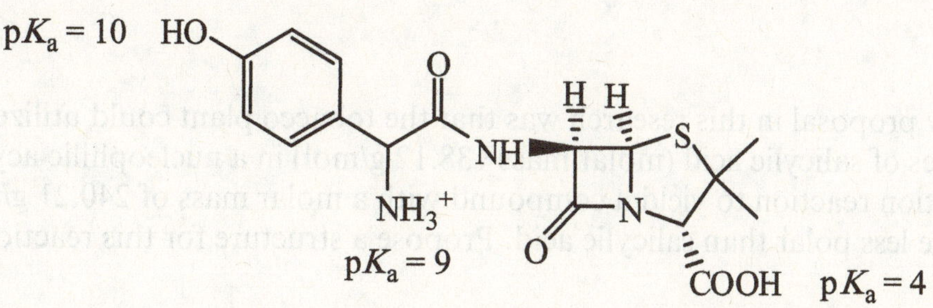

$pK_a = 10$ $pK_a = 9$ $pK_a = 4$

(a) At pH 2.0, which is lower than the pK_a values of all three groups, the three groups will all be found in the protonated form, as shown in the structure.

(b) At pH 5–6, only the carboxylic acid will be ionized, and it will exist as the carboxylate ion.

(c) At pH 11.0, all three functional groups will be deprotonated. The phenol will exist as a phenoxide ion, the ammonium group will exist as an amine (RNH_2), and the carboxylic acid will exist as the carboxylate ion.

14D. Show the compounds that would result if Pyrethrin I and Permethrin were to undergo hydrolysis.

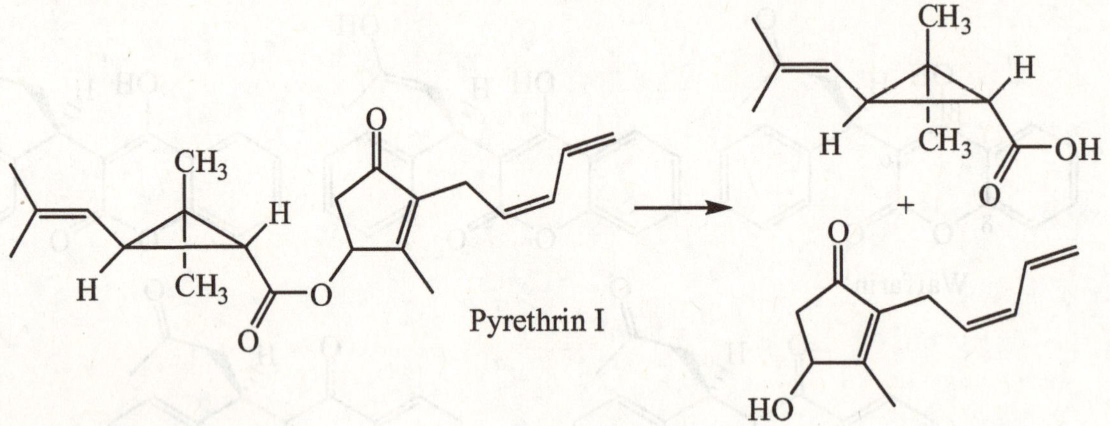

Pyrethrin I

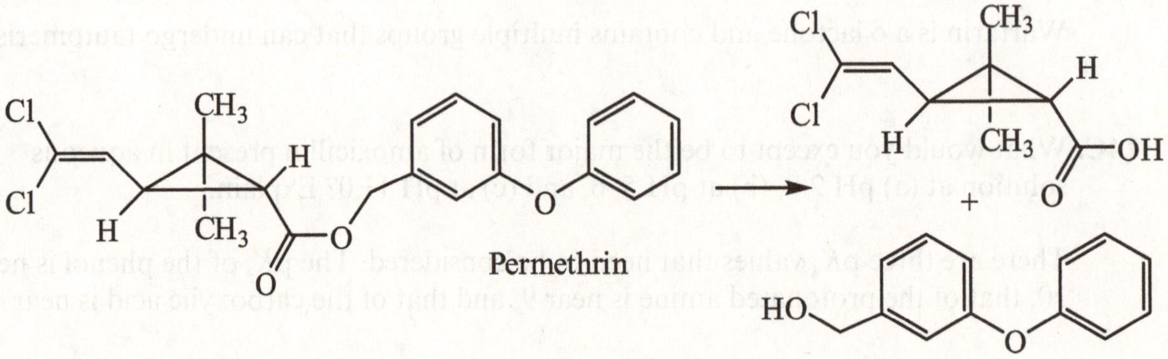

Permethrin

14E. An early proposal in this research was that the tobacco plant could utilize two molecules of salicylic acid (molar mass 138.12 g/mol) in a nucleophilic acyl substitution reaction to yield a compound with a molar mass of 240.21 g/mol that would be less polar than salicylic acid. Propose a structure for this reaction product.

The molar mass of the product is 36.03 g/mol less than the molar masses of the two moles of salicylic acid combined. This difference of 36.03 g/mol can be attributed to the two water molecules that are released upon ester formation.

Quick Quiz

1. The stronger the base, the better the leaving group. *False*. Better leaving groups are those that are more stable and therefore weaker bases.

2. Anhydrides can contain C−O double bonds or P−O double bonds. *True*. Anhydrides made from carboxylic acids contain the former, while those made from phosphoric acid contain the latter.

3. Acid anhydrides react with ammonia and amines without the need for acid or base. *True*. Anhydrides are sufficiently reactive with ammonia and amines in the absence of a catalyst.

4. Derivatives of carboxylic acids are reduced by H_2/M. *False*. Carboxylic acids and their derivatives cannot be reduced by catalytic hydrogenation. Strong hydride-based reducing agents, such as $LiAlH_4$, are required.

5. Aldehydes and ketones undergo nucleophilic acyl substitution reactions, while derivatives of carboxylic acids undergo nucleophilic addition reactions. *False*. The opposite is true.

6. Esters react with ammonia and amines without the need for acid or base. *True*. Although acid or base can accelerate the reaction, the formation of an amide is thermodynamically favorable and can proceed without a catalyst, especially at high temperatures.

7. An acyl group is a carbonyl bonded to an alkyl (R) group. *True*.

8. Hydrolysis is the loss of water from a molecule. *False*. Hydrolysis is a bond-breaking reaction caused by the addition of water.

9. Esters react with water without the need for acid or base. *False*. The hydrolysis of esters requires an acid or base catalyst.

10. Acid anhydrides react with water without the need for acid or base. *True*. Anhydrides are sufficiently reactive towards water, and no catalyst is required.

11. An acid halide can be converted to an amide in one step. *True*. The reaction of an acid halide with ammonia or an amine results in the formation of an amide.

12. An ester can be converted to an acid halide in one step. *False*. To prepare an acid halide from an ester, the ester must first be hydrolyzed to a carboxylic acid. The carboxylic acid then needs to be converted to an acid halide.

13. In the hydrolysis of an ester with base, hydroxide ion is a catalyst. *False*. The hydroxide ion is the nucleophile in the base hydrolysis of an ester. Hydroxide is consumed stoichiometrically and is not a catalyst.

14. Derivatives of carboxylic acids are reduced by $NaBH_4$. *False*. A stronger reducing agent, such as $LiAlH_4$, is needed to reduce carboxylic acids and their derivatives.

15. Acid anhydrides react with alcohols without the need for acid or base. *True*. Like their reaction with water, no acid or base is necessary.

16. Acid halides react with water without the need for acid or base. *True*. Acid halides are the most reactive of all acid derivatives and can react violently even without acid or base.

17. An ester of formic acid reacts with Grignard reagents to form 3° alcohols. *False*. Using an ester of formic would lead to the formation of 2° alcohols. The hydrogen atom bonded to the carbonyl group of the ester would still be present in the alcohol.

18. Acid halides react with ammonia and amines without the need for acid or base. *True*. The highly reactive properties of acid halides make acid or base unnecessary.

19. A cyclic amide is called a lactone. *False*. A cyclic amide is called a lactam. A lactone is a cyclic ester.

20. The reactivity of a carboxylic acid derivative is dependent on the stability of its leaving group. *True*. The more stable the leaving group, the more reactive the acid derivative.

21. Amides react with ammonia and amines without the need for acid or base. *False*. Amides are very stable and require acid or base for reaction to occur.

22. An amide can be converted to an ester in one step. *False*. A carboxylic acid must first be prepared from the hydrolysis of the amide.

23. Amides react with water without the need for acid or base. *False*. Amides are the least reactive of the acid derivatives. Even in the presence of acid or base, the hydrolysis of amides is slow and requires high temperature.

24. Esters react with alcohols without the need for acid or base. *False*. Transesterification reactions generally require an acid or base catalyst.

25. Amides react with alcohols under acidic or basic conditions. *False*. The conversion of an amide to an ester, a more reactive acid derivative, is unfavorable.

26. Esters other than formic acid esters react with Grignards to form ketones. *False*. The reaction of esters with Grignard reagents leads to the formation of alcohols.

27. Acid halides react with alcohols without the need for acid or base. *True*. Like their reaction with water, no acid or base is required.

28. An –OR group attached to a P–O double bond is known as an ester. *True*. Specifically, this type of ester is known as a phosphate ester.

End-of-Chapter Problems

Structure and Nomenclature

14.9 Draw a structural formula for each compound:

(a) Dimethyl carbonate (b) *p*-Nitrobenzamide   (c) Octanoyl chloride

(d) Diethyl oxalate   (e) Ethyl *cis*-2-pentanoate   (f) Butanoic anhydride

(g) Dodecanamide   (h) Ethyl 3-hydroxybuanoate (i) Ethyl benzoate

(j) Benzoyl chloride   (k) *N*-Ethylpentanamide 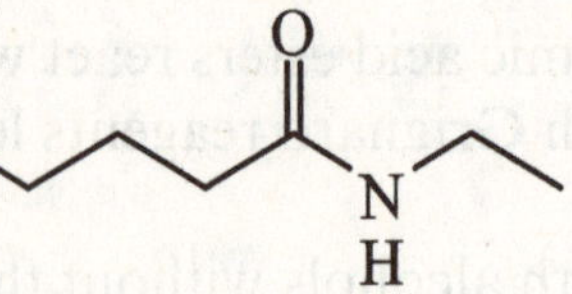   (l) 5-Methylhexanoyl
                                                                     chloride

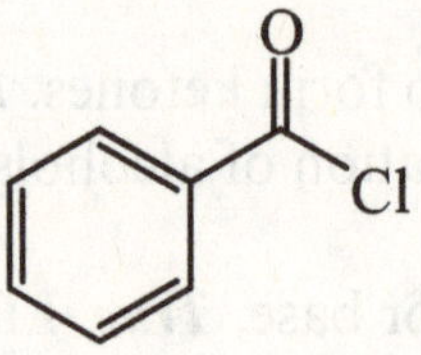

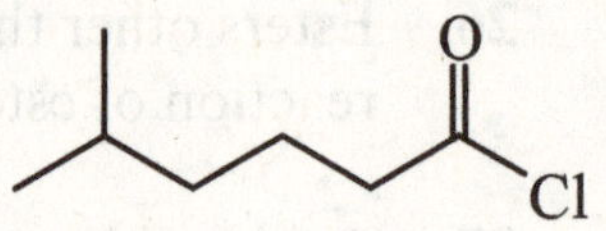

14.10 Write the IUPAC name for each compound:

(a)

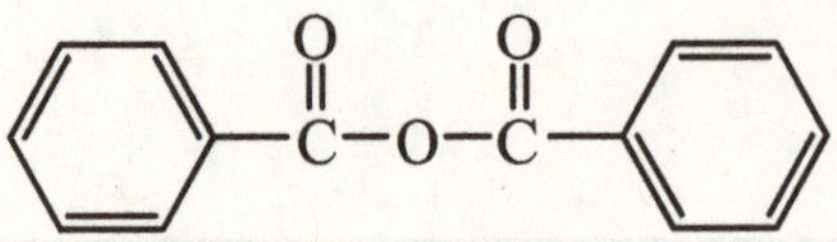

Benzoic anhydride

(b)

$CH_3(CH_2)_{14}COCH_3$

Methyl hexadecanoate

(c)

$CH_3(CH_2)_4CNHCH_3$

N-Methylhexanamide

(d)

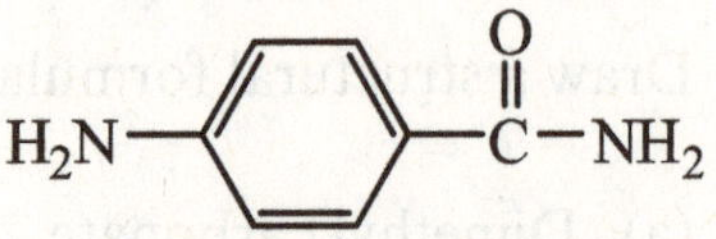

H_2N —C—NH_2

4-Aminobenzamide

(e)

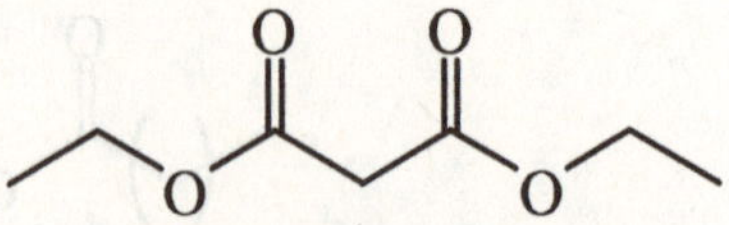

Diethyl propanedioate
(diethyl malonate)

(f)

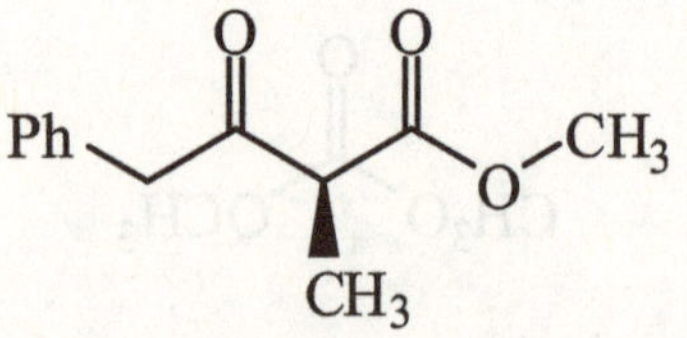

Ph —C—C—O—CH_3
         |
        CH_3

Methyl (*S*)-2-methyl-3-oxo-
4-phenylbutanoate

(g)

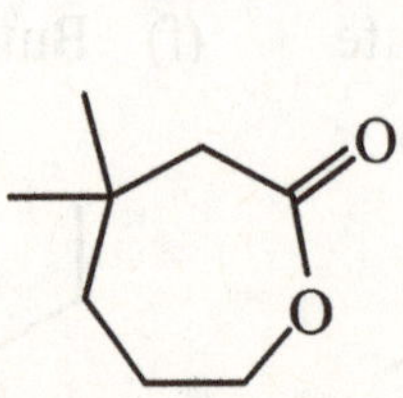

3,3-Dimethyl-6-hexanolactone

(h)

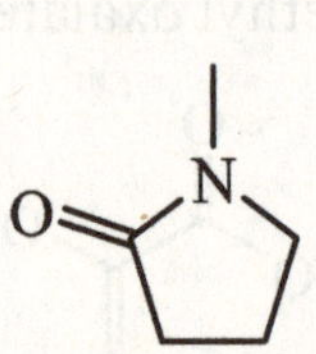

N-Methyl-4-butanolactam

(i)

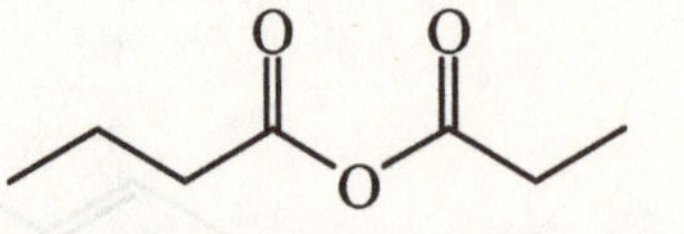

Butyric propionic anhydride

(j)

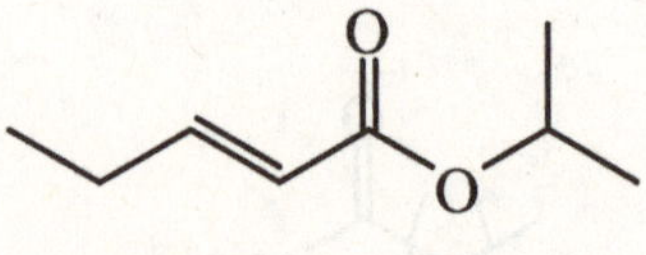

Isopropyl (*E*)-2-pentenoate

(k)

3-Hydroxypentanoyl chloride

(l)

4-Oxopentanoyl chloride

14.11 When oil from the head of a sperm whale is cooled, spermaceti, a translucent wax with a white, pearly luster, crystallizes from the mixture. Spermaceti, which makes up 11% of whale oil, is composed mainly of hexadecyl hexadecanoate (cetyl palmitate). At one time, spermaceti was widely used in the making of cosmetics, fragrant soaps, and candles. Draw a structural formula of cetyl palmitate.

Physical Properties

14.12 Acetic acid and methyl formate are constitutional isomers. Both are liquids at room temperature: one with a boiling point of 32°C, the other with a boiling point of 118°C. Which of the two has the higher boiling point?

Acetic acid is the compound with the higher pointing point. This is because acetic acid, which has an −OH group, can undergo intermolecular hydrogen bonding between the acetic acid molecules. Intermolecular hydrogen bonding between the molecules of methyl formate is not possible, so it has a lower boiling point than acetic acid.

14.13 Butanoic acid (M.W. 88.11 g/mol) has a boiling point of 162°C, whereas its propyl ester (M.W. 130.18 g/mol) has a boiling point of 142°C. Account for the fact that the boiling point of butanoic acid is higher than that of its propyl ester, even though butanoic acid has a lower molecular weight.

Although propyl butanoate has a higher molecular weight and a larger surface area, and thus has more dispersion forces than does butanoic acid, there is no intermolecular hydrogen bonding between the molecules of propyl butanoate. The strength of the hydrogen bonding interactions in butanoic acid is greater than the increase in dispersion forces when going from butanoic acid to its propyl ester.

14.14 Pentanoic acid and methyl butanoate are both slightly soluble in water. One of these has a solubility of 1.5 g/100 mL (25°C), while the other has a solubility of 4.97 g/100 mL (25°C). Assign the solubilities to each compound and account for the differences.

Both compounds are of the same formula and molecular weight. Yet, pentanoic acid is the compound that is more soluble in water because pentanoic acid, which has an −OH group, is capable of acting as a hydrogen bonding donor and acceptor. This is favorable because water can also donate and accept hydrogen bonds. Whereas, methyl butanoate is not capable of acting as a hydrogen bonding donor; with no −OH groups, it can only act as a hydrogen bonding acceptor.

Reactions

14.15 Arrange these compounds in order of increasing reactivity toward nucleophilic acyl substitution:

(1)   (2)   (3)   (4)

The order of reactivity of acid derivatives towards nucleophilic acyl substitution can be rationalized based on the magnitude of the positive charge at the carbonyl carbon (see Problem 13.51). In general, the order of reactivity of acid derivatives, from the least-reactive to the most-reactive, is: amides, esters, anhydrides, and acid chlorides. Thus, the order of increasing reactivity is 3 < 1< 4< 2.

14.16 A carboxylic acid can be converted to an ester by Fischer esterification. Show how to synthesize each ester from a carboxylic acid and an alcohol by Fischer esterification:

14.17 A carboxylic acid can also be converted to an ester in two reactions by first converting the carboxylic acid to its acid chloride and then treating the acid chloride with an alcohol. Show how to prepare each ester in Problem 14.16 from a carboxylic acid and an alcohol by this two-step scheme.

(a)

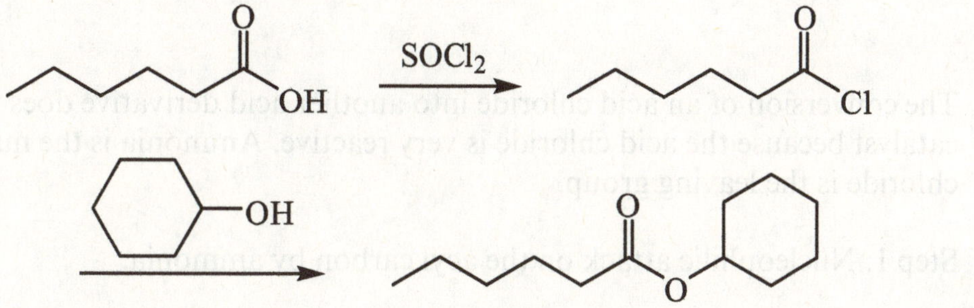

(b)

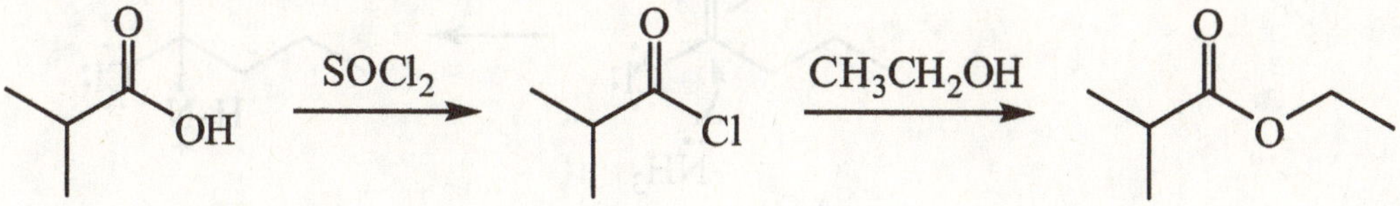

14.18 Show how to prepare these amides by reaction of an acid chloride with ammonia or an amine:

(a)

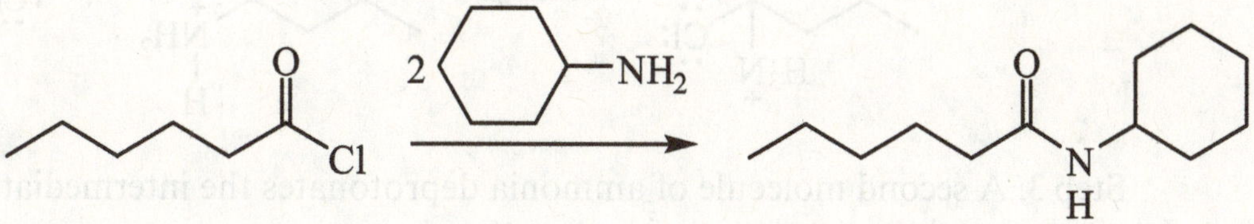

(b)

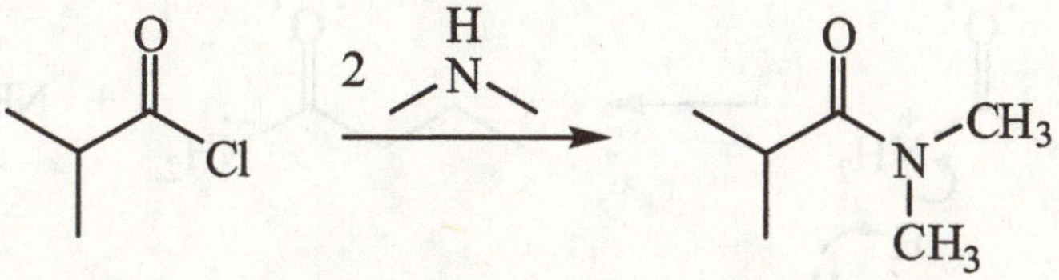

(c)

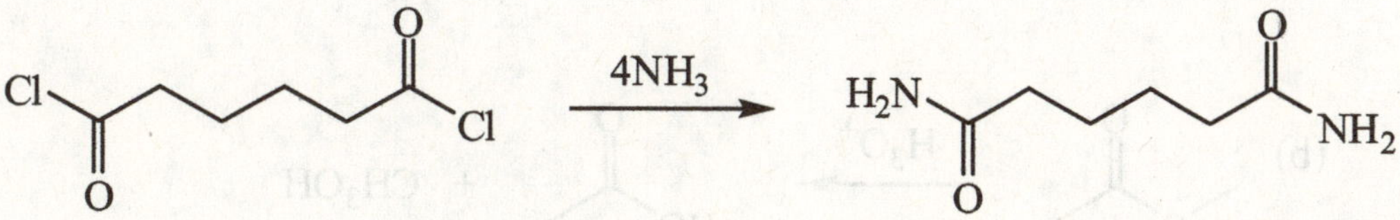

14.19 Balance and write a mechanism for each of the following reactions.

(a)

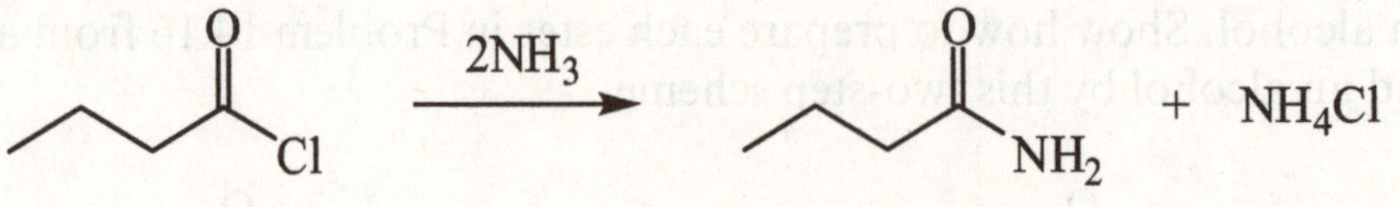

The conversion of an acid chloride into another acid derivative does not require a catalyst because the acid chloride is very reactive. Ammonia is the nucleophile, and chloride is the leaving group.

Step 1: Nucleophilic attack on the acyl carbon by ammonia.

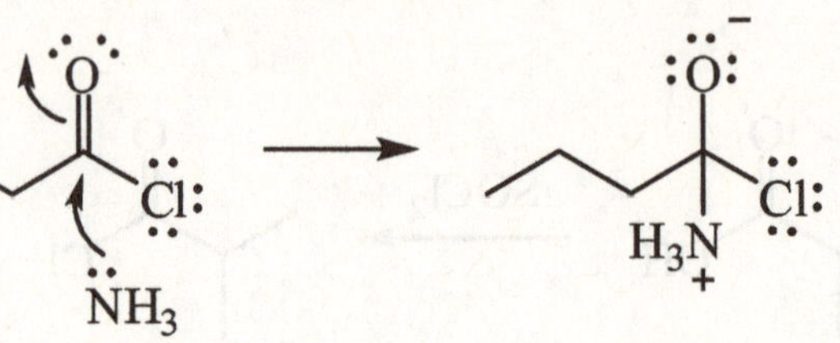

Step 2: Collapse of the tetrahedral intermediate and the loss of chloride.

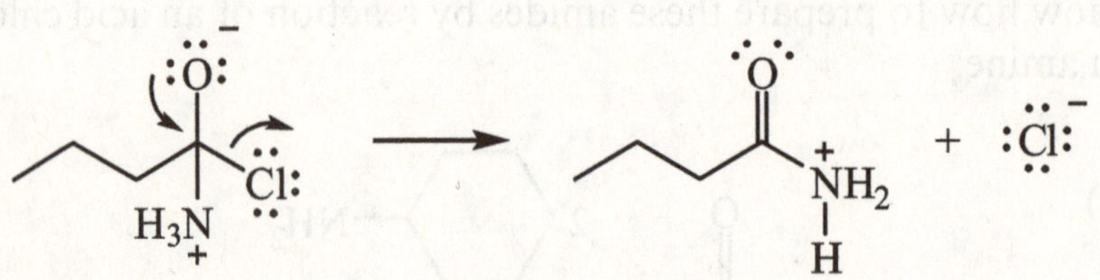

Step 3: A second molecule of ammonia deprotonates the intermediate to form the product and the ammonium ion.

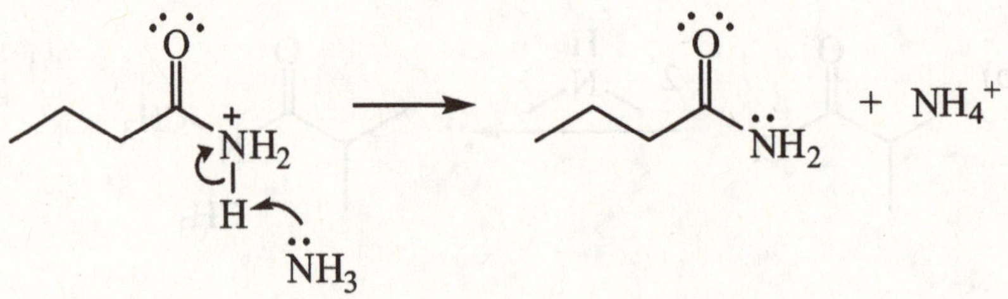

(b)

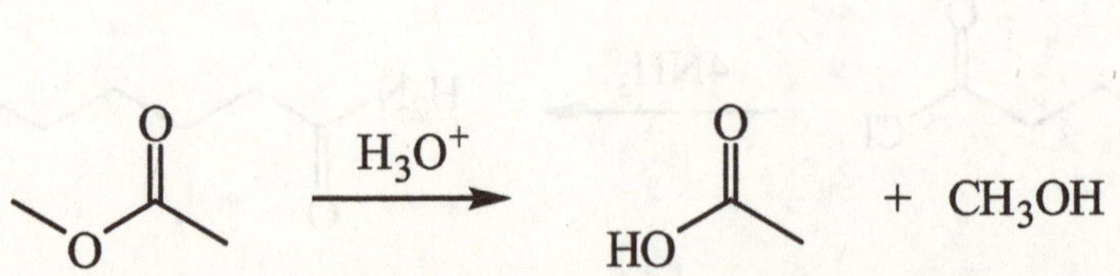

This reaction is mechanistically the reverse of a Fischer esterification. In this reverse reaction, water is the nucleophile and the alcohol is the leaving group.

Step 1: Protonation of the acyl oxygen to form a resonance-stabilized cation.

Step 2: Nucleophilic attack to form a tetrahedral intermediate.

Step 3: Proton transfer from the water portion of the intermediate to the alcohol portion. This converts the alkoxy group into a good leaving group.

Step 4: Departure of the alcohol and deprotonation generates the carboxylic acid and the alcohol, as well as the acid catalyst. Note that either one of the two −OH groups can be deprotonated.

$+ \; CH_3OH \; + \; H_3O^+$

(c)

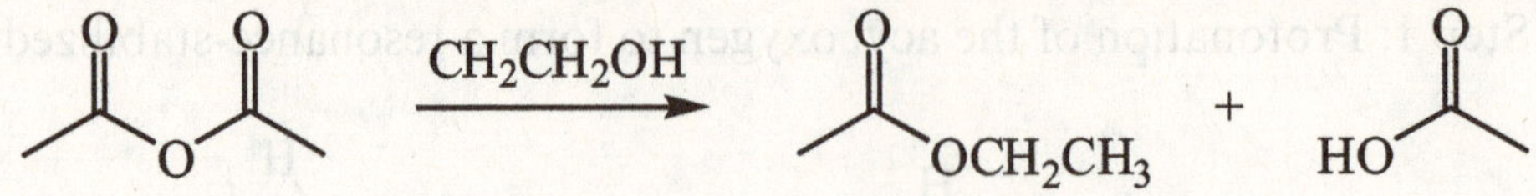

In the absence of an acid catalyst, the reaction proceeds similarly to (a).

Step 1: Nucleophilic attack on the acyl carbon by ethanol.

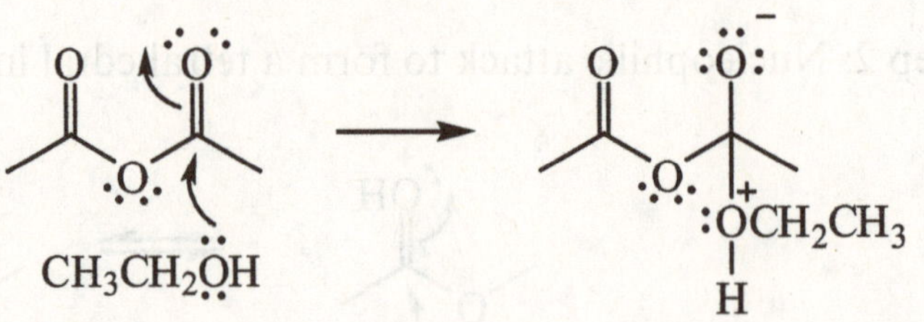

Step 2: Collapse of the tetrahedral intermediate and the loss of acetate.

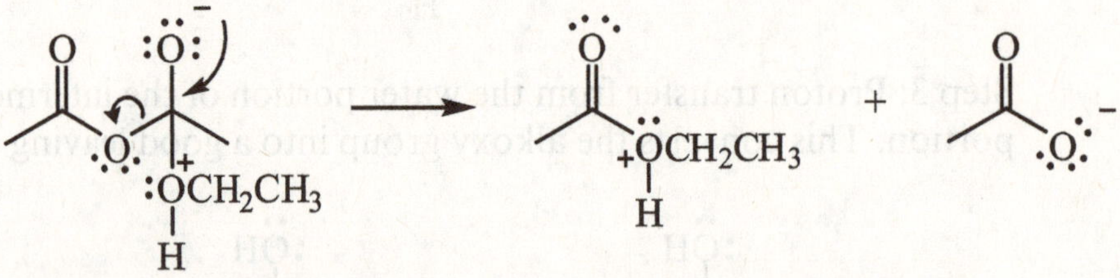

Step 3: Deprotonation of the oxonium ion generates the products.

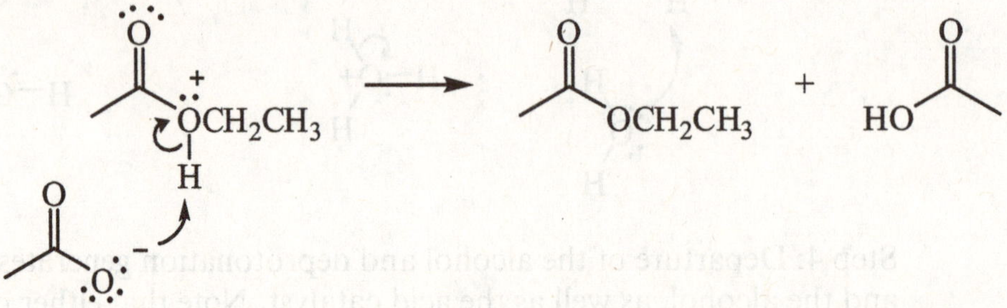

14.20 What product is formed when benzoyl chloride is treated with these reagents?

(a)

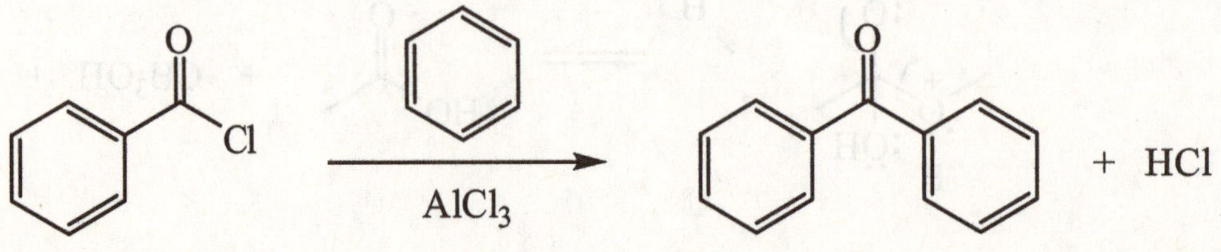

(b)

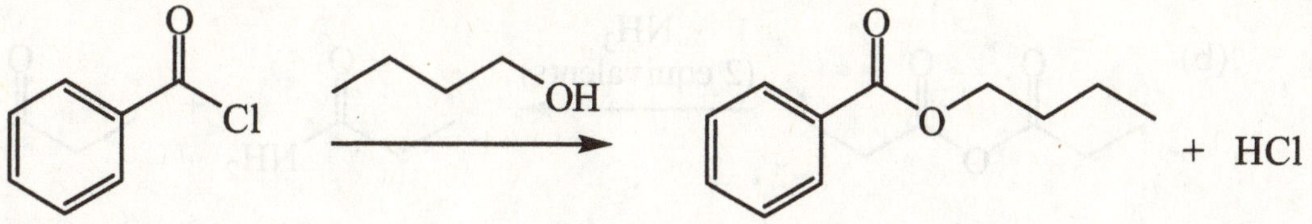

+ HCl

(c)

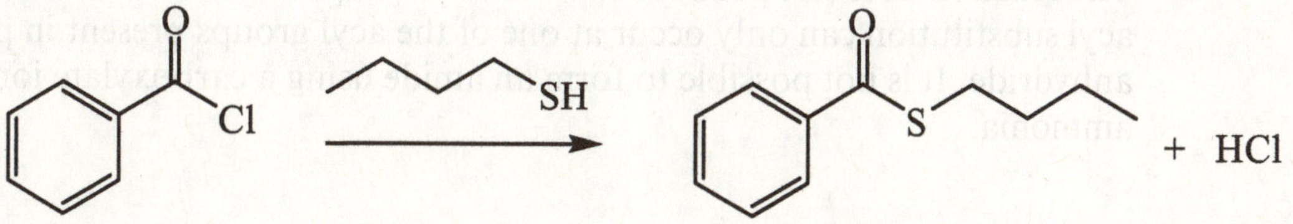

+ HCl

(d)

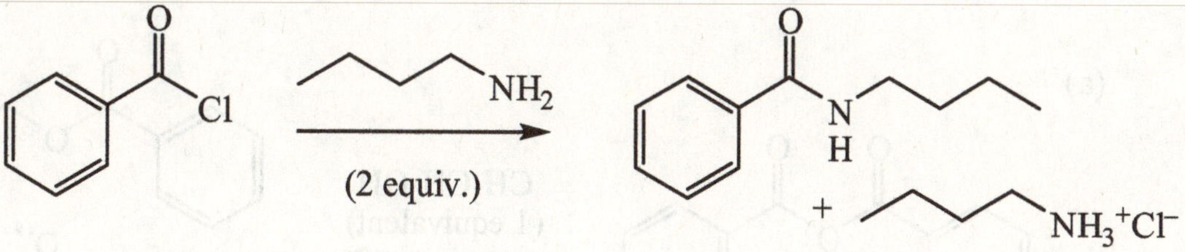

(2 equiv.)

(e)

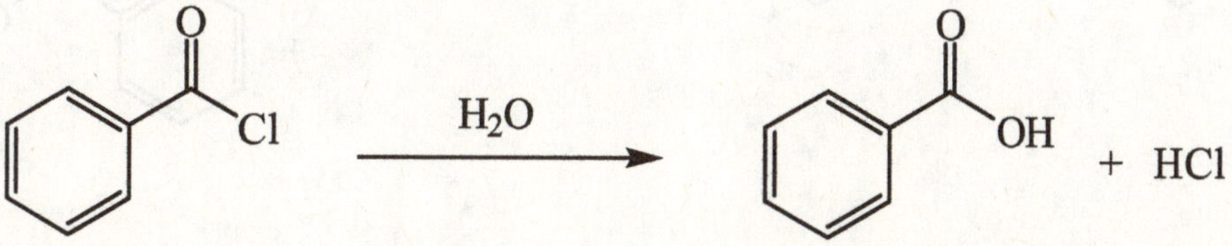

H_2O

+ HCl

(f)

(2 equiv.)

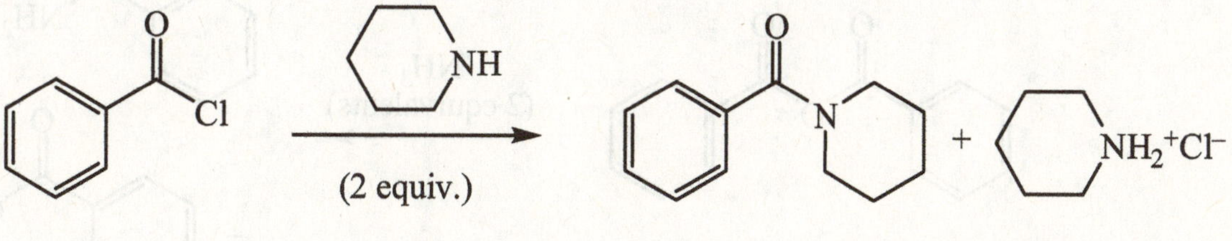

14.21 Write the product(s) of the treatment of propanoic anhydride with each reagent:

(a)

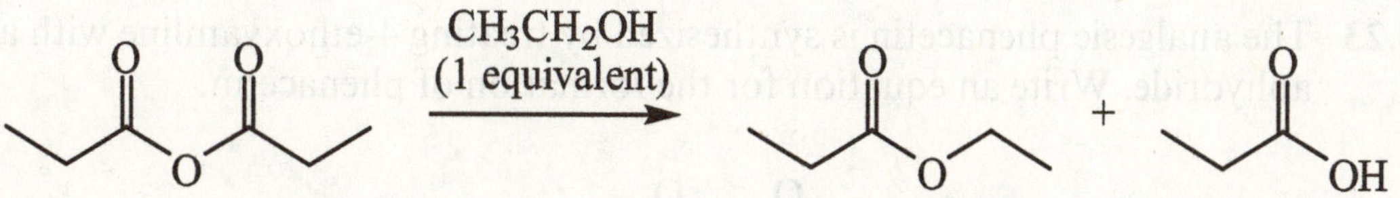

CH_3CH_2OH
(1 equivalent)

(b)

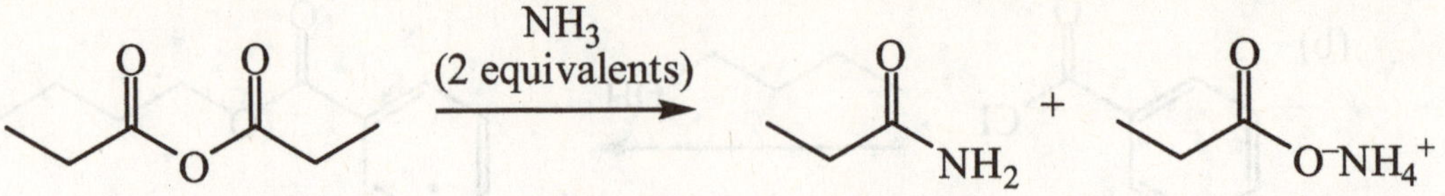

The reaction does not produce two moles of propanamide because nucleophilic acyl substitution can only occur at one of the acyl groups present in propanoic anhydride. It is not possible to form an amide using a carboxylate ion and ammonia.

14.22 Write the product of the treatment of benzoic anhydride with each reagent:

(a)

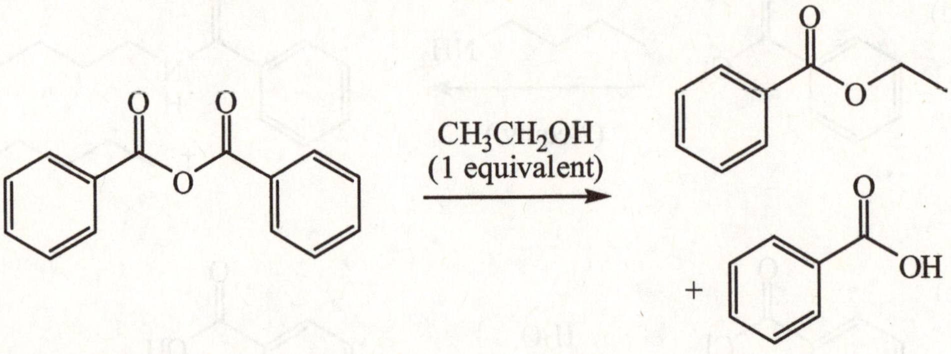

(b)

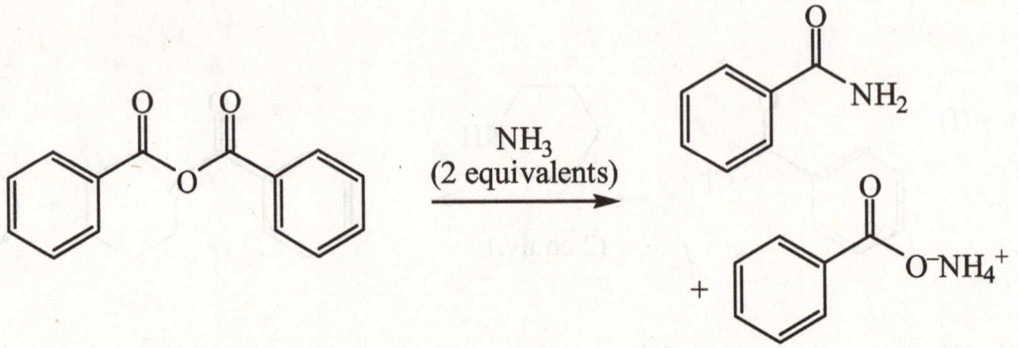

14.23 The analgesic phenacetin is synthesized by treating 4-ethoxyaniline with acetic anhydride. Write an equation for the formation of phenacetin.

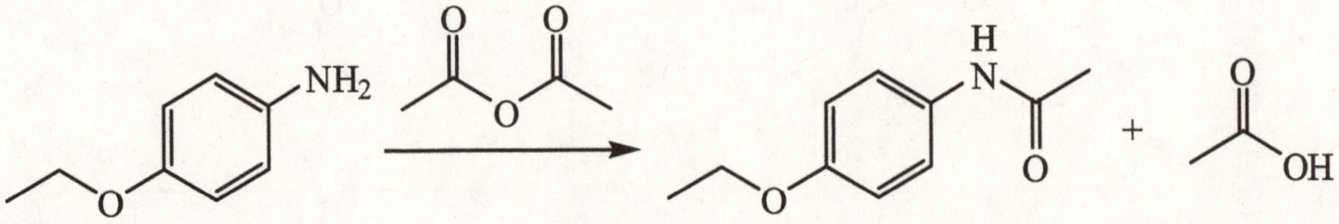

14.24 The analgesic acetaminophen is synthesized by treating 4-aminophenol with one equivalent of acetic anhydride. Write an equation for the formation of acetaminophen. (*Hint*: Remember from Section 7.5A that an –NH₂ group is a better nucleophile than an –OH group.)

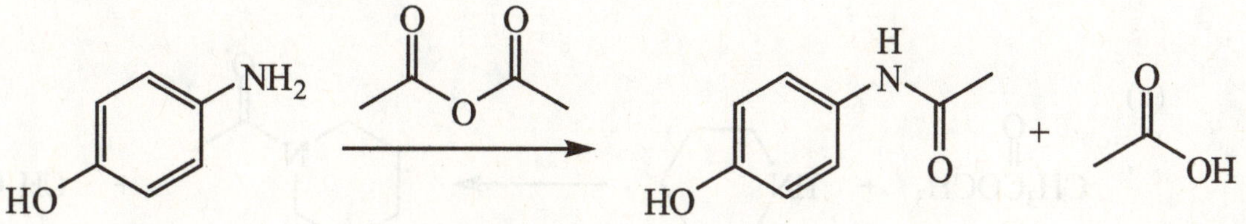

14.25 Nicotinic acid, more commonly named niacin, is one of the B vitamins. Show how nicotinic acid can be converted to ethyl nicotinate and then to nicotinamide:

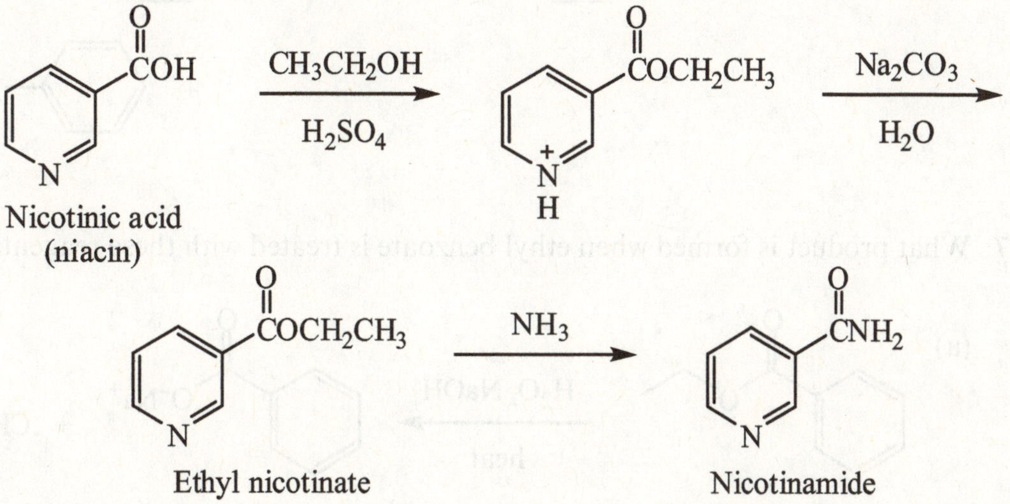

Nicotinic acid
(niacin)

Ethyl nicotinate   Nicotinamide

Under the acidic conditions of a Fischer esterification, the basic pyridine ring will be protonated. Therefore, the product of the first reaction is subsequently treated with a mild base, such as sodium carbonate, to deprotonate the pyridine ring.

14.26 Complete these reactions:

(a)

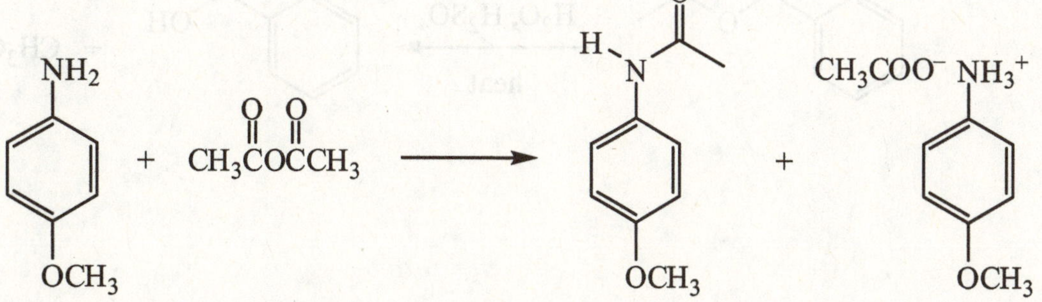

(b)

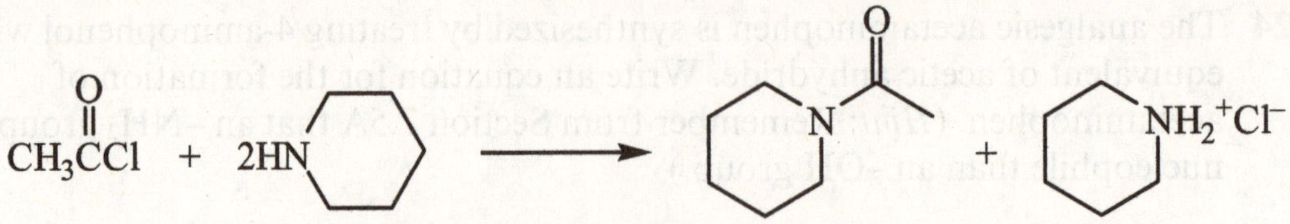

$$CH_3CCl \ + \ 2HN \bigcirc \longrightarrow \bigcirc N-CCH_3 \ + \ \bigcirc NH_2^+Cl^-$$

(c)

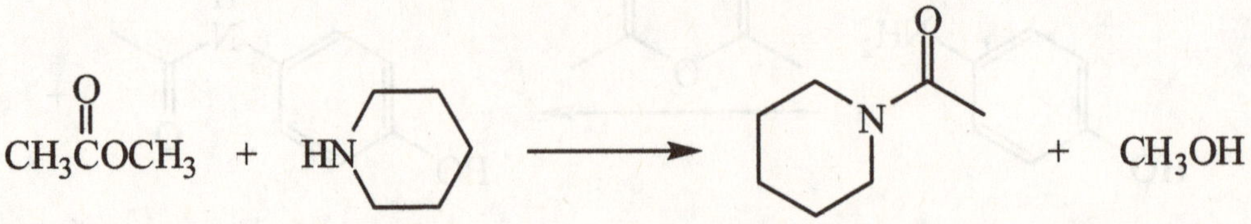

$$CH_3COCH_3 \ + \ HN \bigcirc \longrightarrow \bigcirc N-CCH_3 \ + \ CH_3OH$$

(d)

$$CH_3(CH_2)_5CCl \ + \ \bigcirc -NH_2 \longrightarrow \bigcirc -NHC(CH_2)_5CH_3$$

$$+ \ \bigcirc -NH_3^+Cl^-$$

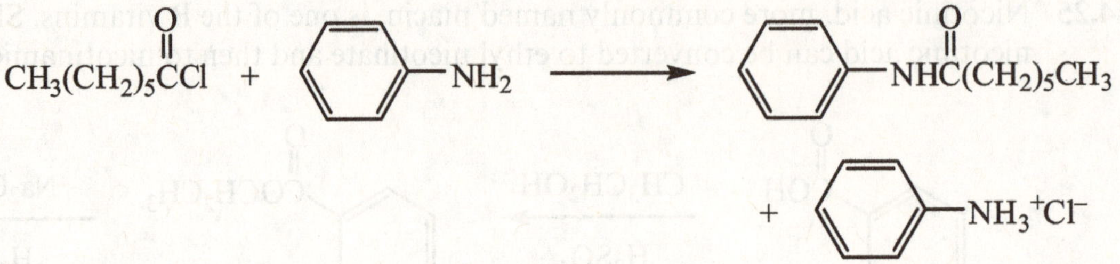

14.27 What product is formed when ethyl benzoate is treated with these reagents?

(a)

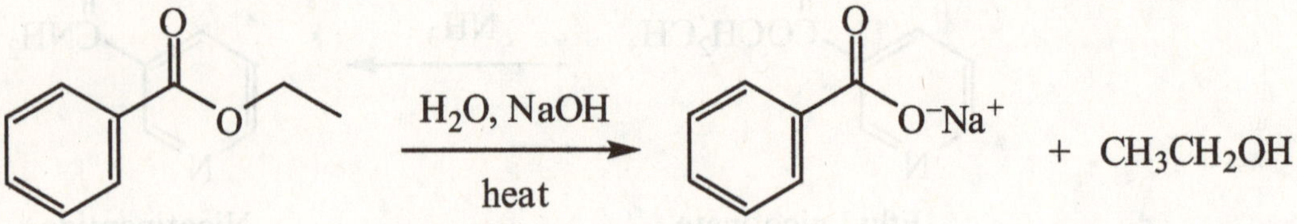

$$\bigcirc \!\!-\!\! \overset{O}{\underset{\parallel}{C}} \!\!-\!\! O \!\!-\!\! \diagup \quad \xrightarrow[\text{heat}]{H_2O, \ NaOH} \quad \bigcirc \!\!-\!\! \overset{O}{\underset{\parallel}{C}} \!\!-\!\! O^-Na^+ \ + \ CH_3CH_2OH$$

(b)

$$\bigcirc \!\!-\!\! \overset{O}{\underset{\parallel}{C}} \!\!-\!\! O \!\!-\!\! \diagup \quad \xrightarrow[\text{2. } H_2O]{1. \ LiAlH_4} \quad \bigcirc \!\!-\!\! \diagup OH \ + \ CH_3CH_2OH$$

(c)

$$\bigcirc \!\!-\!\! \overset{O}{\underset{\parallel}{C}} \!\!-\!\! O \!\!-\!\! \diagup \quad \xrightarrow[\text{heat}]{H_2O, \ H_2SO_4} \quad \bigcirc \!\!-\!\! \overset{O}{\underset{\parallel}{C}} \!\!-\!\! OH \ + \ CH_3CH_2OH$$

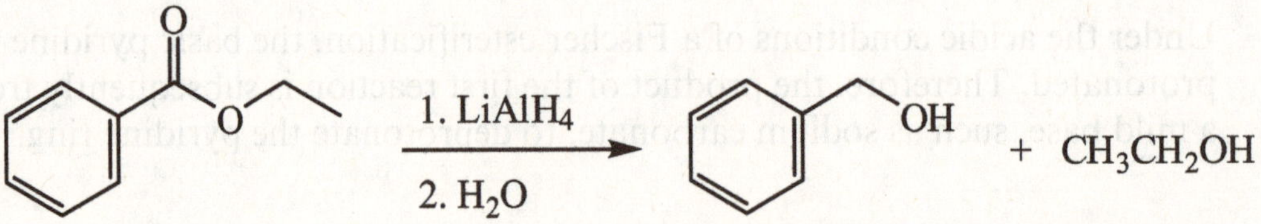

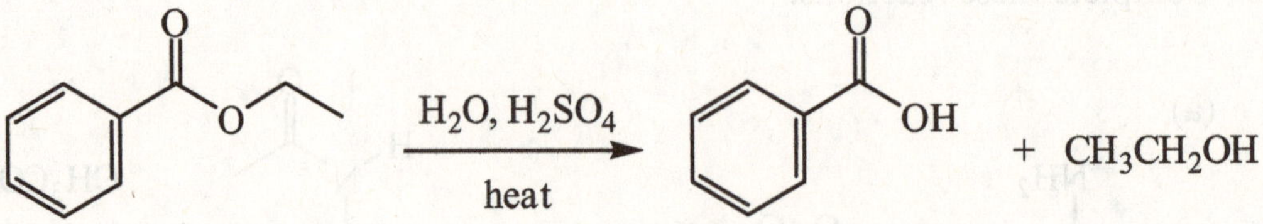

(d)

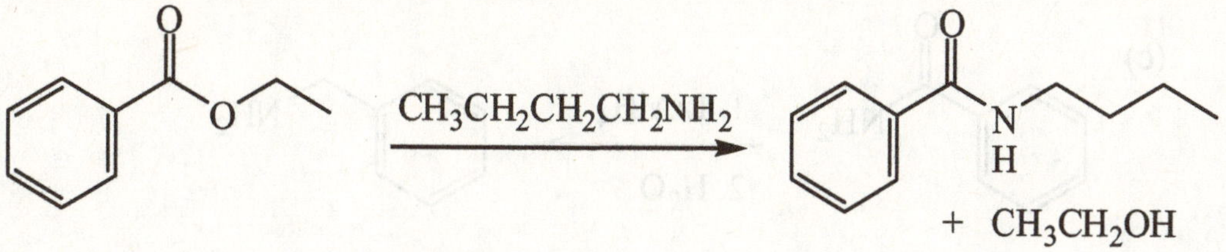
O
O
$CH_3CH_2CH_2CH_2NH_2$
O
N
H
+ CH_3CH_2OH

(e)

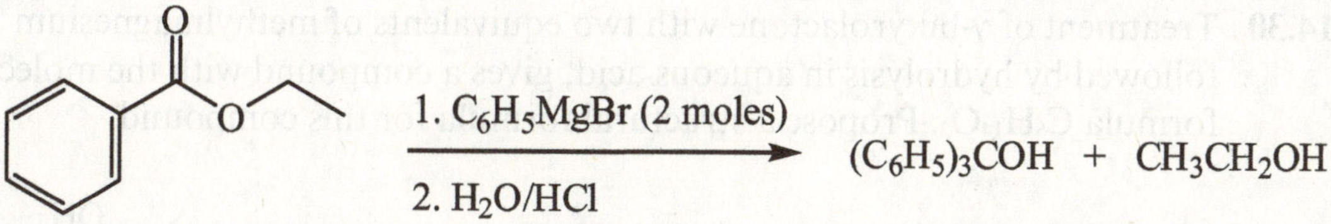
O
O
1. C_6H_5MgBr (2 moles)
2. H_2O/HCl
$(C_6H_5)_3COH$ + CH_3CH_2OH

14.28 Show how to convert 2-hydroxybenzoic acid (salicylic acid) to these compounds:

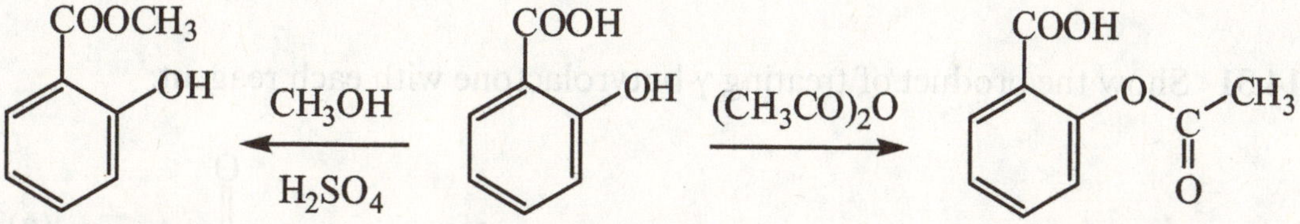
$COOCH_3$
OH
CH_3OH
H_2SO_4
COOH
OH
$(CH_3CO)_2O$
COOH
O
C
CH_3
O

Methyl salicylate (Oil of wintergreen) — Salicylic acid — Acetylsalicylic acid (Aspirin)

In the synthesis of acetylsalicylic acid, it is important to use a reagent that reacts with only the –OH group of salicylic acid. Although acetyl chloride could acetylate the –OH group, it would also acetylate the –COOH group and form an anhydride. By using acetic anhydride as the acetylating agent, only the –OH group is acetylated. The reaction of acetic anhydride with the –COOH group would form another anhydride, but there is no thermodynamic advantage in doing so.

14.29 What product is formed when benzamide is treated with these reagents?

(a)

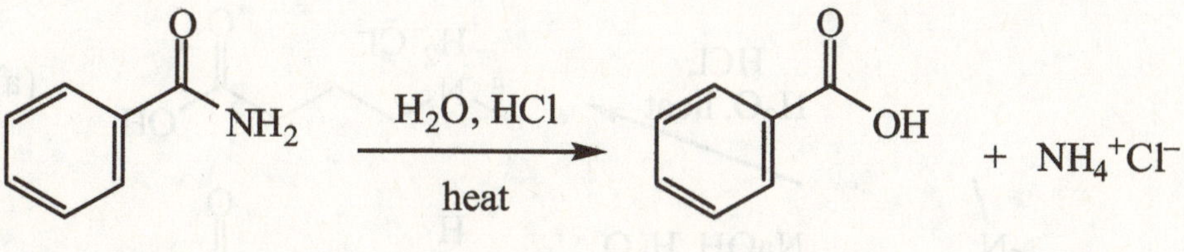
O
NH_2
H_2O, HCl
heat
O
OH
+ $NH_4^+Cl^-$

(b)

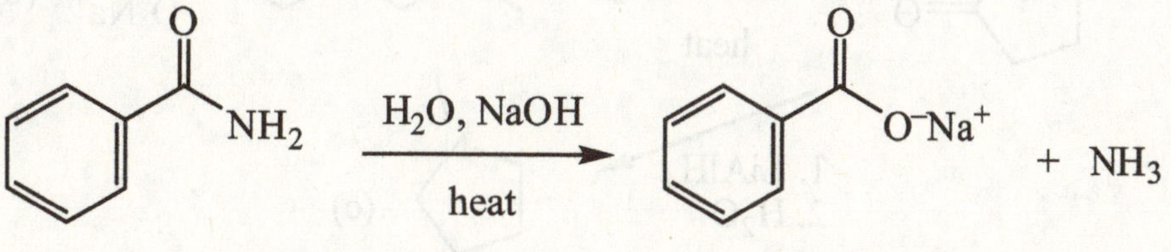
O
NH_2
H_2O, NaOH
heat
O
O^-Na^+
+ NH_3

(c)

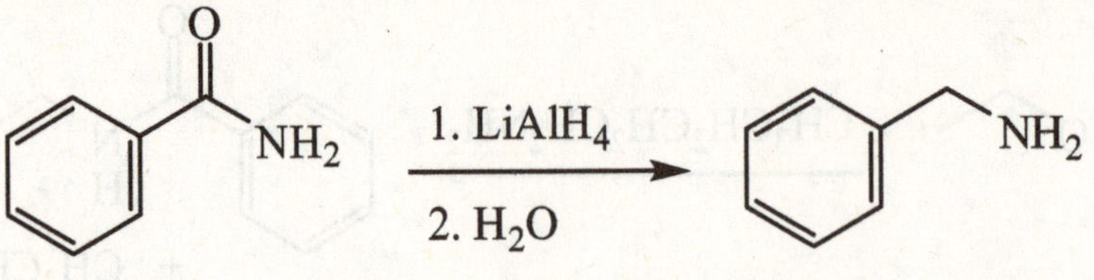

14.30 Treatment of γ-butyrolactone with two equivalents of methylmagnesium bromide, followed by hydrolysis in aqueous acid, gives a compound with the molecular formula $C_6H_{14}O_2$. Propose a structural formula for this compound.

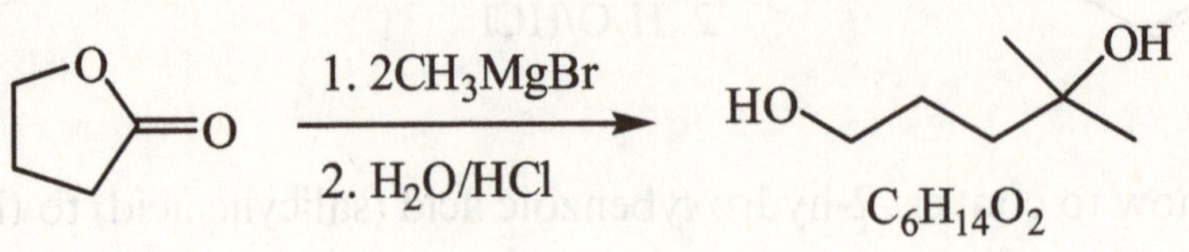

14.31 Show the product of treating γ-butyrolactone with each reagent:

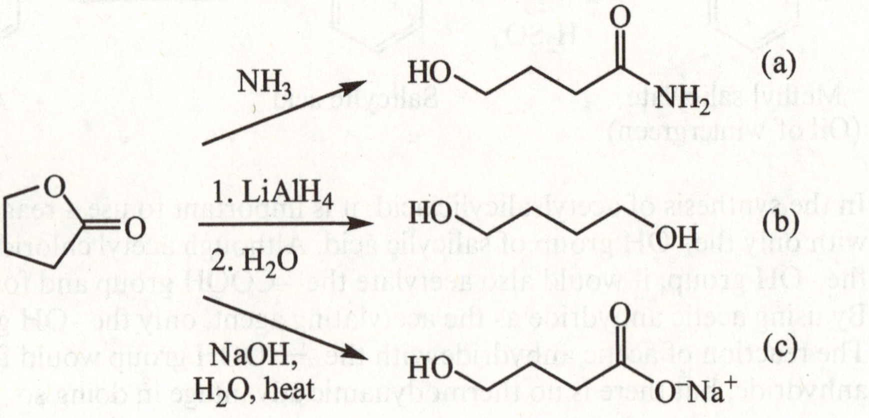

14.32 Show the product of treating *N*-methyl-γ-butyrolactam with each reagent:

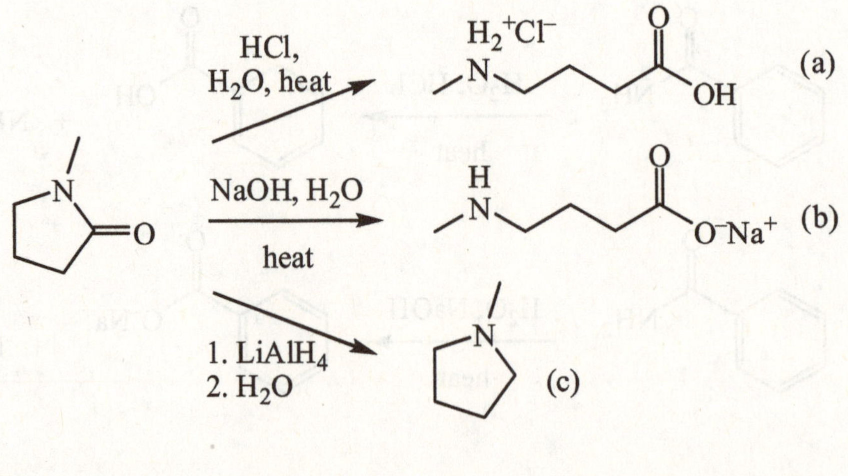

14.33 Complete these reactions:

(a)
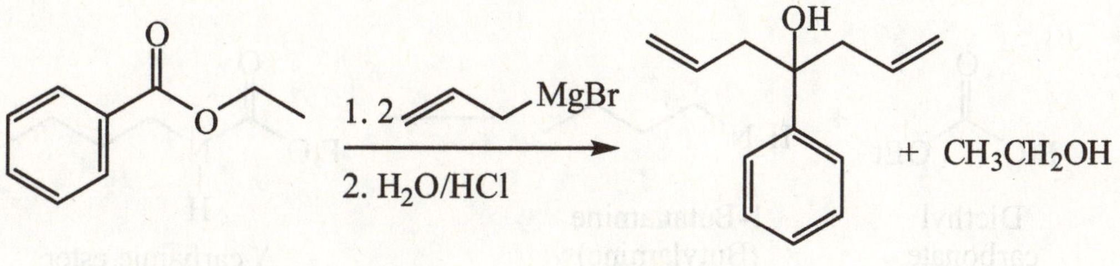
1. 2 MgBr
2. H_2O/HCl
OH
+ CH_3CH_2OH

(b)
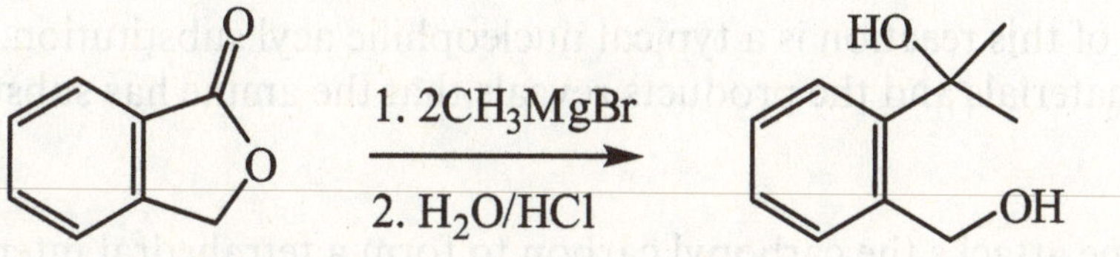
1. $2CH_3MgBr$
2. H_2O/HCl
HO
OH

(c)
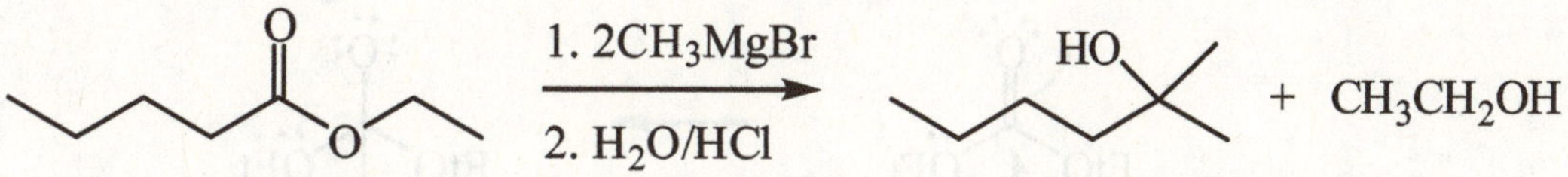
1. $2CH_3MgBr$
2. H_2O/HCl
HO
+ CH_3CH_2OH

14.34 What combination of ester and Grignard reagent can be used to prepare each alcohol?

In each reaction, the alcohol component of the ester (the alkoxy group) can be variable. The alkoxy group is released as the alcohol during the reaction. For example, if an ethyl ester is used, ethanol will be generated.

(a)
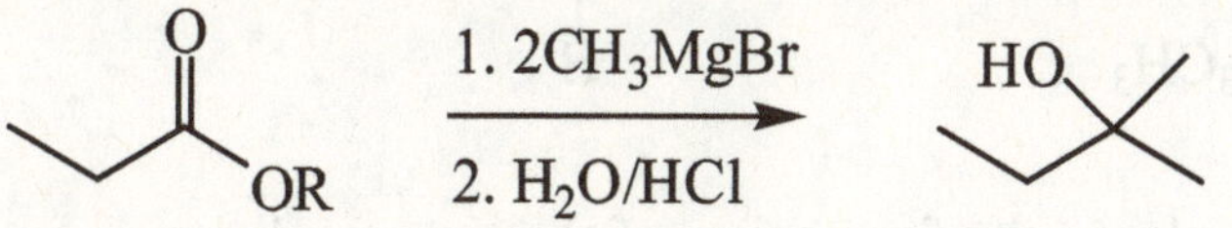
OR
1. $2CH_3MgBr$
2. H_2O/HCl
HO

2-Methyl-2-butanol

(b)
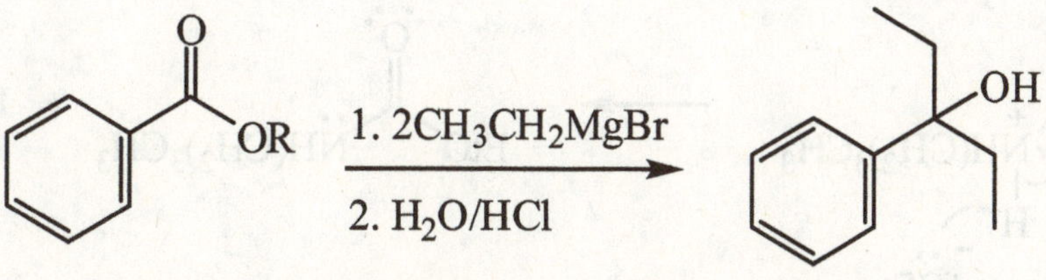
OR
1. $2CH_3CH_2MgBr$
2. H_2O/HCl
OH

3-Phenyl-3-pentanol

(c)
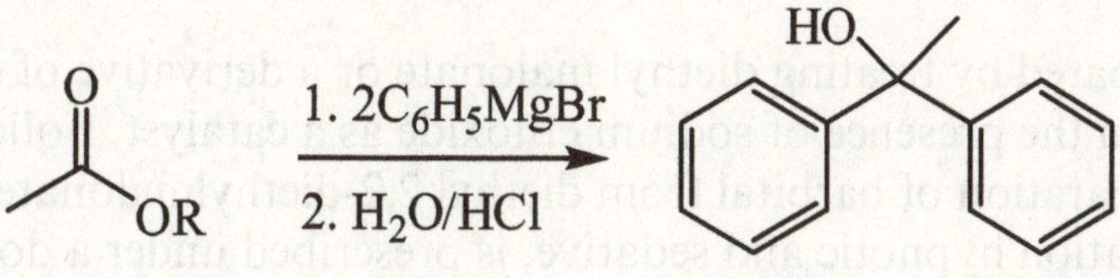
OR
1. $2C_6H_5MgBr$
2. H_2O/HCl
HO

1,1-Diphenylethanol

14.35 Reaction of a 1° or 2° amine with diethyl carbonate under controlled conditions gives a carbamic ester. Propose a mechanism for this reaction.

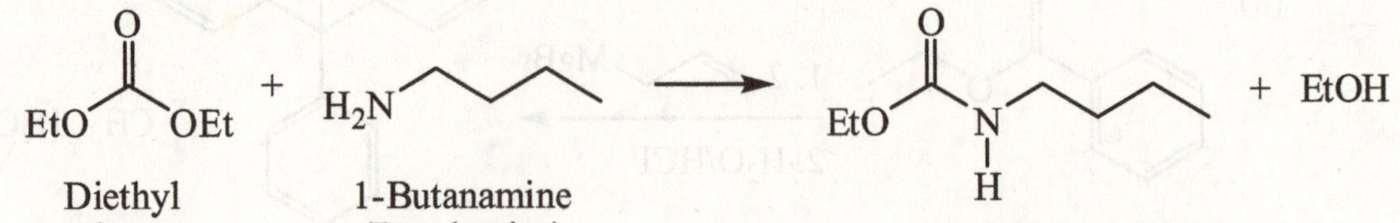

| Diethyl carbonate | 1-Butanamine (Butylamine) | A carbamic ester | + EtOH |

The mechanism of this reaction is a typical nucleophilic acyl substitution. Inspection of the starting materials and the products reveals that the amine has substituted the alcohol.

Step 1: The amine attacks the carbonyl carbon to form a tetrahedral intermediate.

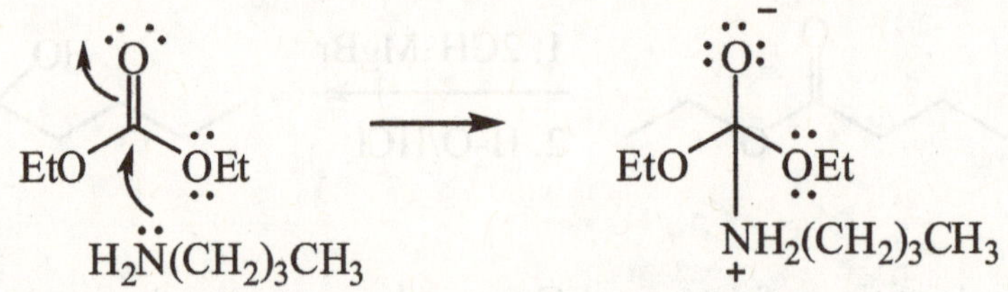

Step 2: Collapse of the tetrahedral intermediate forms a protonated carbamic ester.

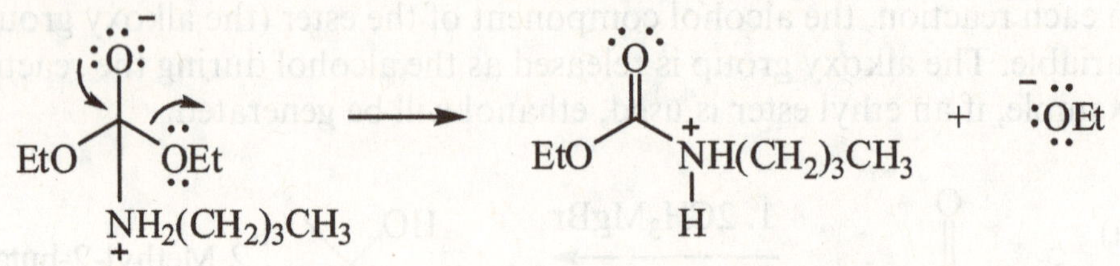

Step 3: Deprotonation results in the formation of the two products.

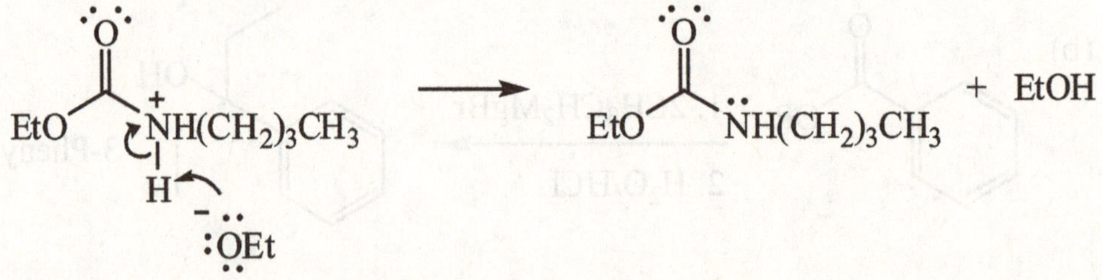

14.36 Barbiturates are prepared by treating diethyl malonate or a derivative of diethyl malonate with urea in the presence of sodium ethoxide as a catalyst. Following is an equation for the preparation of barbital from diethyl 2,2-diethylmalonate and urea (barbital, a long-duration hypnotic and sedative, is prescribed under a dozen or more trade names):

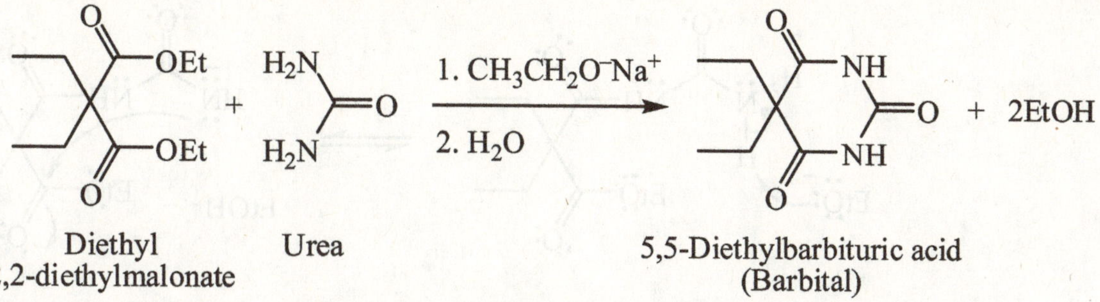

Diethyl   Urea   5,5-Diethylbarbituric acid
2,2-diethylmalonate   (Barbital)

(a) Propose a mechanism for this reaction.

A comparison of the starting materials and the products reveals that the nitrogen
atoms of urea have substituted the two ethoxy groups in diethyl 2,2-diethylmalonate.

Step 1: Under basic conditions, urea is deprotonated to form a good nucleophile.

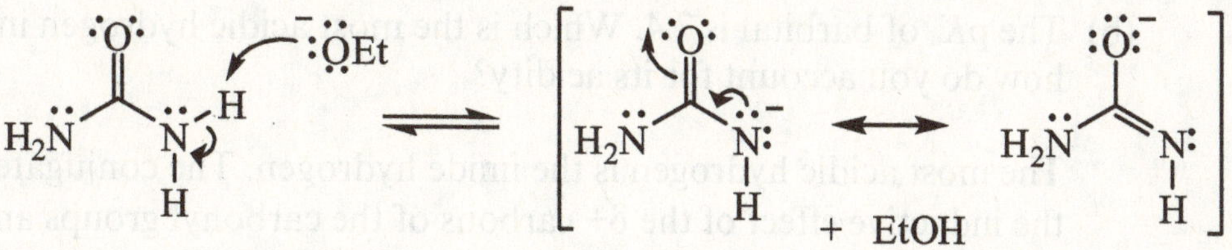

Step 2: The nucleophile attacks diethyl 2,2-diethylmalonate to form a tetrahedral
carbonyl addition intermediate.

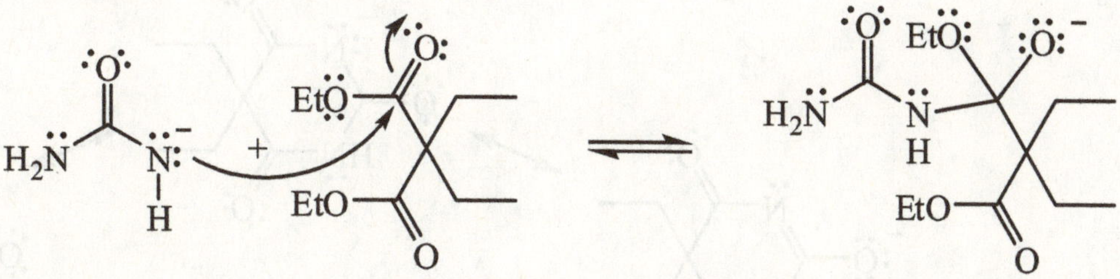

Step 3: Collapse of the tetrahedral intermediate regenerates the carbonyl group
and dispalces ethoxide.

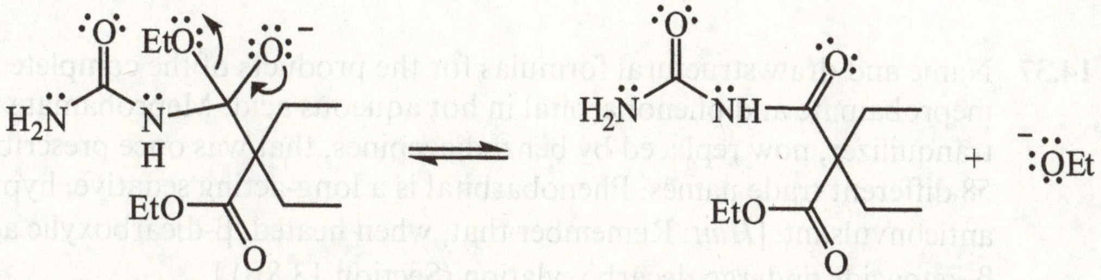

The other nitrogen of the urea portion of molecule then reacts with the remaining
ethyl ester in a manner similar to steps 1 through 3 to give 5,5-diethylbarbituric acid.

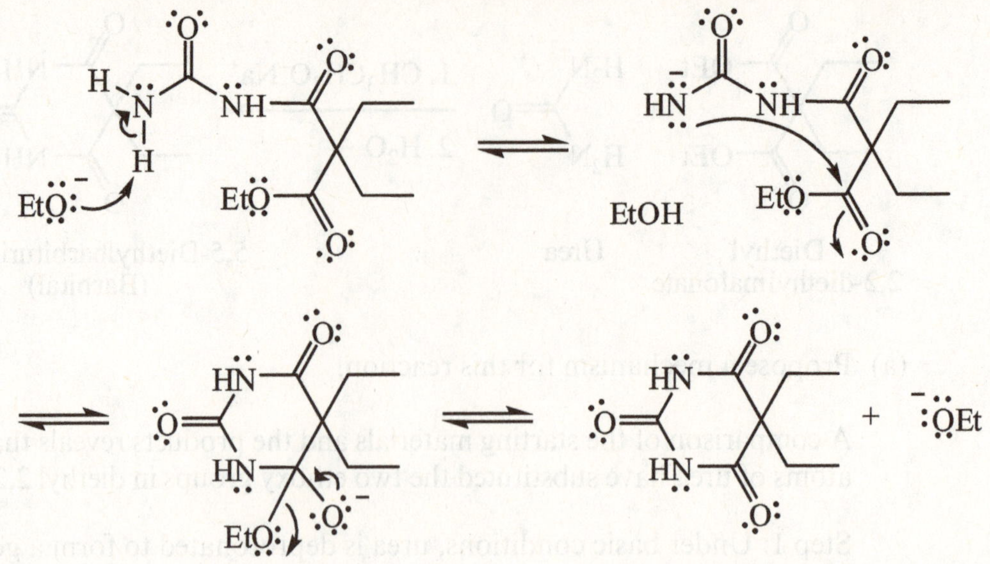

EtOH

EtOH

+ :ÖEt

(b) The pK_a of barbital is 7.4. Which is the most acidic hydrogen in this molecule, and how do you account for its acidity?

The most acidic hydrogen is the imide hydrogen. The conjugate base is stabilized by the inductive effect of the δ+ carbons of the carbonyl groups and by the resonance delocalization of the negative charge to the carbonyl oxygens. Three contributing structures can be drawn for the conjugate base.

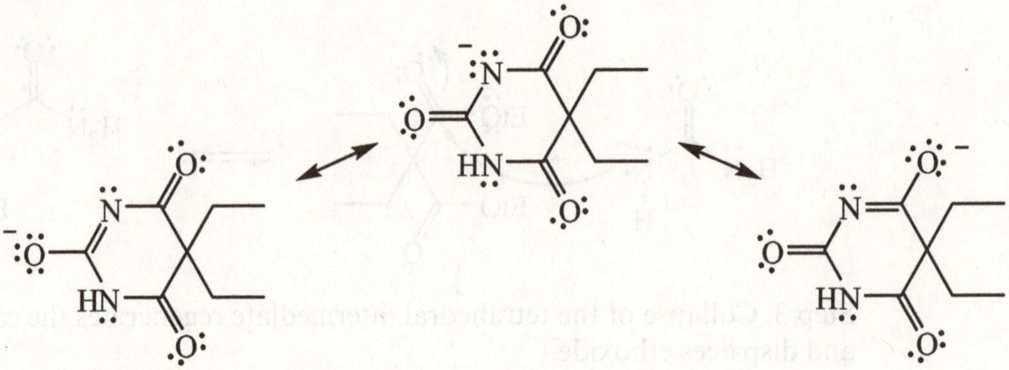

14.37 Name and draw structural formulas for the products of the complete hydrolysis of meprobamate and phenobarbital in hot aqueous acid. Meprobamate is a tranquilizer, now replaced by benzodiazepines, that was once prescribed under 58 different trade names. Phenobarbital is a long-acting sedative, hypnotic, and anticonvulsant. [*Hint*: Remember that, when heated, β-dicarboxylic acids and β-ketoacids undergo decarboxylation (Section 13.8B).]

In both reactions, the hydrolysis reaction forms H_2CO_3, which decomposes to water and CO_2.

(a)

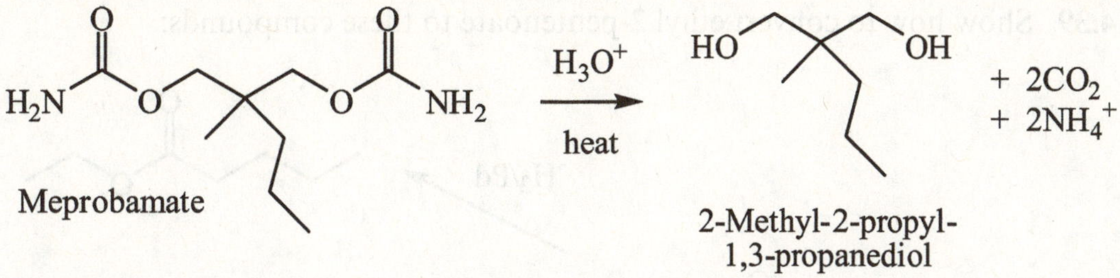
H2N
O
O
O
O
NH2
H3O+
heat
HO
OH
+ 2CO2
+ 2NH4+
Meprobamate
2-Methyl-2-propyl-
1,3-propanediol

(b)

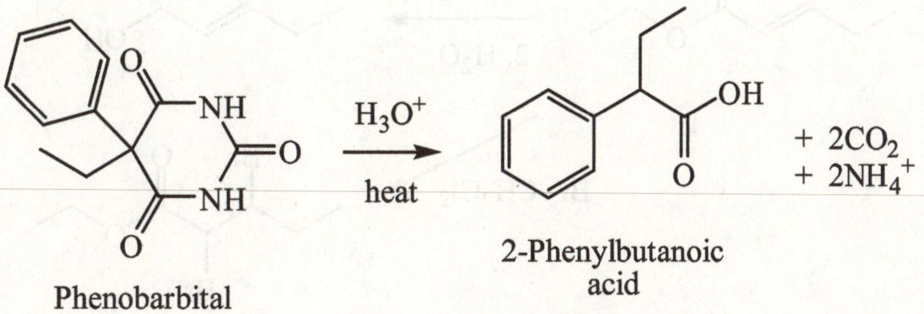
O
NH
O
NH
O
H3O+
heat
OH
O
+ 2CO2
+ 2NH4+
Phenobarbital
2-Phenylbutanoic
acid

Synthesis

14.38 The active ingredient in several common insect repellents *N,N*-Diethyl-*m*-toluamide (Deet) is synthesized from 3-methylbenzoic acid (*m*-toluic acid) and diethylamine. Show how this synthesis can be accomplished.

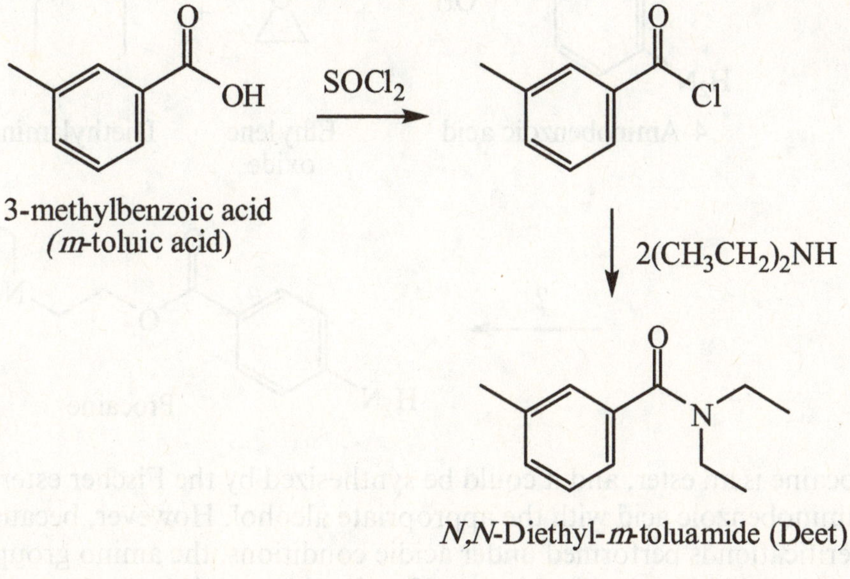
O
OH
SOCl2
O
Cl
3-methylbenzoic acid
(m-toluic acid)
2(CH3CH2)2NH
O
N
N,N-Diethyl-m-toluamide (Deet)

14.39 Show how to convert ethyl 2-pentenoate to these compounds:

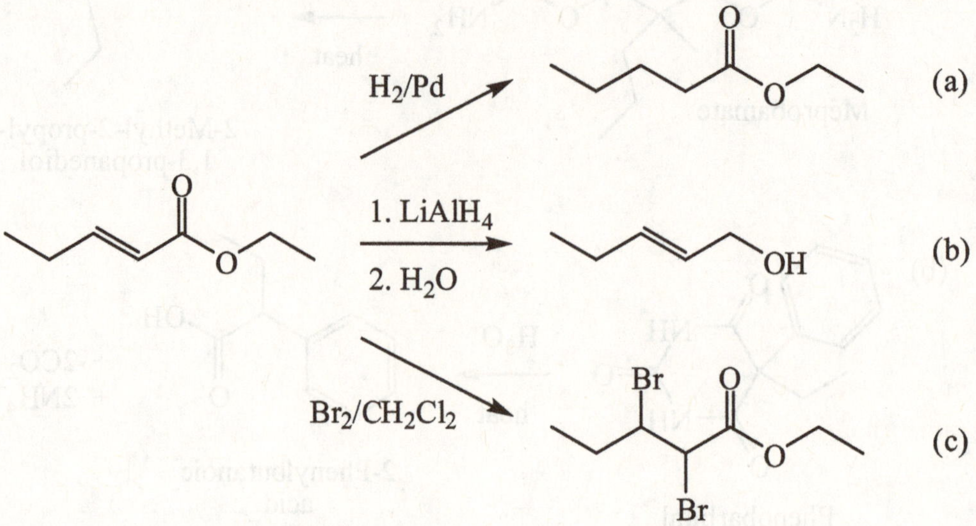

14.40 Procaine (whose hydrochloride is marketed as Novocaine) was one of the first local anesthetics for infiltration and regional anesthesia. Show how to synthesize procaine, using the three reagents shown as the sources of carbon atoms:

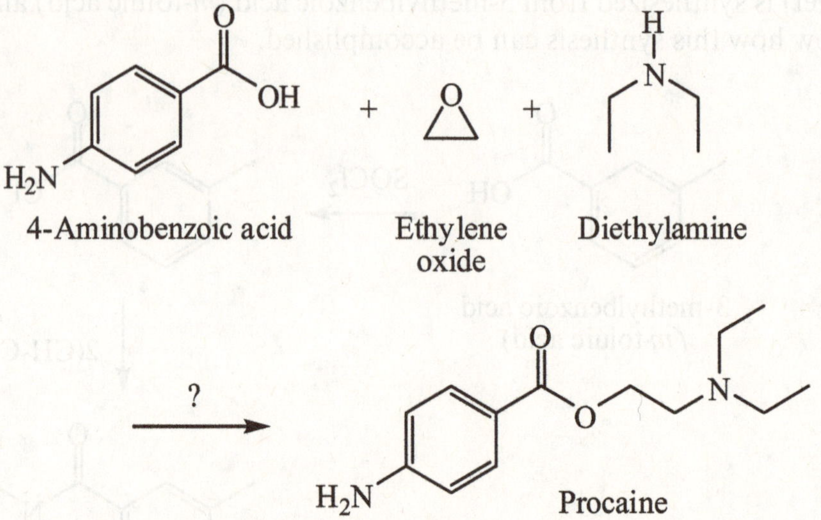

4-Aminobenzoic acid   Ethylene   Diethylamine
                            oxide

Procaine

Procaine is an ester, and it could be synthesized by the Fischer esterification of 4-aminobenzoic acid with the appropriate alcohol. However, because a Fischer esterification is performed under acidic conditions, the amino groups will be protonated. After the Fischer esterification, the product can be treated with a base, such as aqueous NaOH, to generate the free amines.

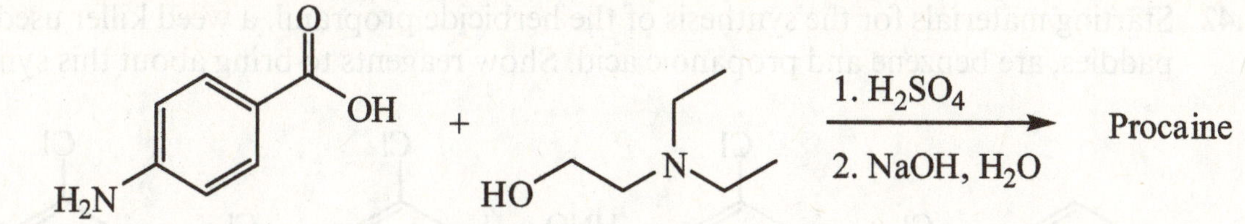

The alcohol used in the Fischer esterification can be synthesized from diethylamine and ethylene oxide.

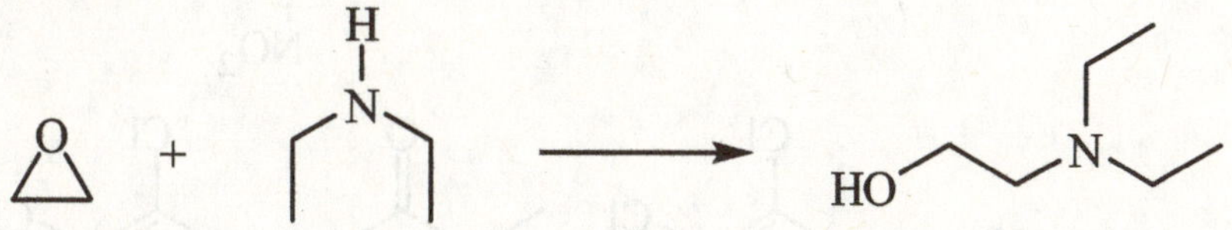

14.41 There are two nitrogen atoms in procaine. Which of the two is the stronger base? Draw the structural formula for the salt that is formed when Procaine is treated with 1 mole of aqueous HCl.

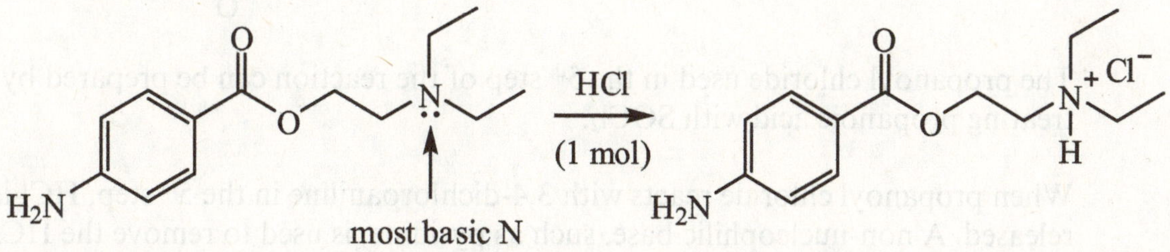

The indicated nitrogen atom, an alkylamine, is more basic than the aniline nitrogen. This is because the lone pair on the aniline nitrogen is stabilized by resonance, which makes the nitrogen atom less basic (recall from Chapter 2 that atoms with lone pairs that are more stable are less basic). When procaine is treated with one equivalent of acid, the strongest base is protonated first.

14.42 Starting materials for the synthesis of the herbicide propranil, a weed killer used in rice paddies, are benzene and propanoic acid. Show reagents to bring about this synthesis:

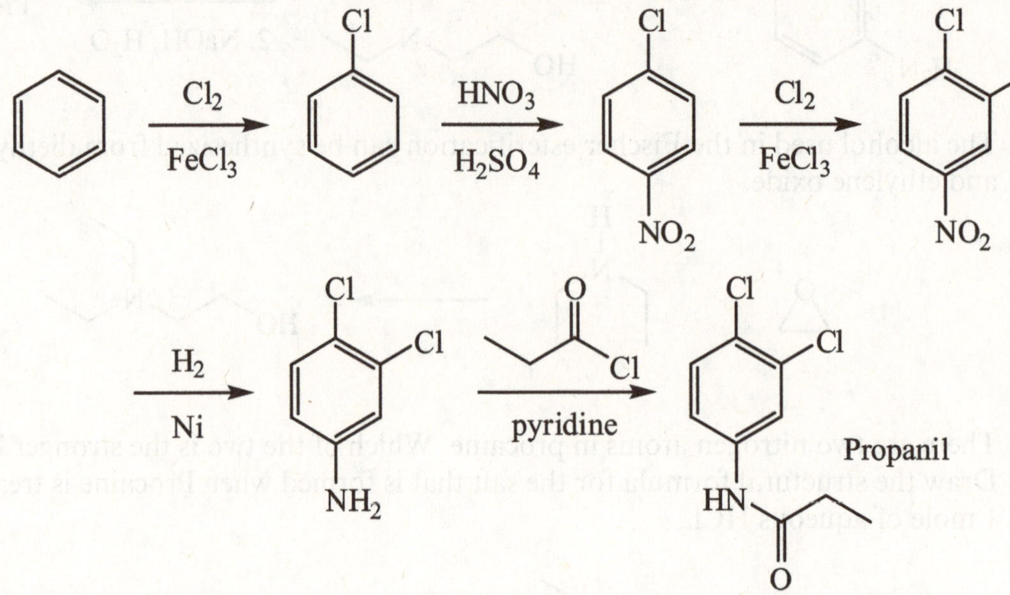

The propanoyl chloride used in the 5th step of the reaction can be prepared by treating propanoic acid with SOCl$_2$.

When propanoyl chloride reacts with 3,4-dichloroaniline in the 5th step, HCl is also released. A non-nucleophilic base, such as pyridine, is used to remove the HCl. If pyridine is not used, two equivalents of 3,4-dichloroaniline would be necessary (one mole for the nucleophilic acyl substitution reaction and the other for the removal of HCl).

14.43 Following are structural formulas for three local anesthetics: Lidocaine was introduced in 1948 and is one of the most widely used local anesthetics for infiltration and regional anesthesia. Its hydrochloride is marketed under the name Xylocaine. Mepivacaine (its hydrochloride is marketed as Carbocaine) is faster and somewhat longer in duration than lidocaine. Articaine is the most widely used local anesthetic in Europe and contains two carboxylic acid derivative groups.

(a) Propose a synthesis of lidocaine from 2,6-dimethylaniline, chloroacetyl chloride (ClCH$_2$COCl), and diethylamine.

The acid chloride group of chloroacetyl chloride is more reactive than the alkyl chloride. As a result, chloroacetyl chloride first reacts by nucleophilic acyl substitution.

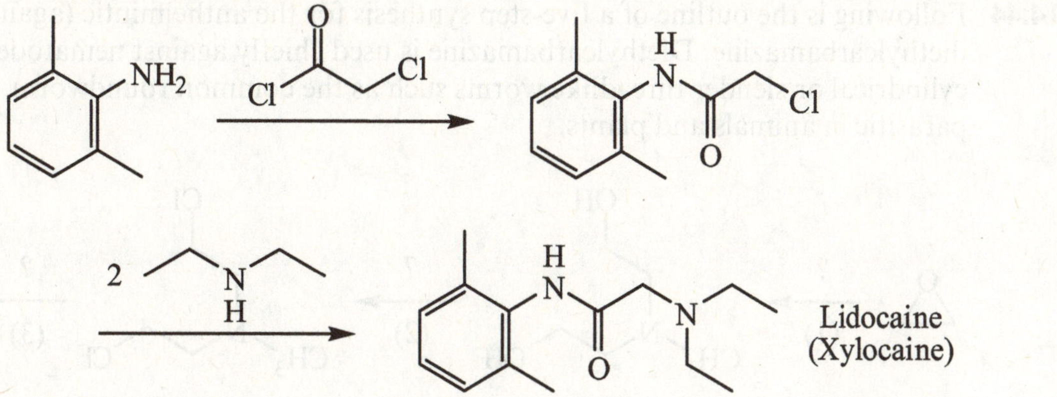

(b) What amine and acid chloride can be reacted to give mepivacaine?

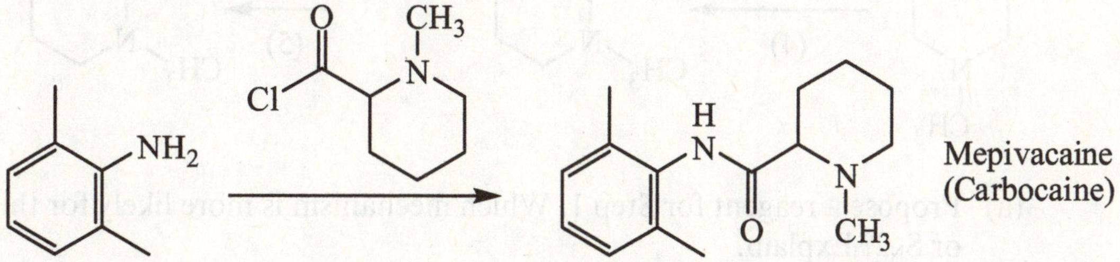

Mepivacaine
(Carbocaine)

(c) Draw the products of acid-catalyzed hydrolysis of both acid derivative groups in articaine.

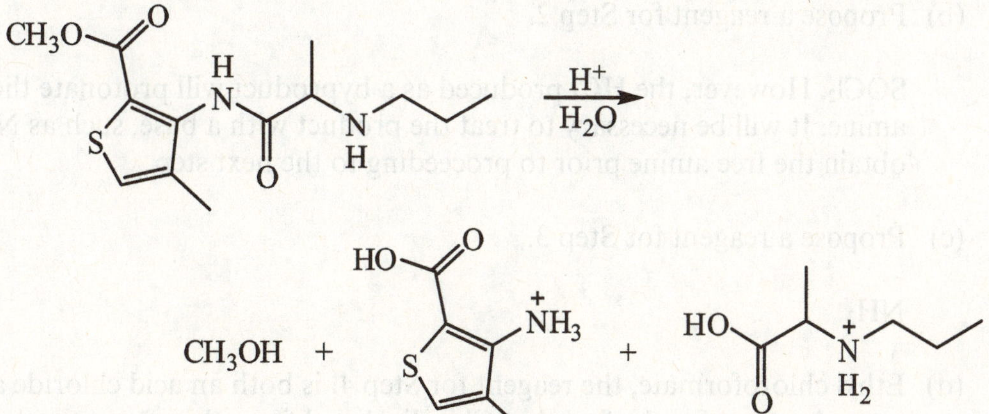

14.44 Following is the outline of a five-step synthesis for the anthelmintic (against worms) diethylcarbamazine. Diethylcarbamazine is used chiefly against nematodes, small cylindrical or slender threadlike worms such as the common roundworm, which are parasitic in animals and plants.

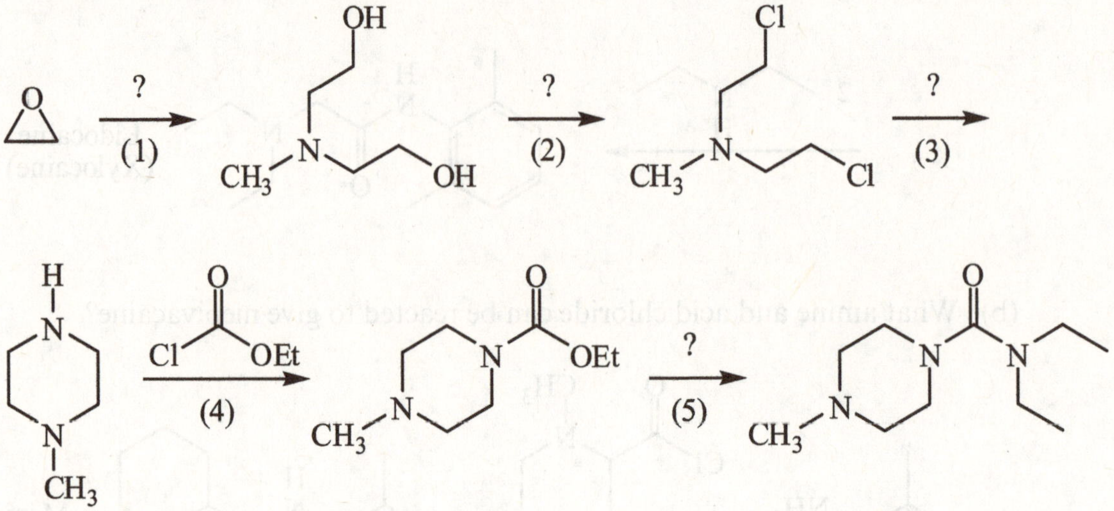

(a) Propose a reagent for Step 1. Which mechanism is more likely for this step; S_N1 or S_N2? Explain.

The reagent for Step 1 is methylamine, CH_3NH_2, which reacts with ethylene oxide by an S_N2 mechanism. Due to the highly strained three-membered ring, ethylene oxide is very reactive towards moderate or good nucleophiles.

(b) Propose a reagent for Step 2.

$SOCl_2$. However, the HCl produced as a byproduct will protonate the basic amine. It will be necessary to treat the product with a base, such as NaOH, to obtain the free amine prior to proceeding to the next step.

(c) Propose a reagent for Step 3.

NH_3

(d) Ethyl chloroformate, the reagent for Step 4, is both an acid chloride and an ester. Account for the fact that Cl is displaced from this reagent rather than OCH_2CH_3.

Chloride is displaced from the reagent because it is a much better leaving group than ethoxide. Recall that leaving groups that are more stable are better leaving groups.

14.45 Following is an outline of a multi-step synthesis for methylparaben, a compound widely used as a preservative in foods. Propose reagents for Steps 1-4.

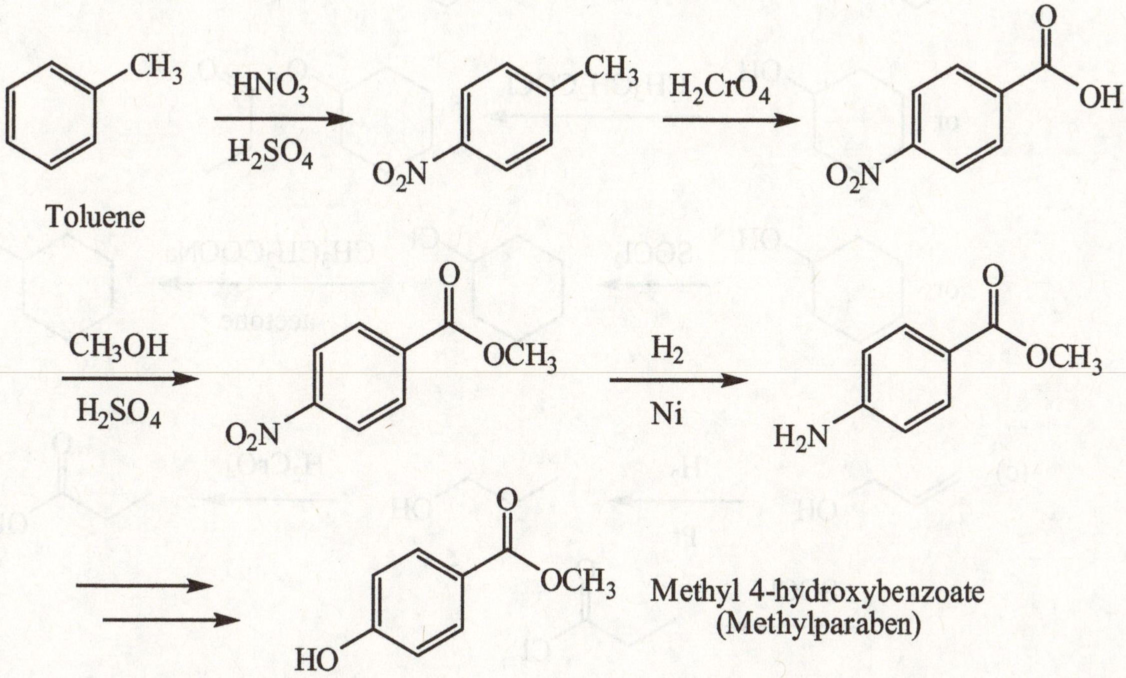

Toluene

Methyl 4-hydroxybenzoate
(Methylparaben)

Chemical Transformations

14.46 Test your cumulative knowledge of the reactions learned thus far by completing the following chemical transformations. *Note: Some will require more than one step.*

(a)

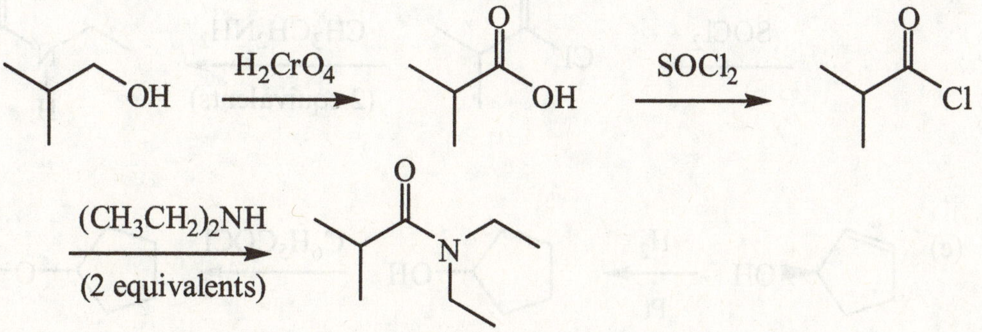

(b)

(c)

(d)

(e)

(f)

(g)

(h)

(i)

(j)

(k)

(l)

(m)

(n)

(o)

(p)

Looking Ahead

14.47 Identify the most acidic proton in each of the following esters:

(a)

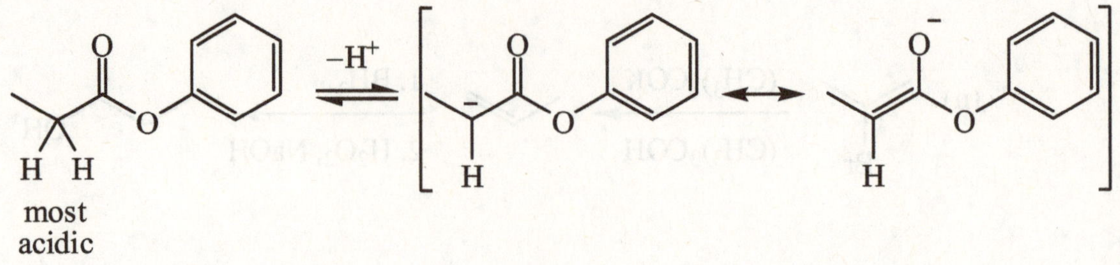

(b)

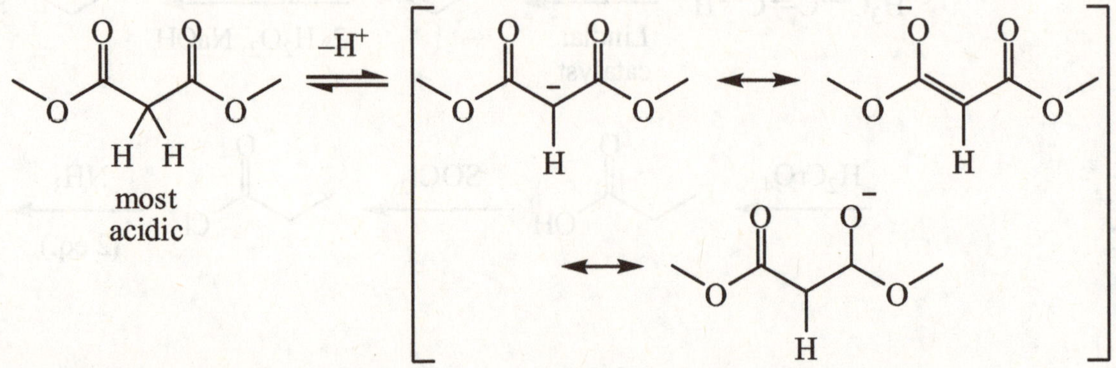

In both compounds, the most acidic hydrogen is the one bonded to the α-carbon (the carbon adjacent to the C=O group). This is because the resulting conjugate base is stabilized by resonance, and the more stable is the conjugate base, the stronger is the acid.

14.48 Does a nucleophilic acyl substitution occur between the ester and the nucleophile shown? Propose an experiment that would verify your answer.

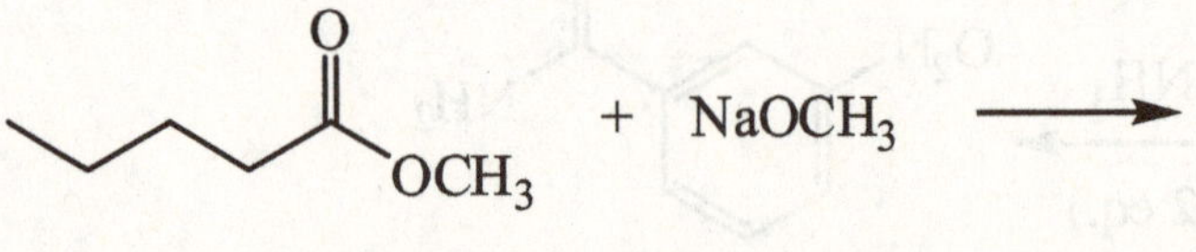

Label the oxygen in sodium methoxide with an isotope such as oxygen-18. If the methoxy group in the ester is replaced by a methoxy group containing oxygen-18, which originated from sodium methoxide, then nucleophilic acyl substitution must have occurred.

14.49 Explain why a nucleophile, Nu, attacks not only the carbonyl carbon, but also the other β-carbon as indicated in the following α,β-unsaturated ester:

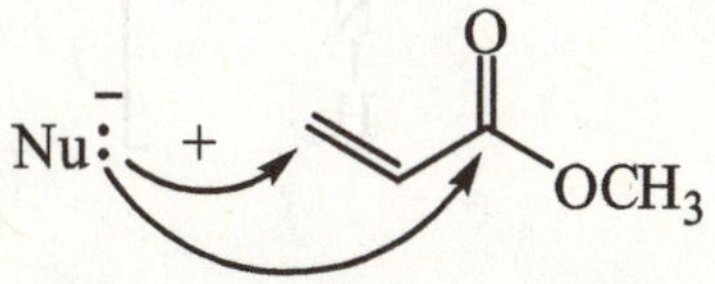

In the α,β-unsaturated ester, the C=C bond is conjugated to the C=O bond. Resonance structures can be drawn to show that the β-carbon is positively charged and therefore electrophilic. Nucleophiles can therefore attack not only the carbonyl carbon, which is positively charged due to the polar carbon-oxygen bond, but also the β carbon.

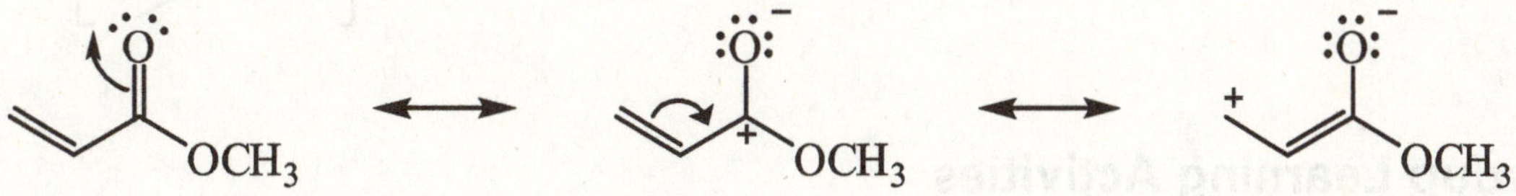

14.50 Explain why a Grignard reagent will not undergo nucleophilic acyl substitution with the following amide:

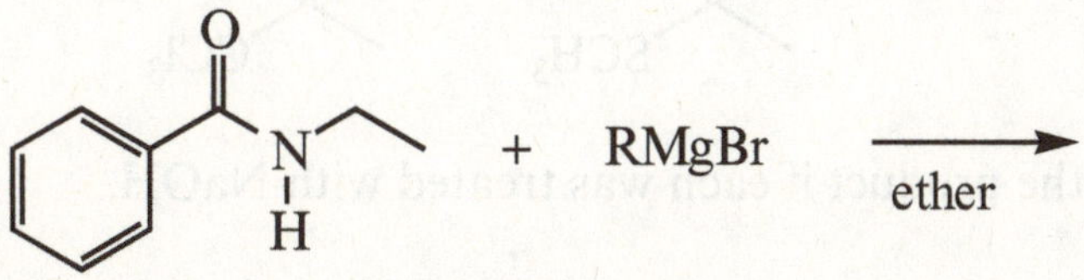

Grignard reagents behave as carbanion equivalents, which are also very strong bases that can be destroyed by acidic protons. The amide proton is sufficiently acidic enough to protonate, and therefore destroy, Grignard reagents.

14.51 At low temperatures, the amide shown below exhibits *cis-trans* isomerism, while at higher temperatures it does not. Explain how this is possible.

Recall that amides are stabilized by resonance and that the best representation of a molecule that is stabilized by resonance is the resonance hybrid. Amides have partial double-bond character between the carbon and nitrogen atoms, and at low temperature, the double bond restricts rotation, giving rise to *cis-trans* isomerism. However, at higher temperatures, there is sufficient energy to overcome the barrier of CN bond rotation.

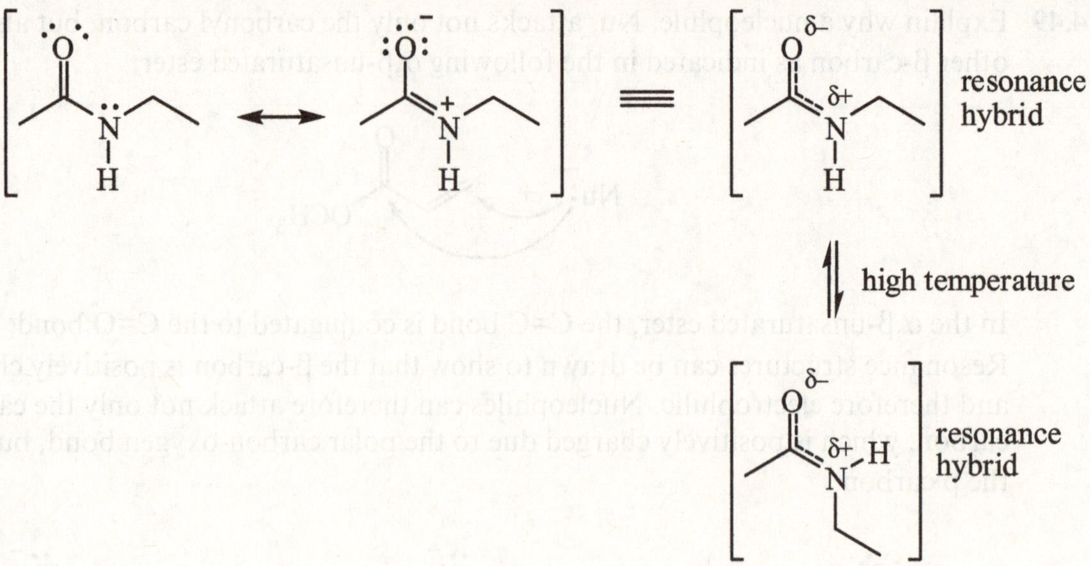

Group Learning Activities

14.52 Following are two compounds that can also undergo nucleophilic acyl substitution. As a group:

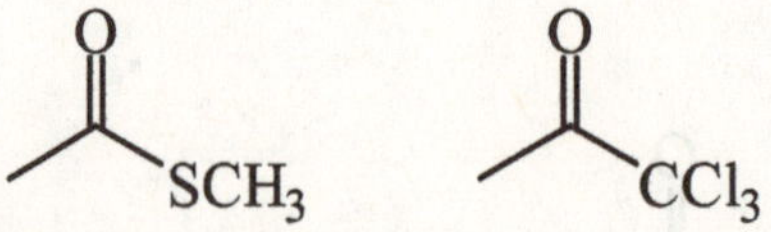

(a) Predict the product if each was treated with NaOH.

> When treated with NaOH, a base, both compounds would form acetate, CH_3COO^-.

(b) Provide a mechanism for each reaction.

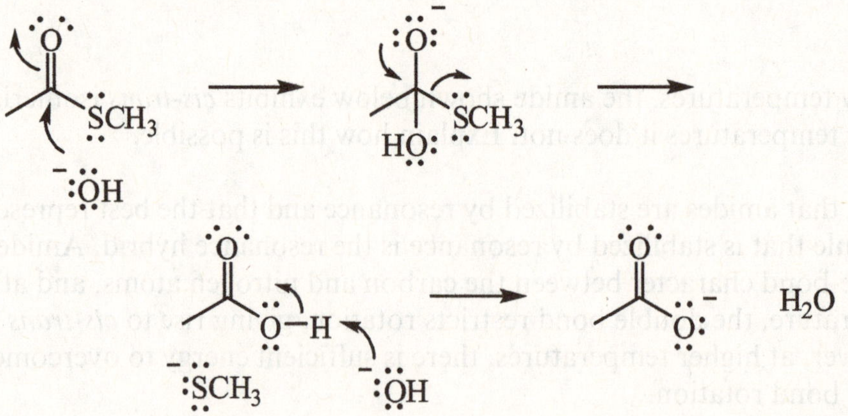

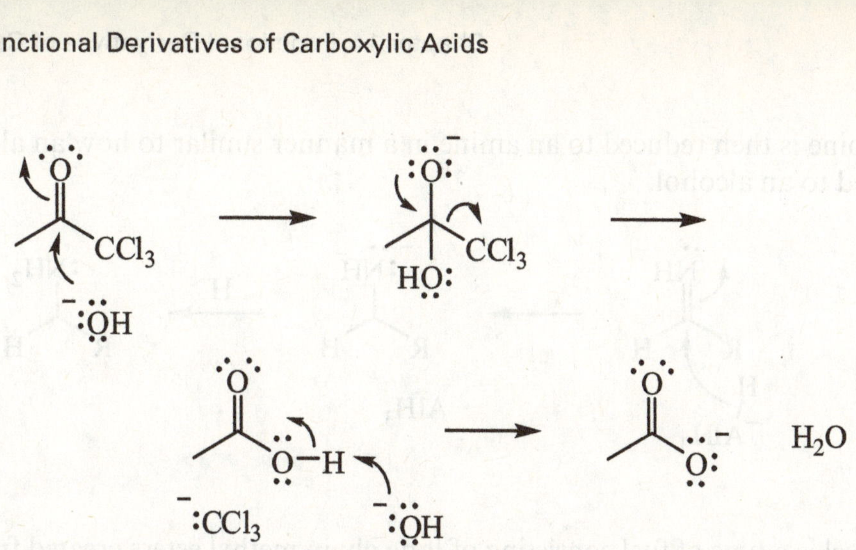

(c) Compare the leaving group ability of ⁻SCH₃ with that of ⁻OCH₃ and that of ⁻CCl₃ with that of ⁻CH₃.

⁻SCH₃ is a better leaving group (more stable) than ⁻OCH₃ because of the larger atomic size of S.

⁻CCl₃ is a better leaving group (more stable) than ⁻CH₃ because of the presence of three inductively withdrawing chlorine atoms.

14.53 The mechanism of the reduction of amides to amines by LiAlH₄ contains many steps. Work as a group to figure out this mechanism. Hint: the carbonyl oxygen is removed as ⁻OAlH₂.

First, the amide is reduced to an imine.

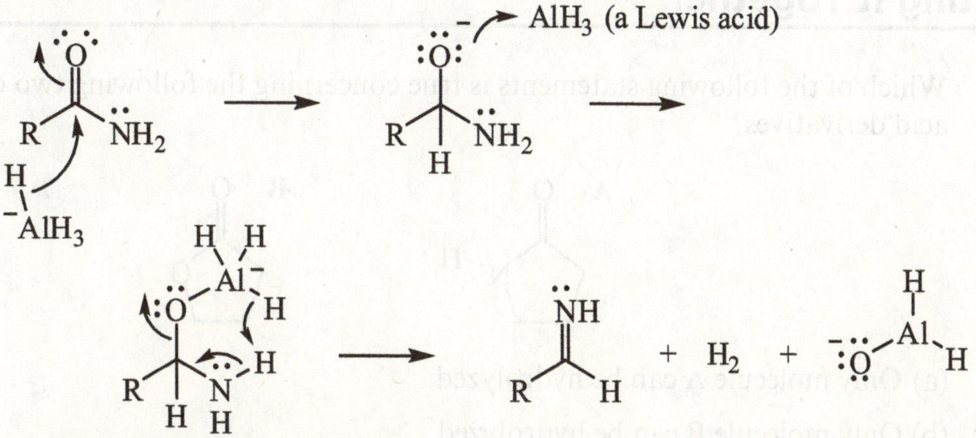

A hydride acts as a base and causes an elimination reaction

The imine is then reduced to an amine in a manner similar to how an aldehyde is reduced to an alcohol.

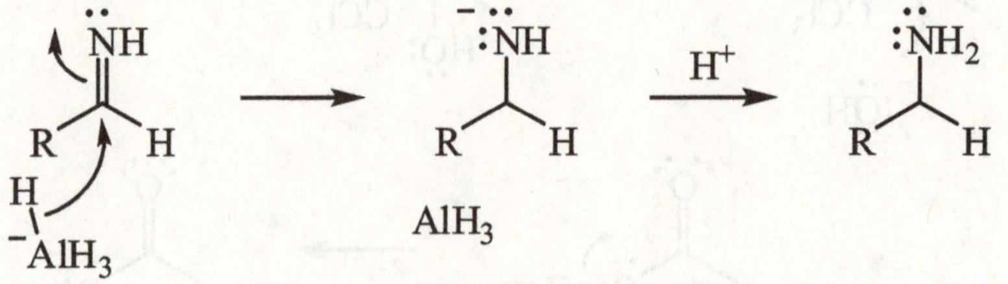

14.54 Biodiesel is a type of fuel consisting of long chain methyl esters created from naturally obtained lipids (*e.g.* vegetable oils), which have the general structure shown below. Propose a reaction or series of reactions that will convert lipids to biodiesel.

The methyl esters of fatty acids can be made by a transesterification reaction using an excess of methanol. The alcohol component (glycerol) of the lipids is replaced by methanol.

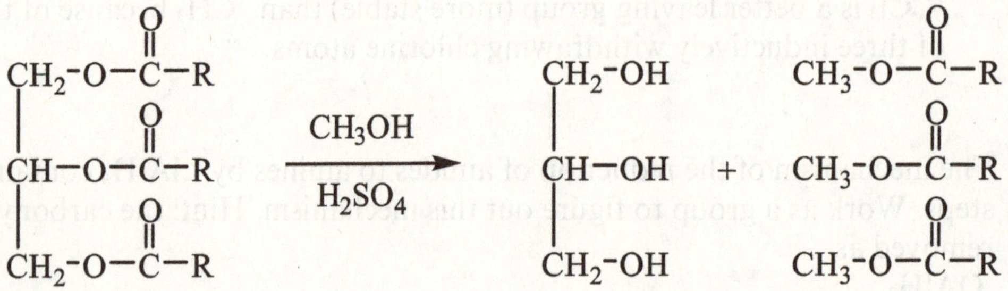

Putting It Together

1.   Which of the following statements is true concerning the following two carboxylic acid derivatives?

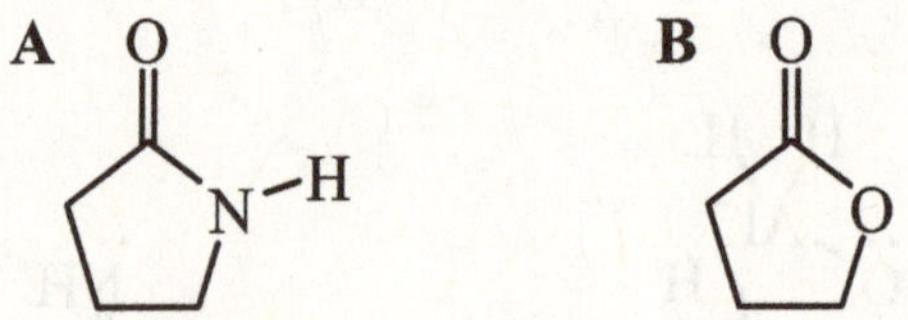

(a) Only molecule **A** can be hydrolyzed.

(b) Only molecule **B** can be hydrolyzed.

(c) Both molecules are hydrolyzable, but **A** will react more quickly than **B**.

(d) Both molecules are hydrolyzable, but **B** will react more quickly than **A**.

(e) **A** and **B** are hydrolyzed at roughly the same rate.

All acid derivatives, including amides and esters, can be hydrolyzed. Esters are more reactive than amides, so **B** is hydrolyzed more quickly.

2.   How many unique reaction products are formed from the following reaction?

$$CH_3CH_2-\overset{\displaystyle O}{\overset{\displaystyle \|}{C}}-CH_2CH_2-\overset{\displaystyle O}{\overset{\displaystyle \|}{C}}-OCH_3 \quad \xrightarrow[H_2O]{H^+}$$

(a) one   **(b) two**   (c) three   (d) four   (e) five

In this reaction, the ester is hydrolyzed. Methanol and 4-oxohexanoic acid are formed.

3.   What sequence of reagents will accomplish the following transformation?

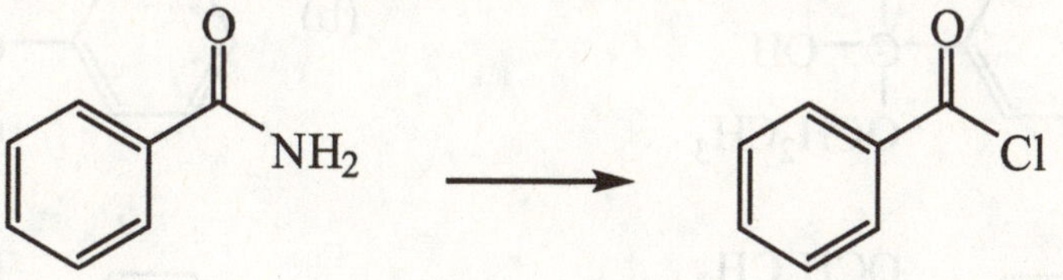

(a) 1. SOCl₂

(b) 1. H₂O₂, SOCl₂

(c) 1. H₂O₂, HCl

(d) 1. H⁺ / H₂O   2. SOCl₂

(e) All of the above

To convert an amide to an acid chloride, the amide must first be hydrolyzed into a carboxylic acid. The carboxylic acid can then be treated with SOCl₂.

4.   Which of the following reactions will not yield butanamide? Select all that apply.

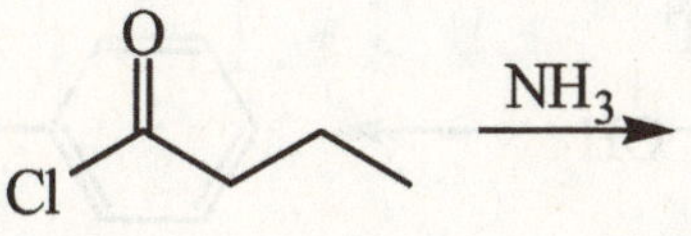

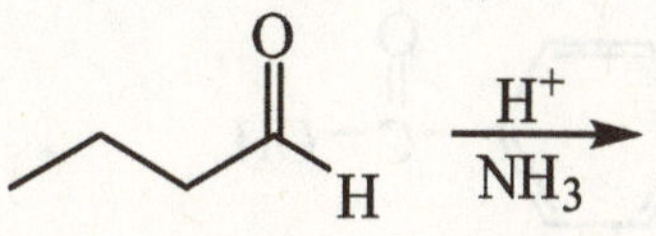

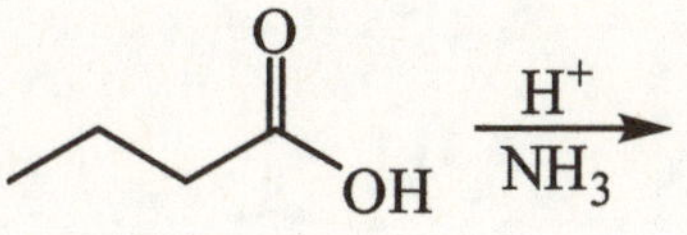

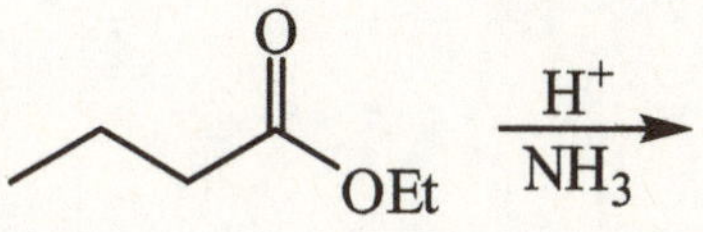

(e)

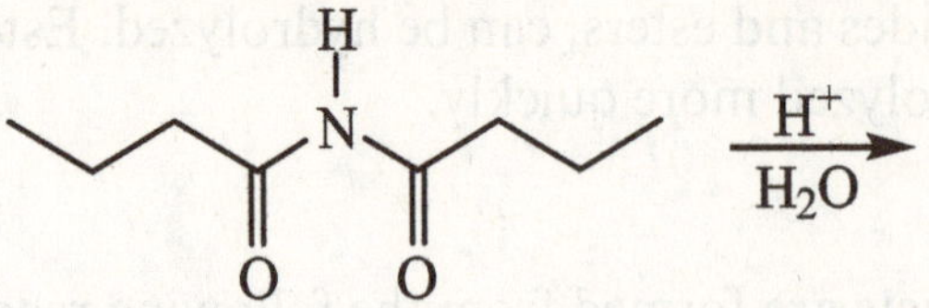

$$\xrightarrow[\text{H}_2\text{O}]{\text{H}^+}$$

Reactions (b), (c), and (e) will not yield butanamide. Reaction (b) involves the reaction of an aldehyde with ammonia, which leads to the formation of an imine. In (c), the combination of a carboxylic acid and ammonia would lead to an acid-base reaction that forms ammonium butanoate (under modest conditions, it is not possible to prepare amides directly from carboxylic acids and amines). The imide in reaction (e) would be hydrolyzed into ammonia and two equivalents of butanoic acid.

5.   Which of the following is the tetrahedral carbonyl addition intermediate (TCAI) for the Fischer esterification of ethanol and benzoic acid?

(a)

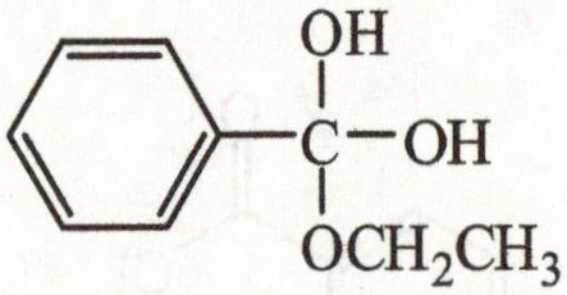

(b)

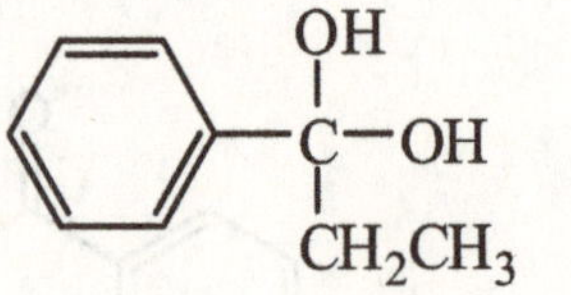

(c)

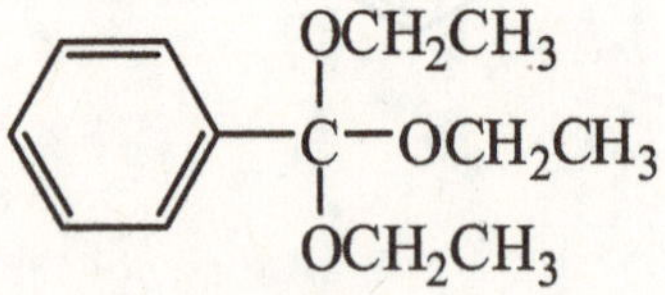

(d)

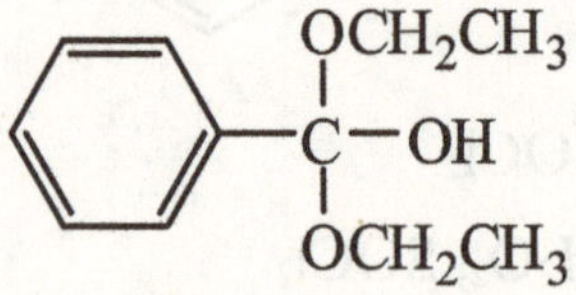

(e)

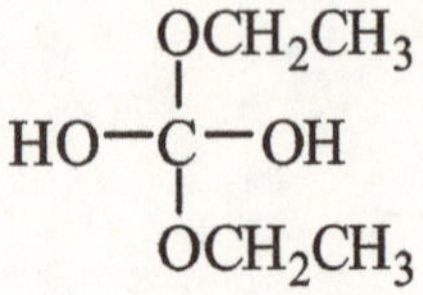

A good strategy for this type of problem is to work through the mechanism. In a Fischer esterification, the carboxylic acid is first protonated by the acid catalyst. Subsequent nucleophilic attack by the alcohol forms a TCAI, which can be deprotonated to give the structure shown in (a).

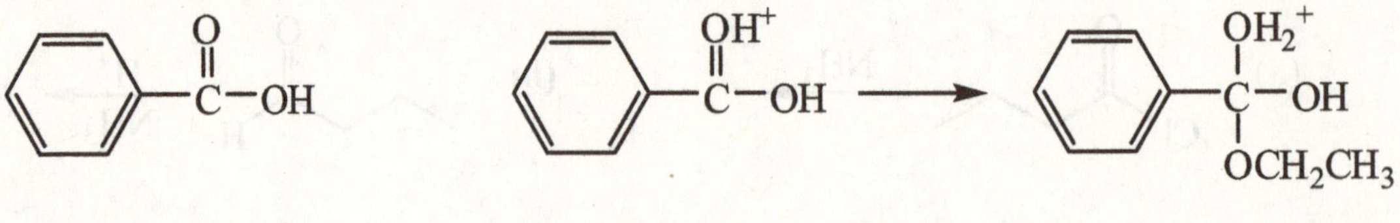

6. Which of the following is the enol intermediate in the decarboxylation of ethylpropanedioic acid?

(a) 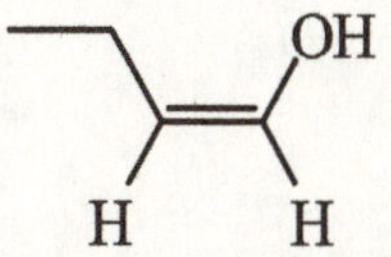   (b) 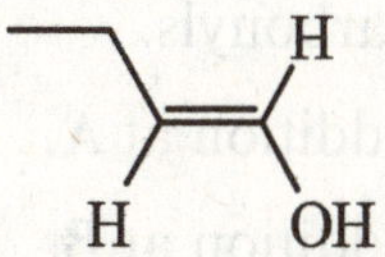   (c)

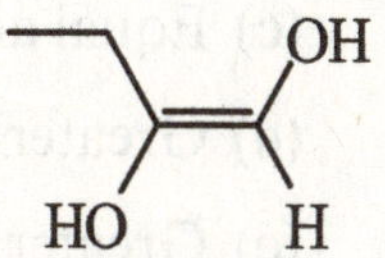

(d) 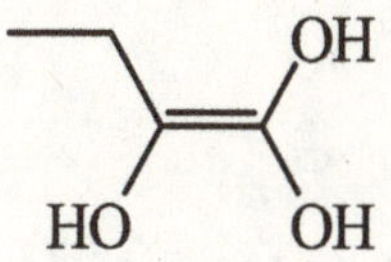   (e)

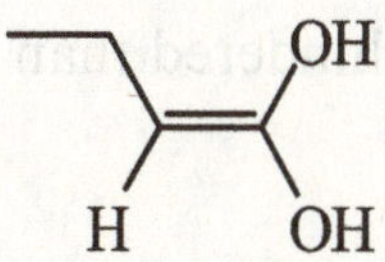

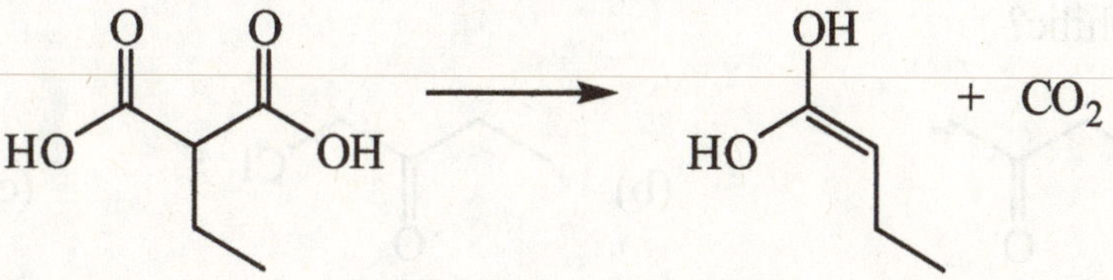

7. Which of the following statements is true concerning the two carboxylic acids shown?

A   B

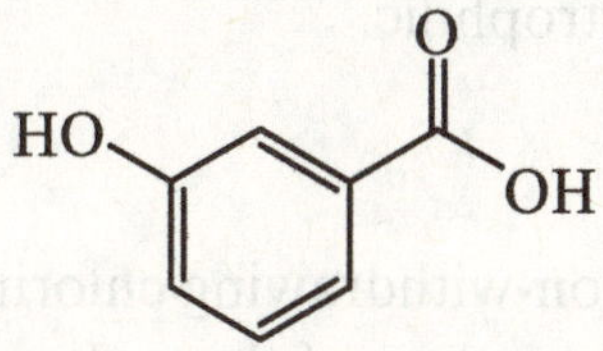

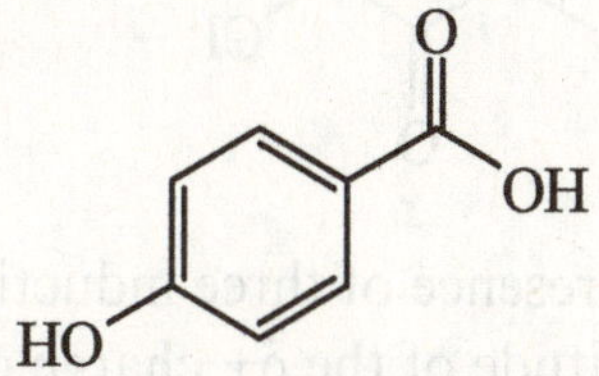

(a) **A** is more acidic than **B** because of an additional resonance effect.

(b) **B** is more acidic than **A** because of an additional resonance effect.

(c) Only the conjugate base of **A** experiences an inductive effect.

(d) Only the conjugate base of **B** experiences an inductive effect.

(e) None of the above.

Both the conjugate bases of **A** and **B** experience the inductive effect of the phenolic – OH.

8. What would be the expected outcome if one equivalent of a Grignard reagent were reacted with the molecule below?

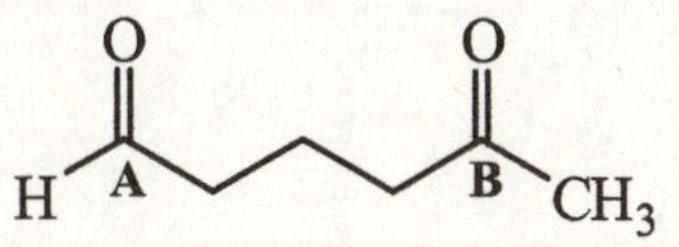

(a) 100% addition at carbonyl **A**.

(b) 100% addition at carbonyl **B**.

(c) Equal addition at both carbonyls.

(d) Greater distribution of addition at **A**.

(e) Greater distribution of addition at **B**.

Carbonyl **A** is less sterically hindered than **B**.

9. Which of the following carbonyl carbons would be considered the most electrophilic?

(a)   (b)   (c)

(d) 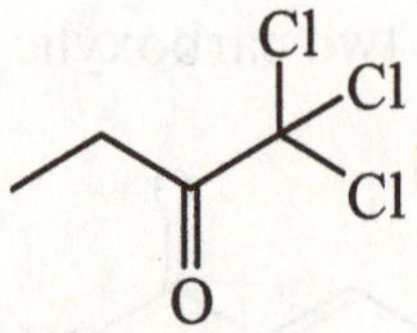   (e) All are equally
                              electrophilic.

The presence of three inductively electron-withdrawing chlorines in (d) increases the magnitude of the $\delta+$ charge on the carbon atom of the carbonyl group.

10. The following reaction will occur as shown:

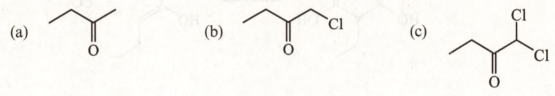

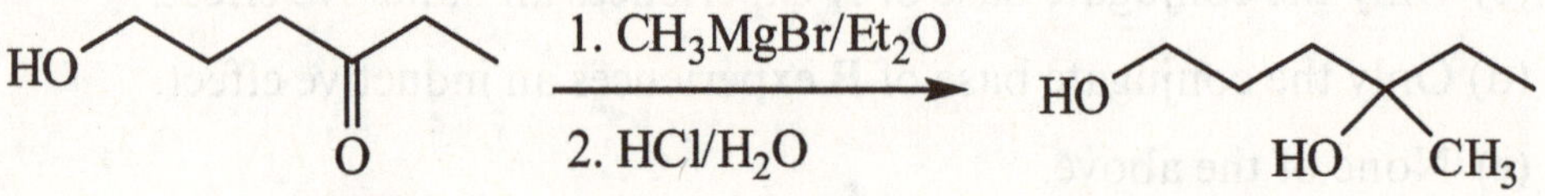

1. CH_3MgBr/Et_2O

2. HCl/H_2O

(a) True   (b) False

Grignard reagents are highly basic and will be destroyed by the alcohol in the reactant.

11. Provide a structure for the starting compound needed to produce the product shown. Then show the mechanism of its formation. Show all charges and lone pairs of electrons in your structures.

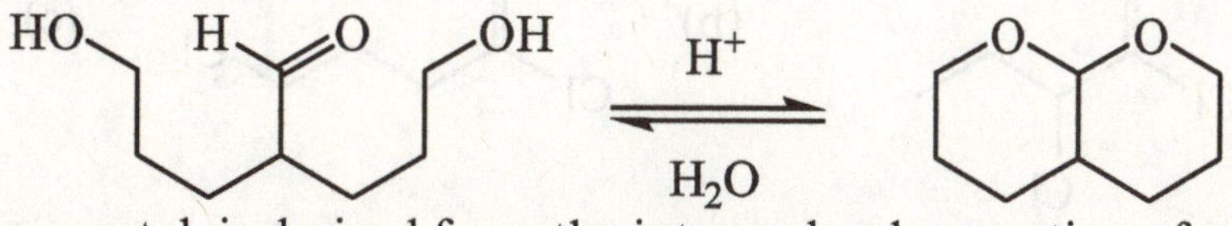

The product, an acetal, is derived from the intramolecular reaction of an aldehyde with and two alcohol groups. The oxonium ions in steps 3 and 7 are deprotonated by H_2O.

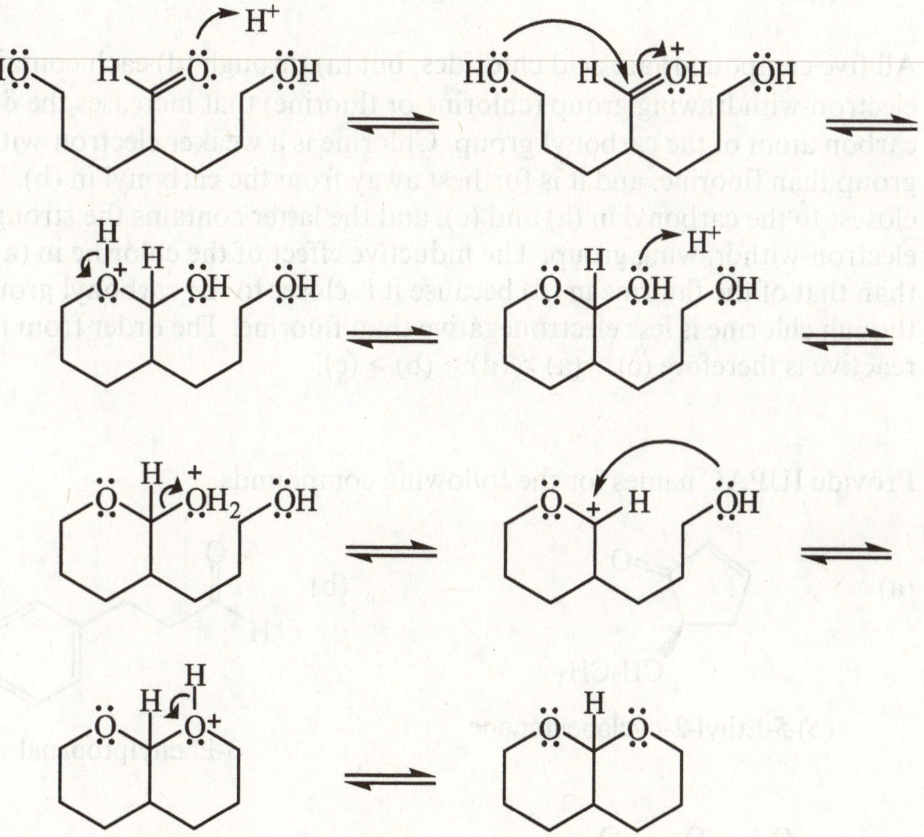

12.   Rank the following from most to least reactive with EtOH. Provide a rationale for your ranking.

(a)

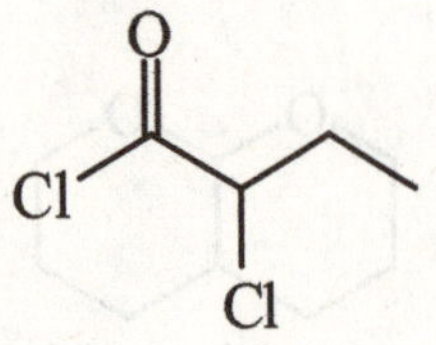

(b)

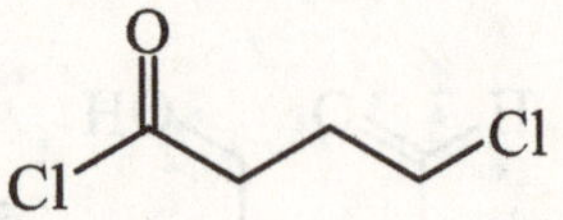

(c)

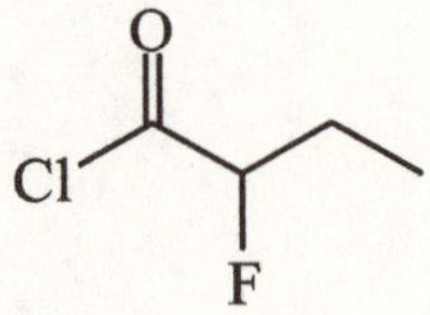

(d)

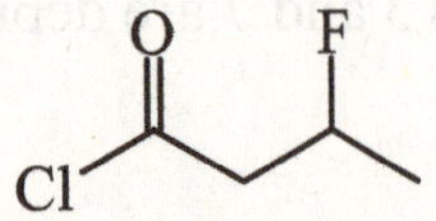

(e)

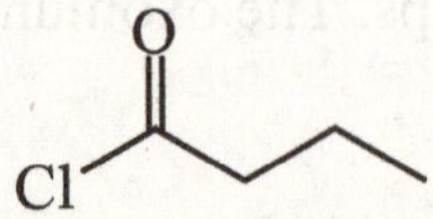

All five compounds are acid chlorides, but (a) through (d) each contain an electron-withdrawing group (chlorine or fluorine) that increases the $\delta+$ charge of the carbon atom of the carbonyl group. Chlorine is a weaker electron-withdrawing group than fluorine, and it is furthest away from the carbonyl in (b). The groups are closest to the carbonyl in (a) and (c), and the latter contains the strongest electron-withdrawing group. The inductive effect of the chlorine in (a) is greater than that of the fluorine in (d) because it is closer to the carbonyl group, even though chlorine is less electronegative than fluorine. The order from most to least reactive is therefore (c) > (a) > (d) > (b) > (e).

13.   Provide IUPAC names for the following compounds.

(a)

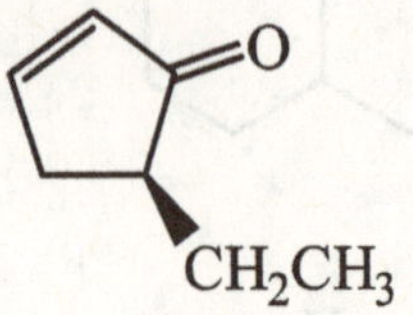

CH₂CH₃

(S)-5-Ethyl-2-cyclopentenone

(b)

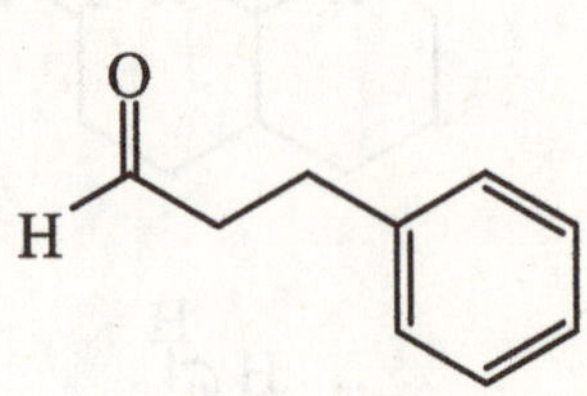

H

3-Phenylpropanal

(c)

H   OH

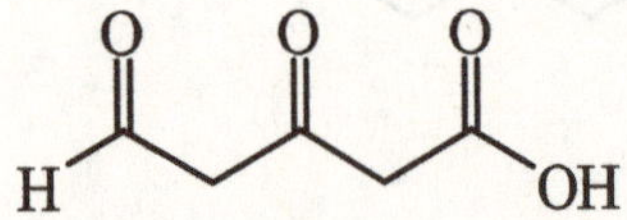

3,5-Dioxopentanoic acid

14.   Provide a mechanism for the following reaction. Show all charges and lone pairs of electrons in your structures, as well as the structures of all intermediates.

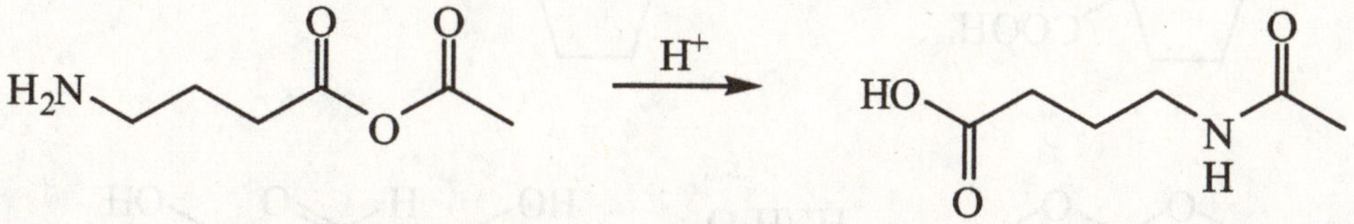

In this reaction, an intramolecular nucleophilic acyl substitution has occurred at the anhydride carbon furthest to the right. The oxonium ions in steps 3 and 6 are deprotonated by H_2O.

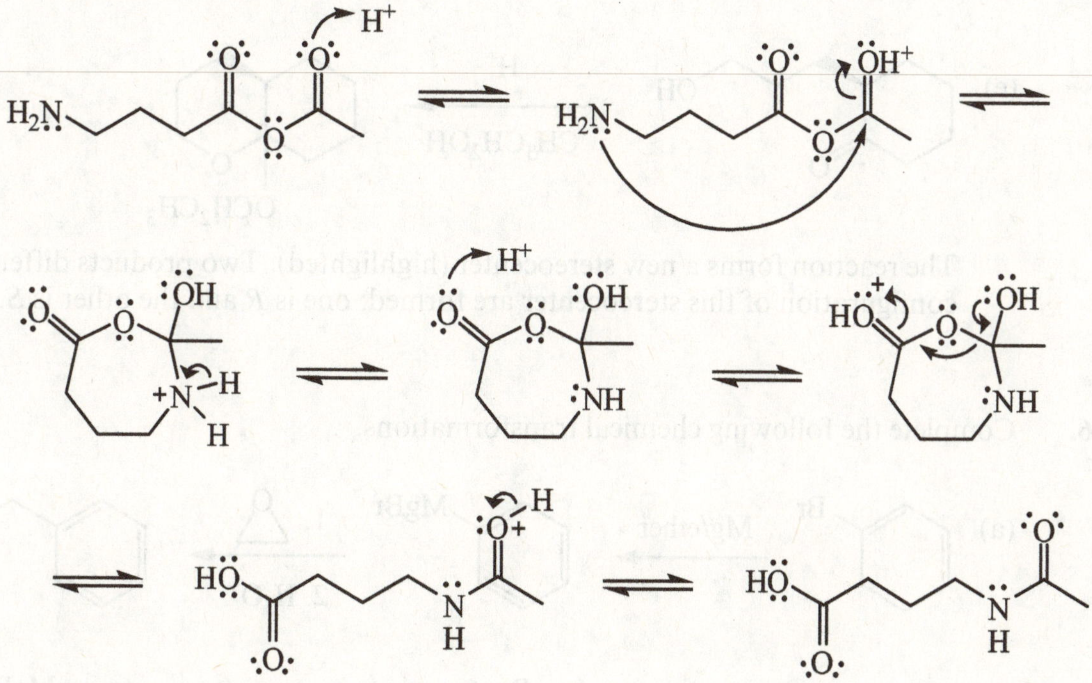

15.   Predict the major product of each of the following reactions.

(a)

(b)

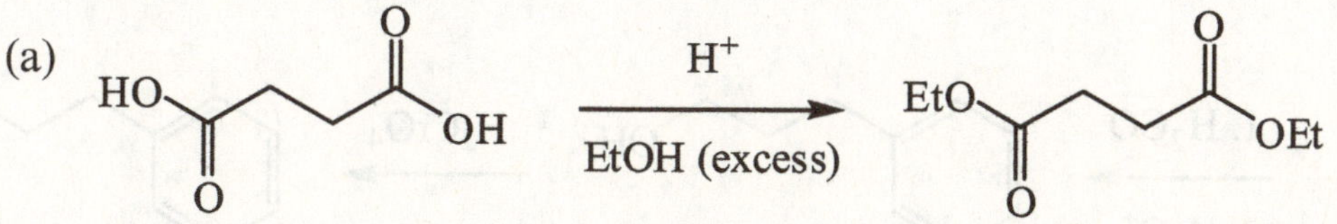

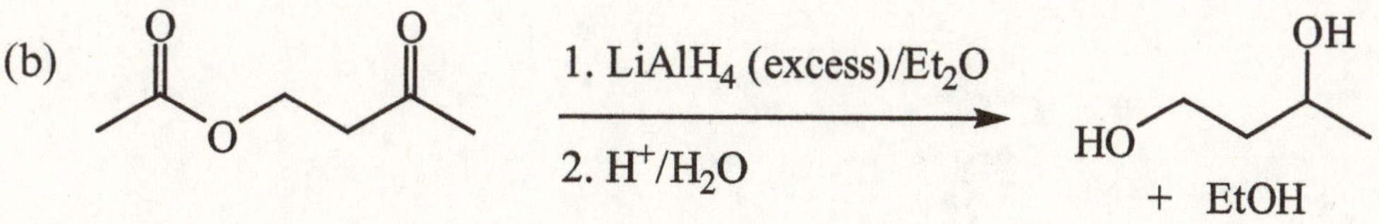

(c)

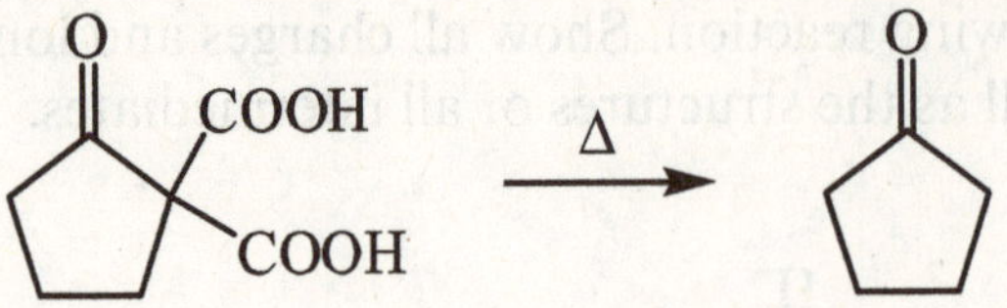

(d)

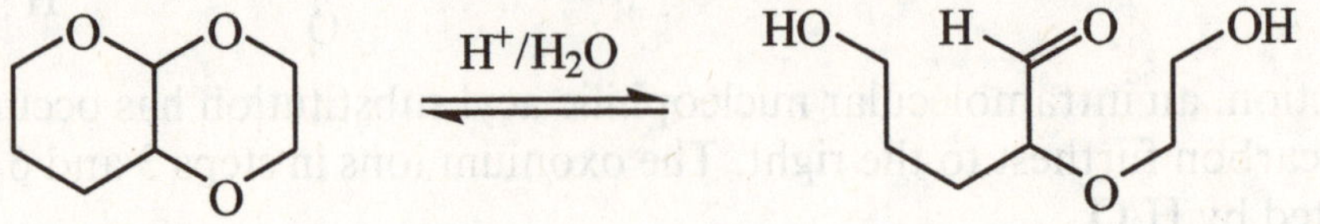

$$H^+/H_2O$$

(e)

$$\xrightarrow[CH_3CH_2OH]{H^+}$$

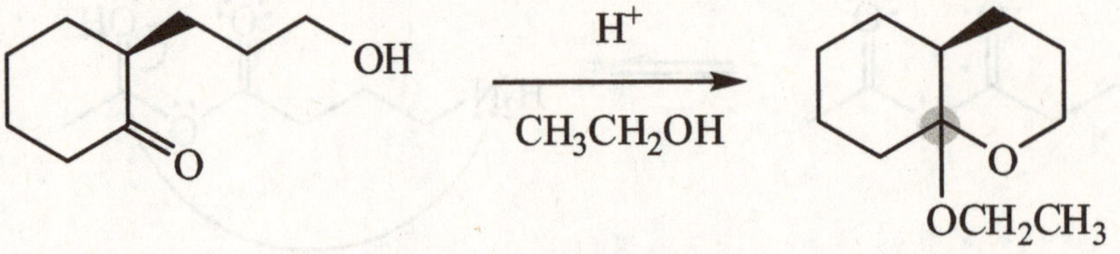

The reaction forms a new stereocenter (highlighted). Two products differing in the configuration of this stereocenter are formed; one is R and the other is S.

16.   Complete the following chemical transformations.

(a)

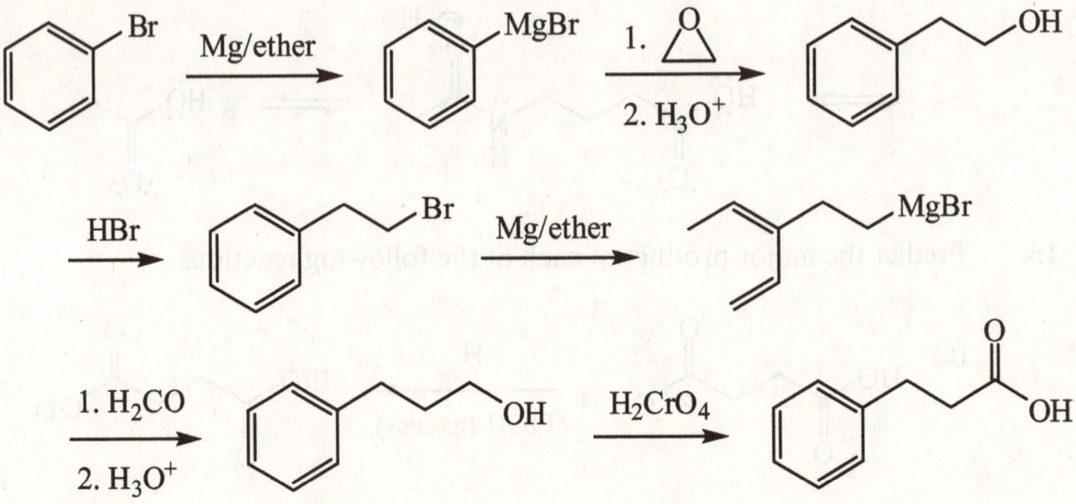

$$\xrightarrow{Mg/ether}$$

$$\xrightarrow[\text{2. } H_3O^+]{\text{1. } \triangle}$$

$$\xrightarrow{HBr}$$

$$\xrightarrow{Mg/ether}$$

$$\xrightarrow[\text{2. } H_3O^+]{\text{1. } H_2CO}$$

$$\xrightarrow{H_2CrO_4}$$

(b)

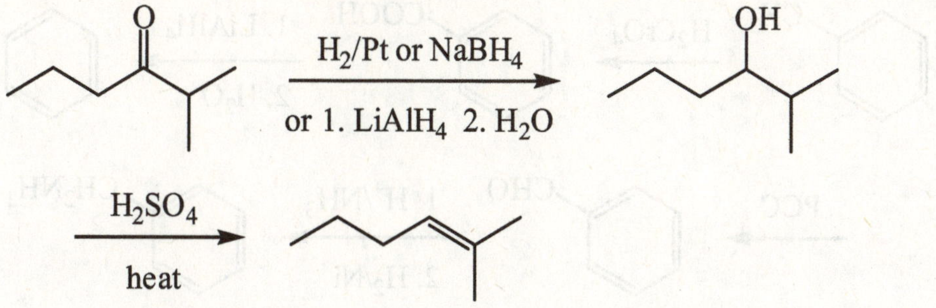

$$\text{ketone} \xrightarrow[\text{or 1. LiAlH}_4 \quad 2.\ H_2O]{H_2/Pt\ \text{or}\ NaBH_4} \text{alcohol}$$

$$\xrightarrow[\text{heat}]{H_2SO_4} \text{alkene}$$

(c)

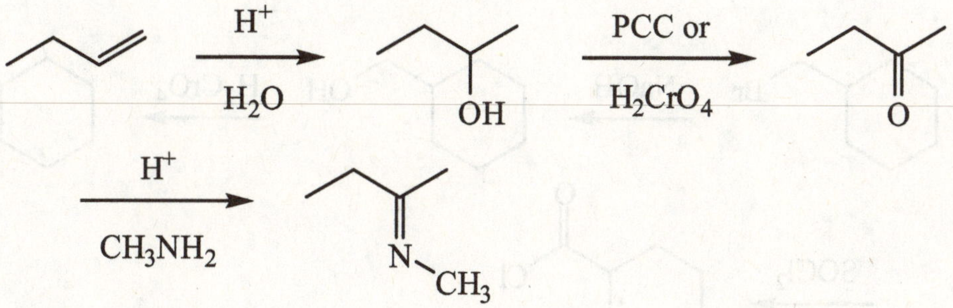

$$\text{alkene} \xrightarrow[H_2O]{H^+} \text{2-butanol} \xrightarrow[H_2CrO_4]{PCC\ \text{or}} \text{ketone}$$

$$\xrightarrow[CH_3NH_2]{H^+} \text{imine (N–CH}_3)$$

(d)

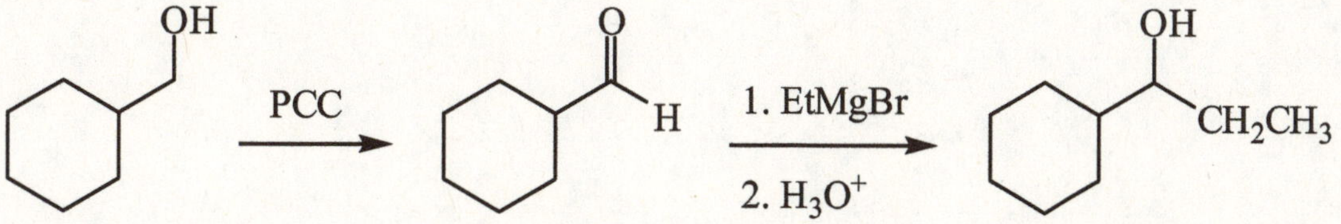

$$\text{cyclohexylmethanol} \xrightarrow{PCC} \text{aldehyde} \xrightarrow[2.\ H_3O^+]{1.\ EtMgBr} \text{alcohol (CH}_2CH_3)$$

(e)

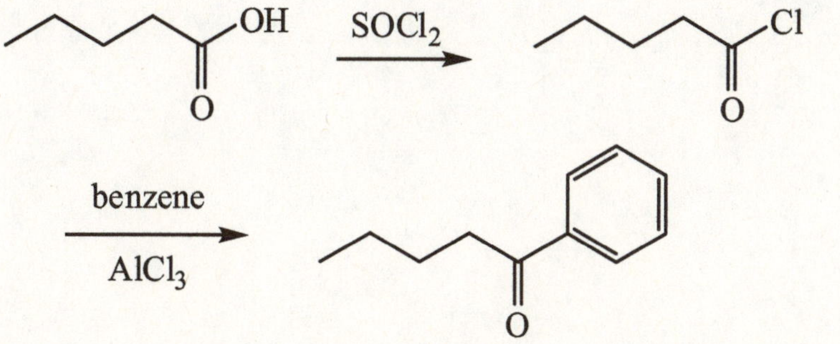

$$\text{carboxylic acid (OH)} \xrightarrow{SOCl_2} \text{acid chloride (Cl)}$$

$$\xrightarrow[AlCl_3]{\text{benzene}} \text{aryl ketone}$$

(f)

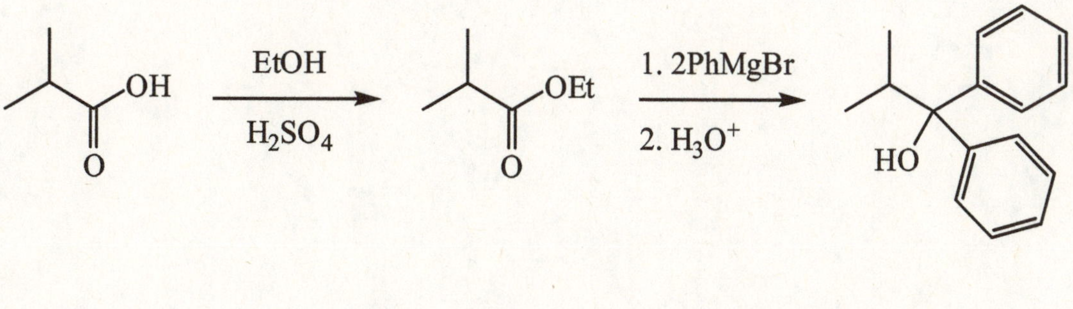

$$\text{carboxylic acid (OH)} \xrightarrow[H_2SO_4]{EtOH} \text{ester (OEt)} \xrightarrow[2.\ H_3O^+]{1.\ 2PhMgBr} \text{tertiary alcohol (HO)}$$

(g)

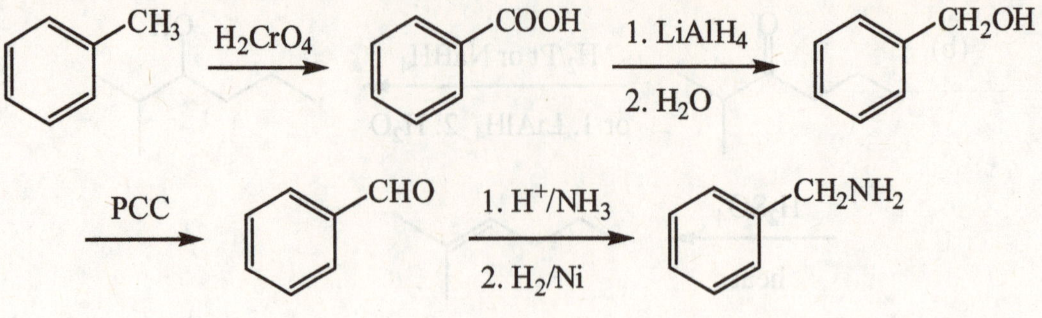

$$\text{C}_6\text{H}_5\text{CH}_3 \xrightarrow{\text{H}_2\text{CrO}_4} \text{C}_6\text{H}_5\text{COOH} \xrightarrow[\text{2. H}_2\text{O}]{\text{1. LiAlH}_4} \text{C}_6\text{H}_5\text{CH}_2\text{OH}$$

$$\xrightarrow{\text{PCC}} \text{C}_6\text{H}_5\text{CHO} \xrightarrow[\text{2. H}_2/\text{Ni}]{\text{1. H}^+/\text{NH}_3} \text{C}_6\text{H}_5\text{CH}_2\text{NH}_2$$

(h)

$$\text{(cyclohexyl)CH}_2\text{Br} \xrightarrow{\text{NaOH}} \text{(cyclohexyl)CH}_2\text{OH} \xrightarrow{\text{H}_2\text{CrO}_4} \text{(cyclohexyl)COOH}$$

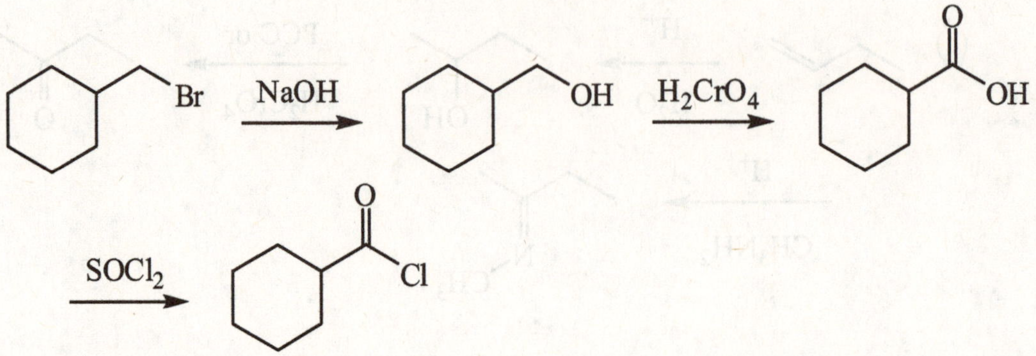

$$\xrightarrow{\text{SOCl}_2} \text{(cyclohexyl)COCl}$$

Chapter 15: Enolate Anions

Problems

15.1 Identify the acidic α-hydrogens in each compound:

(a) 2-Methylcyclohexanone (b) Acetophenone (c) 2,2-Diethylcyclohexanone

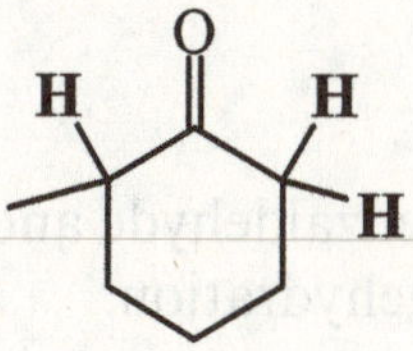

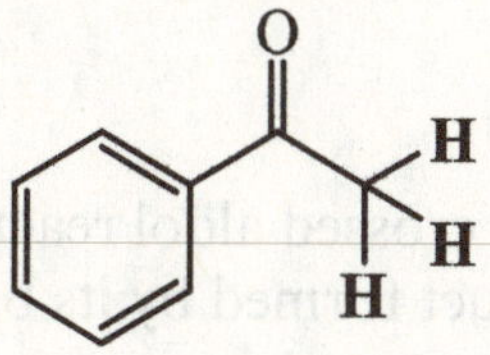

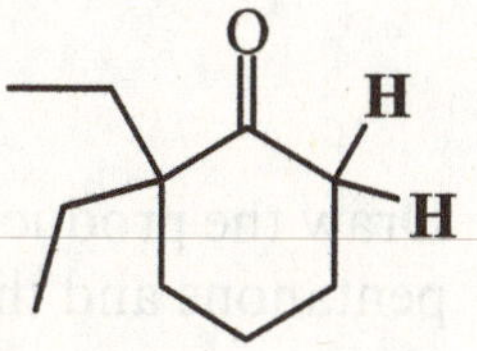

15.2 Draw the product of the base-catalyzed aldol reaction of each compound:

(a)

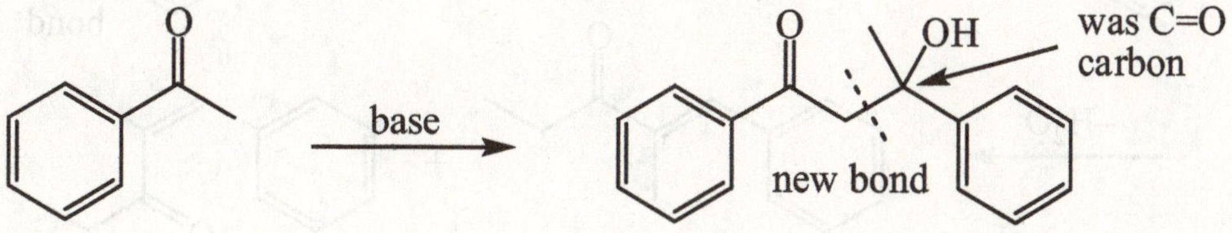

base

was C=O carbon

new bond

(b)

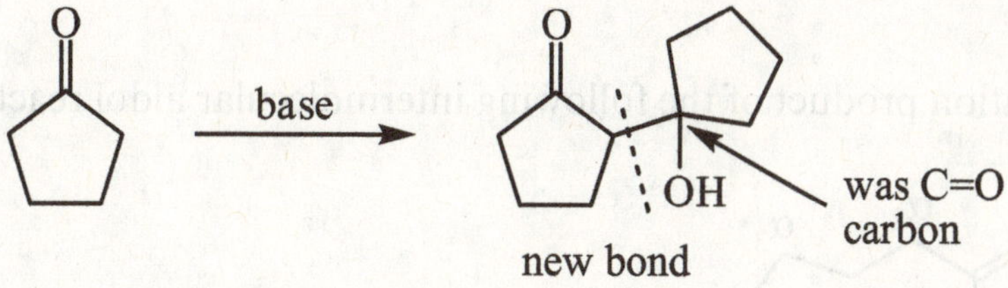

base

was C=O carbon

new bond

(c)

base

was C=O carbon

new bond

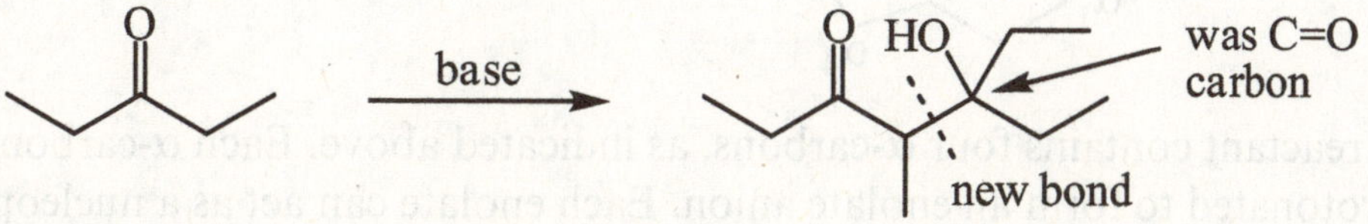

15.3 Draw the product of the base-catalyzed dehydration of each aldol product from Problem 15.2.

(a)

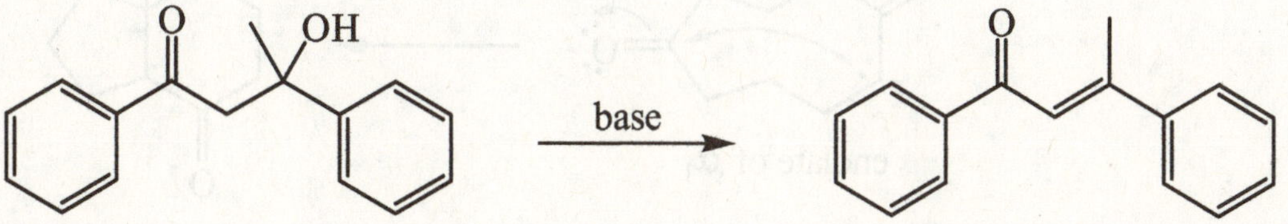

base

(b)

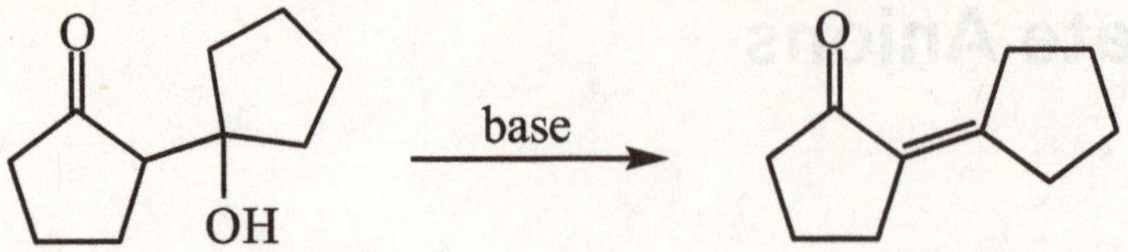

(c)

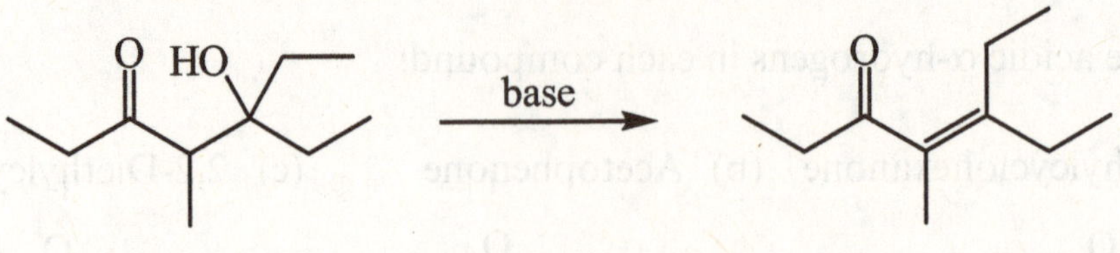

15.4 Draw the product of the crossed aldol reaction between benzaldehyde and 3-pentanone and the product formed by its base-catalyzed dehydration.

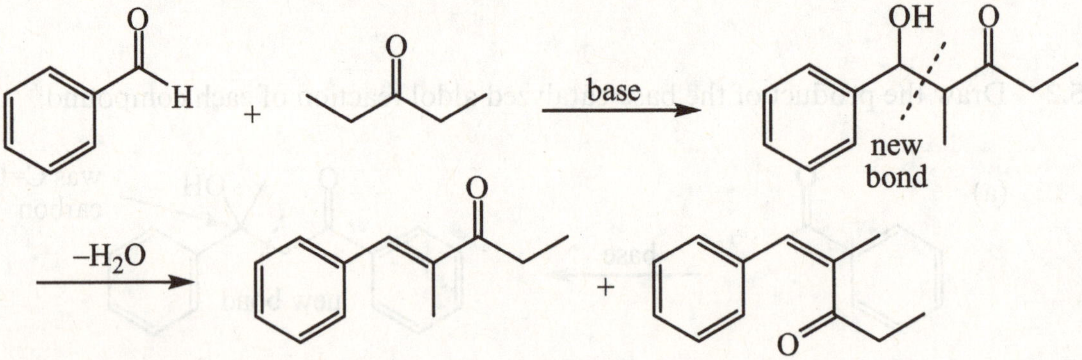

15.5 Draw the dehydration product of the following intermolecular aldol reaction.

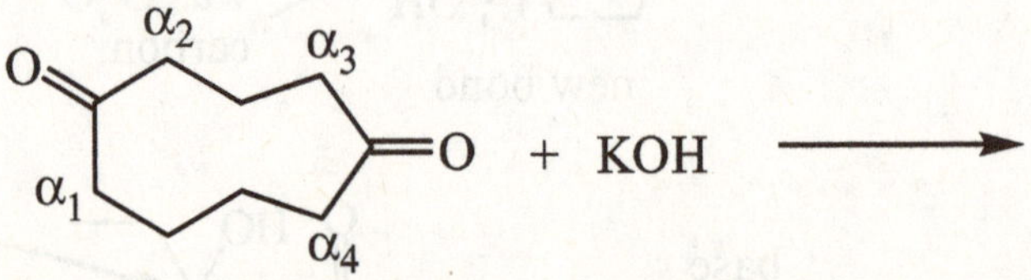

The reactant contains four α-carbons, as indicated above. Each α-carbon can be deprotonated to form an enolate anion. Each enolate can act as a nucleophile and undergo an intramolecular aldol reaction with the other carbonyl group. The four possible enolates and the first step of each aldol reaction are shown below.

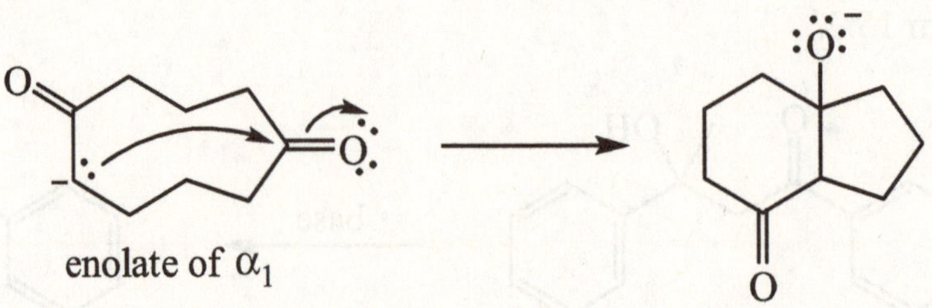

enolate of α_1

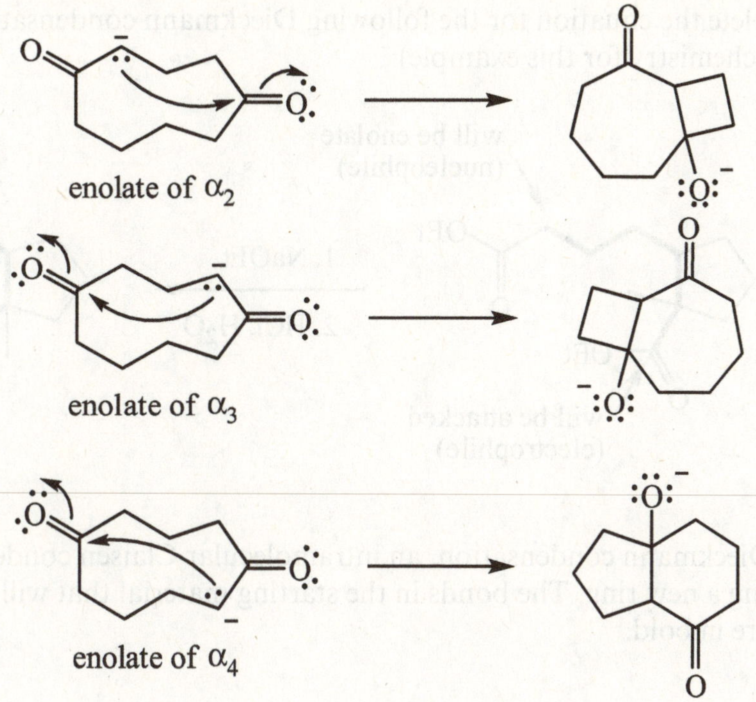

enolate of α_2

enolate of α_3

enolate of α_4

The alkoxide anions formed from the enolates of α_1 and α_4 are identical, and they are also more favorable than those formed from the enolates of α_2 and α_3. Five- and six-membered rings are more favorable than four- and seven-membered rings.

The predominant aldol product is therefore derived from the enolates of α_1 and α_4. Dehydration of β-hydroxyketone yields the α,β-unsaturated ketone below.

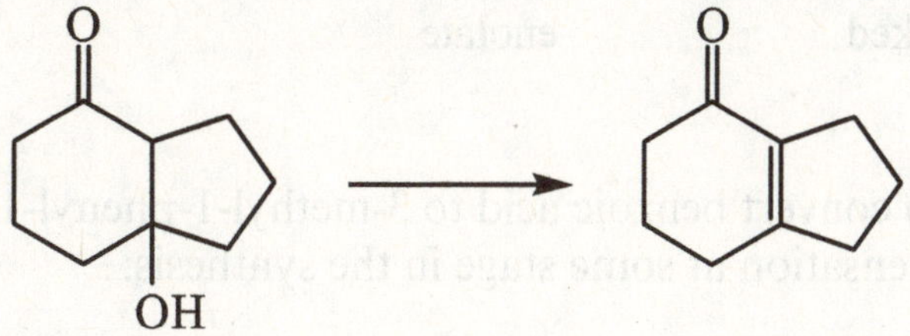

15.6   Show the product of Claisen condensation of ethyl 3-methylbutanoate in the presence of sodium ethoxide.

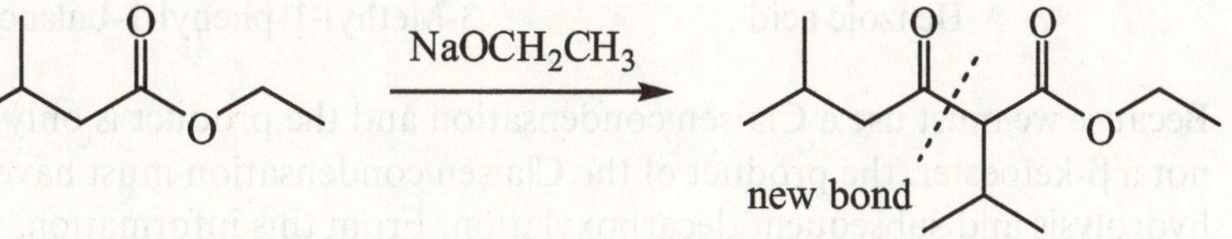

new bond

15.7   Complete the equation for the following Dieckmann condensation (disregard the stereochemistry for this example):

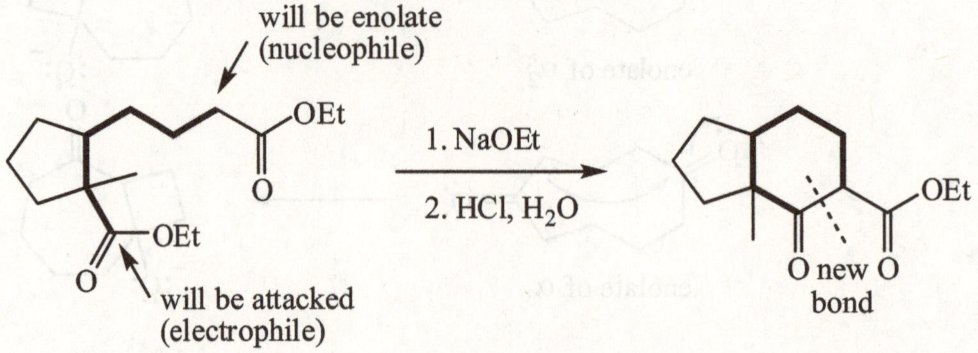

In a Dieckmann condensation, an intramolecular Claisen condensation occurs, forming a new ring. The bonds in the starting material that will be part of the new ring are in bold.

15.8   Complete the equation for this crossed Claisen condensation:

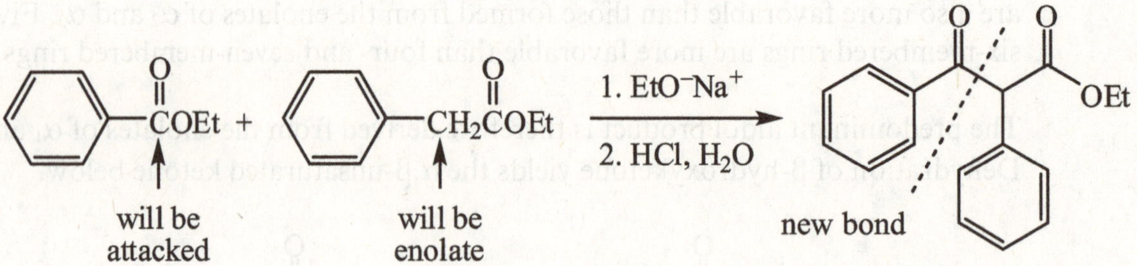

15.9   Show how to convert benzoic acid to 3-methyl-1-phenyl-1-butanone by using a Claisen condensation at some stage in the synthesis:

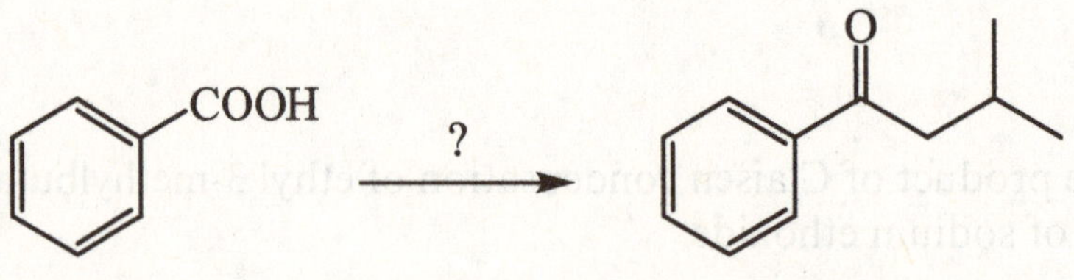

Benzoic acid   3-Methyl-1-phenyl-1-butanone

Because we must use a Claisen condensation and the product is only a ketone and not a β-ketoester, the product of the Claisen condensation must have undergone hydrolysis and subsequent decarboxylation. From this information, we can determine the structure of the initial Claisen condensation product.

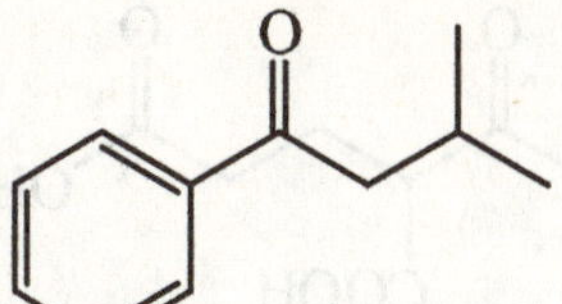

is obtained from the decarboxylation of

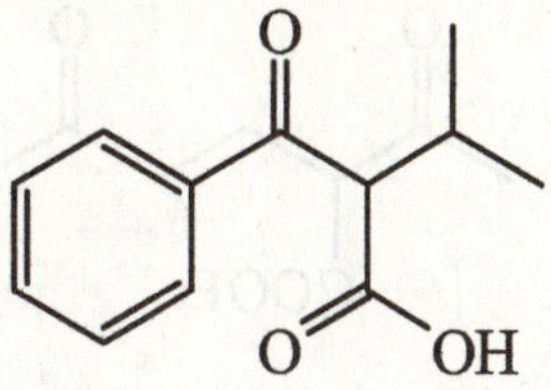

which is in turn obtained from the hydroysis of ester

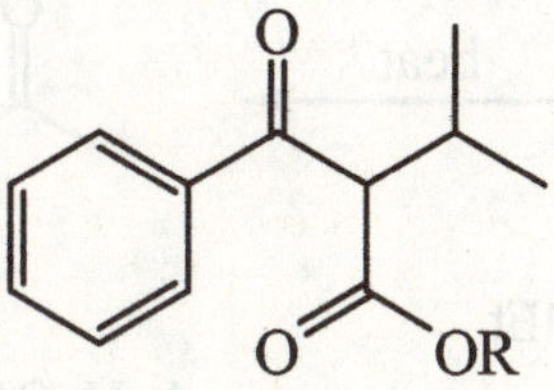

The synthetic scheme is therefore:

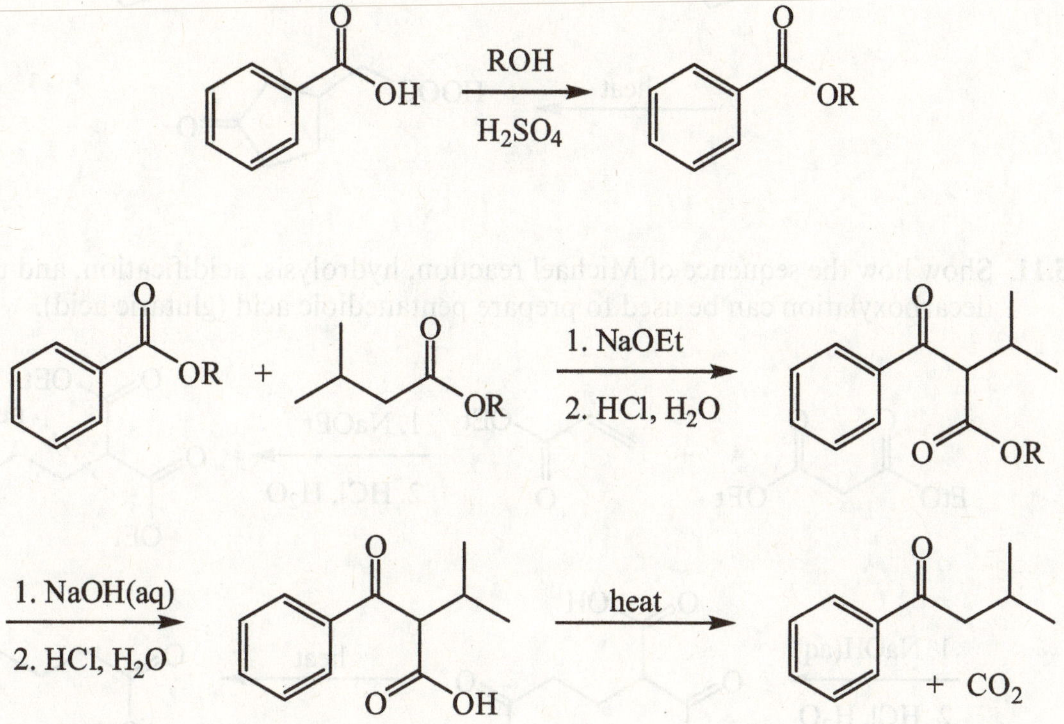

ROH / H₂SO₄ reaction scheme and subsequent steps

15.10 Show the product formed from each Michael reaction in the solution to Example 15.10 after (1) hydrolysis in aqueous NaOH, (2) acidification, and (3) thermal decarboxylation of each β-ketoacid or β-dicarboxylic acid. These reactions illustrate the usefulness of the Michael reaction for the synthesis of 1,5-dicarbonyl compounds.

(a)

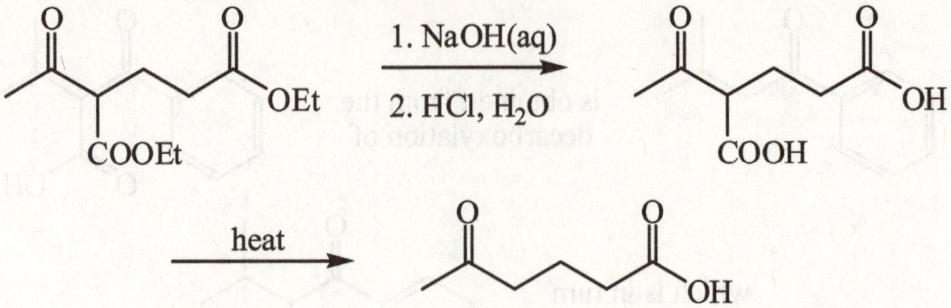

$$\text{(reactant)} \xrightarrow[\text{2. HCl, H}_2\text{O}]{\text{1. NaOH(aq)}} \text{(product)}$$

$$\xrightarrow{\text{heat}} \text{(product)}$$

(b)

$$\xrightarrow[\text{2. HCl, H}_2\text{O}]{\text{1. NaOH(aq)}}$$

$$\xrightarrow{\text{heat}} \text{HOOC}$$

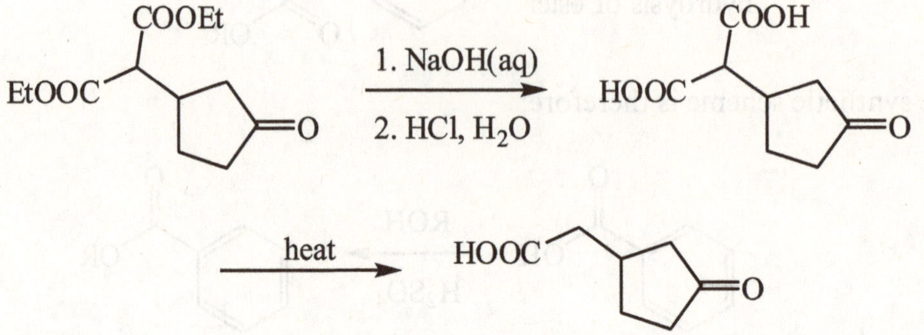

15.11 Show how the sequence of Michael reaction, hydrolysis, acidification, and thermal decarboxylation can be used to prepare pentanedioic acid (glutaric acid).

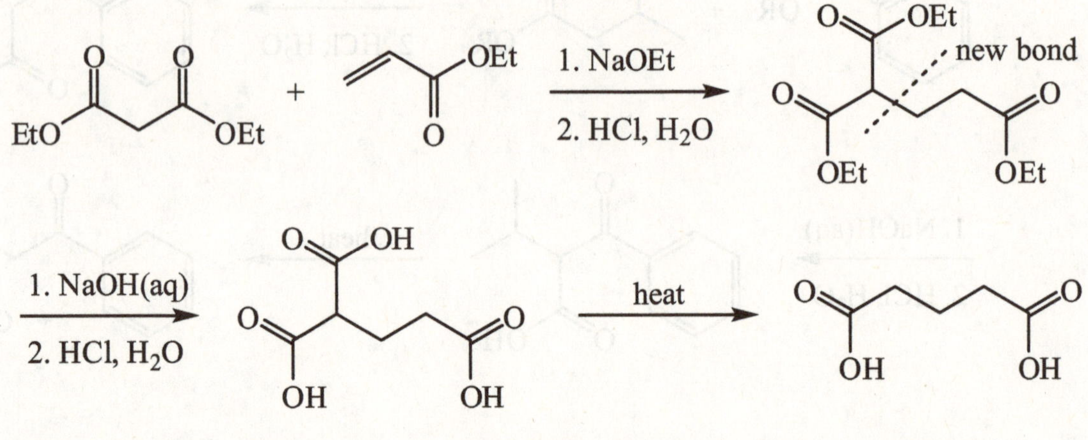

$$\xrightarrow[\text{2. HCl, H}_2\text{O}]{\text{1. NaOEt}} \text{(new bond)}$$

$$\xrightarrow[\text{2. HCl, H}_2\text{O}]{\text{1. NaOH(aq)}} \xrightarrow{\text{heat}}$$

15.12 The product of the double Michael reaction in Example 15.12 is a diester that, when treated with sodium ethoxide in ethanol, undergoes a Dieckmann condensation. Draw the structural formula for the product of this Dieckmann condensation followed by acidification with aqueous HCl.

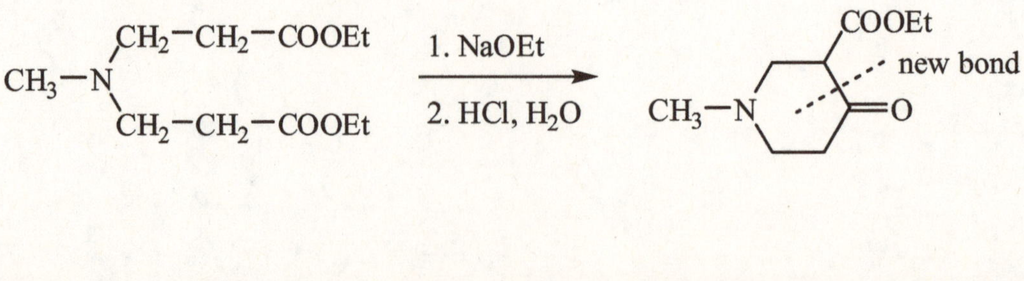

$$\text{CH}_3-\text{N} \begin{array}{l} \text{CH}_2-\text{CH}_2-\text{COOEt} \\ \text{CH}_2-\text{CH}_2-\text{COOEt} \end{array} \xrightarrow[\text{2. HCl, H}_2\text{O}]{\text{1. NaOEt}} \text{new bond}$$

Chemical Connections

15A. Which of the biomolecules in the above reaction scheme could be the product of an aldol reaction?

3-Hydroxy-3-methylglutaryl-CoA (HMG-CoA), which has a β-hydroxy group, could be the product of an aldol reaction. Recall that in an aldol reaction, the carbonyl group that accepts the enolate nucleophile becomes the β-hydroxy group.

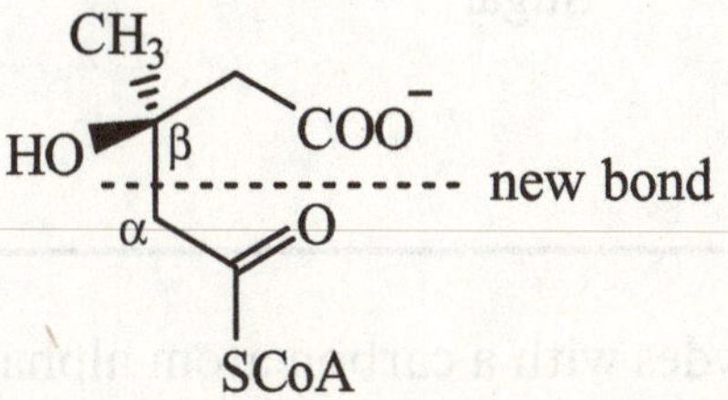

15B. Provide a complete mechanism for the Michael reaction of (b) to produce (c). Using fishhook arrows, provide a mechanism for the rearrangement of (c) to produce (d). The intramolecular molecular Michael addition of (b) first forms a resonance-stabilized enolate anion. Protonation of the enolate at the α-carbon forms (c).

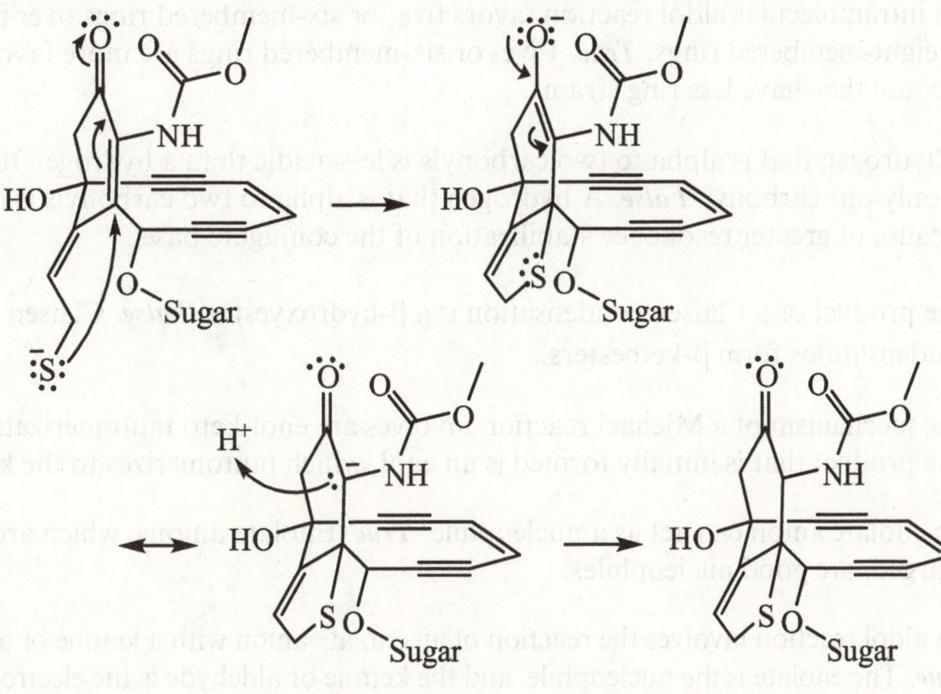

The rearrangement of (c) to (d) occurs as follows:

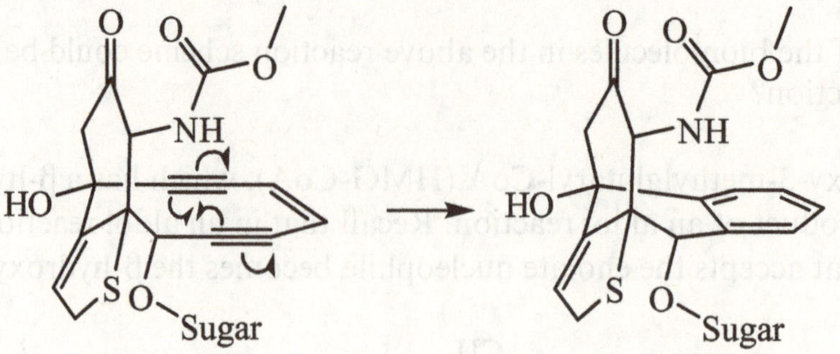

Quick Quiz

1.   All ketones and aldehydes with a carbon atom alpha to the carbonyl group can be converted to an enolate anion by treatment with a catalytic amount of base. *True*. The α-hydrogen is relatively acidic due to resonance stabilization of the conjugate base.

2.   A Dieckmann condensation favors seven- or eight-membered rings over four-, five-, or six- membered rings. *False*. The exact opposite is true.

3.   An intramolecular aldol reaction favors five- or six-membered rings over four-, seven-, or eight-membered rings. *True*. Five- or six-membered rings are more favorable because they have less ring strain.

4.   A hydrogen that is alpha to two carbonyls is less acidic than a hydrogen that is alpha to only one carbonyl. *False*. A hydrogen that is alpha to two carbonyls is more acidic because of greater resonance stabilization of the conjugate base.

5.   The product of a Claisen condensation is a β-hydroxyester. *False*. Claisen condensations form β-ketoesters.

6.   The mechanism of a Michael reaction involves are enol-keto tautomerization. *True*. The product that is initially formed is an enol, which tautomerizes to the keto form.

7.   An enolate anion can act as a nucleophile. *True*. Enolate anions, which are negatively charged, are good nucleophiles.

8.   An aldol reaction involves the reaction of an enolate anion with a ketone or an aldehyde. *True*. The enolate is the nucleophile, and the ketone or aldehyde is the electrophile.

9.   The product of an aldol reaction is a β-hydroxyester. *False*. An aldol reaction typically forms a β-hydroxyaldehyde or a β-hydroxyketone.

10. Aldol reactions and Claisen condensations can be catalyzed by enzymes. *True*. Enzymes can deprotonate the alpha carbon to generate an enolate anion.

11. A crossed aldol reaction is most effective when one of the carbonyl compounds is more reactive toward nucleophilic addition and cannot form an enolate anion. *True*. This minimizes the number of different combinations of products that could be made.

12. Hydrogen atoms alpha to a carbonyl are many times more acidic than vinyl or alkyl hydrogens. *True*. The increased acidity of the alpha hydrogen is due to resonance stabilization of the conjugate base.

13. The Claisen condensation is a reaction between an enolate anion and an ester. *True*. The reaction proceeds via a nucleophilic acyl substitution.

14. The α-hydrogen of an ester is more acidic than the α-hydrogen of a ketone. *False*. The α-hydrogen of a ketone has a lower pK_a (20) than that of an ester (22).

15. An enolate anion is stabilized by resonance. *True*. The negative charge is delocalized over the oxygen and the α-carbon atoms.

16. All carbonyl compounds with an alpha hydrogen can be converted to an enolate anion by treatment with a catalytic amount of base. *False*. The formation of enolate anions using a base is either quantitative or an equilibrium.

17. A crossed Claisen condensation is most effective when one of the carbonyl compounds can only function as an enolate anion acceptor. *True*. Like the crossed aldol, this limits the number of possible combinations of products that can be formed.

18. An enolate anion can participate in a Michael reaction. *True*. Enolate anions are nucleophiles that can add to an α,β-unsaturated carbonyl compound.

19. The product of an aldol reaction can be dehydrated to yield an α,β-unsaturated carbonyl compound. *True*. The α,β-bond is also the carbon-carbon bond made in the aldol reaction.

20. The Michael reaction is the reaction of a nucleophile with the β-carbon of an α,β-unsaturated carbonyl compound. *True*. The β-carbon of an α,β-unsaturated carbonyl compound is electrophilic and can be attacked by an nucleophile.

21. An enolate anion can act as a base. *True*. In addition to being nucleophilic, enolate anions are also basic.

22. The product of a Claisen condensation can be hydrolyzed and decarboxylated to form a ketone. *True*. Claisen condensations form β-ketoacids, which can be hydrolyzed to form β-ketoacids. β-ketoacids decarboxylate upon heating to form a ketone and CO_2.

23. An amine can participate in a Michael reaction. *True*. Amines are nucleophiles that can attack the β-carbon of an α,β-unsaturated carbonyl compound.

End-of-Chapter Problems

The Aldol Reaction

15.13 Identify the most acidic hydrogen(s) in each compound:

The acidity of a hydrogen depends on the stability of the conjugate base. In general, the order of acidity, from most acidic to least acidic, is: carboxylic acid, phenol, alcohol, and α-hydrogen. That is, an enolate anion, which is the conjugate base formed from the deprotonation of an α-hydrogen, is still less stable than an alkoxide ion.

The most acidic hydrogen(s) in each compound is/are indicated in bold.

(a)

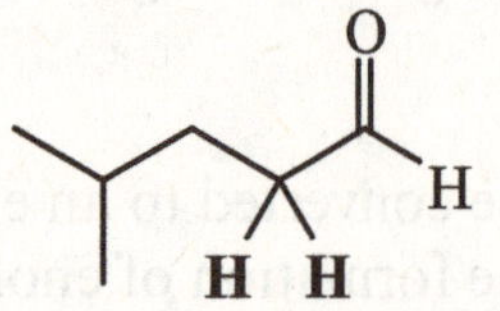

(b)

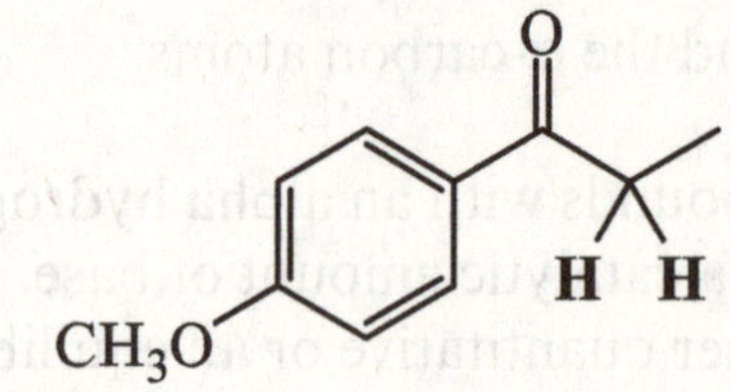

(c)

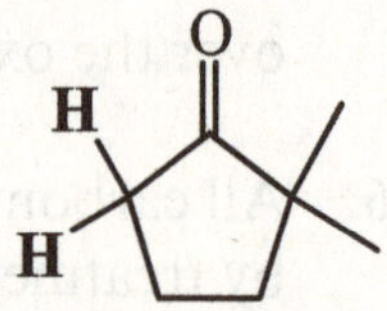

(d)

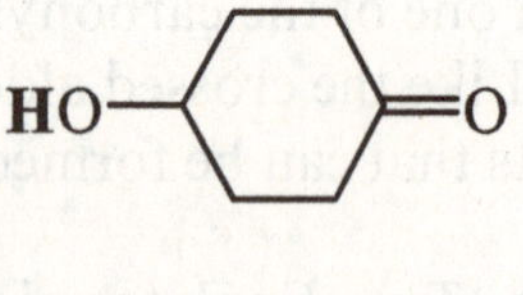

(e)

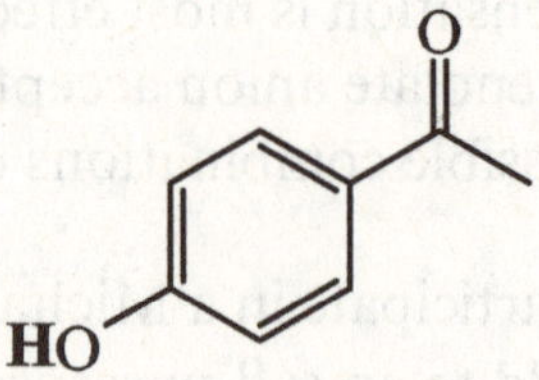

(f)

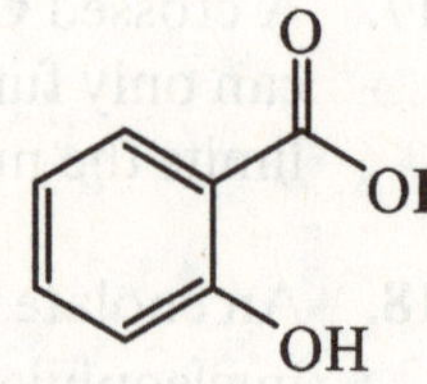

15.14 Estimate the pK_a of the compounds, and in each set arrange them in order of increasing acidity:

$$CH_3CCH_3 < CH_3CHCH_3 < CH_3CCH_2COCH_2CH_3 < CH_3CH_2COH$$

pK_a 20   17   11   5

The approximate pK_a values of the functional groups are: carboxylic acid, 5; α-hydrogen between the two carbonyl groups of a β-ketoester, 11; alcohol, 17; and α-hydrogen of a ketone, 20.

15.15 Write a second contributing structure of each anion, and use curved arrows to show the redistribution of electrons that gives your second structure:

(a)

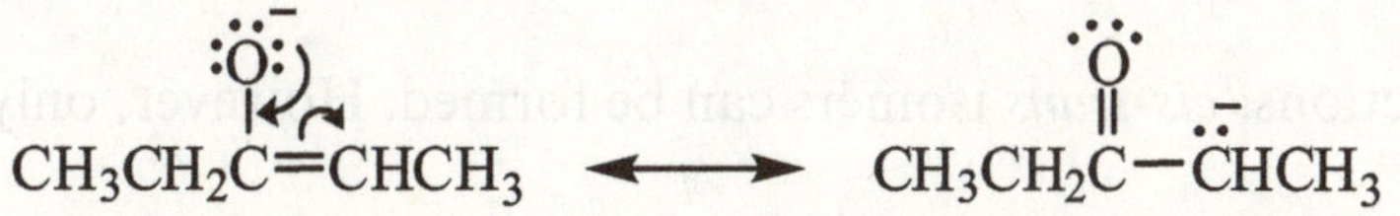

$$CH_3CH_2\overset{:\overset{-}{O}:}{C}=CHCH_3 \quad \longleftrightarrow \quad CH_3CH_2\overset{\overset{\cdot\cdot}{O}}{C}-\overset{\cdot\cdot}{C}HCH_3$$

(b)

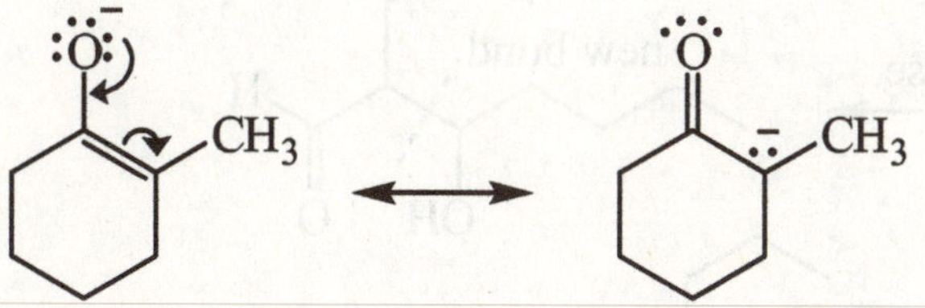

(c)

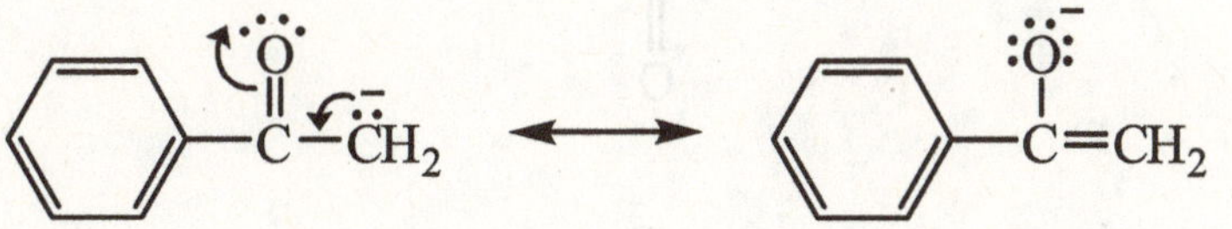

(d)

$$CH_3\overset{:\overset{-}{O}:}{C}=\overset{\overset{\cdot\cdot}{O}}{C}-COCH_2CH_3 \quad \longleftrightarrow \quad CH_3\overset{\overset{\cdot\cdot}{O}}{C}-\overset{:\overset{\cdot\cdot}{O}:^-}{C}=COCH_2CH_3$$
$$H H$$

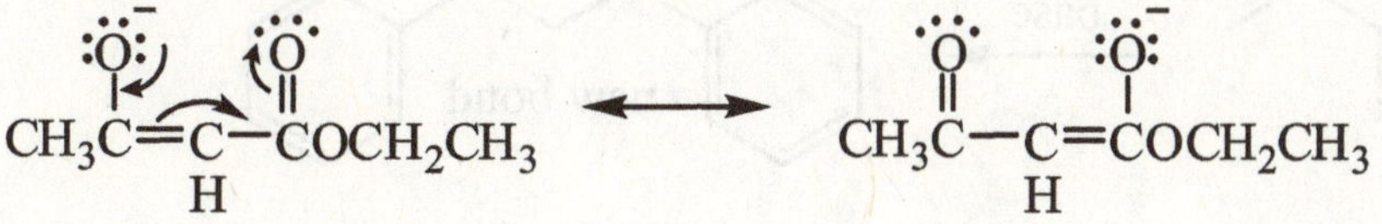

15.16 Treatment of 2-methylcyclohexanone with base gives two different enolate anions. Draw the contributing structure for each that places the negative charge on carbon.

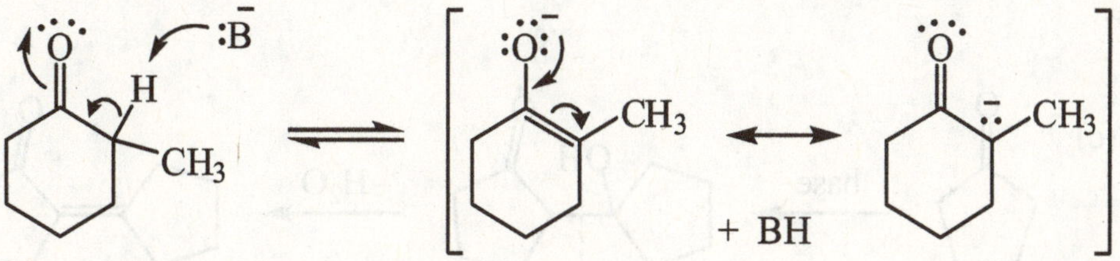

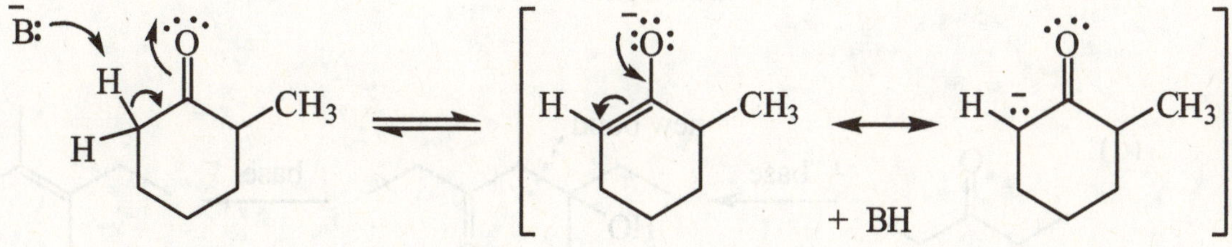

15.17 Draw a structural formula for the product of the aldol reaction of each compound and for the α,β-unsaturated aldehyde or ketone formed by dehydration of each aldol product:

In each of these reactions, *cis-trans* isomers can be formed. However, only one of the isomers is shown.

(a)

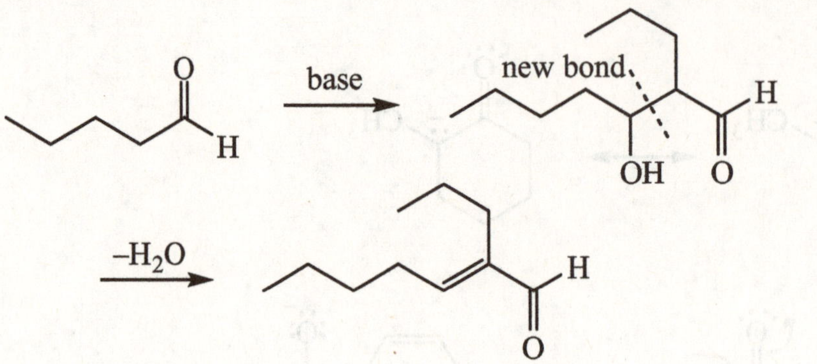

(b)

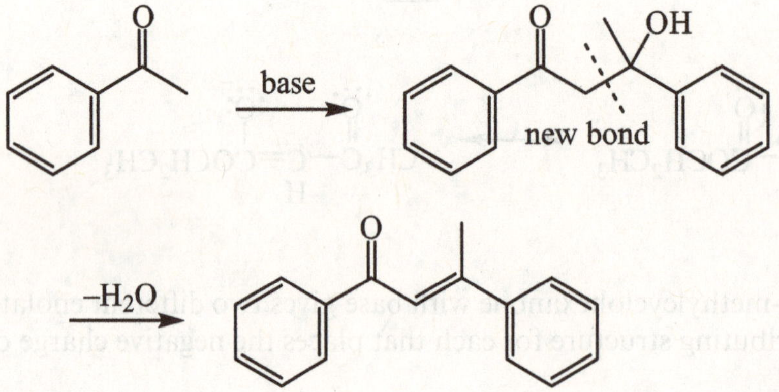

(c)

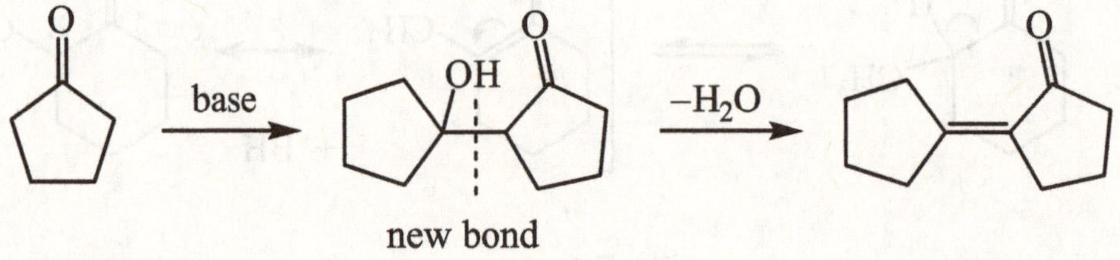

(d)

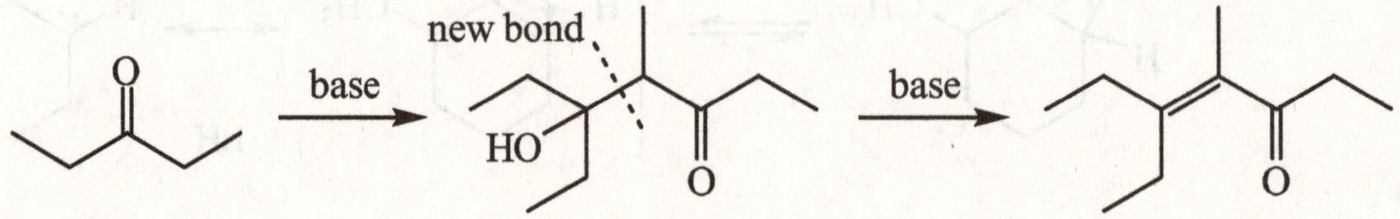

(e)

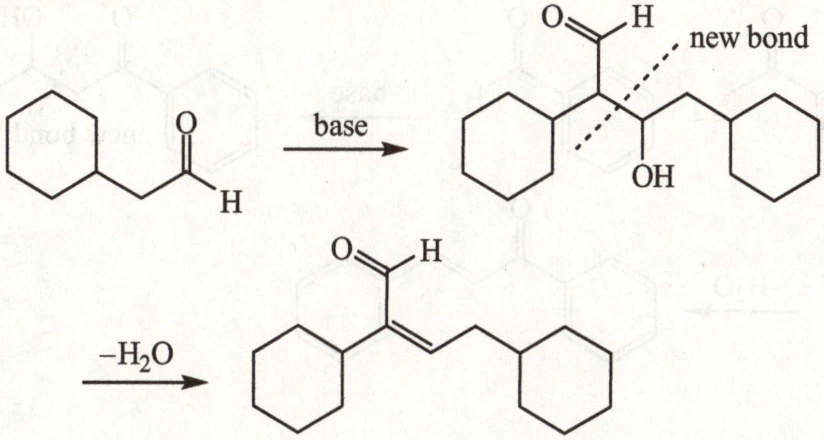

base

–H₂O

(f)

base

–H₂O

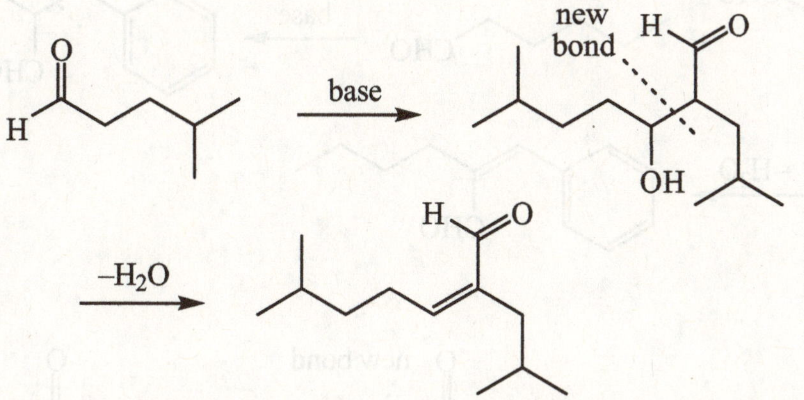

15.18 Draw a structural formula for the product of each crossed aldol reaction and for the compound formed by dehydration of each aldol product:

In each of these reactions, *cis-trans* isomers can be formed. However, only one of the isomers is shown.

(a)

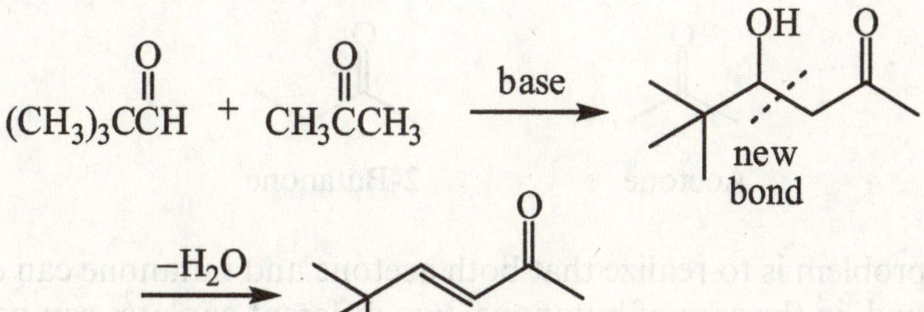

$(CH_3)_3CCH$ + CH_3CCH_3 $\xrightarrow{\text{base}}$

–H₂O

(b)

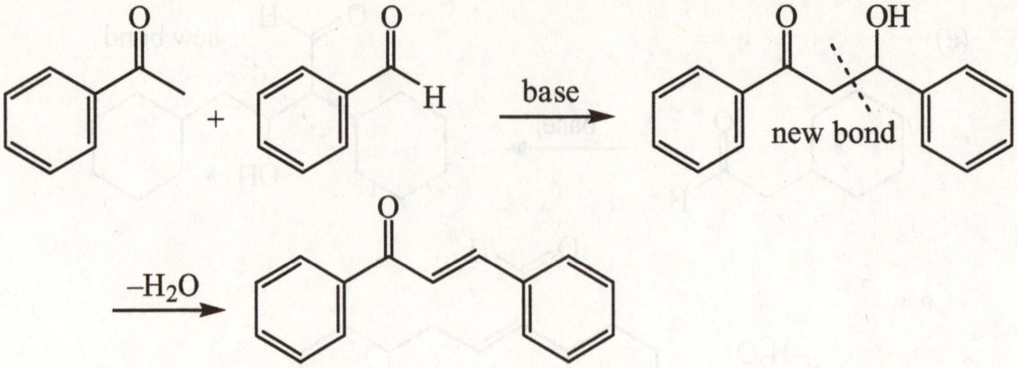

(c)

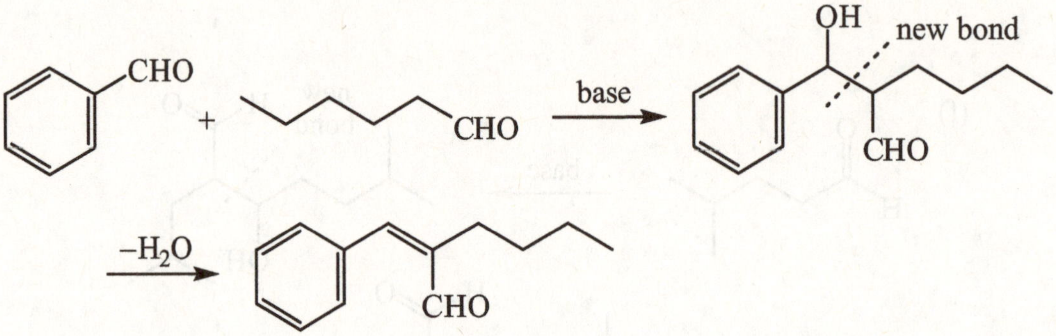

(d)

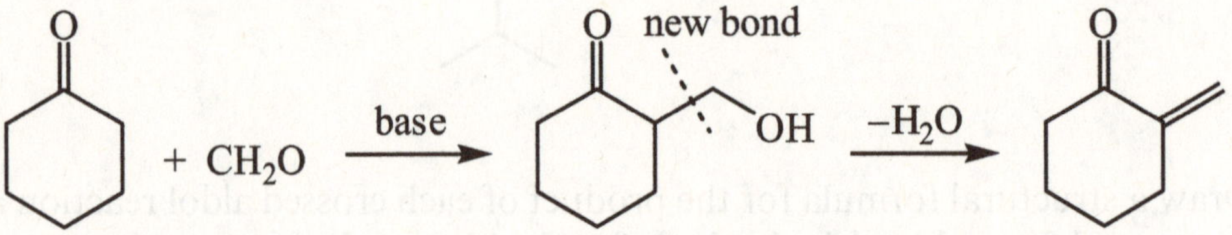

15.19 When a 1:1 mixture of acetone and 2-butanone is treated with base, six aldol products are possible. Draw a structural formula for each.

Acetone   2-Butanone

The key to this problem is to realize that both acetone and butanone can each form enolate anions, and, in the case of butanone, two different enolates can be formed (one at carbon 1 and the other at carbon 3). These enolates are shown below as (a), (b), and (c). Each of these three enolates can then attack acetone or butanone, giving rise to six possible aldol products. In reactions where a new stereocenter is formed in the product, a mixture of R and S stereoisomers results.

(a)

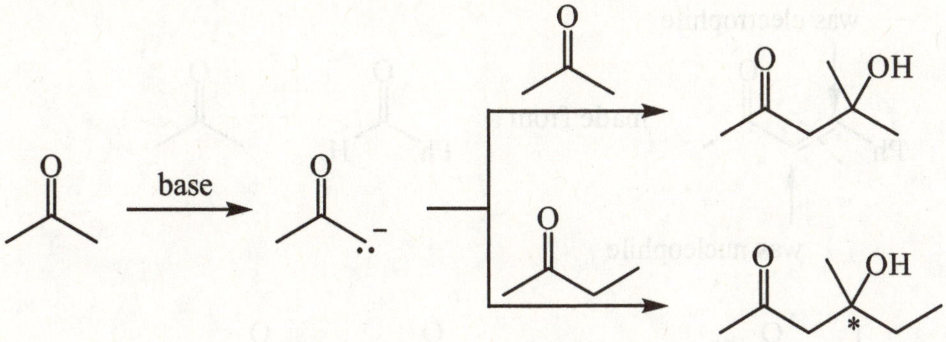

(b)

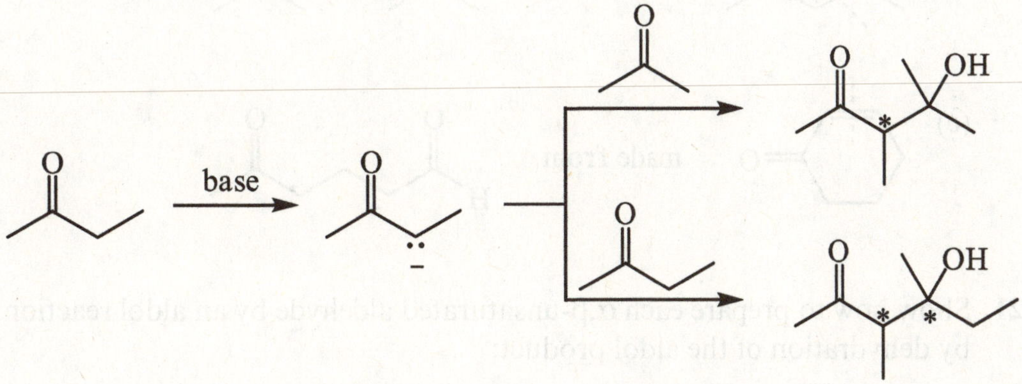

(c)

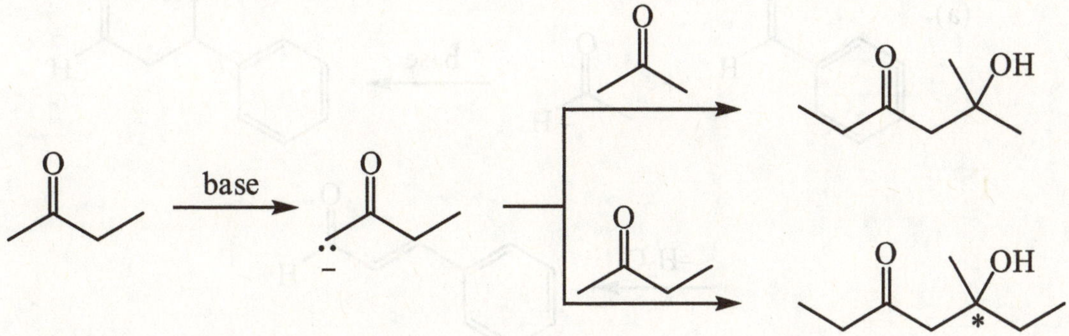

15.20 Show how to prepare each α,β-unsaturated ketone by an aldol reaction followed by dehydration of the aldol product:

Recognize that the β-carbon was formerly the electrophile (the nucleophile acceptor) and that the α-carbon was the nucleophile. During the aldol reaction, the α,β-bond was formed, and the hydroxyl group was bonded to the β-carbon prior to dehydration.

(a)

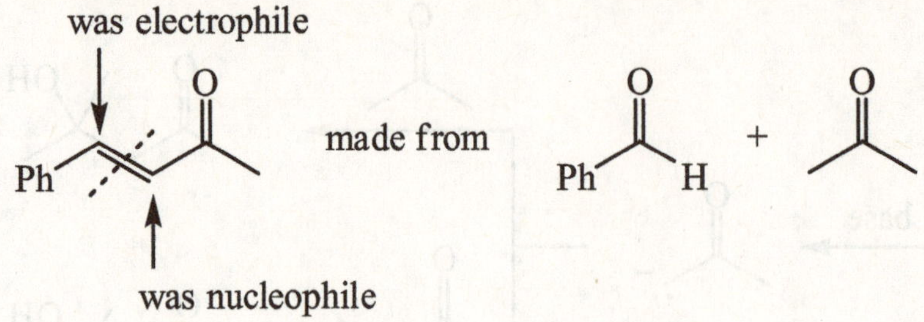
was electrophile
O
Ph
was nucleophile
made from
O
Ph H
+
O

(b)

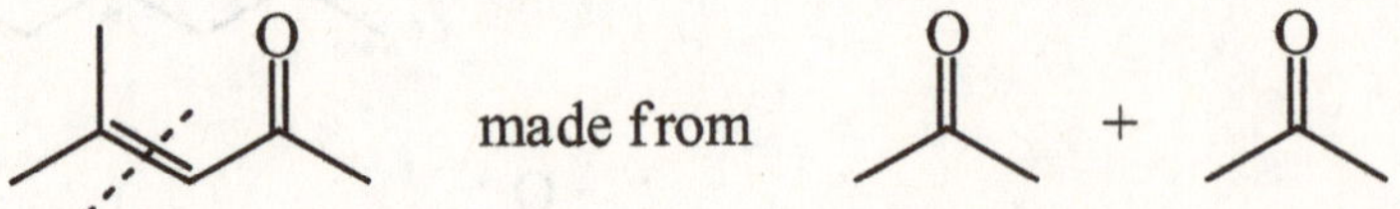
O
made from
O
+
O

(c)

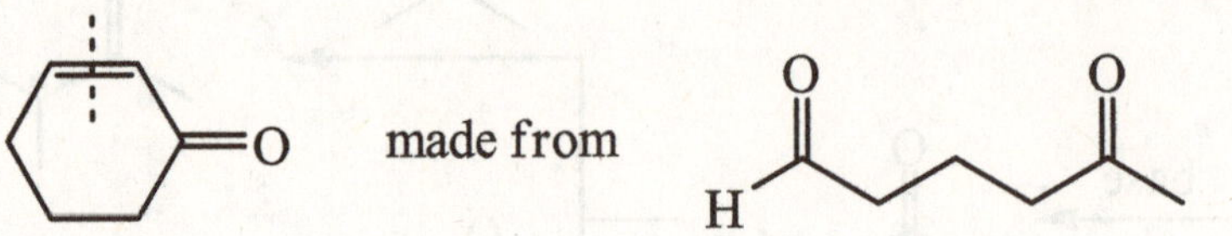
O
made from
O
O
H

15.21 Show how to prepare each α,β-unsaturated aldehyde by an aldol reaction followed by dehydration of the aldol product:

(a)

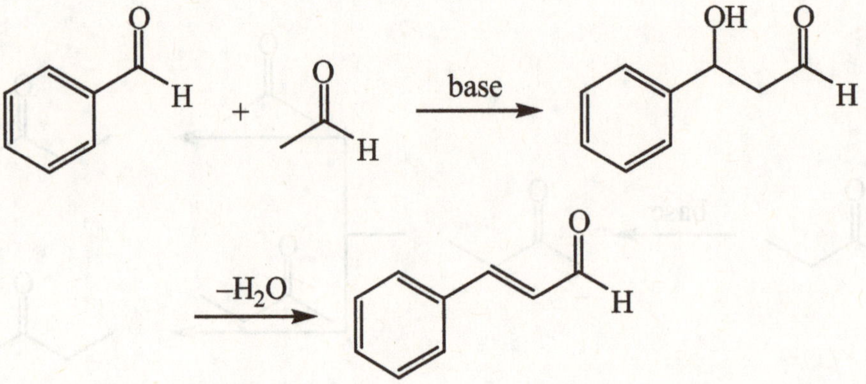
O
H
+
O
H
base
OH O
H
$-H_2O$
O
H

(b)

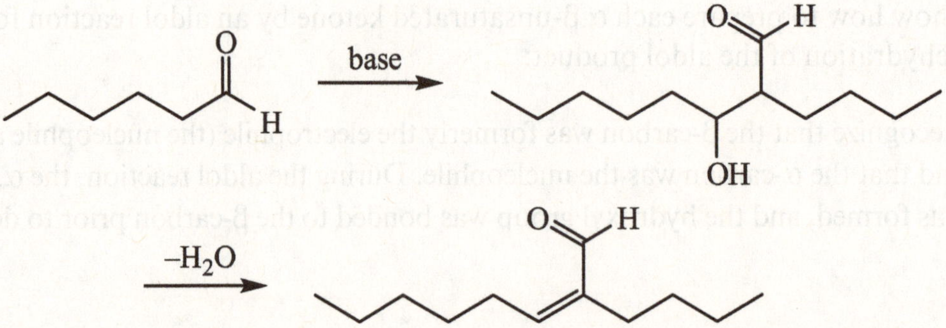
O
H
base
O H
OH
$-H_2O$
O H

15.22 When treated with base, the following compound undergoes an intramolecular aldol reaction, followed by dehydration, to give a product containing a ring (yield 78%). Propose a structural formula for this product.

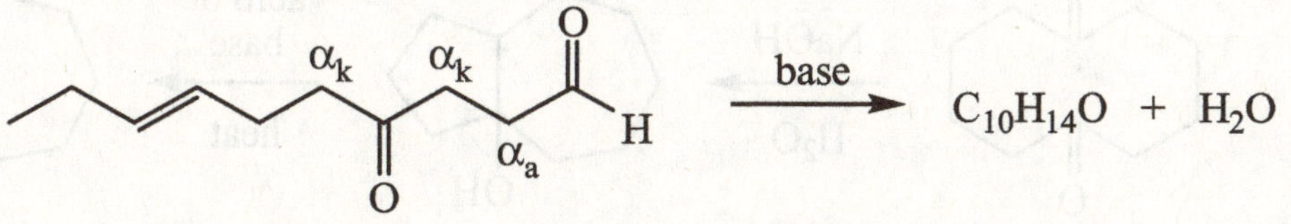

The compound contains three α-carbons, two of which are α to the ketone and one is α to the aldehyde. These are respectively indicated as α_k and α_a. Each of these α-carbons can form an enolate anion, but only one of these (the α_k on the left) will undergo an intramolecular reaction to form a stable five-membered ring. Reaction of the other two would form unfavorable three-membered rings.

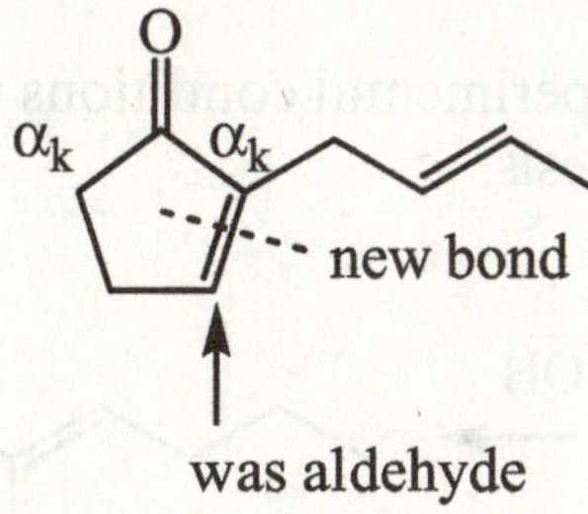

15.23 Propose a structural formula for the compound of molecular formula $C_6H_{10}O_2$ that undergoes an aldol reaction followed by dehydration to give this α,β-unsaturated aldehyde:

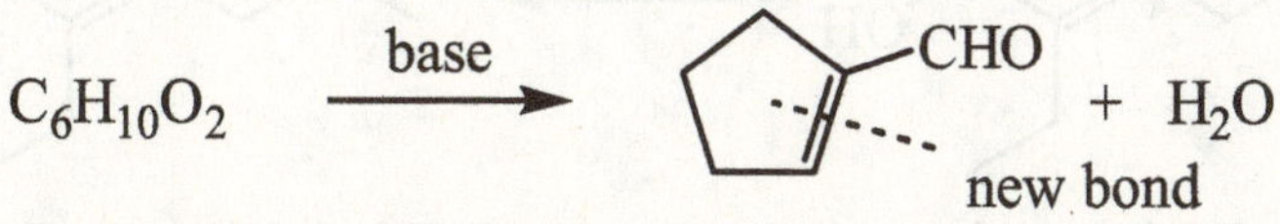

Because the product is cyclic, the reaction must have been an intramolecular aldol reaction. The new bond was formed between the α and β carbons in the product. The starting material that gives rise to this product is hexanedial.

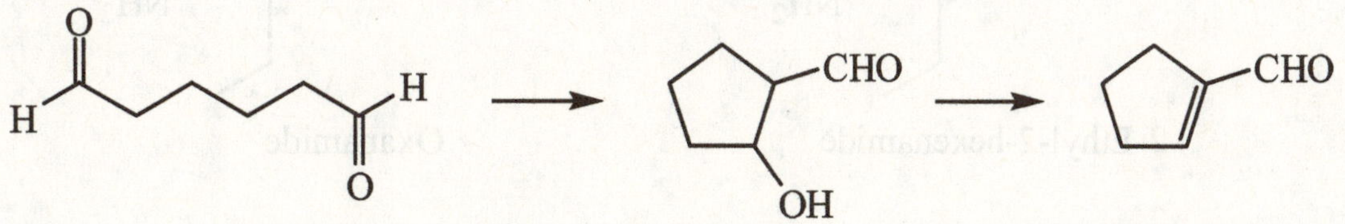

15.24 Show how to bring about this conversion:

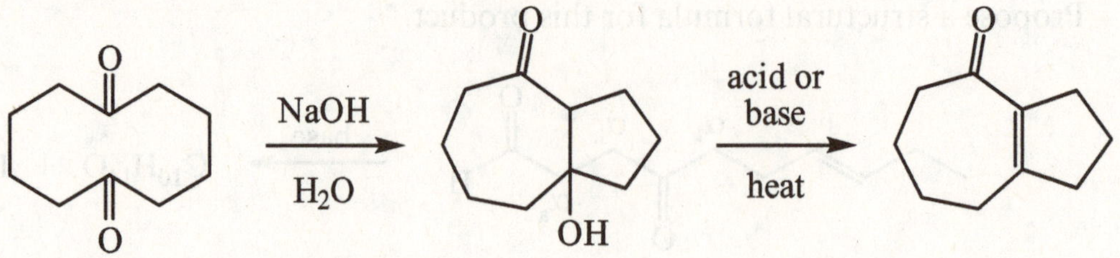
NaOH
H₂O
OH
acid or
base
heat

Due to the extensive symmetry present in the starting material, all four α-carbons are identical and it does not matter which one is used to form the enolate anion. All four will yield the same product.

15.25 Oxanamide, a mild sedative, is synthesized from butanal in these five steps:

(a) Show reagents and experimental conditions that might be used to bring about each step in this synthesis.

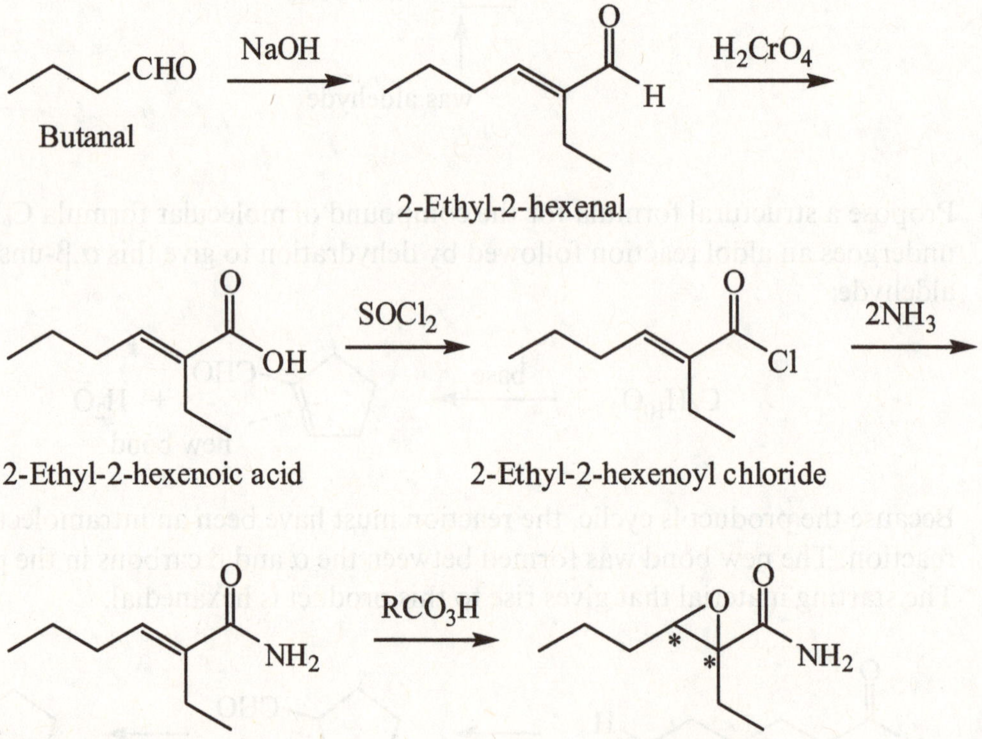
CHO
Butanal
NaOH
2-Ethyl-2-hexenal
H_2CrO_4
2-Ethyl-2-hexenoic acid
$SOCl_2$
2-Ethyl-2-hexenoyl chloride
$2NH_3$
2-Ethyl-2-hexenamide
RCO_3H
Oxanamide

Note that in the fourth step, two moles of NH_3 are required because the HCl byproduct also reacts with NH_3 in a 1:1 ratio to form NH_4Cl.

(b)   How many stereocenters are in oxanamide? How many stereoisomers are possible for oxanamide?

Oxanamide contains two stereocenters, giving rise to four possible stereoisomers. However, if oxanamide is synthesized using this synthetic sequence, only two stereoisomers are produced because the epoxidation reaction is stereoselective.

15.26 Propose structural formulas for compounds A and B:

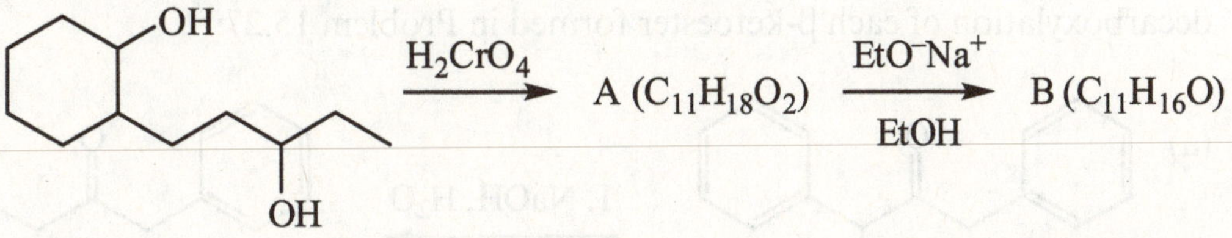

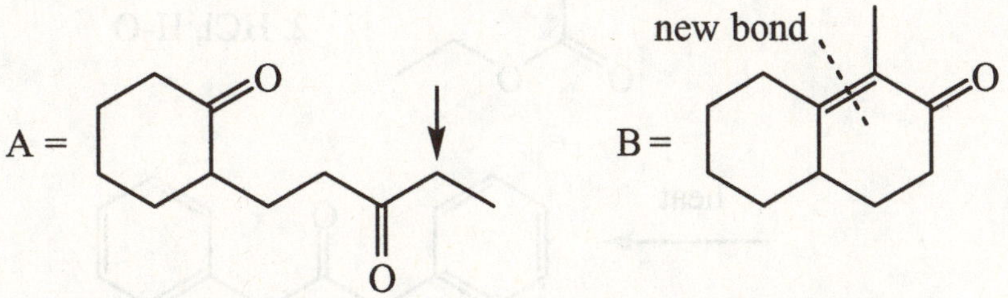

$$\text{(structure)} \xrightarrow{H_2CrO_4} A\ (C_{11}H_{18}O_2) \xrightarrow[\text{EtOH}]{EtO^-Na^+} B\ (C_{11}H_{16}O)$$

A =   (structure)

B =   (structure with "new bond" label)

Only the reaction of the indicated α-carbon in **A** forms a favorable ring size.

The Claisen and Dieckmann Condensations

15.27 Show the product of the Claisen condensation of each ester:

(a)

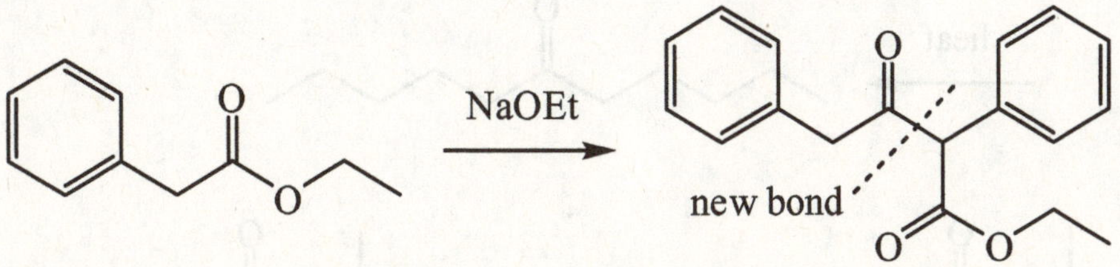

$$\text{(structure)} \xrightarrow{NaOEt} \text{(product with "new bond" label)}$$

(b)

$$\text{(structure)} \xrightarrow{NaOCH_3} \text{(product with "new bond" label)}$$

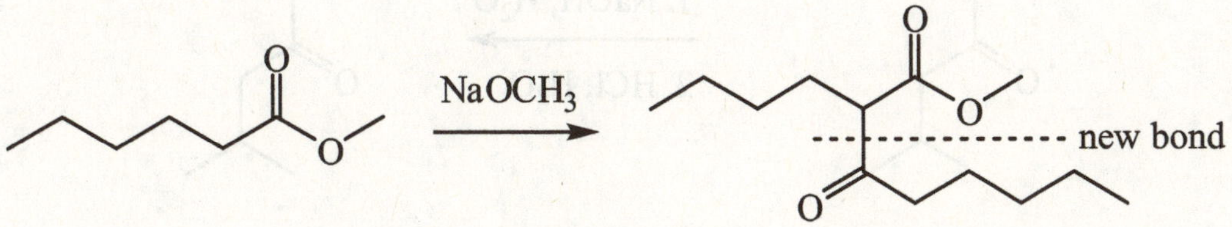

(c)

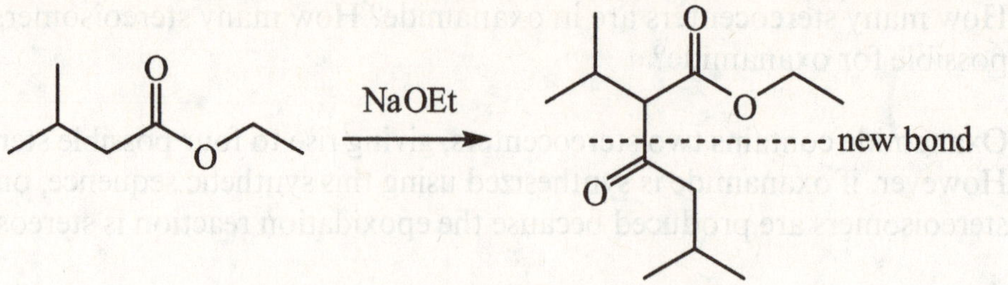

NaOEt → new bond

15.28 Draw a structural formula for the product of saponification, acidification, and decarboxylation of each β-ketoester formed in Problem 15.27:

(a)

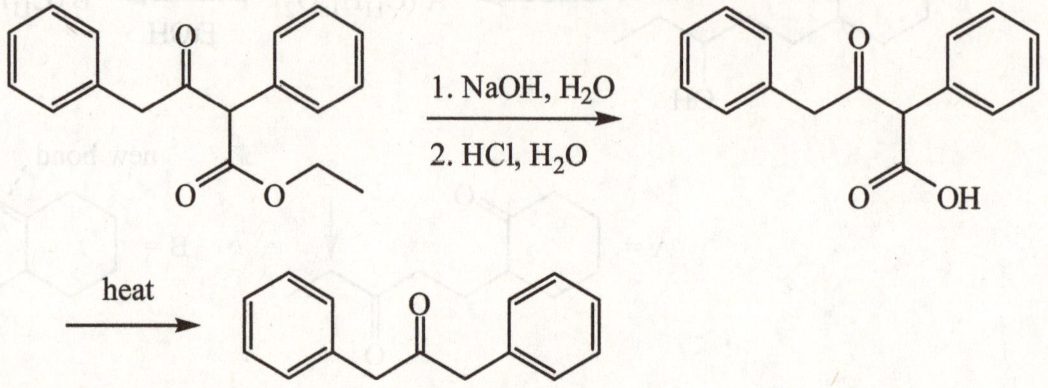

1. NaOH, H$_2$O

2. HCl, H$_2$O

heat

(b)

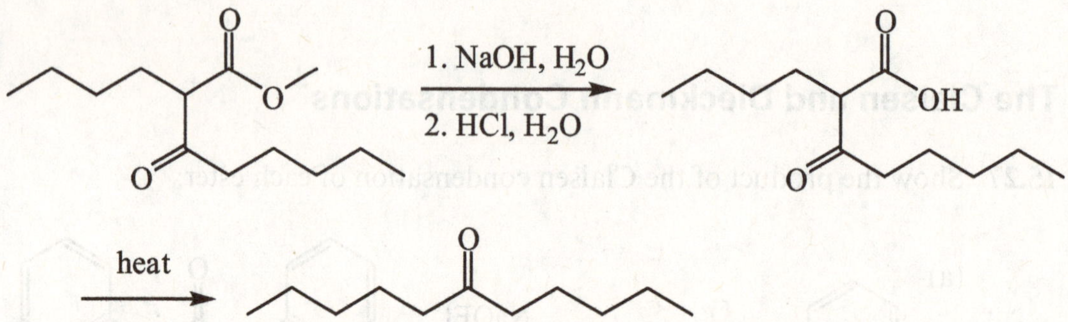

1. NaOH, H$_2$O

2. HCl, H$_2$O

heat

(c)

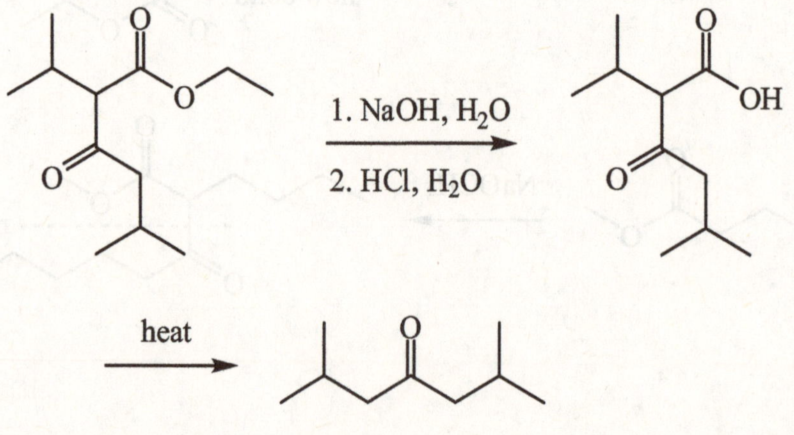

1. NaOH, H$_2$O

2. HCl, H$_2$O

heat

15.29 When a 1:1 mixture of ethyl propanoate and ethyl butanoate is treated with sodium ethoxide, four Claisen condensation products are possible. Draw a structural formula for each product.

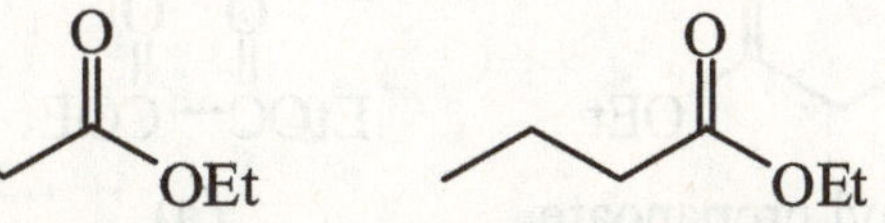

Ethyl propanoate   Ethyl butanoate

The strategy to this problem is similar to that of 15.19. Each of these esters can form an enolate ion, giving a total of two different enolates, which are shown below as (a) and (b). In turn, each one of these enolates can attack ethyl propanoate or ethyl butanoate, giving rise to four product combinations. Each product exists as a racemic mixture.

(a)

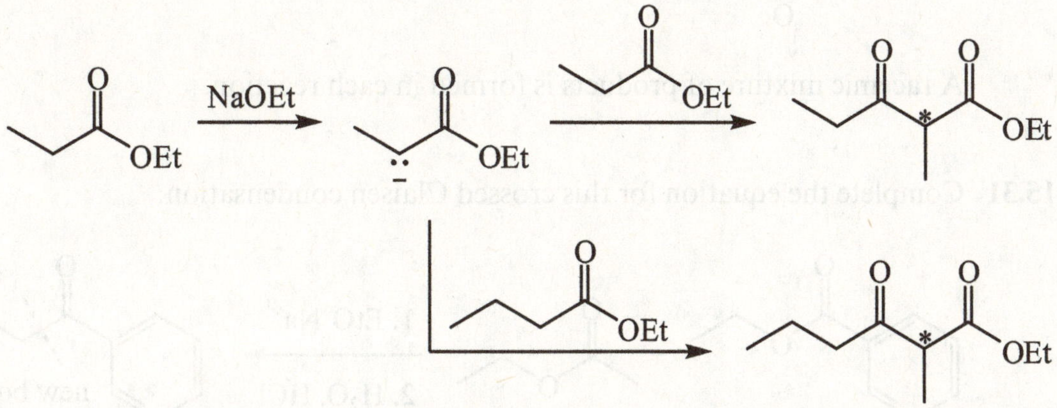

(b)

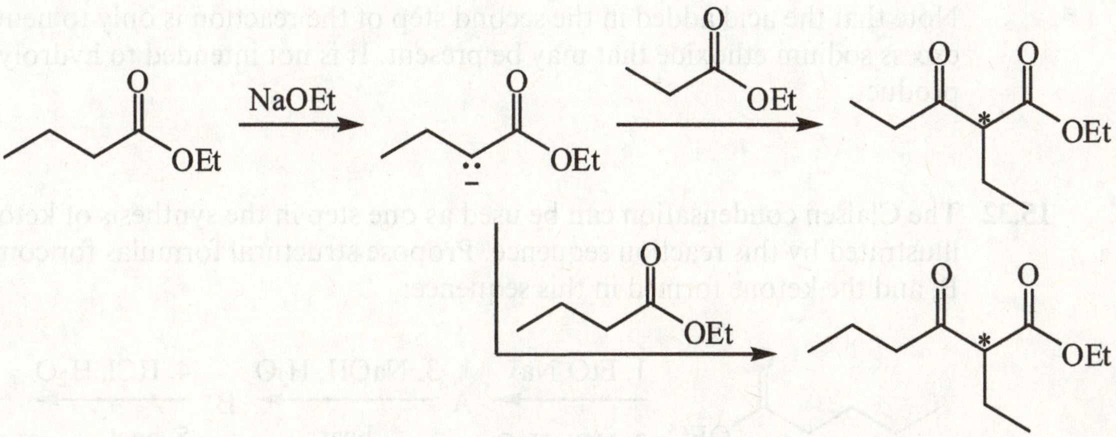

15.30 Draw a structural formula for the β-ketoester formed in the crossed Claisen condensation of ethyl propanoate with each ester:

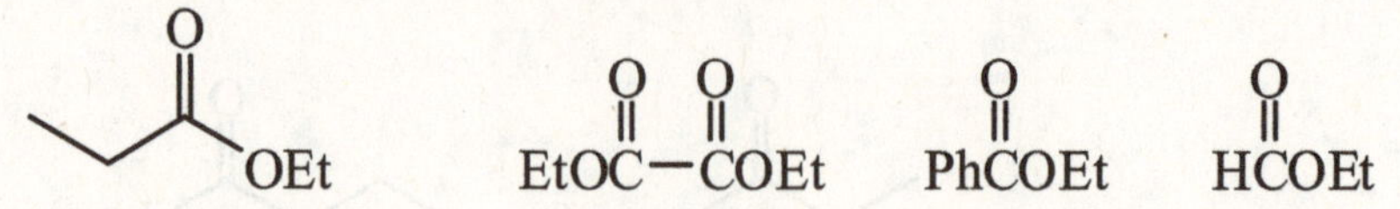
OEt
EtOC—COEt
PhCOEt
HCOEt

Ethyl propanoate (a) (b) (c)

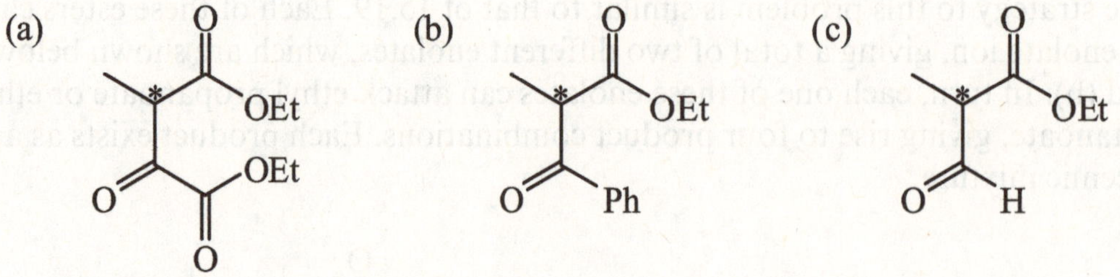
(a)
(b)
(c)
OEt
Ph
H

A racemic mixture of products is formed in each reaction.

15.31 Complete the equation for this crossed Claisen condensation:

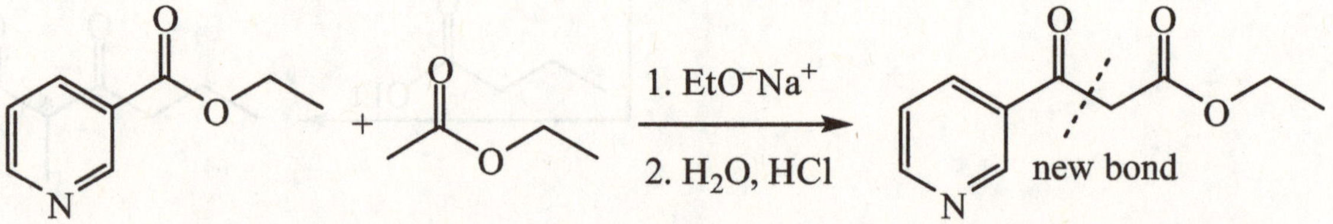
1. EtO⁻Na⁺
2. H₂O, HCl
new bond

Note that the acid added in the second step of the reaction is only to neutralize any excess sodium ethoxide that may be present. It is not intended to hydrolyze the ester product.

15.32 The Claisen condensation can be used as one step in the synthesis of ketones, as illustrated by this reaction sequence. Propose structural formulas for compounds A, B, and the ketone formed in this sequence:

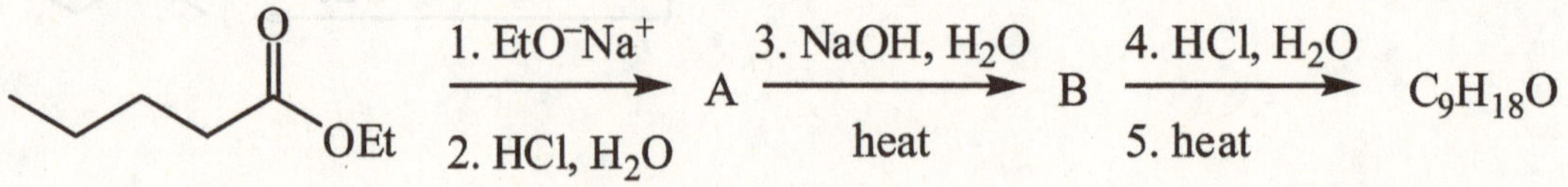
OEt
1. EtO⁻Na⁺
2. HCl, H₂O
A
3. NaOH, H₂O
heat
B
4. HCl, H₂O
5. heat
C₉H₁₈O

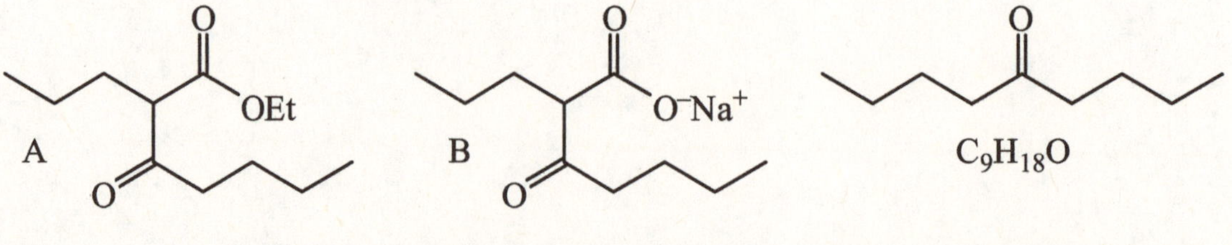
OEt
A
O⁻Na⁺
B
C₉H₁₈O

15.33 Draw a structural formula for the product of treating each diester with sodium ethoxide followed by acidification with HCl (*Hint*: These are Dieckmann condensations.):

(a)

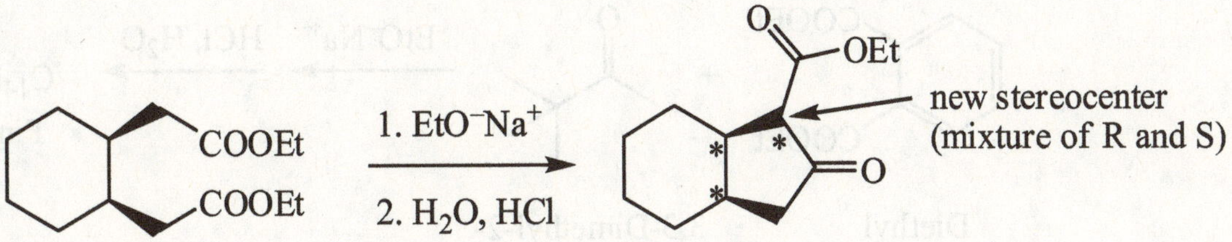

new stereocenter
(mixture of R and S)

1. EtO⁻Na⁺

2. H₂O, HCl

(b)

1. EtO⁻Na⁺

2. H₂O, HCl

racemic
mixture

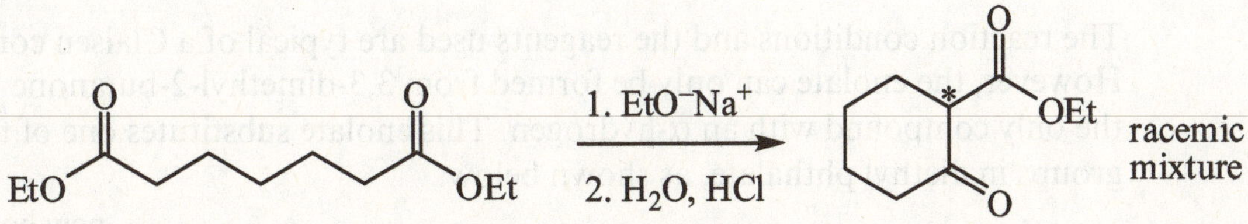

15.34 Claisen condensation between diethyl phthalate and ethyl acetate followed by saponification, acidification, and decarboxylation forms a diketone, $C_9H_6O_2$. Propose structural formulas for compounds A, B, and the diketone:

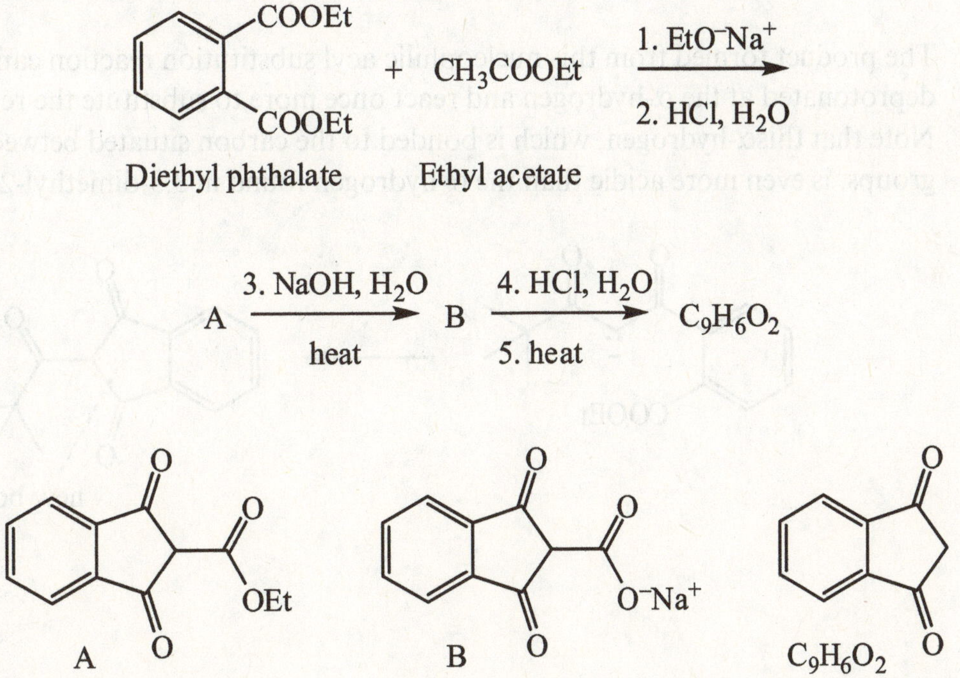

COOEt

+ CH₃COOEt

Diethyl phthalate   Ethyl acetate

1. EtO⁻Na⁺

2. HCl, H₂O

$$A \xrightarrow[\text{heat}]{3.\ NaOH,\ H_2O} B \xrightarrow[\text{5. heat}]{4.\ HCl,\ H_2O} C_9H_6O_2$$

A

B

$C_9H_6O_2$

15.35 The rodenticide and insecticide pindone is synthesized by the following sequence of reactions. Propose a structural formula for pindone.

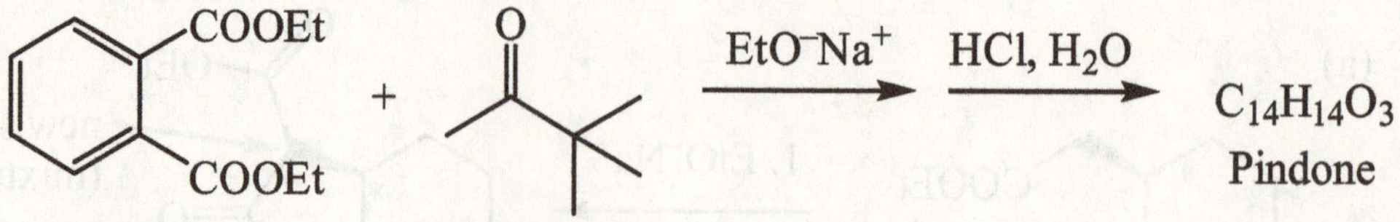
COOEt
COOEt
O
EtO⁻Na⁺
HCl, H₂O
C₁₄H₁₄O₃
Pindone

Diethyl phthalate   3,3-Dimethyl-2-butanone

The reaction conditions and the reagents used are typical of a Claisen condensation. However, the enolate can only be formed from 3,3-dimethyl-2-butanone because it is the only compound with an α-hydrogen. This enolate substitutes one of the ethoxy groups in diethyl phthalate, as shown below.

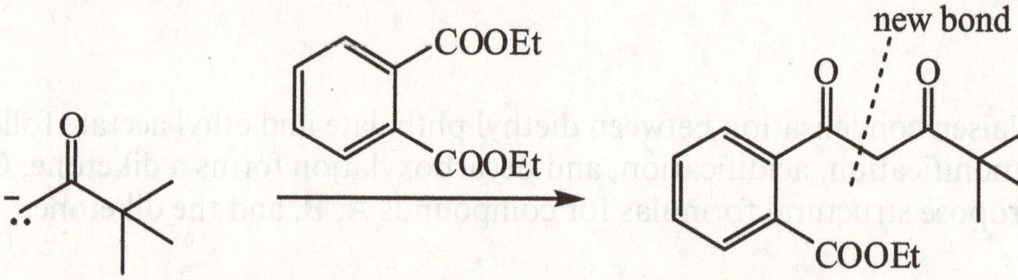
COOEt
COOEt
O
new bond
O
O
COOEt

The product formed from this nucleophilic acyl substitution reaction can be deprotonated at the α-hydrogen and react once more to substitute the remaining ester. Note that thisα-hydrogen, which is bonded to the carbon situated between two C=O groups, is even more acidic than the α-hydrogen found in 3,3-dimethyl-2-butanone.

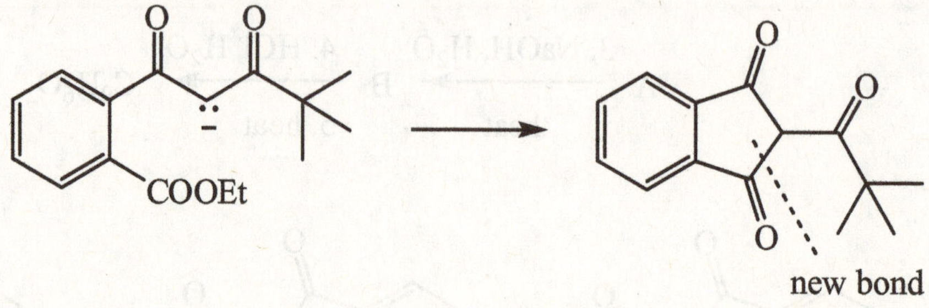
O
O
COOEt
O
O
O
new bond

The Michael Reaction

15.36 Show the product of the Michael reaction of each α,β-unsaturated carbonyl compound:

In reactions that form a new stereocenter, both R and S at each stereocenter are formed.

(a)

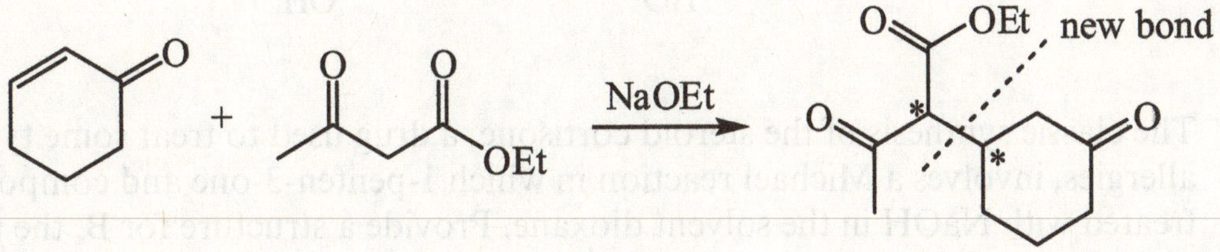

(b)

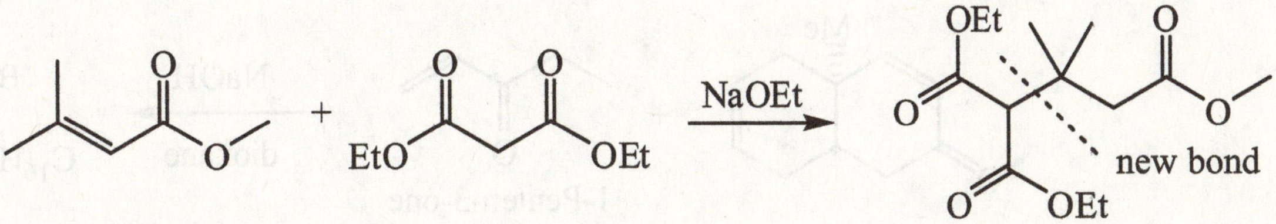

(c)

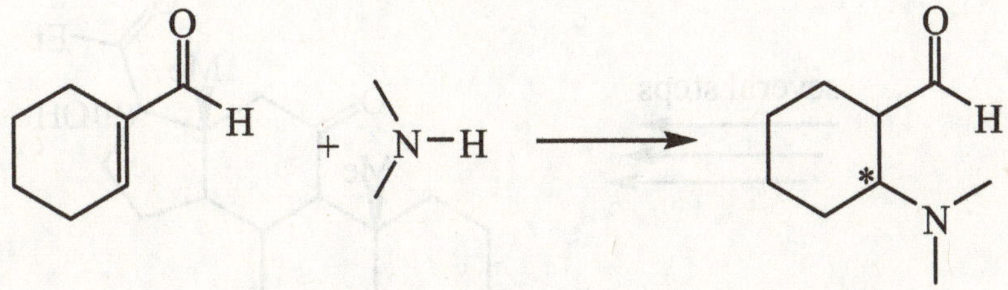

15.37 Show the outcomes of subjecting the Michael reaction products in Problems 15.36a and 15.36b to hydrolysis, followed by acidification, followed by thermal decarboxylation.

(a)

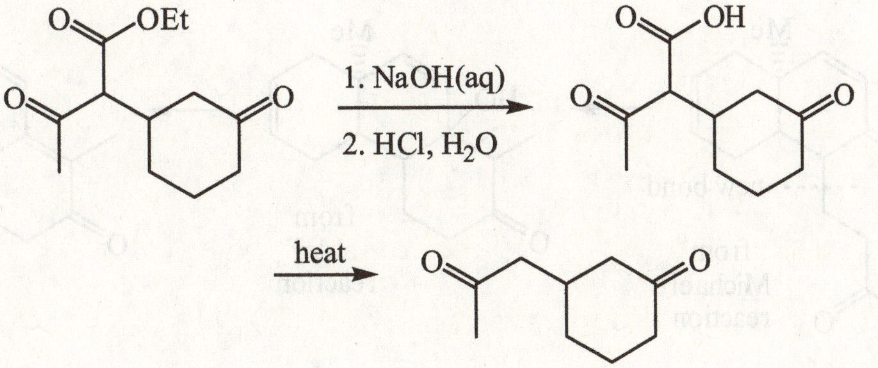

(b)

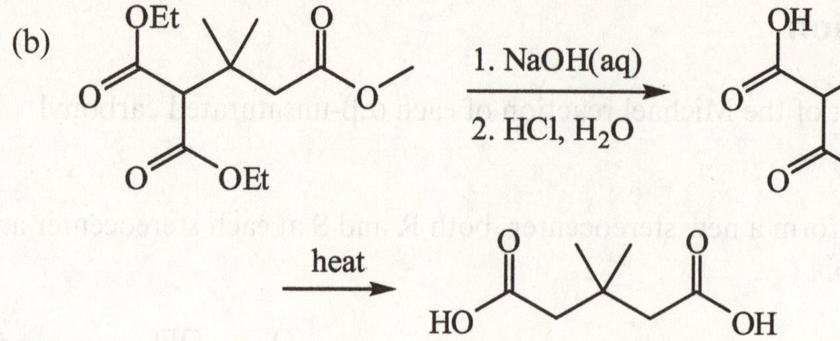

15.38 The classic synthesis of the steroid cortisone, a drug used to treat some types of allergies, involves a Michael reaction in which 1-penten-3-one and compound A are treated with NaOH in the solvent dioxane. Provide a structure for B, the product of this reaction.

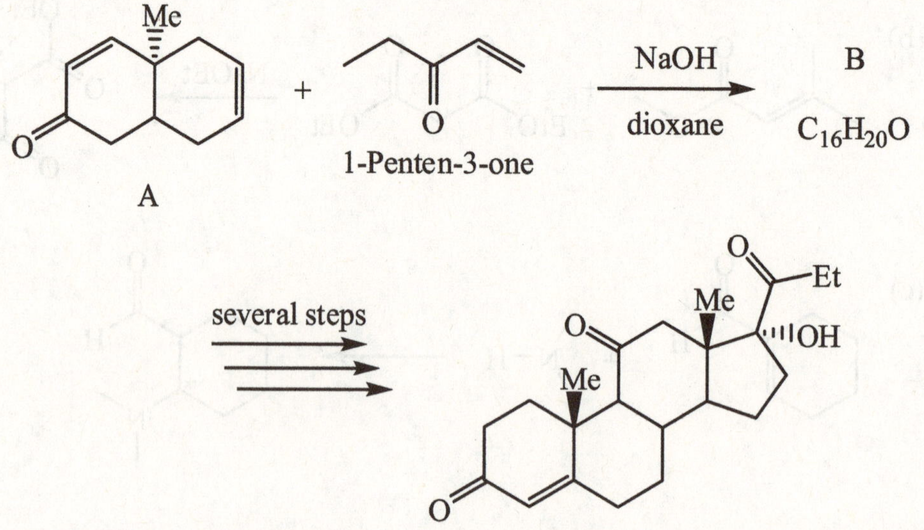

1-Penten-3-one

$$\xrightarrow[\text{dioxane}]{\text{NaOH}} \quad \begin{array}{c} \text{B} \\ C_{16}H_{20}O \end{array}$$

several steps

The molecular formula of **A** suggests that the reaction involves the elimination of water. The compound initially formed after the Michael reaction undergoes an intramolecular aldol reaction to give compound **A**.

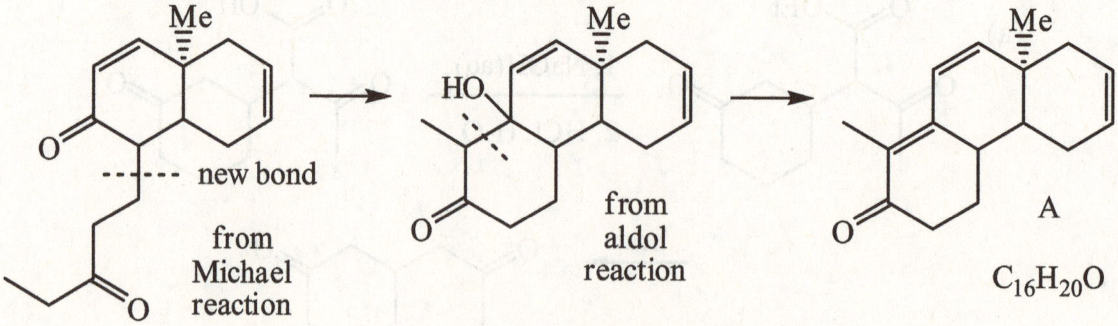

new bond

from Michael reaction

from aldol reaction

A

$C_{16}H_{20}O$

Synthesis

15.39 Fentanyl is a nonopoid (nonmorphinelike) analgesic used for the relief of severe pain. It is approximately 50 times more potent in humans than morphine itself. One synthesis for fentanyl begins with 2-phenylethanamine:

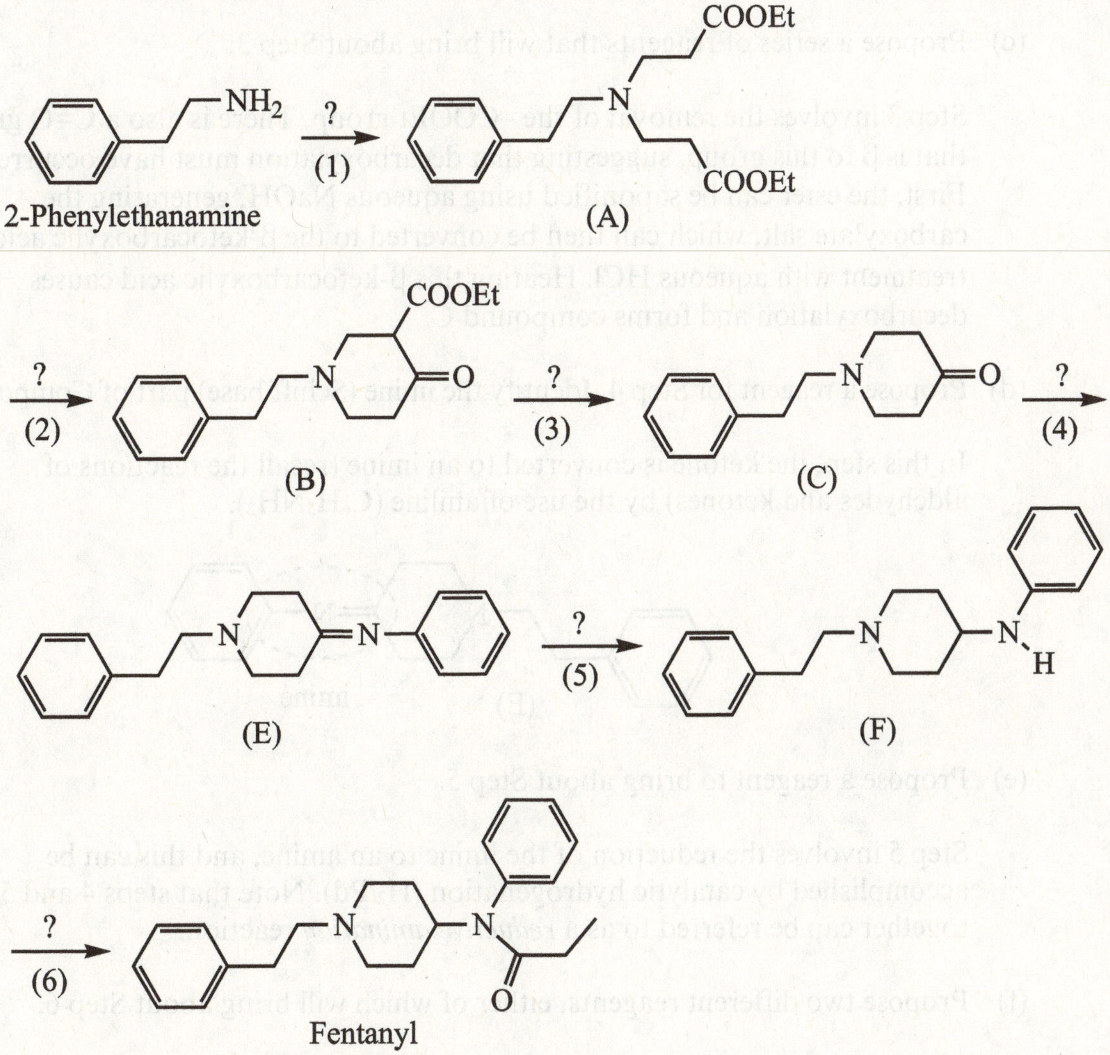

2-Phenylethanamine $\xrightarrow[\text{(1)}]{\text{?}}$ (A)

$\xrightarrow[\text{(2)}]{\text{?}}$ (B) $\xrightarrow[\text{(3)}]{\text{?}}$ (C) $\xrightarrow[\text{(4)}]{\text{?}}$

(E) $\xrightarrow[\text{(5)}]{\text{?}}$ (F)

$\xrightarrow[\text{(6)}]{\text{?}}$ Fentanyl

(a) Propose a reagent for Step 1. Name the type of reaction that occurs in this step.

A suitable reagent is the ester ethyl 2-propenoate, and a Michael addition is performed twice.

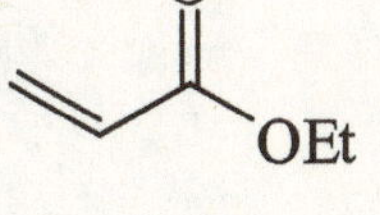

(b) Propose a reagent to bring about Step 2. Name the type of reaction that takes place in this step.

In this step, an intramolecular reaction forms a new ring. The reaction is a Dieckmann condensation and requires the use of a strong base, such as sodium ethoxide, followed by protonation with an aqueous acid, such as HCl in water.

(c) Propose a series of reagents that will bring about Step 3.

Step 3 involves the removal of the –COOEt group. There is also a C=O group that is β to this group, suggesting that decarboxylation must have occurred. First, the ester can be saponified using aqueous NaOH, generating the carboxylate salt, which can then be converted to the β-ketocarboxylic acid by treatment with aqueous HCl. Heating this β-ketocarboxylic acid causes decarboxylation and forms compound C.

(d) Propose a reagent for Step 4. Identify the imine (Schiff base) part of Compound E.

In this step, the ketone is converted to an imine (recall the reactions of aldehydes and ketones) by the use of aniline ($C_6H_5NH_2$).

(e) Propose a reagent to bring about Step 5.

Step 5 involves the reduction of the imine to an amine, and this can be accomplished by catalytic hydrogenation (H_2/Pd). Note that steps 4 and 5 together can be referred to as a *reductive amination* reaction.

(f) Propose two different reagents, either of which will bring about Step 6.

In this step, the amine is converted to an amide, which requires the use of an acid derivative that is more reactive than the amide. There are three possibilities that can be used, and they are the acid chloride, the anhydride, and the ester.

(g) Is fentanyl chiral? Explain.

No. Fentanyl is achiral because it does not have any stereocenters.

15.40 Meclizine is an antiemetic. (It helps prevent or at least lessen the vomiting associated with motion sickness, including seasickness.) Among the names of the over-the-counter preparations of meclizine are Bonine, Sea-Legs, Antivert, and Navicalm. Meclizine can be produced by the following series of reactions:

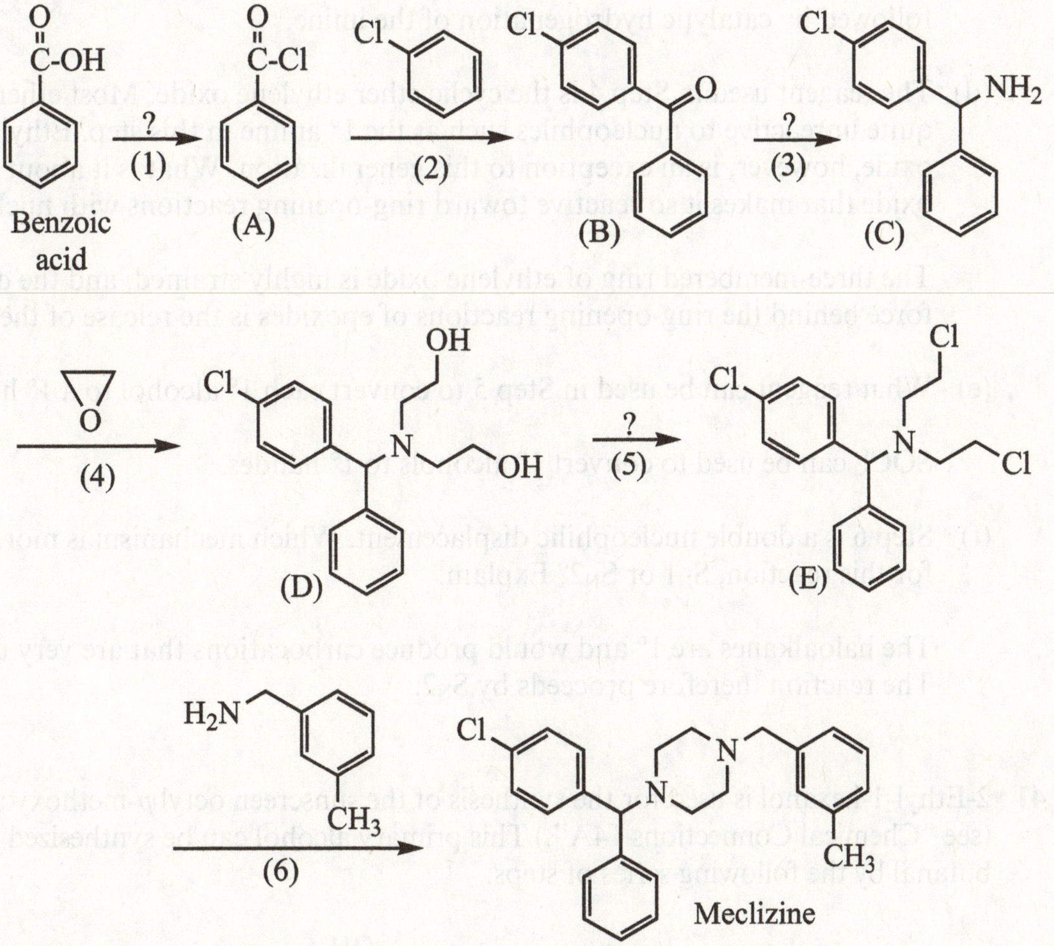

(a) Name the functional group in (A). What reagent is most commonly used to convert a carboxyl group to this functional group?

The functional group is an acid chloride, which is commonly made by treating a carboxyl group with SOCl₂.

(b) The catalyst for Step 2 is aluminum chloride, AlCl₃. Name the type of reaction that occurs in this step. The product shown here has the orientation of the new group para to the chlorine atom of chlorobenzene. Suppose you were not told the orientation of the new group. Would you have predicted it to be ortho, meta, or para to the chlorine atom? Explain.

The reaction in this step is a Friedel-Crafts acylation reaction. The chlorine substituent in chlorobenzene is an ortho-para director.

(c) What set of reagents can be used in Step 3 to convert the C=O group to an −NH₂ group?

The reaction is a reductive amination and can be performed by treating the carbonyl compound with NH₃ in the presence of a catalytic amount of acid, followed by catalytic hydrogenation of the imine.

(d) The reagent used in Step 4 is the cyclic ether ethylene oxide. Most ethers are quite unreactive to nucleophiles such as the 1° amine in this step. Ethylene oxide, however, is an exception to this generalization. What is it about ethylene oxide that makes it so reactive toward ring-opening reactions with nucleophiles?

The three-membered ring of ethylene oxide is highly strained, and the driving force behind the ring-opening reactions of epoxides is the release of the strain.

(e) What reagent can be used in Step 5 to convert each 1° alcohol to a 1° halide?

SOCl₂ can be used to convert 1° alcohols to 1° halides.

(f) Step 6 is a double nucleophilic displacement. Which mechanism is more likely for this reaction, S_N1 or S_N2? Explain.

The haloalkanes are 1° and would produce carbocations that are very unstable. The reaction therefore proceeds by S_N2.

15.41 2-Ethyl-1-hexanol is used for the synthesis of the sunscreen octyl p-methoxycinnamate (see "Chemical Connections 14A".) This primary alcohol can be synthesized from butanal by the following series of steps:

(a) Propose a reagent to bring about Step 1. What name is given to this type of reaction?

Sodium hydroxide can be used to perform this reaction, an aldol reaction.

(b) Propose a reagent for Step 2.

The aldol product can be dehydrated by heating it in NaOH or H_2SO_4.

(c) Propose a reagent for Step 3.

Catalytic hydrogenation using H_2/Pd will reduce both the alkene and the aldehyde.

(d) Following is a structural formula for the commercial sunscreening ingredient octyl-*p*-methoxycinnamate. What carboxylic acid would you use to form this ester? How would you bring about the esterification reaction?

Fischer esterification of 2-ethyl-1-hexanol with *p*-methoxycinnamic acid, in the presence of a sulfuric acid catalyst, generates octyl-*p*-methoxycinnamate.

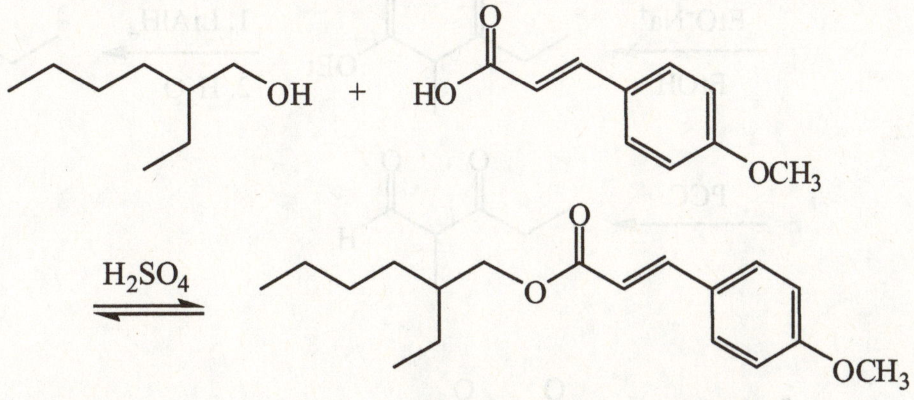

Octyl *p*-methoxycinnamate

Chemical Transformations

15.42 Test your cumulative knowledge of the reactions learned thus far by completing the following chemical transformations. *Note: Some will require more than one step.*

In reactions that form a new stereocenter, both R and S at each stereocenter are formed.

(a)

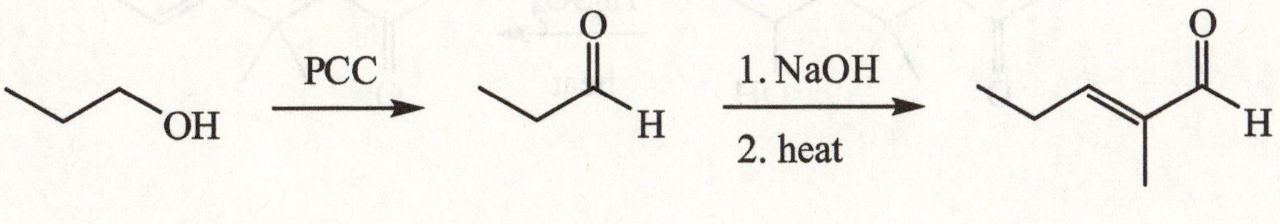

(b)

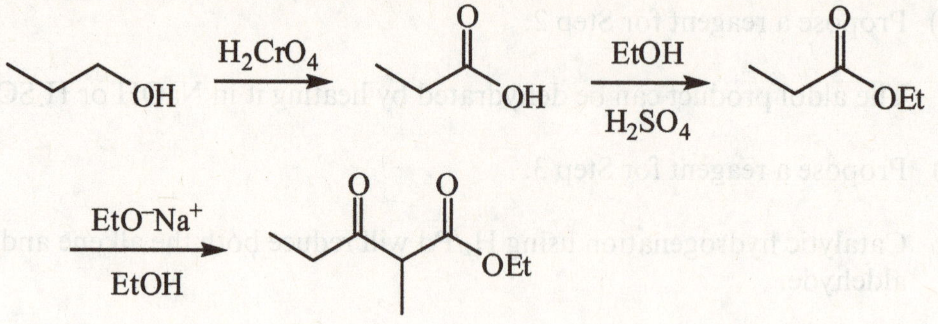

H_2CrO_4

$\xrightarrow{EtOH, H_2SO_4}$

$\xrightarrow[EtOH]{EtO^-Na^+}$

(c)

H_2CrO_4

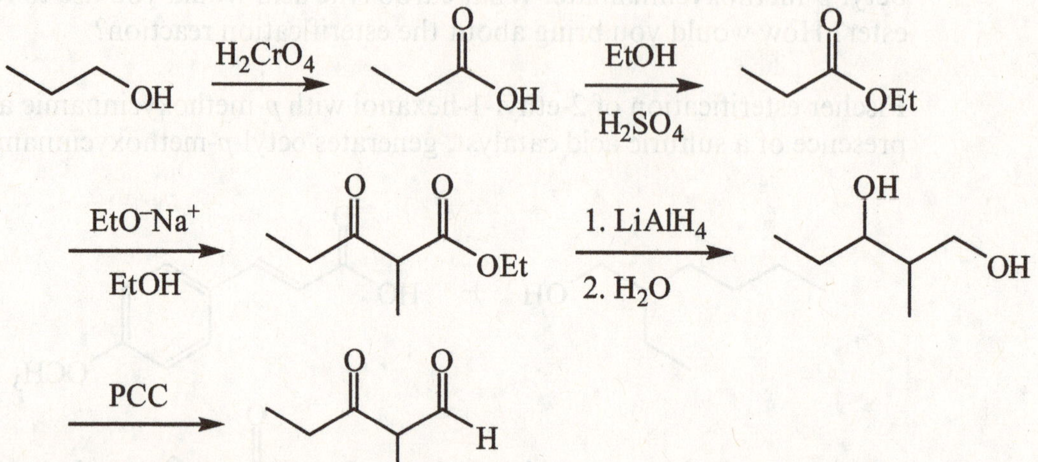

$\xrightarrow[H_2SO_4]{EtOH}$

$\xrightarrow[EtOH]{EtO^-Na^+}$

$\xrightarrow[2.\ H_2O]{1.\ LiAlH_4}$

$\xrightarrow{PCC}$

(d)

$\xrightarrow[EtO^-Na^+,\ EtOH]{}$ (OEt)

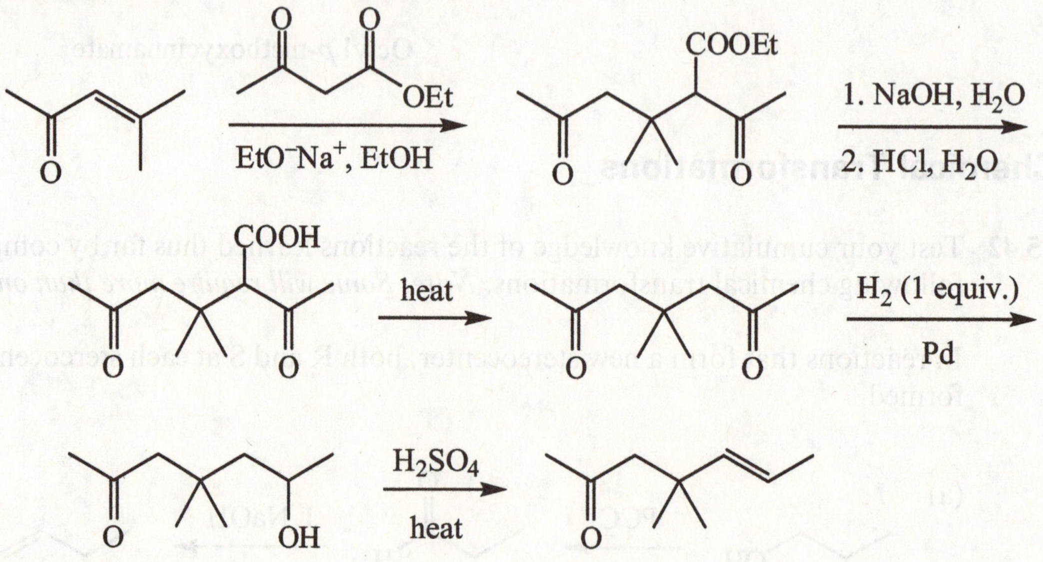

$\xrightarrow[2.\ HCl,\ H_2O]{1.\ NaOH,\ H_2O}$

COOEt

$\xrightarrow{heat}$

$\xrightarrow[Pd]{H_2\ (1\ equiv.)}$

COOH

$\xrightarrow[heat]{H_2SO_4}$

(e)

(f)

(g)

(h)

(i)

(j)

(k)

(l)

(m)

(n)

Looking Ahead

15.43 The following reaction is one of the 10 steps in glycolysis, a series of enzyme-catalyzed reactions by which glucose is oxidized to two molecules of pyruvate. Show that this step is the reverse of an aldol reaction.

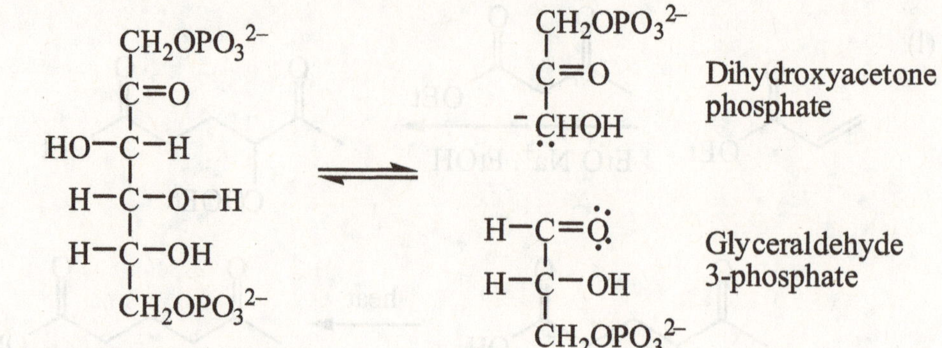

$$
\begin{array}{c}
\text{CH}_2\text{OPO}_3^{2-} \\
| \\
\text{C}{=}\text{O} \\
| \\
\text{HO}{-}\text{C}{-}\text{H} \\
| \\
\text{H}{-}\text{C}{-}\text{O}{-}\text{H} \\
| \\
\text{H}{-}\text{C}{-}\text{OH} \\
| \\
\text{CH}_2\text{OPO}_3^{2-}
\end{array}
\quad \rightleftharpoons \quad
\begin{array}{c}
\text{CH}_2\text{OPO}_3^{2-} \\
| \\
\text{C}{=}\text{O} \\
| \\
{}^-\text{CHOH}
\end{array}
\quad
\begin{array}{c}
\text{Dihydroxyacetone} \\
\text{phosphate}
\end{array}
$$

$$
\begin{array}{c}
\text{H}{-}\text{C}{=}\text{O} \\
| \\
\text{H}{-}\text{C}{-}\text{OH} \\
| \\
\text{CH}_2\text{OPO}_3^{2-}
\end{array}
\quad
\begin{array}{c}
\text{Glyceraldehyde} \\
\text{3-phosphate}
\end{array}
$$

Fructose 1,6-bisphosphate

A helpful strategy is to think about a "forward" aldol reaction between dihydroxyacetone phosphate and glyceraldehyde 3-phosphate to form fructose 1,6-bisphosphate. An important feature of aldol reactions is that they generate β-hydroxyaldehydes or β-hydroxyketones. In the forward reaction, the enolate of dihydroxyacetone phosphate attacks the aldehyde, generating a new carbon-carbon bond. The −OH group (formerly the aldehyde) is β to the C=O in the product.

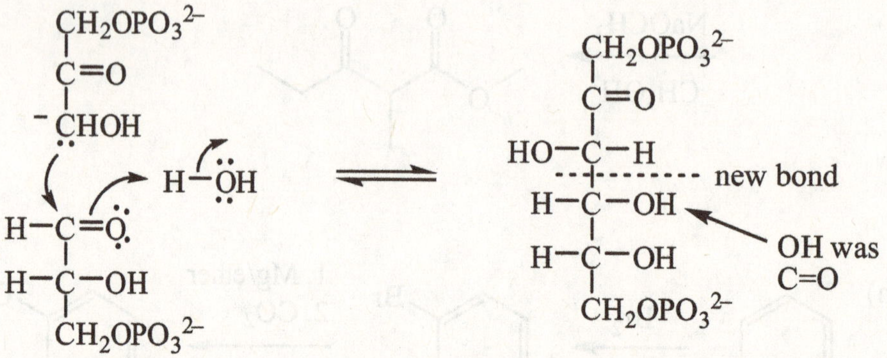

$$
\begin{array}{c}
\text{CH}_2\text{OPO}_3^{2-} \\
| \\
\text{C}{=}\text{O} \\
| \\
{}^-\text{CHOH} \\
\\
\text{H}{-}\text{C}{=}\text{O} \\
| \\
\text{H}{-}\text{C}{-}\text{OH} \\
| \\
\text{CH}_2\text{OPO}_3^{2-}
\end{array}
\quad
\text{H}{-}\text{OH}
\quad \rightleftharpoons \quad
\begin{array}{c}
\text{CH}_2\text{OPO}_3^{2-} \\
| \\
\text{C}{=}\text{O} \\
| \\
\text{HO}{-}\text{C}{-}\text{H} \quad \text{---- new bond} \\
| \\
\text{H}{-}\text{C}{-}\text{OH} \quad \longleftarrow \text{OH was} \\
| \qquad\qquad\qquad\;\; \text{C}{=}\text{O} \\
\text{H}{-}\text{C}{-}\text{OH} \\
| \\
\text{CH}_2\text{OPO}_3^{2-}
\end{array}
$$

In a reverse aldol reaction, the opposite occurs. Deprotonation of the β-hydroxyl group in fructose 1,6-bisphosphate initiates the reverse reaction, breaking the α,β bond. Protonation of the released enolate gives the products.

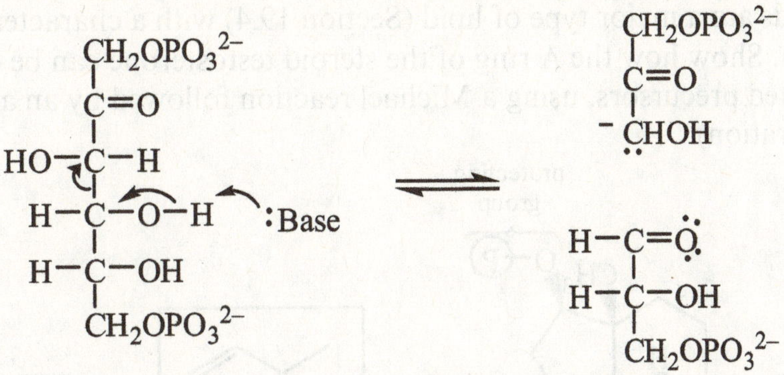

$$CH_2OPO_3{}^{2-}$$
$$C=O$$
$$HO-C-H$$
$$H-C-O-H \quad :Base$$
$$H-C-OH$$
$$CH_2OPO_3{}^{2-}$$

$$\rightleftharpoons$$

$$CH_2OPO_3{}^{2-}$$
$$C=O$$
$$^-CHOH$$

$$H-C=\overset{..}{\underset{..}{O}}$$
$$H-C-OH$$
$$CH_2OPO_3{}^{2-}$$

15.44 The following reaction is the fourth in the set of four enzyme-catalyzed steps by which the hydrocarbon chain of a fatty acid is oxidized, two carbons at a time, to acetyl-coenzyme A. Show that this reaction is the reverse of a Claisen condensation.

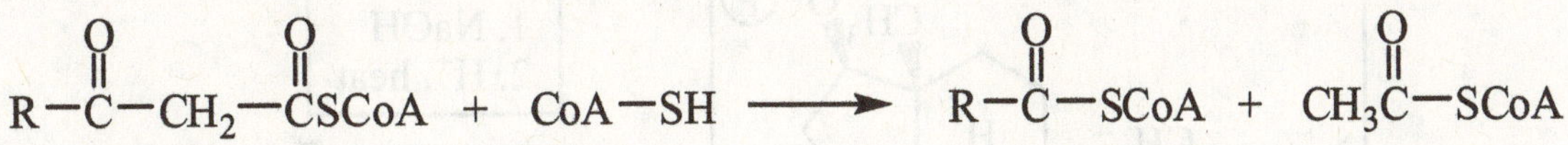

$$\underset{\beta\text{-Ketoacyl-CoA}}{R-\overset{O}{\overset{\|}{C}}-CH_2-\overset{O}{\overset{\|}{C}}SCoA} + CoA-SH \longrightarrow \underset{\text{An acyl-CoA}}{R-\overset{O}{\overset{\|}{C}}-SCoA} + \underset{\text{Acetyl-CoA}}{CH_3\overset{O}{\overset{\|}{C}}-SCoA}$$

Like Problem 15.43, consider a "forward" Claisen condensation that gives a β-ketoacyl-CoA. In this reaction, the enolate anion of acetyl-CoA substitutes the thiol portion of the thioester (the acyl-CoA) via a nucleophilic acyl substitution reaction.

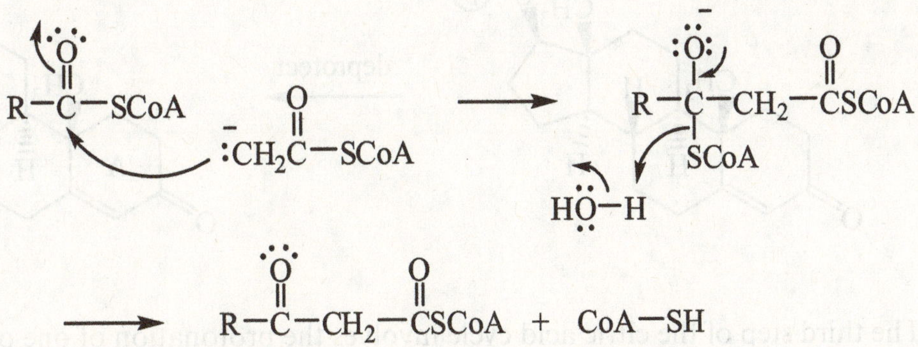

$$R-\overset{\overset{..}{\underset{..}{O}}}{\overset{\|}{C}}-SCoA \quad \overset{O}{\underset{\|}{:CH_2C-SCoA}} \longrightarrow R-\overset{\overset{:\overset{..}{O}:^-}{|}}{\underset{|}{C}}-CH_2-\overset{O}{\overset{\|}{C}}SCoA$$
$$SCoA$$
$$H\overset{..}{O}-H$$

$$\longrightarrow R-\overset{\overset{..}{\underset{..}{O}}}{\overset{\|}{C}}-CH_2-\overset{O}{\overset{\|}{C}}SCoA + CoA-SH$$

In a reverse Claisen condensation, CoAS⁻ is the nucleophile that initiates the reaction and breaks the carbon-carbon bond.

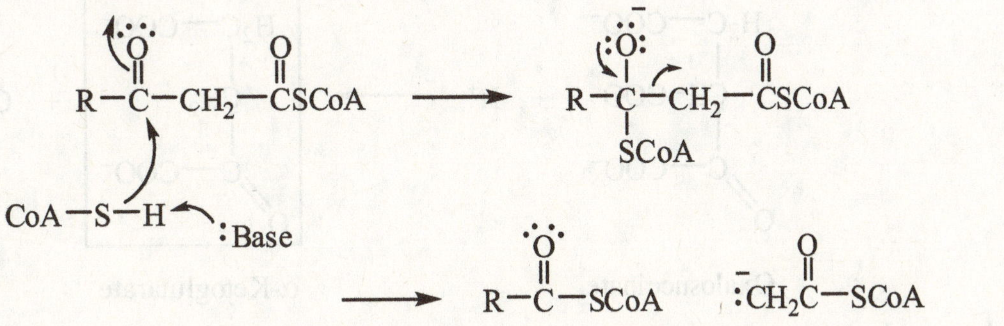

$$R-\overset{\overset{..}{\underset{..}{O}}}{\overset{\|}{C}}-CH_2-\overset{O}{\overset{\|}{C}}SCoA \longrightarrow R-\overset{\overset{:\overset{..}{O}:^-}{|}}{\underset{|}{C}}-CH_2-\overset{O}{\overset{\|}{C}}SCoA$$
$$SCoA$$

$$CoA-S-H \quad :Base$$

$$\longrightarrow R-\overset{\overset{..}{\underset{..}{O}}}{\overset{\|}{C}}-SCoA \quad \overset{O}{\underset{\|}{:CH_2C-SCoA}}$$

15.45 Steroids are a major type of lipid (Section 19.4) with a characteristic tetracyclic ring system. Show how the **A** ring of the steroid testosterone can be constructed from the indicated precursors, using a Michael reaction followed by an aldol reaction (with dehydration):

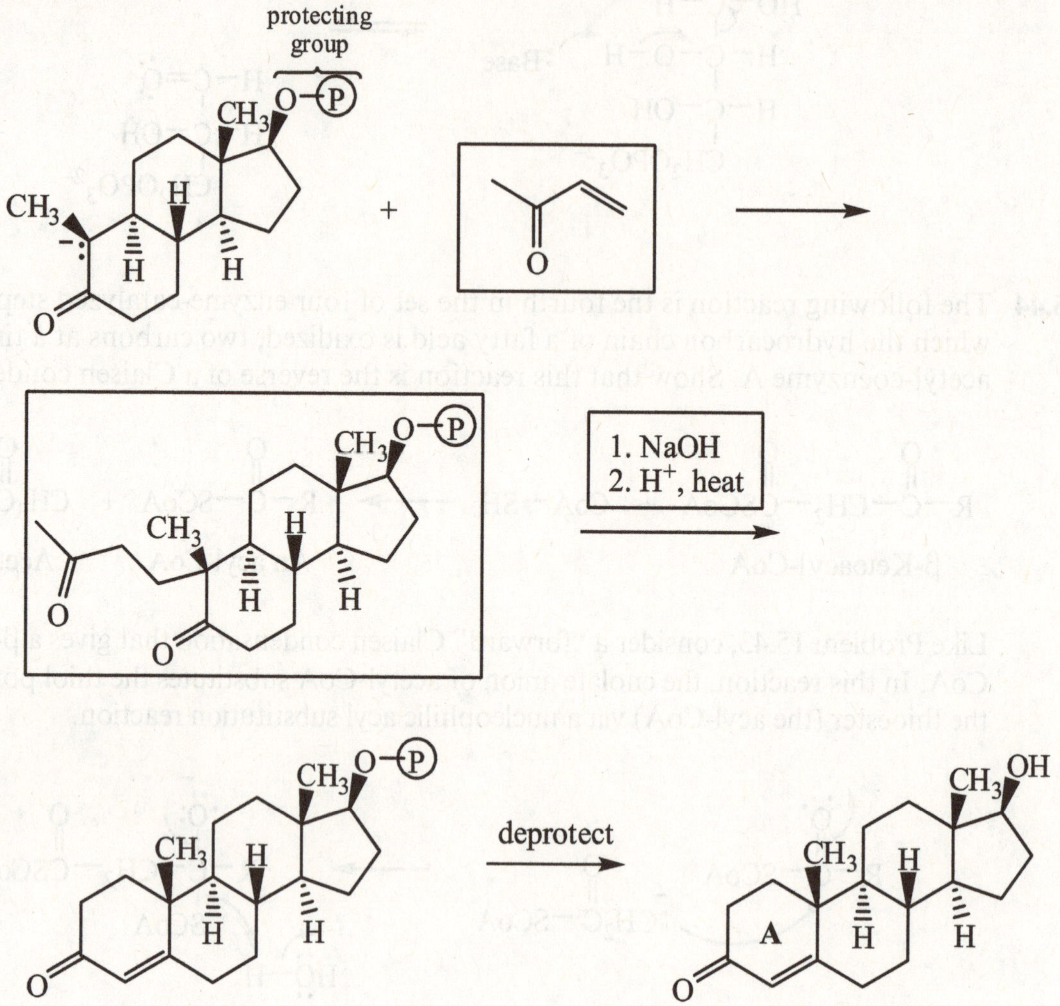

1. NaOH
2. H^+, heat

deprotect

15.46 The third step of the citric acid cycle involves the protonation of one of the carboxylate groups of oxalosuccinate, a β-ketoacid, followed by decarboxylation to form α-ketoglutarate. Write the structural formula of α-ketoglutarate.

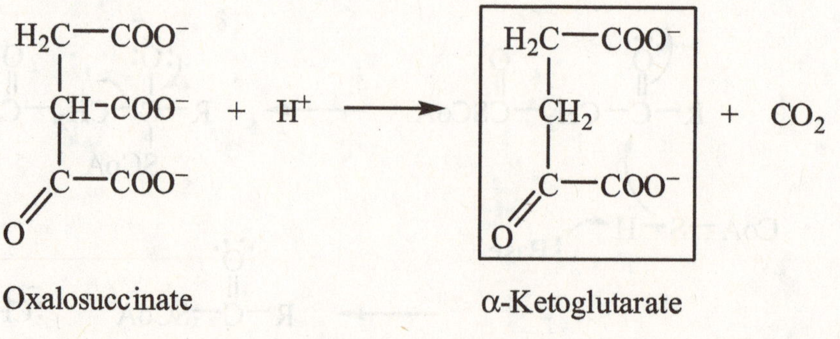

Oxalosuccinate + H^+ → α-Ketoglutarate + CO_2

Group Learning Activities

15.47 Nitroethane has a pK_a of 8.5, which makes it slightly more acidic than ethyl acetoacetate. Acetonitrile has a pK_a of 25, which makes it comparable in acidity to most esters. Account for the acidities of nitroethane and acetonitrile. As a group, decide whether the conjugate bases of these compounds could act similarly to enolates by drawing examples of such reactions.

The conjugate bases of both nitroethane and acetonitrile are stabilized by resonance, but the conjugate base of nitroethane is also stabilized by the electron-withdrawing effect of the positively charged nitrogen atom.

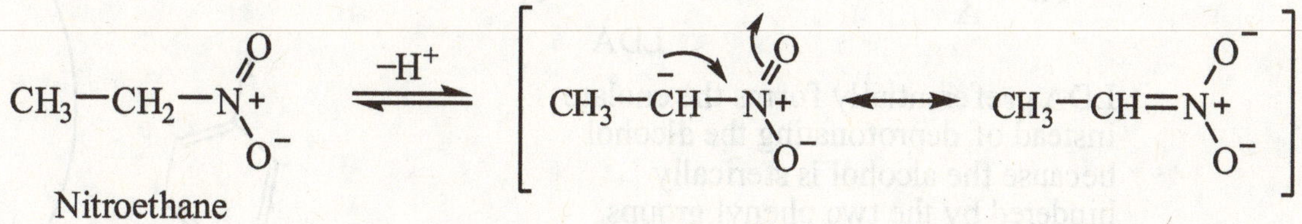

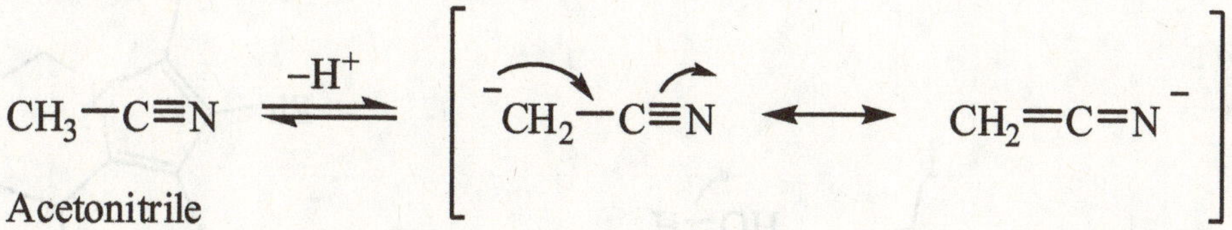

The two conjugate bases are carbanions and can therefore act as nucleophiles in a manner similar to enolates. They could, for example, add to carbonyl groups.

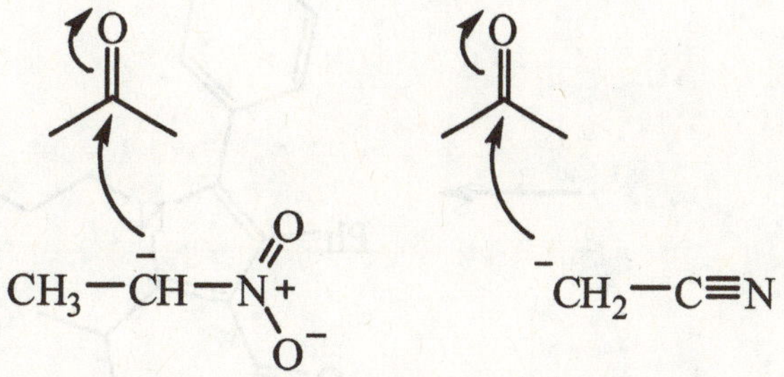

15.48 As a group, discuss why the two compounds shown do not undergo self- Claisen condensation reactions. Refer to Section 15.3A if needed.

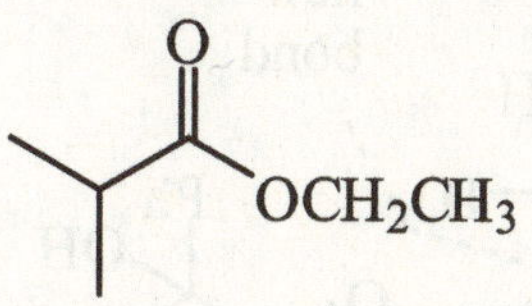

The enolate formed from this compound would be too sterically hindered to act as a nucleophile.

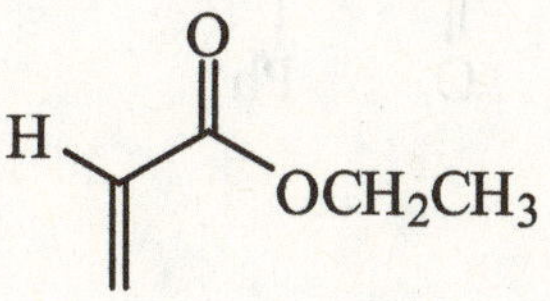

The enolate formed from this compound would undergo a Michael addition with another molecule of the compound.

15.49 One synthesis of atorvastatin, the common drug sold under the trade name Lipitor®
and used to treat high cholesterol, involves the generation of an enolate from **A**
followed by reaction with the fluorinated compound **B**. The enolate is generated
quantitatively by reaction with lithium diisopropylamide (LDA), a very strong
nitrogenous base. As a group, predict the product of the reaction of the enolate of **A**
with **B** and provide a mechanism for the reaction. Also, debate why LDA preferentially
forms the enolate rather than reacting with the more acidic alcohol proton in **A**.

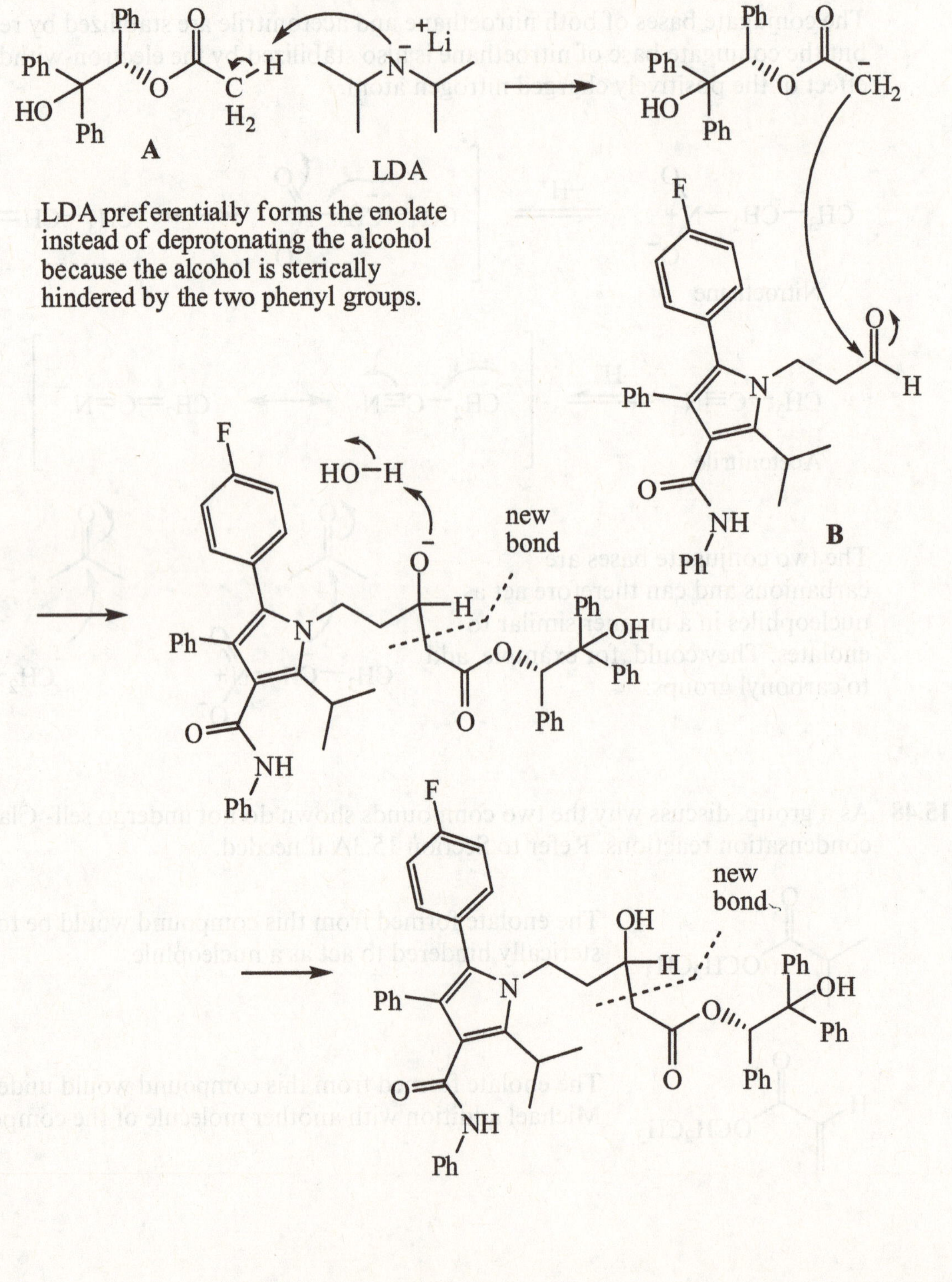

LDA preferentially forms the enolate
instead of deprotonating the alcohol
because the alcohol is sterically
hindered by the two phenyl groups.

Chapter 16: Organic Polymer Chemistry

Problems

16.1   Given the following structure, determine the polymer's repeat unit; redraw the structure, using the simplified parenthetical notation; and name the polymer:

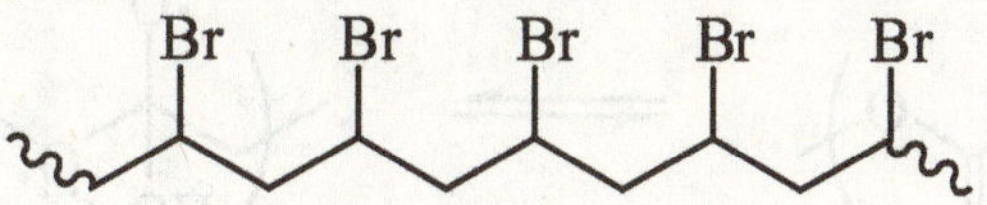

This polymer is derived from bromoethene (vinyl bromide) and is therefore called poly(vinyl bromide).

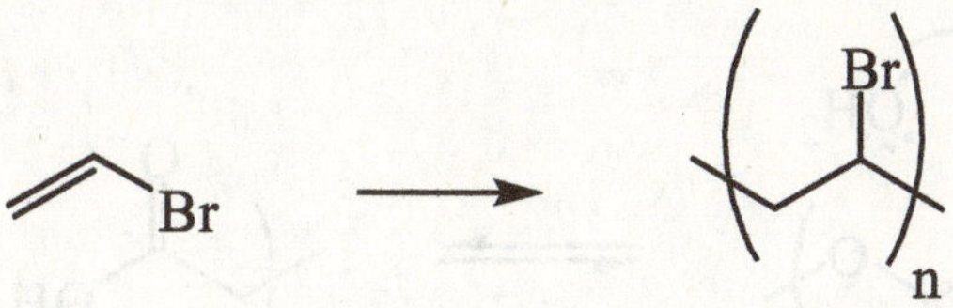

16.2   Write the repeating unit of the epoxy resin formed from the following reaction:

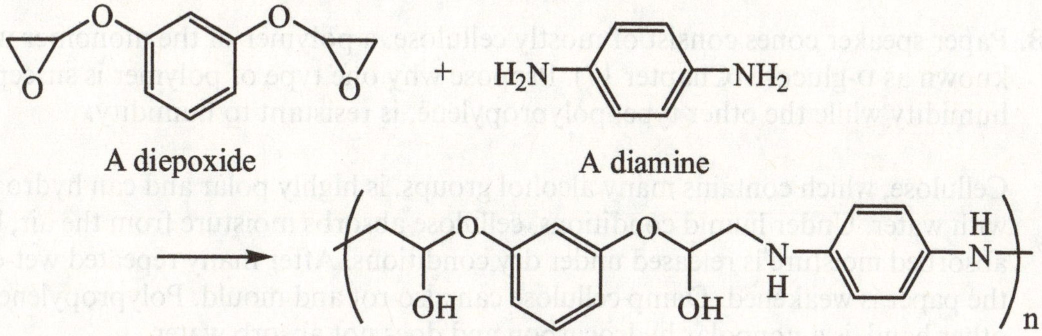

A diepoxide   A diamine

In this reaction, nucleophilic attack of the epoxide by the amine occurs at the carbon atom that is least sterically hindered.

Chemical Connections

16A. Propose a mechanism for the hydrolysis of one repeating unit of the copolymer poly(glycoic acid)-poly(lactic acid).

Each repeating unit of the copolymer consists of one monomer of each of glycolic and lactic acids linked together as an ester. The hydrolysis of the unit is the reverse of a Fischer esterification.

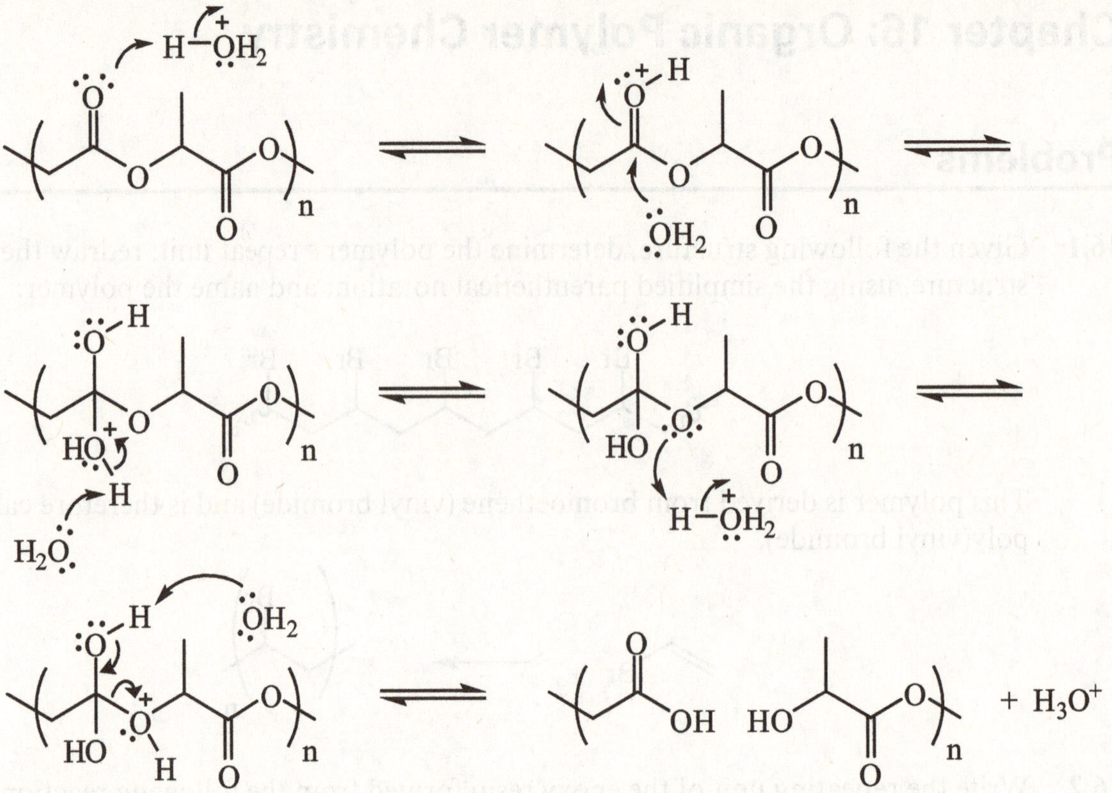

16B. Paper speaker cones consist of mostly cellulose, a polymer of the monomer unit known as D-glucose (Chapter 17). Propose why one type of polymer is susceptible to humidity while the other type, polypropylene, is resistant to humidity.

Cellulose, which contains many alcohol groups, is highly polar and can hydrogen bond with water. Under humid conditions, cellulose absorbs moisture from the air, but the absorbed moisture is released under dry conditions. After many repeated wet-dry cycles, the paper is weakened. Damp cellulose can also rot and mould. Polypropylene, on the other hand, is a nonpolar hydrocarbon and does not absorb water.

Quick Quiz

1. Radicals can undergo chain-growth polymerization. *True*. Chain-growth polymerization reactions generally involve the propagation of radicals.

2. A thermosetting plastic cannot be remelted. *True*. Thermosetting plastics can only be shaped when they are prepared, but they harden irreversibly and cannot be remelted.

3. Chain-transfer reactions can lead to branching in polymers. *True*. The transfer of a radical from the end of the polymer to another location within the polymer leads to branching.

4. Polymers that have low glass transition temperatures can behave as elastomers. *True*. At temperatures above the T_g value, the polymer will be elastic. A polymer with a T_g lower than room temperature will be elastic at room temperature.

5. A highly crystalline polymer will have a glass transition temperature. *False*. Glass transition temperatures are associated with amorphous, not crystalline, polymers.

6. Polymers can be named from the monomeric units from which they are derived. *True*. For example, the polymer polyethylene is prepared from ethylene. Note that the name polyethylene is used even though the polymer does not contain any double bonds.

7. Only compounds that have two or more functional groups can undergo step-growth polymerization. *True*. Two functional groups are needed to attach the monomer units.

8. The propagation step of a radical polymerization mechanism involves the reaction of a radical with another radical. *False*. The reaction of two radicals would lead to chain termination.

9. A radical is a molecule with an unpaired electron and a positive charge. *False*. A radical is any atom or molecule with one or more unpaired electrons. There is no charge requirement.

10. The term *plastics* can be used to refer to all polymers. *False*. *Plastics* are only the polymers that can be molded when hot and retain their shape when cooled.

11. The mechanism of a radical polymerization reaction involves 3 distinct steps. *True*. The three steps are radical initiation, propagation, and termination.

12. Hydrogen bonding will usually weaken the fibers of a polymer. *False*. Just as hydrogen bonding increases intermolecular forces, hydrogen bonding increases the strength of the fibers of a polymer.

13. A secondary radical is more stable than a tertiary radical. *False*. The relative stabilities of radicals are similar to the relative stabilities of carbocations. Tertiary radicals are more stable than secondary radicals.

14. A thermoplastic can be molded multiple times through heating and cooling. *True*. Thermoplastics, but not thermoset plastics, can be reversibly melted, molded, and cooled.

15. Ziegler-Natta Polymerization uses a titanium catalyst. *True*. The titanium catalyzes the polymerization of alkenes under moderate conditions.

End-of-Chapter Problems

Step-Growth Polymers

16.3   Identify the monomers required for the synthesis of each step-growth polymer:

(a)

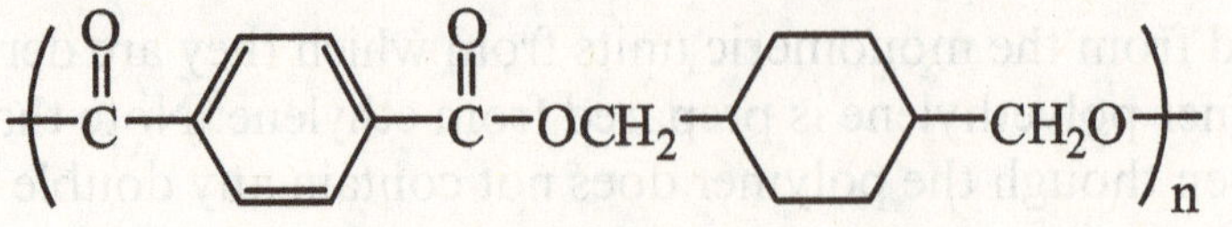

Kodel (a polyester)

Monomers: HO—C(=O)—⟨benzene⟩—C(=O)—OH   +   HOCH₂—⟨cyclohexane⟩—CH₂OH

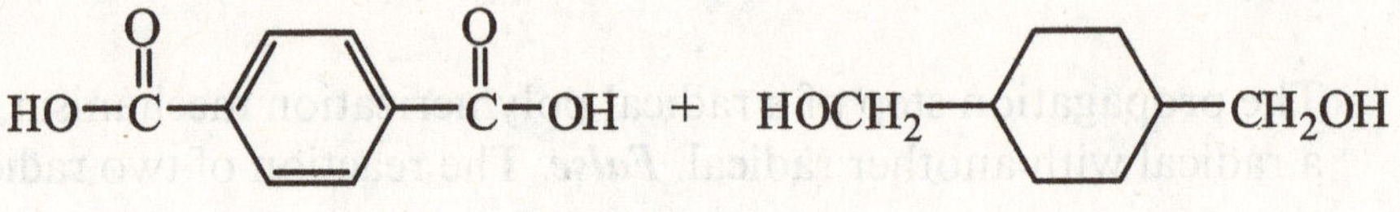

(b)

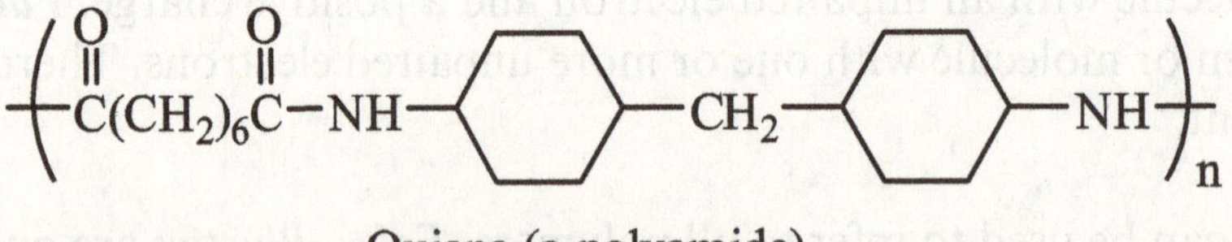

Quiana (a polyamide)

Monomers: HO—C(=O)(CH₂)₆C(=O)—OH   +   H₂N—⟨cyclohexane⟩—CH₂—⟨cyclohexane⟩—NH₂

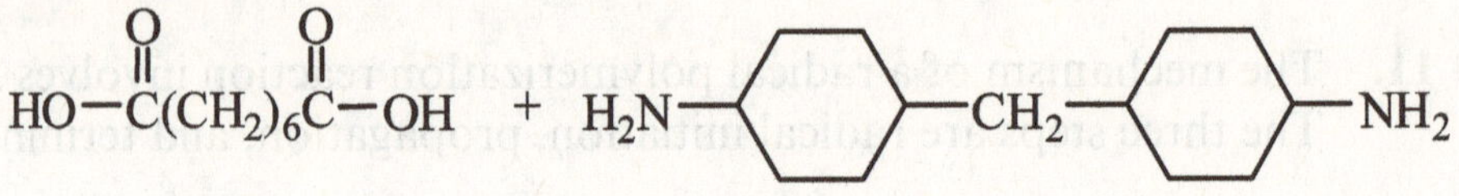

(c)

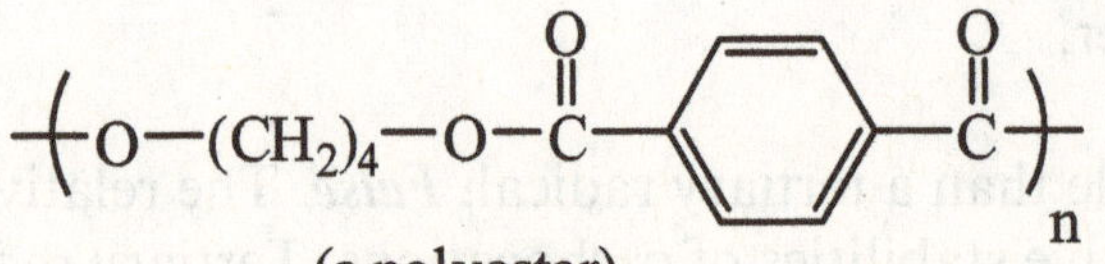

(a polyester)

Monomers: HO(CH₂)₄OH   +   HOOC—⟨benzene⟩—COOH

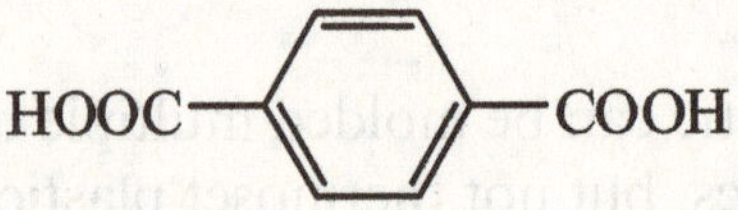

(d)

Nylon 6,10 (a polyamide)

Monomers: HOOC(CH₂)₈COOH   +   H₂N(CH₂)₆NH₂

16.4   Poly(ethylene terephthalate) (PET) can be prepared by the following reaction. Propose a mechanism for the step-growth reaction in this polymerization.

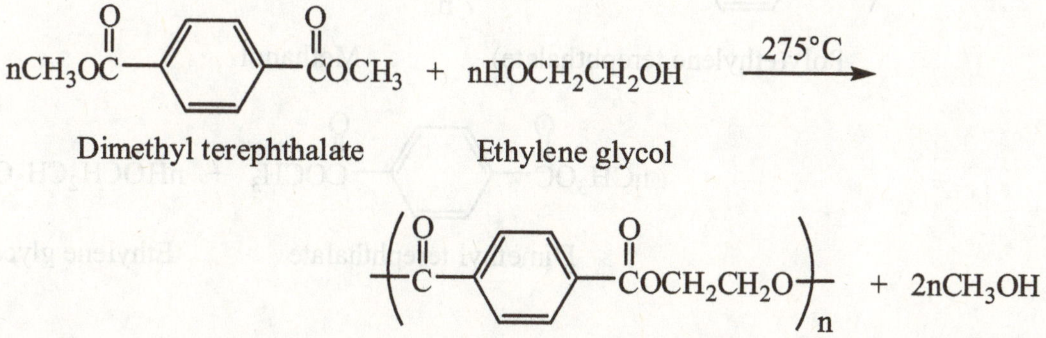

Dimethyl terephthalate   Ethylene glycol

The reaction is a transesterification. Although an acid catalyst is typically used to accelerate the reactions of esters, no catalyst is necessary under the extreme temperature used for the synthesis of PET. The high temperature also causes the methanol to boil off, thus causing the equilibrium reaction to favor the products. The mechanism below shows the formation of one of the ester bonds.

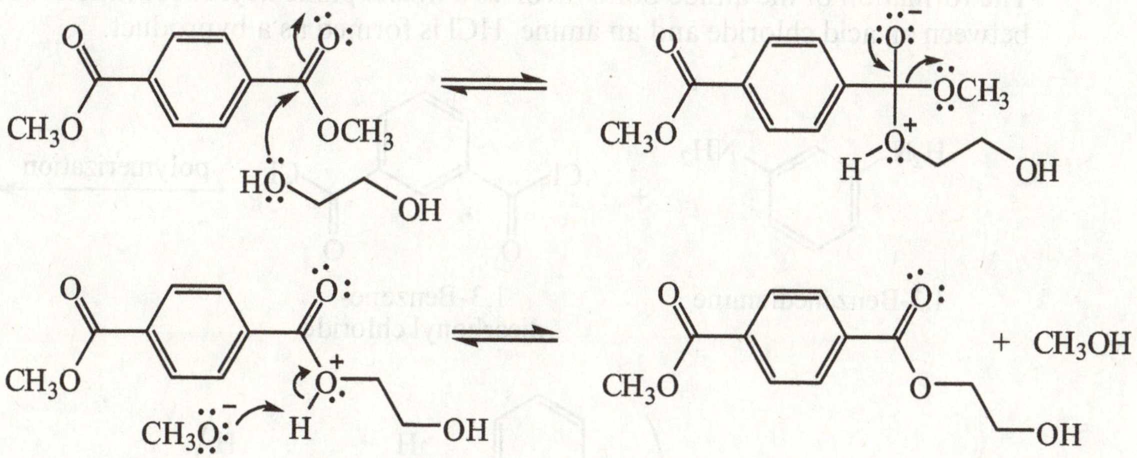

16.5   Currently, about 30% of PET soft-drink bottles are being recycled. In one recycling process, scrap PET is heated with methanol in the presence of an acid catalyst. The methanol reacts with the polymer, liberating ethylene glycol and dimethyl terephthalate. These monomers are then used as feedstock for the production of new PET products. Write an equation for the reaction of PET with methanol to give ethylene glycol and dimethyl terephthalate.

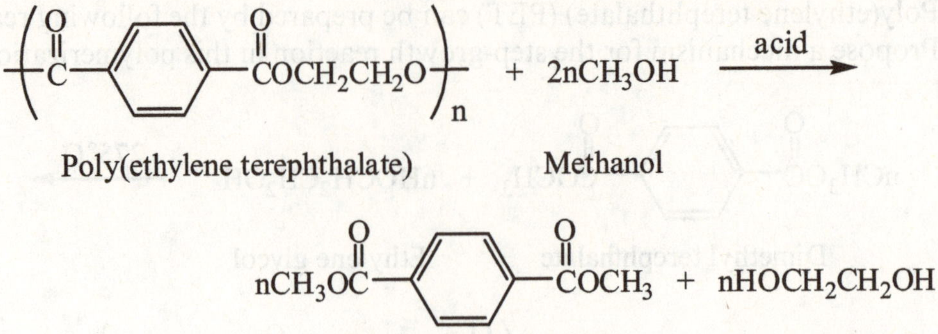

Poly(ethylene terephthalate)   Methanol

nCH_3OC—⬡—$COCH_3$ + $nHOCH_2CH_2OH$

Dimethyl terephthalate   Ethylene glycol

16.6   Nomex is an aromatic polyamide (aramid) prepared from polymerization of 1,3-benzenediamine and the acid chloride of 1,3-benzenedicarboxylic acid. The physical properties of the polymer make it suitable for high-strength, high-temperature applications such as parachute cords and jet aircraft tires. Draw a structural formula for the repeating unit of Nomex.

The formation of the amide bond involves a nucleophilic acyl substitution reaction between an acid chloride and an amine. HCl is formed as a byproduct.

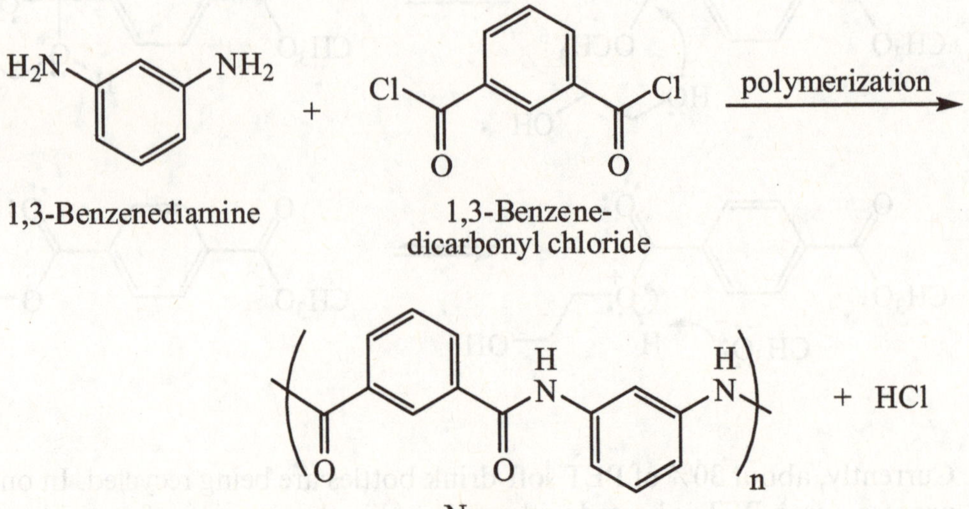

1,3-Benzenediamine   1,3-Benzene-
                                dicarbonyl chloride

Nomex

16.7   Nylon 6,10 [Problem 16.3(d)] can be prepared by reaction of a diamine and a diacid chloride. Draw the structural formula of each reactant.

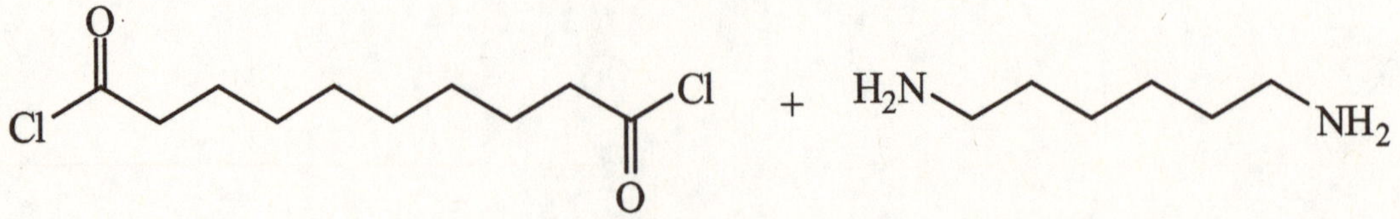

Chain-Growth Polymerization

16.8 Following is the structural formula of a section of polypropylene derived from three units of propylene monomer:

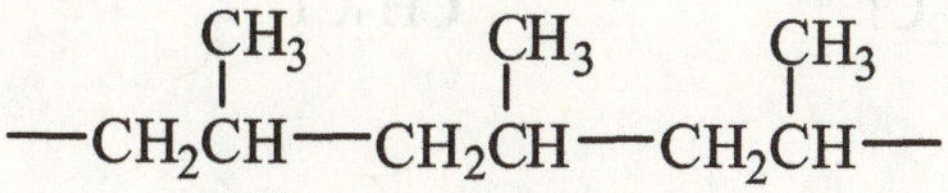

$$\underset{\text{Polypropylene}}{-CH_2CH-CH_2CH-CH_2CH-}$$

(with CH_3 groups on each CH carbon)

Draw a structural formula for a comparable section of:

(a) Poly(vinyl chloride)   (b) Polytetrafluoroethylene (PTFE)

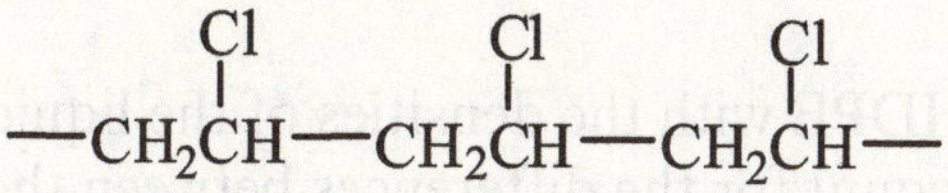

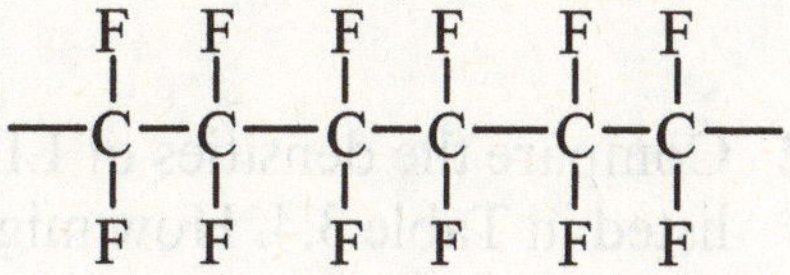

$$-CH_2CH-CH_2CH-CH_2CH-$$
(with Cl on each CH)

$$-\underset{F}{\overset{F}{C}}-\underset{F}{\overset{F}{C}}-\underset{F}{\overset{F}{C}}-\underset{F}{\overset{F}{C}}-\underset{F}{\overset{F}{C}}-\underset{F}{\overset{F}{C}}-$$

(c) Poly(methyl methylacrylate)

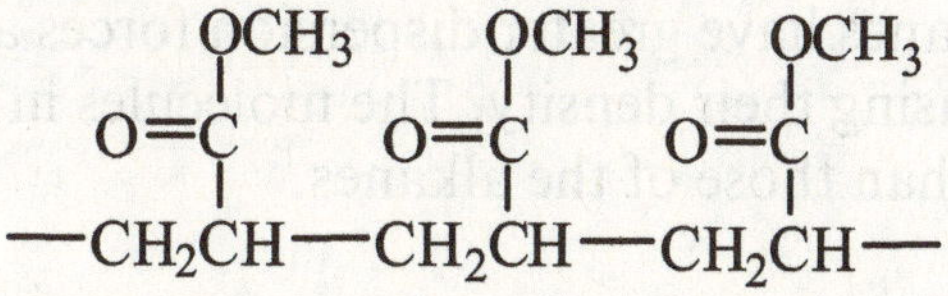

$$-CH_2CH-CH_2CH-CH_2CH-$$
(with $O=C-OCH_3$ on each CH)

16.9 Following are structural formulas for sections of two polymers. From what alkene monomer is each derived?

(a)
$$-CH_2CCH_2CCH_2C-$$
(with Cl above and Cl below each quaternary carbon)

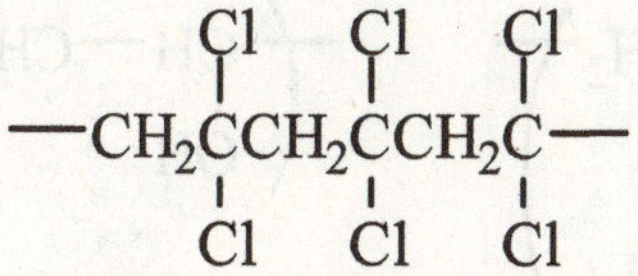

from

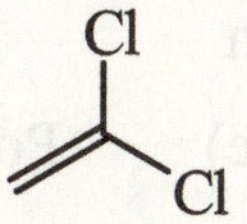

$$\underset{Cl}{\overset{Cl}{C}}=CH_2$$ (1,1-dichloroethene)

(b)
$$-CH_2CCH_2CCH_2C-$$
(with F above and F below each quaternary carbon)

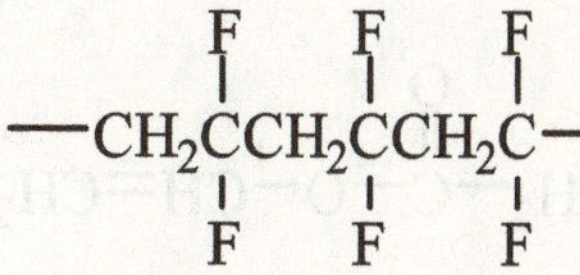

from

$$\underset{F}{\overset{F}{C}}=CH_2$$ (1,1-difluoroethene)

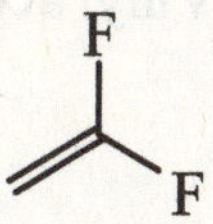

16.10 Draw the structure of the alkene monomer used to make each chain-growth polymer:

(a) $\left(\right)_n$ from (propene)

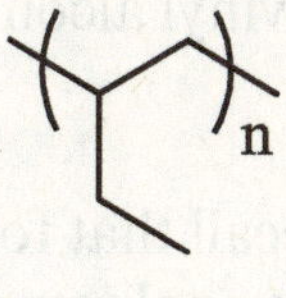

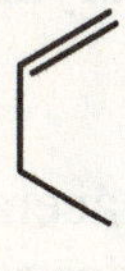

(b) $\left(\right)_n$ (with O-CH_2CH_3 group)   from (ethyl vinyl ether)

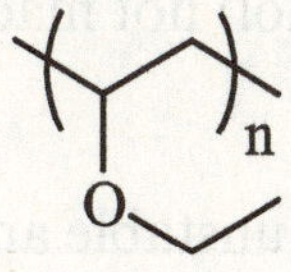

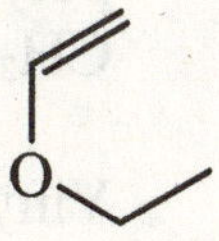

(c)   from

(d)   from

16.11 LDPE has a higher degree of chain branching than HDPE. Explain the relationship between chain branching and density.

When the degree of chain branching increases, the chains are less able to pack together in the solid state. As a result, the density of the polymer decreases.

16.12 Compare the densities of LDPE and HDPE with the densities of the liquid alkanes listed in Table 3.4. How might you account for the differences between them?

LDPE and HDPE, both of which have densities over 0.9 g/mL, are substantially more dense than the liquid alkanes, which range from 0.626 g/mL for pentane to 0.730 g/mL for decane. The larger alkanes have greater dispersion forces and are held together more tightly, thus increasing their density. The molecules in LDPE and HDPE are even more tightly packed than those of the alkanes.

16.13 The polymerization of vinyl acetate gives poly(vinyl acetate). Hydrolysis of this polymer in aqueous sodium hydroxide gives poly(vinyl alcohol). Draw the repeat units of both poly(vinyl acetate) and poly(vinyl alcohol).

Vinyl acetate   Poly(vinyl acetate)   Poly(vinyl alcohol)

16.14 As seen in the previous problem, poly(vinyl alcohol) is made by polymerization of vinyl acetate, followed by hydrolysis in aqueous sodium hydroxide. Why is poly(vinyl alcohol) not made instead by the polymerization of vinyl alcohol, $CH_2=CHOH$?

Vinyl alcohol is unstable and tautomerizes to acetaldehyde. Recall that for most carbonyl compounds, the keto tautomer is more stable than the enol tautomer.

Vinyl alcohol
(enol tautomer)   **Acetaldehyde**
                             (keto tautomer)

16.15 As you know, the shape of a polymer chain affects its properties. Consider the following three polymers. Which do you expect to be the most rigid? Which do you expect to be the most transparent? (Assume the same molecular weights.)

Polymers A and B are expected to be both rigid and opaque polymers. Polymer C is expected to be a more flexible, transparent material. Both of these physical characteristics depend on the degree of crystallinity of the polymer. These three polymer chains all have repeating stereocenters. Both A and B are termed stereoregular; that is, the configurations of the stereocenters repeat in a consistent pattern over the length of the chain. In A, all stereocenters have the same configuration, whereas in B, they alternate between R and S. Because of this stereoregular pattern, molecules of both polymers A and B pack well in the solid state with strong intermolecular interactions between molecules. Polymers A and B therefore have a high degree of crystallinity and are rigid polymers. Polymer C, on the other hand, has a random pattern of stereocenters, and therefore, its chains do not pack as well in the solid state; it has a low degree of crystallinity. The lower is the degree of crystallinity, the more transparent is the polymer.

Looking Ahead

16.16 Cellulose, the principle component of cotton, is a polymer of D-glucose in which the monomeric unit repeats at the indicated atoms. Draw a three-unit section of cellulose.

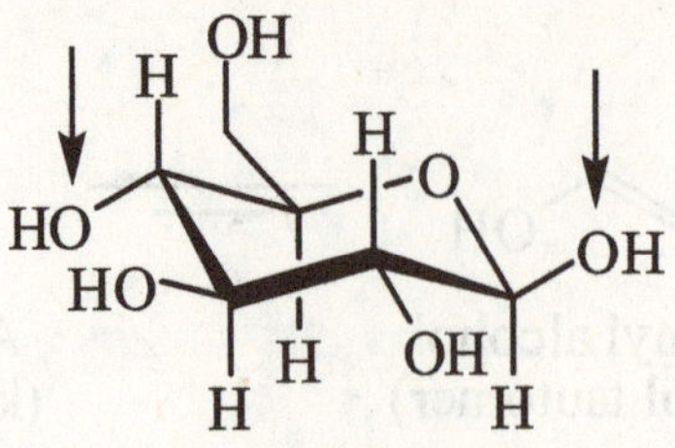

OH

HO
HO

O

OH
OH

D-Glucose

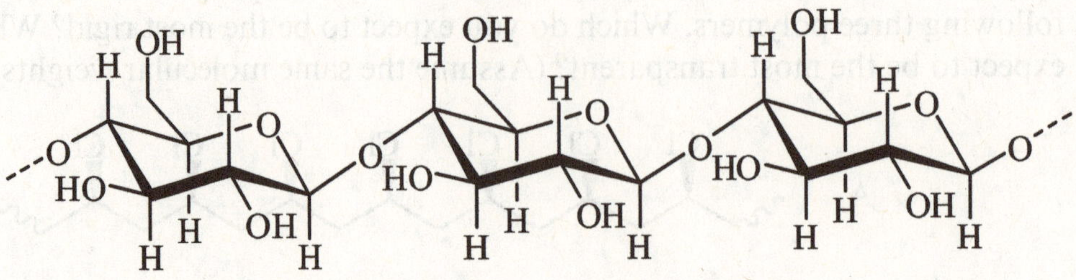

OH

O

HO

OH

O

OH

OH

O

HO

OH

O

16.17 Is a repeating unit a requirement for a compound to be called a polymer?

Yes. Polymers consist of repeating units called monomers.

16.18 Proteins are polymers of naturally occurring monomers called amino acids:

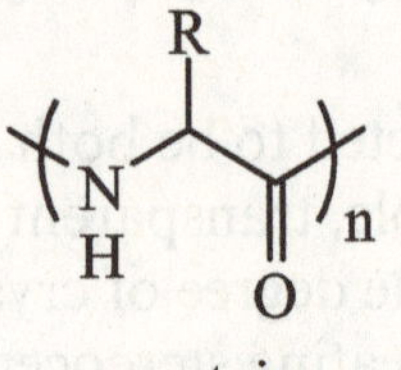

R

N
H

O

n

a protein

Amino acids differ in the types of R groups available in nature. Explain how the following properties of a protein might be affected upon changing the R groups from $-CH_2CH(CH_3)_2$ to $-CH_2OH$:

(a) Solubility in water

The replacement of a $-CH_2CH(CH_3)_2$ with a $-CH_2OH$ allows the R group to undergo hydrogen bonding. Because water is capable of hydrogen bonding, the protein with the $-CH_2OH$ groups is more soluble in water.

(b) Melting point

The protein with the $-CH_2OH$ groups will have higher intermolecular forces (due to increased hydrogen bonding) and will melt at a higher temperature.

(c) Crystallinity

Both the lattice energy and the amount of order within the solid increase with increasing intermolecular hydrogen bonding, so the protein with the $-CH_2OH$ groups will be more cyrstalline.

(d) Elasticity

The protein with the $-CH_2OH$ groups will be less elastic. Due to the protein's higher intermolecular forces, its chains will be held more tightly to each other, reducing the motion of one chain relative to another.

Group Learning Activities

16.19 Only certain kinds of polymers are readily biodegradable, that is only certain types have chemical bonds that are easily broken in the process of composting. Chief among these are polymers that contain ester bonds, because ester bond are readily broken by esterases, microbial enzymes that catalyze the hydrolysis of esters. For this reason, all presently available biodegradable polymers are polyesters. Following are structural formulas for three such biodegradable polyesters. As a group, draw structural formulas and write names for the monomer units present in each. Provide a mechanism for the acid catalyzed hydrolysis of all hydrolyzable bonds in Ecoflex.

(a)

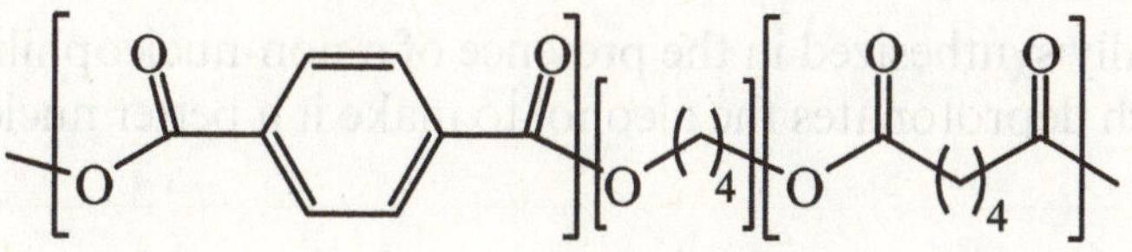

Ecoflex (BASF)

Monomers:

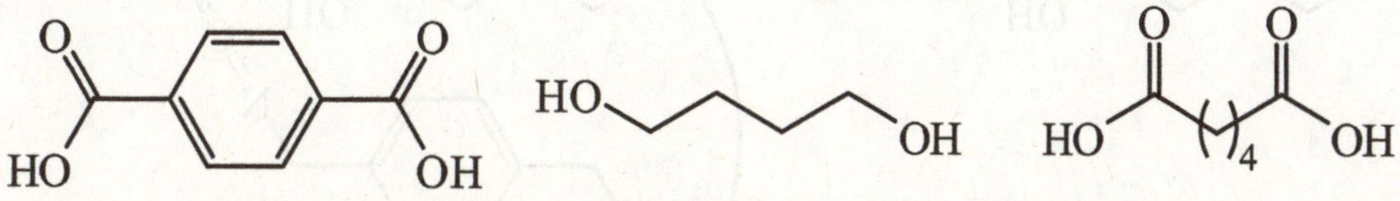

All of the hydroyzable bonds in Ecoflex are esters. The mechanism for the acid hydrolysis of these esters is identical to the one shown in Problem 14.19(b).

(b)

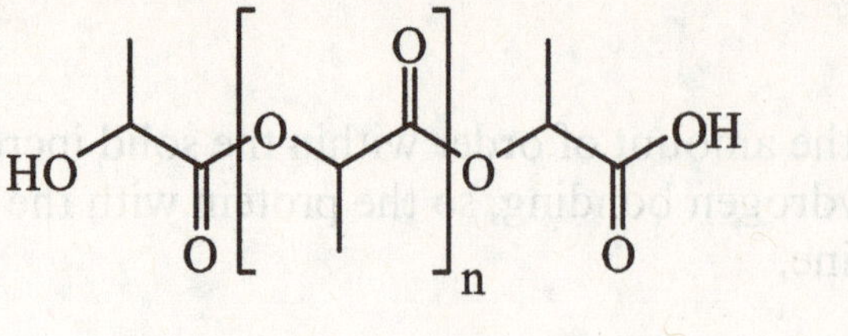

Polylactic Acid (PLA)

Monomer:

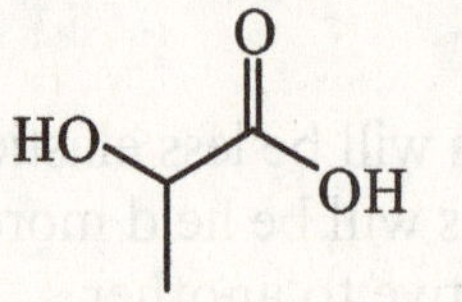

Lactic acid

(c)

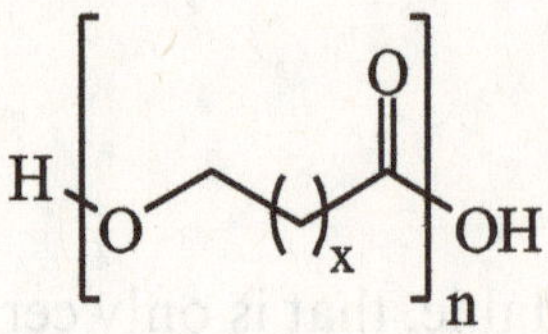

Monomer:

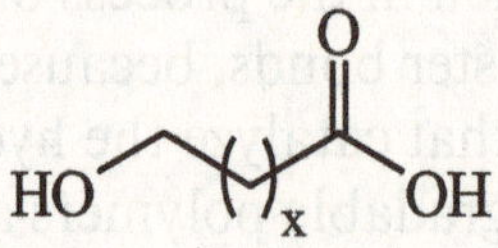

16.20 The two monomers shown below form a polyurethane (Section 16.4D). Propose a structure for the polymer and a mechanism for its formation.

Polyurethanes are typically synthesized in the presence of a non-nucleophilic base, such as a tertiary amine, which deprotonates the alcohol to make it a better nucleophile.

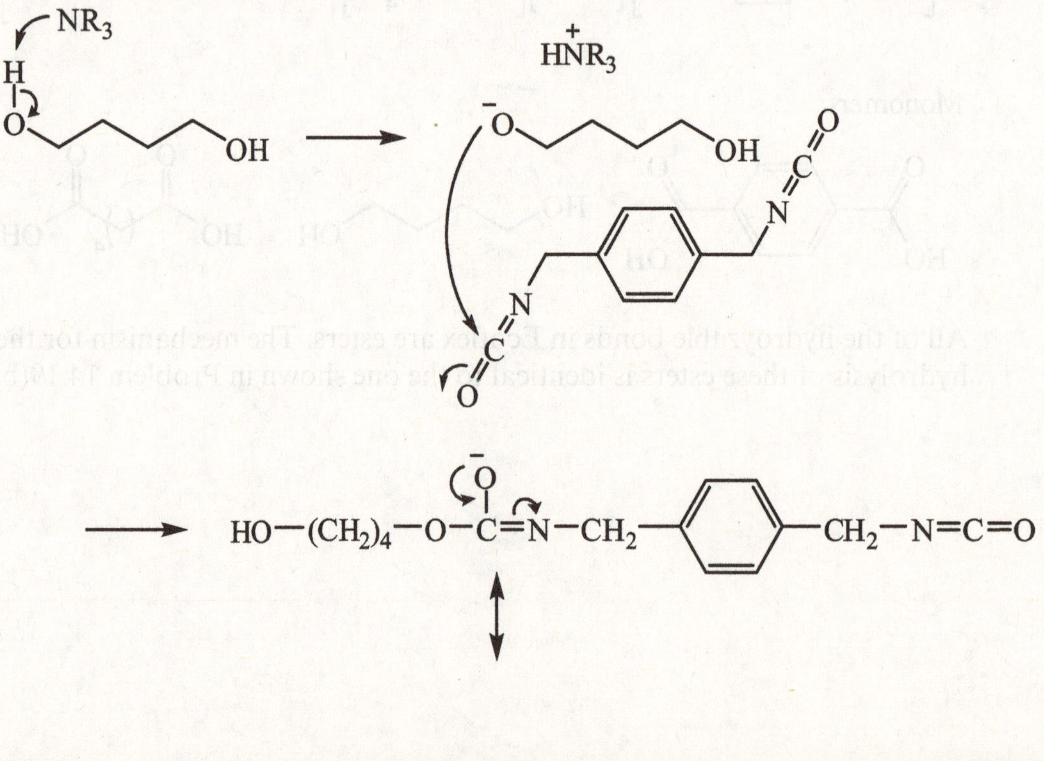

$$\longrightarrow \quad HO-(CH_2)_4-O-C=N-CH_2-\bigcirc-CH_2-N=C=O$$

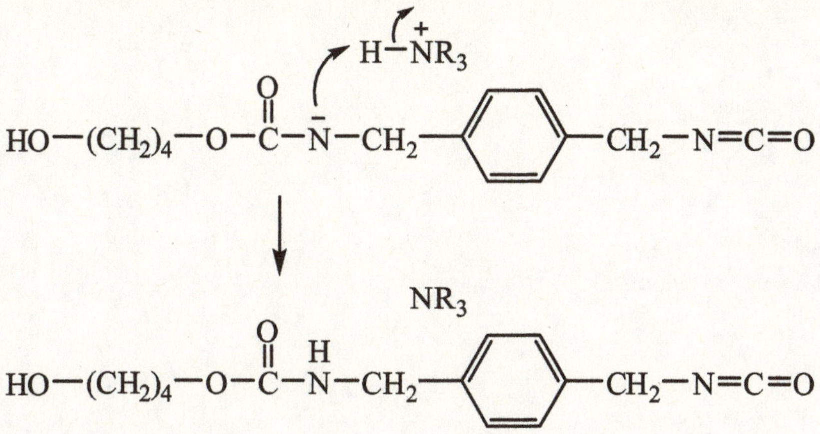

$$HO-(CH_2)_4-O-\overset{\overset{O}{\parallel}}{C}-\overset{-}{N}-CH_2-\!\!\left\langle\!\!\!\bigcirc\!\!\!\right\rangle\!\!-CH_2-N\!\!=\!\!C\!\!=\!\!O$$

H—$\overset{+}{N}R_3$

$$HO-(CH_2)_4-O-\overset{\overset{O}{\parallel}}{C}-\overset{\overset{\overset{NR_3}{}}{\underset{}{}}}{\underset{}{N}}\overset{H}{}CH_2-\!\!\left\langle\!\!\!\bigcirc\!\!\!\right\rangle\!\!-CH_2-N\!\!=\!\!C\!\!=\!\!O$$

The structure of the polymer contains a repeating unit comprised of the two monomers:

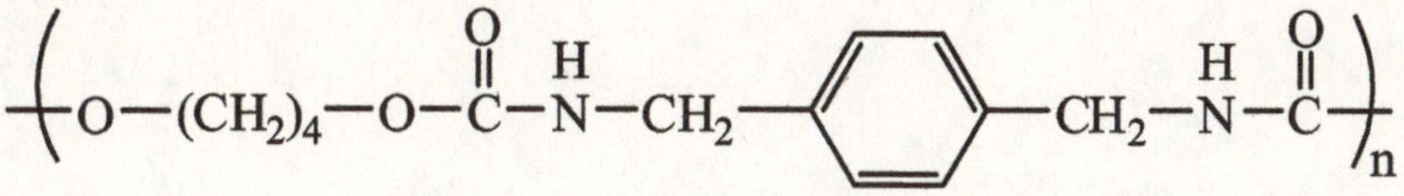

$$\left[\!\!-O-(CH_2)_4-O-\overset{\overset{O}{\parallel}}{C}-\overset{H}{N}-CH_2-\!\!\left\langle\!\!\!\bigcirc\!\!\!\right\rangle\!\!-CH_2-\overset{H}{N}-\overset{\overset{O}{\parallel}}{C}-\!\!\right]_n$$

Chapter 17: Carbohydrates

Problems

17.1 (a) Draw Fischer projections for all 2-ketopentoses.
(b) Which are D-ketopentoses, which are L-ketopentoses, and which are enantiomers?

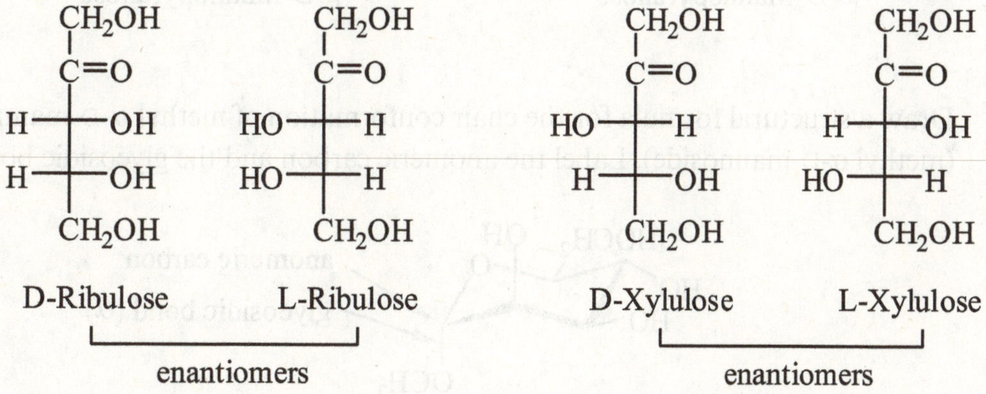

2-Ketopentoses have two stereocenters, so there are a total of four stereoisomers. Two of these are D-ketopentoses and two are L-ketopentoses. Note that when two sugars have different D/L designation, but the same common name, they are enantiomers. The four 2-ketopentoses are also diastereomers of each other.

17.2 Mannose exists in aqueous solution as a mixture of α-D-mannopyranose and β-D-mannopyranose. Draw Haworth projections for these molecules.

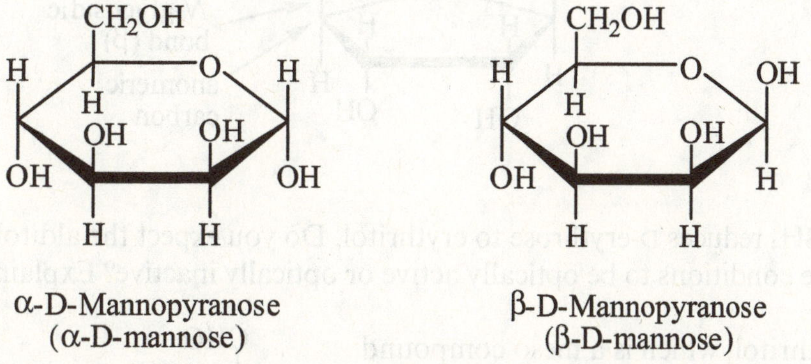

When two sugars have the same D/L designation and the same name, but one is α and the other is β, they differ only at the configuration of the anomeric carbon. The two anomers are diastereomers of each other.

17.3   Draw chair conformations for α-D-mannopyranose and β-D-mannopyranose. Label the anomeric carbon atom in each.

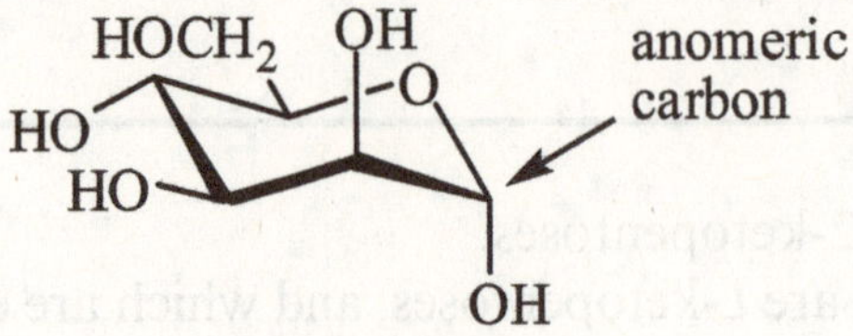

α-D-Mannopyranose 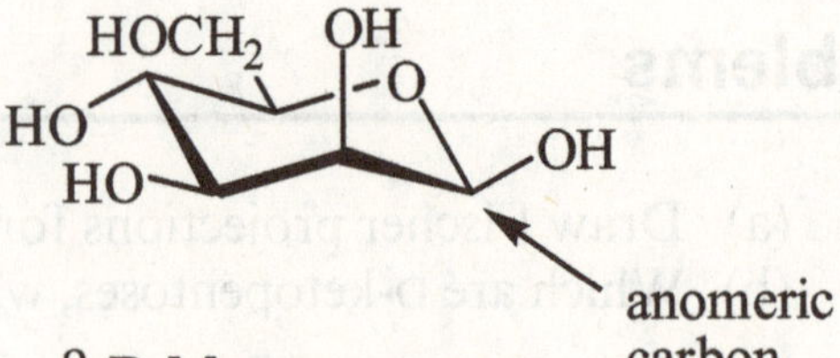   β-D-Mannopyranose

17.4   Draw a structural formula for the chair conformation of methyl α-D-mannopyranoside (methyl α-D-mannoside). Label the anomeric carbon and the glycosidic bond.

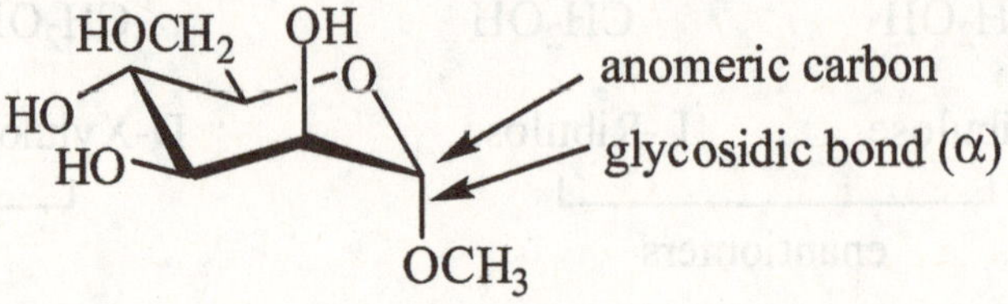

17.5   Draw a structural formula for the β-N-glycoside formed between β-D-ribofuranose and adenine.

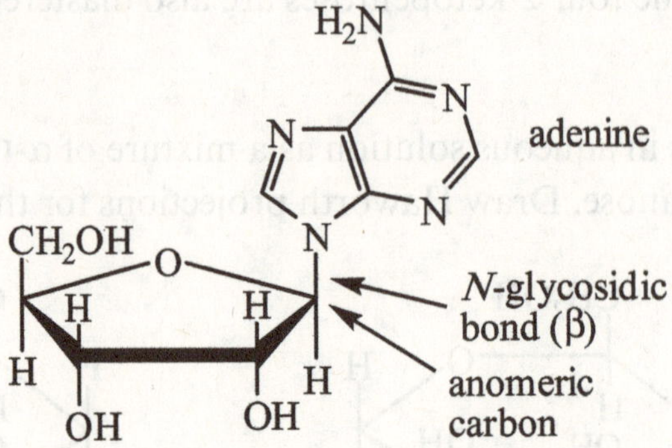

17.6   NaBH₄ reduces D-erythrose to erythritol. Do you expect the alditol formed under these conditions to be optically active or optically inactive? Explain.

Erythritol, which is a meso compound (a compound that has two or more stereocenters but is achiral due to the presence of an internal plane of symmetry), is optically inactive.

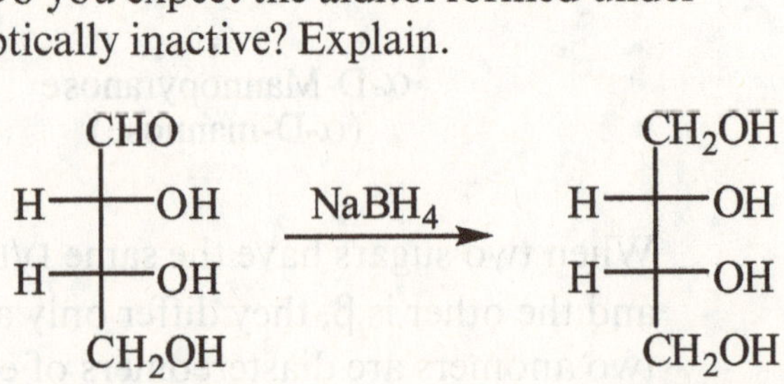

D-Erythrose   Erythritol (meso)

17.7   Draw Haworth and chair formulas for the α form of a disaccharide in which two units of D-glucopyranose are joined by a β-1,3-glycosidic bond.

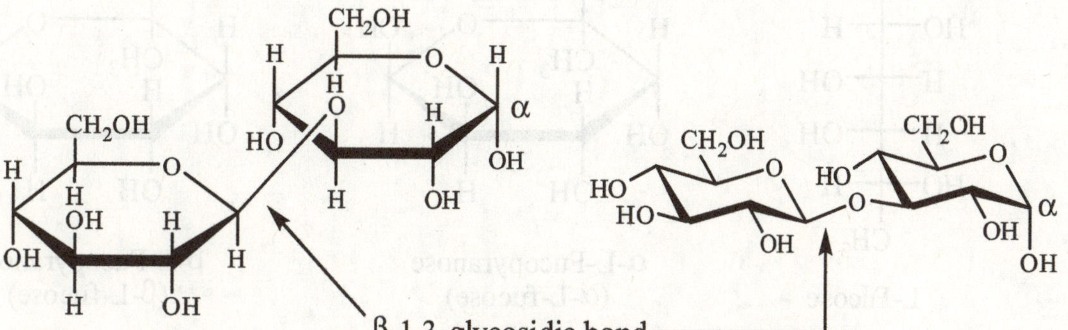

β-1,3-glycosidic bond

Chemical Connections

17A. The following is the structure of the artificial sweetener sucralose. Indicate all the ways in which it differs from sucrose.

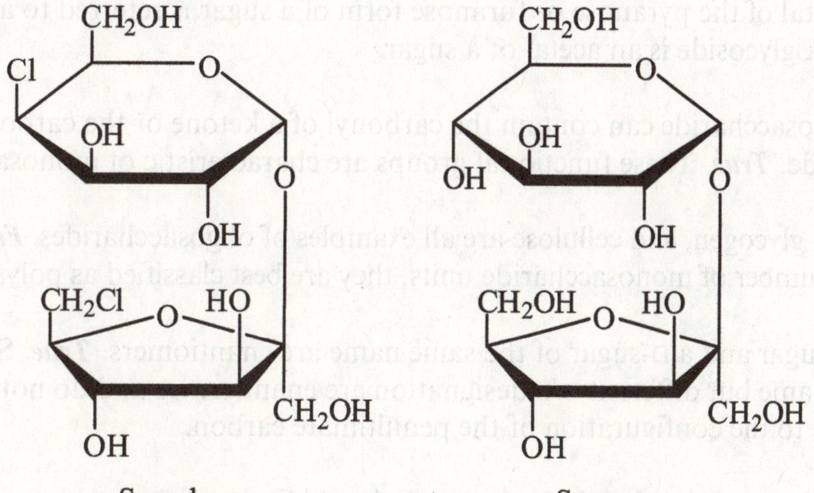

Sucralose   Sucrose

First, two of the hydroxyl groups in sucrose (carbon 4 of glucose and carbon 6 of fructose) are replaced by chlorine atoms. Second, there is stereochemical inversion at carbon 4 of glucose, rendering the glucose portion of sucrose a derivative of galactose.

17B. Draw the two pyranose forms of L-fucose.

L-Fucose

α-L-Fucopyranose
(α-L-fucose)

β-L-Fucopyranose
(β-L-fucose)

An α configuration is assigned when the –OH group bonded to the anomeric carbon is *trans* to the group of the terminal carbon, which is in this case the –CH$_3$ group. If the two groups are *cis* to each other, a β configuration is assigned.

Quick Quiz

1. An acetal of the pyranose or furanose form of a sugar is referred to as a glycoside. *True*. A glycoside is an acetal of a sugar.

2. A monosaccharide can contain the carbonyl of a ketone or the carbonyl of an aldehyde. *True*. These functional groups are characteristic of monosaccharides.

3. Starch, glycogen, and cellulose are all examples of oligosaccharides. *False*. Due to their large number of monosaccharide units, they are best classified as polysaccharides.

4. An L-sugar and a D-sugar of the same name are enantiomers. *True*. Sugars with the same name but different D/L designation are enantiomers and do not just differ with respect to the configuration of the penultimate carbon.

5. Alditols are oxidized carbohydrates. *False*. Alditols are formed by reduction.

6. D-Glucose and D-ribose are diastereomers. *False*. Stereoisomers must be of the same constitutional isomer (same molecular formula but different atom connectivity). The two sugars do not even have the same molecular formula.

7. A pyranoside contains a 5-membered ring. *False*. Pyranosides are based on pyran, a compound with a 6-membered ring.

8. All monosaccharides dissolve in ether. *False*. Monosaccharides are highly polar and are not very soluble in ether.

9. Monosaccharides exist mostly as cyclic hemiacetals. *True*. However, the cyclic hemiacetal is in equilibrium with the open-chain form.

10. A polysaccharide is a glycoside of two monosaccharides. *False*. The compound is a disaccharide. *Poly* is reserved for compounds with many, many monosaccharide units.

11. α and β in a monosaccharide are used to refer to the positions 1 and 2 carbons away from the carbonyl group. *False*. These designations are used to indicate the relative stereochemical configuration of the anomeric carbon.

12. Carbohydrates must have the formula $C_n(H_2O)_n$. *False*. Carbohydrates frequently contain other elements, or are modified via biochemical reactions, and may not have this formula.

13. Mutarotation is the establishment of an equilibrium concentration of α and β anomers of a carbohydrate. *True*. The progress of the establishment of equilibrium can be monitored by measuring the rotation of plane-polarized light.

14. D-Glucose and D-Galactose are diastereomers. *True*. The two monosaccharides are also epimers of each other.

15. Only acyclic carbohydrates that contain aldehyde groups can act as reducing sugars. *False*. Under the appropriate conditions, ketones can isomerize to aldehydes.

16. A methyl glycoside of a monosaccharide cannot act as a reducing sugar. *True*. Glycosides, which are acetals, do not easily revert to hemiacetals and open-chain forms.

17. The penultimate carbon of an acyclic monosaccharide becomes the anomeric carbon in the cyclic hemiacetal form of the molecule. *False*. The carbon that becomes the anomeric carbon is either the aldehyde carbon (carbon 1) or the ketone carbon (carbon 2).

18. A Fischer projection may be rotated 90°. *False*. By convention, a Fischer projection must be written such that the carbon chain runs vertically. When the Fischer projection is rotated 90°, the resulting representation cannot be called a Fischer projection.

End-of-Chapter Problems

Monosaccharides

17.8   What is the difference in structure between an aldose and a ketose? Between an aldopentose and a ketopentose?

An aldose is a monosaccharide with an aldehyde functional group, while a ketose has a ketone group. An aldopentose is a five-carbon monosaccharide with an aldehyde group, and a ketopentose is a five-carbon monosaccharide with a ketone group.

17.9   Which hexose is also known as dextrose?

D-Glucose is also known by the common name dextrose.

17.10 What does it mean to say that D- and L-glyceraldehyde are enantiomers?

D- and L-glyceraldehyde are enantiomers because they are nonsuperimposable mirror images. The stereocenters in the two sugars have opposite configuration.

17.11 Explain the meaning of the designations D and L as used to specify the configuration of carbohydrates.

The two designations refer to the configuration of the penultimate carbon, which is the stereocenter farthest away from the carbonyl group. If the monosaccharide is drawn as a Fischer projection and the −OH group bonded to this carbon is on the right, the D designation is used; likewise, a monosaccharide is L if the −OH group is on the left.

It is worth noting that with the exception of glyceraldehyde, the D and L designations do not predict the direction in which a monosaccharide will rotate plane-polarized light.

17.12 How many stereocenters are present in D-glucose? In D-ribose? How many stereoisomers are possible for each monosaccharide?

D-glucose has four, and D-ribose has three, stereocenters. Accordingly, the number of possible stereoisomers for these monosaccharides are respectively 16 and 8. (Out of the 16 possible stereoisomers of the aldohexoses, only one of these is D-glucose. Similarly, out of the 8 possible stereoisomers of aldopentoses, only one of these is D-ribose.)

17.13 Which compounds are D-monosaccharides and which are L-monosaccharides?

(a)
```
        CHO
    H——OH
   HO——H
    H——OH
    H——OH
        CH2OH
```

(b)
```
        CHO
   HO——H
    H——OH
   HO——H
        CH2OH
```

(c)
```
        CH2OH
        C=O
    H——OH
    H——OH
        CH2OH
```

(d)
```
        CHO
   HO——H
        CH2OH
```

Based on the position of the –OH group on the penultimate carbon, compounds (a) and (c) are D-monosaccharides, while (b) and (d) are L-monosaccharides.

17.14 Draw Fischer projections for L-ribose and L-arabinose.

```
        CHO
    H——OH
    H——OH
    H——OH
        CH2OH
     D-Ribose
```

```
        CHO
   HO——H
   HO——H
   HO——H
        CH2OH
     L-Ribose
```

```
        CHO
   HO——H
    H——OH
    H——OH
        CH2OH
    D-Arabinose
```

```
        CHO
    H——OH
   HO——H
   HO——H
        CH2OH
    L-Arabinose
```

Recall that D- and L-monosaccharides with the same common name are enantiomers of one another. That is, they are mirror images and all the stereocenters must be of the opposite configuration. Thus, L-ribose is the enantiomer of D-ribose, and the same can be said for arabinose. Therefore, start with the D-monosaccharide and draw its mirror image to obtain the L-monosaccharide.

A very common error is to start with the D-monosaccharide and then change only the configuration of the penultimate carbon. While doing so does give an L-monosaccharide, the resulting L-monosaccharide would have a different common name.

17.15 Explain why all mono- and disaccharides are soluble in water.

Mono- and disaccharides are polar compounds and are able to participate in hydrogen bonding with water molecules.

17.16 What is an amino sugar? Name the three amino sugars most commonly found in nature.

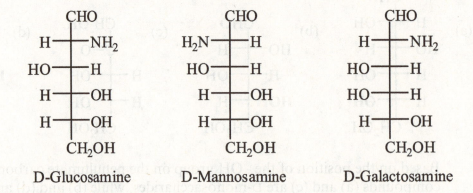

An amino sugar contains an amino group in place of a hydroxyl group. D-Glucosamine is found in cartilage and is a popular nutritional supplement, D-mannosamine is a component of glycoproteins and glycolipids, and D-galactosamine is an important component of blood-type antigens A, B, and AB.

17.17 2,6-Dideoxy-D-altrose, known alternatively as D-digitoxose, is a monosaccharide obtained from the hydrolysis of digitoxin, a natural product extracted from purple foxglove (*Digitalis purpurea*). Digitoxin has found wide use in cardiology because it reduces pulse rate, regularizes heart rhythm, and strengthens the heartbeat. Draw the structural formula of 2,6-dideoxy-D-altrose.

$$
\begin{array}{ccc}
& \text{CHO} & \\
\text{HO} & \!\!\!-\!\!\! & \text{H} \\
\text{H} & \!\!\!-\!\!\! & \text{OH} \\
\text{H} & \!\!\!-\!\!\! & \text{OH} \\
\text{H} & \!\!\!-\!\!\! & \text{OH} \\
& \text{CH}_2\text{OH} &
\end{array}
\qquad
\begin{array}{ccc}
& \text{CHO} & \\
\text{H} & \!\!\!-\!\!\! & \text{H} \\
\text{H} & \!\!\!-\!\!\! & \text{OH} \\
\text{H} & \!\!\!-\!\!\! & \text{OH} \\
\text{H} & \!\!\!-\!\!\! & \text{OH} \\
& \text{CH}_3 &
\end{array}
$$

D-Altrose   2,6-Dideoxy-D-altrose

The Cyclic Structure of Monosaccharides

17.18 Define the term *anomeric carbon*.

In the context of carbohydrate chemistry, an anomeric carbon is the new carbon stereocenter that is created upon the formation of a cyclic structure. With respect to monosaccharides, the anomeric carbon is also the hemiacetal carbon.

17.19 Explain the conventions for using α and β to designate the configuration of cyclic forms of monosaccharides.

The α and β designations refer to the position of the hydroxyl (−OH) group on the anomeric carbon relative to the terminal −CH₂OH group. If both of these groups are on the same side of the ring (*cis*), a β designation is assigned. If the two groups are on the opposite side of the ring (*trans*), the monosaccharide is assigned α.

17.20 Are α-D-glucose and β-D-glucose anomers? Explain. Are they enantiomers? Explain.

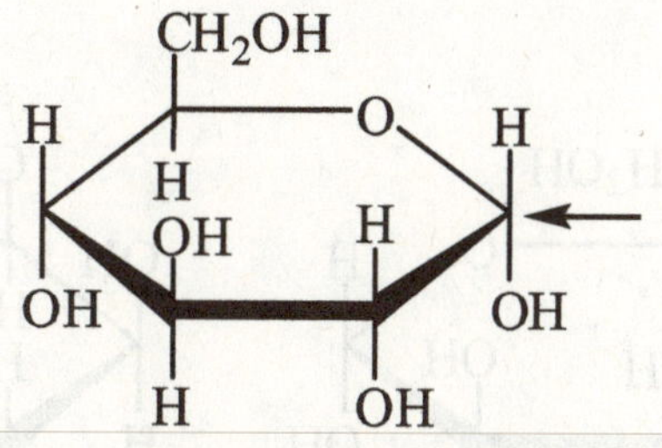

α-D-Glucose 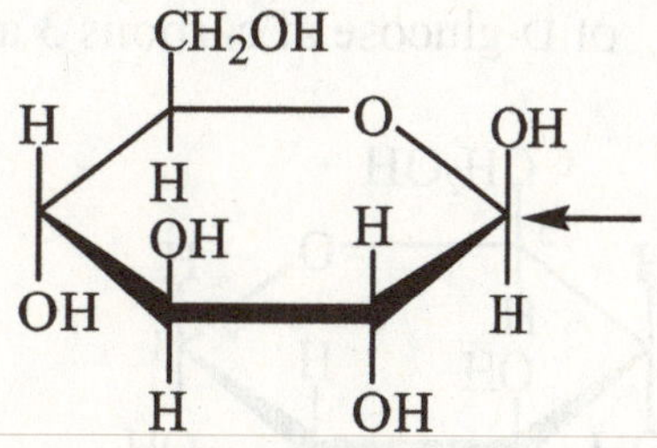   β-D-Glucose

Anomers are compounds that differ only in the configuration of the anomeric carbon (indicated by the arrow), so α-D-glucose and β-D-glucose are anomers. All other stereocenters have the same configuration. Therefore, anomers are diastereomers and not enantiomers. (Recall that two compounds are only enantiomers if all the stereocenters in one compound are opposite to those of the other compound.)

17.21 Are α-D-gulose and α-L-gulose anomers? Explain.

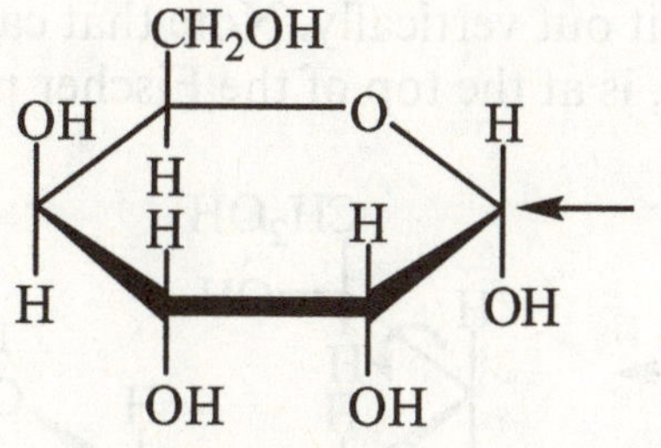

α-D-Gulose 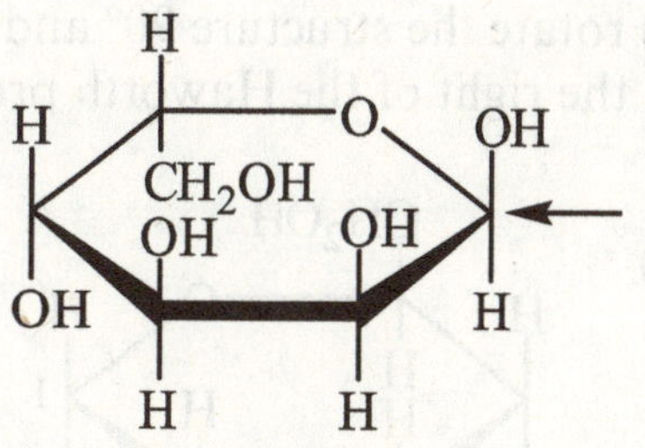   α-L-Gulose

No, because anomers are compounds that differ only in the configuration of the anomeric carbon (indicated by the arrow). Comparing α-D-gulose to α-L-gulose, all of the stereocenters have their configurations inverted. These two compounds are enantiomers.

17.22 In what way are chair conformations a more accurate representation of molecular shape of hexopyranoses than Haworth projections?

Haworth projections show the six-membered rings of hexopyranoses as though they were planar. However, as we have seen with cyclohexane, six-membered rings adopt chair conformations, which are puckered.

17.23 Draw α-D-glucopyranose (α-D-glucose) as a Haworth projection. Now, using only the information given here, draw Haworth projections for these monosaccharides.

(a) α-D-mannopyranose (α-D-mannose). The configuration of α-D-mannose differs from that of α-D-glucose only at carbon 2.

(b) α-D-gulopyranose (α-D-gulose). The configuration of D-gulose differs from that of D-glucose at carbons 3 and 4.

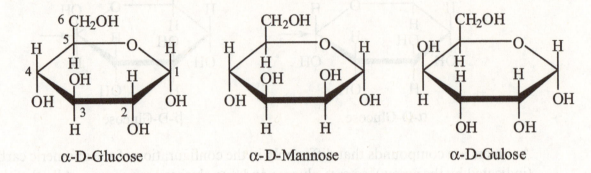

α-D-Glucose   α-D-Mannose   α-D-Gulose

17.24 Convert each Haworth projection to an open-chain form and then to a Fischer projection. Name the monosaccharide you have drawn.

To convert a Haworth projection to a Fischer projection, it is best to (1) revert the hemiacetal to the aldehyde and the alcohol, (2) rotate the bond between carbons 4 and 5 so that the terminal –CH₂OH group is in the horizontal position, and then (3) rotate the structure 90° and stretch it out vertically. Note that carbon 1, which is on the right of the Haworth projection, is at the top of the Fischer projection.

(b)

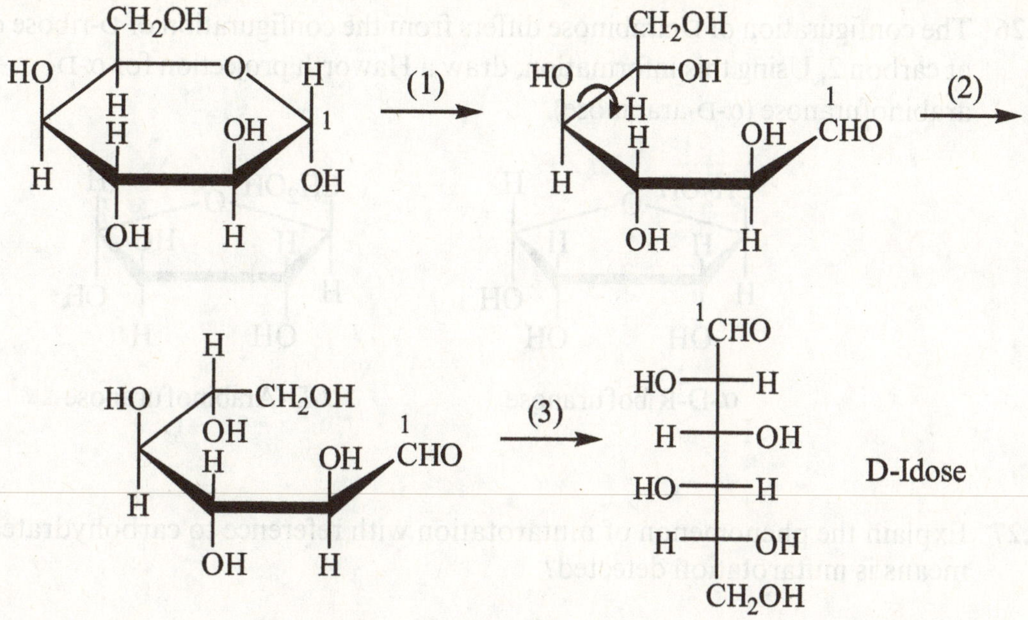

D-Idose

17.25 Convert each chair conformation to an open-chain form and then to a Fischer projection. Name the monosaccharides you have drawn.

(a)

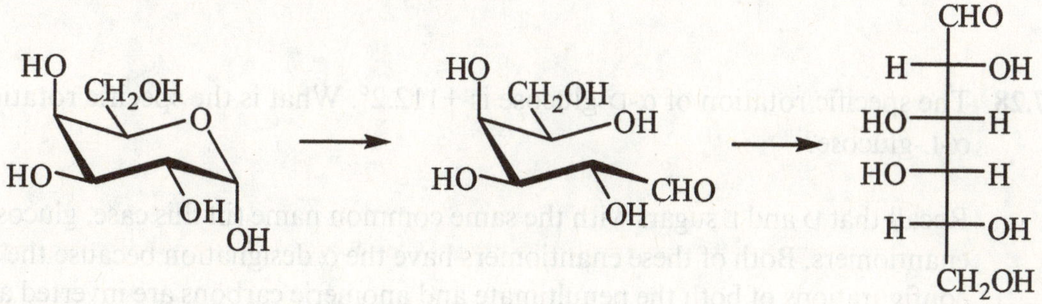

D-Galactose

(b)

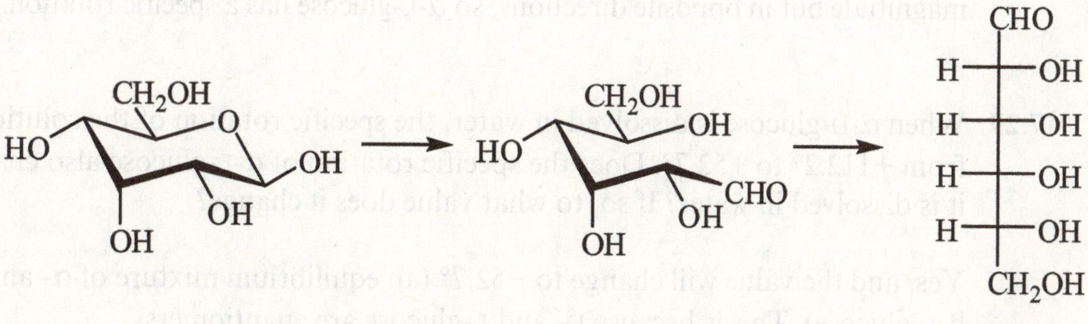

D-Allose

17.26   The configuration of D-arabinose differs from the configuration of D-ribose only at carbon 2. Using this information, draw a Haworth projection for α-D-arabinofuranose (α-D-arabinose).

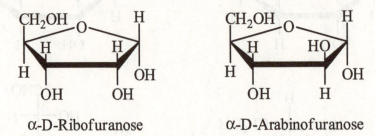

α-D-Ribofuranose   α-D-Arabinofuranose

17.27   Explain the phenomenon of mutarotation with reference to carbohydrates. By what means is mutarotation detected?

Mutarotation involves the formation of an equilibrium mixture of the α and β anomers of a carbohydrate. Either anomer (cyclic hemiacetal) can revert to the open-chain form, which can cyclize again and form the other anomer. Over time, an equilibrium is established. This process can be detected by observing the change in the optical activity of the solution over time.

17.28   The specific rotation of α-D-glucose is +112.2°. What is the specific rotation of α-L-glucose?

Recall that D and L sugars with the same common name (in this case, glucose) are enantiomers. Both of these enantiomers have the α designation because the configurations of both the penultimate and anomeric carbons are inverted and, as a result, the terminal −CH$_2$OH group and the −OH group on the anomeric carbon are *trans* in both enantiomers. Enantiomers rotate plane-polarized light to the same magnitude but in opposite directions, so α-L-glucose has a specific rotation of −112.2°.

17.29   When α-D-glucose is dissolved in water, the specific rotation of the solution changes from +112.2° to +52.7°. Does the specific rotation of α-L-glucose also change when it is dissolved in water? If so, to what value does it change?

Yes, and the value will change to −52.7° (an equilibrium mixture of α- and β-L-glucose). This is because D- and L-glucose are enantiomers.

Reactions of Monosaccharides

17.30 Draw the structural formula for ethyl α-D-galactopyranoside (ethyl α-D-galactose). Label the anomeric carbon and the glycosidic bond.

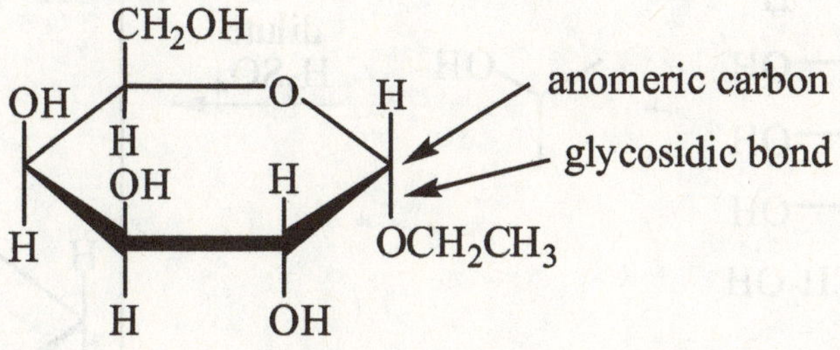

17.31 Draw the structural formula for methyl β-D-mannopyranoside (methyl β-D-mannose). Label the anomeric carbon and the glycosidic bond.

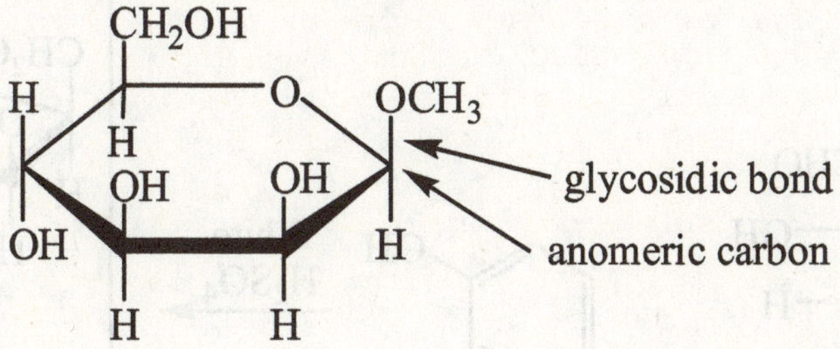

17.32 Show the two possible products of each reaction (refer to Table 17.1). Label the α and β anomers in each reaction.

(a)

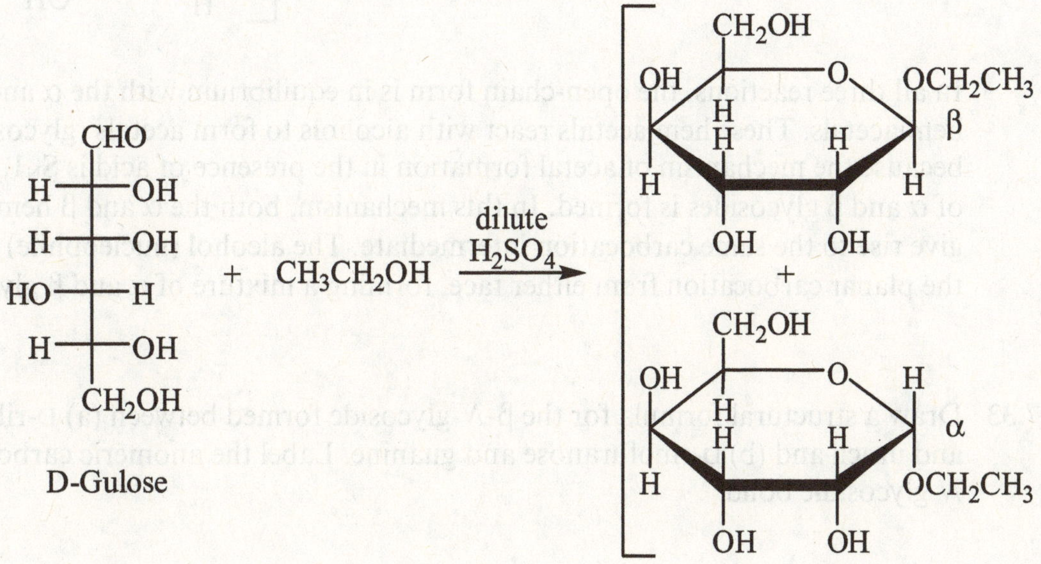

D-Gulose

(b)

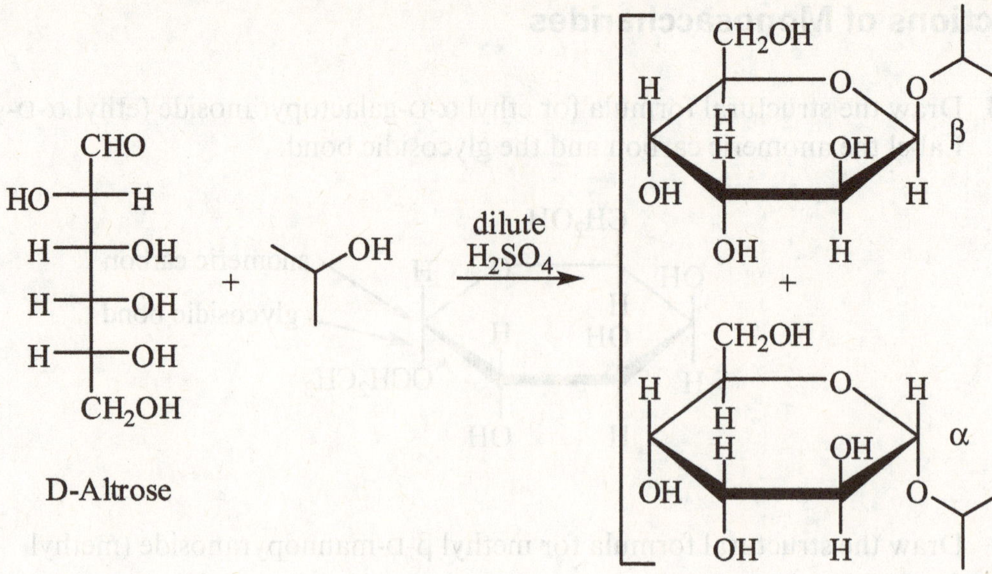

D-Altrose

(c)

D-Xylose

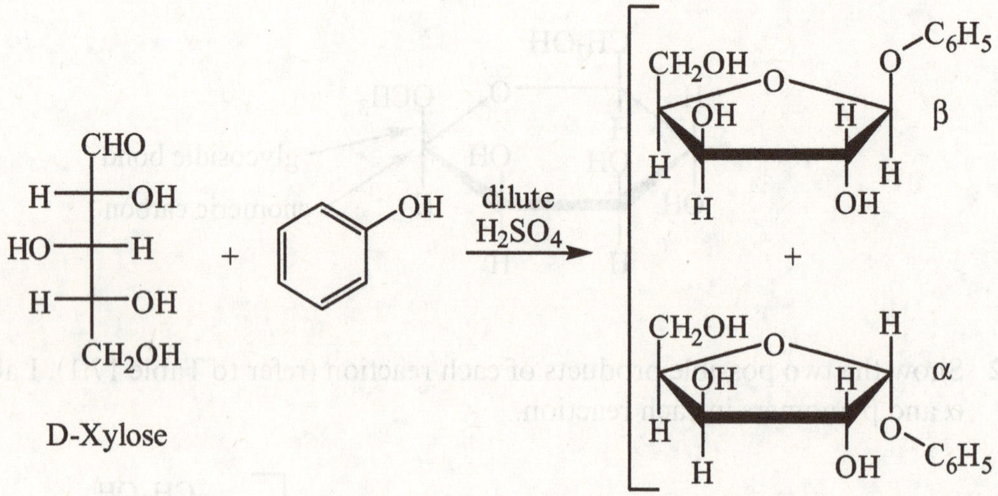

In all three reactions, the open-chain form is in equilibrium with the α and β cyclic hemiacetals. These hemiacetals react with alcohols to form acetals (glycosides), and because the mechanism of acetal formation in the presence of acid is S_N1, a mixture of α and β glycosides is formed. In this mechanism, both the α and β hemiacetals give rise to the same carbocation intermediate. The alcohol (nucleophile) can attack the planar carbocation from either face, forming a mixture of α and β glycosides.

17.33 Draw a structural formula for the β-*N*-glycoside formed between (a) D-ribofuranose and uracil and (b) D-ribofuranose and guanine. Label the anomeric carbon and the *N*-glycosidic bond.

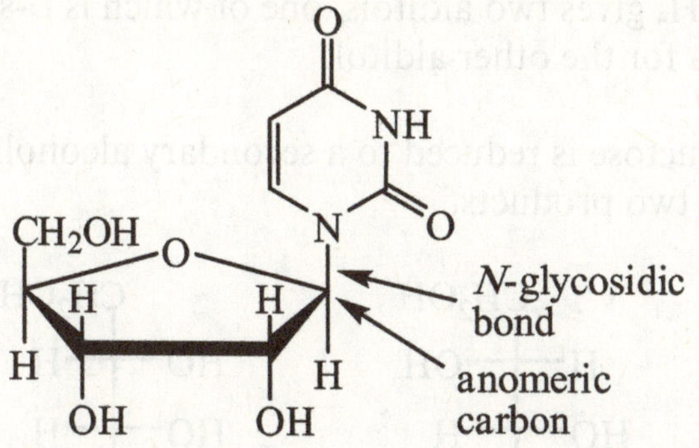

(a) 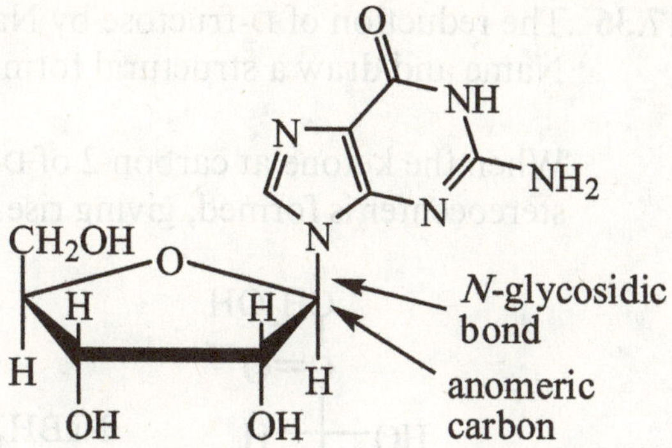   (b)

17.34 Draw Fischer projections for the product(s) formed by reaction of D-galactose with (a) NaBH₄ in H₂O and (b) AgNO₃ in NH₃, H₂O, and state whether each product is optically active or optically inactive.

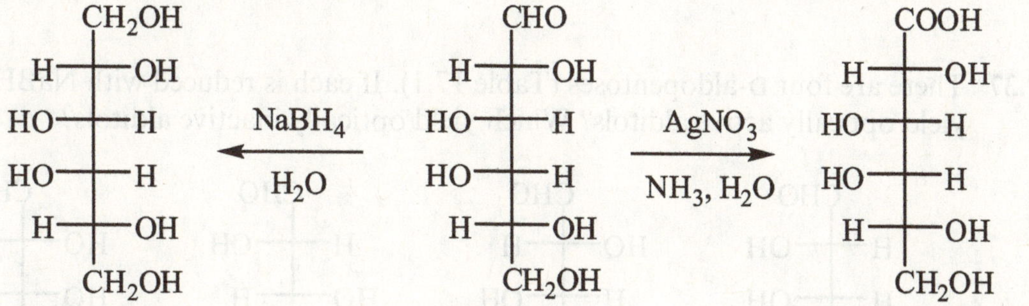

D-Galacitol

optically inactive
(meso)

D-Galactose

D-Galactonic acid

optically active
(chiral)

17.35 Repeat Problem 17.34, but using D-ribose in place of D-galactose.

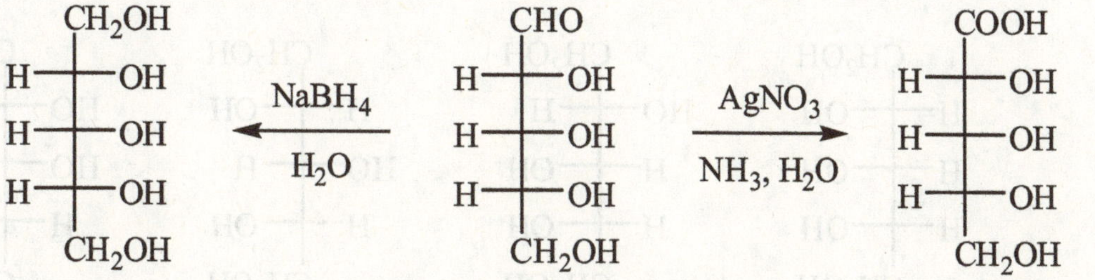

Ribitol

optically inactive
(meso)

D-Ribose

D-Ribonic acid

optically active
(chiral)

17.36 The reduction of D-fructose by $NaBH_4$ gives two alditols, one of which is D-sorbitol. Name and draw a structural formula for the other alditol.

When the ketone at carbon 2 of D-fructose is reduced to a secondary alcohol, a new stereocenter is formed, giving rise to two products.

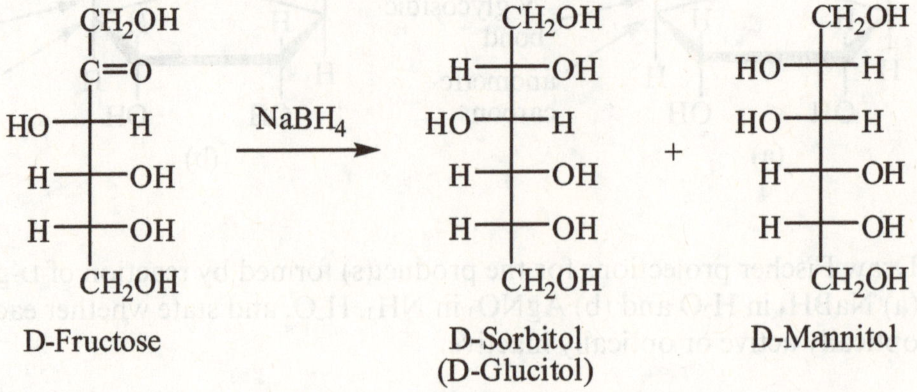

	D-Fructose		D-Sorbitol (D-Glucitol)		D-Mannitol

17.37 There are four D-aldopentoses (Table 17.1). If each is reduced with $NaBH_4$, which yield optically active alditols? Which yield optically inactive alditols?

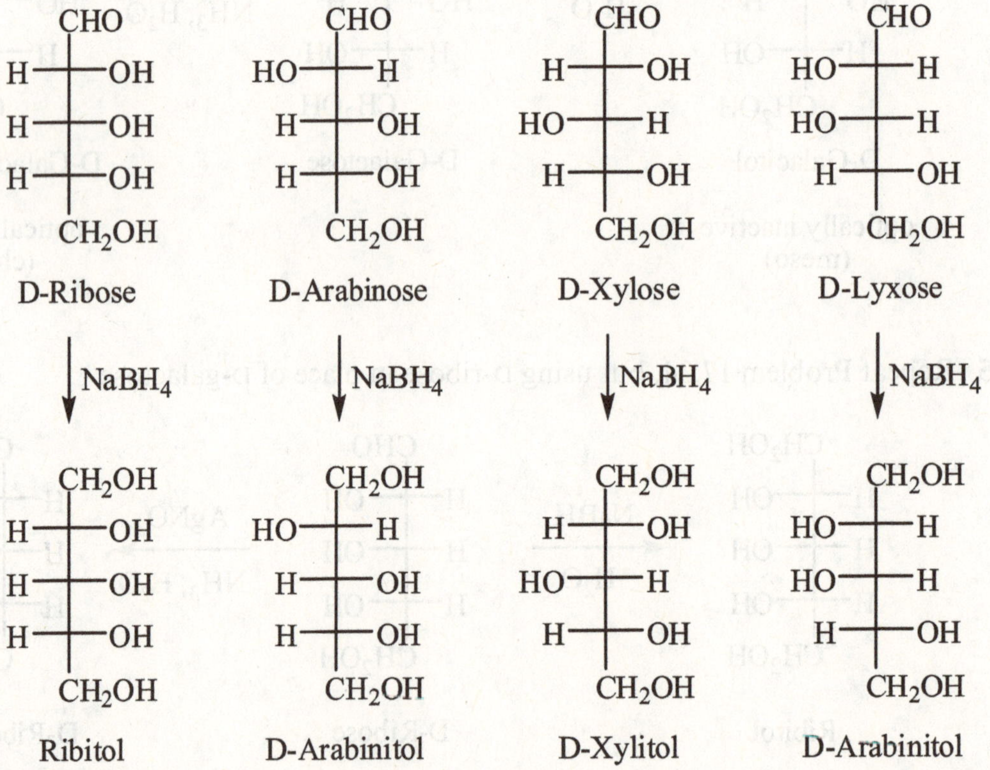

D-Ribose	D-Arabinose	D-Xylose	D-Lyxose

| $\downarrow NaBH_4$ | $\downarrow NaBH_4$ | $\downarrow NaBH_4$ | $\downarrow NaBH_4$ |

| Ribitol | D-Arabinitol | D-Xylitol | D-Arabinitol |

D-Ribose and D-xylose yield meso compounds, which are optically inactive. D-Arabinose and D-lyxose yield the same optically active product, D-arabinitol.

17.38 Account for the observation that the reduction of D-glucose with NaBH₄ gives an optically active alditol, whereas the reduction of D-galactose with NaBH₄ gives an optically inactive alditol.

The reduction product of D-glucose is chiral, but that of D-galactose is achiral (meso). This reaction is historically very significant; prior to the existence of modern instrumentation, it was one of the methods used to differentiate between the two sugars.

$$
\begin{array}{c}
\text{CHO} \\
\text{H}-\text{OH} \\
\text{HO}-\text{H} \\
\text{H}-\text{OH} \\
\text{H}-\text{OH} \\
\text{CH}_2\text{OH}
\end{array}
\xrightarrow{\text{NaBH}_4}
\begin{array}{c}
\text{CH}_2\text{OH} \\
\text{H}-\text{OH} \\
\text{HO}-\text{H} \\
\text{H}-\text{OH} \\
\text{H}-\text{OH} \\
\text{CH}_2\text{OH}
\end{array}
\qquad
\begin{array}{c}
\text{CHO} \\
\text{H}-\text{OH} \\
\text{HO}-\text{H} \\
\text{HO}-\text{H} \\
\text{H}-\text{OH} \\
\text{CH}_2\text{OH}
\end{array}
\xrightarrow{\text{NaBH}_4}
\begin{array}{c}
\text{CH}_2\text{OH} \\
\text{H}-\text{OH} \\
\text{HO}-\text{H} \\
\text{HO}-\text{H} \\
\text{H}-\text{OH} \\
\text{CH}_2\text{OH}
\end{array}
$$

D-Glucose   D-Glucitol (chiral)   D-Galactose   D-Galacitol (meso)

17.39 Which two D-aldohexoses give optically inactive (meso) alditols on reduction with NaBH₄?

Only D-allose and D-galactose will give meso products upon reduction with NaBH₄.

$$
\begin{array}{c}
\text{CHO} \\
\text{H}-\text{OH} \\
\text{H}-\text{OH} \\
\text{H}-\text{OH} \\
\text{H}-\text{OH} \\
\text{CH}_2\text{OH}
\end{array}
\xrightarrow{\text{NaBH}_4}
\begin{array}{c}
\text{CH}_2\text{OH} \\
\text{H}-\text{OH} \\
\text{H}-\text{OH} \\
\text{H}-\text{OH} \\
\text{H}-\text{OH} \\
\text{CH}_2\text{OH}
\end{array}
\qquad
\begin{array}{c}
\text{CHO} \\
\text{H}-\text{OH} \\
\text{HO}-\text{H} \\
\text{HO}-\text{H} \\
\text{H}-\text{OH} \\
\text{CH}_2\text{OH}
\end{array}
\xrightarrow{\text{NaBH}_4}
\begin{array}{c}
\text{CH}_2\text{OH} \\
\text{H}-\text{OH} \\
\text{HO}-\text{H} \\
\text{HO}-\text{H} \\
\text{H}-\text{OH} \\
\text{CH}_2\text{OH}
\end{array}
$$

D-Allose   D-Allitol (meso)   D-Galactose   D-Galacitol (meso)

17.40 L-Fucose, one of several monosaccharides commonly found in the surface polysaccharides of animal cells (Chemical Connections 17B), is synthesized biochemically from D-mannose in the following eight steps:

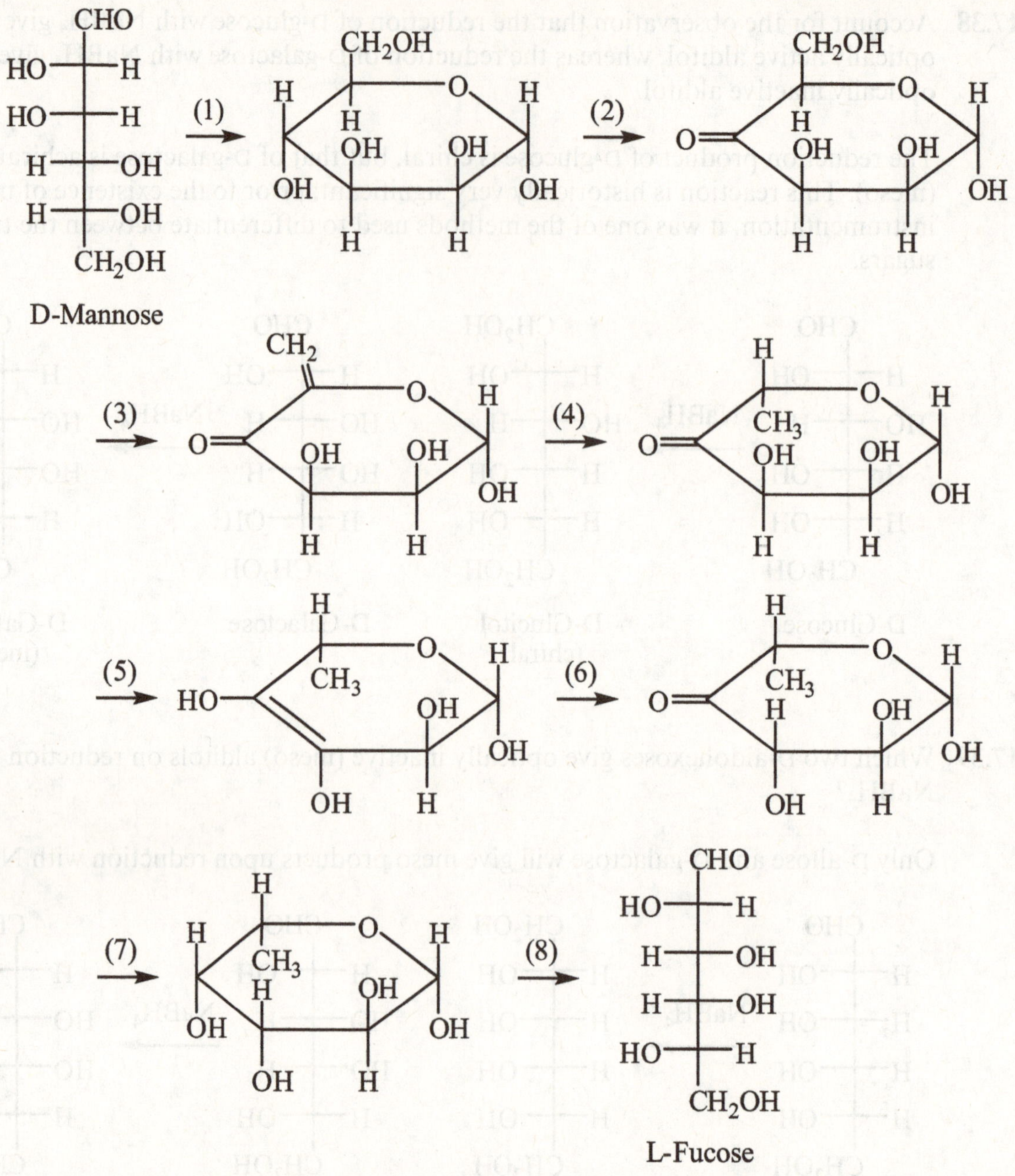

D-Mannose

L-Fucose

(a) Describe the type of reaction (oxidation, reduction, hydration, dehydration, and the like) involved in each step.

(1) Formation of a cyclic hemiacetal
(2) Oxidation of a secondary alcohol to a ketone
(3) Dehydration of an alcohol to form an alkene
(4) Reduction of an alkene to an alkyl group
(5) Tautomerism from the keto form to the enol form
(6) Tautomerism from the enol form to the keto form
(7) Reduction of a ketone to a secondary alcohol
(8) Opening of a cyclic hemiacetal to form an aldehyde and an alcohol

(b) Explain why this monosaccharide, which is derived from D-mannose, now belongs to the L-series.

> In the Fischer projection of D-mannose, the hydroxyl group bonded to the penultimate carbon (carbon 5) is on the right. The chirality of this carbon is lost in step (3), and when it is regenerated in (4), it is inverted to the L-series.

Disaccharides and Oligosaccharides

17.41 Define the term *glycosidic bond*.

> A glycosidic bond, also known as a glycosidic linkage, is the bond formed between the anomeric carbon of a glycoside (an acetal carbon of a carbohydrate) and an –OR group.

17.42 What is the difference in meaning between the terms *glycosidic bond* and *glucosidic bond*?

> A glucosidic bond is a specific type of glycosidic bond. While the anomeric carbon of a glycosidic bond could be that of any glycoside (an acetal of any sugar), the anomeric carbon of a glucosidic bond must be that of a glucoside (an acetal of glucose).

17.43 Do glycosides undergo mutarotation?

> No. Glycosides, which are acetals of sugars, are stable in neutral and alkaline conditions and are not in equilibrium with their open-chain forms. α- and β-glycosides do not form an equilibrium mixture. Under strongly acidic conditions, glycosides would hydrolyze.

17.44 In making candy or syrups from sugar, sucrose is boiled in water with a little acid, such as lemon juice. Why does the product mixture taste sweeter than the starting sucrose solution?

> Sucrose is a disaccharide consisting of D-glucose and D-fructose. These two monosaccharides are linked by a glycosidic bond that is hydrolyzed under acidic conditions. The resulting 1:1 mixture of D-glucose and D-fructose has a sweeter taste than sucrose alone.

17.45 Which disaccharides are reduced by NaBH$_4$?

(a) Sucrose   (b) Lactose   (c) Maltose

In order for a sugar to be reduced by NaBH$_4$, which is capable of reducing a carbonyl group to an alcohol, the sugar must contain at least one anomeric carbon that is in equilibrium with the open-chain form. For this to be the case, the sugar must have a hemiacetal. Maltose and lactose both contain a monosaccharide that is a hemiacetal, but sucrose does not. Both of the anomeric carbons in sucrose are glycosides (acetals).

17.46 Draw Haworth and chair formulas for the β form of a disaccharide in which two units of D-glucopyranose are joined by a β-1,4-glycosidic bond.

The "β-1,4" designates the configuration of the anomeric carbon at the glycosidic bond that connects the anomeric carbon, via an oxygen, to carbon 4 of the other sugar. The anomeric carbon of the other sugar is a hemiacetal in the β configuration. As a whole, this disaccharide has a hemiacetal and is thus a reducing sugar and can mutarotate. However, its ability to reduce and mutarotate is due exclusively to the anomeric carbon that is the hemiacetal and not to the anomeric carbon of the glycosidic bond.

17.47 Trehalose is found in young mushrooms and is the chief carbohydrate in the blood of certain insects. Trehalose is a disaccharide consisting of two D-monosaccharide units, each joined to the other by an α-1,1-glycosidic bond.

Trehalose

(a) Is trehalose a reducing sugar?

No. In order for a sugar to be a reducing sugar, it must have at least one anomeric carbon that is in equilibrium an open-chain aldehyde. That is, the sugar must have at least one hemiacetal that is an aldehyde.

(b) Does trehalose undergo mutarotation?

No. Sugars that do not contain at least one hemiacetal do not undergo mutarotation. Both anomeric carbons in trehalose are glycosides (see Problem 17.43).

(c) Name the two monosaccharide units of which trehalose is composed.

A good strategy to this problem is to first realize that the monosaccharide at the bottom of trehalose is not the drawn the usual way. Typically, monosaccharides are drawn with the oxygen atom of the ring in the top-right corner of the ring. The best approach is to redraw the bottom monosaccharide as shown below (build a model if necessary). Both monosaccharides are D-glucose.

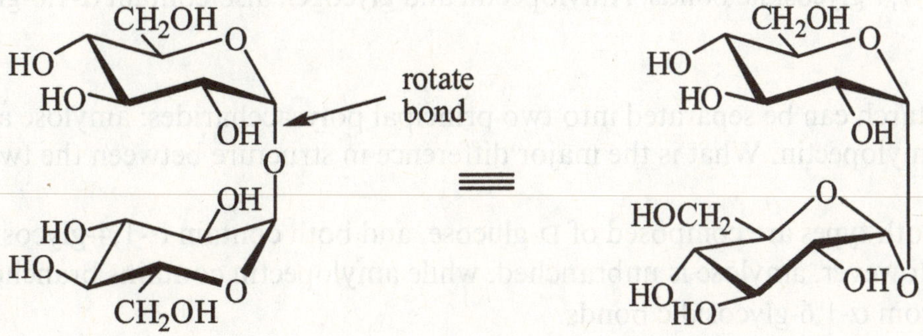

17.48 Hot-water extracts of ground willow bark are an effective pain reliever. Unfortunately, the liquid is so bitter that most persons refuse it. The pain reliever in these infusions is salicin. Name the monosaccharide unit in salicin.

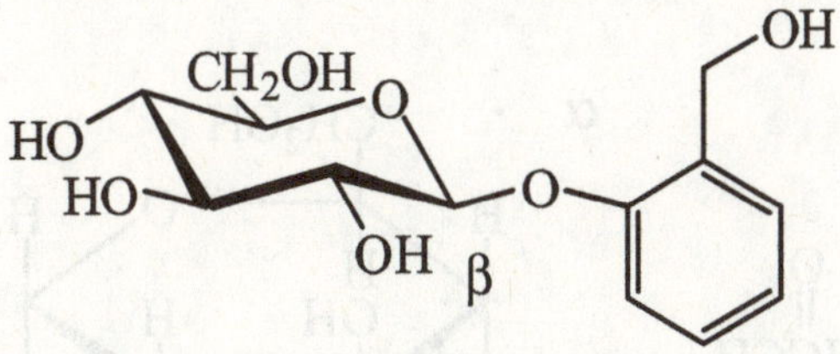

The monosaccharide in salicin is D-glucose, and it is connected by a β glycosidic bond.

Polysaccharides

17.49 What is the difference in structure between oligosaccharides and polysaccharides?

An oligosaccharide is a short polymer of about 6 to 10 monosaccharides. While there is no definite rule, polysaccharides generally contain more than 10 monosaccharides.

17.50 Name three polysaccharides that are composed of units of D-glucose. In which of the three polysaccharides are the glucose units joined by α-glycosidic bonds? In which are they joined by β-glycosidic bonds?

Three polysaccharides that are composed of D-glucose are cellulose, starch (which includes amylose and amylopectin), and glycogen. The glucose units in cellulose are joined by β-1,4-glycosidic bonds, while those in starch and glycogen are joined by α-1,4-glycosidic bonds. Amylopectin and glycogen also contain α-1,6-glycosidic bonds.

17.51 Starch can be separated into two principal polysaccharides: amylose and amylopectin. What is the major difference in structure between the two?

Both types are composed of D-glucose, and both contain α-1,4-glycosidic bonds. However, amylose is unbranched, while amylopectin contains branches that result from α-1,6-glycosidic bonds.

17.52 A Fischer projection of *N*-acetyl-D-glucosamine is given in Section 17.2E.

(a) Draw Haworth and chair structures for the α- and β-pyranose forms of this monosaccharide.

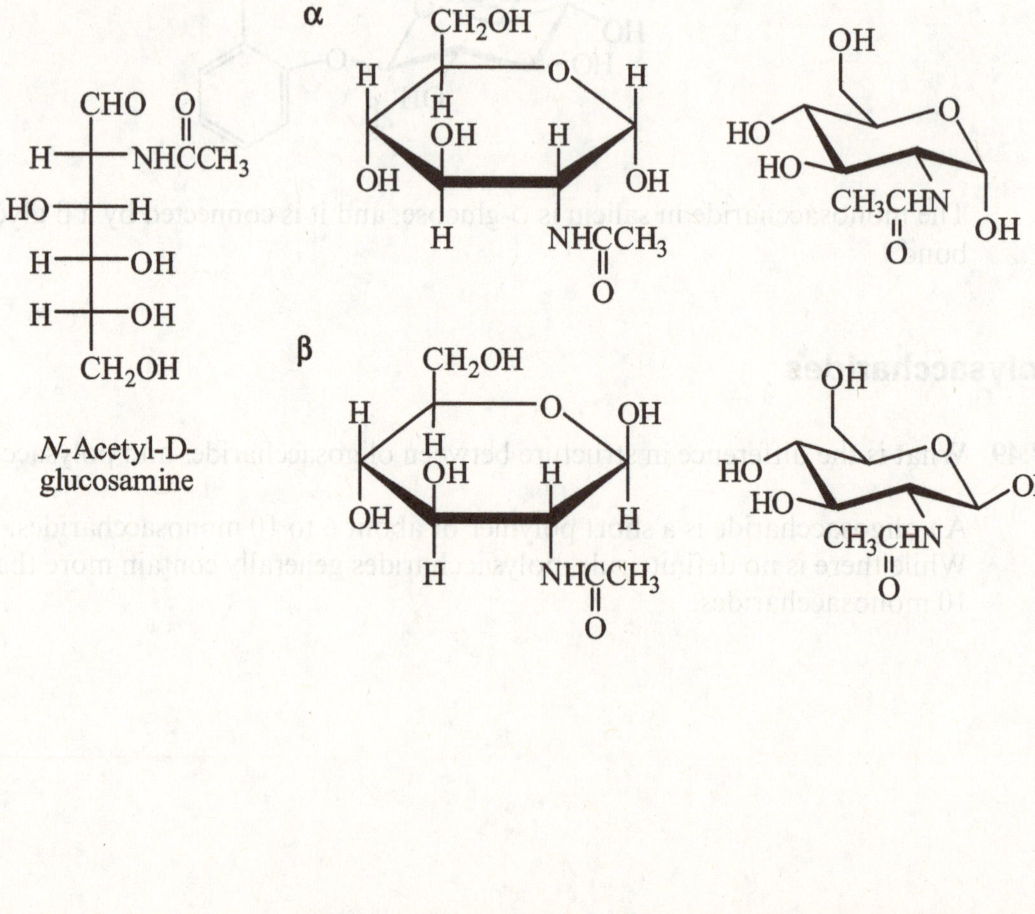

N-Acetyl-D-glucosamine

(b) Draw Haworth and chair structures for the disaccharide formed by joining two units of the pyranose form of *N*-acetyl- D-glucosamine by a β-1,4-glucosidic bond. If your drawing is correct, you have the structural formula for the repeating dimer of chitin, the structural polysaccharide component of the shell of lobster and other crustaceans.

Although the anomeric carbon of the hemiacetal is shown in the structures as the α-configuration, it can mutarotate and form an equilibrium mixture of the α and β anomers.

17.53 Propose structural formulas for the repeating disaccharide unit in these polysaccharides (see Section 17.4D for the treatment of uronic acids):

(a) Alginic acid, isolated from seaweed, is used as a thickening agent in ice cream and other foods. Alginic acid is a polymer of D-mannuronic acid in the pyranose form joined by β-1,4-glycosidic bonds.

D-Mannuronic acid

(b) Pectic acid is the main component of pectin, which is responsible for the formation of jellies from fruits and berries. Pectic acid is a polymer of D-galacturonic acid in the pyranose form joined by α-1,4-glycosidic bonds.

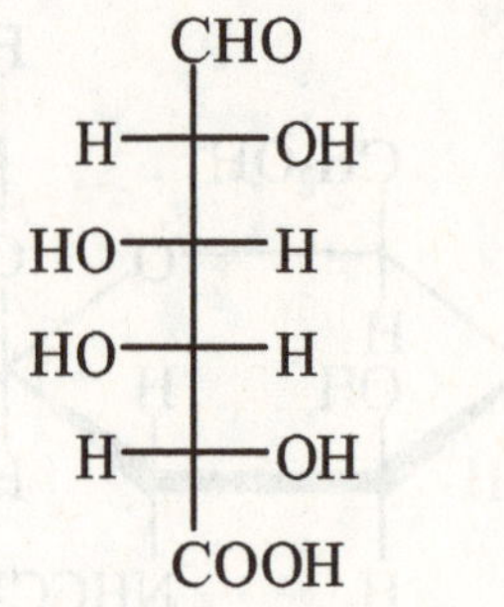

CHO
H——OH
HO——H
HO——H
H——OH
COOH

D-Galacturonic acid

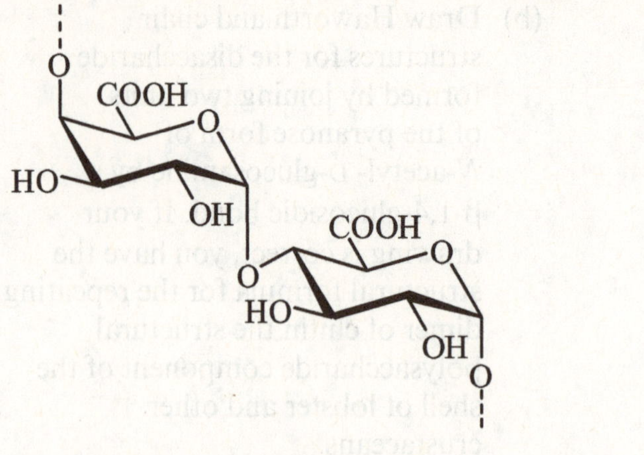

17.54 On the left is a Haworth projection and on the right is a chair conformation for the repeating disaccharide unit in chondroitin 6-sulfate. This biopolymer acts as a flexible connecting matrix between the tough protein filaments in cartilage and is available as a dietary supplement, often combined with D-glucosamine sulfate. Some believe this combination can strengthen and improve joint flexibility.

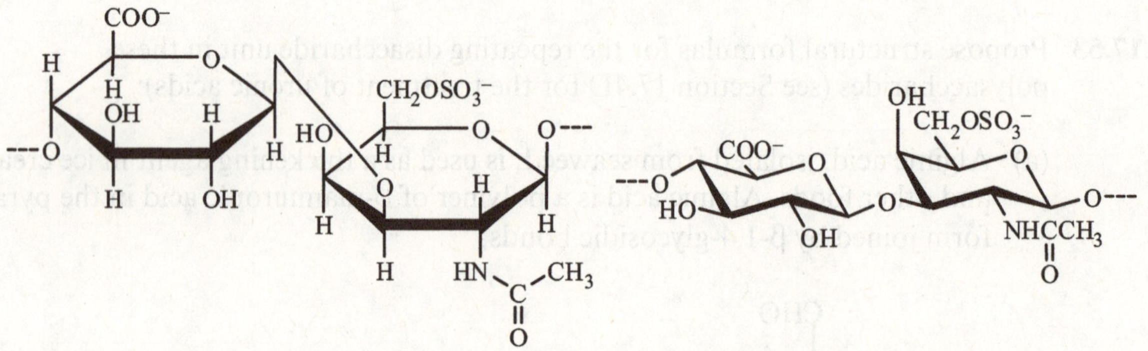

(a) From what two monosaccharide units is the repeating disaccharide unit of chondroitin 6-sulfate derived?

The repeating units in chondroitin 6-sulfate are from D-glucuronic acid (shown in the ionized form) and N-acetyl-D-galactosamine 6-sulfate.

(b) Describe the glycosidic bond between the two units.

The anomeric carbon of D-glucuronic acid is linked to N-acetyl-D-galactosamine 6-sulfate via a β-1,3-glycosidic bond.

17.55 Certain complex lipids are constantly being synthesized and decomposed in the body. In several genetic diseases classified as lipid storage diseases, some of the enzymes needed to decompose the complex lipid are defective or missing. As a consequence, the complex lipids accumulate and cause enlarged liver and spleen, mental retardation, blindness and in certain cases early death. At present no treatment is available for these diseases. The best way to prevent them is genetic counseling. Some of them can be diagnosed during fetal development.

The following is the structure of the lipid that accumulates in Fabray's disease. The genetic defect in this case is that the enzyme α-galactosidase is either missing or defective. This enzyme catalyzes the hydrolysis of glycosidic bonds formed by α-D-galactopyranose.

(a) Name the three hexoses present in this lipid.
(b) Describe the glycosidic bond between each.

D-Galactose   D-Glucose   D-Glucose

(c) Would you expect this molecule to be soluble or insoluble in water? Explain.

With the two large, nonpolar hydrocarbon chains on the right of the molecule, the molecule is expected to be relatively insoluble in water.

Looking Ahead

17.56 One step in glycolysis, the pathway that converts glucose to pyruvate (Section 21.3), involves an enzyme-catalyzed conversion of dihydroxyacetone phosphate to D-glyceraldehyde 3-phosphate. Show that this transformation can be regarded as two enzyme-catalyzed keto-enol tautomerizations (Section 12.8).

$$
\begin{array}{ccc}
\underset{\substack{| \\ \text{C}=\text{O} \\ | \\ \text{CH}_2\text{OPO}_3{}^{2-}}}{\text{CH}_2\text{OH}} & \xrightarrow[\text{catalysis}]{\text{enzyme}} & \underset{\substack{\text{H} \longrightarrow \text{OH} \\ | \\ \text{CH}_2\text{OPO}_3{}^{2-}}}{\text{CHO}}
\end{array}
$$

Dihydroxyacetone   D-Glyceraldehyde
phosphate   3-phosphate

The first tautomerization involves the conversion of the hydroxyketone to an
enediol. This enediol is both the enol of the ketone and the enol of the aldehyde.
Subsequent tautomerization of the enediol into the aldehyde forms the product.
Note that when the enediol tautomerizes to the aldehyde, a stereocenter is formed;
however, because the reaction is enzyme-catalyzed, the enzyme ensures that only
D-glyceraldehyde is made.

$$
\begin{array}{ccccc}
\underset{\substack{\text{CH}_2\text{OH} \\ | \\ \text{C}=\text{O} \\ | \\ \text{CH}_2\text{OPO}_3{}^{2-}}}{} & \rightleftharpoons & \underset{\substack{\text{H} \diagdown \text{C} \diagup \text{OH} \\ \| \\ \text{C}-\text{OH} \\ | \\ \text{CH}_2\text{OPO}_3{}^{2-}}}{} & \rightleftharpoons & \underset{\substack{\text{CHO} \\ | \\ \text{H} \longrightarrow \text{OH} \\ | \\ \text{CH}_2\text{OPO}_3{}^{2-}}}{}
\end{array}
$$

enediol

17.57 One pathway for the metabolism of glucose 6-phosphate is its enzyme-catalyzed
conversion to fructose 6-phosphate. Show that this transformation can be regarded
as two enzyme-catalyzed keto-enol tautomerizations.

Similar to Problem 17.56, the conversion involves the formation of an enediol
intermediate, except in the reverse direction. The aldehyde tautomerizes to the
enediol, which then tautomerizes to the ketone.

$$
\begin{array}{ccccc}
\underset{\substack{\text{CHO} \\ | \\ \text{H}-\text{C}-\text{OH} \\ \text{HO} \longrightarrow \text{H} \\ \text{H} \longrightarrow \text{OH} \\ \text{H} \longrightarrow \text{OH} \\ \text{CH}_2\text{OPO}_3{}^{2-}}}{} & \rightleftharpoons & \underset{\substack{\text{H} \diagdown \text{C} \diagup \text{OH} \\ \| \\ \text{C}-\text{OH} \\ \text{HO} \longrightarrow \text{H} \\ \text{H} \longrightarrow \text{OH} \\ \text{H} \longrightarrow \text{OH} \\ \text{CH}_2\text{OPO}_3{}^{2-}}}{} & \rightleftharpoons & \underset{\substack{\text{CH}_2\text{OH} \\ | \\ \text{C}=\text{O} \\ \text{HO} \longrightarrow \text{H} \\ \text{H} \longrightarrow \text{OH} \\ \text{H} \longrightarrow \text{OH} \\ \text{CH}_2\text{OPO}_3{}^{2-}}}{}
\end{array}
$$

D-Glucose 6-phosphate   enediol   D-Fructose 6-phosphate

17.58 Epimers are carbohydrates that differ in configuration at only one stereocenter.

(a) Which of the aldohexoses are epimers of each other?

There are a number of aldohexoses that are epimers of each other. For example, D-altrose, D-glucose, D-gulose, and L-talose are epimers of D-allose; the four monosaccharides each differ from D-allose at one (and only one) stereocenter.

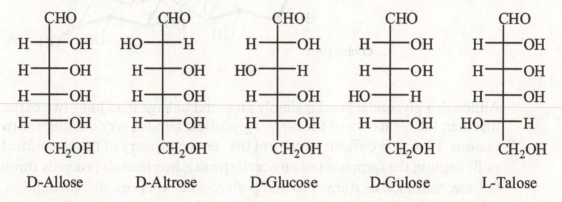

D-Allose   D-Altrose   D-Glucose   D-Gulose   L-Talose

(b) Are all anomer pairs also epimers of each other? Explain. Are all epimers also anomers? Explain.

Epimers differ in the configuration of one stereocenter other than the anomeric carbon, so pairs of anomer are not epimers of each other. For example, α- and β-D-glucose are anomers of each other and cannot be classified as epimers. Epimers are not anomers, because anomers differ in only the configuration of the anomeric carbon.

17.59 Oligosaccharides are very valuable therapeutically and are especially difficult to synthesize, even though the starting materials are readily available. Shown is the structure of globotriose, the receptor for a series of toxins synthesized by some strains of *E. coli*. From left to right, globotriose consists of an α-1,4-linkage of galactose to galactose which is part of a β-1,4-linkage to glucose. The squiggly line indicates that the configuration at that carbon can be α or β. Suggest why it would be difficult to synthesize this trisaccharide, for example by first forming the galactose-galactose glycosidic bond and then forming the glycosidic bond to glucose.

Globotriose

Although a glycosidic bond is simply an acetal linkage than joins two carbohydrates together, the formation of the *correct* glycosidic bond is very difficult for two reasons. First, it is difficult to control the stereochemistry of the glycosidic bond (α or β) because the formation of an acetal (from a hemiacetal) proceeds through an S_N1 mechanism; a mixture of α- and β-glycosidic bonds results. In addition, it is difficult to form only the desired 1,4-linkage, because any one of the hydroxyl groups located on carbons 2, 3, 4, and 6 could be used to form the bond. Therefore, for each glycosidic bond formed, we would obtain a mixture of 1,2-, 1,3-, 1,4-, and 1,6-linkages of both α- and β-stereochemistry. Clearly, this is not going to give the desired product in very good yield.

Group Learning Activities

Solutions are only provided for activities that are not open-ended.

17.61 Discuss how nature settled upon the D form of carbohydrates as the sole stereoisomeric form in living systems. Use the internet to learn about different scientific theories to this question and debate the merits of each.

There are several different theories, but recent research (*Org. Biomol. Chem.*, 2012, **10**, 1565-1570) has shown that under prebiotic conditions, L-amino acids can catalyze the formation of D-sugars. This may suggest that living organisms adapted to use the D-sugars that were already present in the environment.

17.62 Work as a group to provide the mechanism for the reaction shown below. Hint: each step is a mechanistic pattern we have covered in this or previous chapters.

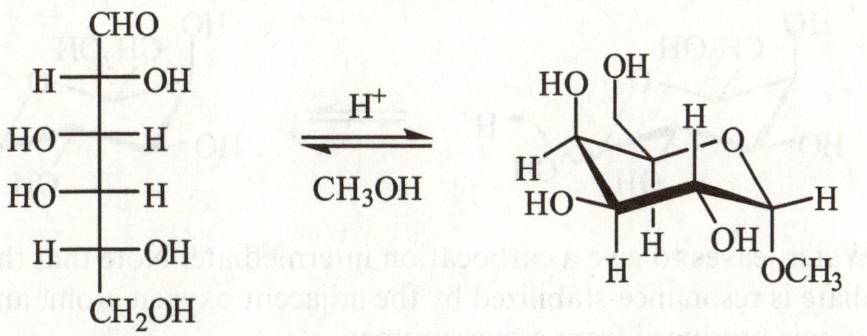

Examination of the product suggests that the starting material underwent intramolecular cyclization via a nucleophilic addition reaction to form a hemiacetal, which then underwent substitution to form an acetal.

Step 1: Protonation of the carbonyl group to form an electrophilic intermediate. (Note that the protonated carbonyl group is resonance-stabilized.)

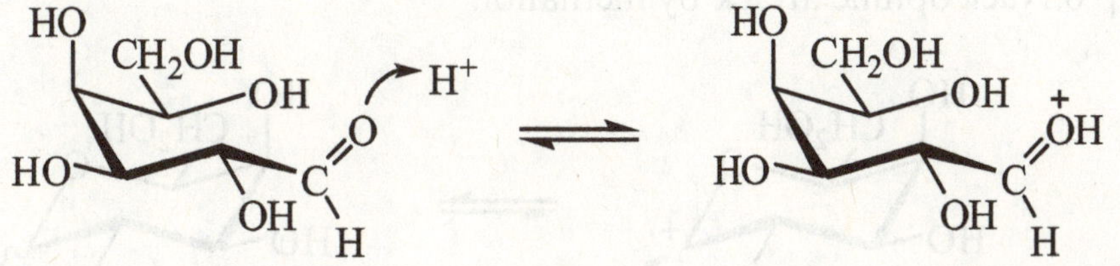

Step 2: The hydroxyl group bonded to the penultimate carbon attacks the protonated carbonyl group, forming a protonated hemiacetal. Note that a mixture of anomers is formed (the wavy bond indicates a mixture of configurations at the respective carbon).

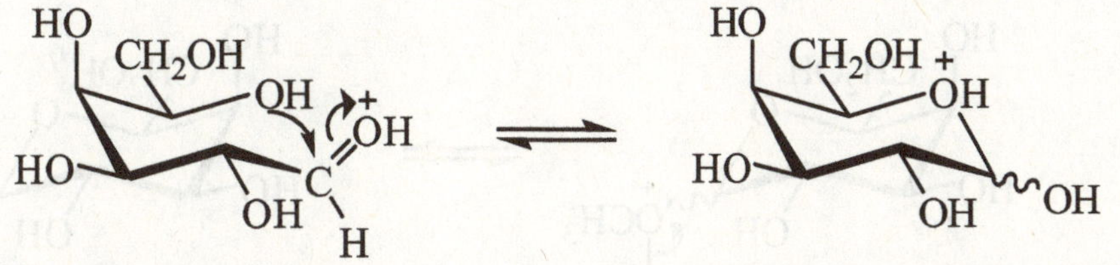

Step 3: Deprotonation of the oxonium ion by methanol generates the hemiacetal.

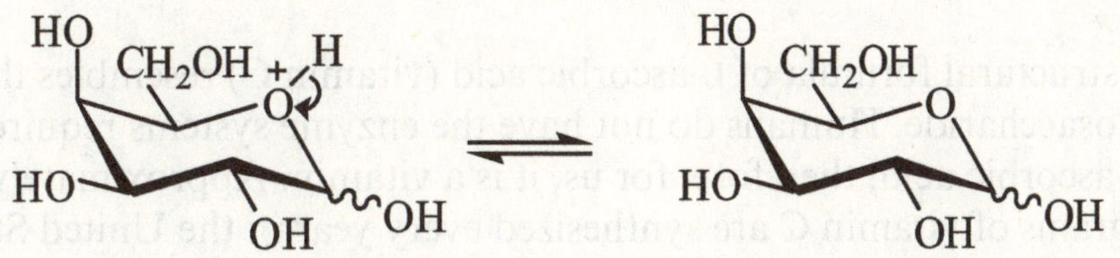

Step 4: The subsequent steps are characteristic of an S_N1 substitution. The hydroxyl group of the hemiacetal is protonated, forming a good leaving group.

Step 5: Water leaves to give a carbocation intermediate. Note that this carbocation intermediate is resonance-stabilized by the adjacent oxygen atom, and that the same carbocation is produced from either anomer.

Step 6: Nucleophilic attack by methanol.

Step 7: Deprotonation of the oxonium ion results in the final product.

(mixture of anomers)

17.63 The structural formula of L-ascorbic acid (vitamin C) resembles that of a monosaccharide. Humans do not have the enzyme systems required for the synthesis of L-ascorbic acid; therefore, for us, it is a vitamin. Approximately 66 million kilograms of vitamin C are synthesized every year in the United States. Ascorbic acid contains four hydroxyl groups. With your group, determine which –OH group is most acidic debate the merits of your selection. Recall from Section 2.5 that various structural features can enhance the stability of a conjugate base, thus increasing the acidity of the original acid.

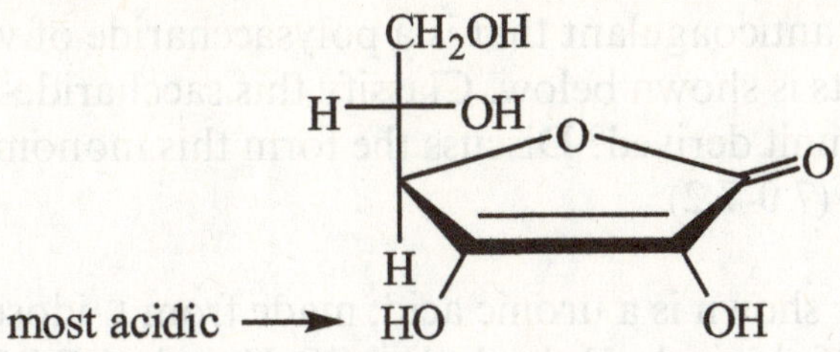

CH₂OH

H——OH

H

most acidic ⟶ HO   OH

**L-Ascorbic Acid
(vitamin C)**

The conjugate base formed by the deprotonation of the most-acidic proton is stabilized by three contributing structures that delocalize the negative charge over two oxygen atoms and one carbon atom. The pK_a of this proton is about 4.1.

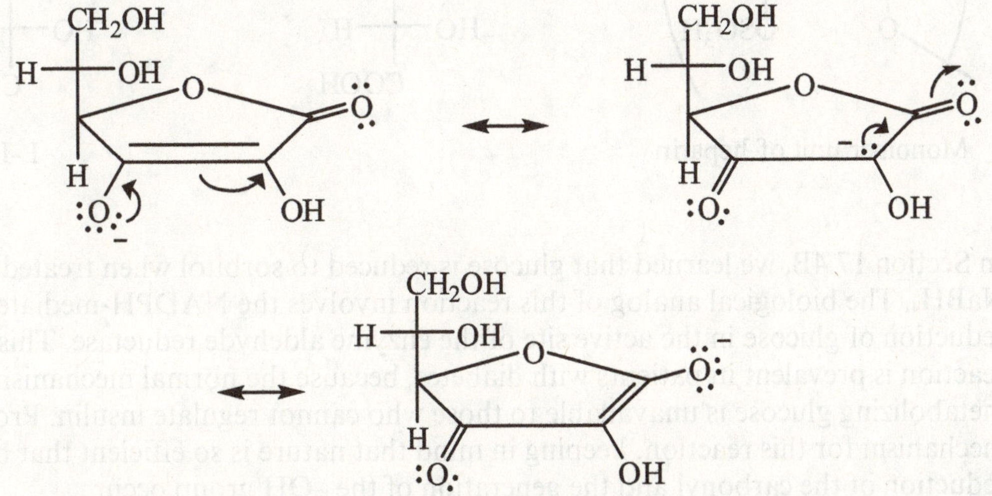

CH₂OH

H——OH

H

:O:

OH

⟷

CH₂OH

H——OH

H

:O:

OH

CH₂OH

H——OH

H

:O:

OH

⟷

Ascorbic acid has another acidic –OH group, but it is less acidic because the conjugate base is stabilized by only two contributing structures that delocalize the negative charge over one oxygen atom and one carbon atom. This lesser degree of resonance stabilization of the conjugate base makes the proton less acidic; its pK_a is about 11.8.

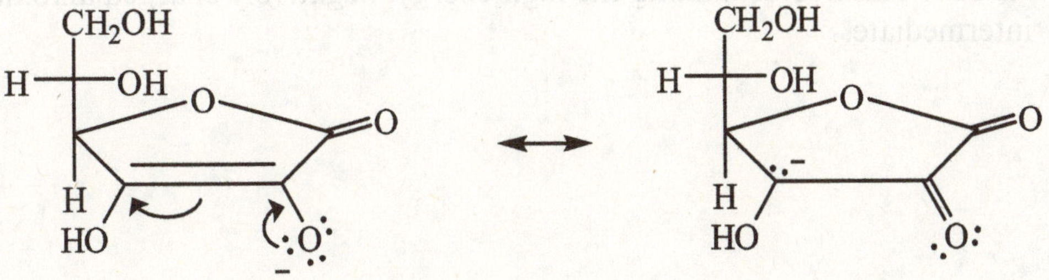

CH₂OH

H——OH

H
HO

:O:

⟷

CH₂OH

H——OH

H
HO

:O:

17.64 Heparin is an anticoagulant that is a polysaccharide of varying chain size. One of its monomer units is shown below. Classify this saccharide-based unit. From which hexose is this unit derived? Discuss the form this monomer unit would assume at biological pH (7.0–7.2).

The monomer shown is a uronic acid, made from L-idose, that has been sulfated at carbon-2. At biological pH, both the $-SO_3H$ and $-COOH$ groups would be negatively charged.

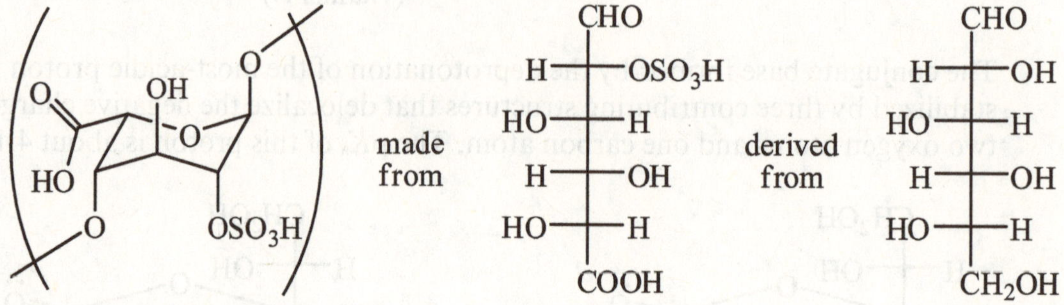

Monomer unit of heparin   L-Idose

17.65 In Section 17.4B, we learned that glucose is reduced to sorbitol when treated with NaBH$_4$. The biological analog of this reaction involves the NADPH-mediated reduction of glucose in the active site of the enzyme aldehyde reductase. This reaction is prevalent in patients with diabetes, because the normal mechanism of metabolizing glucose is unavailable to those who cannot regulate insulin. Propose a mechanism for this reaction, keeping in mind that nature is so efficient that both the reduction of the carbonyl and the generation of the –OH group occur simultaneously. As a group, discuss how nature is often a much better synthetic chemist than synthetic chemists!

In most enzyme-catalyzed reactions that involve the addition of a nucleophile to a carbonyl group, the oxygen of the carbonyl group is simultaneously protonated by a nearby acid source (the side chain of an amino acid) as the nucleophile attacks the carbon. This avoids making the high-energy, negatively charged alkoxide intermediate.

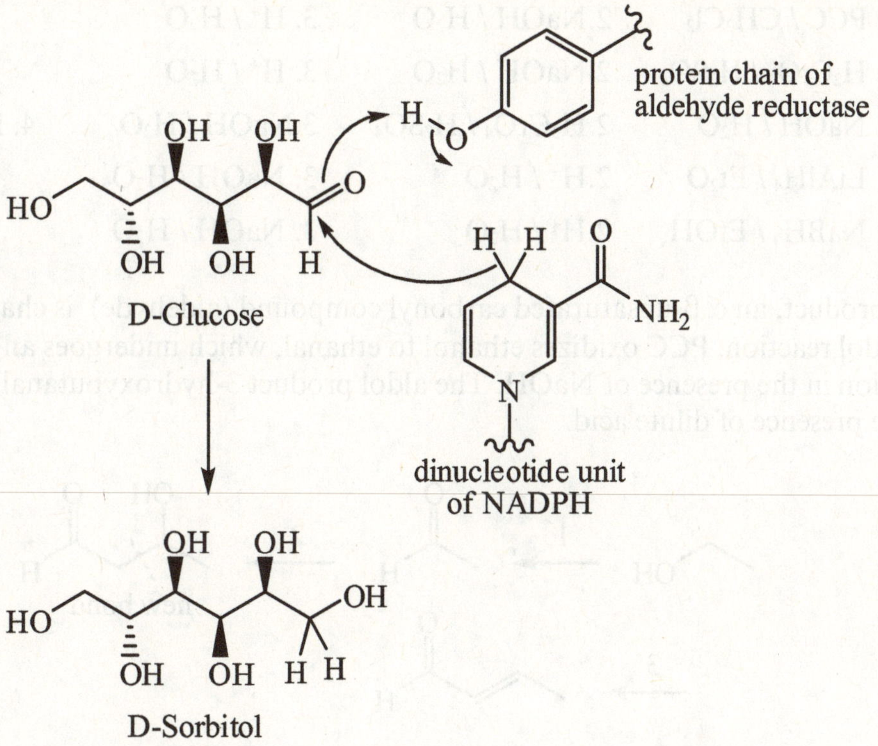

D-Glucose

protein chain of
aldehyde reductase

dinucleotide unit
of NADPH

D-Sorbitol

Putting It Together

1. Which carbon on β-maltose will be oxidized by Tollens' reagent?

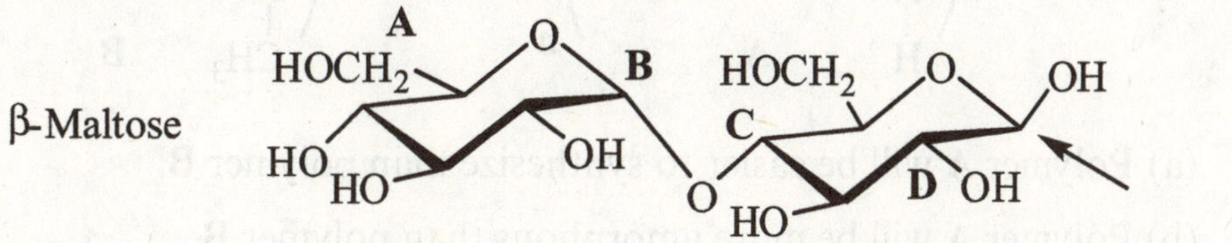

β-Maltose

(a) **A**   (b) **B**   (c) **C**   (d) **D**   **(e)** None of the above

Tollens' reagent will oxidize aldehyde carbons, and none of the above will be
oxidized. Rather, the carbon indicated by the arrow will be oxidized; it is a
hemiacetal and is in equilibrium with the open-chain aldehyde form.

2. What sequence of reagents will accomplish the following transformation?

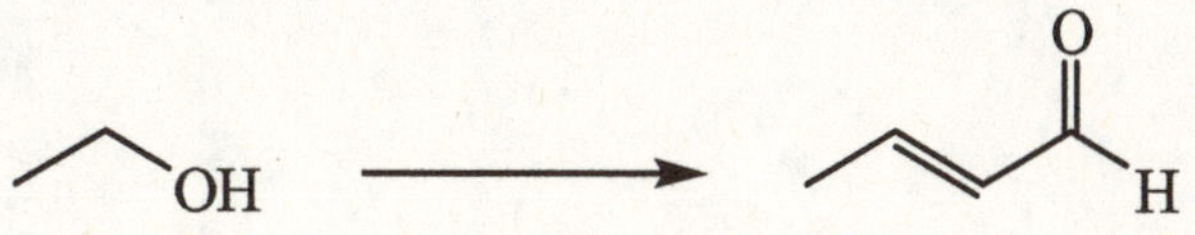

(a) 1. PCC / CH$_2$Cl$_2$ 2. NaOH / H$_2$O   3. H$^+$ / H$_2$O

(b) 1. H$_2$CrO$_4$ / H$_2$SO$_4$ 2. NaOH / H$_2$O   3. H$^+$ / H$_2$O

(c) 1. NaOH / H$_2$O   2. H$_2$CrO$_4$ / H$_2$SO$_4$ 3. NaOH / H$_2$O   4. H$^+$ / H$_2$O

(d) 1. LiAlH$_4$ / Et$_2$O   2. H$^+$ / H$_2$O   3. NaOH / H$_2$O

(e) 1. NaBH$_4$ / EtOH   2. H$^+$ / H$_2$O   3. NaOH / H$_2$O

The product, an α,β-unsaturated carbonyl compound (aldehyde), is characteristic of an aldol reaction. PCC oxidizes ethanol to ethanal, which undergoes an aldol reaction in the presence of NaOH. The aldol product 3-hydroxybutanal dehydrates in the presence of dilute acid.

3.   Assuming that the following two polymers are manufactured in similar ways, which of the following statements is *true*?

(a) Polymer **A** will be easier to synthesize than polymer **B**.

(b) Polymer **A** will be more amorphous than polymer **B**.

(c) Polymer **A** will be weaker than polymer **B**.

(d) Polymer **A** will have a higher T_g than polymer **B**.

(e) None of the above.

Compared to polymer **B**, polymer **A** contains an N–H group that is capable of hydrogen bonding. As a result, polymer **A** is expected to be more ordered (the chains can align themselves to maximize hydrogen bonding), stronger, and have a higher T_g value.

4.   Which of the following carbohydrates does *not* undergo mutarotation?

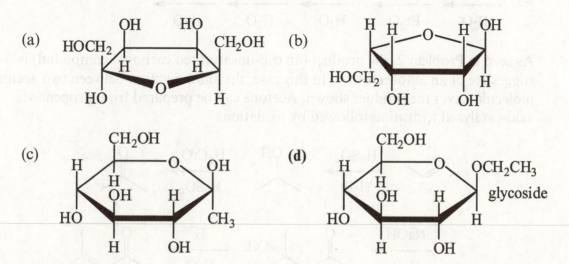

(e)   All of the above

For a carbohydrate to be able to undergo mutarotation, the anomeric carbon needs
to be a hemiacetal. Unlike glycosides (acetals of sugars), which are much more
stable, hemiacetals are in equilibrium with their open-chain forms.

5.   What sequence of reagents will accomplish the following transformation?

(a) $\xrightarrow[\text{H}_2\text{SO}_4]{\text{HBr}}$ $\xrightarrow[\text{H}_2\text{O}]{\text{H}_2\text{CrO}_4}$ $\xrightarrow{\text{NaOH}}$

(b) $\xrightarrow[\text{H}_2\text{O}]{\text{NaOH}}$ $\xrightarrow[\text{H}_2\text{O}]{\text{H}^+}$

(c) $\xrightarrow[\text{H}_2\text{SO}_4]{\text{H}_2\text{CrO}_4}$ $\xrightarrow[\text{H}_2\text{O}]{\text{NaOH}}$ $\xrightarrow[\text{H}_2\text{O}]{\text{H}^+}$

(d) $\xrightarrow[\text{H}_2\text{O}]{\text{H}_2\text{SO}_4}$ $\xrightarrow[\text{H}_2\text{SO}_4]{\text{H}_2\text{CrO}_4}$ $\xrightarrow[\text{H}_2\text{O}]{\text{NaOH}}$ $\xrightarrow[\text{H}_2\text{O}]{\text{H}^+}$

(e) $\xrightarrow[\text{H}_2\text{O}]{\text{H}_2\text{SO}_4}$ $\xrightarrow[\text{Et}_2\text{O}]{\text{LiAlH}_4}$ $\xrightarrow[\text{H}_2\text{O}]{\text{H}^+}$ $\xrightarrow[\text{H}_2\text{O}]{\text{NaOH}}$ $\xrightarrow[\text{H}_2\text{O}]{\text{H}^+}$

As seen in Problem 2, the product (an α,β-unsaturated carbonyl compound) is suggestive of an aldol reaction. In this case, the aldol reaction between two acetone molecules gives the product shown. Acetone can be prepared from propene via acid-catalyzed hydration followed by oxidation.

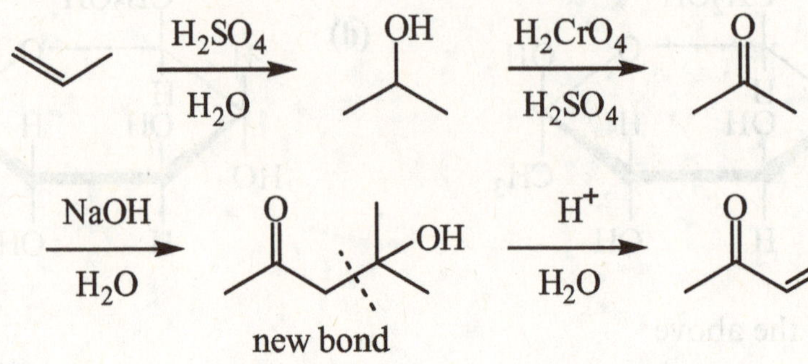

6.   Select the most likely product of the following reaction.

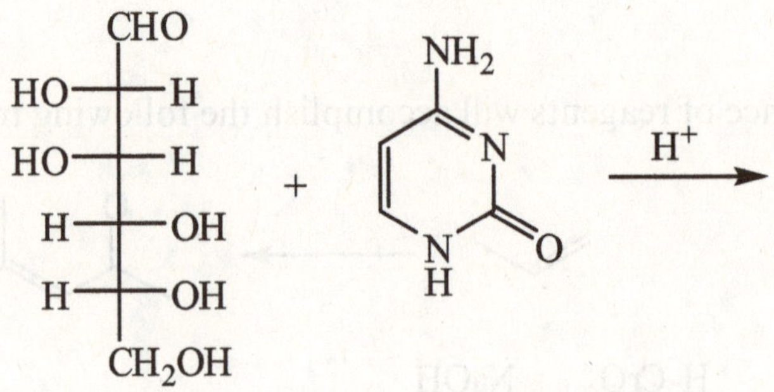

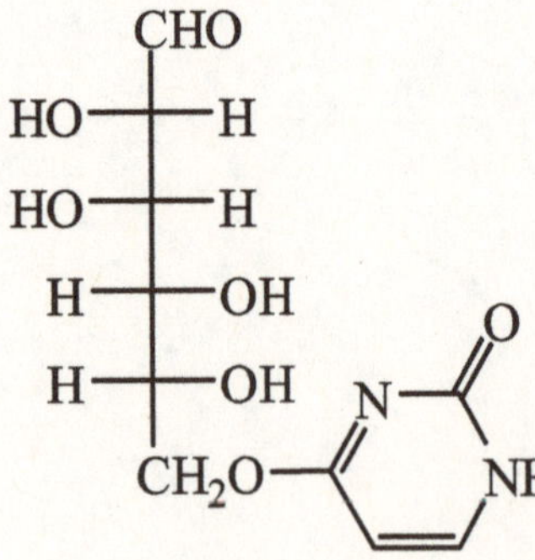

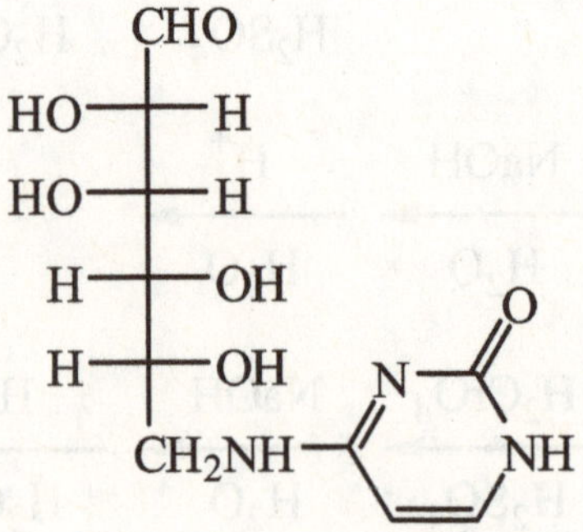

(c)

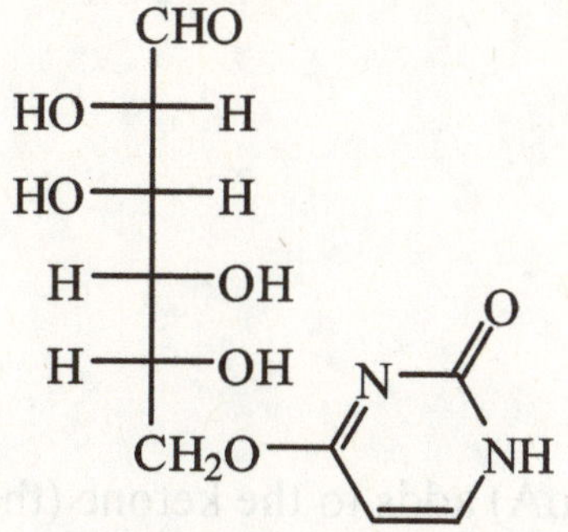

(d)

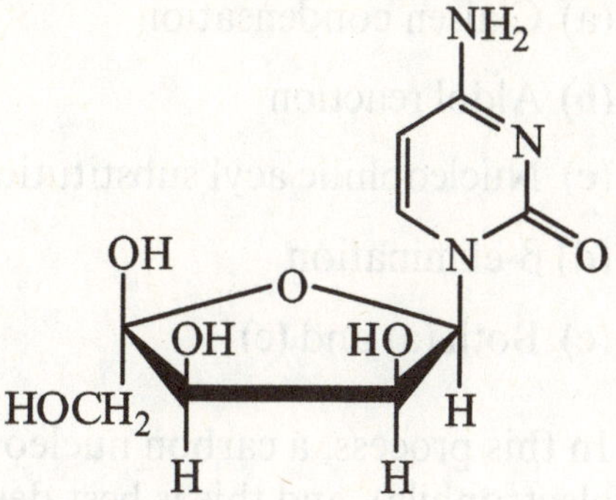

(e)   None of the above

The most likely product is the *N*-glycoside of D-mannose, as shown below.

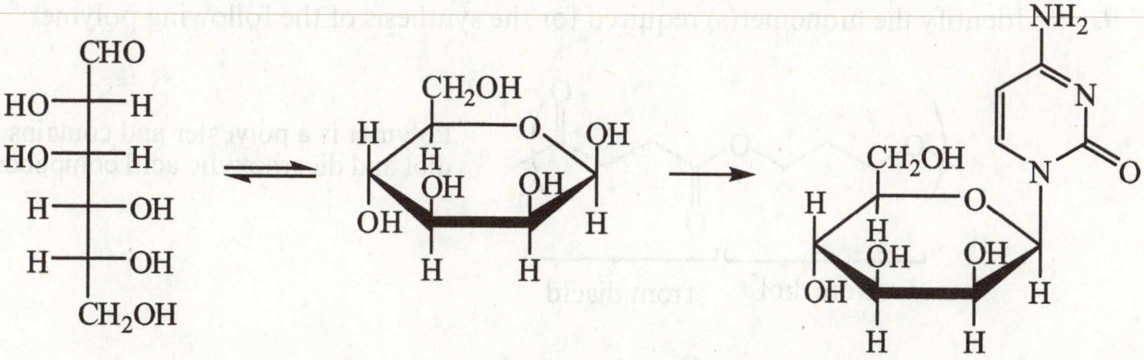

7.   How many glycosidic bonds exist in the following polysaccharide?

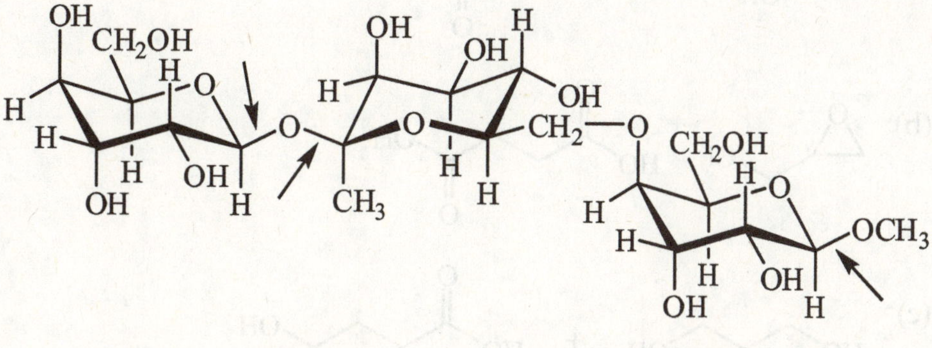

(a) one   (b) two   (c) three   (d) four   (e) five

8.   Which of the following best classifies the following biological process?

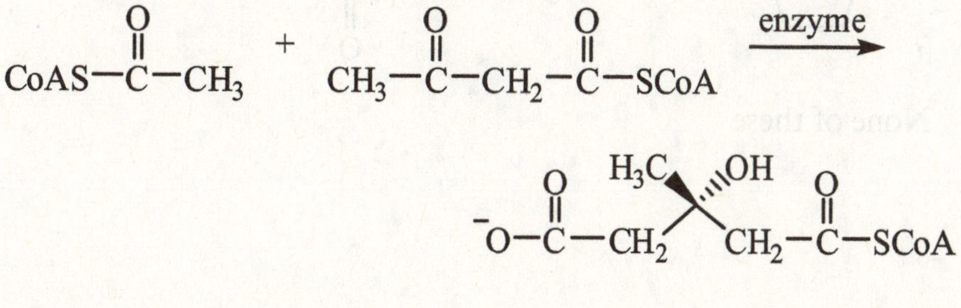

(a) Claisen condensation

(b) Aldol reaction

(c) Nucleophilic acyl substitution

(d) β-elimination

(e) Both (a) and (c)

In this process, a carbon nucleophile (acetyl CoA) adds to the ketone (the electrophile), and this is best described as an aldol reaction. A thioester is also hydrolyzed.

9. Identify the monomer(s) required for the synthesis of the following polymer:

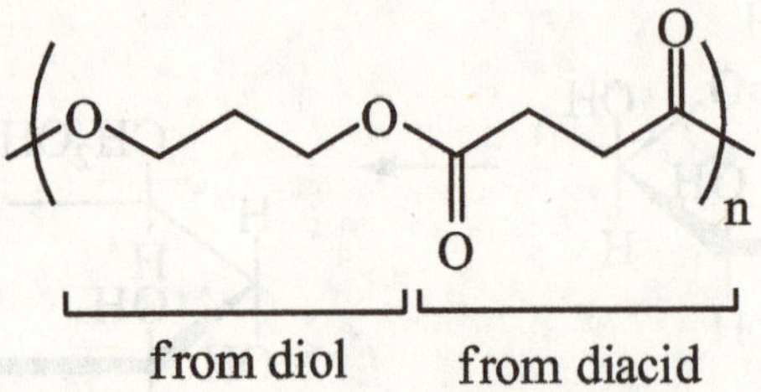

Polymer is a polyester and contains diol and dicarboxylic acid components

(a)

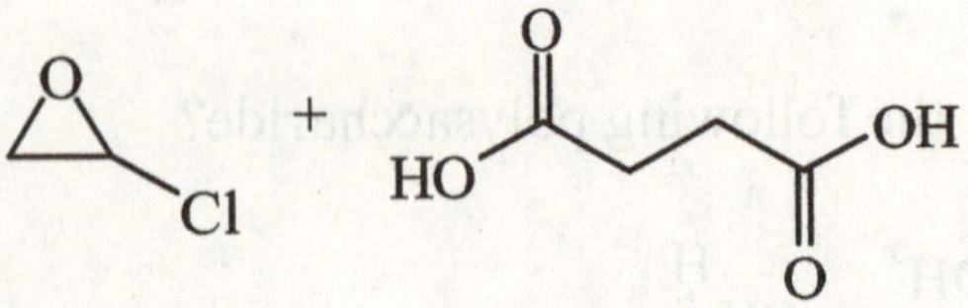

(b)

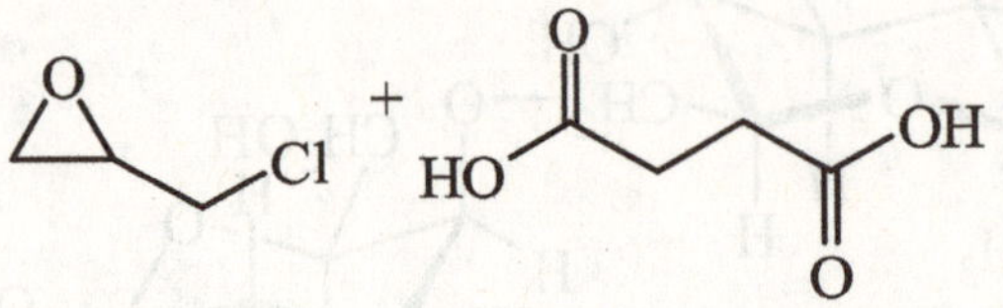

(c)

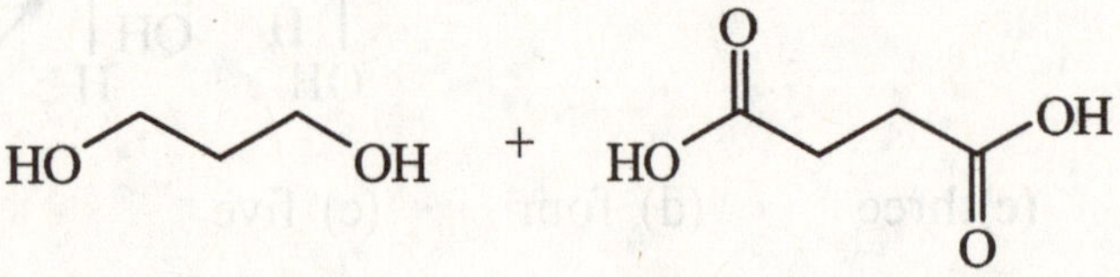

(d)

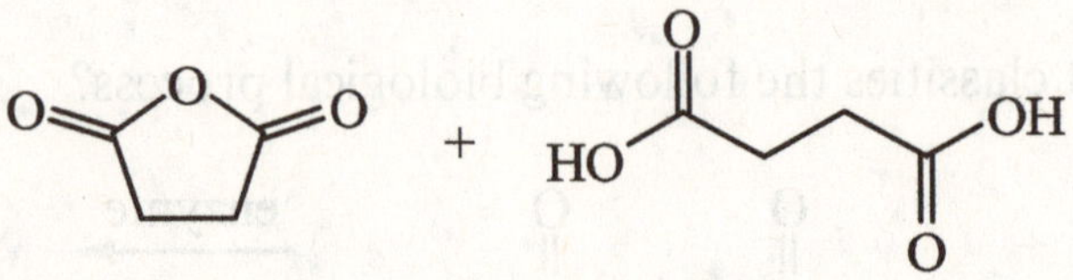

(e) None of these

10.   An unknown carbohydrate is placed in a solution of NaBH₄/EtOH. After isolating
the product, it was discovered that no alditol products were formed. Which of the
following carbohydrates could be the unknown?

(a)

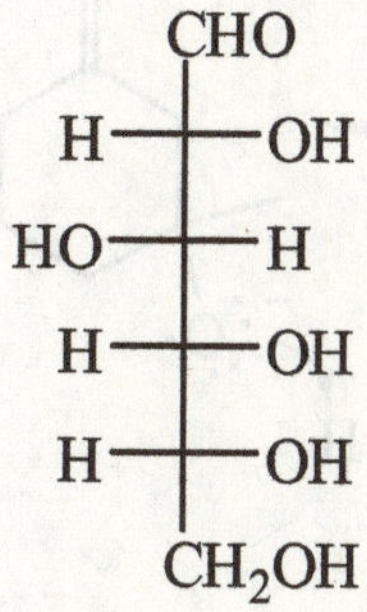

```
        CHO
   H ——— OH
  HO ——— H
   H ——— OH
   H ——— OH
        CH₂OH
```

(b)

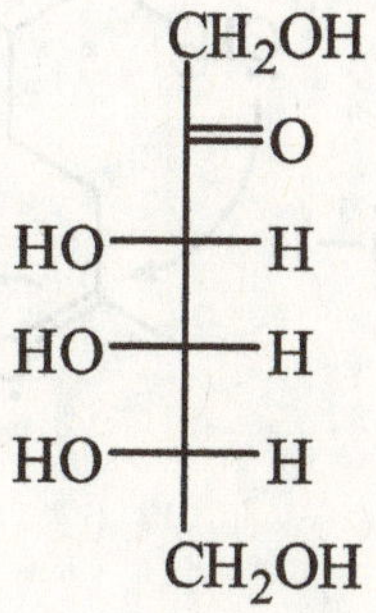

```
        CH₂OH
        ═O
  HO ——— H
  HO ——— H
  HO ——— H
        CH₂OH
```

(c)

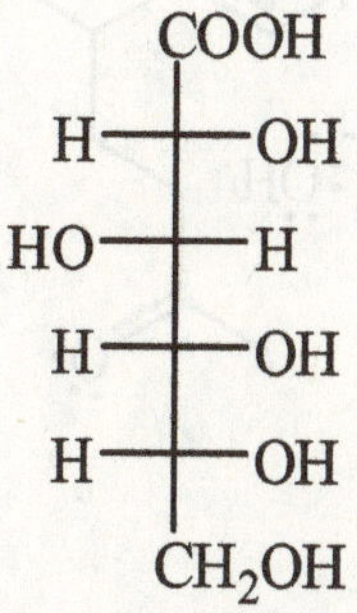

```
        COOH
   H ——— OH
  HO ——— H
   H ——— OH
   H ——— OH
        CH₂OH
```

(d)

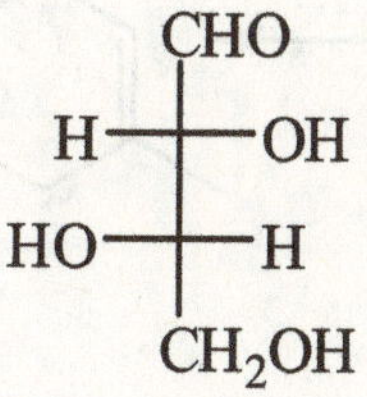

```
        CHO
   H ——— OH
  HO ——— H
        CH₂OH
```

(e)

```
        CHO
  HO ——— H
  HO ——— H
  HO ——— H
  HO ——— H
        CH₂OH
```

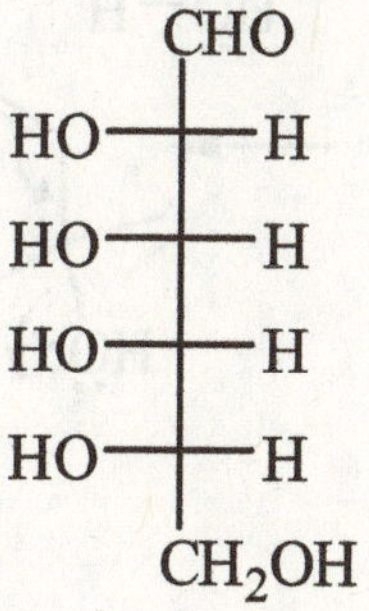

Alditols are sugars with
only alcohol groups.
NaBH₄ is a relatively mild
reducing agent and will
only reduce aldehydes and
ketones to alcohols.
Carboxylic acids, as in (c),
are not reduced.

11.   Each of the products shown can be made by the reaction indicated under the arrow.
Provide a structure for the starting compound(s) needed to produce the product
shown. Then show the mechanism of its formation. Show all charges and lone pairs
of electrons in your structures.

(a)

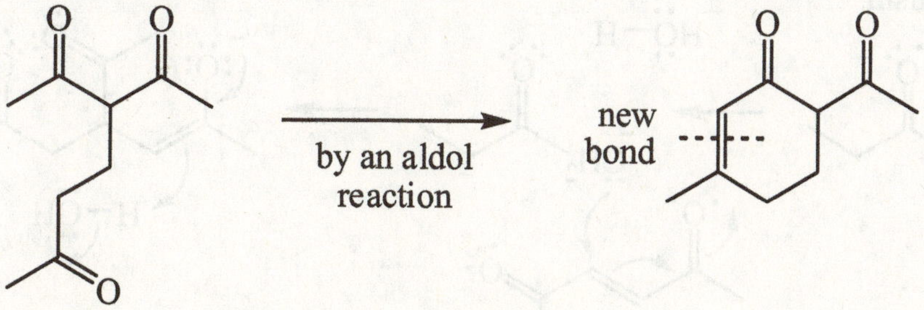

by an aldol
reaction

new
bond

Mechanism:

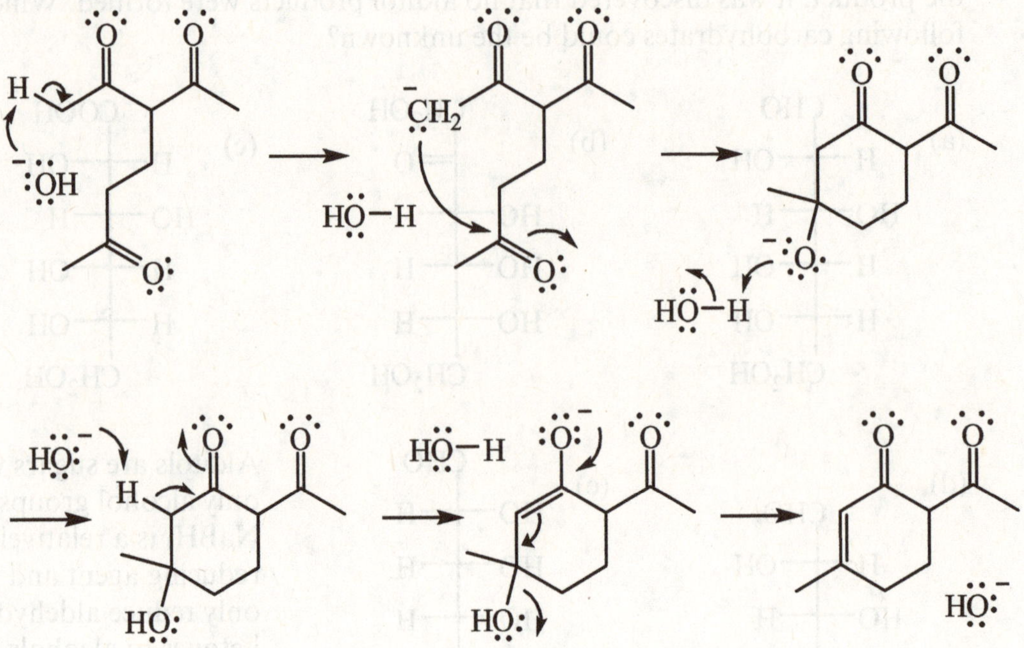

(b)

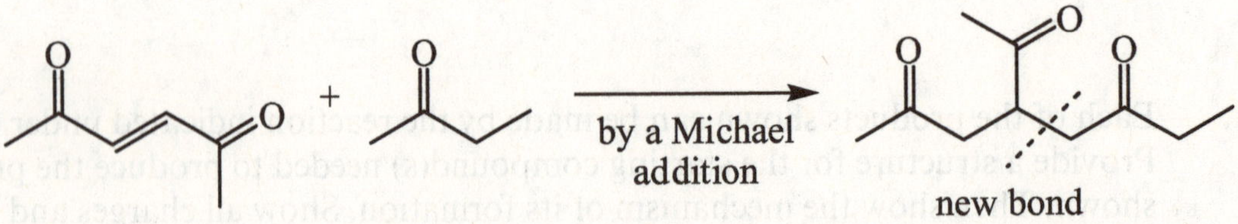

Mechanism:

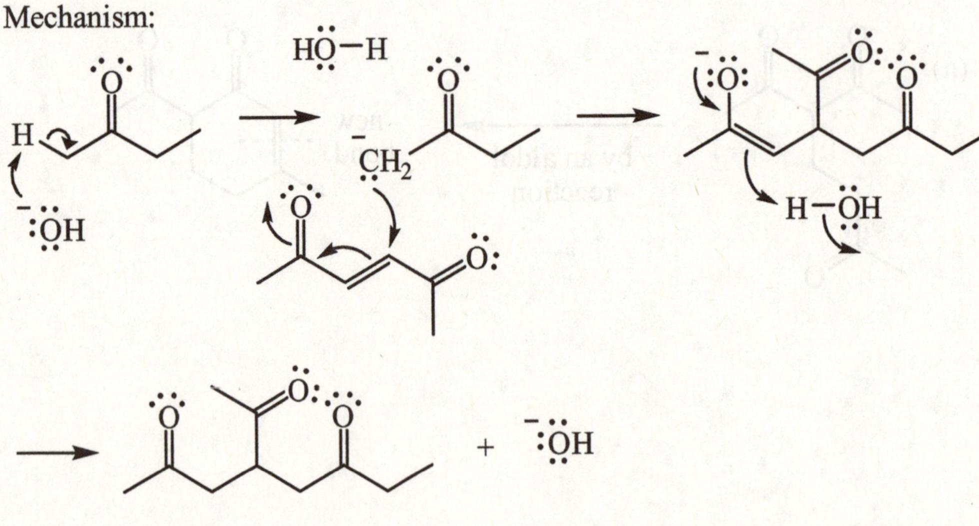

12.   NAD⁺ is a coenzyme found in all living cells. It acts to carry electrons during biological reactions.

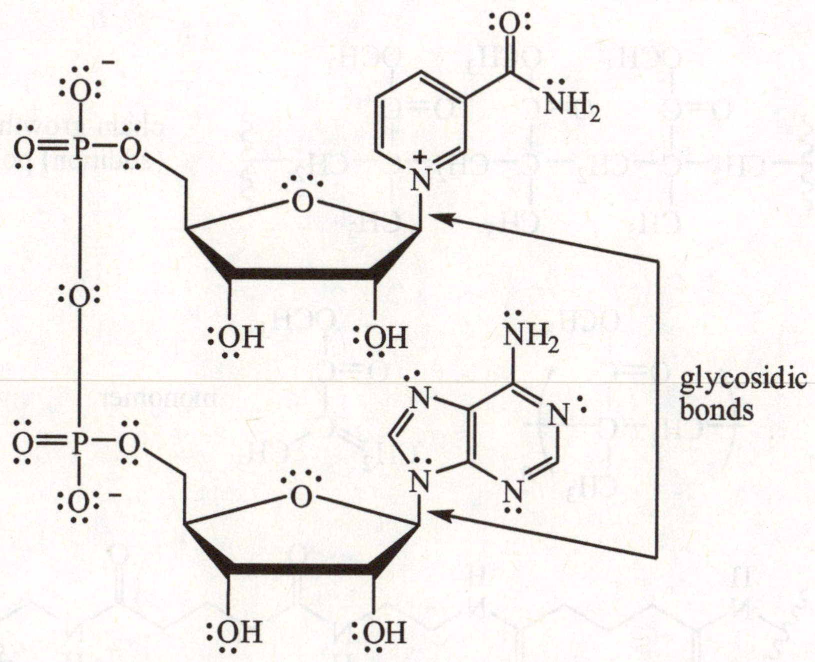

glycosidic bonds

(a) Assign formal charges to the atoms in NAD⁺.

(b) Is NAD⁺ a reducing sugar?

No. Both anomeric carbons are glycosides (acetals), which are not reducing.

(c) Identify the glycosidic bonds in NAD⁺.

(d) From what naturally occurring sugar(s) is NAD⁺ derived?

Both carbohydrate units in NAD⁺ are derived from D-ribose.

(e) Is NAD⁺ an α sugar, a β sugar, or both?

Only β sugars are found in NAD⁺.

(f) Does NAD⁺ undergo mutarotation?

No. Anomeric carbons that are glycosides do not undergo mutarotation.

13. Redraw each polymer using the notation in which parentheses are placed around the repeating unit. Identify each as chain-growth or step-growth polymers and identify the monomers required for its synthesis.

(a)

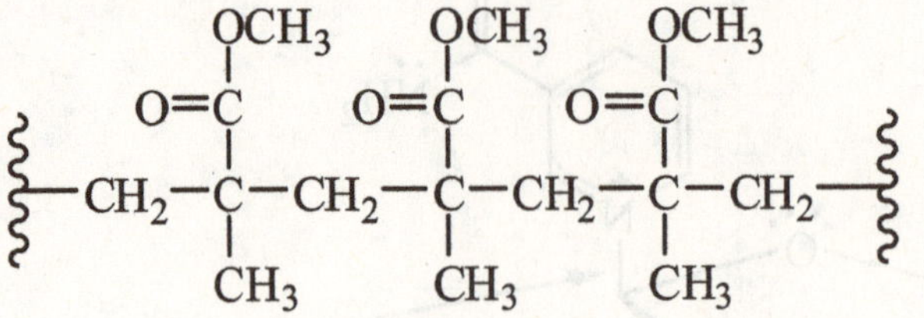

chain-growth
(addition) polymer

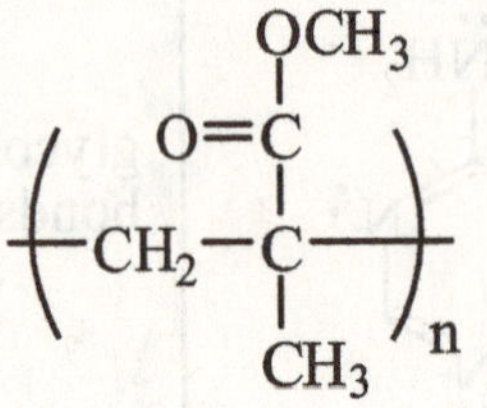

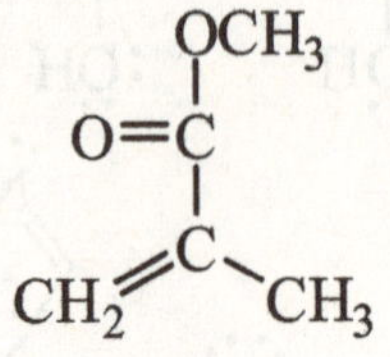

monomer

(b)

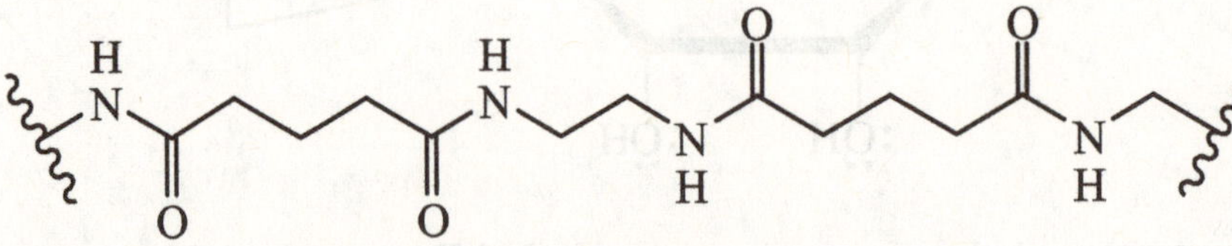

step-growth (condensation) polymer

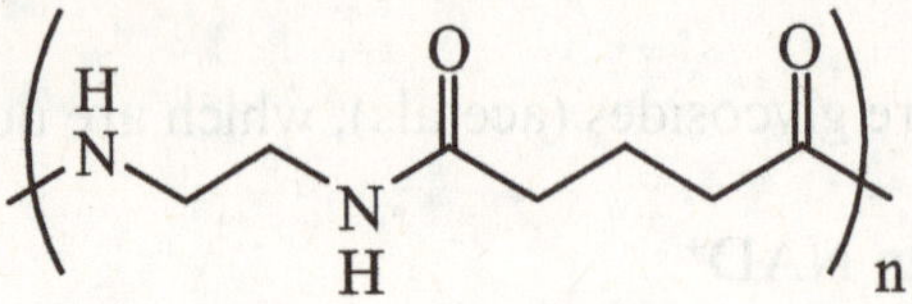

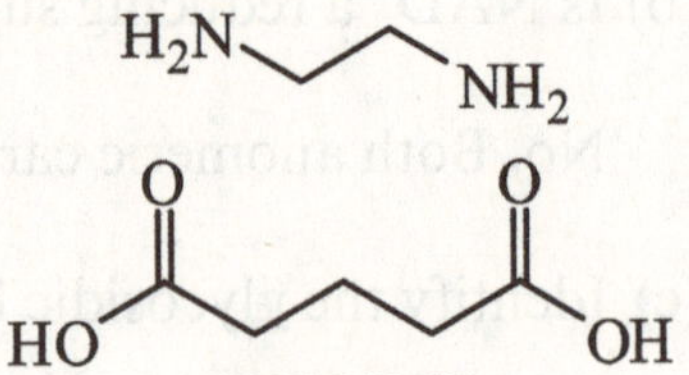

monomers

(c)

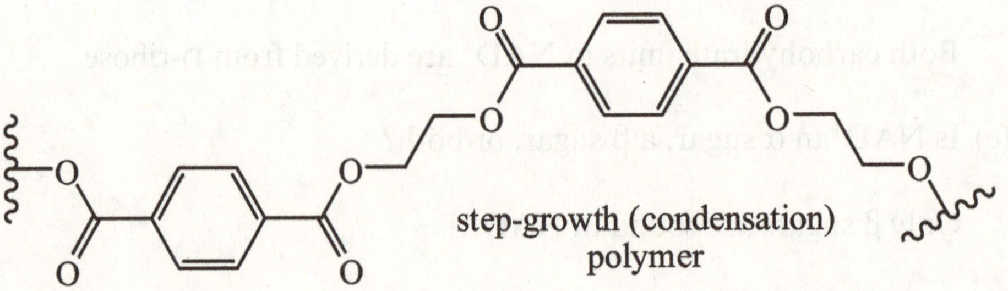

step-growth (condensation)
polymer

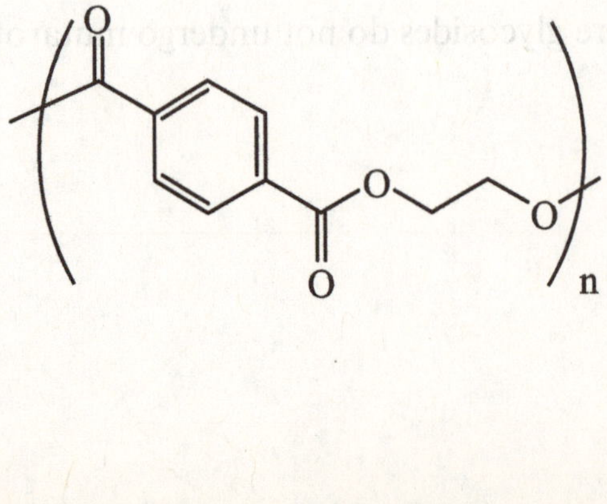

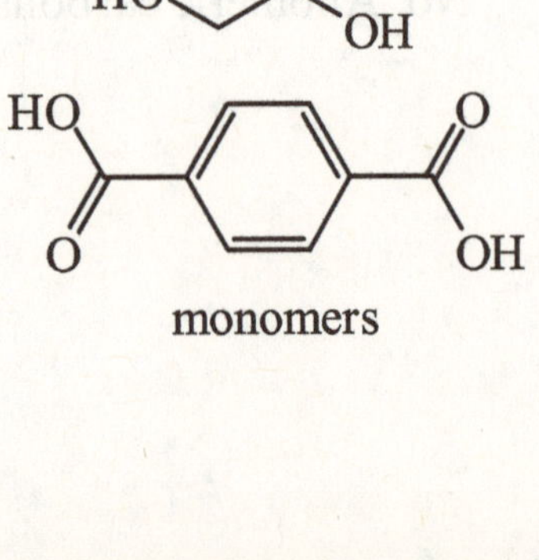

monomers

14. Provide a mechanism for the following series of reactions. Show all charges and lone pairs of electrons in your structures as well as the structures of all intermediates.

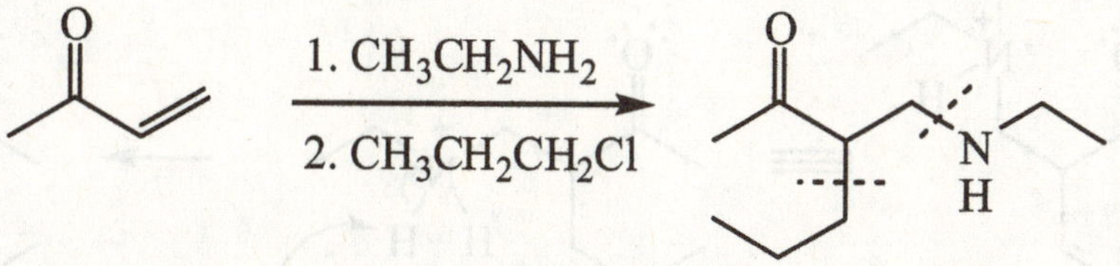

The dashed lines indicate where new bonds have formed. The starting material first undergoes a Michael addition where ethylamine adds to the β-carbon. In the second part of the reaction, a propyl group is added to the α-carbon.

Step 1: Addition of ethylamine to the β-carbon, forming a resonance-stabilized enolate.

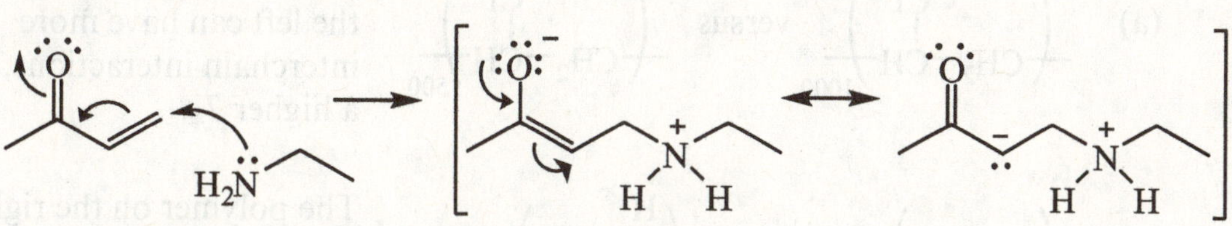

Step 2: Formation of the enol and deprotonation of the nitrogen.

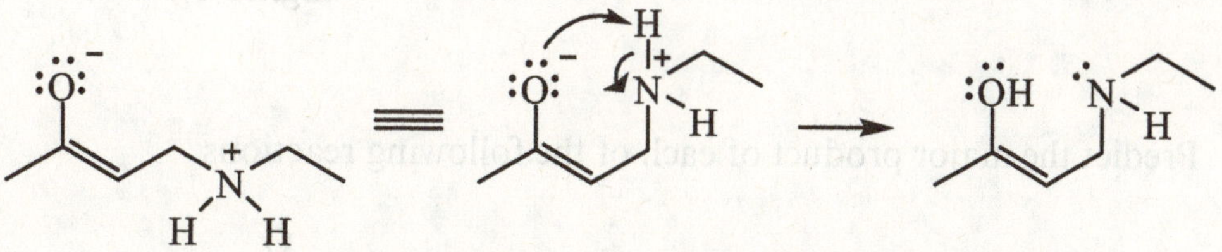

Step 3: After the addition of chloropropane to the mixture, the nitrogen deprotonates the enol, and a nucleophilic aliphatic substitution follows.

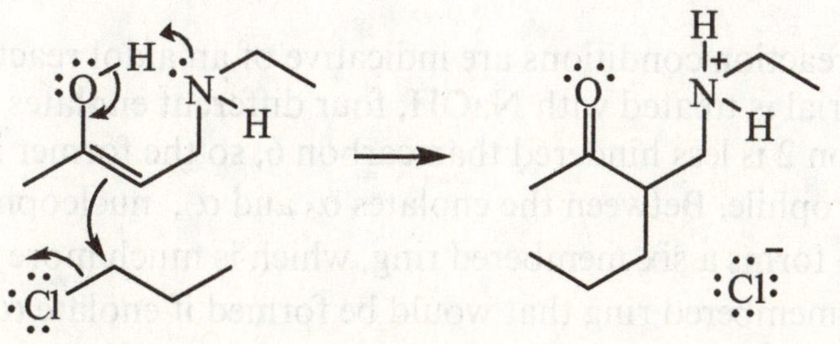

Step 4: Deprotonation of the ammonium ion to generate the product.

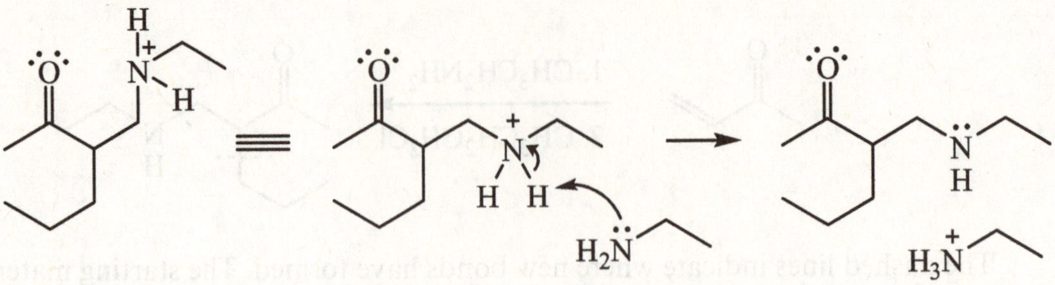

15. Select the polymer from each pair that would have the higher glass transition temperature and provide an explanation for selection.

(a) 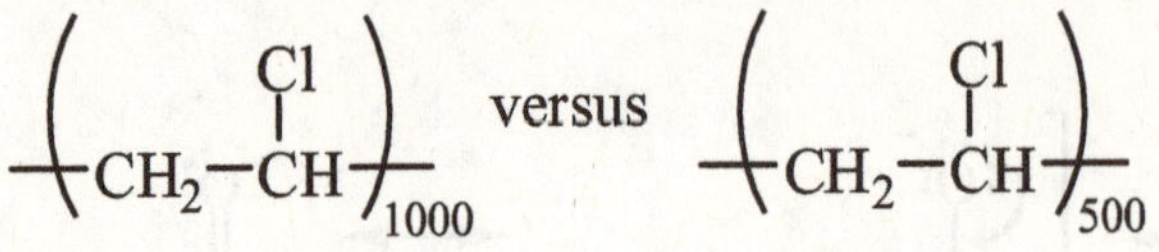   versus

The longer polymer chain on the left can have more interchain interactions, so it has a higher T_g.

(b) 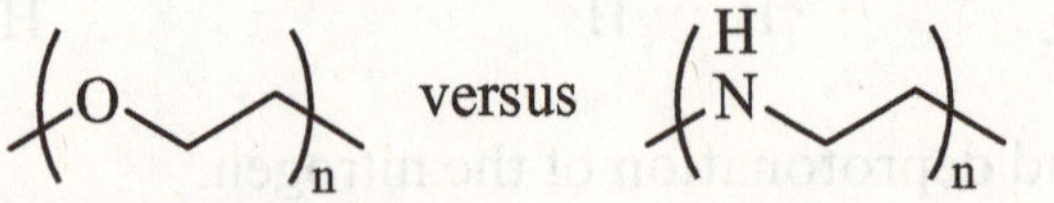   versus

The polymer on the right can undergo intermolecular hydrogen bonding and has a higher T_g.

16. Predict the major product of each of the following reactions.

(a)

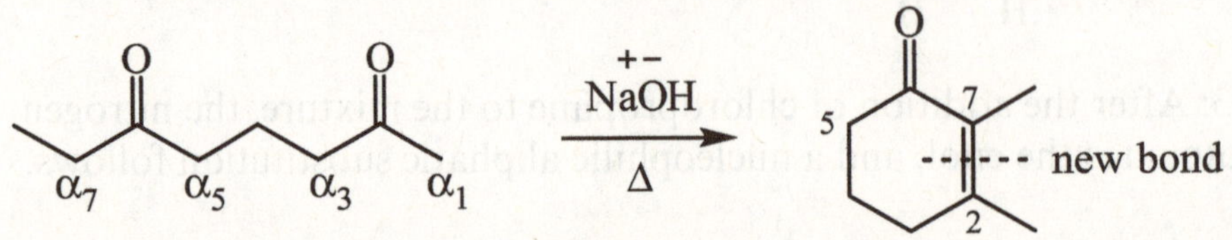

The reaction conditions are indicative of an aldol reaction. When the starting material is treated with NaOH, four different enolates can be formed. However, carbon 2 is less hindered than carbon 6, so the former is most likely to act as the electrophile. Between the enolates α_5 and α_7, nucleophilic attack of carbon 2 by α_7 forms a six-membered ring, which is much more favorable than the four-membered ring that would be formed if enolate α_5 attacked carbon 2.

(b)

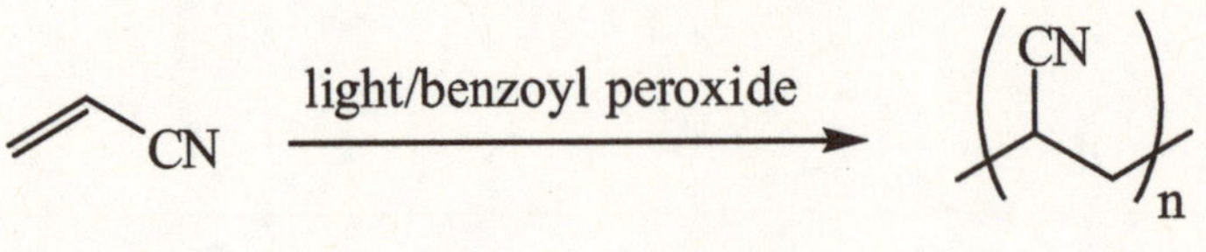

(c)

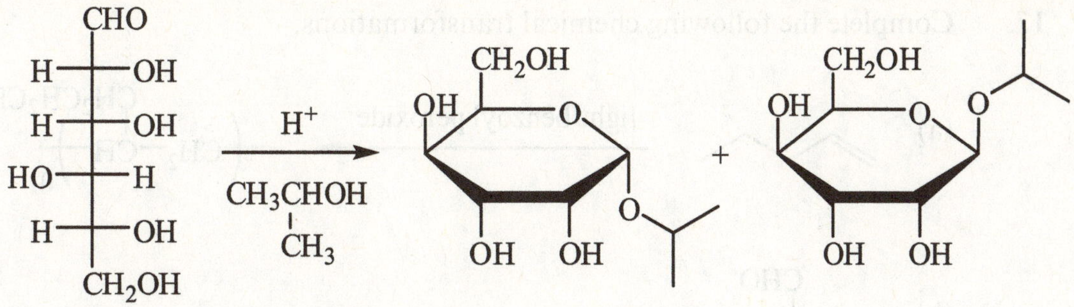

Fischer projection:
CHO
H—OH
H—OH
HO—H
H—OH
CH₂OH

$\xrightarrow[\substack{CH_3CHOH \\ | \\ CH_3}]{H^+}$

Two pyranose products with CH₂OH, OH, O-isopropyl groups, and OH OH substituents, joined by +

(d)

$$\underset{\substack{| \\ O=C \\ | \\ OH}}{HO-\overset{O}{\overset{||}{C}}-CH_2-CH-\overset{O}{\overset{||}{C}}-CH_3} \xrightarrow{heat} HO-\overset{O}{\overset{||}{C}}-CH_2-CH_2-\overset{O}{\overset{||}{C}}-CH_3$$

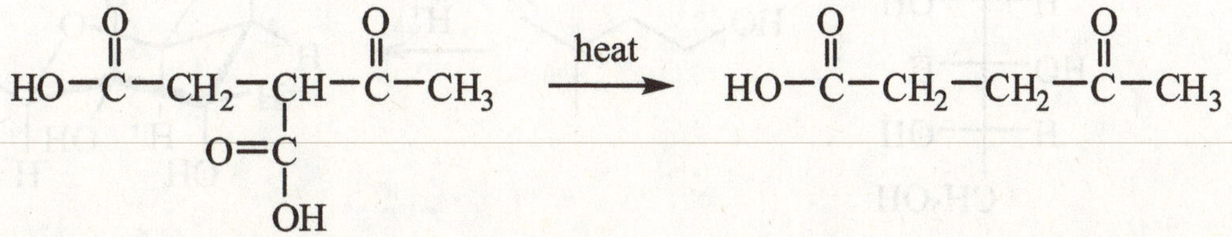

Only the carboxylic acid in the center can decarboxylate because the decarboxylation of a carboxylic acid requires a β-keto group.

(e)

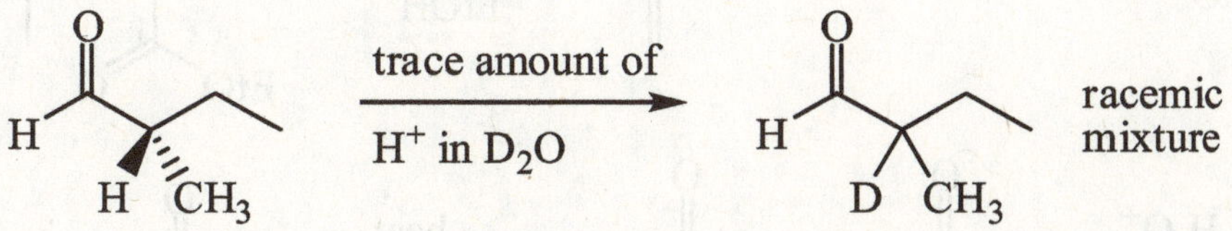

$\xrightarrow[\substack{H^+ \text{ in } D_2O}]{\text{trace amount of}}$

racemic mixture

Keto-enol tautomerism is accelerated by the presence of acid. When the enol is formed from the starting material, the chiral center is destroyed. Reprotonation (using D^+ from HD_2O^+) can occur from either face of the enol, forming a racemic mixture.)

(f)

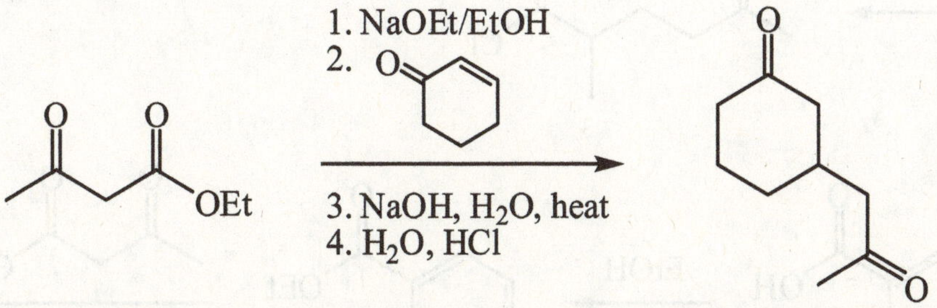

1. NaOEt/EtOH
2. (2-cyclohexenone)

3. NaOH, H₂O, heat
4. H₂O, HCl

Treatment of the starting material with sodium ethoxide in ethanol (Step 1) generates the enolate, which subsequently reacts with 2-cyclohexenone via a Michael addition (Step 2). Step 3 hydrolyzes the ethyl ester, forming a carboxylate salt, which is neutralized by the acid in Step 4 to give a β-ketoacid that decarboxylates.

17.   Complete the following chemical transformations.

(a)

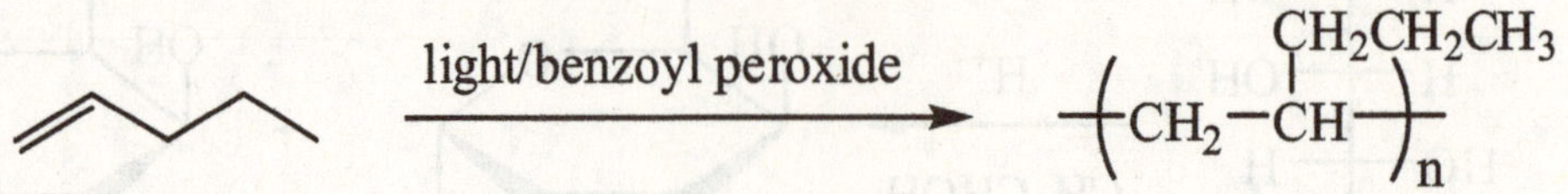

light/benzoyl peroxide

$$\left(CH_2-CH\right)_n$$ with $CH_2CH_2CH_3$ substituent

(b)

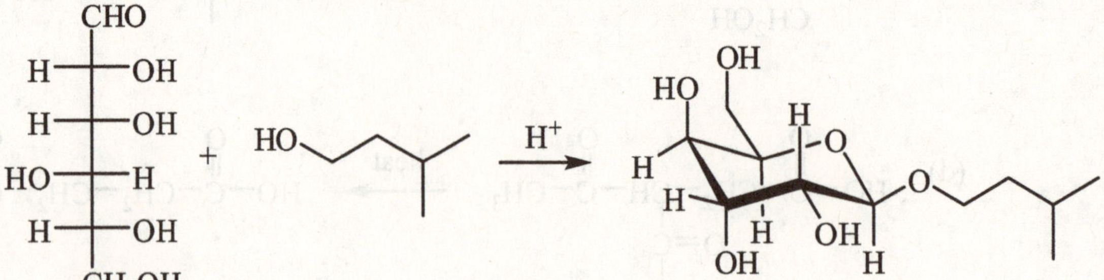

CHO
H—OH
H—OH
HO—H
H—OH
CH₂OH

+ HO—CH₂CH₂CH(CH₃)₂

$\xrightarrow{H^+}$

(c)

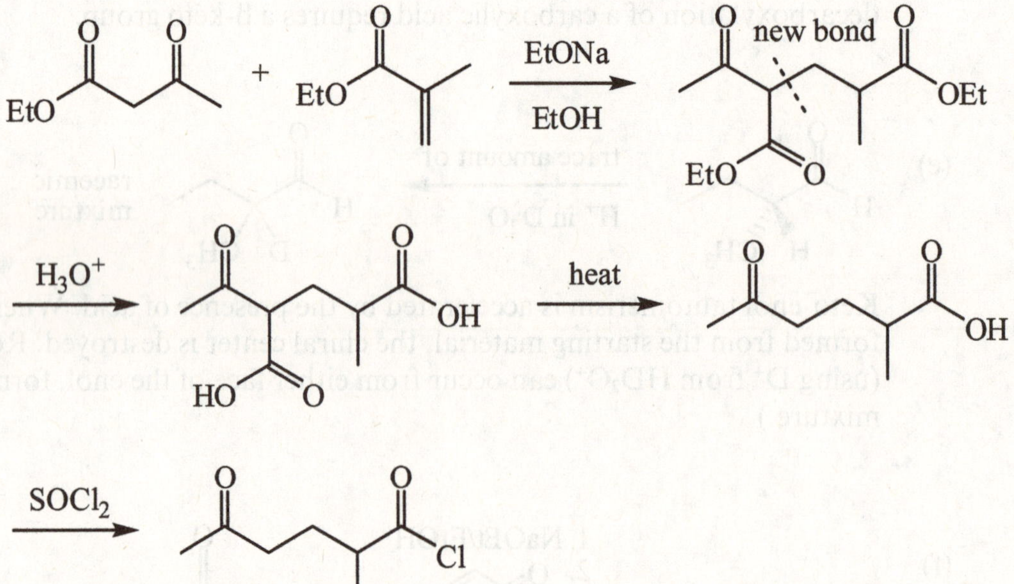

+ methacrylate ester $\xrightarrow[\text{EtOH}]{\text{EtONa}}$ (product with new bond)

$\xrightarrow{H_3O^+}$ (diacid/keto product)

$\xrightarrow{\text{heat}}$ (keto acid)

$\xrightarrow{SOCl_2}$ (acid chloride)

(d)

benzoic acid $\xrightarrow[\text{H}_2\text{SO}_4]{\text{EtOH}}$ ethyl benzoate $\xrightarrow[\text{EtONa/EtOH}]{}$ (with ethyl acetoacetate OEt)

$\xrightarrow[\text{2. heat}]{\text{1. H}_3\text{O}^+}$ (product)

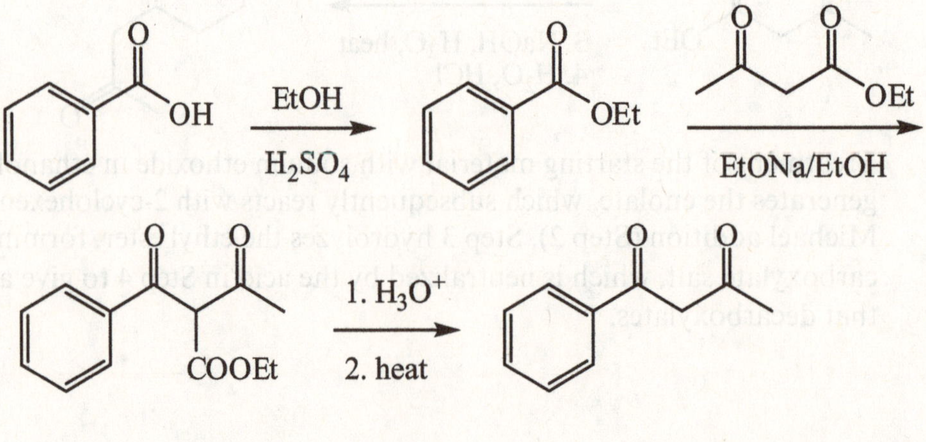

(e)

(f)

Chapter 18:Amino Acids and Proteins

Problems

18.1   Of the 20 protein-derived amino acids shown in Table 18.1, (a) which contain no stereocenter and (b) which contain two stereocenters?

Of the 20 amino acids, only glycine does not contain a stereocenter. Isoleucine and threonine each contain two stereocenters.

18.2   The isoelectric point of histidine is 7.64. Toward which electrode does histidine migrate during paper electrophoresis at pH 7.0?

If the pH is identical to the pI of the amino acid, the amino acid would have no net charge. When the pH is lower than the pI of amino acid, the amino acid is positively charged and will migrate towards the negative electrode.

18.3   Describe the behavior of a mixture of glutamic acid, arginine, and valine during paper electrophoresis at pH 6.0.

The pI values for glutamic acid, arginine, and valine are respectively 3.08, 10.76, and 6.00. At pH 6.0, arginine is positively charged and will migrate towards the negative electrode, glutamic acid is negatively charged and will migrate towards the positive electrode, and valine is uncharged and will remain at the origin.

18.4   Draw a structural formula for Lys-Phe-Ala. Label the *N*-terminal amino acid and the *C*-terminal amino acid. What is the net charge on this tripeptide at pH 6.0?

pH 6.0 is higher than the pK_a of the carboxyl group but lower than the pK_a of an ammonium group. The carboxyl group is therefore deprotonated to form a carboxylate, and the ammonium groups are protonated. The net charge is +1.

18.5   Which of these tripeptides are hydrolyzed by trypsin? By chymotrypsin?

(a) Tyr-Gln-Val   (b) Thr-Phe-Ser   (c) Thr-Ser-Phe

Trypsin catalyzes the hydrolysis of peptide bonds formed from the carboxyl groups of basic amino acids, with a preference for lysine and arginine; none of the three peptides satisfy these criteria. Chymotrypsin catalyzes the hydrolysis of peptide bonds formed from the carboxyl groups of phenylalanine, tyrosine, and tryptophan, so it will cleave between the Tyr and Gln residues in (a) and between the Phe and Ser residues in (b). Although Phe is present in (c), the carboxyl group of Phe is not involved in a peptide bond. (Recall that unless specified, the *N*-terminal is on the left side, and the *C*-terminal is on the right side.)

18.6   Deduce the amino acid sequence of an undecapeptide (11 amino acids) from the experimental results shown in the following table:

Experimental Procedure	Amino Acids Determined from Procedure
Amino Acid Analysis of Undecapeptide	Ala, Arg, Glu, Lys$_2$, Met, Phe, Ser, Thr, Trp, Val
Edman degradation	Ala
Trypsin-Catalyzed Hydrolysis	
Fragment E	Ala, Glu, Arg
Fragment F	Thr, Phe, Lys
Fragment G	Lys
Fragment H	Met, Ser, Trp, Val
Chymotrypsin-Catalyzed Hydrolysis	
Fragment I	Ala, Arg, Glu, Phe, Tyr
Fragment J	Lys$_2$, Met, Ser, Trp, Val
Treatment with Cyanogen Bromide	
Fragment K	Ala, Arg, Glu, Lys$_2$, Met, Phe, Thr, Val
Fragment L	Trp, Ser

From the Edman degradation data, the *N*-terminal residue of the peptide is Ala. From amino acid analysis data, there is only one Ala in the peptide. Ala is found in Fragment E, and because it is one of the products of trypsin hydrolysis, Arg must be the *C*-terminal of Fragment E. The sequence of Fragment E must be Ala-Glu-Arg, and because it contains Ala, it is also first three amino acids of the undecapeptide.

Fragment G, which is just a single Lys residue, suggests that there must be either two lysine residues, or one arginine and one lysine, adjacent to each other. Yet, the chymotrypsin results indicate that Lys$_2$ and Arg are in separate fragments, so there must be two adjacent Lys residues (one Lys adjacent to an Arg would not have been cleaved by chymotrypsin). Up to now, the sequence information that we have is:

Ala-Glu-Arg-(unknown)-Lys-Lys-(unknown)

From the chymotrypsin data, the C-terminal of Fragment J is Trp. Because there is only one Trp in the undecapeptide, Fragment L, which contains only two amino acids, must be Ser-Trp. Furthermore, because Fragment L was obtained from CNBr cleavage, we must have the sequence Met-Ser-Trp. These three residues are present in Fragment H, which must therefore be Val-Met-Ser-Trp. Thus, the data we have up to now are:

Ala-Glu-Arg-(unknown)-Lys-Lys-(unknown)

(unknown)-Val-Met-Ser-Trp-(unknown)

We know that the two Lys residues are adjacent to each other, so we can deduce the sequence of Fragment J as Lys-Lys-Val-Met-Ser-Trp, giving the sequence below.

Ala-Glu-Arg-(unknown)-Lys-Lys-Val-Met-Ser-Trp-(unknown)

From the chymotrypsin data, the C-terminal of Fragment I is Phe. Fragment I has the amino acids Ala, Glu, and Arg, and was earlier determined to be Ala-Glu-Arg, plus Thr. Fragment I must therefore be Ala-Glu-Arg-Thr-Phe. The data we now have are:

Ala-Glu-Arg-Thr-Phe-(unknown)

Ala-Glu-Arg-(unknown)-Lys-Lys-Val-Met-Ser-Trp-(unknown)

Combining these sequences gives us the overall sequence, with Ala as the N-terminal and Trp as the C-terminal.

Ala-Glu-Arg-Thr-Phe-Lys-Lys-Val-Met-Ser-Trp

18.7 At pH 7.4, with what amino acid side chains can the side chain of lysine form salt bridges?

At pH 7.4, the side chain of lysine is positively charged because the pK_a of the ammonium group is higher than 7.4. The side chain can only form salt bridges with negatively charged side chains, which are those of aspartic and glutamic acids. The side chains of these acids have pK_a values lower than 7.4 and are therefore negatively charged at this pH.

Chemical Connections

18A. Draw two pentapeptide strands of polyalanine and show how they hydrogen-bond to each other.

Quick Quiz

1. The isoelectric point of an amino acid is the pH at which the majority of molecules in solution have a net charge of -1. *False*. The isoelectronic point is the pH at which the majority of the molecules have no net charge.

2. Proteins can protect an organism against disease. *True*. In animals, skin, which is made of protein, acts as a physical barrier. Antibodies, which recognize foreign substances and infectious agents, are also made of proteins.

3. Titration of an amino acid can be used to determine both the pK_a of its ionizable groups and its isoelectric point. *True*. pK_a values are found from the midpoint of each neutralization reaction, while the isoelectronic point can be determined from the pK_a values.

4. Hydrogen bonding, salt bridges, hydrophobic interactions, and disulfide bonds can each be categorized as stabilizing factors in a protein. *True*. These all contribute to the maintenance of secondary, tertiary, and quaternary structure.

5. A polypeptide chain is read from its *C*-terminal end to its *N*-terminal end. *False*. The opposite is true. A polypeptide chain is read, from left to right, from its *N*-terminal end to its *C*-terminal end.

6. The majority of naturally occurring amino acids are from the D-series. *False*. Most naturally occurring amino acids are L-amino acids.

7. The amino group of an α-amino acid is more basic than the amino group of an aliphatic amine. *False*. The amino group of an α-amino acid is slightly less basic than that of an aliphatic amine.

8.   Lysine contains a basic side chain. *True*. The side chain of lysine contains an amino group.

9.   Electrophoresis is the process of creating a synthetic protein. *False*. Electrophoresis is a method used to separate amino acids or proteins.

10. A peptide bond exhibits free rotation at room temperature. *False*. Peptide (amide) bonds, as a result of resonance, have partial double-bond character and cannot freely rotate.

11. In Edman degradation, a polypeptide is shortened one amino acid at a time using the reagent phenyl isothiocyanate. *True*. The *N*-terminal amino acid is cleaved from the polypeptide one at a time.

12. Phenylalanine contains a polar side chain. *False*. The side chain of phenylalanine has a phenyl ring and is therefore hydrophobic.

13. The side chain of arginine shows enhanced basicity because of resonance stabilization of the ion that results after protonation. *True*. The increased stability of the conjugate acid makes the side chain more basic.

14. α-helices and β-pleated sheets are examples of the tertiary structure of a protein. *False*. They are examples of secondary structure.

15. The majority of amino acids in proteins are β-amino acids. *False*. Most amino acids in proteins are α-amino acids.

16. Proteins can act as catalysts in chemical reactions. *True*. Enzymes are catalytic proteins.

17. In electrophoresis, species with a net negative charge will move toward the negative electrode. *False*. Negatively charged species migrate towards the positive electrode.

18. All naturally occurring amino acids are chiral. *False*. Glycine, one of the twenty essential amino acids, is achiral.

19. Cyanogen bromide, trypsin, and chymotrypsin each act to cleave peptide bonds at specific amino acids. *True*. CNBr cleaves at methionine, trypsin cleaves at basic amino acids, and chymotrypsin cleaves at hydrophobic aromatic amino acids.

20. The carboxyl group of an α-amino acid is more acidic than the carboxyl group of an aliphatic carboxylic acid. *True*. This is due to the electron-withdrawing effect of the positively charged α-ammonium group.

21. The amino acid sequence of a protein or polypeptide is known as its secondary structure. *False*. Sequence information is contained in primary structure.

22. The 20 common, naturally occurring amino acids can be represented by both three- and one-letter abbreviations. *True*.

23. The side chain of histidine shows enhanced basicity because of the electron-withdrawing inductive effects that stabilize the ion that results after protonation. *False*. The conjugate acid is stabilized by resonance and not by induction.

24. The quaternary structure of a protein describes how smaller, individual protein strands interact to form the overall structure of a protein. *True*. Quaternary structure refers to the aggregation of individual protein molecules.

25. An amino acid with a net charge of +1 is classified as a zwitterion. *False*. A zwitterion contains internal charge but has no net charge.

26. The side chain in serine is polar and can undergo hydrogen bonding. *True*. The side chain of serine contains a hydroxyl group.

End-of-Chapter Problems

Amino Acids

18.8 What amino acid does each abbreviation stand for?

(a) Phe	(b) Ser	(c) Asp	(d) Gln
Phenylalanine	Serine	Aspartic acid	Glutamine
(e) His	(f) Gly	(g) Tyr	(h) Trp
Histidine	Glycine	Tyrosine	Tryptophan

18.9 The configuration of the stereocenter in α-amino acids is most commonly specified using the D,L convention. The configuration can also be identified using the *R,S* convention (Section 6.3). Does the stereocenter in L-serine have the *R* or the *S* configuration?

The D,L convention used for amino acids is derived from D- and L-glyceraldehyde, where the most oxidized carbon is placed at the top of the Fischer projection. Amino acids are treated similarly, except the configuration is determined by the position of the α-amino group (D = right, L = left).

The D,L convention is entirely different from the *R,S* convention, which is based on the assignment of group priorities. A very common error is to assume that D corresponds to *R*, and L to *S*. Because the two conventions are based on different principles, this assumption is not always true. It is best to determine *R,S* from the actual structure of the amino acid regardless of its D,L designation. It may be helpful to redraw the Fischer projection as a line-angle diagram or to build a model.

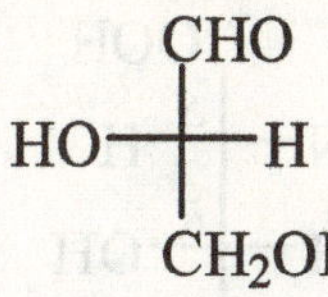

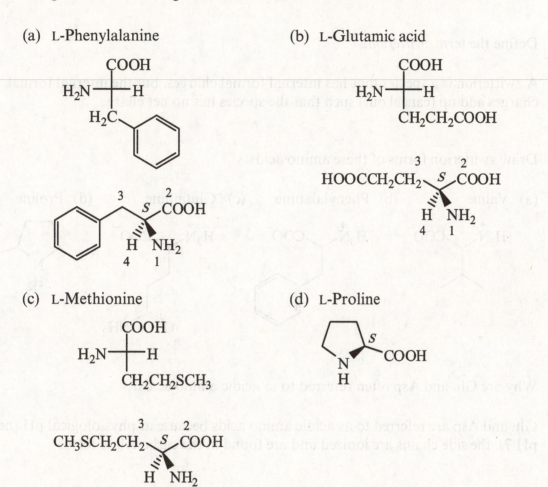

L-glyceraldehyde 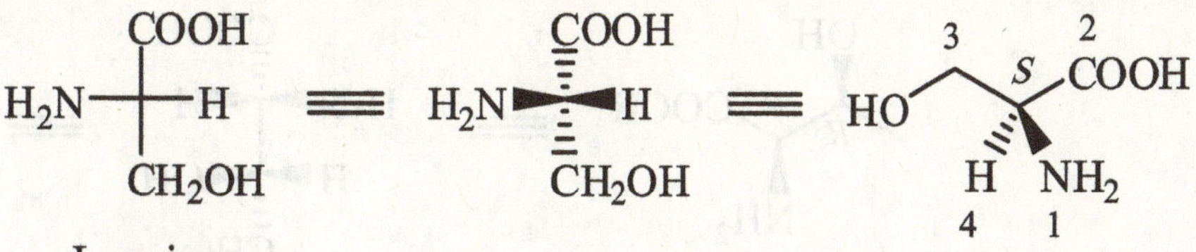   L-serine

18.10 Assign an *R* or *S* configuration to the stereocenter in each amino acid:

(a) L-Phenylalanine

(b) L-Glutamic acid

(c) L-Methionine

(d) L-Proline

18.11 The amino acid threonine has two stereocenters. The stereoisomer found in proteins has the configuration 2*S*,3*R* about the two stereocenters. Draw a Fischer projection of this stereoisomer and also a three-dimensional representation.

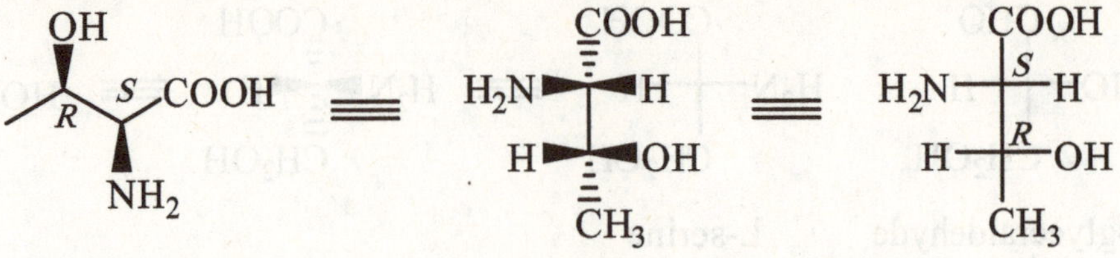

For this problem, it may be easier to start with a line-angle diagram, with the proper stereocenters, followed by redrawing the diagram as a Fischer projection.

18.12 Define the term *zwitterion*.

A zwitterion is a species that has internal formal charges, but the internal formal charges add up (cancel out) such that the species has no net charge.

18.13 Draw zwitterion forms of these amino acids:

(a) Valine   (b) Phenylalanine   (c) Glutamine   (d) Proline

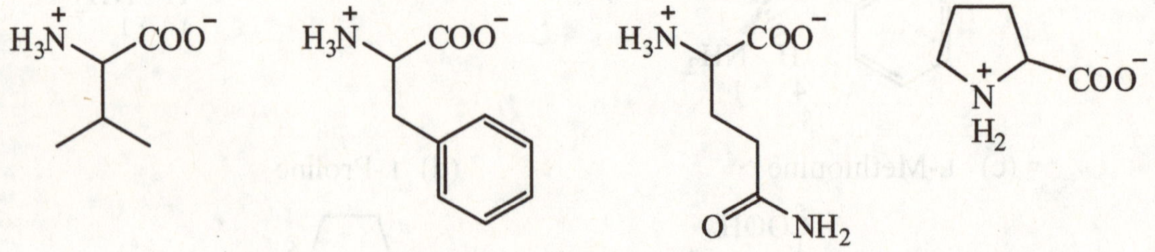

18.14 Why are Glu and Asp often referred to as acidic amino acids?

Glu and Asp are referred to as acidic amino acids because at physiological pH (near pH 7), the side chains are ionized and are found in the carboxylate form.

18.15 Why is Arg often referred to as a basic amino acid? Which two other amino acids are also referred to as basic amino acids?

Arg is referred to as a basic amino acid because at physiological pH, its side chain is protonated and is positively charged. Lys and His are also basic amino acids.

18.16 What is the meaning of the alpha as it is used in α-amino acid?

The α refers to the fact that the amino group is bonded to the α-carbon, which is the carbon directly adjacent to the carboxyl group.

18.17 Several β-amino acids exist. A unit of β-alanine, for example, is contained within the structure of coenzyme A (Section 21.1D). Write the structural formula of β-alanine.

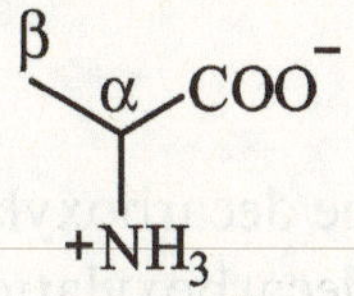

α-Alanine 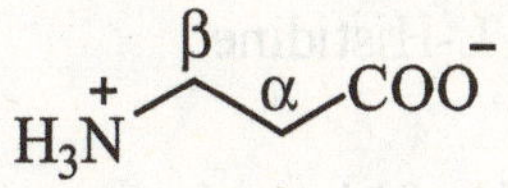   β-Alanine

18.18 Although only L-amino acids occur in proteins, D-amino acids are often a part of the metabolism of lower organisms. The antibiotic actinomycin D, for example, contains a unit of D-valine, and the antibiotic bacitracin A contains units of D-asparagine and D-glutamic acid. Draw Fischer projections and three-dimensional representations for these three D-amino acids.

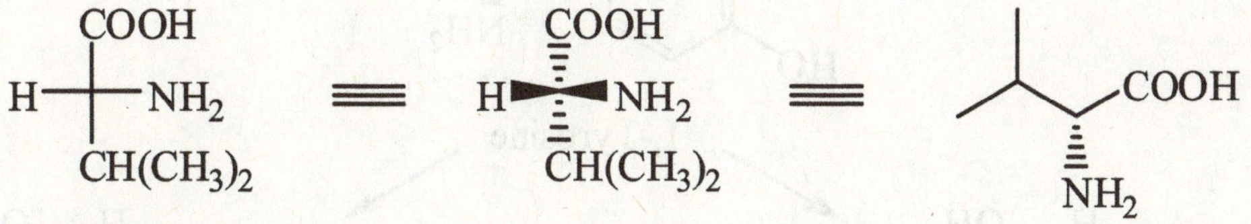

D-Valine

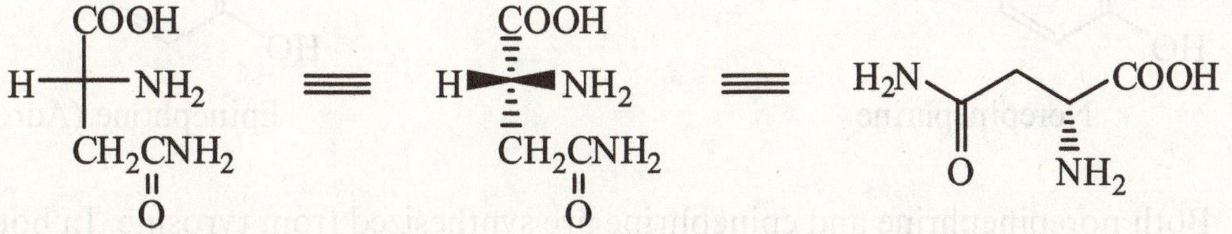

D-Asparagine

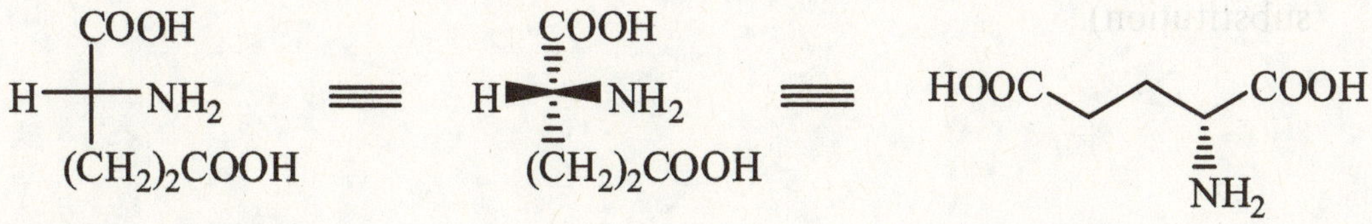

D-Glutamic acid

18.19 Histamine is synthesized from one of the 20 protein-derived amino acids. Suggest which amino acid is the biochemical precursor of histamine, and the type of organic reaction(s) (e.g., oxidation, reduction, decarboxylation, nucleophilic substitution) involved in its conversion to histamine.

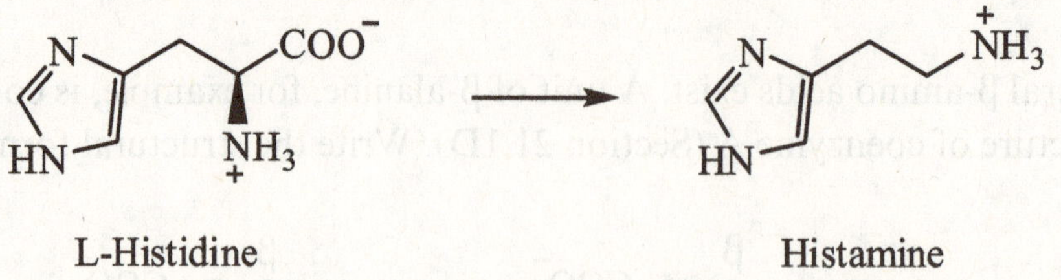

L-Histidine → Histamine

The biosynthesis of histamine occurs via the decarboxylation of histidine. Although a β-keto group is not present in histidine, decarboxylation still occurs because the impressive enzyme and coenzyme involved in the reaction can mimic a β-keto group.

18.20 Both norepinephrine and epinephrine are synthesized from the same protein-derived amino acid. From which amino acid are they synthesized, and what types of reactions are involved in their biosynthesis?

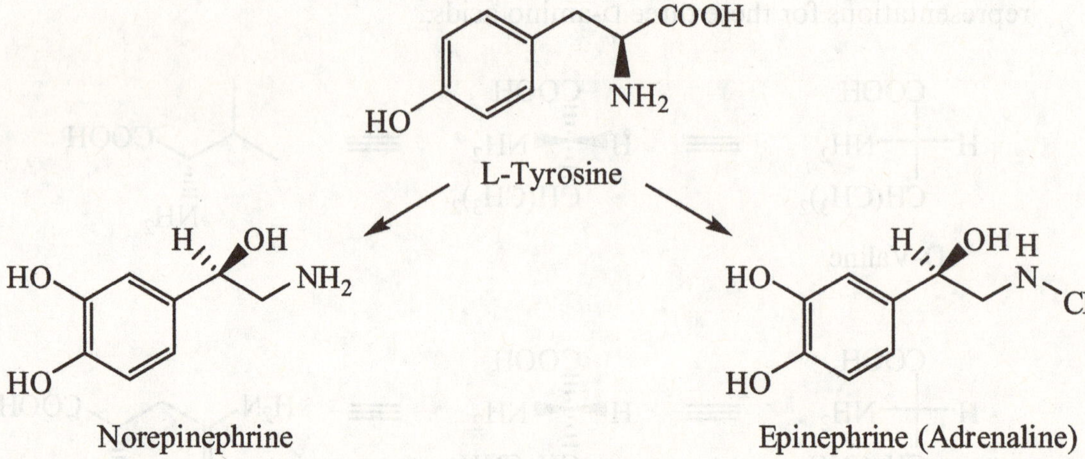

L-Tyrosine

Norepinephrine   Epinephrine (Adrenaline)

Both norepinephrine and epinephrine are synthesized from tyrosine. In both cases, the biosynthesis involves a decarboxylation, hydroxylation of the aromatic ring ortho to the original phenolic –OH group, and hydroxylation of the benzylic carbon atom (oxidation). The amino group of epinephrine is also methylated (nucleophilic substitution).

18.21   From which amino acid are serotonin and melatonin synthesized and what types of reactions are involved in their biosynthesis?

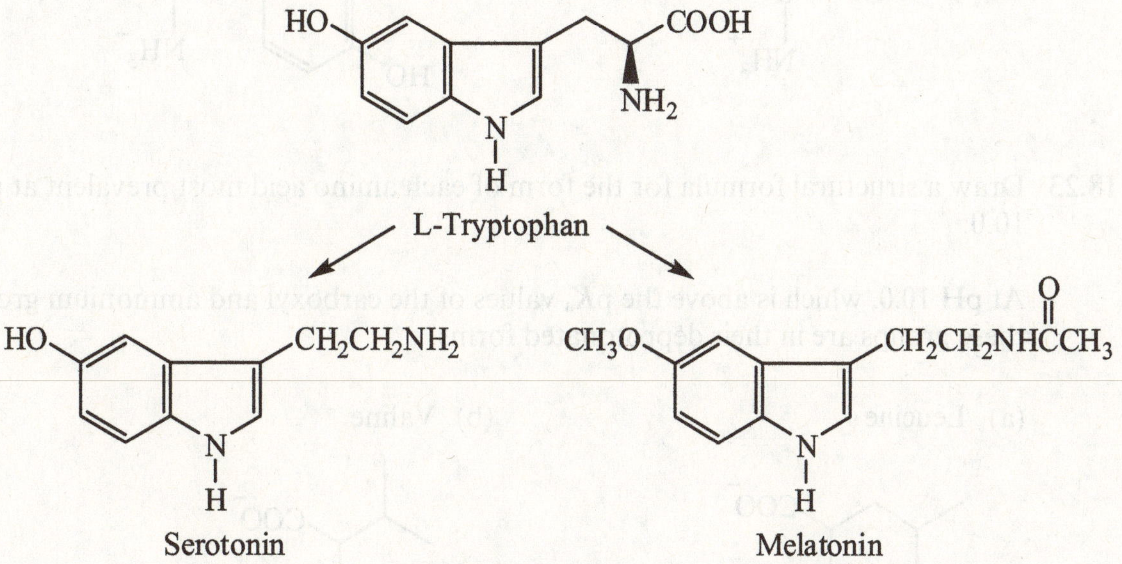

L-Tryptophan

Serotonin

Melatonin

Serotonin and melatonin are both synthesized from tryptophan. In both cases, the biosynthesis involves a decarboxylation and an oxidation (hydroxylation) of the aromatic ring. In the case of melatonin, there are two more reactions: after the oxidation of the aromatic ring, the hydroxyl group is methylated (nucleophilic substitution) and the amino group is acetylated (nucleophilic acyl substitution).

Acid-Base Behavior of Amino Acids

18.22   Draw a structural formula for the form of each amino acid most prevalent at pH 1.0:

At pH 1.0, which is below the pK_a values of the carboxyl, ammonium, and guanidinium groups, all of these groups are protonated.

(a)   Threonine

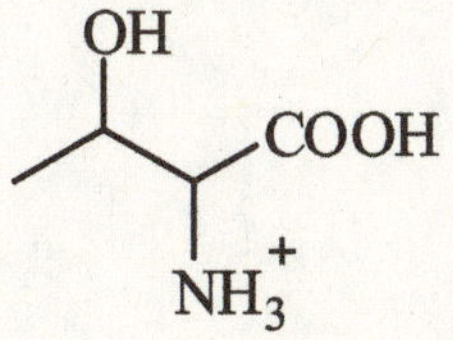

(b)   Arginine

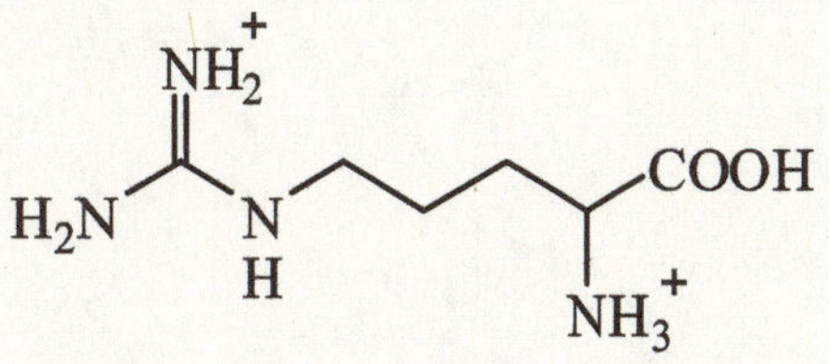

(c) Methionine

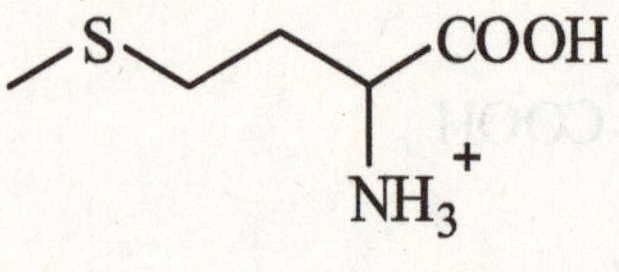

(d) Tyrosine

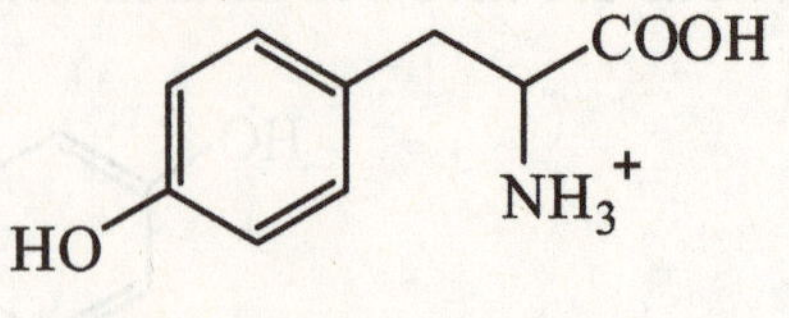

18.23 Draw a structural formula for the form of each amino acid most prevalent at pH 10.0:

At pH 10.0, which is above the pK_a values of the carboxyl and ammonium groups, these groups are in their deprotonated forms.

(a) Leucine

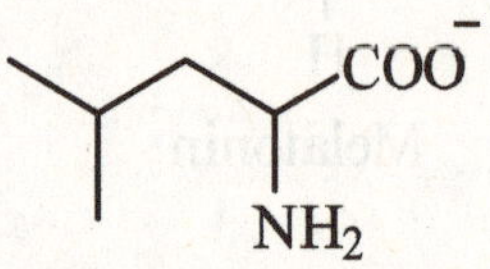

(b) Valine

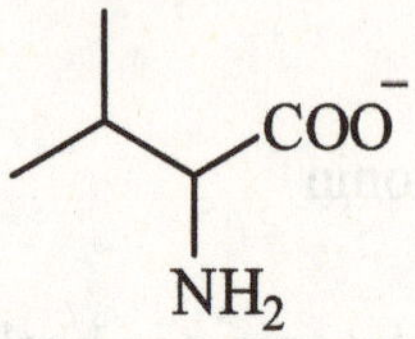

(c) Proline

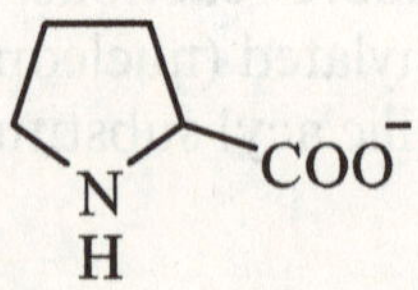

(d) Aspartic acid

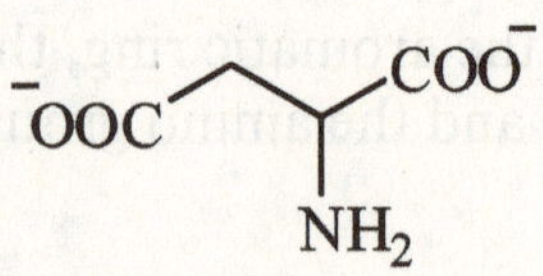

18.24 Write the zwitterion form of alanine and show its reaction with

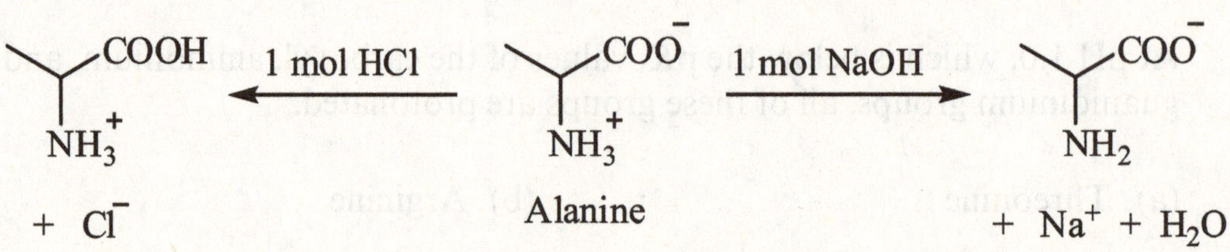

Alanine

18.25 Write the form of lysine most prevalent at pH 1.0, and then show its reaction with the following (consult Table 18.2 for pK_a values of the ionizable groups in lysine):

When fully protonated lysine is treated with NaOH, the order of deprotonation corresponds to the order of the decreasing acidity of the groups. The most-acidic group is deprotonated first, and the least-acidic group is deprotonated last.

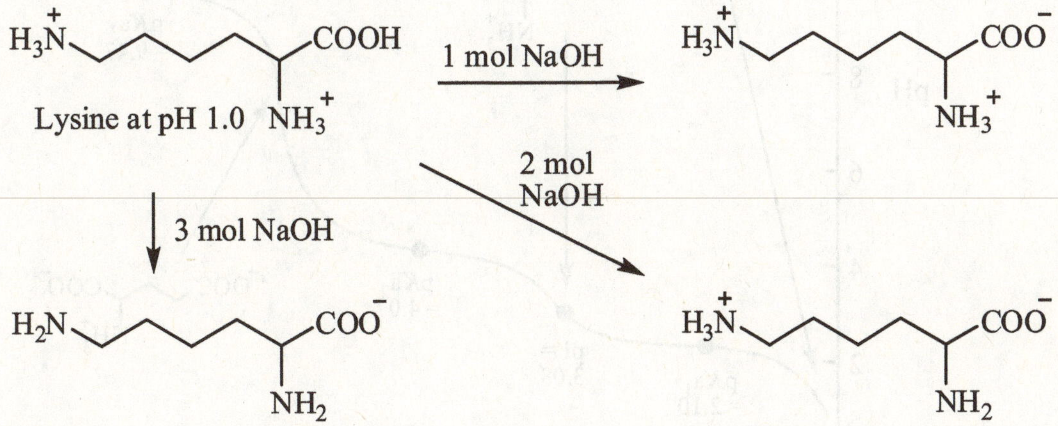

18.26 Write the form of aspartic acid most prevalent at pH 1.0, and then show its reaction with the following (consult Table 18.2 for pK_a values of the ionizable groups in aspartic acid):

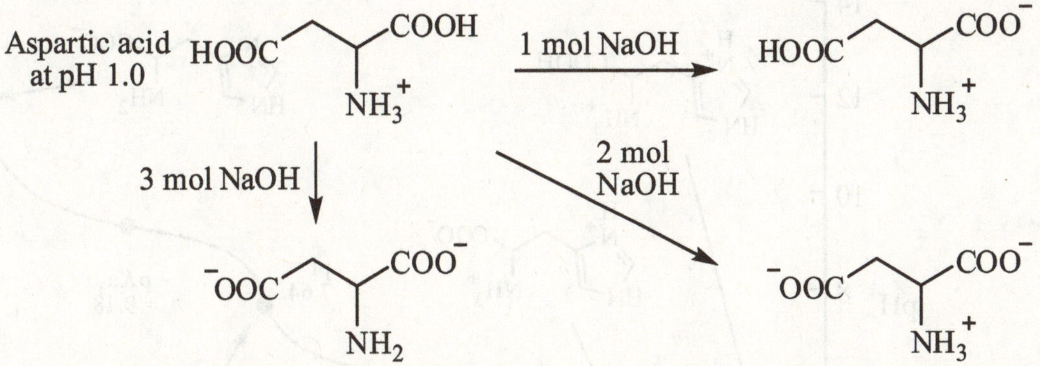

Note that the most-acidic carboxyl group is the one closest to the positively charged ammonium group; this carboxyl group experiences the greatest inductive effect.

18.27 Given pK_a values for ionizable groups from Table 18.2, sketch curves for the titration of:

(a) Glutamic acid with NaOH

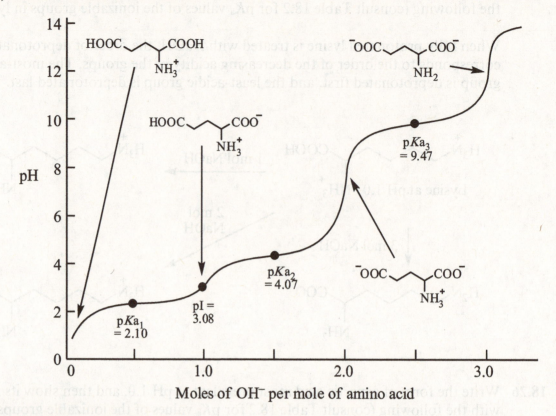

(b) Histidine with NaOH

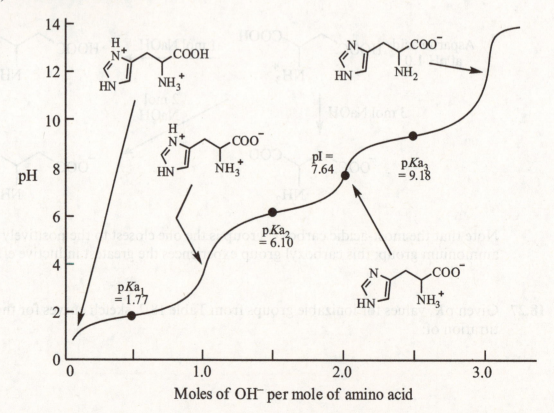

18.28 Draw a structural formula for the product formed when alanine is treated with the following reagents:

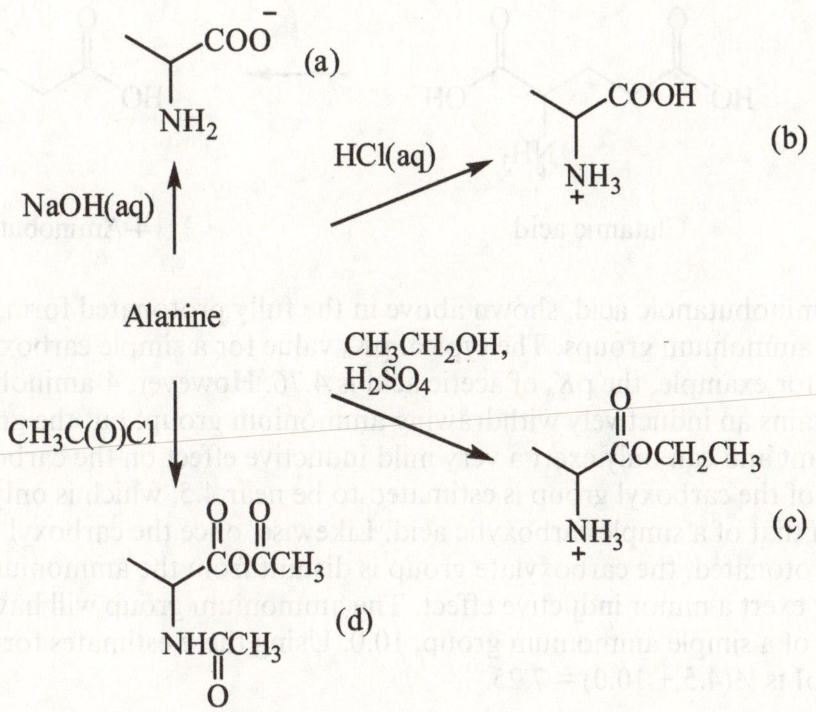

18.29 Account for the fact that the isoelectric point of glutamine (pI 5.65) is higher than the isoelectric point of glutamic acid (pI 3.08).

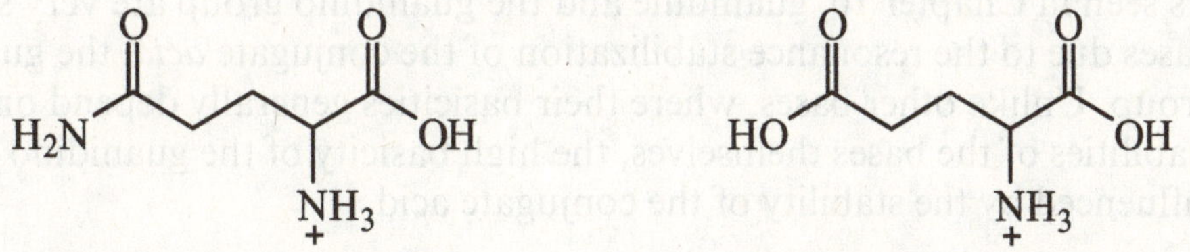

Glutamine (pI = 5.65)   Glutamic acid (pI = 3.08)

The side chain of glutamine contains an amide functional group, which is not basic or acidic. On the other hand, the side chain of glutamic acid contains a carboxyl group.

When the pH is equal to the pI of glutamic acid (3.08), the amino acid has no net charge. For this to occur, one of the carboxyl groups must be deprotonated. The pI for glutamic acid therefore falls between the pK_a values of the two carboxyl groups, which are 2.10 and 4.07.

With glutamine, its pI value is determined by the pK_a values of the only two ionizable groups, the carboxylic acid (pK_a 2.17) and the α-amino group (pK_a 9.03).

18.30 Enzyme-catalyzed decarboxylation of glutamic acid gives 4-aminobutanoic acid (Section 18.2D). Estimate the pI of 4-aminobutanoic acid.

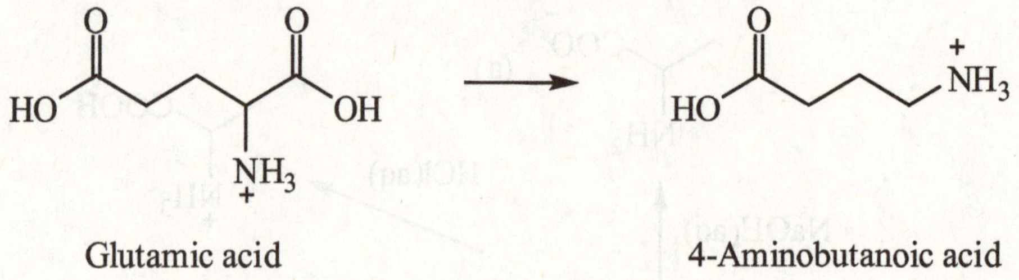

Glutamic acid   4-Aminobutanoic acid

4-Aminobutanoic acid, shown above in the fully protonated form, contains carboxyl and ammonium groups. The typical pK_a value for a simple carboxylic acid is near 4.8; for example, the pK_a of acetic acid is 4.76. However, 4-aminobutanoic acid contains an inductively withdrawing ammonium group, but the group is relatively distant and can only exert a very mild inductive effect on the carboxyl group. The pK_a of the carboxyl group is estimated to be near 4.5, which is only slightly lower than that of a simple carboxylic acid. Likewise, once the carboxyl group is deprotonated, the carboxylate group is distant from the ammonium group and can only exert a minor inductive effect. The ammonium group will have a pK_a close to that of a simple ammonium group, 10.0. Using these estimates for the pK_a values, the pI is $\frac{1}{2}(4.5 + 10.0) = 7.25$.

18.31 Guanidine and the guanidino group present in arginine are two of the strongest organic bases known. Account for their basicity.

As seen in Chapter 10, guanidine and the guanidino group are very strong amine bases due to the resonance stabilization of the conjugate *acid*, the guanidinium group. Unlike other bases, where their basicities generally depend on the relative stabilities of the bases themselves, the high basicity of the guanidino group is greatly influenced by the stability of the conjugate acid.

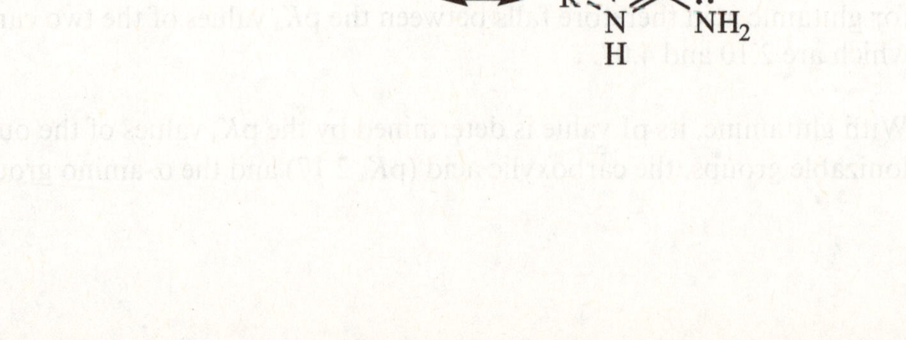

18.32 At pH 7.4, the pH of blood plasma, do the majority of protein-derived amino acids
bear a net negative charge or a net positive charge?

The majority of the amino acids have pI values between 5 and 6, so they are
negatively charged at pH 7.4. The only exceptions are the three basic amino acids
(Arg, His, and Lys).

18.33 Do the following compounds migrate to the cathode or to the anode on
electrophoresis at the specified pH?

(a) Histidine at pH 6.8   (b) Lysine at pH 6.8
(c) Glutamic acid at pH 4.0   (d) Glutamine at pH 4.0

When the pH is lower than the pI of the amino acid, the amino acid has a net
positive charge and migrates toward the negative electrode (cathode). Conversely,
when the pH is higher than the pI of the amino acid, the amino acid has a net
negative charge and migrates toward the positive electrode (anode). His, Lys, and
Gln have pI values of 7.64, 9.74, and 5.65, respectively, and they will migrate
towards the cathode under the stated conditions. Glu has a pI of 3.08 and will
migrate towards the anode at pH 4.0.

(e) Glu-Ile-Val at pH 6.0   (f) Lys-Gln-Tyr at pH 6.0

A good approach is to determine the charges of the *N*-terminal, *C*-terminal, and the
side chains. At pH 6.0, the *N*-terminal will be positively charged because the pK_a of
the α-ammonium group is higher than 6.0. The *C*-terminal will be negatively
charged because the pK_a of the α-carboxyl group is less than 6.0. With (e), the
carboxylic acid side chain of Glu will be negatively charged at pH 6.0, so (e) has an
overall negative charge and will migrate towards the anode. With (f), the amino side
chain of Lys will be positively charged at pH 6.0, so (f) has a net positive charge and
will migrate towards the cathode.

18.34 At what pH would you carry out an electrophoresis to separate the three amino
acids in each mixture?

(a) Ala, His, Lys

The pI values for Ala, His, and Lys are 6.11, 7.64, and 9.74, respectively. If
electrophoresis is performed at pH 7.64, His would have no net charge and will
not move. Ala would be negatively charged, and Lys positively charged, and
will respectively migrate to the positive and negative electrodes.

(b) Glu, Gln, Asp

The pI values for Glu, Gln, Asp are 3.08, 5.65, and 2.98, respectively. If electrophoresis is performed at pH 3.08, Glu would have no net charge and will not move. Asp would be negatively charged, and Gln positively charged, and will respectively migrate to the positive and negative electrodes.

(c) Lys, Leu, Tyr

The pI values for Lys, Leu, and Tyr are 9.74, 6.04, and 5.63, respectively. If electrophoresis is performed at pH 6.04, Leu would have no net charge and will not move. Tyr would be negatively charged, and Lys positively charged, and will respectively migrate to the positive and negative electrodes.

18.35 Examine the amino acid sequence of human insulin (Figure 18.13), and list each Asp, Glu, His, Lys, and Arg in this molecule. Do you expect human insulin to have an isoelectric point nearer that of the acidic amino acids (pI 2.0−3.0), the neutral amino acids (pI 5.5−6.5), or the basic amino acids (pI 9.5−11.0)?

Insulin contains and equal number of acidic amino acids (4 Glu) and basic amino acids (2 His, 1 Lys, and 1 Arg). Therefore, its isoelectric point must be at a pH where all four acidic groups are deprotonated (pH >4) and all four basic groups are protonated (pH < 6). The pH of insulin should therefore fall between 5.5−6.5, which is similar to that of a neutral amino acid.

Primary Structure of Polypeptides and Proteins

18.36 If a protein contains four different SH groups, how many different disulfide bonds are possible if only a single disulfide bond is formed? How many different disulfides are possible if two disulfide bonds are formed?

Think about the possible combinations that are possible! For reference, we will refer to the four different SH groups as A, B, C and D.

If only one disulfide bond is formed, we could have any *one* of the following six possible combinations: A−B, A−C, A−D, B−C, B−D, and C−D. However, if two disulfide bonds are formed, we could have any *two* of the possible six combinations listed above, without using any SH group twice. With two disulfide bonds, only three sets of two are possible: A−B and C−D, A−C and B−D, and A−D and B−C.

18.37 How many different tetrapeptides can be made if

(a) The tetrapeptide contains one unit each of Asp, Glu, Pro, and Phe?

Because we are determining the number of possible sequences that contain one unit of each of the four amino acids, this problem involves permutations. For that reason, students in most biology, biochemistry, and chemistry programs need to take math courses! At the first position (the *N*-terminal), any one of the four amino acids can be used. At the second position, there are only three possible choices, as the tetrapeptide must contain one of each of the four amino acids. At the third position, there are two choices. Finally, at the fourth position (the *C*-terminal), there is only one choice. The number of possible tetrapeptides is 4 factorial ($4! = 4 \times 3 \times 2 \times 1 = 24$). All 24 of these tetrapeptides are constitutional isomers of each other.

(b) All 20 amino acids can be used, but each only once?

Using the same logic, except with 20 amino acids, the number of possible tetrapeptides would be $20 \times 19 \times 18 \times 17 = 116,280$.

18.38 A decapeptide has the following amino acid composition:

$$Ala_2, Arg, Cys, Glu, Gly, Leu, Lys, Phe, Val$$

Partial hydrolysis yields the following tripeptides:

$$Cys\text{-}Glu\text{-}Leu + Gly\text{-}Arg\text{-}Cys + Leu\text{-}Ala\text{-}Ala + Lys\text{-}Val\text{-}Phe + Val\text{-}Phe\text{-}Gly$$

One round of Edman degradation yields a lysine phenylthiohydantoin. From this information, deduce the primary structure of the given decapeptide.

The Edman degradation indicates that Lys is the *N*-terminal amino acid. Using this information, the sequence of the peptide can be determined by overlapping the sequences of the tripeptides (look for common residues). Note that overlaps are possible because we are always starting with many, many molecules of the original compound and never just one single molecule of the original compound.

```
        Lys-Val-Phe
            Val-Phe-Gly
                Gly-Arg-Cys
                    Cys-Glu-Leu
                        Leu-Ala-Ala
        ─────────────────────────────────
        Lys-Val-Phe-Gly-Arg-Cys-Glu-Leu-Ala-Ala      overall sequence
```

18.39 Following is the primary structure of glucagon, a polypeptide hormone of 29 amino acids. Glucagon is produced in the α-cells of the pancreas and helps maintain the blood glucose concentration within a normal range. Which peptide bonds are hydrolyzed when glucagon is treated with each reagent?

1   5   10   15
His-Ser-Glu-Gly-Thr-Phe-Thr-Ser-Asp-Tyr-Ser-Lys-Tyr-Leu-Asp-Ser-Arg-Arg-

                           20   25   29
                Ala-Gln-Asp-Phe-Val-Gln-Trp-Leu-Met-Asn-Thr

Glucagon

(a) Phenyl isothiocyanate

Phenyl isothiocyanate, the reagent used in the Edman degradation, cleaves only the *N*-terminal amino acid. The bond cleaved between residues 1 and 2.

(b) Chymotrypsin

Chymotrypsin cleaves peptide bonds on the carboxyl side (*C*-terminal side) of Phe, Tyr, and Trp. It therefore cleaves the bonds between residues 6 and 7, 10 and 11,
13 and 14, 22 and 23, and 25 and 26.

(c) Trypsin

Trypsin cleaves at the carboxyl side of the two highly basic amino acids, Arg and Lys. Cleavage occurs between residues 12 and 13, 17 and 18, and 18 and 19.

(d) Br-CN

Cyanogen bromide cleaves at the carboxyl side of Met, so between 27 and 28.

18.40 A tetradecapeptide (14 amino acid residues) gives the following peptide fragments on partial hydrolysis. From this information, deduce the primary structure of the given polypeptide. Fragments are grouped according to size.

Pentapeptide Fragments
Phe-Val-Asn-Gln-His
His-Leu-Cys-Gly-Ser
Gly-Ser-His-Leu-Val

Tetrapeptide Fragments
Gln-His-Leu-Cys
His-Leu-Val-Glu
Leu-Val-Glu-Ala

Phe-Val-Asn-Gln-His
Gln-His
His-Leu-Cys
His-Leu-Cys-Gly-Ser   Leu-Val-Glu-Ala
Gly-Ser-His-Leu-Val
His-Leu-Val-Glu

Phe-Val-Asn-Gln-His-Leu-Cys-Gly-Ser-His-Leu-Val-Glu-Ala   overall sequence

18.41 Draw a structural formula for each of the following tripeptides: marking each peptide bond, the *N*-terminal amino acid, and the *C*-terminal amino acid:

(a) Phe-Val-Asn

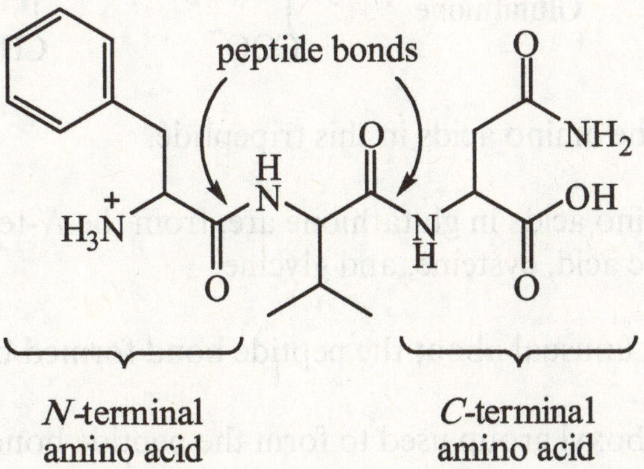

(b) Leu-Val-Gln

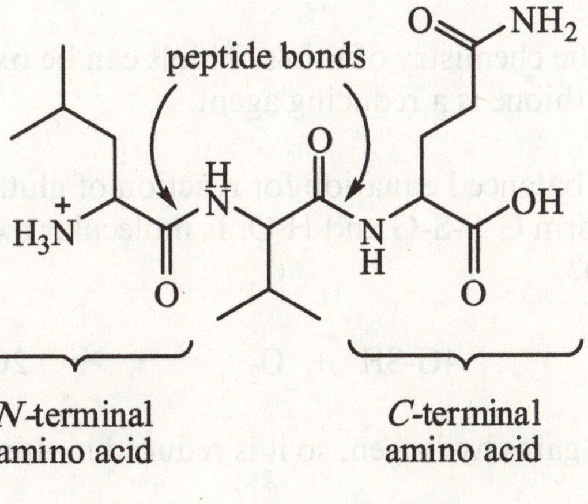

18.42 Estimate the pI of each tripeptide on Problem 18.41.

None of the side chains in Phe-Val-Asn and Leu-Val-Gln are acidic or basic, so the pI is determined by only pK_a values of the ammonium group at the N-terminal and the carboxyl group at the C-terminal. The pK_a of the ammonium group in Phe is 9.4 and that of the carboxyl group in Asn is 4.8, so the pI of Phe-Val-Asn is estimated to be ½(9.4 + 4.8) = 7.2. The pK_a of the ammonium group in Leu is 9.76 and that of the carboxyl group in Gln is 4.8, so the pI of Leu-Val-Gln is estimated to be ½(9.76 + 4.8) = 7.3.

18.43 Glutathione (G-SH), one of the most common tripeptides in animals, plants, and bacteria, is a scavenger of oxidizing agents. In reacting with oxidizing agents, glutathione is converted to G-S-S-G.

$$
\underset{\underset{\text{COO}^-}{|}}{\overset{+}{\text{H}_3\text{N}}\text{CHCH}_2\text{CH}_2\overset{\overset{\text{O}}{\|}}{\text{C}}\text{NH}\underset{\underset{\text{CH}_2\text{SH}}{|}}{\text{CH}}\overset{\overset{\text{O}}{\|}}{\text{C}}\text{NHCH}_2\text{COO}^-}
$$

Glutathione

(a) Name the amino acids in this tripeptide.

The amino acids in glutathione are, from the N-terminal to the C-terminal, glutamic acid, cysteine, and glycine.

(b) What is unusual about the peptide bond formed by the N-terminal amino acid?

The carboxyl group used to form the peptide bond is not that of the α-carboxyl group but rather that of side chain.

(c) Is glutathione a biological oxidizing agent or a biological reducing agent?

Recall the chemistry of thiols. Thiols can be oxidized to form disulfide bridges, so glutathione is a reducing agent.

(d) Write a balanced equation for reaction of glutathione with molecular oxygen, O_2, to form G-S-S-G and H_2O. Is molecular oxygen oxidized or reduced in this reaction?

$$4\text{G-SH} + O_2 \longrightarrow 2\text{G-S-S-G} + 2H_2O$$

Oxygen gains hydrogen, so it is reduced in this reaction. Glutathione is oxidized.

18.44 Following is a structural formula for the artificial sweetener aspartame.

(a) Name the two amino acids in this molecule.

The amino acids in aspartame are aspartic acid (*N*-terminal) and phenylalanine (*C*-terminal). The carboxyl group of the *C*-terminal is esterified as a methyl ester.

Aspartame

(b) Estimate the isoelectric point of aspartame?

The only two acidic or basic functional groups are the α-ammonium group and the carboxylic acid side chain, both of which are part of Asp. The pK_a values of the ammonium group and the carboxylic acid are 9.82 and 3.86, respectively, resulting in a pI value of 6.84.

(c) Draw structural formulas for the products of hydrolysis of aspartame in 1 M HCl.

The two functional groups in aspartame that are susceptible to acid hydrolysis are the ester and the amide. Recall that esters are more reactive than amides, so the methyl ester would be hydrolyzed first. Depending on the temperature and the length of the reaction, it is also possible to hydrolyze the amide bond, releasing Asp and Phe.

+ CH_3OH

Three-Dimensional Shapes of Polypeptides and Proteins

18.45 Examine the α-helix conformation. Are amino acid side chains arranged all inside the helix, all outside the helix, or randomly?

All the amino acid side chains extend outside of the helix. There is no room inside the helix for the side chains because the diameter of the helix is relatively narrow. When we examine nucleic acids in the next chapter, we'll see that the α-helix of DNA is much wider and that all of the nucleobases are inside the helix.

18.46 Distinguish between intermolecular and intramolecular hydrogen bonding between the backbone groups on polypeptide chains. In what type of secondary structure do you find intermolecular hydrogen bonds? In what type do you find intramolecular hydrogen bonding?

Intermolecular hydrogen bonds are those formed between two polypeptide chains or molecules, while intramolecular hydrogen bonds are those formed between groups within a single chain or molecule. Because secondary structure refers the conformation of a single molecule, no intermolecular hydrogen bonds exist in secondary structure. Intramolecular hydrogen bonds between the NH and C=O groups are found in both α-helices and β-sheets.

18.47 Many plasma proteins found in aqueous environment are globular in shape. Which of the following amino acid side chains would you expect to find on the surface of a globular protein, in contact with the aqueous environment, and which would you expect to find inside, shielded from the aqueous environment? Explain.

(a) Leu   (b) Arg   (c) Ser   (d) Lys   (d) Phe

Amino acids that are polar, acidic, and basic will prefer to be on the surface of the protein, in contact with the aqueous environment to maximize hydrophilic and hydrogen-bonding interactions; these amino acids include Arg, Ser, and Lys. Nonpolar amino acids will prefer to avoid contact with the aqueous environment and turn inward to maximize hydrophobic interactions; these include Leu and Phe.

Looking Ahead

18.48 Some amino acids cannot be incorporated into proteins because they are self-destructive. Homoserine, for example, can use its side-chain −OH group in an intramolecular nucleophilic acyl substitution to cleave the peptide bond and form a cyclic structure on one end of the chain. Draw this structure and explain why serine does not suffer the same fate.

Homoserine undergoes a cyclization to form a five-membered ring (a γ-lactone), as shown below. The same reaction with serine would form a four-membered ring (a β-lactone), which is much more strained, and higher in energy, than a five-membered ring.

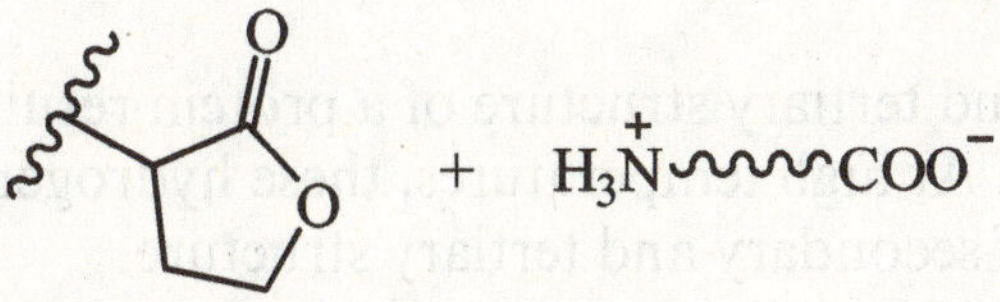

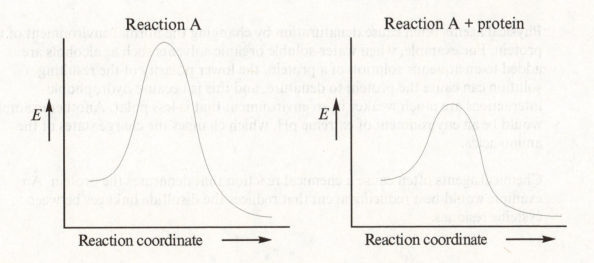

18.49 Would you expect a decapeptide of only isoleucine residues to form an α-helix? Explain.

A decapeptide of only isoleucine residues is not expected to form an α-helix. First, the numerous bulky *sec*-butyl side chains would destabilize an α-helix. Second, if this α-helix were in an aqueous environment, all of the side chains, which are nonpolar, would be on the outside of the helix and exposed to the aqueous environment.

18.50 Which type of protein would you expect to bring about the following change?

In the presence of the protein, the energy difference (ΔH) between the starting materials and the products does not change, but the activation energy is lower. These results are consistent with the protein being a catalyst, hence an enzyme.

Group Learning Activities

18.51 Heating can disrupt the 2° and 3° structure of a protein. Apply what you know about intermolecular forces and discuss the chemical processes that could occur upon heating a protein.

The secondary and tertiary structure of a protein result from the formation of hydrogen bonds. At high temperatures, these hydrogen bonds are broken, leading to the disruption of secondary and tertiary structure.

18.52 Enzymes are examples of proteins. Discuss why enzymes lose their catalytic activity at higher than physiological temperatures?

In order for an enzyme to function, it must be in its proper conformation. Conformation includes both secondary and tertiary structure, and these are disrupted at high temperatures.

18.53 Denaturation is the loss of secondary, tertiary, and quaternary structure of a protein by a chemical or physical agent and the resulting loss of function. The previous two problems revealed that heat can cause denaturation. As a group, discuss other physical or chemical agents that could cause denaturation and explain the processes that would affect denaturation.

Physical agents often cause denaturation by changing the normal environment of the protein. For example, when water-soluble organic solvents such as alcohols are added to an aqueous solution of a protein, the lower polarity of the resulting solution can cause the protein to denature, and this is because hydrophobic interactions are much weaker in an environment that is less polar. Another example would be an environment of extreme pH, which changes the charge states of the amino acids.

Chemical agents often cause a chemical reaction that denatures the protein. An example would be a reducing agent that reduces the disulfide linkages between cysteine residues.

Chapter 19: Lipids

Problems

19.1   (a)   How many constitutional isomers are possible for a triglyceride containing
one molecule each of palmitic acid, oleic acid, and stearic acid?

Three constitutional isomers are possible, and they differ in the fatty acid that is
esterified to carbon 2 of glycerol.

$$CH_2-palmitate \qquad CH_2-palmitate \qquad CH_2-stearate$$
$$CH-oleate \qquad CH-stearate \qquad CH-palmitate$$
$$CH_2-stearate \qquad CH_2-oleate \qquad CH_2-oleate$$

(b)   Which of the constitutional isomers that you found in part (a) are chiral?

All three constitutional isomers are chiral, with each containing one stereocenter.

Chemical Connections

19A. How is a lysolecithin able to act as a detergent?

The structure of a lysolecithin resembles that of a soap. Lysolecithins have one
hydrophobic "tail" and one polar "head." The head consists of a phosphorylated
glycerol.

19B. Determine the E/Z configuration of the carbon-carbon double bond in tamoxifen.
What is the role of progesterone and similar compounds in contraceptive pills?

The carbon-carbon double bond in tamoxifen has an E configuration.

The compounds in contraceptive pills, such as progesterone and its related analogues,
are members of the estrogen family. Estrogen hormones control the menstrual cycle,
and these specific compounds inhibit ovulation and therefore act as a contraceptive.

Quick Quiz

1.   Prostaglandins are formed in response to certain biological triggers. *True*. Prostaglandins
are not stored but are synthesized in response to triggers such as inflammation.

2. A fatty acid consists of a carboxyl group on the end of a long hydrocarbon chain. *True*. As a result, from a chemical perspective, a fatty acid is simply a long-chained carboxylic acid.

3. Vitamin D is synthesized from a steroid. *True*. Vitamin D is synthesized from 7-dehydrocholesterol, a steroid.

4. All steroids have the same tetracyclic ring system in common. *True*. The tetracyclic structure contains three six-membered rings and one five-membered ring.

5. Fatty acids with high melting points are called oils, while fatty acids with low melting points are called fats. *False*. There are two reasons why this statement is false. First, oils and fats refer to triglycerides and not the fatty acids themselves. Second, triglycerides with high melting points are called fats, while those with low melting points are called oils.

6. Steroids are achiral molecules. *False*. Steroids contain numerous stereocenters.

7. Synthetic detergents are based on fatty acids where the carboxylate group has been replaced by a sulfonate group. *True*. Detergents are sulfonates, and soaps are carboxylates.

8. According to the fluid-mosaic model, proteins in lipid bilayers are stationary and cannot move around. *False*. The membrane is *fluid* and the components embedded in the membrane are free to move around. Yet, the membrane is also heterogeneous from one location to another, hence the term *mosaic*.

9. Phospholipids organize into micelles in aqueous solution. *False*. Due to the presence of two hydrophobic "tails," phospholipids form lipid bilayers.

10. The greater the degree of unsaturation in a fatty acid, the higher is its melting point. *False*. The exact opposite is true. Unsaturation decreases melting point.

11. The function of both low- and high-density lipoproteins is to transport cholesterol throughout the body. *True*. The former is known as "bad cholesterol," and the latter as "good cholesterol."

12. A fatty acid that has been deprotonated and coordinated with a sodium ion can act as a soap. *True*. Soaps are sodium (or alkali metal) salts of fatty acids.

13. Bile acids differ from most other steroids in the number of their rings. *False*. Bile salts are made from steroids and retain the same number of rings.

14. Treatment of a triglyceride with aqueous base followed by acidification yields glycerol and up to three fatty acids. *True*. Each triglyceride contains three (hence the prefix *tri*) fatty acids esterified to glycerol.

15. Vitamin A can be synthesized from β-carotene. *True*. For that reason, β-carotene is also known as provitamin A (*pro* refers to the fact β-carotene is a precursor to vitamin A).

16. In fatty acids with C−C double bonds, the *trans* form of each double bond predominates. *False.* The carbon-carbon double bonds in naturally occurring fatty acids are almost always in the *cis* form.

End-of-Chapter Problems

Fatty Acids and Triglycerides

19.2 Define the term *hydrophobic*.

Hydrophobic literally means "water fearing" or "water hating." Hydrophobic species are relatively nonpolar and do not dissolve in water.

19.3 Identify the hydrophobic and hydrophilic region(s) of a triglyceride.

Each triglyceride contains three hydrophobic regions (the hydrocarbon chains of the fatty acids) and three hydrophilic regions (the ester groups).

Note that although a triglyceride contains hydrophilic regions, the extremely large hydrophobic regions render the triglyceride water-insoluble. This is why fats and oils do not dissolve in water.

19.4 Explain why the melting points of unsaturated fatty acids are lower than those of saturated fatty acids.

Saturated fatty acids, which are more linear in structure, are able to pack more closely in the solid state, thus increasing the amount of intermolecular dispersion forces that hold the molecules together. The higher are the intermolecular forces, the higher is the melting point (also see Problem 4.35).

19.5 Oleic acid has a melting point of 16°C. If you were to convert the *cis* double bond to a *trans* double bond, what would happen to the melting point?

The conversion of the double bond in oleic acid from *cis* to *trans* would increase the melting point of the compound. The *trans* fatty acid is relatively linear and can pack together much better than the *cis* fatty acid, which has a "kinked" structure (see Problem 4.35 for more details).

19.6   Which would you expect to have a higher melting point, glyceryl trioleate or glyceryl trilinoleate?

Oleic acid contains one *cis* double bond, while linoelic acid contains two *cis* double bonds. Oleic acid therefore has a higher melting point than linoleic acid, and this trend also applies to the triglycerides. Glyceryl trioleate has a higher melting point.

19.7   Which animal fat has the highest percentage of unsaturated fatty acids? Which plant oil has the highest percentage of unsaturated fatty acids?

The animal fat that contains the highest percentage of unsaturated fatty acids is actually human fat, while with plant oil, it is olive oil. This is one of the reasons why olive oil is better for your health than are many other oils.

19.8   Draw a structural formula for methyl linoleate. Be certain to show the correct configuration of groups about each carbon-carbon double bond.

19.9   Explain why coconut oil is a liquid triglyceride, even though most of its fatty acid components are saturated.

Despite having a large percentage of saturated fatty acids, coconut oil is a liquid triglyceride because a very high percentage of the saturated fatty acids have a low molecular weight. Recall that when compounds are of lower molecular weight, they have weaker dispersion forces and therefore lower melting points.

19.10 It is common now to see "contains no tropical oils" on cooking-oil labels, meaning that the oil contains no palm or coconut oil. What is the difference between the composition of tropical oils and that of vegetable oils, such as corn oil, soybean oil, and peanut oil?

Tropical oils primarily consist of saturated fatty acids, while non-tropical vegetable oils primarily consist of unsaturated fatty acids, which are better for your health.

19.11 What is meant by the term *hardening* as applied to vegetable oils?

Hardening refers to the catalytic hydrogenation of the C=C bonds in vegetable oil. By increasing the percentage of saturation in the oil, the melting point is increased, often to the point where the oil is a solid a room temperature. In that sense, "hydrogenated vegetable oil" is harder than non-hydrogenated vegetable oil.

An unintended consequence of the hydrogenation process is that under the reaction conditions used, some of the *cis* C=C bonds isomerize to the *trans* configuration. These "*trans* fats" have gained immense attention in the past few years.

19.12 How many moles of H_2 are used in the catalytic hydrogenation of one mole of a triglyceride derived from glycerol, stearic acid, linoleic acid, and arachidonic acid?

The hydrocarbon chain of stearic acid is saturated, while those of linoelic and arachidonic acids contain 2 and 4 C=C bonds, respectively. One mole of such a triglyceride will require 6 moles of H_2 for complete reduction to a saturated triglyceride. Note that the C=O groups of the esters are unaffected by catalytic hydrogenation.

19.13 Characterize the structural features necessary to make a good synthetic detergent.

A good synthetic detergent should have a long, hydrophobic chain and a very hydrophilic head group (either ionic or very polar). In addition, it should not form insoluble precipitates with the ions that are commonly found in hard water, such as Ca^{2+}, Mg^{2+}, and Fe^{2+}.

19.14 Following are structural formulas for a cationic detergent and a neutral detergent. Account for the detergent properties of each.

$$CH_3(CH_2)_6CH_2\overset{\overset{\displaystyle CH_3}{|+}}{\underset{\underset{\displaystyle CH_2C_6H_5}{|}}{N}}CH_3 \quad Cl^- \qquad\qquad HOCH_2\overset{\overset{\displaystyle HOH_2C}{|}}{\underset{\underset{\displaystyle HOH_2C}{|}}{C}}CH_2O\overset{\displaystyle O}{\overset{\displaystyle \|}{C}}(CH_2)_{14}CH_3$$

Benzyldimethyloctylammonium chloride   Pentaerythrityl palmitate
(a cationic detergent)   (a neutral detergent)

Both detergents contain a long, hydrophobic tail that is attached to a very hydrophilic head. The cationic detergent has a positively charged head, while the neutral detergent has an uncharged head that is very polar (multiple –OH groups). When placed in water, both of these compounds will form micelles and act as detergents.

19.15 Identify some of the detergents used in shampoos and dishwashing liquids. Are they primarily anionic, neutral, or cationic detergents?

Although the composition of a shampoo or a dishwashing liquid obviously depends on the brand and the formulation, the detergents are usually anionic and are typically sodium salts of alkylbenzene sulfonates or alkyl sulfates. The most common is an alkyl sulfate known as sodium dodecyl sulfate (also known as SDS, sodium lauryl sulfate, or sodium laureth sulfate). Next time you're in the shower, check your shampoo bottle!

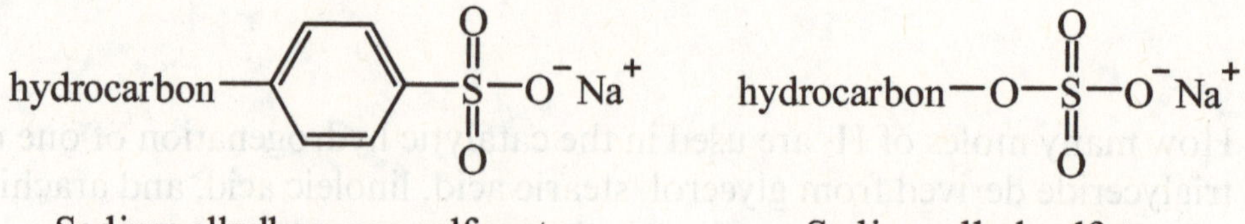

Sodium alkylbenzene sulfonate   Sodium alkyl sulfate

19.16 Show how to convert palmitic acid (hexadecanoic acid) into the following:

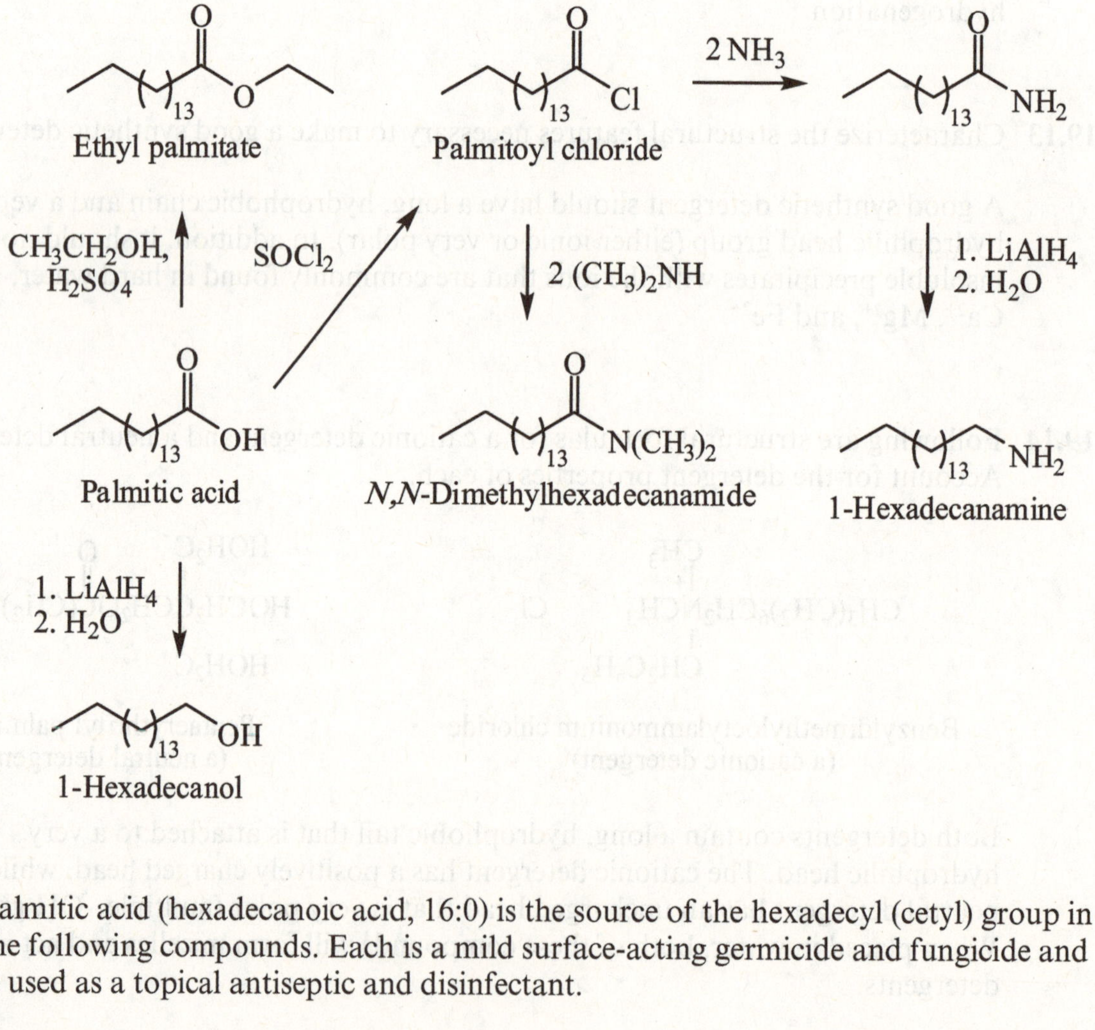

Ethyl palmitate

Palmitoyl chloride

NH₂

$CH_3CH_2OH,$
H_2SO_4

$SOCl_2$

$2 (CH_3)_2NH$

1. $LiAlH_4$
2. H_2O

Palmitic acid

N,N-Dimethylhexadecanamide

1-Hexadecanamine

1. $LiAlH_4$
2. H_2O

1-Hexadecanol

19.17 Palmitic acid (hexadecanoic acid, 16:0) is the source of the hexadecyl (cetyl) group in the following compounds. Each is a mild surface-acting germicide and fungicide and is used as a topical antiseptic and disinfectant.

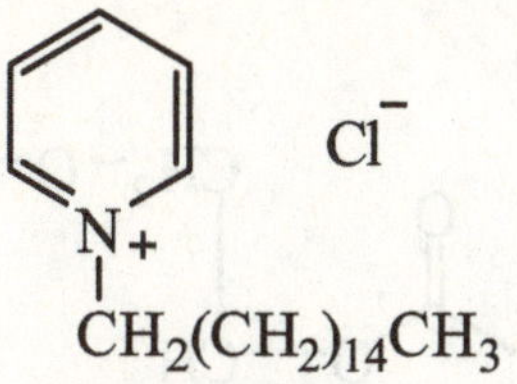

Cetylpyridinium chloride 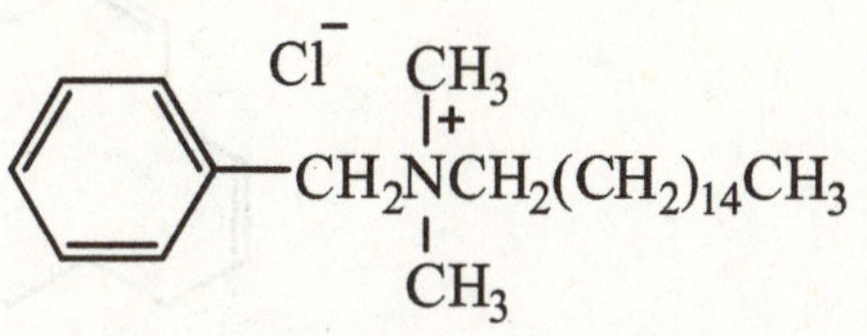   Benzylcetyldimethylammonium chloride

(a) Cetylpyridinium chloride is prepared by treating pyridine with 1-chlorohexadecane (cetyl chloride). Show how to convert palmitic acid to cetyl chloride.

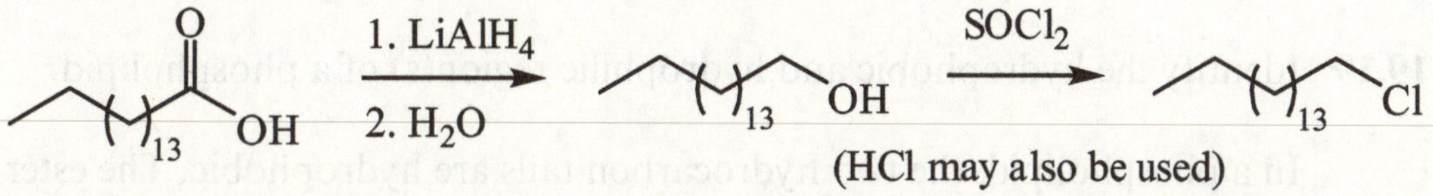

(HCl may also be used)

(b) Benzylcetyldimethylammonium chloride is prepared by treating benzyl chloride with N,N-dimethyl-1-hexadecanamine. Show how this tertiary amine can be prepared from palmitic acid.

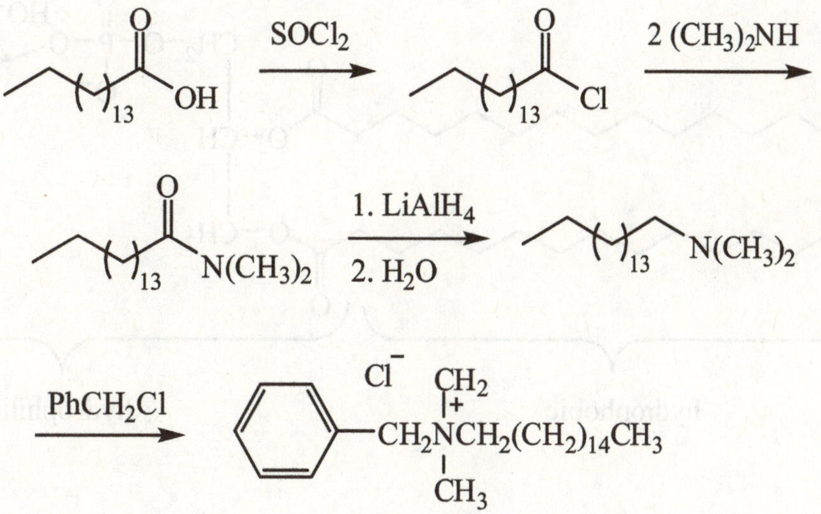

Phospholipids

19.18 Draw the structural formula of a lecithin containing one molecule each of palmitic acid and linoleic acid.

Phospholipids contain a glycerol backbone, two esters of fatty acids, and one phosphate ester. Lecithin is a type of phospholipid and contains choline that is esterified to the phosphate.

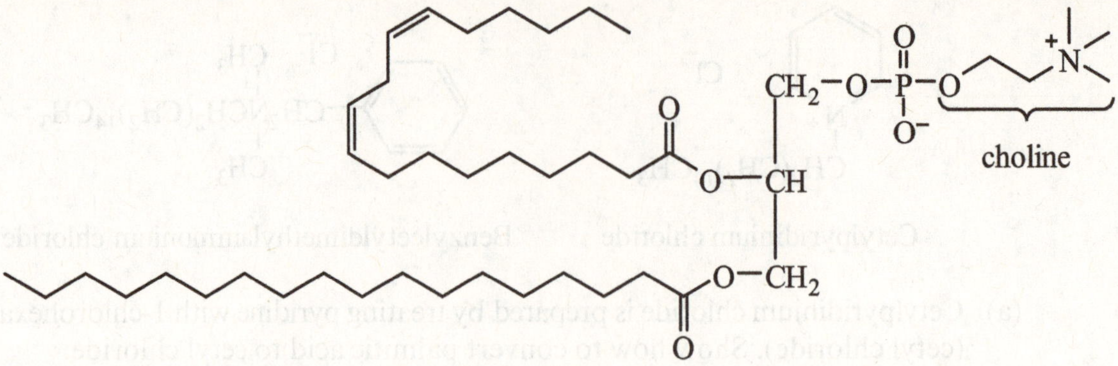

19.19 Identify the hydrophobic and hydrophilic region(s) of a phospholipid.

In a phospholipid, the two hydrocarbon tails are hydrophobic. The ester and the phosphate groups (as well as any other charged or polar groups that are esterified to the phosphate) are hydrophilic. These are indicated using phosphatidylinositol as an example.

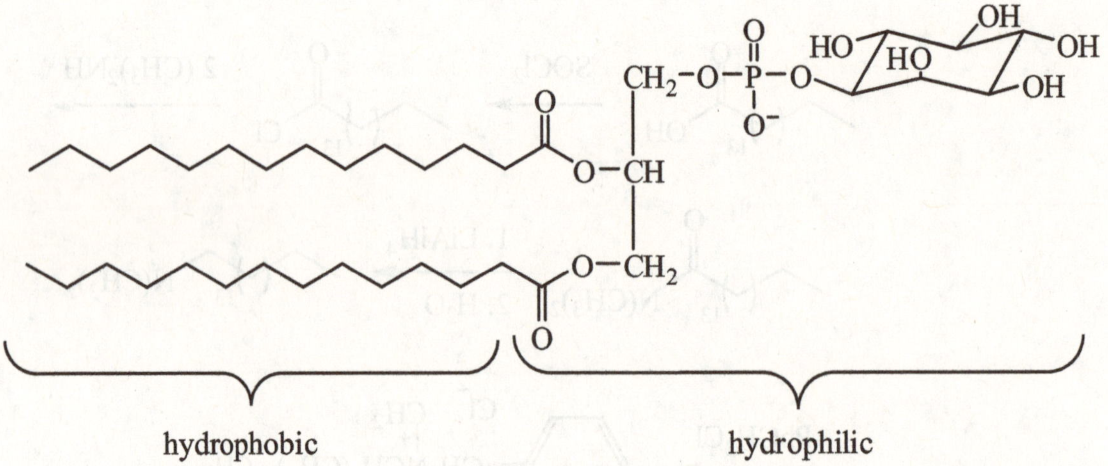

19.20 The hydrophobic effect is one of the most important noncovalent forces directing the self-assembly of biomolecules in aqueous solution. The hydrophobic effect arises from tendencies of biomolecules (1) to arrange polar groups so that they interact with the aqueous environment by hydrogen bonding and (2) to arrange nonpolar groups so that they are shielded from the aqueous environment. Show how the hydrophobic effect is involved in directing

(a) The formation of micelles by soaps and detergents.

Soap and detergent molecules contain a single hydrophobic tail and a very hydrophilic head (usually ionic, but can also be a set of polar groups). The supramolecular structure (structures that are held together by noncovalent forces) that these molecules adopt is that of a micelle. The tails aggregate inside the micelle,

where they are shielded from water, while the heads are located on the outside so
that they can interact with the aqueous environment.

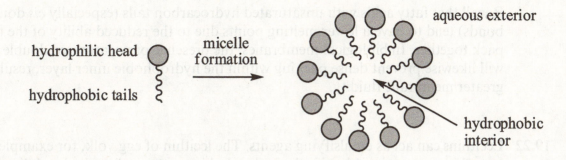

(b) The formation of lipid bilayers by phospholipids.

Unlike soap and detergent molecules, phospholipids contain *two* hydrophobic
tails and a hydrophilic head. The extra space required by the extra tail results in
the formation of a lipid bilayer. The heads are on both sides of the bilayer and are
exposed to water. The tails are in between the layers and shielded from water.

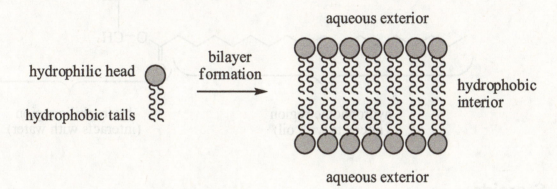

(c) The formation of the DNA helix.

In the DNA double helix, the relatively
hydrophobic bases are stacked inside the helix.
This allows the bases to interact by
hydrophobic interactions and also shields them
from the aqueous environment. The sugar-
phosphate backbone, which is hydrophilic, is
located on the outside of the helix and can
interact with the aqueous environment.

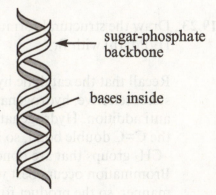

19.21 How does the presence of unsaturated fatty acids contribute to the fluidity of biological membranes?

Recall that fatty acids with unsaturated hydrocarbon tails (especially *cis* double bonds) tend to have a lower melting points due to the reduced ability of the tails to pack together. In biological membranes, the presence of unsaturated double bonds will likewise prevent dense packing within the hydrophobic inner layer, resulting in greater membrane fluidity.

19.22 Lecithins can act as emulsifying agents. The lecithin of egg yolk, for example, is used to make mayonnaise. Identify the hydrophobic part(s) and the hydrophilic part(s) of a lecithin. Which parts interact with the oils used in making mayonnaise? Which parts interact with the water?

hydrophobic region   hydrophilic region
(interacts with oil)   (interacts with water)

Steroids

19.23 Draw the structural formula for the product formed by treating cholesterol with (a) H$_2$/Pd; (b) with Br$_2$.

Recall that the catalytic hydrogenation and the bromination of an alkene are stereoselective; hydrogenation occurs by syn addition, while bromination occurs by anti addition. Hydrogenation will preferentially occur on the least hindered side of the C=C double bond, so hydrogen is added from the side opposite to the –OH and –CH$_3$ groups that are bonded to ring A and the junction of rings AB, respectively. Bromination occurs best when the two bromine groups are added in a *trans* diaxial manner, so the product formed has the –Br groups axial and *trans* to each other.

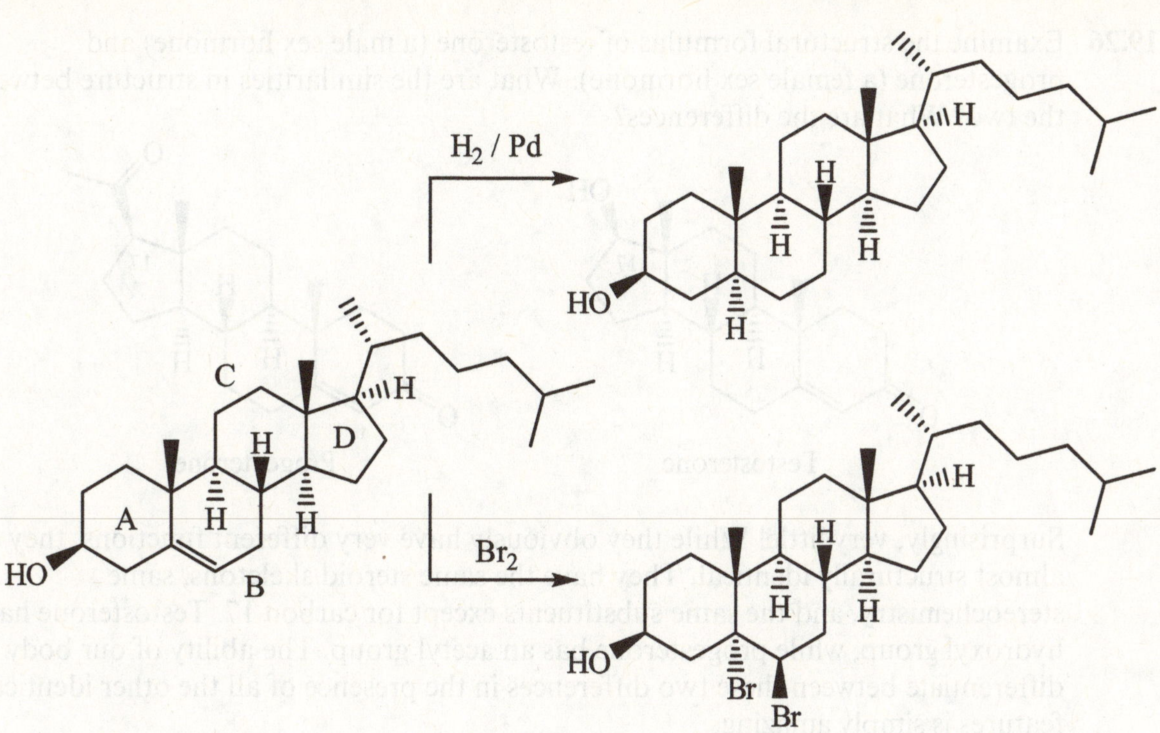

19.24 List several ways in which cholesterol is necessary for human life. Why do many people find it necessary to restrict their dietary intake of cholesterol?

Although there is a plenty of bad press about cholesterol, cholesterol is essential for human life. It is a component of biological membranes, where it helps to maintain the proper fluidity of the membrane. Cholesterol is also an important biosynthetic precursor to a variety of steroid hormones. Despite these essential functions, the consumption of too much cholesterol increases the risk of cardiovascular disease. Note that cholesterol is also biosynthesized by the body, and if blood cholesterol levels cannot be controlled by diet alone, there are pharmaceutical drugs available to modulate its biosynthesis.

19.25 Both low-density lipoproteins (LDL) and high-density lipoproteins (HDL) consist of a core of triacylglycerols and cholesterol esters surrounded by a single phospholipid layer. Draw the structural formula of cholesteryl linoleate, one of the cholesterol esters found in this core.

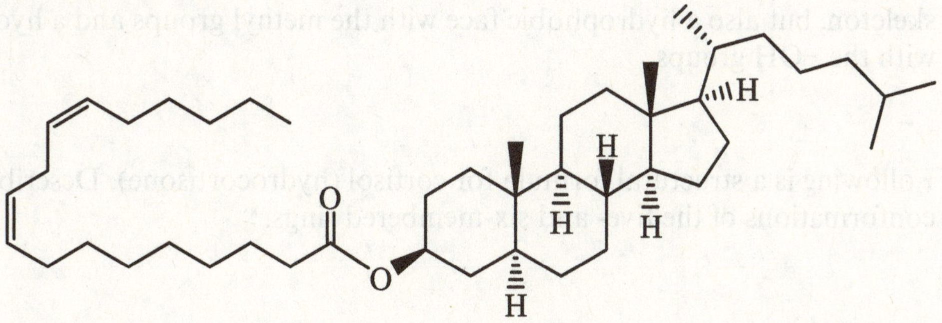

19.26 Examine the structural formulas of testosterone (a male sex hormone) and progesterone (a female sex hormone). What are the similarities in structure between the two? What are the differences?

Testosterone   Progesterone

Surprisingly, very little! While they obviously have very different functions, they are almost structurally identical. They have the same steroid skeletons, same stereochemistry, and the same substituents except for carbon 17. Testosterone has a hydroxyl group, while progesterone has an acetyl group. The ability of our body to differentiate between these two differences in the presence of all the other identical features is simply amazing.

19.27 Examine the structural formula of cholic acid, and account for the ability of this bile salt and others to emulsify fats and oils and thus aid in their digestion.

Choic acid

Cholic acid is able to act as an emulsifier, just like lecithin, because it contains both hydrophobic and hydrophilic groups. The hydrocarbon rings of the steroid skeleton are hydrophobic, while the carboxylate and hydroxyl groups are hydrophilic.

It is interesting to note that all three –OH groups point in the same direction, and the two methyl groups in the other direction. As a result, cholic acid has not only a hydrophilic head with a carboxylate group and a hydrophobic tail with a steroid skeleton, but also a hydrophobic face with the methyl groups and a hydrophilic face with the –OH groups.

19.28 Following is a structural formula for cortisol (hydrocortisone). Describe the conformations of the five- and six-membered rings.

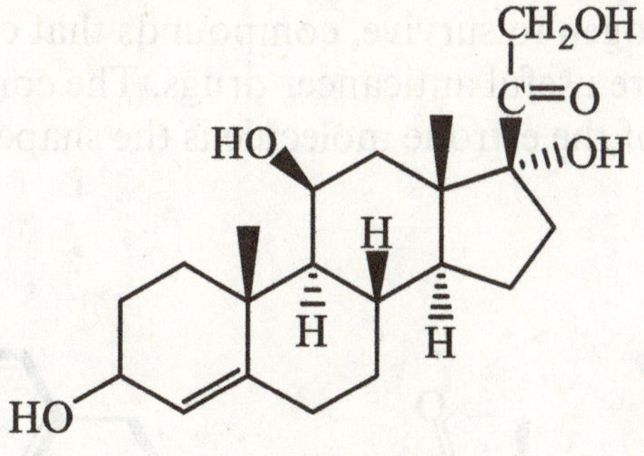

**Cortisol
(Hydrocortisone)**

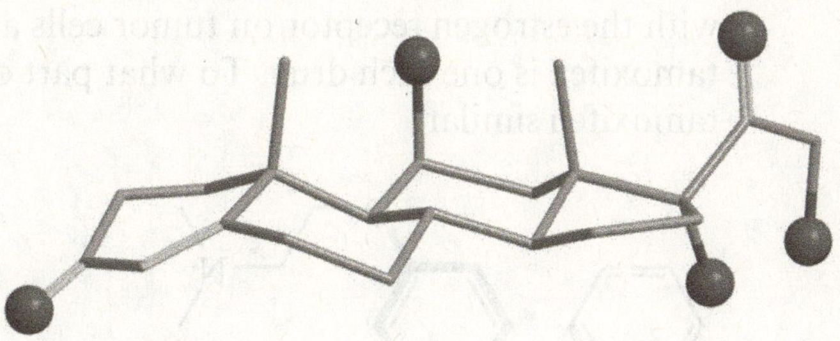

As seen in Chapter 3, polygonal depictions of cyclic compounds do not reveal much information about their conformation, so it is necessary to examine the ball-and-stick model that is provided. The ball-and-stick model shown above has been redrawn with the hydrogens omitted for clarity. The two six-membered rings are chairs, and the five-membered cycloalkane ring is an envelope.

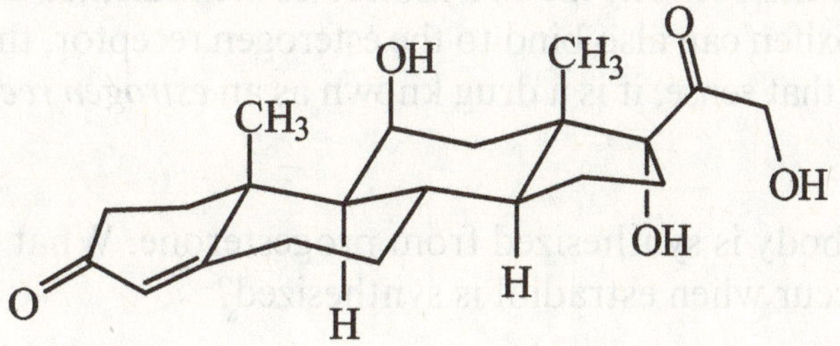

19.29 How does the oral anabolic steroid methandrostenolone differ structurally from testosterone?

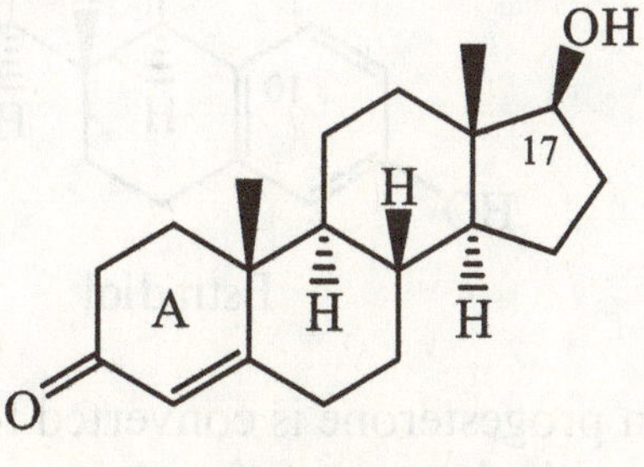

Testosterone 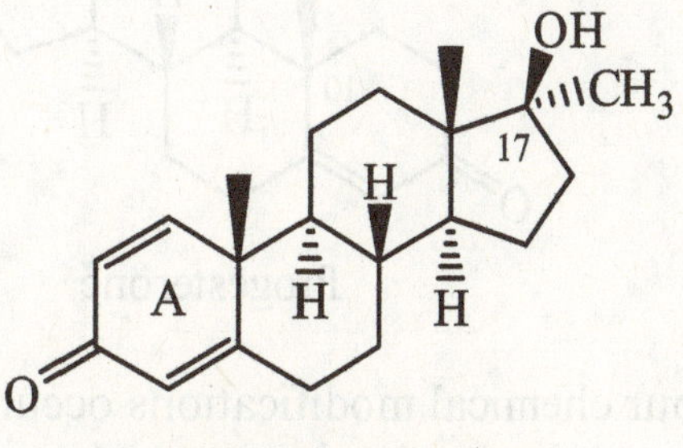   **Methandrostenolone**

Like testosterone and progesterone (Problem 19.26), methandrostenolone and testosterone are structurally very similar. Methandrostenolone has an extra carbon-carbon double bond in ring A and a methyl substituent on carbon 17. These two seemingly innocent changes result in a muscle-growth agent that is illegal in many countries.

19.30 Because some types of tumors need an estrogen to survive, compounds that compete with the estrogen receptor on tumor cells are useful anticancer drugs. The compound tamoxifen is one such drug. To what part of the estrone molecule is the shape of tamoxifen similar?

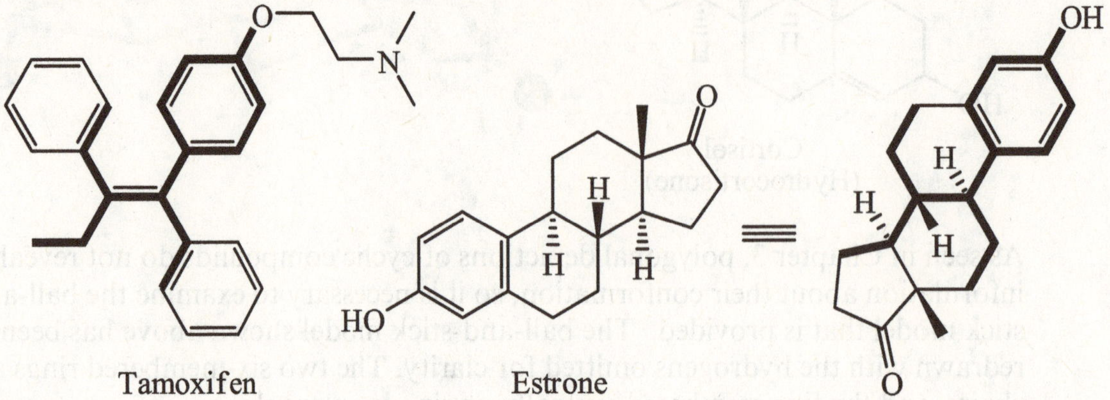

Tamoxifen   Estrone

Structural similarities between the two molecules are indicated in bold. Because of these similarities, tamoxifen can also bind to the esterogen receptor, thus preventing estrogen from binding. In that sense, it is a drug known as an *estrogen receptor antagonist*.

19.31 Estradiol in the body is synthesized from progesterone. What chemical modifications occur when estradiol is synthesized?

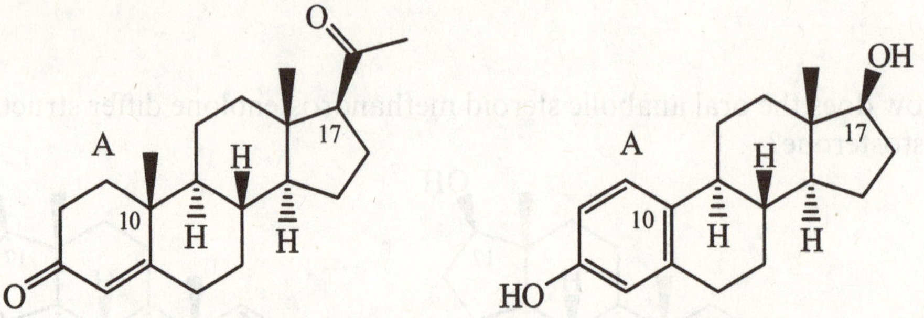

Progesterone   Estradiol

Four chemical modifications occur when progesterone is converted to estradiol: Ring A is converted into an aromatic ring, the ketone of ring A is converted into a phenol, the methyl group on carbon 10 is removed, and the acetyl group on carbon 17 is replaced by a hydroxyl group.

Prostaglandins

19.32 Examine the structure of PGF$_{2\alpha}$ and

(a) identify all stereocenters.
(b) identify all double bonds about with *cis-trans* isomerism is possible.
(c) state the number of stereoisomers possible for a molecule of this structure.

PGF$_{2\alpha}$ contains five stereocenters, which are each marked with an asterisk, and two double bonds for which *cis-trans* isomerism is possible. Each stereocenter can be *R* or *S*, and each double bond can be *cis* or *trans*, so $2^7 = 128$ stereoisomers are possible.

19.33 Following is the structure of unoprostone, a compound patterned after the natural prostaglandins (Section 19.5). Rescula, the isopropyl ester of unoprostone, is an antiglaucoma drug used to treat ocular hypertension. Compare the structural formula of this synthetic prostaglandin with that of PGF$_{2\alpha}$.

Unoprostone differs from PGF$_{2\alpha}$ as follows: The double bond between carbons 13 and 14 is reduced, carbon 15 is oxidized from an alcohol to a ketone, and there are two extra carbons (21 and 22) at the end of the molecule.

19.34 How does aspirin, an anti-inflammatory drug, prevent strokes caused by blood clots in the brain?

Aspirin is an inhibitor of cyclooxygenase, the enzyme that converts arachidonic acid to prostaglandin G$_2$. PGG$_2$ is a biosynthetic precursor to many other prostaglandins, including thromboxane A$_2$ (TXA), the prostaglandin that causes the aggregation of platelets (blood clots). The inhibition of PGG$_2$ biosynthesis lowers the level of TXA.

Fat-Soluble Vitamins

19.35 Examine the structural formula of vitamin A, and state the number of *cis-trans* isomers possible for this molecule.

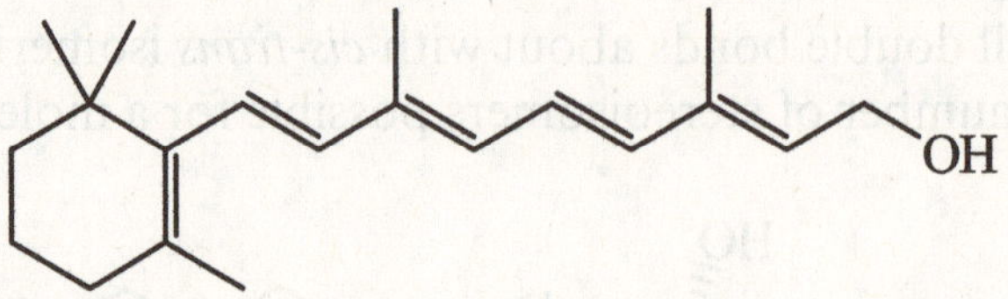

Vitamin A contains four double bonds that can exhibit *cis-trans* isomerism. These are the double bonds found in the acyclic chain; the double bond that is a part of the six-membered ring cannot exist as a *trans* alkene with respect to the ring skeleton (recall Chapter 4). Therefore, $2^4 = 16$ *cis-trans* isomers are possible for Vitamin A.

19.36 The form of vitamin A present in many food supplements is vitamin A palmitate. Draw the structural formula of this molecule.

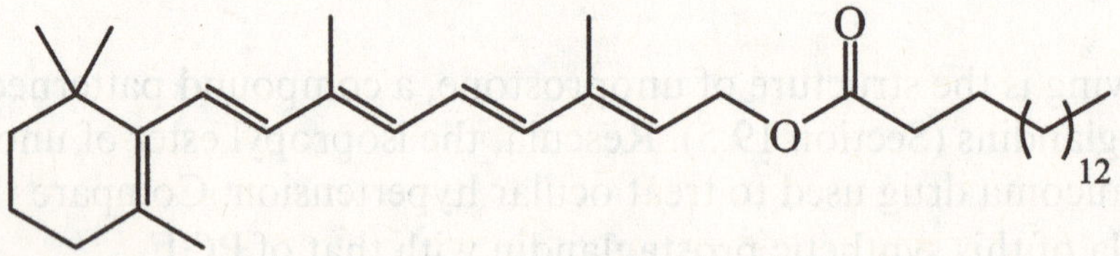

When vitamin A palmitate is ingested, the ester is hydrolyzed to release vitamin A. Notice that both vitamin A and its palmitate ester are very hydrophobic, so both of these are soluble in fats and oils. Milk is often supplemented the palmitate ester.

19.37 Examine the structural formulas of vitamin A, 1,25-dihydroxy-D$_3$, vitamin E, and vitamin K$_1$. Do you expect them to be more soluble in water or in dichloromethane? Do you expect them to be soluble in blood plasma?

All of these vitamins are extremely hydrophobic, so they are not expected to be very soluble in water or blood plasma. However, they will be much more soluble in dichloromethane, which is much less polar than water.

Looking Ahead

19.38 Here is the structure of a glycolipid, a class of lipid that contains a sugar residue. Glycolipids are found in cell membranes. (a) Which part of the molecule would you expect to reside on the extracellular side of the membrane? (b) Which monosaccharide unit is present in this glycolipid?

hydrophilic   hydrophobic

HO

CH$_2$OH

HO   O

O

HN   (CH$_2$)$_n$CH$_3$

OH   O

(CH$_2$)$_{12}$CH$_3$

OH

A glycolipid

The glycolipid has a hydrophilic region (the monosaccharide D-galactose) that is expected to reside on the extracellular side of the membrane. The extremely hydrophobic region (the hydrocarbon chains) is expected to be embedded in the membrane.

19.39 How would you expect temperature to affect fluidity in a cell membrane?

The fluidity of a cell membrane is expected to increase when temperature increases. This is because when temperature increases, so does molecular motion. This is true for all compounds and substances; for example, butter is hard at low temperature, but as it warms up, it becomes softer and softer until it eventually melts.

19.40 Which type of lipid movement is most favorable in cell membranes? The darker portion represents the polar head group of the lipid. Explain.

Movement A is most favorable, because B would require the polar head group of the lipid to pass through the nonpolar part of the lipid bilayer.

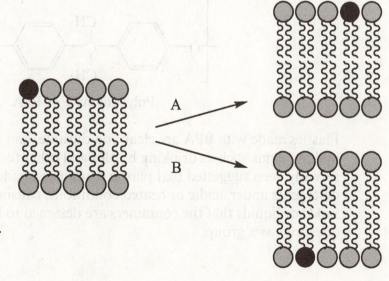

19.41 Aspirin works by transferring an acetyl group to the side chain of the 530th amino
acid in the protein prostaglandin H_2 synthase-1. Draw the product of this reaction:

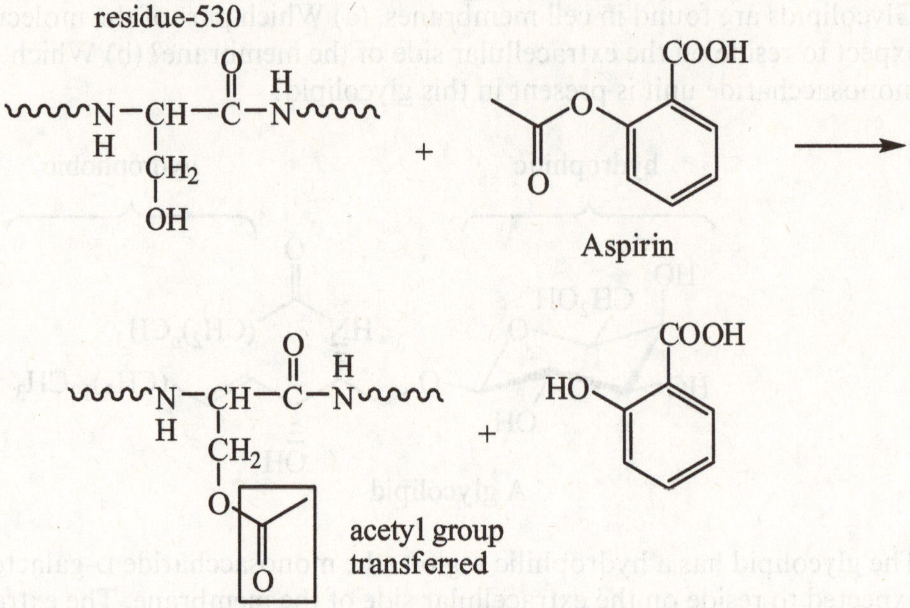

residue-530

Aspirin

acetyl group
transferred

Notice that this reaction is simply a nucleophilic acyl substitution reaction. The next
time you take aspirin, you should appreciate that the reactions you have learned in
Chapter 14 are being used to reduce your pain and inflammation!

Group Learning Activities

19.42 Bisphenol A, commonly known as BPA, has been widely used as a monomer to
make polycarbonates (Section 16.4C).

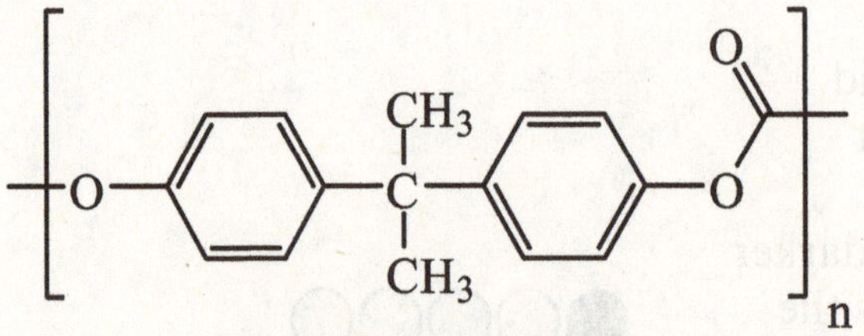

Polycarbonate of BPA

Plastics made with BPA are clear and durable, and for this reason, are ideal for
making items such as drinking bottles and other food containers. However, it has
recently been suggested that plastic containers made with BPA can react with strong
detergents under acidic or heated conditions, causing BPA to leach into the very
foods or liquids that the containers are designed to hold. Discuss the following
questions as a group:

(a) Provide a structure for BPA by drawing the mechanism for the acid-catalyzed
     hydrolysis of its polycarbonate.

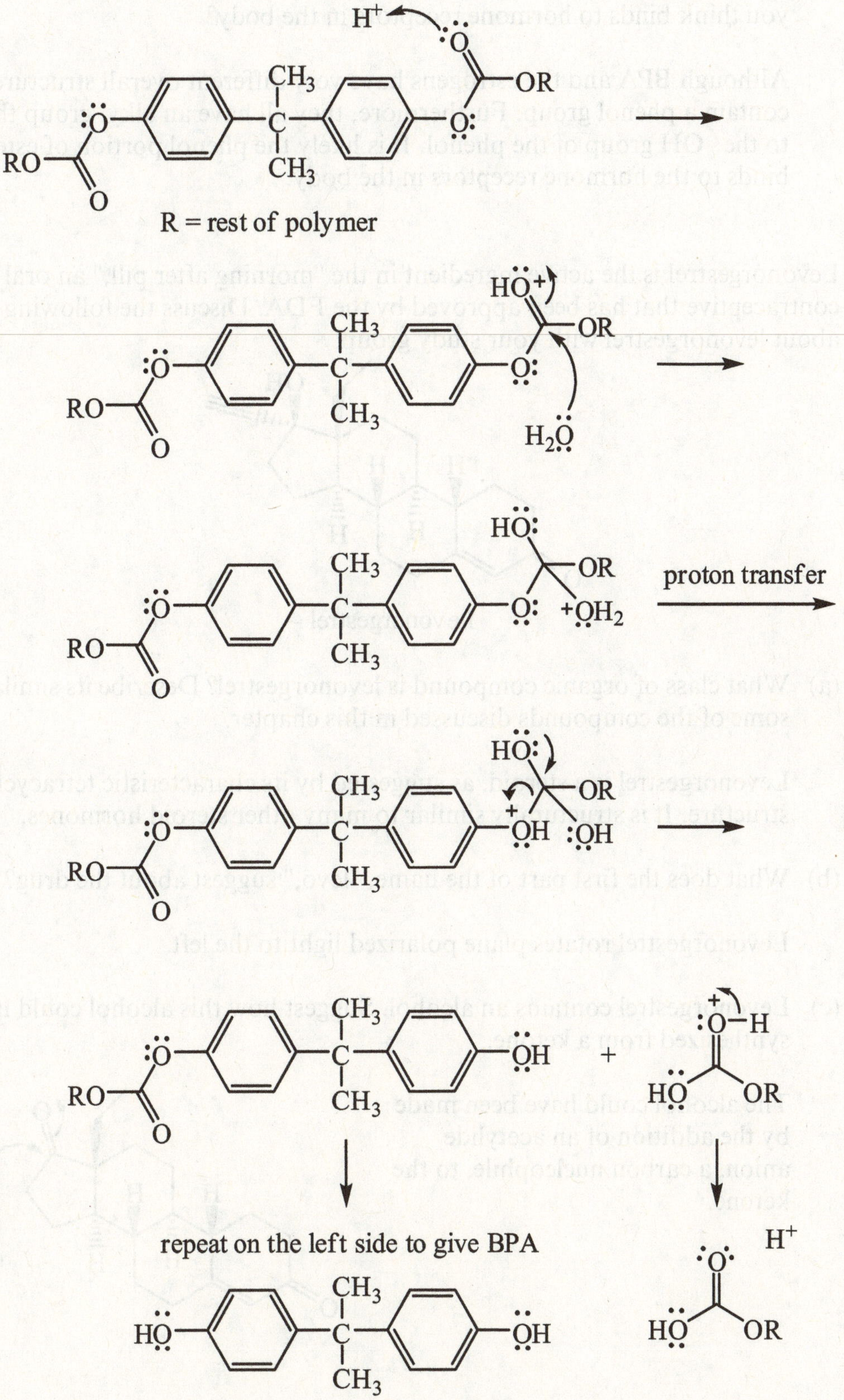

R = rest of polymer

proton transfer

repeat on the left side to give BPA

(b) BPA acts as an endocrine disruptor by mimicking estrogen in the body. Compare the structure of BPA to the three estrogens in Chemical Connections 19B. How are they different? How are they similar? Which part of estrogen do you think binds to hormone receptors in the body?

Although BPA and the estrogens have very different overall structures, they all contain a phenol group. Furthermore, they all have an alkyl group that is *para* to the –OH group of the phenol. It is likely the phenol portion of esterogen that binds to the hormone receptors in the body.

19.43 Levonorgestrel is the active ingredient in the "morning after pill," an oral contraceptive that has been approved by the FDA. Discuss the following questions about levonorgestrel with your study group:

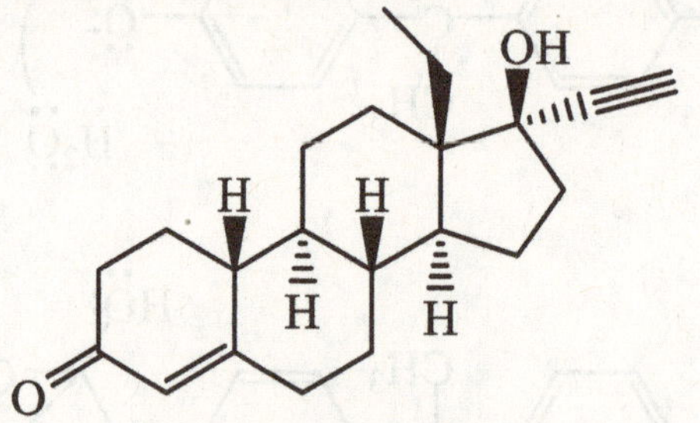

Levonorgestrel

(a) What class of organic compound is levonorgestrel? Describe its similarities with some of the compounds discussed in this chapter.

Levonorgestrel is a steroid, as suggested by its characteristic tetracyclic structure. It is structurally similar to many other steroid hormones.

(b) What does the first part of the name, "levo," suggest about the drug?

Levonorgestrel rotates plane polarized light to the left.

(c) Levonorgestrel contains an alcohol. Suggest how this alcohol could have been synthesized from a ketone.

The alcohol could have been made by the addition of an acetylide anion, a carbon nucleophile, to the ketone.

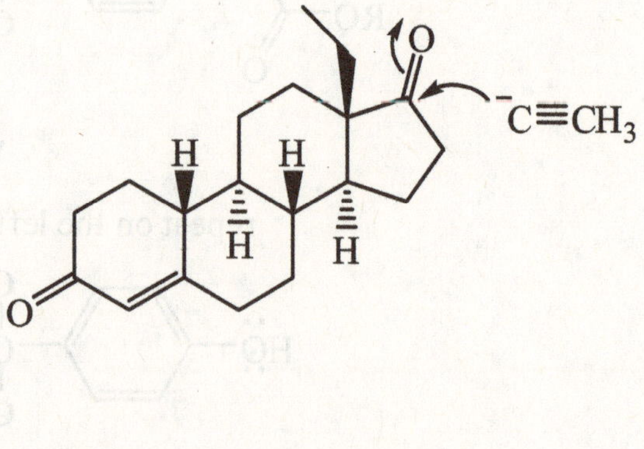

(d) Levonorgestrel contains an α, β-unsaturated ketone. Suggest how this part of the compound could have been synthesized from a Michael reaction followed by an aldol reaction.

Working backwards, the α, β unsaturated ketone in levonorestrel can be prepared from the dehydration of the corresponding β-hydroxyketone.

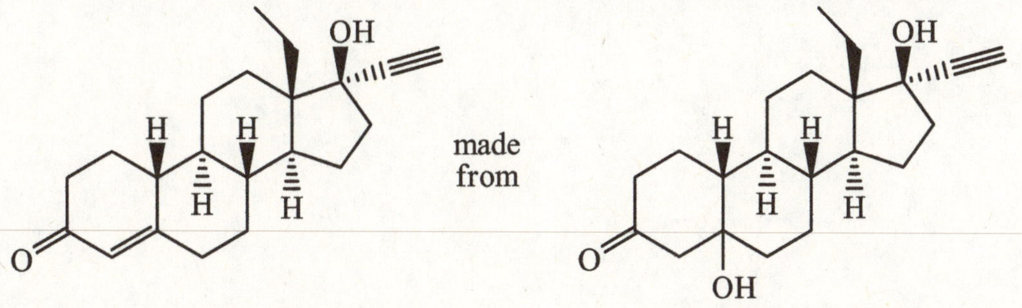

The β-hydroxyketone can be synthesized by an aldol reaction.

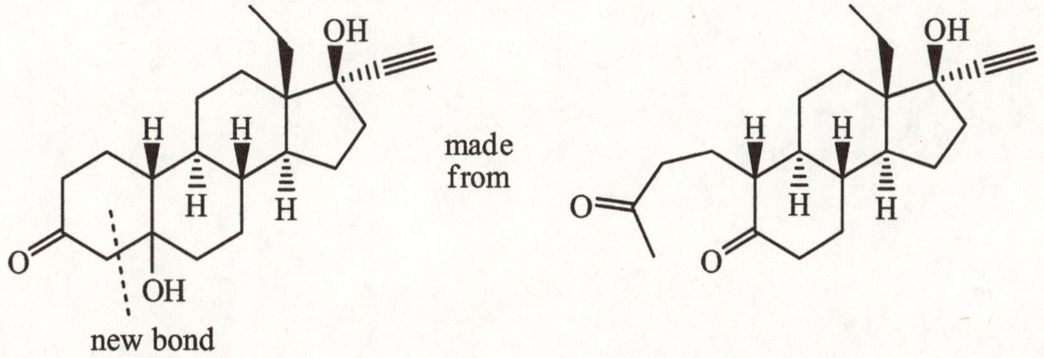

The starting material for the aldol reaction can be made by a Michael addition.

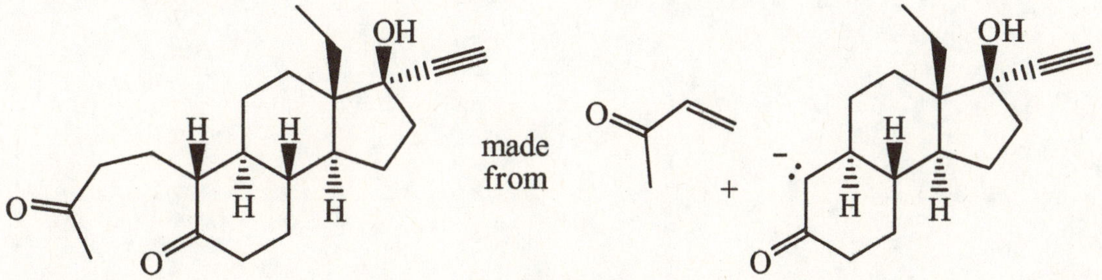

Chapter 20: Nucleic Acids

Problems

20.1 Draw a structural formula for 2'-deoxythymidine 3'-monophosphate.

The "prime" after each of the numbers refers to the carbon atoms present in the sugar.

Unless otherwise specified, nucleotides are assumed to be derived from D-ribose, and the *N*-glycosidic linkage is β.

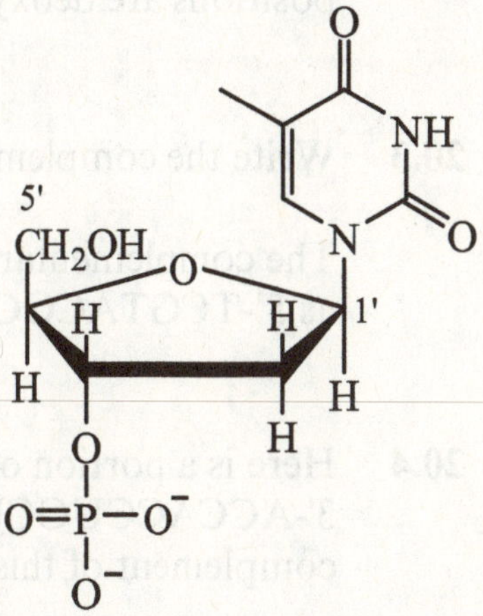

20.2 Draw a structural formula for the section of DNA that contains the base sequence CTG and is phosphorylated at the 3' end only.

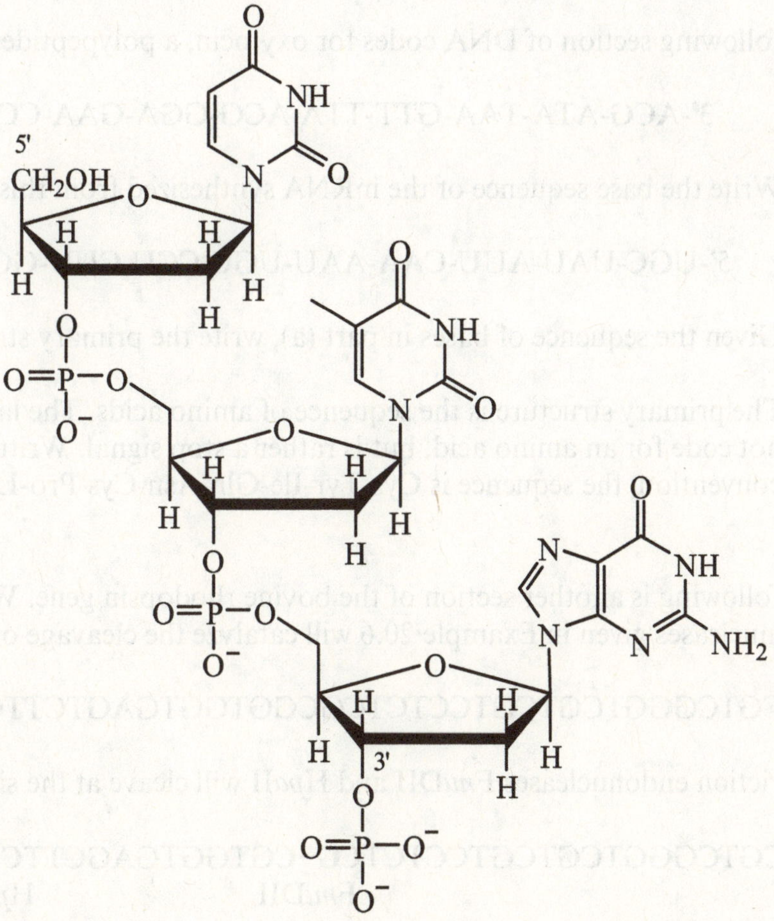

By convention, sequences for nucleic acids are written in the 5'-to-3' direction, from left to right, unless otherwise indicated. This is similar to the convention used for proteins, where the left is the *N*-terminal and the right is the *C*-terminal. These conventions may seem arbitrary, but they actually correspond to the directions in which nucleic acids and proteins are biologically synthesized. In DNA, the 2' positions are deoxygenated.

20.3   Write the complementary DNA base sequence for 5'-CCGTACGA-3'.

The complementary sequence is 3'-GGCATGCT-5', which could also be written as 5'-TCGTACGG-3'.

20.4   Here is a portion of the nucleotide sequence in phenylalanine tRNA: 3'-ACCACCUGCUCAGGCCUU-5'. Write the nucleotide sequence of the DNA complement of this sequence.

Remember that the base uracil (U) in RNA is the complement to adenine (A) in DNA. The complementary DNA sequence is 5'-TGGTGGACGAGTCCGGAA-3'.

20.5   The following section of DNA codes for oxytocin, a polypeptide hormone:

3'-ACG-ATA-TAA-GTT-TTA-ACG-GGA-GAA-CCA-ACT-5'

(a)   Write the base sequence of the mRNA synthesized from this section of DNA.

5'-UGC-UAU-AUU-CAA-AAU-UGC-CCU-CUU-GGU-UGA-3'

(b)   Given the sequence of bases in part (a), write the primary structure of oxytocin.

The primary structure is the sequence of amino acids. The last codon UGA does not code for an amino acid, but is rather a stop signal. Written according to convention, the sequence is Cys-Tyr-Ile-Gln-Asn-Cys-Pro-Leu-Gly.

20.6   The following is another section of the bovine rhodopsin gene. Which of the endonucleases given in Example 20.6 will catalyze the cleavage of this section?

5'-ACGTCGGGTCGTCGTCCTCTCGCGGTGGTGAGTCTTCCGGCTCTTCT-3'

Restriction endonucleases F*nu*DII and H*pa*II will cleave at the sites indicated below.

5'-ACGTCGGGTCGTCGTCCTCT**CG**−**CG**TGGTGAGCTT**C**−**CGG**CTCTTCT-3'
                        F*nu*DII   H*pa*II

Chemical Connections

20A. Draw the triphosphate of acyclovir as it would exist in solution at pH 7.4.

Like ATP, the triphosphate of acyclovir has four ionizable hydrogens. The first three have pK_a values less than 5.0, while the fourth has a pK_a of about 7.0. At pH 7.4, the first three are fully ionized, while over 50% of the fourth hydrogen is ionized.

20B. Explain the basis of separation for the bands shown on the DNA fingerprinting gel.

In the separation of DNA by polyacrylamide gel electrophoresis, the DNA is separated on the basis of the size of the fragment. Smaller (shorter) fragments migrate faster through the gel and are found further away from the origin.

Quick Quiz

1. Endonucleases are used to make DNA strands radioactive for sequencing. *False*. Endonucleases are enzymes that cleave DNA strands, sometimes at specific sequences.

2. A strand of RNA is the product of transcription. *True*. Transcription refers to the synthesis of RNA from a DNA template.

3. There are the same number of hydrogen bonds between a G-C base pair as there are between an A-T base pair. *False*. A G-C base pair has three hydrogen bonds, while an A-T base pair has only two hydrogen bonds.

4. There are multiple types of RNA, each serving a different function in a cell. *True*. The main types of RNA are ribosomal, messenger, and transfer RNA.

5. The dideoxy method of DNA sequencing uses a nucleoside triphosphate that is deoxygenated at both the 5' and 3' positions of the pentose ring. *False*. Only the 3' position is deoxygenated.

6. The primary structure of DNA is the sequence of its bases and is always read from the 5' end to the 3' end. *True*. This direction also happens to be the direction in which the biosynthesis of DNA occurs.

7. A primer is a short section of single-stranded DNA used in the synthesis of DNA. *True*. DNA polymerase adds nucleotides to the 3' end of an existing strand, so a primer is required.

8. There is an equal number of each type of nucleoside (A, T, G, and C) in the DNA of humans. *False*. The number of A must equal the number of T, and the number of G must equal the number of C, but the number of A and T, and G and C, do not need to be equal.

9. The four heterocyclic amine bases of DNA are uracil, cytosine, guanine, and adenine. *False*. Thymine, not uracil, is found in DNA.

10. A-DNA is the most common form of DNA in living systems. *False*. B-DNA is the most common form.

11. Histones complex with DNA mainly through hydrogen bonding interactions. *False*. The complexation of histones with DNA mainly occurs by electrostatic interactions.

12. The ionizable oxygen atoms of monophosphate, diphosphate, and triphosphate groups are all protonated at biological pH. *False*. At biological pH, the groups are deprotonated.

13. Messenger RNA (mRNA) molecules carry the amino acids necessary for protein synthesis. *False*. mRNA molecules carry the sequence information of the protein to be synthesized. Amino acids are carried by transfer RNA (tRNA).

14. A codon is the amino acid that is produced by a three-letter sequence of DNA. *False*. Codons are found in mRNA, not DNA.

15. There are an equal number of purines and pyrimidines in the DNA of humans. *True*. Human DNA is double-stranded, and each base pair contains a purine and a pyrimidine.

16. A nucleoside is a compound containing a pentose bonded to a heterocyclic aromatic amine base by a β-*N*-glycosidic bond, where a molecule of phosphoric acid has been esterified with a free hydroxyl of the pentose ring. *False*. No phosphate is present in nucleosides, which contain only a sugar and a nitrogenous base.

17. The double helix is an example of the tertiary structure of DNA. *False*. The double helix is the secondary structure of DNA. Tertiary structure refers to supercoiling.

18. The carbohydrate ring in DNA is deoxygenated at the 3' position. *False*. In DNA, it is the 2' position that is deoxygenated.

End-of-Chapter Problems

Nucleosides and Nucleotides

20.7 Write the names and structural formulas for the five nitrogen bases found in nucleosides.

Notice that all five of these nucleobases are aromatic compounds and that with each, a set of tautomers can be drawn.

Purines

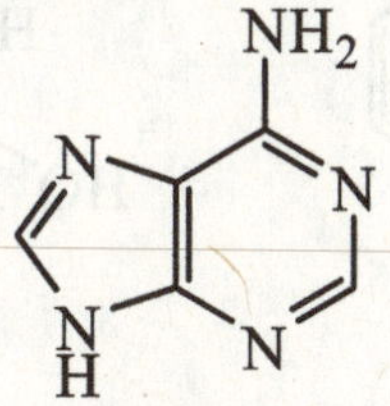

Adenine

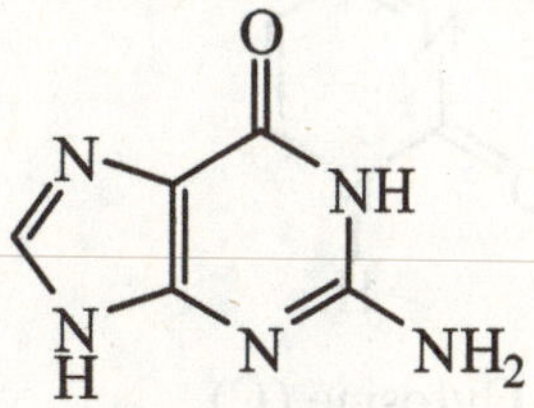

Guanine

Pyrimidines

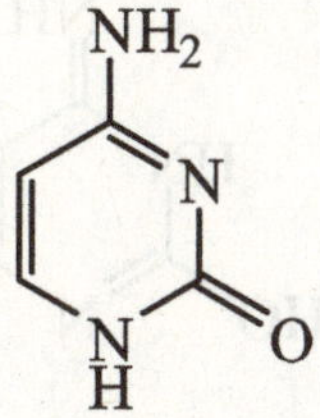

Cytosine

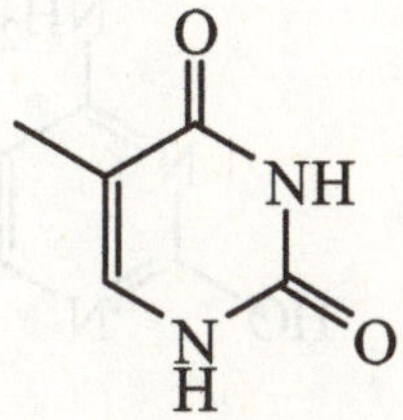

Thymine

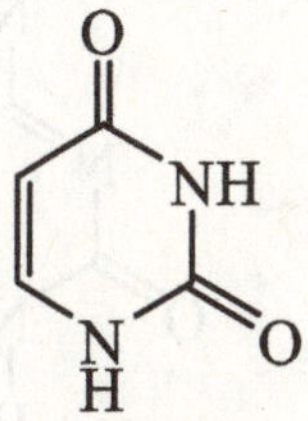

Uracil

20.8 Write the names and structural formulas for the two monosaccharides found in nucleosides.

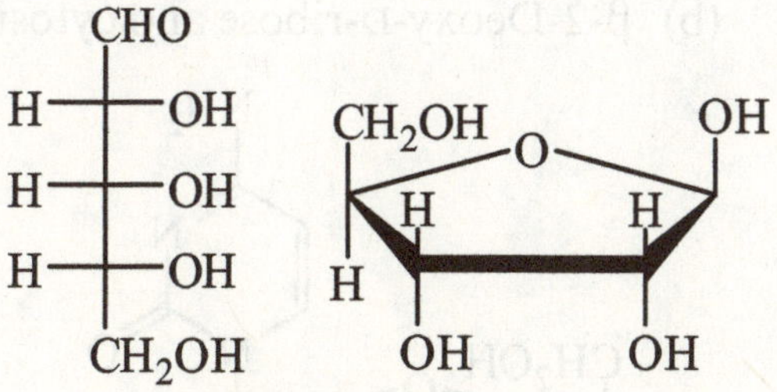

D-Ribose
(found in RNA)

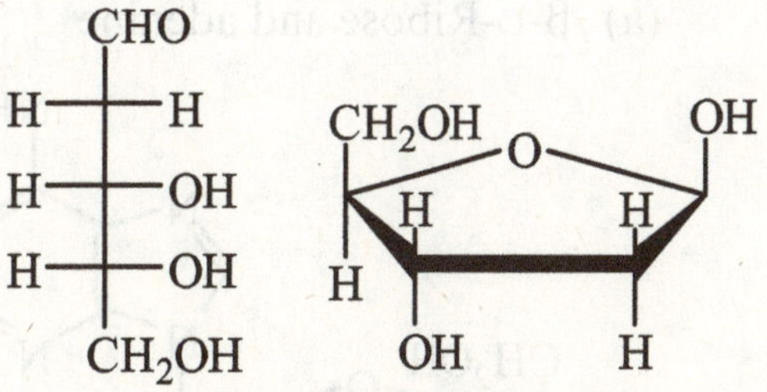

2-Deoxy-D-ribose
(found in DNA)

20.9 Explain the difference in structure between a nucleoside and a nucleotide.

A nucleoside consists of only a nucleobase and either ribose or 2-deoxyribose. On the other hand, a nucleotide is a phosphorylated nucleoside. Nucleotides therefore contain a nucleobase, either ribose or 2-deoxyribose, and one or more phosphates.

20.10 Following are structural formulas for cytosine and thymine. Draw two additional tautomeric forms for cytosine and three additional tautomeric forms for thymine.

Recall that tautomerism involves the interconversion between enol and keto forms. With respect to nucleobases, many tautomers are possible. A good approach is to examine one keto-enol pair at a time. Each of C and T has three additional tautomeric forms.

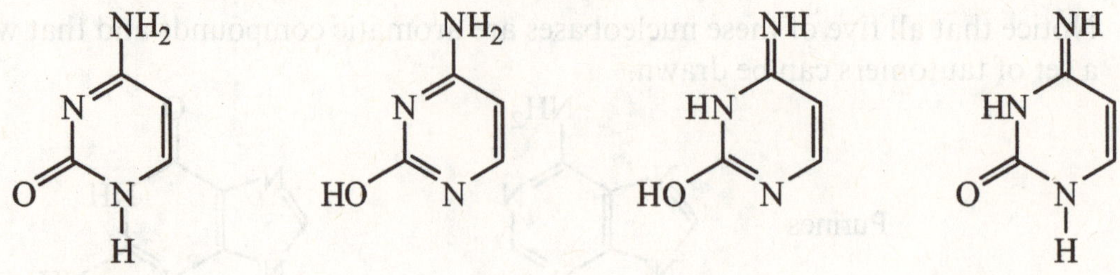

Cytosine (C)

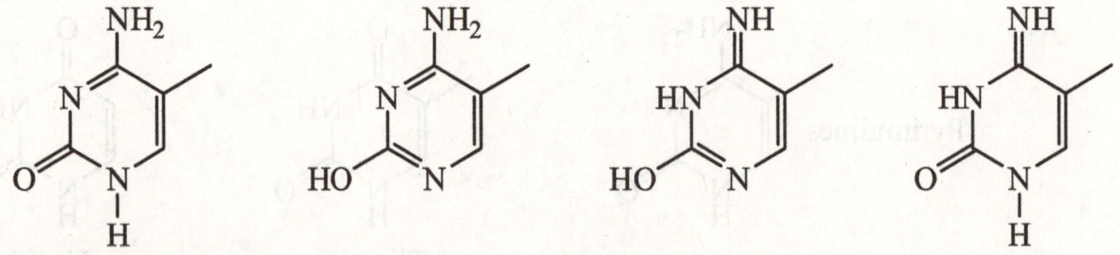

Thymine (T)

20.11 Draw a structural formula for a nucleoside composed of:

(a) β-D-Ribose and adenine

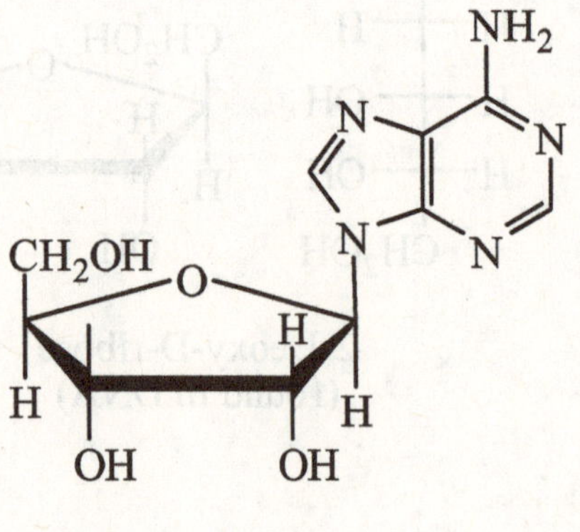

(b) β-2-Deoxy-D-ribose and cytosine

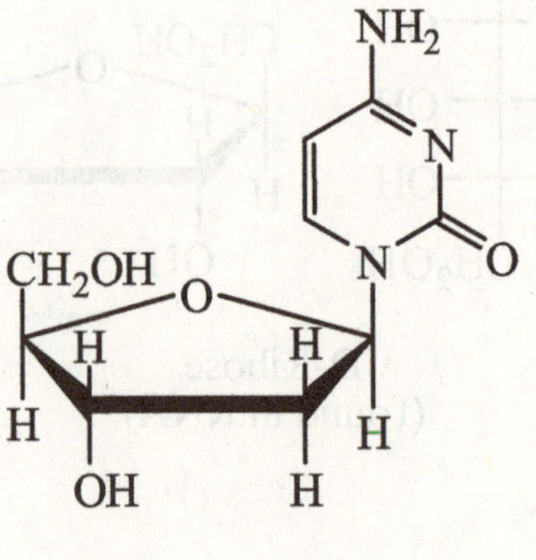

20.12 Nucleosides are stable in water and in dilute base. In dilute acid, however, the glycosidic bond of a nucleoside undergoes hydrolysis to give a pentose and a heterocyclic aromatic amine base. Propose a mechanism for this acid-catalyzed hydrolysis.

The acid-catalyzed of hydrolysis of the *N*-glycosidic bond is no different than the hydrolysis of acetals, except the leaving group is an amine instead of an alcohol. Overall, the reaction proceeds by an S_N1 mechanism. To enhance your learning, compare the similarities between the hydrolysis of nucleosides and the hydrolysis of acetals! (For convenience, we'll examine the hydrolysis of adenosine, the mechanism of which is similar for all other nucleosides.)

First, protonation of the nucleobase turns it into a good leaving group. With purines, protonation does *not* occur at the nitrogen participating in the glycosidic bond because the lone pair of that nitrogen is involved in the aromatic sextet; that nitrogen is less basic than the other nitrogen that is protonated, as shown below.

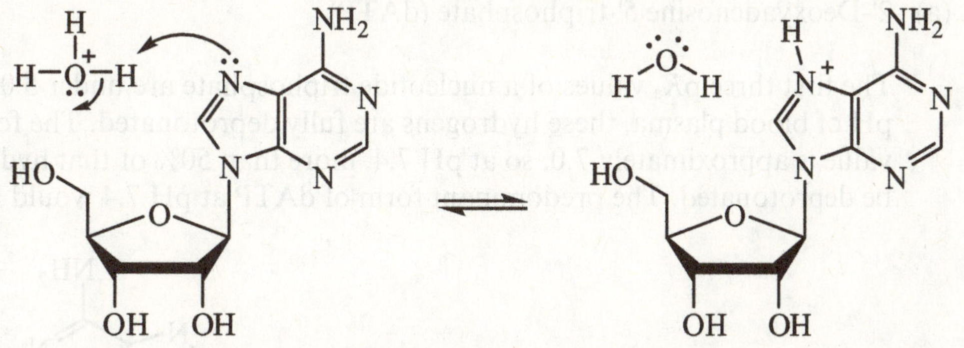

Departure of the leaving group (one of the tautomers of adenine) results in a carbocation that is stabilized by resonance. Recall that carbocations are sp^2-hybridized and are planar.

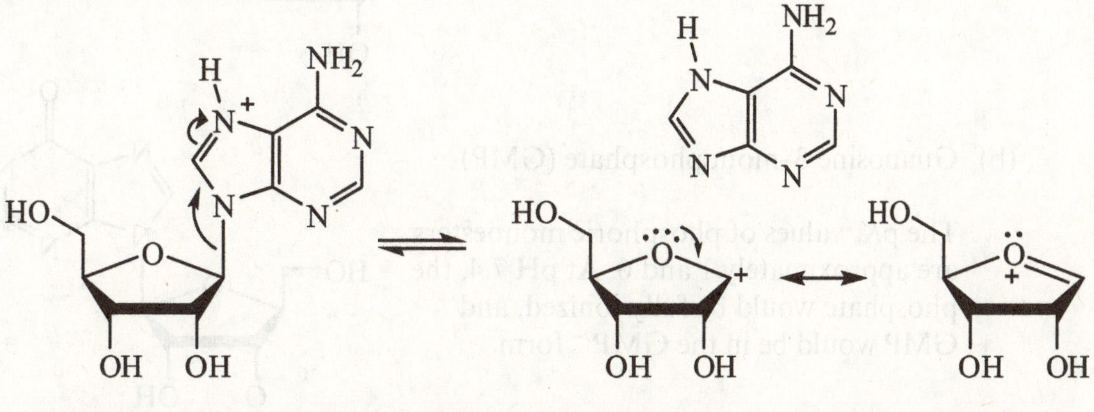

Nucleophilic attack of the carbocation by water generates a mixture protonated of α- and β-anomers of D-ribose. Subsequent deprotonation results in a cyclic hemiacetal. Note that hemiacetal pentose is also in equilibrium with its open-chain form.

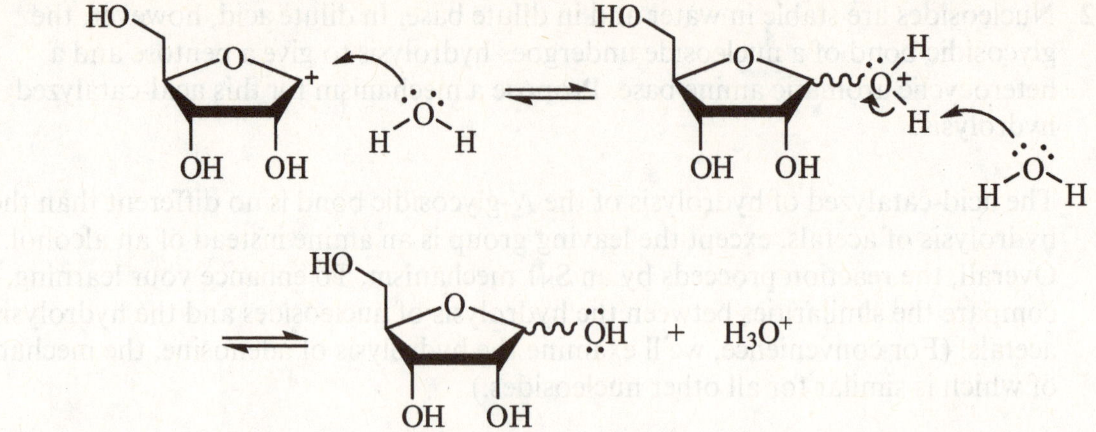

20.13 Draw a structural formula for each nucleotide and estimate its net charge at pH 7.4, the pH of blood plasma:

(a) 2'-Deoxyadenosine 5'-triphosphate (dATP)

The first three pK_a values of a nucleotide triphosphate are under 5.0, so at the pH of blood plasma, these hydrogens are fully deprotonated. The fourth pK_a value is approximately 7.0, so at pH 7.4, more than 50% of that hydrogen would be deprotonated. The predominant form of dATP at pH 7.4 would be dATP^{4-}.

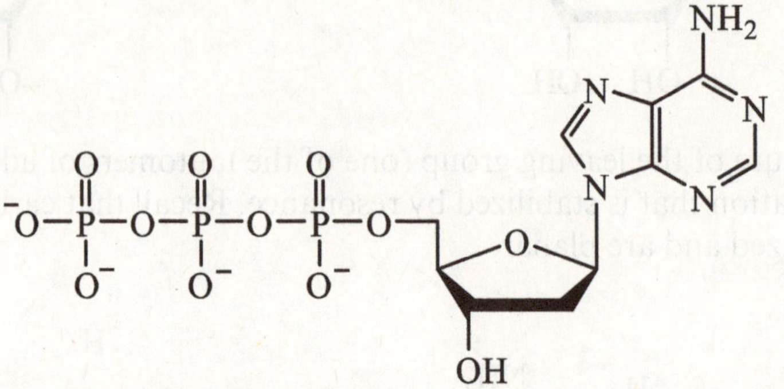

(b) Guanosine 3'-monophosphate (GMP)

The pK_a values of phosphoric monoesters are approximately 1 and 6. At pH 7.4, the phosphate would be fully ionized, and GMP would be in the GMP^{2-} form.

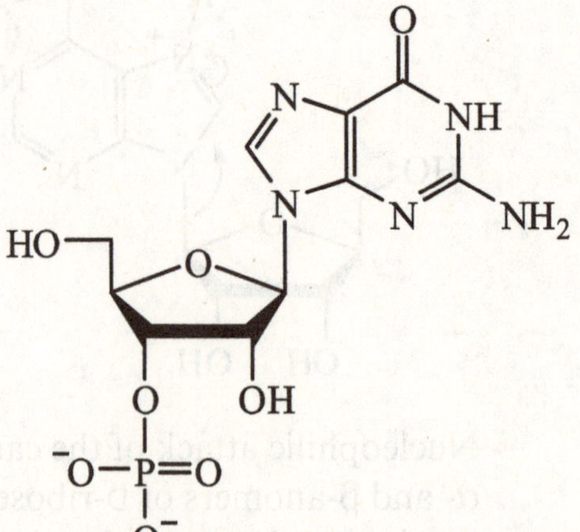

(c)   2'-Deoxyguanosine 5'-diphosphate (dGDP)

The first two pK_a values of a nucleotide diphosphate are 5.0, so those hydrogens are fully deprotonated. The third pK_a is approximately 6.7, so at pH 7.4, that hydrogen is mostly deprotonated. The predominant form of dGDP at pH 7.4 would be $dGTP^{3-}$.

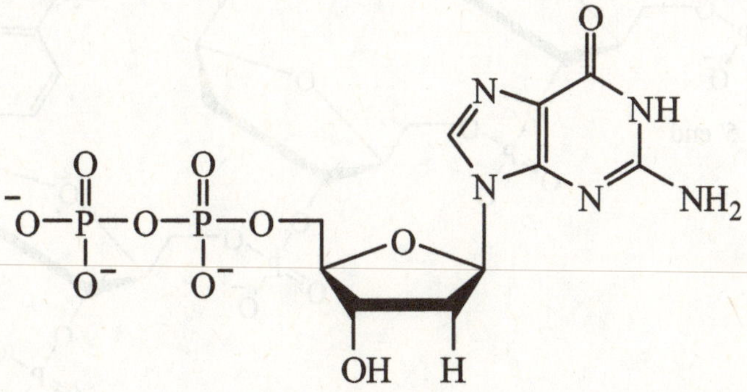

The Structure of DNA

20.14 Why are deoxyribonucleic acids called acids? What are the acidic groups in their structure?

Deoxyribonucleic acids are called acids because the phosphodiester groups of the backbone are acidic. At biological pH, the groups are fully deproneated and found in their negatively charged, conjugate base form.

20.15 Human DNA contains approximately 30.4% A. Estimate the percentages of G, C, and T and compare them with the values presented in Table 20.1.

A pairs with T, so human DNA must also contain 30.4% T. Thus, A and T comprise 60.8% of human DNA. The remaining 39.2% must be equally split between 19.6% G and 19.6% C, because G pairs with C. These values agree very well with the experimental values found in Table 20.1.

20.16 Draw a structural formula for the DNA tetranucleotide 5'-A-G-C-T-3'. Estimate the net charge on this tetranucleotide at pH 7.0. What is the complementary tetranucleotide to this sequence?

At pH 7.0, the phosphate groups are fully ionized, as shown in the structure below. The complementary tetranucleotide is 3'-T-C-G-A-5'.

20.17 List the postulates of the Watson-Crick model of DNA secondary structure.

- DNA consists of a two, antiparallel strands of polynucleotide that are coiled in a right-handed manner and arranged about the same axis to form a double helix.
- The nucleobases project inward towards the axis of the helix and are always paired in a very specific manner, A with T and G with C. (By projecting the bases inwards, the acid-labile *N*-glycosidic bonds are protected from the surrounding environment.)
- The base pairs are stacked with a spacing of 3.4 Å between them.
- There is one complete turn of the helix every 34 Å (ten base pairs per turn).

20.18 The Watson-Crick model is based on certain experimental observations of base composition and molecular dimensions. Describe these observations and show how the Watson-Crick model accounts for each.

Chargaff discovered that regardless of the organism, the percentage of A is always equal to that of T, and the percentage of G is always equal to that of C, even though the amounts of A and G can differ from organism to organism. This phenomenon is fully explained by Watson and Crick's base-pairing model. (As you've learned in introductory chemistry, the analysis of chemical composition is very important!)

The X-ray diffraction data of Franklin and Wilkins suggested that DNA had a certain periodicity and thickness, both of which are explained by the double-helical structure of Watson and Crick.

20.19 List the compositions, abbreviations, and structures of the various nucleotides in DNA.

All nucleotides in DNA are based on 2-deoxyribose that is phosphorylated at C-5, and C-1 of the sugar is linked to the nucleobase via an *N*-glycosidic bond. The nucleotides are 2-deoxyadenosine 5'-monophosphate (dAMP), 2-deoxythymidine 5'-monophosphate (dTMP), 2-deoxyguanosine 5'-monophosphate (dGMP), and 2-deoxycytidine 5'-monophosphate (dCMP).

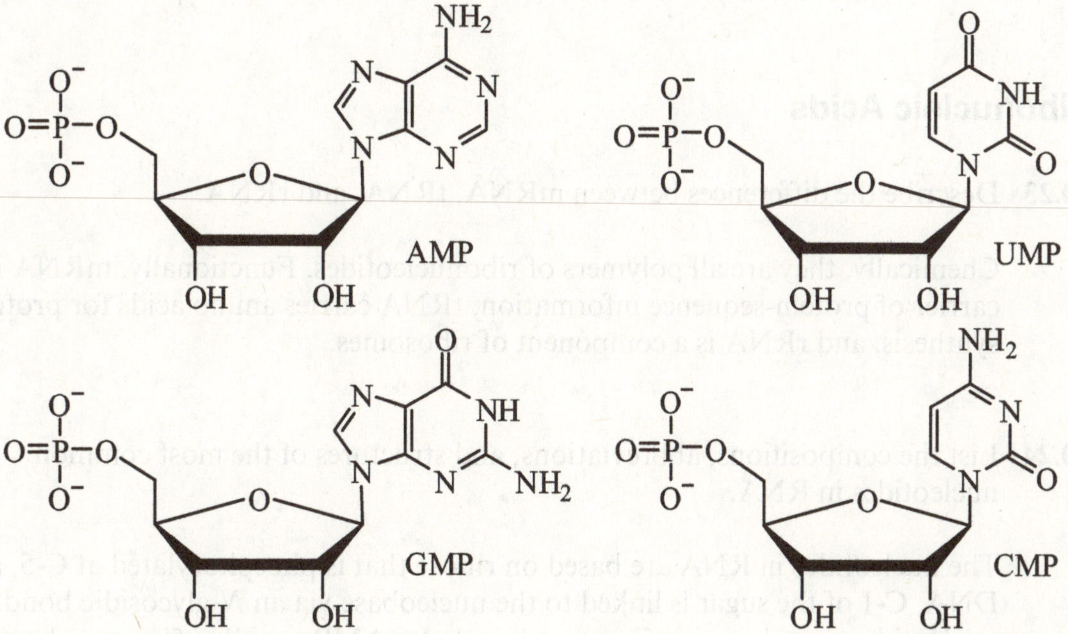

20.20 What is meant by the term "complementary bases"?

In DNA, the bases are complementary in that A always pairs with T, and G always pairs with C. That is, A and T, and G and C, complement each other.

20.21 Discuss the role of the hydrophobic interactions in stabilizing double-stranded DNA.

Recall that in the structure of the double helix, the nucleobases, which are hydrophobic, are pointed inwards. This minimizes their contact with water on the outside of the helix and also allows them to stack via hydrophobic interactions.

20.22 In terms of hydrogen bonding, which is more stable, an A-T base pair or a G-C base pair?

The stability of a base pair, or the resistance of the base pair towards unpairing, is determined by the strength of the hydrogen-bonding interactions between the two bases. An A-T base pair is held together by two hydrogen bonds, while a G-C base pair is held together by three hydrogen bonds, thus making the G-C pair more stable.

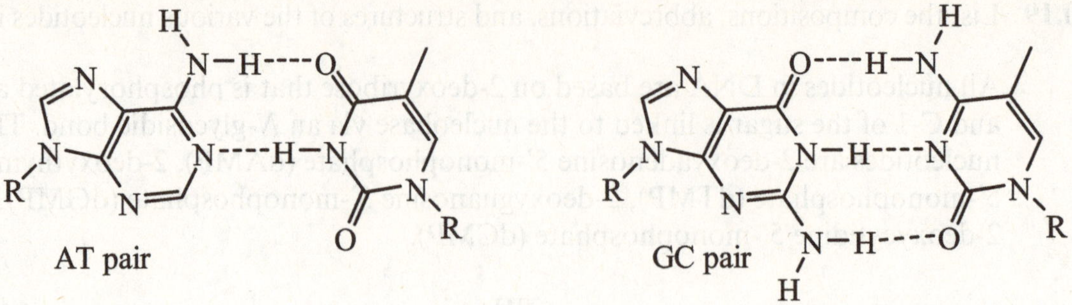

AT pair

GC pair

Ribonucleic Acids

20.23 Describe the differences between mRNA, tRNA, and rRNA.

Chemically, they are all polymers of ribonucleotides. Functionally, mRNA is a carrier of protein-sequence information, tRNA carries amino acids for protein synthesis, and rRNA is a component of ribosomes.

20.24 List the compositions, abbreviations, and structures of the most common nucleotides in RNA.

The nucleotides in RNA are based on ribose that is phosphorylated at C-5, and like DNA, C-1 of the sugar is linked to the nucleobase via an *N*-glycosidic bond. The nucleotides are adenosine 5'-monophosphate (AMP), uridine 5'-monophosphate (UMP), guanosine 5'-monophosphate (GMP), and cytidine 5'-monophosphate (CMP).

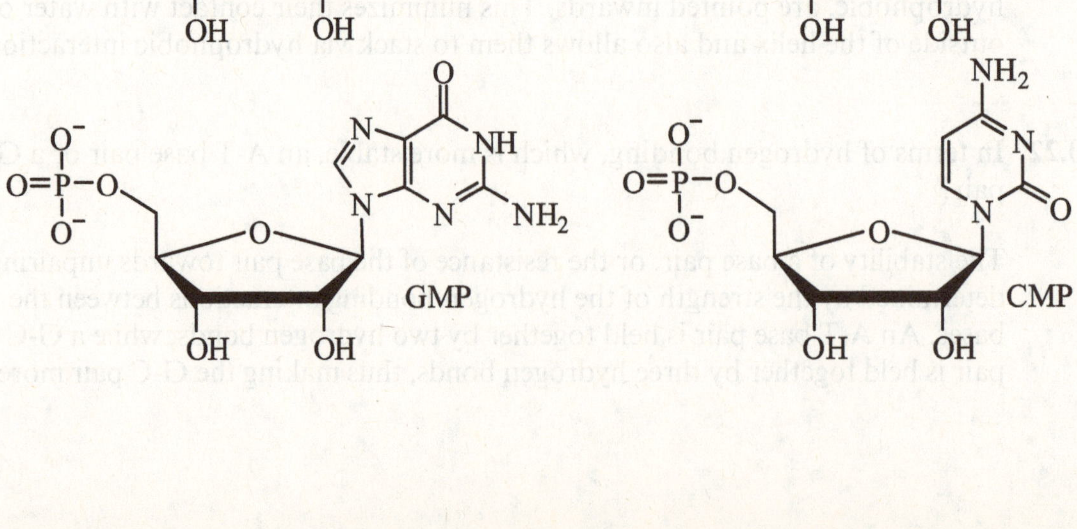

AMP

UMP

GMP

CMP

20.25 Compare the degree of hydrogen bonding in the base pair A-T found in DNA with that in the base pair A-U found in RNA.

The only difference between T and U is the absence of a methyl group in U. The absence of this methyl group has no impact on hydrogen bonding.

AT pair   AU pair

20.26 Compare DNA and RNA in these ways:

(a) Monosaccharide units

DNA contains 2-deoxy-D-ribose, while RNA contains D-ribose. In both DNA and RNA, the nucleobases are linked via *N*-glycosidic bonds in the β-configuration.

(b) Principal purine and pyrimidine bases

Both DNA and RNA use the purines A and G and the pyrimidine C. DNA uses the pyrimidine T while RNA uses the pyrimidine U.

(c) Primary structure

Primary structure refers to the sequence of nucleotides. DNA consists of A, T, G, and C, while RNA consists of A, U, G, and C. Unlike proteins, where there are 20 different amino acids, there are only 4 different nucleobases in nucleic acids.

(d) Location in the cell

In eukaryotic organisms, which include animals, plants, and fungi, DNA is found inside the nucleus and RNA is found in the cytoplasm. Whereas, prokaryotes (bacteria) do not have a nucleus, so both DNA and RNA are found in the cytoplasm.

(e) Function in the cell

In cells, DNA is a long-lived carrier of all genetic information. RNA can also carry information (as in mRNA), but this information is only the information pertaining to one protein sequence. Protein synthesis also needs tRNA, which carries amino acids, and rRNA, which is part of the ribosome and acts as a ribozyme (RNA with catalytic activity). Note that in some viruses, such as HIV, RNA is the carrier of genetic information, but viruses are not considered to be cells.

20.27   What type of RNA has the shortest lifetime in cells?

mRNA has the shortest lifetime in cells, usually on the order of a few minutes or less. This short lifetime allows the cell to have excellent control over how much protein is synthesized at any one time.

20.28   Write the mRNA complement of 5'-ACCGTTAAT-3'. Be certain to label which is the 5' end and which is the 3' end of the mRNA strand.

The mRNA complement is 3'-UGGCAAUUA-5'.

20.29   Write the mRNA complement for 5'-TCAACGAT-3'.

The mRNA complement is 3'-AGUUGCUA-5'.

The Genetic Code

20.30   What is the genetic code?

The genetic code refers to the sequence of the nucleotides. Each sequence of three consecutive nucleotides (one unit) codes for a particular amino acid.

20.31   Why are at least three nucleotides needed for one unit of the genetic code?

There are 20 amino acids that are specified by the genetic code. With three nucleotides, 64 different sequence combinations are possible; for each nucleotide in the unit, there are A, T, G, and C to choose from. If only two nucleotides were present in each unit, only 16 combinations would be possible.

20.32   Briefly outline the biosynthesis of proteins, starting from DNA.

First, the DNA template is used to synthesize mRNA (a process known as transcription), tRNA, and rRNA. rRNA is contained in ribosomes, which catalyze the synthesis of proteins by "reading" the mRNA (a process known as translation) and using tRNAs that are carrying specific amino acids.

20.33 How does the protein-synthesizing machinery of the cell know when a genetic sequence is complete?

Three codons, UAA, UAG, and UGA, are known as the "stop codons" and they indicate that the primary sequence of the protein is complete.

20.34 At elevated temperatures, nucleic acids become denatured; that is, they unwind into single-stranded DNA. Account for the observation that the higher the G-C content of a nucleic acid, the higher the temperature required for its thermal denaturation.

DNA normally exists as a double-stranded helix. In order to form single-stranded DNA, the hydrogen-bonding interactions between the two strands must be overcome. G-C base pairs each contain three hydrogen bonds, compared to just two in each A-T base pair. Naturally, more energy (higher temperature) is required to break the G-C base pair.

20.35 Write the DNA complement of 5'-ACCGTTAAT-3'. Be certain to label which is the 5' end and which is the 3' end of the complement strand.

The DNA complement is 3'-TGGCAATTA-5'.

20.36 Write the DNA complement for 5'-TCAACGAT-3'.

The DNA complement is 3'-AGTTGCTA-5'.

20.37 What does it mean to say that the genetic code is degenerate?

The genetic code is degenerate because there are certain amino acids that are coded for by more than one codon. For example, all of AUU, AUC, and AUA code for the same amino acid, isoleucine. However, it is important to note that each codon codes for only one single amino acid.

20.38 Aspartic acid and glutamic acid have carboxyl groups on their side chains and are called acidic amino acids. Compare the codons for these two amino acids.

Interestingly, the codons for these two acidic amino acids all begin with GA. The codons for Asp are GAU and GAC, while those for Glu are GAA and GAG. Even more interesting is that for Asp, the third nucleotide in both codons is a pyrimidine, and similarly, for Glu, the third nucleotide is a purine.

20.39 Compare the structural formulas of the aromatic amino acids phenylalanine and tyrosine. Compare also the codons for these two amino acids.

Both Phe and Tyr are structurally similar, except the latter contains a hydroxyl group on the aromatic ring. The codons for Phe are UUU and UUC, while those for Tyr are UAU and UAC; the codons for the two amino acids differ only in the second position.

20.40 Glycine, alanine, and valine are classified as nonpolar amino acids. Compare their codons. What similarities do you find? What differences?

Glycine	GGU	GGC	GGA	GGG	GGX
Alanine	GCU	GCC	GCA	GCG	GCX
Valine	GUU	GUG	GUA	GUG	GUX

All of these three amino acids each have four codons. All codons start with G, and in each case, the first two nucleotides are identical for a given amino acid. As a result, the third nucleotide (X) in each codon is irrelevant. By now, you've probably noticed that there are trends between the structural (and chemical) properties of the amino acids and the sequences of the codons!

20.41 Codons in the set CUU, CUC, CUA, and CUG all code for the amino acid leucine. In this set, the first and second bases are identical, and the identity of the third base is irrelevant. For what other sets of codons is the third base also irrelevant, and for what amino acid(s) does each set code?

As seen in Problem 20.40, the last base in the codons for Gly, Ala, and Val is irrelevant. Other codons in which the third base is irrelevant include those for Arg (CGX), Pro (CCX), and Thr (ACX).

20.42 Compare the amino acids coded for by the codons with a pyrimidine, either U or C, as the second base. Do the majority of the amino acids specified by these codons have hydrophobic or hydrophilic side chains?

The majority of the amino acids that have a pyrimidine in the second position of their codons are hydrophobic; these amino acids include Phe, Leu, Ile, Met, Val, Pro, and Ala. Only Ser and Thr, which are hydrophillic, contain a pyrimidine in the second position.

20.43 Compare the amino acids coded for by the codons with a purine, either A or G, as the second base. Do the majority of the amino acids specified by these codons have hydrophilic or hydrophobic side chains?

With the exception of Trp and Gly, all codons with a purine in the second position code for polar, hydrophilic side chains.

20.44 What polypeptide is coded for by this mRNA sequence?

$$5'\text{-GCU-GAA-GUC-GAG-GUG-UGG-}3'$$

This mRNA sequence codes for the peptide Ala-Glu-Val-Glu-Val-Trp, as written according to standard convention (*N*-terminal on the left side).

20.45 The alpha chain of human hemoglobin has 141 amino acids in a single polypeptide chain. Calculate the minimum number of bases on DNA necessary to code for the alpha chain. Include in your calculation the bases necessary for specifying termination of polypeptide synthesis.

Each amino acid requires one codon (three nucleotides). Therefore, $3 \times 141 = 423$ bases are required for the amino acids alone, plus another three for the stop codon, giving a total of 426 bases.

20.46 In HbS, the human hemoglobin found in individuals with sickle-cell anemia, glutamic acid at position 6 in the beta chain is replaced by valine.

(a) List the two codons for glutamic acid and the four codons for valine.

Glu   GAA GAG
Val   GUA GUG GUU GUA

(b) Show that one of the glutamic acid codons can be converted to a valine codon by a single substitution mutation; that is, by changing one letter in one codon.

Both codons for Glu can converted to two Val codons by changing the central A residue in the middle to a U. This seemingly tiny change is detrimental to the proper function of hemoglobin – think about the chemical differences between the side chains of Glu and Val.

Additional Problems

20.47 Two drugs used in the treatment of acute leukemia are 6-mercaptopurine and 6-thioguanine. Note that, in each drug, the oxygen at carbon 6 of the parent molecule is replaced by divalent sulfur. Draw structural formulas for the enethiol (the sulfur equivalent of an enol) forms of 6-mercaptopurine and 6-thioguanine.

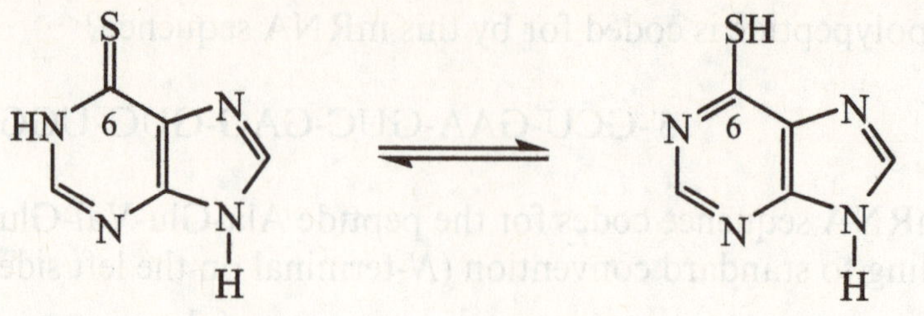

6-Mercaptopurine

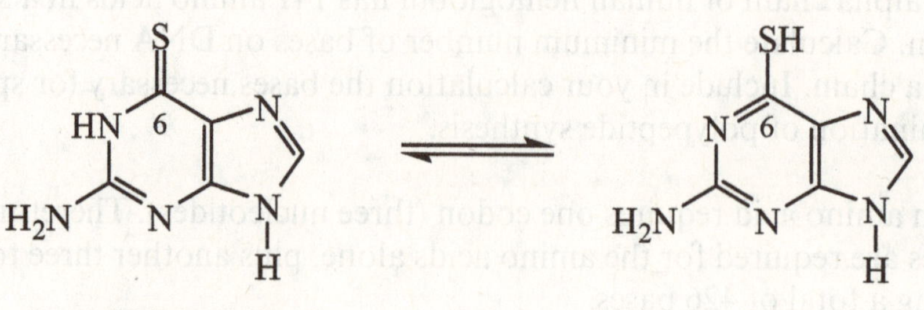

6-Thioguanine

20.48 Cyclic-AMP, first isolated in 1959, is involved in many diverse biological processes as a regulator of metabolic and physiological activity. In this compound, a single phosphate group is esterified at both the 3' and 5' hydroxyls of adenosine. Draw a structural formula of cyclic-AMP.

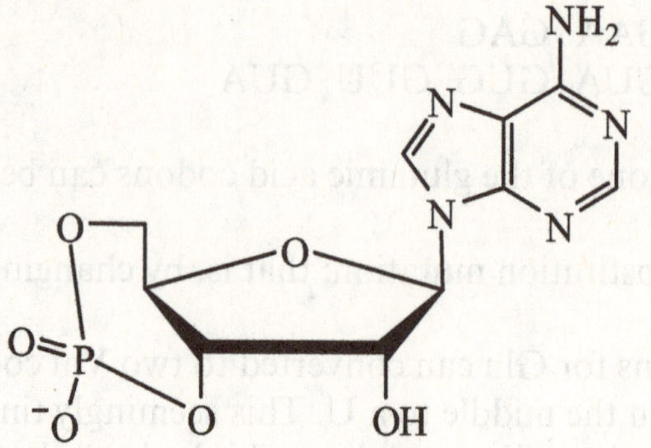

20.49 Compare the α-helix of proteins and the double helix of DNA in terms of:

(a) The units that repeat in the backbone of the polymer chain.

In the α-helices of proteins, the repeating units are amino acids that are linked by peptide (amide) bonds. Whereas, the repeating units in DNA are 2'-deoxy-D-ribose linked via 3',5'-phosphodiester bonds. The backbones of proteins and DNA are thus very different, but this is no surprise given their different biological functions.

(b) The projection in space of substituents along the backbone (the R groups in the case of amino acids, purine and pyrimidine bases in the case of double-stranded DNA)relative to the axis of the helix.

The R groups in the α-helices of proteins point outward from the helix, whereas the nucleobases in the DNA double helix point inward and away from the aqueous environment of the cell.

20.50 The loss of three consecutive units of T from the gene that codes for CFTR, a transmembrane conductance regulator protein, results in the disease known as cystic fibrosis. Which amino acid is missing from CFTR to cause this disease?

The loss of three consecutive T residues (TTT) from the gene would result in the loss of the codon AAA on the complementary strand of mRNA. The codon AAA codes for Lys, which is the missing amino acid in CFTR in cystic fibrosis.

20.51 The following compounds have been researched as potential anti-viral agents. Suggest how each of these compounds might block the synthesis of RNA or DNA.

(a) Cordycepin (3'-deoxyadenosine)

3'-Deoxyadenosine mimics the adenosine nucleoside that is used to synthesis RNA. However, it is missing the 3'-OH group, so it acts as a chain terminator; RNA elongation cannot continue unless a 3'-OH group is present.

(b) 2,5,6-Trichloro-1-(β-D-ribofuranolsyl)benzimidazole

The trichlorinated benzimidazole fragment present in this compound, an RNA nucleoside analog, mimics a purine base. This compound likely interferes with RNA polymerase, the enzyme that transcribes RNA from DNA.

(c) 9-(2,3-Dihydroxypropyl)adenine

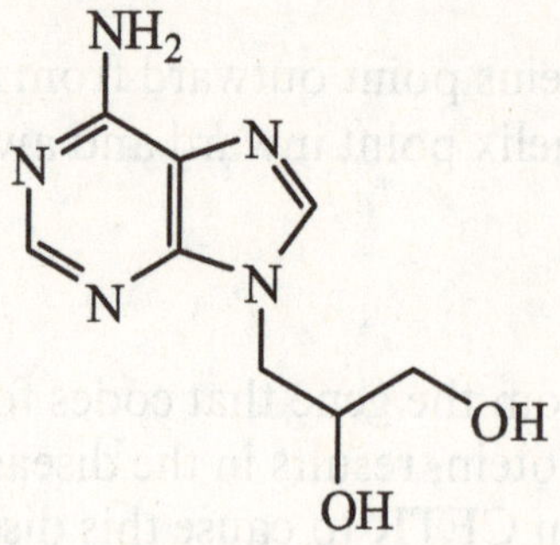

Adenosine

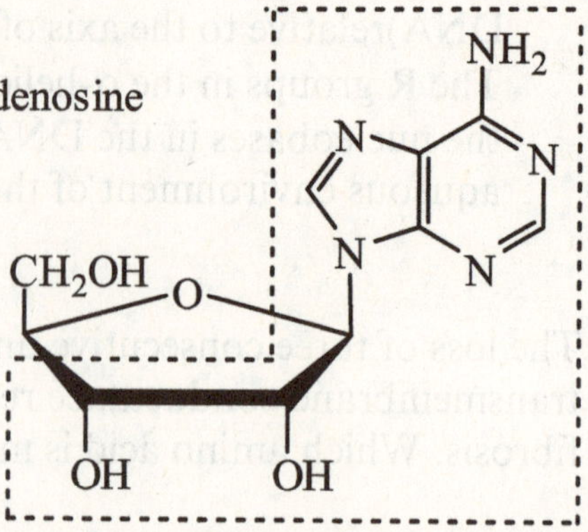

9-(2,3-Dihydroxypropyl)adenine is an analog of adenosine and likely interferes
with the enzymes involved in nucleic acid synthesis.

20.52 Name the type of covalent bond(s) joining monomers in these biopolymers.

(a) Polysaccharides   Glycosidic bonds (linkages)
(b) Polypeptides   Peptide (amide) bonds
(c) Nucleic acids   Phosphate ester bonds

20.53 The ends of chromosomes, called telomeres, can form unique and nonstandard
structures. One example is the presence of base pairs between units of guanosine.
Show how the guanine base can pair with another via hydrogen bonding.

Guanine

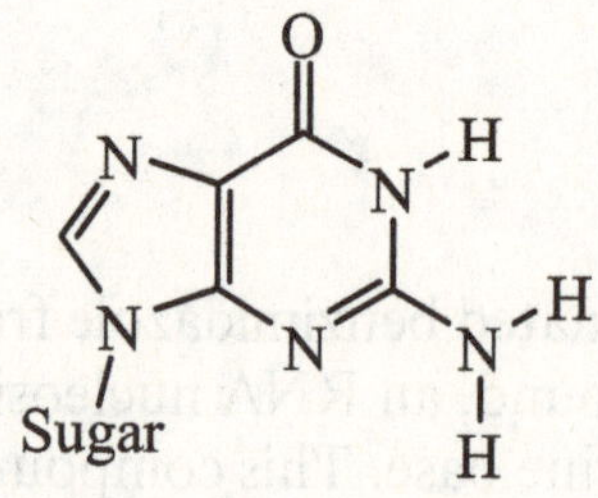

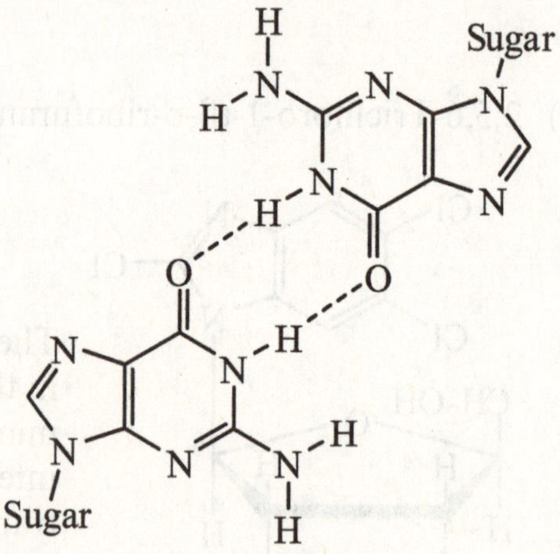

20.54 One synthesis of zidovudine (AZT) involves the following reaction (DMF is the solvent *N,N*-dimethylformamide). What type of reaction is this?

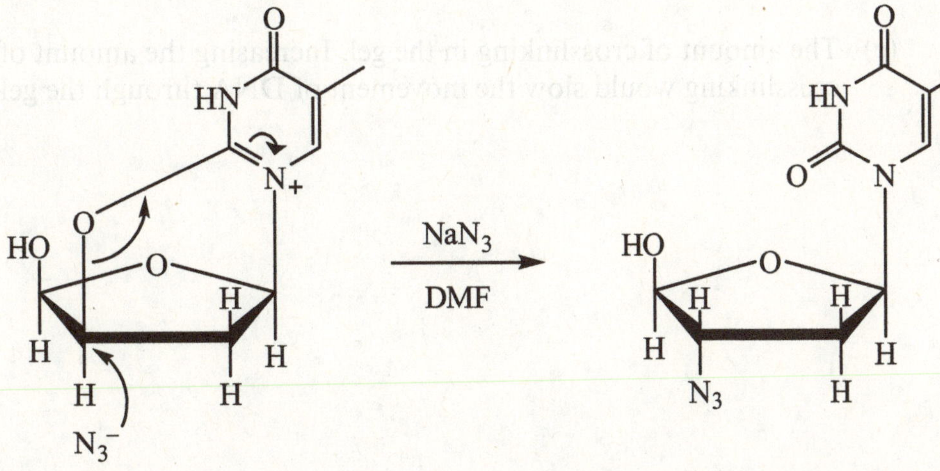

Notice the details of this reaction: azide (N_3^-), a powerful nucleophile, replaces the –OR group on the 3' carbon; an inversion of stereochemistry occurs at this carbon; and DMF is a polar, aprotic solvent. All of these details are supportive of an S_N2 reaction.

Group Learning Activities

20.55 Using your knowledge of the 1° and 2° structure of DNA and the principles learned thus far, discuss the following questions with your study group:

(a) Would you expect DNA to be more soluble in water or in ethanol? Why?

DNA is an ionic compound and is expected to be more soluble in water than in ethanol. Water is more polar than ethanol.

(b) If DNA were placed a solution of water and ethanol, how might changing the ratio of the two solvents affect the 2° structure of DNA?

Increasing the amount of ethanol in the solution is expected to increase the amount of A-DNA and decrease the amount of B-DNA. Water is required for B-DNA to maintain its structure.

(c) In Section 20.5, we saw how electrophoresis can be used to sequence DNA. How might changes in the following conditions affect how quickly or slowly a strand of DNA moves through the gel?

(a) The voltage applied across the gel. Increasing the voltage across the gel increases the charge difference between the two ends of the gel. This larger charge difference increases the rate at which DNA moves through the gel.

(b) The pH of the medium in the gel. The pH affects the overall charge of DNA. At low pH, more phosphate groups would be protonated and therefore neutrally charged, causing DNA to move slower through the gel.

(c) The amount of crosslinking in the gel. Increasing the amount of crosslinking would slow the movement of DNA through the gel.

Chapter 21: The Organic Chemistry of Metabolism

Problems

21.1   Under anaerobic (without oxygen) conditions, glucose is converted to lactate by a metabolic pathway called anaerobic glycolysis or, alternatively, lactate fermentation. Is anaerobic glycolysis a net oxidation, a net reduction, or neither?

$$C_6H_{12}O_6 \xrightarrow[\text{glycolysis}]{\text{anaerobic}} 2CH_3\overset{\displaystyle OH}{\underset{}{C}}HCOO^- + 2H^+$$

Anaerobic glycolysis is neither a net oxidation nor a net reduction. There is no change in the number of hydrogen or oxygen atoms when glucose is converted to two equivalents of lactic acid. Furthermore, the half-reaction written above does not require electrons on either side for charge balance, so the reaction is neither oxidation nor reduction.

21.2   Does lactate fermentation result in an increase or a decrease in blood pH?

Lactate fermentation results in the production of lactic acid, which is fully ionized to the lactate and H_3O^+ under at pH 7.4. As a result of H_3O^+, blood pH decreases.

Quick Quiz

1.   Fatty acids are metabolized through the process of β-oxidation. *True*. The metabolism of fatty acids involves the oxidation of the β-carbon, hence the <u>name</u> β-oxidation.

2. NAD^+, NADH, FAD, and $FADH_2$ are coenzymes that undergo oxidation and reduction during metabolism. *True*. NAD^+ and FAD are oxidizing agents, while NADH and $FADH_2$ are reducing agents.

3.   An end product of glycolysis is pyruvate. *True*. Although pyruvate is subsequently converted into other compounds, those reactions are not part of glycolysis.

4.   The starting point of the citric acid cycle involves acetyl CoA-SH, which can be formed from carbohydrates, triglycerides, or proteins. *True*. The citric acid cycle is a central metabolic pathway, and it uses the acetyl CoA-SH formed from the metabolism of different groups of biomolecules.

5.   The conversion of a $-CH_2-CH_2-$ unit to $-CH=CH-$ by FAD is a reduction reaction. *False*. Hydrogens have been removed, so the reaction is an oxidation reaction.

6. Lactate fermentation allows pyruvate to be metabolized under aerobic conditions. *False*. Fermentation is an anaerobic process.

7. All carbohydrates are directly metabolized through the process of glycolysis. *False*. Glycolysis only pertains to the metabolism of glucose. All other carbohydrates must first be converted to glucose before glycolysis can occur.

8. Keto-enol tautomerism is an important reaction in glycolysis. *True*. The conversion of phosphoenolpyruvate to pyruvate, which generates ATP, involves a keto-enol tautomerism.

9. Coenzyme A is used in metabolism to store and transfer hydroxyl groups. *False*. Coenzyme A is a carrier of acetyl groups, not hydroxyl groups.

10. Coenzyme A contains a terminal $-OH$ group, which forms a new bond to acetyl groups in the degradation of monosaccharides, fatty acids, glycerol, and amino acids. *False*. Coenzyme A contains a terminal thiol group that forms a thioester with an acetyl group.

11. Alcoholic fermentation is one possible means of pyruvate metabolism. *True*. Yeasts and a few other organisms ferment pyruvate to ethanol.

12. ATP, ADP, and AMP are involved in the storage and transfer of adenine groups. *False*. They are involved in the storage and transfer of phosphate groups. The phosphate groups are linked as phosphoric anhydrides.

13. NAD^+ acts to oxidize both hydroxyl and carbonyl groups. *True*. NAD^+ is a very versatile biological oxidizing agent that functions as a hydride acceptor.

End-of-Chapter Problems

Glycolysis

21.3 In many enzyme-catalyzed reactions, a group on the enzyme functions as a proton donor. List several amino acid side chains that could function as proton donors.

Side chains that commonly function as proton donors are those of the acidic amino acids, which are aspartic acid and glutamic acid, as well as those of the conjugate acids of histidine, lysine, and arginine. Less frequently, the side chains of serine and cysteine can act as proton donors.

21.4 In many other enzyme-catalyzed reactions, a group on the enzyme surface functions as a proton acceptor. List several amino acid side chains that might function as proton acceptors.

Side chains that might function as proton acceptors are those of the basic amino acids, which are histidine, lysine, and arginine. In addition, the conjugate bases of the side chains of aspartic acid, glutamic acid, serine, and cysteine can also be proton acceptors.

21.5 Name one coenzyme required for glycolysis. From what vitamin is the coenzyme derived?

One coenzyme required for glycolysis (the oxidation steps) is nicotinamide adenine dinucleotide (NAD^+), and it is derived from the vitamin niacin.

21.6 Number the carbons of glucose 1 through 6. Which carbons of glucose become the carboxyl groups of the two pyruvates?

Carbons three and four of glucose become the carboxyl groups of the two pyruvates.

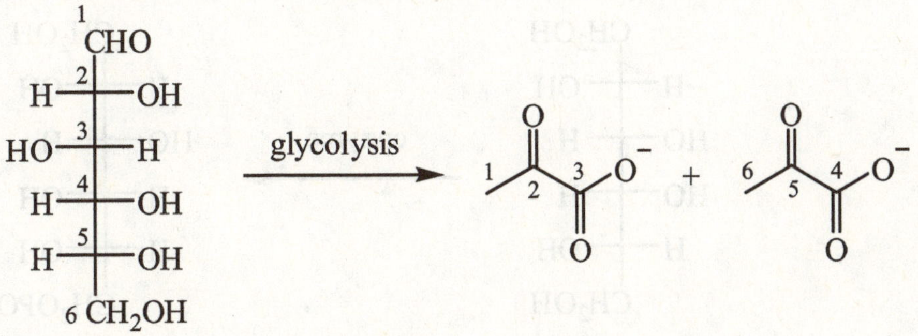

21.7 How many moles of lactate are produced from 3 moles of glucose?

In anaerobic fermentation, one mole of lactate is formed from one mole of pyruvate. Each mole of glucose forms two moles of pyruvate. Thus, three moles of glucose produces six moles of lactate.

21.8 Although glucose is the principal source of carbohydrates for glycolysis, fructose and galactose are also metabolized for energy.

(a) What is the main dietary source of fructose? Of galactose?

Surcrose, a disaccharide consisting of glucose and fructose, is the main dietary source of fructose. The main dietary source of galactose is lactose, a disaccharide of glucose and galactose.

(b) Propose a series of reactions by which fructose might enter glycolysis.

$$
\begin{array}{ccc}
\begin{array}{c}
\text{CH}_2\text{OH} \\
| \\
\text{C}=\text{O} \\
\text{HO}-\!\!\!-\text{H} \\
\text{H}-\!\!\!-\text{OH} \\
\text{H}-\!\!\!-\text{OH} \\
| \\
\text{CH}_2\text{OH}
\end{array}
&
\xrightarrow{\text{phosphorylation}}
&
\begin{array}{c}
\text{CH}_2\text{OH} \\
| \\
\text{C}=\text{O} \\
\text{HO}-\!\!\!-\text{H} \\
\text{H}-\!\!\!-\text{OH} \\
\text{H}-\!\!\!-\text{OH} \\
| \\
\text{CH}_2\text{OPO}_3^{2-}
\end{array} \\
\text{D-Fructose} & & \text{D-Fructose 6-phosphate}
\end{array}
$$

Fructose is phosphorylated to form fructose 6-phosphate, which can enter glycolysis at the third reaction, where it will then be converted to fructose 1,6-bisphosphate.

(c) Propose a series of reactions by which galactose might enter glycolysis.

$$
\begin{array}{ccc}
\begin{array}{c}
\text{CH}_2\text{OH} \\
\text{H}-\!\!\!-\text{OH} \\
\text{HO}-\!\!\!-\text{H} \\
\text{HO}-\!\!\!-\text{H} \\
\text{H}-\!\!\!-\text{OH} \\
| \\
\text{CH}_2\text{OH}
\end{array}
&
\xrightarrow{\text{enzyme}}
&
\begin{array}{c}
\text{CH}_2\text{OH} \\
\text{H}-\!\!\!-\text{OH} \\
\text{HO}-\!\!\!-\text{H} \\
\text{H}-\!\!\!-\text{OH} \\
\text{H}-\!\!\!-\text{OH} \\
| \\
\text{CH}_2\text{OPO}_3^{2-}
\end{array} \\
\text{D-Galactose} & & \text{D-Glucose}
\end{array}
$$

Galactose can be converted to glucose, which are epimers of each other, by the enzyme-catalyzed epimerization of the hydroxyl group at carbon 4.

21.9 How many moles of ethanol are produced per mole of sucrose through the reactions of glycolysis and alcoholic fermentation? How many moles of CO_2 are produced?

Sucrose is first hydrolyzed into two six-carbon monosaccharides, glucose and fructose. Each of these enters glycolysis to give two moles of pyruvate, so a total of four moles of pyruvate are formed from one mole of sucrose. Each mole of pyruvate is fermented into one mole of ethanol and one mole of CO_2. Accordingly, each mole of sucrose forms four moles of ethanol and four moles of CO_2.

21.10 Glycerol that is derived from the hydrolysis of triglycerides and phospholipids is also metabolized for energy. Propose a series of reactions by which the carbon skeleton of glycerol might enter glycolysis and be oxidized to pyruvate.

Glycerol can be converted to dihydroxyacetone phosphate, an intermediate in glycolysis, by enzyme-catalyzed phosphorylation and oxidation.

Glycerol

Dihydroxyacetone phosphate

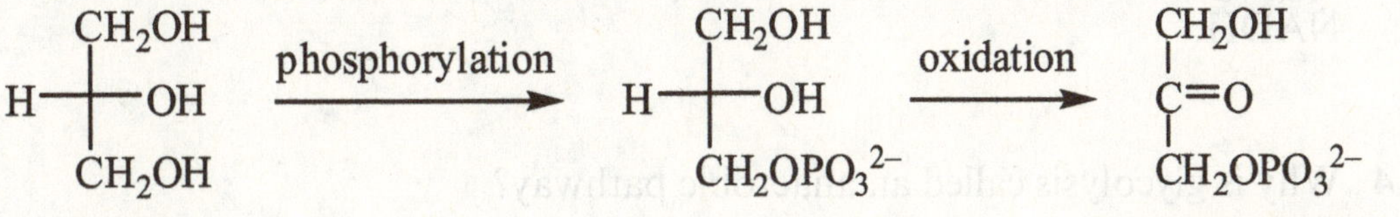

21.11 Write a mechanism to show the role of NADH in the reduction of acetaldehyde to ethanol.

NADH functions as a hydride (H⁻) donor. The hydride nucleophile is accepted by the carbonyl group of acetaldehyde, and this step is concomitant with the protonation of the carbonyl group by a proton donor present in the enzyme's active site.

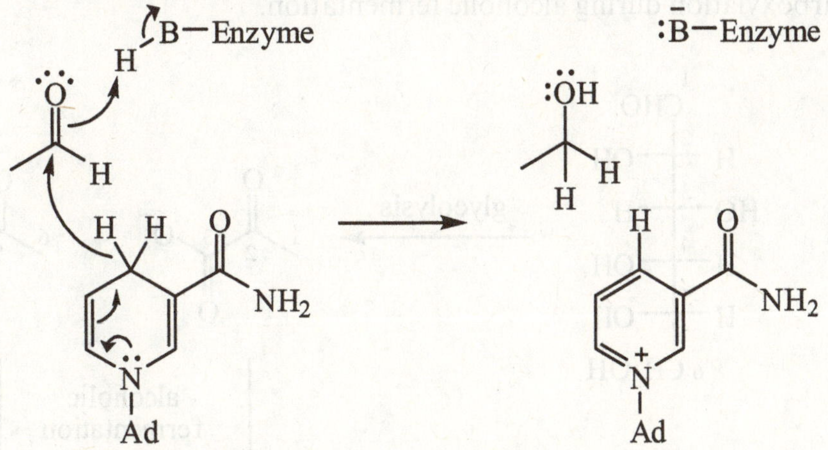

21.12 Ethanol is oxidized in the liver to acetate ion by NAD^+.

(a) Write a balanced equation for this oxidation.

$$CH_3CH_2OH + 2NAD^+ + H_2O \rightarrow CH_3COO^- + 2NADH + 3H^+$$

(b) Do you expect the pH of blood plasma to increase, decrease, or remain the same as a result of metabolism of a significant amount of ethanol?

The pH is expected to decrease because the metabolism of ethanol produces H^+.

21.13 When pyruvate is reduced to lactate by NADH, two hydrogens are added to pyruvate: one to the carbonyl carbon, the other to the carbonyl oxygen. Which of these hydrogens is derived from NADH?

As seen in Problem 21.11, NADH is a hydride donor. Hydride is a strong nucleophile and reacts with the electrophilic carbon of the carbonyl group. The hydrogen that is added to the carbonyl carbon is therefore the one derived from NADH.

21.14 Why is glycolysis called an anaerobic pathway?

Glycolysis is called an anaerobic pathway because no oxygen is involved. From an evolutionary perspective, glycolysis was likely used by organisms that existed during the first billion or so years of life on Earth, during which time there was no oxygen in the air.

21.15 Which carbons of glucose end up in CO_2 as a result of alcoholic fermentation?

As seen in Problem 21.6, carbons C3 and C4 of glucose end up as the carboxylate carbons of pyruvate. These same two carbons are removed as CO_2 through decarboxylation during alcoholic fermentation.

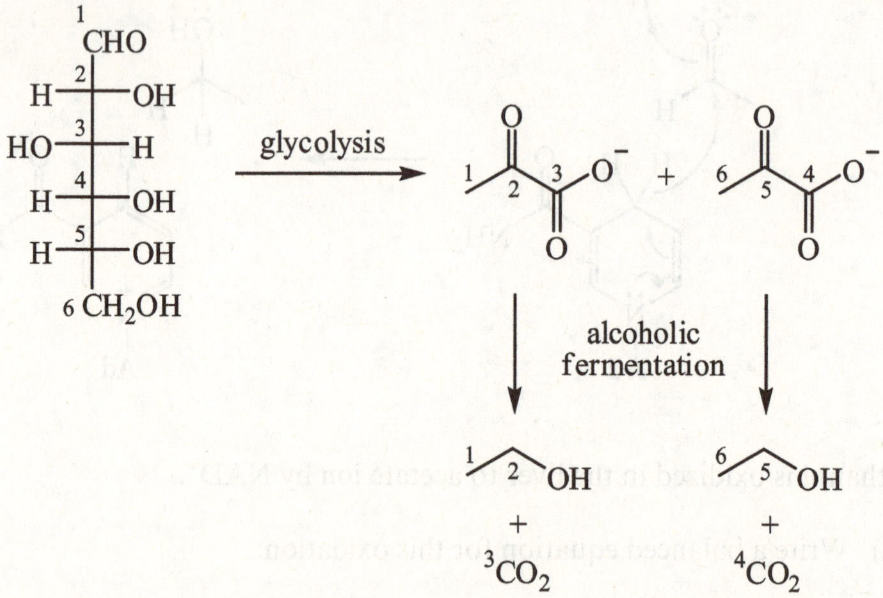

21.16 Which steps in glycolysis require ATP? Which steps produce ATP?

Reactions 1 and 3 of glycolysis (see Figure 21.1) consume two equivalents of ATP for per glucose. Reactions 7 and 10 produce four equivalents of ATP for per glucose.

β-Oxidation

21.17 Write structural formulas for palmitic, oleic, and stearic acids, the three most abundant fatty acids.

Palmitic acid (C$_{16}$)

Stearic acid (C$_{18}$)

Oleic acid (C$_{18}$)

21.18 A fatty acid must be activated before it can be metabolized in cells. Write a balanced equation for the activation of palmitic acid.

The activation of a fatty acid involves its conversion to a fatty acid-AMP mixed anhydride, which is subsequently converted to the CoA thioester.

$$CH_3(CH_2)_{14}COO^- + ATP + CoASH \rightarrow$$
$$CH_3(CH_2)_{14}COSCoA + AMP + P_2O_7^{4-} + H^+$$

21.19 Name three coenzymes necessary for β-oxidation of fatty acids. From what vitamin is each derived?

The three coenzymes and the vitamins from which they are derived are FAD (riboflavin), NAD$^+$ (niacin), and coenzyme A (pantothenic acid). All three of these coenzymes contain the heterocyclic aromatic amine nucleobase adenosine.

21.20 We have examined β-oxidation of saturated fatty acids, such as palmitic acid and stearic acid. Oleic acid, an unsaturated fatty acid, is also a common component of dietary fats and oils. This unsaturated fatty acid is degraded by β-oxidation, but, at one stage in its degradation, requires an additional enzyme named enoyl-CoA isomerase. Why is this enzyme necessary, and what isomerization does it catalyze? (*Hint:* Consider both the configuration of the carbon-carbon double bond in oleic acid and its position in the carbon chain.)

The carbon-carbon double bond in oleic acid is *cis*. However, the carbon-carbon double bonds involved in β-oxidation are of the *trans* configuration. After three rounds of β-oxidation, enoyl-CoA isomerase performs a *cis*-to-*trans* isomerization reaction so that β-oxidation can continue.

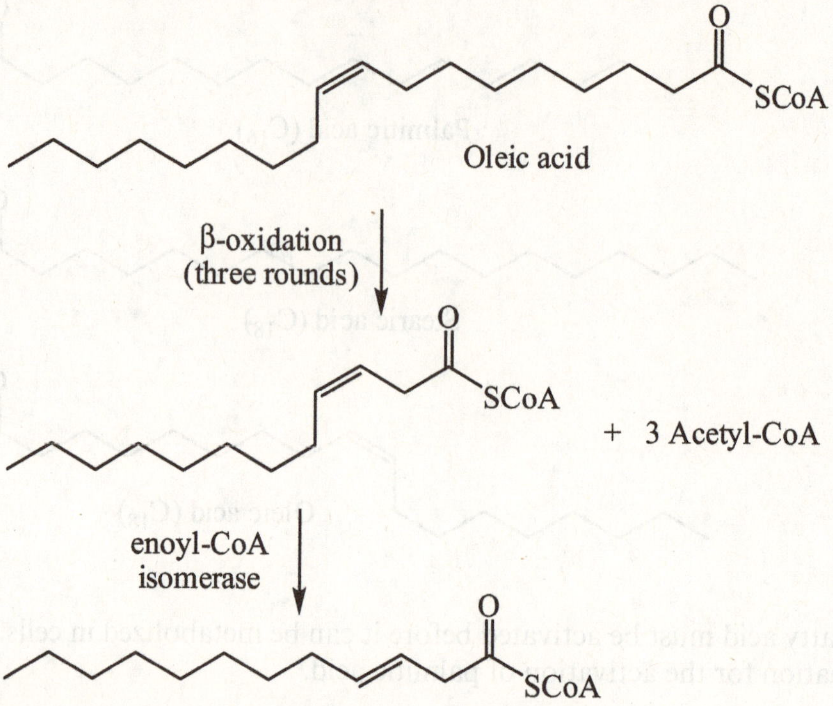

Oleic acid

β-oxidation
(three rounds)

+ 3 Acetyl-CoA

enoyl-CoA
isomerase

Citric Acid Cycle

21.21 What is the main function of the citric acid cycle?

The main function of the citric acid cycle is to oxidize the two carbon atoms of the acetyl group that is present in acetyl-CoA into carbon dioxide. This oxidation generates the reduced coenzymes that are subsequently used for electron transport and oxidative phosphorylation, which allows the aerobic production of ATP.

21.12 Which steps in the citric acid cycle involve

(a) The formation of new carbon-carbon bonds

The first step the aldol reaction of oxaloacetate with acetyl-CoA forms a new carbon-carbon bond.

(b) The breaking of carbon-carbon bonds

Steps 3 and 4 are decarboxylation reactions, each of which breaks a carbon-carbon bond. These two steps also involve oxidation reactions.

(c) Oxidation by NAD⁺

Steps 3, 4, and 8, which are respectively the oxidations of isocitrate, α-ketoglutarate, and malate, involve NAD⁺.

(d) Oxidation by FAD

FAD is involved in the oxidation of succinate, Step 6.

(e) Decarboxylation

Steps 3 and 4

(f) The creation of new stereocenters

Steps 1, 2, and 6 create new stereocenters. Step 1 is the aldol reaction of acetyl-CoA with oxaloacetate. Step 2 is an isomerization reaction that involves tautomerism. Step 6 adds water to fumarate to form malate.

21.23 What does it mean to say that the citric acid cycle is catalytic – that is, that it does not produce any new compounds?

The citric acid cycle is catalytic in that none of the intermediates involved in the cycle are destroyed or created in the net reaction. Both carbon atoms of acetyl-CoA are converted to carbon dioxide. NADH and FADH₂ are respectively generated from NAD⁺ and FAD.

Additional Problems

21.24 Review the oxidation reactions of glycolysis, β-oxidation, and the citric acid cycle, and compare the types of functional groups oxidized by NAD⁺ with those oxidized by FAD.

NAD⁺ is generally used for the oxidation of hydroxyl groups to carbonyl groups, while FAD is typically used for the oxidation of alkyl groups to alkenes.

21.25 The *respiratory quotient* (RQ), used in studies of energy metabolism and exercise physiology, is defined as the ratio of the volume of carbon dioxide produced to the volume of oxygen used:

$$RQ = \frac{\text{Volume } CO_2}{\text{Volume } O_2}$$

(a) Show that RQ for glucose is 1.00. (*Hint:* Look at the balanced equation for the complete oxidation of glucose to carbon dioxide and water.)

$$C_6H_{12}O_6 + 6O_2 \longrightarrow 6CO_2 + 6H_2O$$

The oxidation of glucose into six moles of CO_2 requires six moles of O_2.

(b) Calculate RQ for triolein, a triglyceride of molecular formula $C_{57}H_{104}O_6$.

$$C_{57}H_{104}O_6 + 80O_2 \longrightarrow 57CO_2 + 52H_2O \quad RQ = 0.71$$

(c) For an individual on a normal diet, RQ is approximately 0.85. Would this value increase or decrease if ethanol were to supply an appreciable portion of caloric needs?

$$C_2H_6O + 3O_2 \longrightarrow 2CO_2 + 3H_2O \quad RQ = 0.67$$

The RQ of ethanol is 0.67, so an individual's RQ would decrease if ethanol were to supply an appreciable portion of caloric needs.

21.26 Acetoacetate, β-hydroxybutyrate, and acetone are commonly referred to within the health sciences as "ketone bodies," in spite of the fact that one of them is not a ketone at all. All are products of human metabolism and are always present in blood plasma. Most tissues (with the notable exception of the brain) have the enzyme systems necessary to use ketone bodies as energy sources. Ketone bodies are synthesized by the enzyme-catalyzed reactions shown below. Describe the type of reaction involved in each step.

Step 1 is a Claisen condensation. Step 2 is an aldol reaction followed by the hydrolysis of one of the two thioesters. Step 3 is a reverse aldol reaction. Step 4 is a decarboxylation reaction. Step 5 involves the reduction of a ketone.

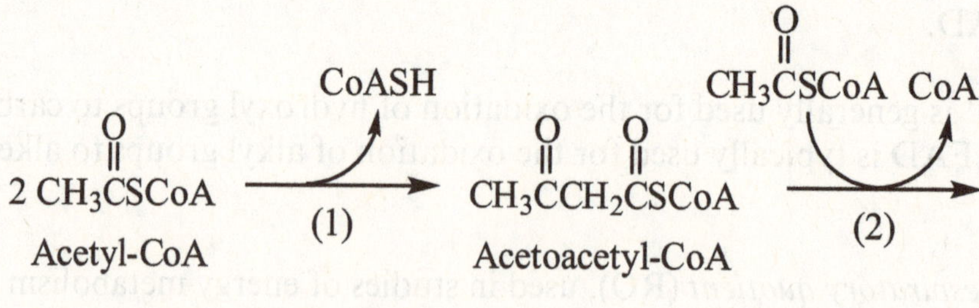

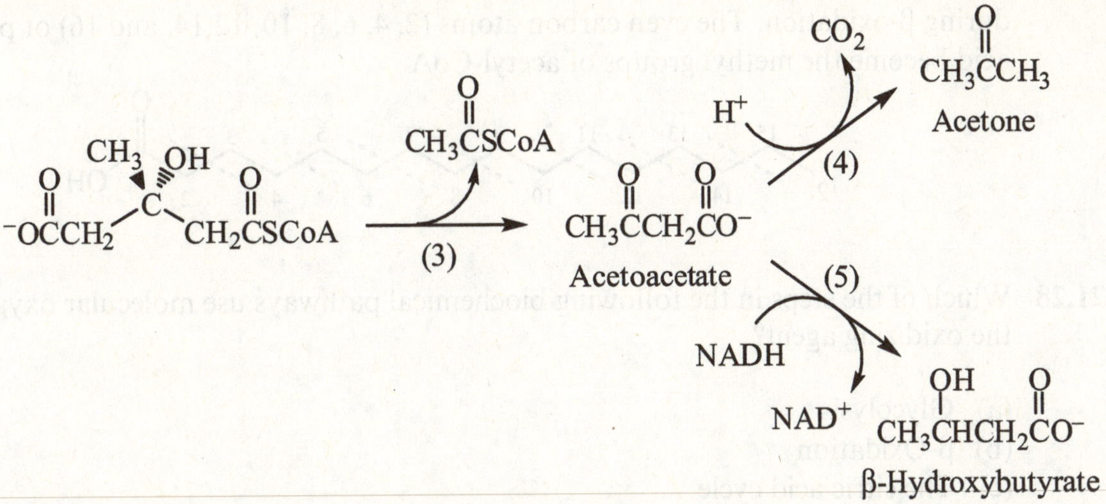

21.27 A connecting point between anaerobic glycolysis and β-oxidation is the formation of acetyl-CoA. Which carbon atoms of glucose appear as methyl groups of acetyl-CoA? Which carbon atoms of palmitic acid appear as methyl groups of acetyl-CoA?

As seen in Problem 21.6, C3 and C4 of glucose become the carboxylate groups of the two pyruvates. During oxidative decarboxylation, they are removed, and C1 and C6 of glucose become two methyl groups of acetyl-CoA.

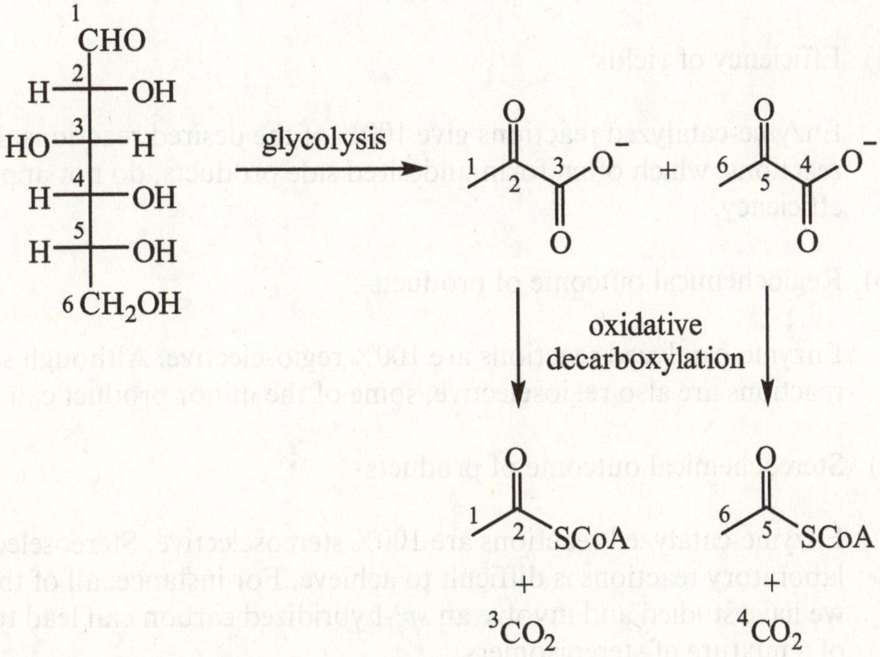

Palmitic acid undergoes β-oxidation and subsequent cleavage (indicated by the dashed lines) to produce acetyl-CoA. Odd-numbered carbon atoms are oxidized during β-oxidation. The even carbon atoms (2, 4, 6, 8, 10, 12,14, and 16) of palmitic acid become the methyl groups of acetyl-CoA.

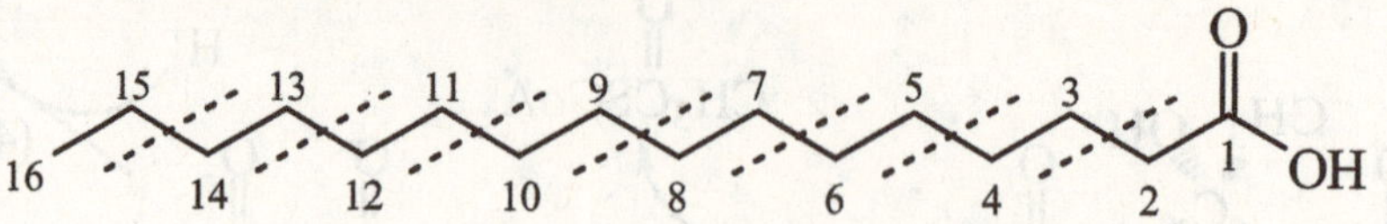

21.28 Which of the steps in the following biochemical pathways use molecular oxygen as the oxidizing agent?

(a) Glycolysis
(b) β-Oxidation
(c) The citric acid cycle

None of these biochemical pathways use molecular oxygen as the oxidizing agent. Glycolysis uses NAD^+ as the oxidizing agent. β-Oxidation and the citric acid cycle both use NAD^+ and FAD. Oxygen is used only during electron transport and oxidative phosphorylation.

Group Learning Activities

21.29 Compare biological (enzyme-catalyzed) reactions to laboratory reactions in terms of

(a) Efficiency of yields

Enzyme-catalyzed reactions give 100% of the desired reactions. Laboratory reactions, which often form undesired side products, do not approach this efficiency.

(b) Regiochemical outcome of products

Enzyme-catalyzed reactions are 100% regioselective. Although some laboratory reactions are also regioselective, some of the minor product can be formed.

(c) Stereochemical outcome of products

Enzyme-catalyzed reactions are 100% stereoselective. Stereoselectivity in laboratory reactions is difficult to achieve. For instance, all of the reactions that we have studied and involve an sp^2-hybridized carbon can lead to the formation of a mixture of stereoisomers.

21.30 Comment on the importance of stereochemistry in the synthesis of new drugs.

Enzymes and receptors, which interact with drugs, are chiral. An important property of enantiomers is that they behave differently in the presence of other chiral molecules. Therefore, one enantiomer of a drug may be biologically active and give the desired effects, while the other enantiomer may be inactive or even toxic. With respect to the laboratory synthesis of drugs, it is desirable to use advanced synthetic methods that are stereoselective and produce just one enantiomer instead of a racemic mixture.

21.31 Of the functional groups that we have studied, which are affected by the acidity of biological environments (biological pH)?

The functional groups that are affected by the acidity of biological environments are amino groups, carboxyl groups, and phosphate groups. Their degree of protonation or deprotonation is dependent on the pH of the environment.

21.32 Can you think of any aspect of your day to day life that does not involve or is not affected by organic chemistry? Explain.

All biological processes involve the reactions of organic chemistry. The combustion of fossil fuels that release energy for heat, electricity, and transportation involve the oxidation of organic molecules. The manufacture of household goods, pharmaceutical drugs, materials, and consumer electronics involves the use of organic compounds. To that end, the list of organic reactions that are necessary for everyday life is exceptionally large. Organic chemistry, traditionally defined as the study of carbon compounds, is only natural to life on Earth. After all, we are a carbon-based life form.